Fiddler Crabs of the World

OCYPODIDAE: GENUS *UCA*

PRINCETON UNIVERSITY PRESS · PRINCETON, NEW JERSEY

Fiddler Crabs of the World

OCYPODIDAE: GENUS *UCA*

by Jocelyn Crane

Publication of this book has been aided by a
subvention from the New York Zoological Society

Library of Congress Cataloging in
Publication Data will be found
on the last printed page of this book.

Printed in the United States of America by
Princeton University Press
This book has been composed in Linotype Times Roman

Preface

This survey of fiddler crabs results from work in both field and laboratory. It emphasizes the relation of striking form to complex behavior. The approach is comparative and the viewpoint that of a student of evolutionary biology.

The systematic revision and proposed phylogeny arise from morphological comparisons aided by evidence from social behavior, biogeography, and ecology. While a microscope indoors still arbitrates details, binoculars on a mudflat give better evolutionary insights.

Contributions to the subject from other fields await future research. No geneticist has worked on *Uca*. Physiologists, while they have already illumined a number of problems, have not yet provided enough comparative material to help systematists. Finally, in the present state of our knowledge the approaches of numerical taxonomy appear both unsuitable and impractical.

Wholly apart from the interest of their characteristics, fiddler crabs form an excellent group for comparative study. Their genus ranges throughout the warmer parts of the world, divides more or less acceptably into about 90 species and subspecies, is readily accessible to field observations, and can often be collected in series with relative ease.

It is hoped that this contribution will prove useful to three groups of workers: to those interested in the taxonomy of the genus, to the increasing number of physiologists employing *Uca* as an experimental animal, and to students of comparative ethology and evolution. Meanwhile, as this inquiry is handed on, the fiddlers continue to challenge us with surprising puzzles and elusive facts.

Contents

Part One. Systematic Section

Part Two. Toward an Evolutionary Synthesis

Indexes

List of Figures

List of Maps

List of Plates

Acknowledgments

One warm December morning William Beebe, John Tee-Van, and I were walking along the shore of a small and lonely bay in Costa Rica, binoculars at the ready. The white beach was suitably fringed with green and out in the turquoise water three pelicans were fishing. It was obviously the kind of place you dream about when it is cold up north, and gray, and you are waiting for a bus. Suddenly, on a muddy patch ahead, we saw a crowd of tiny creatures, each brandishing a great red claw that flashed in the sun. We froze motionless. "What . . . ?" I asked, and Will said shortly, "Fiddlers, of course," shocked that an assistant of his didn't know. And that, more than thirty years ago, was really the beginning of this book. I was hooked on fiddler crabs.

So, first of all, my appreciation goes to two people who are not here any more to receive it—William Beebe, Director of the New York Zoological Society's Department of Tropical Research, and John Tee-Van, for long Dr. Beebe's chief associate and later General Director for the Zoological Society of the Bronx Zoo and the New York Aquarium. For many seasons we and the rest of our group continued to study animal life in the American tropics and to work up the results at home. At intervals the fiddlers took time from other parts of my job, both on field trips made alone and in our laboratory at the Bronx Zoo. Yet Will and John were tireless, practical encouragers of my crab-watching at the expense of their own convenience. Always, Will's enthusiasm for living beings and for work—his, mine, everybody's—was endless, and his queries beat a rhythm through the years. "*Why* do fiddlers wave? . . . *Why* do they have those claws? . . . WHY . . . ?" Meanwhile I enjoyed the advantage of John's example in learning the details of making field trips run. He taught me, for instance, how to write a packing-list some forty pages long, when to rewrite the budget, when to stop trusting a taxonomic key, and what to do with a developing tank that is stuck in the heat with the film half in.

Another person who was important throughout much of this study was the late Fairfield Osborn, President of the New York Zoological Society. Dr. Osborn's brand of enthusiastic prodding was notable. Now and then he made me go to scientific meetings for which I did not want to take the time, and which sometimes changed my thinking; such was the one on ritualization, organized in 1965 by Sir Julian

Huxley. And from time to time memoranda turned up: "That last outline was fine. Now, how about a chapter?"

To two institutions in particular go my thanks. First, of course, is my parent organization, the New York Zoological Society. Through its officers and trustees it supported our department's research for decades before there were any government grants to help. Even when aid for special studies, such as work on the fiddler crabs, became available, the Society still provided our essentials—staff salaries, operating funds, and special allocations, along with the vital gifts of time and trust and understanding. The second institution is the National Science Foundation, which supported this study with grants extending from 1955 to 1970. The book was not planned, in fact, until the first grant gave me the means to work at last with fiddlers in Malaysia and Singapore. Foundation funds met the costs of all the principal crabbing trips that followed, which I made alone as time could be managed. These awards also provided for other needs that ranged from artists' fees to instruments and film. Throughout our association I especially appreciated the friendly cooperation of the staff in the Divisions of Systematic Biology and of Psychobiology.

Other organizations have contributed financial aid and a wide variety of other kinds of help during periods of the work. The National Geographic Society provided generous funds both for field work and for a summer in European museums. Through the Scripps Institution of Oceanography, University of California, I became a member of the Alpha Helix Expedition of 1969 to New Guinea, a trip that resulted in the clarification of questions that had puzzled me for years. Special thanks go also to several industrial firms which, with pleasing incongruity, gave a hand to the fiddler work. In Venezuela, the Creole Petroleum Corporation acted as a generous godfather on four expeditions of the Zoological Society; the fiddlers enjoyed the benefits, along with the spiders, fish, butterflies, and birds, with all of which we were at the time principally concerned. Similarly, at our station in Trinidad the Alcoa Steamship Company gave us aid of many kinds, which proved incidentally to help the study of the crabs. Finally, the Suriname Bauxite Maatchappij was our good friend on recurrent field trips to Guyana and Surinam.

Certain zoologists in diverse fields have been exceedingly helpful to me throughout this study. Each knows the job has been under way, but not one has read any part of the text and is therefore not at all to blame for any errors of fact or judgment. It is rather through their own published contributions and their friendly encouragement of my attempt to deal with fiddlers as a whole that I have felt the force of their aid. They are Alfred E. Emerson, Donald R. Griffin, Konrad Lorenz, Ernst Mayr, Waldo L. Schmitt, and Nikolaas Tinbergen. I'll comment here on only two of them. Dr. Schmitt, when he was director of the National Museum at the Smithsonian, suggested the wide-angle view in the first place: "Go on—why stop with those chaps on the Pacific coast? That's only *one* piece of ocean. . . . Tackle the lot!" And then there is Don Griffin, my husband; although not responsible for anything visible in these pages, he has certainly contributed a benign effect on them through his endless patience while I finished the job—at the cost of lots of tidiness in living.

Anyone who works on a systematic study of any group depends with confidence on the cooperation of fellow specialists in museums and universities and on the custodians of their libraries. My gratitude goes especially to Fenner L. Chace, Jr., of the National Museum, Smithsonian Institution, who has helped me for years by lending material from the museum's outstanding collection and receiving the collections I have made, as well as providing laboratory space as needed. He, along with these others, each of whom numbers crustaceans among his interests, has advised me on particular questions, procedures, or sections, but none of them is in the least responsible for the result. Their aid has also ranged from locating elusive specimens to providing welcomes in far-off places. These fellow enthusiasts are Rudolf Altevogt, Dorothy E. Bliss, Richard Bott, Jacques Forest, John S. Garth, Isabella Gordon, Danièle Guinot, Heinrich-Otto von Hagen, Lipke B. Holthuis, Teresita Maccagno, Don C. Miller, William Macnae, Lejeune P. Oliveira, Ernst S. Reese, Tsumei Sakai, Michael Salmon, Michael W. F. Tweedie, J. Verwey, and Torben Wolff.

In the libraries I have especially appreciated the knowledge and helpful spirit, in equal parts, of Hazel Gay and Mary Wissler at the American Museum of Natural History in New York. Without their continuing cooperation, and the use of their institution's superlative resources, this book could not have been written. During briefer periods I have also enjoyed the aid of Sonia Mirsky at the library of The Rockefeller University, and of Marie Gabrielle Madier at the Bibliothèque Centrale du Muséum National d'Histoire Naturelle de Paris.

Toward the people who helped most immediately to put this volume together I feel a special warmth. Julie C. Emsley provided most of the drawings,

working with me at our Trinidad field station. Frances Waite Gibson made almost all the others, keeping up with the book's progress in both Trinidad and New York, while several other skilled artists contributed the rest. The figures made by each are listed under the heading Illustration Credits, below.

With this kind of a job, as anyone who has been through it knows, a drawing and the writing of the associated text demand a close collaboration between the scientific artist and the author. Even when a zoologist is his own artist, he must often ask himself what he "really sees." When two people, artist and author, are involved, as in this book, the artist depends on the zoologist for plans and checks, while the zoologist learns and profits from the artist's trained eye. In this particular study our arguments have sometimes been cheerfully strenuous as we took turns peering through the microscope—and very often the artist proved right.

For photographers, small crustaceans make trying subjects. The specimens are inconveniently three-dimensional with steep slopes; it is accordingly difficult to bring salient details into simultaneous focus. The beasts also have far too many legs. Most of the plates illustrating the systematic section result from the persistent and ingenious efforts of two photographers, Russ Kinne and William Meng. The full list of specialists who provided the illustrations is given on p. xxiii, and my appreciation goes to each.

Of special aid was Michael Flinn. A physicist by profession, he spent his time at our Trinidad field station on acoustic studies of bats, katydids, and fiddlers, recording with equal verve ultrasonic trills from the jungle air and deeper sounds from the mud by the shore. For this study he secured fine material, and for all he taught me I am most grateful.

My close friend and assistant in dealing with numerous details was Kathleen Campbell, whose skilled help in Trinidad and New York proved invaluable; I especially appreciated her persistent energy in analyzing film and in tracking stray references. Eleanora d'Arms also worked most ably with the film during a more recent season. In preparing the thousands of crabs resulting from the field trips for their permanent home in the National Museum, I was unusually fortunate in having Helen Miller's aid in the Bronx Zoo laboratory. For her efficiency in that long task go special thanks; they also go to her and to Rosanne Blair for typing the final drafts of the manuscript.

Recently I have been enjoying, with gratitude, the help of Joanna Hitchcock and Frank Mahood of the Princeton University Press, who are carrying the principal weight of guiding this volume through publication. Their skill is a pleasure to watch.

Finally, it would be wonderful to include by name, followed by the appropriate accolade, every one of the other people who has been important in helping this study on its way. Some have been close friends for decades; others I met briefly when they made my work easier, or pleasanter, or even possible. One of them dug crabs with me in Java on a hot noon in the midst of Ramadan, when the rest of the town was trying to sleep through the daily fast. Another opened his office on a Sunday in Australia to fix a camera. One, at a dinner in the Hotel Manila, proffered from a brown paper bag a quart jar full of fiddlers. Unfortunately, the list of kindnesses is impossibly long. So I shall ask two special friends, who have played large and unlikely roles in the making of this book, to take my warmest thanks on behalf of all the rest—A. Parks McCombs and Helen Damrosch Tee-Van.

ILLUSTRATION CREDITS

Figures

Julie C. Emsley: 5-7, 10-40, 45-49, 53-85, 87-89, 92-94, 96-101.
P. Gaillard: 8, 9.
Frances Waite Gibson: 1-3, 41, 44, 50-52.
Edward R. Ryan: 4.
Pauline Thomas: 86, 95.
Dorothy F. Warren: 90, 91.

Maps

Julie C. Emsley, Frances Waite Gibson, Pauline Thomas, J. Crane.

Photographs

Toshio Asaeda: 44 *E*.
J. Crane: 45 *A-F*, 48 *A-C*, 49, 50 *A-B*.
P. Gaillard: 2 *E-F*, 5 *E-F*, 7 *I-J*, 9 *A-D*, 38 *I-J*, 40 *A-F*.
Russ Kinne: 1 *E-H*, 2 *A-D*, 3 *A-H*, 4 *A-H*, 5 *A-D*, 6 *A-H*, 7 *A-H*, 8 *A-D*, 9 *E-H*, 10 *A-D* and *G-J*, 11 *A-D*, 12 *A-H*, 13 *A-H*, 14 *A-H*, 16 *A-H*, 18 *A-D*, 38 *E-H* and *K-L*, 39 *A-F*.

A. Veenstra, through William Macnae: 8 *E*.
William Meng: 21 *A*, 22 *C-H*, 23 *C-D*, 24 *D-H*, 25 *A-D*, 26 *A-D*, 27 *A-D*, 29 *C-D*, 30 *C-I*, 31 *C-D* and *G-H*, 32 *A-H*, 33 *A-D*, 34 *C-D* and *G-H*, 35 *E-H*, 36 *A-D* and *G-H*, 37 *A-H*, 41 *C-H*, 42 *A-D* and *G-H*, 43 *A-D* and *G-H*.
Margaret Sweeting: 1 *A-D*, 11 *E-H*, 15 *A-D* and *G-J*, 17 *A-D*, 21 *B-D*, 23 *E-G*, 24 *A-C*, 28 *A-H*, 29 *A-B*, 30 *A-B*, 31 *A-B* and *E-F*, 34 *A-B* and *E-F*, 35 *A-D*, 36 *E-F*, 41 *A-B*, 42 *E-F*, 43 *E-F*.
Natica Waterbury: 17 *E-H*, 18 *E-H*, 19 *A-H*, 20 *A-H*, 21 *E-H*, 22 *A-B*, 23 *A-B*, 25 *E-H*, 26 *E-H*, 27 *E-H*, 29 *E-H*, 33 *E-H*, 38 *A-D*.
M. Woodbridge Williams: 44 *B* (© National Geographic Society).
Photographic Service of the British Museum (Natural History): 5 *G-H*, 10 *E-F*, 15 *E-F*.

During the final stages of preparing the figures, maps and photographs for publication, the Illustration Service of The Rockefeller University provided appreciated help in many details.

Fiddler Crabs of the World

OCYPODIDAE: GENUS *UCA*

General Introduction

Any human being who finds a suitable piece of warm shore, sits down, keeps quiet, and watches fiddler crabs must be impressed. Each adult male has one of his claws longer than his body and, when conditions are right, he wields it vigorously in threat and courtship. Sometimes he links claws with another male, in either a strenuous fight or ceremonial encounter.

Even when he is not socially occupied a fiddler is active, providing only that the tide is out, the day warm, and, preferably, the sun shining. He feeds by straining bits of organic matter from the surface of the shore. The small claw, specialized for lifting mud or sand, rhythmically brings pinches to the mouthparts. Here food is separated from the mineral matter which, accumulating outside, is wiped off now and then and dropped to form a growing line of pellets. Except under special conditions every crab feeds near the burrow in which he has passed the preceding hours of high tide.

A burrow is the center of a fiddler's life. Even when the water's edge is far away, he rushes down the hole when startled or when driven by heat or dehydration. He repairs it as the tide approaches, scraping out with his legs large packets of earth which he sometimes flings inches away. If a crab does not, at the moment, have a burrow of his own, he takes active measures to acquire one before the water reaches him, either finding an empty hole or ousting another crab.

Some tropical fiddlers change color strikingly before they start the day's display, brightening to shining white with scarlet claws. Others stay vividly spotted with blue. Most, while otherwise dull, at least have the great claw patched with color and tipped with white.

These brachyuran crabs are without doubt among the best of all invertebrates for comparative study. They are generally considered an evolutionary apex of crustaceans, where they form the well-marked genus *Uca*. In this contribution they are divided into 62 species. For the biologist one of their great advantages is a combination of prevalence and accessibility. Fiddlers thrive around the world in the muddier parts of the tropics; several forms reach even to New England and Japan. All the species are largely diurnal, active at low tide, often sympatric, and always gregarious. Their morphological specializations

alone show a challenging variety; at the latest count 84 structures on the large claw are devoted to combat, and as many as 40 of these occur in a single species. This claw sometimes attains more than two fifths of the entire weight of the crab. A fiddler's repertory of social components is equally remarkable. For example, in pages to come 14 different threat postures and motions are counted, 16 methods of sound production, 13 components of combat, and 18 of waving display.

People have been writing about fiddler crabs for more than 300 years. Marcgrave recognizably described two Brazilian forms in 1648, long before Linnaeus was born; the German naturalist even remarked on their color and habitat, and mentioned that one kind was eaten locally. Rumphius, who spent most of his life in the East Indies, first described display in fiddlers. "During ebb tide," he wrote, "it waves its larger claw strenuously and continuously, as though it wanted to call people, and when one comes to it, it hides in the sand." He listed names for it in Malay, Dutch, and Latin, all signifying "calling crab." His account was published in 1705, several years after his death. Linnaeus (1758) picked up the old description and gave this species officially the name *vocans*—the name we use today. About a hundred years later Fritz Müller was in Brazil, enthusiastically accumulating material for his *Facts and Arguments for Darwin*, published in English translation in 1869; here he gave in a footnote the first statement on color change in fiddlers, having found that one species changed rapidly from white to dark when it was caught and held. His lively report on color change and courtship appeared in 1881. Meanwhile Darwin, in touch with Müller, referred several times to fiddler crabs in *The Descent of Man* . . . (1871).

Alcock (1892), another fine naturalist, watched fiddler crabs in India and gave an excellent sketch of their habits. After describing one use of the large claw in courting, he continued: "The second function, as a fighting weapon, becomes apparent when, in the general tournament, one of the rival males approaches too close to the other. The great claw is then used as a club, the little creatures making savage, back-handed sweeps at each other. . . . I did not actually see the rival males seize each other in the conflict, but I have no doubt that they do so, for on going over the field of action I saw several freshly dismembered chelae lying on the mud. So that the chela is probably used as a shears as well as a club." Alcock's tentative conclusion always disturbed me since, in years of observing fiddlers, I have never seen a claw detached in a fight. Finally, in 1943, when M. R. Raut was watching whimbrels near Bombay, he saw the birds catching fiddlers. Each whimbrel seized a crab by its large claw, "lifted it up into the air and then sharply jerked its head." The fiddler promptly broke away from its claw and fell to the ground, the appendage remaining in the bird's beak. The bird then dropped the claw and picked up the crab before it could escape. Raut also reported that the place was littered with inedible claws. When, long after its publication, I read Raut's account it made a satisfying sequel to Alcock's original observation: a man who was bird-watching solved a crab-watcher's problem, apparently without knowing there was a puzzle.

To the early naturalists it seemed clear that both waving and bright colors were concerned primarily with direct sexual selection. As Alcock put it: "From prolonged watching I feel convinced that the waving of the claw by the male is a signal of entreaty to the female, and I think that no one can doubt that the claw of the male has become conspicuous and beautiful in order to charm the female." Although the full explanation does not now seem to be a simple matter of clubbing rivals and charming mates, the idea of the basic ambivalence of this appendage is strongly upheld in the present contribution.

Dr. Alcock was representative of many of the enthusiastic naturalists to whom we owe not only the early collections of fiddler crabs but also the first observations on their behavior. Some of the men were physicians in distant colonies, others were captains of merchant ships. Sometimes they staffed government survey vessels. Accounts of their voyages make it easy to understand why puzzles turn up when old collections are consulted.

One such round-the-world trip was undertaken by the French vessel "Uranie," sailing on a government mission in 1817. The expedition aimed to circle the earth, survey the seas, describe the human beings, and collect the fauna and flora. The ship's company made good progress for more than two years, rounded the Horn in safety, and headed at last for home. Then they piled up on a submerged rock off Patagonia.

The expedition's leader and captain, M. le Chevalier de Freycinet, seems to have been both efficient and exuberant. At the beginning of the voyage he smuggled his wife, disguised as a sailor, on board for the trip, to the Admiralty's later shock. In the narrative he quoted Byron from time to time and used exclamation marks with freedom. After the ship-

wreck he wrote: "Continuous vicissitudes—such is the life of the sailor! Dangers forgotten as soon as passed—such are men!" Somehow they rescued stores, gear, and all the scientific specimens. Finally they bought another ship and sailed home. It does seem likely that a crab label could have been displaced that night they hit the rock. (De Freycinet, 1825-1829; Rouch, 1953.)

In the early days of taxonomic study it was not often possible for systematists to work with one another's collections, which in any case were meager. Among the predictable results are the sometimes extensive synonymies that appear in the present revision. In addition the characters important for distinguishing species were still wholly uncertain. Consequently *Uca*'s taxonomy very early reached a state of chaos. A number of capable systematists from time to time attacked the group on a regional basis; their work made it possible to distinguish species in restricted areas. Of these workers the most ambitious and successful was Mary J. Rathbun, who organized the numerous American species of the genus in her monograph *The Grapsoid Crabs of America* (1918.1). She was the first to recognize the taxonomic value of a number of *Uca*'s characters. Much of the morphological work in the current study rests on the foundation she built.

The form of the present systematic section is the result of four basic aims that carry roughly equal weight. The first is to clarify the man-made taxonomy. The second is to provide means of species identification for workers with the group, regardless of their special interest. The third aim is to describe in sufficient detail the tuberculation, ridges, and other armature so that, even though the uses of these structures are not yet fully understood, ethologists in the future will find a morphological groundwork already in place. Finally, a continuing effort is made to seek out and include characteristics of apparent importance in the group's evolution, whether they be morphological or behavioral. Unfortunately not enough is yet known of comparative physiology in *Uca* to include more than hints of physiological characters in the systematic treatment. Even more regrettably, we remain completely ignorant of the genetics of the genus.

The second part of the book consists of general chapters on biogeography, ecology, morphology, and behavior. Here the aim has been to survey the group as a whole, the emphasis being both on its general phylogeny and on its development of social behavior.

Throughout the progress of the study a major effort was made to link knowledge of the living crabs with that of their morphological specializations. Toward that end the practical work was divided into three categories. The first consisted of field trips, made to all the major regions inhabited by *Uca*. Information on the localities visited and amount of time spent in each is given in Table 24 (p. 662). Field time throughout most of the trips was divided between observation and cinematography. In recent years tape recordings have been made and, most recently, videotape apparatus is proving useful. After 1957, when the importance of the semi-lunar cycle in social behavior became apparent, trips were planned to permit work at key points at optimal tides. The longer itineraries were built around stays ranging from about ten days to a month or more in important places, with shorter stops in less strategic areas.

Second, supplementary observations on living crabs were made in terraria, here termed crabberies, both out-of-doors on the West Indian island of Trinidad and indoors in New York. Altogether 27 species have been carried to the crabberies from Asia, the eastern Pacific, and the Caribbean. Through work with the captive crabs valuable supplementary data were obtained, chiefly on waving display and sound production. Because of artifacts of captivity, no display timings were recorded on these individuals, nor are observations on social behavior included here that were made only on captive crabs.

Finally, interspersed among the field trips was work on preserved specimens in the laboratory and in museums. The alternations of behavior work with the morphological study of collections proved stimulating to both aspects of the program. For example, for years I casually watched the fighting of male fiddlers in the field, paying the combats little attention, since they seemed so much alike throughout the genus. Meanwhile in the laboratory I often had to describe the ridges and tubercles that make patterns on the claws. By manipulating specimens I sometimes tried, always without success, to fit the most puzzling structures into positions suitable for stridulation against another part of the fiddler's anatomy. When the idea of the use of the structures during combat eventually arrived, I was not watching living crabs on a tropical mudflat. Instead, the explanation appeared on a winter afternoon in the Bronx Zoo laboratory, when I was describing a long-dead *festae* through a microscope and trying simply to finish my stint and go home. Suddenly it seemed that those tubercles on the claw might somehow be used in ceremonial fighting. Perhaps in linking chelae males played a kind of "duet." So when I next went to Trinidad I checked the idea out on the first good tide. The only surprise then was that the explanation had not been obvious to everyone who ever watched a fiddler fight during the past hundred years.

In this survey of a genus with many species, the traditional approach of an evolutionary biologist appears to be the only one feasible. The approaches here have accordingly followed the paths mapped out in particular by Huxley (1942), Lorenz (1941), Simpson (1953, 1961), Tinbergen (1951, 1953), Mayr, Linsley, and Usinger (1953), and Mayr (1963, 1969).

After years of intermittent preparation, the accumulated material appears finally to be fitting together. Naturally unknown answers and unfinished lines of evidence still dangle in tantalizing plenty. I hope and believe that other people, as they carry on from here, will also thoroughly enjoy the search.

Part One · Systematic Section

KEYS: Appendix B, p. 615.

DENDROGRAMS: Figs. 96-101.

CONVENTIONS, ABBREVIATIONS, AND GLOSSARY: p. 678.

Introduction

The first paragraphs below concern procedures used in this section. These comments are followed by an annotated list of the topics that will be covered. The section ends with comments on style.

CATEGORIES AND THEIR TREATMENT

Every taxonomist who revises a group of animals must ask questions that do not have definite answers. Often the basic query is merely whether to use a particular category above or below the level of a species. For instance, will the division of a complex genus, such as *Uca*, into subgenera be helpful to workers or an inefficient complication?

A more important example concerns the designation of subspecies. It is an old problem. Species have biological reality, while the other taxonomic categories are man-made conveniences. Species can be defined, once we know enough about them. The trouble is that we must often base our systematic decisions on inadequate material. Especially in these difficult cases we usually have little or no acquaintance with the animal in its habitat. These uncertainties have, of course, been prevalent in the present study of fiddler crabs.

What, in particular, is the best treatment for allopatric populations made up of individuals that are both similar and neighboring, yet each a recognizable member of one form or the other? Material is almost always insufficient for reliable statistical analysis. More important, we do not know whether forms interbreed, or whether they could or would do so if populations were naturally coincident. Often no field work has yet been done in the most likely area of coincidence. In only three species of *Uca* have I found good evidence of hybridization in such cases; here I could with confidence designate the subgroups as subspecies. Yet, in a number of other forms the designations of subspecies rather than species seem also to be clearly the more reasonable on the basis of the morphological and behavioral evidence we have in hand. The resulting divisions are of practical importance for future progress, since specialists in other fields understandably tend to disregard subspecific names; sometimes, too, both taxonomists and non-taxonomists do not consider the concept of subspecies as valid or even useful. Nevertheless it is now clear in *Uca* that populations of the same species in different geographic localities do differ in ways that

will prove to be of increasing interest to biologists.

What, again, is the best taxonomic procedure when the subjects are similar forms occurring on the two sides of the Isthmus of Panama? This barrier has kept related fiddler populations apart for at least two, and more probably five, million years. Several pairs of these forms seem to have evolved at slower rates than, for instance, some forms of *Uca* in the Indo-Pacific. Obviously the populations cannot, at this geological moment, interbreed; obviously, too, they must by now differ genetically; they do, in fact, show distinctions in minor ways. A taxonomist perhaps should automatically consider each a full species; yet if he does so some of his other decisions will appear particularly illogical.

Fortunately we may look forward to improvements on any current classification as soon as genetical programs, planned studies in the field, adequate samplings, physiological techniques, and judicious statistical analyses begin to produce results. Meanwhile it seems to me that the most helpful interim arrangement of the genus should be strongly hierarchical, even though the degrees of relationships cannot yet be proved. I have therefore used subdivisions based on varying degrees of morphological and behavioral similarity. In addition to species, the categories found serviceable in *Uca* consist of formal subspecies, informal superspecies, and formal subgenera. In addition, another informal term, *alliance*, is used to cover one or more species in a superspecies along with the other species that appear also to be close relations.

ALLOPATRIC CATEGORIES

Pragmatic criteria of subspecies and species among allopatric groups of *Uca*, where we have no data concerning hybridization, are used here as follows:

Subspecies usually differ from one another in only about one to three morphological characters that are readily apparent to the practiced worker; these distinctions include any characters on the gonopod and major cheliped. Sometimes only one character is diagnostically reliable; behavioral and habitat differences are usually minimal, although exceptions occur.

Allopatric species differ from one another in about two or three readily apparent morphological characters in addition to those of gonopods and major chelipeds; except in the subgenus *Minuca*, the gonopods show clear distinctions, while differences are also discernible in waving display. In *Deltuca* and *Minuca* the major chelipeds are similar among allopatric species; in *Celuca* distinctions are more apparent.

Evolution in *Uca* being the complex process it obviously is, any attempt to pin down degrees of differences for application throughout the genus can only be approximately satisfactory, as workers using the above criteria will swiftly agree.

The informal category superspecies is useful for assembling under a single name two or more allopatric species. Except for occasional zones of coincidence, each of these species occupies a different range and each appears, through a number of morphological characters, to be more closely related to one or more other members of its superspecies than to other species in the genus. In the table of contents and in headings for species and subspecies in the text, superspecies are indicated by the name or initial, enclosed in brackets, of one of the species in the group.

ALLIANCES AND LOCAL ALLIANCES

In a few localities, especially in the tropical eastern Pacific, members of superspecies are found in close sympatric association with other species that appear to be equally or almost as closely related as are the allopatric members of the group. These associated species are here termed members of the *local alliance*, while the entire related group, composed of the members of the superspecies and the members of the local alliances, are termed an *alliance*; sometimes the name of a species is used with the term *and its allies*, or *and the members of its local alliance*. The only large group of species to which these terms apply is that of the American superspecies *crenulata* and its relations in the subgenus *Celuca*.

SUBGENERA

The last kind of group used in this study is the subgenus, once more a formal designation. During the first half of the work on this contribution it seemed inadvisable to stress feeble homogeneities by dividing the genus formally into subgenera. Such divisions usually have an added disadvantage: either they include intermediate forms, thereby obscuring relationships, or else monospecific subgenera have to be erected for the interesting misfits. Characteristically, such a monospecific group cannot be adequately defined on the subgeneric level because of the lack of species for comparison; in any case, it is often insufficiently differentiated for the standing accorded it. Nevertheless, in this study the convenience of subgeneric handles came to seem essential.

The nine subgenera described are helpful both in indicating relationships on a global basis and in breaking up the genus into more manageable units. As an inspection of the diagnoses, descriptions, the key, and Table 1 will show, their morphological boundaries are something less than concise, while the selection of subgeneric characters has been difficult. The comments under the descriptions of the genus and of the individual subgenera will, it is hoped, demonstrate the reasons for these subgeneric divisions.

Unlike the treatment of species, described below, non-morphological categories are curtailed in the descriptions of subgenera, since they form the subjects of most of the chapters in Part Two. A special section on *Relationships* follows the description of each subgenus; in addition, two of them, *Uca* and *Minuca*, end with comments on the only fossils known in the genus.

ANNOTATED LIST OF TOPICS CONSIDERED UNDER SYSTEMATIC HEADINGS

Following short introductions, the species are treated under the topics given below and in corresponding order.

Subspecies are described as briefly as is practicable, usually with emphasis on morphological characters. Unlike the procedure in species descriptions, their order is governed by their importance or convenience for diagnosis, not by consistency among species. Data on measurements, color, and social behavior of subspecies are usually included under the appropriate headings in descriptions of the species in order to facilitate comparisons; exceptions occur where the material was too extensive for effective presentation under single headings.

Since zoeae and megalopa in *Uca* have been identified in only five species, descriptions and discussions of these stages are omitted. The following references are pertinent: Hyman (1920, 1922), Herrnkind (1968.2), and Feest (1969).

Morphology

Diagnosis

The characters given are of two kinds. The most general, especially the breadth of the front, are for the reorientation of workers who come directly to a page after using one of the regional keys. Second, several characters are given which at least in combination separate the species from sympatric and nearby allopatric relations.

These characters are purely those of convenience and do not necessarily appear in the order in which they will be included, often with amplified descriptions, under the detailed treatment of the morphology.

Description

MALE

The selection of characters of course varies with the subgenus. Their locations are illustrated in Figs. 1-4, 42-44, and 58, and are included in the introduction to the keys (p. 614) and Glossary. Among the subgenera, the species of *Deltuca* and *Celuca*, the first and the last, are described in most detail as the least and most specialized both morphologically and in social behavior. Especially in *Celuca*, morphological details are described where the use is not yet known, particularly in the armature of the major cheliped. Thus the descriptions contain not only material useful in the identification of a crab in the hand, but also material that is used in discussing relationships; finally, some of the details are expected to be useful only to future ethologists.

On the other hand, some characteristics are treated very briefly, even though they can be expected to prove of importance in future work of various kinds. The reason, of course, was lack of time combined with either great variation, or difficulty in making efficiently rapid examinations, or both. The most obvious structures that have received short shrift are the spoon-tipped setae on the second maxilliped, the gill on the third, details of armature on the eight ambulatories (which vary not only from leg to leg but from one side to the other) and, finally, the external form of the spermatophore, which almost certainly can provide excellent specific characters.

FEMALE

Although the topic *Description* is followed by the subheading *Male*, most characters will be found to apply to both sexes. Unless there is a statement to the contrary, the female, except for differences noted under that heading in the subgenus and genus, shares with the male all characters that are not of obvious primary and secondary sexual significance. Only exceptions and special characters are listed for each species under the subheading *Female*.

Measurements

Methods of measurement appear in Fig. 4. Although many carcinologists have used the carapace breadth as the standard measurement of size, this custom has not been followed here; length seems to be the more satisfactory dimension from a biological point of view. Heterogony strikingly affects many parts of fiddler crabs beside the major claw, and carapace breadth, next to the claw size and proportions, shows the greatest relative change with growth. If this dimension were selected as the size standard, the growth ratio of the claw to carapace would reflect the heterogonic values less clearly. In each species, with the exception of *rapax*, measurements are given on only a few specimens. The measurements selected are carapace length and breadth in both sexes, plus the lengths of propodus and major dactyl in males. These dimensions not only give an idea of normal and unusually large adult sizes, but they automatically indicate the principal length proportions of the major cheliped. In addition, they show the extreme ratio of claw to body size attained in the material at hand. This information may help future workers interested in heterogony in their selection of material.

The appearance of the initials J.C. in a table of measurements indicates that I made the measurement, usually on a type-specimen, to complete information omitted in the published description.

Morphological Comparison and Comment

The section is often divided between practical suggestions for identifying the species and remarks on its relationships to other members of the subgenus. Most of the dendrograms (Figs. 96-101) resemble the aerial view of a tree more than the usual lateral aspect. The reasons for the design reflect, of course, both our lack of fossils and the close interrelations of subgenera and species. These apparently basic and varied relationships are indicated by the open bases of the branches, and, in the dendrogram of subgenera, the black lines proceeding from some of the bases toward non-adjacent groups. Superspecies are shown by the grouping of species in solid rings; local alliances are similarly circled by unshaded rings.

Color

The basis of each description is the color of males during waving display. Often the description is confined to this phase, because nonwaving males as well as females and young frequently do not show distinctive hues and patterns. Throughout the genus individual color variation is prevalent, while some paling or brightening of color by displaying males is frequent. Populations in different parts of the range often differ in their coloration, either slightly or strikingly. Therefore in systematic work color cues from field notes should be used with caution unless one is thoroughly familiar with the species and its sympatric associates.

Social Behavior

When data are available the following topics are treated under each species: waving display, precopulatory behavior, chimney or other construction activities, acoustic behavior, and combat. The indexes to names and subjects will often provide other references to the species appearing in Part Two. Definitions of behavior components and references to their descriptions in the text are given in the Glossary.

Range

A summary is given under this heading, along with a statement concerning marginal records that are open to question. More detailed localities will be found under the heading References and Synonymy at the end of the account of each species. Additional records are given under the species name in Appendix A: Material Examined. In maps with dots showing records of occurrence, only those localities are included where I have checked the material; this was essential in most Indo-Pacific forms where the systematics were especially confused and many of the recorded specimens could not be located.

Biotopes

A short résumé of the habitats is given for use in conjunction with Figs. 21-23 and Table 10.

Sympatric Associates

Only the most usual associates are listed. In localities especially rich in food the list can often be considerably lengthened. Examples are given under suitable species. Comments on degrees of closeness in sympatry are given on p. 690.

Material Resulting from Field Work

This résumé is supplemented by the data given in Appendix A: Material Examined.

Type Material and Nomenclature

Not all types that are known to be missing or that have not been found are replaced by a newly designated lectotype or neotype. Where no apparent taxonomic confusion or question exists at present, this step does not seem to be necessary, or, in view of the ultimate fragility of specimens, desirable. Lectotypes from among older collections, particularly with specimens from more than one locality all labeled "types," seem more useful; the best example is that of the two males from different localities described by Milne-Edwards as *Gelasimus dussumieri*; they proved to belong to two species, one new and one already described. Lost types and indeterminate material are discussed following the last species in the systematic section (p. 322). The index to scientific names lists all names, whether in current use, in synonymy, discarded, or of indeterminate significance.

References and Synonymy

Under this heading two classes of intended omissions occur. First are occasional citations that not only show taxonomic confusion but also lack special historical or other interest. Second are recent references wholly devoted to physiological studies; comments on this class of omission appear on p. 448.

The omissions in the first group are based on erroneous synonymies, on material belonging to more than one species, or both, and, furthermore, not described in such a way that useful attempts can be made to disentangle them. Almost all of these references are to contributions that surveyed the genus toward the end of the nineteenth century. At that time it was impossible for the authors to examine more than a fraction of the scattered material that had been recorded; their synonymies and descriptions consequently often reflected the abbreviated descrip-

tions and inadequate drawings of their predecessors. Among the references that are sometimes omitted for these reasons are, for example, Kingsley, 1880.1, and Ortmann, 1897, as well as some of the references they included in their synonymies.

References to illustrations are included if they meet any of the following conditions: First, they accompany a type description. Second, the illustrations, most of them recent, appear helpful from a taxonomic viewpoint; these citations are often given only in general terms, such as "fig. of claw," where no confusions exist and the author under the species heading gives the full reference. Third, the illustrations accompany early references of historical interest. Omitted figures or plate numbers refer to inaccurate sketches and those illustrating characteristics now known to be of little or no taxonomic importance. Future students of a particular species or group will consult the references for themselves; meanwhile it is certain that the great majority of older illustrations do not include, or else they picture inexactly, the characters that are now known to be of importance both taxonomically and functionally, such as, particularly, those used in acoustic and combat behavior.

Finally, the annotations accompanying the references attempt to give in extremely brief form the content of each citation. Where the page number, with or without a reference to an illustration, is followed only by a geographic locality, the author gives no taxonomic description or discussion; he is, however, recording new material; he may also give a list of references in such a record and arrange them without comment according to his conception of the synonymies of the names included. If the reference is based only on the work of previous investigators and does not record new material, that fact is stated following the citation.

Where the word *taxonomy* is included in the annotation, the reference comments, whether briefly or in a detailed description, on the morphology of this material, and often on its systematic position.

Any of the rarely occurring comments on color, habitat, or, in the older references, habits are included in the annotation, unless the author merely uses a word or phrase so general as to be unhelpful, such as *intertidal zone* or *mangroves* in groups where the association can today be taken for granted.

At the end of some annotations appears the abbreviation for an institution, as given on p. 678, followed by "!", the symbol indicating that I have examined some, at least, of the material included in the reference cited. Since, in older collections and references, it is often impossible to determine whether or not particular specimens were those described in a given publication, I have omitted this symbol except in examples where I felt sure beyond reasonable doubt that the reference cited was actually based on the indicated material.

COMMENTS ON STYLE

The generic name *Uca*, or its abbreviation *U.*, is usually omitted in this study before a species name. Since almost all the crabs mentioned in the book are members of this genus, and since *Uca* is used whenever other genera are also being discussed, this disregard of taxonomic convention seems warranted.

More questionable may be my occasional use of a species name, or the phrase "this species," as though it were a living animal capable of activity. An example is the sentence, "This species, when pursued, runs faster than *inversa*." To be traditionally acceptable it should be phrased more precisely, as in the form, "In this species the crabs, when pursued, run faster than do individuals in *U. inversa*." Such problems of usage were rare in the days when taxonomic names were used principally in morphological descriptions of dead specimens. At present, since so much of this book concerns behavior, it seems to me desirable to avoid circumlocutions whenever this practice does not obscure the meaning.

In the selection and spelling of geographic names occasional conflicts arose between familiar usage and recent nomenclature. In statements of range, the forms selected are in general in accordance with those used in current leading American and British atlases and hydrographic charts, as well as on large-scale maps, in English, that are available in some countries. Exceptions occur when a name that has been officially discarded in recent years probably remains more familiar to most English-speaking readers than does the current name; an example is my use of Celebes for Sulawesi. In Appendix A, however, an effort has been made to use the contemporary names of political entities with the more familiar name following, in brackets. The practice of using brackets in similar fashion is also used where both names are at present current, as in the Nansei [Ryukyu] Islands. In the sections entitled "References and Synonymy" the names used in the publication cited are retained; for example, Taiwan is not substituted for Formosa; in foreign names, English versions are used, as in the use of China for "Chine," when no doubt of the author's intention is apparent.

FAMILY **OCYPODIDAE** ORTMANN, 1894

Reference: Ortmann, 1894.2: 700.

Brachygnathous crabs with orbits occupying almost entire anterior border of carapace, the front being usually narrow and somewhat deflexed; outer orbital margins often incomplete; eyestalks slender. Palp of third (external) maxilliped articulating at or near antero-external angle of merus; exognath usually slender and often more or less concealed; the maxilliped usually completely covers the large buccal cavity. One cheliped in male often enlarged. Male abdomen narrow.

Amphibious, burrowing crabs confined largely to the tropics of the world, where they occupy a variety of habitats on shore and in estuaries. Most are gregarious and some genera have complex patterns of social behavior. The family comprises about a dozen genera of which the best known include *Macrophthalmus, Scopimera, Ilyoplax, Heloecius, Ocypode,* and *Uca.*

SUBFAMILY **OCYPODINAE** DANA, 1851

Reference: Dana, 1851: 248.

Carapace deep, roughly quadrilateral but sometimes almost pentagonal, the regions usually indistinct; front narrow, usually a lobe curved slightly downward between the eyestalks; flagellum of antenna small, folding obliquely or almost vertically; interantennular septum broad; external maxillipeds completely enclosing buccal cavity; exognath always at least partly visible, its flagellum present or absent. Afferent branchial opening, thickly fringed with setae, between bases of 2nd and 3rd pairs of ambulatories. Female gonopore on 3rd segment of sternum.

One cheliped enlarged in males, often remarkably; chelipeds in females equal or somewhat unequal. Merus of chelipeds and ambulatories always distinctly three-sided at least proximally, there being one margin dorsally and two ventrally. The merus of the major cheliped, however, is rotated toward the rear, so that the dorsal margin is directed postero-dorsally and accordingly is termed the *postero-dorsal margin* in the descriptions; similarly the margin that is homologous with the antero-ventral margins of the ambulatories is rotated upward and becomes the antero-dorsal margin, while the homologously ventral surface faces forward, becoming functionally and taxonomically the *anterior surface.*

Carpus and manus of ambulatories have one dorsal and two ventral margins, occasionally with the addition of an antero-lateral or postero-lateral ridge, short or long; one or more of all these margins, as well as those of the meri, may be indistinct or missing, especially in males; ambulatory dactyls always have six margins marked by ridges. Carpus, propodus, and dactyl of both chelipeds showing a wide variety in shape, armature, and relative size.

The subfamily's two genera, *Ocypode* and *Uca,* are the only cosmopolitan genera in the family, excluding the rare and marginal genus *Euplax.* On American shores as well as in the Indo-Pacific the Ocypodinae are so well known that they almost always have local names; in English the species of *Ocypode* are recognized as *beach crabs* or *racing crabs,* while those of *Uca* are the calling crabs or *fiddlers.*

GENUS *UCA* LEACH, 1814

MORPHOLOGY

Diagnosis

Brachygnathous crabs differing from all other genera in the enormous size of the male's major cheliped, the minor cheliped and both in the female being minute. *Uca* differs additionally from *Ocypode*, the other genus in the subfamily, in having smaller eyes on longer stalks, longer antennae, and shorter ambulatories, while it lacks a stridulating ridge on the ischium of the major cheliped.

Description

With the characteristics of the subfamily.

MALE

Carapace. Body deep. Carapace subquadrilateral to almost hexagonal, slightly wider than long, never flattened. Front deflexed, moderately to very narrow (about 2.5 to 20 times in breadth of carapace), spatulate to widely arched in shape; a small, median, distal notch always present in frontal margin. Orbits sometimes straight, but usually directed obliquely backward. Antero-lateral angles usually pointed and slightly produced, sometimes a right angle; variable, often dissimilar on major and minor sides. Antero-lateral margins ranging from absent to moderately long; converging, parallel, or, rarely, slightly diverging. Postero-dorsal margins moderately converging, of varying lengths and degrees of distinctness, rarely extending almost to posterior edge of carapace above base of 4th ambulatory; usually marked by a raised line only, sometimes beaded or spinulous, rarely tuberculate. One or two postero-lateral striae, one above the other, sometimes present behind and below postero-dorsal margin. Lateral margins slightly to moderately converging. Vertical lateral margin ranging from almost absent, to developed ventrally only, to strong throughout. Surface of carapace usually smooth, rarely partly setose or tuberculate, especially posteriorly or on sides; except for H-form depression, regions never deeply demarcated.

Orbit usually considerably wider than diameters of eye and its stalk. Upper orbital margin more or less sinuous, often composed of two distinct edges, sometimes serrate or beaded, with or without a space between them which, when present, forms the eyebrow; latter often deflexed; upper marginal edge continues internally as a raised line and proceeds around the front, sometimes as a slender margin, sometimes as

a thickened border. External angle of orbit without a raised margin, but often with one or more separated tubercles on its outer edge; a narrow channel, free of tubercles, always present at its lower, posterior corner. Lower orbital margin with crenellations weak to strong, always strongest near outer angle; orbital floor with or without a mound, ridge, or tubercles. Suborbital regions, behind crenellations, usually tuberculate, sometimes naked, often covered with setae which may or may not be in the form of dense pile. Pterygostomian regions usually setose, often tuberculate, never altogether naked and smooth.

Anterior Appendages. Antennal flagellum not hidden beneath front; antenna large compared with that in *Ocypode*. Eye subterminal on stalk, the extreme end of which, usually flattened, curves partway around the eye's tip; rarely the stalk projects in a short or long style beyond the eye. Buccal cavity broader than long. Third maxilliped with ischium ranging from flat, grooved, and moderately setose to convex and smooth, its gill form ranging from large with many books to vestigial with none. Second maxillipeds with spoon-tipped setae ranging from few to numerous, and of varying form.

Major Cheliped. Enormous, strikingly different from the minor, and with a wide diversity of form, relative sizes of segments, and occurrence of ridges, tubercles, and similar specializations. *Ischium* without a stridulating ridge, but sometimes with minute tubercles. *Merus* in adults always projecting beyond carapace; postero-dorsal margin often lacking except proximally; antero-dorsal margin (in homology with ambulatories, the antero-ventral margin) always marked strongly or weakly by a ridge, crest, tubercles, or spines, especially distally; ventral margin (homologously the postero-ventral) always with one or more rows of serrations, tubercles, or rugosities; posterior and dorsal surfaces usually partly rugose, or with tuberculate striae, or tuberculate, at least dorso-distally; anterior surface (homologously the ventral) flat and smooth, finely granulate, or, rarely, partly tuberculate, especially dorso-distally.

Carpus with dorsal surface flat or rounded, with or without rugosities, its edge usually bent toward inner side, with or without one or more teeth or small tubercles, its anterior surface flat and smooth, sometimes with a blunt, oblique ridge, which is rarely armed with one or more tubercles, large or small.

Outer *manus* ranging from rough with large tubercles to a surface covered with tubercles so minute that the surface appears macroscopically smooth. Tubercles usually largest near upper margin, sometimes also enlarged near base of pollex; lower proximal region (heel) often rounded and protruding to various degrees; dorsal margin more or less bent over toward inner side (palm), sometimes broad and flattened, always tuberculate in various patterns, sometimes further marked off externally by a submarginal groove. Predactyl area, at upper, distal end of manus, almost always set off by a predistal groove and distal blunt ridge, the latter either smooth or tuberculate. Lower distal outer manus, near base of pollex, with or without a depression, which may be large or small, and tuberculate, smooth or pilous. Ventral margin usually marked by a row of beaded tubercles, sometimes further set off as a low carina by an external groove at its base; rarely with a second ridge, submarginal, smooth or tuberculate, continuing distally into pollex.

Palm always with an upper proximal depression, the carpal cavity, with its distal boundary sometimes well marked, sometimes incomplete. An oblique ridge, tuberculate and prominent with few exceptions, runs from lower distal edge of carpal cavity to lower margin of palm near pollex base; sometimes it almost parallels the ventral margin; sometimes it is strongly oblique; in some species it continues up around the cavity's anterior margin. One or two tuberculate ridges, roughly parallel to base of dactyl, on distal part of palm; the more proximal is confluent ventrally with inner row of tubercles along upper pollex margin. Center of palm tuberculate to smooth. Lower distal part of palm with a hollow, behind pollex base, ranging from large and deep, with a trifid trench, to very shallow. Lower proximal part of palm almost smooth except, in one subgenus, for differing patterns of stridulating tubercles.

Pollex and *dactyl* usually much longer than manus, their shape and the armature of gape varying greatly even within populations. Each finger with or without a long external groove running almost its full length; each tuberculate outside, or rarely pitted, usually only proximally. Ventral margin of pollex sometimes with a continuation of the arrangement of tubercles, sometimes carinate, that starts on ventral margin of manus. Dorsal margin of dactyl usually flattened proximally, practically always rough with tubercles, which continue for some distance along margin and may reach halfway to tip; among the proximal tubercles a short groove is usually deeply marked; rarely it continues far along outer upper side of dactyl, paralleling the lateral groove. Dactyl never shorter than pollex; its curvature is often greater, and it often tapers to a slender tip which curves down beyond pollex. Gape narrow to wide, the fingers not in contact except distally; upper margin of pollex and lower of dactyl basically with three rows each of tubercles

or teeth; any row may be missing, or present only in a portion of the gape; one or more tubercles, most often a part of the central row, are usually enlarged, sometimes strikingly so, in each margin; in some species part of either margin, rimmed with tubercles, projects into gape, sometimes forming a forceps-shaped tip at end of claw; tip of pollex often bifid or trifid, through grouping of enlarged distal tubercles.

Minor Cheliped. Very small, even in comparison with ambulatories. *Merus* with dorsal margin armed weakly or strongly with tubercles, spinules, or continuations of rugosities from postero-dorsal surface. Antero-ventral margin unarmed. Postero-ventral margin serrate or tuberculate, the tubercles sometimes in form of sharp and slender spinules. Postero-dorsal surface almost always somewhat roughened with tubercles, spinules, or rugosities, sometimes weakly and only on upper distal part. Anterior and ventral surfaces smooth; dorso-distal margin at least irregular, sometimes formed of a regular row of minute tubercles; a parallel ridge, similarly armed, usually present predistally on dorsal surface. *Carpus* with or without a subdorsal ridge on outer side. A longitudinal stria, the mano-pollex ridge, is almost always conspicuous; it starts at some point on lower, outer manus and continues part way along lower half of pollex. Dactyl with or without a faint, broad, longitudinal ridge; sometimes also with a shallow groove above it and a faint second ridge above the groove. Pollex and dactyl usually clearly longer than manus. Gape ranging from very narrow, with the tubercles or serrations of the prehensile edges in contact, to wide, with the fingers arching to meet only distally. Tips horny, often expanded, meeting perfectly or nearly so, almost always with spoon-like excavations on inner sides. Margins of manus and fingers with setae, ranging from sparsely scattered to thick fringes; outer manus and palm with a few setae only; inner subdistal portions of fingers with fringes of setae forming a single, basket-like tuft when tips of fingers are in contact; they are located along the lower and upper margins of the excavated ends of pollex and dactyl respectively, especially distally.

Ambulatories. All always well developed, but shorter and the segments often relatively deeper than in *Ocypode*. Second leg is longest; 3rd leg next longest; 4th leg shortest and most slender. All segments beyond ischium moderately compressed antero-posteriorly. Merus often expanded dorso-ventrally, all three margins marked by at least a blunt edge, usually armed at least dorsally with serrations directed distally, rarely with separated tubercles; serrations on dorso-distal margin usually distinct; postero-ventral margin lower than antero-ventral, and its armature stronger. Posterior side of merus, especially in dorsal half, often with scattered tubercles or vertical, tuberculate striae;

they are most numerous on 2nd and 3rd legs, often entirely missing on 4th. Similar tubercles or striae rarely and sparsely present antero-dorsally on first two or three meri.

Anterior carpus, and sometimes the distal merus, the manus, or both of 1st leg, rarely with a ridge or a row of tubercles (some *Celuca* only). Posterior carpus of 2nd and 3rd legs sometimes similarly equipped. Dorsal margins of carpus and manus of two or more legs sometimes with one or two ridges, the intervening dorsal surface sometimes roughened. Manus and dactyl fringed variously with setae; pile sometimes present on any of the segments, especially close to the joints.

Abdomen. Seven segments except in a few species of the subgenus *Celuca*; in these two or more segments are fused.

Gonopod. All parts showing a wide range of form and proportion.

FEMALE

The outstanding characteristics of females throughout the genus are as follows: first, the absence of a large cheliped, the cheliped on the two sides being minute and similar; second, the stronger armature of carapace and ambulatories; and, third, the relatively greater volume of their carapace in comparison with that of males of equal length. This capacity results from some or all of these characteristics: more transverse orbits, less converging sides, higher arching, and more shallow grooves between regions. (In species where the males are often exposed to desiccation the carapaces are similarly shaped, so that in this particular these males appear feminized.)

In more detail, females differ from males in the following secondary sexual characteristics: Carapace less wide, often more arched. Orbits less oblique, except in species where male orbits are almost transverse; antero-lateral angles similar to each other. Antero-lateral margins, where absent or very short in male, present or longer. Dorso-lateral margins less converging; these margins and, when present, postero-lateral striae, sometimes longer and stronger, the ridging being higher and the elements of its armature—beading, serrations, or tubercles—larger; often beading replaces unarmed ridging or serrations replace beading. Granules in rows or clusters, or tuberculate striae in this postero-lateral area sometimes present in female only; entire dorsal and postero-lateral surfaces of carapace sometimes granulate or even tuberculate, in contrast to its smoothness in male. Vertical lateral margins stronger, the beads usually larger. Carapace regions less well marked; pile, when present, stronger in female, sometimes present in female only. Eyebrow usually wider; teeth on floor of orbit when present more numerous and

more regular, sometimes present in female only; suborbital crenellations sometimes larger and more extensive; setae sometimes thicker and more widely distributed in suborbital, pterygostomian, and adjacent lateral regions. Both chelipeds very small, closely similar to minor cheliped of male, although sometimes with distinctions among gape teeth and in ridging of carpus and dactyl. Meri of ambulatories broader, their marginal serrations or tubercles well developed, larger than in male; posterior spinules, tubercles, or tuberculate striae always stronger; carpus and dorsal manus often rougher, the ridges being stronger, or distinguishable where wholly absent in male; any or all ambulatory segments more often pilous, sometimes densely so.

Gonopore. Often in a depression; with or without an elevated rim, either simple or sculptured and surrounding the pore either wholly or in part; with or without one or more tubercles, simple or sculptured, adjoining it. This characteristic is highly subject to differentiation among subspecies but relatively stable within populations. Spermatophores sometimes distort the gonopore's appearance and can be difficult to distinguish from a possible tubercle on the margin.

Size

Small to moderate, in comparison with crabs of other groups, the apparent size always increased by disproportionate growth of major cheliped. Carapace length of smallest known male that is apparently adult: 4.5 mm, propodus 11 mm, in (*Celuca*) *batuenta*; measurements of the largest: (*Uca*) *maracoani insignis* with carapace length 31.5 mm, propodus 94 mm; (*Afruca*) *tangeri* length 33 mm, propodus 105 mm. Females usually smaller than males, about equal in some species but never attaining greater lengths than their males. Size and growth in the genus are discussed on p. 449.

Color

Color changes before waving display, sometimes to polished white, are prevalent among males in several subgenera. Red or orange tints or tones are frequent and persistent on major chelipeds. These and all other color families occur in the genus on carapace and ambulatories, very roughly in species-specific patterns but with wide variation within and among populations. The general discussion of color begins on p. 466.

SOCIAL BEHAVIOR

Visual display involving conspicuous motions of the major cheliped is known to occur in all species that have been sufficiently observed; similar behavior may be presumed for the few species that are either unknown in life or have not been observed during display. Stridulation is also a genus characteristic, although far less is known about it; usually sound production is employed principally if not exclusively in agonistic situations, but some species certainly have incorporated one or more of its components into their courtship patterns. Territoriality that centers closely around an intertidal burrow is well developed. Intermale combat is not only prevalent throughout the genus, as has long been known, but has recently proved to be highly ritualized and to consist of a variety of components employing numerous special structures of the major cheliped. In some species chimneys, pillars, or semi-domes are erected beside the burrows, made of substrate; at present little is known of their functions. Social behavior is described and discussed in Part Two, Chapters 5 and 6.

RANGE

Uca is known from all continents except Antarctica. Its most northern limits are about 34° N in Japan and 42° N on the northern shore of Cape Cod in the United States. The most southern records are about 32° in South Africa and Australia, and about 35° in Uruguay. Details are given in Part Two, Chapter 1, in Tables 7, 8, and 9, and on Map 1.

BIOTOPES

All *Uca* are intertidal, but their burrows are not necessarily either covered by the water or exposed every day. Substrates range from mud to sand with a minimum of silt. Mangroves or other forms of vegetation are usually close, although no *Uca* occur regularly in deep or constant shade. Preferred salinities range from fully marine to practically, or temporarily, fresh; tolerance to change is high. Chapter 2 in Part Two, Table 10, and Figs. 21-23 give further information.

RELATIONSHIPS

Although more closely related to *Ocypode* than to any other genus in the family, *Uca* is somewhat less well adapted to withstand desiccation, while no fiddler has developed the capacity for speed that characterizes the racing crabs. On the other hand the complexity and diversity of social patterns developed in *Uca*, along with the accompanying morphological specializations, are not approached by those of *Ocypode*. Fig. 57 illustrates examples of the gonopods in other genera of ocypodids for comparison with those of *Uca*; Fig. 33 shows the form of the maxillipeds in *Ocypode*, and Figs. 76, 77, and 81 the gills in *Mac-rophthalmus* and *Ocypode*. Resemblances to the corresponding organs in some of the species of *Uca* are apparent.

In the present study, *Uca* is divided into nine subgenera. This arrangement continues the work started by Bott in 1954, when he erected two subgenera between which he distributed a group of species from the eastern Pacific. His nominative subgenus, *Uca*, was based on those American species characterized by extremely narrow fronts. For his other subgenus, *Minuca*, he selected as the type-species *Uca mordax*, which is characterized by a front exceptionally wide. In the descriptions that follow, Bott's conception of the subgenus *Uca* remains unchanged. The species he placed in *Minuca*, however, are here distributed among three other subgenera: *Minuca* Bott, *Boboruca* subgen. nov., and *Celuca*, subgen nov. These four subgenera include all American forms. In one of the subgenera, *Celuca*, are placed in addition two Indo-Pacific species. Of the five remaining subgenera, all new, one is erected for an eastern Atlantic species; the other four are confined to the Indo-Pacific.

The apparent relationships are indicated in the dendrogram of the subgenera (Fig. 96); the principal distinctions on which they are based are summarized in Table 1.

The subgenus *Deltuca* is considered the least specialized among the nine; in general, members of this subgenus depend more on continuously damp substrates than most other species; they are relatively inactive and devote little time to social behavior; waving displays are simple and mating uncomplicated, always taking place on the surface. Morphologically these characteristics are reflected in relatively flat carapaces and only moderate diversity and specialization in armature for both stridulation and combat.

The second subgenus, *Australuca*, appears to be very close to *Deltuca*; it differs principally in its slightly stronger armature and tube-tipped gonopods in all species; similar gonopods appear fully formed in only one species of *Deltuca*. Waving displays and mating activities in *Australuca* show a slight trend toward the greater energy, higher reach, lateral direction, and underground mating developed in greater degrees by other subgenera. These species have also adapted to a wider range of habitats than have those in *Deltuca*.

In both morphology and behavior *Thalassuca* appears closer to *Deltuca* than to other subgenera, particularly than to the specialized *Minuca* and *Celuca* of America. This close affinity is shown especially in the shape of the front, gonopods, armature, basic waving display, and mating behavior. Its intermediate position is indicated by details of armature on the ambulatories and, in the waving display, by the more frequent occurrence, duration, and occasional trend toward laterality. The basic affinities with the non-Indo-Pacific groups seem to be chiefly with *Afruca* and *Uca*, although in the cheliped armature there are resemblances to that of *Minuca*. These apparent relationships are indicated on the dendrogram by the directions of the lines emerging from the base of *Thalassuca*'s branch.

The fourth subgenus, *Amphiuca*, occupies a median position between the three Indo-Pacific groups listed above and the five remaining groups that appear to have evolved in the eastern Pacific and the Atlantic. The shape of the front in *Amphiuca* is basically that of a broad-fronted American *Uca*, as is the lack of long grooves on the major chela; yet details of the palmar armature are those of Indo-Pacific groups; the armature of the ambulatories is intermediate. The form of the gill on the 3rd maxilliped resembles especially that of *Afruca*. The waving display is almost a perfect intermediate between the flexed vertical waves of *Deltuca* and the unflexed, more spacious motions of New World species, yet the low level of social activity usually rivals that of *Deltuca*. The dendrogram indicates *Amphiuca*'s intermediate position; no base lines are directed toward other subgenera, except for a short one aimed toward *Deltuca*.

The fifth subgenus, *Boboruca*, is erected for the American species *thayeri*, which like *Amphiuca* occupies an intermediate position. The shape of its front and the armature of its ambulatories are essentially characteristic of *Minuca* and *Celuca*, but some of the armature on the major cheliped resembles that of *Deltuca*. Its lethargy and the form of its waving display are also similar to those characteristics in the Indo-Pacific subgenera. Its closest affinities must remain inconclusive, as indicated on the dendrogram.

The sixth and seventh subgenera, *Afruca* and *Uca*, appear more closely related than any others previously characterized, and hence are shown arising from a common stem. In spite of different fronts, which are moderately narrow in the eastern Atlantic *Afruca* and extremely narrow in the American *Uca*, a number of distinctions are held in common, ranging from details of the setae on the 2nd maxillipeds and the armature of the major cheliped to components of social behavior. Their affinities, more clearly than in most subgenera, are with the Indo-Pacific *Thalassuca*, as shown especially in the shape of the chelipeds in some of the species and in the general form of the gill on the third maxilliped; in addition, *Afruca* has a waving display closely similar to those of several species of *Minuca*.

The last two subgenera, *Minuca* and *Celuca*, like *Afruca* and *Uca*, have more in common with each other than with any of the other groups. Some species in each are better adapted to a semi-terrestrial life,

are more active, and have a more varied social repertory than any of the other groups. The two subgenera share similar forms of front and complex details of armature. Nevertheless they differ significantly from each other. *Minuca* adapts to a semi-terrestrial way of life in a manner different from that of *Celuca*. Most *Celuca* have thick bodies and sometimes a highly reflectant display coloration, which probably helps both in withstanding heat and avoiding desiccation. *Minuca*, on the other hand, while its species also leave the low-tide areas of marine shores, have specialized by adapting to conditions found in brackish swamps; when these dry out the crabs aestivate underground; long periods of inactivity, sometimes extending to weeks or months, are the rule in this group. In contrast most *Celuca* have to aestivate, if at all, only during semi-monthly neap tides. When *Minuca* become socially active, however, they are extremely so, waving daily for long periods, unlike species in *Deltuca* and related subgenera, where waving is sporadic. In brief, in adaptation to a definitely littoral life and in social activity, *Minuca* is closer to *Celuca* than to any other subgenus; when social activity alone is concerned, however, it is rivaled by both *Afruca* and *Uca*. On the dendrogram lines from the base of the *Minuca* branch are also directed toward *Deltuca* and *Thalassuca*, in order to emphasize distant affinities with these Indo-Pacific stocks, as shown particularly in the gonopod shapes of some of the species of *Minuca*.

When the morphological and behavioral specializations of *Celuca* are viewed as a whole, its preeminent position is clear. Unquestionably it represents the apex of the evolution of the genus. This culmination shows in its adaptations to a semi-terrestrial life of high activity; it shows with equal clarity in the structural and functional specializations connected with social patterns at once complex and sustained.

Its adaptations toward drier and less muddy biotopes include the trend toward broad, semi-cylindrical carapaces, and toward gaping small chelae that effectively hold scoops of more solid substrate. In both these particulars a number of *Celuca* contrast with most other species in the entire genus; the uncommon exceptions—adapted for aestivation, hibernation, or sandier substrates—never reach the morphological extremes found in *Celuca*.

Associated with longer periods in the air is the increase of time for social activities and associated elaboration of structure and function. These take the form of high diversity of armature used in sound production and combat, and of the development of non-ambivalent display components used only during courtship. In the elaboration of acoustic equipment the most obvious examples are the high development of suborbital crenellations and of stridulating mechanisms on the major palm and first leg. Some species of *Celuca* also, unlike those in other subgenera, incorporate acoustic elements wholly into the fabric of waving display. Finally, in this subgenus construction activities reach their apex.

Relationships among the subgenera will be considered further in Part Two, Chapter 7.

The three fossils known in the genus are discussed on pp. 127 and 157.

REFERENCES AND SYNONYMY

Genus *Uca* Leach, 1814

Uca

Leach, 1814: 430. Type description. Type-species: *Cancer vocans major* Herbst, 1782 = *Uca* (*Uca*) *major* (Herbst, 1782) in present study, p. 136.

Leach, 1815: 323. Brief description.

Rathbun, 1897.1: 154. Nomenclature, explaining the need for recognizing *Uca* Leach, 1814, as the valid name for the genus, replacing *Gelasimus* Latreille, 1817, the name that had been in general use for the preceding 80 years.

Stebbing, 1905: 39. Nomenclature, including a refutation of Desmarest's (1825) criticism of the acceptance of *Uca* Leach, 1814, as the valid name for the genus; taxonomy.

Rathbun, 1918.1: 374. Taxonomy.

Barnard, 1950: 89. Taxonomy.

Holthuis, 1962: 239, 240, 245, 246, 251. Nomenclature of the genus and type-species; rejection of *Uca* Latreille, 1819.

Gelasimus

Latreille, 1817.2: 517. Type description. Type-species: *Cancer vocans* Linnaeus, 1758 = *Gelasimus vocans* as figured by Milne-Edwards, ?1836: Pl. 18, Fig. 1 = *Uca* (*Thalassuca*) *vocans* in present study, p. 85; p. 67: habits (no original observations).

Latreille, as pointed out by Holthuis in a personal communication, did not designate a type-species, although he listed three species in his description of the genus, including *G. maracoani*, which has heretofore been cited as the type-species. Holthuis found that Milne-Edwards, above, instead is responsible for the designation of the type-species, since his contribution is the first to deal with the matter, including in its title the phrase "edition accompagnée de planches gravées, representant les types de tous les genres." Latreille starts his account with the words "GELASIME, *Gelasimus* (Buffon). Genre de crustacés . . ." Like Stebbing (1905), I have been unable to trace the use of the word by Buffon.

Desmarest, 1825: 122. Taxonomy and nomenclature, including objections to the establishment of *Uca* Leach as the valid name for the genus; general habits (no original observations).

de Man, 1891: 20. Key to 15 Indo-Pacific species.

Alcock, 1900: 350. Taxonomy; habits.

Acanthoplax

Milne-Edwards, 1852: 151. Type description. Type-species: *Acanthoplax insignis* Milne-Edwards, 1852 = *Uca* (*Uca*) *maracoani insignis* (Milne-Edwards, 1852) in present study, pp. 146, 147, and 149.

I. *DELTUCA* SUBGEN. NOV.

Typus: *Gelasimus forcipatus* Adams & White, 1848

(Indo-Pacific: East Africa to Fiji; Japan to Western Australia)

PLATES 1-9.
GONOPOD DRAWINGS: FIGURES 61, 62 (part).
DENDROGRAMS: FIGURES 96, 98.
MAPS 2, 18, 19.
TABLES 1, 8, 10, 12, 14, 19, 20.

MORPHOLOGY

Diagnosis

Uca with front narrow, antero-lateral margins short to absent, eyestalks slender, postero-lateral striae absent at least in males. Major manus outside rough with tubercles that are usually largest near pollex base; oblique ridge on palm never continued upward around carpal cavity; longitudinal grooves along outside of both pollex and major dactyl. Postero-dorsal surface of ambulatory meri with simple tubercles, not with tuberculate striae vertically arranged. Gill on 3rd maxilliped small, without books. Females with rare exceptions have a pair of enlarged teeth in gape of at least one chela.

Description

MALE

Carapace. Front narrow, narrowest between eyestalk bases, its minimum breadth subequal to, rarely 1.5 times, basal breadth of erected eyestalk. Central area of front between raised margins is a groove narrow to broad, its margins parallel or subparallel. Orbits straight to strongly oblique; antero-lateral angles always acute, often produced, especially on major side; antero-lateral margins lacking or short, except in *arcuata*, usually finely serrate or beaded; dorso-lateral margins distinctly marked, except in *dussumieri*, by raised lines which are strongly diverging, except in *arcuata*; this margin usually anteriorly spinulate and posteriorly beaded, or beaded throughout. Postero-lateral striae absent. Vertical lateral margin usually not reaching antero-lateral margin, usually beaded. Carapace profile in lateral view little arched, except in *arcuata*. Suborbital region practically or wholly naked. Pterygostomian region little convex, moderately rugose; strongly setose.

Eyebrows moderate to short, narrow and, except in *arcuata*, vertical; one or both margins serrate, finely granulate or beaded. Orbit with lower margin rolled more or less outward, somewhat sinuous, with or without tubercles, granules, a blunt ridge or mound on its floor; suborbital margin practically entire, the crenellations absent toward inner angle and otherwise feeble except, sometimes, for enlargement near outer angle. This angle, along with outer edge of orbital floor and lower, outer margin of the antero-lateral angle, makes a continuous angle, obtuse or acute; part of its sharp margins are sometimes microscopically serrate but never toothed or crenellate. Eyestalks very slender in middle sections, the eyes in life clearly of larger diameter than the stalks, sometimes notably so even in preservative.

Second Maxilliped. Spoon-tipped hairs, few to moderate in numbers.

Third Maxilliped. Ischium with central portion almost flat; no median groove but a slight central depression subdistally; setae short, sparse. Gill small to vestigial, the books never distinct.

Major Cheliped. Merus with postero-dorsal margin well defined throughout in smaller specimens; in larger individuals margin is rounded at least distally, except in some *demani*; a conspicuous crest or large tooth never present on antero-dorsal margin; dorsal surface always smooth, slightly concave.

Carpus with dorsal surface flat, coarsely tuberculate, its anterior margin armed proximally with one to several enlarged tubercles; posterior surface slightly convex, smooth or sparsely rugose.

Manus: Outer surface covered with coarse tubercles, those near base of pollex largest; heel not notably convex; upper margin of manus strongly bent over toward inside, marked by a low ridge which is more or less flattened, bounded externally by a submarginal groove and, above and adjacent to the groove, by at least 2 rows of tubercles which are most numerous, and usually largest, distally near base of dactyl. No smooth, shallow triangular depression on lower distal outer manus, near pollex base, although a small tuberculate pit is sometimes present at beginning of the long pollex furrow (see below). Lower margin of manus with a distinct carina, marked by one row of tubercles, ending on proximal part of pollex.

Palm with carpal cavity extending distally almost to upper part of dactyl base, although diminishing in breadth and depth; no beaded margin around cavity's upper distal part; oblique ridge always present, always tuberculate, extending from near distal part of carpal cavity's lower margin obliquely down to ventral margin at pollex base; no tubercles ever continue upward around anterior margin of carpal cavity. Tuberculate ridges at base of gape and dactyl highly variable intraspecifically and in large specimens sometimes vestigial. Tubercles of lower part of proximal ridge, however, practically always strong. Center of palm strongly convex with or without distinct tubercles. Lower distal part of palm with a deep depression, trifid, its distal arm longest and encroaching on pollex base, its dorsally directed arm shortest.

Pollex and *dactyl* each with a long furrow externally, running along middle of side throughout most of its length; the pollex furrow in several species is indistinct in some individuals; sometimes a 2nd, short pollex furrow is present proximally, below it. Dactyl usually with a 2nd furrow traceable subdorsally, usually short, near dactyl base only; in several species it continues to parallel the median furrow distinctly throughout most of its length. Proximal part of dactyl dorsally tuberculate. Both pollex and dactyl always subequal, usually almost straight, the dactyl never arching well beyond and below a contrastingly straight pollex; extreme tip of each simple, tapering, that of pollex directed up, that of dactyl, down. Gape and gape teeth various and variable, but the 3 basic rows of teeth are usually traceable proximally on pollex, those of center row being often reduced in number but enlarged; supplementary rows sometimes present; on distal half of pollex the median row usually clearly continues, while on distal part of dactyl the inner row so continues. In both pollex and dactyl enlarged subdistal teeth often combine with distal inner concavities to form a forceps-like tip to chela; sometimes only dactyl thus equipped. Outer distal part of pollex sometimes with a submarginal carina.

Minor Cheliped. Merus with dorsal and postero-ventral margins serrate, tuberculate or spinulate; antero-ventral margin smooth and rounded; postero-dorsal surface with scattered spinules; anterior and ventral surfaces smooth; dorso-distal margin either with slight irregularities, or a series of denticles. Carpus with subdorsal ridge absent or faintly represented by minute bumps, pits or both. Mano-pollex ridge short, never starting proximal to middle half of manus, often ending in proximal half of pollex. Dac-

tyl sometimes with a faint, broad, longitudinal ridge, rarely also with a shallow groove and faint second ridge above it; dactyl longer than manus, gape moderate to narrow, its edges finely and evenly serrate. Upper palmar setae in a close-set row or, rarely, in a tuft.

Ambulatories. Meri expanded, sometimes conspicuously, except 4th leg of *dussumieri*. Armature weak throughout subgenus, best developed on 2nd and 3rd legs. Merus with dorsal row of serrations largest; subdistal transverse row rudimentary; serrations on dorso-distal margin distinct; antero-ventral ridge strongest proximally, sometimes absent on 1st leg; antero-ventral serrations, if any, chiefly distal, often separated; postero-ventral ridge present only in distal three-quarters or less, strongest distally; serrations small, often absent, on 4th leg; posterior surface with scattered spinules, not tuberculate striae, on 1st, 2nd and 3rd legs; sometimes almost absent. Carpus with dorsal ridges present but often weak, with or without intervening rugosities; ridge on posterior carpus of 2nd and 3rd legs present or absent. Manus with ridges similar to those of carpus but weaker.

Gonopod. Various.

FEMALE

With the characteristics of females in the genus (p. 17). However, unlike females in some other subgenera, only in *acuta* is the carapace clearly more arched than in males; in all the suborbital crenellations are not larger and more extensive; setae are never thicker or more widely distributed on suborbital, pterygostomian, and adjacent lateral regions, except for several species in which a few setae are scattered on floor of orbit and below suborbital crenellations, these areas being naked in males.

The principal characteristics, in comparison with *Deltuca* males, are as follows:

Carapace. Granulate, the roughness extending partway over postero-lateral regions. In the position of postero-lateral striae in other subgenera (below and behind end of dorso-lateral margin) one or more tubercles are present in most species; these are arranged either in a cluster or in a vertical or horizontal line of granules, which do not arise from a raised line (stria).

Cheliped. Merus more expanded, its marginal and posterior armature stronger. No subdorsal ridge, such as is sometimes present in males, on outer carpus. Ridge on outer dactyl distinguishable more often than in male. Gape proximally sometimes slightly wider; often with 2 enlarged teeth, one near middle of pollex, the other opposing it on dactyl; in the same population these teeth may be altogether absent, or present on one or both sides. Other teeth in gape always small; except in *acuta* they are minute or absent proximal to enlarged teeth.

Ambulatories. Merus of at least 4th leg in two species with felt-like pile usually bordering at least postero-ventral margins. In addition to the stronger marginal and posterior armature on all meri characteristic of *Uca* females, in *Deltuca* a posterior subdorsal row of spinules or small tubercles is conspicuous, especially on 2nd and 3rd legs; in addition, a curved row of postero-dorsal subdistal spinules is pronounced in females, plus the usual series of minute denticles or irregularities found on the distal margin in both sexes. Carpus with posterior tuberculate ridge strongly developed at least on 2nd leg, usually also on 3rd; these ridges are far less subject to variation within populations and on the two sides of individuals than in males.

Size

Moderate to large.

Color

Display lightening absent. In fact, in contrast to their occasional brilliant white or yellow in some other subgenera, mature *Deltuca* males are often darker and duller than the females or young, tending to lack both bright colors and striking markings. In all only the lower distal part of the outer major manus is always distinctively colored, usually in contrast to adjacent regions; it ranges from purplish, red, or orange to ochraceous or yellow brown, while at least the tips of the chela are pale, usually clear white. An exception to the prevailing dullness of males is *urvillei*, the blue carapace of which is brightest in displaying individuals.

In contrast to these larger males, juvenile *Deltuca* of both sexes and sometimes mature females are both more contrastingly marked and vividly colored; white or bright blue spots occur in several species on a dark carapace and on the legs, especially the meri of the last pair; in several species red or yellow bands sometimes cross the carapace anteriorly, while red, yellow, blue, or white at times marks the orbits and other regions of the anterior aspect of the carapace, the maxillipeds, minor chelipeds in both sexes, and anterior face of the major merus; the ambulatories are sometimes solidly blue or red.

Several species have color peculiarities that are of diagnostic value in the field. In both sexes the adults of *rosea* are peculiarly translucent, its carapace and appendages variously pearly gray to deep rose. In *d. demani* the general hue is also characteristic; it is the only Indo-Pacific form, except for some populations of *chlorophthalmus*, in which the pervading hue

is dark crimson or purplish red; the remaining carapace reds all have a scarlet or orange red cast. In the Singapore and Sarawak areas, at least, it is possible promptly to distinguish *dussumieri* from *forcipata* by color alone; the *dussumieri*, except for large males, always show some markings of bright blue; the spot on the lower outer major manus of males is ochraceous yellow or brownish; in contrast, *forcipata* shows scarlet markings except in large males, which are distinguished by a distinct purplish patch on the outer manus; in addition, in *forcipata* (and rarely in its allopatric, *coarctata*) the eyestalks are sometimes vivid scarlet, a hue which in Singapore and Sarawak extends even to the globular eyes.

Post-megalopal stages, where known, are brilliant, solid colors—white, yellow, yellow with orange legs, scarlet, or blue.

SOCIAL BEHAVIOR

Waving display is less developed than in any other subgenus. In several species it consists only of a low, vertical, up-and-down motion (waving component 1), with no specializations except for the occasional grouping of waves into a series. In several other species there is a tendency to semi-oblique waves (component 3), the claw being directed slightly outward at a narrowly acute angle to the front of the body. Except in *demani*, the upper margin of the dactyl reaches scarcely or not at all above the tips of the eyes. Again with the exception of *demani*, the carapace is raised on the legs little or not at all. When the waves are in series, the crab often makes several waves of small or diminishing amplitude in quick succession. Sometimes the cheliped is raised in a series of upward jerks (component 2); the jerking waves are usually confined to displays of high intensity and are often interspersed with non-jerking waves that reach even a lower level. Except for the development of jerks, high intensity specializations of the wave itself are absent, including lateral displays of all kinds. Except again for *demani*, even leg-stretches (component 10), among the motions often associated with waving in the genus, are rare or absent. Display is sporadic and largely unpredictable in most species; it often occurs while the crabs are moving about, sometimes away from their burrows.

In no species is mating always or even usually preceded by waving, although no crabs have been seen to copulate that were not waving during the same period of low tide. The male, with or without display, approaches the female and copulates close to her burrow, after stroking her carapace as usual with his ambulatories before turning her into position. Tapping with the ambulatories and plucking motions with the minor cheliped have been observed less often.

Chimneys sometimes surround the burrows in about half the species in the subgenus, including all those in the superspecies *coarctata*. Most construction in some seems to be by mature females; in others chimneys seem to characterize, instead, juveniles of both sexes. In no observed population were they numerous.

The study of acoustic behavior and combat in this subgenus has scarcely begun. The presumably acoustic components so far observed on the surface and performed by single crabs are the major-merus-rub, minor-merus-rub, leg-wag, major-merus-drum, major-manus-drum, and leg-stamp (acoustic components 1, 2, 5, 7, 9, 11). Tape recordings resulting from the introduction of an intruder into a burrow have been secured in 2 species. Filmed components of ritualized combat include only the manus-rub, dactyl-slide, and heel-and-hollow (components 1, 6, 11).

RELATIONSHIPS

Comparisons with other groups will be found on p. 18 in connection with the discussion of the subgeneric dendrogram; bases for superspecies are given on p. 10. Within the subgenus the members of the superspecies *acuta* appear, at least in the simplicity

of their waving displays and low development of social activity, to be less specialized than the other species. In ability to withstand desiccation and in form and prevalence of waving activity, *demani* is clearly slightly in advance of the rest of the subgenus, although without outstanding morphological characteristics. On the other hand, the members of the superspecies *coarctata* show an almost perfect progression in gonopod structure; these organs range from the basic ocypodid form, showing large flanges and a prominent inner process, to the flangeless, projecting tube of *urvillei* which occurs in 3 other subgenera; none of the other morphological characters and waving displays in this superspecies show such progressive specialization, although the northern species, *arcuata*, has developed the increased depth characteristic of forms exposed to desiccation during hibernation.

NAME

Deltuca: From the Greek capital letter, *delta*, with its successive derivations of triangle and river delta. The proposed name, therefore, signifies "fiddlers of the delta," in reference to the usual habitat of all these species of *Uca*.

1. *UCA (DELTUCA) [ACUTA] ACUTA* (STIMPSON, 1858)

(Central Indo-Pacific)

PLATE 1.
FIGURES 61 *A, B*; 90 *A, B*; 98.

MAP 2.
TABLES 8, 10, 12, 19, 20.

INTRODUCTION

A northern and a southern form are here viewed as subspecies of *Uca acuta*; one ranges along the coast of China, while the other is known from Singapore and Borneo. Both are mud-livers found near river mouths on banks and flats. Males of some populations in both subspecies have the major manus red, while the carapace in both sexes of the southern form is strikingly reticulated with yellow and black. This form, *a. rhizophorae*, is also distinguished by having the simplest waving display known in the genus. The entire performance consists only of fast, slight lifts of the flexed cheliped, repeated in series and never reaching above the eye.

The southern form was rare near Singapore, its northern limit, according to both Tweedie's experience and my own. In contrast, it was the dominant ocypodid near the mouth of the Kuching River, Sarawak, where the small crabs covered some of the muddy banks and adjacent flats by the thousand. Farther upriver they were replaced by *U. forcipata*, there being a wide area of intermingling.

The new material here referred to *acuta acuta* was all collected in Hong Kong, not far from Macao, the type locality for Stimpson's *Gelasimus acutus*. From a distance these fiddlers resemble young *arcuata*, having similar dark carapaces and red claws. Stimpson's types of *acutus* were unfortunately among those lost in the Chicago fire, along with others from the region. The consequent confusion has been striking, even in a group notable for its taxonomic uncertainties. Specimens I have examined which were published under the name of *acutus*, all taken farther to the south and west, proved to be principally *rosea* and *forcipata*. On the other hand, specimens from China, clearly belonging to the same species as the present examples from Hong Kong, are labeled variously *demani, arcuata,* and *manii*.

MORPHOLOGY

Diagnosis

A single, long groove running laterally almost entire length of major dactyl, the subdorsal groove being short; chela tip forceps-like, because of an opposable series of enlarged tubercles, but not strikingly so. Gonopod with anterior flange large, posterior rudimentary; inner process minute, scarcely reaching base of flange. Floor of orbit without tubercles, although occasionally with a few, minute, irregular granules and always with a prominent blunt ridge or mound on inner third. Carapace with fronto-orbital margin only slightly oblique, almost straight in comparison with that of *rosea*; antero-lateral margins well developed.

Description

With the characteristics of the subgenus *Deltuca* (p. 21).

MALE

Carapace. Frontal groove wide or narrow, its sides parallel or slightly convergent. Fronto-orbital margins slightly oblique; antero-lateral angles acute to varying degrees, moderately produced; antero-lateral margins short but definite, finely or not at all serrate, moderately converging, then turning sharply to form strongly converging dorso-lateral margins which are beaded or smooth. Vertical lateral margin extending almost to dorsal surface, beaded throughout except, in some individuals, in upper portion. Eyebrow traceable through more than half of orbit; upper margin strong, and either finely beaded or smooth; lower margin indistinct, granulate. Suborbital margin less sinuous than usual in *Deltuca*; crenellations present on outer half of orbit only, separate, broad and spatulate, not continuing externally beyond angle. Tubercles on floor of orbit absent or, rarely, barely traceable as irregular fine granules; a blunt ridge or mound running up to one-third width of orbit on its floor, at inner side behind margin.

Major Cheliped. Merus with a row of tubercles, slightly enlarged distally on antero-dorsal margin; distally with one or two short, irregular rows of tubercles on anterior surface; latter surface with other small tubercles, especially near ventral margin, variable in distribution and size; ventral row of tubercles single or compound. Tubercles of outer manus relatively small compared with other *Deltuca*, notably large only near pollex base; smaller tubercles extending proximally onto pollex and dactyl. Inner surface of palm between ridges covered with small, low tubercles. Depression at pollex base a longitudinal trench rather than clearly trifid. Outer side of dactyl with subdorsal groove short, lined with tubercles.

In gape, a median pollex tooth and several proximal teeth on dactyl usually enlarged in brachychelous individuals; subdistally the edges of dactyl and pollex become suddenly closely apposed, the margins being straight, not concave; they are furnished with corresponding series of slightly enlarged teeth, the most proximal pair in the group largest, giving a distal forceps-like formation which is distinct but not extreme. Outer distal pollex without submarginal carina.

Minor Cheliped. Merus with serrations of dorsal margin well separated, those of postero-ventral less so. Carpal and dactyl ridges present or absent.

Ambulatories. Merus: Antero-ventral ridge on 1st leg absent proximally, represented distally by separated fine serrations; corresponding serrations on 2nd, 3rd, and 4th legs weak or moderate. *Carpus:* Dorsal ridges weak or moderate; posterior ridges present or absent.

Gonopod. Inner process small, scarcely projecting onto horny surface of flange base; anterior flange large, curving moderately forward, its breadth greater than diameter of pore; posterior flange very small; thumb short, subdistal, barely reaching base of flange.

FEMALE

Gonopore externally with a raised rim.

Measurements (in mm)

	Length	Breadth	Propodus	Dactyl
U. acuta rhizophorae				
(Singapore)				
Largest male	11.0	19.0	36.0	26.0
Moderate male	10.0	16.0	26.0	20.0
Larger female				
(ovigerous)	9.0	14.0	–	–
Smaller female				
(ovigerous)	8.5	14.0	–	–
U. acuta acuta				
Largest male (Amoy)	12.0	22.0	34.0	23.0
Neotype (Hong Kong)	9.0	16.0	22.5	13.5
Female (Hong Kong)	8.5	15.0	–	–

Morphological Comparison and Comment

U. acuta differs from all other Indo-Pacific narrow-fronts except *urvillei* in the minute size of the gonopod's inner process; from its closest relation, *rosea*,

as well as from *dussumieri* in having one, not two, long lateral furrows on major dactyl; from *demani*, *coarctata*, and *urvillei* by the absence of a row of tubercles on orbit's floor; from *arcuata* in the combination of a large anterior with a minute posterior flange on gonopod and in the much smaller size of adults, young *arcuata* having typically short fingers on the major cheliped; from *forcipata* and *coarctata* by the well-developed flanges on gonopod; from narrow-fronted crabs of other subgenera in having long grooves on both dactyl and pollex. Armature is weaker than in *rosea*.

The subspecies *acuta rhizophorae* stands out among its sympatric associates through a combination of its small size, reticulated carapace, and slender claw, with a series of slightly enlarged predistal teeth, diminishing in size, in both pollex and dactyl.

U. acuta acuta differs from its sometimes close sympatric, young *arcuata*, as follows. First, the gonopod's anterior, not posterior, flange is large even in young specimens, while the posterior flange is virtually absent. Second, the ventral margin of the 4th merus is almost straight, especially in the distal half, not distinctly convex as in *arcuata*. Third, the manus is more slender and the fingers do not have the juvenile shortness of *arcuata* of similar size. From *dussumieri*, the 3rd and last species of *Deltuca* now thought to occur in north China, *a. acuta* differs in having only one, not two, distinct long grooves on the dactyl, as well as in the obvious differences in the gonopod, and in the broad ambulatory meri. Large, leptochelous examples show a slight tendency for the subdorsal groove on the dactyl to extend in extremely shallow, shadowy form more distally, always just below the dorsal margin; in each of these examples the series of enlarged subdistal tubercles of both dactyl and pollex remain in the otherwise edentulous gape; the resemblance to a forceps is therefore far more striking, as in many *a. rhizophorae*, than in the brachychelous examples, which are always well provided with gape tubercles throughout; these opposed subdistal tubercles serve as a convenient key for avoiding preliminary confusion of *a. acuta* with *dussumieri* in specimens with the trace of a second dactyl groove described above.

Color

Display lightening absent; carapace sometimes conspicuously marbled with light and dark; major manus white to red.

SOCIAL BEHAVIOR

Information on waving display, precopulatory behavior, and acoustic behavior is available for *a. rhizophorae* only (p. 27).

RANGE

Singapore, northwest Borneo, and coast of mainland China north to Fu-chou [Foochow]. No records between Borneo and Hong Kong.

BIOTOPES

Muddy banks, both steep and shelving, near mouths of streams and rivers; also on protected flats in deltas; always near vegetation. (Biotopes 12, 13.)

SYMPATRIC ASSOCIATES

In the tropics its usual associates are *forcipata* and, especially, *dussumieri*. In Hong Kong it occurs sympatrically with young *arcuata*.

MATERIAL RESULTING FROM FIELD WORK

(The complete list of specimens examined is given in Appendix A, p. 592.)

Observations and Collections. Singapore, Sarawak: near Santobong. Hong Kong.

Film. Singapore.

TYPE MATERIAL AND REFERENCE

Uca acuta acuta (Stimpson, 1858)

Stimpson's type of *Gelasimus acutus* is not extant. Type-locality: Macao.

The present material from Hong Kong is referred to this species, and one of the series selected as a neotype, for the following reasons. The virtually straight sides of the front are notable, as is the mound on the floor of the orbit, these characters being the most distinctive mentioned in the type description (Stimpson 1858, 1907). The rather short dactyl of the holotype is in agreement with its small size. To Stimpson, unacquainted with a wide variety of Indo-Pacific *Deltuca*, the moderately produced antero-lateral angles and moderately converging sides or dorso-lateral margins, or both, appeared extreme, as described; so they are in comparison with those of *vocans*, *chlorophthalmus*, and *lactea*, the species familiar to him. The proximity of Hong Kong to Macao, the type-locality, prompted the selection of the neotype from Hong Kong material.

NEOTYPE. In Smithsonian Institution, National Museum of Natural History, Washington. Cat. no. 137665. Hong Kong: Kowloon (Castle Peak area). Male. Measurements on p. 26.

Uca acuta rhizophorae Tweedie, 1950

HOLOTYPE. Male. Sarawak, near Santobong. Length 17.2 mm. Described as *Uca rhizophorae*. (BM !)

Uca (Deltuca) [acuta] acuta rhizophorae Tweedie, 1950

(Singapore and Sarawak)

MORPHOLOGY

With the characteristics of the species.

Groove of front with sides convergent distally, narrower than in *a. acuta*. Suborbital granules altogether absent. Carapace sides little converging. Major cheliped with fingers tending toward leptochelous form and manus tubercles externally small for group. Armature in general weaker than in *a. acuta*.

Color

Carapace in both sexes dark, strikingly marbled with a paler tint, ranging from olive green to yellow or light brown, especially antero-dorsally. Major manus white in Singapore but bright orange red in large males in Sarawak; pollex brown, dactyl white in both localities.

SOCIAL BEHAVIOR

Waving Display

Wave vertical, low, the dactyl tip maximally reaching distal end of eye; cheliped held obliquely at highest reach, so that mano-carpal joint is much lower than tip of chela; cheliped not touching ground between waves in a series. Jerks absent. Body scarcely or not at all raised. Walking frequent both during waving and between series. Waves very fast, at rate of 3 to 6 waves per second; in series of 4 to 7 waves; often 6 to 8 series with only several seconds' pause between. (Waving component: 1; timing elements: Table 19, p. 656.)

Precopulatory Behavior and Copulation

Three apparently complete copulations were seen on surface, close to female's burrow. One was preceded by waving during walking; the other two were prefaced by no waving whatever, although both males had been waving some minutes earlier. Before each of the three males turned a female into position, he stroked and patted her carapace with his ambulatories, but made no plucking motions. Several other approaches reached the stroking stage close to a female's burrow but then broke off, as the female stiffened her ambulatories on one side and thrust them into the air in a typical legs-out (agonistic component 14).

Acoustic Behavior

During field observations I believed the ground was forcefully touched at the end of each wave. Film analyses have since shown that the cheliped does not, in fact, touch the ground in any of the approximately 20 waves, in both threat and courtship, where the point is clear. Leg-wagging was both observed in the field and filmed (component 5).

Uca (Deltuca) [*acuta*] *acuta acuta* (Stimpson, 1858)

(Hong Kong north to Fu-chou [Foochow])

MORPHOLOGY

With the characteristics of the species.

Differs as follows from *a. rhizophorae*. Groove of front broader with sides parallel, not at all converging distally; sides of carapace converging more; carapace slightly more arched, the regions less well defined. Suborbital granules detectable, although few are usually clear. Major cheliped with fingers tending toward brachychelous, not leptochelous, form, although moderately leptochelous examples occur; tubercles on outer manus larger. Gonopod with tube and flanges slightly longer. Armature in general stronger than in *a. rhizophorae*.

Color

Non-displaying males: Major cheliped with at least manus bright red. (Observation casual.)

REFERENCES AND SYNONYMY

Uca (Deltuca) acuta (Stimpson, 1858)

TYPE DESCRIPTION. See under *U. (D.) acuta acuta*, below.

Uca (Deltuca) acuta rhizophorae Tweedie, 1950

Uca rhizophorae

TYPE DESCRIPTION. Tweedie, 1950.1: 357; Fig. 7 a-c. Borneo: Sarawak: "mangrove and nipah swamp by the river below Kuching near Santobong." (BM !)

  Tweedie, 1954: 118. (Distribution note only, type-locality.)

  Crane, 1957. Singapore; Sarawak. Preliminary classification of waving display.

Uca (Deltuca) acuta acuta (Stimpson, 1858)

Gelasimus acutus

TYPE DESCRIPTION. Stimpson, 1858: 99. Macao (off east coast of China).

  Stimpson, 1907: 105; Pl. 14, Fig. 3. (Description of holotype, slightly expanded and in English.)

Uca demani (not of Ortmann)

  Gee, 1925: 165 (part). China (Fukien): Amoy. (USNM !)

  Kellogg, 1928: 356. China (Fukien): Amoy. (USNM !)

Uca arcuata (not *Gelasimus arcuatus* de Haan)

  Shen, 1937.2: 309 (part). China (Fukien): Amoy. (USNM ! and BM !)

2. *UCA (DELTUCA) [ACUTA] ROSEA* (TWEEDIE, 1937)

(Tropical Indo-Pacific)

PLATE 2 *A-D.*
FIGURES 24 *A-C*; 25 *A-C*; 30 *A, B*; 32 *A-C*; 35 *A, B*;
 45 *J-LL*; 46 *A*; 47 *A, B*; 48 *A, B*; 61 *E*; 98.

MAP 2.
TABLES 8, 10, 12, 19, 20.

INTRODUCTION

On the western coast of Malaya this pinkish, fragile-looking fiddler lives in sheltered mud close to mangroves. Its populations occur both near stream mouths and, in delta country, in drainage ditches; the water is probably almost always of low salinity, although subject to the usual wide range. At Penang a vigorous population occurred in close proximity to similarly thriving groups of *forcipata* and *triangularis*. Its waving display, poorly known, is wholly vertical and includes jerks.

Ranging around most of the Bay of Bengal, this species is undoubtedly the western form of *acuta*. In the older collections it was identified variously as *acuta* and *manii* (a name here synonymized with *forcipata*).

MORPHOLOGY

Diagnosis

Two grooves running almost entire length of major dactyl; chela tip forceps-like, because of an opposable series of enlarged tubercles. Gonopod with anterior and posterior flanges both broad; inner process broad, well developed. Floor of orbit without a mound, tubercles or granules. Orbits strongly oblique. Antero-lateral carapace margins absent, the angles acute and produced; dorso-lateral margins strongly converging. Meri of ambulatories much enlarged.

Description

With the characteristics of the subgenus *Deltuca* (p. 21).

MALE

Carapace. Frontal groove wide, its sides subparallel. Fronto-orbital margins strongly oblique, antero-lateral angles strikingly produced and acute, scarcely directed forward; antero-lateral margins absent; dorso-lateral margins very strongly converging; margins from the angle anteriorly finely serrate, posteriorly beaded. Vertical lateral margin extending more than halfway to dorsal surface, beaded throughout. Cuticle notably thin, translucent, and fragile. Eyebrow traceable through more than half of orbit, both margins strong and tuberculate. No tubercles or granules on orbital floor, but a low, smooth, variable ridge distinguishable. Suborbital crenellations small throughout, separate only in outer half.

Major Cheliped. Merus with tuberculate antero-dorsal margin equipped distally with only several slightly enlarged, clustered tubercles, less developed than in any other species of *Deltuca*; ventral margin proximally with rugosities extending onto anterior surface; otherwise equipped with a single row of strong tubercles, starting anteriorly just beyond rugosities but continuing directly on margin. Tubercles on outer manus relatively small compared with most *Deltuca* but slightly larger than in *acuta* and *demani*; smaller tubercles extend variably far out on pollex and dactyl. Inner surface of palm between ridges covered with small, low granules. Depression at pollex base a longitudinal trench rather than clearly trifid. Outer side of dactyl with subdorsal groove long, paralleling lateral groove through about two-thirds of dactyl's length. In gape a median tubercle usually enlarged near middle of both pollex and dactyl in brachychelous individuals; distal forceps-like formation distinct. Outer distal pollex without submarginal carina.

Minor Cheliped. Merus with dorsal and postero-ventral margins serrate, the dorsal serrations separated. Carpal ridge present, faint. No ridges or groove on dactyl.

Ambulatories. Merus: All notably expanded. Antero-ventral ridge on 1st leg absent proximally; distal serrations, non-contiguous, on 1st, 2nd, and 3rd legs; absent on 4th and from postero-ventral ridge on 4th leg. *Carpus:* 1st and 2nd legs each with 2 dorsal ridges present, the anterior the stronger, surface between spinous; 3rd and 4th legs each with anterior ridge only; all ridges minutely serrate. Posterior ridge, finely serrate, present on 2nd and 3rd legs.

Gonopod. Inner process well developed, flat; both flanges broad, the anterior slightly the broader; both much wider than pore; thumb short, subdistal, reaching base of flange.

FEMALE

The shape of the carapace closely resembles the form of the male.

Measurements (in mm)

	Length	Breadth	Propodus	Dactyl
Largest male	16.0	27.5	41.0*	23.0
Holotype male (Tweedie)	15.0	25.6	45.0	29.0
Moderate male	12.0	20.5	30.0	19.0
Largest female	10.5	17.5	–	–
Ovigerous female	9.0	16.0	–	–

* Estimated projection, since pollex regenerated.

Morphological Comparison and Comment

Uca rosea differs from *acuta*, its closest relation, in having two furrows, not one, on the dactyl, a large posterior flange and inner process, strongly oblique orbits, and no mound or ridge on the orbit's floor.

It differs from *forcipata* in the large gonopod flanges, in having two grooves, not one, on the dactyl, and in the more strongly produced antero-lateral angles and more converging dorso-lateral margins; from *dussumieri* and *arcuata* in the strongly oblique orbits and absence of antero-lateral margins; from *demani* in having two grooves on the dactyl and no tubercles on the orbit's floor; from *coarctata* and *urvillei* in the latter two characters as well as in having flanges. It differs from narrow-fronted crabs of other subgenera in having long grooves on both dactyl and pollex combined with distinct dorso-lateral margins on the carapace.

Its differences from *acuta* appear to be unquestionably on the species level. Its range is not known to coincide at all with that of its nearest allopatric relation, *acuta rhizophorae*, except for two young specimens taken by Tweedie (1937) in Singapore and a few *a. rhizophorae* taken by each of us in other parts of the same island. A thorough comparison of their displays there, or possibly in Pontianak, would be rewarding, particularly if any populations could be found living in close sympatry.

Color

Both sexes, including rare, displaying males: Carapace and eyestalks usually pale turquoise-gray, almost translucent, giving an effect unlike that of any other *Uca*; carapace sometimes partly or entirely pink, but still retaining the striking translucence. The pink, when only partial, appears on epibranchial regions and sides of carapace. Major cheliped and ambulatories usually pink. Very young crabs bright yellow, sometimes orange, not at all translucent; in a subsequent stage the carapace is whitish, the legs pink.

SOCIAL BEHAVIOR

Waving Display

Only 2 high-intensity waves seen and filmed, both by a courting male at Penang. Wave vertical, low, the cheliped tip raised slightly above tip of eye. Jerks present, each wave being divided by unequal nodes, 5 or 6 on upstroke, 3 on downstroke; each jerk shows a slight regression of wave direction: those on way up lower cheliped slightly, while those on way down re-elevate it. Jerks were absent at low intensity and the wave was lower, not reaching base of eyestalks; jerks absent and wave lower also by the single fully displaying male observed when he was rapidly following the female between jerking waves. Body scarcely or not at all raised. Waving rate: the waves lasted 3.7 and 4.1 seconds respectively. (Waving components: 1, 2; timing elements: Table 19, p. 656.)

Acoustic Behavior

Leg-wagging (sound component 5). A male beside his own burrow was threatening a small male *triangularis*, moving from a lateral-stretch to a moderate high-rise (agonistic components 10, 13). The leg-wagging, occurring in the middle of the high-rise, certainly included the rubbing of meri against one another, and perhaps also the rubbing of the same segments against the vertical lateral margin in a leg-side-rub (component 6). Components filmed; no data on sound production.

RANGE

Northeast India; Burma; west coast of Malay Peninsula; Sumatra; Borneo.

BIOTOPES

Muddy stream banks, sloping or steep, near mouths and adjacent protected flats, always near vegetation and sometimes partly shaded. (Biotopes 12, 13.)

MATERIAL RESULTING FROM FIELD WORK

(The complete list of specimens examined is given in Appendix A, p. 592.)

Observations and collections. West coast of Malaya: Malacca; Negri Selambang: young on yellowish, clayey banks of drainage ditches emptying into river; Penang. August.

Film. Penang.

Type Material and Nomenclature

Gelasimus roseus Tweedie, 1937

HOLOTYPE. In British Museum (Natural History), London. Male. Malaya: Port Swettenham. Measurements on p. 30. (!)

References and Synonymy

Uca (*Deltuca*) *rosea* (Tweedie, 1937)

Gelasimus roseus

TYPE DESCRIPTION. Tweedie, 1937: 145. Malaya: Port Swettenham. Singapore: Jurong R. (BM !)

Gelasimus acutus (not of Stimpson)

de Man, 1887.1: 113. Mergui Arch. (Leiden !)
de Man, 1891: 21, 30 (part). Locality not given.

Taxonomy.

de Man, 1895: 573. Malaya: Penang. Sumatra: Atjeh. Borneo: Pontianak.

Uca acuta (not *Gelasimus acutus* Stimpson)

Lanchester, 1900.1: 753. Malacca.
Rathbun, 1909: 114 (part). Taxonomy, in con-nection with description of *U. manii*; no records of *rosea* from Siam included.

Rathbun, 1910: 322 (part). The preceding comments apply.

Uca manii

Pearse, 1932: 292. Ganges delta.
Pearse, 1936: 353. Ganges delta. (USNM !)
Chopra & Das, 1937: 422. Mergui Arch.

Uca rosea

Tweedie, 1954: 118. West coast of Malaya. Distribution.
Crane, 1957. Malaya: Penang. Preliminary classification of waving display.

3. *UCA (DELTUCA) DUSSUMIERI* (MILNE-EDWARDS, 1852)

(Indo-Pacific)

PLATES 2 *E-F*; 3.
FIGURES 5; 8 *A*; 9 *A*; 27 *A-C*; 34 *A*; 35 *C*, *D*; 36 *A*; 37 *A*, *B*; 38 *E-H*; 45 *A-DD*; 46 *B*; 52 *A-BB*; 54 *A-C*; 56 *A*; 60 *A*, *B*; 61 *F*, *G*, *I*; 72 *A-CC*; 74; 81 *C*; 92; 98.

MAP 18.
TABLES 8, 10, 12, 14, 19, 20, 23.

INTRODUCTION

From the point of view of a waiting human being, this bluish fiddler seems especially likely to sit indefinitely on the mud without giving any information whatever on the form of his waving display. It is true that when tides are low in the morning a flourishing population may display with energy. Yet even at high intensity these waves look like mere waggings of the claw, intermittent and unimpressive in comparison with the tireless, spacious circlings of *lactea* nearby.

U. dussumieri is probably the most abundant and certainly the most widely ranging species of its subgenus, being found near mangroves from eastern India to New Caledonia and north China. Its usual sympatric associate is a member of the superspecies *coarctata* and in any tropical swamp color differences usually distinguish most individuals, *dussumieri* being marked with blue, while *forcipata* or *coarctata* is to some degree patterned with red.

MORPHOLOGY

Diagnosis

Two distinct grooves running most of length of major dactyl. No tubercles on floor of orbit. Gonopod with anterior flange large, sometimes represented chiefly by a spinous tubercle; posterior flange small; inner process broad, non-spinous. Carapace with anterior margin almost straight; antero-lateral margins present but short and variable. Merus of last ambulatory in male notably slender, its dorsal margin straight; last merus in female bordered conspicuously with pile along ventral margin, except in northwest Australia. Eye not notably larger in diameter than adjacent eyestalk.

Description

With the characteristics of the subgenus *Deltuca* (p. 21).

MALE

Carapace. Frontal groove moderate to extremely narrow; fronto-orbital margins practically straight; antero-lateral angles acute but little produced; antero-lateral margins short but definite, finely serrate or beaded except in some individuals; dorso-lateral margins anteriorly smooth or finely beaded, posteriorly absent or very weak. Vertical lateral margin distinct and beaded in lower one-third to two-thirds, of variable height within subspecies. Eyebrow poorly marked; upper margin beaded; lower margin indicated by a row of fine granules, usually double internally, sometimes dying out halfway to antero-lateral angle, sometimes extending almost to the angle. Suborbital crenellations low but separate and truncate, the outer angle often slightly projecting, rectangular, entire. No ridges or mound on floor of orbit.

Major Cheliped. Merus with antero-dorsal margin marked by tubercles except proximally, in a single row, increasing in size regularly or abruptly near the distal end; antero-distal surface also with a few small tubercles; latter surface also with scattered granules concentrated chiefly near ventral margin, their number, size, and arrangement variable. Ventral margin with close-set tubercles, usually in an irregularly compound row. Inner surface of palm between ridges either finely granulate or covered with flat tubercles. Outer side of pollex base with or without a short submarginal furrow proximally, below the usual longitudinal furrow. Outer side of dactyl with subdorsal furrow long, paralleling lateral furrow through most of its length; dactyl dorsally and ventral margin of pollex each tuberculate in its proximal half or more, while their outer surfaces are somewhat granulate, at least proximally. In gape, an enlarged median or submedian tooth on the pollex is characteristic in brachychelous individuals. No notable forceps-like formation distally. Outer distal pollex without submarginal carina.

Minor Cheliped. Merus with a few spinules on dorsal margin; postero-ventral margin either serrate or with separated spinules. Carpal and dactyl ridges scarcely indicated or absent. Palmar setae in a short row.

Ambulatories. Merus: Little expanded; on 4th leg dorsal and ventral margins about parallel. Antero-ventral margins with ridges on 1st and 2nd virtually absent; a few minute serrations distally; 3rd with ridge and serrations weak; 4th with ridge moderate but serrations barely indicated or absent. Postero-ventral margins with ridges in distal two-thirds to three-quarters; small serrations present in at least distal half of 1st, 2nd, and 3rd legs, present or absent on 4th. *Carpus*: 1st leg dorsally with posterior ridge, 2nd with anterior and posterior, 3rd and 4th with anterior only; ridges slightly beaded or finely spinulous; dorsal surface between ridges on 2nd sometimes with fine tubercles; posterior ridge absent or present on 2nd or 3rd, sometimes on major only; faintly tuberculate or smooth.

Gonopod. Inner process broad and flat; anterior flange large, sometimes with a large, dark tubercle or spine representing its anterior edge; posterior to the spine the flange joins the tube only subdermally or nearly so, as a low, horny ridge; posterior ridge small; thumb short, practically distal.

FEMALE

Merus of last ambulatory broader than in male but more slender than that of other *Deltuca*, its dorsal margin being straight. Except in *d. capricornis*, the postero-ventral margin is covered with short, dense pile, usually firmly attached, although sometimes missing partly or wholly from one side; in *d. spinata* the serrate edge of the segment is wholly buried in pile; pile sometimes also present on third merus. Gape of chela with 1 pair of enlarged teeth present or absent, except in *d. spinata*, where it is always absent on both sides.

Morphological Comparison and Comment

Uca dussumieri differs notably from *acuta* and *forcipata* in having two grooves, not one, running almost full length of dactyl; from *acuta* and *rosea* in having greatly enlarged tubercles all over outer manus, instead of only near pollex base; from *rosea* in having the fronto-orbital border practically straight, usually with a distinct antero-lateral margin, instead of oblique with sides converging immediately behind antero-lateral angle; from *demani*, *coarctata*, and *urvillei* in the lack of tubercles on floor of orbit; from *arcuata* in the enlargement of the anterior, not posterior, gonopod flange and from *forcipata*, *coarctata*, and *urvillei* by having a large flange, instead of flanges vestigial to absent. Differs from narrow-fronted *Uca* of other subgenera in having long grooves on both dactyl and pollex. Females of *dussumieri spinata* and *d. dussumieri* differ from those of all species except *acuta* in having a postero-ventral border of pile on merus of last ambulatory.

Along the southern and western margins of the

Philippines, in Tawi-Tawi, Joló, and northeast Palawan, the local *dussumieri* showed characters with tendencies intermediate between those of *d. dussumieri* and *d. spinata*. Notably, the anterior flange of the gonopod tended to be slightly produced anterodistally, as in *d. spinata*, while the pile usually present on the female's 3rd merus in *d. dussumieri* was lacking, as often in *d. spinata*. The frontal furrow in all remained typical of *d. dussumieri*. Specimens from Nokota Bay, on the southwest coast of Palawan, are close enough to *d. spinata* to be here so identified.

Some male *d. dussumieri* in the Philippines superficially resemble *acuta rhizophorae* in having a few separated tubercles in the gape of the major chela which are distally enlarged.

Although I have not been able to distinguish morphologically specimens here referred to *d. spinata* from temperate north China from those taken in tropical Thailand, Malaysia, and elsewhere, it seems highly unlikely that further material will not provide distinctions. We have no data on the appearance in life or behavior of members of temperate populations.

Color

The subspecies share four color characteristics. First, blue occurs at least in young crabs to varying degrees on carapace and ambulatories. Second, pale spots, ranging from turquoise to white, are typical of young crabs, excluding post-megalopal stages; their location and persistence are highly variable; spots sometimes remain in mature females, particularly on the merus of the 4th ambulatory, but not in mature males. Third, the lower major manus, at least, is never blue and never white; it ranges instead from pale yellow, orange, or red to dull ocher or cream. Finally, large individuals are always darker and duller than smaller ones; dullest of all are large males.

The differences among subspecies are minor. Blue is most prevalent and persistent in *spinata*; spotting is most highly developed in *capricornis*. The manus in the young is yellow, apricot, or orange in *capricornis* and *spinata*, but bright orange red in *dussumieri*. A few females of *dussumieri*, seen only in New Caledonia, had the carapace reddish and the minor chelipeds bright red; the rest of the population showed the usual coloration. Very small crabs in *spinata* are entirely brilliant cobalt blue, without spots, but white in *capricornis*; crabs of similarly small size in *dussumieri* were not identified in life.

The following description covers all subspecies.

Adult displaying male. Carapace dorsally without spots, although white sometimes persists on ptery-gostomian or other regions of frontal aspect. Eyestalks translucent gray. Major cheliped with merus rarely orange or ochraceous, inside or out; lower part of outer manus usually dull ochraceous, less often apricot to creamy, the color typically extending onto pollex; upper manus and dactyl creamy to white; tips of fingers white. Minor cheliped and ambulatories dark.

Adult female. Carapace and ambulatories as in male except as follows. The blue cast of the carapace is often pronounced even in large individuals; some of the pale spots of the young often persist into maturity, particularly the single large spot on the merus of each 4th ambulatory; 3rd maxilliped sometimes strikingly white, as may be other aspects of the frontal view; ambulatories sometimes bright blue as in young.

Young. Smallest crabs, to a length of about 6 mm, unspotted, brilliant cobalt blue to white. Later, large single pale spots, ranging from turquoise to white, are distributed singly on any number of the following general regions: front, gastric, hepatic, cardiac, branchial, orbital, suborbital, pterygostomian, 3rd maxillipeds, minor cheliped merus; anterior parts of merus, carpus and manus of first 3 ambulatories; posterior merus of 4th ambulatory. Eyestalks translucent yellowish or greenish. Major cheliped always with tips of fingers white, otherwise sometimes entirely yellow or cream; more often only lower manus is colored, whether yellow, apricot or orange red.

SOCIAL BEHAVIOR

Waving Display

Wave in all subspecies strictly vertical, held close against buccal region, with no tendency toward a semi-unflexed wave; wave low, the dactyl tip not reaching above eye and often, even at maximum intensity, not beyond eyestalk.

Depending on the subspecies, the waves are either in series, the waves of each series being clear of the ground, or else each wave is compound, consisting of several jerks, each successively higher, of increasing length, without regression of wave direction, and with a usually unbroken downstroke.

Body scarcely or not at all raised. At least in *d. dussumieri*, waving occurs during walking, both by aggressive wanderers and by courting males. Waving rate: single waves: about 2.5 to 8 per second; compound (jerking) waves: 0.5 to 2.5 per second. (Waving components: 1, 2, and, uncommonly and weakly, 10; timing elements: Table 19, p. 656.)

Remarks. There seems to be no question but that the jerking wave of *d. dussumieri* is derived from the serial wave of *d. capricornis*, and that the higher held serial wave of *d. spinata* is also a derivation of the *capricornis* form, but with the specialization in another direction, toward increased height through the slight raising of the body at high intensities. As is usual in the genus, at low intensities non-serial waves occur throughout all subspecies, while non-jerking display is then an additional characteristic of *d. dussumieri*.

Precopulatory Behavior

Male approaches female, with or without waving. Stroking of female's carapace with male ambulatories and plucking motions of her epibranchial regions with his minor cheliped seen in *d. capricornis*. No copulations observed.

Acoustic Behavior

In each subspecies, unreceptive females responded to courting males with high-rises (agonistic component 13), often accompanied by leg-wagging (sound component 5), with some of the details showing clearly in several film sequences. As usual there is no particular order for the rubbing by various ambulatories; the motions definitely include rubbing the tuberculate, posterior parts of the meri with the strong dorsal serrations of the next merus to the rear. Rarely leg-stamps (sound component 11) accompanied leg-wagging. Although in the rejection posture the 4th ambulatories are often crooked high up against the carapace, no signs of stridulating by or against these appendages have been noted.

A feeding male of *d. spinata* performed a typical major-manus-drum (sound component 9) in the middle of a strong lateral-stretch (agonistic component 10), directed toward a female digging nearby. A male *d. capricornis* performed a major-merus-rub (sound component 1), the antero-distal part of the merus rubbing an indeterminate point close below antero-lateral angle. Surface components filmed only. Underground sounds were recorded on tape in New Caledonia.

Combat

The following components were filmed: manus-rub and dactyl-slide (components 1, 6).

Chimney Construction

Chimneys erected by this species were found only in north-central New Guinea. Some, at least, were made by females although they were mounted by both sexes. Both construction and courtship activities occurred only for several days close to the time of new moon during both June and July, the months of observation.

RANGE

The species has been reported a number of times from western India, east Africa, and Madagascar, both in the literature and on labels attached to museum specimens. Each of these specimens that I have been able to examine has proved, with one exception, to be an unmistakable example of *U. urvillei*. The exception is a male in the Rijksmuseum van Natuurlijke Historie, Leiden, where it forms one of a group under no. 1243, labeled "Pollen v. Dam, Nossy Faly"; this is the material reported by Hoffmann (1874) from Madagascar; as mentioned by de Man (1891), he figures the *dussumieri* male; it is unmistakably an example of *dussumieri spinata*. Crosnier (1965, p. 110), working recently in Madagascar, was unable to locate other specimens of *dussumieri*. In the present contribution the locality record of the exceptional specimen is regarded as questionable.

The subspecies are distributed as follows. *U. dussumieri capricornis*: known only from Broome and the Monte Bello Is., Australia; several specimens reported in the literature and labeled in collections as *dussumieri* from northwest Australia proved to be examples of *U. coarctata flammula*. *U. dussumieri spinata*: Bay of Bengal, Sumatra, and other parts of Indonesia; northwest Palawan in the Philippines; Thailand; mainland of China. *U. dussumieri d.*: Sumatra and other parts of Indonesia; Territory of Papua and New Guinea; northeast Australia south to Moreton Bay; New Caledonia; Solomon Islands; Caroline Islands; Philippines; apparently rare in Nansei [Ryukyu] Islands. A male labeled "Toan Atoll, Tuamotu Archipelago" was almost certainly collected in the Philippines (USNM 93283, part; see p. 41).

BIOTOPES

Muddy stream banks and protected flats, near mouth, near mangroves. Not found as far upstream as members of the superspecies *coarctata*. An evidently rare occurrence in the Nansei (Ryukyu) Islands seems to be confined to Ishiyaki Island, the northern limit of mangroves (Macnae, 1968). (Biotope 12.)

SYMPATRIC ASSOCIATES

Although all three subspecies of *dussumieri* live at lower levels of the shore than corresponding representatives of the superspecies *coarctata*, mingling is frequent, and their close sympatric association is one of the most typical sights of Indo-Pacific mangrove communities. In the Philippines *demani* is another frequent associate; in Sarawak it is *acuta rhizophorae*, and along the west coast of Malaya *rosea*

sometimes intermingles. See also the discussion of the patchwork distribution of subspecies in the East Indies, under the corresponding headings concerning *forcipata* and *coarctata* (pp. 51 and 54).

MATERIAL RESULTING FROM FIELD WORK

(The complete list of specimens examined is given in Appendix A, p. 592.)

Observations and Collections. U. dussumieri capricornis. Australia: Broome. *U. dussumieri spinata.* West coast of Malaya; Sarawak; Singapore; Java, near Surabaja. *U. dussumieri dussumieri.* New Caledonia: neighborhood of Dumbéa River, near Nouméa; Philippines: Zamboanga; west coast, Gulf of Davao; New Guinea: north-central coast, near Madang.

Although a determined search was made for specimens of *dussumieri* around Semarang, the type-locality of the species, none was found. The explanation very likely lies in the change in habitat, due to building of fish ponds and cutting of mangroves. The only *Uca* apparent was *U. bellator*, which was represented by large and flourishing populations; normally this species occurs farther back from the shore than *dussumieri*, and often in much less saline water.

Films. Motion pictures were made at each of the major observation localities listed.

Sound Recordings. New Caledonia.

TYPE MATERIAL AND NOMENCLATURE

Uca dussumieri (Milne-Edwards, 1852)

LECTOTYPE. In Muséum National d'Histoire Naturelle, Paris. Listed by the museum as a "type non specifié." Male, selected by J. Crane as the lectotype, one of two males listed by Milne-Edwards (1852) in the type description of *Gelasimus dussumieri*. This specimen, from Samarang, has the following label: "*Gelasimus dussumieri*. M. Kunaraht   Samarang." Dried; relaxed for study. Measurements in mm: length 14; breadth 23.5; propodus 43. The specimen is in poor condition, but its specific characters are unmistakable, as are the identifying details of the gonopod and the narrow frontal groove, which are typical of *d. dussumieri*.

The second male listed by Milne-Edwards has the type-locality designated in the description and on the label as "Malabar." This specimen proved to be an example of *U. urvillei* (p. 60).

Uca dussumieri capricornis subsp. nov.

HOLOTYPE. In Smithsonian Institution, National Museum of Natural History, Washington. Male, cat. no.

137675. Western Australia. Collected by J. Crane. Measurements on this page.

Named in reference to its occurrence near the Tropic of Capricorn.

Uca dussumieri spinata subsp. nov.

HOLOTYPE. In Smithsonian Institution, National Museum of Natural History, Washington. Male, cat. no. 137677. Singapore. Collected by J. Crane. Measurements on this page.

Named in reference to the spinous projection on the gonopod.

Type of *Gelasimus dubius* Stimpson, 1858. Loo Choo (Ryukyu or Nansei) Islands. Type destroyed in Chicago fire. The brief description of the apparently unique type agrees well with the appearance of a young *dussumieri*, which has the orbits more oblique than do *arcuata* of similar size; they would probably appear strikingly oblique to Stimpson, whose acquaintance with Indo-Pacific narrow-fronts may well have been limited to *vocans* and his own small holotype of *acuta*. In Professor Sakai's collection were specimens referred to *Uca dubia* from the same islands. Examination showed them to be characteristic examples of the tropical subspecies of *dussumieri*, *d. dussumieri*, as were those similarly labeled *dubia* from the Palau Islands.

Other material referred to Stimpson's species from the north China mainland and elsewhere, deposited in several museums, turned out to be, variously, specimens of *acuta acuta*, *dussumieri spinata*, and *arcuata*. It is proposed here that *G. dubius* be considered a synonym of *U. d. dussumieri*.

Uca (Deltuca) dussumieri capricornis subsp. nov.

(Western Australia: Broome and Monte Bello Is.)

MORPHOLOGY

With the characteristics of the species.

Distal border of anterior flange of gonopod strongly concave, the anterior tip protracted. Front with marginal ridges well separated, leaving median groove distinct. Subdorsal groove of major dactyl shallow; supramarginal groove at pollex base well developed, extending along proximal part of pollex. Male with dorsal margin of last ambulatory merus definitely convex, slightly broader than in other subspecies. Female with last ambulatory merus not bordered ventrally with pile, except occasionally for a trace near proximal end. A pair of slightly enlarged teeth in female chela present or absent.

Measurements (in mm)

	Length	Breadth	Propodus	Dactyl
Holotype male	15.0	24.0	38.0	26.0
Largest male	18.0	27.0	45.0	29.5
Moderate female	13.0	18.0	–	–
Largest female	15.5	23.0	–	–

SOCIAL BEHAVIOR

Waving Display

With the characteristics of the species.

Jerks absent. Waves serial, with 3 to 11 waves in each series, the usual number being about 6. Often several or more series occur without more than a momentary pause. No trace of drumming on ground. Rate of waving from about 2.5 to about 8 waves per second.

Uca (Deltuca) dussumieri spinata subsp. nov.

(From northeast India through Malaya; sporadically in East Indies and Philippines to China)

MORPHOLOGY

With the characteristics of the species.

Distal border of anterior flange of gonopod very low proximally, close to canal, connecting with the greatly enlarged, tuberculate, distal strut. In large specimens the flattened dark tip of this flange is often the only distal structure of the gonopod easily visible without dissecting away setae and flesh. Front with submarginal ridges well separated, leaving a median groove narrow but distinct. Subdorsal groove of major dactyl well developed in proximal half or throughout. Female with last ambulatory merus bordered conspicuously with pile along postero-ventral margin; pile also sometimes similarly present on 3rd merus, but only distally. No enlarged teeth in either chela of female. A definite boss, sometimes tuberculate, on female carapace behind dorso-lateral margin.

Measurements (in mm)

	Length	Breadth	Propodus	Dactyl
Holotype male (Singapore)	20.0	33.0	61.5	47.0
Largest male (Foochow)	24.0	38.0	77.0	65.0
Moderate female, ovigerous (Sarawak)	15.0	24.5	–	–
Largest female, ovigerous (Foochow)	20.5	30.0	–	–

SOCIAL BEHAVIOR

Waving Display

With the characteristics of the species.

Jerks absent. Waves serial, with 3 to 8, usually 4 to 5 waves in each series without cheliped being brought close to ground; sometimes cheliped is slightly unflexed at the point of highest reach, and the waves, of very low amplitude, made there. At highest intensity the body is sometimes raised in front on the first two straightened ambulatories; sometimes, instead, the middle ambulatories on the major side are straightened, giving extra elevation to the waving cheliped. Waving rate: about 3.3 to 6.7 waves per second.

Uca (Deltuca) dussumieri dussumieri (Milne-Edwards, 1852)

(Indonesia; east Australia; New Guinea; Philippines; certain islands in west Pacific)

MORPHOLOGY

With the characteristics of the species.

Anterior flange of gonopod complete, the anterior strut not especially strengthened. Front with median groove absent or nearly so, its marginal ridges being almost or entirely fused in front's midline, but with the line of juncture always clearly marked. Subdorsal groove of major dactyl well developed in proximal half or throughout. Female with last ambulatory merus bordered conspicuously with pile along postero-ventral margin; pile usually similarly present on 3rd merus. A pair of enlarged teeth almost always present in both chelae of female. A definite boss, sometimes tuberculate, on female carapace behind dorso-lateral margin.

Measurements (in mm)

	Length	Breadth	Propodus	Dactyl
Moderate male (Zamboanga)	13.0	21.0	34.0	24.0
Largest male	17.0	27.0	58.0	48.5
Female (ovigerous and largest) (Zamboanga)	14.5	22.0	–	–
Smallest ovigerous female (Joló)	8.5	12.5	–	–
Mating pair (Madaum)				
male*	9.0			
female	10.5		–	–

* The male's cheliped was of juvenile form. He waved with high intensity, then mated.

SOCIAL BEHAVIOR

Waving Display

With the characteristics of the species.

Jerks present at high intensity, 2 to 4 on upstroke; almost always none, rarely 1, on downstroke. In series of up to 5 waves in threat, 8 in courtship. Waving rate of complete waves, including jerks: about 2.5 to less than 0.5 wave per second.

In Zamboanga jerks were not seen after 0940, or in the young at any time. Then the display was simply an up-and-down vertical wave, apparently indistinguishable from that of *coarctata coarctata.*

REFERENCES AND SYNONYMY

Uca (Deltuca) dussumieri (Milne-Edwards, 1852)

See under *U. (D.) dussumieri dussumieri*, below.

Uca (Deltuca) dussumieri capricornis subsp. nov.

Uca dussumieri capricornis

TYPE DESCRIPTION. P. 36. Western Australia: Broome.

Uca (Deltuca) dussumieri spinata subsp. nov.

Gelasimus dussumieri

de Man, 1887.1: 108. Mergui: Kisseraing I. Taxonomy.

Ortmann, 1894.2: 755 (part). Java; Singapore.
de Man, 1895: 576 (part). Malaya: Penang. Taxonomy. (Amsterdam !)
Alcock, 1900: 361 (part). Mergui, Andamans, Nicobars, Bimlipatam. Taxonomy.
de Man, 1902: 486 (part). East Indies. Taxonomy.
Tweedie, 1937: 141. Malaya: Singapore. (Raffles !)

Uca dussumieri

Nobili, 1899.3: 516 (part). Sarawak. Taxonomy.
Lanchester, 1900.1: 753. Color. Singapore.
Tesch, 1918: 39 (part). Labuan.
Gee, 1925: 165. China: Macao, China Bay.
Shen, 1940: 232. Hong Kong.
Tweedie, 1950.1: 356, fig. Labuan; Sarawak. Taxonomy; ecology. (Raffles !)

Crane, 1957: Singapore; Sarawak. Preliminary classification of waving display.

Uca dussumieri spinata

TYPE DESCRIPTION. P. 36. Singapore.

Uca (Deltuca) dussumieri dussumieri (Milne-Edwards, 1852)

Gelasimus dussumieri

TYPE DESCRIPTION. Milne-Edwards, 1852: 148 (part). Java: Semarang [Samarang]. (Paris !)

Milne-Edwards, A., 1873: 274. New Caledonia. Taxonomy. (Paris !)

Kingsley, 1880.1: 145. Taxonomy.

de Man, 1892: 306. Celebes. Taxonomy. (Leiden !)

Ortmann, 1894.2: 755 (part). Philippines: Mindanao.

de Man, 1895: 576 (part). East Indies: Pontianak. Taxonomy. (Leiden !)

Alcock, 1900: 361 (part). Indonesia: Bimlipatam. Taxonomy.

de Man, 1902: 486 (part). East Indies. Taxonomy. (Amsterdam !)

Gordon, 1934: 12. Indonesia. Taxonomy.

Gelasimus dubius

Stimpson, 1858: 99. Type description. (Type not extant.) Loo Choo Islands [Ryukyu or Nansei].

Kingsley, 1880.1: 141. Description after Stimpson.

Stimpson, 1907: 104, fig. Taxonomy, in English, of type material.

Sakai, 1936: 170. Palao Is. Color. (Yokohama !)

Sakai, 1939: 621, fig. Palao Is. Taxonomy.

Sakai, 1940: 47. Ryukyu Is. (Yokohama !)

Uca dussumieri

Nobili, 1899.3: 516 (part). Sumatra. Taxonomy. (Torino !)

Schenkel, 1902: 578. Celebes: Kema. Taxonomy.

Nobili, 1903.1: 22. Borneo: Samarinda. Taxonomy.

Grant & McCulloch, 1906: 20, 26. Australia: Port Curtis. Taxonomy; habitat.

Roux, 1917: 614. New Guinea: Merauke.

Estampador, 1937: 543. Philippines: Joló; Cebu; Negros; Negros Orientale.

Crane, 1957. Philippines. Preliminary classification of waving display.

Estampador, 1959: 100. Records from same islands as in 1937.

Macnae, 1966: 77, 79. Australia: Queensland (Thursday I. to Moreton B). Color; ecology.

Uca dubia

Miyake, 1936: 511. Ryukyu Is.: Miyara, mangrove swamp.

Miyake, 1938: 109. Palao Is.: Ngardok; Babelthaob.

Miyake, 1939: 222, 241. Palao Is.

4. *UCA (DELTUCA) DEMANI* ORTMANN, 1897

(Indo-Pacific: Philippines, Indonesia, Northwest Australia)

PLATE 4; 46 *F.*
FIGURES 61 *C, D, H*; 90 *E, F*; 98.
MAP 2.
TABLES 8, 10, 12, 19, 20.

INTRODUCTION

In the southern Philippines *demani* stands out among other *Deltuca* both in color, which is largely a dark red, and in waving display. For the first time in this arrangement of the species of *Uca*, the pattern includes a raising of the body high on the stretched legs—a character otherwise found only in subgenera showing more advanced social behavior. In another indication of social development, *demani* often waves without interruption in continuous series for minutes at a time. This characteristic I have found in no other *Deltuca*.

In Zamboanga *demani* was common rather high up on open banks of the town's tidal drainage ditches and even in the shade under houses raised on stilts. The few individuals seen outside of villages, near Davao, were also on higher ground, well back from the edge of a protected tidal flat.

The species as understood in this contribution ranges in a north-south band from northwest Australia to the northern Philippines.

In collections and in the literature the names most often interchanged with *demani* are *acuta, dussumieri, forcipata,* and *coarctata.* In addition, a species described from the southern Philippines, *zamboangana,* is a synonym of *demani.*

MORPHOLOGY

Diagnosis

A long row of tubercles on orbit's floor behind margin (very rarely, in large males only, and in the unique Australian specimen, vestigial). Major chela without a large projection, hook-like or triangular, on either pollex or dactyl; subdorsal groove on dactyl usually short, never more than half dactyl's length. Fronto-orbital margin practically straight; dorso-lateral margins strongly convergent. Gonopod with anterior flange wider than posterior, strongly curving away from pore; pore diameter wide; inner process distally broad, rounded. Female gonopore in Philippine specimens trapezoidal, with a small tubercle and a low rim.

Description

With the characteristics of the subgenus *Deltuca* (p. 21).

MALE
Carapace. Frontal groove wide, its sides subparallel; fronto-orbital margins practically straight; antero-lateral angles acute, produced, their upper external margin serrate; antero-lateral margins short to absent, curving gradually into dorso-lateral margins which converge strongly; anteriorly these are strongly ridged, sometimes faintly beaded, but die out before level of mid-cardiac region. Vertical lateral margin weak, finely or not at all beaded, dying out variably well below dorsal surface. Upper margin of eyebrow strong and beaded; lower marked by a single or irregularly double row of close-set granules, dying out in external half. Lower margin of orbit little rolled out and little sinuous, at least on minor side. Suborbital crenellations low but well marked and extending almost full width of orbit; lower external corner of orbit bluntly right-angled; external to this, margin is finely tuberculate all the way to the narrow channel leading to the orbit and dividing suborbital margin from lower edge of antero-lateral angle. A long row of tubercles, usually close-set and regular, along floor of orbit just behind marginal crenellations; in *d. demani* they are better developed than in any other species in the genus, ranging in number from about 18 to 22 or more; in exceptional, large specimens they are low and indistinct; in the other two subspecies they are less numerous and sometimes vestigial; in all they usually occupy about the middle three-fifths to two-thirds of orbital floor; in some individuals the row has one or two gaps; the row on the minor side is usually the better developed, and the tubercles almost always reach maximum development in immature individuals.

Major Cheliped. Merus with antero-dorsal margin proximally with rugosities, distally with an irregular row or small cluster of tubercles; ventral margin with a row, usually compound and irregular, of small, close-set tubercles. Tubercles rather small compared with some other *Deltuca*; similar tubercles extend, becoming progressively smaller, partway or entirely along the lengths of pollex and dactyl; young with tubercles better developed than large crabs. Oblique ridge inside palm low for the subgenus, its tubercles relatively small. Inner surface of palm covered with small tubercles, close-set, but so low that the surface appears almost smooth. Depression at pollex base clearly trifid, especially in large specimens. Outer side of dactyl with subdorsal groove short to moderate, never more than half dactyl's length; the dactyl's long, lateral groove relatively shallow and indistinct, sometimes practically absent, especially in larger specimens; at its strongest it extends clearly throughout most or all of the dactyl's length; sometimes, in contrast, it is necessary to blot off the preservative and manipulate the dactyl under the lamp to detect traces of the groove. Pollex groove also shallow; sometimes it is slightly more distinct than is long groove on dactyl. Tubercles of prehensile edges low and rounded except subdistally; here two series are slightly enlarged, but the fingers can meet only at their extreme tips.

Minor Cheliped. Merus with a few spinules on dorsal margin; postero-ventral margin with weak, separated serrations, better described as minute, oblique rugosities; carpal and dactyl ridges almost or completely absent. Palmar setae in a tuft.

Ambulatories. Merus: 1st, 2nd, and 3rd moderately expanded, 4th slightly. Antero-ventral ridge on 1st leg scarcely indicated, distally with several widely separated serrations; 2nd, 3rd, and 4th with ridges strong proximally, serrations indicated distally. Postero-ventral sides ridged in distal two-thirds of all; small serrations at least on distal half on 1st, 2nd, and 3rd; serrations practically absent on 4th. *Carpus*: Marginal and dorsal armature well developed; 1st, 2nd, and 3rd legs with antero-dorsal and postero-dorsal ridges, 4th with antero-dorsal; all ridges serrate and the spaces between them rugose. Posteriorly, 2nd leg on major side only with a weak tuberculate ridge and 1st, 2nd, and 3rd legs slightly rugose.

Gonopod. Inner process broad and flat; anterior flange larger than posterior, strongly curved. Pore larger than in other members of the subgenus, except some of the superspecies *coarctata*; overlap of components of distal part of tube are clearly visible, unlike the usual obscurity in *dussumieri* and the superspecies *acuta*. Thumb short, practically distal.

FEMALE

Gonopore roughly trapezoidal in shape, the base being external; a small tubercle on posterior margin; additionally with a low rim, sometimes interrupted laterally. Meri of posterior ambulatories broad. No enlarged teeth in chela. As in the male the great

length of the row of tubercles on orbital floor is diagnostic, but, in contrast to their relative strength in other species, the female's tubercles are sometimes smaller and fewer in number than those of males. Material is insufficient for a satisfactory comparison in either *d. demani* or *d. typhoni*. In the 3rd subspecies, *d. australiae*, the female is unknown.

Measurements (in mm) (both sexes)

	Length	Breadth	Propodus	Dactyl
demani australiae				
Holotype male (Broome)	15.0	25.0	40.0	27.0
demani typhoni				
Holotype male (Manila)	17.0	28.0	38.0	21.0
demani d. (Zamboanga)				
Largest male	25.5	41.0	73.0	49.0
Moderate male	18.0	28.0	43.0	29.0
Moderate female	13.0	21.5	–	–

Morphological Comparison and Comment

Uca demani differs most obviously from *dussumieri* and *forcipata*, and from (*Thalassuca*) *tetragonon* and (*Thalassuca*) *vocans*, in the presence of tubercles on the orbital floor. It differs from *coarctata* and (*Australuca*) *bellator* in the well developed flanges on gonopod, as well as in the absence of a large tooth on pollex or dactyl. The tubercles of the major manus are smaller than in other *Deltuca*, while the eye diameter, in contrast particularly to its large size in *forcipata* and *coarctata*, is little greater than that of the stalk. The trapezoidal shape of the female's gonopore is a useful diagnostic character; this structure, combined with the unusual extent and regularity of the tubercles on the orbit's floor, make the identification of even young females relatively easy.

The subspecies are differentiated by the relative development of tubercles on the orbit's floor and by the forms of the gonopod and gonopore.

Color

Known only for the subspecies *d. demani* (p. 42).

Social Behavior

Waving Display and Acoustic Behavior

Known only for the subspecies *d. demani* (p. 42).

Range

U. demani ranges roughly along a north-south axis from northwest Australia to the north Philippines. The scanty material available is distributed as follows: *d. australiae*: Broome, Australia; *d. demani*: eastern Java, Sumbawa, Celebes, and, in the south

Philippines, the island of Mindanao; *d. typhoni*: Philippines, from Negros to Luzon. Three specimens (USNM 93283, part), labeled "Toan Atoll, Tuamotu Archipelago," certainly were not collected there; mixed with them were specimens of two other forms found in the Philippines, *d. dussumieri* and *forcipata*; the name given on the original label was *dussumieri*.

BIOTOPES

(Known only for *d. demani* on Mindanao.)

Open flats of muddy sand to mud in delta lagoon; also muddy stream banks near mouth and protected flats, near vegetation. In general nearer high tide levels than low. (Biotopes 11, 12.)

SYMPATRIC ASSOCIATES

(Known only for *d. demani* on Mindanao.)

In rich localities *d. demani* occurs in close sympatry with *d. dussumieri* and *c. coarctata*.

MATERIAL RESULTING FROM FIELD WORK

(The complete list of specimens examined is given in Appendix A, p. 593.)

Observations and Collections. Philippine Islands: Mindanao: Zamboanga; west coast, Gulf of Davao: Malalag and the neighborhood of Sasa.

Film. Same.

TYPE MATERIAL AND NOMENCLATURE

Uca demani

LECTOTYPE. In Rijksmuseum van Natuurlijke Historie, Leiden; no. 1257. Name proposed by Ortmann, 1897, for 2 of de Man's specimens published under the name of *G. forcipatus*, of which this is one (de Man, 1891). This male in Leiden selected by J. Crane as the lectotype. It is labeled: "Celebes *Uca demani* (Ortmann) = *G. forcipatus* de Man." In alcohol. Excellent condition. See References and Synonymy, p. 42. (!)

Associated Material. The 2nd male referred to by Ortmann, was recorded from Sumbawa by de Man in 1892; it is deposited in the Zoölogisch Museum, Amsterdam. (!)

Uca demani australiae subsp. nov.

HOLOTYPE. In Smithsonian Institution, National Museum of Natural History, Washington. Cat. no. 64250. Broome, Australia. Collected by E. Mjoberg, identified by M. J. Rathbun on label as *Uca demani*. Measurements on p. 40. (!)

Named in reference to its southern distribution. (From the Latin noun *auster*, "the south.")

Uca demani typhoni subsp. nov.

HOLOTYPE. In Smithsonian Institution, National Museum of Natural History, Washington. Cat. no. 43041. Manila, Philippine Is. Collected by A. S. Pearse and identified by him on label as *Uca demani*. Measurements on p. 40. (!)

Named in reference to its habitat in a land often subject to typhoons. (From the Greek noun *typhon*, "a stormy wind"; "a whirlwind.")

Type Material of *Uca zamboangana* Rathbun, 1913. In Smithsonian Institution, National Museum of Natural History, Washington. 6 males from Zamboanga, Philippine Is. (Cat. no. 43207). There appears to be no reason to distinguish this material from the specimens from Celebes and Sumbawa. (!)

Uca (Deltuca) demani australiae subsp. nov.

(Northwest Australia)

MORPHOLOGY

(From holotype male, the unique specimen.)

With the characteristics of the species.

Tubercles on floor of orbit poorly developed, being indistinct and few in comparison with those of *d. demani* and *d. typhoni*; an accurate count is impossible, since some can be considered either tubercles or merely irregularities on the supporting ridge. Antero-lateral margin short but distinct on major side; corresponding region of minor side missing.

Gonopod with anterior flange very broad, strongly curved and high, its distal edge projecting beyond those of pore and posterior flange, its outer corner projecting most. Pore diameter less than breadth of anterior flange but wider than the tapering distal end of the narrow posterior flange which does not project beyond it. Inner process broadly rounded, ending proximal to pore, covering not more than posterior half of distal part of tube and none of anterior flange.

Uca (Deltuca) demani typhoni subsp. nov.

(Northern Philippines)

MORPHOLOGY

With the characteristics of the species.

Tubercles on floor of orbit well developed, but

usually less numerous than in *d. demani*; their range is from about 12 to about 18.

Gonopod very similar to that of *d. australiae*, with the anterior flange even wider and more strongly curving, and very high; its most distal section is sometimes more median than in *d. australiae* in which the outer angle projects most; its entire distal edge and angles are slightly rounded, almost truncate. Posterior flange narrow and tapering as in *d. australiae*. Inner process strongly bent away from both anterior process and pore region, neither of which it overlies.

Uca (Deltuca) demani demani Ortmann, 1897

(Indonesia; southern Philippines)

MORPHOLOGY

With the characteristics of the species.

Tubercles on floor of orbit almost always strong and numerous, ranging from about 18 to 22 or more.

Anterior flange wide and strongly curved as in other subspecies but diminishing rapidly in height away from pore; dissection is sometimes needed to reveal its full extent. Posterior flange similar but less curved. Pore exceptionally wide, almost equal to flange breadth. Inner process not strongly curved.

Color

Mature crabs of both sexes with carapace, at brightest, pale, dull red; sometimes dark with variable pale blotches anteriorly. Eyestalks rose to red. Major cheliped ranging from rose red through dull purplish red to true purple, often including fingers. Minor chelipeds and ambulatories rose red to pale violet. In immature individuals the prevailing color is rose red rather than purplish; half-grown young sometimes with carapace and legs translucent grayish, but major cheliped dark red. Individuals of all sizes in Zamboanga tended toward rose red rather than purple or violet tones, while the opposite was true in populations near Davao.

SOCIAL BEHAVIOR

Waving Display

Wave vertical, tending toward a semi-unflexed wave, the cheliped being held at an acute angle with respect to buccal region. Tips of chela extending above eye, and, at highest intensity, mano-carpal joint raised even higher. Jerks absent. Body raised throughout a series of waves (high intensity, near Davao, Philippines) or raised with each wave (medium intensity, Zamboanga, Philippines). At high intensity at least the 2nd or 3rd ambulatories, or both, on both sides, held off ground, but not in contact with each other or with other legs. No seriality apparent. Waving rate: slightly more than 1 per second. (Waving components: 1, 3, 10, 11; timing elements: Table 19, p. 656.)

Unlike *acuta rhizophorae*, *rosea*, and *dussumieri*, this subspecies partly unflexes the cheliped, raises it relatively high above the eyes at the peak of the wave, simultaneously elevates the body, and often displays from a raised point on the substrate. These characteristics all are frequent in species with more complex displays.

Acoustic Behavior

Minor-merus-rub (sound component 2): The rub is apparently against sides of carapace and outer suborbital region, as well as mutual rubbing with 1st ambulatory; latter also appears to rub side of carapace in a leg-side-rub (component 6). Leg wagging (component 5) questionably occurs of 2nd and 3rd ambulatories on major side, as a male waves when partly down burrow; this may be merely a sample of abortive leg-waving, which is part of the regular waving display. Components filmed; no data on sound production.

REFERENCES AND SYNONYMY

Uca (Deltuca) demani Ortmann, 1897

TYPE DESCRIPTION. See under *U. (D.) demani demani*, below.

Uca (Deltuca) demani australiae subsp. nov.

TYPE DESCRIPTION. P. 41.

Uca (Deltuca) demani typhoni subsp. nov.

TYPE DESCRIPTION. P. 41.

Uca (Deltuca) demani demani Ortmann, 1897

Gelasimus forcipatus (not of Adams & White)
    de Man, 1891: 32. Celebes. Taxonomy. One of

2 specimens for which Ortmann, 1897, below, proposed the name *demani*. (Leiden, no. 1257 !)

de Man, 1892: 306. Sumbawa: Baii van Bima. Taxonomy. The 2nd specimen for which Ortmann, 1897, proposed the name *demani*. (Amsterdam !)

Uca demani

TYPE DESCRIPTION. Ortmann, 1897: 349. New name as indicated above for 2 of de Man's specimens.

Ward, 1941: 3. Philippines: west coast, Gulf of Davao.

Uca zamboangana

Rathbun, 1913: 615; Pl. 74. Philippines: Zamboanga. Type description. (USNM 43307 !)

Estampador, 1937: 544. Philippines: Zamboanga. Record of type material.

Estampador, 1959: 102. Same comment as above.

Ward, 1941: 3. Philippines: west coast, Gulf of Davao.

Crane, 1957. Philippines: west coast, Gulf of Davao. Preliminary classification of waving display.

5. *UCA (DELTUCA) [COARCTATA] ARCUATA* (DE HAAN, 1835)

(Subtropical and warm temperate Indo-Pacific: China to Japan)

PLATE 5 *A-F.*   MAP 19.
FIGURES 8 *C*; 9 *C*; 61 *J*; 98.   TABLES 8, 10, 12, 20.

INTRODUCTION

Uca arcuata stands out among other *Deltuca* in its northern distribution, which lies quite clearly in a band that includes both mainland and islands from Hong Kong to Japan.

Toward its northern limit the crab hibernates. In northwest Taiwan I found its breeding season well under way in late April, while a month later in south Japan waving had not begun. This breeding pattern, showing local adaptations to temperature, is also characteristic of temperate zone fiddlers in the western Atlantic. The rhythm in *arcuata* seems to be partly endogenous, judging by the behavior of specimens flown to the crabberies in Trinidad, West Indies. After their arrival in early June they displayed regularly until October. In spite of frequent observations, no more waving was seen until the following spring, although the crabs fed actively and tended their burrows. Then, in mid-April, the several surviving males recommenced vigorous display and continued almost daily for several months. The last individual disappeared late the following fall, 18 months after capture. The great reduction, if not cessation, of waving during the months of a northern winter was obviously not connected with temperature, since Trinidad lies in the equatorial zone. Other possible factors, drought and rainfall, do not appear to be involved, since no seasonal shifts coincided with the crabs' behavior change; in Trinidad the dry season starts in January and ends in May.

The Japanese, with a combination of their usual enterprise and taste for seafood, have introduced canned *arcuata* to the local market. The label shows a large, red, unmistakable, male *Uca*—the only medium in which I have ever seen a fiddler commercially depicted. Fortunately for the welfare of the crabs, the industry is small and the biologists of Kyushu University are devoted conservationists. Only one additional fiddler, *U. tangeri*, seems to be regularly eaten but only in Europe. On waterfront streets in the Algarve of Portugal I have sampled the claws, freshly caught, boiled, and enjoyed promptly by the citizens as a snack or appetizer (p. 118).

In collections and the literature the names *arcuata*, *forcipata*, *dussumieri*, *acuta*, and their synonyms have been confused in the identification of mainland specimens. Examples from the tropics occasionally labeled *arcuata* always prove to be members of other species.

MORPHOLOGY

Diagnosis

A single groove running most of length of major dactyl. Gonopod with anterior flange rudimentary, posterior flange large; inner process broad, non-spinous, straight. No tubercles behind margin on floor of orbit; traces only of indistinct granules, rare in male, frequent in female. Carapace with anterior margin almost straight; antero-lateral margins always well developed. Merus of 4th ambulatory widely expanded in both sexes, its dorsal margin clearly convex. Female: gonopore oval, with a low rim except antero-internally, highest postero-internally.

Description

With the characteristics of the subgenus *Deltuca* (p. 21).

MALE

Carapace. Strongly arched, unlike the other species in the subgenus. Tip of front little bent downward, compared with other species in the subgenus; frontal groove moderate, its sides parallel; fronto-orbital margins practically straight; antero-lateral angles little produced, almost rectangular; antero-lateral margins long for the subgenus, scarcely or not at all converging, margins strong, equipped with minute tubercles; these margins curve inward to form the moderately converging dorso-lateral margins, well marked but scarcely or not at all beaded and ending at about level of mid-cardiac region. Vertical lateral margin distinct only in the basal third or a little more, not beaded. Eyebrow oblique instead of vertical, well marked, the upper margin with minute tubercles, the lower by a row of granules, dying out more than halfway to antero-lateral angle. Suborbital crenellations low, rounded, increasing moderately toward orbit's lower outer angle, which does not project; beyond this point two or more smaller crenellations, less often minute granules, sometimes fill in entire outer margin below lower edge of antero-lateral angle. Granules on floor of orbit usually absent, sometimes faintly indicated; a blunt ridge in same position, most prominent slightly internal to mid-point of floor, dying out externally.

Major Cheliped. Merus with antero-dorsal margin with tubercles present or absent in proximal half, distally in a regular row, all small, or enlarged only at segment's end where one or two are sometimes larger, or several small ones may be united longitudinally; close by on anterior distal surface is a small cluster of tubercles; ventral margin with an irregular row of moderate and small tubercles. Center of palm with indications of large tubercles, but so flat that the surface feels smooth and appears so except when appropriately lighted; surface also with a few minute granules. Depression at pollex base clearly trifid. Pollex and dactyl marginally and externally with large tubercles only proximally; remainder of length granulous; subdorsal furrow of dactyl short. In gape a median tooth usually enlarged near middle of pollex; one or more smaller enlarged teeth in dactyl's proximal half. Forceps-like tip of chela present, but not as pronounced as in *coarctata* or *forcipata.* No submarginal carina on outer distal part of pollex.

Minor Cheliped. Merus with a few spinules on dorsal margin; postero-ventral margin with a single row of separated tubercles; carpal ridge and dactyl groove and ridges absent or faintly traceable. Palmar setae in a row.

Ambulatories. Merus: All notably expanded. Antero-ventral ridge virtually absent on 1st leg, weak on 2nd, 3rd, and 4th, which are equipped only with a few minute, separated serrations, often practically absent. Postero-ventral ridges all notably strong, serrations weak or moderate. *Carpus:* 1st leg with postero-dorsal ridge only; 2nd with antero-dorsal and postero-dorsal; 3rd and 4th with antero-dorsal only; serrations well developed except on 4th; dorsal areas rugose. Posterior ridge faintly indicated on 2nd leg of minor side only.

Gonopod. Inner process broad, flat, straight; anterior flange narrow, posterior flange notably broad, forming a combination found in no other species and distinct even in very young crabs. Gonopore smaller in diameter than width of posterior flange, which is only slightly curved; overlapped margins of gonopore distinguishable. Thumb of moderate length, subdistal.

FEMALE

No bumps, clustered tubercles or tuberculate striae on carapace behind postero-dorsal margins. Minute granules usually detectable on orbital floor. Gape of minor cheliped with a pair of teeth slightly enlarged at least on one side. Merus of 4th ambulatory strikingly broad, its dorsal margin convex throughout; no pile along the segment's postero-ventral margin.

Gonopore oval, slightly longer than wide on the obliquely antero-posterior axis; rimmed except antero-internally; rim highest around postero-internal angle; sometimes with a 2nd, smaller rise, almost a tubercle, opposite at antero-outer angle. Gonopore depression in sternum shallow.

Measurements (in mm)

	Length	Breadth	Propodus	Dactyl	Depth of manus at dactyl base
Largest male	23.0	37.0	68.0	45.0	
Moderate male lepto-chelous	19.5	33.0	58.0	42.0	18.0
brachy-chelous	19.5	33.0	52.0	33.0	20.0
Largest female (ovigerous) (Taiwan)	17.0	28.0	–	–	–
Moderate ovigerous female (Amoy)	14.5	24.5	–	–	–

Morphological Comparison and Comment

U. arcuata differs from all the narrow-fronted *Uca* occupying in part the same range in having the posterior, not anterior, flange of the gonopod well developed. It differs additionally from them as follows: From *dussumieri spinata* in having the angle of the wider flange (regardless of its identity) low and indistinct, instead of produced and pointed; major dactyl with 1, not 2, long grooves; female gonopore with rim highest postero-internally, not antero-externally, if it projects at all; female chela always with a pair of enlarged teeth on at least one side, instead of being always without enlarged teeth; female merus of last ambulatory without a submarginal band of pile on lower, posterior margin; finally, in both sexes, merus of last ambulatory very broad, its dorsal margin convex throughout. From *dussumieri dussumieri* it differs as above except, first, that the differences between the distal angles of the wider gonopod flanges are not so strikingly different, since in *d. dussumieri* the angle is not produced; second, female *d. dussumieri,* as in *arcuata,* have a pair of enlarged teeth on at least one side; a helpful additional difference in both sexes lies in the submarginal ridges of the front, which are well separated in *arcuata* but close together or virtually confluent in *d. dussumieri.* From *acuta acuta* it differs in both sexes in having the antero-lateral margins of the carapace distinct instead of practically or wholly absent; in the absence in both sexes of a mound or blunt ridge on the orbit's floor; in the female's last ambulatory merus, which lacks submarginal pile on its ventral posterior surface; and, finally, in the female's gonopore, which does not lack a rim on its inner margin. From all species of the subgenus *Thalassuca* it differs in the gonopod tip proportions, broad and convex merus of the last ambulatory, weak crenellations on suborbital margin, and slender claw on minor cheliped.

From narrow-fronted Indo-Pacific *Uca* found in other regions, *arcuata* differs as follows: From its closest relations (*forcipata, coarctata,* and *urvillei*) in the presence of well-developed gonopod flanges, which also distinguish it from all members of the subgenus *Australuca*. The wide posterior flange also sets it off from the other members of the subgenus *Deltuca, rosea* and *demani,* not previously mentioned.

Color

Adults of both sexes with carapace dark brown or, sometimes, dull maroon; sometimes with a light or dark band across anterior part, between antero-lateral margins. Eyestalks always dark. Major manus dull yellowish, at least on lower outer portion; tips of fingers white. Young with varying amount of red; major manus always red, at least on outer side; sometimes entire cheliped red except for usual white fingertips; ambulatories often red, as are sometimes the orbits, buccal region, and minor cheliped.

The "Van Dyck Red reticulations" mentioned by Sakai (1938) were not seen by me in living or freshly dead specimens; in formalin, however, darkish reticulations appear on a lighter ground. The red may possibly be an effect of alcohol on fresh specimens, or perhaps a phase characteristic of the breeding season in Japan.

SOCIAL BEHAVIOR

Waving Display

Wave vertical, high for the subgenus; at highest intensity the entire chela, as far as lower part of pollex, reaches above eye. Jerks absent. Upstroke much longer than downstroke. Body raised in a leg-stretch with each wave; during highest reaches, only the minor side is raised so that the finger tips are pointed up; even at moderate intensity the upper margins of manus and dactyl, held horizontally rather than tilted, extend slightly above eye level. During one threat situation the wave combined with a warding off threat motion to give the effect of a semi-unflexed wave (see acoustic behavior, below). At low intensities, during feeding, the wave is of very small amplitude and much faster, with the cheliped not touching ground. No trace of seriality. Rate of regular, high intensity waving ranging from about 0.3 to 1.2 waves per second. The rapid, low-intensity waves were at the rate of between 1.5 and 3.3 to the second. (Waving components 1, 3 [weak], 10.)

Acoustic Behavior

A questionable example secured of a major-merus-rub during a low-intensity threat sequence, in which a wave was combined with a lateral threat as described above. The rub appears to be of the antero-dorsal part of the merus against an indeterminate point below the antero-lateral angle. (Acoustic component 1.) Filmed; no data on sound production.

Chimney Construction

Chimneys erected around burrows by both sexes, young as well as adults, in both Hong Kong and Taiwan. None were seen in Japan. In Hong Kong a young female 10 mm long built a chimney 75 mm high, bringing the material from a distance of 3 or 4 inches as usual; during the same period, she brought loads of substrate from time to time from inside the burrow, carrying them also to a distance before dropping them.

RANGE

East coast of Asia from the Gulf of Tonkin north to Korea; Hong Kong; Taiwan; Japan north to Fukuoka, on Kyushu; formerly extending at least to the Kii Peninsula on Honshu.

BIOTOPES

On mud flats and low banks near mouths of large rivers; immature specimens occur on tidal flats well to the sides of actual river mouths. (Biotopes 12, 13, 16.)

SYMPATRIC ASSOCIATES

Judging by collections from the north China mainland on deposit in museums, *arcuata*'s regular associates in the area are *dussumieri spinata* and *acuta acuta*. In Hong Kong I saw no mature *arcuata*; the young collected were mingled with mature *a. acuta* on a sheltered tidal flat; others were seen on a higher flat above the zone frequented by *vocans*; it was not possible to travel upriver, beyond the borders of the colony, to seek populations of large individuals. At Tamsui in Taiwan the young occurred in a location similar to that at Hong Kong; displaying and large adults were found only upstream from the river mouth, although still in strongly brackish water; here they were the only *Uca*. Adults lived in a similar situation close to Ariake Bay, Kyushu, Japan; no other species of *Deltuca* reaches as far north as Kyushu.

MATERIAL RESULTING FROM FIELD WORK

(The complete list of specimens examined is given in Appendix A, p. 594.)

Observations and Collections. Hong Kong: young only; May. Tamsui, Taiwan: young and displaying adults, including one ovigerous female; late April. Ariake Bay, Kyushu, Japan: no display seen, but weather rainy; late May. Living males and females brought from Ariake Bay to Trinidad, West Indies.

Film. Hong Kong; Tamsui.

TYPE MATERIAL AND NOMENCLATURE

Uca arcuata (de Haan, 1835)

LECTOTYPE. In Rijksmuseum van Natuurlijk Historie, Leiden. One male, selected by J. Crane from a group of 4 males, one female, numbered "243" and labeled: "*Uca arcuata* (de Haan) (= *Gelasimus arcuatus* de Hn. Type) v. Siebold 4 m., 1 f. Japon."

Associated Material. Above specimens, carapace lengths 14-22 mm; propodus of largest, 64 mm. Material in alcohol, in excellent condition. (!)

In the same collection: no. 244 with label "*Uca arcuata* de Haan Japon m. v. Siebold," consisting of a jar of medium to large specimens. No. 245 with label "*Uca arcuata* (de Haan) v. Siebold F. Japon": 1 female. Finally, dried specimens, designated "Types." (!)

Type Material of *Gelasimus brevipes* Milne-Edwards, 1852. Examined in Muséum d'Histoire Naturelle, Paris. Listed by Museum as "types non spécifiés," all dried. One 15 mm male relaxed for study. The relative size and form of the major cheliped agree with some of those of similar size of *arcuata* from Hong Kong and elsewhere, while the gonopod and other specific characters are indistinguishable from those of *arcuata*. *G. brevipes* is therefore here synonymized with *U. arcuata*. (!)

REFERENCES AND SYNONYMY

Uca (Deltuca) arcuata (de Haan, 1835)

Gelasimus arcuatus

TYPE DESCRIPTION. De Haan, 1835: 53; Pl. 7, Fig. 2. Japan. (Leiden !)
Milne-Edwards, 1852: 146. Taxonomy (brief).
Kingsley, 1880.1: 143. Taxonomy. Refs. to date.
Miers, 1880: 309. Taxonomy.
de Man, 1891: 28. Japan. Taxonomy.
Ortmann, 1894.2: 755. Japan.
Ono, 1959. Japan: Fukuoka, in estuary of Tatara River. Ecology.

Gelasimus brevipes

Milne-Edwards, 1852: 146. China. Type description. (Paris !)
Stimpson, 1907: 106. Comparison with *G. acutus*.
Gee, 1925: 165. China.

Gelasimus arcuatus var. forcipatus

Miers, 1880: 309. Taxonomy.

Uca arcuata

Parisi, 1918: 93. Japan: Wakanoura, Hondo, Tamsui on Formosa.

Kellogg, 1928: 354. China: Foochow.
Gordon, 1931: 528. China: Amoy.
Shen, 1932: 273; figs. N. China: Shantung Peninsula. Taxonomy (brief), color, habitat.
Sakai, 1934: 320. Japan: Nagasaki.
Takahasi, 1935: 78. Formosa. Habits.
Kamita, 1935: 61, 69. W. Korea: Yellow Sea.
Serène, 1937: 76. Indochina: Tonkin.
Shen, 1937.1: 184. North China: southern coast of Shantung Peninsula, salt marshes. Taxonomy.
Shen, 1937.2: 309. North China. Check list only.
Sakai, 1939: 619 (part). Japan: Kii Peninsula, Nagasaki, Ariake Bay.
Sakai, 1940: 42. Japan. Distribution.
Shen, 1940: 231. Hong Kong: Repulse Bay, Wong Chuk Hang, Kowloon (Shatin, Ts'ue, Wan).
Lin, 1949: 26. Formosa.
Miyake, 1961: 175. Japan: Kyushu: Ariake Bay. Local use, salted, for food.
Ono, 1965. Japan: Kyushu: Fukuoka, in Tatara-Umi River estuary. Ecology; feeding in relation to morphology of mouthparts.

6. *UCA (DELTUCA) [COARCTATA] FORCIPATA* (ADAMS & WHITE, 1848)

(Tropical west-central Indo-Pacific)

PLATES 5 *G-H*; 6 *A-D*; 46 *A*.   MAP 19.
FIGURES 26 *B*; 38 *A-D*; 61 *K*; 98.   TABLES 8, 10, 19, 20, 23.

INTRODUCTION

A crab characteristic of the backwaters of river deltas, *forcipata* extends upstream a little beyond the limits of the mangroves.

In living crabs the large, spherical eyes and their stalks are sometimes bright scarlet. This characteristic, along with other markings of red or purple, is helpful in distinguishing *forcipata* from *dussumieri spinata*, its frequent sympatric associate among the mangroves near shore; on *dussumieri* red and purplish shades are absent while blue is prevalent. An additional difference among females is *forcipata*'s habit of building chimneys around their burrows.

Prompt identification of living crabs is necessary, since so much remains to be learned of their social patterns, which, unlike those of most other subgenera, usually unfold at uncertain intervals that try human patience. When at last *forcipata* suddenly waves or undertakes a brief, lethargic courtship, the protagonists are usually hard to catch. Although slow to rouse to social actions they show the speed of all fiddlers in dropping down their burrows when alarmed; in adult *forcipata* the holes are often deep and lead through a tangle of mangrove roots. When there is time to focus, binoculars can contribute reliably to field identification. In particular they show whether the major dactyl has the single long groove of *forcipata* or the two grooves of *dussumieri*. Another character observable in the field is *forcipata*'s strong "forceps" at the end of the claw. This effect is formed by a large, hook-like, tuberculate projection into the gape that arises just before the tips of dactyl and pollex. We do not yet know how this elaborate specialization is used, although presumably it functions during certain components of combat, still to be observed.

MORPHOLOGY

Diagnosis

Major chela with a strong, tuberculate, hook-like projection distally on both dactyl and pollex, giving a forceps-like appearance. A single groove running most of length of major dactyl. Tubercles behind margin on orbit's floor altogether absent. Gonopod with flanges obsolescent, the posterior slightly more distinct than anterior, widest basally, practically absent distally; pore much wider than either flange where they adjoin it distally; inner process flattened but distally pointed. Carapace with orbits distinctly oblique and with its sides converging strongly posteriorly. Eyes notably larger in diameter than stalks even in preserved specimens. Merus of 4th ambulatory moderately expanded. Female gonopore triangular, partially rimmed, with the highest point at postero-inner angle.

Description

With the characteristics of the subgenus *Deltuca* (p. 21).

MALE

Carapace. Frontal groove of variable width, widest distally; never so narrow that its sides are almost in contact. Fronto-orbital margins oblique; antero-lateral angles moderately to strongly acute and produced; antero-lateral margins short to absent, being convergent immediately behind antero-lateral angle, and, where distinct at all, soon curving into the more strongly converging dorso-laterals; entire margin from tip of antero-lateral angle to end of dorso-laterals at level of mid-cardiac region, strongly ridged and with small projections; these range in decreasing size from small separate serrations near the angle to beading posteriorly. Vertical lateral margin well developed and beaded throughout, ending just below dorsal surface. Eyebrow vertical, well marked, its upper margin marked by small tubercles continuing full width of orbit; lower margin formed of minute granules in a single or compound row, usually in two distinct rows internally, the continuing row ending just external to middle of orbit. Suborbital crenellations low, rounded or flattened, close-set, increasing regularly in size toward orbit's outer angle; latter projects little or not at all; decreasing slightly, the crenellations continue around the angle along entire outer, lower edge of orbit. No trace of tubercles or granules on floor of orbit; blunt ridge in same locality scarcely developed.

Major Cheliped. Merus with a single row of tubercles on antero-dorsal margin, increasing somewhat in size toward distal end, where the last one or two tubercles are sometimes bicuspid; an irregular cluster of tubercles, highly variable in location and extent, on antero-distal surface; ventral margin with an irregular row of tubercles; adjacent, submarginally on posterior surface, are several other rows of smaller tubercles. Center of palm with minute, scattered granules. Depression at pollex base clearly trifid. Pollex and dactyl marginally and externally with tubercles and fine granules extending practically to their tips; included are pollex tubercles in a regular row along proximal third of ventral margin, and just below gape. Outer side of dactyl with subdorsal furrow very short. Outer side of pollex also with a short furrow proximally, below regular outer furrow and above ventral margin. In gape, an enlarged submedian tooth, often compound, on pollex; at least one enlarged tooth on dactyl's proximal half. Forceps-like tip of chela strongly developed, the proximal tooth of the forceps series in both pollex and dactyl notably enlarged. No submarginal carina on outer distal part of pollex.

Minor Cheliped. Merus with dorsal and postero-ventral margins with well developed, separated serrations; carpal ridge faintly discernible; dactyl groove and ridges absent. Palmar setae in a row.

Ambulatories. Merus: All notably expanded. Antero-ventral and postero-ventral ridges well developed in usual relative strengths; serrations on antero-ventrals relatively strong and close-set except on 1st leg where they are wide apart; serrations on postero-ventrals well separated on 1st, regular and scarcely separated on 2nd and 3rd, minute on 4th. *Carpus:* 1st, 2nd, and 3rd legs with well developed anterior and posterior ridges, 4th leg with anterior only; all ridges serrated; strongest ridges and serrations are the posterior on 1st and 2nd and anterior on 3rd; area between ridges definitely rugose. Posterior ridge strong and tuberculate on 2nd and 3rd legs, major and minor sides.

Gonopod. Inner process flat and moderately broad but with tip pointed; both flanges virtually absent distally beside pore; proximally they are traceable, the posterior being the broader; both are in the form of thickened struts. Pore very wide, its overlapped margins distinguishable.

FEMALE

No bumps, clustered tubercles or tuberculate striae on carapace behind postero-dorsal margins. No tubercles or granules on floor of orbit. Gape of minor cheliped with a pair of enlarged teeth at least on one side and usually on both. Merus of 4th ambulatory moderately enlarged, its dorsal margin clearly convex throughout; no pile along the segment's pos-

tero-ventral margin. Gonopore triangular, the angles being directed internally, posteriorly and antero-externally. Rim present except on each side of inner apex; highest part at posterior angle; a secondary peak at antero-outer angle. Gonopore depression in sternum shallow.

Measurements (in mm)

	Length	Breadth	Propodus	Dactyl
Largest male (Singapore)	20.0	33.0	54.0	37.0
Moderate male				
(Penang) leptochelous	14.0	23.0	34.0	23.0
(Penang) brachychelous	17.0	26.0	34.5	20.0
Largest female				
(Negri Sembilan)	15.0	24.0	–	–
Ovigerous female				
(Negri Sembilan)	13.0	20.0	–	–

Morphological Comparison and Comment

U. forcipata shares a complete forceps-like development on the chela only with *rosea*, its frequent sympatric associate on the west coast of Malaya. From *rosea*, as well as from *acuta*, *dussumieri*, and *demani*, it is easily distinguished by its gonopod, which has the flanges vestigial and the genital opening large; it also differs from *rosea* and *dussumieri* in having one, not two, long grooves on the major dactyl, and from *demani* in lacking tubercles on the orbit's floor. The meri of the last ambulatories are much broader than in *dussumieri*. It differs from 2 members of its superspecies, *coarctata* and *urvillei*, in having vestigial flanges on the gonopod, instead of none, and a corresponding absence of a tubular distal canal, and in having the inner process flat, not spinous; also, there is a complete absence of tubercles on the orbit's floor. The female's gonopore, which lacks a rim on its outer side, differs distinctly enough from those of the species named above. *U. forcipata* differs from the remaining member of the superspecies, *arcuata*, its neighbor to the north, in the wide genital opening combined with vestigial flanges, *arcuata* having a narrower opening and a broad posterior flange; *forcipata* also has narrower ambulatories than *arcuata*.

The length of the antero-lateral margins as usual in this group varies considerably within populations. In *forcipata* a slight geographical trend is apparent. The individuals with these margins best developed were taken in the northern part of the species range, closest to that of *arcuata*, on the west coast of Palawan; those with the shortest margins occurred in western Malaya.

Another somewhat variable characteristic is the width of the vestigial posterior flange on the gonopod; fortunately, the amount of variation is never such that *forcipata* can be confused with any other species.

Color

The outstanding color characteristics of this species are as follows: a sporadic distribution of bright red, except on large males; the occasional prevalence of bright red eyestalks and eyes; the reliable presence of a purplish spot on lower, outer part of major manus in mature males; the occasional marbling of the carapace with yellowish.

Adults of both sexes with carapace ranging from plain dark to solidly and finely mottled or marbled with yellow, cream, or pale yellowish gray, especially antero-dorsally. Mature females sometimes additionally with a wide border of bright scarlet or rusty red across full width of antero-dorsal part of carapace, while some or all of the following regions may also be partly or wholly bright red: pterygostomian regions, ischium of 3rd maxilliped, both chelipeds, all ambulatories; the latter appendages sometimes pale. Mature male with a spot or patch on lower outer manus, often extending onto pollex, ranging from clear purple to dull purplish red or bluish brown; upper part of manus usually pale brownish; chela yellowish or white, at least distally. Both sexes and all ages with eyestalks, or the globular eyes, or both, often bright red; whether this coloration is restricted to certain populations, or to individuals at certain seasons, is not known. Young crabs of both sexes have more red than do adult females, in varying and irregular amounts and positions, including often a red spot on major merus. Early post-megalopal stages are entirely red.

SOCIAL BEHAVIOR

Waving Display

Only a few individuals were observed while they were performing courtship displays at high intensity. Wave vertical; chela reaching at most scarcely beyond level of eye; 3 jerks on upstroke, 1 on downstroke. In between these jerking displays were single, short waves of small amplitude, characteristic of lower intensities; unlike their equivalents in the high intensity displays of *coarctata*, these short waves did not diminish successively in height; instead, they were very similar to those found in *arcuata* (p. 46). During low intensity and during a filmed example of moderately intense threat, the carapace was raised in a leg-stretch of intermediate height, while the cheliped was slightly raised and lowered, without jerks, 3 to 4 times in a series; during each series the manus did not touch the ground. Upstrokes in both kinds of waves longer than downstrokes. In the very young the cheliped appeared to bounce against the ground in very fast, short displays; since no films were secured showing whether the ground was actually hit, these juvenile

displays are not included in the table of acoustic components. Waving rate: high intensity, jerking waves, more than 0.5 wave per second; moderate intensity, simple waves, more than 1 wave per second. (Waving components: 1, 2, and, at low intensity, 11; timing elements: Table 19, p. 656.)

Precopulatory Behavior

A number of incomplete courtships were observed, most of them not preceded by waving display. Instead, each time a male simply stopped feeding and moved toward a nearby female, attempting to climb on her as usual as she sat beside her own burrow, and stroking her carapace with his ambulatories; no tapping or plucking motions were seen. Twice the female was on top of a chimney surrounding her hole. Courtships were not confined to females with chimneys.

Chimney Construction

Chimneys, about the height of mature females, occurred sporadically, and were apparently confined to females' burrows, although the evidence is scanty. One female while excavating placed some of the substrate on the chimney's already high rim, then carried the rest of the load 6 inches away.

RANGE

U. forcipata is the characteristic representative of its superspecies only in the western part of the Malay Peninsula, Singapore, Sarawak, and Thailand. South, east, and north of that area it occurs sporadically in a general range otherwise inhabited by *U. coarctata coarctata*; these localities are in Borneo, Sumatra, Java, Celebes, and the Philippines. Two specimens (USNM 93283, part), labeled "Tuan Atoll, Tuamotu Archipelago," were probably taken in the Philippines (see p. 41).

BIOTOPES

Muddy banks, sloping or steep, near mouths of streams and rivers, and associated protected flats and swamps. Always near vegetation and sometimes partly shaded. Characteristically the species extends farther upstream or in mangroves back from the shore than any other *Deltuca* within its range, with the exception of its consuperspecific *coarctata* and, sometimes, *rosea*; it is not, however, confined to those less saline regions. (Biotopes 12, 13, 14.)

SYMPATRIC ASSOCIATES

The most common associate of *forcipata* in Malaysia is *dussumieri spinata*; in Indonesia, beyond this heartland of its range, *d. spinata* also is found with it, instead of *d. dussumieri*, in the restricted areas where *forcipata* sometimes replaces *c. coarctata*. From personal field experience I know only one place, the neighborhood of Surabaja in Java, where both forms occur in this fashion, outside their regular ranges; there I saw and collected only *d. spinata* and *forcipata*; other localities have come to light in museum collections. I have not yet found evidence, except perhaps in the central Philippines, that both *forcipata* and *coarctata* occur in actual close sympatry; a field study of the populations in Negros Occidentalis, Samar, and Panay should yield most interesting results. The odd distribution pattern seems certainly to be connected with the geological history of the Sunda Shelf area and orogenic activity in adjacent regions.

U. forcipata is also sometimes closely sympatric with *rosea* on the west coast of the Malay Peninsula, as well as, in Sarawak and Singapore, with *acuta rhizophorae*.

MATERIAL RESULTING FROM FIELD WORK

(The complete list of specimens examined is given in Appendix A, p. 594.)

Observations and Collections. Malaysia: Penang and Negri Sembilan; observations but no collections in Sarawak. Singapore. Indonesia: Near Surabaja. Philippines: Palawan.

Film. Singapore.

TYPE MATERIAL AND NOMENCLATURE

Uca forcipata (Adams & White, 1848)

HOLOTYPE. In British Museum (Natural History), London. One male, labeled: "*Gelasimus forcipatus* Adams & White, 1848 Holotype Borneo 44.106 447b Note: Specimen 447a having disappeared I select this as holotype. (W. T. Calman 15/12/11)." Specimen dried, in good condition. Relaxed for study. Measurements in mm: length 10.5, breadth 17, propodus 23.

In spite of its rather small size, the specimen is characteristic of *forcipata* as described in this contribution. Although the locality "Borneo" could apply equally well either to *forcipata* or to *coarctata* .*coarctata*, the gonopod, orbital floor, tuberculation of the outer major manus, and armature of the major chela's gape all agree with specimens from Penang and Singapore rather than with populations of *coarctata coarctata* from Fiji, Australia, and parts of the Philippines where the two species do not adjoin.

Type material of *Uca manii* Rathbun, 1909: Lem Ngob, Thailand. The series deposited in both Copenhagen and the USNM (no. 39714) were examined, including a male of the latter species compared by Calman with the holotype of *forcipata*. I agree with Calman on the synonymy of the type series of *manii* with *forcipata*.

REFERENCES AND SYNONYMY

Uca (Deltuca) forcipata (Adams & White, 1848)

Gelasimus forcipatus

NAME ONLY. White, 1847: 36. Borneo.

TYPE DESCRIPTION. Adams & White, 1848: 50. (BM !)
? de Man, 1892: 306. Sumbawa. Taxonomy.
Milne-Edwards, 1852: 147. Copy of type description.
Kingsley, 1880.1: 142 (part). Taxonomy.

Gelasimus acutus (not of Stimpson, 1858)
de Man, 1891: 21, 30 (part); no locality given. Taxonomy.
de Man, 1892: 306. ? Sumatra & Celebes. Taxonomy.
de Man, 1895: 573. Part of queries. (Amsterdam !)

Lanchester, 1900.1: 753. Singapore. Habits. Color.

Uca acuta (not *Gelasimus acutus* of Stimpson)
Nobili, 1903.1: 21. Borneo: Samarinda. Taxonomy. (Torino !)
Maccagno, 1928: 19. Nobili specimen. Taxonomy.

Gelasimus rubripes
Estampador, 1937: 545 (part). Philippines: Zamboanga. Questionable locality; see Appendix A: Material Examined. (BM 84.31 !) No new material.

Uca manii
Rathbun, 1909: 114. Siam: Lem Ngob. Mangroves. Type description. (Copenhagen !; USNM !)
Rathbun, 1910: 322. Siam. Mangroves.
Tweedie, 1937: 143. Pleopod fig. Malaya.
Suvatti, 1938: 74. Siam: Lem Ngob.
Tweedie, 1950.1: 356. Labuan & Sarawak. Taxonomy and Ecology.
Crane, 1957. Singapore. Preliminary classification of waving display.

52

7. *UCA (DELTUCA) [COARCTATA] COARCTATA* (MILNE-EDWARDS, 1852)

(Tropical central and eastern Indo-Pacific)

PLATES 6 *E-H*; 7; 8; 46 *C-D*.
FIGURES 6, 9 *B*; 26 *A*; 29 *A*; 31 *A*; 36 *B*; 37 *C*; 38 *M-T*; 46 *C*; 62 *A-D*; 81 *D*; 82 *H*; 92; 98.
MAP 19.
TABLES 8, 10, 12, 14, 19, 20, 23.

INTRODUCTION

Wherever it occurs, this species is the most colorful crab in sight. It is usually marked conspicuously with scarlet and often with white and blue as well, the combination being set off by a black carapace.

In *coarctata*, a consuperspecific of *forcipata*, the chela is equipped only on the dactyl with a hook-like projection, the upper half of the forceps-like specialization found in complete form in *forcipata*. In the eastern subspecies of *coarctata* the dactyl's projection is especially strong, with the pollex end always clearly slender. This combination distinguishes the crab at once from *forcipata*, an occasional inhabitant of the same general region. As far as we yet know, these species are never actual sympatric associates. The point needs particular attention in the central Philippines, where the two species have been taken on the adjacent islands of Panay and Negros, the populations almost facing each other across a narrow strait.

MORPHOLOGY

Diagnosis

Major chela with a tuberculate, hook-like projection distally on dactyl only, strong or weak; any tuberculate projection on pollex is submedian, not distal. A single groove running most of length of major dactyl. On orbit's floor is a short line of tubercles or of granules which in one subspecies are sometimes missing. Gonopod with flanges altogether absent, the pore being at the end of a short to moderate tube; inner process spinous. Carapace with orbits distinctly oblique and with its sides converging strongly posteriorly. Eyes notably larger in diameter than stalks even in preserved specimens. Merus of 4th ambulatory moderately expanded. Female gonopore with a large tubercle on postero-outer margin.

Description

With the characteristics of the subgenus *Deltuca* (p. 21).

MALE

Carapace. Frontal groove of variable width and length; always narrow but never so narrow that its sides are almost in contact; sides always parallel or subparallel; fronto-orbital margins moderately to strongly oblique; antero-lateral angles moderately to strongly acute and produced; antero-lateral margins practically or completely absent, being convergent immediately behind antero-lateral angle, the dorso-lateral margins in effect starting at the angle; the margin has several fine serrations starting immediately behind angle; behind these the margin, never strongly ridged, is armed only by minute beads, well separated and decreasing in size posteriorly to its end at level of mid-cardiac region. Vertical lateral margin weak but distinct and finely beaded through about three-fourths of way to dorsal surface. Eyebrow very narrow in depth, vertical or even directed posteriorly to form part of orbit's ceiling; well marked, its upper margin with small tubercles continuing almost to antero-lateral angle; lower margin short, ending slightly but variably beyond middle of orbit; formed of minute granules, in a double or irregularly compound row internally, regularly single beyond. Suborbital crenellations low, flattened, and indistinct except externally, where they are somewhat larger, rounded and separated; orbit's outer angle projects slightly as a right or obtuse angle; several weak crenellations or irregularities present beyond angle along outer, lower edge of orbit. A short row of granules or small tubercles present or absent on floor of orbit just behind margin; their distribution, size, and number highly variable, both within and among populations; a short low, blunt ridge distinguishable in same region.

Major Cheliped. Merus with a row of tubercles on antero-dorsal margin all small except distally, where several are enlarged, the distal being sometimes bi- or tricuspid; two or more tubercles, highly variable in number, size and arrangement, on antero-distal surface; ventral margin with a row of small tubercles; adjacent, submarginally on posterior surface, are several other rows, less regular, of smaller tubercles, while the entire posterior surface is covered more densely than usual in the subgenus with small tubercles. Tubercles of outer manus smaller, especially in *c. coarctata*, than in the other species of the superspecies, and are scarcely larger at base of pollex than elsewhere on manus. Center of palm with minute scattered granules. Depression at pollex base clearly trifid. Outer side of dactyl with subdorsal furrow very short, indistinct. Pollex and dactyl externally practically smooth except proximally. In gape, pollex of brachychelous individuals with an enlarged submedian tooth; dactyl with one enlarged tooth proximal to middle. Forceps-like tip of chela always strongly developed on dactyl, sometimes on pollex as well; pollex, however, usually tapers to its upturned tip without any enlarged teeth subdistally. No submarginal carina on outer distal part of pollex.

Minor Cheliped. As in *forcipata* (p. 49), except that dactyl groove and ridges are sometimes distinguishable.

Ambulatories. Merus: As in *forcipata*, except that serrations are much smaller on both antero-ventral and postero-ventral margins. *Carpus*: 1st leg with postero-dorsal ridge only; 2nd with antero- and postero-dorsal; 3rd and 4th with antero-dorsal only; all ridges weakly serrate, the dorsal surfaces slightly rugose. Posterior ridge present on 2nd leg only, or absent.

Gonopod. Inner process spinous, either not appressed closely to tubular end of shaft or bent anteriorly and closely appressed; pore wide, its overlapped margins distinguishable; thumb of moderate length, subdistal or with tube extending well beyond it; flanges absent.

FEMALE

Carapace usually with a few slightly enlarged tubercles, more or less in linear formation, extending posteriorly between end of postero-dorsal margins and base of last ambulatory. A short row of tubercles always present on floor of orbit in one subspecies, represented if at all by granules in the other; their number and variability about as in males. Gape of minor cheliped with a pair of enlarged teeth at least on one side and usually on both. Merus of 4th ambulatory moderately enlarged, its dorsal margin clearly convex throughout; no pile along the segment's postero-ventral margin. Gonopore marked conspicuously with a large tubercle that takes up the entire postero-outer margin and overhangs the orifice; remainder of pore without sharp angles, either rimless or with a low rim at least anteriorly and antero-internally. Gonopore depression in sternum small, ranging from slight to moderate in depth.

Measurements (in mm)

The two subspecies usually differ in size, *c. coarctata* being smaller in the best known populations, in the Philippines and Fiji, although large individuals were taken in Cairn, northeast Australia. The following examples illustrate more typical differences:

	Length	Breadth	Propodus	Dactyl
coarctata coarctata				
Largest male (Zamboanga)	14.0	26.0	43.0	31.0
Moderate male (Zamboanga)	12.0	20.5	27.0	19.0
Large female (Zamboanga)	13.0	20.0	–	–
Large ovigerous female (Fiji)	11.0	17.0	–	–
Holotype female of *U. mearnsi* (see p. 55) (Davao)	14.2	21.2	–	–
coarctata flammula				
Large male (longest claw) (Broome)	21.0	34.5	69.0	54.0
Longest male (Broome)	23.5	36.5	67.0	49.0
Moderate male (holotype) (Darwin)	16.0	26.0	40.5	27.0
Largest female (Broome)	25.0	38.0	–	–
Moderate female (Darwin)	14.5	22.0	–	–

Morphological Comparison and Comment

Uca coarctata differs as follows from the species with which it associates sympatrically, or with which its range may sometimes coincide: from *dussumieri* and *demani* by the tubular gonopod, without flanges, and by the concave, hook-like structure on the dactyl; additionally from *dussumieri* by always having tubercles or at least granules on the orbital floor and one, not two, long grooves on the major dactyl: additionally from *demani* by the relatively poor development of tubercles on orbital floor; from *forcipata* by having the hook-like structure developed on the dactyl alone, by the absence of vestigial flanges on the gonopod and by the reliable presence of tubercles on the orbit's floor in all parts of the general range of both species where their distributions approach each other. The female's gonopore in *coarctata* differs from that of all other *Deltuca* in having a large tubercle on the postero-outer margin.

U. coarctata's shorter gonopod tube and shorter row of tubercles on orbital floor set it off from *urvillei* in the west; its lack of flanges distinguish it at once from the remaining members of *Deltuca—acuta, rosea,* and *arcuata*—all of which have other ranges, as well as from species of the subgenus *Thalassuca*.

Possible confusion with species of the subgenus *Australuca* is avoided through the shortness of the gonopod tube, the relatively poor development of tubercles on the orbital floor, the different shape of both front and minor chelae, the lack of a large tuberculate projection on the major pollex, and the presence of strong grooves on both dactyl and pollex.

The eyes, even in preservative, are notably large in comparison with the stalk, as in other members of the superspecies *coarctata* and in *rosea*.

Color

A vivid species, notable for striking markings that always include red and often one or more additional colors ranging from white to yellow and blue. Carapace basically black; lower major manus red. Aside from these two characteristics, variation is extensive among individuals in single populations. Unlike some other *Deltuca*, males, females, and immature crabs are similarly colored, with little tendency for the male to be less strikingly marked. Details of color ranges are given in the subspecies descriptions.

SOCIAL BEHAVIOR

Waving Display

Wave vertical, with some tendency toward a semi-unflexed wave. Upper edge of dactyl reaching at most, and rarely, scarcely above tip of eye. Jerks present on upstroke only and only at high intensities. The waves are then in series, a relatively high, jerking wave being followed by one or more lower waves of diminishing height, with jerks weak or absent. Up-strokes always longer and more variable in duration than downstrokes. At highest intensity the body is sometimes raised slightly in front on the ambulatories and tilted down posteriorly, through bending of the last pair of legs. Waving rate: 0.3 to 1.2 per second. (Waving components: 1, 2, 3 and, weakly, 10; timing elements: Table 19, p. 656.)

Precopulatory Behavior

Male approaches female, with or without waving. For the single observed exception, where, between normal copulations by her burrow, the female approached the male, see the section on unusual behavior, p. 506. Stroking of female's carapace by male ambulatories, and plucking motions from it by his minor cheliped, usual in both subspecies.

Chimney Construction

Occurrence of chimneys sporadic, most examples being very small, and their inhabitants juveniles of either sex measuring less than 10 mm in length. Only one chimney was seen in *c. flammula*, the Western Australian subspecies, and one in Fiji, the eastern

boundary of *c. coarctata*. In the Philippines, on the contrary, chimneys were fairly common.

Acoustic Behavior

Leg-wagging (component 5) is the only component noted at the surface. Films only. Underground sounds were recorded on tape in Fiji.

Combat

Manus-pushes and manus-rubbing (component 1) filmed.

RANGE

U. coarctata ranges from Sumatra to the Fiji Islands and from northwest Australia to the northern Philippines. It is absent from the Asian mainland. The two subspecies are distributed as follows. *U. c. coarctata*: Sumatra and other parts of Indonesia; Philippines; parts of western New Guinea; eastern New Guinea; northeast Australia south to Moreton Bay; New Caledonia; Fiji. *U. coarctata flammula*: northwest Australia and parts of western New Guinea. One male (USNM 19662), labeled "Tahiti," was almost certainly collected elsewhere. Two other locality records are undoubtedly erroneous: Odessa (see p. 55) and near Lissa, in the Adriatic (Stossich, 1877: 190).

BIOTOPES

As in *forcipata* (p. 50). (Biotopes 12, 13, 14.)

SYMPATRIC ASSOCIATES

In rich localities in the Philippines, *c. coarctata* occurs often in close sympatry with *d. dussumieri* and *d. demani*, although it occupies in general higher levels of the shore. Elsewhere it is either the only *Deltuca*, as in Fiji, or else all the large populations occupy stream banks in the more inland reaches of the mangroves, as does *c. flammula* near Darwin and Broome, where no other *Deltuca* is found. As usual in *Uca*, however, there are sympatric zones of coincidence close to shore, particularly with *d. dussumieri* or *d. capricornis*, depending on the subspecies of *coarctata* involved. For the possible close association of *c. coarctata* with *forcipata*, and their patchwork distribution in Indonesia and the Philippines, see under *forcipata*, p. 51. The known distribution of *c. coarctata* and *c. flammula* in the western part of New Guinea indicates a similar situation.

MATERIAL RESULTING FROM FIELD WORK

(The complete list of specimens examined is given in Appendix A, p. 594.)

Observations and Collections. U. c. coarctata: Australia: Queensland: Gladstone. New Caledonia: neighborhood of Dumbéa River, near Nouméa. Fiji: Viti Levu. Territory of Papua and New Guinea: near Madang. Philippines: Mindanao: Zamboanga; west coast, Gulf of Davao; Luzon: near Manila. *U. c. flammula*: Australia: Darwin; Broome.

Film. Fiji; Philippines: Zamboanga and neighborhood of Sasa. Australia: Darwin.

Sound Recordings. Fiji.

TYPE MATERIAL AND NOMENCLATURE

Uca coarctata (Milne-Edwards, 1852)

LECTOTYPE *and Associated Material*. In Muséum National d'Histoire Naturelle, Paris. One male, selected by J. Crane from a group listed by the museum as "types non specifiés," with the following label: "*Gelasimus coarctatus*, Edw. M. Nordmann. Odessa." The box contained 2 males and one female, all dried; they were relaxed for study. The individual designated the lectotype has the following measurements in mm: length 13; breadth 23; propodus 35; dactyl 21. The gonopod drawn (Fig. 9 *B*) is from the lectotype; the cheliped photographed (Pl. 7, *I* and *J*) is from the 2nd male, with a carapace measuring 11 mm. All 3 specimens are in poor condition, and at least one was apparently caught when newly molted. Some of the lectotype's legs are wired on. The locality given, Odessa, is clearly an impossible habitat for this or any other *Uca*, because of the early date at which the Black Sea region was blocked off from Tethys. Yet the specific characters of the species and even of the subspecies are intact, and, whatever the explanation of the label's error, it occurred before Milne-Edwards published his description.

The identity of the small female associated with the 2 males, and of a 2nd small female, legless, mounted upside down in another box, and similarly labeled, should remain unsettled.

Uca coarctata flammula subsp. nov.

HOLOTYPE. In Smithsonian Institution, National Museum of Natural History, Washington. Male, cat. no. 137676. Australia: Darwin. Collected by J. Crane. Measurements on p. 53.

Named in reference to its striking red markings. (From the Latin noun *flammula*, "a little flame.")

Type Material of *Uca rathbunae* Pearse, 1912.1. Apparently not extant. Two males (no. 43040) from Manila, Philippine Islands, the type-locality, are deposited in the Smithsonian Institution, National Museum of Natural History, Washington, collected and identified by Dr. Pearse and presented by him to the museum. I have found no trustworthy characters by which this Philippines form of the superspecies *coarctata* can be distinguished from examples of *c. coarctata* occurring elsewhere, either from examination of these and other specimens from the Philippines, or from the type description and figures. (!)

Holotype of *Uca mearnsi* Rathbun, 1913. In Smithsonian Institution, National Museum of Natural History, Washington. A large female, the unique specimen (no. 43383) from Davao, Philippine Islands. This specimen is only questionably referred to *U. coarctata*, since the gonopore is apparently distorted by a large spermatophore and the orbits are practically straight. In making the tentative identification, the sex and size of the specimen were taken into account in disregarding the unusual direction of the orbits. More material in an adequate size range will be needed for certainty in synonymizing *U. mearnsi*. Measurements on p. 53. (!)

Type Material of *Uca ischnodactylus* Nemec, 1939. In Field Museum, Chicago, from the Fiji Islands. Type not seen. Paratype (no. 99261) in Smithsonian Institution, Washington. (!) I am also familiar through field work and collections with the Fijian representative of the superspecies *coarctata*. There appears to be no reason for distinguishing it by any formal designation from more western populations of *coarctata coarctata*.

Uca (*Deltuca*) [*coarctata*] *coarctata coarctata* (Milne-Edwards, 1852)

(Indonesia and the Philippines south to New Caledonia and east to the Fiji Islands)

MORPHOLOGY

With the characteristics of the species.

Differs as follows from *coarctata flammula*. Carapace with orbits distinctly oblique. Tubercles always present on floor of orbit. Major dactyl with tuberculate, hook-like, distal projection large and well formed. Genital pore at end of a short tube, the inner process not bent anteriorly.

Color

In spite of complex patterns and great variability, the general impression given by this subspecies compresses to a dark crab with light spots, red claw, and scattered accents of red, white, blue, and sometimes yellow.

Mature crabs of both sexes are similar and differ little from the young, except that as usual in *Deltuca* the female tends to retain more spots longer than does the male; unlike *forcipata* and *dussumieri*, however, adult males are about as brightly colored as females and young. The most frequently occurring patterns are as follows.

Carapace antero-dorsally sometimes with a broad band across its entire width, ranging from shining white through pale yellow to orange or scarlet; sometimes a white spot occurs behind the band. More often there is no band, but only one or more spots, ranging from polished, pure white to cerulean blue; when at their maximum development one spot occurs on each large region of the carapace except the hepatic—namely, at base of antero-lateral angle and on gastric, branchial, cardiac, and intestinal regions. A frequent arrangement shows one spot each on gastric and cardiac regions plus the branchial pair, giving a diamond formation. Any or all of the following anterior regions of the carapace may be white, blue, or scarlet: around antero-lateral angle, orbits, suborbital and pterygostomian regions, and 3rd maxillipeds. Major and minor chelipeds sometimes both largely orange red, inside and out; outer, lower, major manus and at least base of pollex always bright scarlet; upper manus, dactyl, and distal pollex usually polished white. Ambulatories usually brownish, sometimes bluish; white spots present or absent in a variety of combinations, but frequently with a single, large spot on posterior side of each 4th merus.

Post-megalopal young, up to a length of about 6 mm, in Zamboanga translucent yellow with orange legs; since a male was reared to the point where mature characters were developed in the Trinidad crabbery, there was no question of identity, in spite of the wealth of related species in the Philippines; young of similar color filmed among adults of the species near Davao. In Fiji corresponding stages were entirely white; identification in the latter locality, where the young were locally abundant, was also reliable since the only other narrow-fronts reaching the islands are *Thalassuca*; at corresponding sizes their fronts and orbital margins are unmistakable.

SOCIAL BEHAVIOR

Waving Display

During high-intensity courtship, in both the Philippines and in Fiji, 2 to 4 jerks occurred on the upstroke of the primary wave in a wave complex, followed by the usual series of several secondary waves, each of which diminished in height. In the Philippines the secondary waves were much stronger than in Fiji, with the 1st, or the 1st and 2nd, sometimes also showing jerks. In the Philippines the entire series of secondary, diminishing waves was always a conspicuous part of the display; several of these wave complexes were usually included in a series.

At low and moderate intensities jerks were absent, the waves then being more or less equal in height and very low. At intermediate intensities a jerking, primary wave was sometimes followed by several of these low waves of equal height, rather than by diminishing waves.

Waving rate from 0.3 to 1.2 waves per second; jerking waves about 1 per second (Table 19, p. 656.)

Uca (*Deltuca*) [*coarctata*] *coarctata flammula* subsp. nov.

(Northwest Australia; parts of west New Guinea)

MORPHOLOGY

With the characteristics of the species.

Differs as follows from *coarctata c.* Carapace with orbits less oblique. Tubercles on orbit's floor represented, if at all, by scattered minute granules. Major dactyl with tuberculate, hook-like distal projection weak, sometimes indistinct except by contrast with other species. Distal tube of gonopod longer, the inner process bent anteriorly.

Color

General color strikingly black and scarlet orange.

Both sexes with carapace black, sometimes with a pair of spots, round or oval, white to pale orange, located on branchial regions or nearer the center of dorsal area; sometimes the spots are narrow, in the form of two short, longitudinal stripes. Orbits, 3rd maxillipeds, most of underparts and, in particular, both chelipeds and all ambulatories brilliant scarlet orange, except for white tips of major chela; even they are sometimes washed with pale orange.

SOCIAL BEHAVIOR

Waving Display

Waves regularly serial, each combining a relatively high jerking wave with subsequent waves of diminishing height in which jerks are questionable or clearly absent. Each series with 3 of the high, always jerking waves, combined with 2 or 3 diminishing low waves. The high waves all have 3 to 5 jerks on upstroke only; the low waves may have up to 3 indistinct upward jerks. Waving rate of highest jerking waves from 0.4 to 0.7 wave per second.

REFERENCES AND SYNONYMY

Uca (Deltuca) coarctata
(Milne-Edwards, 1852)

TYPE DESCRIPTION. See under *U. (D.) coarctata coarctata*, below.

Uca (Deltuca) coarctata coarctata
(Milne-Edwards, 1852)

? "A small *Luzone Crab* with a large fight claw."
Petiver, 1767: Pl. 78, Fig. 5. Philippines. [See p. 326.]

Gelasimus coarctatus

TYPE DESCRIPTION. Milne-Edwards, 1852: 146. "Odessa"; see p. 55. (Paris !)
Heller, 1863: 100. Copy of type description.
A. Milne-Edwards, 1873: 272; fig. New Caledonia.
Haswell, 1882: 93. Listed in catalogue of Australian Crustacea, but no localities given. Taxonomy.
de Man, 1891: 21, 31. Moluccas; Ponape. Taxonomy.
? de Man, 1892: 308. Celebes. Taxonomy.
Nobili, 1899.2: 273. Andai, Borepata.
Nobili, 1899.3: 517. Lelemboli. Taxonomy.
Tweedie, 1937: 143. Sumatra: Simaloe I. Brief taxonomy. (BM !)

Uca arcuata (not *Gelasimus arcuatus* de Haan)

? Grant and McCulloch, 1906: 20. Northeastern Australia: Port Curtis. (This synonymy is suggested only on the basis of locality.)

Uca rathbunae

Pearse, 1912.1: 91. Philippines. Type description.
Pearse, 1912.2: 113ff. Philippines. Habits.
Ward, 1941.3. Philippines: Mindanao: west coast, Gulf of Davao.
Crane, 1957. Philippines. Preliminary classification of waving display.

? *Uca mearnsi*

Rathbun, 1913: 616, Pl. 75. Philippines: Mindanao: Davao. Type description. (USNM !)
Estampador, 1959: 102. Listing of type.

Uca coarctata

Gordon, 1934: 11. Arroe Is. (Strait of Manoembaii). Taxonomy. (BM !)
Macnae, 1966: 85, 89. Australia: Queensland (south to Moreton Bay). Color; ecology.

Uca ischnodactylus

Nemec, 1939: 107; Fig. 1. Fiji: Suva. Type description.
Crane, 1957. Fiji. Preliminary classification of waving display.

Uca (Deltuca) coarctata flammula
subsp. nov.

Gelasimus arcuatus (not of de Haan)

Haswell, 1882: 92. Northwestern Australia: Port Darwin. Taxonomy.

Gelasimus dussumieri (not of Milne-Edwards)

Haswell, 1882: 93 (at least part). Northwestern Australia: Port Darwin.

Uca forcipata (not *Gelasimus forcipatus* Adams & White)

Rathbun, 1914.1: 661 (part). Western Australia: Monte Bello Is. Taxonomy. (USNM 46348 !; BM 713.1.2 !)
? Miyake, 1939: 222, 241. Micronesia: Palau Is.; Caroline Is. (Ponape).

Uca dussumieri (not *Gelasimus dussumieri* Milne-Edwards)

Rathbun, 1924.2: 8. Western Australia: Broome. Taxonomy. (USNM 56419 !, 56420 !)

Uca coarctata flammula

TYPE DESCRIPTION. P. 56.

8. *UCA (DELTUCA) [COARCTATA] URVILLEI* (MILNE-EDWARDS, 1852)

(Tropical and subtropical western Indo-Pacific)

PLATE 9.
FIGURES 7; 8 *B, D*; 9 *D, E*; 27 *G, H*; 38 *U-X*;
   62 *E*; 75; 98.

MAP 19.
TABLES 8, 10, 12, 14, 19, 20, 23.

INTRODUCTION

This large blue fiddler is almost certainly the only *Deltuca* that reaches Africa. It lives characteristically in mangroves along streams well back from the shore, as do all the other members of its superspecies. In *urvillei*, however, its niche lies close to low-tide levels.

In Natal and in Cape Province *urvillei* is exposed seasonally to chilly weather, although not to the extent of that tolerated, through hibernation, by its northern consuperspecific, *arcuata*, in Japan. Since both species, without apparent morphological differences, occur also in much warmer climates, comparative study of their range of physiological adaptations to temperature would be of special interest.

MORPHOLOGY

Diagnosis

A single groove running most of length of major dactyl. On orbit's floor is a long line of well-developed tubercles. Genital opening at end of a long tube. Carapace with anterior margin practically straight; antero-lateral angles acute and produced. Female gonopore without tubercle on rim, which is low but distinct on three sides and absent internally.

Description

With the characteristics of the subgenus *Deltuca* (p. 21).

MALE

Carapace. Frontal groove of variable width, always short and narrow; sides parallel or converging distally, never almost in contact. Fronto-orbital margins practically straight; antero-lateral angles strongly produced and acute; antero-lateral margins moderately long in larger specimens, convergent throughout their variable length before curving into the more strongly convergent dorso-laterals; antero-laterals with separated serrations throughout; dorso-laterals with ridge distinct, finely beaded, the beads diminishing in size posteriorly. Vertical lateral margin well developed, minutely beaded throughout, ending just below dorsal surface. Eyebrow vertical except internally, where it is obliquely horizontal; upper margin marked by tubercles, smaller and set farther apart than serrations of antero-lateral margin, dying out near antero-lateral angle, absent close to front; lower margin formed of small, distinct, close-set tubercles, in two rows internally, the continuing row ending slightly external to middle of orbit. Suborbital crenellations are abruptly larger, sometimes shallowly bifid or flattened; angle somewhat projecting, almost rectangular; minute crenellations or irregularities continuing around angle along most of outer, lower edge of orbit. A long row of distinct tubercles always present on floor of orbit behind margin; although highly variable in number, size and extent, they generally occupy about the middle half or two-thirds of orbital breadth, and total about 9 to 16; they arise from an irregular blunt ridge.

Major Cheliped. Merus with a row of tubercles on antero-dorsal margin, all small except distally, where several are enlarged, one or more of them sometimes bicuspid; one to a cluster of smaller tubercles, highly variable, on antero-distal surface; ventral margin with up to several irregular rows of small tubercles; additional rows of smaller ones adjacent on posterior surface which, more dorsally, has irregular, oblique, curving, compound rows of granules or incipient rugosities. Center of palm with minute scattered granules. Depression at pollex base relatively broad and shallow, but trifid character apparent. Outer side of dactyl with subdorsal furrow very short, indistinct; on pollex in larger specimens a short, proximal groove sometimes present below regular lateral groove. Pollex and dactyl externally practically smooth except proximally. In gape, pollex of brachychelous individuals with an enlarged, submedian tooth; dactyl with enlarged teeth submedially in brachychelous, subdistally in leptochelous individuals. Forceps-like tip of chela scarcely or not at all indicated. No submarginal carina on outer distal part of pollex.

Minor Cheliped. As in *forcipata* (p. 49), except that lower dactyl ridge is always distinct, although upper ridge and intervening groove are weak or absent.

Ambulatories. As in *coarctata.*

Gonopod. Inner process very short, moderately flat and narrow but not notably spinous, not reaching free base of external tube of shaft; flanges absent, the pore at tip of an elongate, strongly calcified tube, its overlapped margins distinguishable throughout its length; thumb usually of moderate length, reaching base of tube, but sometimes reduced or vestigial and sometimes much enlarged, reaching halfway to tip of tube.

FEMALE

Carapace with an irregular band of enlarged tubercles extending posteriorly between end of posterodorsal margin and base of last ambulatory, often including one or two tuberculate striae or lumps. A long row of tubercles on floor of orbit, in general more numerous than in male and often more regular. Gape of minor cheliped with a pair of slightly enlarged teeth at least on one side. Merus of 4th ambulatory very wide, its dorsal margin clearly convex throughout; no pile along the segment's posteroventral margin. Gonopore without a tubercle on the rim, although a sharp, tubercle-like projection protrudes from the pore to varying extents in all the females at hand; all are from an actively breeding population in Pemba, and almost certainly the projections are spermatophores; the necessary precise dissections have not been made; rim of gonopore low but definite, present on posterior, outer and anterior margins but absent along the internal edge of the rounded pore. Gonopore depression in sternum small, ranging from slight to moderate in depth.

Measurements (in mm)

	Length	Breadth	Propodus	Dactyl
(All specimens from Pemba)				
Largest male	21.0	34.0	61.0	46.0
Moderate male	15.5	25.5	36.5	23.0
Largest female (ovigerous)	16.5	27.0	–	–
Smallest ovigerous female	11.5	19.5	–	–

Morphological Comparison and Comment

U. urvillei differs notably from *dussumieri*, with which it has been most frequently confused, in having a long, slender, flangeless tube forming the distal end of the gonopod, instead of distinct flanges and no projecting tube; in having one, not two, long grooves on major dactyl; in the row of tubercles on

floor of orbit; in the broad ambulatory meri; and in the absence of a rim on the gonopore's inner margin. It differs clearly from *coarctata* and *forcipata*, its nearest consuperspecifics to the east, in the much longer gonopod tube, in the long row of strong tubercles on floor of orbit, and in the straight anterolateral margin, the orbits being scarcely or not at all oblique; in addition the female's gonopore lacks the large marginal tubercle of *coarctata*, but has a distinct outer rim that is absent in *forcipata*.

The form of the gonopod tube alone sets it off strongly from all other species of *Deltuca*. The parallel sides to the frontal groove, strongly tuberculate manus, lack of a triangular tooth on distal part of pollex, form of the minor chelae, and presence of a groove on major dactyl distinguish it easily from any member of the subgenus *Australuca* with which it might be confused, their gonopods being similar. Differs from species in *Thalassuca* in having long grooves on both dactyl and pollex, expanded meri on the ambulatories, and a tubular gonopod.

Color

Males and young largely a characteristic blue, ranging from soft, clear, violet blue to steel blue, quite different from the vivid cobalt found particularly in Malayan *dussumieri*, or of the greenish blue in African *chlorophthalmus*.

Mature males with carapace, buccal region, dorsal parts of ambulatory meri and visible underparts entirely blue except that postorbital and antero-lateral margins of carapace are sometimes white. Major cheliped largely ochraceous to apricot brown, with a patch of clearer, brighter apricot usually present on outer, lower manus, sometimes extending onto pollex; major dactyl often white, the distal pollex less frequently. Females darker than males, unmarked, except for chelipeds which in small individuals are sometimes entirely bright red. Young males often with pale and dark blotches on the blue. Post-megalopal stages of both sexes white.

SOCIAL BEHAVIOR

Waving Display

Display among the simplest. Wave vertical, low, the upper edge of dactyl at maximum reaching to or barely above tip of eye. Jerks absent. Waves in series of 3, uncommonly 4, the manus not touching the ground between waves; series unpredictably far apart. Crabs often walk during displays. Waving at rate of slightly more to slightly less than 1 wave per second. (Waving component 1; timing elements: Table 19, p. 656.)

Precopulatory Behavior

Numerous coverings of female by male were seen, the large majority not preceded by waving. Several males followed females briefly partway down the latter's burrows. No copulations observed.

Chimney Construction

Chimneys plentiful, apparently only around female burrows. Twice, however, when a male had reached the precopulatory stage and followed a female partly down her burrow, he added substrate from below to the chimney's rim; several times a female behaved similarly, though construction was usually with material brought from a distance. Both sexes, in excavating, otherwise took the material from underground a few inches from the burrow, as usual, slinging it in a single direction.

Acoustic Behavior

A major-merus-rub (acoustic component 1) occurred just before a combat. A major-merus-drum (component 7) accompanied an attempt by a male to dig out another male, the appendages of the minor side being thrust down a neighboring burrow. Leg-wags (component 5) were common in females; twice they took place when unreceptive females stood in high-rise positions (threat component 13) on their chimneys during the approach of males. Females, also during high-rises, apparently performed leg-stamps (component 11) in confronting other females. Films only; no data on sound production.

Combat

The following components were filmed: manus-rub, dactyl-slide, heel-and-hollow (components 1, 6, 11).

RANGE

East coast of Africa from Giumbo, Somalia, to Cape Province, South Africa (mouth of Umtata R.); Madagascar; Karachi, Pakistan; western India.

BIOTOPES

In mangrove mud, sometimes partly shaded; well back from the sea but near low tide levels. (Biotopes 8, 12.)

SYMPATRIC ASSOCIATES

Since no other *Deltuca* are known to share the range, *urvillei* has no sympatric associates in the strict sense of the term. Two species of other subgenera, *chlor-* *ophthalmus* and *lactea*, characteristically occur within sight of *urvillei*, but on slightly higher ground; *chlorophthalmus* and *urvillei* often mingle along their lower and upper borders respectively.

MATERIAL RESULTING FROM FIELD WORK

(The complete list of specimens examined is given in Appendix A, p. 595.)

Observations and Collections. Tanzania: Pemba, Zanzibar, and Dar-es-Salaam; August. Seen but not collected in Mozambique: Inhaca I.; September.

Film. Pemba.

TYPE MATERIAL AND NOMENCLATURE

Uca urvillei (Milne-Edwards, 1852)

LECTOTYPE. In Muséum National d'Histoire Naturelle, Paris. Listed by the museum as a "type non specifié." Male, selected by J. Crane from a box containing this specimen and 2 females. Label inside box: "*Gelasimus urvillei* Edw. M. M. Quoy & Gaimard Vanikoro"; label on box: "MM. Quoy & Gaimard." Specimens dried; in poor condition. Male alone relaxed for study. Measurements in mm: length 11, breadth 18.5, propodus 17, dactyl 11.5.

Although the specimen is small, with the proportions of the major cheliped indicating immaturity, the specific characters, including the unmistakable gonopod, are all clearly those of the *Deltuca* characteristic of East Africa. Yet "Vanikoro," the locality given on the label and in Milne-Edwards' type description, was a stop in the New Hebrides on the *Astrolabe* expedition under Dumont d'Urville. It seems clear that labels were confused before completion of Milne-Edwards' manuscript.

Associated Material. The identity of the 2 females associated with the lectotype is unsettled.

Type Material of *Gelasimus dussumieri* Milne-Edwards, 1852. The Malabar specimen listed by Milne-Edwards in his description of *G. dussumieri* is another specimen of *U. urvillei* as defined in the present contribution. His Semarang specimen was designated on p. 35 as the lectotype of *U. dussumieri*. The gonopod of the dried Malabar specimen, when the latter was relaxed, again is unmistakable; although the thumb is short, it is apparently broken. It is possible that in western India the thumb is shorter than in African populations; specimens in the British Museum from Karachi, however, show thumbs of standard length. The left gonopod is more damaged than the right, except for the spinous inner process which shows well only on the left. Carapace length 17 mm.

References and Synonymy

Uca (*Deltuca*) *urvillei* (Milne-Edwards, 1852)

Gelasimus urvillei

TYPE DESCRIPTION. Milne-Edwards, 1852: 148. "Vanikoro" [New Hebrides]. Locality name unquestionably erroneous. (Paris !)

Kingsley, 1880.1: 145. Description from type description.

Ortmann, 1894.1: 59. East Africa: Lindi, Dar-es-Salaam. Color; habitat.

Alcock, 1900: 362 (part). Pakistan: Karachi. Taxonomy.

de Man, 1891: 34. Madagascar: Nossy-Faly.

Gelasimus arcuatus (not of de Haan)

Krauss, 1843: 39. South Africa: Natal Bay.
Stebbing, 1905: 40. South Africa.
Stebbing, 1910: 327. Annotated references.
Stebbing, 1917: 15. Natal. Taxonomy.

Gelasimus dussumieri (not of Milne-Edwards 1852, Semarang specimen)

Milne-Edwards, 1852: 148 (part). India: Malabar specimen only (Paris !)

Hilgendorf, 1869: 84; fig. East Africa: Zanzibar. Taxonomy.

A. Milne-Edwards, 1868: 71. East Africa: Zanzibar.

Hoffmann, 1874: 17; fig. (part). Madagascar: Nossy-Bé.

de Man, 1880: 68 (part). Madagascar. Taxonomy.

Lens & Richters, 1881: 423. Madagascar. Taxonomy.

Pfeffer, 1889: 30. East Africa: Zanzibar. Habitat.

de Man, 1891: 20, 26 (part). East Africa. Taxonomy.

Gelasimus acutus (not of Stimpson)

Alcock, 1900: 360 (part). Pakistan: Karachi. Taxonomy.

Uca dussumieri (not *Gelasimus dussumieri* of Milne-Edwards, 1852, Semarang specimen)

Lenz, 1910: 559. East Africa: Zanzibar; Pemba. Brief taxonomy.

Maccagno, 1928: 17 (part). Somaliland: Giumbo.

Vatova, 1943: 24. Somaliland.

Chapgar, 1957: 510; Pl. 14. Western India: Kolak; Umarsadi. Taxonomy.

Uca urvillei

Vatova, 1943: 24; fig., photo. Somaliland: Giumbo. Taxonomy.

Barnard, 1950: 93; fig. South Africa. Taxonomy.

Fourmanoir, 1953: 90. Madagascar: Near Canal de Mozambique. Color; ecology.

Day & Morgans, 1956: 277; 305. South Africa: Durban Bay. Ecology.

Macnae & Kalk, 1958: 40, 67, 125. Mozambique: Inhaca I. Color; general behavior; ecology.

Macnae, 1963: 3, 7, 23. Mozambique (Inhaca I.) to Cape Province, South Africa (mouth of Umtata R.). Ecology.

Crosnier, 1965: 110; figs. Madagascar.

II. *AUSTRALUCA* SUBGEN. NOV.

Typus: *Gelasimus bellator* Adams & White, 1848

(Tropical Indo-Pacific: West Java to New Guinea; Philippines to Queensland and Western Australia)

PLATES 10-12.
GONOPOD DRAWINGS: FIGURE 62 (part).
DENDROGRAMS: FIGURES 96, 97.
MAPS 3, 4.
TABLES 1, 8, 10, 12, 19, 20.

MORPHOLOGY

Diagnosis

Uca with front narrow, differing most clearly from the related subgenus *Deltuca* as follows. One postero-lateral stria always represented at least by a pro-tuberance, instead of occurring only rarely in males and occasionally in females; a ridge, often with tuber-cles or granules, always, not sometimes, present on orbit's floor. Major cheliped with merus having an antero-dorsal crest, but no distal enlarged tooth; manus outside moderately to very smooth, never roughened by large, well-separated tubercles; its largest tubercles near dorsal margin, never around pollex base; pollex without a long lateral furrow and, in distal half, with a large, triangular projection or a single large tubercle; major dactyl never with a sub-distal hook or projection, so that the chela never has a forceps-like tip. Minor cheliped in both sexes with teeth almost always unusually large for the genus, with one or more on each finger further enlarged. Gonopod tip always, not sometimes, a produced tube. (Table 1.)

Description

With the characteristics of the genus (p. 15), and with the following differences from *Deltuca*.

MALE

Carapace. Tip of front pointed, or only minutely ex-cavate, instead of clearly bifid; front's central de-pression starting farther back from tip, the margins diverging moderately to widely, instead of being sometimes parallel or nearly so. Fronto-orbital mar-gins moderately oblique, never strongly or scarcely. Antero-lateral margins always distinct, instead of sometimes; a single postero-lateral stria always pres-ent, represented variously by a short, raised line, a tubercle surmounting a small protuberance or only

by a protuberance; in *Deltuca* the homologous struc-ture rarely occurs at all.

Eyebrow extremely narrow to practically absent, in general less well marked than in *Deltuca*, although longer; always strongly vertical. Floor of orbit al-ways, not sometimes, with a ridge, often tuberculate or granulate. Suborbital crenellations weak to absent except near outer angle, where they are almost al-ways at least present, sometimes fairly strong.

Major Cheliped. Merus with antero-dorsal margin always with a strong distal crest, never found in *Del-tuca*, with or without tubercles or serrations. Carpus with dorsal margin strongly bent over, with three or more, not one or two distinct tubercles; sometimes, instead, with a crest; dorsal surface sometimes almost smooth, at other times tuberculate almost as in *Del-tuca*. Outer manus always smoother than in *Deltuca*, with tubercles small and close-set or, in some *bella-tor*, of moderate size but never large and well sepa-rated; in comparison with *Deltuca*, therefore, the surface appears almost or strikingly smooth; tuber-cles near pollex base little or not at all larger than those in the vicinity, in contrast to *Deltuca*, while those near dorsal margin are always the largest on the manus. Tuberculate carina of ventral margin of manus sometimes continued distally almost entire length of pollex, where it is marked by an external crease along its upper margin. Oblique ridge inside palm almost lacking in *polita*, otherwise strong, as throughout *Deltuca*; depression on lower distal part of palm large, but more shallow than in *Deltuca* and never clearly trifid. Pollex externally never with the long lateral furrow characteristic of *Deltuca*. Dactyl externally always with 1 long furrow, sometimes shallow, running almost its entire length, never with 2 long ones as sometimes found in *Deltuca*; proximal subdorsal furrow present or absent, the tubercles of the area continuing partway along dactyl. In gape near or beyond middle of pollex the middle row of teeth follows the edge of a triangular projection, large except in *polita*, that culminates in an enlarged tubercle; distal to the tooth the projection's margin is straight or concave, ending in the slender, rounded, curved-up pollex tip. A short, subdistal keel sometimes present close to gape inside pollex and, less often and weaker, a corresponding keel inside dactyl; keel continuous or discontinuous with inner row of gape tubercles; unlike *Deltuca* such a keel is never found outside pollex tip. Unlike *Deltuca*, subdistal part of dactyl never has a hook-like structure or other projection.

Minor Cheliped. Propodus large, equal in length at least to distance from antero-lateral angle to outer base of antenna; fingers long; gape narrow to moderate; teeth always well developed, usually stronger than elsewhere in the genus, 1 pair of opposing teeth always enlarged except in one subspecies of *bellator* (Brisbane area).

Ambulatories. Meri of first 3 pairs sometimes moderately wide, with dorsal margins strongly convex; those of 4th legs scarcely or not at all enlarged. Spinules on posterior surfaces few and weak to absent.

Gonopod. End of shaft always projecting, tubular; pore large, terminal; flanges absent. Inner process various, ranging from spinous and closely applied to shaft, to projecting, curved and even broad and somewhat thickened. Thumb usually a vestigial shelf or nubbin, arising well below base of projecting shaft; rarely short, thick, reaching just above shaft's base.

FEMALE

With the characteristics of the females in the genus (p. 17). A pair of enlarged teeth in gape of cheliped except in some individuals of 2 subspecies of *U. Australuca bellator*. The larger teeth in the gape and more slender meri of the 4th ambulatories distinguish females most easily from sympatric members of *Deltuca*. Other distinctions between the subgenera, in characters showing little sexual dimorphism, are given above, in the subgeneric description of the male.

Size

Small to medium.

Color

Display lightening almost or entirely absent. Red brown, dark red, and pink occur on major cheliped. Yellow phase, blue and green absent.

SOCIAL BEHAVIOR

Waving display moderately prevalent during waving periods and at moderate rates. Wave vertical but sometimes tending to semi-lateral. No jerks. No high-intensity courtship components. Copulation usually at the surface with male approaching female, but infrequently female is pushed down male's burrow and he follows. Waving display usually precedes copulation. No construction activities. Combat components unknown. Acoustic components unknown except leg-wagging observed and filmed.

RELATIONSHIPS

Comparisons with other groups will be found on p. 18 in connection with the discussion of the subgeneric dendrogram. Within the subgenus both *seismella* and *polita* in morphology are further removed from the general pattern found in *Deltuca* than is *bellator*. The differences, in different directions, are apparent in the smoother cheliped of *polita*, along with its overall pale color, and in the shape of the major cheliped in *seismella*. In addition, *seismella*'s waving display is far more active than any other known in the Indo-Pacific, being analogous to those of some *Celuca*, such as *batuenta* and *saltitanta*.

NAME

Australuca: From the Latin adjective *australis*, "southern." The proposed name therefore signifies "southern fiddlers," since this subgenus is largely confined to that hemisphere.

9. *UCA (AUSTRALUCA) BELLATOR* (ADAMS & WHITE, 1848)

(Central Indo-Pacific)

PLATES 10; 11.
FIGURES 11; 29 *B*; 46 *D*; 62 *G-J*; 81 *N*; 90 *C, D*; 97.
MAP 3.
TABLES 2, 8, 10, 12, 19, 20.

INTRODUCTION

Four closely related forms of *Uca*, described in the past under various specific names, are here considered as subspecies of the single species, *bellator*. Only one of them occurs outside Australia, where it is probably restricted to a triangle with angles formed by the Philippines, Borneo, and western New Guinea. The several forms show very minor morphological differences, principally in the armature of the major cheliped and lower orbit. Unlike *vocans* and *lactea*, the subspecies are not known to intermingle. Material, however, remains scanty.

In the form of its waving display, *bellator* shows clearly one aspect of the intermediate character of the subgenus. The wave ranges in form from strictly vertical, especially at low intensities, through a usual semi-unflexed movement, to a complete, lateral-straight wave, uncommonly at high intensities. Lateral-circular waves, however, characteristic only of the socially most advanced subgenera, apparently do not occur.

In one subspecies the male sometimes pushes the female ahead of him down his own burrow, instead of following her and mounting on the surface, in the manner characteristic of Indo-Pacific narrow-fronts. This courtship behavior appears broadly intermediate between surface mating and the pattern characteristic of socially advanced subgenera, where the male almost always attracts the female to his burrow and precedes her below ground.

None of the waving displays has been observed in more than one or two populations except in *b. bellator*, while in all the subspecies combat and threat behavior remain practically unknown. A comparative study of social behavior throughout the range of the species would certainly be rewarding.

MORPHOLOGY

Diagnosis

Australuca with a projection on gape of major cheliped beyond middle of pollex, almost always triangular and denticulate, rarely merely an enlarged tubercle. Major merus with crest tuberculate (except in *b. minima*); palm with oblique ridge and its tubercles strong. Small cheliped in both sexes without enlarged distal teeth in gape. Female without patches of pile on carpus and manus of 4th ambulatory; gonopore always with some structure on rim.

Description

With the characteristics of the subgenus (p. 62).

MALE

Carapace. Front slightly wider than basal breadth of erected eyestalk. Antero-lateral angle acute, slightly produced; antero-lateral margins short to moderate, always converging, forming dorso-lateral margins with a gradual turn; these margins distinct or obsolescent, but never tuberculate. Postero-lateral stria weak, short, direction variable. Eyebrow very narrow, vertical or nearly so, variable. Floor of orbit strongly rolled out, always with a mound but with or without tubercles or granules; suborbital crenellations present or absent, except near external angle where they are always apparent.

Major Cheliped. Antero-dorsal crest on merus always tuberculate (except in *b. minima*). Outer surface of carpus and manus with tubercles well developed for the subgenus except in *longidigita*; palm with oblique ridge strong, its tubercles more or less regular; distal ridge at dactyl base always with distinct tubercles. Pollex with or without a ventral marginal carina and associated crease. Subdorsal furrow of dactyl present or absent. Subdistal carina inside pollex long; except in *b. minima*, clearly marked by tubercles or beading except at tip. Fingers long, slender, compressed. Base of pollex slightly if at all deeper than corresponding part of dactyl; a triangular projection strongly marked, its apex always beyond middle of gape and usually surmounted by an enlarged tubercle. An enlarged tooth on dactyl often present opposite apex of pollex projection. Gape narrow throughout, almost or entirely lacking distal to the projection apex.

Minor Cheliped. Gape narrow; teeth regular or slightly irregular, not graduate; a single pair of opposed, enlarged teeth present or absent.

Gonopod. All elements varying slightly with the subspecies.

Measurements. Small to moderately small *Uca*, the largest (*b. bellator*) attaining a known maximum length of 12 mm, the smallest (*b. minima*) less than 7 mm, the smallest Indo-Pacific form known in the genus.

FEMALE

Carapace without tubercles on dorso-lateral margin, although this marginal line is always strong and rarely minutely beaded. Tubercles on orbital floor present or absent. Minor cheliped with or without a pair of enlarged, opposing teeth in gape, but never with an enlarged tooth distally; dentition as a whole closely similar to that of male. Manus and carpus of 2nd, 3rd and 4th ambulatories with or without posterior spinules, never with a marked, close-set patch on proximal part of segment; no patches of pile in these areas on 4th ambulatory. Rim of gonopore always with some form of structure, always either externally or postero-externally.

Morphological Comparison and Comment

The key on p. 624 distinguishes *bellator* from the other two members of the subgenus *Australuca*. The combination of a large projection on the major pollex and a long, tubular tip on the gonopod should quickly distinguish this species from Indo-Pacific narrow-fronts of other subgenera.

Color

Displaying males: Display whitening usually partly developed, never enveloping entire crab. Carapace dark with pale markings or vice versa. Major cheliped often with shades of red or orange on proximal segments; manus characteristically brown to orange brown; fingers white. Crab's anterior aspect usually marked with white, or it is wholly white, including minor cheliped; ambulatories dark, with or without white markings on posterior surfaces of last two legs, rarely with anterior white on first legs. Females similar to males; uncommonly with red on cheliped and first legs.

SOCIAL BEHAVIOR

Waving Display

Wave vertical, semi-unflexed or lateral straight. Jerks absent. Tips of chela reach well above eyes during displays of moderate intensity, at least during all waves in a series except the last two or three; these final motions are sometimes diminishing waves of low amplitude. Minor cheliped sometimes makes

synchronous motion. Body raised slightly or not at all with each wave. Legs not raised during display, except to take occasional steps to one side or the other. Display at rate of about 1 wave per second, with or without a pause between waves. (Component nos. 1, 3, 4, 9, 10, 14; timing elements in Table 19, p. 656.)

Precopulatory Behavior

Apparently principally on the surface, prefaced by stroking, as in *Deltuca*. Herding, however, is better developed than in any other species where it is known, since the male sometimes actually nudges a partly responsive female toward the mouth of his burrow and pushes her down the hole, after which he follows at once (p. 498).

Acoustic Behavior

Three components so far known, all in one subspecies (p. 67).

RANGE

Labuan; Indonesia; Philippines; western New Guinea; tropical and subtropical Australia; 1 record from Nicobar Islands in the Indian Ocean.

BIOTOPES

Sheltered flats of mud or muddy sand inside mouths of streams and rivers, sometimes close to the sea and sometimes upstream as far as the limits of mangroves. (Biotope nos. 12, 14, 15.)

SYMPATRIC ASSOCIATES

The more seaward populations are sometimes mingled with stray individuals of the other *Australuca*, *seismella* and *polita*. Other subgenera are represented chiefly by *coarctata*, although most of each population of this *Deltuca* are assembled nearby and mingle only marginally. Both (*Thalassuca*) *vocans* and (*Celuca*) *lactea* are concentrated in more marine habitats and occur with *bellator* only marginally if at all, or through adventitious individuals. Farther upstream the known populations of *bellator* are pure cultures.

MATERIAL RESULTING FROM FIELD WORK

(The complete list of specimens examined is given in Appendix A, p. 595.)

Observations and Collections. U. b. bellator: Indonesia: Djawa: Semarang and near Surabaja. Philippines: Mindanao: Zamboanga; west coast, Gulf of Davao; Palawan: Puerto Princesa; Luzon: near Manila. *U. b. signata*: Australia: Queensland: in and near Gladstone. *U. b. longidigita*: Australia: Queensland: near mouth of Brisbane R. *U. b. minima*: Australia: Darwin.

Film. Indonesia: Semarang; Philippines: west coast, Gulf of Davao and near Manila. Australia: near Gladstone; mouth of Brisbane R.; Darwin.

TYPE MATERIAL AND NOMENCLATURE

Uca bellator (Adams & White, 1848)

HOLOTYPE. In British Museum (Natural History), London. Label: "*Gelasimus bellator.* Holotype. Philippine Ids. 43.6. Cuming Coll." Measurements in mm: length 12.5; breadth 21; propodus 41. Specimen in good condition. (!)

Uca bellator signata (Hess, 1865)

TYPE-SPECIMEN. Zoologisch Institut, Göttingen. Label: "53b. *Gelasimus signatus* Hess. 1889. Pöhl. Australien." Inner label (tied on ambulatories): "3665." Measurements in mm: length 12; propodus 29. Specimen in good condition. (!). Data in catalogue: "a. Sydney. 1864. Schütte. b. Australien. 1889. Capt. Pöhl Ibbry." Specimen "a" is missing; Sydney highly improbable as the site of capture.

Uca bellator longidigita (Kingsley, 1880)

HOLOTYPE of subspecies. Not present in Academy of Natural Sciences at Philadelphia or found elsewhere.

Uca bellator minima subsp. nov.

HOLOTYPE. In Smithsonian Institution, National Museum of Natural History, Washington, no. 137668. Australia: Darwin. Collected by J. Crane.

Named in allusion to its being the smallest of the subspecies in the species. (L. *minimus,* adj.: least.)

Gelasimus signatus var. *angustifrons* de Man, 1892. A good series was deposited by de Man at various times in Leiden, collected in several parts of Java. Cat. no. 1538 refers to a single male, with the following label, written (*fide* Holthuis) in de Man's handwriting: "*Uca signata* Hess var. *angustifrons* de Man. Dr. J. Semmelink. Batavia. 1882." Measurements in mm: length 10; propodus 25. If this form were not, as in this study, synonymized with *U. bellator bellator*, this specimen could have served well as neotype, no designated type having been found.

Ortmann (1897: 350) included both *Gelasimus bellator* and *G. signatus* in the synonymy of *G. forceps* (see p. 323).

Uca (Australuca) bellator bellator (Adams & White, 1848)

(Philippines; Labuan; Indonesia; western New Guinea; one record from Nicobar Islands)

MORPHOLOGY

With the characteristics of the species.

Gonopod: Inner process straight and tumid except for flattened tip, extending only partway to tip of tube against which it is appressed; thumb thick and short but well formed and reaching slightly beyond exposed base of tube. Gonopore with a well-developed postero-external tubercle, neither high nor sharp but extensive. Orbital floor with strong tubercles on at least one side in both sexes. Suborbital crenellations in male small but distinct and regular, variable in size and extent. Major cheliped: merus with antero-dorsal crest low, straight except for slight distal expansion, tuberculate throughout; pollex with ventral marginal carina and crease well developed; subdistal carina inside pollex long and minutely beaded; dactyl without subdorsal furrow. Minor cheliped: in both sexes teeth are similar, being unequal, moderate sized, largest near middle, serrate; in females and some males a single pair, further enlarged and opposed, occurs at or just beyond middle. Female ambulatories: no patches of close-set spinules on 3rd and 4th carpus and manus, but often a sparse scattering on 3rd manus.

Measurements (in mm)

	Length	Breadth	Propodus	Dactyl
(All specimens from Semarang material)				
Largest male	12.0	19.0	33.5	23.0
Moderate male	10.0	15.5	26.5	16.5
Largest female (ovigerous)	10.0	14.5	–	–
Smallest ovigerous female	6.5	10.0	–	–

Color

Displaying males: carapace brown with extremely variable markings of white, blue or both; very rarely white with dark markings. Within single populations the range extends from a single, white, transverse band across anterior third of dorsal surface to white speckles mixed with blue spots that are distributed both dorsally and laterally. Major cheliped: merus often red to orange brown or dull yellow; carpus sometimes similar; lower manus brown to orange brown, orange, or white; upper manus and fingers white. Eyestalks sometimes yellowish green. Orbits uncommonly rimmed with greenish or blue; maxillipeds, suborbital and pterygostomian regions, minor chelipeds, and ambulatories all dark with highly variable white spots; of these the most common and constant in position is a single large spot on posterior merus of 4th leg; similar spots only uncommonly found on posterior merus, or on carpus and manus, of 3rd leg; apparently never occurring on 1st or 2nd; anterior surface of 1st legs, however, are sometimes wholly white. Females with carapace and appendages similar to those of male, except that eyestalks and anterior surface of merus on 1st ambulatory are infrequently bright red.

Social Behavior

Waving Display

Wave vertical, ranging to a semi-unflexed wave at high intensities. In each series 2 to 8 relatively high waves, usually about 3, are followed by about 2 diminishing waves of low amplitude. Body raised and lowered with each wave.

Although, as stated already in the description of waving display in the species, the ambulatories are not raised as a part of the waving display, in wholly agonistic situations the 4th legs, with the meri often spotted with white, are sometimes kicked out, either toward the opponent in males, or toward a male by an unreceptive female. At these times no contact of the 4th legs with either the 3rd or with the carapace is made. Functional herding of female by male occasional (see below).

Precopulatory Behavior

The usual final phases of courtship seem to consist, as in *Deltuca*, of approaching female with or without waving, followed by stroking and then mating at the surface. However two examples were seen, one in the Philippines and the other in Java, where the male, after display to a somewhat inattentive female and an attempt to stroke her at a distance of some inches from his burrow, ended the visible courtship by nudging her toward his burrow using chiefly the legs of his minor side, and literally pushing her below. He then, with his minor side already partly inside the burrow's mouth, in one of the two examples, vibrated his major cheliped in a major-merus-rub that shows distinctly in the filmed sequence, before he too vanished beneath the surface.

Acoustic Behavior

The following components have been observed in the field and in films: major-merus-rub (component no. 1; see also above description of its use in courtship); leg-wagging (5); leg-side-rub (6).

Uca (Australuca) bellator signata (Hess, 1865)

(Eastern Australia: Tropical Queensland)

Morphology

With the characteristics of the species.

Gonopod: Inner process not tumid, curving, ending far below tip of tube to which it is closely appressed throughout; thumb represented only by a shelf, its edge transverse, not slanting. Gonopore with a small, distinct tubercle located clearly on outer side of rim, not near its posterior part. Orbital floor in both sexes with granules or tubercles. Suborbital crenellations in male small but distinct and regular. Major cheliped: merus with antero-dorsal crest low, thick, practically straight, with irregular tubercles; pollex with ventral carina and crease present; subdistal carina inside pollex beaded; dactyl with a short subdorsal furrow. Minor cheliped: teeth in both sexes small, mostly erect; a distinctly enlarged pair opposed near middle absent in males, at least sometimes present in females (chelae in material missing or detached). Female ambulatories: no patches of spinules on 3rd and 4th carpus and manus, or weakly present on entire upper half of 3rd manus.

Measurements (in mm)

	Length	Breadth	Propodus	Dactyl
(Both from Gladstone)				
Largest male	9.5	15.5	21.0	13.5
Largest female	8.5	11.5	–	–

Color

Displaying males: Carapace rarely fully white; usually dark with or without paler mottlings or with white spots on branchial regions only. Major cheliped: merus red, pink or reddish orange; carpus dull; outer manus brown to reddish orange; fingers white. Minor cheliped similar to major. Ambulatories dark, not spotted.

Social Behavior

Waving Display

Wave vertical to semi-unflexed, sometimes a full lateral-straight wave, the cheliped being completely unflexed. Diminishing waves present. Body raised and lowered with each wave, which last more than 1 second each, with a pause of equal length or more between waves of a series.

Acoustic Behavior

Leg-wagging (component no. 5).

Uca (*Australuca*) *bellator longidigita* (Kingsley, 1880)

(Eastern Australia: known only from neighborhood of Brisbane)

MORPHOLOGY

With the characteristics of the species.

Gonopod: inner process not tumid, slightly curved, tip reaching variably close to end of tube to which it is closely appressed throughout; thumb small but distinct, ending below exposed base of tube. Gonopore with no trace of a tubercle, but with external half of rim very slightly raised, its edge smooth. Orbital floor in both sexes smooth, without tubercles or granules. Suborbital crenellations absent in male, or, near external angle, with vestiges. Major cheliped: merus with antero-dorsal crest weakly tuberculate or serrate, the extreme distal portion slightly expanded and convex; outer manus smoother than in other subspecies; pollex with ventral carina and crease absent; pollex long, minutely tuberculate; dactyl with subdorsal furrow absent. Minor cheliped: teeth are clearly larger in female, but in both sexes of moderate size, somewhat serrate, and with a single pair enlarged and opposed slightly distal to middle. Female ambulatories: patches of weak spinules present or absent on 3rd and 4th carpus and manus.

Measurements (in mm)

	Length	Breadth	Propodus	Dactyl
(All specimens from near Sandgate, near Brisbane)				
Largest male	10.5	18.0	31.5	23.5
Largest female (ovigerous)	9.5	19.0	–	–
Smallest ovigerous female	8.0	12.0	–	–

Color

Displaying males: carapace dull, inconspicuously marbled with paler. Appendages dull except for major manus and chela, which are usually white, with the manus, instead, being sometimes pale blue green.

SOCIAL BEHAVIOR

Waving Display

(Observed briefly; not filmed.) Wave semi-flexed, sometimes almost unflexing fully in a lateral-straight wave. Diminishing waves apparently absent. Waving at the rate of about 1 wave per second; no perceptible pause between waves; a large number forming each series.

Uca (*Australuca*) *bellator minima* subsp. nov.

(Northwest Australia: known only from Darwin)

MORPHOLOGY

With the characteristics of the species.

Gonopod: inner process long, slender, straight, not appressed to terminal tube; thumb absent, represented by a slanting shelf. Gonopore: about as in *bellator*, with an extensive postero-external tubercle. Orbital floor with the usual mound merging in external half into a linearly narrow ridge, its edge in male either smooth or with a few minute granules, in female always with tubercles. Suborbital crenellations in male minute, even near external angle.

Major cheliped: merus with antero-dorsal crest large and convex, but smooth; pollex with ventral marginal carina and weak associated crease; subdistal carina inside pollex without tubercles; dactyl without subdorsal furrow. Minor cheliped: with a pair of enlarged, opposed teeth absent in male, sometimes present near middle in female; teeth small, mostly erect. Female ambulatories: no patches of spinules on 3rd and 4th carpus and manus.

Measurements (in mm)

	Length	Breadth	Propodus	Dactyl
Largest male (holotype)	6.8	10.0	15.5	9.5
Moderate male	5.0	8.0	9.5	5.5
Largest female	5.5	11.0	–	–

Color

Displaying males: Carapace in some individuals wholly white. Major cheliped: Outer merus and carpus reddish brown; inner merus orange red; manus, pale orange to white; fingers white. Maxillipeds, suborbital area, and pterygostomian regions white. Ambulatories dark.

SOCIAL BEHAVIOR

Waving Display

Wave vertical, apparently not attaining the unflexed, lateral-straight form. Cheliped raised slowly, brought down rapidly. Diminishing waves absent. Waving at rate of about 1 per second; waves usually occur singly, but sometimes several are grouped in a series.

REFERENCES AND SYNONYMY

Uca (Australuca) bellator
(Adams & White, 1848)

TYPE DESCRIPTION. See under *U. (A.) bellator bellator*, below.

Uca (Australuca) bellator bellator
(Adams & White, 1848)

Gelasimus bellator

NAME ONLY. White, 1847: 36. Philippines.

TYPE DESCRIPTION. Adams & White, 1848: 49. (BM !)

White, 1848: 85. No new material.

Milne-Edwards, 1852: 146. Taxonomy. No new material.

Kingsley, 1880.1: 138 (part). Mention of White's specimen.

Gelasimus signatus var. angustifrons

de Man, 1891: 38. Type description. (Leiden: part !)

Gordon, 1934: 12. Dutch East Indies. Taxonomy. (BM !)

Uca signata

Roux, 1917: 614. Western New Guinea: Merauke. Record.

Estampador, 1937: 543. Philippines: Panay. Local distribution.

Estampador, 1959: 101. Philippines: Panay. Local distribution.

Verwey, 1930: 172ff.; 199ff. Java. Ecology. Mouthparts. Behavior.

Crane, 1957. Philippines. Preliminary classification of waving display.

Uca signata var. angustifrons

Maccagno, 1928: 25. Western New Guinea: Merauke. Taxonomy. (Torino !)

Uca angustifrons

Tweedie, 1950.1: 357. Borneo: Labuan. Taxonomy.

Tweedie, 1954: 118. Note on restricted local distribution in relation to Sunda Shelf.

Uca (Australuca) bellator signata
(Hess, 1865)

Gelasimus signatus

TYPE DESCRIPTION. Hess, 1865: 146; fig. Australia: Sydney; "Australien." (Göttingen, part !)

? Kingsley, 1880.1: 138 (part). Specimen listed from Australia (not extant).

Kingsley, 1880.1: 146. Translation of type description.

Haswell, 1882: 93. No new material. Considers Sydney a doubtful locality record.

Miers, 1884: 236. Australia: Port Curtis (BM !); Swan River.

de Man, 1891: 35. Eastern Australia. Taxonomy. (Leiden !)

Ortmann, 1894.2: 756. Eastern Australia. Record (in Strassburg Museum, received from Godeffroy Museum).

Uca signata

Crane, 1957. Australia: Gladstone. Preliminary classification of waving display.

Uca bellator

Macnae, 1966: 77, 79. Australia: Queensland (Thursday I. to Port Curtis). Ecology.

Uca (Australuca) bellator longidigita
(Kingsley, 1880)

Gelasimus longidigitum

TYPE DESCRIPTION. Kingsley, 1880.1: 144. Australia: near Brisbane: Moreton Bay.

Uca longidigitum

Macnae, 1966: 79, 80. Australia: Queensland (Moreton Bay). Ecology.

10. *UCA (AUSTRALUCA) SEISMELLA* SP. NOV.

(Australia)

PLATES 12 *A-D*; 46 *E*.   MAP 4.
FIGURES 12; 46 *E*; 62 *K*; 97.   TABLES 8, 10, 12, 19, 20.

INTRODUCTION

This Australian *Uca* is perhaps the most unmistakable of all fiddler crabs, whether watched in action on a mudbank or examined briefly with a lens. During its waving display the small crab shakes all over as the large claw waggles up and down. Only the neotropical *saltitanta* gives a similar impression. It seems fitting to give the proposed new species a name signifying "little earthquake."

In both sexes the teeth of the small cheliped, always large and serrate, are characteristic, as described below. Unfortunately we have no clue to the uses of this formidable armature, whether in feeding, in stridulation, or in both.

MORPHOLOGY

Diagnosis

Australuca with large, triangular teeth in gape of small cheliped, always with the two opposing distal teeth much larger than the rest; characteristic of both sexes. Major pollex with a long, low, triangular projection on upper edge; ventral marginal pollex ridge present; palm with oblique ridge strong; major merus with a large crest, its edge smooth or nearly so. Suborbital margin smooth except occasionally near outer angle. Female with dorso-lateral margin tuberculate; carpus and manus of last ambulatory each with a patch of pile posteriorly.

Description

With the characteristics of the subgenus (p. 62).

MALE

Carapace. Front little or not at all wider than basal breadth of erected eyestalk. Antero-lateral angle rectangular, not produced; antero-lateral margins scarcely or not at all converging, forming dorso-lateral margins in a sharply angular turn; these margins, although variable, are sometimes marked by raised lines scarcely longer than antero-lateral margins, continuing posteriorly as a blunt ridge, sometimes interrupted and thus forming a series of widely spaced, blunt tubercles. Postero-lateral stria strong and longitudinal. Eyebrow very narrow, vertical. Floor of orbit strongly rolled out and with a blunt ridge, but without tubercles or granules; no crenellations on suborbital margin except for several small irregularities, sometimes present near one or both outer angles at level of eyeball tip.

Major Cheliped. Antero-dorsal crest on merus high and convex in portion beyond level of antero-lateral angle but without tubercles, the edge being either smooth, slightly uneven, or with a few minute serrations. Outer surface of carpus and manus almost smooth, although less so than in *polita*; the tubercles or granules are very distinct, but low and well separated. Palm with oblique ridge strong, its tubercles nearly regular; distal ridge at dactyl base weak, variable. Pollex with a strong, ventral, marginal carina and associated crease; subdorsal furrow of dactyl absent; both fingers compressed, covered by minute granules externally and near margins on inner side. Pollex in general deeper than dactyl because of a long, triangular projection on gape, its base occupying entire pollex edge except tip; its apex at middle of pollex low but marked by a tubercle. Subdistal carina inside pollex tip weak, its proximal portion represented by a regular row of close-set, minute tubercles that are never continuous with inner row of gape tubercles; dactyl with a similar carina, usually weaker and with a shorter row of tubercles. Dactyl gape usually with two enlarged teeth, one near base, the other beyond level of tubercle surmounting pollex projection.

Minor Cheliped. Manus and base of pollex notably deep, fingers longer than manus, gape practically absent. About 8 (range 7 to 10) teeth in each finger, occupying entire gape, triangular, usually directed obliquely distad, sometimes almost upright, smallest proximally. After the sometimes irregular proximal one or two minute teeth, the size in each finger increases in a regular gradation at least to the 2nd tooth before the distal, the predistal pair being sometimes reduced. Each tooth of the distal pair is invariably much larger than any others in the series and forms the crab's most useful diagnostic character.

Ambulatories. Merus of first 3 pairs moderately enlarged.

Gonopod. Inner process long, strong, spinous, closely appressed to tube and extending to level of pore;

tube itself moderately long, its terminal diameter large; thumb a vestigial shelf near base of projecting tube.

FEMALE

Minor cheliped about as in male. Carpus and manus of 2nd, 3rd and 4th ambulatories each strongly spinulous posteriorly, those of the fourth having in addition a patch of pile overlaying part of the spinules. Gonopore without marginal structures.

Measurements (in mm)

	Length	Breadth	Propodus	Dactyl
Largest male	9.5	14.5	22.5	15.5
Holotype male	7.5	13.0	18.0	12.0
Largest female (ovigerous)	9.0	13.5	–	–
Smallest ovigerous female	7.0	10.0	–	–

Morphological Comparison and Comment

This very distinct species lacks close relations. Through its possessing characters regarded here as of subgeneric rank, however, it clearly belongs in the subgenus *Australuca*. These characters combine a narrow front with a tubular gonopod tip, a strong postero-lateral stria, strong dentition on minor chela, absence of a groove on the major pollex and presence of a large protuberance on its prehensile edge.

Color

Display whitening absent. Carapace dark brown to mottled gray. Major cheliped: merus brown to dull yellow. Manus at least on distal half and entire chela shiny white, conspicuous against the usually dark background of both the crab's body and the mud. The palm, never visible during display, was grayish in freshly caught specimens. Ambulatories dark, sometimes faintly banded. Females with carapace and appendages often pale gray except in many individuals for white markings on chelipeds and on anterior surfaces of 1st and 2nd ambulatories.

SOCIAL BEHAVIOR

Waving Display

Wave partly vertical, partly semi-unflexed. Major cheliped at beginning of a series is raised wholly flexed almost as far as base of eyestalks, then partly unflexed and flexed again, in the same plane, obliquely up and out, in an arc of narrow amplitude. At most each wave extends slightly above eye. The usual rate of the upper-level, vibrating waves is about 5 per second, but unusually and briefly 8; the range in a series is from 4 to 6; at no time during a series

does the manus or pollex touch the ground. At the end of each series, however, the claw is brought fully down into rest position and at least several seconds usually elapse between series. During display the minor cheliped apparently is held motionless and the body is not raised. (Component nos. 1, 3; timing elements in Table 19, p. 656.)

Acoustic Behavior

Three components are known: the minor-merus-rub in both sexes (component no. 2), leg wag (5), and major-merus-drum (7). Contrary to its appearances in the field, film analysis shows that the major manus does not come in contact with the ground during waving display.

RANGE

Known only from 3 localities in Australia: Broome, on the northwest coast, Darwin in the Northern Territory, and Gladstone, Queensland.

BIOTOPES

Muddy banks near mouths of streams, often steep and usually near mangroves. (Biotope nos. 12, 13.)

SYMPATRIC ASSOCIATES

Very rarely marginally mingled with *bellator* or, more often, with populations of this species of *Australuca* concentrated nearby. Usually associated on the same mudbank with (*Deltuca*) *coarctata* the appropriate subspecies of which usually occupies more upper levels of the bank while *seismella* occurs typically somewhat lower.

MATERIAL RESULTING FROM FIELD WORK

(The complete list of specimens examined is given in Appendix A, p. 596.)

Observations and Collections. Australia: Broome; Darwin; Gladstone.

Film. Darwin.

TYPE MATERIAL AND NOMENCLATURE

Uca (Australuca) seismella sp. nov.

HOLOTYPE. In Smithsonian Institution, National Museum of Natural History, Washington. Male, cat. no. 137666. Type-locality: Australia: Northern Territory: Darwin. Measurements above.

Named in allusion to its shaking during waving display. (From the Greek noun for "earthquake" from verb for "to shake," plus diminutive.)

11. *UCA (AUSTRALUCA) POLITA* SP. NOV.

(Australia)

PLATE 12 *E-H*. MAP 4.
FIGURES 10; 31 *B*; 62 *F*; 92; 97. TABLES 8, 10, 12, 20.

INTRODUCTION

This species, for which the name *Uca polita* is proposed, shows the active, semi-unflexed wave characteristic of *Australuca*. It has, however, none of the speed of *seismella*, or the more complete lateral-straight wave-form attained by some *bellator*. The largest species in the subgenus, *polita* is conspicuous among all Australian fiddlers for the rose pink that often suffuses the claw.

Except for *seismella*, it is the only *Uca* found on both the eastern and western coasts of Australia that does not appear to warrant the erection of subspecies.

MORPHOLOGY

Diagnosis

Australuca with outer carpus and manus notably smooth even for the subgenus; pollex near middle of gape with a single enlarged tooth, but no triangular projection and no ventral marginal ridge; base of pollex not notably wider than base of dactyl; oblique ridge inside palm low, blunt with weak tuberculation. Ridge on floor of orbit blunt, without tubercles or denticles. Minor cheliped in both sexes with an opposing pair of enlarged teeth beyond middle of gape, followed distally by small teeth. Female without tubercles on dorso-lateral margins, without patches of pile on carpus and manus of last ambulatories, without generally distributed spinules on these segments of last 3 ambulatories, and without any structure on gonopore rim.

Description

With the characteristics of the subgenus (p. 62).

MALE

Carapace. Front slightly wider than basal breadth of erected eyestalk. Antero-lateral angle rectangular or acute and slightly projecting; antero-lateral margins scarcely to moderately convergent; dorso-lateral margins marked by raised lines of variable length. Postero-lateral stria a weak protuberance. Eyebrow not distinct. Floor of orbit, in inner half only, with a blunt ridge, without denticles or tubercles; weak crenellations or irregularities present to a variable ex-

tent in external half of suborbital margin, never extending to extreme outer angle and usually confined to region below non-erected eyeball.

Major Cheliped. Merus with antero-dorsal crest distally minutely serrate; outer carpus and manus, except near dorsal margin of manus, unusually smooth. A large, indistinct, shallow depression outside base of pollex, notable only in comparison with its absence elsewhere in the subgenus. Palm with oblique ridge low, its tubercles small and irregular; distal ridge at base of dactyl weak and highly variable. Pollex without a ventral, marginal carina and crease in adult, but the structure is indicated proximally in young; depth of pollex at base scarcely or not at all deeper than base of dactyl except in young. Dactyl with subdorsal furrow absent. Pollex without a triangular projection near middle of gape in adult, but with a strong tooth in that area even in leptochelous examples; in the young a broad, low, triangular structure is usually present instead of the tooth, as in adult *Australuca* of other species. Subdistal keel inside pollex weakly indicated if at all. Dactyl with a similarly enlarged tooth opposite that of pollex, as well as several smaller ones in proximal region.

Minor Cheliped. Pollex proximally clearly much deeper than corresponding part of dactyl. Gape beyond middle with two large, opposing teeth; rest of gape both proximally and distally with minute denticles and serrations, irregular and variable, but with several in proximal part of dactyl usually slightly enlarged.

Ambulatories. Meri of first 3 pairs moderately wide, their dorsal margins strongly convex; spinules on posterior surfaces few and weak to absent.

Gonopod. Inner process slim, spine-like, closely appressed to tube and ending below pore; thumb represented by a round nubbin well below base of projecting tube.

FEMALE

Carapace without tubercles on dorso-lateral marginal lines. Cheliped with gape armed about as minor in male, but dentition often stronger and pollex base deeper, at least on one side. Second, 3rd, and 4th ambulatories without a generally distributed covering of spinules on posterior surface of each carpus and manus; no pile on same segments of 4th ambulatory. Gonopore without tubercles or other structures on margin.

Measurements (in mm)

	Length	Breadth	Propodus	Dactyl
Largest male	16.5	25.0	41.0	27.0
Holotype male	14.5	22.5	35.0	21.0
Largest female (ovigerous)	19.0	21.5	–	–

Morphological Comparison and Comment

This moderately large *Uca* is easily distinguished from the several subspecies of *bellator*, its closest relation, by at least two of the following characters: the lack of a triangular, tuberculate tooth in distal half of major pollex; lack of tubercles or denticles on floor of orbit in both sexes; female without any structure on gonopore rim. The dentition of the male's minor chela and of both in the female, each being equipped with only one enlarged tooth, not at its distal end, bears no resemblance to the strong, serial serrations of *seismella*, the last species in *Australuca*. The smoothness of the major carpus and manus differentiates *polita* at once from the remaining Australian narrow-fronts.

The dorso-lateral surface of the manus appears smoother in specimens from Broome than in those from Gladstone, the tubercles in the former examples being fewer and smaller.

Color

Displaying males: Display whitening partial. Carapace ranging from uniformly dark to, rarely, white. The usual phases are dull bluish with white marblings or mottlings, especially anteriorly, or yellowish with dark to white markings. Major cheliped: Merus and carpus, inside and out, reddish rufous to pale orange. Outer manus and chela largely or completely rose pink in most specimens; in some individuals, especially larger males in Broome, both manus and chela pale partly or wholly to white. Third maxillipeds, suborbital areas, and pterygostomian regions often variously marked with white. Minor cheliped with manus and chela usually pink. Ambulatories banded light and dark, sometimes with white spots on posterior meri.

SOCIAL BEHAVIOR

Waving Display

Wave semi-unflexed, ranging almost to a full lateral-straight. Jerks absent. At beginning of each wave,

mero-carpal joint is raised, the chela tips then touching or almost touching ground; from this position they are unflexed, partly or wholly, up and out, the chela tips at most reaching clearly above eyes. At high intensities the cheliped is not fully flexed between waves. Minor cheliped usually does not make a corresponding motion. Carapace little or not at all raised between waves, and legs are not raised, except in making steps during a wave to one side or the other. Waving is at the rate of less than 1 display per second, and displays continue without pauses between up to a maximum of about 12 in a series (component nos. 3, 4, ?9, ?10).

Precopulatory Behavior

Several males were observed stroking females in the usual fashion. In each example the male approached the female without waving.

Acoustic Behavior

The sole component known was observed only in a filmed sequence. It is a major-merus-rub (component no. 1).

RANGE

Known in Australia from Nicol Bay and Broome in Western Australia and, in Queensland, from Thursday Island south to Moreton Bay.

BIOTOPES

Bayshores on flats of muddy sand or mud, the substrate being characteristically in a thin layer over a bed of conglomerate, usually near or among shoots of pioneer mangroves; uncommon on muddy banks of streams bordered with mangroves. (Biotope nos. 4, 5, 13.)

SYMPATRIC ASSOCIATES

Uncommonly individuals were observed among populations of (*Australuca*) *seismella* and (*Deltuca*) *coarctata*. The more marine groups occur among (*Deltuca*) *dussumieri* and near (*Thalassuca*) *vocans* which frequents levels nearer low tide.

MATERIAL RESULTING FROM FIELD WORK

(The complete list of specimens examined is given in Appendix A, p. 596.)

Observations and Collections. Australia: Broome; Gladstone.

Film. Broome.

Type Material and Nomenclature

Uca (Australuca) polita sp. nov.

HOLOTYPE. In Smithsonian Institution, National Museum of Natural History, Washington. Male, cat. no. 137667. Type-locality: Australia: Queensland: Gladstone. Measurements on p. 73.

Named in reference to the exceptionally smooth surface of the major cheliped's outer manus. (From Latin *polire*, "to polish.")

Reference and Synonymy

Uca (Australuca) polita sp. nov.

Uca unnamed sp. (pink claw)

Macnae, 1966: 85, 89. Australia: Queensland (south to Moreton Bay). Color; ecology.

III. *THALASSUCA* SUBGEN. NOV.

Typus: *Cancer tetragonon* Herbst, 1790

(Tropical and subtropical Indo-Pacific: East Africa to the Tuamotus; Red Sea to Natal; Nansei [Ryukyu] Islands to northern New South Wales and central Western Australia)

PLATES 13, 14.
GONOPOD DRAWINGS: FIGURES 63, 64.
DENDROGRAMS: FIGURES 96, 99.

MAPS 4, 20.
TABLES 1, 8, 10, 12, 14, 19, 20.

MORPHOLOGY

Diagnosis

Uca with front narrow, antero-lateral margins short to absent; eyestalks slender; postero-lateral striae absent; no tubercles or other irregularities on orbital floor; suborbital margin erect, not rolled out, its crenellations always distinct and usually strong. Major manus outside rough with moderate to large tubercles; oblique ridge on palm never continued upward around carpal cavity; a long, lateral furrow sometimes present outside pollex, never on major dactyl. Serrations of minor cheliped absent or few and weak. Gonopod always with at least one flange well developed, without a projecting, distal tube; entire terminal portion often twisted. Gill on 3rd maxilliped with up to about 11 books in some individuals of each species. Female never with an enlarged tooth in cheliped gape.

Description

With the characteristics of the genus (p. 15).

MALE

Carapace. Front as in *Deltuca*, but margins of central depression never parallel, always diverging posteriorly. Antero-lateral margins absent to moderately short; dorso-lateral margins well-marked to absent, neither margin serrate or beaded, posteriorly moderately to slightly converging; postero-lateral striae absent; sides of carapace little converging, practically straight in posterior two-thirds; vertical lateral margins ranging from distinct throughout to dorsally absent. Carapace profile in lateral view little arched, except in *formosensis*. Suborbital and pterygostomian regions nearly naked to moderately setose; scarcely tuberculate.

Eyebrow when present vertical or oblique, always narrow. Lower margin of orbit not rolled outward but erect, variously sinuous, never with tubercles, granules, ridge, or mound on its floor; suborbital

margin always with strong, close-set, truncate crenellations that extend across entire orbit, rarely excepting the extreme inner angle. Eyestalks slender; diameter of eyes little or not at all greater than that of stalks.

Second Maxilliped. Spoon-tipped setae numerous.

Third Maxilliped. Ischium nearly flat to slightly convex, with or without a slight distal depression; little setose. Gill with up to 11 distinct books but sometimes vestigial.

Major Cheliped. Similar to *Deltuca*, except as follows: oblique ridge inside palm sometimes weak; a large, shallow depression with tubercles reduced in number outside base of pollex; the depression sometimes continued along outer pollex as a lateral furrow; major dactyl without a long outer furrow; row of tubercles on ventral margin of manus continued onto pollex; as in *Deltuca*, however, there is no carina on pollex. Shape of chela and tuberculation of gape widely various, even within subspecies and when among leptochelous or brachychelous individuals of the same population.

Minor Cheliped. Similar to *Deltuca* except that serrations are almost or wholly absent. Gape moderate.

Ambulatories. Meri wide to narrow; armature weak; postero-dorsal tubercles when present always small, usually scattered, sometimes on vertical striae.

Gonopod. At least one flange always well developed; inner process always broad, distinct, sometimes thickened; gonopod tip never, therefore, appears simple and tubular to a casual inspection; thumb well developed, nearly distal but always ending clearly below flange base. Torsion sometimes strong.

FEMALE

With the characteristics of the females in the genus (p. 17).

Size

Medium to large.

Color

The résumé below applies only to *tetragonon* and *vocans*, the color of *formosensis* in life being unknown.

Display lightening sometimes extreme, sometimes absent. Young never more vividly colored than adults. At least lower distal part of outer major manus always distinctively colored, ranging from ochraceous through orange to red. Carapace, where display whitening is not attained, ranging from dull gray to bright blue, often spotted, sometimes passing through a yellow, predisplay phase before whitening. Ambulatories various, ranging from dull gray or brown to bright red.

Social Behavior

Waving display basically of vertical type; serial characteristics present or absent; at high intensities semilateral elements sometimes present accompanied by elevation of carapace. Special high intensity movements lacking. Male usually but not always displays when approaching female. Copulation on surface near female's burrow.

Subsurface vibration has been recorded and surface stridulation observed and recorded in one species (*vocans*). Some combat components are fairly well known in *vocans*.

No construction activities by males have been observed. Some females may erect chimneys.

Relationships

Comparisons with other groups will be found on p. 18, in connection with the discussion of the subgeneric dendrogram. Within the subgenus, *tetragonon* is clearly closer to the less specialized species in *Deltuca* than is *vocans*, judging by the morphology of the major cheliped and gonopod, as well as by the more restricted time spent in waving display. Morphologically, at least, *formosensis* occupies an intermediate position.

Name

Thalassuca: From the Greek noun *thalassa*, "the sea"; the entire name therefore signifies "sea fiddlers," in reference to the fact that members of this subgenus often live on less sheltered shores than is usual among *Uca*, sometimes thrive on oceanic islands, and always burrow close to low-tide levels.

12. *UCA (THALASSUCA) TETRAGONON* (HERBST, 1790)

(Tropical Indo-Pacific)

PLATE 13.   MAP 4.
FIGURES 37 *D*; 63 *A*, *B*; 81 *F*; 82 *E*; 99.   TABLES 8, 10, 12, 14, 19, 20.

INTRODUCTION

A striking red and blue species, *tetragonon* is the most marine of all the genus and has the widest range. From Africa to Tahiti, it often thrives on shores with no more protection than a barrier reef and never lives near muddy river mouths. Although *chlorophthalmus* shares a similar range, that species shows geographical distinctions and the subspecies occur in different biotopes.

In contrast, the populations of *tetragonon* show no differences in the material at hand that encourage either the proposal of subspecies or the discernment of clines.

The populations are locally uncommon, the ecological requirements being obviously more restricted than in many other species. The most likely places to search are islands, either offshore or in mid-ocean, and always in regions rich in living coral. The burrows, always near low-tide level, often end deep in a layer of dead coral covered by silt and sand; sometimes they slant beneath encrusted shells and stones. I have never found these crabs, nor have they been reported, on small atolls such as the Cook Islands; they probably need in their substrate the debris from a rather rich onshore vegetation, and perhaps variations from a wholly marine salinity. Good examples of islands that characteristically support *tetragonon* are Raiatea in the Societies and Green Island, off Massawa in the Red Sea.

A physiological study of the zoeae would doubtless be illuminating. It seems likely that this species will prove to have the ability to delay metamorphosis for a long time, further development being triggered by stimuli near more or less suitable shores.

MORPHOLOGY

Diagnosis

Thalassuca with major pollex and dactyl normally rounded, with furrows absent or, if faintly traceable, not visible distal to middle of length; oblique ridge inside palm indistinct, without enlarged tubercles; major merus with a large, sharp tooth at distal end of its antero-dorsal margin. Gonopod with no noteworthy torsion of tip or inner process; posterior flange slightly larger than anterior; pore large and easily seen. Carapace with orbits oblique, antero-lateral margins short, ill-defined. Minor cheliped with gape scarcely longer than manus. Female: gonopore without marginal tubercles or ridges; an oblong patch of pile on postero-lateral part of carapace.

Description

With the characteristics of the subgenus (p. 75).

MALE

Carapace. Frontal groove of variable width anteriorly even in same population, but usually moderately wide, with the sides diverging rapidly posteriorly, and continuing to form upper margin of eyebrows; fronto-orbital margins with orbits moderately oblique, little sinuous; antero-lateral angles slightly produced, acute; antero-lateral margins short, converging from the angles, usually distinguished from dorso-lateral margins by a slight bend in the line of granules demarcating the margins. Carapace not highly arched, the branchial chambers not notably elevated. Eyebrow narrow, almost vertical, usually traceable to near outer angle of orbit. Suborbital crenellations strong, distinct, extending from extreme outer angle of orbit to its inner boundary; crenellations in outer half well separated from each other; near inner boundary they are smaller and close-set but well formed; lower side of antero-lateral angle with apex sometimes serrate beneath.

Major Cheliped. Antero-dorsal margin of merus without a crest, but with an irregular row of tubercles and a large, sharp tooth subdistally; ventral margin with several irregular rows of fine granules; postero-dorsal margin rounded with transverse rugosities of similar fine granules, extending slightly postero-laterally; merus otherwise smooth. Carpus externally with granules so small that the segment is almost smooth; its dorsal margin with a small proximal crest marked by one to several small tubercles. Tubercles of outer manus larger than those of carpus but small compared with those of *vocans* or the subgenus *Deltuca*; they are slightly enlarged around base of pollex, and absent both from a submarginal region along base of gape and lower part of dactyl base, and to various degrees from center of the large, shallow depression at base of pollex. This depression is of variable extent, even within the same population; always triangular, it extends along the pollex itself ending with its attenuated acute angle more than one-third the distance to pollex tip; hence it lies in the usual position of the pollex groove found in *Deltuca* and *Australuca*. Oblique ridge inside palm low and blunt, not marked with enlarged tubercles; at base of gape the proximal of the two ridges has large, distinct tubercles; opposite dactyl base however it diverges widely from the vertical and proceeds up and back as small tubercles merging with those covering the palm; distal ridge at dactyl base represented only by numerous similar small tubercles, not in rows. Palm with a moderate depression at pollex base. Pollex and dactyl normally rounded, slender and tapering, rather short in proportion to manus, when compared with *vocans* or many other *Uca*. Lower margin of pollex and upper margin of dactyl normally variable, often arched, never straight. Pollex and dactyl practically always without a trace of grooves externally, except for the forward extension of the triangle at base of pollex; very rarely in an oblique light a trace shows of one or two faint, very shallow depressions, subdorsally or both subdorsally and laterally, near dactyl's base; the tubercles are never interrupted here and the surface unevenness is not apparent except in a detailed inspection. Pollex and dactyl both covered externally with small tubercles about as large as those on upper half of manus; smaller tubercles or granules usually almost cover the inner surfaces as well. Gape often with two slightly enlarged tubercles on pollex at middle and beyond; and with several similar or smaller enlarged tubercles scattered along middle section of dactyl; no trace of a forceps-like tip to chela.

Minor Cheliped. Fingers short, scarcely longer than manus; base of pollex unusually broad; serrations feeble, few, variable; gape slight. No continuous fringes of setae except for subdistal tuft.

Ambulatories. Meri slightly expanded, about midway between the very broad segments of *formosensis* and their slenderness in *vocans*; tuberculate striae well developed in comparison with *vocans*.

Gonopod. Inner process broad, not swollen, not twisted, lying against distal end of shaft in normal position; anterior flange slightly narrower than posterior, which itself is narrower than diameter of the conspicuous pore. Thumb of moderate size, not nearly reaching base of flange.

FEMALE

Except for the usual secondary sexual differences, the carapace and appendages closely resemble those of the male. Carapace with a small, oblong patch of pile above base of last ambulatory; firmly attached, it is rarely missing and develops before maturity. Margin of gonopore smooth.

Measurements (in mm)

	Length	Breadth	Propodus	Dactyl
Largest male (Zanzibar)	21.0	34.0	49.0	36.0
Largest male (Tahiti)	21.5	33.5	55.0	36.0
Moderate male (Massawa)	15.5	23.0	31.0	17.5
Largest female (Massawa)	23.0	32.0	–	–
Largest female (ovigerous) (Zanzibar)	17.5	25.0	–	–
Largest female (ovigerous) (Bora Bora)	22.0	30.0	–	–
Smallest ovigerous female (Massawa)	12.5	18.0	–	–
Smallest ovigerous female (Tahiti)	14.5	20.0	–	–

Morphological Comparison and Comment

U. tetragonon differs markedly from the other species in the subgenus, *formosensis* and *vocans*, in the lack of unusual compression of pollex and dactyl, in the lack of a strong groove on pollex; in the absence of a pronounced, tuberculate ridge inside palm; and in the short fingers of minor cheliped. It differs additionally from *formosensis* in the uncontorted inner process of gonopod, with larger anterior flange; in having a single large tooth antero-dorsally on distal end of major merus; in the oblique orbits and short, indistinct antero-lateral margins. It differs additionally from *vocans* in the less converging carapace sides and slightly expanded ambulatory meri. The relatively simple gonopod tip, lacking torsion, differs sharply from that of all subspecies of *vocans* except *v. borealis* and *v. pacificensis*. The female is easily distinguished from that of *formosensis* and *vocans* by the distinctive patch of pile on postero-lateral carapace and by the gonopore's smooth edge; additional differences are given in the discussion of *vocans* (p. 86).

U. tetragonon differs from all species of the subgenus *Deltuca* in the strong crenellations throughout lower margin of orbit, in the virtual absence of grooves on dactyl and pollex, and the form and position of the female's patch of pile.

It differs from most species of the subgenus *Australuca* in the absence of tubercles on orbital floor and absence of an enlarged tooth in minor chela, and from all in the lack of a produced tubular tip to the gonopod.

The species differs from *chlorophthalmus* in the much narrower front, which is expanded anteriorly, in the rough cheliped, in the presence of a large tooth distally on major merus, and in the wholly different gonopod, with subterminal thumb and distinct flanges, the terminal end of the shaft not being produced.

As indicated in the introduction, any subdivisions of the species do not seem to be warranted on the basis of the available material. The species, as usual in widely ranging forms, is variable; allometry causes the inevitable complications; gonopods show minimal differences; collections from all localities are small.

Color

The brilliant coloration of this crab is characteristic in at least two regions almost half a world apart: the Red Sea and the Society Islands. Here display coloration shows a turquoise blue carapace with white spots and coral red appendages.

In more detail the color is as follows: Carapace often dark, finely marbled at least in male and especially anteriorly with buffy, yellow, green or green blue. Large females often have a posterior pair of large, round, blue spots, one just above the base of each posterior ambulatory; middle-sized females are often marked with bluish reticulations on gray backs. As display lightening develops, the carapace shows a ground color of marine blue to turquoise, with transverse markings or spots of pale blue to white; pale buffy anterior markings may persist. All-white carapaces are not attained. No red ever appears on dorsal part of carapace.

Pterygostomian regions and 3rd maxillipeds blue or red; sides of carapace, sternum, and abdomen usually with red markings. Eyestalks dark to mottled pink to yellowish. Both chelipeds and all ambulatories bright red, usually orange red, at least in larger displaying males, and usually in an entire population at an active time of the day. Outer surface of major cheliped sometimes paler than inner surface. Lower outer manus sometimes with a red spot distinct from the surrounding paler color, but sometimes not standing out from the suffusing red of the appendage; this spot has been mentioned in old descriptions as persisting in preserved specimens. Major chela pink to white.

SOCIAL BEHAVIOR

Waving Display

Wave vertical, but with cheliped held consistently, and brought to rest between waves of a series, at an acute angle to the front. Wave low, the tip of chela rarely reaching tip of eye; at this height the cheliped is unflexed, the mero-carpal joint being held low. Jerks absent. Waves strongly serial, 3 to 10 waves in each series in both western and eastern parts of Indo-Pacific region. Body very slightly raised during each wave so that it is at most level with mero-carpal joints

of ambulatories. Walking during waving is usual, and males without territories sometimes wave. Waving rate: more than 2 waves per second. (Waving components: 1, 3, 9, 10; timing elements: Table 19, p. 656.)

Precopulatory Behavior

Males on Bora Bora and at Massawa seen to approach females, with or without waving, and cover them on the surface.

Chimney Construction

Some females in east Africa inhabited chimneys, to which they at least added material, brought as usual from some distance away. Once in Zanzibar a male followed such a female through the chimney into her burrow.

Acoustic Behavior

At the surface major-merus-rubs and leg-wags were observed and filmed (components 1 and 5). Underground sounds were recorded on tape in Massawa (biotope no. 4).

Combat

A single brief combat was filmed in Zanzibar. A heel-and-hollow (component 11) preceded the forceful end.

RANGE

U. tetragonon ranges throughout the Indo-Pacific from East Africa to the Tuamotus, from the Sinai Peninsula to Madagascar, and from Wake Island to subtropical Australia.

BIOTOPES

Lowest tidal levels, often barely protected by a strait or reef from open ocean. Often among conglomerates of mixed coral and shell, on mud or muddy sand, sometimes with underlying dead coral. Further comment in the introduction to this species. (Biotope nos. 4, 5.)

SYMPATRIC ASSOCIATES

In Massawa and Zanzibar *U. tetragonon* and *vocans* were found mingling, as described on p. 88. On islands of the tropical Pacific *chlorophthalmus crassipes* (subgenus *Amphiuca*), the only other *Uca* found east of Samoa, often lives adjacent to *tetragonon* on the next higher biotope; the two populations sometimes overlap.

MATERIAL RESULTING FROM FIELD WORK

(The complete list of specimens is given in Appendix A, p. 596.)

Observations and Collections. Ethiopia: Eritrea: Massawa. Zanzibar. Philippines: Tawi Tawi. British Samoa: Apia. Society Islands: Bora Bora; Raiatea; Tahiti.

Film. Ethiopia: Massawa. Zanzibar. Tahiti.

TYPE MATERIAL AND NOMENCLATURE

Uca tetragonon (Herbst, 1790)

TYPE of *Cancer tetragonon* Herbst. Apparently not extant. Originally in Herbst collection, Berlin.

NEOTYPE. In Muséum National d'Histoire Naturelle, Paris. One male, selected by J. Crane from a group of 3 males and 3 females, all dried and in good condition; they were distributed among 3 boxes, 1 male and 1 female in each, from 3 localities; each box was marked "Type," both on the inside label and on the bottom of the box. The specimen selected is from Box 1 and labeled, both inside and on the bottom, in full, as follows: "*Gelasimus tetragonon* Herbst. Type. Egypte."

The other specimens are labeled as follows: Box 2: "*Gelasimus tetragonon* Herbst. Type. A. Milne Edwards det. Collection A.M. Edwards 1903. Nouvelle Caledonie." Box 3: "*Gelasimus tetragonon* Herbst. Type male, female. 792.66. Iles Sandwich."

All the specimens are characteristic examples of *U. tetragonon.*

The handwriting on the inside labels of all the boxes is the same, and different from that on the boxes' undersides, which were obviously labeled by another person. Herbst (1790) ends his type description with the words: "Das Vaterland ist mir unbekannt." It seems that A. Milne-Edwards probably selected the above 6 specimens in the Paris museum to stand in place of Herbst's presumably missing type. H. Milne-Edwards does not comment on these specimens in his account of *tetragonon* (1852: 147). Since no designation was made of a single neotype, the male from Egypt is now suggested for this role.

The Hawaiian ("Iles Sandwich") locality on the label in the third box may well be an error for Bora Bora, as is apparently the case with similarly labeled specimens of *chloropthalmus.* While it is of course possible that both these species formerly occurred in Hawaii, it seems more likely that a substitution of place names has occurred (see p. 597 and Edmondson, 1933: 270).

Type Material of *Gelasimus duperreyi* Guérin, 1826. Not found in Muséum National d'Histoire Naturelle, Paris, or in Academy of Natural Sciences, Philadelphia. Synonymy of this name with *tetragonon* based on type illustration (Pl. 1) and my complete lack of success in finding any specimens throughout *tetragonon*'s wide range which are not referable to the single form, *tetragonon.* Milne-Edwards (1852), Kingsley (1880.1), and Alcock (1900) also synonymized them.

Type Material of *Gelasimus variatus* Hess, 1865. Examined in museum at Göttingen. Three males in 2 vials, each with outer label on red paper, indicating type material. Cat. no. 52a. Outer label: "*Gelasimus variatus.* Schütte 1864. Sydney." Inner label: "*Gelasimus variatus* identisch mid *Gelasimus tetragonon* (Herbst) Rüpp. Teste Dr. J. G. De Man 12/4 1886." Contents: 2 males; measurements in mm: length 15, propodus 31; length 15, propodus 28. Cat. no. 52b. Outer label: "*Gelasimus variatus.* Hess. Schütte 1876. Australien." (No inner label.) Measurements in mm: length 17; propodus 41. My examination of the specimens leads me wholly to agree with de Man's conclusion on no. 52a; *G. variatus* is therefore here synonymized with *U. tetragonon.* Milne-Edwards (1837: 52) in a footnote to his account of *G. tetragonon* ends the list of references with the entry: "*Gelasima variegata*, Latr. Coll. du Mus. (fem.)." This specimen was not found in the Muséum National d'Histoire Naturelle at Paris when I examined the collection in 1959.

Type Material of *Gelasimus tetragonon* var. *spinicarpa* Kossmann, 1877. Red Sea. In the Rijksmuseum van Natuurlijk Historie, Leiden, under cat. no. 1493, is a single male with the following label: "*Uca tetragonon* (Herbst) var. *spinicarpus* Kossm. Kossmann 1880. Roode Zee." Measurements in mm: length 12; propodus 23. Examination shows this crab to be a young example of *Uca vocans*, leptochelous, with the claw probably regenerated. On the basis of this specimen, apparently identified by Kossmann and presented by him to the museum, the variety is here referred to the synonymy of *U. vocans.*

Type Material of *Gelasimus desjardinii* Guérin, MS. Mauritius: Not found in Muséum National d'Histoire Naturelle, Paris. Identification of this unpublished form made on the basis of dried specimens from Mauritius in the Academy of Natural Sciences, Philadelphia. Kingsley (1880) also referred this material to *tetragonon.*

Note on the spelling of *tetragonon.* Several of the authors cited below under References and Synonymy used the spelling *tetragonum.* No distinction is made in the references between the two endings, there being no apparent justification for the -*um* termination. Some other authors listed under *Uca tetragonon* give the species name the feminine ending *a*, treating it as an adjective. This practice is also disregarded in the list, the original name, *tetragonon*, being considered a substantive in apposition, not an adjective in agreement.

REFERENCES AND SYNONYMY

Uca (Thalassuca) tetragonon
(Herbst, 1790)

Cancer marinus, minor, vociferans

Seba, 1758 and 1761. Vol. 3: 48; Pl. 19, Fig. 15 (a female).

Cancer tetragonon*

TYPE DESCRIPTION. Herbst, 1790: 257; atlas, Pl. 20, Fig. 110 (after Seba).

Ocypode tetragona**

Olivier, 1811: Vol. 8: 418. East Indies. No new material.

Goneplax tetragonon

Latreille, 1817.1: 17. Reference to illustration of type of *Cancer tetragonon*.

Gelasimus duperreyi

Guérin, 1829: Pl. 1, Figs. 2, 2 *A*. Bora Bora. Illustrations in lieu of type description.
Dana, 1852: 317. Tongatabu; Upolu. Taxonomy.

Gelasimus tetragonon

Rüppell, 1830: 25; Pl. 5, Fig. 5; Pl. 6, Fig. 20. Red Sea: near Massawa. Taxonomy; color; brief habitat and habits.
Milne-Edwards, 1837: Vol. 2: 52. Red Sea; Mauritius. Taxonomy.
Guérin, 1838: 10 (part). Synonymizes *G. duperreyi*, represented in Guérin, 1829, by Pl. 1, Figs. 2 and 2 *A*, with *G. tetragonon*. He also synonymizes with *G. tetragonon* another species, *G. affinis*, which is represented by Fig. 3 on the same plate and illustrates a major claw. For comments on the claw's identity and the taxonomic status of *G. affinis*, see p. 322.
White, 1847: 36.
Milne-Edwards, 1852: 147. Red Sea; Tongatabou. Taxonomy.
Heller, 1865: 37. Red Sea; Nicobar Is.
Hilgendorf, 1869: 84. Zanzibar. Taxonomy.

Hoffmann, 1874: 16. Madagascar: Nossy-Bé. Taxonomy.
Kossmann, 1877: 52. Red Sea.
Kingsley, 1880.1: 143. Taxonomy.
Miers, 1886: 243. Tahiti; Arrou Is. Taxonomy.
de Man, 1887.2: 353. Noordwachter I.
de Man, 1891: 24. Tahiti; Samoa. Taxonomy.
Ortmann, 1894.2: 754. Tahiti; Salanga.
Whitelegge, 1898: 138. Ellice Is.; Funafuty I. Taxonomy.
Alcock, 1900: 357. Andaman Is. Taxonomy. First use of spelling *tetragonum*.
Sewell, 1913: 344. Southeast Burma: Tavoy I. Habitat; color.
Bouvier, 1915: 303. Seychelles: Grand Port. Taxonomy.

Gelasimus variatus

Hess, 1865: 146. Australia: Sydney. Type description.
Kingsley, 1880.1: 154. Description from type description.

Gelasimus tetragonon var. *spinicarpa*

Kossmann, 1877: 52.

Uca tetragonon

Ortmann, 1897: 348. Tongatabou.
Doflein, 1899: 193. Sinai Peninsula.
Lanchester, 1900.1: 754. Singapore.
Lanchester, 1902: 549. Malaya: Trengganu.
Rathbun, 1902.2: 123. Maldive Is.
Nobili, 1906.2: 151. Red Sea: Ilot Ente Ara. Taxonomy; color.
Nobili, 1906.3: 313. Red Sea. Taxonomy.
Rathbun, 1907: 26. Society Is.: Bora Bora. Friendly Is.: Tongatabou. Gilbert Is.: Tarawa.
Borradaile, 1907: 66. Seychelles: Mahé. Chagos Arch.: Barachbis, Diego Garcia, Chagos. Specimens in mangrove swamp at Mahé. [This atypical habitat for the species needs checking.]
Nobili, 1907: 408. Polynesia: Rikitea; Gatavake Kirimino. Taxonomy.
Laurie, 1915: 416. Red Sea.
Tesch, 1918: 39. East Indies: Bay of Bima, Sumbawa, Karakelang, Talaud Is., Kur I., west of Kei Is. Taxonomy.

---

*Not *Cancer tetragonus* of Fabricius, 1798: 341.
**Not *Ocypoda tetragona* of Bosc, 1802: 198, or Bosc, edited by Desmarest, 1830: 251.

Gravier, 1920: 472. Madagascar.

Balss, 1924: 15. Red Sea.

Edmondson, 1925: 59. Central Pacific: Wake I. (Bishop !)

Maccagno, 1928: 22; Text Figs. 6, 7, 8. Red Sea: Isole Key; Massawa. Persian Gulf. New Guinea. Taxonomy. (Torino !, except specimens from Persian Gulf.)

Edmondson, 1933: 270; in 1946 edition: 311. Not found in Hawaii. Brief taxonomy.

Balss, 1938: 75. Gilbert Is.; Marshall Is.

Estampador, 1937: 543. Philippines.

Holthuis, 1958: 52. Red Sea: Sinai Peninsula (Abu Zabad). Carapace breadth 33 mm.

Miyake, 1959: 223, 242. Gilbert Is.; Marshall Is.

Estampador, 1959: 100. Philippines.

Forest & Guinot, 1962: 70. Distribution.

Uca duperreyi

Ward, 1939: 14. Tongareva (Penryn). Taxonomy.

13. *UCA (THALASSUCA) FORMOSENSIS* RATHBUN, 1921

(Subtropical Indo-Pacific: northwest Taiwan)

PLATE 14 *A-D*.   MAP 4.
FIGURES 63 *C*; 99.   TABLE 8.

INTRODUCTION

This elusive species appears to be known from less than a dozen preserved specimens. I was unable to find it in northwest Taiwan at the type-locality in its only known habitat. According to Takahasi (1935), it lives on muddy shores near low tide level in the neighborhood of a river.

The species obviously belongs with *tetragonon* and *vocans* in the subgenus *Thalassuca*; its differences from both, equally obviously, are of specific rank. It seems likely that *formosensis* is the northern, allopatric representative of *tetragonon*, a species that has not been recorded in this latitude. Because of the scanty material, it is not listed here as an allopatric species under a superspecies, *tetragonon*.

MORPHOLOGY

Diagnosis

Thalassuca with major pollex and dactyl flattened, furrows very faint or absent; oblique ridge inside palm low but with its tubercles moderately large and fairly regular; major merus with a cluster or row of small, sharp tubercles at distal end of antero-dorsal margin. Gonopod with anterior flange rudimentary, posterior flange large; inner process broad, thickened, strongly curved forward. Carapace with orbits almost straight; antero-lateral margins well developed. Minor cheliped with fingers much longer than manus. Merus of first 3 ambulatories broad; that of 4th moderate but in male dorsally almost straight. Female: dorso-lateral margins strongly beaded; gonopore with outer, posterior quarter of margin raised; no pile on postero-lateral part of carapace; 4th ambulatory with merus convex.

Description

With the characteristics of the subgenus (p. 75).

MALE

Carapace. Tip of front expanded; groove with anterior margin narrowly U-shaped. Fronto-orbital margin practically straight; antero-lateral angles little produced, almost rectangular; antero-lateral margins strongly developed, scarcely converging, sometimes even diverging slightly on major side; then curving in to form convergent, well-marked, dorso-lateral margins. Carapace highly arched, the branchial chambers strongly tumid. Eyebrow broad internally, narrow externally, almost vertical, traceable to middle of orbit or beyond. Suborbital crenellations low; distinct, except sometimes at internal angles.

Major Cheliped. Merus with a low, sharply tuberculate, antero-dorsal crest, highest distally, where it sometimes ends in a cluster of small, sharp, unequal tubercles; in one of the type specimens the distal tubercles remain in a single line surmounting the crest; ventral margin of merus with two rows, almost regular, of similar tubercles; postero-dorsal margin rounded and indistinct, rugose with irregular small, blunt tubercles. Carpus with dorsal margin proximally crested, its tubercles continuing those on outer surface. Tubercles of outer manus not larger than those on carpus, low, well separated; those on lower half sometimes in a faintly reticulate pattern; at pollex base a large, slightly depressed, non-triangular area with its tubercles much smaller than those nearby. Oblique ridge inside palm low but definite with distinct tubercles in an almost regular row; proximal ridge at base of dactyl well developed and tuberculate; distal ridge variable, with the tubercles strong or weak, sometimes irregular. Central palm almost smooth, the granules being minute and well-separated. A moderate depression inside palm at pollex base. Fingers flattened, moderately slender, pollex lower margin and dactyl's upper one practically straight, not arched. Pollex smooth except for a low ridge near outer lower margin, marked proximally by tubercles that start on lower manus and die out before end of pollex; against upper side of this ridge runs a slight crease; both crease and ridge essentially form a submarginal rim along pollex edge, unlike the lateral furrow of *vocans* and *Deltuca*, and are not continuous with the depression at pollex base. Dactyl smooth inside and out except for tubercles proximally and dorsally; subdorsal furrow moderate; laterally a faint furrow is traceable only in middle region. Tubercles of gape very small except for an enlarged tooth in pollex and one or two in dactyl, all proximal to middle; distal to this point gape is almost lacking, its edges meeting through most of its length; no trace of a forceps-like tip to chela.

Minor Cheliped. Fingers much longer than manus. Gape extremely narrow, practically absent in distal two-thirds where alone minute serrations occur; horny tip broad, obliquely spatulate.

Ambulatories. All meri wide compared with other *Thalassuca,* the 2nd and 3rd notably so; nevertheless dorsal margin of 4th is almost straight.

Gonopod. Inner process long, broad, thick; almost its entire visible portion is bent obliquely forward, at right angles, so that its tip lies across base of pore; anterior flange very narrow; posterior flange broad and truncate, little curved, much wider than pore which, however, has a moderately large diameter. Thumb of moderate length, not nearly reaching base of flange.

FEMALE

Antero-lateral and, especially, dorso-lateral margins strongly beaded. No pile on postero-lateral carapace. All ambulatory meri broad, more or less convex dorsally, including 4th. Gonopore with entire postero-external quarter of rim elevated but not tubercular in form.

Measurements (in mm)

	Length (Rathbun)	Breadth (Rathbun)	Propodus	Dactyl
Holotype male	18.4	28.8	61.0	47.5
Smaller type specimen, male	17.0	27.6	56.0	40.0
Moderate male	15.5	25.0	39.0	26.5
Largest female	18.0	28.5	–	–

Morphological Comparison and Comment

U. formosensis differs clearly from the other two species in the subgenus in having small, spinous tubercles, not a single large spine or tooth, antero-dorsally on distal end of major merus, in the broader merus of all ambulatories in both sexes and in the thickened and twisted inner process of the gonopod. Differs additionally from *tetragonon* in the flattened major pollex and dactyl and in having distinct and regular

tubercles on oblique ridge inside palm, while the female lacks pile on posterior part of carapace, and has a beaded dorso-lateral margin. Differs additionally from the sympatric *vocans borealis* in having posterior, not anterior, flange of gonopod the larger. Differs from *arcuata,* also found in the same locality, in the flattened major pollex and dactyl, virtually without furrows, in the small size of tubercles on outer manus, in the thick and twisted inner process of gonopod, and in the form of lower orbital margins, which lacks a blunt ridge behind edge, this margin being erect, as in all *Thalassuca,* not rolled out as is usual in *Deltuca.*

With its straight orbits, distinct antero-lateral margins, arched profile, and tumid branchial regions the carapace is similar to that of *arcuata* and other species subjected to periodic desiccation (p. 451).

SOCIAL BEHAVIOR

Combat

"In those species which are small in population the fighting habits are generally well developed and this is most remarkable in *Uca formosensis*" (Takahasi, 1935).

RANGE

Northern Taiwan from central part of west coast to extreme northeast.

BIOTOPES

According to Takahasi (1935), *formosensis* occurs on muddy beaches near the mouth of the Tamsui River, both in clayey mud with *arcuata* and in the "vegetation" zone, associating with *vocans* and *arcuata.*

TYPE MATERIAL

Uca formosensis Rathbun, 1921

TYPE MATERIAL. In Smithsonian Institution, National Museum of Natural History, Washington. Holotype, cat. no. 54472. Formosa: Taichu: Rokko. One additional male marked "type" from the type-locality. Measurements in mm above. (!)

REFERENCES AND SYNONYMY

Uca formosensis

TYPE DESCRIPTION. Rathbun, 1921.3: 155. Formosa. (!)
    Takahasi, 1935: 78. Formosa. Habits.

Gelasimus formosensis

    Sakai, 1939: 620. Formosa. Taxonomy; habitat.
    Sakai, 1940: 58. Distribution (Formosa only).
    Lin, 1949: 26. Formosa.

14. *UCA (THALASSUCA) VOCANS* (LINNAEUS, 1758)

(Tropical Indo-Pacific)

PLATES 14 *E-H*; 47 *A*.
FIGURES 38 *I-L*; 56 *B*; 60 *C-E*; 64 *A-FF*; 92; 99.
MAP 20.
TABLES 3, 8, 10, 12, 14, 19, 20, 22.

INTRODUCTION

This Indo-Pacific fiddler is undoubtedly one of the most abundant in the world. Because of its exposed habitat, *vocans* is also one of the easiest to find and observe. Populations thrive in the sandy mud along the edges of protected bays. Although the crabs are most numerous near river mouths, freely surging tides seem essential. This species lives closer to low-tide levels than does *lactea*, its usual associate, although the populations are often mixed in the middle levels.

Throughout most of its wide range *vocans* can be recognized promptly by the yellow to orange red color on the lower manus and pollex. Displaying crabs in some populations change color strikingly and completely, shifting from gray to yellow to pure white in a fashion similar to some species in the American subgenus *Uca*.

For many years *vocans* was best known as *marionis* and its variety, *m. nitida*. These two forms, fully sympatric, represent merely strong examples of a claw dimorphism common in the genus. Holthuis (1959) resurrected the appropriate older name used by Linnaeus, who in turn was preserving the name given by Rumphius in 1705.

The species divides readily into six subspecies, which occasionally show the usual signs of mingling in the Sunda region. The forms are distinguished primarily by morphological details of gonopods and gonopores, by visual display rhythms, and, sometimes, as mentioned above, by cheliped color. As in the subspecies of *dussumieri* and *lactea*, the most specialized form morphologically and behaviorally is the one occurring in the crowded Philippines-Java axis. The least specialized morphologically is the subspecies found in subtropical Hong Kong and west Taiwan; behaviorally, the least specialized is the subspecies in northwest Australia, in which the waving display is strongly serial. These forms, especially in places where they seem to hybridize, offer superb opportunities for further study.

MORPHOLOGY

Diagnosis

Thalassuca with major pollex and dactyl flattened, the dactyl especially notably broad, without furrows except for basal traces subdorsally; pollex with deep outer furrow in basal two-thirds; ridge inside palm high, its tubercles strong; tubercles on outer margin of palm large, particularly near depression at pollex base; major merus with a large, sharp tooth at distal end of its antero-dorsal margin. Gonopod usually with entire tip, or some of its elements, twisted; in subspecies without twisting anterior flange is wider than posterior. Carapace in male with orbits oblique, antero-lateral as well as dorso-lateral margins absent or nearly so. Minor cheliped with gape much longer than manus. Merus of all ambulatories slender. Female: gonopore with marginal tubercle or other structures; no pile on postero-lateral part of carapace; dorso-lateral margin strongly beaded.

Description

With the characteristics of the subgenus (p. 75).

MALE
Carapace. As in *tetragonon* (p. 77), except for the more converging carapace sides in most subspecies and frequent absence both of an antero-lateral margin and of any line of granules demarcating a dorso-lateral margin.

Major cheliped. Merus as in *tetragonon*, except that the postero-dorsal ridge distally often is distinctly crested, with or without tubercles; a second, small tooth, or short low crest, sometimes present at base of large subdistal tooth on antero-dorsal margin. Carpus dorsally with a single medium-sized small tooth or low crest culminating in the tooth; outer surface of carpus completely smooth except in dorsal and distal parts, where there are granules or small tubercles, strongly developed only in the most northern and eastern subspecies. Tubercles of outer manus large, but varying with the subspecies, largest on lower half of manus, absent from shallow, triangular depression at pollex base. In *vocans* this depression is extremely large in all subspecies, one angle reaching proximal to middle of manus, another almost to the lower edge of dactyl, and the third, as in *tetragonon*, well out on pollex, whence in *vocans* it continues still further as a groove, usually pronounced and always distinguishable; depression and groove sometimes with pile, easily detached. Oblique ridge inside palm high, thin, sharp, except in *borealis* where it is neither thin nor especially high; in all subspecies it is usually crowned with close-set tubercles; proximal ridge at dactyl base lower but similar, highest opposite ventral margin of dactyl; distal ridge obsolescent or absent. A moderately deep depression inside palm at pollex base. Palm between and on ridges covered with small to fine tubercles or granules. Pollex and dactyl strongly compressed and notably broad, even in the most leptochelous specimens; base of dactyl particularly broad. Except in *borealis* and, weakly, in *vomeris*, no trace of grooves on dactyl except for one or two short subdorsal depressions among the tubercles; pollex groove described above in connection with depression at pollex base. Pollex and dactyl practically smooth externally except for minute granules near gape and for small tubercles, proximally and dorsally, on dactyl. Both edges of gape notably sinuous and uneven, the pollex often having two large triangular projections margined with fine tubercles, one occurring near gape's middle, the other subdistally. These projections, characteristic of Milne-Edwards' *marionis nitidus*, occur in all populations known to me except in the subspecies *dampieri* and *vomeris*, where the median projection seems never to appear; in all others this striking pollex represents merely the broad form of chela characteristic of all *Uca*, but especially notable in Indo-Pacific narrow-fronted species, including *vocans* (see p. 459 and Fig. 38). No trace of a forceps-like tip on chela.

Minor Cheliped. Fingers clearly longer than palm, serrations absent or nearly so, gape varying with the subspecies from moderate to broad.

Ambulatories. Fifth merus always slender; others slender to slightly expanded. Armature weak even for this subgenus, with marginal serrations small, subdorsal spinules minute and posterior surfaces almost or wholly unarmed.

Gonopod. Highly various among the subspecies, but scarcely variable within them. These intrasubspecific differences appear to be confined to minute details of proportion, as in individual flanges or thumb shape. Inner process always broad and distally flat; sometimes its base is greatly swollen and overhangs the

flanges and pore; sometimes the entire structure twists toward outer side of shaft. Flanges always present, sometimes also showing torsion, but toward the inner side, opposite from the direction of twist by inner process; either anterior or posterior flange the broader, depending on subspecies. Pore with diameter small, always smaller than breadth of the wider flange. Thumb various, ranging from short and not reaching base of flange, to long and reaching beyond its base; often twisted in accordance with the torsion of the inner process.

FEMALE

Except for the usual secondary sexual differences, the carapace and appendages closely resemble those of the male. No pile on postero-lateral region of carapace. Gonopore margin always with one or more tubercles or thickened ridges.

Measurements (in mm)

Maximum length about 16; additional data included in descriptions of subspecies.

Morphological Comparison and Comment

U. vocans differs markedly from the other species in the subgenus, *formosensis* and *tetragonon*, in the distinct furrow extending along most of pollex, in the high, usually thin, oblique ridge inside manus, and in the notably slender meri of the ambulatories. Differs additionally from *formosensis* in having a single large tooth antero-dorsally on distal end of major merus, in the more oblique orbits, and in the lack of marked antero-lateral margins. Differs additionally from *tetragonon* in the more converging carapace and, in most of its subspecies, in the twisted tip of the gonopod.

Differs from all species of the subgenus *Deltuca* in the presence of strong crenellations across entire breadth of suborbital margin and in the absence of a distinct groove running most of dactyl's length.

Differs from *Australuca* and *Amphiuca* in ways exactly similar to the differences listed under *tetragonon* (p. 78).

Female *vocans* differ clearly from some females of the subgenus *Deltuca* and all *Australuca* in lacking a large tooth in the cheliped. Differ from females of *tetragonon* in having marginal structures along the edge of the gonopore. Differ additionally from *tetragonon* as follows: no pile on postero-lateral region of carapace; orbits less oblique; cheliped manus and chela less deep with fingers longer, their dorsal and ventral margins always regularly fringed with long setae.

Morphological Bases of Subspecies (pp. 90ff.; Table 3; Fig. 62). The subspecies are characterized distinctly by differences in gonopods and gonopores; no differences between or within populations have been found of a kind to make questionable the subspecific identification of an individual through its gonopod alone; these characters remain distinct even in zones of apparent hybridization.

Secondary distinctions that set off one or more subspecies consist of tubercle size on outer manus, height of oblique ridge on palm, shape and grooving of dactyl, absence in all individuals of median *nitida* projection, and presence of dorso-lateral carapace margin in male. Other differences occur principally in males, but are too variable to be of much taxonomic use; they include eyebrow width, shape of frontal groove, and development of antero-lateral margins. In all subspecies the females and young males have eyebrows and dorso-lateral margins well developed, while antero-lateral margins are distinct.

In contrast to the form of the gonopod, the most distinct of the secondary subspecific characters, and sometimes also the female's gonopore, show strongly mixed characteristics in zones of subspecific overlap. In both New Guinea and the South Philippines, where *pacificensis* extends westward to mingle rarely with *vomeris* and *vocans*, respectively, the gonopod remains clearly *pacificensis*, a relatively simple structure, and the gonopore usually also as in *pacificensis*. The dactyl's breadth, markedly wide in *pacificensis*, appears intermediate, and the dorso-lateral marginal line, wholly absent in males of *vomeris* and *vocans*, is always distinct as in *pacificensis*, though almost always shorter than farther east. In the rare specimens with gonopods like *hesperiae* that have been taken in Malaya and Singapore, secondary differences are too slight and the material too rare to make comment desirable at this time.

Female gonopores are not oriented in conformity with the torsion of the gonopod but rather are adapted to it through the varying development and location of marginal ridges and tubercles (p. 465).

Color

Characteristic of all subspecies in all phases is an orangeish patch on the lower manus and base of pollex. The only exceptions are rare specimens attaining complete display white. The spot ranges from dull ochre in individuals not in full display color through bright orange or flaming orange red to pale yellow in specimens otherwise brilliantly white. Display white of at least the carapace and usually the underparts is developed throughout the range in both sexes, but in most populations many non-white individuals display. The buccal and pterygostomian re-

gions often whiten before the rest of the carapace. During the whitening process the major cheliped is usually yellow except for the deeper colored manus and pollex base. The ambulatories whiten only rarely, often brightening only from chocolate to purple; in females the anterior faces of the meri often are bright purplish red, as are the minor chelipeds in both sexes. On Inhaca Island, Mozambique, yellow carapaces and chelipeds were first reported by Gordon (1958); this yellow phase also occurs occasionally in Fiji. There is a clear similarity in all these display and pre-display colors to those of *stylifera* among the members of the American subgenus *Uca*, and in cheliped characteristics to *heteropleura* in the same group.

In contrast, in northwest Australia pre-display color is characterized by a carapace blue with white spots, shifting to white with blue and turquoise spots, and finally, as usual, to white. The pre-display phase in Zanzibar is dull with pale gray spots. In Singapore the lightening commonly commences with horizontal pale streaks. Throughout the geographical range of the species the non-displaying phase varies from dull gray to dark brown and is often mottled.

In the observed populations of Fiji, New Caledonia, eastern Australia, northwest Australia, and, rarely, of Zanzibar, the proximal major dactyl, regardless of the degree of display white attained, ranges from violet to pale pink, rarely continuing to white; in contrast, in all observed populations in Taiwan, Hong Kong, the Philippines, northern New Guinea, Malaysia, west India, and east Africa, save for the noted Zanzibar exceptions, the major dactyl is clear yellow, often continuing to white. In Zanzibar and New Guinea this violet pink versus yellow distinction does not coincide with the prevailing dactyl color of the local subspecies.

SOCIAL BEHAVIOR

Waving Display

Wave vertical, often tending toward a semi-unflexed wave, depending on intensity and on subspecies. Upper edge of dactyl reaching at most slightly above the tip of eye. Jerks absent. Waves non-serial except at middle and high intensities in northwest Australia (*v. dampieri*) and, in at least the New Caledonia population of *v. vomeris*, at middle intensity only. Upstrokes ranging from much longer than downstroke to an equal duration. Carapace with each wave markedly raised at least anteriorly being simultaneously depressed behind; at highest intensity the entire body is sometimes raised. Walking during waving is usual, and males without territories sometimes wave. Waving rate ranges widely from more than 2 waves per second to 1 wave every 2 seconds. (Waving components: 1, 3, 9, 10, 14; timing elements: Table 19, p. 656.)

Precopulatory Behavior

Male approaches female, with or without waving, and the usual stroking and plucking motions follow. In both Zanzibar and New Guinea I have observed copulations with ovigerous females, while the prevalence of spermatophores in the gonopores of these egg-bearing individuals indicates that it is a usual type of behavior. See also p. 504. Herding of immature females occurred occasionally in Fiji and New Guinea (p. 503).

Acoustic Behavior

At the surface two components have been observed and filmed: major-merus-rubs, with circling motion and leg-wags (nos. 1 and 5). Underground sounds have been recorded on tape in Massawa, New Caledonia and Fiji (Pl. 47).

Combat

The following components have been filmed: manus-rubs and heel-&-hollows (nos. 1 and 11), as well as forceful manus-pushes. In New Guinea, manus-rubs were more prevalent than manus-pushes, and manus-&-pollex tappings in the course of a rubbing component were prevalent. During 1969 in Fiji and New Guinea several other components were observed but not filmed; these were the interlace (no. 9), a component probably ubiquitous in the genus, and two others hitherto unknown, the dactyl-along-pollex-groove and the subdactyl-&-suprapollex saw (nos. 14, 15). Each of these last two components is of special interest since it involves morphological peculiarities of the *vocans* pollex.

RANGE

U. vocans ranges from East Africa to Samoa, from the Red Sea to Natal, and from Okinawa to subtropical Australia. Although reported from Japan, it now seems to be extinct there (*fide* Sakai). The six subspecies to be distinguished are distributed as follows. *U. vocans borealis*: northeast China; northwest Taiwan; Hong Kong. *U. v. pacificensis*: islands of western Pacific north to Guam and Marshalls, east and south to Fiji and Samoa; in the west it apparently hybridizes with the local subspecies in three regions, namely New Guinea, Amboina, and the southern Philippines. *U. v. vomeris*: New Caledonia; eastern Australia regularly south to Trial Bay; also

present in a sheltered locality near Sydney; eastern New Guinea. *U. v. dampieri*: northwest Australia south to Broome. *U. v. hesperiae*: east Africa from the Red Sea to the northern part of Cape Province, western India, Ceylon, Burma, rare and perhaps hybridizing in Malaya and Singapore. *U. v. vocans*: Malaysia, Indonesia, Philippines, Nansei (Ryukyu) Islands.

BIOTOPES

Unshaded sandy mud to mud along lower tide levels of protected bays. (Biotopes 6 B, 8.)

SYMPATRIC ASSOCIATES

In Massawa and Zanzibar *U. vocans* and *tetragonon* were found mingling in restricted areas; in each of these examples *vocans* occupied its usual biotope, while *tetragonon*, its consubgeneric, was on a shore less exposed than usual. In each area of overlap, however, the substrate included or was adjacent to encrusted stones and old coral, such as are characteristic of *tetragonon*'s typical habitat.

The usual associate of *vocans* throughout its range is *lactea*, of the subgenus *Celuca*, a species typical of a slightly higher biotope (6 A). The two species frequently share a wide intermediate strip.

MATERIAL RESULTING FROM FIELD WORK

(The complete list of specimens examined is given in Appendix A, p. 597.)

Observations and Collections. U. vocans borealis: Taiwan: near Taipei. Hong Kong. *U. v. pacificensis*: Samoa. Fiji. Northeast New Guinea: near Madang (apparently hybridizing with *U. v. vomeris*). Philippines: various localities on Mindanao (apparently hybridizing with *U. v. vocans*). *U. v. dampieri*: northwest Australia: Darwin; Broome. *U. v. vomeris*: east Australia: near mouth of Brisbane R.; New Caledonia: near Nouméa; northeast New Guinea: near Madang. *U. v. hesperiae*: Ethiopia: Eritrea: Massawa. Zanzibar. Mozambique: Inhaca I. Ceylon: Negombo (near Colombo). India: Kerala: Ernakulam (not collected). Malaya: Penang. Singapore. *U. v. vocans*. Malaya: Penang. Singapore. Indonesia: Java: near Surabaja. Philippines: Tawi Tawi; Joló; Zamboanga; Gulf of Davao (Sasa and Madaum); Palawan I.; near Manila.

Film. Hong Kong. Fiji. Australia: Darwin; Broome. New Caledonia: near Nouméa. Ethiopia: Massawa. Zanzibar. Philippines: Zamboanga; Gulf of Davao; Palawan I.; near Manila.

Sound Recordings. Fiji. New Caledonia. Ethiopia.

TYPE MATERIAL AND NOMENCLATURE

Uca vocans (Linnaeus, 1758)

TYPE. Rumphius' specimen, the basis of Linnaeus' type description of *Cancer vocans*, is not extant. See References & Synonymy, p. 93.

NEOTYPE. *Uca vocans vocans*: In Smithsonian Institution, National Museum of Natural History, Washington. 1 male, cat. no. 137673. Type-locality: Madaum, Mindanao, Philippine Is. Collected by J. Crane. Measurements on p. 92.

Uca vocans borealis subsp. nov.

HOLOTYPE. In Smithsonian Institution, National Museum of Natural History, Washington. 1 male, cat. no. 137669. Type-locality: Hong Kong. Collected by J. Crane. Measurements on p. 90.

Named in allusion to its northern distribution. (Genitive of Latin adjective from the Greek noun *Boreas*, "the north wind.")

Uca vocans pacificensis subsp. nov.

HOLOTYPE. In Smithsonian Institution, National Museum of Natural History, Washington. 1 male, cat. no. 137670. Type-locality: Suva, Fiji Islands. Collected by J. Crane. Measurements on p. 90.

Named in reference to its distribution in the Pacific Ocean.

Uca vocans dampieri subsp. nov.

HOLOTYPE. In Smithsonian Institution, National Museum of Natural History, Washington. Male, cat. no. 137671. Darwin, Australia. Collected by J. Crane. Measurements on p. 91.

Named in honor of Dampier, the explorer.

Uca vocans hesperiae subsp. nov.

HOLOTYPE. In Smithsonian Institution, National Museum of Natural History, Washington. Male, cat. no. 137672. Zanzibar, Tanzania. Collected by J. Crane. Measurements on p. 92.

Named in reference to its being the most western of the subspecies. (Genitive of Greek adjective for "evening," "the west," and, later, "Hesperia," "Land of the West," then Italy and Spain.)

Type of *Gelasimus marionis* Desmarest, 1823 (1825). Philippines. Apparently not extant. Syn., on basis of the illustration (Pl. 13, Fig. 1), *fide* Holthuis, 1959. I agree.

Type of *Gelasimus cultrimanus* Adams & White, 1848. Philippines. In British Museum (Natural History). Measurements in mm: length 20; propodus 53. Examined and found identity with *U. vocans* un-

questionable. (!) Examination made before the erection of subspecies.

Type of *Gelasimus nitidus* Dana, 1851. Fiji Is. (Viti). Not in Smithsonian Institution, National Museum of Natural History, Washington. The illustrations and fuller description provided by Dana in 1852 leave no doubt as to the identity of his *G. nitidus* with *U. vocans*, as the species is understood in the present study.

Dana made no reference either in the type description or in his work of 1852 to a fossil crab described by Desmarest in 1817 as *Goneplax nitidus* and in 1822 as *Gelasima nitida*; the same fossil specimen was referred to by Milne-Edwards, 1837, as *Gelasimus nitidus*. Milne-Edwards in 1852 said nothing of the fossil reference but commented on Dana's recently published description of the material from Fiji. No doubt Dana selected the name *nitidus* independently, using the phrase "carapax nitidus" in the type description, just as Desmarest earlier selected it in reference to the "carapace . . . d'un noir luisant" of his fossil crab (p. 324).

If Dana's type material (which presumably would be clearly referrable to *U. vocans pacificensis*) were extant, and if Desmarest's fossil crab were also extant and proved after all not to be a member of the genus *Uca*, the name *nitidus* would have been maintained in place of the subspecies name *pacificensis* proposed here. Since neither Desmarest's fossil specimen nor Dana's material appears, however, to have survived, it seems best to avoid the use of the name *nitida* in the current nomenclature of *Uca* observed in this study. (See also p. 459.)

Type of *Uca marionis* var. *vomeris* McNeill, 1920. Rivermouth, Queensland, Australia. ? In National Museum, Sydney. In present study given subspecific rank as *Uca vocans vomeris*.

Status of name *excisa*, proposed by Nobili, 1906. In his paper on the crustaceans of the Red Sea, Nobili (1906.3: 314) suggests the use of the name *excisa* in the following paragraph, at the end of his comments on the taxonomy and synonymy of "*Uca marionis* var. *nitida* Dana [sic]: Le *G. cultrimanus* dans le sens de Kingsley et de Ortmann est identique avec le *G. nitidus* Dana. Ce dernier nom serait donc le nom de cette espèce ou variété, mais comme il y a déjà un *Gelasimus nitidus* Desmarest, espèce fossile, je propose pour cette forme le nom d'*excisa*."

For several reasons the name is here considered to have no taxonomic standing. Dana was not the author of any of the names mentioned, all of which are here regarded as synonyms of *vocans*. Furthermore, neither type material nor any locality was mentioned by Nobili, and no illustration was provided.

If the name had been based definitely on Dana's Fijian material, the name *excisa* could stand by priority as a subspecific name in place of *pacificensis*; the wording, however, does not appear to be sufficiently precise to warrant such a procedure.

Uca (Thalassuca) vocans borealis subsp. nov.

(Hong Kong to northwest Taiwan)

MORPHOLOGY

With the characteristics of the species.

Gonopod practically without torsion; anterior flange longer and broader than posterior; inner process not tumid. Gonopore with postero-external ridge strongly developed; anterior ridge virtually absent. Major cheliped: oblique ridge on palm lower than usual in the species, always with strong, regular tubercles; pollex with or without the two projections, median and subdistal, of a weak "*nitida*" form; dactyl with faint traces of two grooves externally near base, strongest in young; dactyl in young slightly deeper than opposite part of pollex. Carapace: dorsolateral marginal line absent in large males, weakly present in young; frontal groove variable, sometimes shorter than distal projection of front beyond it, apex usually bluntly angled, never widely U-shaped; male eyebrow developed internally only, narrow; branchial regions notably elevated, a characteristic of this subspecies alone. Minor cheliped with gape narrower than adjacent part of pollex.

Measurements (in mm)

	Length	Breadth	Propodus	Dactyl
Holotype male (Hong Kong)	12.5	19.0	26.0	17.0
Largest male (Taiwan)	17.0	26.0	47.0	32.0
Female (ovigerous and largest) (Hong Kong)	11.0	15.0	–	–
Smallest ovigerous female (Hong Kong)	6.5	10.0	–	–

Color

Dull in comparison with other subspecies, even including the displaying crabs of Hong Kong; in May they appeared to be in full breeding condition, almost all the females taken having spermatophores and, in addition, egg masses. Few carapaces were white, the major manus was marked only with dull ochraceous yellow, and most legs were dark.

SOCIAL BEHAVIOR

Waving Display

Vertical. Non-serial. At highest intensity cheliped extends well above eye and carapace is raised on all legs during each wave. At lower intensities only anterior carapace is raised and cheliped scarcely reaches tip of eye. Rate of waving from about 0.4 to 1.4 waves per second.

Uca (Thalassuca) vocans pacificensis subsp. nov.

(Tropical west Pacific)

MORPHOLOGY

With the characteristics of the species.

Gonopod similar to that of *borealis* but with slight torsion; anterior flange broader than posterior; inner process broad and flat throughout. Gonopore differing from that of *borealis* in the somewhat greater development of the postero-external ridge, the anterior ridge being virtually absent in both subspecies. Major cheliped: pollex usually with the two strong projections of the "*nitida*" form; basal half of major dactyl notably broader than corresponding part of pollex, a characteristic in the adult unique to this subspecies. Carapace: dorso-lateral marginal line in male distinct; front variable but its median groove is not short, and has the tip almost always pointed, never widely U-shaped; a narrow eyebrow present in male. Minor cheliped variable but gape usually narrower than pollex.

Measurements (in mm)

	Length	Breadth	Propodus	Dactyl
Holotype male (Fiji)	12.0	17.0	24.5	16.0
Largest male (Samoa)	14.0	22.0	37.5	25.0
Female (ovigerous and largest) (Samoa)	12.0	17.0	–	–
Smallest ovigerous female (Fiji)	8.0	11.0	–	–

Color

Pre-display phase sometimes yellow. Major dactyl usually violet to pale pink.

SOCIAL BEHAVIOR

Waving Display

Vertical. Non-serial. At highest intensity cheliped extends well above eye. Waving rate highly variable; see Table 19.

Uca (Thalassuca) vocans dampieri subsp. nov.

(Northwest Australia)

MORPHOLOGY

With the characteristics of the species.

Gonopod with moderate torsion; posterior flange broader than anterior, but both are narrow; base of inner process strongly tumid; thumb very short, set far down on shaft, its tip not nearly reaching base of flange. Gonopore similar to *borealis* and *pacificensis*, except that anterior ridge is present as a small projection. Major cheliped: outer manus with tubercles smaller than in other subspecies; pollex with lower margin practically straight; a single triangular projection, not two, on its upper margin, always in distal half, often small and subdistal or absent, especially in specimens from Broome. Carapace: dorsolateral marginal line of carapace absent in adult males, as is the eyebrow except, rarely, for a short, internal indication; frontal groove with sides strongly convergent, the apex curving or bluntly angled.

Measurements (in mm)

	Length	Breadth	Propodus	Dactyl
Holotype male (Broome)	12.0	19.0	28.5	17.0
Largest male (Darwin)	15.0	22.5	41.5	27.0
Female (ovigerous and largest) (Darwin)	14.0	20.5	–	–
Smallest ovigerous female (Darwin)	8.0	11.0	–	–

Color

Pre-display phase usually blue with white spots lightening to white with turquoise spots. Major dactyl sometimes pink.

SOCIAL BEHAVIOR

Waving Display

Vertical, to a more literal extent than any of the other subspecies, the cheliped not usually pushing outward at an acute angle, although this semi-lateral movement sometimes occurs at highest intensity. Strongly serial at middle and high intensities. In courtship body elevated with each approach wave, though cheliped reaches only slightly above eye even at high intensity, not at all during low. Five to 8 waves in each series; rate of waving 0.2 to 1.8 waves per second.

Three degrees of intensity show particularly well in this subspecies. Lowest intensity: non-serial, fast-est, up- and downstrokes nearly equal, cheliped little raised, body not elevated. Medium intensity: serial, cheliped raised higher but still not reaching above eye, and body still not elevated. High intensity: serial, slowest, upstroke always longer than down, dactyl tip reaching above eye tip, chela directed out at a slight angle, body slightly raised with each wave.

Uca (Thalassuca) vocans vomeris McNeill, 1920

(New Caledonia, Eastern Australia, New Guinea)

MORPHOLOGY

With the characteristics of the species.

Gonopore stout throughout in relation to its length, in comparison with the slenderness of other subspecies, in particular *pacificensis* and *vocans*; torsion strong, the inner process and its tumid base, subdistal part of canal, and thumb all being twisted far around to the shaft's external side, while the flanges flanking the canal's orifice are twisted somewhat in the opposite direction; the normally posterior (now external) flange is wider than the anterior. Gonopore with postero-external ridge much reduced while anterior ridge is well developed. Major cheliped: pollex usually with a single triangular projection, subdistal; dactyl with distinct traces of two grooves. Carapace: dorso-lateral marginal line absent in male; front unexpanded, frontal groove widely triangular.

Measurements (in mm)

	Length	Breadth	Propodus	Dactyl
Largest male (Sandgate, Brisbane)	17.5	27.0	46.5	32.0
Largest female (N.S.W.)	12.0	19.5	–	–

Color

Major dactyl in New Caledonia and near Brisbane lavendar to pink; in northern New Guinea, pale lemon to white (see p. 87).

SOCIAL BEHAVIOR

Waving Display

Vertical. Non-serial except in New Caledonia, as noted below. Wave high-reaching, the dactyl's tip reaching well above that of eye, the body elevated with each wave. At medium intensity near Nouméa, but not on New Guinea's north coast, the wave is definitely serial, similar to that found in northwest

Australia; uniquely, however, in these *vomeris* the series is made up of 1 high wave followed by about 3 diminishing waves. In both localities at lowest intensity the wave, as usual in the species, is non-serial, fast and very low. Unfortunately the subspecies was not displaying in Queensland during my September stay.

Uca (Thalassuca) vocans hesperiae subsp. nov.

(West Africa to Singapore)

MORPHOLOGY

With the characteristics of the species.

Gonopod similar to that of *vomeris*, but torsion even more extreme; thumb shorter and more distally placed; anterior flange (internal through torsion) vestigial; a deep hollow bounded distally by the flange-and-orifice structure on one side and the oppositely twisted inner process on the other. Gonopore closely similar to that of *vomeris*, but anterior ridge larger. Pollex often of "*nitida*" form with two large, triangular projections. Carapace: dorso-lateral marginal line vestigial or absent in male; frontal groove narrowly triangular, the apex pointed; eyebrow distinct at least internally, often complete, always very narrow. Minor chela with gape notably narrow for the species.

Measurements (in mm)

	Length	Breadth	Propodus	Dactyl
Holotype male (Zanzibar)	14.5	23.0	36.5	25.0
Largest male (Negombo, Ceylon)	16.5	25.0	39.0	25.0
Largest claw (Zanzibar)	15.5	25.5	44.0	33.0
Female (ovigerous and largest) (Zanzibar)	15.0	22.0	–	–
Smallest ovigerous female (Zanzibar)	10.0	13.5	–	–

Color

Pre-display phase sometimes yellow, more often gray with pale spots. Major dactyl rarely violet.

SOCIAL BEHAVIOR

Waving Display

Vertical. Non-serial. Dactyl reaches only a little above eye, although anterior part of body is raised high during each wave, the posterior part being depressed. At high intensity, when the object of display

is not straight in front of the displaying crab, the cheliped usually waves semi-laterally, at a wider acute angle than is characteristic of other subspecies. Rate of waving from about 0.3 to 2 waves per second. Upstroke sometimes only half length of downstroke, but usually 1.5 to 3 times longer; rarely the strokes are equal.

Uca (Thalassuca) vocans vocans (Linnaeus, 1758)

(Malaya to Nansei [Ryukyu] Islands)

MORPHOLOGY

With the characteristics of the species.

Gonopod similar to those of *vomeris* and particularly *hesperiae*; differs from the latter chiefly in having the homologously anterior flange (now inner) larger than the posterior, instead of vestigial; in this character alone it agrees with *borealis* and *pacificensis*; it differs additionally from *vomeris* in the much deeper anterior hollow, made through the curvature of the flange bases and of the inner process, and in the much higher anterior (now inner) flange; entire gonopod more slender than that of *vomeris*. Gonopore similar to *vomeris* and *hesperiae*, except that the anterior ridge is more curved; western populations with a more definite margin posteriorly and externally, and often with a small postero-external tubercle; a similar one also sometimes postero-internally. Pollex often of the "*nitida*" form, the two triangular projections being large. Carapace: dorso-lateral margins absent in male; frontal furrow narrow, tapering; tip of front often expanded; eyebrow almost always present, narrow but complete; some of suborbital teeth always bicuspid. Minor cheliped with gape usually wider than adjacent part of pollex.

Measurements (in mm)

	Length	Breadth	Propodus	Dactyl
Neotype (Madaum)	15.0	22.0	38.5	26.0
Moderate male (Madaum)	12.5	14.5	27.0	18.5
Largest male (Singapore)	16.0	25.0	44.0	29.0
Largest male broken claw (Zamboanga)	16.5	26.0	–	–
Smallest ovigerous female (Zamboanga)	7.0	10.0	–	–
Largest ovigerous female (Singapore)	15.0	21.5	–	–
Largest female (Singapore)	15.5	22.5	–	–
Mating pair (Singapore)				
male	9.0	13.0	14.0	8.0
female	14.0	19.5	–	–

Color

Predisplay phase sometimes yellow. Major dactyl yellow to white.

SOCIAL BEHAVIOR

Waving Display

Vertical. Non-serial. Low, the upper dactyl edge maximally reaching slightly above eye. Anterior part of body slightly elevated with each wave, while posterior is further depressed. Rate of waving from about 0.5 to 2.0 waves per second; upstrokes from 1.1 to 7 times longer than downstrokes.

REFERENCES AND SYNONYMY

Uca (Thalassuca) vocans (Linnaeus, 1758)

The references to this species are not distributed among the subspecies because of apparent hybridization of the forms in parts of the range (see p. 87).

Cancer vocans

Rumphius, 1705, 1740, 1741: 14; Pl. 10, Fig. E; 1711, pl. only, without text (only 1st ed. seen; others *fide* Holthuis, 1959). East Indies, probably Amboina. Description; color; habits, including first account in literature of waving behavior in *Uca*. See p. 3.

Petiver, 1713: 2 (caption only); Pl. 9, Fig. 5 (redrawn from Rumphius).

TYPE DESCRIPTION. Linnaeus, 1758: 626. From Rumphius' description and illustration.

Linnaeus, 1767: 1041. "C. brachyurus, thorace quadrato inermi, chela altera magna. . . . *Habitat in Indiis, latitans sub saxis, prehendens digitos occurrentes cum dolore. Variat chela majore manuum.*" Confused synonymy.

Fabricius, 1775: 401 (part). "In Indiis." Minimal description; confused synonymy.

Fabricius, 1787: 318 (part). Minimal description.

Fabricius, 1798: 340 (part). Minimal description.

Olivier, 1811: Vol. 6: 157 (part). Confused synonymy; no new material; paraphrases Rumphius' account of habits.

Ocypode vocans

Bosc, 1802: 198 (part). No new material. Acceptable description and references, but gives "Amerique meridionale" as habitat.

Latreille, 1801-1802: 45. "Aux Indes." Discussion of earlier synonymies.

Goneplax vocans

Latreille, 1817.1: 17. Brief mention of large claw motion "as if it wished to make signals."

Gonoplax vocans

Lamarck, 1818: 254. "Habite l'Océan indien."

Brief Latin diagnosis. Common name: "Rhombille appellant."

Henschell, 1838: 204. Reference only cited to Rumphius, 1705, Book I, Pl. 10, Fig. E. (First part of Henschell's volume consists of an excellent biography of Rumphius, in Latin.)

Gelasimus vocans

Desmarest, 1825: 123 (part). Apparently no new material. Brief description.

Milne-Edwards, 1852: 145. Java; Malabar. Taxonomy.

Stimpson, 1858: 99. Hongkong; Riu Kiu Is.: Okinawa. Habitat.

Heller, 1865: 37. Nicobars.

Hilgendorf, 1869: 83. Zanzibar. Taxonomy.

A. Milne-Edwards, 1873: 272. New Caledonia. Taxonomy.

Hoffmann, 1874: 16. Madagascar: Nossy-Faly, Nossy-Bé. Taxonomy.

Miers, 1879.2: 488. Rodriguez. Taxonomy.

de Man, 1880: 67. E. Indies; Madagascar. Taxonomy.

Richters, 1880: 155. Fouquets. [Not seen.]

Miers, 1880: 308. Taxonomy.

Lenz & Richters, 1881: 423. Madagascar. Taxonomy.

Haswell, 1882: 92. Louisade Arch.: Woodlark Is. Northwestern Australia: Port Darwin. Taxonomy.

Miers, 1886: 242. Fiji Is.; Arrou Is. Taxonomy.

Walker, 1887: 110. Singapore.

de Man, 1887.2: 353. East Indies: Amboina. Brief taxonomy.

de Man, 1891: 23. Sumatra; Fiji Is.; Marataia. Taxonomy.

de Man, 1892: 305. Celebes; Macassar; Pare-Pare. Taxonomy.

Aurivillius, 1893: 26. Mauritius: Port Louis; East Indies. Amphibious characteristics; morphology; ecology.

de Man, 1895: 572. East Indies: Malacca. Taxonomy.

Stimpson, 1907: 104. Taxonomy. Habitat.

Gelasimus marionis

Desmarest, 1825: 124; Pl. 13, Fig. 1. Philippines: Manila. Type description.

? Milne-Edwards, 1837: 53. Taxonomy. (No locality; synonymized by Milne-Edwards, 1852, with *G. perplexus*; specimen not extant.)

Milne-Edwards, 1852: 145. India: Malabar. Taxonomy.

Hoffmann, 1874: 15. Madagascar: Nossy-Faly. Taxonomy.

Miers, 1880: 308. Malaya. Taxonomy.

Kingsley, 1880.1: 141. No new record. Taxonomy.

de Man, 1880: 67. E. Indies. Taxonomy.

Ortmann, 1894.2: 754. New Guinea: Kaiser-Wilhelm-Land. Taxonomy.

Alcock, 1900: 359. Andamans. Taxonomy.

de Man, 1902: 487. E. Indies: Ternate; Halmahera. Taxonomy.

Lenz, 1910: 559. Madagascar: Gulf of Tulear. Taxonomy.

Raj, 1927: 148. Gulf of Manaar: Krusadei Island. Taxonomy.

Tweedie, 1937: 143. Malaya; East Indies: Ternate; Halmahera. Taxonomy.

Chopra, 1937: 422. Mergui Arch.: Palaiow on east coast of Doung I. Taxonomy.

Shen, 1940: 232. Hong Kong: Wong Chuk Hang.

Chapgar, 1957: 509, Pl. 13. West India: Bombay. Taxonomy.

Gelasimus nitidus (not *Goneplax nitida* Desmarest, 1817, *Gelasima nitida* of Desmarest, 1822, or *Gelasimus nitidus* of Milne-Edwards, 1837; see p. 324).

Dana, 1851: 248. Fiji Is. Type description (Latin).

Dana, 1852: 316; Pl. 19, Fig. 5 a-d. No new material. Description (Latin and English).

Milne-Edwards, 1852: 147. Taxonomy.

Thallwitz, 1892: 42. E. Indies: Celebes: Manado; Ternate; Togian; Tamini; Java: Madura; Timur; S. New Guinea. Taxonomy.

Stebbing, 1921: 16. South Africa: Durban. Taxonomy.

Gelasimus cultrimanus

White, 1847: 35. Philippines. (No description.)

Adams & White, 1848: 49. Philippines. Type description. (!)

White, 1848: 85. No new material.

Kingsley, 1880.1: 140. Taxonomy.

Cano, 1889: 233. Philippines: Manila.

Ortmann, 1894.1: 58, 67. E. Africa: Lindi; Kilwa; Dar-es-Salaam. Taxonomy; ecology.

Ortmann, 1894.2: 752. Record of museum specimens from Indian Ocean, Singapore, Philippines, New Guinea, Fiji, Samoa.

Gelasimus vocans var. *cultrimanus*

de Man, 1891: 23. Taxonomy (reducing *cultrimanus* to a variety).

Uca cultrimana

Ortmann, 1897: 348. No new records. Taxonomy.

Doflein, 1899: 192. South Seas; Fiji Is.; Indian Ocean. Taxonomy.

Nobili, 1899.2: 272. Australo-Malaysia; Ternate; Andai. Taxonomy.

Nobili, 1899.3: 516. Sumatra: Siboga. Taxonomy.

Borradaile, 1900: 595. Ellice Is.: Funafuti; Fiji Is.: Rotuma.

Pesta, 1913: 58, Text Fig. 3. Samoa. Taxonomy.

Raben, 1934: 428. Sumatra: Padang. Respiratory system (illustrated).

Estampador, 1937: 545. Philippines.

Estampador, 1959: 103. Philippines.

Gelasimus marionis nitidus

Alcock, 1900: 360. Andamans, Nicobars, Coromandel and Malabar coasts.

de Man, 1902: 487. East Indies. Taxonomy.

Raj, 1927: 148. Ceylon: Gulf of Manaar. Taxonomy.

Tweedie, 1937: 143. Malaya. Taxonomy.

Sakai, 1940: 32. Japan: Distribution.

Shen, 1940: 232. Hong Kong: Wong Chuk Hang, Kowloon, Nagau Chi Wan.

Lin, 1949: 27. Taiwan.

Chapgar, 1957: 510, Pl. 13. West India. Taxonomy.

Uca vocans

Lanchester, 1900.1: 754. Malaya: Malacca; Singapore. Taxonomy. Habits.

Parisi, 1918: 92. Indo-Pacific.

Holthuis, 1959.1: 115; Pl. 10, Fig. 1. Taxonomy (incl. discussion of synonymy). Photo of Rumphius' illustration.

Holthuis, 1962: 240, 246.

Macnae, 1966: 85, 86, 89. Australia: Queensland and New South Wales. Ecology.

Uca marionis

Nobili, 1906.3: 314. Red Sea.

Nobili, 1907: 408. Polynesia: Samoa (Apia). Taxonomy.

Pearse, 1912.2: 114. Philippines. Habits.

Laurie, 1915: 416. Red Sea.

Tesch, 1918: 38. Celebes: Dongala; Kwandang Bay; Kara elang; Talaut I. Taxonomy.

McNeill, 1920: 105; fig. Australia. Taxonomy.

Gee, 1925: 165. Hong Kong: Chin Bey.

Maccagno, 1928: 23. Samoa. Taxonomy.

Ward, 1928: 243. Australia. Color. Habits.

Musgrave, 1929: 342. Australia: Queensland; New South Wales.

Verwey, 1930: 177. Java. Ecology; physiology; behavior.

Estampador, 1937: 544. Philippines: Luzon; Maytunig; Riza; Panay.

Sakai, 1939: 625. Japan: Misaki; Nagasaki; Miyako Is.

Ward, 1941: 3. Philippines: Mindanao (west coast, Gulf of Davao).

Chace, 1942: 202. Tanganyika Territory: Lindi.

Buitendijk, 1947: 280. Malaya: Port Dickson.

Barnard, 1950: 90; fig. South Africa. Taxonomy.

Altevogt, 1955.1, 1955.2, 1956.1, 1956.2, 1957.1, 1957.3: India: near Bombay. Morphology, ecology, and behavior; the early references are under the heading of *Uca marionis nitidus*, but included here for completeness in listing the series.

Day & Morgans, 1956: 276, 305. South Africa: Durban Bay. Ecology.

Crane, 1957. Indo-Pacific. Preliminary classification of display.

Macnae & Kalk, 1958: 41, 67, 125. Mozambique: Inhaca I. Color; general behavior; ecology.

Estampador, 1959: 102. Philippines. Local distribution.

Macnae, 1963: 3, 23. Mozambique (Inhaca I.) to Union of South Africa (mouth of Umngazana River, in northern Cape Province). Ecology.

Uca marionis: *nitida* and variety *nitida*

Nobili, 1903.1: 21. Borneo: Samarinda.
Nobili, 1906.3: 314. Red Sea. Taxonomy.
Pearse, 1912.2: 114. Philippines. Habits.

Laurie, 1915: 416. Red Sea.

Maccagno, 1928: 24. Moluccas. Taxonomy.

Gordon, 1931: 528. Hong Kong.

Sakai, 1934: 319. Japan: Nagasaki.

Balss, 1938: 75. Fiji Is.

Takahasi, 1935: 78. Taiwan. Habits.

Sakai, 1936: 170. Palao Is.

Miyake, 1936: 511. Riu Kiu Is.: Miyara. Mangrove swamp.

Estampador, 1937: 544. Philippines: Luzon; Manila; Rizal Prov.; Malabon.

Miyake, 1939: 190, 223, 242; Pl. 16, Fig. 1. Micronesia: Palau Is.; Cardina Is.

Tweedie, 1950.1: 356. Labuan.

Altevogt, 1955.1ff.: See under *Uca marionis* for full list of references in series.

Estampador, 1959: 102. Philippines.

Uca marionis forma *excisa*

Nobili, 1906.3: 314. [See p. 89.]

Uca marionis var. *vomeris*

McNeill, 1920: 106. Type description.

Rathbun, 1926.1: 177. Queensland, Australia: Rivermouth.

Dakin, Bennett & Pope, 1954: 187. Australia: east coast south to Pittwater, Broken Bay, N.S.W. Ecology. General habits: 187-88 (part).

Uca vomeris

? Dakin, Bennett & Pope, 1954: Pl. 46.

Uca marionis excisa

Sankarankutty, 1961: 113. Bay of Bengal: Andaman and Nicobar Is.

IV. *AMPHIUCA* SUBGEN. NOV.

Typus: *Gelasimus chlorophthalmus* Milne-Edwards, 1837

(Tropical and subtropical Indo-Pacific: East Africa to the Marquesas; Red Sea to Natal; Nansei [Ryukyu] Islands to Friendly Islands; apparently absent from Australia)

PLATES 15, 16.   MAPS 5, 6, 7.
GONOPOD DRAWINGS: FIGURES 68 (part), 69 (part).   TABLES 1, 8, 10, 12, 14, 19, 20.
DENDROGRAMS: FIGURES 96, 99.

MORPHOLOGY

Diagnosis

Uca with front moderately wide; antero-lateral margins always distinct, short to moderate, posteriorly rounded; no ridge, mound, or tubercles on orbital floor; suborbital margin with crenellations minute or absent except near outer angle. Major manus outside with very small, close-set tubercles except dorsally; dorsal margin externally with a beaded keel; a depression with definite edges outside pollex base; tuberculate predactyl ridges on palm strongly divergent; no long furrows on pollex or dactyl. Posterior merus of minor cheliped without tubercles in vertical rows. Gill on 3rd maxilliped large, with distinct books. Female never with an enlarged tooth in cheliped gape.

Description

With the characteristics of the genus (p. 15).

MALE

Carapace. Front intermediate compared with most other subgenera, much wider than in *Deltuca* and other narrow-fronts, but not approaching the extremes found in *Minuca* and some *Celuca*; in general shape it agrees with the latter two groups, being narrowest below eyestalk bases, its breadth between them twice or more basal breadth of erected eyestalk, in front of which it tapers gradually to a broadly rounded tip; frontal breadth measured between inner eyebrow margins contained from about 3.5 to about 4.5 times in carapace width between tips of antero-lateral angles. Orbits moderately to strongly oblique. Antero-lateral angles variously acute, slightly to moderately produced on major side but scarcely or not at all on minor; antero-lateral margins short to moderate, converging, curving gradually, without sharp angles, into the moderately converging dorso-lateral margins; these margins range from strong to vestigial. One pair of postero-lateral striae present or absent. Vertical lateral margins either strong, reaching

antero-lateral margins, or dorsally absent. All carapace margins non-tuberculate, sometimes weakly beaded, usually unarmed. Carapace profile in lateral view moderately arched except in Hong Kong specimens of *chlorophthalmus*, where the arching is strong. Suborbital region at least in inner half notably flat and almost or completely naked. Pterygostomian region convex and thickly setose.

Frontal margin always raised, broad, unbroken. Eyebrows well developed, moderately broad, obliquely vertical. Lower margin of orbit rolled slightly outward, somewhat sinuous; no tubercles, distinct ridge or mound on its floor, but in some specimens of some populations a trace of elevation, with a few, irregular, minute granules, near inner corner; suborbital margin with crenellations either absent or small and close-set, except near external angle. Eyestalks moderately slender, not fitting orbits closely, their diameter scarcely less than that of eyes.

Second Maxilliped. Spoon-tipped setae moderately numerous.

Third Maxilliped. Ischium almost flat with a median longitudinal groove or convex without groove; setae sparse to almost absent. Gill large, but ranging in shape from that of a typical, flat, tapering gill to hand-like with dactyloid projections; number of distinct books highly variable, from vestigial to more than a dozen on each side of median vane. For more details on these structures in *Amphiuca*, see Chapter 3, p. 469.

Major Cheliped. Merus variously armed, in about the range found in *Deltuca*, but sometimes with a large crest distally on antero-dorsal margin, instead of a small crest or row of tubercles.

Carpus externally almost smooth with bent-over dorsal margin edged with a series of small tubercles, sometimes indistinct.

Outer manus appearing almost smooth in comparison with this segment in *Deltuca* and *Thalassuca*, although not as smooth as in many *Celuca* and *Minuca*; surface covered with very small, well separated tubercles, clearly larger on upper manus near the slightly bent-over, thickened, dorsal margin; this roughened margin is marked along its outer edge by a row of tubercles, often bead-like, with a minute groove along their outer base; the latter is practically continuous with the wider, short, subdorsal groove among the tubercles on proximal part of dactyl. A depression, large or small, outside base of pollex. Ventral margin of manus with a tuberculate carina, with or without a groove above it and always ending at pollex base.

Palm with proximal, lower part covered by small tubercles ranging down to minute granules. Oblique ridge present or absent; when present it is moderately high, broad, blunt, with or without a few large tubercles surmounting it. No tubercles ever continue upward around the gently sloping anterior region (there is no margin) of carpal cavity; this cavity sometimes traceable distally below dorsal margin as a shallow depression; no beaded margin around cavity's upper distal part. Inner surface of palm granulate or smooth, almost flat. Proximal ridge inside dactyl base always widely separated from distal ridge in lower portion and generally diverging more from the distal as it proceeds dorsally; both ridges, but especially the proximal, with well-developed tubercles except in some large specimens.

Both pollex and dactyl lacking long grooves on outer surfaces, or, in *inversa inversa* only, with a faint trace on dactyl, clearest distally. Subdorsal furrow on proximal dactyl short but distinct, surrounded by tubercles. Dactyl slightly longer and more convex than pollex, its tip curving strongly downward slightly beyond pollex. Outer surfaces of pollex and dactyl usually covered with minute tubercles, and with their lower and upper margins, respectively, minutely spinulous; extreme tips of both spinulous. All three rows of gape tubercles represented at least proximally in both pollex and dactyl; an enlarged submedian tooth on pollex almost constant; other enlargements various. No submarginal carina on outer distal part of pollex.

Minor Cheliped. Merus with dorsal margin armed with transverse, sparse, weakly sinuous rugosities; postero-ventral margin similar or with a single line of small, strong tubercles; antero-ventral margin and anterior surface smooth or practically so; posterior surface smooth except for scattered granules occurring chiefly proximally and distally. Carpus with subdorsal ridge represented by minute tubercles and pits. Mano-pollex ridge short, starting about middle of manus or more distally and extending about two-thirds of way to dactyl tip. Dactyl much longer than manus. Gape moderately narrow to wide; a few weak serrations present or absent. Upper palmar and other setae sparse.

Ambulatories. Meri wide to moderately narrow; dorsal armature consisting of small tubercles, often close-set, flattened, and slanting distad, sometimes overlapping in scale-like formation; postero-dorsal tubercles usually on vertical striae, fairly numerous and strong on first 3 legs; ventral armature very weak.

Gonopod. Distinct flanges present or absent; inner process flattened, spinous, thin, scarcely visible; pore at end of a moderate to long tube, or flanked by broad flanges; thumb short to absent; when present arising far down shaft.

FEMALE

With the characteristics of the females in the genus (p. 17).

Gonopores sometimes with a portion of the rim elevated, but without tubercles in the known forms (female unknown in *inversa sindensis*).

Size

Small to moderate.

Color

Display whitening poorly developed to absent. Carapace blue or green blue with white or red. Major cheliped with shades or tints of red. Ambulatories dark to red. Yellow and orange absent.

Social Behavior

Wave semi-unflexed to almost lateral, rapid at high intensities. No special courtship components. Little time devoted to waving display, but territoriality strongly developed. Copulation at surface. Chimneys sometimes constructed in one of the two species. Combat and acoustic behavior too little known for general comment, except that ritualized combat is highly developed and, during restricted periods, prevalent.

Relationships

Comparisons with other groups will be found on p. 19, in connection with the discussion of the subgeneric dendrogram. The two species recognized seem to have reached comparable levels of specialization both in adaptation to a semi-terrestrial life and in social behavior.

Name

Amphiuca: From the Greek prefix *amphi*, "on both sides." In combination with *Uca*, the proposed name may be translated "ambiguous fiddlers," in reference to their equivocal position between the morphologically and socially less advanced subgenera of the Indo-Pacific and those more highly developed centering in America.

15. *UCA (AMPHIUCA) CHLOROPHTHALMUS* (MILNE-EDWARDS, 1837)

(Tropical Indo-Pacific)

PLATES 15; 46 *B*.
FIGURES 13; 14; 15; 26 *C*; 31 *C*; 37 *H, I*; 39 *A, B*; 56 *C*;
60 *L, M*; 68 *A, B*; 81 *G*; 82 *G*; 83 *A*; 99.

MAPS 5, 6.
TABLES 8, 10, 12, 14, 19, 20.

INTRODUCTION

Uca chlorophthalmus is one of those species, prevalent in the Indo-Pacific, that combine vivid color with persistent lethargy. Nevertheless its morphology and display both show characteristics intermediate between those of the Indo-Pacific narrow-fronts and the broad-fronts of the world.

In both males and females the colors are unfading scarlet red, blue, blue green, and white, usually in a striking mixture, although sometimes the entire crab is scarlet. An individual, either male or female, often lives in the same burrow for days or weeks. The waving periods seem regularly to be confined to two or three days of spring tides, with only a few individuals ever waving at the same time. Except in the final stages of courtship, no crab waves for more than a minute or so without a long rest period that may last until the tide comes in. Combats are not only restricted to the same brief hours as waving but seldom occur even then.

In the literature *chlorophthalmus* has been described under the additional names of *crassipes* (here kept as the name of the eastern subspecies), *gaimardi, latreillei, splendidus, pulchellus,* and *novaeguineae*. On the basis of the material examined and of field observations near the extremes of the range, it seems reasonable to refer all populations to a single species, dividing them among two subspecies. Of these, one occurs in East Africa and Mauritius and the other east to the Marquesas.

MORPHOLOGY

Diagnosis

Uca with front moderately broad, ambulatory meri broad. Major cheliped with a small triangular depression outside pollex base; no distal crest on merus and no large tooth near tip of dactyl. Dorso-lateral and vertical lateral carapace margins strong and complete; one postero-lateral stria. Lower margin of minor cheliped merus with a row of strong tubercles. Gonopod with pore arising at end of a long tube; flanges vestigial to absent. Female with meri of middle ambulatories posteriorly armed with tubercles on vertical striae near dorsal margin and with separate tubercles on lower part of segment, an area bent abruptly forward.

Description

With the characteristics of the subgenus (p. 96).

MALE

Carapace. Orbits strongly oblique. Antero-lateral margins short; dorso-lateral margins always distinct, ending opposite middle of cardiac region. On each side a strong postero-lateral stria below and parallel to posterior end of dorso-lateral margin, above base of fourth ambulatory. Vertical lateral margins strong, reaching antero-lateral margins.

Major Cheliped. Merus without a distinct crest on antero-distal margin. Depression outside pollex base small, strongly marked, roughly triangular, often pubescent. Palm with oblique ridge clearly present, its few enlarged tubercles restricted to region near the low apex, but continuing proximally along lower margin of carpal cavity almost to articulation with carpus. Upper carpal cavity not extending distally below dorsal margin as a narrow trench, but sometimes confluent with a general depression on upper palm. Center of palm little convex, finely granulate. Tip of pollex irregularly bifid through the enlargement of a subdistal tooth arising from median row. No enlarged, triangular tooth subdistally on dactyl.

Minor Cheliped. Merus with postero-ventral margin armed with a single row of strong, small tubercles. Gape narrow or moderate, with a few weak serrations.

Ambulatories. Merus wide or moderately so; dorsal margins of last three strongly convex.

Gonopod. Flanges absent; pore at end of a projecting tube, moderate to long; thumb short but always distinct; lengths of both tube and thumb highly variable, even within populations.

FEMALE

Meri of 2nd and 3rd ambulatories with lower, posterior surfaces near ventral margin bent forward and conspicuously tuberculate, the tubercles in these localities separate, not set on vertical striae; remaining ambulatory armature as usual stronger than in male but closely similar.

Measurements (in mm)

Small to moderate, attaining a maximum length of 14.5 mm. Details in descriptions of subspecies.

Morphological Comparison and Comment

The well-marked triangular depression outside the major pollex base is almost always the best single diagnostic character throughout the range; in one of the two type-specimens of *novaeguineae* alone is it absent; often, however, it is small and, if not emphasized by pile, can be inconspicuous. In Africa in both sexes the short ambulatories with broad meri distinguish *c. chlorophthalmus* easily from the other two local broad-fronted *Uca, i. inversa* and *lactea annulipes*, and the differences in leg breadth hold between local subspecies of *chlorophthalmus* and *lactea* throughout the Indo-Pacific. Where *chlorophthalmus* coincides in range with *triangularis*, the more slanting orbits of *triangularis* and characteristic armature of its small cheliped, again in both sexes, ensure easy identification of the two species. Details of tuberculation on the major manus, along with the morphology of gonopores and gonopods, give, as usual, unmistakable identifications, in spite of the great variability of the gonopod in *chlorophthalmus*.

The differences between the eastern and western forms of *chloropthalmus* seem, now that fairly large samples from various populations have been compared, to show distinctions of no more than subspecific rank, being restricted to minor differences in gape and pollex breadth and of major cheliped, breadth of ambulatory meri, length of projecting distal tube and of thumb on gonopod, and details of gonopore. The differences in depth of manus and length of fingers between most specimens in the two subspecies, remarked on by several investigators in discussing under various names the western and eastern forms, would almost certainly prove to be reliable differences when investigated statistically. At present the material is wholly inadequate for that treatment, while the usual difficulties are caused by homochelous and heterochelous dimorphism, as well as by growth rate differences; it is therefore necessary to defer investigation of these characteristics. Unfortunately material from the eastern Indian Ocean, Malaysia, and Indonesia is so scanty that no light is shed on whether intermediate forms exist.

Figs. 13-14 show how the gonopod tips vary, both among and within populations. If the largest specimens available from Hong Kong had been the only ones examined, Stimpson's name, *splendidus*, would certainly have been kept to designate at least a subspecies, since the gonopod tips were strikingly longer than in other examples and the arching and other details of the carapace were distinctive. When I collected and examined elsewhere several smaller specimens from Hong Kong, they were not distinctive and it was apparent that the usual variation was responsible; in addition, two specimens from Cocos-Keeling had the gonopod tips similarly produced.

There seems to me to be no question of the synonymy of the other specific names involved. I agree wholly with Forest & Guinot (1961) concerning the synonymy of the types in Paris, except for preferring to consider the eastern and western forms as subspecific instead of specific rank. Details are given under the headings on Type Material (p. 100) and References and Synonymy (p. 102).

Color

Color differences are not at all correlated with the ranges of the two subspecies.

Displaying males: Display whitening absent. Color of individuals persisted unchanged throughout day and night in the field, as well as during transportation to Trinidad, and for almost a year thereafter in the crabberies. Carapace usually dark blue to bluish green, spotted with pale blue, blue green, or white; exceptions are the populations in the Philippines and Hong Kong, in which both carapace and major cheliped are entirely scarlet red. Major cheliped usually entirely scarlet red; in a few individuals red is confined to a wash of rose red on manus, while intermediate forms occur; chela pink to white. Eyestalks often pale green yellow. Maxillipeds, suborbital areas, and pterygostomian regions sometimes spotted with white, except when carapace is entirely red. Ambulatories usually scarlet red; at least on Tahiti, however, they are bright blue with pale spots. Adult females: carapace and ambulatories as in males of same population, except as follows. In New Guinea the carapace is wholly bright red, as in males in the Philippines and Hong Kong, while the corresponding New Guinea males have the spotted blue carapace found throughout the rest of the species-range. Finally, in Tahiti many females have an anterior band of red on carapace. Young with carapace and legs usually dark.

SOCIAL BEHAVIOR

Waving Display

Wave semi-unflexed to lateral-straight. Jerks absent. Tips of chela reaching scarcely or not at all above eyes. Minor cheliped makes motions, more or less in synchrony with the major's wave. Body scarcely or not at all raised during waves, which are usually grouped in series of 3 to 7 or more. Display rate ranging from about 1 to about 2 waves per second, which within a series follow one another without pauses. (Component nos. 3, 4, 9; timing elements in Table 19, p. 656.)

In northern New Guinea a burrow holder, which had been established for some days within several inches of the burrow of a particular female, was observed to wave as he moved toward his own hole, after an absence, from a distance of at least three feet; at the time no other male or female was at the surface within 12 feet—far beyond the range of his attention. This anecdotal observation perhaps may prove, when further supported, to be another indication of a behavioral similarity of *Amphiuca* to that of Indo-Pacific narrow-fronted crabs, rather than to members of the subgenera *Minuca* and *Celuca*.

Precopulatory Behavior

About a dozen examples have been seen in various parts of the species' distribution, and almost as many more in the Trinidad crabberies. Males approach females, usually but not always waving. They often climb on the female's chimney. After the usual plucking and stroking of the female's carapace, copulation takes place either there or down on the ground. Several times, however, when a male climbed a chimney, the female promptly went underground and the male followed.

Chimneys

Apparently only large females build chimneys. The structures were observed in East Africa, the Philippines, and Tahiti. As with waving display, their construction is closely associated with spring tides, but not with every one; the rhythm persisted even in individuals from the Philippines brought to Trinidad and kept on wholly different tidal schedules.

Acoustic Behavior

Only leg-wagging has been observed in the field and filmed (component no. 5).

Combat

Two combats only were seen, one of which was filmed. One occurred in East Africa, the other in New Guinea. Both consisted chiefly of supraheel-rubs, otherwise known only in (*Celuca*) *pugilator*. Associated taps were present in the African example, but absent in New Guinea. The New Guinea combat was preceded by a manus-rub. Both examples were fully ritualized. (Component nos. 1, 13.)

RANGE

Indo-Pacific, from eastern Africa to the Marquesas; north to the Nansei [Ryukyu] Islands; south to Union of South Africa (Cape Province) and Friendly Islands; apparently absent from Australia. The western subspecies (*c. chlorophthalmus*) occurs in Africa and Mauritius, while the eastern form (*c. crassipes*) is found throughout the rest of the range.

BIOTOPES

Characteristically, not close to the open sea but near high-tide levels on the muddy banks and flats of mangrove estuaries. On smaller islands in the South Pacific, however, in the absence of mangroves *chlorophthalmus* lives closer to the open shore, usually on muddy sand, again near high-tide levels and close to the mouths of streams or rivulets; often the only shelter on these islands is formed by a small indentation in the coast, with a barrier reef offshore. In all habitats spring tides alone reach some of the burrows. (Biotope nos. 5, 6, 11, 12, 14.)

SYMPATRIC ASSOCIATES

In East Africa (*Deltuca*) *urvillei* is the only associate of *chlorophthalmus*; occurring on slightly lower levels of the same biotope, it mingles on its upper margins with *chlorophthalmus*. On South Pacific islands (*Thalassuca*) *tetragonon* replaces *urvillei* in its altitudinal relation to *chlorophthalmus*. In rich areas in the Philippines, as at Zamboanga, *chlorophthalmus* lives among *coarctata*, *demani*, and occasional *dussumieri*, all members of the subgenus *Deltuca*; this close association is not characteristic.

MATERIAL RESULTING FROM FIELD WORK

(The complete list of specimens examined is given in Appendix A, p. 598.)

Observations and Collections. Uca chlorophthalmus crassipes: Hong Kong; Zamboanga, Philippine Is.; near Madang, New Guinea; near Nouméa, New Caledonia; Suva, Fiji Is.; near Apia, Samoa; Bora Bora, Raietea, and Tahiti, Society Is. *Uca chlorophthalmus c.*: Pemba I., Zanzibar I., and near Dar-es-Salaam, Tanzania; Inhaca I., Mozambique.

Film. Zamboanga, Madang, Tahiti, and Pemba.

TYPE MATERIAL AND NOMENCLATURE

Uca chlorophthalmus (Milne-Edwards, 1837)

TYPE. In Muséum National d'Histoire Naturelle, Paris. 1 dried male, labeled as follows: "*Gelasimus chlorophthalmus* Latr. M. Mathieu. Ile de France.

Anc. collection sèche. 1958." (This date represents the year during which D. Guinot was working on the collection; she considered it a "Type non specifié," personal communication.) (See Forest & Guinot, 1961.) In his type description, apparently of a single specimen, Milne-Edwards gives the length as "5 lignes," the locality "l'Ile-de-France," and the information, in his customary footnote, "Latreille, Coll. du Muséum." The identity of the specimen is unmistakable as an example of *U. chlorophthalmus* as here described, even though the crab is in fragments except for the major manus and pollex, frontal and orbital regions, and a posterior ambulatory. The minor chela appears at first sight to have a wider gape than is characteristic; examination, and comparison with similar individuals in other collections, show that the breadth is an incidental effect of drying. Propodus length in mm 19.5.

Uca chlorophthalmus crassipes
(Adams & White, 1848)

LECTOTYPE of the subspecies. In British Museum (Natural History), London. A single male (cat. no. 436) was selected by Calman, 1911, as holotype of *G. crassipes*—the only extant specimen of the two listed by White (1847), from same locality, given as "Philippine Ids. (Siquejor)." Label: "*Gelasimus crassipes* n.s. Holotype (selected 1911 WTC). Philippine Ids. 902. 43b." Measurements in mm: length 11; breadth 16; propodus 22; dactyl 14. The pollex was probably regenerated at an early age; only slight irregularity is shown. Although at one time the specimen was obviously dried, it has been relaxed, is in good condition, and is an excellent example of the subspecies. (!)

Type Material of *Gelasimus gaimardi* (Milne-Edwards, 1852). In Muséum National d'Histoire Naturelle, Paris. Consists of a "type non specifié," a male, dried, with the following measurements in mm: length 12; propodus 30.5. The label and Milne-Edwards both give the locality as "Tongatabou." The label in addition states "Mm. Quoy & Gaimard." The specimen is altogether characteristic of the form here referred to *c. crassipes*. The only discrepancy is that the "Blossom," the vessel on which Mm. Quoy and Gaimard traveled, never went to Tongatabou (Tonga Is.). (!)

Type Material of *Gelasimus latreillei* (Milne-Edwards, 1852). In Muséum National d'Histoire Naturelle, Paris. Consists of a "type non specifié," a male, dried, with the following measurements in mm: length 14; propodus 31. The label and Milne-Edwards both give the locality as "Bora Bora." Label also states "M. Duperrey; anc. collection sèche; 1958." This date, as in the case of *G. chlorophthal-*

mus, represents the year during which D. Guinot was working on the collection. (See Forest & Guinot, 1961.) (!)

Type Material of *Gelasimus splendidus* Stimpson, 1858. Type, from Hong Kong, destroyed in Chicago fire. Synonymy of this species with *U. chlorophthalmus crassipes* is on the basis of the type description, the illustration in Stimpson, 1907, specimens in the Smithsonian Institution, National Museum of Natural History, Washington (which were referred to it by Rathbun), and specimens I collected; all material is from Hong Kong.

Type Material of *Gelasimus pulchellus* Stimpson, 1858. Type, from Tahiti, destroyed in Chicago fire. Synonymy of this species with *U. chlorophthalmus crassipes* is based on the type description, the illustration in Stimpson, 1907, and my personal acquaintance with the *Uca* fauna of Tahiti and nearby islands.

Type of *Uca amazonensis* Doflein, 1899. 1 male, labeled in full as follows, is in the collections of the Museo di Zoologia della Università di Torino: "*Uca amazonensis*, Dofl. Tipo. Teffè—Amazoni. Da F. Doflein 1901." The specimen is a characteristic example of *U. c. chlorophthalmus*. Obviously, the type-locality as given on the label and in Doflein's description is erroneous. Carapace length 15.5 mm; propodus 45 mm. (!) See also p. 322, including note on von Hagen's recent location of type-specimens in Munich.

Type Material of *Uca novaeguineae* Rathbun, 1913. In Smithsonian Institution, National Museum of Natural History, Washington. 2 males, cat. no. 6374. New Guinea. (!)

Uca (*Amphiuca*) chlorophthalmus crassipes (Adams & White, 1848)

(Malaya; Indonesia; Philippine Islands; western and central Pacific)

MORPHOLOGY

With the characteristics of the species.
Minor cheliped with gape moderate, subequal to or scarcely narrower than adjacent part of pollex. Meri of 3rd and 4th ambulatories only moderately broad, the maximum breadth of 3rd merus on minor side being clearly less than half the length. Gonopod tube relatively long and slender in most specimens, sometimes extremely so, in comparison with that of *c. chlorophthalmus*. Gonopore set in a roughly triangular pit, without marginal elevations.

Measurements (in mm)

	Length	Breadth	Propodus	Dactyl
Largest male (Hong Kong)	14.5	23.0	34.0	24.0
Moderate male (Tahiti)	12.0	18.0	27.0	18.0
Largest female (Marshalls)	14.0	21.0	–	–
Largest female (ovigerous) (Tahiti)	11.0	17.0	–	–
Smallest ovigerous female (Zamboanga)	7.0	10.0	–	–

Uca (Amphiuca) chlorophthalmus chlorophthalmus (Milne-Edwards, 1837)

(East Africa; Indian Ocean)

MORPHOLOGY

With the characteristics of the species.

Minor cheliped with gape clearly narrower than adjacent part of pollex. Meri of both 3rd and 4th ambulatories notably broad, the dorsal and ventral margins of third strongly convex; maximum breadth of third merus on minor side well over half the length. Gonopod tube relatively short and broad when compared with that of *c. crassipes*. Gonopore with postero-inner margin slightly raised and uneven; other margins depressed.

Measurements (in mm)

	Length	Breadth	Propodus	Dactyl
Largest male (Pemba)	11.0	19.0	29.0	18.5
Moderate male (Zanzibar)	10.0	17.5	23.0	14.0
Largest female (Pemba)	12.5	19.5	–	–
Ovigerous female (unique) (Pemba)	9.0	14.0	–	–

REFERENCES AND SYNONYMY

Uca (Amphiuca) chlorophthalmus (Milne-Edwards, 1837)

TYPE DESCRIPTION. See under *U. (A.) chlorophthalmus chlorophthalmus*, below.

Uca (Amphiuca) chlorophthalmus crassipes (Adams & White, 1848)

Gelasimus crassipes

NAME ONLY: White, 1847: 36. Philippines. (No description.)

TYPE DESCRIPTION. Adams & White, 1848: 49. Philippines: Siquejar. (BM !)

White, 1848: 84. No new material.

Kingsley, 1880.1: 146. Description from Adams & White.

Cano, 1889: 92, 232. Manila. Taxonomy.

Gelasimus gaimardi

Milne-Edwards, 1852: 114; Pl. 4, Fig. 17, 17a. Tongatabou. Type description. (The upper fig., labeled "17" on the plate, represents this subspecies; cf. Milne-Edwards, 1852, under *c. chlorophthalmus*.) (Paris !)

Heller, 1865: 38. Tahiti. (Identity checked at Munich by Forest & Guinot.)

Kingsley, 1880.1: 150. Description from Milne-Edwards.

de Man, 1891: 39. Samoa; East Indies: Banda Sea, Amboina. Taxonomy.

Thallwitz, 1892: 44. East Indies: Manado, Celebes; Ternate; Togien Is.; Philippines: Ilo-Ilo. Taxonomy.

Sakai, 1939: 617; Pl. 104, Fig. 3; Text Fig. 92. Formosa; Bonin Is.

Sakai, 1940: 47. Distribution in Japan.

Shen, 1940: 232. Hong Kong: Wong Chuk Hang.

Lin, 1949: 26. Taiwan.

Gelasimus latreillei

Milne-Edwards, 1852: 114; Pl. 4, Fig. 20, 20a. Bora Bora. Type description. (Paris !)

A. Milne-Edwards, 1873: 274. New Caledonia.

Brocchi, 1875: 73. No locality given. Brief statement on male appendages; fig. without details. No taxonomy.

Kingsley, 1880.1: 152; Pl. 10, Fig. 31. Philippines. Taxonomy.

Ortmann, 1894.2: 757. Samoa.

Ortmann, 1897: 353. Taxonomy.

Balss, 1922.1: 142. Annam; Ryukyu Is. (Miyako).

Estampador, 1937: 545. Philippines.

Sakai, 1939: 618. No new record.

Sakai, 1940: 47. No new record. Geographical distribution.

Gelasimus splendidus

Stimpson, 1858: 99. Hong Kong. Type description. Habitat.

Kingsley, 1880.1: 149. Description from Stimpson.

Stimpson, 1907: 106; Pl. 14, Fig. 2. Habitat. Taxonomy, in English, of type.

Gelasimus pulchellus

Stimpson, 1858: 100. Tahiti. Type description. Habitat.

Stimpson, 1907: 107; Pl. 13, Fig. 1. Color. Taxonomy, in English, of type.

Gelasimus chlorophthalmus

Kingsley, 1880.1: 151 (part). Moluccas: Bourou I. (one of 2 specimens from Guérin's collection at Philadelphia Academy; 2nd is a *lactea*). (!)

Cano, 1889: 92 and 234. Amoy. Taxonomy. (Identification questionable.)

de Man, 1891: 41. New Caledonia. Taxonomy.

de Man, 1902: 484; Pl. 19, Fig. 4. East Indies: Ternate. Taxonomy.

Uca gaimardi

Ortmann, 1897: 353. Taxonomy.

Doflein, 1899: 192. Fiji Is.: Kandavu. Taxonomy.

Nobili, 1899.2: 274. Australo-Malaysia: Pt. Moresby; Hula.

Nobili, 1907: 408. Polynesia: Apia-Samoa.

Rathbun, 1907: 26. Society Is.: Tahiti, Bora Bora. Caroline Is.: Kusaie.

Pesta, 1911: 55; Pl. 3, Fig. 3; Text Fig. 2. Samoa: Upolu. Taxonomy.

Pearse, 1912.2: 114. Philippines. Habits.

Tesch, 1918: 39, 40. West coast of Flores; Labuan, Badjo. Taxonomy.

Sendler, 1923: 22. West Carolines: Yap.

Maccagno, 1928: 30. Samoa; Kei, Jule, Hula. Taxonomy.

Gordon, 1931: 528. Hong Kong.

Miyake, 1939: 222, 241. Caroline Is.: Kusaie.

Tweedie, 1950.2: 127; Text Fig. 4c. Cocos-Keeling Is. Ecology. Taxonomy.

Crane, 1957. Society Is. Preliminary classification of display.

Forest & Guinot, 1961: 140; Text Figs. 140-145, 153, 156a, b. Tahiti. General taxonomic discussion.

Forest & Guinot, 1962: 70. Distribution.

Uca chlorophthalmus

Nobili, 1907: 408. Polynesia: Taravao-Tahiti. Taxonomy. (Torino !)

Uca novaeguineae

Rathbun, 1913: 617. New Guinea. Type description. (USNM !)

? Miyake, 1938: 110, part. Palau Is. (Part or all may be *Uca triangularis*.)

? Miyake, 1939: 223, 242. Palau Is.; Caroline Is.: Kusaie. (Part or all may be *Uca triangularis*.)

Uca pulchella

Parisi, 1918: 93. Bonin Is.: Misaki.

Uca latreillei

Balss, 1922.1: 142. Loo Choo [Ryukyu] Is.: Miyako I.

Estampador, 1959: 103. Distribution in Philippines.

Uca splendidus

Gee, 1925: 165. Hong Kong.

Uca (Amphiuca) chlorophthalmus chlorophthalmus (Milne-Edwards, 1837)

Gelasimus chlorophthalmus

TYPE DESCRIPTION. Milne-Edwards, 1837: 54. Mauritius. (!)

MacLeay, 1838: 64. South Africa. [No specific localities given.]

Krauss, 1843: 40. South Africa. No new material.

Guérin, 1829-1843: Vol. 2: Pl. 4, Fig. 3. Vol. 3: 7 (caption to Pl. 4, Fig. 3). Mauritius.

White, 1847: 36. (Listed.)

Milne-Edwards, 1852: 114; Pl. 4, "Fig. 19" in text and caption, "17" on plate. Ile-de-France (Mauritius). [On plate there are 2 drawings labeled "17," the upper of an orbit, the lower of a claw; the specimen from Mauritius, if either, can be identified as the lower, the example drawn being somewhat immature, but having the proportionately higher (broader) manus and shorter fingers characteristic of the form recognized here as a subspecies. There is no indication that the illustration is intended to represent the type-specimen. Cf. under *gaimardi* Milne-Edwards, 1852, p. 102.]

Hilgendorf, 1869: 85 (part). Zanzibar; Mozambique; Mascarenes. Taxonomy.

Hilgendorf, 1879: 803 (part). Taxonomy.

Kingsley, 1880.1: 151 (part). No new records (cf. same ref. under *U. c. crassipes*).

Bouvier: 1915: 301. Mauritius: Port Louis, Grand Port, Chaland.

Uca amazonensis

Doflein, 1899: 193. Teffè—Amazon River. Type description. See p. 322.

Maccagno, 1929: 27; Text Fig. 14. Remarks on a type-specimen, now deposited in Torino; resemblance to *U. gaimardi* mentioned. See p. 101. (!)

Uca chlorophthalmus

Stebbing, 1910: 327. Annotated references.

Barnard, 1950: 95. South Africa. Taxonomy.

Fourmanoir, 1953: 89. Madagascar: near Canal de Mozambique. Color; ecology.

Day & Morgans, 1956: 277, 305. South Africa: Durban Bay. Ecology.

Macnae & Kalk, 1958: 39, 41, 67, 125. Mozambique: Inhaca I. Color; general behavior; ecology.

Forest & Guinot, 1961: 141 (part). Taxonomy.

Macnae, 1963: Mozambique (Inhaca I.) to South Africa (mouth of Umtata River in northern part of Cape Province). Local distribution and ecology.

Uca gaimardi

Vatova, 1943: 24. Italian Somaliland: Giumbo. Taxonomy.

16. *UCA (AMPHIUCA) INVERSA* (HOFFMANN, 1874)

(Western Indo-Pacific)

PLATE 16.
FIGURES 39 *G*; 69 *I, J*; 99.
MAP 7.
TABLES 8, 10, 12, 14, 19, 20.

INTRODUCTION

Uca inversa, like *chlorophthalmus*, shows characteristics of both morphology and waving display that are intermediate between those of Indo-Pacific narrow-fronts and the remaining subgenera of *Uca*.

In contrast to the situation in *chlorophthalmus*, individual *inversa* devote a relatively large amount of time to waving display, while strictly agonistic behavior also appears more prevalent. Unfortunately the social behavior of the African form is known only in outline, while the eastern subspecies has been observed in life during part of a single low tide, and then before it was recognized. The use of the large tooth at the tip of the major dactyl in *i. inversa*, doubtless in combat, is still to be learned. Perhaps it makes an efficient instrument for tapping in the combat component termed the supraheel-tap.

In East Africa *inversa* ranges from the Red Sea to Natal, through an impressive range of climates (p. 442). Although no morphological differences are apparent among the limited collections examined, it seems that physiological differences would certainly emerge with study. Furthermore, this African subspecies would be an excellent form in which to search for behavior differences among populations, particularly in droving patterns and in those of combat, both of which have given hints that they would repay attention.

MORPHOLOGY

Diagnosis

Uca with front moderately broad. Major cheliped with a distinct distal crest on merus and, in African specimens, with a very large tooth near tip of dactyl. Dorso-lateral and vertical lateral carapace margins weak, in part vestigial to absent; no postero-lateral striae. Lower margin of minor cheliped merus with small, weak rugosities. Gonopod with broad flanges flanking the pore. Female (in subspecies where known) without tubercles or other specializations on lower, posterior surfaces of ambulatory meri above ventral margins.

Description

With the characteristics of the subgenus (p. 96).

MALE

Carapace. Orbits moderately to strongly oblique. Antero-lateral margins moderate; dorso-lateral margins vestigial to absent. No postero-lateral striae. Vertical lateral margins weak, not reaching antero-lateral margins.

Major Cheliped. Merus with a definite crest on antero-distal margin, either minutely serrate or practically entire. Depression outside pollex base large, roughly triangular but with boundaries indistinct; not pubescent. Palm without oblique tuberculate ridge, the structure represented only by a low, broad convexity, either smooth or continuing the general distribution of minute tubercles or granules that cover lower proximal surface of palm. Upper part of carpal cavity extending distally either as a trench or represented by a distal depression somewhat separated by a convex area from the cavity itself. Center of palm smooth. A large triangular tooth present or absent subdistally on dactyl.

Minor Cheliped. Merus with postero-ventral margin armed only with sparse, weakly spinous, transverse rugosities. Gape moderate to wide, always equal to or wider than breadth of adjacent part of pollex; a few serrations present distally or absent.

Ambulatories. Merus moderately wide to narrow.

Gonopod. Flanges present, both broad; tube not projecting beyond them; thumb short or absent.

Measurements

Small to moderate crabs, attaining a known length of 13.5 mm.

Morphological Comparison and Comment

The western form of *U. inversa* can be distinguished at once from the two other broad-fronted fiddlers, *chlorophthalmus* and *lactea*, by the large hook on the tip of the major dactyl. Both the western and eastern (Karachi) subspecies have a well-developed crest on the major merus and lack postero-lateral striae. See also under *U. chlorophthalmus*, p. 99.

The small chelipeds in *inversa* are unusually large in both sexes, but this characteristic has not yet been systematically investigated. When making a preliminary sorting of mixed females, or males with the major chelipeds missing, this difference between *inversa* and *lactea* is sufficient to be a practical aid.

Color

Displaying males: Display whitening absent to partial. Major cheliped marked with orange red to pink.

SOCIAL BEHAVIOR

Waving Display

Wave obliquely vertical, semi-unflexed or lateral straight. Jerks absent. Tips of chela reaching scarcely or not at all above eyes. Minor cheliped makes motions, either in synchrony with the major's wave or more frequently. Body not raised, or scarcely raised, during waving except when waving accompanies walking. Ambulatories raised during display in the eastern subspecies. Herding known in the African subspecies. Display usually at rate of less than 1 second per wave, with a pause between waves ranging from short to long. (Component nos. 1, 3, 4, 9, 10 [weak], 12, 14; timing elements in Table 19, p. 656.)

Acoustic Behavior

Leg-wagging in *i. inversa* is the only component that has been observed and filmed (component 5).

Combat

The following three components have been filmed, all only in *i. inversa* at Zanzibar. Manus-rubs which, unusually, are in rapidly vibrating series (component no. 1). Supraheel-taps, without rubs; these taps were also extremely rapid and in series (no. 13). Dactyl-along-pollex-groove; this component was unusual in that the dactyl's motion started proximally, in the hollow at pollex base, and proceeded distally in each of the several examples known, instead of proceeding proximally from the pollex tip (no. 14). Interfemale combat was observed several times in Zanzibar, the ambulatories being, as usual, interdigitated and vibrated. On Inhaca Island, in mid-September, four days after full moon during a morning low tide, wandering and fighting was at a peak I have rarely seen in other species. Waving display was also common by both aggressive wanderers and, less often, by neighboring burrow-holders; no interest was shown

in females. The nights were still cool, with temperatures around 13° C to 17° C although clear days, including the morning of the observation, maintained surface temperatures in the sun of 29° C to 31° C.

RANGE

Western Indo-Pacific. *U. i. inversa*: East Africa from the approximate latitude of Massawa in the Red Sea to Natal; islands near the coast (in the Red Sea; Sokotra; Zanzibar; Madagascar; Inhaca); Arabia, from Aden and Mukalla. *U. i. sindensis*: Known only from Karachi, Pakistan.

BIOTOPES

Salt flats usually at or near highest tide levels, sometimes covered only by spring tides, and always partly cut off from the open sea; sometimes on the flat banks of brackish streams, or on flats in the upper reaches of mangrove estuaries; substrate usually muddy sand. (Biotope nos. 11, 12.)

SYMPATRIC ASSOCIATES

A few populations of (*Celuca*) *lactea annulipes* have been found associated with those of *inversa*. In both subspecies one or two populations mingled more than marginally.

MATERIAL RESULTING FROM FIELD WORK

(The complete list of specimens examined is given in Appendix A, p. 599.)

Observations and Collections. U. i. inversa: Massawa, Eritrea; Zanzibar; Inhaca I., Mozambique. *U. inversa sindensis*: Karachi, Pakistan.

Films. Motion pictures were made in Zanzibar and Karachi.

TYPE MATERIAL AND NOMENCLATURE

Uca inversa (Hoffmann, 1874)

LECTOTYPE. In Rijksmuseum van Natuurlijk Historie, Leiden. One male, selected by J. Crane from a group of 3 males, of which this is the largest, numbered 251 and labeled: "*Uca inversa* (Hoffm.) (= *Gelasimus inversus* Type Hoffm.) Nossi-Faly." Fragile condition, with major cheliped detached and most ambulatories missing. There is, however, no question of its identity, the diagnostic characteristics, except for those of the gonopod, being clearly apparent. In view of its fragility it was not removed from jar. Size typical of well-grown males in this species.

Associated Material. The 2 additional males are both small, with major chelipeds and some of the ambulatories missing. Their identity, since they were not removed for gonopod examination, necessarily remains slightly questionable.

De Man (1891) reported that this material then consisted of 3 specimens in poor condition, just as is true today. He also states their indigenous name, in Madagascar, is "Cava Tangena."

Type Material of *Gelasimus smithii* Kingsley, 1880. A dried male, cat. no. 3003, in the Academy of Natural Sciences, Philadelphia, is labeled under the specific name of *inversa*, from the Wilson collection, with the locality given as Natal. (!) No specimen labeled *smithii* was found. Kingsley, under the heading *Gelasimus inversus*, states that he never saw a copy of Hoffmann's description of *inversus*; his description and illustration of *smithii* show that the name has been rightly synonymized with *U. inversa* by several subsequent authors, including Barnard (1950).

Uca inversa sindensis (Alcock, 1900)

TYPE MATERIAL of the subspecies. *Gelasimus inversus* var. *sindensis* Alcock, 1900, from Karachi, is clearly described by its author. The 30 specimens were in the collections of the Indian Museum, Calcutta. I have not been able to trace their present location; they are not in the British Museum (Natural History), London.

Uca (Amphiuca) inversa inversa (Hoffmann, 1874)

(Red Sea to Natal; Madagascar)

MORPHOLOGY

With the characteristics of the species.

Major cheliped with upper part of carpal cavity extending distally as a distinct trench; dactyl with a strikingly large, subdistal, triangular tooth, sometimes minutely serrate. Minor cheliped with gape much wider than breadth of adjacent part of pollex; serrations absent. Ambulatory meri narrow, the dorsal margins of 3rd and 4th relatively straight. Gonopod thumb absent. Gonopore with postero-outer margin corneous and very slightly raised, but not tuberculate.

Measurements (in mm)

	Length	Breadth	Propodus	Dactyl
Largest male (Mozambique)	13.5	24.0	37.5	30.0
Moderate male (Zanzibar)	9.5	16.0	24.0	15.0
Largest female (Zanzibar)	9.5	15.0	–	–

Color

Displaying males: Display whitening absent or incomplete. Carapace black to blue with white transverse markings, often consisting of fine or coarse marblings. Frequently present and, when so, diagnostic in the field, is a posterior pair of large white spots or bars, usually oval and transverse; rarely another, similar median pair is divided in the midline by black. In Zanzibar the white was sometimes more extensive, so that the effect was of a white crab with blue markings. Major cheliped with merus, outside and in, and palm orange red; outer manus pink, the color sometimes extending onto fingers which are otherwise white. Eyestalks (Inhaca) yellowish. Anterior merus of minor cheliped and of 1st and 2nd ambulatories red (Inhaca); fingers of minor cheliped white; ambulatories otherwise dark. Females similar to males, except that white on posterior part of carapace usually is not broken by black in midline. Minor chelipeds sometimes (Inhaca) with propodus and dactyl blue in some specimens.

SOCIAL BEHAVIOR

Waving Display

Described from films made in Zanzibar and observations made in Zanzibar and on Inhaca; waving at Massawa was practically absent during the observation periods. Wave at low intensities obliquely vertical, the cheliped being only slightly raised; at high intensities, the wave can be termed a lateral straight, although the cheliped is rarely unflexed much farther than the point where the manus makes a right angle with merus and carpus, only rarely reaching level of tips of eyes, and more rarely above them. Minor cheliped often observed making 3 to 4 corresponding motions to every wave of major; this motion was not filmed; since the field work was done before the recognition of the probable stridulating component termed minor-merus-rubs (component no. 2), no definite statement is possible; the films show only an occasional, single corresponding motion by the minor. Carapace raised scarcely to not at all during display, except during walking; then, as usual, it is held well above ground during and between waves. Ambulatories not raised during display. Herding noted occasionally in Zanzibar, the male displaying, facing away from female, sometimes pushing against her with his posterior legs and carapace, and running as she ran. Only single males behaved in this fashion toward single females, and there was no attempt to urge a female toward a particular burrow. I do not know whether the females were immature (p. 503). High intensity courtship display unknown.

Uca (Amphiuca) inversa sindensis (Alcock, 1900)

(Karachi, Pakistan)

MORPHOLOGY

With the characteristics of the species.

Major cheliped with a predactyl depression somewhat separated by a convexity from upper part of carpal cavity; outer manus, palm, and fingers smoother than in *i. inversa*; dactyl without a predistal triangular tooth. Minor cheliped with gape subequal to breadth of adjacent part of pollex; weak serrations present, chiefly in distal part. Ambulatory meri moderately wide, the dorsal margins of 3rd and 4th clearly convex. Gonopod thumb present. (Female unknown.)

Measurements (in mm)

Largest male: length 8.0; breadth 16.0; propodus 16.0; dactyl 10.5.

Color

Displaying males: display whitening well developed, except for pink or pinkish orange on major palm.

SOCIAL BEHAVIOR

Waving Display

Known only from a brief and unsatisfactory film strip. Wave obliquely vertical to lateral straight, the major cheliped not fully unflexed. Minor makes corresponding motion. Several ambulatories on minor side raised in turn during display.

REFERENCES AND SYNONYMY

Uca (Amphiuca) inversa (Hoffmann, 1874)

TYPE DESCRIPTION. See under *U. (A.) inversa inversa*, below.

Uca (Amphiuca) inversa inversa (Hoffmann, 1874)

Gelasimus chlorophthalmus (not of Milne-Edwards)

Hilgendorf, 1869: 85; 1878 (1879): 803. Mozambique. Taxonomy.

Gelasimus inversus

TYPE DESCRIPTION. Hoffmann, 1874: 19; Pl. 4, Figs. 23-26. Madagascar: Nossy-Faly.

Kingsley, 1880.1: 155. Record of type description only.

de Man, 1891: 44. Madagascar: Nossy-Faly. Taxonomy.

Ortmann, 1894.1: 59. Tanganyika: Lindi; Dar-es-Salaam.

Alcock, 1900: 355. States that specimens in the Indian Museum, Calcutta, from Madagascar and the Red Sea are different from those from Karachi, which he then describes as a variety, *sindensis* (see below).

Gelasimus smithii

Kingsley, 1880.1: 144. Natal. Type description.

Uca inversa

Ortmann, 1897: 351. No new record. Taxonomy.

Pocock, 1903: 213, 216. Gulf of Aden: Sokotra I.: Hadibu and Haulaf; Abd-el-Kuri I.

Nobili, 1906.2: 151. Red Sea: Massawa. Taxonomy.

Nobili, 1906.3: 312. Red Sea. Taxonomy.

Stebbing, 1910: 328. South African records, none new.

Laurie, 1915: 416. Red Sea.

Balss, 1924: 15. Red Sea.

Colosi, 1924: 32. Somaliland.

Maccagno, 1928: 26. Specimens from Red Sea deposited in Torino museum. Taxonomy.

Rathbun, 1935.2: 27. Kenya: Gongoni.

Chace, 1942: 202. Tanganyika Territory: Lindi.

Vatova, 1943: 25. Lists published records; no new data.

Barnard, 1950: 94. South African records. Taxonomy.

Holthuis, 1958: 53. Sinai Peninsula: Ras Muhammed; Shora el Monpata; Abu Zabad.

Macnae & Kalk, 1958: 38, 67, 125. Mozambique: Inhaca I. Color; general behavior; ecology.

Macnae, 1963: 23. Mozambique: Inhaca I.

Uca (Amphiuca) inversa sindensis (Alcock, 1900)

Gelasimus inversus var. *sindensis*

TYPE DESCRIPTION. Alcock, 1900: 356. West Pakistan: Karachi.

V. *BOBORUCA* SUBGEN. NOV.

Typus: *Uca thayeri* Rathbun, 1900 (Tropical and subtropical coasts of America)

PLATE 17.
GONOPOD DRAWINGS: FIGURE 73.
DENDROGRAMS: FIGURES 96, 99.

MAP 11.
TABLES 1, 9, 10, 12, 19, 20.

MORPHOLOGY

Diagnosis

Large *Uca* with front of intermediate width; pile prevalent at least on ambulatories, often on carapace; 3rd and 4th ambulatory meri wide; major manus appearing smooth, never rough with large tubercles; fingers usually long and slender, never broad and flat; eyes never with styles; gonopods with subequal flanges, large thumb, little torsion; female gonopore without marginal tubercle.

Description

With the characteristics of the genus (p. 15).

MALE

Carapace. Front barely of moderate width, narrowest below eyestalk bases, its breadth between them about twice or less width of erected eyestalk at its base, in front of which the halves of the distal margin converge to the usual slightly excavate tip; frontal breadth contained about 5 to almost 6 times in that of carapace. Orbits almost straight; antero-lateral angles approximately right angles, scarcely or not at all projecting. Antero-lateral margins of moderate length, well defined, either slightly diverging, parallel, or slightly converging, curving obliquely inward, or, occasionally, turning in at an obtuse angle to form the strong dorso-lateral margins; these margins are usually conspicuously slightly concave, stopping opposite middle of cardiac region; dorso-lateral and, usually, antero-lateral margins finely and often indistinctly, beaded. Two pairs of postero-lateral striae, paralleling dorso-lateral margins, the members of each well separated from the adjacent margin and from each other; upper stria always long, strong and clearly beaded, lower much shorter, sometimes vestigial. Vertical lateral margin strong, complete, beaded. Carapace profile in lateral view little arched. Suborbital region concave, practically naked except along its dorsal margin; pterygostomian region little convex, finely rugose, setose.

Frontal margin always raised, broad, unbroken.

Eyebrows long and wide, oblique, sharply marked both dorsally and ventrally by beaded margins, the ventral being stronger. Orbit with lower margin scarcely rolled outward, slightly sinuous; without tubercles, granules, ridge, or mound on its floor. Suborbital margin with moderate-sized crenellations throughout, only slightly larger and farther apart near and beyond outer angle, continuing laterally to postero-lateral channel. Ventral margin of antero-lateral angle usually blunt, rarely sharp, always unarmed. Diameter of eyestalks moderate, that of the eyes in life distinctly larger than stalks, though not to the extent found in *Deltuca*.

Second Maxilliped. Spoon-tipped setae few to moderate.

Third Maxilliped. Ischium with central portion almost flat, its central groove represented only by a small, shallow depression near anterior margin. Gill minute, without books.

Major Cheliped. Merus with postero-dorsal margin undefined throughout, antero-dorsal blunt, armed with weak rugosities which distally are little stronger, and represent only a continuation of the general armature of postero-dorsal surface; proximo-dorsal surface smooth; ventral margin bluntly angled, with multiple, irregularly oblique rows of small tubercles; merus postero-ventrally and anteriorly practically smooth.

Carpus weakly rugose only postero-dorsally and dorsally. Bent-over dorsal margin slightly rugose; it and anterior surface each with or without one or more small to moderate tooth-like tubercles.

Outer *manus* appearing moderately smooth; surface covered with small to very small tubercles, the sizes mixed and largest on upper half, near the slightly bent-over dorsal margin. This margin, thin except distally where it is thick and tuberculate, is marked along its outer edge by a row of beads with a minute groove along their outer base; the latter is practically continuous with the short subdistal groove among the tubercles on proximal part of dactyl. No depression outside pollex base. Ventral margin of manus with a carina, beaded to tuberculate, with a narrow groove above it on its outer side, both ending at pollex base.

Palm with proximal, lower part smooth or, ventrally only, minutely tuberculate. Oblique ridge of moderate height, the part distal to apex slightly higher; its tubercles distinct, of similar size, several of them continuing up beyond apex around anterior margin of carpal cavity; sometimes others, weak, occur farther up, where the carpal cavity merges gradually, without a margin, into the finely tuberculate, nearly flat, central and upper palm; a very short, variable beaded margin extending from dorsal margin down along extreme antero-dorsal part of carpal cavity. Proximal and distal ridge inside dactyl base both well developed, with tubercles distinct even on distal ridge, parallel or with the long proximal ridge diverging slightly in its dorsal portion. Hollow inside pollex base moderate, its boundaries not sharply marked.

Both *pollex* and *dactyl* lacking long grooves on outer surfaces. Subdorsal furrow on proximal dactyl well developed, surrounded by tubercles. Both fingers ranging from moderate in length and breadth to very long and slender; dactyl almost straight in proximal two-thirds, curving down distally beyond pollex tip. Gape narrow. Outer surfaces of pollex and dactyl smooth except, sometimes, for small tubercles proximally, especially on dactyl. Lower margin of pollex smooth or proximally with tubercles; upper margin of dactyl tuberculate throughout roughly proximal half. All three rows of gape tubercles represented in both pollex and dactyl, one or more rows being indistinct only distally if anywhere; no tuberculate teeth or strikingly enlarged tubercles, although the usual enlarged tubercles—one submedian on pollex and two on dactyl—are present in most individuals.

Minor Cheliped. Merus with dorsal margin and posterior surface armed with moderately large, transverse rugosities; postero-ventral margin with a single line of tubercles, small or large; antero-ventral margin and anterior surface smooth or practically so except sometimes for a row of small tubercles, proximally only, on margin.

Carpus with subdorsal ridge finely and usually only partially beaded; ridge and adjacent groove strong. Mano-pollex ridge strong with trace of beading, starting about middle of manus and traceable about two-thirds of way to pollex tip; upper distal half of pollex with another, submarginal ridge, faintly beaded, just below gape teeth. Dactyl much longer than manus, its proximal two-thirds subdorsally with a ridge and, below and adjacent to it, a moderate to weak groove; sometimes a second ridge, proximally only, adjacent to and below groove. Gape narrow proximally, very narrow to absent in distal half or more; this section armed with moderately strong serrations.

Ambulatories. Meri all wide, those of 3rd and 4th legs notably so, their dorsal margins convex; dorsal armature consisting of low, close-set, transverse, spinulous rugosities; postero-dorsal tubercles small, on striae which are vertical near margin, sometimes curved or horizontal toward middle of posterior surface and always sparse to absent lower down; these tubercles and marginal armature largely concealed in pile, particularly in the young; ventral margins distinct, the posterior usually finely tuberculate and the anterior beaded, but one or both margins sometimes smooth, especially on first 2 ambulatories and in large specimens; these margins are always free of pile. Anterior surfaces of all segments of 1st ambulatory smooth (except, on merus, for the usual short row of predistal tubercles continued down from the dorsal and postero-dorsal regions).

Gonopod. Flanges present, large, subequal in breadth, confluent for a slight distance beyond and external to pore; inner process flattened, spinous, thick, scarcely visible; thumb large but arising far down shaft and ending below, or scarcely reaching, base of flange; torsion slight.

Pile. In adult males virtually confined to armature of minor merus and of ambulatories; highly variable and easily detached. In young prevalent on carapace as a velvety pubescence.

FEMALE

With the characteristics of the females in the genus (p. 17). Plentiful velvety pubescence on carapace and dense pile on legs persists in this sex throughout life, although it is most dense in the very young; in large females it is sometimes scanty, especially on carapace regions that are subject to abrasion. Gonopore set in a small depression and lacking tubercles on margin.

Size

Large.

Color

No display whitening. Carapace and appendages dark, sometimes with dark red or dark orange red, sometimes brighter on major cheliped.

SOCIAL BEHAVIOR

Waving display obliquely vertical with jerks. No special components during high intensity courtship. Little time devoted to waving. Male attracts female into his burrow. Chimneys constructed. Combat and threat behavior poorly known, apparently never prevalent.

Relationships

Comparisons with other groups will be found on p. 19 in connection with the discussion of the subgeneric dendrogram. The subgenus is treated here as monospecific.

Name

Boboruca: From the Greek *borborus*, "mud," *borborodes*, "muddy," the first *r* being elided for euphony. The name is proposed in reference to the frequent clinging of the biotopic mud to these *Uca*, the "muddy fiddlers."

17. *UCA (BOBORUCA) THAYERI* RATHBUN, 1900

(Tropical eastern Pacific; tropical and subtropical western Atlantic)

PLATE 17.
FIGURES 46 *K*; 56 *E*; 60 *H*, *I*; 73 *A*, *B*; 81 *I*; 82 *I*; 99.

MAP. 11.
TABLES 9, 10, 12, 19, 20.

INTRODUCTION

As mentioned above in the description of the subgenus *Boboruca*, certain characteristics of *Uca thayeri* suggest a closer kinship with Indo-Pacific subgenera than with the American groups. Most of them are concerned with social behavior, the unusual aspects being the infrequency of *thayeri*'s waving display; the hours of waving, with their concentration near sunrise and sunset; the almost vertical wave, without special behavior during high intensity courtship; the low general level of social activity; and the female's habit of constructing chimneys. In contrast, the male's attraction of the female to his own burrow, as well as morphological details such as most of the armature of the palm, are characteristics of American *Uca*.

The two forms for which the subgenus *Boboruca* has been erected are here reduced from species rank to that of subspecies, even though they have been divided by the Isthmus of Panama for some five million years. As in (*Minuca*) *vocator*, this course seems the only logical one to take, in view of the extreme similarity of *t. thayeri* and *t. umbratila*, both in morphology and in social behavior. Bott (1954: 163) first recognized the morphological closeness of forms in the two oceans.

Although *t. umbratila* has been only briefly observed in the field during a time of its almost complete inactivity, a group of specimens brought from western Costa Rica to Trinidad lived for months in the crabberies. It is there that its display and chimney building were seen, and compared with the activities of *t. thayeri*.

MORPHOLOGY

Diagnosis

As for the subgenus *Boboruca*, p. 109.

Description and Discussion

With the characteristics of the subgenus.

As indicated in the introduction to the species, the differences between the Atlantic and Pacific forms are so slight that the maintenance of two species does not seem warranted. In brief, the morphological distinctions consist in the Pacific chiefly of slightly stronger armature on the major cheliped and a more brachychelous claw; in the Atlantic, an exaggeratedly leptochelous form is usual among large specimens. In display, general habits, and habitat no distinctions are apparent. Because no other material is referable to this subgenus, and because of the similarity of the current forms, no characters are here singled out from the subgeneric description as of specific value, the differences being given below under the subspecies. So few adult specimens of *t. umbratila* have been taken that even most of the selected characters may prove to be unreliable.

Color

Displaying males: Display whitening absent. Carapace brown to orange brown, sometimes with faint reticulations. All appendages brown to orange, the major manus always brightest, being dull rufous to orange, with the color often extending to tip of chela.

SOCIAL BEHAVIOR

Waving Display

Wave practically vertical at low intensities, otherwise semi-unflexed. Jerks present, the major cheliped usually making between 2 and 4 during elevation but none on the descent. At its highest, chela tips reach well above eyes. Minor cheliped usually makes a synchronous motion. Carapace scarcely raised during display. Legs remain on the ground. Steps are not usually taken from side to side. Waving is at the rate of about 0.5 wave per second, and single waves occur at indefinite intervals ranging from several seconds to many minutes. (Component nos. 2, 3, 9; timing elements in Table 19, p. 656.)

As in many Indo-Pacific narrow-fronts, *thayeri*'s display period is short and its times of greatest and sometimes only activity are during the late afternoon and early morning, when low tide falls at those hours. As in all other *Uca*, however, occasional individuals in peak display phase wave at wholly atypical times of the day.

Precopulatory Behavior

After waving that is little or not at all faster than routine display, the male attracts the female to his own burrow and precedes her below.

Chimneys

Apparently only females that are in or approaching breeding condition build chimneys. The structures are large and usually symmetrically formed. Females dug out of them in Trinidad proved often to be ovigerous. Males often display from the wide rims, but seem never to construct their own. In southern Florida male (*Minuca*) *rapax* sometimes display in the same position on top of a *thayeri* chimney.

Acoustic Behavior

In Trinidad sounds were heard in both field and laboratory, when the crabs were below ground and intruders had been introduced. Von Hagen (1967.2) reports having recorded sound production on that island, but details are not yet available.

RANGE

Tropical and subtropical western Atlantic (*t. thayeri*) and tropical eastern Pacific (*t. umbratila*).

BIOTOPES

Deep mud on sloping banks of mangrove-bordered estuaries and streams, often partly shaded. In the Pacific *t. umbratila* was collected and observed somewhat farther upstream than the Atlantic *t. thayeri*. (Biotope nos. 12, 13, 14.)

SYMPATRIC ASSOCIATES

In the Atlantic *t. thayeri* often mingles closely with (*Celuca*) *cumulanta*. Marginally its usual distribution coincides with that of (*Uca*) *maracoani* on the tidal levels slightly below it and closer to open water, and, along its upper border, with that of (*Minuca*) *rapax*.

MATERIAL RESULTING FROM FIELD WORK

(The complete list of specimens examined is given in Appendix A, p. 600.)

Observations and Collections: Western Atlantic. United States: Florida: St. Augustine, Miami. Guatemala: Puerto Barrios. Guadeloupe. Trinidad and Tobago. Venezuela: Aragua: Turiamo; Delta Amacuro: Pedernales. Brazil: Pernambuco: Recife; Bahia: São Salvador; Rio de Janeiro. *Eastern Pacific:* Costa Rica: Puntarenas; Golfito; Ballenas Bay. Panama: Near Old Panama; Balboa (Canal Zone).

Film. Trinidad; Recife; Rio de Janeiro.

Sound Recordings. Trinidad.

TYPE MATERIAL AND NOMENCLATURE

Uca thayeri Rathbun, 1900

TYPE MATERIAL. In Smithsonian Institution, National Museum of Natural History, Washington. 7 males, 1 female, cat. no. 23753. Cabedello, mouth of Rio Parahyba do Norte, Brazil. June 20, 1899. Collected by A. W. Greeley. Received as gift from Branner–Agassiz expedition. On mangroves. Measurements in mm of largest: length 17; breadth 25; propodus 42.5; dactyl 34. (!)

See also note on *Gelasimus palustris*, p. 324.

Uca thayeri umbratila Crane, 1941

HOLOTYPE of *Uca umbratila*. In Smithsonian Institution, National Museum of Natural History, Washington. 1 male, no. 138132 (formerly New York Zoological Society no. 381, 129); Puntarenas, Costa Rica. Measurements in mm: length 17.5; breadth 29; propodus 48. (!)

Additional type material. In same institution. Paratypes: 1 young male, 1 young female, no. 79407 (formerly New York Zoological Society no. 4118); Balboa (Canal Zone). 1 female, no. 138133 (formerly New York Zoological Society no. 381130; Puntarenas, Costa Rica. (!)

Type Material, including holotype, of *Uca thayeri zilchi* Bott, 1954. In Forschungsinstitüt Senckenberg, Frankfurt a. M. (!) I have compared this material with examples of *U. t. umbratila* brought with me to Frankfurt, and they are undoubtedly all representatives of the same form. The major claw of the holotype has been regenerated, certainly from an early age, since it is well formed but too short for the carapace length; its smoothness, approaching that of *t. thayeri*, has been found to be characteristic of other species when the original claw was broken off when the crab was small.

Uca (Boboruca) thayeri umbratila Crane, 1941

(Tropical eastern Pacific from
El Salvador to Panama)

MORPHOLOGY

Carapace: Front-orbital margin notably straight, even in young; adults of both sexes lacking or practically lacking dorsal pile. Major cheliped: Carpus usually with several similar, close-set tubercles on dorsal margin and always with a single large tooth on its inner surface; manus with outer tubercles and those of center palm relatively large; short ridge curving down from dorsal margin around upper carpal cavity present or absent, when present less well marked than in *t. thayeri*, formed of separate tubercles rather than contiguous beads. Fingers shorter and wider than in *t. thayeri* and gape narrower, being less wide than adjacent part of dactyl. Gonopore of largest female only with postero-outer portion of margin slightly raised, but not enough to appear as a distinct tubercle.

Measurements (in mm)

	Length	Breadth	Propodus	Dactyl
Largest male				
(Rio Abajo, Panama)	19.0	29.0	49.5	32.0
Holotype male				
(Costa Rica)				
(dactyl imperfect)	17.5	29.0	48.0	–
Largest female (paratype)				
(Costa Rica)	19.5	30.0	–	–

Uca (Boboruca) thayeri thayeri Rathbun, 1900

(Subtropical and tropical western Atlantic from northern Florida, at Sarasota and St. Augustine, to Estado São Paulo, Brazil)

MORPHOLOGY

Carapace: Fronto-orbital margin very slightly oblique, even in large specimens, when compared with *t. umbratila*. Some pile almost always persists dorsally on carapace in adults, especially in female. Major cheliped: Carpus with row of tubercles on dorsal margin and tooth on inner margin present or absent; the two structures appear independently, outer manus and center palm appearing relatively smooth, the tubercles being smaller than in *t. umbratila*; short ridge curving down from dorsal margin around upper carpal cavity very distinct, its crest clearly beaded as characteristic of the subgenera *Celuca* and *Minuca*; fingers in adult males always long and slender; in large specimens the dactyl is more than twice length of manus and the pollex is sinuous, its distal end curving downward; gape wider than adjacent part of dactyl.

Measurements (in mm)

	Length	Breadth	Propodus	Dactyl
(All from Trinidad specimens)				
Large male				
(longest claw)	18.5	30.0	56.0	43.0
Longest male				
(broken claw)	19.5	31.0		
Moderate male	15.0	24.0	36.5	26.0
Largest female	19.0	27.5		
Largest ovigerous				
female	17.5	26.5	–	–
Smallest ovigerous				
female	12.5	17.5	–	–

REFERENCES AND SYNONYMY

Uca (Boboruca) thayeri Rathbun, 1900

TYPE DESCRIPTION. See under *U. (B.) thayeri thayeri*, below.

Uca (Boboruca) thayeri thayeri Rathbun, 1900

Cicie Ete

Marcgrave de Liebstad, 1648: 185; fig. Brazil. Color. Description.

Uca thayeri

TYPE DESCRIPTION. Rathbun, 1900.3: 134; Pl. 8, Figs. 1 and 2. Brazil: Rio Parahyba do Norte at Cabedello. (USNM !)

Rathbun, 1902.1: 7. Puerto Rico. Taxonomy.

Rathbun, 1918.1: 406; text figs.; photos. Puerto Rico, Jamaica. Brazil: Victoria; Natal; Rio Parahyba do Norte; Plataforma (Bahia); São Matheos. Taxonomy. (USNM part !)

Crane, 1957. Trinidad; Brazil. Preliminary classification of display.

Gerlach, 1958: 672. Brazil: Estado São Paulo (near Cananéia). Ecology.

Holthuis, 1959.2: 275; text figs. of gonopod; photos. Suriname: shore of Matappica Canal near mouth.

Barnwell, 1963. Illus. Brazil. Endogenous daily and tidal rhythms of melanophore and motor activity.

Miller, 1965. U.S.A. Morphology, physiology, and ecology in relation to distribution.

Salmon, 1967: 451. Florida. Distribution.

von Hagen, 1967.2. Trinidad. Tape recording secured. (Preliminary statement.)

Chace & Hobbs, 1969: 216; text fig. of gonopod. West Indian distribution stated: Cuba, Jamaica, Puerto Rico, Guadeloupe, Curaçao. Taxonomy.

Warner, 1969.1: 382-85. Jamaica. Ecology.

von Hagen, 1970.1: 226. Material recorded from Colon, Panama to Sta. Catarina, Brazil. New West Indian record: Hispaniola. Distribution refs. stated.

Uca (Boboruca) thayeri umbratila Crane, 1941

Uca umbratila

TYPE DESCRIPTION. Crane 1941.1: 181; Text Fig. 7;

Pl. 7, Fig. 34. Costa Rica: Puntarenas and Ballenas Bay; Panama (Canal Zone): Balboa. (USNM !)

Uca thayeri thayeri

Bott, 1954: 163; Text Fig. 5; Pl. 15, Figs. 5 a, b. El Salvador: El Triunfo. (Frankfurt !)

Uca thayeri zilchi

Bott, 1954: 164; Text Fig. 6; Pl. 15, Figs. 6a, b. El Salvador: La Herradura; La Union. Type description. (Frankfurt !)

VI. *AFRUCA* SUBGEN. NOV.

Typus: *Gelasimus tangeri* Eydoux, 1835

(Eastern Atlantic from southern Europe to Angola)

PLATE 18 *A-D*.   MAP 8.
GONOPOD DRAWINGS: FIGURE 63 (part).   TABLES 1, 10, 12, 14, 19, 20.
DENDROGRAMS: FIGURES 96, 99.

MORPHOLOGY

Diagnosis

Front moderate tending toward narrow. Carapace and pterygostomian region covered with large, well-separated tubercles; dorso-lateral margins extending back almost to posterior carapace border; a large, sharp tubercle on orbital floor near inner corner; spoon-tipped setae on 2nd maxilliped with a pointed process at base of each spoon. Major cheliped: manus rough; fingers flattened, the pollex broader.

Description

With the characteristics of the genus (p. 15).

MALE

Carapace. Front barely of moderate width, narrowest below eyestalk bases, its breadth between them about twice basal breadth of erected eyestalk, each half of distal margin obliquely converging to the usual slightly excavate tip. Orbits straight or scarcely oblique; antero-lateral angles almost right angles, scarcely projecting; antero-lateral margins short, sometimes almost absent on major side, parallel or slightly diverging, turning obliquely inward almost at once as the moderately converging, dorso-lateral margins; the latter continue almost to posterior edge of carapace; here they are stronger and curve slightly inward. Postero-lateral striae absent. Vertical lateral margins strong, complete, beaded. Carapace profile in lateral view strongly convex but not nearly semi-cylindrical. Regions all unusually deeply marked for *Uca* except for hepatic-branchial boundary, which is absent; regional central areas otherwise separately tumid and tuberculate, especially on branchial and cardiac regions. Area between dorso-lateral and lateral margins slightly tuberculate in large specimens. Suborbital region tuberculate; pterygostomian region strongly convex and tuberculate.

Dorsal margin of orbit single, only the upper (posterior) part ridged; below this ridge it is usually tuberculate, sometimes almost smooth, and always merging without a definite lower boundary into the orbit's smooth, inner surface. Orbit with lower margin not rolled outward. Orbital floor near inner corner with a large, sharp tubercle, flattened laterally; external to the tubercle is a mound, close to it but separate, that is covered with a variable arrangement of granules and small tubercles. Suborbital crenellations strong throughout, stopping externally just before lowest part of outer, orbital trench. Diameter of eyestalks moderate, that of eyes in life not strikingly greater.

Second Maxilliped. Spoon-tipped setae numerous, each spoon with a basal pointed process.

Third Maxilliped. Ischium smooth, convex except for a shallow, submarginal groove along inner edge. Gill large; on ventral side a large lobe, a large, thumblike process, and, between them 4 to 6 dactyloid books.

Major Cheliped. Ischium with a ventral tubercle. *Merus* with a large tooth distally on antero-dorsal margin; ventral margin tuberculate; postero-dorsal surface almost smooth and dorsally approximately flat, the postero-dorsal margin absent in distal half but represented by tuberculate, transverse striae. *Carpus* with dorsal margin tuberculate.

Outer *manus* rough with large tubercles, largest both near dorsal margin and in a broad, variable band starting behind gape and continuing, after an angular turn, downward to ventral margin; depression outside pollex slight and smooth. Ventral margin of manus with a tuberculate carina continuing on pollex.

Palm with oblique ridge strong, continuing upward along lower part of distal edge of carpal cavity; latter extending almost to upper base of dactyl, none of its margin being beaded; both ridges at base of dactyl usually irregularly tuberculate, diverging; depression inside pollex base slight; central area of palm almost smooth.

Pollex and *dactyl* long, flattened, moderately slender but not strongly tapering; pollex broader, somewhat concave proximally on upper edge, widest near middle; dactyl with lower edge almost straight, the upper curving little more than pollex, not extending beyond pollex and sometimes shorter. Both fingers without long, lateral grooves, although dactyl's submarginal proximal groove is traceable for half or more of dactyl's length. Gape narrow, almost absent in distal half. Rows of tubercles on both prehensile edges complete except distally, where the median row swerves outward and displaces outer row, leaving a smooth area between it and the inner row; all tubercles low and rounded, especially proximally; none enlarged except occasionally one slightly larger near base of dactyl and, almost always, another at widest (median) part of pollex; median row giving rise to them in each finger at that point usually is slightly raised, in a weakly triangular formation.

Minor Cheliped. Appendage long. Merus somewhat rugose, tuberculate externally and along postero-ventral margin. Carpus and manus weakly tuberculate, sometimes pitted; smooth in small adults; a small tubercle on carpus on inner distal side below dorsal margin; manus broad, shorter than dactyl; mano-pollex ridge strong, starting slightly proximal to middle of manus, running moderately near ventral margin, serrate. Pollex base broad; dactyl with one subdorsal ridge well developed, flanked by a groove both above and below; second ridge weak. Gape narrow, less than breadth of adjacent part of pollex, gradually decreasing distally; serrations few and weak, sometimes absent. Upper palmar setae in a tuft.

Ambulatories. Meri of moderate breadth, those of the 4th leg narrow, all with dorsal margins practically straight. Marginal armature strong. Vertical, tuberculate striae present on posterior margins.

Gonopod. Basically of the *Thalassuca* form, with flanges, a large inner process, and a subdistal thumb, the pore not being at the end of a produced tube.

FEMALE

With the characteristics of the females in the genus (p. 17) except as follows. Carapace not more arched and scarcely narrower; carapace regions only

slightly less marked; carapace tubercles more homogenous than in male, extending all the way to the smooth posterior margin, which is bounded anteriorly by a tuberculate edge; tubercles of carapace sides usually more extensive. Orbits less oblique than in similar sized males only in smaller specimens, since the corresponding male characteristics are already extreme. Antero-lateral angles more acute; tubercles on orbital floor somewhat larger on the whole than in male but not more regular.

Size

Large to very large.

Color

Slight display lightening but never to polished white. Major cheliped largely with grayed yellowish or dull orange. All parts of the crab including ambulatories often with shades of purple, usually dull. True red, blue, and blue green absent.

SOCIAL BEHAVIOR

Plentiful time devoted to waving display. Wave strongly lateral-straight, the circular component being usually weak. One special component, the curtsy, present in high intensity courtship. Female follows displaying male down his burrow with the usual exceptional surface copulations or attempted copulations. No construction activities. Down-pointing component included in threat postures.

RELATIONSHIPS

Comparisons with other groups will be found on p. 19, in connection with the discussion of the subgeneric dendrogram. Subgenus considered monospecific.

NAME

Afruca: Derived from the root of the Latin adjective, *africus*; hence, "the African fiddler," since the group is largely confined to that continent.

18. *UCA (AFRUCA) TANGERI* (EYDOUX, 1835)

(Tropical to warm temperate eastern Atlantic)

PLATE 18 *A-D*.
FIGURES 27 *D-F*; 37 *E*; 45 *E-II*; 46 *F*; 63 *D*; 81 *E*; 82 *F*; 99.

MAP 8.
TABLES 10, 12, 14, 19, 20.

INTRODUCTION

Uca tangeri holds a number of distinctions that are rare or unique among fiddlers. It is the only member of the genus found in Europe and probably the only one that lives in west Africa. Its range extends farther from north to south—a distance, measured around the bulge of the continent, of more than 9,500 kilometers—than that of any other *Uca*. In spite of the lack of competition from other fiddlers, it maintains a complex waving display unchanged from one end of this vast strip to the other; whatever the function of waving patterns in helping among sympatric populations to maintain species barriers, in *tangeri* this need obviously does not exist; hence the species provides clear-cut evidence that the intraspecific functions of waving are important.

A further interest lies in its courtship behavior. In particular, it is one of the several fiddlers, all of them in non-tropical regions, known to court females actively at night through using sound production instead of waving display.

Again, droving is apparently better developed in *tangeri* than in any other *Uca* except southern populations of *pugilator*. In Angola, during a week's observation in September, I saw large groups of both sexes of *tangeri*, none of them in territorial or display phases, surging to and fro, feeding at the tide's edge. In Spain, von Hagen (1962) observed similar droving, but here it was closely connected with tidal phases and ecological conditions. In Portugal I saw the pattern only once and the group was small. These differences are further considered in the general account of droving (p. 478).

The species is found throughout an almost full spectrum of *Uca* biotopes, with groups within the same population distributing themselves from mud to sand, and from areas exposed to air during only brief periods daily to the upper margins of beaches, the distribution being on the basis of sex, age, and physiological phase.

On both morphological and behavioral evidence, *tangeri* appears to be more closely related to American species, in particular to those of the subgenus *Uca*, than to other fiddlers. It probably holds the size record for the genus, in spite of close runners-up in America within the related group; the largest individual *tangeri* encountered in this study measures 33 mm long, 47 mm across, and has a claw 105 mm long from the base of the manus to the pollex tip.

Two final distinctions of the species are man-made. It is the only fiddler that is regularly savored as an hors-d'œuvre and the only one which receives protection through a policy of conservation. In Portugal during the late afternoon fishermen sometimes bring large, fresh, fiddler claws to waterfront shops; there they are boiled to an attractive pink, spread in rows on trays, and sold to passersby who nibble as they stroll. The rest of each crab meanwhile continues his life back on his native flat, where he was tossed after a fisherman detached his claw. There the fiddler is expected to grow another morsel suitable for harvest. Reports vary on the effect of this practice in relation to the numbers of crabs. The species is certainly becoming rarer in Spain (Altevogt, 1959), and as early as 1939 Vilela, a concerned Portuguese zoologist, wrote a paper deploring the reduction of local populations. Fortunately, in 1959 I found the fiddler still abundant in selected areas along the Algarve coast of that country, while regenerating chelae appeared to be no more numerous than in other species. We have as yet no information on the effect of claw loss on the success of courtship—a problem that certainly deserves attention.

MORPHOLOGY

Diagnosis

As for the subgenus, p. 116.

Description

With the characteristics of the subgenus.

MALE

Carapace. Postero-dorsal and lateral margins beaded throughout with close-set small tubercles. Carapace naked, covered except in regional grooves by well-separated, large, rounded tubercles which are largest over hepatic-branchial and cardiac regions, but absent over front and antero-lateral angles and weak to absent on intestinal region and near posterior margin; small specimens relatively smooth. Mound on orbital floor surmounted by highly variable small tubercles and granules, none of them nearly as large as the sharp protuberance near by. The latter arises farther inside the orbit than the mound, and the depressed eyestalk, much more slender than the orbit, fits behind and against it; the mound, which lies nearer the suborbital margin, is not at all in contact with eyestalk. Tubercles of pterygostomian region similar to large ones dorsally on carapace. Pilous setae absent from suborbital, pterygostomian, and buccal areas; longer setae scanty.

Second Maxilliped. Spoon-tipped setae with about 3 pairs of large lateral lobes and a basal pointed process, short and broad, almost a right angle in profile.

Major Cheliped. Merus with antero-dorsal margin finely granulate, the granules curving down proximally on anterior surface and ending distally in a large, blunt tooth fringed with long setae; on or slightly above ventral margin anteriorly a row of large, separated tubercles proximally curves from the edge up toward proximal end of granules from antero-dorsal margin; below or behind this row of large tubercles is a second row of smaller ones, closer together, confined to margin and submargin, sometimes irregular and compound on a thick ridge. Carpus with several enlarged marginal tubercles, variable, on bent-over dorsal edge and a very large, spinous tubercle on middle of inner surface. Dorsal manus margin double-edged, the inner with sharp, separated tubercles, the outer with smaller, rounded tubercles set closer together; ventral margin with a low carina, set with small tubercles that continue unbroken along pollex to its tip. A few additional tubercles inside pollex just above lower margin.

Ambulatories. Each merus with dorsal ridge armed with serrations, largest, roughest, broadest, and most irregular on 2nd leg, next on 1st, weakening and becoming regular on 3rd; finally, on 4th, the edge is thin, the serrations being both small and regular. Ventral margins with both edges of all legs equipped with moderate, well-separated, tubercular serrations, those of anterior series always somewhat smaller and more closely set. Anterior series largest on major 1st leg, decreasing in size regularly on the other 3 legs, but becoming close-set and numerous; vestigial on 4th leg. Posterior series about equally large on 1st and 2nd legs, reduced on 3rd, marked on 4th by a beaded ridge; the beads well separated, no great differences between major and minor side armatures, except on antero-ventral series of 1st leg. Posterior striae: variable but best developed on 2nd and 3rd legs, in vertical, tuberculate, long rows, numbering about 9 to 11, interspersed with shorter, less regular

striae; all confined to upper half of merus; less developed on 1st leg, absent from 4th; 1st and 2nd meri with a few striae near upper margin anteriorly.

Ambulatory carpus on major side dorsally flattened and rugose on 1st through 3rd legs, the area bounded by serrate edges anteriorly and posteriorly; on minor side carpus dorsally smoother; on 1st leg only posterior edge distinctly sharp and granulate; on 2nd, both edges; on 3rd, only the anterior.

Gonopod. Right flange strongly developed, extending far beyond pore, which opens at its side; left flange very low, almost lacking; inner process broad, thick except for distal tongue, which lies partly against left flange, is not curved forward, and ends just below level of pore; thumb well developed, not reaching base of flange.

FEMALE

Gonopore. Orifice slit-like, external to a large, somewhat flattened tubercle. Tubercle with apex antero-inner, the rest of its surface marked by several longitudinal creases.

Measurements (in mm)

(See also p. 449.)

	Length	Breadth	Propodus	Dactyl
Largest male (AMNH specimen 5933, part; Africa)	33.0	47.0	109.0	75.5
Largest male (Faro) (claw broken)	28.0	37.5		
Largest male (Loanda) (claw broken)	26.0	38.5		
Smallest displaying male (near Faro)	13.5	18.5	25.0	15.0
Largest female (Faro)	31.0	40.0	–	–
Largest ovigerous female (Faro)	28.5	35.5	–	–
Smallest ovigerous female (Faro)	17.0	23.0	–	–

Color

Displaying males. No display whitening. Carapace dorsally and on sides purplish to neutral gray to grayed orange yellow, or pale yellowish gray. Major cheliped with merus and carpus dark to purple to dull orange; entire manus and chela ranging to pure white, but often with lower outer manus and outer pollex dull orange or yellowish. Eyestalks yellowish; maxillipeds, suborbital areas, and pterygostomian regions often reddish purple. Minor cheliped sometimes yellowish to grayish white. Ambulatories purple to dull orange except for reddish purple often on anterior surface of meri. In large specimens narrow, distal bands at segment ends of both chelipeds and of some or all ambulatories occasionally bright orange or red.

The proportions of dull orange and purple vary greatly, without respect to size, sex, or phase. For example, within each of a number of local populations in South Portugal some displaying individuals were almost entirely dull orange and a few altogether purple, except for the white major manus and chela; these observations were made during June, at the height of the breeding season. In Angola in September the color range was the same as in Portugal, but dull gray carapaces were more prevalent and the paler orange yellow tones rare. In a geographically intermediate population in Lagos, Nigeria, during several days of rainy weather the color was dull but comparable to that in Portugal and Angola, the carapace being variable dull violet to purplish and the appendages in general dull yellowish with orange at the joints; one small group of inactive individuals had violet carapaces speckled with white.

Neither in Portugal, Nigeria, nor Angola were clear hues seen, the tones of yellow, orange, reddish purple, and purple always being strongly grayed. Nevertheless the color families and distribution were similar to those of several members of the subgenus *Uca*, particularly *stylifera* and *heteropleura* in the eastern Pacific.

Color records by Altevogt (1959) and von Hagen (1962) made in southern Spain agree well with the observations summarized above.

Social Behavior

Waving Display

Wave lateral-circular except at highest intensity courtship, when it becomes a lateral-straight. Jerks absent. Waves in series, usually numbering no fewer than about 12, and often totalling more than 60; at low intensities there are sometimes short pauses between waves of single series, the duration of each low intensity wave being about a second or somewhat less. At high intensities the rate is faster; whether in courtship or agonistic behavior the cheliped is then not brought to the ground, the wave starting and ending with the cheliped partly unflexed, the chela pointing forward. During a series the body remains raised on the extended ambulatories, none of which is lifted from the ground during display, except in the course of high intensity courtship. At this time and only then each wave is preceded by a curtsy, through the usual flexing of the ambulatories. Sometimes, when the male is displaying toward an attentive female behind him, his posterior legs are strongly bent while the anterior ones remain extended; the carapace consequently is almost vertical, as in the display of at least several other species having the curtsy component. (Waving components: 4, 5, ?9, 11, 13; timing elements in Table 19, p. 656.)

Precopulatory Behavior

Altevogt (1959) and von Hagen (1962) both report that in Spain *tangeri* males not only attract females down their burrows by waving display and sound production, but also, very often, mate at the surface both by day and at night. These observers were careful to count as examples only pairs that were actually in copulating position, including the arrangement of their abdominal flaps. In Portugal, Nigeria, and Angola, on the other hand, I observed or filmed a total of more than 20 apparently normal, diurnal courtships through the stage in which the female followed the male down his own burrow; each time the following was preceded by the male's high intensity waving display, as usual in American crabs. In contrast, I only rarely saw pairs in position at the surface, and of these daytime coverings of females I could see clearly in three examples that, although the female was not in an otherwise non-receptive posture, still her abdominal flap remained tightly closed, so that the male's gonopods were not in contact with her gonopores; in each of these pseudo-pairings, as well as in several others in which the abdomens were either invisible or not clearly seen, the male appeared from his general behavior, including failure to wave and lack of a burrow, to be a typical aggressive wanderer (p. 691); hence none of these males was presumably in condition for successful copulation. I did not observe *tangeri* at night. Perhaps the diurnal copulations above ground in this and other species in which the female usually follows the displaying male by day down his own burrow are characteristic of nocturnal courtships, the behavior occasionally becoming temporally displaced into daylight. The fact remains that the frequency of surface pairings in southern Spain was much higher than in the populations familiar to me, even when pseudo-copulations are counted.

Another unusual characteristic of the population in Spain was the prevalence of attraction of males as well as females down the burrows of waving males (von Hagen, 1962, p. 690). In this and other species I have observed such behavior only exceedingly rarely; in each case where I was able to identify the phase of a second male descending a burrow after the burrow-holder had already gone underground, the following male was in one of three categories. Either he had recently been dispossessed of this same burrow (usually by a wanderer when he was engaged in combat nearby) and was attempting to resume possession of the burrow, or a wandering male was investigating the burrow, or, finally, a male of any kind was startled into retreat down the nearest hole. None of these obvious explanations would of course explain von Hagen's observations, which were apparently numerous. The behavior appears wholly atypical of the species in several other places, as does the prevalence of surface copulations. Whether the explanation of both phenomena will prove to be due to ecological, seasonal, or still other factors at present unknown, they will certainly be of special interest. The subject is further considered in the general account of precopulatory behavior on p. 500.

The sound recordings of Altevogt (1962) and von Hagen (1962) also showed that when a male entered the hole, after attracting a female close to his burrow, he produced the "long whirl" described in the next section. Judging from my own observations, this drumming during the day takes place so far within the burrow that the motion is out of sight; this is unlike the occurrence of a corresponding component in the courtships of some other species, such as *pugilator* and *cumulanta*, where the drumming occurs clearly at the burrow's entrance.

Acoustic Behavior

Altevogt (1959, 1961, 1962) and von Hagen (1962) described and recorded two varieties of sounds that occurred both by day and at night in the courtship of *tangeri* populations in Spain. Both were made in the same way, by striking the lower margin of the major manus against the substrate; the investigators called the two sounds the *kurze Wirbel* ("short whirl") and *lange Wirbel* ("long whirl"); both are covered in the present contribution by the term major-manus-drum (sound component 9). In *tangeri* the differences consist in the number of vibrations in a series, which range from 1 to 3 in the short whirl and from 7 to 12 in the long. The investigators found that the short form serves to attract a female close to the burrow of the signaling male at night, in place of waving display; this short whirl is also used occasionally during daylight when a male and female are hidden from each other by physical obstructions; each short whirl proved to be as long as a single wave performed at the same temperature, so that there was a temporal equivalence to the visual signal. The long whirl, they found, is used in all high intensity courtships when the female is close to the burrow and the male inside it; it is also used, they report, by a male on top of an inhabited burrow, where it acts as "an irresistible signal for the occupant to emerge." If the crab is a male, combat may follow; if a female, copulation at the surface sometimes occurs.

My own records consist wholly of field observations and films, acoustic recordings not having been secured. They concerned only the long whirl and were gathered during daylight only in Portugal, Nigeria, and Angola. These data differ from the series in Spain, summarized above, principally in the situations that elicited the whirl, which will hereafter be termed for conformity the major-manus-drum or drumming. These sound components were infrequent and, with one atypical exception, were always produced only during agonistic encounters between males. I never once saw or filmed drumming during courtship in this species except in the following abortive case photographed in Angola. Its equivalent has been observed in other species. In this example an aggressive wanderer, with the short chela of an immature crab, climbed on the carapace of a much larger female. As she sat with her legs extended stiffly in the non-receptive posture, he plucked at her carapace with his minor chela, making the irregular, incomplete feeding motions of displacement behavior that are sometimes incorporated into the precopulatory pattern as carapace-feeding. From time to time he attempted to turn her into mating position but did not succeed. Finally he hit her carapace repeatedly and vigorously with the lower edge of his claw in a typical major-manus-drum (8 raps), then climbed down and wandered away.

As noted in the preceding section on precopulatory behavior, major-manus-drumming, in the apparently normal, high intensity courtships with which I am familiar during daylight, did not occur close enough to the burrow mouth to be visible.

Another filmed sequence shows a use of major-manus-drumming in intermale behavior. Photographed in Angola, it depicts the take-over of a burrow from a display-phase male by a slightly smaller, aggressive wanderer. The wanderer first attempted, unsuccessfully, to dig the burrow-holder out, using the major cheliped. He then suddenly withdrew his claw and knocked it sharply on the ground beside the burrow in a major-manus-drum (6 raps). The burrow-holder at once emerged and went away in submissive posture, eventually taking over an empty burrow. Meanwhile the former wanderer emerged from the hole he had just usurped and promptly started to wave. Except for the drumming, no threat, much less combat, occurred. We can only speculate that the former burrow-holder was close to the end of a display phase, while the wanderer was ready to enter one. Again, I have seen this particular use of drumming in *pugilator* and *lactea*.

Following a later series of observations in Morocco and Spain, Altevogt (1964) published a preliminary report, with sonogram, on sound production during waving through contact of the pollex with the ground, the instrument being the same as during major-manus-drumming. The sound consists, however, of only single raps, occurring at the low point of many waves. The sounds, it seems likely from the investigator's account, occurred during the high intensity waving-display components termed curtsies in the present study. My own films made in Angola and Portugal show that at least during the highest intensities the cheliped did not usually touch the ground between waves. Perhaps we have here another most interesting example of behavior differences among populations that will repay detailed study in the future.

A consideration of acoustic behavior in *tangeri* in comparison with that of other species is given on p. 481f.

In addition to the major-manus-drum (component 9) the following acoustic components have been filmed: leg-wag (5), leg-side-rub (6), major-merus-drum (7), and leg-stamp (11), all in agonistic situations only.

Combat

The following components were filmed: manus-rub (1) and interlace (9), the latter preceded by down-pointing (agonistic posture 2).

RANGE

From the south coasts of Spain and Portugal to Baia dos Tigres, at approximately lat. 16° S. Absent from the Mediterranean. In 1960 the species still occurred near the mouth of the Guadalquivere River (Altevogt, 1961, 1962; von Hagen, 1962). In 1959 I found it in all suitable places along the south coast of Portugal, from São Antonio do Réal in the east to Portamão in the west. I did not find it, nor was there any report of it, on the Atlantic coast of Portugal, even in highly suitable terrain. Altevogt, who earlier believed it, after a search, to be extinct in Tangiers, the type-locality, finally succeeded in 1963 in locating a small population near the mouth of Oued el Melaleh, about 5 km east of town (Altevogt, 1965.1). In Angola it was abundant in 1957 at various localities close to Loanda, which was the most southern point I reached.

BIOTOPES

Near mouths of streams, rivers, and their deltas, and on flats beyond their mouths in sheltered bays. Substrate ranging, even in the same locality, from almost pure mud to sand. The species range extends both north and south of the limits of mangroves; in purely tropical localities mangroves are often close to the crabs, as at Lagos, Nigeria, although populations have not been found in their shade. Burrows occur sometimes, however, among clumps of marsh grass. In south Portugal, where large rivers are absent, I did not find *tangeri* farther than about 1.5 km upstream, and usually they were much closer to the sea. The largest specimens and populations there were on

fully marine flats near Faro; in Angola the optimum habitat lay out of town, on a sheltered bay, with a full range of biotopes between rocky headlands. (Biotope nos. 4, 5, 6, 8, 9, 11, 12, 14, 16.)

MATERIAL RESULTING FROM FIELD WORK

(The complete list of specimens examined is given in Appendix A, p. 600.)

Observations and Collections. South coast of Portugal from São Antonio do Réal to Portamão; Lagos, Nigeria; Loanda, Angola.

Films. Motion pictures were made in all these localities.

TYPE MATERIAL AND NOMENCLATURE

Uca tangeri (Eydoux, 1835)

Type material of *Gelasimus tangeri*. In the collections of the Academy of Natural Sciences, Philadelphia, is a single male from the type-locality. One of its two labels, apparently much the older, reads as follows: "Gelasimus tangeri Eydoux Mag. Zool (type) I. tanger. 151 Guérin's colln." The second label reads: "Gelasimus tangieri (151 Guérin) Isle tangier Dr. Wilson," along with two male symbols and, in pencil, the words "type" and "Eydoux." The specimen's measurements in mm are as follows: Length (approximate, since posterior edge of carapace is broken), 23; breadth 33; propodus 64; dactyl 44. Except for the minor damage mentioned, this specimen is in good condition. Its present number in the Academy's collection is 9-3028 (!)

Kingsley (1880.1: 153) in listing the material he examined of *Gelasimus tangieri* wrote: "Tangier! Guérin's Collection (Eydoux' Types) . . . (Phila. Acad.)." Rathbun (1918.1: 387) concludes her reference to the type description: "type-locality Tanger; type in Mus. Phila. Acad. Sci.)."

In the Muséum National d'Histoire Naturelle de Paris are deposited a number of other specimens from the type-locality (!), four of which are listed by the Muséum as "types non spécifiés," and consist of three males and one female. One male, No. 3478-1, in a box labeled "Gelasimus tangerii Eyedoux Tanger," measures as follows in mm: Length 33; breadth 45; propodus 103; dactyl 66 (slightly shorter than pollex, probably because of regeneration at an early age).

Type of *Gelasimus perlatus* Herklots 1851. Boutry. In Rijksmuseum van Naturlijk Historie, Leiden. (!) I agree with several previous workers that *perlatus* is not specifically distinct from more northern examples.

Type of *Gelasimus cimatodus* Rochebrune, 1883. Senegambia. Reported by Rathbun (*loc. cit.*) to be in Paris museum; not found there by me in 1959. I agree with Rathbun that *cimatodus* should be synonymized with *tangeri*.

Type of *Uca tangeri* var. *platydactylus* Monod, 1927.

Not found by me in Paris museum, 1959. The variation described appears to come within the general range occurring in single populations.

Uca tangeri var. *matandensis* Monod, 1928. *Nomen novum* for *U. t.* var. *platydactylus*, the latter name being preoccupied.

REFERENCES AND SYNONYMY

Uca (Afruca) tangeri (Eydoux, 1835)

Note. This species has been the subject of a large number of relatively recent contributions concerned wholly with its biology. For the convenience of workers in this subject, the references are grouped at the end of the section, p. 124. In each of them the species is referred to as *Uca tangeri*.

Gelasimus tangeri

TYPE DESCRIPTION. Eydoux, 1835: cl. 7; Pl. 14 (colored). Morocco: Tangiers.

Milne-Edwards, 1852: 151. Cadiz; coasts of Morocco. Brief taxonomy.

Heller, 1863: 101. No new record. Brief taxonomy.

Kingsley, 1880.1: 153. New records listed as "West Africa! (Duchaillu); (?) Bahia! E. Wilson (Phila. Acad.)." Taxonomy. (See p. 327.)

Miers, 1881: 262. Senegambia: Gorée Island. Taxonomy.

Hilgendorf, 1882: 24. Senegambia.

Osorio, 1887: 226. Barre du Dande, Benguela, Lobito.

Osorio, 1888: 186, 190. L'Île du Prince.

Osorio, 1889: 133. Iles St. Thomé, du Prince; Praia Almoxarife.

Osorio, 1891: 46. Ile St. Thomé: Iogo-Iogo.

Aurivillius, 1893: 34. Portugal. Morphology; ecology; amphibious characteristics.

Ortmann, 1894.2: 760. Liberia.

Baudouin, 1903: 341. Spain. Autotomy, regeneration; use as food for man.

Osorio, 1906: 150. Angola: Loanda.

Baudouin, 1906: 1. Spain. General account.

Nobre, 1931.1: 179. Portugal. Taxonomy; local distribution.

Nobre, 1931.2: 100. Portugal. Taxonomy; color; local distribution.

Nobre, 1936: 58. Portugal. Local distribution; use as food for man.

Gelasimus perlatus

Herklots, 1851: 6. Near Boutry. Type description.

Milne-Edwards, 1852: 151. No new records. Taxonomy.

Hilgendorf, 1879: 806. Liberia; Chinchoxo; Loanda.

de Man, 1879: 66. Gulf of Guinea. Suggests synonymy of *perlatus* with *tangeri*.

Kingsley, 1880.1: 153. Guinea. Taxonomy.

Studer, 1882: 13. West Africa. Taxonomy.

Aurivillius, 1893: 31. Cameroons. Morphology; ecology; amphibious characteristics.

Aurivillius, 1898: 9. Cameroons: Bibundi.

Johnston, 1906: 862. Liberia.

Gelasimus cimatodus

Rochebrune, 1883: 171. Senegambia: "Les Deux Mamelles & côte de Maringouins; pointes des Cameaux." Type description.

Uca tangeri

Ortmann, 1897: 356. (Spelling: *tangieri.*) Saw Eydoux's specimen.

Rathbun, 1900.1: 276. West Africa. Taxonomy.

Baudouin, 1906.1: Annual claw harvest; cooking and sale in Andalusia; regeneration.

Bouvier, 1906: 187. Mauritania: Casanda Bay (Baie du Levrier).

Nobili, 1906.1: 317. Spanish Guinea. Taxonomy.

Rathbun, 1918.1: 387; photos. Dakar; Monrovia and vicinity; Baya R., Elmina, Ashantee; Chinchoxo; Loanda; mouth of Kwilu R.

Rathbun, 1921.1: 465. Belgian Congo. Taxonomy. Habits.

Balss, 1922.2: 80. Distribution.

Roux, 1927: 238. Gabon: Port Gentil.

Monod, 1927: 612; text figs. of tubercle development with growth, pterygostomian regions, feeding positions. Cameroons. Habits, with special reference to feeding and food.

Maccagno, 1928: 33; text fig. Congo region (Principe I., S. Thomé, Bissau).

Monod, 1932: 538. Mauritania. Species abundant near Port Etienne.

de Miranda y Rivera, 1933.1: 55; color pls. Southwest Spain. Old local records.

de Miranda y Rivera, 1933.2: southwest Spain. Habits; capture; distribution.

Werner, 1938: 134; habitat photo. Morocco. Brief, general habits.

Vilela, 1939: 177. Portugal. Reduction of *Uca* population through harvesting of claws.

Vilela, 1949: 65. Portuguese Guinea: local records.

Monod, 1956: 399; Text Figs. 559, 560. West Africa.

Rossignol, 1957: 86, 119. French Equatorial Africa, near Pointe Noire. Brief color note.

Guinot & Ribeiro, 1962: 67. Angola: Luanda to Baia dos Tigres.

Altevogt, 1965.1: 31; one text fig. Occurrence in type-locality (near Tangiers).

Uca tangeri var. *platydactylus*

Monod, 1927: 619. Cameroons. Type description.

Uca tangeri var. *matadensis*

Monod, 1928: 252. *Nomen novum* proposed for *U. t. platydactylus*, since latter name preoccupied.

SPECIAL LISTING. CONTRIBUTIONS TO THE BIOLOGY OF *Uca tangeri*

Monod: 1923. Morocco. Feeding.

1927. Cameroons. Food and feeding. (See also regular listing, above.)

Hediger: 1933.1. Morocco. Behavior.

1933.2. Morocco. Behavior.

1934. Morocco. Escape behavior. (In a general paper.)

Crane: 1957. West Africa. Preliminary classification of waving display.

Altevogt: 1959. Spain. Ecology and ethology.

1962. Spain. Acoustical behavior.

1963.1. Spain. Responses to polarized light.

1963.2. Spain. Experiments on learning behavior.

1964.1. Spain. Drumming and waving.

1964.2. Spain. 3 films on feeding and social behavior.

1965.2. Spain. Orientation methods.

1968. Spain. Text associated with film.

1969.2. Spain. "Bubbling" as a means of regulating temperature.

1970. Spain. Form and function of acoustic signals.

Altevogt & von Hagen: 1964. Spain. Orientation.

von Hagen: 1961. Spain. Nocturnal activity.

1962. Spain. Ethology, including acoustic behavior; post-larval development.

Gunther: 1963. Spain. Distribution and ecology.

Korte: 1966. Spain. Experiments on vision.

Jansen: 1970. Spain. Physiology and ecology of "posing."

VII. SUBGENUS *UCA* LEACH (*SENSU* BOTT, 1954)

Typus: *Ocypode maracoani* Latreille, 1802-1803. (East and west coasts of tropical America)

PLATES 18 *E-H*; 19-21.   MAPS 9, 10.
GONOPOD DRAWINGS: FIGURE 65.   TABLES 1, 9, 10, 12, 14, 19, 20.
DENDROGRAMS: FIGURES 96, 99.

MORPHOLOGY

Diagnosis

Uca of moderate to large size. Front extremely narrow, its distal end abruptly somewhat wider; eyestalks strikingly slender; eye on major side sometimes with a terminal style, unique in the genus; lower edge of eyebrow absent; no elevation on orbital floor. Major cheliped with oblique ridge sometimes continued up around carpal cavity; no beaded ridge along upper edge of cavity, which always continues unimpeded toward dactyl base; no long furrow on pollex or dactyl; pollex with ventral carina. Spoon-shaped tips on setae of 2nd maxilliped each with a proximal projection. Gill on 3rd maxilliped large, with books.

Description

With the characteristics of the genus (p. 15).

MALE

Carapace. Front exceedingly narrow, narrower than base of erected eyestalks, spatulate, the distal portion beyond eyestalk bases being distinctly broader than proximal. Frontal groove with margins usually parallel and almost fused in midline, not diverging until near posterior border of basal segment of eyestalk, but sometimes with margins diverging slightly starting at anterior end; groove never continuing close to edge of front which is notably smooth and rounded. Orbits scarcely oblique. Antero-lateral margins short, sometimes almost lacking; postero-dorsal margins usually distinct almost to postero-lateral angle of carapace; postero-lateral striae absent. Vertical lateral margins complete except occasionally in extreme dorsal portions, and in *princeps monilifera* where these margins are entirely absent. Carapace profile in lateral view and pterygostomian regions little convex except in *p. monilifera*; suborbital and pterygostomian regions always non-tuberculate and only sparsely setose.

Eyebrow region with no definite margin marking lower border. Orbit with lower margin not rolled outward; no tubercles, granules, ridge, or mound on orbit's floor; suborbital margin with crenellations moderate to strong, of varying widths, always continued almost to inner angle and, continuing around outer angle, to orbit's lateral channel; orbit on major side, along with the corresponding eyestalk, often strikingly longer than on minor side. Eyestalks very slender, especially in middle; eyes even in life not much larger in diameter than adjacent part of stalk.

Ocular Styles. A style, short or long, present or absent at distal end of cornea on major side only, being a prolongation of the portion of the stalk that continues along inner margin of cornea. Except in *stylifera*, styles never present on all males of any population, usually on rare individuals only, and more often in young than in adults.

Second Maxilliped. Spoon-tipped setae with narrow projections at bases of each spoon-like distal enlargement, unique in genus except for *U. (Afruca) tangeri*.

Third Maxilliped. Gill well developed, similar to those of *Thalassuca* and *Amphiuca*.

Major Cheliped. Ischium with or without a ventral tubercle. Merus with an antero-dorsal crest, a large tooth, or both, the structure being tuberculate, serrate, or smooth; ventral margin almost always tuberculate, sometimes variably rugose; postero-dorsal surface rounded and sparsely to moderately rugose, granulate, or tuberculate; variable, being in some individuals of all species practically smooth. *Carpus* with dorsal margin proximally or entirely tuberculate; postero-dorsal surface with small tubercles except in *princeps monilifera* and *ornata*; ventral margin smooth (except for occasional extensions of posterior small tubercles) except in *major* where larger, sharp tubercles are present.

Outer manus rough with large, coarse tubercles, always somewhat larger near dorsal margin and sometimes, in addition, near pollex base; below dorsal margin they sometimes form one or more definite rows, horizontal or oblique, that are separated by a smooth area or definite groove from a marginal series of one or more somewhat irregular rows of tubercles; the most distal of these tubercles is sometimes enlarged and similar in size to the one at the apex of the oblique ridge on palm (see below); one or more individual tubercles in some species overhang dorsal margin, directed toward palm; sometimes one or more large tubercles arise, instead, on distal part of palm itself, just below dorsal margin. *Outer pollex* with a large, very shallow depression at its base, tuberculate or punctate, the margins being indefinite. Ventral margin of manus and pollex as well as dorsal margin of manus described above, and dorsal margin of dactyl with tubercles that are strong and regular except for occasional obsolescence near tips of fingers, the tubercles usually surmounting a more or less distinct ridge that is sometimes flanked on outer side by a slight groove.

Palm with oblique ridge high, the tubercles usually well developed but tending toward obsolescence in large specimens; an enlarged tooth always present and persistent regardless of size at junction of ridge with carpal cavity; except in *heteropleura* and *ornata* tubercles of the ridge continue along distal margin of cavity to near base of dactyl; no beaded margin around cavity's upper distal part; extreme upper part of cavity always extends distally, unblocked by tubercles or elevation of predactyl area, open to base of dactyl; only one predactyl ridge, with proximal row of tubercles strong and fairly regular; distal ridge and row represented if at all by clustered tubercles or a few irregularities; a moderate depression inside palm at pollex base; center area of palm smooth.

Pollex and *dactyl* both compressed, usually broad, never markedly slender, the tip of dactyl never curving down beyond tip of pollex. A tuberculate, supra-marginal ridge starts on distal part of outer manus more or less close to ventral margin and continues along pollex; except in *major* it almost reaches pollex tip; a narrow, adjacent groove is distinct along its upper side, and, in *major*, continues beyond it; often the groove supports pile; in distal half of pollex the ridge either is still separate from margin, closely adjoins it, or (*maracoani* and *ornata*) merges with it; in *major*, where the ridge stops well before tip, the groove continues and supports plentiful pile. Pollex and dactyl both without long, outer grooves (excluding any very narrow grooves adjoining marginal and submarginal tuberculate ridges). Outer side of pollex smooth, granulate, weakly tuberculate, or punctate; when punctate the depressions sometimes are associated with irregular rugosities and are often filled with pile which may extend beyond them. Dactyl outside proximally tuberculate except in *ornata*, with or without a weak subdorsal groove, and, in *heteropleura* and *stylifera*, with a second groove faintly indicated proximally above central margin; remainder of outer surface finely tuberculate to smooth; inside dactyl proximally a similar subdorsal groove is sometimes developed, flanked below by a blunt ridge which is

sometimes tuberculate; the groove's location is closely adjacent to distal extension of carpal cavity. Gape narrow, usually present only proximally; tuberculation of prehensile edges various.

Minor Cheliped. Appendage long, the propodus roughly as long as erectile part of adjacent eyestalk plus cornea; armature weak; setae long but often abraded. Merus with serrations, sometimes vestigial, on dorsal margin; postero-ventral margin smooth to weakly serrate, or, in *maracoani* and *ornata*, with large tubercles, widely spaced, variable in number and rarely absent; rare individuals with smooth margins occur; antero-ventral margin smooth and rounded; postero-dorsal surface with scattered sharp granules; dorso-distal margin usually with a regular series of minute granules, but variable within species, the row being sometimes short, vestigial, or absent. Carpus with subdorsal ridge weak to absent; mano-pollex ridge except in *major* weak, setose, starting slightly proximal to middle of manus, running close to ventral margin and continuing throughout most of pollex close to the finely granulate or beaded pollex margin from which it is separated only by a narrow groove. Dactyl with the two subdorsal longitudinal ridges and intervening groove usually distinct but weak. Dactyl always very long, sometimes more than twice length of manus; gape narrow, being always less wide than adjacent part of either pollex or dactyl, present in *princeps* only proximally; serrations moderate (*princeps* only) to vestigial. Upper palmar setae in a close-set row.

Ambulatories. Meri of ambulatories never greatly expanded, that of 4th being always slender. Armature about as in *Deltuca*, being weak except on postero-ventral margin of merus in *ornata*. In this subgenus alone of American subgenera the armature of posterior surfaces of meri is confined to scattered spinules, except for traces in several species of striae or rugosities; tuberculate striae are absent. No rugosities intervening between the weak carpal ridges.

Gonopod. Showing large interspecific differences; in less specialized species it is basically of the *Thalassuca* type, with flanges present, inner process either broad and flat or fleshy, and thumb subdistal. In the most specialized forms the parts are variously exaggerated, displaced, reduced, or omitted; torsion is sometimes extreme.

FEMALE

With the characteristics of the females in the genus (p. 17).

Size

Moderate to very large.

Color

Display whitening sometimes maximally developed in 3 of the 6 species, absent in the others. Yellow and orange phases present before whitening. Carapace sometimes with purplish tones. Tones of red and orange on major cheliped usual in all species, sometimes associated with purple. Purple, purplish red, and dull orange occurring frequently on ambulatories. Blue and blue green absent, except for a bluish tinge appearing rarely on gray carapaces.

SOCIAL BEHAVIOR

Wave strongly lateral, both straight and circular, sometimes at high intensities developing into overhead circling. Plentiful time devoted to waving display. No special components confined to high intensity courtship, except that ritualized feeding from the female's carapace often accompanies the customary plucking or stroking motions of the male. Female normally follows male down his burrow, with the usual exceptional copulations and, especially, attempted copulations at the surface.

RELATIONSHIPS

Comparisons with other groups will be found on p. 19 in connection with the discussion of the subgeneric dendrogram. Within the subgenus *Uca*, *princeps* appears to be the least specialized, as shown by both gonopod and major cheliped, while *maracoani* and *ornata* are most so in their morphological attainments of shape and armature, as well as by their waving displays and large size.

Between *princeps* on the one hand and *maracoani* and *ornata* on the other, *heteropleura*, *major*, and *stylifera* appear clearly intermediate. Their points of greatest interest are their highly plastic gonopods. These appendages, in spite of the resemblances of the species among themselves in other particulars, show great diversity along with exaggerations of one or more parts; these differences occur, it seems probable, in response to the varied demands of sympatry.

The relationships of these median species among themselves and to the other members of the sub-genus are clearly not simple. Of the trio, *heteropleura* in the Pacific appears to be the least specialized and closest to *princeps*, also a Pacific form; the indications lie both in the relatively unexaggerated shape and armature of the cheliped and in the presence of a similar, anomalous projection on the gonopod. In the Atlantic, *major* seems, in spite of a wholly dissimilar gonopod, to be closer to *heteropleura* than to the Pacific *stylifera*; hence *heteropleura* and *major* are here termed allopatric species, although their resemblance to each other is not as striking as between the members of most other pairs of notably similar species in the two oceans. Finally, it seems that *stylifera* is least closely related to the other two; it alone has the broad and pitted pollex found to stronger degrees in *maracoani*, *ornata*, and (*Afruca*) *tangeri*; again, its ocular style occurs more constantly and attains a greater length than in any other species; finally its gonopod, with an attenuated tip and extreme torsion, resembles that of *major* only superficially, due principally to the reduction of flanges common to both species.

FOSSIL SPECIMEN

Rathbun described, from a male lacking both chelipeds, a fossil species from the Pliocene of southern California, *Uca oldroydi*. I agree with her conclusion that it is "near *monilifera*." The carapace, sternum, and abdomen are in good condition, while some details of the ambulatories' armature are discernible. Although the distal expansion of the front is minimal, being perhaps damaged, it is clearly of the American not Indo-Pacific form. Again, American rugosities, not simple tubercles, show on the merus and carpus of an ambulatory. Measurements in mm: carapace length 15; breadth about 28. This great width recalls that of *Macrophthalmus*, but it seems that crushing has probably affected the proportions. The type data are as follows: Rathbun, 1926.2: 29; Pl. 7, Figs. 1 and 2. From Deadman I., near San Pedro, Los Angeles County. C., Pliocene. Holotype originally at Stanford University; now in Smithsonian Institution, National Museum of Natural History, Washington, no. 1371025. (!)

19. *UCA (UCA) PRINCEPS* (SMITH, 1870)

(Subtropical eastern north Pacific and tropical eastern Pacific)

PLATES 18 *E-H*; 19 *A-D.*   MAP 9.
FIGURES 54 *E*; 65 *A, C*; 99.   TABLES 9, 10, 20.

INTRODUCTION

One of the most interesting observations that has been made on the biology of *Uca princeps* is the fact that waving display in one population is different in details of posture and in timing from that of several others. Comparison of specimens from Panama City, Panama, and from Puerto Bolivar, Ecuador, the two localities where I have observed displaying males, shows no morphological indication whatever that two subspecies are concerned. Several differences between the populations, aside from their wave forms, were noticed. First, the displaying crabs seen in Panama, although of mature proportions, were much smaller than those in Ecuador; second, the Panamanian crabs did not develop the striking display whitening or even yellowing found in the Ecuadorian group; and, third, young *princeps* collected in Panama did not have the styles on the eyes characteristic of the young in Ecuador. My description of waving in the Panama group was published in 1941; the observations in Ecuador were made in 1944 and have remained until now unpublished. Meanwhile Peters (1955) noted that a population he studied and filmed in El Salvador displayed in a fashion quite different from my description of the Panamanian crabs, while von Hagen (1968), after working with *princeps* in northern Peru, again found differences. Both von Hagen and Peters were apparently observing large males, similar in size to those I watched in Ecuador, and their descriptions agree well with my own made in that locality.

U. princeps at its maximum size is one of the largest species in the genus. Accordingly, the most likely explanation of these differences between displays in small, Panamanian crabs and in large individuals elsewhere seems to be that large crabs eliminate or alter components which would be awkward to perform as the crabs become heavier and the claw, through inexorably continuing allometric growth, becomes not only more unwieldy but probably more difficult to balance. For example, in small crabs the body is raised and lowered with every wave, as the ambulatories repeatedly unflex and flex. Furthermore, in small crabs the waves have greater amplitude, are faster, and are separated by shorter pauses. All of these characteristics may be expected in crabs needing to expend less energy to combat gravity.

Suggestive is the fact that two other species of *Uca*, (*Afruca*) *tangeri* and (*Minuca*) *mordax*, show precisely similar differences when the waves of very large individuals are compared with those of small but mature members of the population. An additional fact, possibly related to the problem, is that although both *ornata* and *maracoani*, both of them also unusually heavy members of the subgenus *Uca*, display vigorously when mature at all sizes, they show a wave form that puts the cheliped high overhead, above the center of the body; there it stays, well balanced, as it windmills through a long series of high-intensity waves. These same two species also perhaps derive extra support from their preferred substrate, which is sticky mud, into which they often sink slightly, sometimes bending their ambulatory dactyls inward; the resultant flat surfaces, braced by surrounding mud, perhaps aids both balance and support. In contrast, *princeps* usually displays on firm, muddy sand and has not been observed to stand on flexed dactyls.

Since in all rich localities *princeps* is more or less closely sympatric with approximately the same group of species, including all three other species of the subgenus *Uca* that commonly occur in similar biotopes in the eastern Pacific, no reasons for local development of courtship differences are apparent. Indeed the existence of these striking differences in wave form and timing add one more perplexity to the unsolved questions concerned with species-specific characteristics and the functions of waving display.

All the biological notes on *princeps*, both in this introduction to the species and below, refer only to *p. princeps*, since the northern form, here regarded as a subspecies, *p. monilifera*, has not been observed in life.

MORPHOLOGY

Diagnosis

Members of the narrow-fronted subgenus *Uca* with fingers of major cheliped slender and tapering or moderately deep, but always without pits on outer pollex. Short ocular style occasional, but only in young. Carapace with sides of frontal groove parallel, almost contiguous, no large tubercles on dorso-lateral margins; suborbital crenellations large throughout, not narrow and pointed near outer angle. Gonopod tip blunt, with a large, anterior, subdistal projection. Female with pile along dorsal margin of 4th ambulatory merus; gonopore with a large marginal tubercle externo-posteriorly.

Description

With the characteristics of the subgenus (p. 125).

MALE

Carapace. Dorsal profile little arched except in the northern subspecies, where the branchial areas, especially, are expanded. Frontal groove very narrow, its raised lateral margins parallel and almost contiguous; antero-lateral margins short; beaded, weakly serrate, or smooth; converging from the angles and usually distinguished from dorso-lateral margins, when they are present, on minor side only and then only by a slight bend in margin. Dorso-lateral margins present or absent; when present they are distinctly beaded or granulate but never strongly tuberculate or spinous. Vertical lateral margins present or absent. Suborbital crenellations large throughout, those near orbit's outer angle not abruptly narrower than those of central portion and not pointed. Lower margin of antero-lateral angles on both sides smooth.

Ocular Styles. Sometimes present in small males, especially in populations from Ecuador and Peru.

Major Cheliped. Ischium with a ventral tubercle. Antero-dorsal margin of merus with a tuberculate crest running its full length, slightly higher in middle than at ends; the tubercles are all small, although distally either slightly enlarged, more regular or both. Carpus with upper surface finely tuberculate or smooth. Palm with oblique ridge continued around anterior margin of carpal cavity; tuberculation of the single ridge at base of gape and dactyl highly variable, even within populations, ranging from a single strong, straight row to a cluster of tubercles at dactyl's base. Pollex and dactyl externally smooth in large crabs, granulate in young; always moderately to strongly compressed and either slender and tapering or very deep, depending on subspecies. Gape moderately wide in proximal half but with edges practically articulating throughout distal half through elevation of that portion of pollex. Dactyl with a single enlarged tooth in middle row slightly proximal to middle of segment; dactyl's inner row of tubercles commencing well beyond base of gape and consisting in its proximal part of large, well-separated tubercles diminishing distally in size.

Minor Cheliped. Gape absent in distal half or more, the prehensile edges being not only in contact but with the serrations better developed than in the remainder of the subgenus.

Ambulatories. Marginal armature of merus notably weak.

Gonopod. Both flanges usually present, scarcely curved, the anterior much wider than posterior, the latter being at best narrower than diameter of pore and, in large specimens from Ecuador, vestigial to absent. Inner process a well-developed, flat spine extending over the short tube in normal position. Thumb discernible but vestigial. Torsion slight. The most conspicuous structure on the gonopod is a large, firm, flattened projection below base of tube, arising below junction of tube with anterior flange; distal edge concave, antero-distal angle rounded or acute, where anterior edge curves to meet shaft. The projection's homologies are indeterminate and, save for the occurrence of a similar structure in *heteropleura*, it is unique in the genus; possibly it corresponds to a displaced version of the large, flattened tubercle in *U. (Deltuca) dussumieri spinata*, but in the latter the tubercle is clearly a semi-detached portion of the anterior flange and there is nothing in the present species to indicate a similar origin; in Ecuadorian specimens, at least, it could more readily be considered a strongly curved, atypical posterior flange, in which case the narrow structure here considered the posterior flange would when present be interpreted merely as an unusual strut on the surface of the tube.

FEMALE

No large, separated tubercles marking dorso-lateral margins of carapace. Merus and, sometimes, carpus and proximal part of manus of 4th ambulatory with pile on posterior surface adjoining dorsal margin, longest distally. Gonopore with a large marginal tubercle externo-posteriorly, depressible in large specimens because of an elastic basal area; it seems clear, however, in the specimens examined that the tubercle is not part of an inserted spermatophore (see p. 465).

Measurements (in mm)

	Length	Breadth	Propodus	Dactyl
*Uca princeps p.**				
Largest males, type series				
(Smith measurements)	24.1	41.1	64.0	
	23.4	39.8	71.4	
Largest male				
(Ecuador)	23.0	38.5	63.5	44.0
Largest female				
(Ecuador)	14.5	23.5	–	–
Uca princeps monilifera				
Male (USNM 67735)	29.5	49.0	80.0	55.0
Holotype male				
(MCZ 1578)	28.0	45.0	70.0	49.0
Female (USNM 67735)	28.0	42.5	–	–

*Larger specimens formerly in the collection consisted of a male from Golfito, length 24.5, a female from the same locality, 19.5, and an ovigerous female, same locality, length 14.5. Small displaying males near Balboa measured around 15 in length.

Morphological Comparison and Comment

Uca princeps shares only with *heteropleura* the presence of a large, anterior, subdistal projection on the gonopod. It shares with *major, maracoani,* and *ornata* the shape of the frontal groove, its sides being parallel and almost contiguous instead of diverging. In the breadth and compression of the major pollex and dactyl, the subspecies *p. monilifera* approaches the exaggerated shape found in *maracoani* and *ornata*; the armature and general form of the entire crab, however, indicates its close relationship to *p. princeps,* although the armature of carapace and ambulatories in both sexes is, in the northern form, even weaker than in *p. princeps.*

The collection of additional material in Mexico, particularly between Guaymas and Manzanillo, could yield material of much interest. Two males (USNM 57066) from Empalme, Sonora, have the characteristics of *p. princeps,* although the type material of *p. monilifera* came from nearby Guaymas. In addition, a male (USNM 20654) from San Blas, Tepic Territory (now Nayarit), has the gonopod projection similar to that in *p. monilifera*; the carapace tuberculation, and that of females in the same lot, is characteristic of *p. princeps.* The ambiguity is sufficient to suggest hybridization, in spite of the fact that the major cheliped is missing.

Color

Displaying males (*p. princeps*): display white complete in Ecuador except for bright orange manus; this stage is preceded there by cream, and that, in turn, by bright yellow, the phase shown by the majority of displaying males during the April period of observation. In Panama the color at brightest was as follows. Carapace purplish gray with bluish posterior margin. Major cheliped: outer side of merus proximally orange, distally dark purple; carpus, same; manus with lower two-thirds orange, upper third and chela white. Inner side, ischium light orange; merus, carpus, and chela white; palm orange buff. Eyestalks green yellow. Maxillipeds, suborbital areas, and pterygostomian regions white. Minor cheliped and ambulatories purplish gray except for anterior meri; these are sometimes orange on the major side and white on the minor, as in some Ecuadorian crabs that later in the day achieve display white. Females: carapace and legs with or without whitening, as in males of same populations.

An account of the color changes occurring in populations of Ecuadorian *princeps,* both daily and during growth, is given by Crane [1944].

SOCIAL BEHAVIOR

Waving Display

(See also introduction to the species.)

Panama (Crane's observations only [1941.1]; no films secured). Small specimens, about 15 mm long, actively courting. Wave lateral, straight to slightly circular. Jerks absent. With every wave carpus is raised, so that chela hangs obliquely down, and from that position the propodus is unflexed up and out, then brought without pause more rapidly into rest position; during unflexing the dactyl is raised slightly from the pollex. Minor cheliped makes a motion synchronous with the wave. Body raised high with each wave. Ambulatories remain on ground except to run a few fast steps to one side or the other. Waving at rate of about 1 wave every three-fourths of a second; a pause of similar length between waves. (Component nos. 4, 5, 9, 10, modified 14; timing elements in Table 19, p. 656.)

Ecuador (Crane's observations only, unpublished; no films secured). Large specimens. Differs from Panamanian display in maintenance of elevation of carpus during a series of waves, during which the manus and chela never touch the ground, but when at rest are held at carpal level, parallel to substrate; body kept raised on extended legs throughout a series (component 11), not raised and lowered at every wave; waving about one-third slower, at the rate of about 1 wave per second. Pauses between waves much longer, ranging from about 2 to 5 seconds. At maximum intensity carapace posteriorly depressed on bent ambulatories. The crab usually mounts some small elevation close to his burrow before starting to wave.

Precopulatory Behavior

No complete courtships were seen, either in Panama or Ecuador, although waving was often directed toward apparently attentive females. Twice in Panama, males seized females after strenuous waving displays and tried unsuccessfully to drag them down a burrow. In each case the burrow was that of the courting male. Grip used not noted [Crane, 1941.1].

RANGE

Eastern Pacific from Santo Domingo, San Bartolomé Bay, Lower California, and Gulf of California to Peru. One subspecies (*p. monilifera*) is known only in the Gulf, from San Felipe and the mouth of the Colorado River to Guaymas, while the other (*p. princeps*) occurs elsewhere.

BIOTOPES

The subspecies *p. princeps* occurs on salt marshes largely cut off from the sea, and on sheltered muddy sand and sandy mud beaches, near mangroves and sometimes among their rhizophores. In Panama displaying individuals were on the lower edges of muddy sand beaches, while non-displaying crabs fed farther out from shore, on almost enclosed mudflats. (Biotope nos. 6, 8, 11.)

SYMPATRIC ASSOCIATES

Occasionally closely associated with each of the other Pacific members of the subgenus *Uca*, except only rarely and atypically with *ornata*, a species that characteristically inhabits deep, open mudflats.

MATERIAL RESULTING FROM FIELD WORK

(The complete list of specimens examined is given in Appendix A, p. 601.)

Observations and Collections (U. princeps princeps only). Costa Rica: Golfito; Ballenas Bay. Canal Zone: Balboa. Ecuador: Puerto Bolivar.

TYPE MATERIAL AND NOMENCLATURE

Uca princeps (Smith, 1870)

COTYPES. *Gelasimus princeps*: In Museum of Comparative Zoology, Harvard University, Cambridge, Massachusetts. Nine males (cat. no. 5813) and 3 males (cat. no. 5814), all from or close to Corinto, Nicaragua, the type-locality. Measurements on p. 129. (!)

Uca princeps monilifera Rathbun, 1914

HOLOTYPE. *Uca monilifera*: Also in the Museum of Comparative Zoology (above). One male, cat. no. 1578. Type-locality: Guaymas, Mexico. "From Capt. C. P. Stone; USNM; rec'd. 1859." Measurements in mm: length 28; breadth 45; propodus 70; dactyl 49. (!)

Additional type material. In Museum of Comparative Zoology: 7 males, cat. no. 8272, labeled from Guaymas, Mexico, "out of type lot." In Smithsonian Institution, National Museum of Natural History, Washington, also from Guaymas, Mexico: one male, cat. no. 22180, labeled paratype, "received from the Museum of Comparative Zoology from no. 1578." (!)

Uca (Uca) princeps princeps (Smith, 1870)

(Tropical eastern Pacific)

MORPHOLOGY

With the characteristics of the species.

Carapace. Antero-lateral and dorso-lateral margins distinctly beaded or granulate in both sexes continuing almost to posterior edge of carapace. Vertical lateral margin always clearly marked and beaded at least in lower part; upper portion sometimes interrupted or weak and, close to antero-lateral margin, absent.

Major Cheliped. Merus with tuberculate crest moderately high. Carpus with postero-dorsal surface tuberculate. Outer manus with tubercles numerous, not differing much in size among themselves, extending to gape and on proximal part of dactyl but absent (except in young, where they are very small) from outside of pollex. Dactyl and pollex not notably deep. Lower manus and pollex with supramarginal, tuberculate ridge distinct from marginal ridge throughout, although close to it distally. Prehensile edge of pollex with distal part of median row having some slightly enlarged, blunt teeth, the proximal being both largest and, sometimes, bifid or irregularly tuberculate.

Gonopod. Anterior projection smaller than in *p. monilifera*, its antero-external angle acute.

Uca (Uca) princeps monilifera Rathbun, 1914

(Subtropical eastern Pacific: Gulf of California)

MORPHOLOGY

With the characteristics of the species.

Carapace. Arching of dorsal profile higher than in *p. princeps*, with branchial regions especially expanded. Antero-lateral margins indistinct, either unarmed or weakly serrate; dorso-lateral margins in male absent; in female faint, with irregular minute granules, the most posterior enlarged; vertical lateral margins in male absent, in female weak.

Major Cheliped. Merus with tuberculate crest higher than in *p. princeps*, and more regularly arched. Carpus with postero-dorsal surface smooth. Outer manus with few tubercles, widely spaced, not extending onto sides of either dactyl or pollex in the few specimens, all large, that are known. The outstanding characteristic of the subspecies is the great depth and compression of both pollex and dactyl, the widest part of each being an area just beyond middle of their length; both are notably concave internally near their proximal ends, while the pollex has a concave region externally near its tip, the extreme tip being curved outward. The supramarginal tuberculate ridge along lower part of pollex merges distally with the ventral ridge. These ventral and supramarginal ridges on the pollex, the outer prehensile rows of both pollex and dactyl, and the dorsal tuberculate ridge of the pollex all combine to give a strong, raised, tuberculate border to all the outer edges of the chela, the effect being more conspicuous than in *p. princeps*. Prehensile

edge of pollex, unlike that of *p. princeps*, without notably enlarged tubercles.

Ambulatories. Armature of merus especially weak, the margins and posterior surfaces in males being usually practically smooth and in females with small tubercles only on some margins. No young available for examination.

Gonopod. Thumb and anterior projection slightly larger than in *p. princeps*, the antero-external angle of projection being rounded. The two gonopods, however, are so closely similar that it seems reasonable to consider *monilifera* a subspecies of *princeps* in this subgenus where the gonopods differ widely even among closely related species.

REFERENCES AND SYNONYMY

Uca (Uca) princeps (Smith, 1870)

TYPE DESCRIPTION. See under *U. (U.) princeps princeps*, below.

Uca (Uca) princeps princeps (Smith, 1870)

Gelasimus platydactylus (not of Milne-Edwards, 1837, which is a synonym of *Uca major* [Herbst, 1782])

? Saussure, 1857: 362. Mexico: Mazatlan. Brief taxonomy.

Gelasimus princeps

TYPE DESCRIPTION. Smith, 1870: 120; Pl. 2, Fig. 10; Pl. 3, Figs. 3-3c. Nicaragua: Corinto. (MCZ !) (Not Pl. 2, Fig. 8, questionably identified by Smith as *princeps*; see this study under *stylifera*.)

Smith, 1871: 91. Record of type description.

Lockington, 1877: 146. West coast, Lower California: San Bartolomé Bay. Taxonomy. Habitat: "Holes under rocks at low tide." Material not extant.

Uca princeps

Rathbun, 1911: 550. Peru. Brief taxonomy; local distribution (by R. E. Coker).

Rathbun, 1918.1: 382; photos. Abreojos Point, Lower California, to Peru. Taxonomy.

Rathbun, 1924.4: 376. Gulf of California: Balandra Bay at Carmen I. (lat. 26° N).

Maccagno, 1928: 16; text fig. (claw). Mexico: Salina Cruz. Taxonomy.

Crane, 1941.1: 170; text figs. Costa Rica. Canal Zone. Taxonomy, color, waving display, habitat.

Crane, 1944. Ecuador. Color changes in the field.

Bott, 1954: 158; photos. El Salvador. Taxonomy (part).

Peters, 1955: 436; text figs. Salvador. Behavior; morphology; ecology.

Crane, 1957. Panama; Ecuador. Preliminary classification of waving display.

Altevogt & Altevogt, 1967: E 1269. Film of waving display.

von Hagen, 1968.2: 436; text figs. Peru. Taxonomy, color, waving display.

Altevogt, 1969.1: 238; text figs. Behavior and gonopod structure as isolating mechanisms.

Uca (Uca) princeps monilifera Rathbun, 1914

Eurychelus monilifer

Agassiz, L. (MS label.)

Uca monilifera

TYPE DESCRIPTION. Rathbun, 1914.2: 126; Pl. 9. Mexico: Guaymas. (MCZ !; USNM !)

Rathbun, 1918.1: 380; photos. Gulf of California: mouth of Colorado R. (Montague I.). Taxonomy.

Maccagno, 1928: 15; text fig. (claw). Mexico. Taxonomy.

20. *UCA (UCA) [MAJOR] HETEROPLEURA* (SMITH, 1870)

(Tropical eastern Pacific)

PLATE 19 *E-H*.
FIGURES 24 *D, E*; 32 *J, K*; 37 *F*; 40 *A, B*; 54 *D*;
  65 *D*; 82 *D*; 99.

MAP 10.
TABLES 9, 10, 12, 19, 20.

INTRODUCTION

Among the six members of the subgenus, *heteropleura* is the smallest, the most vigorous in waving display, and the most abundant. The bright orange on its lower manus and pollex, combined with its general appearance, flattened claw, size, and habitat, all make this species a close parallel to one of the Indo-Pacific narrow-fronts, (*Thalassuca*) *vocans*. The displays of the two forms, however, are wholly unlike, each showing characteristics typical of its own part of the world (p. 533).

MORPHOLOGY

Diagnosis

Members of the narrow-fronted subgenus *Uca* with fingers of major cheliped moderately slender and the pollex not pitted. Short ocular style occasional in adults and young. Carapace with sides of frontal groove diverging; suborbital crenellations broad and truncate in central portion but narrow and pointed near outer angle; in males only at least one tubercle beneath base of antero-lateral angle, above external orbital channel. Gonopod tip slender, with a large anterior projection at base of projecting tube. Female without pile on merus of last ambulatory; gonopore without tubercle, slightly curved; setae on chela margins moderately long, numerous, close-set.

Description

With the characteristics of the subgenus.

MALE

Carapace. Frontal groove narrow, its sides diverging gradually backward from its pointed anterior end; eyebrows wholly absent, the dorsal margins of orbits being thin, usually beaded but sometimes practically smooth; antero-lateral angles scarcely to slightly produced, directed either outward or obliquely forward, sometimes fully rectangular. Antero-lateral margins short and granulate, weakly serrate, or practically smooth, converging from the angles, distinguished from dorso-lateral margins only by a wide angle or

slight bend that is usually absent on major side; dorso-lateral margins little or not at all raised but always marked throughout by distinct granules or small tubercles that continue almost to posterior margins. Vertical lateral margins present and finely beaded in lower part but weak to absent in upper. Suborbital crenellations lower than in other species of the subgenus except *stylifera*, being broad and truncate in central section but narrow and pointed, sometimes almost spinous, well separated close to outer angle and beyond it on orbit's lower, outer margin. Lower margin of antero-lateral angle on major side sometimes tuberculate on inner side of its fringing setae; always at its lower (posterior) end, just above external orbital channel and therefore opposite end of suborbital armature, there is a relatively large tubercle, or a cluster or short row of several small tubercles; this structure always lies outside the aforesaid row of setae.

Ocular Styles. Present in some males, about as long as cornea.

Major Cheliped. Ischium with ventral tubercle present or absent. *Merus* with distal end of antero-dorsal margin having a large, flattened tooth or narrow crest, edged with variable denticulations; proximally the merus margin is edged only with small tubercles. Dorsal margin of *carpus* with a short proximal row of tubercles. *Outer manus* with tubercles largest near dorsal margin and around the slight hollow at pollex base; smaller tubercles about size of those on carpus occur inside this hollow and extend, unbroken, up along edge of gape and throughout lengths of both pollex and dactyl. *Palm* with oblique ridge ending at carpal cavity; above that level ridge is represented by a scattering of faint granules; one or usually two pointed tubercles, comparable in size to the usual enlarged tubercle that ends the oblique ridge at its junction with carpal cavity, occur submarginally near upper distal end of carpal cavity, the proximal being always the larger. Pollex with supramarginal ventral ridge conspicuous, farther removed from ventral margin in pollex' proximal half than in *princeps*, but distally close to marginal row; pile usually present in at least distal part of the narrow groove adjoining pollex submarginal ridge. *Pollex* and *dactyl* both moderately compressed, not notably deep except for pollex in large individuals; tapering; dactyl usually with traces of a short, proximal groove above its ventral margin, as well as the usual weak, subdorsal groove; pollex and dactyl each in distal half with a tooth enlarged in middle row, that on dactyl being more proximal; margins of gape beyond these teeth concave; dactyl's inner row of tubercles commencing well beyond base of gape and consisting in its proximal part of small, moderately close-set, rounded tubercles, the most proximal being abruptly the largest and the rest diminishing gradually in size.

Ambulatories. Merus with marginal serrations relatively well developed on postero-ventral margins.

Gonopod. Anterior and posterior flanges each represented by a narrow strut along each side of tube. Inner process vestigial, apparently represented by a brush of hair-like setae on slight thickening at base of projecting terminal tube. Latter is bent almost at right angles to main shaft, and distally shows slight torsion toward rear. Thumb absent. At base of tube is a large, stiff, flattened projection, directed anteriorly, similar to that found in *princeps* (p. 129), and of equally uncertain derivation.

FEMALE

Unlike male, female has no tubercle beneath base of antero-lateral angle; the suborbital crenellations, however, are smallest and narrowest near outer angle, similar to those of male. No pile on merus of last ambulatory. Gonopore without tubercle, a slightly curved slit.

Measurements (in mm)

	Length	Breadth	Propodus	Dactyl
Largest male (Ecuador)	15.5	24.0	38.0	22.0
Moderate male (Panama)	11.0	17.5	24.0	13.5
Largest female (Panama)	9.0	14.0	–	–
Ovigerous female (Panama)	6.5	(damaged)	–	–

Morphological Comparison and Comment

U. heteropleura is unique in having the suborbital crenellations abruptly narrow and pointed near outer angle, with one or more tubercles beneath base of antero-lateral angle. It shares only with *princeps* the presence of a large, anterior, subdistal projection on gonopod, and only in *stylifera* do the sides of the frontal groove also diverge.

From time to time during the preparation of this

study it has seemed to me that *heteropleura* must be synonymous with *stylifera*, the displaying members of a *heteropleura* population perhaps indicating a form of paedomorphism in which a majority of individuals never reached full breeding condition. Most of the few morphological differences—except for the gonopod—between the two could be readily explained as growth differences, while the fact that I have never, during limited observation periods, seen males that have attracted the attention of females was also suggestive. Finally, the energetic display of *heteropleura*, when compared with the slower movements found in *stylifera*, suggested the similar differences found between the displays of smaller and larger *princeps*. In spite of these earlier doubts, however, it now seems clear that *heteropleura* and *stylifera* are distinct species, although closely related sympatric forms, just as, for example, (*Celuca*) *beebei* and (*Celuca*) *stenodactylus*.

Color

Displaying males: Display whitening often developed but confined to carapace. A yellow phase precedes the white at Puerto Bolivar, Ecuador. Major cheliped: Outer merus, carpus, and upper half of manus purple; lower half of merus persistent dark reddish brown, providing a good recognition character at a distance in the field even among non-displaying males; lower half of manus and proximal carpus bright orange; rest of chela white inside and out; upper inner side of merus and carpus orange, their lower parts purple; palm orange. Minor cheliped and ambulatories brown to gray. Displaying males often do not develop white carapaces, remaining dark. Females dark brown to brownish gray.

SOCIAL BEHAVIOR

Waving Display

Wave lateral; sometimes straight at low intensities, circular at high. Jerks absent. At beginning of each wave carpus is raised, the manus and chela pointing obliquely down, not touching ground; cheliped then unflexed laterally, after which, except during low intensity straight display, the chela is raised maximally before being flexed abruptly down into position. Dactyl slightly raised and held parallel to pollex during each wave. Minor cheliped makes synchronous motion. During each wave body is raised high on middle 2 ambulatories, which are stretched vertically to such a degree that the shorter first and last legs leave the ground. The legs otherwise are not raised except often between waves in taking a few steps from side to side. Waving is at the rate of about 0.5 to more than 1 second per wave, with a pause between waves in a series about half as long; more than 100 waves sometimes occur in a single series. (Component nos. 4, 5, 9, 10; timing components in Table 19, p. 656.)

Acoustic Behavior

Leg-wagging (component no. 5) has been filmed.

RANGE

Tropical eastern Pacific from the Gulf of Fonseca (El Salvador) to Puerto Pizarro in northern Peru.

BIOTOPES

Tidal mudflats close to shore, along more or less open bays, not close to mouths of large rivers; displaying populations usually adjoining these flats on the lower levels of muddy sand beaches. (Biotope nos. 6, 8.)

SYMPATRIC ASSOCIATES

A few of the closely related species (*Uca*) *stylifera* sometimes share the same shore, as does (*Uca*) *princeps*. Although (*Uca*) *ornata* also lives on mudflats, *heteropleura* usually does not mingle with it, since the preferred habitat of the present species is farther from river mouths on somewhat firmer ground. Two members of the subgenus *Celuca* often mingle closely with *heteropleura*—*oerstedi* and *beebei*.

MATERIAL RESULTING FROM FIELD WORK

(The complete list of specimens examined is given in Appendix A, p. 601.)

Observations and Collections. Costa Rica: Golfito. Panama and Canal Zone: Bahia Honda; Panama City (Bellavista, Old Panama); Balboa. Ecuador: Puerto Bolivar.

Film. Old Panama.

TYPE MATERIAL AND NOMENCLATURE

Uca heteropleura (Smith, 1870)

HOLOTYPE of *Gelasimus heteropleurus*. In Museum of Comparative Zoology, Harvard University, Cambridge, Massachusetts. One male, cat. no. 5819. Alcohol. Gulf of Fonseca, Salvador. Collected by J. A. McNiel. Received from Peabody Academy of Science, Nov. 1885. Measurements in mm: length 15.8 (Smith); breadth 25 (Smith); propodus 32 (Smith); dactyl 20 (J.C.) (!)

REFERENCES AND SYNONYMY

Uca (Uca) heteropleura (Smith, 1870)

Gelasimus heteropleurus

TYPE DESCRIPTION. Smith, 1870: 118; Pl. 2, Fig. 7; Pl. 3, Figs. 2-2b. Gulf of Fonseca: El Salvador. (MCZ !)

Smith, 1871: 91. Record only; no new material.

Lockington, 1877: 147. Record only; no new material.

Kingsley, 1880.1: 139. No new material. Taxonomic comments on type.

Uca heteropleura

Rathbun, 1918.1: 385; Pl. 161, Figs. 1-4 (all drawn after Smith). No new material. Taxonomic description after Smith.

Crane, 1941.1: 171; text fig. of minor cheliped. Costa Rica; Panama. Taxonomy, color, waving display, habitat.

Holthuis, 1954: 162. No new material; record only.

Crane, 1957. Panama; Ecuador. Preliminary classification of waving display.

von Hagen, 1968.2: 435; text figs. Peru. Taxonomy; waving display; habitat.

Uca stylifera (not *Gelasimus styliferus* Milne-Edwards)

Bott, 1954: 159 (part); not text fig. nor plate. Cat. no. 1884 (Frankfurt) (*fide* von Hagen, 1968.2: 436); Alcadia de Triunfo, El Salvador.

21. *UCA (UCA) [MAJOR] MAJOR* (HERBST, 1782)

(Tropical western north Atlantic)

PLATE 20 *E-H*.   MAP 10.
FIGURES 65 *F*; 99.   TABLES 9, 10, 12, 20.

INTRODUCTION

Uca major is a moderately large fiddler that seems to be uncommon. Only scattered individuals have been observed and collected from single localities, and apparently no large population of the species has ever been reported.

Its closest relations appear to be *heteropleura* and *stylifera* in the eastern Pacific. The problem of the contrasting gonopods in these species is considered on p. 127.

The species for many years was known by the name of its synonym, *heterochelos*.

MORPHOLOGY

Diagnosis

Large, narrow-fronted *Uca* with fingers of major chela tapering, not broad and flat and not touching except distally. Differs from *U. heteropleura* most distinctly as follows. Ocular styles apparently absent. Frontal groove very narrow with sides parallel, practically contiguous; suborbital crenellations largest near outer angle, not centrally in male; no tubercle beneath base of antero-lateral angle. Gonopod tip without anterior projection at base of tube, which is extremely produced and twisted. Female with pile on last ambulatory along dorsal and ventral margins of merus, dorsally on carpus, and, sometimes, on manus; gonopore, although without a distinct tubercle, has the outer rim raised.

Description

With the characteristics of the subgenus (p. 125). Differs as follows from *heteropleura*.

MALE

Carapace. Frontal groove as in *princeps*, being very narrow, its raised lateral margins parallel and almost or wholly contiguous; dorsal margins of orbits thicker than in *heteropleura*; antero-lateral angles always acute and more produced, directed obliquely forward, the upper (anterior) part of the angle's lower margin finely serrate on major side; antero-lateral margins very short, tuberculate or serrate, even less distinct from dorso-lateral margins than in *hetero-*

pleura; dorso-lateral margins always granulate at least anteriorly, instead of throughout as in *heteropleura* or at least posteriorly as in *stylifera*; central and posterior tubercles are few or lacking in the largest specimens. Vertical lateral margins complete even at dorsal end. Suborbital crenellations large, largest close to outer orbital angle, not centrally; the several truncate crenellations, if any, are near the angle. No tubercles on lower side of antero-lateral angle above external orbital channel.

Ocular Styles. Absent from known specimens. None of those examined, however, have been very young.

Major Cheliped. Ischium with ventral tubercle absent, or only indicated by an enlargement of angle. *Merus* with antero-dorsal margin with a large crest running its entire length; proximally it is low, irregularly and variably shaped and almost entire; highest in middle whence it curves downward slightly to its end; distal half strongly tuberculate or serrate. *Carpus* with dorsal margin tuberculate throughout most of its length; one to three tubercles also present along lower margin. Tubercles of *outer manus* slightly more homogenous in size than in *heteropleura*. *Palm* with oblique ridge continuing strongly around anterior margin of carpal cavity; no submarginal tubercles below dorsal margin. Supramarginal ridge on pollex dying out beyond middle, although the groove along the ridge's upper margin continues, slanting slightly upward almost to pollex tip; groove, including its distal extension, supporting short pile which, beyond end of tuberculate ridge, is much better developed than in related species, sometimes filling in entire space between top of groove and ventral marginal ridge; in the best-developed examples the pile in this distal region is very long; in some specimens, as in other species, all pile is virtually lacking, apparently through abrasion. Pollex slender for the subgenus, even in large specimens, usually with a few minute punctae among the granules. *Dactyl* outside without definite trace of a lower proximal groove, and its subdorsal groove is rudimentary; inside dactyl the proximal subdorsal groove is well developed, its lower edge marked by a short line of tubercles. Middle row of tubercles on prehensile edge of pollex with a tubercle beyond middle only slightly enlarged, never in the form of a large, triangular tooth; a slightly enlarged, more proximal tubercle on dactyl is even less developed; the long distal margins of prehensile edges beyond these two are not at all concave (excluding the frequent slight curve of extreme tip of each finger).

Minor Cheliped. Mano-pollex ridge well developed, clearly stronger than in the other species of the subgenus, scarcely setose in the specimens examined.

Ambulatories. Some of the spinules near dorsal margin of posterior sides of meri are arranged on rudimentary striae.

Gonopod. More similar to that of *stylifera* than of *heteropleura*, the anterior projection being absent, but with *stylifera*'s characteristics exaggerated. Entire tip very elongate, twisted to such an extent that the positions of the very narrow flanges are reversed from the normal, the morphologically posterior flange, which projects beyond the pore, being approximately anterior; in contrast, the shorter, morphologically anterior, flange is distally in a posterior position.

FEMALE

As in *stylifera*, the suborbital crenellations in this sex are not a reliable character. Tips of chela clearly pointed, not obliquely truncate; setae on chela short and scanty, similar to those in *stylifera*, rather than in *heteropleura*. Tubercles on dorso-lateral margins present or absent. Merus of 4th ambulatory with pile on posterior surface adjoining both dorsal and ventral margins; above ventral margin it is absent proximally; carpus of same leg with pile on dorsal surface only; pile sometimes additionally present proximally on dorsal surface of manus. Gonopore without a distinct tubercle but with parts of the outer and inner rims slightly and unevenly raised.

Measurements (in mm)

	Length	Breadth	Propodus	Dactyl
Largest male (Trinidad)	20.0	34.0	52.5	34.0
Largest female (Bahamas)	15.0	25.0	–	–

Morphological Comparison and Comment

U. major alone among the members of the subgenus lacks ocular styles at all ages, so far as is known; it should be remembered however that this species is uncommon, and no large collections have been available for examination. The species is also unique in having the suborbital crenellations largest near outer angle, and in the extreme elongation and torsion of the gonopod's tip.

The male is easily distinguished from that of *maracoani*, the other Atlantic *Uca* with a very narrow front, by its normally tapering fingers on the major cheliped and the elongate tip to the gonopod; of similar diagnostic use in the female is the submarginal pile on the 4th ambulatory merus.

Color

Displaying males: display whitening usually strongly developed over entire carapace and major cheliped except for occasional purple markings in any of the following positions: in a band across posterior margin of carapace, or on merus, carpus, or manus of major cheliped; in Venezuela proximal part of pollex was bright purple red in the few examples seen but pure white in Trinidad. Eyestalks green yellow. Orbits, antennal area, and suborbital regions sometimes purplish. Minor cheliped with manus and chela usually white. Ambulatories with purple at least on anterior sides of meri; often all ambulatories are entirely purple or purplish brown even when the crab is otherwise completely polished white. Females dull brown to yellowish white except for yellow eyestalks and white on carpus, manus, and chela of each cheliped.

SOCIAL BEHAVIOR

Waving Display

(Known only from a few examples in Venezuela and Trinidad; films not secured.) Wave at low intensity semi-unflexed, approaching the vertical; more often, at higher intensities, straight to narrowly circular, the cheliped being unflexed obliquely upward. Jerks absent. Manus and chela not brought all the way down to contact with the ground between waves. Usually only front of body raised with each wave, through straightening of the more anterior ambulatories; posterior legs meanwhile are further bent, depressing posterior part of carapace; as each wave ends, the crab settles abruptly; sometimes at high intensities body is held moderately high during a series. Minor cheliped makes synchronous motion during wave. Ambulatories kept on ground during waving except that during moderate to high intensity rapid steps are usually taken to one side or the other, often toward the side of the major cheliped. Wave highly variable in duration; often lasting about 2 seconds, the unflexing and flexing each taking about 1 second without a pause at peak. At low intensities indefinitely long pauses occur between waves, but at higher intensities waves sometimes occur in long series with pauses between waves imperceptible. (Components 3, 4, 5, 9, 10, 11.)

Acoustic Behavior

Major-merus-suborbital rubbing (component no. 1) observed a number of times after each wave in agonistic situations. Von Hagen (1967.2) reports having made a tape-recording of sounds produced by this species.

RANGE

Tropical western Atlantic, known with certainty from the West Indies, Panama, Venezuela, and French Guiana. Reported once from Mexico (Kingsley, 1880.1), but no material has been found and the record should probably be considered questionable. Occurrence in Brazil (Seba, 1758 and 1761) also unproved (see under type material). The range in the West Indies consists of the following localities: Chace & Hobbs (1969) listed the Bahamas (Bimini, San Salvador), Cuba, Jamaica, Puerto Rico, Saint Croix, and Guadeloupe; von Hagen (1970) added Haiti, on the basis of specimens he found in the Hamburg Museum, and Trinidad, where I also found the species to be present but rare.

BIOTOPES

Muddy sand to sandy mud, usually on salt flats almost cut off from open water and well above low tide levels; near but not among mangroves. (Biotope nos. 6, 11.)

SYMPATRIC ASSOCIATES

Sometimes *major* occurs among lower, feeding (not displaying) populations of (*Minuca*) *rapax*. It does not seem to live among populations of *maracoani*, the other Atlantic species of the subgenus, although in one area familiar to me in Trinidad the two species live within a hundred yards of each other with no physical barriers between them, excluding the differences in their preferred substrates.

MATERIAL RESULTING FROM FIELD WORK

(The complete list of specimens examined is given in Appendix A, p. 601.)

Observations and Collections. Trinidad. Venezuela: Turiamo (Aragua).

TYPE MATERIAL AND NOMENCLATURE

Uca major (Herbst, 1782)

LECTOTYPE. Fig. in Seba (1761), labeled "*Cancer Uka una, Brasiliensis*," selected by Holthuis (1962) as lectotype of species. Brazil as the type-locality is in question, since many localities cited by Seba have proved to be incorrect; also, the species, except for that single record, has not been recorded south or east of Cayenne. I agree with Chace & Hobbs (1969: 214) that "As the identity of the species is not in question at the present time, there seems to be no justification for designating a neotype, apparently the only means by which the type-locality can now be corrected."

Type Material of *Ocypoda heterochelos* Lamarck, 1801. Lamarck cited only the figures of Seba and Herbst (1782, drawn after Seba), without a description, except the phrase "n. *Cancer vocans* Lin."

Type Material of *Gelasimus platydactylus* Milne-Edwards, 1837. In Muséum National d'Histoire Naturelle, Paris. Two specimens listed by the museum as "types non specifiés." One found, male, dried, with the following label: "*Gelasimus platydactylus* Latr. Cayenne." Carapace and major cheliped in good condition, other appendages not. However, there is no question but that this is the specimen illustrated by Milne-Edwards, 1852: Pl. 3, Fig. 2. (!)

Type of *Gelasimus grangeri* Desbonne in Schramm, 1867. In Muséum National d'Histoire Naturelle, Paris. Listed by the museum as "type non specifié." Guadeloupe. Synonymized already by Rathbun (1918.1: 381) with *heterochelos*, the name in turn synonymized by Holthuis (1962) with *major*. Specimen (female) dried, in poor condition, but recognizable as *major*. (!)

REFERENCES AND SYNONYMY

Uca (*Uca*) *major* (Herbst, 1782)

Cancer Uka una, Brasiliensis

Seba, 1758 and 1761: 44; Pl. 18, Fig. 8. Brazil (locality questionable; see above). External anatomy; habitat; habits.

Cancer vocans major

TYPE DESCRIPTION. Herbst, 1782: 83; Pl. 1, Fig. 11 (after Seba). Herbst's own material apparently from Bahamas.

Ocypoda heterochelos

Lamarck, 1801: 150. Cayenne. Reference only cites figures of Seba and Herbst.

Cancer Uca (not *Cancer vca* (*uca*) Linnaeus, 1767)

Shaw & Nodder, 1802: Color Pl. 588 (after Seba), with accompanying text, headed "The Uka Crab" (*Cancer Uca* in index). No new material.

Ocypode heterochelos

Olivier, 1811: Vol. 8: 417 (part). Brazil. Brief, confused taxonomy.

Uca una

Leach, 1814: 430. Refers to Seba and Herbst. This reference and date constitute the type description of the genus *Uca*.

Leach, 1815: 323. Further statement of referral of *Cancer vocans major* as figured by Herbst to *Uca una*.

Gelasimus platydactylus

Milne-Edwards, 1837: 51. Cayenne. Type description. (Paris !)

Milne-Edwards, 1852: 144; Pl. 3, Figs. 2, 2a, 2b. No new material.

Gelasimus grangeri

Desbonne in Schramm, 1867: 45. Guadeloupe. Type description. (Paris !)

Gelasimus heterocheles

Kingsley, 1880.1: 137 (part). ? Eastern Mexico; Jamaica. Taxonomy.

Young, 1900: 271 (part). Atlantic localities only. Taxonomy.

Uca heterochela

Rathbun, 1897.2: 27. Taxonomy, in reference to Seba and Herbst.

Uca heterochelos

Rathbun, 1918.1: 381; photos. Bahamas, Cuba, Jamaica. Taxonomy.

Rathbun, 1924.1: 156. Panama: Colon.

Crane, 1957. Venezuela. Preliminary classification of waving display.

Uca major

Holthuis, 1962: 240, 245, 246. Resurrection of Herbst's name. Specimen, not extant, figured by Seba selected as lectotype.

von Hagen, 1967.2. Trinidad. Tape recordings secured. (Preliminary statement.)

Chace & Hobbs, 1969: 213; text fig. Distribution in West Indies.

von Hagen, 1970.1: 224. Distribution in West Indies.

von Hagen, 1970.4; text figs. Trinidad. Adaptations to a particular intertidal level.

22. *UCA (UCA) STYLIFERA* (MILNE-EDWARDS, 1852)

(Tropical eastern Pacific)

PLATE 20 *A-D*.　　　MAP 10.
FIGURES 29 *C*; 65 *E*; 99.　　TABLES 9, 10, 20.

INTRODUCTION

When in full display color a male *Uca stylifera* is one of the most striking fiddlers in the world, the shiny white of his carapace being set off by an orange and yellow cheliped and bright purple legs. The yellow phase that precedes the white seems to human eyes no less notable. This large species, although uncommon, would be a convenient subject for experimental work on color changes and their relation, if any, to social behavior.

Comments on styles will be found on p. 455.

Uca stylifera and *heteropleura* appear to be so closely related and are so closely sympatric that they are here considered to form an alliance. Two probable species barriers exist in the forms of the gonopod and the movements and timing of waving display. It must, however, be kept in mind that another member of the same subgenus, *princeps*, shows two different forms of display, apparently depending on the size of the crab (pp. 128, 130).

MORPHOLOGY

Diagnosis

Differs from *U. heteropleura* most distinctly as follows. Major pollex faintly pitted. Ocular style always present, in adults usually as long as peduncle. Suborbital crenellations near center little wider than those near outer angle in male. Gonopod tip slender, without anterior projection at base of projecting tube. Female with gonopore strongly curved or medially angular; setae on chela margins short and few.

Description

With the characteristics of the subgenus (p. 125). Differs as follows from *heteropleura*.

MALE

Carapace. Dorso-lateral margin in larger specimens anteriorly weak to absent, with the granules small and widely spaced; more posteriorly the margin always persists while the granules are larger and more closely set; in the largest specimens the margin is confined to the area close to the posterior border, is well ele-

vated and surmounted with close-set granules enlarged to a size now better termed tubercles. Central suborbital crenellations less broad than in *heteropleura*, while the outer ones, although narrow tubercles widely set, are not so contrastingly small and sharp as in *heteropleura*. Almost always without tubercles on lower side of antero-lateral angle above external orbital channel, although adjacent part of lower margin of angle flares outward.

Ocular Styles. Present on all males, usually longer than peduncle in adults, sometimes with soft, flat expansions near tip; von Hagen (1968) reports some with short styles in his Peruvian material.

Major Cheliped. Ischium with a ventral tubercle. Merus with a low crest, weakly serrate to smooth, on antero-dorsal margin, not a flattened tooth or very short crest as in *heteropleura*; instead, in *stylifera* the crest runs throughout the segment's length, but proximally is merely a low ridge; it then increases gradually to attain moderate height near distal end; posterodorsal surface of merus almost smooth. Outer manus with tubercles not specially enlarged near pollex base. Palm with oblique ridge continuing around anterior margin of carpal cavity; 2 to 5 submarginal tubercles below dorsal margin, near anterior end of carpal cavity, relatively smaller than in *heteropleura*. Pollex moderately deep in comparison with most *heteropleura*, its outer side punctate in at least proximal half; rest of its outer surface, as well as that of dactyl, finely granulate except, in large specimens, for indistinct rugosities on pollex and, in all, for the usual proximal tubercles on upper dactyl. Middle row of tubercles on pollex gape with a large, triangular tooth, sometimes tuberculate, near middle or slightly proximal to a smaller, enlarged tooth on dactyl—not, as in the corresponding structure in *heteropleura*, distal to it; subdistal margins of gape not clearly concave. Occasional variations in both species however make these distinctions on the gape unsuitable for a key or diagnosis.

Ambulatories. 2nd and 3rd meri more expanded than in *heteropleura*, but with their armature slightly weaker.

l

Gonopod. No trace of anterior projection. Distal part of tube projecting farther and with its torsion greater than in *heteropleura.*

FEMALE

Suborbital crenellations, unlike those of male, become smaller toward outer angle, somewhat as in *heteropleura* but usually less abruptly; in any case in this sex they are not a good distinguishing character between the two species. Setae on chela short and scanty in comparison with those in *heteropleura,* those on upper margin of dactyl shorter than dactyl's breadth. Gonopore more strongly curved than in *heteropleura,* its margin's central portion angular in larger specimens.

Measurements (in mm)

	Length	Breadth	Propodus	Dactyl
Largest male (Corinto) (claw broken)	17.5	29.0		
Large male (Corinto)	16.0	24.5	31.5	20.5
Male (largest claw) (Golfito)	15.0	23.5	35.0	23.0
Largest female (Golfito)	14.0	23.0	–	–
Ovigerous female (Golfito)	12.5	20.0	–	–

Morphological Comparison and Comment

In *U. stylifera* alone do all the males except the youngest have well developed styles; in adults they are longer than in any other member of the subgenus. The species shares only with *heteropleura* the divergent sides of the frontal groove. Finally, it shares only with *maracoani* and *ornata* the pitting on the major pollex; in *stylifera,* however, these minute depressions are faint.

Color

Displaying males: Display whitening fully developed on carapace, including its lateral and anterior aspects; white does not extend to the appendages, except for several small areas; the exception to white on the carapace is a purple band bordering its posterior edge. White phase preceded seasonally, daily, or both by clear yellow, at least in Puerto Bolivar, Ecuador, and in Panama City, Panama. Major cheliped: outer merus and carpus orange yellow; manus orange or, rarely, rose pink, its tubercles white and with white near upper margin; outer pollex orange; outer dactyl white. Inner merus, carpus, upper palm, and dactyl pale yellow to pale orange; lower palm and inner pollex orange. Eyestalks and eyes greenish yellow; styles yellow. Minor cheliped yellow except for or-

ange tips for chela. Ambulatories pinkish violet to bright purple except as follows: first 3 legs on major side with anterior merus often orange; that of first 2 on minor side often white.

Females dull, the carapace purplish gray to grayish white; chelipeds largely white; ambulatories dark purple.

SOCIAL BEHAVIOR

Waving Display

(No films secured)

Wave lateral, apparently chiefly straight with little or no circularity. Jerks absent. At beginning of a series merus and carpus are raised and extended straight out at side, the manus and chela being unflexed forward from them at right angles and the whole cheliped held clear of the ground; this is the lowest position attained in the course of a single series. From here, manus and chela are raised slowly in a motion of narrow amplitude and promptly returned to the lower position. During waving the dactyl is slightly raised, parallel to pollex, and the minor cheliped sometimes makes a synchronous motion. The body is held fairly high throughout a series, although during a wave the crab sometimes stretches the first 2 pairs of ambulatories so that anterior part of carapace is tilted up with them while posterior portion is lowered on the bending of the last 2 pairs of legs. This seesaw motion is sometimes suddenly inserted in the midst of a series, or all waves in a series may include the motion. The ambulatories are kept on the ground except to make several steps in any direction during a wave and during high intensity courtship to raise and lower one middle leg on each side, not necessarily the corresponding leg, simultaneously; in the succeeding waves the members of the 2 pairs of middle legs alternate with one another, not in a wholly regular fashion. Display is at the rate of about 1 wave every three-fourths second; waves are repeated without pause to a total of 30 to 50 in a series. Pauses between series may be as short as 5 seconds. (Component nos. 3, 4, 9, modified 11.)

Precopulatory Behavior

Males and females come inshore from the mudflats to muddy sand beaches to display and court. Apparent copulation, preceded by high intensity display and by stroking of the female, was seen once at the surface. Although numerous high intensity displays directed toward apparently attentive females were observed, only one female followed a male down his burrow. Further details are given in the general discussion of precopulatory behavior, p. 504.

RANGE

Tropical eastern Pacific from Pacific coast of El Salvador to northern Peru.

BIOTOPES

As in *heteropleura* (p. 135). Sometimes closer to low-tide levels.

SYMPATRIC ASSOCIATES

As in *heteropleura* (p. 135).

MATERIAL RESULTING FROM FIELD WORK

(The complete list of specimens examined is given in Appendix A, p. 601.)

Observations and Collections. Nicaragua: Corinto; Costa Rica: Golfito. Canal Zone and Panama: Balboa and Panama City. Ecuador: Puerto Bolivar.

TYPE MATERIAL AND NOMENCLATURE

Uca stylifera (Milne-Edwards, 1852)

LECTOTYPE. In Muséum National d'Histoire Naturelle, Paris. Listed by the museum as one of two "types non specifiés." One specimen found, male, dried, with the following label: "*Gelasimus styliferus* M. Edw. M. Th. Bell. Guayaquil." Also has "Guayaquil" written on major dactyl. Designated here as lectotype. Measurements in mm: length 20; breadth 34; propodus 50.5; dactyl 35. Eyestalk on major side missing, as well as minor cheliped, 1st left ambulatory, 3rd right, and dactyl of 4th right. Otherwise in good condition. Milne-Edwards' illustration (1852, Pl. 3, Fig. 3a) shows a short style and mentions "un petit stylet" in his description (p. 144); his drawing of the major cheliped unmistakably depicts *stylifera*, not *heteropleura*, and shows well the characteristics of the type. Specimen not relaxed for gonopod examination, there being no confusion apparent in specimen, locality, or name. (!)

Type Material of *Gelasimus heterophthalmus* Smith, 1870. In Museum of Comparative Zoology, Harvard University, Cambridge, Massachusetts. 4 male cotypes from the Gulf of Fonseca. Already synonymized by Rathbun (1918.1) with *stylifera*; I agree. (!)

REFERENCES AND SYNONYMY

Uca (Uca) stylifera (Milne-Edwards, 1852)

Gélasime platydactyle

Milne-Edwards, 1836: Pl. 18, Fig. 1a (not *Gelasimus platydactylus* Milne-Edwards, 1837).

Gelasimus styliferus

TYPE DESCRIPTION. Milne-Edwards, 1852: 145; Pl. 3, Figs. 3, 3a. Ecuador: Guayaquil. (Paris !)
Smith, 1870: 118. No new material.
Kingsley, 1880.1: 139. No new material. Brief taxonomy.

Gelasimus heterophthalmus

Smith, 1870: 116; Pl. 2, Figs. 6, 6a; Pl. 3, Figs. 1-1b. Gulf of Fonseca. Type description. (MCZ !)
Smith, 1871: 91. No new material.
Lockington, 1877: 147. No new material.
Kingsley, 1880.1: 139. No new material. Brief taxonomy.

Gelasimus sp.

Smith, 1870: 126; Pl. 2, Fig. 8. Nicaragua: Corinto. (2 females, MCZ !)

Uca platydactyla var. *stylifera*

Ortmann, 1897: 347 (part). No new material.

Uca stylifera

Rathbun, 1918.1: 383; photos. Costa Rica: Puntarenas. Taxonomy.
Pesta, 1931: 180. Costa Rica.
Crane, 1941.1: 171. Nicaragua; Costa Rica; Panama. Taxonomy; color; waving display; copulation; habitat.
Crane, 1944. Panama; Ecuador. Color change in the field.
Holthuis, 1954: 162. No new material.
Crane, 1957. Panama; Ecuador. Preliminary classification of waving display.
Bott, 1954: 159; Text Fig. 3 (gonopod); Pl. 15, Fig. 3a-b (photos). (One specimen listed, Frankfurt cat. no. 1884, referred to *U. heteropleura* by von Hagen, 1968.2. Illustrated specimens are characteristic examples of *stylifera*.) El Salvador: Alcadia de Triunfo. Taxonomy.
Altevogt & Altevogt, 1957: E 1268. Film of waving display.
von Hagen, 1968.2: 433; text figs. (style; gonopod). Peru: Puerto Pizarro. Taxonomy; color; habitat.
Altevogt, 1969.1: 238; text figs. Peru. Behavior and gonopod structure as isolating mechanisms.

23. *UCA (UCA) MARACOANI* (LATREILLE, 1802-1803)

(Tropical western Atlantic and tropical eastern Pacific)

PLATES 21 *A-D*; 44 *B*; 45 *B*.
FIGURES 28; 33 *A-FF*; 36 *C*; 37 *G*; 34 *B*; 53 *A, B*;
  55; 56 *D*; 60 *F, G*; 65 *B*; 78; 79; 80; 81 *H*;
  82 *C*; 83 *C*; 84; 88; 89; 94; 99.

MAP 9.
TABLES 4, 9, 10, 12, 14, 19, 20.

INTRODUCTION

Uca maracoani and a related species, *ornata*, are both notable for large size and a heavy claw with broad, flat fingers. Their displays are equally unusual. At high intensity the great claw is lifted above the body and balanced high overhead, the divergent fingers pointing toward the sky; then, while the whole cumbersome appendage revolves, the chela describes wide circles for long seconds at a time.

It is interesting that two other, less specialized members of the subgenus, *princeps* and *stylifera*, in their high intensity display closely resemble the low intensity simplicity of the more specialized *ornata* and *maracoani*.

In Trinidad the behavioral phases leading to waving display and courtship in *maracoani* were worked out in the crabberies, following field observations on this and other species (p. 505).

A very similar Pacific form, previously known with some confusion as *Uca insignis*, is treated here as a subspecies of *maracoani*, a species name hitherto restricted by taxonomists to the Atlantic form.

MORPHOLOGY

Diagnosis

Members of the narrow-fronted subgenus *Uca* with both fingers of the major cheliped strikingly deep and compressed; pollex conspicuously pitted, dactyl deeper than pollex, its upper border strongly arched; antero-dorsal margin of merus with separated, large tubercles or blunt spines. No large, separated tubercles on dorso-lateral margin of carapace in male, except for a single one at posterior end. Gonopod tip with tube not produced; inner process tumid and extending to pore; no anterior process. Female with vertical lateral margin of carapace strong and either granulate or tuberculate; posterior margin of carapace without a row of conspicuous tubercles; surface of carapace moderately rough (Atlantic) to almost smooth (Pacific); suborbital crenellations distinct near outer angle; gonopore with a large, outer tubercle surrounded on three sides by a deep depression;

a patch of pile covering posterior part of carapace sides present (Pacific) or absent (Atlantic). See also Table 4, p. 638.

Description

With the characteristics of the subgenus.

MALE

Carapace. Branchial and pterygostomian regions with branching, vein-like striae present or absent. Frontal groove very narrow as in *princeps* and *major*, its raised lateral margins parallel and almost contiguous. Eyebrows, as characteristic of the subgenus, not provided with lower margins, but in *maracoani* the orbits' prominent upper margins run behind the most anterior extensions of the carapace, so that the smooth ceilings of the orbits appear to be rolled back; antero-lateral angles always acute and moderately produced; antero-lateral margins short but definite, granulate or tuberculate, straight, and curving in more abruptly than usual in the subgenus on both sides to continue as the dorso-lateral margins. These dorso-lateral margins range from distinct and granulate, rarely with several enlarged granules or small tubercles along their length, to obsolescent or absent; never with long, blunt, spinous tubercles along the margin itself, although such a tubercle sometimes occurs close to its anterior end while a large tubercle of variable shape, sometimes more or less compound, always ends the marginal area close to postero-lateral edge of carapace, even in specimens where the margin itself is obsolescent or absent. Vertical lateral margins beaded or granulate and usually extending almost or entirely to antero-lateral margins. Suborbital crenellations usually large, their form highly variable but with their edges, regardless of breadth, usually more or less curved, sometimes truncate; center of margin sometimes entire; in large specimens crenellations near outer angle always distinct, regardless of size, not more or less fused. No tubercles underneath base of antero-lateral angle above external orbital channel; entire lower margins of antero-lateral angles smooth.

Ocular Styles. Absent, except rarely, in young *m. maracoani*, where they are about as long as cornea. No young specimens of *m. insignis* available for examination.

Major Cheliped. **Ischium** with a ventral tubercle. **Merus** with a row of well-separated tubercles or blunt spines numbering from 5 to 9, running entire length of antero-dorsal margin and largest distally, where they clearly surmount a small crest. Dorsal margin of *carpus* tuberculate only proximally. **Manus** extremely short, broader (higher) than long; entire outer surface thickly covered with tubercles, well-separated, large, and semi-spherical or bluntly pointed; below dorsal margin they are sometimes aligned in several obliquely vertical rows below the usual, submarginal, smooth space which is not in this species a groove; dorsal margin with a single row of blunt tubercles, one of which, usually the most distal, is large and outstanding; none of them however is directed inward toward palm and no additional tubercles arise from submarginal surface of palm itself. **Palm** with oblique ridge strong to obsolescent, continued by a row of low, flat tubercles, sometimes obsolescent, along anterior margin of carpal cavity; predactyl ridge broad and high with tubercles usually irregular and unusually variable. **Dactyl** and **pollex** attaining more than three times length of manus, both fingers being broad, strongly compressed, and narrower proximally than near middle; dactyl much wider than pollex in middle and subdistal portion, the dorsal margin being highly arched; dorsal margin of dactyl and ventral of pollex bent slightly inward, the inner surface of each being in addition concave; finally, near distal end the pollex curves outward. Also involved in these complex bendings of the chela are the prehensile edges: the narrow gape is present proximally only, formed by a slight, proximal concavity at least in the pollex; in the Pacific form, *m. insignis*, the proximal prehensile edge of the dactyl also is concave or, at least, slants downward from the dactyl's base; in their distal two-thirds the prehensile edge of dactyl overlaps that of pollex externally as far as chela tip; there the fingers can hook firmly together through small, pointed, curved, terminal teeth, that of pollex then falling outside the dactyl's. Pollex with outer surface punctate except proximally where are continued the tubercles of the manus, and except on its submarginal area below prehensile edge which is covered with very small tubercles and granules. Dactyl with outer surface entirely covered with similar fine armature or else smooth, depending on the subspecies, except for small tubercles which are always present proximally. Pollex with supramarginal tuberculate ridge arising only near distal end of manus, marked there by a row of well-separated tubercles and on pollex by small, close-set tubercles separated from marginal row only by a narrow, smooth strip which dies out distally as the two rows become contiguous; a narrow groove adjoins dorsal side of supramarginal row only. Pile, never dense, usually distinguishable in this groove as well as in the punctae and, often, among tubercles of outer manus. Tuberculation of lower and upper margins of pollex and dactyl, respectively, weak to obsolescent in distal halves. Dactyl without traces of proximal grooves, either inside or out. Prehensile edge of pollex with median row having a large, triangular crest, edged with flat tubercles, slightly proximal to middle; sides of crest finely tuberculate and granulate. Dactyl with median row absent except proximally, where it lies close to outer row; outer row with a slightly enlarged tooth opposite triangular crest of pollex; inner row absent proximally as usual, or represented by clustered granules or small tubercles; in contrast to *heteropleura* and its relations, the most proximal regular tubercle of inner row is not enlarged but merely forms the first of a well-developed series that usually do not diminish in size until about middle of prehensile edge.

Minor Cheliped. **Merus** with postero-ventral margin usually with a few large widely spaced tubercles, reduced in some individuals to one or two, rarely absent.

Ambulatories. Armature of merus notably weak, except for presence in some individuals of a few wide-spaced, short tubercles on postero-ventral margins of 1st-3rd legs; their occurrence is highly variable even within populations and on the two sides of the same crab; serrations of dorsal margins more or less concealed in pile; posterior surfaces slightly and variably rugose; they show little tendency to parallel arrangement, and the sparse spinules only occasionally arise from them. Dorsal surface of carpus densely pilous.

Gonopod. Inner process tumid, extending to pore; latter at end of tube which along with anterior flange is strongly bent and somewhat twisted; flange long and narrow, extending beyond pore; posterior flange absent. Thumb absent. Anterior projection absent.

FEMALE

Vein-like striae on carapace present or absent; dorsal surface scarcely to moderately rough with tubercles and small granules except on cardiac region and, usually, gastric and intestinal regions at least near midline; their occurrence is however very variable within populations and in the only two specimens seen (including the type) of the Pacific subspecies the surface is smoother than in any of the numerous Atlantic examples. Dorso-lateral margin armed with tubercles, small and close set (Atlantic) or larger and separated (Pacific), the most posterior being slightly enlarged. Posterior margin of carapace never with large, close-set, rounded tubercles. Vertical lateral margin strong throughout, armed with granules (Atlantic) or separated tubercles (Pacific) and bordered with pile. Posterior sides of carapace covered with pile (Pacific) or not (Atlantic). Setae of chela moderately long, close-set, forming distinct fringes when not abraded. Last ambulatory with pile at least subdorsally on posterior merus and sometimes also above its postero-ventral margin which has long tubercles present (Pacific) or absent (Atlantic); a long tubercle also present on ischium. Gonopore with a large, outer tubercle overhanging a steep-sided depression that surrounds it on three sides—anteriorly, internally, and posteriorly.

Measurements (in mm)

	Length	Breadth	Propodus	Dactyl
Uca maracoani m.				
Largest male (Georgetown)	25.0	42.5	66.0	51.0
Male (largest claw) (Brazil)	23.5	34.0	68.0	53.0
Moderate male (Georgetown)	21.5	32.0	50.0	39.5
Largest female (ovigerous) (Brazil)	23.5	33.0	–	–
Smallest ovigerous female (Georgetown)	18.5	26.5	–	–
Uca maracoani insignis				
Largest male (Ecuador)	31.5	46.5	94.0	77.0
Female (Ecuador)	28.0	36.5	–	–

Morphological Comparison and Comment

In the subgenus *Uca*, *maracoani* is unique in the strong development of large tubercles or blunt spines on the antero-dorsal margin of the major merus. It shares only with *ornata* the extreme breadth and characteristic shape of the major pollex and dactyl, combined with a short manus; a comparable exaggeration is approached but not nearly attained by *princeps monilifera*. *U. maracoani* also shares with *ornata* the presence of conspicuous pits on the pollex; similar but less-developed pitting is present in *stylifera* and is strongly present in the West African species, (*Afruca*) *tangeri*.

The differences between the Atlantic and Pacific forms here discussed under the species heading *maracoani* are trivial in comparison with differences among other species in the subgenus. They are therefore regarded as subspecies, *U. m. maracoani* in the Atlantic, and *U. maracoani insignis* in the Pacific. The subspecies are compared with each other and with the closely related species, *ornata*, in Table 4. The other *Uca* with forms in the two oceans that are considered subspecies rather than species are (*Boboruca*) *thayeri* and (*Minuca*) *vocator*.

From an evolutionary point of view it is interesting that the female of *m. insignis* shares the strong armature of carapace margins and ambulatory meri with *ornata* but the form of the gonopore with *m. maracoani*. The ranges of *m. insignis* and *ornata* coincide.

Color

Displaying males: Display whitening absent. Carapace grayish purple; anterior margins of eyebrows sometimes orange. Major cheliped brown to gray except as follows: outer manus and pollex tinged with orange or dull rufous, rarely brighter red orange; dactyl outside and inside pale, tinged with purplish or orange. Pterygostomian and adjacent regions sometimes purple. Minor cheliped and ambulatories dark to dull orange. Even at height of display the crabs remain partly coated with mud.

SOCIAL BEHAVIOR

Waving Display

U. m. maracoani: Wave at low intensities lateral-straight, the cheliped being unflexed obliquely up and out, then returned directly to position, without pause and without circling. At high intensity the wave is wholly circular, although in a form shared only by *m. insignis* and *ornata*, the cheliped being lifted above the body, the fingers pointed upward, and repeated circles described in the air, the figure being flat and parallel to the ground, instead of the obliquely vertical circles usual in lateral circular waves. In the remaining characteristics, low and high intensity displays are similar: major dactyl is held widely separated from pollex throughout; minor cheliped makes a synchronous motion with every wave; body is held high off ground on extended ambulatories throughout a series; one or more ambulatories on each side are meanwhile lifted in typical leg waves, perhaps aiding in balance. During display the crab also often makes a few steps from side to side. Waving at low intensity is at the rate of about 1 wave every several seconds; at high each wave is completed in about 1.0 to 1.5 seconds. (Component nos. 4, 8, 9, 11, 12; timing elements in Table 19, p. 656.)

U. m. insignis: Similar, but not known in detail.

Precopulatory Behavior

Male attracts female to mouth of his own burrow and follows her underground. In none of the frequent examples seen where males covered females briefly at the surface did actual copulation take place. Either the abdominal flaps of one or both crabs were clearly not extended, or the crabs never reached copulatory position. In each of these cases an aggressive wanderer was making the unsuccessful overtures. (Observation on *m. maracoani* only.)

Acoustic Behavior

Leg-stamps seen repeatedly in the field and filmed (component no. 11). Leg-wags (component no. 5) occurred once following a combat; they were performed by both crabs. Sounds were recorded in the laboratory in response to intruders introduced in burrows, but their origin remains unknown. (Observations on *m. maracoani* only.) In 1967.2 von Hagen reported that he also made sound recordings in Trinidad.

Combat

Only the manus rub (component no. 1) has been filmed of the ritualized components. Irregular forceful combat, once ending in a complete upset of one opponent, was observed and filmed. Each time it was preceded, as in *tangeri, ornata, inaequalis,* and sometimes *pugilator,* by down-points (threat component no. 2), from which position the chelae engaged. Dactyl-hooks also accompanied combat, the dactyls of the 1st or 1st and 2nd legs of one crab bending and hooking behind the corresponding dactyl of the opponent. It is not known whether this action, found also in *ornata,* tends to throw the other crab off balance, or whether the armatures on these segments are engaged in stridulation. (Observations on *m. maracoani* only.)

RANGE

Western Atlantic from Venezuela and Trinidad to Brazil (Estado São Paulo). Single West Indian records from Jamaica (Sloane) and Santo Domingo (specimen in AMNH). Eastern Pacific from El Salvador to northern Peru; possibly formerly known from Chile (p. 147).

BIOTOPE

In Atlantic always near low-tide levels on muddy substrates, close to mangroves. The flats may be along edges of bays with gentle tides, on flats inside the more marine portions of mangrove estuaries, or on flat banks near mouths of streams (biotope nos. 8, 9, 12); in Pacific on salt flats at higher levels, the substrate being sandy mud. (Biotope no. 11.)

SYMPATRIC ASSOCIATES

Apparently not typically mingling with *major,* the other Atlantic member of the subgenus *Uca* (p.

136). Often *(Boboruca) thayeri* and *(Celuca) cumulanta* occupy adjacent patches of mud, usually at slightly higher levels, or slightly farther inshore. In the Pacific found with *(U.) ornata* (von Hagen); the few I collected (Ecuador) were farther inland than others of the subgenus *Uca,* among *(Minuca) vocator ecuadoriensis.*

MATERIAL RESULTING FROM FIELD WORK

(The complete list of specimens is given in Appendix A, p. 601.)

Observations and Collections. U. maracoani m.: Trinidad: Port of Spain. Eastern Venezuela: Pedernales. Guyana: Georgetown and vicinity. Brazil: São Salvador. *U. maracoani insignis:* Ecuador: Puerto Bolivar.

Film. Trinidad. Ecuador: Puerto Bolivar.

Sound Recordings. Trinidad.

TYPE MATERIAL AND NOMENCLATURE

Uca maracoani (Latreille, 1802-1803)

LECTOTYPE. In Muséum National d'Histoire Naturelle, Paris. Listed by the museum as a "type non spécifié" and believed (1959) by D. Guinot (personal communication) to be, along with 3 other males, Latreille's original specimens. These specimens, dried but in good condition, were in two boxes, each containing two specimens, with identical labels, as follows: "*Gelasimus maracoani* Margr. M. Leprieur. Cayenne." Milne-Edwards (1837) first listed their habitat as Cayenne, and his usual note, "(C.M.)," indicates that at that time they were in the collections of the museum. The specimen selected by J. Crane as lectotype is the only one in almost perfect condition, it being marred only by a slightly damaged carapace but with all characters, except the concealed gonopods, clearly characteristic of the Atlantic subspecies. Not relaxed. Measurements in mm: length 27; breadth 39; propodus 73; dactyl 58. (!)

Marcgrave's (1648) original specimens presumably came from Brazil; Latreille (1802-1803) gave the type-locality as "Le continent de l'Amerique Meridionale." Since Latreille, through his use of a post-Linnaean binomial, stands as the author of the species, and since the material in the Paris museum appears to be authentically his, it has seemed appropriate to designate the lectotype as described, with Cayenne as the type-locality.

Uca maracoani insignis (Milne-Edwards, 1852)

HOLOTYPE. In Muséum National d'Histoire Naturelle, Paris. Listed by the museum as a "type non spécifié." Apparently the unique female, and therefore automatic holotype, described by Milne-Edwards. The illustrations accompanying his 1854 paper are characteristic, as is the specimen, of *U. m. insignis* and not of *U. ornata*, the very similar sympatric species. The label is as follows: "*Acanthoplax insignis* Edw. M. Gay Chili." Measurements in mm: length 24; breadth 31. Spines on both sides of carapace well developed, 8 on left dorso-lateral margin, 9 on right; dorsal surface of carapace only slightly roughened, in comparison with that of *ornata*. Vertical lateral margin with a series of well-developed, small spines and tubercles diminishing in size ventrally, one of the best diagnostic characters for females of this subspecies. 2 to 4 spines along posterior margins of all ambulatories. Specimen dried; not relaxed. "Chili" almost certainly erroneous as the type-locality (see p. 438). (!)

Type of *Gelasimus armatus* Smith, 1870. In Museum of Comparative Zoology, Harvard College, Cambridge, Massachusetts. One male, cat. no. 5816. Characteristic of *m. insignis*, not *ornata*. Measurements in mm: length 25.2; breadth 35.5; propodus 60; dactyl 45.6 (Smith). Gulf of Fonseca, Salvador. Collected by J. A. McNiel. Received from Peabody Academy of Sciences November, 1885. (!)

Uca (Uca) maracoani maracoani (Latreille, 1802-1803)

(Tropical western Atlantic)

MORPHOLOGY

With the characteristics of the species.

MALE

Carapace. Vein-like striae of branchial and pterygostomian regions present or absent. Dorso-lateral margins always clearly marked by ridges with close-set granules; they are best developed in specimens from Guyana, where they are not only slightly larger than elsewhere but usually interspersed anteriorly on major side with still larger structures that may be called small tubercles; in all specimens throughout the subspecies range the margins on both sides near posterior angle of carapace end in a single, relatively large tubercle the apex of which is sometimes incised to form two or three beads.

Major Cheliped. Tubercles of outer manus somewhat more thickly set than in *m. insignis*. Palm with oblique ridge not tending to obsolescence. Dactyl with lower margin not slanting or excavate proximally; outer surface finely tuberculate.

Ambulatories. Meri with the slight rugosities of posterior surfaces, when present, not confined to subdorsal regions.

Gonopod. Differs from that of *m. insignis* chiefly in the slightly more tumid inner process; anterior flange bent slightly less, its torsion beginning more distally, and its tip scarcely projecting beyond pore.

FEMALE

Tubercles of dorso-lateral margin small and close set, with pile on each side of margin. No patch of pile posteriorly on carapace; vertical lateral margin armed with close-set granules; pile present or absent on merus of 4th ambulatory supramarginally along postero-ventral margin; no long, separated tubercles on postero-ventral margins of ambulatories.

Uca (Uca) maracoani insignis (Milne-Edwards, 1852)

(Tropical eastern Pacific)

MORPHOLOGY

With the characteristics of the species.

MALE

Carapace. Vein-like striae on branchial and pterygostomian regions always present. Dorso-lateral margins faint or absent, never with more than one tubercle anteriorly, always large and spinous, and with another close to postero-lateral margin of carapace; the anterior tubercle is sometimes absent and there are no traces of other tubercles along the path of the margin before the posterior tubercle which is always present and is never incised at the apex to form two or three beads.

Major Cheliped. Tuberculation of outer manus somewhat sparser than in *m. maracoani*, but variable, tending toward that of *ornata*. Palm with oblique ridge tending to obsolescence, especially in large specimens, more than in either *m. maracoani* or *ornata*. Dactyl with its lower margin proximally notably sloping or concave; outer surface sometimes as smooth as in *ornata*, although without traces of the vermiculations or punctae found in that species, and sometimes finely tuberculate as in *m. maracoani*.

Ambulatories. As in *m. maracoani*, except that rugosities on posterior sides of meri are practically absent; when indicated they are in the subdorsal area only, as in *ornata*.

Gonopod. As in *m. maracoani*, with these minor exceptions: inner process slightly less tumid; anterior flange bent slightly more than in some *m. maracoani*; additionally its torsion begins more proximally and its tip projects slightly farther beyond pore.

FEMALE

(From a single female, from Ecuador, which agrees with von Hagen's [1968] description.)

Vein-like striae of branchial and pterygostomian regions present. Carapace scarcely roughened with tubercles and granules; tubercles of dorso-lateral margins moderately large and separated, about as in most females of *U. ornata*; vertical lateral margin armed with similar, but slightly smaller, separated tubercles; a large area of pile on sides of carapace near posterior margin; merus of last ambulatory without pile above its postero-ventral margin; long, spinous, blunt tubercles present on postero-ventral margins of all ambulatories.

REFERENCES AND SYNONYMY

Uca (Uca) maracoani
(Latreille, 1802-1803)

TYPE DESCRIPTION. See under *U. (U.) maracoani maracoani*, below.

U. (U.) maracoani maracoani
(Latreille, 1802-1803)

Maracoani

Marcgrave de Liebstad, 1648: 184; text fig. Brazil. Description; including color.

? Cancer palustris cuniculos sub terra agens

Sloane, 1725: 260. Jamaica: "by Passage-Fort." Gives "*Maracoani*. Margr. p. 184 ed. 1648" as reference, followed by: "This crab agrees in everything to the Description of *Marcgrave*." Habitat in marshes and among mangroves. (See p. 324.)

Ocypode maracoani

TYPE DESCRIPTION. Latreille, 1802-1803: 46. Cayenne. (Paris !)

Gelasimus maracoani

Latreille, 1817.2: 519. No new material. New genus, *Gelasimus*. (See p. 20.)
Desmarest, 1825: 123. Apparently no new material. Taxonomy.
Milne-Edwards, 1852: 144; Pl. 3, Figs. 1, 1a, 1b. Taxonomy.
Dana, 1852: 318. Brazil: Rio de Janeiro. Taxonomy.
Smith, 1869.1: 35 (part). Brazil.
Brocchi, 1875: 73; text fig. of gonopod (no morphological details shown).
Kingsley, 1880.1: 136 (part). Western Atlantic material only, including that from Bahia, Brazil.
Cano, 1889: 92, 231. Brazil: Pernambuco.
Ortmann, 1894.2: 756. "Antillen"; Surinam.
Aurivillius, 1893: 35. Brazil. Description of amphibious characteristics.

Young, 1900: 270 (part); text fig. Western Atlantic material only. Taxonomy.

Gelasima Maracoani

Latreille, 1818: 3 (consists only of caption); Pl. 296, Fig. 1. (Not Pl. 275, nor its caption on p. 1 erroneously applied to Pl. 274; Pl. 275 is, instead, a copy of Seba's illustration of the form now referred to *U. major*.)

Gonoplax maracoani

Lamarck, 1818: 254. "Habite l'Amérique méridionale." Brief Latin diagnosis. Common name: "Rhombille maracoan."

Uca maracoani

Rathbun, 1900.3: 134. Brazil: Natal; Rio Parahybo do Norte at Cabedello.
Moreira, 1901: 52. Brazil: northern part and Rio de Janeiro.
Rathbun, 1918.1: 378. Photos. Cayenne. Brazil: Maranhão, Natal, Plataforma, Porto Seguro. Taxonomy.
Oliveira, 1939.1: 123. Brazil: Rio de Janeiro. Taxonomy; color; habitat.
Crane, 1943.2: 35; text fig. Venezuela. Taxonomy; waving display; habitat.
Crane, 1957. Western Atlantic. Preliminary classification of waving display.
Crane, 1958. Trinidad. Social behavior.
Gerlach, 1958.1: 672. Brazil: Estado São Paulo (near Cananéia: Rio Perequé). Ecology.
Holthuis, 1959.2: 260. Surinam. Taxonomy.
Guinot-Dumortier, 1959: 515; Text Fig. 17a, b. French Guiana: Cayenne.
Barnwell, 1963; text figs. Brazil. Endogenous daily and tidal rhythms of melanophore and motor activity.
von Hagen, 1967.2. Trinidad. Tape recordings secured. (Preliminary statement.)
von Hagen, 1970.1. Distribution in West Indies.
von Hagen, 1970.4; text figs. Trinidad. Adaptations to a particular intertidal level.

Uca (Uca) maracoani insignis
(Milne-Edwards, 1852)

Acanthoplax insignis

TYPE DESCRIPTION. Milne-Edwards, 1852: 151. Chili. (Paris !)

Milne-Edwards, 1854: 162; Pl. 11, Figs. 1, 1a, 1b. Further description of the type.

Gelasimus armatus

Smith, 1870: 123; Pl. 2, Fig. 5; Pl. 3, Figs. 4, 4a, 4b, 4c. Gulf of Fonseca. Type description. (MCZ !)

Smith, 1871: 91. Record of type. No new material or further description.

Gelasimus insignis

Smith, 1870: 126. Taxonomy.

Uca insignis

? Rathbun, 1911: 550. Peru. Brief taxonomy; local distribution (by R. E. Coker).

Porter, 1913: 317. Not found by this author in Chile.

Rathbun, 1918.1: 385-387 (part: description of male only); Pl. 161, Figs. 9, 10, 11, 12, all of which are from drawings published by Smith in the type description of *G. armatus*.

Bott, 1954: 156; Pl. 14, Fig. 1a, 1b (photos). El Salvador. Taxonomy.

Altevogt & Altevogt: 1967: E 1288. Film of waving display.

von Hagen, 1968.2: 442; Text Figs. 14e (gonopod), 15b (photo), 16b (photo), 20. Peru: Puerto Pizarro. Taxonomy; waving display; habitat.

Altevogt, 1969.1: 238; Text Figs. Peru. Behavior and gonopod structure as isolating mechanisms.

24. *UCA (UCA) ORNATA* (SMITH, 1870)

(Tropical eastern Pacific)

PLATE 21 *E-H*.
FIGURES 26 *E*; 31 *D*; 39 *C, D*; 46 *G*; 65 *G*; 99.
MAP 10.
TABLES 4, 9, 10, 12, 19, 20.

INTRODUCTION

Because of the broad fingers of its claw, *Uca ornata* probably attains a greater weight than any other fiddler, although in dimensions it resembles its close relation, *maracoani insignis*, and is surpassed by (*Afruca*) *tangeri*.

Both its habitat and the form of its display are probably adaptations connected with the lopsided distribution of its weight. When mature it occurs, at least in Panama when breeding, on the inshore edges of mudflats which are so moist that at most low tides the surface remains semi-liquid ooze. From this substrate the crabs may derive some support during their vigorous display. The situation recalls those herbivorous dinosaurs which were apparently confined by their weight to semi-aquatic habitats.

The display of *ornata* is a somewhat exaggerated version of *maracoani*'s performance, the great claw circling high overhead above the body of the crab.

MORPHOLOGY

Diagnosis

Differs from *U. maracoani* of both subspecies most distinctly as follows: Male: Antero-dorsal margin of major merus with close-set, small, rounded tubercles and a single, distal, contrastingly large, crest-like tooth, not with a row of separated, large, spinous tubercles; major dactyl externally smooth, not granulate. Carapace with dorso-lateral margins almost always with a row of separated tubercles, either all large or some vestigial. Gonopod tip slender, the inner process not tumid, the tube not projecting beyond the long, narrow flanges. Female: Vertical lateral margin of carapace vestigial; carapace dorsally very rough with tubercles and granules, with a row of rounded tubercles across its posterior border; suborbital crenellations partly fused near outer angle; no patch of pile on posterior part of sides of carapace. Gonopore without a large tubercle but with a long, slightly raised rim directed antero-posteriorly. (See also Table 4, p. 638.)

Description

With the characteristics of the subgenus (p. 125).

MALE

Carapace. Branchial and pterygostomian regions without vein-like striae or with faint traces only. Upper margin of orbit erect, scarcely rolled back; antero-lateral margins less definite than in *maracoani*, usually not clearly separated from dorso-laterals, with or without one or more large, spinous tubercles; dorso-lateral margins marked not by close-set granules but by widely-spaced, large, sharp tubercles, some of them with the intervening spaces indicated only by a faint, very low ridge, smooth except, sometimes, for the widely spaced, low bumps of vestigial large tubercles; a large tubercle, sometimes compound, as in *m. maracoani*, always present close to posterior angle of carapace, and in one exceptional specimen of *ornata* from Panama it is the only one on dorso-lateral margin; in this peculiarity alone this example resembles *maracoani insignis*. The number of tubercles, including the posterior and the one or more sometimes present on antero-lateral margin, range from 2 to 13 on each side. The number and spacing of tubercles on the two sides are rarely the same and the larger number may be on either major or minor side; the usual range is 7 to 10; 1 to 4 of the penultimate tubercles are sometimes vestigial. Vertical lateral margin faint to absent in dorsal half. Suborbital margins little or not at all incised, giving the broadest crenellations, appearing partly fused, close to each orbit's outer angle, not in middle of margin. Young not pilous.

Major Cheliped. Merus with antero-dorsal margin armed throughout entire length except distally only by close-set, very small, rounded tubercles, usually scarcely more than beading and sometimes obsolescent either predistally or even throughout entire margin; they are set on a low ridge that is sometimes indistinct; occasionally one to several tubercles in proximal part are slightly enlarged; distally the margin ends in a strikingly large tooth, compressed, triangular, and pointed, the apex sometimes curved so that the point is directed distally. *Carpus* with postero-dorsal surface smooth, its distal portion as well as ventral part of carpus often densely pilous. Outer manus with tubercles fewer, more widely spaced and lower than in *maracoani*; those of dorsal margin also few, between 2 and 5, widely separate, sometimes all about equally outstanding from the dense pile of upper outer manus and its margin but usually with the one or two most distal projecting beyond the others. *Palm* with oblique ridge well developed but ending at carpal cavity and represented if at all along cavity's anterior border by indistinct rugosities or by one or two vestigial tubercles. *Pollex* with outer surface uneven, but not strongly tuberculate; grooves and punctae present, variable; *dactyl* with outer surface entirely smooth except for traces of vermiculations and, sometimes, minute punctae, all discernible in cross-lighting. Pile on pollex denser than in *maracoani*, always prevalent in grooves and punctae, as on outer manus. Prehensile edge of dactyl with median row even closer to external row than in *maracoani* and with inner row extending throughout length, forming a straight series of small tubercles that are close set and similar except at proximal end, where the first few are smaller, more widely spaced and, sometimes, imperfectly aligned; these are followed by several slightly enlarged tubercles, those succeeding them distally diminishing regularly to merge with the rest of the series.

Ambulatories. Postero-ventral tubercles on merus, while variable in number and position, are much larger than in *maracoani*, almost always represented on all legs of at least first three pairs by at least one or two elongate tubercles resembling blunt spines; posterior surfaces of meri on the other hand are usually practically smooth, with the rudimentary rugosities found often on *maracoani* scarcely represented and confined to the subdorsal surface, while spinules are very sparse to absent.

Gonopod. Inner process not at all tumid; instead it is a flat, moderately narrow, translucent structure arising from setose tip of fleshy covering of tube; latter is moderately projecting and the inner process overlays it all the way to terminal pore. Tube flanked closely by a narrow anterior and vestigial posterior flange, both of them ending at pore. Curvature of tube moderate, torsion slight, beginning only subdistally.

FEMALE

Dorsal surface of carapace rough with tubercles and granules, as in *maracoani*, but almost always much stronger, approaching their appearance in (*Afruca*) *tangeri*; dorso-lateral margins with 8 to 11 large, widely separated tubercles, without pile, approximately as in female *m. insignis*, the posterior tubercle not especially enlarged; posterior border of carapace with a row of strong, rounded tubercles, moderately close set, numbering about 10 or 12. Vertical lateral margin vestigial, distally absent, without pile or large tubercles, armed only proximally and then only with granules or fine beading. Sides of carapace posteriorly not covered with pile. Suborbital crenellations partly fused near outer angle. A subdorsal border of pile posteriorly on 4th ambulatory merus, as in *maracoani*, but none supramarginally along postero-ventral margin, which is armed, along with the same margin in the other ambulatories, with 1 to 7 (usually 3 or 4) blunt, spinous tubercles (in addition to a spine on each ischium), highly variable in number and position, as in *m. insignis*; dorsal margin of carpus with traces of pile along its posterior side. Gonopore with tubercle lower and more elongate than in *maracoani*, the depression less deep and extending laterally less far both anteriorly and posteriorly.

Measurements (in mm)

	Length	Breadth	Propodus	Dactyl
(All specimens from Panama)				
Largest male	31.5	50.5	87.0	67.5
Moderate male	29.5	45.0	67.0	52.5
Largest female (ovigerous)	30.0	42.0	–	–
Smallest ovigerous female	26.0	37.0	–	–

Morphological Comparison and Comment

The male of *ornata* is unique in the subgenus in having, almost always, large separated tubercles on the postero-dorsal margins of carapace, instead, as in *maracoani*, of only posteriorly. Only *maracoani* shares with *ornata* the extreme characteristics of the major pollex, although the shape is approached by that of *princeps monilifera*. Pits on the pollex are also equivalent to those in *maracoani*; similar structures occur, in weak form, in *stylifera* and strongly in (*Afruca*) *tangeri*. They seem to be closely associated, as usual, with the presence of pile and, probably, algae (p. 466). Female *ornata* also have the dorso-lateral carapace margins strongly armed with separated, spinous tubercles, but they share this characteristic with females of *maracoani insignis*; as usual in the armature of carapace and legs of female *Uca*, it is stronger and less variable than in males. In most female *ornata* the dorsal surface of the carapace is so well provided with tubercles that the appearance is somewhat similar to that of *tangeri*, even though the tubercles in the American species are neither as large nor as extensive as in the African form.

Further comparisons are given in the account of *maracoani* (p. 144).

Color

Displaying males: Display whitening in Panama extremely rare, confined to carapace; observed by von Hagen (1968.2) in Peru, Ecuador or both. Carapace in Panama occasionally pinkish or yellow, usually purple or rose red; eyebrows sometimes orange. Major cheliped with propodus and dactyl entirely brown to dull orange or purplish red. Eyestalks yellow. Ambulatories proximally, at junctions between segments or in entirety, purplish red or darker. The crab is rarely free of mud even during full display; then usually only the dactyl is clean, the color of the propodus and carapace remaining obscured.

Social Behavior

Waving Display

At high intensity closely similar in all details to that of *maracoani*. At low intensity, however, the cheliped instead of giving a lateral-straight wave circles in the same general pattern as at high, but the cheliped is not extended so far up and the diameter of the circles, as well as the speed at which they are described, is not so great; finally, the circles tend to more of a normal tilt, instead of being parallel to the ground.

In detail, the high intensity display is as follows. Major cheliped with manus and chela thrust straight above crab; dactyl separated as far as possible from pollex so that the claw is widely "open"; the tips then describe a wide circle for up to several minutes at a time without pause, although most series are shorter. In a single series I counted a maximum of 186 circular waves, each made at the rate of slightly less than one circle per second. The minor cheliped often makes a corresponding motion. Body is held high off ground on extended ambulatories throughout a series, one or more legs on each side being raised and lowered in typical leg-waves, their meri not touching; as in *maracoani*, this may be partly concerned in balance, rather than strictly an embellishment of display that gives the effect of greater size. (Component nos. 4, 8, ?9, 11, 12; timing elements in Table 19, p. 656.)

Precopulatory Behavior

Females have been seen twice to follow males down burrows. Three pairs were dug up in shared burrows in Panama; since courting was active at the time, they may merely have been startled into joint descent by my approach.

Acoustic Behavior

Major-merus-rubs, leg-wags, minor-chela-taps and leg-stamps were observed and filmed (component nos. 1, 5, 10, 11).

Combat

Several forceful combats were filmed but the details are unclear. As described under *maracoani* (p. 146), down-points (agonistic posture no. 2) and leg-hooks initiated the combats. Stamping on turned-under dactyls occurred between rounds, as well as repetitions of the leg-hooks.

Range

Tropical eastern Pacific from Central America to northern Peru.

Biotope

Open mudflats on bay shores beside and inside the mouths of large streams and rivers; uncommonly on almost landlocked mudflats not reached by neap tides, so that at these times the crabs aestivate; mangroves seem always to be nearby; young sometimes occurring on stream banks slightly farther from sea than usual. (Biotope nos. 8, 11; young on 9, 12.)

Sympatric Associates

None when adult, except sometimes on their inshore boundaries, *saltitanta*.

Material Resulting from Field Work

(The complete list of specimens is given in Appendix A, p. 602.)

Observations and Collections. Mouth of Abajo R., near Old Panama, Panama City, R.P.; young from La Boca, near mouth of Panama Canal, Balboa, C.Z.

Film. Same locality.

Type Material and Nomenclature

Gelasimus ornatus Smith, 1870

HOLOTYPE. In Museum of Comparative Zoology, Harvard University, Cambridge, Massachusetts. One female. Cat. no. 5817. Type-locality: west coast of Central America. Collected by J. A. McNiel. Received from Peabody Academy of Sciences, November 1885. (!)

Rathbun (1918.1) synonymized this species, represented by the holotype only, with *U. insignis*, this female being the only member of that sex included in the Pacific specimens of these close-related species available to her for examination. Her males, on the other hand, probably are all specimens of *insignis*, here considered the Pacific subspecies of *U. maracoani*.

I base my identification of male *ornata* on field

observations and collections made personally near Panama City. Not only did I not find any specimens of *maracoani insignis* in the area, but I watched individual females, later captured, attracted by the high intensity courtship of the males; two of these females had reached the stage of following displaying males underground. In addition, three other burrows proved to be inhabited by pairs; see, however, comment on p. 152.

In Ecuador the few specimens of *m. insignis* I observed were in a different habitat from that frequented by *ornata* in Panama—salt flats far back from open water, subject to desiccation at neap tides, and with the substrate sandy mud rather than mud; von Hagen's (1968) specimens of similar southern populations of both species, however, were closely sympatric at high levels. Obviously further ecological observations are needed.

Comments on the type in Paris of *Acanthoplax insignis* are given on p. 146.

Type Material of *Uca pizarri* von Hagen, 1968. Holotype male in Rijksmuseum van Natuurlijke Historie, Leiden. Cat. no. D 23061. Measurements in mm from type description: length 27.3; breadth 40.2. Additional measurements by J. Crane: propodus 55; dactyl 44. The propodus was regenerated, doubtless at an early age, as shown by its normal shape combined with small size and atypical tuberculation. Type-locality: Puerto Pizarro, Peru. Also in Leiden: 1 paratype male (cat. no. D 23062); breadth in mm 43.3; Guayaquil, Ecuador. (!)

REFERENCES AND SYNONYMY

Uca (Uca) ornata (Smith, 1870)

Gelasimus ornatus

TYPE DESCRIPTION. Smith, 1870: 125; Pl. 2, Figs. 9-9a; Pl. 3, Figs. 5-5c. West coast of Central America. (MCZ !)

Uca insignis (not *Acanthoplax insignis* Milne-Edwards)

Rathbun, 1918.1 (part): 385; Pl. 161, Figs. 5, 6, 13, 14, 15. Part of description referring to female only (p. 387); figs. are from drawings published by Smith in the type description of *G. ornatus*.

Crane, 1941.1: 173. Panama. Brief taxonomy of young.

Crane, 1957. Panama. Preliminary classification of waving display.

Uca pizarri

von Hagen, 1968.2: 439; Text Figs. 14d, 15a (photo), 16a (photo), 20. Puerto Pizarro, Peru; Guayaquil, Ecuador. Type description. (Leiden !)

VIII. SUBGENUS *MINUCA* BOTT, 1954

Typus: *Gelasimus mordax* Smith, 1870

(East and west coasts of America)

PLATES 22-28.
GONOPOD DRAWINGS: FIGURES 66, 67, 68 (part).
DENDROGRAMS: FIGURES 96, 100.

MAPS 10, 11, 12, 13, 14.
TABLES 1, 9, 10, 12, 14, 19, 20.

MORPHOLOGY

Diagnosis

Very small to large *Uca* with front wide to very wide; except in *pygmaea* and *panamensis* antero-lateral margins long and curving posteriorly, not sharply angled. Eyestalks short, their diameter large, almost filling orbits; suborbital crenellations small, often obscured by pile except near outer angle. No special armature on lower proximal palm and on anterior surface of carpus of 1st major ambulatory. Major manus with tubercles on upper half moderate, never very small; except in *brevifrons* carpal cavity with beaded edge on upper margin, the cavity never extending distally. Small cheliped always with gape narrow, minutely serrate. Gonopod with flanges well developed, never ending in a projecting tube, thumb large and subdistal. Female gonopore always with a tubercle, although minute in *burgersi* and *mordax*. Segments of male abdomen never partly fused.

Description

With the characteristics of the genus. Related closely to *Celuca*, which is described in detail beginning on page 211. Differences are as follows.

MALE

Carapace. General shape of front as in *Celuca* but usually wider; measured between eyebrow margins its breadth is contained from about 4 to 2.5 times in carapace width between tips of antero-lateral angles, instead of from 6 to, unusually, about 3. Orbits more than slightly oblique only in *pygmaea*, where the backward slant is strong. Antero-lateral margins differ as follows: longer, except in *pygmaea*, than in most *Celuca*; except in *panamensis*, slightly convex rather than straight or concave, and characteristically slightly divergent, the widest part of carapace being usually behind, not between, antero-lateral angles, while the antero-lateral margins continue into dorso-laterals with a rounded turn, without forming the sharp angle characteristic of most *Celuca*. Upper postero-lateral stria sometimes long and strong, ex-

tending anteriorly below dorso-lateral margins. Vertical lateral margins always complete and strong. Carapace profile in lateral view moderately to strongly arched but never semi-cylindrical. Antero-lateral part of carapace usually with irregular, low and indistinct rugosities which are never tuberculate but are sometimes replaced by scattered low tubercles and are sometimes continued farther back, internal to dorso-lateral margins; patterned pile present on carapace only in *vocator*. In about half the species the postorbital groove is strongly marked, rather than moderately to not at all. Frontal margin always complete, instead of rarely weak or absent; sometimes with faint and variable traces of beading on its inner (posterior) edge. Margin of posterior carapace also at times faintly beaded on anterior edge, continuing the usual beading of lateral margin beyond postero-lateral corner of carapace. Suborbital region flat to convex, setae thicker and more numerous than in most *Celuca*, on both suborbital and pterygostomian regions.

Orbital floor never tuberculate; submarginal setae sometimes thick and continuing over inner part of margin to merge with those of suborbital region, obscuring the weak crenellations, instead of being, as in *Celuca*, set in a single row; suborbital crenellations present throughout margin, sometimes extending around outer margin and even onto lower margin of antero-lateral angle; in no species, however, do they reach the high development shown in some *Celuca*.

Eyestalk and eye almost fill orbit, except near suborbital and outer orbital margins. Stalk always short and thick and, in these characters, *Minuca* includes the most extreme examples in the genus; stalk about equal to eye in diameter.

Second Maxilliped. Few spoon-tipped hairs.

Third Maxilliped. Gill vestigial.

Major Cheliped. Merus: Ventral margin except in *panamensis* and *pygmaea* with two distinct rows of tubercles, not contiguous; the anterior is always the more regular, slightly supramarginal, with the proximal end starting higher, then descending obliquely to parallel the second row.

Carpus: Postero-dorsal surface more often with tubercles mixed with or replacing the rugosities, than strictly rugose; sometimes practically smooth; degree and type of roughness highly variable.

Manus: Outer surface usually covered with small to moderate tubercles, rarely with the minute tubercles characteristic of most *Celuca*. While always largest as in that subgenus in upper part of manus, they differ in never showing patterns of arrangement, such as reticulations or vertical rugosities, and in never having more than an extremely narrow band of minute tubercles, or a smooth area, close to subdorsal groove; latter groove, except in *panamensis*, always present, strong and full length of the well-marked, raised, tuberculate outer edge of dorsal margin. Unlike many *Celuca*, margin is never distally broad, triangular, or strongly flattened, never has specialized arrangements of the tubercles, and never is practically smooth. Instead, it continues beyond carpal cavity as a relatively narrow, slightly convex ridge, little or not at all broadened distally; the surface is equipped only with a few small tubercles, low and unspecialized. Lower distal outer manus at pollex base always flattened as in *Celuca*, but with the depression, except in *pygmaea* and *subcylindrica*, weaker than in most *Celuca*. Except in *panamensis* a ventral marginal keel is always distinguishable and always flanked by an outer groove; both, however, stop at pollex base, where they are strongest, instead of sometimes continuing onto pollex; tubercles and beading only occasionally surmounting or replacing keel instead of being usually present. Heel always well rounded; in *panamensis* it projects postero-laterally to an extent unique in the genus.

Palm: Lower triangular area differs from that of *Celuca* in having minute tubercles at least adjacent to oblique ridge, rather than on the area's lower part; again unlike *Celuca* the tubercles are never larger proximally and near ventral margins than elsewhere and never are specialized in arrangement or shape. Central area of palm always at least partly tuberculate, instead of usually smooth; no definite depression ever in upper part of center palm, below predactyl region. Carpal cavity with proximal part of lower portion sometimes so low that it lacks a margin, instead of being always defined throughout. Upper part of cavity, except in *brevifrons*, without an extension distally into the flattened predactyl area, being cut off sharply by the downward curving, strongly beaded inner edge of dorsal margin; this barrier which occurs occasionally and then often incompletely in *Celuca*, in *Minuca* is incomplete only in *subcylindrica*. Proximal predactyl ridge either subparallel to the adjacent groove, as in *Celuca*, or diverging from it, the ridge then being directed proximo-dorsally.

Proximal ridge always stronger than distal ridge, which may be variably obsolescent.

Pollex and *dactyl*: Differ as follows from those of *Celuca*: Tip of dactyl, when closely apposed to that of pollex, never lies against inside of pollex tip, but always touches the tip itself as it curves slightly beyond and below in exactly the same plane; pollex tip except in *panamensis* and *subcylindrica*, is always roughly trifid, instead of being usually simple and only occasionally bifid or trifid. Pollex and dactyl of similar breadths, instead of pollex usually being clearly wider; only in *pygmaea* is pollex somewhat triangular. Long external grooves present in *brevifrons* and *subcylindrica*. Outer, proximal, subdorsal groove on dactyl always well developed except in *brevifrons* and *panamensis*, where it is respectively weak and absent. A distal keel occurs only in *subcylindrica*, on pollex only.

Minor Cheliped. As in *Celuca*, except as follows. Antero-ventral margin of merus, instead of being smooth and rounded, is distinct at least proximally and marked at least faintly with minute tubercles, spinules, or beading; this same margin is flanked supramarginally on anterior surface at its proximal end by a cluster of minute spinules or, sometimes, by a distinct, short row of spinules or tubercles; the anterior surface also is sometimes slightly roughened, at least proximally, by various means; all of these meral characteristics are subject to individual variation within populations. Carpal ridge and mano-pollex ridge both always strongly developed, not sometimes short or weak; carpal ridge often clearly beaded. Except in *panamensis*, gape always, instead of sometimes, extremely narrow with middle portion clearly serrate, the edges almost or wholly in contact when tips are apposed.

Ambulatories. As in *Celuca*, except that the ventral margins of merus are completely unarmed, instead of weakly so, except in *panamensis* and *subcylindrica*. Pile in general more prevalent than in any *Celuca*, especially on carpus and manus; extent in several species controlled by presence or absence of underlying, very minute, sharp granules, and of taxonomic importance. No special armature on anterior surface of 1st carpus or other segments.

Abdomen. All segments distinct.

Gonopod. Basically of the *Deltuca-Thalassuca* type. Flanges always well developed, the anterior the wider, never confluent beyond pore; pore never at end of a produced tube; inner process strong, ranging from broad and expanded when it almost overhangs pore, to flat and moderately narrow, although never spinous. Thumb always well developed. Torsion absent to slight, never extreme.

FEMALE

With the characteristics of females in the genus (p. 17) and of the subgenus *Celuca* (p. 215), when the following comments are taken into account. Otherwise as in male, as described above.

The female in *Minuca* differs in secondary characters most definitely from the male in having both ventral margins of the ambulatories always distinct and, at least on part of the 2nd and 3rd legs, armed with serrations or tubercles; in *Celuca* these margins are armed weakly but distinctly even in males. In some individuals of most species a longitudinal ridge on anterior surface of at least 1st carpus; although seldom tuberculate, it is otherwise similar to the armature in males of some *Celuca*. In general the armature of the carapace is relatively stronger in comparison with that of the males in *Minuca* than in *Celuca*, particularly in the anterior and posterior beading; in no species, however, are the carapace differences between the sexes as pronounced as in some other subgenera. Gonopore of *Minuca* with one to three tubercles, unlike *Celuca*, where tubercles are absent, except in one subspecies; in two species of *Minuca*, however, *burgersi* and *mordax*, the single tubercle is very small.

Size

Small to large.

Color

Display whitening strongly attained in one subspecies only, *g. galapagensis*, where it is preceded by a bright yellow phase. Minor lightening of carapace and major cheliped apparent in a number of other *Minuca*. Red and orange frequent on major cheliped; less often these hues appear on ambulatories and, in one subspecies, *vocator ecuadoriensis*, dark red extends to the carapace. Green and blue green are confined to the crab's anterior regions and occur only in some populations of two species, *rapax* and *pugnax*.

SOCIAL BEHAVIOR

Wave lateral-straight to lateral-circular, the cheliped usually unflexing obliquely upward, often with jerks. One special courtship component, the curtsy, present in several species. Tempo of waving often slow; in no *Minuca* is the speed of waving attained that occurs in a number of *Celuca*, although in jerking displays direct comparisons are invalid since a jerking display represents compound single waves (p. 524). Nevertheless even when a display consists only of a simple wave it is characteristically slower than in most *Celuca*. Waving display prevalent in both popula-

tions and individuals during periods of social activity, but these periods are often sharply limited by drought or temperature. Male normally attracts female into his burrow, even attempted matings, much less completed copulations, being rare at the surface during daylight. In the temperate zone, however, courtship in two species occurs at night as well as during the daytime; in these forms the incidence of nocturnal surface pairings may be high. Chimney construction prevalent, chimneys being apparently sometimes constructed only by females but sometimes also serving as a display perch for males. Hoods are formed regularly, apparently by displaying males, in *minax* but rudimentary ones were made in a single population of *pugnax*. Sound production reaches moderate diversity and complexity. Combat, at least in Trinidad populations of *rapax* and probably *burgersi*, is more thoroughly ritualized—the elements of force being more often apparently excluded—than in any other *Uca* where the patterns are known.

RELATIONSHIPS

Nine of the twelve species included in the subgenus *Minuca* appear to be closely related. This majority is treated in the paragraphs that follow; the three aberrant species are discussed in the section's final paragraph.

One species, *vocator*, is composed of populations of great similarity, even though they occur in both the Atlantic and Pacific; the groups in each ocean are therefore given only the rank of subspecies; their similarities include not only superficial details of armature and pile, but a close resemblance in gonopods which, in the form of the inner process and the presence of a unique subdistal tooth, set off the species from its close relations; the gonopod of this species, in fact, is most similar to those of certain *Deltuca* in the Indo-Pacific.

The eight other closely related species divide with considerable justification into two superspecies, each containing three geographically distinct species and an additional allied species; each of these two allies shows geographical coincidence with one member of a superspecies.

The first superspecies, *minax*, is composed of *mordax* and *minax*, respectively in the tropical and temperate Atlantic, and *brevifrons*, in the tropical Pacific; these three species all show the widest fronts in the genus, highly arched carapaces adapting them well to desiccation during drought or cold, similarities of armature on the major cheliped, including diverging predactyl ridges (which are also shared by *vocator*), and a toleration for almost fresh water; in fact each of the three reaches its highest population size and includes the largest individuals far from the mouths of rivers or in swamps of minimal tidal influence. A fourth species that is closely allied to the superspecies, and in particular to *mordax*, is *burgersi*, discussed further in the account of that species (p. 169). The gonopods of this superspecies and its ally are exceedingly similar to one another and to those of the second superspecies, described below, as well as minutely variable in similar ways; they are, therefore unreliable as taxonomic tools.

The second superspecies, *galapagensis,* groups three similar species. These are *galapagensis* in the tropical Pacific, *rapax* in the tropical and subtropical Atlantic, and *pugnax* in the subtropical and temperate Atlantic. Each is divided here into two subspecies. All of these forms are strikingly similar morphologically. In contrast to members of the superspecies *minax*, the species are confined to habitats of relatively high salinity; with the exception of one subspecies (*g. galapagensis*) that aestivates in very dry flats, the carapaces are not so highly arched as is usual in the superspecies *minax* and the fronts are less wide throughout; the similarity of the gonopods within and between the two groups has already been mentioned. Within the superspecies all of the species share fine details of armature of the major cheliped; these consist principally of a high oblique ridge on the palm with tubercles often in a partly double row, parallel predactyl ridges, and a predistal crest on the pollex distinctively formed. Although reliable gonopod differences doubtless exist in all these forms, adequate statistical studies of variations will be needed to determine them; thereafter only devoted use both of high power lenses and micrometer scales can take advantage of the information. A fourth species, *zacae*, appears to be closely allied to this superspecies; it shares the Pacific range of *galapagensis*.

The three remaining species included in *Minuca* are *pygmaea, subcylindrica,* and *panamensis.* Each is highly and diversely specialized. The peculiarities of *pygmaea* are associated with its paedomorphic characteristics, which appear in one or more species in most of the other subgenera. This species is little known either in life or preservative, and its affinities to the rest of the subgenus will doubtless become certain as soon as more information and material become available; meanwhile it is placed among its more probable relations, *vocator* and the superspecies *minax*, on the left of the dendogram. The other two aberrant species, *subcylindrica* and *panamensis,* could perhaps equally well be considered members of the less homogeneous subgenus *Celuca.*

The basic resemblance of the gonopods, however, to those characteristic of *Minuca* seems to be sufficient reason for the inclusion of the species here, particularly since our knowledge of their habitats is sufficient to explain most of their structural oddities.

Fossil Specimens

The holotype and unique specimen of *Uca hamlini* Rathbun, 1926, is a fossil claw from the Pliocene of southern California. Its visible characters are more similar to those of the members of the subgenus *Minuca* than to those of species composing other current subgenera. The visible part of the specimen consists chiefly of the outer side of the right major propodus and dactyl. Characteristics of the armature are distinct, in spite of cracks and other damage. The manus tubercles are small, as in living *Minuca* and unlike the large tubercles found in the subgenus *Uca* or the usually smaller ones in *Celuca*. The manus at first view appears abnormally long; further examination shows that the outer manus has almost certainly been detached and pushed somewhat proximally, so that distally part of the internal aspect of the palm (that is, a portion never exposed in an intact crab) is visible. The fingers are also damaged, although some blunt gape tubercles are distinct, all resembling closely those that occur in species of *Minuca* today. Measurements in mm: propodus more than 30 mm, dactyl ca. 24 mm. The type data are as follows: Rathbun, 1926.2: 30; Pl. 8, Fig. 1. From Third St. tunnel, Los Angeles. Pliocene. Collected by Homer Hamlin, U.S. Geological Survey, 1901. Holotype in Smithsonian Institution, National Museum of Natural History, Washington, no. 353372. (!)

Another fossil specimen seems to me to be unsuitable for referral, at least at present, even to a subgenus. It consists of a single dactyl from an ambulatory and was referred by Rathbun, 1918, to *Uca macrodactylus* (Milne-Edwards & Lucas). Thanks to our present expanded knowledge of the genus, it seems clear that the morphology of dactyls is on the one hand similar, and on the other has not yet been investigated in enough detail, to distinguish the small interspecific differences. Measurement: length ca. 6 mm. Reference: Rathbun, 1918.2: 177; Pl. 64, Fig. 7. From Panama, Canal Zone; Pleistocene. Specimen in Smithsonian Institution, National Museum of Natural History, Washington, no. 324251. (!) The taxonomic history of the name *macrodactylus* is discussed below, on p. 186.

25. *UCA (MINUCA) PANAMENSIS* (STIMPSON, 1859)

(Tropical eastern Pacific)

PLATE 22 *A-D*. MAP 12.
FIGURES 46 *H*; 66 *G*; 100. TABLES 9, 10, 12, 20.

INTRODUCTION

Uca panamensis is the only fiddler adapted by both morphology and behavior to a stony habitat. It occurs only at the ends of beaches, where sand so often gives way to rocky headlands. The fiddler depends on stones and small rocks for protection, dodging around and beneath them, or freezing motionless in their shelter for minutes at a time. Its variable colors, ranging widely from light to dark, are always in general accord with the local substrate. The carapace is unusually flat for a *Uca*, in apparent adaptation to sheltering under stones. Burrows, insofar as they have been sampled, are shallow and very temporary; sometimes *panamensis* does not dig at all before the tide comes in; the crab simply shelters under a rock like the numerous xanthids nearby.

In this environment, where the sight of waving individuals would usually be restricted by intervening stones, probably lies the explanation for the small amount of waving that seems to be typical of this species. A further hint exists in the exaggerated posterior extension of the heel of the major manus; presumably it could act as an effective instrument in major-manus-drumming against the ground. Unfortunately this species has not been observed in life since our knowledge of acoustic behavior in *Uca* began to be assembled.

The stiff brushes of setae on the tips of the small chelae scrape algae from the stones. These minute plants form the principal food of all *panamensis* that have reached a moderate size. Only the young and occasional adults sift organic matter from the sand in typical *Uca* fashion.

MORPHOLOGY

Diagnosis

Major manus with proximal outer end strikingly thick and projecting; oblique ridge on palm vestigial, being smoothly rounded and without tubercles. Minor cheliped in both sexes with gape wide and with a long, distal brush of stiff, bristle-like setae on each finger. Front broad, about one-third width of carapace; antero-lateral angles acute and produced.

The posterior extension on the major manus and the striking brushes of setae on the small chelae distinguish this species from all others in the genus.

Description

With the characteristics of the subgenus *Minuca* (p. 154).

MALE

Carapace. Front contained about 3 times in carapace breadth. Carapace shape antero-laterally atypical of the subgenus, being widest between antero-lateral angles which are acute and slightly produced; in addition, the margins converge slightly from the angle before turning inward, sometimes in an almost angular turn. In addition to the usual postero-lateral striae, very short, weak, additional striae, of varying size, number, and position, occur on sides, behind the vertical lateral margins; posteriorly the roughening takes the form of single spinules. Suborbital crenellations strong throughout; the outer angle of the orbit flares anteriorly, so that the crenellations project; the angle has a smooth, vertical outer border, the crenellations recommencing on a low level and continuing posteriorly along outer edge of orbit and onto lower edge of antero-lateral angle.

Major Cheliped. Merus on ventral margin with a single row of tubercles, not two, or a row of weak, transverse rugosities; sometimes margin is almost smooth. Manus with outer surface marble smooth or with very minute tubercles on lower part, and always with small, low tubercles extending from upper part of heel area obliquely dorso-distally across upper third; subdorsal groove absent; dorsal margin little bent over, broad, convex, not set off from either outer manus or palm, tubercles of marginal area being continuous with those of both surfaces; the only exception is a wavy, linear row of the same tubercles that can sometimes be distinguished in proximal half of marginal area, in the region normally part of outer edge and groove; ventral margin also smoothly rounded, with or without faint traces of a line of minute tubercles in distal half, both keel and groove being absent; heel notably enlarged, extending posteriorly in a rounded protuberance that prevents full lateral extension of manus. Palm with lower triangle uniformly covered throughout by very minute tubercles; central area smooth, the only general roughening of entire palm being near dorsal margin, where the small tubercles of outer manus and dorsal margin die out; tubercles altogether lacking on the low, rounded, indistinct, oblique ridge, except for the very minute tubercles of lower triangle which die out in this area; carpal cavity small, its lower margin lacking, its distal edge marked with small tubercles, the beading of its upper edge minute; predactyl ridges weak. Pollex and dactyl with tubercles of prehensile edges low, weakly differentiated; tip of pollex blunt, not appearing trifid; dactyl with proximal subdorsal groove absent and with proximal dorsal tubercles minimal.

Minor Cheliped. Manus rounded, almost as broad as long and two-thirds as thick. Fingers only slightly longer than palm, broad and thick, gaping moderately to the corneous tips which articulate well; prehensile edges without serrations. Entire lower distal end of pollex and upper distal of dactyl covered with a thick brush of long bristles; these continue backward on both fingers, inside and out, as the usual oblique rows of relatively scanty setae.

Ambulatories. Ventral margins of meri serrate for at least part of anterior and posterior margins; strength of serrations variable among populations. Pile absent.

Gonopod. Anterior flange much broader than posterior. Inner process broad, thick except distally, closely applied to and extending beyond tip of posterior flange. Thumb slender, ending well below base of flange.

FEMALE

All armature notably strong, for a female of this subgenus; noteworthy characteristics are the tuberculation of the hepatic and branchial regions, the strength of the unusual short striae of the carapace sides, and the distinct beading of the posterior edge of frontal margin, eyebrows, and anterior edge of posterior margin of carapace. Gonopore with an external tubercle directed antero-internally.

Measurements (in mm)

	Length	Breadth	Propodus	Dactyl
Largest male	13.5	19.5	30.0	19.0
Moderate male	11.0	16.5	21.5	14.5
Largest female	14.5	19.5	–	–
Largest ovigerous female	10.0	13.3	–	–
Smallest ovigerous female	7.0	10.2	–	–

Morphological Comparison and Comment

In spite of its unusual characteristics, *panamensis* is basically a typical member of the subgenus *Minuca*, the front being very broad and the gonopod of the general form found in *vocator* and its relations. The exceptionally smooth manus and chela hint that actual combat may be strongly reduced, while the large heel, as well as a moderate development of armature on the merus and suborbital regions, suggest that this species will prove to depend strongly on acoustic signaling in agonistic behavior as well as in courtship.

Color

Display whitening probably absent, although some individuals are moderately pale. The color of this species definitely varies with the prevailing color of the surrounding sand and stones. Populations on light sand and among pale stones are usually yellowish buff or grayish white, but individuals are sometimes paler. Crabs on dark volcanic sand among dark stones, on the other hand, range through various shades of brown, dull green, and gray; they are often speckled or marbled with lighter or darker, and sometimes with dark red. Major cheliped frequently lighter than rest of crab.

SOCIAL BEHAVIOR

Waving Display

(Incompletely known; only about five examples have been observed; no films were secured.)

Wave apparently always lateral-straight, the major cheliped being extended diagonally outward and upward, then flexed again in the same plane. Minor cheliped motionless; body not raised; ambulatories remain on ground; no steps are taken. Waving is at the rate of about 1 wave to the second. The examples seen all displayed from the tops of stones. (Component no. 4.)

RANGE

Gulf of Fonseca, El Salvador, to Payta, Peru.

BIOTOPES

This species is the only one in the genus found close to the ends of sheltered, sandy beaches, among the stones that often mark the change to rocks and tidepools just beyond. Rarely found also in the adjacent tidepools under stones; this location appears to be accidental, resulting from a hurried escape. (Biotope no. 4 approximately; see above; rarely, no. 2.)

SYMPATRIC ASSOCIATES

None.

MATERIAL RESULTING FROM FIELD WORK

(The complete list of specimens examined is given in Appendix A, p. 602.)

Observations and Collections. Nicaragua: Gulf of Fonseca, near Potosi R.; Corinto. Costa Rica: Port Parker; Culebra Bay; Piedra Blanca Bay; Ballenas Bay; Uvita Bay; Golfito. Panama: Honda Bay; Panama City. Colombia: Gorgona I. Specimens observed and examined, but not collected, at the following additional localities: Costa Rica: Parida I., Cedro I., both in Gulf of Nicoya.

TYPE MATERIAL AND NOMENCLATURE

Uca panamensis (Stimpson, 1859)

Stimpson's type of *Gelasimus panamensis* is not extant. Type-locality: Panama. Since no taxonomic confusion has yet arisen, it seems neither necessary nor desirable to designate a neotype.

REFERENCES AND SYNONYMY

Uca (Minuca) panamensis
(Stimpson, 1859)

Gelasimus panamensis

TYPE DESCRIPTION. Stimpson, 1859: 63. Panama.
   Smith, 1870: 139; Pl. 4, Fig. 5. Taxonomy of type.
   Kingsley, 1880.1: 150. Gulf of Fonseca. Taxonomy.
   Cano, 1889: 92, 235. Gulf of Panama. Taxonomy.

Uca panamensis

   Nobili, 1901.3: 49. Ecuador: Flamenco I. Taxonomy.
   Rathbun, 1918.1: 412; photos. Gulf of Fonseca: El Salvador. Costa Rica: Puntarenas. Panama: Taboga I. and elsewhere. Peru: Payta.

   Maccagno, 1928: 40. Ecuador: I. Flamenco. Taxonomy of Nobili's material.
   Crane, 1941.1: 204; text fig. (minor chela). Nicaragua; Costa Rica; Colombia. Taxonomy; general habits; habitat. (USNM !)
   Garth, 1948: 60. Colombia: Humboldt Bay (AMNH !); Limon Bay.
   Holthuis, 1954.2: 163. El Salvador (no new records).
   Bott, 1954: 162; text fig. (gonopod); photos. Gulf of Fonseca: El Salvador.
   Crane, 1957. Panama. Preliminary classification of waving display.

Uca galapagensis (not of Rathbun).

   Boone, 1927: Text Fig. 97 (part). Cocos I. (eastern Pacific). Specimen illustrated in lower half of Text Fig., designated *galapagensis*, is example of *panamensis*. (!)

26. *UCA (MINUCA) PYGMAEA* CRANE, 1941

(Tropical eastern Pacific)

PLATE 22 *E-H*.   MAP 11.
FIGURES 66 *E*; 100.   TABLES 9, 10.

INTRODUCTION

A very small species with paedomorphic characteristics, *Uca pygmaea* is known from few specimens and has not been observed alive before capture.

MORPHOLOGY

Diagnosis

Front wide; orbits extremely oblique. Major pollex with a deep, narrowly triangular furrow supraventrally at its base; palm without oblique, tuberculate ridge, the area being granulate; manus greatly tumid; no backwardly directed heel. Gonopod with anterior flange projecting clearly beyond posterior flange; inner process moderately thick and broad.

Description

With the characteristics of the subgenus *Minuca* (p. 154).

MALE

Carapace. Front only moderately wide in the subgenus, contained about 4 times in width of carapace between antero-lateral angles, which are strongly acute; orbits extremely oblique; antero-lateral margins almost lacking; dorso-lateral margins strongly converging. Upper postero-lateral stria long; lower absent. Postmarginal part of orbital floor and entire suborbital region densely setose except along outer half of suborbital margin, where the well-developed crenellations stand free of setae and continue around outer edge of orbit.

Major Cheliped. Ventral margin of merus with a single row of sharp tubercles, those near distal end enlarged and spinous. Manus appearing swollen, being very thick, and scarcely or not at all longer than wide; outer surface roughly tuberculate in dorsal half; at outer base of pollex a conspicuous furrow, narrowly triangular, the apex distal; ventral margin proximally a finely beaded keel, otherwise formed of large, low discrete tubercles; beads and tubercles graduated throughout, largest at pollex base. Palm without oblique ridge, or with a rudimentary indication of a ridge, the entire area very convex, rough

in dorsal part with large tubercles that become gradually smaller ventrally; proximal ridge at base of dactyl diverging from distal, which is obsolescent. Pollex and dactyl deep and thick, the pollex appearing distinctly triangular; prehensile edge of pollex has, unusually close to its proximal end, a very large tooth, adjacent to one or two others, somewhat smaller. Dactyl little or not at all longer than manus.

Minor Cheliped. Palm broad and thick.

Ambulatories. Meri moderately broad; legs without pile.

Gonopod. Flanges of similar breadth, both strongly concave externally, forming together with the intervening canal and pore almost a semicircle; anterior flange with its entire anterior edge projecting well beyond inner process, pore, and posterior flange. Inner process moderate in size and thickness, covering only part of the posterior flange and leaving anterior flange wholly exposed. Thumb large, arising near base of anterior flange but ending well below latter's tip.

Measurements in mm: Holotype: length 5.7; breadth 8.4; propodus 10.8. Twelve male paratypes: length 4.0 to 5.5.

FEMALE
Unknown.

Morphological Comparison and Comment

This species appears to be most closely related to *U. zacae.* It differs in the extreme obliqueness of the orbit, in the deeper, tumid palm with coarse granulation on its inner surface, and in details of the gonopod. It resembles *(Celuca) argillicola* in the general shape of the carapace combined with the lack of an oblique ridge on the palm, but differs clearly in the wider front, narrow meri on ambulatories and form of gonopod.

In spite of the small size and paedomorphic form of the males in the type series, the only known specimens, the well-formed details of the gonopods indicate that the specimens are almost or wholly mature.

Range

Golfito, Costa Rica, to Buenaventura, Colombia.

Biotope

Muddy banks of fresh-water stream. (Biotope no. 15.)

Type Material and Nomenclature

Uca pygmaea Crane, 1941

HOLOTYPE. In Smithsonian Institution, National Museum of Natural History, Washington. Cat. no. 137419 (formerly New York Zoological Society cat. no. 381,110). Golfito, Costa Rica. 12 additional males, all designated paratypes and all from the type-locality, deposited in the same institution, cat. no. 137420 (formerly NYZS cat. no. 381, 111 part).

References

Uca (Minuca) pygmaea Crane, 1941

Uca pygmaea

TYPE DESCRIPTION. Crane, 1941.1: 174; Text Fig. 4b (minor cheliped); Pl. 1, Fig. 1 (photo); Pl. 2, Fig. 4 (photo). (USNM !)
    Garth, 1948: 61. Colombia: Humboldt Bay.

27. *UCA (MINUCA) VOCATOR* (HERBST, 1804)

(Tropical eastern Pacific and tropical western Atlantic)

PLATES 23; 24 *A-D*.   MAP 13.
FIGURES 16; 66 *A, B, C, D*; 100.   TABLES 9, 10, 12, 20.

INTRODUCTION

Like *Uca thayeri* in the subgenus *Boboruca* and *maracoani* in *Uca*, the two forms of *vocator* are divided by the Isthmus of Panama and yet are so similar that there seems to be no justification for giving them the rank of separate species.

The fragility of the characteristic pile on the carapace has led to confusions of identification on both coasts. As in (*Amphiuca*) *chlorophthalmus*, variations in the form of the gonopod are prevalent, particularly in the Pacific. Since differences occur, as in *chlorophthalmus*, even within the same populations, further allopatric subdivisions are not indicated, at least on the basis of the present limited material.

MORPHOLOGY

Diagnosis

Front very broad; profuse pile, in a pattern, on carapace of both sexes, but easily detached; traces practically always remain on hepatic and branchial regions. Major palm with oblique ridge obsolescent, the apex very low, tubercles small and irregular, sometimes absent; proximal predactyl ridge not parallel to adjacent groove, but diverging from it throughout. Ambulatories in male only with pile profuse, even present on ventral half of mani, but easily detached, often partly or wholly missing. Gonopod with inner process broad, tumid, twisted, closely appressed against entire width of flanges; a tubercle near base of anterior edge of anterior flange, usually obscured by setae. Gonopore with a large tubercle surrounded on three sides by a strong, uneven rim.

Description

With the characteristics of the subgenus *Minuca* (p. 154).

MALE

Carapace. Front very wide, contained less than 3 times in breadth of carapace between antero-lateral angles. Upper postero-lateral striae long and strong. Pile in a characteristic pattern of patches and slender curving ovals, extending at maximum over most of carapace, though in varying thickness; it is most distinct on hepatic regions but present in all depressions and along all lateral and postero-lateral margins and ridges. Easily removed by abrasion, it is partly or wholly missing on many specimens; beneath the pile the carapace is microscopically pitted. Eyebrows ranging from strongly depressed to almost vertical. Crenellations of suborbital margin small but well formed, especially externally; they are, however, often almost hidden in pile and setae.

Major Cheliped. Manus with tubercles of upper outer surface very small. Palm with oblique ridge low and blunt, not marked by special tubercles in a single row; small, low tubercles, slightly larger than those on outer surface, cover entire area of slope and crest of ridge and continue, enlarged, over center palm; any or all of these tubercles sometimes vestigial or absent, especially in large specimens, but often also in middle-sized individuals; on apex and, sometimes, immediately above and below it, several enlarged tubercles may or may not be clustered; at base of dactyl only proximal ridge is well developed and with distinct tubercles; it diverges widely from the vertical and proceeds up and back, merging with the enlarged tubercles of center palm.

Ambulatories. Meri of 2nd and 3rd ambulatories slightly enlarged, their dorsal margins more or less convex. Pile maximally present dorsally, anteriorly and posteriorly on all legs on all segments except dactyls, and, in addition, on ventral side of manus; the pile is, however, easily dislodged, and is probably naturally scanty or sometimes lacking ventrally in northern Pacific populations.

Gonopod. Anterior flange slightly to greatly broader than posterior; a tubercle, ranging from very small to large, sometimes spiniform, close to base of anterior edge of flange, often obscured by setae. Inner process broad, truncate, and in large specimens slightly tumid, covering all of posterior and at least half of anterior flanges, the coverage increasing with growth; the process is a little twisted anteriorly and in larger specimens curves distally beyond edge of flanges, to which it is closely appressed. Thumb large, reaching well beyond flange base.

FEMALE

Suborbital crenellations not stronger than in male.

Pile on ambulatories in Pacific subspecies sparse compared with that of male and usually absent from ventral surface of manus; in Atlantic subspecies ambulatory pile altogether absent.

Gonopore with a large, central tubercle, surrounded externally, posteriorly and internally by a strong, projecting rim. This rim is deeply incised externally, so that its outer section appears as a small tubercle; the posterior section of the rim joins the internal section at an angle marked by a crease. Anterior edge of gonopore smooth, especially in the Pacific subspecies, or with a very low, straight rim.

Measurements (in mm)

	Length	Breadth	Propodus	Dactyl
U. vocator vocator				
Largest male (Trinidad)	17.0	24.5	40.0	28.0
Moderate male (Trinidad)	14.0	21.0	32.0	22.5
Largest female (ovigerous) (Venezuela)	13.5		–	–
Smallest ovigerous female (Venezuela)	8.2		–	–
U. vocator ecuadoriensis (all from Guayaquil, Ecuador)				
Largest male	17.0	25.0	32.0	25.0
Largest claw (body missing)			39.0	26.0
Moderate male	14.0	19.0	31.0	21.0
Largest female (ovigerous)	15.5	21.5	–	–
Smallest ovigerous female	11.5	15.0	–	–

Morphological Comparison and Comment

U. vocator differs from all other *Uca* with very broad fronts in the presence of widespread patches of pile, maximally over entire carapace, and in the presence, on the gonopod, of a tubercle or spine near base of anterior flange. It differs further from all these broad-fronted *Uca* except *panamensis* in the breadth of the gonopod's inner process, which covers the posterior and more than half the anterior flanges. It differs additionally from sympatric *Minuca* on both Atlantic and Pacific coasts as follows: from *burgersi, mordax,* and *brevifrons* by the low apex of the weak ridge inside palm; from *pygmaea* by the lack of a triangular furrow outside pollex base; and from *galapagensis* and *rapax* by the divergence inside major palm, up and back, of the proximal ridge at dactyl's base from the vertical line of the base itself. The most noticeable differences between the two subspecies are confined to minor distinctions in the armature of one edge of the major merus, the presence or absence of sparse pile on the female ambulatories, and details of the gonopod tip, concerning breadth of one flange and size and variability of the tubercle near its base.

Color

Displaying and non-displaying males: Display whitening absent. In the Pacific subspecies, *v. ecuadoriensis*, the carapace is usually marbled with yellowish and dark brown, or with gray and black; in the salt flats near Guayaquil some rose red individuals were collected, none of them displaying. In the Atlantic subspecies, *v. vocator*, displaying individuals are usually dull brownish or grayish, but sometimes tinged with yellowish or reddish. The brightest individuals, none of which were waving, in the large population at the mouth of the San Juan River, Venezuela, had the carapace rufous orange above, with major manus and chela yellow; in Trinidad the carapace was usually dull, attaining a yellowish buff in only several displaying individuals in one population; in all Trinidadian populations noted the major manus was tinged with yellow to pale orange; in at least one Trinidad stream the orange was confined to the dry season, between February and May. In both the Atlantic and Pacific, in fact, the major chelipeds are often the only non-monochromatic part of the crab, usually showing some yellowish or reddish, especially on the manus, while the fingers are usually white. Carapace and ambulatories of females as in males of the populations.

SOCIAL BEHAVIOR

Waving Display

(Similar in the two subspecies, but not well known in either.)

Wave lateral-straight to narrowly circular, the cheliped usually unflexing obliquely upward, sometimes almost directly laterally. Jerks present but so faintly indicated that they are sometimes difficult to count; in well-accented examples they number 4 to 6 on unflexing followed, usually after an intervening pause, by 4 to 6 more jerks on the descent. Cheliped not brought all the way to ground between waves at high intensities; at these times the pause at peak may be omitted. Minor cheliped usually inactive. Body raised anteriorly as its posterior part is tilted down, through bending of ambulatories. Duration at low intensity about 3 seconds including a 1 second pause at wave's peak; at high intensity each wave, the peak pause being absent, lasts at least 2 seconds. Pauses between waves in a series range from practically none to 1 second. (Component nos. 4, 5, 6, 10.) (Cf. von Hagen, 1970.3.)

Chimney Construction

Large, well-constructed chimneys were found in the population at the mouth of the San Juan River, Venezuela, and only there. As usual, only a small proportion of the population had erected chimneys. A number of pairs of contiguous structures were seen, with a large male and female in each pair; sometimes the individuals sat in rest position on the structure's broad lip, or partially leaned against its base. The majority of chimneys were separated from neighboring structures and seemed to belong to large females that were often ovigerous. No waving males were seen in this population, which was visited only once, in April. The largest chimneys measured about 63 mm high by 50 mm across the widest part, those of smaller crabs being in proportion.

It is possible that some very large mounds are constructed by members of the Pacific subspecies, the occupants of which have not yet been collected and identified.

Acoustic Behavior

Two usual components were observed and tape-recorded in Trinidad, in both field and laboratory. These were the major-merus-rub and major-manus-drum (component nos. 1, 9). In addition membrane vibrations (no. 13) were observed and tape-recorded in the laboratory (p. 484). This is the only species in which this very distinct response has been surely detected. Patterns of striae in the membrane at the bases of both chelipeds are clearly visible in life and in fresh dead specimens; they do not however preserve well in either formalin or alcohol.

Tape recordings were made of the species in Trinidad by von Hagen also (1967).

RANGE

Western Atlantic: From Tampico in Mexico at least to the state of Rio de Janeiro in Brazil, including Guatemala, British Honduras, Venezuela, Guyana, and Surinam, as well as in the West Indies, Santo Domingo, Puerto Rico, Guadeloupe, Dominica, Tobago, and Trinidad. According to Luederwaldt (1919.2), the species was taken farther south, in Iguape in the State of Santa Caterina; his specimens have not been reexamined, however, and perhaps should be referred instead to *mordax*, *burgersi*, or *rapax*. Eastern Pacific: From San Blas (Tepic) in Mexico to Puerto Pizarro in Peru.

BIOTOPES

In partly shaded mud, near mouths of streams on flat banks, close to mangroves, sometimes well back from streams, or rivers in damp mud among tall man-

groves and ferns; as usual, however, with *Uca* habitats, sunlight does reach the mud at some times of the day. Sometimes the species occurs close to or slightly beyond the upstream limit of mangroves, among grasses. (Biotope nos. 9, 12, 14.)

SYMPATRIC ASSOCIATES

Sometimes *vocator* mingles marginally with other *Minuca* including *burgersi*, which typically lives in habitats both more saline and more open than does *vocator*; occasionally individuals are found in populations of *rapax* in the Atlantic and *galapagensis* in the Pacific but this habitat is not their usual one. I have not observed *vocator* in association with either *mordax* or *brevifrons*.

MATERIAL RESULTING FROM FIELD WORK

(The complete list of specimens examined is given in Appendix A, p. 602.)

Observations and Collections. Western Atlantic: Guatemala: Puerto Barrios. Puerto Rico: San Juan. Venezuela: near Maracaibo (Zulia); mouth of San Juan River and at Pedernales (Monagas). Trinidad. Guyana: Georgetown. *Eastern Pacific*: Nicaragua: San Juan del Sur. Costa Rica: Negritos I.; Golfito. Ecuador: Puerto Bolivar.

Film. Pedernales and Golfito.

Sound Recordings. Trinidad.

TYPE MATERIAL AND NOMENCLATURE

Uca vocator (Herbst, 1804)

As stated by Holthuis (1959.2: 273), "Unfortunately, the type specimen of *Cancer vocator* is no longer extant, as Dr. H.-E. Gruner of the Berlin Zoologisches Museum was so kind to inform me (1 December 1958, in litt.)." Holthuis, on the basis of Herbst (1804) p. 1, Pl. 59, Fig. 1, decided that there was little doubt but that *Cancer vocator* is identical with *Uca murifecenta* Crane, 1943, and that a long series of specimens collected in Surinam should be referred to the same species. I agree without reservation with Holthuis' conclusion. Herbst's type locality: "Das Vaterland ist Amerika."

From the Surinam series Holthuis selected a neotype of *U. vocator* (Herbst), as follows:

NEOTYPE. In Rijksmuseum van Natuurlijke Historie, Leiden. Cat. no. D 12329. Type-locality: shore of Suriname River near "Purmerend" plantation, Leonsberg, N. of Paramaribo, Surinam (1 April 1957), L. B. Holthuis no. 1208. Carapace breadth 26 mm. (!)

Type Material, of *Uca murifecenta*, Crane, 1943. In Smithsonian Institution, National Museum of Natural History, Washington: holotype: no. 137424 (formerly New York Zoological Society cat. no. 42167); paratypes: no. 137425 (formerly New York Zoological Society cat. no. 42417). Type-locality: near mouth San Juan R., Venezuela. Holotype illustrated by photos (Crane, 1943.1, Pl. 1, Figs. 1, 2; gonopod of non-type material [Text Fig. 1d, e, f] shows neither the spine at flange base nor the flanges themselves adequately). (!)

Uca vocator ecuadoriensis Maccagno, 1928

TYPES. 2 young males from Esmeraldas, Ecuador, in Museo di Torino. Described as *Uca ecuadoriensis* by Maccagno from material considered by Nobili (1897) to be a variety of *Gelasimus vocator*. Carapace lengths about 4 and 7 mm. There seems to be no need for selecting one of these immature examples as a lectotype for the subspecies; the juvenile cheliped is well illustrated by Maccagno (1928: 49, Fig. 32). (!)

Type Material of *Uca schmitti* Crane, 1943. In Smithsonian Institution, National Museum of Natural History, Washington, cat. no. 80451 (holotype male); type-locality San Blas, Tepic Territory (= Nayarit), Mexico; also 22305 (3 male paratypes from type-locality); illustrated by photos in Pl. 1, Figs. 1 and 2 in Crane, 1943.1: 31-32. Additional paratypes: 1 young male (USNM 137421, formerly New York Zoological Society cat. no. 381116) from San Juan del Sur, Nicaragua; 1 male (USNM 137423, formerly New York Zoological Society cat. no. 381117) from Negritos I., Costa Rica; 1 male (USNM 137422, formerly New York Zoological Society cat. no. 381118), from Golfito, Costa Rica. In Museum of Comparative Zoology, Harvard University, Cambridge, Massachusetts, cat. no. 5892, part (paratype male), from Acapulco, Mexico. The gonopod illustrated in Crane, 1943.1, Text Fig. 1d, c, f) was done from non-type material, depicts proportions found in young rather than adults, and does not show the spine at base of anterior flange, the existence of which had not then been discovered. In all 4 specimens from the type-locality the spine is notably long and slender. (!)

Type Material of *Uca lanigera* von Hagen, 1968. In Rijksmuseum van Naturalijke Historie, Leiden: holotype male, cat. no. 23049, and paratypes, cat. no. D 23050. This material from Puerto Pizarro, Peru, falls within the range of variation of the Pacific subspecies, established from other material. Von Hagen's Text Fig. 10a, b (p. 423), illustrating gonopod tips from Peru and Guatemala, respectively (the latter identified as *U. schmitti*), show two of the characteristic variables—the size of the spine at

flange base, and the length and breadth of the inner process; the latter is strongly dependent on the size of the crab. The carapace pile in Text Fig. 5 (p. 413) shows the pattern at its best preserved. (!)

Uca (Minuca) vocator vocator (Herbst, 1804)

(Tropical western Atlantic)

MORPHOLOGY

With the characteristics of the species.

Major Cheliped. Antero-dorsal margin of merus marked by a row of parallel rugosities, obliquely set, low and tuberculate.

Ambulatories. Pile absent in female.

Gonopod. Anterior flange much broader than posterior; tubercle near anterior flange small, almost adjoining base.

Gonopore. Anterior rim of gonopore usually absent, sometimes vestigial.

Uca (Minuca) vocator ecuadoriensis Maccagno, 1928

(Tropical eastern Pacific)

MORPHOLOGY

With the characteristics of the species.

Major Cheliped. Antero-dorsal margin of merus marked by a row of small, sharp tubercles or serrations, their apices directed distally.

Ambulatories. Pile sparsely present in female, including sometimes on lower surface of manus; often absent, apparently due to abrasion; in northern populations usually missing on lower surfaces of manus and carpus even in males.

Gonopod. Differs from that of *vocator vocator* in having anterior flange only slightly wider than posterior; differs further in having tubercle at base of anterior flange larger and exceedingly variable. The range is from small to large in basal diameter, and from short to long and sometimes very slender. It is usually longer than in *v. vocator* and sometimes separated farther from flange's base. Considerable variability occurs within populations and appears only partly associated with the size of the crab; in the limited material at hand I have not found reliable geographic differences.

Gonopore. Anterior rim of gonopore present at least externally, but poorly developed.

REFERENCES AND SYNONYMY

Uca (Minuca) vocator (Herbst, 1804)

TYPE DESCRIPTION. See under *U. (M.) vocator vocator*, below.

Uca (Minuca) vocator vocator (Herbst, 1804)

Cancer vocator

TYPE DESCRIPTION. Herbst, 1804: 1; Pl. 59, Fig. 1. "Amerika."

Goneplax vocator

Latreille, 1817.1: 17. Reference to illustration of type of *Cancer vocator*.

Uca mordax (not *Gelasimus mordax* Smith)

? Rathbun, 1902.1: 7. Puerto Rico. Taxonomy.

Rathbun, 1918.1: part, at least including Pl. 134, Figs. 3, 4 (photos). British Honduras: Belize. (Re-identification on basis of pile apparent on carapace; in agreement with suggestion of Holthuis, 1959.3 and Chace & Hobbs, 1969. One male from Belize included under USNM 21373 (!) and listed by Rathbun proves to be an example of *vocator* and is included in Appendix A. The same catalogue number is given in the caption to Pl. 134. Accordingly, there seems to be no doubt remaining on the identity of the specimen illustrated.

Uca murifecenta

Crane, 1943.2: 38; Text Figs. 1d, e, f; Pl. 1, Figs. 1, 2. Type-locality: near mouth of Rio San Juan, Venezuela; also from other Venezuelan localities. Type description. Color; general habits. (USNM !)

Uca vocator

? Moreira, 1901: 52. Brazil: northern part and near Rio de Janeiro. Taxonomy.

? Luederwaldt, 1919.1: 370, 384, 398. Brazil: Santos. Brief taxonomy.

? Luederwaldt, 1919.2: 435. Brazil: State of São Paulo south to Iguape.

? Luederwaldt, 1929: 54. Brazil: State of São Paulo: São Sebastião I.

Holthuis, 1959.3: 269; Text Figs. 66 (after Herbst), 67 (ambulatories); Pls. 14, Fig. 1 and 15,

Fig. 1 (photos). Selection of neotype from Surinam; taxonomy; habitat. (Leiden !)

von Hagen, 1967.2. Trinidad. Tape recordings secured. (Preliminary statement.)

Chace & Hobbs, 1969: 217; Text Figs. 73g, h, i, j, and 74. Dominica. Distribution; taxonomy; color; habitat.

von Hagen, 1969; text figs. Trinidad. Eggs taken by poecilid fishes and grackles.

von Hagen, 1970.1: 225. West Indian distribution.

von Hagen, 1970.3: 238; text figs. Trinidad. Display and courtship in relation to a habitat where vegetation obscures vision; use of models; sound production; design for crabbery.

von Hagen, 1970.4; text figs. Trinidad. Adaptation to a particular intertidal level.

Uca (Minuca) vocator ecuadoriensis Maccagno, 1928

Gelasimus vocator (Herbst) var.

Nobili, 1901.3: 49. Ecuador: Esmeraldas. Taxonomy.

Uca ecuadoriensis

TYPE DESCRIPTION. Maccagno, 1928: 49; Text Fig. 32. Described from Nobili's material (above). Stands as type description of the subspecies. (Torino, part !)

Uca mordax (not *Gelasimus mordax* Smith)

Rathbun, 1918.1: 391 (part). Mexico: Tepic Territory: San Blas. Includes type material of *U. schmitti*. (USNM !)

Crane, 1941.1: 176 (part); Text Figs. 2, 3, 4e. Includes type and other material of *U. schmitti*. (USNM !)

Uca schmitti

Crane, 1943.1: 31; Text Fig. 1; Pl. 1, Figs. 1, 2. Mexico: Tepic Territory: San Blas (holotype and 3 paratypes); other material from Mexico, Nicaragua, Costa Rica. Type description. (USNM !)

Garth, 1948: 60. Colombia: Humboldt Bay.

Uca lanigera

von Hagen, 1968.2: 421; Text Figs. 2, 3, 10a. Peru. Type description. Color. (Leiden !)

28. *UCA (MINUCA) BURGERSI* HOLTHUIS, 1967

(Tropical and subtropical western Atlantic)

PLATE 24 *E-H*.
FIGURES 26 *F*; 31 *H*; 54 *G*; 66 *F*; 100.
MAP 12.
TABLES 9, 10, 12, 20.

INTRODUCTION

Until its recent description by Holthuis, specimens of this widespread and very common species, *Uca burgersi*, were usually identified as *U. mordax*. Although minor morphological differences and distinct ecological preferences were sometimes apparent, the characteristics were all so variable that there was overlap in at least parts of the ranges, and no division into two species seemed practical. The difficulties extended then, and still do, even to their waving displays, which in the details so far known are indistinguishable; even the morphology of the gonopods is such that the two forms cannot be decisively identified by that means; their geographical ranges coincide throughout the latitudes inhabited by one of them, the other extending farther to both north and south; finally, during the entire extent of the coincidence, there are no definite ecological barriers between the two, displaying adults occasionally being found in the closest sympatry. Fortunately, Holthuis discovered a single character—a minor difference in pile distribution on the ambulatories—that extends to all known individuals of the two species. In addition, even in the several instances I have found of fully sympatric associations there is no trace apparent of hybridization; furthermore, in each of these areas the *local* populations of each form are otherwise quite easily distinguished from each other by additional characters, particularly by details of palm armature. These facts call for specific status for each form. When their physiologies and ethologies become better known, the puzzling aspects of their evolution and the maintenance of their present barriers will no doubt be clarified.

Throughout the West Indies *U. burgersi* is probably the most common fiddler, its only possible numerical rival being *rapax*, a member of the same subgenus. In populations that include waving individuals, the two are easily distinguished in the field by the display alone; the cheliped of *burgersi* usually makes only 3 or 4 jerks during its rise, while that of *rapax* customarily makes at least 8 and usually more; in both species individuals making occasional displays having more or fewer jerks than are character-

istic of their species are always in a small minority and easily discounted by the observer. In addition, the major manus, and sometimes the carapace, of some individual *burgersi* in most populations are pink to bright red, while no tints or tones of red are found in *rapax*, the manus usually ranging from grayed yellowish to a dull orange.

While he was observing *burgersi* on the Caribbean island of St. Martin, Dr. P. W. Hummelinck (personal communication, with photographs) found some individuals moving about and even displaying when completely submerged in the shallow water of a storage tank for seawater, their movements in and out of it being unimpeded.

One hot July noon on the nearby island of St. Thomas, I found a somewhat similar situation. Here a mangrove-edged lagoon, practically enclosed, included a broad and muddy beach. At low tide fiddler burrows were everywhere. But I saw only one crab and, upon digging carefully, could find none in the burrows. Then I looked out to the water's edge and saw the crabs in thousands, clinging motionless in clusters to stones and driftwood; most of the crabs were fully submerged. When I took my shoes off to wade I understood the reason; the mud was hot enough to fry the proverbial egg, while the water itself was only tepid. The crabs proved to be mostly immature specimens and small adults of both sexes; none was ovigerous and none was waving. When disturbed each fiddler merely dropped to the bottom and scuttled to the nearest stone or another piece of wood; with a little care a handful could be scooped up at once. This kind of seaward migration, which I have seen in no other fiddler, seemed obviously to be a very local means of dealing with excessive heat. The shores of the same lagoon adjoining the broad stretch of muddy beach were exactly similar except that the mangroves came close to the water, leaving only a narrow strip of mud. Even though this mud was also unshaded, it was conspicuously cooler than that of the broad beach. Here, although no crabs were on the surface, the burrows proved to be inhabited, these fiddlers having escaped the midday heat in the usual way by sitting at the bottom of their holes.

Morphology

Diagnosis

Front very broad. Major cheliped with proximal pre-dactyl ridge on palm diverging strongly from adjacent groove at least in lower part. 2nd and 3rd ambulatories in both sexes without pile on ventral side of manus. No pile on hepatic and branchial regions. Oblique, tuberculate ridge on major palm moderately developed, but tubercles small and not in a single row; upper center palm with large tubercles. No crest of graduated tubercles, diminishing distally, on outer tip of pollex. Gonopod with inner process not tumid, narrow, leaving entire anterior flange uncovered. Gonopore with a single, small tubercle.

Description

With the characteristics of the subgenus *Minuca* (p. 154).

MALE

Carapace. Front contained about 3 times in carapace breadth between antero-lateral angles. Upper postero-lateral stria long and strong. No pile on hepatic and branchial regions. Only distal edge of eyebrow is beaded. Inner as well as outer suborbital crenellations not usually obscured by setae and pile.

Major Cheliped. Manus with tubercles of upper, outer surface small. Inner edge of dorsal margin sometimes marked with a distinct row of tubercles. Ventral, marginal keel with surmounting tubercles usually developed, although they are sometimes absent and the keel and adjacent groove themselves may be obsolescent; sometimes the keel is traceable only proximally, being continued to pollex base as a faint row of irregular tubercles. Palm with oblique ridge well developed, the apex always highest, covered with small tubercles that usually extend partway up distal margin of carpal cavity beyond apex; sometimes a single row is distinguishable along crest of ridge, especially near apex; sometimes no linear order is apparent. Near oblique ridge, center palm is more or less minutely tuberculate, moderately high throughout, usually flat or sometimes with a slight depression between the distinct distal border of carpal cavity and the moderate hollow at base of pollex. Upper part of center palm usually with a group of large, distinct, rounded tubercles with which merges the ventral end of the strongly down-curving beaded ridge which, as usual, forms dorsal margin of carpal cavity; the tubercles extending up from apex of oblique ridge may or may not reach this cluster of center palm tubercles. Predactyl area distinctly concave, but not set sharply off, by differences in plane, from the tuberculate center palm. At dactyl base, the proximal ridge always has well-developed tubercles, set almost contiguously, as on a beaded ridge, at least in its median region; the ridge always diverges widely from the vertical at the level of the dactyl's lower edge; then it usually either curves gradually in its upper portion back toward the dactyl base or, often, parallels the base in an almost or entirely vertical line; in some individuals it continues to diverge, however, in its initial proximo-dorsal direction, throughout its length.

Ambulatories. Meri of 2nd and 3rd ambulatories scarcely enlarged, their dorsal margins slightly convex to almost straight. Pile present only on dorsal surfaces of carpus and manus of first 3 legs; it is always altogether absent from ventral margins, including that of manus.

Gonopod. Anterior flange much broader than posterior. Inner process variable in width and length, usually not completely covering anterior flange and usually projecting minutely beyond pore; anterior flange scarcely or slightly produced. Thumb well developed, but does not reach base of flange.

FEMALE

Ambulatories with distribution of pile as in male, including its complete absence on lower margin of manus. Suborbital crenellations not stronger than in male. Gonopore with posterior margin slightly raised, variable, but not forming a definite tubercle. Meri of 3rd and 4th ambulatories scarcely expanded, their dorsal margins almost as straight as in male.

Measurements (in mm)

	Length	Breadth	Propodus	Dactyl
Largest male (Fortaleza, Brazil)	13.0	19.0	33.5	23.5
Moderate male (Fortaleza, Brazil)	10.0	14.0	17.0	10.0
Largest female (Fortaleza, Brazil)	13.5	18.5	–	–
Largest (ovigerous) female (Br. Honduras)	15.0	20.0	–	–
Small ovigerous female (Br. Honduras)	10.5	14.0	–	–

Morphological Comparison and Comment

U. burgersi and *mordax* are so closely similar in morphology, as well as in behavior and habitat, that each would undoubtedly be considered of only subspecific rank if their ranges did not overlap geographically. Nevertheless they do overlap extensively, from Guatemala to northern Brazil, while one of them, *burgersi*, extends from Florida to Rio de Janeiro. Furthermore, in the field I have twice encountered them where they were closely sympatric, yet proved morphologically as distinct as ever; each was in its usual habitat and, in one case, motion picture records were made of both forms, which show minor differences in display. These localities were the north coast of Trinidad, at Blanchisseuse, and Puerto Barrios, Guatemala. In each case, *burgersi* was found on a sheltered, sandy mud, tidal flat, while *mordax* was on the bank of an adjacent stream, near its mouth.

For many years the two forms have been confused both in the literature and in collections. More recently several field workers, including L. B. Holthuis (1959.2), became convinced that the form now known as *burgersi*, and characteristic of the West Indies, is a separate species. At the time I agreed. Later, however, it seemed to me that the two forms were so much more similar than any other pairs of apparently justified specific rank in the genus that some other solution should be found. In 1967 Holthuis, with plentiful material, examined the entire question. His evidence, along with my own examination of a large collection at hand, has convinced me of the specific distinctness of the two forms. Accordingly they are being given specific rank in this contribution.

Their similarities, however, remain challenging: they would undoubtedly form an ideal pair of species for morphological, behavioral, and ecological comparisons, as well as from evolutionary points of view. In general, *burgersi* is a paedomorphic form in comparison with *mordax*; the fingers of the major cheliped are usually slightly shorter than in *mordax*, the manus usually deeper, and pile less developed on the ambulatories. Above all, the armature has juvenile characteristics: the suborbital crenellations are not covered with pile and setae; the ventral margin of the major cheliped in general is better armed; tubercles of the major palm do not tend to obsolescence; the center of the palm is little sculptured, whereas in well-developed examples of *mordax* three large, low areas are separated by two high surfaces.

Holthuis, working primarily with material from the Netherlands West Indies, gives 7 principal differences between *mordax* and *burgersi*. When these characteristics were checked with specimens from other parts of the West Indies, from Mexico and Central America, and from South America it was found that only one—the absence of pile on the lower side of the manus on the two middle ambulatories—held throughout the range. It is true that when a number of examples of a single population are compared, the great majority will share most or all of these characteristics. However, with the single exception given, none of the 7 characteristics occurs in every specimen in a population, and some specimens show as many as 3 or 4 characters typically found only in the other species. In addition to the ventral pubescence mentioned, the most reliable characters in moderate-sized

specimens are as follows: first, the stronger, unobscured suborbital crenellations of *burgersi* (unreliable in young *mordax*); second, the smaller, well-separated tubercles on the proximal predactyl ridge; third, the rudimentary distal margin of carpal cavity above apex, with the well-marked trench connecting it with the depression at pollex base.

There seems to be no basis on which to erect subspecies of *burgersi*, nor is there evidence of a cline. For example, in single populations as far apart as the Bahamas and Rio de Janeiro proximal predactyl ridges occur that curve strongly away from the dactyl base and then proceed parallel with the base above the initial divergence; in populations between, the ridge usually, but not always, curves throughout. Both *burgersi* and *mordax* would furnish excellent material for statistical work on the morphological differences among populations; better yet, as usual, would be a determined attack on genetics.

Color

Display lightening little developed; no white ever apparent. Some members of both sexes sometimes with carapace bright pink, or red, or marbled with these hues, regardless of whether the males are displaying; pink sometimes extends to bases of ambulatory meri; carapace and ambulatories in both sexes however are usually brownish or gray. Major cheliped rarely with entire merus, outer carpus, and outer manus red, the palm then being grayish and fingers white; usually only the outer manus is reddish, its shade ranging from bright red to light pink or pinkish orange, the color often grayed; fingers with at least their tips white.

SOCIAL BEHAVIOR

Waving Display

Wave lateral, circular. Jerks present but variable within populations; no differences indicated from present knowledge of geographic variations in range or average numbers of jerks. The most usual in tropical populations seem to be 3 to 4 on cheliped's rise and 2 or 3 on its descent, but the range is from 2 to 6 on the rise and 2 to 4 on the descent. The example given by Salmon (1967) of a Florida specimen showed 5 jerks on the ascent and 4 on the descent. No pause at peak and almost none between waves. Minor cheliped makes a corresponding motion. Body held high on extended ambulatories during waves of a single series. One or 2 ambulatories, especially on minor side, raised during display. Any special characteristics of high intensity courtship have not yet become evident. Duration of waves ranging from about 5 to about 8 seconds, the usual duration being between 6 and 7. (Component nos. 5, 6, 9, 11, 12.)

Acoustic Behavior

Bubbling (component no. 12, p. 484) was seen, heard, and tape-recorded in Trinidad, both in the field and laboratory. It was performed strictly in response to the experimentally forced entry of a conspecific of either sex into its burrow. Salmon (1967) obtained oscillograms of leg-wagging (component no. 5) of males at night at the entrances to their burrows. Von Hagen (1967.2) in a preliminary statement also reports obtaining recordings of sound production in Trinidad.

RANGE

Western tropical and subtropical Atlantic from Daytona on the east coast of Florida in the United States, Mexico, Central America, the West Indies, and South America as far as São Paulo, Brazil. No records from the Gulf of Mexico except near Progreso, Yucatan. Possibly reaches western Africa (p. 327).

BIOTOPES

Sheltered flats or sloping banks of mud or muddy sand, near mouths of streams, or, more often, along the shores of lagoons and estuaries, usually near mangroves, sometimes on edges of coconut groves; characteristically well sheltered from open seawater, although always exposed to strong tidal influence; sometimes the mouths of the burrows are covered only at spring tides; never found upstream near mangrove limits. In Florida *burgersi* occurred in greatest numbers associated with *U. rapax* above the high-tide level and in the shade of mangrove thickets (Salmon, 1967: 450). (Biotope nos. 9, 11, 12.)

SYMPATRIC ASSOCIATES

Often found close to populations of *rapax*, although the latter usually occur nearer the open sea, probably in more saline situations, and the species usually mingle, if at all, only peripherally. The few small groups of *mordax* found close to *burgersi* occurred slightly more inland, along drainage ditches in Trinidad and Guatemala, the more numerous *burgersi* here occurring along small, almost enclosed lagoons between the ditches and the seashore.

MATERIAL RESULTING FROM FIELD WORK

(The complete list of specimens examined is given in Appendix A, p. 603.)

Observations and Collections. United States: Fort Lauderdale (Florida). Guatemala: Puerto Barrios. West Indies: Puerto Rico, St. Thomas, Guadeloupe,

Barbados, Tobago, Trinidad. Venezuela: Turiamo (Aragua). Brazil: São Luiz; Fortaleza; São Salvador; Rio de Janeiro.

TYPE MATERIAL AND NOMENCLATURE

Uca burgersi Holthuis, 1967

HOLOTYPE. In Rijksmuseum van Natuurlijke Historie, Leiden. Cat. no. Crust. D. 23012. 1 male, length 12 mm, breadth 18 mm. Type-locality: Plantage Knip, Westpunt, Curaçao, Netherlands Antilles; 15 December 1956; leg. A.C.J. Burgers and L. B. Holthuis (L. B. Holthuis no. 1023). (!)

Type Material of *Gelasimus affinis* Streets, 1872. Not now in Academy of Natural Sciences, Philadelphia, but syntypes in Leiden. Holthuis (1967: 53) discusses the history of this name as follows; I agree with his conclusions. "These specimens forming Lot 753 of the Philadelphia Museum, were labelled 'Uca vocator v. Martens.' It is evident that when Kingsley studied the *Uca* material of the Philadelphia Museum for his 1880 revision of the genus, he removed Streets's original label of *Gelasimus affinis* and replaced it by his own new label bearing the name *Gelasimus vocator.* Kingsley (1880: 147) namely placed *Gelasimus affinis* Streets in the synonymy of *Gelasimus vocator* (Herbst), incorrectly using the author's name Von Martens for the latter species. Rathbun (1918: 391) included *Gelasimus affinis* Streets with some doubt in the synonymy of *Uca mordax* (Smith) and indicated the types as being not extant. Kingsley's changing of the labels of the type lot evidently led her astray.

"The rediscovery of Streets's types removed any doubt about the identity of his *Gelasimus affinis,* and consequently I used that name in my papers of 1959. Subsequently, however, I found that the name *Gelasimus affinis* Streets, 1872, is preoccupied by *Gelasimus affinis* Guérin, 1829. Guérin's name appears only in the legend of Plate 1 of the Crustacean part of Duperrey's 'Voyage autour du monde . . . sur la corvette . . . La Coquille . . . ,' which was published in 1829 (see Holthuis, 1961). In his text, which was published as late as 1838, Guérin placed his *Gelasimus affinis* in the synonymy of *Gelasimus tetragonon* (= *Uca tetragonon* [Herbst]), where it has been left by all subsequent authors. There can be no doubt that, although *Gelasimus affinis* Guérin, 1829, is a junior subjective synonym of *Cancer tetragonon* Herbst, 1790, it is an available name, and being a senior primary homonym of *Gelasimus affinis* Streets, 1872, invalidates the latter name. As no other name is available for Streets's species, the new name *Uca burgersi* is now proposed here for it." See also present study, p. 322.

References and Synonymy

Uca (Minuca) burgersi Holthuis, 1967

Gelasimus affinis

Streets, 1872: 131 (not *Gelasimus affinis* Guérin, 1829; see p. 322, present contribution). West Indies: St. Martin.

Gelasimus vocator (not *Cancer vocator* Herbst)

Kingsley, 1880.1: 147 (part).

Uca minax (not *Gelasimus minax* LeConte, 1855)

Rathbun, 1897.2: 27. Jamaica: Kingston Harbor. (USNM !)

Uca mordax (not *Gelasimus mordax* Smith, 1870)

? Pearse, 1916: (part) 532, 554. Colombia: Santa Marta, at mouth of Manzanares River. Habitat. (Synonymy suggested on basis of habitat.)

Rathbun, 1918.1: 391 (part). West Indian records (part). ? Liberian record. (Not Text Fig. 166, nor Pl. 134, Figs. 3, 4.) (USNM, part !)

Beebe, 1928: 59-66. Haiti. Courtship; habitat. (USNM !)

Maccagno, 1928: 46 (part). West Indies records.

Oliveira, 1939.1: 138. Rio de Janeiro, Brazil. Taxonomy; habitat.

Crane, 1957 (part). Western Atlantic. Preliminary classification of waving display.

Salmon, 1967. U.S.A.: Florida. Local distribution; courtship, including waving display and sound production.

Uca affinis

Holthuis, 1959.2: 76. West Indies.

Holthuis, 1959.3: 265, 266. Suriname. Taxonomy (distinguished from *U. mordax*).

von Hagen, 1967.2. Trinidad. Tape-recordings secured. (Preliminary statement.)

Uca burgersi

TYPE DESCRIPTION. Holthuis, 1967: 52. Curaçao. (Leiden !)

Salmon & Atsaides, 1968.2: 634. Mention of sound production, in reference to Salmon, 1967.

Chace & Hobbs, 1969: 207; Text Figs. 70, 71a-d. Dominica. Taxonomy; color; ecology.

von Hagen, 1970.4. Trinidad. Adaptations to a particular intertidal level.

29. *UCA (MINUCA) [MINAX] MORDAX* (SMITH, 1870)

(Tropical western Atlantic)

PLATE 25 *A-D*.   MAP 12.
FIGURES 67 *F*; 100.   TABLES 9, 10, 12, 20.

INTRODUCTION

Although large populations of *Uca mordax* apparently live only far upstream in tropical rivers, a few individuals sometimes mingle with or adjoin populations of *burgersi* and *rapax* in lagoons and estuaries close to the open sea.

The species has the broadest front in the entire genus, a fact which is a reminder that we still do not have any idea of the adaptive value of wide fronts, since their increase, particularly in *Minuca*, is necessarily at the expense of the length of the eyestalks (p. 452).

The puzzling relationship of *mordax* to *burgersi* is reviewed as a whole on p. 168ff.

MORPHOLOGY

Diagnosis

Very close to *U. burgersi* (p. 168), from which it can be invariably distinguished only by the persistent presence of pile completely covering in both sexes the manus of 2nd and 3rd ambulatories.

Description

As in *burgersi* except for the following characteristics. As discussed on p. 170, none of them, with the exception of a detail of pile distribution, always occurs in every specimen.

MALE

Carapace. Inner suborbital crenellations usually very small and largely or wholly obscured by pile and setae on orbit's floor and on suborbital region; outer crenellations clearly visible, but usually slightly smaller than in *burgersi*, with both pile and setae more profuse. These distinctions apply poorly to smaller specimens, or when individuals of *mordax* are compared to those of *burgersi* that live in habitats muddier than usual.

Major Cheliped. Manus with ventral, marginal keel usually weaker than in *burgersi*, more often obsolescent, and with tubercles on it usually found only in the young. Palm with tubercles of oblique ridge never in clearly linear order, largely lacking in many large specimens; they always end in a small cluster slightly beyond the high apex and do not clearly continue up around distal edge of carpal cavity, although this edge sometimes has one or two small separated tubercles. The edge in moderate and large specimens, however, is always extremely low, almost absent, and marks the proximal end of a deep trench which extends disto-ventrally across the center palm to end in the depression at pollex base; in small specimens, even with fingers already longer than the palm, the trench is scarcely to be distinguished from the lesser depression found in *burgersi*. Center palm tubercles, in trench, numerous, few, or absent. The tubercles of upper palm are usually not as large as in *burgersi*; the ventral end of the beaded edge curving down from dorsal margin in many specimens turns sharply distally; it then extends across upper margin of the group of upper palm tubercles as a distinct short row, rather than merging with the group at its ventral end, as in *burgersi*; when present, this is an excellent character for the identification of *mordax*, but it is often absent or indistinct. Predactyl area excavate, particularly in large specimens, bent at a sharp angle to center palm, so that, in effect, the submarginal region of palm is distally partly flattened; in smaller specimens this character is unreliable. At base of dactyl, the proximal ridge always has tubercles smaller and more separated than those characteristic of *burgersi* (but not invariably found in that species), while in typical *mordax* the ridge diverges from the vertical throughout its length; again, exceptions occur, for in some individuals the ridge curves or is straight, after the initial divergence, much as in *burgersi*.

Ambulatories. Pile always present, strongly attached, on ventral margin of manus at least on 2nd and 3rd legs; this is the only character which the present material shows to be completely reliable, in distinguishing *mordax* from *burgersi*. In juvenile crabs pile develops on dorsal surface of manus before it appears ventrally. The distribution of pile in *mordax* in detail is as follows, in specimens where it is apparently undamaged and of maximum distribution: on first 3 legs, dorsal margins of meri and carpi, and all surfaces of mani; 4th leg with pile on same surfaces, but less extensive.

Gonopod. Inner process sometimes narrower and more pointed than in *burgersi*, while anterior flange does not project so far. As in all related species, however, these characters are variable in both *burgersi* and *mordax*.

FEMALE

Ambulatories with pile on ventral margin of manus appearing as reliably as in males. Suborbital crenellations little or not at all stronger than in males, so that it forms a usually good supplementary character distinguishing *mordax* from *burgersi*. Two groups of small tubercles dorsally on carapace, one behind each antero-lateral angle. Gonopore with a tubercle, very small but always distinct.

Measurements (in mm)

	Length	Breadth	Propodus	Dactyl
Largest male (Caripito)	16.5	25.5	50.5	35.5
Largest male cotype (Belém)	16.5	24.5	44.0	32.0
Moderate male (Caripito)	14.0	21.0	39.0	27.0
Largest female (ovigerous)	15.5	23.0	–	–
Smallest ovigerous female (Pedernales)	8.4	13.0	–	–

Morphological Comparison and Comment

The close similarities between *mordax* and *burgersi* have already been discussed (p. 170), as have its distinctions from *vocator*, summarized on p. 164 as well as in the key (p. 631), and the diagnoses. From *rapax*, the only other species in its range with which it might be confused, *mordax* is as usual notable for the pile in both sexes on the ventral side of the manus on the 2nd and 3rd ambulatories; in addition, the front is broader and, on the major palm, the proximal predactyl ridge diverges from the adjacent groove instead of running clearly parallel to it throughout.

From *minax* and *brevifrons*, the temperate Atlantic and tropical Pacific members of the superspecies, *mordax* and *burgersi* in the tropical Atlantic differ principally in the following general characteristics: degree of delicacy of integument and slenderness of appendages, *mordax* and *burgersi* occupying an intermediate position between the fragile, leptochelous *brevifrons* and the heavy-set *minax*; obliqueness of orbits, which are more slanted in *mordax* and *burgersi* than in either of the other species; details of armature of major manus; and degree of pilosity of ambulatories, there being most in *mordax* and least in *brevifrons*.

Color

Displaying males: Display whitening sometimes developed slightly, never strikingly. Carapace ranging from gray to yellowish gray to grayish white. Major cheliped: Usually yellowish except for white fingers. Minor cheliped with white fingers. Ambulatories brownish except, sometimes, for pale bluish on anterior sides of meri and sometimes of other segments. Females brownish.

SOCIAL BEHAVIOR

Waving Display

(Observed but not filmed at Caripito, Venezuela; observed and filmed, but at low intensities only, in rainy weather, at Puerto Barrios, Guatemala.)

Wave lateral, circular at least at high intensities. Jerks present, numbering 3 to 4 during rise and an equal number during descent. No pause at peak. Minor cheliped makes a corresponding motion during each wave. Body held high during a series of waves, and during each wave one or two ambulatories of either or both sides are raised and kicked outward at wave's apex. Duration of single waves about 5 seconds at low intensities but much faster, as little as 2 seconds, at highest courtship intensity; at these times each wave is followed by rapid vibrations of the 3rd, or 2nd and 3rd, ambulatories on each side, not distinguishable in the field notes from the acoustical leg-wagging now familiar in many species; the observed vibrations occurred definitely in connection with high intensity courtship, not agonistic behavior. Normally there is no pause between waves of a series; at highest intensity, however, a single wave, followed by these vibrations, is in turn followed by a period of motionless posing with the cheliped pointed straight overhead. These high intensity elements were seen during a number of observation periods at Caripito, the populations apparently being then, between February and April, at the peak of their breeding season. Unfortunately the very few individuals that have been observed waving since then were all in very small groups near the sea and displayed at low intensity only; none of the few resultant films and notes distinguish them adequately from the corresponding stages of *burgersi*. (Component nos. 5, 6, 9, 11, 12.)

RANGE

Tropical western Atlantic. Because of frequent confusions with *vocator*, *burgersi*, and *rapax*, *mordax* is known with certainty only from the following locations. All are continental, not West Indian, with the exception of Trinidad, which is geologically a portion of South America and where, in any case, the species is rare. Guatemala: Puerto Barrios; Nicaragua: Greytown; Colombia: Turbo; Venezuela; Trinidad: in northern and northeastern swamps; Guyana: Georgetown; Surinam; Brazil: several localities from Belém to Estado São Paulo, near Cananéia. (This southernmost record *fide* Gerlach, 1958.1; his material was identified by Bott.)

BIOTOPES

Only two large populations have been found, both on mudflats along the shores of large, tidal rivers; the largest groups were above the levels of mangroves where the water is practically or wholly fresh. One was at Belém, the type locality, and the other at Caripito, Venezuela. The few other records are all from near the mouths of rivers, and only a few individuals were found in each. It seems likely that the young during the swimming stages are not usually pelagic and that the majority may even develop far upstream. (Biotope nos. 11, 12, 14, and, especially, 15.)

SYMPATRIC ASSOCIATES

In optimum habitats upriver, *mordax* is the only *Uca*. Its occasional occurrence near the sea is usually in association with *burgersi* and *rapax*. In Puerto Barrios, Guatemala, *mordax*, *burgersi*, and *rapax* all occurred at one end of the town, but the small population of *mordax*, which included several displaying males, was farthest inland; that of *burgersi* came next, and then, closest to the bay, were large numbers of *rapax*, the three species maintaining in miniature their usual ecological relations to one another.

MATERIAL RESULTING FROM FIELD WORK

(The complete list of specimens examined is given in Appendix A, p. 604.)

Observations and Collections. Guatemala: Puerto Barrios. Venezuela: San Juan R., at Caripito and near its mouth (Monagas); Pedernales (Monagas). Trinidad: near Blanchisseuse (north coast); near mouth of Oropouche R. (east coast). Guyana: Georgetown. Surinam: near mouth of Marawyne R. Brazil: Belém.

Films. Puerto Barrios.

TYPE MATERIAL AND NOMENCLATURE

Uca mordax (Smith, 1870)

TYPE MATERIAL of *Gelasimus mordax*. In Museum of Comparative Zoology, Harvard University, Cambridge, Massachusetts. Cotypes: no. 5882; 7 males, 5 females. Type-locality: canals at Pará (= Belém), Brazil. October 1858. Collected by Caleb Cook. Received from Peabody Academy of Sciences, November 1885. Measurements on p. 174. (!)

REFERENCES AND SYNONYMY

Uca (Minuca) mordax (Smith, 1870)

Gelasimus mordax

Smith, 1869.1: 35. Pará, Brazil. Name only.

TYPE DESCRIPTION. Smith, 1870: 135; Pl. 2, Fig. 3; Pl. 4, Figs. 4, 4a. Canals at Pará (= Belém), Brazil. (MCZ !)

Uca minax (not *Gelasimus minax* LeConte)

? Pearse, 1916 (part): 532, 554. Colombia: Santa Marta; specimens from Punta Guesa. Habitat. (Synonymy suggested on basis of habitat.)

Uca mordax

Rathbun, 1918.1: 391 (part); Text Fig. 166 (claw after Smith); not Pl. 34, Figs. 3, 4. Not most localities. (USNM, part !)

Crane, 1943.1: 31; Text Figs. 1a-c; Pl. 1, Figs. 3, 4. Taxonomy.

Crane, 1943.2: 37: Eastern Venezuela. Taxonomy; color; waving display; habitat. (USNM !)

Crane, 1957 (part). Guatemala, Venezuela. Preliminary classification of waving display.

Gerlach, 1958.1: 672. Brazil: Estado São Paulo (near Cananéia).

Holthuis, 1959.3: 262; Text Figs. 64a, b, c; Pl. xiv, Fig. 2; Pl. xv, Fig. 2. Suriname (many localities). Taxonomy (including distinctions from *Gelasimus affinis* Streets, described by Holthuis 1967 as *Uca burgersi*); habitat.

Barnwell, 1963. Illus. Brazil. Endogenous daily and tidal rhythms of melanophore and motor activity.

von Hagen, 1967.2. Trinidad. Tape-recordings secured. (Preliminary statement.)

von Hagen, 1970.4. Trinidad. Adaptations to a particular intertidal level.

30. *UCA (MINUCA) [MINAX] MINAX* (LE CONTE, 1855)

(Temperate western north Atlantic)

PLATE 25 *E-H*. MAP 12.
FIGURES 67 *D*; 81 *K*; 100. TABLES 9, 10, 12, 20.

INTRODUCTION

This large fiddler, *Uca minax*, is more completely confined to the temperate zone than any other member of the genus, since even the ranges of *pugnax*, *pugilator*, and *arcuata* extend into the subtropics of Florida and China. For this and other reasons *minax* would repay more special study than it has yet received. For example, the crabs seem to be as well adapted to fresh water as their close allopatric relations, *mordax*, in the rivers of the tropical Atlantic mainland, and *brevifrons* in the tropical Pacific; yet, of the three, only the integument of *brevifrons* is particularly fragile, while that of *minax* is notably thick and tough. Of special interest, as usual, would be a study of its social behavior in comparison with that of other temperate or partly temperate members of the genus, including *tangeri*. All of these species so far studied, through the work of Altevogt (1962, 1964.1) and von Hagen (1962), as well as that of Salmon and his associates (e.g. Salmon & Stout, 1962; Salmon, 1967; Salmon & Atsaides, 1968.2 and refs.), show that these northern populations have nocturnal breeding behavior well developed, with acoustical components, and sometimes special acoustical behavior takes place by day in localities where the vision of individuals is obscured by vegetation.

Such a study will not be easy in *minax*. The populations are most widespread in large marshes and it is there, accordingly, that the social structure can be expected to be most characteristic of the species. These areas of course are usually not conveniently explored except by boat, and a boat, it has always seemed to me, is to be avoided whenever possible for fiddler study. At worst it limits personal movement and at best it is a nuisance that has to be tied up, anchored, or otherwise dealt with at crucial moments of observation. Again, because of intervening vegetation, the smaller movements of the crabs during waving display and, doubtless, of sound production are often obscured. Furthermore, except at intervals controlled by rhythms so far undetermined, individuals seem to devote almost as little time to waving as do some of the Indo-Pacific *Deltuca*. The large size of *minax* is another characteristic that interferes with ethological study in such a lethargic species, since territories are larger than in, for example, *pugnax*, so

that even with good visibility fewer displaying males can be observed at once; finally, the breeding season, as is unfortunately always the case in temperate latitudes, is frustratingly short.

Because the species is still available and even plentiful conveniently near to a number of biological stations on the eastern seaboard of the United States, the outlook is hopeful that some of the desirable ethological and physiological work on this species may soon be undertaken with the requisite enthusiasm. There is considerable need for haste, since great marshes still wholesome enough for *minax* are particularly subject to all the usual hazards—pesticide runoffs from adjacent agricultural lands, pollution from factories upstream, and drainage for reclamation.

MORPHOLOGY

Diagnosis

Front extremely broad; eyebrows almost vertical, widest part at least equal to diameter of adjacent part of depressed eyestalk, as measured through its smaller dimension. Major cheliped with outer upper half of manus coarsely tuberculate; proximal ridge of palm diverging very slightly in a straight or curved line from adjacent groove at base of dactyl; oblique ridge not especially high beside carpal cavity; center palm variably tuberculate. Merus of ambulatories slightly enlarged. Gonopod with inner process narrow; posterior flange almost as wide as anterior. Female with tubercles on carapace near antero-lateral angles.

Description

With the characteristics of the subgenus *Minuca* (p. 154). Closest to *burgersi*, *mordax*, and *brevifrons*. In addition to its much larger size, it differs as follows from *burgersi*.

MALE

Carapace. Front contained clearly less than 3 times in carapace breadth between antero-lateral angles; orbits scarcely oblique. Eyebrow almost vertical, short, and equal to or exceeding smaller dimension of adjacent part of depressed eyestalk, its inner margin strongly concave.

Major Cheliped. Manus with tubercles of entire upper half of outer surface large and outstanding, more so than in any other *Minuca.* In large specimens the subdorsal area is strongly bent and flattened. Ventral marginal keel and groove always distinct, although its tubercles are obsolescent or absent in large examples. Palm with oblique ridge surmounted by a more or less regular row of large tubercles, always well developed, the apex not sharply higher than rest of ridge; tubercles, with uncommon individual exceptions, ending abruptly on apex, not continued upward around carpal cavity. Center palm near oblique ridge smoother than in *burgersi,* but tubercles of its upper part similar and usually conspicuous. At dactyl base, proximal ridge diverges only very slightly from adjacent groove; this diversion is usually a straight line, rarely a curve; the straightness in most resembles that of most *mordax,* although the tubercles of *minax* are stronger; unlike those of *burgersi* they are never so nearly contiguous as to appear beaded. Pollex with a variable, outer, subdistal crest about as in the superspecies *galapagensis,* but in addition with the most distal tooth on inner row enlarged.

Ambulatories. Traces of pile dorsally on meri. Otherwise as in *burgersi,* pile on ventral margins of mani being absent.

Gonopod. Posterior flange wider than in *burgersi* and slightly produced.

FEMALE

Dorsal part of carapace near each antero-lateral angle with a patch of tubercles varying in size but always larger than the minute granules covering other regions. Gonopore with only a very small posterior tubercle; the dome-like end of the spermatophore, when present, protrudes externally, filling the cavity and resembling a large tubercle.

Measurements (in mm)

	Length	Breadth	Propodus	Dactyl
Large male (brachychelous)	23.0	33.0	50.0	33.5
Moderate male (leptochelous)	18.0	27.0	42.0	33.0
Large female	22.0	30.0	–	–

Morphological Comparison and Comment

U. minax shares with only two other fiddlers the long stretch of American coastline from north Florida to Cape Cod. It is usually distinguished at once from both of them, *(Minuca) pugnax* and *(Celuca) pugi-*

lator, by its much larger size, broader front, and shorter, wider eyebrow. In addition, *minax* differs from *pugnax* in the extent of the oblique ridge on major palm, which does not continue up around carpal cavity. From *pugilator* it differs in both sexes by the presence of pile on the 1st, 2nd, and 3rd legs, on the dorsal parts of each carpus and manus; additionally, the major palm in *minax* has a well-developed oblique ridge but lacks a ridge along outer pollex above ventral margin.

The tuberculation of the distal part of the pollex and the form of the proximal predistal ridge show characteristics intermediate between those of the superspecies, *minax* and *galapagensis.* Otherwise the affinities are with *burgersi, mordax,* and *brevifrons.*

On the Gulf coast male *minax* can be distinguished at once from those of *rapax longisignalis* by the absence of pile on the lower surfaces of the ambulatory carpi and mani; *minax* in addition is, so far as known, everywhere a larger species and the front is broader.

Color

Displaying males: Display whitening moderately developed, the lightest color of carapace ranging from grayish white to dull yellowish white; frontal and eyebrow regions sometimes dull orange; middle of cardiac region sometimes with a red spot; major cheliped entirely grayish orange to dull yellowish white except for distal and proximal margins of segments, especially at junction of carpus with manus, which are usually edged narrowly with red; minor cheliped with carpus, manus, and chela white; ambulatories dark. Female similar to male.

SOCIAL BEHAVIOR

Waving Display

Wave lateral, weakly circular. Cheliped on rise usually makes 1 to 3 jerks, sometimes none; on the descent jerks are always present and number about 3 to 6, usually 4; the direction of the wave is often outward, with little rise; during a series, cheliped does not return to ground. During each wave minor cheliped makes a corresponding motion that is unusually vigorous for this appendage and often even includes jerks; very often the motion is strikingly unsynchronized with that of the major, descending, for example, as the major rises. During a series body is raised on extended anterior ambulatories, while the posterior legs are bent, especially when the display is directed toward an individual behind the waving fiddler; the carapace is thus presented almost vertically and the high reach of the cheliped—never at these times unflexed outward—increased. One or both middle ambulatories are often raised on each side and waved but not, unlike high intensity in *mordax*, vibrated. A single series of displays may continue for minutes at a time, but the time between series is indefinite and usually prolonged. Any high intensity components of advanced courtship remain unknown. Duration of a single wave about 2 to 3 or more seconds. (Component nos. 5, 6, 9, 11, 12.)

Hood Construction

During the summertime *minax* builds semi-circular hoods beside the entrance to a burrow. Long known as "ovens" to naturalists and fishermen, they have not yet been described in detail.

RANGE

Temperate western Atlantic along the eastern and southern coasts of the United States, from the southwest shore of Cape Cod (Silver Springs) to northeast Florida, and on the coast of the Gulf of Mexico from Yankeetown in northwest Florida (Salmon, 1967) to Louisiana. A few specimens from the Gulf states, including Texas, have been referred in the past to *minax*, usually tentatively (e.g., Rathbun, 1918.1), which proved on reexamination to be examples of *rapax longisignalis* or *pugnax virens*. The single locality record (MCZ 5900) from Key West is questionable.

BIOTOPES

Marshes threaded by streams, both close to the sea and farther inland, the water ranging from brackish to fresh but always subject to tidal influence; the burrows occur both in the muddy banks of the marsh creeks and among vegetation, *Spartina* and *Salicor-nia*, and freshwater herbs on top of the banks. (Biotope nos. 14, 15, 16.)

SYMPATRIC ASSOCIATES

None of the other temperate American *Uca* occur so far inland, in almost or completely fresh water, and it is in such marshes that the largest populations congregate. Occasionally, however, *minax*, especially when immature, is found in higher salinities scattered among *pugnax* and even, wholly atypically, *pugilator*. In certain juxtapositions of sheltered marsh and adjacent inlet with shores of both mud and sand, the three northern species sometimes live close together in moderate numbers, but with little or no peripheral mingling.

MATERIAL RESULTING FROM FIELD WORK

(The complete list of specimens examined is given in Appendix A, p. 604.)

Observations and Collections. Connecticut; New York (Long Island); New Jersey.

Film. New Jersey (kindness of Dr. Don C. Miller).

TYPE MATERIAL AND NOMENCLATURE

Gelasimus minax LeConte, 1855

TYPE MATERIAL. In Academy of Natural Sciences, Philadelphia: cat. no. 3581. 2 males, 1 female. Dried. Poor condition. Since at present the identity and characteristics of the species are not in question, there seems to be no need for selecting a lectotype, or, possibly, neotype. The specimens are labeled "Dennis Creek, N.J.," although LeConte's type-locality is given as Beesley's Point. (!)

REFERENCES AND SYNONYMY

Uca (Minuca) minax (LeConte, 1855)

Note. U. minax is one of the species of *Uca* that has been used occasionally in laboratory investigations of endogenous rhythms. Key references appear on p. 448.

Gelasimus minax

TYPE DESCRIPTION. LeConte, 1855: 403. U.S.A.: New Jersey. (Philadelphia !)

Smith, 1870: 128; Pl. 2, Fig. 4; Pl. 4, Figs. 1, 1a, 1b. Taxonomy.

Verrill, 1873: 337, 467, 545. U.S.A.: Massa-chusetts. General (not local) distribution; basic ecology and habits (no display or combat).

Verrill & Smith, 1874: 43, 173, 251. A reissue of Verrill, 1873.

Kingsley, 1878: 321. U.S.A.: Georgia. Local habitat.

Gelasimus palustris

? Stimpson, 1859: 62 (part). Probably most of records in U.S.A. Taxonomy; habitat.

Uca vocator, var. *minax*

Ortmann, 1897: 353. Taxonomy.

Uca minax

Rathbun, 1902.1: 583, 585. Key. Habitat noted to extend up rivers to fresh water.

Rathbun, 1905: 1. Northeast U.S.A. Local distribution.

Rathbun, 1918.1: 389; Pl. 137. Eastern and southern U.S.A. Taxonomy. (USNM !)

Hay & Shore, 1918: 451. U.S.A. North Carolina. Taxonomy; color; ecology; habits, including hood construction and breeding season.

Hyman, 1920. Illus. U.S.A.: North Carolina. Postembryological development.

Hyman, 1922. Illus. U.S.A.: North Carolina. Habits, including spawning.

Pearse, 1928: 231ff. U.S.A.: North Carolina. Ecology; tests on toleration of desiccation and fresh water.

Maccagno, 1928: 48; Text Fig. 31 (claw). U.S.A.: Virginia. Taxonomy. (Torino !)

Gray, 1942. U.S.A.: Maryland. Color; ecology; habits.

Crane, 1943.3: 220; Text Fig. 1b. U.S.A.: northeastern states. Taxonomy; color; waving display.

Crane, 1957. U.S.A.: New Jersey. Preliminary classification of waving display.

Teal, 1958. U.S.A.: Georgia. Local distribution; ecology; experiments on tolerances for different foods, temperatures, and salinities, as well as on substrate preferences; comparisons with sympatric species.

Teal, 1959. U.S.A.: Georgia. Respiration and its relation to ecology.

Miller, 1961. U.S.A.: North Carolina. Feeding mechanisms and adaptations.

Miller, 1965. U.S.A.: Morphology, physiology, and ecology in relation to distribution.

Salmon, 1965. U.S.A.: North Carolina. Waving display and sound production in courtship.

Salmon, 1967. U.S.A.: Florida. Local distribution.

31. *UCA (MINUCA) [MINAX] BREVIFRONS* (STIMPSON, 1860)

(Tropical eastern Pacific)

PLATE 26 *A-D*.
FIGURES 24 *F-H*; 30 *E, F*; 32 *G-I*; 34 *C*; 35 *E, F*; 36 *D*;
   37 *J*; 45 *M-OO*; 46 *I*; 47 *C, D*; 48 *C, D*; 66 *H*; 100.

MAP 13.
TABLES 9, 10.

INTRODUCTION

Once in Costa Rica I saw a fiddler up a tree. She was sitting at eye level on a branch, her coral red carapace set off by the glossy green of a jungle vine behind her. Her perch was surrounded by forest, no stream was close by, and the shore was at least a kilometer away. She was the only tree-climbing *Uca* I ever saw. The fact that the forest had been soaked the night before by the first heavy rains of the season helps explain the crab's location, as does the preference of *Uca brevifrons* for water apparently wholly fresh. One small group was collected more than 5 kilometers from the coast, in hilly country, far above tidal influence.

The fragility of the carapace and appendages in this species may be due to the low mineral content of the water, the structure of the integument in the genus—unlike that of pseudothelphusids, for example—being perhaps poorly adapted for non-marine habitats.

The species' social behavior remains wholly to be investigated.

MORPHOLOGY

Diagnosis

Front very broad. Major cheliped with proximal predactyl ridge on palm diverging from distal ridge; oblique tuberculate ridge vestigial in lower portion, but very high at apex; upper, distal beaded ridge that usually bounds carpal cavity vestigial to absent, the cavity continuing as a trench into predactyl area; distal half of pollex with a series of enlarged teeth. No pile dorsally on carapace and very sparse on ambulatories which, in male, are notably slender. Gonopod without tooth or spine at base of anterior flange, which is very wide on the outer side, due to torsion, and strongly concave. Gonopore usually with one small tubercle on anterior lip.

Description

With the characteristics of the subgenus *Minuca* (p. 154). Closest to *burgersi*, *mordax*, and *minax*. The great height of its oblique ridge on palm, the obso-

lescent beaded edge bounding upper carpal cavity, the enlarged serial tubercles on pollex, the slender legs, and minimal pile make it less similar to these Atlantic species of the superspecies than they are to each other. In more detail, it differs as follows from *burgersi*.

MALE

Carapace. Eyebrow more strongly bent in proximal portion, though not as much as in *minax*; the distal part is slightly concave and projecting, bringing into prominence the strong beading of the edge; proximal margin a smooth ridge, as in the other members of the superspecies, but stronger, while the postorbital depression is deeper than usual. Postero-dorsal margins of carapace relatively weak, as are the postero-lateral striae and vertical lateral margin of carapace. Floor of orbit and suborbital region in large specimens with profuse setae which nearly obscure the weak inner suborbital crenellations, and among which the stronger external group only partly emerge; in this character *brevifrons* approaches *mordax*, rather than *burgersi* or *minax*.

Major Cheliped. Keel and groove along ventral margin of manus obsolescent. Palm with oblique ridge low near pollex, but elsewhere very high, narrow, highest at apex, and, except near pollex, surmounted by a single row of tubercles; these tubercles continue barely over peak of apex, stopping abruptly at or before end of ridge at carpal cavity; distal margin of carpal cavity low and blunt; center palm smooth and flat or slightly convex between carpal cavity and hollow at base of pollex; upper part of carpal cavity with low tubercles, usually less distinct than in *burgersi*. Predactyl area consisting of an elongate depression, smooth or with scattered tubercles, its breadth diminishing distally, completely confluent with upper carpal cavity, as in some *Celuca*; beaded inner edge of proximal dorsal margin, so characteristic of *Minuca* and *Celuca*, is either absent in *brevifrons* or represented by a short minutely beaded edge which does not curve downward; more distal portion of inner, dorsal margin not distinctly bounded, there being no trace of the row of tubercles above the predactyl area, often present in *burgersi* and others of the superspecies. Proximal ridge at dactyl base diverging from distal in an approximately straight line, the degree of divergence being less than in *burgersi* and *mordax*, but slightly more than in *minax*; tubercles smaller than in most *burgersi*, never contiguous. Pollex in distal half with about 6 to 8 enlarged tubercles, well separated; an equivalent row, less enlarged, usually present in distal part of dactyl; these are in addition to the usual two teeth that often divide the dactyl into thirds. Dactyl with proximal subdorsal groove usually either short or weak, the dorsal tubercles poorly developed; inner surface of dactyl with a faint, broad depression; a similar one on pollex is shorter, being a continuation of the depression at base of pollex.

Ambulatories. Meri notably more slender than in *burgersi* and the related species. Pile much sparser, though traceable in same areas as *burgersi*, and as in that species wholly absent from lower side of manus.

Gonopod. Tip strongly bent over, especially in large individuals. Torsion moderate, so that anterior flange in position is slightly the more posterior. Inner process somewhat tumid proximally, especially in large specimens, similar to its appearance in *vocator* but not distally truncate; instead, it tapers distally to a rounded tip which alone is freely moveable; process does not project beyond pore and does not cover anterior flange. No tooth or spine, characteristic of *vocator*, at base of flange. Anterior flange very wide, strongly concave on the side directed posteriorly, and distally truncate; truncation more or less oblique, the highest part being adjacent to pore and extending beyond it little or not at all; posterior flange very narrow, its free side diminishing distally in width, to end against pore in an edge either obliquely truncate or pointed. Thumb in large specimens in contact with inner process; in all specimens it reaches flange base or somewhat farther.

FEMALE

Carapace dorsally with an elongate group of tubercles bordering each antero-lateral margin; merus of 4th ambulatory very slender, its dorsal margin straight. Gonopore usually with one small tubercle on anterior lip. When spermatophore present, two conspicuous tubercles project, the larger posterior, the anterior small and slender.

Measurements (in mm)

	Length	Breadth	Propodus	Dactyl
Largest male	20.0	28.0	55.5	41.0
Moderate male	15.5	23.0	30.5	22.0
Largest female	19.0	27.0	–	–
Ovigerous male	15.0	20.5	–	–

Morphological Comparison and Comment

The absence of the down-curving, beaded edge subdorsally on the major palm is the most useful, single character for diagnosis, readily separating male *brevifrons* from other Pacific *Minuca*. A general comparison with the Atlantic relations of *brevifrons* is given on p. 174.

Color

Two males and one female, alive and recently caught in different localities, were dissimilar. Males: Carapaces, respectively, dark brown and reddish brown with fine, black marblings; major cheliped brown to orange pink, one or both fingers orange, finger tips white; upper ambulatories like carapace. Female: carapace and upper parts of appendages bright coral red (orange red) except for white finger tips.

RANGE

Tropical eastern Pacific from Todos Santos Bay, Lower California, to Darien in Panama.

BIOTOPES

Muddy and clayey banks of freshwater and brackish streams, close to water level; also, as described in the introduction to the species, taken above tidal levels and, once, in wet forest, 1.5 meters above the ground and many meters from the nearest stream. (Biotope nos. 14, 15.)

MATERIAL RESULTING FROM FIELD WORK

(The complete list of specimens examined is given in Appendix A, p. 604.)

Observations and Collections. Mexico: Puerto Angeles. Costa Rica: Port Parker; Gulf of Nicoya; Uvita Bay; Golfito; Parida I. (Observations minimal.)

TYPE MATERIAL AND NOMENCLATURE

Gelasimus brevifrons Stimpson, 1860

HOLOTYPE. Not extant. Type-locality: Lagoon at Todos Santos near Cape San Lucas, Lower California.

Additional type material. In Museum of Comparative Zoology, Harvard University, Cambridge, Massachusetts: cat. no. 1332: 1 female, length 17 mm from "type lot," from Smithsonian Institution; Cape San Lucas, Lower California; collected by John Xantus. Specimen fragile, but shows described characteristics of female *brevifrons* well, including those of gonopore with attached spermatophores. (!)

Type of *Uca brevifrons* var. *delicata* Maccagno, 1928. In Regio Museo Zoologico di Torino. One male, from Rio Lara, Darien, Panama. The lepto-chelous form of the major cheliped, illustrated by Maccagno in Text Fig. 33, falls within the range of variation found in the present material. (!)

REFERENCES AND SYNONYMY

Uca (*Minuca*) *brevifrons* (Stimpson, 1860)

Gelasimus brevifrons

TYPE DESCRIPTION. Stimpson, 1860: 292. Mexico: Cape San Lucas, Lower California.

Smith, 1870: 131. Taxonomy of type.

Lockington, 1877: 147. Mexico: Magdalena Bay, Lower California.

Pesta, 1931: 180. Costa Rica.

Gelasimus vocator (not of Herbst)

Nobili, 1897: 3. Panama: Darien: Rio Lara. (Torino !)

Uca brevifrons

Holmes, 1904: 308; Pl. 35, Figs. 1-5. Mexico: San José del Cabo, Lower California.

Rathbun, 1918.1: 395; Pl. 139 (photo). Eastern Pacific. Taxonomy. (USNM !)

Crane, 1941.1: 177 (part); Text Figs. 4f, 5; Pl. 7, Fig. 35. Mexico, Costa Rica, Panama. Taxonomy; color; habitat. (Not 1 male, cat. no. 381,120, from San Juan del Sur, Nicaragua nor 1 male, cat. no. 381,122 part, from Port Parker, C.R.) (USNM !)

Garth, 1948: 60. Panama: Piñas Bay. (AMNH !)

Holthuis, 1954.1: 41. El Salvador. Habitat.

Holthuis, 1954.2: 162. El Salvador.

Uca brevifrons var. *delicata*

Maccagno, 1928: 51; Text Fig. 33. Type description of new variety from specimen referred by Nobili (1897) to *Gelasimus vocator*. (Torino !)

32. *UCA (MINUCA) [GALAPAGENSIS] GALAPAGENSIS* RATHBUN, 1902

(Tropical eastern Pacific)

PLATE 26 *E-H*.   MAP 14.
FIGURES 67 *A*, *B*; 100.   TABLES 9, 10, 12, 19, 20.

INTRODUCTION

Uca galapagensis has long been better known under the early name of *macrodactylus*. As with a number of species in its subgenus, its morphological idiosyncrasies have led to cumulative confusions. On the one hand, its variability and changes during growth invited nomenclatural splitting. On the other, valid species distinctions, as well as bases for subspecies, appear minor and difficult to recognize without series of specimens which were usually unavailable to workers in the group; these conditions led to confusions with other species.

U. *galapagensis* is morphologically so similar to *rapax*, its close allopatric relation in the tropical Atlantic, that single examples of the two species, even when examined under the microscope, can easily be confused if the investigator is not right in the middle of extended work on the group. The similarities also include general characteristics of waving display, absence of structure-building activities, and selection of biotopes. The relationship is in fact undoubtedly so close that there would be justification for considering the two forms as transisthmian subspecies, as has been done with *thayeri*, *maracoani*, and *vocator*. The wide distribution, variability, abundance, and familiarity of *rapax*, along with its continuing taxonomic difficulties, seem to make this course undesirable in the present state of our knowledge. In this contribution, therefore, with a certain lack of logic full specific rank will be maintained for *galapagensis* and *rapax*.

The limited information now at hand shows clear differences in the waving displays of the two subspecies of *galapagensis* here recognized. A comparison of these displays with those of related *Uca* in each locality can be expected to show that, as in some other allopatric groups sufficiently well known, the display differences distinguish each subspecies from the displays of its close sympatrics, which differ somewhat in the two ranges. Our knowledge of the species, as well as that of its close relations, is still insufficient for further useful discussion of this problem. Concentration on the *Minuca* in only several localities would certainly lead to excellent results, even if waving display alone were adequately explored. If combat behavior were simultaneously investigated, we would have a good chance of understanding some

of the most basic factors in the evolution of this taxonomically irritating subgenus.

MORPHOLOGY

Diagnosis

Front only moderately wide, clearly less than one-third carapace width. Major cheliped with proximal and distal ridges inside dactyl base parallel, the proximal not diverging; oblique tuberculate ridge on palm strong throughout, usually with tubercles in a double row for a short distance toward middle; apex high; carpal cavity cut off from predistal area by a down-curving, beaded margin; several distal tubercles on pollex forming a small crest. Gonopod with inner process narrow. Carapace without pile dorsally.

It differs from its closest relations, *rapax* and *pugnax*, both in the Atlantic, as follows: from *rapax* in having the center palm tuberculate to various extents but never almost smooth, in the continuation of the tubercles of the oblique ridge upward around carpal cavity, in having postorbital groove on carapace moderately, not scarcely, developed; and in having upper margin of merus on 2nd ambulatory not strongly convex. From *pugnax* it differs in having the oblique ridge on palm with a high apex, the postorbital groove on carapace moderately strong, and the upper margin of 2nd ambulatory not practically straight in the middle. Finally the female differs from that of the other two species in having no tubercle beside the gonopore.

Description

With the characteristics of the subgenus *Minuca* (p. 154).

MALE

Carapace. Front contained about 3.5 times in carapace breadth between antero-lateral angles. Orbits scarcely to moderately oblique and lateral margins little to moderately convergent. No pile on antero-lateral region, but sometimes a few small tubercles. Eyebrow broader than in related Atlantic species. Beading of distal margin only of eyebrow distinct, that of proximal edge being faint to absent. Suborbital crenellations well formed throughout, not obscured by the sparse setae; pile absent.

Major Cheliped. Manus with tubercles of upper, outer surface usually moderately enlarged, but sometimes smaller and fewer in large specimens. Palm always with oblique ridge high and marked with large tubercles. The row is clearly linear, but most specimens have an irregularity toward middle of ridge, usually a slight break with the proximal part of the row continuing above the distal part, so that they overlap for the space of one to several tubercles, forming a short double row; apex sometimes strikingly high, sometimes slightly lower than the adjacent, more distal tubercles; always, however, apex, ridge, and tubercles are notably strong; tubercles on apex either single or in a small cluster; they continue, reduced in size, variable, and usually well separated upward around distal edge of carpal cavity; tubercles of center palm moderate in size and number, variable, but never large and crowded, sometimes almost lacking in large specimens; beaded inner edge of dorsal margin curving strongly downward, clearly cutting off carpal cavity from predactyl area, merging with tubercles of center palm; proximal predactyl ridge parallel to adjacent groove, not at all divergent, tubercles of both ridges strong except in some large specimens, those of proximal ridge closely set but not contiguous. Pollex tip with the usual two distal teeth enlarged, one from outer row, one from inner; however, no third tooth, subdistal to them and marking end of median row, gives the trifid effect found in most *Minuca*; instead, near its end the median row in effect swerves outward, the external row being there interrupted and absent proximal to the small distal tooth; the median row now consists of 3 to 4 much enlarged tubercles, the most proximal always largest except where there are 4 in the series; then usually the next to the proximal is largest; all are conspicuous, contiguous or nearly so, and give the effect of a toothed crest.

Ambulatories. Meri of 2nd and 3rd ambulatories moderately enlarged, the dorsal margins slightly convex to almost straight. Pile present or absent on dorsal surfaces of carpus and manus of first 3 legs; it is always altogether absent from ventral margins, including that of manus.

Gonopod. Inner process not at all tumid, narrow, little tapering, the tip rounded, reaching almost to or just beyond pore, not at all covering anterior flange, usually lying more or less obliquely against posterior flange and canal. Anterior flange either slightly broader or slightly narrower than posterior, not strongly curved, its distal edge ranging from produced and spinous to truncate; no tooth at base of flange. Thumb usually reaching to or beyond base of flange, but in smaller specimens not attaining it.

FEMALE

Pile as in males. Gonopore rim without tubercle.

Measurements (in mm)

	Length	Breadth	Propodus	Dactyl
galapagensis herradurensis (Panama)				
Longest male	16.0	23.0	33.0	
Male with largest claw	14.5	23.0	40.0	
Large male	14.5	22.0	36.0	
Largest female	10.0		–	–
galapagensis galapagensis (Puerto Bolivar)				
Largest male	14.5	22.0	(claw broken)	
Large males	14.0	21.0	37.0	
	13.0	20.0	30.0	
(Galapagos, fide Garth, 1946)				
Largest male	14.5	22.1	38.6	28
Largest female	13.5	19.6	–	–

Note. The above figures indicate that, contrary to Bott's suggestion, the relative breadths of the carapace in the two forms are similar, when crabs of comparable lengths are considered.

Morphological Comparison and Comment

The first paragraph under the heading Diagnosis summarizes the points by which *galapagensis* can most readily be distinguished from other broad-fronted species from the eastern Pacific, particularly its closest Pacific relations—*vocator*, *brevifrons*, and *zacae*. The second paragraph indicates the apparent scarcity of reliable morphological differences between *galapagensis* and the Atlantic members of its superspecies—*rapax* and *pugnax*.

It is of interest to note that the differences between the two subspecies of *galapagensis*, as will be apparent under the descriptions of the two forms, are slightly more extensive than those between *galapagensis* and its transisthmian allopatric, *rapax*. Furthermore, the northern subspecies, *g. herradurensis*, appears even more similar to *rapax* than does *g. galapagensis*.

Apparently no long series of specimens from the Galapagos has yet been preserved. Such a collection would be of special interest because of the evidence given below that some individuals from those islands show resemblances to both subspecies. Also, the Galapagos specimens examined were all smaller than those on the mainland, although the major chelipeds showed proportions characteristic of larger crabs in Ecuador.

Of the two subspecies, *g. herradurensis* has a more juvenile form, in particular with the carapace less arched, and less adapted therefore for water conservation. The crabs I collected in both localities in Ecuador came from a much drier habitat than did those in Panama and Costa Rica, and in Ecuador alone their disappearance underground during neap tides was established.

The form of the gonopod in this species, as is often the case in *Minuca*, makes an unsatisfactory taxonomic character, particularly when only several specimens are available from a given population. Not only is the gonopod variable, but the distal part of the anterior flange, which is diagnostic in distinguishing the subspecies, seems subject to breakage, possibly even during copulation.

A principal source of taxonomic confusion in the past has been the degree of obliqueness in the orbits, those in juveniles being markedly slanting, while in large adults the orbits and fronts make an almost straight transverse line. From time to time in the literature, therefore, the young have not only been considered different species or subspecies from adults but they have been confused with *U. zacae*. Differences between young *galapagensis* and *zacae* are listed on p. 207.

Color

Display whitening moderately to strongly developed, sometimes preceded by an orange phase.

SOCIAL BEHAVIOR

Waving Display

The following account is derived, first, from von Hagen's analysis (1968.2: 420) of a film made by R. Altevogt at Guayaquil, Ecuador; from my field observations made in Puerto Bolivar, Ecuador; and, third, from my field observations and films made in Golfito, Costa Rica. Also at hand for comparison is a film made by I. Bauman in the Galapagos; except as noted below details requisite for special comment do not show in the low intensity waves of the filmed example. For details of subspecific differences in display see pp. 187, 188, and 658 (Table 20).

Wave lateral, weakly circular, the circularity often apparent only at high intensities. Jerks present on both rise and descent of cheliped except at lowest intensity in Puerto Bolivar, where none was detected; in some of the field observations in Costa Rica the jerks were barely detectable to the eye, but all showed up clearly during film analyses; in Bauman's Galapagos film they are scarcely indicated. Jerks on the cheliped's rise ranged from 1 to 2 in the southern subspecies to 4 to 8 in the north, with fewer on the descent. A strong pause, of variable duration, at apex of wave in the southern subspecies is present in all intensities; at highest intensity the cheliped tip vibrates during the pause (von Hagen). In the north-

ern subspecies there is no pause at apex, but, instead, a definite double peak with jerks between them. In the Galapagos film there is no trace either of a double peak, a pause, or vibration at the apex. In neither subspecies is the cheliped brought to ground between waves of a series except, in the southern form, at high intensities where it lands in front of usual position and returns to place through several small, jerky motions (von Hagen). Minor cheliped makes a corresponding motion not well synchronized with wave and sometimes (Crane films and observations) showing 2 circular motions of the minor cheliped to 1 wave of the major; these motions probably involve sound production, as noted below under acoustic behavior. Body held high during waves of a series; sometimes only the anterior ambulatories are fully extended so that the carapace is held almost vertically. Except at lowest intensities 1 to 3 ambulatories on each side are raised high, but not in simultaneous motions, on each side with each wave; at high intensities their activity is such, although not in a regular pattern, that a rocking effect is achieved; this action is quite different from a curtsy, the common component attained through bending of the ambulatories; meanwhile several steps are usually taken from side to side, further enhancing the complexity of the display. Curtsy component absent. Duration of a single wave about 1 to 2 seconds. (Component nos. 5, 6, 9, 11, 12; timing elements in Table 19, p. 656.)

Acoustic Behavior

Analyses of films made in Golfito, Costa Rica, showed the following components: major-merus-rub (component 1), minor-merus-rub (2), and leg-wagging (5). All were closely associated with some of the waving displays previously described, at times when the individuals were showing signs either of agonistic behavior toward another male or of awareness of disturbance, apparently caused by the proximity of the camera. These signs included above all frequent displacement feeding and displacement cleaning.

RANGE

Tropical eastern Pacific, certainly from Los Blancos, El Salvador, to Puerto Grande, north Peru; Galapagos Islands. Single, questionable specimens also taken in Mexico, at Guaymas in Sonora, and at Tenacatita Bay in Jalisco. Perhaps formerly occurring in Chile.

BIOTOPES

Mud flats near mouths of streams, near mangroves; also, farther upstream, on flats behind fringing mangroves; sometimes on large flats of muddy sand, flooded only at spring tides and inhabited by no other species of *Uca*. Garth (1946) writes of populations in the Galapagos: "*U. galapagensis* Rathbun thrives equally well in a pinkish muck at Charles Island or a red-orange gumbo at South Seymour Island. The Academy Bay specimens, living in gray mud, attain the greatest size." (Biotope nos. 11, 12, 14.)

Sympatric Associates

Other *Minuca* with which *galapagensis* regularly associates consist of *vocator ecuadoriensis* and *zacae*. In Costa Rica it was occasionally found with two *Celuca, latimanus,* and *argillicola,* although the latter occurred in nearby shade. See also Garth's observations (1946) made in the Galapagos (preceding paragraph).

Material Resulting from Field Work

(The complete list of specimens examined is given in Appendix A, p. 605.)

Observations and Collections. Nicaragua: Corinto; Panama: Old Panama; Ecuador: Puerto Bolivar; Guayaquil (El Salado).

Films. Costa Rica: Golfito. Specimens filmed in the field, transported to and identified in Trinidad laboratory, but died and were lost in crabberies. Identification was definite.

Type Material and Nomenclature

Uca galapagensis Rathbun, 1902

HOLOTYPE. In Smithsonian Institution, National Museum of Natural History, Washington. Cat. no. 22319. Individual distinguished by label: "TYPE." Type-locality: Indefatigable I., Galapagos. Measurements in mm: length 13; breadth 19.5; propodus 34; dactyl 24.5. (!) Additional type material, under same number in same jar: 5 males, lengths 8-13. (!)

Type Material of *Gelasimus macrodactylus* Milne-Edwards & Lucas, 1843. In Muséum National d'Histoire Naturelle, Paris. The two cotypes examined by Rathbun (1918.1) of *G. macrodactylus* were not found by me in 1959. Several boxes, including specimens considered by the museum to be "types non-marqués" (!), contained dried crabs in very poor condition, all of them labeled "*Gelasimus macrodactylus* Edwards & Lucas," and one of them had in addition "M. d'Orbigny-Valparaiso," the latter being the type-locality of the species. This specimen, however, does not have the characteristics here ascribed to *galapagensis*. In addition, the correctness of the

type-locality is questionable, since there is no later record of the occurrence of *macrodactylus* in Chile, except for Porter (1913); Porter's material is apparently not extant. The extremely poor condition of the Paris specimen mentioned makes it highly desirable to disregard it as a possible lectotype. I therefore agree with von Hagen (1968.2, 1968.3) who proposes to discard the name *macrodactylus* officially through the International Commission on Zoological Nomenclature, and has given his reasons in full; they include his conclusion that the "Valparaiso" specimen in Paris is probably a representative of the Atlantic species *rapax*. The suppression of the name *macrodactylus* now makes it possible to refer a number of reports of *macrodactylus* to *galapagensis*, a well-described species represented by a type in good condition.

Uca galapagensis herradurensis Bott, 1954

HOLOTYPE. In Forschungsinstitut Senckbenberg, Frankfurt a. M. One male, cat. no. 1865; La Herradura, El Salvador. Measurements in mm: length 12; breadth 20. (!)

Additional type material (listed in type description). Paratypes: cat. no. 2135; La Herradura. Juvenile paratype: cat. no. 1866; Alcaldia de Triunfo.

Type Material, including holotype, of *Uca macrodactyla glabromana* Bott, 1954. In Forschungsinstitut Senckenberg, Frankfurt a. M. Holotype male, cat. no. 1842; Los Blancos, El Salvador; measurements in mm: length 10; breadth 14. (!) This specimen is a young example of *U. g. herradurensis*, a conclusion with which von Hagen, who examined the material independently, agrees (1968.1). The paratype material (!) includes examples of *U. zacae*.

Uca (Minuca) [galapagensis] galapagensis herradurensis Bott, 1954

(El Salvador to Panama; ? Mexico)

Morphology

Differs from *g. galapagensis* as follows. Carapace: Orbits slightly more oblique, sides of carapace and dorso-lateral margins more converging; sides of front slightly more converging; carapace less arched, the regions more clearly defined; antero-lateral region usually with a few small tubercles; in the very distinct H-form depression, pile always is found in at least part of the grooving; eyebrow slightly narrower. Major Cheliped: Apex of oblique ridge never as high as in large specimens of *g. galapagensis*; highest part of ridge almost always, instead of sometimes, slightly distal to apex. Ambulatories: Pile always clearly present on dorsal parts of carpus and manus of first 3 legs, in both sexes. Gonopod: Anterior flange with all of distal margin except inner edge always more or less produced beyond pore, sometimes strikingly so, when it may narrow strongly, its outer margin being thickened and darkly pigmented, so that it appears, in a superficial examination, to be a long spine. In some individuals the prolongation is less, but it is always more than the slight projection of the outer angle described under *g. galapagensis*. Posterior flange usually narrower than the anterior and never projecting beyond it; never strongly flattened.

Atypical Individuals. One example from Corinto, Nicaragua, has a gonopod with a strongly truncate anterior flange; in other respects it is fairly characteristic of the present subspecies (NYZS 381128, part); since the projections are very brittle, it seems possible that this truncation is unnatural. On the other hand, Bott's two specimens from El Salvador referred by him to *U. galapagensis* may be regarded either as strays from the Galapagos, or as intermediates. The sample collected in Panama is the only large population sample known; not one of these individuals has atypical gonopods, an ambiguous carapace or lacks pile on the legs.

In contrast to these ambiguous specimens from El Salvador and Nicaragua, an example from Eden Island, Galapagos (USNM: formerly New York Zoological Society cat. no. 2442, part) seems clearly intermediate. On the one hand, the carapace is that characteristic of *g. galapagensis*; on the other, the gonopod has a strongly projecting edge to the anterior flange, and ambulatory pile, both characteristic of *g. herradurensis*. The specimen is included under the subspecies heading of *g. galapagensis*, in accordance with the usual distribution of the subspecies.

Color

Displaying males: At Golfito, Costa Rica, in August full display whitening was not observed, the lightest individuals being entirely pale buff. Females duller. No orange phases were present.

Social Behavior

Waving Display

Differs from that of *g. galapagensis* in larger number of jerks, in presence of 2 peaks of the wave, each reaching an equally high point, the peaks being separated by a slight descent of the chela; in the absence of vibration during the cheliped's pause at the apex; and in the absence of vibration of the cheliped as it

is brought into rest position from a more forward landing spot; in fact at moderate to high intensity the cheliped is not at all returned to the ground between waves. Numbers of jerks are as follows: on cheliped's rise, 5 to 8; between the 2 peaks at apex: 2 on descent, followed by an unbroken return to apex; jerks on final descent (completing the wave) ranging from none, giving an unbroken descent, to 4.

Uca (Minuca) [galapagensis] galapagensis galapagensis Rathbun, 1902

(Colombia to Peru; Galapagos Islands; ? formerly Chile)

Morphology

Differs from *g. herradurensis* as follows: Carapace: In all larger specimens, the orbits are scarcely oblique, the sides and dorso-lateral margins converge less as do the sides of the front, and the carapace clearly arches more, with the regions faintly or not at all indicated; in moderate-sized and juvenile individuals the differences are far less apparent. Eyebrow broader, at least in larger individuals. Major Cheliped: Apex of oblique ridge sometimes, in the largest specimens, strikingly high, more outstanding than in any other species in the genus and the highest point of the ridge; in many smaller specimens however, including the holotype, the apex is slightly lower than the adjacent part of the ridge and not remarkable, in comparison with other species in the superspecies and elsewhere. Ambulatories: No pile ever present on carpus and manus. Gonopod: Most characteristically both anterior and posterior flanges are sharply truncate, the anterior at about the level of the pore, the posterior slightly beyond it; the posterior is often notably flat and a little wider than anterior. As in *herradurensis*, however, there is a fairly wide range of variation; the most usual, and the only one that might be confusing, is the frequent very slight projection of the outer distal corner of the anterior flange; the direction is as much outward as distal and it would be unremarkable were it not for the pigmentation of this corner, which is abruptly dark as in the large projection of *herradurensis*.

Atypical specimens are mentioned above, under the description of *g. herradurensis*.

Color

Both displaying and non-displaying males achieved display whitening of the carapace and ambulatories at both Guayaquil and Puerto Bolivar in April, even though few individuals were displaying at the time. The palest crabs were brilliant white except for the entire major cheliped; this appendage always ranged from bright reddish orange to yellow orange except for the fingers which were white for more or less of their lengths; sometimes only the tip of the dactyl was white. The dullest individuals of both sexes were highly variable, the appendages usually brown and the carapace brown mottled with orange or else entirely dull orange. From this phase the change in both males and females progressed to bright orange followed by a dull buff or cream phase before true white was attained; lightening of carapace sometimes preceded and sometimes followed that of ambulatories; as usual many individuals remained parti-colored or wholly dull. Females with manus and chelae of both chelipeds apparently always white.

SOCIAL BEHAVIOR

Waving Display

Differs from that of *g. herradurensis* in having no more than one or 2 jerks, on either the rise or descent of the cheliped, usually on both at higher intensities, and in the lack of 2 peaks at the wave's apex. Additionally, as shown in von Hagen's analysis of high intensity display (*loc. cit.*), this subspecies differs in two other ways from the northern form: first, the cheliped vibrates during the pause at the apex of the wave; second, the cheliped is lowered all the way to the ground with each wave but to a position in front of the usual rest position; from there the claw is brought back toward the mouthparts through several small jerky motions.

REFERENCES AND SYNONYMY

Uca (Minuca) galapagensis Rathbun, 1902

TYPE DESCRIPTION. See under *U. (M.) galapagensis galapagensis*, below.

Uca (Minuca) galapagensis galapagensis Rathbun, 1902

Gelasimus macrodactylus

Milne-Edwards & Lucas, 1843: 27; Pl. 11, Figs. 3, 3a. "Côtes de Valparaiso." Type description.
Nicolet, 1849: 165. Chile. Taxonomy; color.
Milne-Edwards, 1852: 149. Taxonomy.
Smith, 1870: 128. Record of type only.
von Hagen, 1968.3: 60. Proposed suppression of specific name, *macrodactylus*, H. Milne-Edwards & Lucas, 1843, as published in the combination *Gelasimus macrodactylus*, under plenary powers of International Commission on Zoological Nomenclature.

Gelasimus annulipes (not of Milne-Edwards)

Kingsley, 1880.1: 148 (part). *G. macrodactylus* Milne-Edwards & Lucas synonymized with *G. annulipes* Milne-Edwards, 1837.

Uca macrodactyla

Nobili, 1901.4: 49. Colombia: Puntilla di S. Elena, Tumaco.
Maccagno, 1928: 37; Text Fig. 22 (photo of claw). No new record. Taxonomy of Nobili's material.
? Porter, 1913: 316. Chile: Quintero (north of Valparaiso; collected in 1896).
Garth, 1957: 106. No new record.

Uca galapagensis

TYPE DESCRIPTION. Rathbun, 1902.3: 275; Pl. 12, Figs. 1, 2. Galapagos Is.: Indefatigable. (USNM !)
Rathbun, 1911: 550. Peru. Brief taxonomy; local distribution (by R. E. Coker).
Rathbun, 1918.1: 403; Text Fig. 167; Pl. 142. Galapagos Is.: James; Indefatigable; South Seymour. Peru: Puerto Grande, Rio Zarumilla. Taxonomy. (USNM !)
Rathbun, 1924.1: 155. Galapagos Is.: Indefatigable; James; South Seymour.
Boone, 1927: 273 (part); Text Fig. 97 (upper part). Galapagos Is.: South Seymour; James; Eden. Not specimen from Cocos I., illustrated in lower half of Text Fig. 97, which should be referred to *U. panamensis*. Specimen in upper half of Text Fig. 97 is young; it was referred by Crane (1941.1: 176) to *macrodactyla*, the species name now suppressed. Taxonomy. (USNM !)
Sivertson, 1934: 20. Galapagos.
Hult, 1938: 14. Galapagos. Measurements; local distribution.
Crane, 1941.1: 176. No new records. Taxonomic comments on material listed by Boone, 1927.
Garth, 1946: 515; Pl. 87, Figs. 3, 4. Galapagos Is.: Charles I. (Flamingo Lagoon and east of Postoffice Bay); Indefatigable I. (Academy Bay and Conway Bay); South Seymour I. Taxonomy; habitat.
von Hagen, 1968.2: 415; Text Figs. 7, 8a, 8b, 17. Ecuador: Guayaquil; Peru: Puerto Pizarro. Taxonomy; color; display.

Uca galapagensis galapagensis

Bott, 1954: 166; Pl. 16, Fig. 8a, b. Peru. Taxonomy.
Bott, 1958: 209. Galapagos. Taxonomy.

Uca macrodactyla macrodactyla

Bott, 1954: 167 (part). No new material. Taxonomy.

Uca (Minuca) galapagensis herradurensis Bott, 1954

Uca macrodactylus

Rathbun, 1918.1: 404 (part); Pl. 143. ? Mexico: Guaymas (MCZ !). Nicaragua: Corinto; Costa Rica: Santo Domingo. Taxonomy; food of owl (*Ciccaba nigrolineata*); mangrove habitat. (USNM !)

? Rathbun, 1918.2: 177. Pleistocene of Panama. Fossil. See p. 157. (USNM !)

Crane, 1941.1: 178. ? Mexico: Tenacatita Bay. Nicaragua: Corinto. (Not young Galapagos specimen from Boone's 1927 material discussed on p. 179.) Taxonomy. (USNM !)

Uca galapagensis herradurensis

TYPE DESCRIPTION. Bott, 1954: 166; Text Fig. 9; Pl. 16, Fig. 9a, b. El Salvador: La Herradura. (Frankfurt !)

Barnwell, 1968.1. Costa Rica. Chromatophoric responses to light and temperature. Taxonomic note on pp. 221-22.

Uca macrodactyla macrodactyla

Bott, 1954: 167 (part). No new material. Taxonomy.

Uca macrodactyla glabromana

Bott, 1954: 168 (part); Text Fig. 10; Pl. 16, Fig. 10a, b. El Salvador: Los Blancos and Puerto El Triunfo. (Frankfurt !)

Uca herradurensis

von Hagen, 1968.2: 417. Discussed during his treatment of *U. galapagensis*; considers *U. g. herradurensis* Bott of specific, not subspecific rank. No new records.

190

33. *UCA (MINUCA) [GALAPAGENSIS] RAPAX* (SMITH, 1870)

(Tropical and subtropical western Atlantic)

PLATES 27 *A-D*; 45 *C-F*.
FIGURES 52 *C-DD*; 54 *F*; 67 *C*; 86; 91 *E, F*; 100.

MAP 14.
TABLES 9, 10, 11, 12, 14, 15, 16, 17, 18, 19, 20.

INTRODUCTION

A waving male *Uca rapax* stands out among other fiddlers on any shore in his range by the numerous small jerks dividing the raising of his large cheliped. Almost always they number at least 8; sometimes they reach more than 30. Nothing else about his appearance is striking. His carapace never changes to shining white; his major claw is never colorful, the lower manus achieving at most a rather dull orange; his size is small to medium; and the tempo of the display, especially at low intensities on chilly mornings, has a soporific effect on the watching ethologist. In fact its slowness is a distinction, since single waves last longer under these conditions than in any other species in the genus; the longest waves, in Rio de Janeiro during early December, at temperatures around 22 C, lasted up to 13 seconds each.

Another characteristic less than stimulating to the worker is the fact that *rapax* is one of the few tropical fiddlers that has a prolonged period of non-breeding, that is, excluding any lunar or semi-lunar rhythms. Even in localities where the tides cover the burrows regularly during the dry season, so that the fiddlers can feed and hence do not aestivate, the populations almost always pass through a non-waving period during two or three of the months having little or no rain. This appears to be true throughout the West Indies as well as in adjacent parts of the mainland; here from February through April waving is so rare that field work on the species' social behavior should not be scheduled. These fiddlers are also inactive in southern Florida in December, and presumably throughout the colder weather farther north.

Yet in less obvious characteristics members of the species are among the most remarkable in the genus. They are undoubtedly the most abundant neotropical fiddler and the most adaptable to varied biotopes. As with several other *Uca* that are outstanding in extent of range, numerical abundance, and ecological tolerance, a study of their biological success would be illuminating.

Perhaps their most outstanding characteristics, undoubtedly related closely to the fact of their success, are those most obviously related also to their systematics. This species and its allopatric relations, in fact, form one of those Jeckyll-and-Hyde groups that

delight an evolutionary biologist and give nightmares to a practicing taxonomist. Fortunately these days most workers tend to be at least a little of both, so people concerned with the group are unlikely to be unhappy all the time.

The most widespread form, here termed a subspecies, *Uca r. rapax*, appears to be wholly recalcitrant to further taxonomic subdivisions; it extends, as do several other western Atlantic *Uca*, from southern Florida to the Tropic of Capricorn at Rio de Janeiro. Like most such successful forms this one is highly variable in details of its morphology; no evidence of clines has yet appeared.

The other forms all have lesser ranges. In this study the group, forming the superspecies *galapagensis*, is considered to be composed of three closely related species—*galapagensis* in the Pacific, *rapax* chiefly in the tropical and subtropical Atlantic, mentioned above, and *pugnax* in the temperate regions of the north Atlantic. In each species two subspecies are recognized. As already pointed out in the Introduction to *galapagensis* (p. 183) the species is morphologically so close to *rapax* that logically the two forms should be considered subspecies.

In the Atlantic a zone of coincidence occurs in east Florida where *pugnax* and *rapax*, living on the same shore, apparently do not interbreed. Unfortunately for the taxonomist, the species characteristics of each are exceedingly variable, being often shared in various parts of the ranges where no coincidence occurs. As a result firm specific distinctions are, once again, difficult to erect.

The two Gulf coast species, *Uca longisignalis* and *U. virens*, recently proposed by Salmon & Atsaides (1968.1) largely on behavioral and color characteristics, are here regarded as subspecies of *rapax* and *pugnax* respectively. Again, firm morphological distinctions are virtually absent, while the behavioral and color characteristics crop up in closely similar or apparently identical forms somewhere in the ranges of the other species and subspecies of the superspecies; yet at present we have no evidence of interbreeding in the zones of overlap.

Non-morphological peculiarities in the superspecies giving food for thought concerning the close relationships of these rather fluid forms may be itemized as follows:

1. In the Pacific, *U. g. galapagensis* has the rare waving display component of a pause at the apex of the wave as a regular part of the display; in the Atlantic *r. rapax*, a pause occurs frequently, but not as a regular component, in various parts of its range.

2. *U. galapagensis herradurensis* shows no pause at the apex, but a double peak there in every wave; this is an even rarer component than the pause. A similar double peak occurs very rarely in far separated populations of *r. rapax*.

3. *U. g. galapagensis* during high intensity courtship performs leg-wagging stridulation on the surface of the ground, a component extremely rare in the courtship of other species; populations of *r. rapax* in Florida perform similarly, but do not do so farther south.

4. In northwest Florida and adjacent states, *r. longisignalis* in display jerks very feebly, but this is true also of some typical populations of *r. rapax*.

5. In Texas, *pugnax virens* is characterized by relatively strong jerking and a color pattern not found in populations on Florida's east coast and adjoining states; yet stronger jerks and color similar to those in Texas are found in northern populations of *p. pugnax*.

6. The outstanding color characters of *rapax longisignalis* in northwest Florida and nearby areas differ from those of Florida populations of *r. rapax*; yet in Venezuela (Turiamo) rare individuals of *r. rapax* occur with coloring apparently very similar to that of *longisignalis*.

Some of these points will be discussed in Chapter 7, in the section on sympatry.

If one were being altogether logical and if it were not for two locations—Yankeetown in northwest Florida and Crescent Beach on that state's east coast, where two forms coincide without interbreeding—the most acceptable procedure would be to call the entire superspecies a single species, *pugnax*, divided into 6 subspecies (*galapagensis, herradurensis, rapax, longisignalis, pugnax,* and *virens*). Because of two considerations I am not proposing this course. First, more work, both ethological and morphological, with larger study collections available, is needed, particularly at the zones of coincidence. Second, the convenience of biologists appears to be of sufficient importance to avoid yet another change of names in forms widely known and currently the subjects of active research. This is particularly desirable since work is so obviously demanded on the group's genetics which, one excellent day, will certainly illuminate the problem. After all, parts of the genus *Drosophila* have caused similar difficulties and their solutions advance steadily, with large rewards for general biology. *Uca* will take a little longer—but optimism is in order.

Morphology

Diagnosis

Distinguishing characteristics from related broad-fronted *Uca*, except from the two members of its superspecies, as in the first paragraph of the diagnosis for *U. galapagensis* (p. 183). From these closest relations, *galapagensis* in the Pacific and *pugnax* in the Atlantic, *rapax* is distinguished most reliably in the following respects: From both in having the center of major palm almost smooth with small granules at most, even in small specimens, and not clearly tuberculate; and in having the tubercles of the oblique ridge only rarely continuing up around carpal cavity, and then only weakly and for a short distance. Additionally, from *galapagensis* only, in having postorbital groove on carapace very shallow and in having a tubercle on female gonopore. Additionally, from *pugnax* only, in having merus of second ambulatory broad, even in males, with its upper margin usually slightly convex throughout, but sometimes almost straight in the middle; eyebrow slanting, not vertical; apex of oblique ridge higher, with the ridge tubercles larger and more regular. The subspecies in the Gulf of Mexico, *r. longisignalis*, shares almost all these characters; it differs most distinctly in having the eyebrow nearly vertical and scanty pile on lower margins of some segments of at least the middle ambulatories; the armature of the palm is more variable than in *r. rapax*, and should be considered unreliable as a diagnostic aid.

Description

With the characteristics of the subgenus. Extremely close to its Pacific allopatric, *galapagensis*, particularly to *g. herradurensis*; the similarity is greater than to its Atlantic allopatric, *pugnax*; see discussion (p. 193). Table 11 gives an idea of the variability of the tropical subspecies, *r. rapax*, throughout its wide range.

MALE

Carapace. Front contained about 3.5 times in carapace breadth between antero-lateral angles, its sides converging more than in related Atlantic species and the tip therefore appearing slightly narrower in most individuals; both the shape and breadth of front, however, vary, and it is distinctly broader in *r. longisignalis*. Orbits moderately oblique, dorso-lateral and lateral margins moderately convergent. No pile on antero-lateral region, but scattered, small tubercles or granules detectable in some individuals of all populations; pile in H-form depression present or absent. Eyebrow of moderate width, almost always intermediate between the clearly wider eyebrow of the Pacific species, *galapagensis*, and the usually narrower one of the Atlantic *pugnax*; in *r. rapax* it is also clearly less depressed than in *p. pugnax*, but the relationships are reversed between the subspecies in the Gulf of Mexico. Beading of its distal margin always better developed than that of proximal, but latter is distinctly beaded in the larger specimens of some populations, unlike the character in related species. Suborbital crenellations well formed throughout, not obscured by the sparse setae; pile absent.

Major Cheliped. Manus with tubercles of upper, outer surface moderately enlarged and only rarely reduced, and then only in large specimens, in size and number. Palm always with oblique ridge high and marked with large tubercles; the degree of regularity of their arrangement is, however, extremely variable both within and among populations; in general small crabs with claws of mature shape show the most regular rows, with well-formed tubercles throughout, there being sometimes, as in *galapagensis*, a slight break and overlap in the middle, proximal to which the tubercles are slightly the largest and the ridge highest just distal to the apex. Some larger individuals also maintain this regular formation, with ridge and apex relatively high and tubercles relatively regular, so that the entire structure is indistinguishable from that found in characteristic examples of the Pacific *g. herradurensis*. Other individuals in *rapax* show a breakdown in the distal part of the ridge, which is then very low, with tubercles small and not at all in linear formation, while the apex and adjacent ridge remain high and covered with abruptly large tubercles; these form a single or double row before the apex and a cluster on the apex itself; sometimes several very large tubercles, well separated from each other, occur along the middle of the ridge. When numerous individuals from a series of populations are compared, however, two points are clear. First, the oblique ridge and its apex in *rapax* as a rule are slightly less high than in *g. herradurensis*, while its tubercles in most individuals are far less regular. Second, the apex in most *rapax* is higher and the tubercles more regular than in *pugnax*. When compared with the ridges of its three tropical, sympatric forms, *vocator*, *burgesi*, and *mordax*, as well as with the northern *minax*, both ridge and apex are high and its tubercles large and regular.

Similar variability is shown in the tuberculation around the distal edge of the carpal cavity and of the center palm. In many individuals the oblique ridge stops abruptly at or immediately over the crest of the apex, at the edge of the carpal cavity; in most, several tubercles mark the cavity's borders, always non-contiguous and often well separated; when present they merge with the small tubercles of the center palm. The latter group are usually minute and well separated; in at least some individuals in all populations they are practically lacking; as usual in this species, exceptions are frequent: a few individuals have large tubercles near the end of the downcurved part of the beaded edge of dorsal margin; a few have the distal border of the carpal cavity bordered with an almost continuous row of tubercles, while the tubercles of the upper palm are as large and numerous as in *g. herradurensis*, from which in this character it would then be impossible to distinguish *rapax*. The great majority of individuals, however, in all populations of *rapax r.* have far fewer tubercles, and smaller ones, both around the distal part of the carpal cavity and on the center palm, than occur either in the related *galapagensis* or *pugnax*, or in its tropical sympatrics, *vocator*, *burgersi*, and *mordax*. Their relative smoothness is characteristic also of small specimens. In *rapax longisignalis* these characters are not so reliable.

Proximal ridge at base of dactyl with tubercles always strong and close set, but very rarely (as noted only in part of a Miami population) almost contiguous in its lower part, as is characteristic of *burgersi*; there is never any trace of proximal divergence from the distal ridge, although the upper tubercles of the latter sometimes slant toward the dactyl, as in related species; sometimes the entire row of distal tubercles is obsolescent, particularly in large specimens.

In shape and proportions of manus and fingers, *rapax* closely resembles *galapagensis*, although, as pointed out by Tashian and Vernberg (1958) they are distinct from those of *pugnax*. In *rapax* the pollex and dactyl are definitely shorter in comparison with the manus, thicker, and less tapering. In addition, the distal part of the pollex tends to curve upward in *rapax*, in contrast to its straightness, or even downward curve, in *pugnax*. The caveats concerning variability that are usual in this species apply in these differences also. Since populations vary widely in lengths at which their "normal" morphological proportions are attained, and because of leptochelous and brachychelous tendencies, the selection of specimens for meristic series must be made with special care. Again, although most individuals show a slightly upcurved pollex, it is not apparent in all. For example, it is prevalent in almost all individuals from Puerto Barrios, Guatemala, but weak or absent in many from Tobago and from Rio de Janeiro, particularly in those of large size. The differences in slenderness of dactyl and pollex and the degree of taper are apparent to the eye throughout all populations of the two species, although the necessarily rigorous series of measurements to establish it have not been undertaken.

Predistal crest on pollex in many individuals is distinct as in *U. galapagensis* (p. 184), but it is far more variable.

Ambulatories. Meri of 2nd and 3rd ambulatories moderately enlarged, the dorsal margins slightly convex to almost straight. The breadth of the meri is similar to that found in *galapagensis*, but greater than in *pugnax*; the segments are clearly wider than in its tropical sympatrics *vocator*, *burgersi*, and *mordax*. Dorsal bristles of manus and carpus less numerous, less stout and less stiff than in *pugnax*. Pile always present on dorsal surfaces of carpus and manus of first 3 legs; it is always altogether absent from ventral margins, including that of manus, in *r. rapax*; scanty pile present on ventral surface of some segments in *r. longisignalis*.

Gonopod. Inner process of moderate width, usually straight, sometimes slightly curved; anterior flange clearly wider than posterior, its distal margin always produced, but to a variable extent.

FEMALE

Gonopore. Tubercle on margin of gonopore apparently always present.

Measurements (in mm)

	Length	Breadth	Propodus	Dactyl
r. rapax				
Largest male (Tobago)	21.0	32.0	63.0	45.5
Moderate male (Tobago)	10.5	16.5	22.0	15.0
Largest female (Santo Domingo)	17.0	27.0	–	–
Largest ovigerous female (Tobago)	14.5	22.0	–	–
Smallest ovigerous female (San Juan)	11.0	17.5	–	–
r. longisignalis				
Male (holotype) (Ocean Springs)	13.0	21.0	37.0	28.0
Female (Ocean Springs)	12.0	16.0	–	–

Morphological Comparison and Comment

The close resemblance of *rapax* to its transisthmian allopatric, *galapagensis*, has already been discussed (p. 183). Series of preserved *rapax* can usually be easily distinguished from series of the northern allopatric species, *pugnax*, but as usual in the subgenus the characteristics are so superficial and there is so much variation within and between populations that individuals atypical enough to fit the description of the opposite species can usually be found without difficulty. Although Tashian & Vernberg (1958) found differences in the proportions of the chelipeds in the particular populations with which they worked in Florida, in the zone of coincidence they discovered, these differences do not always hold good in other populations in the north and south. I found no evidence of regular differences among tropical populations, or of clines.

Recent complications in the taxonomy of the two species have arisen as a result of the description by Salmon & Atsaides (1968.1) of two new species of *Uca*, *longisignalis* and *virens*, from the Gulf coast of the United States. Although these authors considered both to be most closely related to *pugnax*, I am treating them here as subspecies of *rapax* and *pugnax*, respectively. This decision is unsatisfactory, because it is not based on personal acquaintance with the new forms, except for a few preserved specimens, and because "good" morphological characters are apparently extremely scarce. The authors have described excellent behavioral differences, as will later be discussed. There is a slight coincidence of ranges in north Florida, between apparently *rapax rapax*, as here understood, and *r. longisignalis*; the material is unfortunately insufficient to furnish evidence of interbreeding. See details of differences under the subspecies.

Color

Displaying males: Display whitening of carapace poorly to moderately developed; anterior part of carapace and eyestalks green blue in populations of *rapax longisignalis* and in rare individuals in the tropics of *r. rapax*, especially at Turiamo, Venezuela. Cheliped about as light as carapace in both subspecies except for outer manus which, at least on lower part, ranges from dull, grayed, greenish yellow (approximately ochraceous) to tints of orange (often described in field notes as apricot), sometimes extending on to pollex; red tints and tones always absent; fingers white, at least at tips. Minor cheliped with usually manus and always fingers white, except at Turiamo, Venezuela, where sometimes blue. Ambulatories usually remain somewhat darker than carapace. Females attain about the same degree of grayish white as males.

SOCIAL BEHAVIOR

Waving Display

Wave lateral, weakly circular, at low intensities often practically straight. Jerks always present, on both rise and descent of major cheliped, more numerous on the rise than in any other species in the genus, and highly variable, largely regardless of geographical locality. The range of numbers is, on the rise, between 6 and 34 and on the descent from 2 to 8; the total number of jerks therefore can and very rarely does, reach 42, which was attained in one crab filmed in Rio de Janeiro, Brazil and another in Puerto Rico. The highest numbers of jerks are at lowest intensities; at high intensities where curtsies are included, with or without the acoustical component of stamping, the jerks on the rise range from 6 to 12, those on the descent from 2 to 4, and the total number of jerks from 8 to 16. Comparably low numbers of jerks are also recorded in populations of *r. rapax* studied in southern Florida by Salmon (1967), where the total ranged from 7 to 12 and of *r. longisignalis* described by Salmon & Atsaides (1968.1) in northwest Florida and Mississippi, with totals between 8 and 15; in the latter, northern populations the jerks were less distinctly marked than in the south. None of the individuals in these Florida populations, even at low intensities, reached the high number of jerks found in the south. In waves showing the highest number of jerks (6 to 8) on the descent, the last several are usually made after the cheliped reaches its low point, which is usually not in contact with the ground, somewhat in front of its usual rest position, and the cheliped jerks back into place in front of the mouthparts; a similar motion sometimes occurs in the Pacific with the closely allopatric form, *U. g. galapagensis*, observed by von Hagen (1968.2). At high intensities the cheliped often does not nearly reach the ground after the final jerk, being simply flexed into rest position barely before the onset of the curtsy, with or without stamping. A brief pause at peak of wave is present or absent, similar to but shorter than that also found in *U. g. galapagensis* (*loc. cit.*, and this study, p. 184). Very rarely the wave includes a double peak at the apex, as in *U. galapagensis herradurensis* (p. 184). Minor cheliped usually makes a corresponding motion, but without jerks. During a series the body is held high, except for the curtsy component of high intensity courtship, and any combinations on the first 3 ambulatories on each side are kicked up and out in leg-waves in which the meri are definitely not brought into contact, unlike sound-producing leg-wags. The curtsy component is always a characteristic of high intensity courtship and not of agonistic behavior; as usual it occurs close to the entrance of the male's burrow when the attention of a female has been aroused but, at least in tropical populations, waving does not necessarily stop with the first curtsy, the curtsy rather forming, with or without associated stamping, the final component of a wave, to be followed almost at once by another in the same series. In Florida the curtsy, usually followed by others, takes place at the end of the series of rapid waves of courtship and immediately preceding the male's descent into his burrow; it apparently always includes sound production through stamping and, at night, by leg-wagging (see under Acoustic Behavior, below). Duration of single waves, at low intensity with many jerks, up to 13 seconds; the slowest of all were filmed in Rio de Janeiro on cloudy mornings in early December when waving was just beginning, an hour before low tide; the surface temperature ranged from 20° to 23° C. Waves almost as slow, however, lasting up to 12 seconds in *r. rapax*, with totals of between 20 and 34 jerks, have been filmed on clear hot afternoons, at surface temperatures between 28° and 30° C, in both Venezuela and Puerto Rico. Duration of waves of high intensities about 2 to almost 6 seconds. (Component nos. 4, 5, 6, 9, 11, 12, 13, 14; timing elements summarized in Table 19, p. 656.)

Precopulatory Behavior

Female follows male into his burrow. No records of surface copulations, except for the usual occasional abortive attempts by aggressive wanderers. No records of nocturnal mating have yet been published. Salmon (1967) reported herding motions by single males of females in populations in south Florida; he reports that the males sometimes approached females as much as a meter away. It will be interesting to learn if the females in question are immature in these instances as in at least two other species (p. 503).

Acoustic Behavior

In Florida Salmon (1967) and Salmon & Atsaides (1968.1, 1968.2) made recordings of sound components in *rapax* during both diurnal and nocturnal courtships. The diurnal sounds were produced close to and just within the burrow mouths by the stamping component of curtsies made during high intensity waving. Another sound occurred when the male was farther within the burrow, as well as when a female was introduced when such a male was already within. Salmon describes the diurnal sound production as follows: "The males stopped waving when the females came within 5 to 10 cm of their burrows. The major chela was held flexed just above the carapace. While in this posture sounds were produced when the second and third pairs of ambulatories were raised and lowered in unison against the substrate, 2 to 6 times in a series. With each blow of the legs against the surface, the body was lowered in steps from an initially raised position almost to the substrate. A series of movements lasted from 1 to 3 sec. The male might remain at the same spot between consecutive series of ambulatory movements, but if the female moved closer he would dash toward and sometimes just inside his burrow. The sounds resulting from the movements were sharp pulses of short duration (30 to 50 msec), each pulse corresponding to one contact between the ambulatories and the substrate. But when the males moved inside their burrows the pulses contained a low-frequency component which extends pulse durations to 120 to 200 msec. . . ." During the evening, "Sound-producing males of *U. rapax* were either several millimetres inside the burrow or near the entrance with their ambulatories protruding from the burrow opening. The sounds consisted of two to eight pulses produced in a series, each composed of low frequencies with fundamentals between 170 and 200 Hz. The sounds were similar to those heard during the day from males inside their burrows. Conspecific crabs, as well as other species, elicited an increase in rate and intensity of sound production from males after contact with their ambulatories. Four of ten males showed these responses when the substrate 1 cm from the burrow was scratched lightly with a twig. Several other males, silent at the time, began producing sounds when a few grains of sand were pushed into their burrows.

"The sounds were produced by movements of the ambulatories similar to those produced during the day. . . . During the first one to three pulses, the ambulatories were lowered to the substrate, raised and rapidly vibrated in a dorso-ventral plane, then lowered again. In the remaining pulses of the sound the movements were identical but *the ambulatories did not touch the surface* [italics mine]. The low frequency components were always correlated with ambulatory vibration. The major cheliped was not moved during sound productions . . ." (Salmon 1967, pp. 453-54). The authors provide data on variation in time of onset and cessation of the sounds and in acoustical properties, as well as comparisons with sounds produced in particular by *speciosa*. In the subspecies *rapax longisignalis* Salmon & Atsaides (1968.1: 287) describing nocturnal sound production report that, as with the two subspecies of *pugnax*, *r. longisignalis* "produced sounds during nocturnal low tides for periods of up to several hours . . . the sounds were produced by movements of the ambulatory legs; each sound was composed of several pulses produced in a series, . . . the sounds containing 5-14 pulses, and successive sounds in a series were each separated by long intervals (8-10 seconds). . . . The signals . . . were all produced by lone males."

No other detailed study of *rapax* sounds has yet appeared. In 1968.1 von Hagen published a preliminary statement that he had secured recordings of the species in Trinidad, but his results are not yet available.

In the course of the present study observations of sound-producing mechanisms through films made diurnally throughout much of the range of *rapax* were made, as well as a few recordings, also in Trinidad, both by day and night. As with the Florida populations, leg-wagging produced sounds, as in the phrase in italics on the preceding page. However, leg-stamping was only rarely seen to be included in the numerous curtsies of high-intensity courtships that have been observed and photographed. This element was always either in an agonistic situation at the surface or in response to an intruder, regardless of sex, introduced into the burrow by day or night. This is fully in accord with leg-wagging throughout the genus, except in some *mordax* during high intensity courtship. Another component has been heard and recorded, by day and night, in Trinidad *rapax* when an intruder is introduced and sometimes spontaneously; this I have termed "honking"; it is unmistakable and we have as yet no idea how it is produced. (Component nos. 5 and 11, plus "honking.")

Combat

Combat behavior in *rapax* has been more closely examined than in any other species. It was described in detail for a population in Trinidad in Crane (1967); it is reviewed, with components added, in the present study, in comparison with that of other species, in Chapter 5, beginning on p. 485. (Component nos. 1, 2, 4, 6, 7, 9, 12.)

RANGE

Tropical, subtropical, and warm temperate western Atlantic from the south coast of the United States along the Gulf of Mexico and Florida's eastern shores, then continuing south on the coasts of Mexico, Central America, the West Indies, north and east coasts of South America as far as Cananéia, Estado São Paulo, Brazil. One form, here viewed as a subspecies, *longisignalis*, is known in the warm temperate part of the range, from Matagorda Bay to Yankeetown; near Yankeetown Salmon & Atsaides (1968.1) found it occurring sympatrically with *r. rapax*. Throughout the rest of the species' range there seems to be no justification for further subdivision.

BIOTOPES

Typically, in the tropics, on sheltered flats of mud or sandy mud in the vicinity of mangroves, in similar substrate edging lagoons, in river deltas, and along the streams and rivers themselves on flat banks close to their mouths; also on muddy sand, especially on tropical salt flats covered only at spring tides. Even displaying parts of the populations often in the partial shade of the edges of mangroves. Occasionally the substrate includes so little silt that it appears almost fully sandy, especially in food-rich localities, such as around pneumatophores of some mangrove associations, or near fishermen's landings. (Biotope nos. 6, 9, 11, 12.)

SYMPATRIC ASSOCIATES

When a full spectrum of local species of *Uca* occurs on a tropical shore, distributed in typical stratification, the species arrange themselves as follows, from those nearest the sea and in highest salinities to those farthest inland, including the banks of water courses: (*Celuca*) *leptodactyla*, (*Uca*) *maracoani*, (*Boboruca*) *thayeri*, (*Celuca*) *cumulanta*, (*Uca*) *major*, and all members of the subgenus *Minuca*—*rapax*, *burgersi*, *vocator*, and *mordax*. Such a diagrammatic distribution of course occurs very rarely both because of the complex configurations of shores and estuaries, and the still unknown requirements of the different species in regard to the compositions of their general biotopes. Three or more species are almost always missing—most often *leptodactyla, major*, and *mordax*—from any locality.

U. rapax, the most abundant of the nine species if the range is viewed as a whole, is most often closely sympatric with another *Minuca, burgersi*. In areas rich in food, such as strips of shore where fish or mollusks are cleaned or coconuts opened, pigs tethered, or close to huts and outhouses straggling at the ends of towns, the two species sometimes mingle in almost equal numbers; these mixed populations appear always to be small, though they usually include fully adult, strongly displaying individuals. In large, normal populations, however, the species mingle, if at all, peripherally, *rapax* usually being found on the more seaward flats and less often in almost landlocked lagoons. On the other hand, large populations of very large *rapax*, apparently never including active wavers and with the individuals probably approaching senility, are often found farther from the sea than the more active groups of both *rapax* and *burgersi*; in these cases they are not up rivers or streams, but in higher, drier locations, open or partly shaded on the inland edges of mangrove swamps, their burrow mouths reached if at all by spring tides only. In these localities, *rapax* is the only *Uca*.

Crescent Beach, on the east coast of Florida, near Daytona, marks the northern limit of *rapax*. Here *rapax* and *pugnax*, members of the same superspecies, are closely sympatric on the same stretch of beach, but at different levels; *rapax* is the higher, in sandier locations (Tashian & Vernberg, 1958).

For additional remarks on the associations of *rapax* see the corresponding sections of *major, burgersi, vocator, pugnax*, and *cumulanta*.

MATERIAL RESULTING FROM FIELD WORK

(The complete list of specimens examined is given in Appendix A, p. 605.)

Observations and *Collections*. U.S.A.: Fort Lauderdale, Miami. Guatemala: Puerto Barrios. West Indies: Puerto Rico; St. Thomas; Guadeloupe; Martinique; Barbados; Curaçao; Tobago; Trinidad. Colombia: Cartagena. Venezuela: Maracaibo and neighborhood; Puerto Cabello; Turiamo; Puerto La Cruz; Pedernales. Guayana: Georgetown. Surinam: near Paramaribo and near mouth of Marawyne R. Brazil: São Luiz, Fortaleza, Recife, São Salvador, Rio de Janeiro.

Films. Miami, Puerto Barrios, Puerto Rico, St. Thomas, Trinidad, Puerto Cabello, Pedernales, Paramaribo, São Luiz, Fortaleza, Recife, São Salvador, Rio de Janeiro.

Sound Recordings. Trinidad.

TYPE MATERIAL AND NOMENCLATURE

Uca rapax (Smith, 1870)

TYPE of *Gelasimus rapax* Smith, 1870. Formerly in Peabody Museum, Yale University, New Haven; from Aspinwall, near Colon, Atlantic coast of Panama; presumably checked by Rathbun (1918.1: 397); not now extant.

? Type Material of *Uca salsisitus* Oliveira, 1939. ? In Instituto Oswaldo Cruz. Male type: cat. no. 30; 4 male and female paratypes: cat. nos. 171, 172. Type-locality: Guanabara Bay, Rio de Janeiro, Brazil. Carapace lengths not given. The male specimens are either young, have the claw regenerated, or both. Synonymy is on the basis of the photograph of the dorsal view (Pl. 11, Fig. 57) and of the drawing of the claw's inner view (Pl. 8, Fig. 43). The latter shows a non-divergent proximal ridge at the dactyl's base, instead of the divergent ridge characteristic of the sympatric *burgersi* and *mordax*. The 4th ambulatory shown in Pl. 3, Fig. 11, and Pl. 8 is slender for mature *rapax*, but, again, the size of the specimen is not known. In Pl. 8 there are suggestions of ambulatory pile which make it possible that *salsisitus* will prove to be synonymous with *vocator*, but in the ambulatory drawings (Pl. 3) no pile is shown. The female abdomen in Pl. 7, Fig. 40 is that of a mature individual, as indicated by the shape and scale, but this character does not aid in species identifications. Finally the text does not state which specimen(s) were used for the illustrations. Examination of the type material may show that more than one species is included.

Type Material of *Uca pugnax brasiliensis* Oliveira, 1939. ? In Instituto Oswaldo Cruz, Rio de Janeiro. Male type, cat. no. 191; paratypes; no. 299. Type-locality: Ilha do Pinheiro, Rio de Janeiro, Brazil. The distinctions in the specimens described are within the range of variation of *Uca rapax* in the type-locality.

Uca rapax longisignalis Salmon & Atsaides, 1968

HOLOTYPE. In Smithsonian Institution, National Museum of Natural History, Washington. *Uca longisignalis*: One male, cat. no. 121599. Paratype males, 5, cat. no. 122204. Type-locality: Ocean Springs. Measurements on p. 193. (!)

Uca (*Minuca*) [*galapagensis*] *rapax rapax* (Smith, 1870)

(Subtropical and tropical western Atlantic. Northern boundaries: in Mexico, Tamaulipas; in west Florida, Yankeetown; in east Florida, Crescent Beach. Southern boundary in Brazil, Cananéia)

MORPHOLOGY

With the characteristics of the species.

In both sexes, *r. rapax* differs from *r. longisignalis* in having the eyebrow only slightly inclined and the front slightly narrower. Additionally, the male differs in the absence of pile on lower margins of merus, carpus, and manus of 2nd and 3rd ambulatory. Finally, the female differs in the presence of a distinct tubercle on the gonopore and the absence of pile on lower margins of ambulatory meri.

The subspecies differs from *p. pugnax*, with which it occasionally coincides on Florida's east coast, chiefly in the slightly, not strongly, inclined eyebrows, a smoother major palm with apex of oblique ridge higher and ridge tubercles larger and more regular, and in having merus of 2nd ambulatory broader, with its upper margin usually slightly convex throughout, but sometimes almost straight in middle. In the particular populations with which they worked, Tashian & Vernberg (1958) found useful differences in the proportions of the major chelipeds, the manus in *rapax* being somewhat deeper and the fingers shorter; these differences are unreliable as subspecies or species distinctions.

In distinguishing members of the majority of populations, all fully tropical, of this subspecies from sympatric species, the following characters used in combination seem to be the most reliable and practical: on major palm, non-divergent predactyl ridges and a well-developed, oblique, tuberculate ridge; pollex with a distinct subdistal crest; broad ambulatory meri, showing conveniently in the convex dorsal margin of the 4th; no pile on lower margins of 2nd and 3rd ambulatory mani; female with a single, well-formed tubercle on posterior margin of gonopore. See also comparative comments in treatments of *vocator*, *burgersi*, and *mordax* (pp. 164, 168, 173).

Uca (Minuca) [galapagensis] rapax longisignalis Salmon & Atsaides, 1968
(Northern part of Gulf of Mexico from Matagorda Bay, Texas to Yankeetown, Florida)

MORPHOLOGY

With the characteristics of the species.

The differences between this form and the tropical *r. rapax* are as follows: in both sexes, eyebrow strongly inclined (although, as usual, less so in females) and front wider (widest in females). In the male, pile that is scanty but distinct is present on lower surfaces of merus, manus, and carpus of at least 2nd and 3rd ambulatories; it is arranged along anterior and posterior borders of each segment and is persistent enough to make it a reliable diagnostic character—the only one I have found; the only specimens examined in which it is lacking are some examples received from Yankeetown, from which all pile and setae had been previously removed for special purposes. In the female, of which I have seen only three individuals, pile is also present, but on the meri only, again along anterior and posterior ventral margins, of 1st, 2nd, and 3rd legs; all have the gonopore with antero-inner margin of lip slightly raised, but not in the form of a tubercle.

From sympatric species along the Gulf, *minax* and *pugnax virens*, *r. longisignalis* is again most clearly distinguished by the pile on the ventral margins of the ambulatories, as described above, and, additionally, from *pugnax virens* alone, by the strongly vertical eyebrows; subdistal crest on pollex characteristically stronger than in the other two species, but experience is needed in comparing them. Even in females of *longisignalis* the front is narrower and the ambulatory meri wider than in female *minax*, while the eyebrows, although also strongly bent, are narrower than in *minax*.

Examination of the types and of additional material in the Smithsonian Institution in Washington has led to the following conclusions on the characters used in the type description. On the major cheliped these characters include an unbroken tuberculate ridge inside the carpus; in 4 males from Ocean Springs, Mississippi, this series is interrupted, as in *p. virens*; in *r. rapax* from southern Florida and elsewhere and in *minax* it is sometimes interrupted and sometimes not. The degree of tuberculation inside the palm and details of the oblique ridge prove to be similarly unreliable characters because of a wide range of variation; similarly, the differences cited in the minor chelipeds of *p. virens* and *r. longisignalis* are undependable. In most, but not all, the examples I have seen of *r. longisignalis*, the anterior flange of the gonopod projects somewhat less than that of most specimens of *p. virens*; again, the degree of projection proves to be variable and overlapping in the two forms.

Although *r. longisignalis* as far as known is a smaller species than *minax*, the several females examined are confusingly similar to those of the larger species, especially since few specimens of *minax* from Gulf Coast populations seem to exist, and they may well prove to have local distinctions. Large, well-preserved series of all three of the local *Minuca* are urgently needed.

REFERENCES AND SYNONYMY

Uca (*Minuca*) *rapax* (Smith, 1870)

TYPE DESCRIPTION. See under *U.* (*M.*) *rapax rapax*, below.

Uca (*Minuca*) *rapax rapax* (Smith, 1870)

Gelasimus rapax

TYPE DESCRIPTION. Smith, 1870: 134; Pl. 2, Fig. 2; Pl. 4, Fig. 3. Atlantic coast of Panama: Aspinwall.

Aurivillius, 1893: 34. West Indies: St. Barthélemy. Morphology; amphibious characteristics; ecology.

Gelasimus minax (not of LeConte)

Nobili, 1897: 3. Venezuela: Puerto Cabello. (Synonymy *fide* Maccagno, 1928: 45.)

Uca pugnax rapax

Rathbun, 1902.1: 7. Puerto Rico. Taxonomy, reducing *G. rapax* to a subspecies of *U. pugnax*.

Rathbun, 1918.1: 397; Pl. 140 (photos). U.S.A.: Florida: Miami, Key West. West Indies: Cuba; Haiti; Puerto Rico; St. Thomas; Jamaica, Curaçao; Trinidad. Panama: Colon. Colombia: Sabanilla. Venezuela: Puerto Cabello. Brazil: Pernambuco (= Recife), Plataforma, Bahia (= São Salvador), Maruim, Itabapuana. Taxonomy; habitat. (Not specimens listed from Alabama and Texas.) (Part USNM !)

Rathbun, 1920: 342. Curaçao. Local distribution.

Rathbun, 1924.3: 19. Curaçao. Local distribution.

Maccagno, 1928: 45; Text Fig. 29 (claw). Curaçao. Venezuela: Puerto Cabello. Taxonomy.

Oliveira, 1939.1: 134. Brazil: Rio de Janeiro. Taxonomy; habitat.

Crane, 1943.2: 40; Text Figs. 1a, b, c. Western and eastern Venezuela. Taxonomy; color; waving display. (!)

Crane, 1957. Subtropical and tropical western Atlantic. Preliminary classification of waving display; in Table 2 its characteristics are included under the species name, *pugnax*.

? *Uca salsisitus*

Oliveira, 1939.1: 131; Pl. 3, Figs. 7-15; Pl. 4, Figs. 19-21; Pl. 5, Figs. 23, 24; Pl. 7, Fig. 40; Pl. 8, Figs. 43, 44; Pl. 11, Figs. 57, 58. Brazil: Rio de Janeiro. Type description. (See p. 196.)

Uca pugnax brasiliensis

Oliveira, 1939.1: 136; Pl. 6, Figs. 29-32; Pl. 7, Fig. 36; Pl. 10, Fig. 56; Pl. 11, Fig. 57; Pl. 12, Figs. 59, 60. Brazil: Rio de Janeiro. Type description.

Oliveira, 1939.2. Brazil: Rio de Janeiro. Ecology.

Uca pugnax (not *Gelasimus pugnax* Smith)

Gmitter & Wotton, 1953: 271; Text Figs. 16-18 (all captioned under the name *Uca pugnax rapax*). West Indies: St. Thomas. Brief, general taxonomy.

Uca rapax

Tashian & Vernberg, 1958. Illus. U.S.A.: eastern Florida. Taxonomy (specific distinctness of *rapax* and *pugnax* restored); color; display; habitat.

Gerlach, 1958.1: 672. Brazil: Estado São Paulo (near Cananéia). Ecology.

Holthuis, 1959.3: 266; Text Figs. 64d-f, 65; Pl. 14, Figs. 4-6; Pl. 15, Fig. 3. Surinam. Taxonomy; habitat.

Barnwell, 1963. Illus. Brazil. Endogenous daily and tidal rhythms of melanophore and motor activity.

Miller, 1965. U.S.A. Morphology, physiology, and ecology in relation to distribution.

Vernberg & Costlow, 1966. Development of right or left handedness.

Crane, 1967. Illus. West Indies: Trinidad. Combat and its ritualization.

Salmon, 1967: 450. Illus. U.S.A.: Florida. Local distribution; display; sound production.

Altevogt & Altevogt, 1967: E 1291. Film of waving display.

von Hagen, 1967.1. Illus. West Indies: Trinidad. Kinaesthetic orientation to burrow.

von Hagen, 1967.2. West Indies: Trinidad. Taperecordings secured. (Preliminary statement.)

Salmon & Atsaides, 1968.2. Illus. U.S.A. Included in their general paper on visual and acoustical signaling in *Uca*.

Chace & Hobbs, 1969: 214; Text Figs. 73a, b. West Indian distribution: Bimini I.; Cuba; Jamaica; Hispaniola; Puerto Rico; St. Thomas; St. Croix; Antigua; Guadeloupe; Providencia; Bonaire; Curaçao; Los Roques; Trinidad. Taxonomy; habitat.

Warner, 1969: 382-85. West Indies: Jamaica. Ecology.

von Hagen, 1970.1: 226. Caribbean distribution. Taxonomy.

von Hagen, 1970.4. Illus. West Indies: Trinidad. Adaptations to a particular intertidal level.

Uca (*Minuca*) *rapax longisignalis* Salmon & Atsaides, 1968

Gelasimus vocator (not *Cancer vocator* Herbst)

Herrick, 1887: 44; Pl. 5, Fig. 4 (outer major claw). U.S.A.: Alabama: Mobile B. Brief taxonomy; synonymizes *G. pugnax* Smith with *vocator*; attributes *vocator* to von Martens.

Uca longisignalis

TYPE DESCRIPTION. Salmon & Atsaides, 1968.1: 279; Text Figs. 1-4, 6, 7. U.S.A.: Florida: Yankeetown; Mississippi: Ocean Springs; Louisiana: Cameron. Color; waving display; sound production.

Salmon & Atsaides, 1968.2. Illus. U.S.A. No new material. Included in this general paper on visual and acoustical signaling in *Uca*.

34. *UCA (MINUCA) [GALAPAGENSIS] PUGNAX* (SMITH, 1870)

(Temperate and subtropical western north Atlantic)

PLATE 27 *E-H*.   MAPS 10, 14.
FIGURES 39 *H*; 46 *J*; 67 *E*; 81 *J*; 100.   TABLES 9, 10, 12, 14, 19, 20.

INTRODUCTION

A small fiddler with an often lethargic display, *Uca pugnax* is the most abundant of its genus on the eastern coast of the United States. From Cape Cod to northern Florida it often associates more or less closely with (*Celuca*) *pugilator*, although *pugnax* occurs typically on muddy substrates, while *pugilator* lives on sandier strips of shore. Throughout these latitudes waving members of the two species can be distinguished in the field, even at a considerable distance, by the slower, less frequent display of *pugnax*, the waving always showing distinct traces of jerks. The degree and directions of circularity in the waves are also different; that of *pugnax* is much less broadly circular and of the usual form, the cheliped moving obliquely-out-and-up-then-in-and-down, instead of the reverse wave (up-and-out-obliquely-down-and-in) characteristic of *pugilator*. In addition, color differences are an additional help to fiddler-watchers unfamiliar with the species along the same coasts; except in the extreme southern part of its range, *pugnax* during the summer can be relied on to show at least some individuals with blue or blue green fronts, eyestalks, or both, while a member of the same species never has a purple or red spot on the carapace, as do many individuals of *pugilator*; finally, but least helpfully, many individual *pugilator* pale to a color more nearly approaching display white during display than do *pugnax*.

The third species of *Uca* sharing the same, temperate range with *pugnax* is *minax*. Because it lives characteristically when adult in marshes with waters of very low salinity, close sympatry in the field with *pugnax* rarely occurs. The wave of the much smaller *pugnax* is much more frequent and its jerks are slight, in comparison with the uncommon wave of *minax* with its very strong jerks; in addition, *minax* in summer usually shows distinctive, narrow red rings around the segment joints of the major cheliped.

Two species considered closely related to *pugnax* were described by Salmon & Atsaides, 1968.1. One is here termed a subspecies of *pugnax*, *U. p. virens*, and the other of *rapax*, *U. r. longisignalis*. The first, *virens*, originally reported only from Mississippi and Texas, has since been found in old collections from Mexico; the second, *longisignalis*, is known at present from Texas, Louisiana, Mississippi, and northwest Florida. The authors report that in *virens* the anterior one-third of its carapace, except for a white anterior border, is green in males, while *longisignalis* (which overlaps in range with *virens* and lives in close sympatry with it in Mississippi), has the front bright turquoise in both sexes, blending into a blue green band more posteriorly.

Both forms of *pugnax*, like other northern species, produce sounds in the evening; these differ from those both of each other and of the two subspecies of *rapax*.

The introduction to the preceding species (*rapax*, p. 190) includes remarks on the relationships of *pugnax*; the general section on sympatry (p. 529) in Chapter 7 discusses peculiarities of the zone of overlap with *rapax* on the eastern coast of Florida.

MORPHOLOGY

Diagnosis

Distinguishing characteristics from related broad-fronted *Uca*, except from the two members of its superspecies, are as in the first paragraph of the diagnosis for *galapagensis*, with allowance for relative weakness of oblique ridge on major palm. From these closest relations, *galapagensis* in the Pacific and *rapax* in the Atlantic, *pugnax* is distinguished most reliably in the following respects: From both in having merus of 2nd ambulatory comparatively slender, its dorsal margin always practically straight except at each end; in having extreme apex of oblique ridge on major palm low, with its tubercles relatively small and irregular; and, in the Atlantic subspecies only, in having the eyebrow slanting so steeply it is practically or wholly vertical. Additionally, from *galapagensis* only, in having postorbital groove on carapace very shallow and in having a tubercle beside female gonopore. Additionally, from *rapax* only, in having center of major palm clearly tuberculate, never almost smooth, and in the definite continuation of tubercles of oblique ridge upward around carpal cavity.

Description

With the characteristics of the subgenus. Very close to its Atlantic allopatric, *rapax*, from which it differs most reliably in the following respects.

MALE

Carapace. Sides of front converging less, the tip therefore appearing slightly broader in most individuals than in *rapax*; eyebrow usually narrower, and, in the Atlantic form only, more depressed than in the Atlantic subspecies of *rapax*, often practically vertical and so facing forward; in the Gulf of Mexico these relationships of the eyebrows are reversed; proximal margin of eyebrow apparently never beaded, instead of often so in large specimens of *rapax*.

Major Cheliped. Palm with oblique ridge almost always weaker than in *rapax*, being both lower as a whole, with the apex usually slightly lower than adjacent part of ridge, the tubercles being relatively small and often in an irregularly compound row, especially in distal part of ridge where they are sometimes obsolescent; traces of the distinct break and overlap of rows of tubercles near the ridge are often discernible, especially when the observer is familiar with the corresponding structures in *rapax* and *galapagensis*; the ridge variations are further discussed in association with the major cheliped of *rapax*, p. 192. Beyond apex the ridge tubercles are always, instead of rarely, continued distinctly, although for a variable distance, upward around carpal cavity. Center of palm always covered with tubercles that mingle proximally with those of the upward continuation of oblique ridge; usually they are of moderate to large size for the area, but sometimes so small that their diameter relative to the space is well within the range of this character in *rapax*, a species noted for the fineness of any granulation of its palm. Tip of pollex with a small, outer, dentate crest that is usually weaker and with its tubercles less closely spaced than in *rapax* and *galapagensis*; it is nevertheless clearly a homologous structure derived, as in the other two species, from the displaced distal end of the median row of the pollex' prehensile edge.

Ambulatories. Meri, although moderately widened throughout, are less so than in *rapax*; dorsal margin of 2nd merus practically straight except proximally, and distally where it curves down to the segment's ends.

Gonopod. Inner process apparently always straight, unlike *rapax*, where it is sometimes slightly curved.

FEMALE

Gonopore. A small tubercle indicated in the Atlantic form, absent in the Gulf of Mexico.

Measurements (in mm)

	Length	Breadth	Propodus	Dactyl
(All from *p. pugnax*, Long Island; measurements of *p. virens* are comparable)				
Largest male	12.0	19.0	33.0	23.5
Largest claw (detached)	–	–	37.0	25.0
Moderate male	10.0	16.0	25.0	10.0
Largest female	13.0	18.0	–	–
Moderate female	9.0	13.0	–	–

Morphological Comparison and Comment

As in other closely related species of *Minuca*, the few morphological characters that present themselves for taxonomic use are extremely variable, both within and among populations. Accordingly, identifications can be made with confidence only after a certain amount of experience with preserved specimens; even then it is more than usually important that series be examined. In most areas, of course, the ranges of these particular species do not coincide, and when displaying populations are observed they are easily identified.

A most interesting point concerns the chief differences between *rapax* and *pugnax*. When the several characters are looked at as a whole, it is apparent that in *rapax* both the palm and the pollex tip are distinguished by armature facilitating regularity of ridge-rubbing in performance of the heel-and-ridge component during combat. In *pugnax*, on the other hand, the tuberculate palm, low ridge with irregular rows of small tubercles, and the weak pollex crest seem conducive to a less precise juxtaposition of palm and pollex. A detailed comparison of heel-and-ridge performance in these two species has yet to be attempted.

Diagnostic differences between *pugnax pugnax* and *minax* are summarized on p. 177; between *pugnax pugnax* and *pugilator* on p. 225. These species are the only *Uca* known to occur on the Atlantic coast north of Florida.

Color

Displaying males: Display whitening poorly or not at all developed, the carapace usually brown, sometimes lightening to pale gray at least on branchial regions on individuals that have dried in the sun more thoroughly than usual; no purple or red spot ever present on carapace; anterior region of carapace, eyestalks, and anterior parts of 3rd maxillipeds often blue to blue green (turquoise) in specimens from Massachusetts to St. Augustine; the color appears, strongly or weakly, on any or all of these areas; in St. Augustine it was slight and in few individuals (Crane, unpublished); farther south Salmon (1967) found it absent. In Texas and Mississippi Salmon & Atsaides

(1968.1, in the description of *virens*) found the anterior one-third of carapace, except for an anterior border, bright green in males. Major cheliped lightens to various degrees of light brownish or yellowish, the manus being dull yellowish orange to yellowish white in the brightest examples in the northern states mentioned, with fingers always white or nearly so; proximal and distal ends of segments not red. Manus and dactyl of minor cheliped white. Ambulatories dark, banded with darker. Females similar to males, including development of blue and blue green, but not as strongly as in male.

SOCIAL BEHAVIOR

Waving Display

Lateral, weakly circular, at low intensity almost straight. Jerks almost always distinguishable but very variably so and very weak in comparison with those of *minax* and *rapax*; in all areas they are strongest in northern and western (Mississippi and Texas) parts of range and weak to practically absent near the zone of overlap with *rapax* in eastern Florida; jerks ranging from 3 to 14 on the cheliped's rise and from 1 to 7 on its descent. A pause at apex of wave, longer than that at end, present or absent, variable in occurrence even within populations. Minor cheliped makes a roughly corresponding motion. Body held raised throughout waves of a series; one or more ambulatories kicked out during display, but in the leg-waves (waving component 12) the meri do not touch as in sound-producing leg-wagging (sound component 5). Curtsy present during high intensity courtship between waves, the cheliped held flexed during the component, which is often or always accompanied by stamping (see under Acoustic Behavior, below). Minimal herding sometimes takes place; Salmon (1967) observed this component in areas where waving was partly obscured by vegetation. (Component nos. 5, 6, 9, 11, 12, 13, 14; timing elements in Table 19, p. 656.)

Precopulatory Behavior

Female follows male down his burrow after high intensity waving display with stamping. At night females are attracted by sound production achieved by the males through motions of the ambulatories, both with leg-wagging, the meri vibrating in intermittent contact as usual in agonistic behavior (component 5), and with stamping (11) (Salmon, 1967; Salmon & Atsaides, 1968.1, 1968.2).

Hood Construction

Hood-building has been observed only on Long Island, at Port Jefferson, and there the habit was rarely

and minimally developed. In these examples the male's burrow was always on the side of a small elevation, such as a clump of weed; over the entrance hung a small, roughly formed half-dome, which I first interpreted as a chance result of pushing up the soft substrate during excavation; however, in several examples the male added as I watched to the ceiling of the structure from outside the burrow, in the fashion characteristic of the builders of turrets and hoods in *Celuca*.

Acoustic Behavior

Salmon & Atsaides (1968a) found that both subspecies of *pugnax* made sounds through movements of the ambulatories at night for periods of up to several hours. "Each sound . . . was composed of several pulses produced in a series. In [*U. p. pugnax*] the sounds were usually composed of 2-3 pulses and intervals between successive sounds that averaged about 7 seconds. In [*U. p. virens*] most sounds contained 2 pulses, but the intervals between successive sounds were about 3 seconds. . . . The signals described above were all produced by lone males." The stamping component was of lower frequency than that of the leg-wagging. Films of *pugnax* made during the daytime at St. Augustine and on Long Island also show stamping during the leg-*waving* of courtship, and, in agonistic situations, leg-wagging without stamping. (Components 5, 11.) (See also general remarks in Chapter 5, under Acoustic Behavior, p. 482.)

Combat

The following components are known, all observed on the south shore of Cape Cod: Manus-rub, subdactyl-&-subpollex-slide, dactyl-slide, upper-&-lower-manus-rub, interlace, heel-&-ridge. (Component nos. 1, 4, 6, 7, 9, 12.)

RANGE

Temperate western north Atlantic along the eastern and southern coasts of the United States from Massachusetts, on the northwest coast of Cape Cod at Provincetown (lat. 42° N) to the state of Vera Cruz, Mexico. The Atlantic subspecies (*p. pugnax*) extends as far south as Daytona Beach on Florida's east coast; the other subspecies (*p. virens*) is known from Ocean Springs to the neighborhood of Coatzacoalcos, slightly north of lat. 18° N.

BIOTOPES

Salt marshes and other sheltered shores close to bays, inlets, and estuaries, the substrate ranging from mud to sandy mud and muddy sand. (Biotope nos. 6, 12, 16.)

SYMPATRIC ASSOCIATES

On the Atlantic coast sometimes associated marginally with (*Minuca*) *minax* or (*Celuca*) *pugilator* (pp. 178 and 226); in the zone of overlap in Florida with (*Minuca*) *r. rapax* (Vernberg & Tashian, 1953; Salmon, 1967); in Ocean Springs, Mississippi, with *r. longisignalis* (Salmon & Atsaides, 1968.1), where the two species occur within sight of each other (Salmon, personal communication).

MATERIAL RESULTING FROM FIELD WORK

(The complete list of specimens examined is given in Appendix A, p. 606.)

Observations and Collections. United States: Massachusetts, on Cape Cod, near Cotuit; Connecticut; New York on Long Island; New Jersey, near Brigantine; Florida, near St. Augustine.

Films. Long Island; St. Augustine.

TYPE MATERIAL AND NOMENCLATURE

Uca pugnax (Smith, 1870)

TYPE MATERIAL. In Peabody Museum, Yale University, New Haven. Cat. no. 1060; West Haven. Measurements of extant specimen in mm: length 11; breadth 16.5; propodus 24; dactyl 19. (!) The other type-specimens have not survived. The extant specimen has the right cheliped closely similar to and almost as large as the left, the propodus measuring 20 and the dactyl 15. The anomaly was described by Smith (1870: 133), who noted in addition: "The specimen which was examined while alive was very active and used both hands with equal facility." This specimen is also the subject of a preliminary note by Smith (1869.2), headed "A Fiddler-crab with Two Large Hands," in which he remarks: "It does not appear to differ from the common *Gelasimus palustris* except in the right cheliped. . . ." (For comments on *palustris*, see p. 324.)

Uca virens Salmon & Atsaides, 1968

HOLOTYPE. In Smithsonian Institution, National Museum of Natural History, Washington. Cat. no. 121598. Type-locality: Port Aransas. Measurements in mm: length 12; breadth 21; propodus 34; dactyl 24.5. (!)

Additional type material. Same locality: paratype male; cat. no. 122205. (!)

Uca (Minuca) [galapagensis] pugnax pugnax (Smith, 1870)

(From the northeast coast of Cape Cod to Daytona Beach)

MORPHOLOGY

With the characteristics of the species.

In both sexes, *p. pugnax* differs from *pugnax virens* in having the eyebrow almost always strongly inclined and somewhat narrower; and in never having the anterior margin of the front appearing truncate; female gonopore with posterior edge slightly raised, and a minute tubercle present or absent.

The distinctions between *p. pugnax* and *r. rapax* in their zone of coincidence are given on p. 197.

From the other two *Uca* occurring north of Florida, *p. pugnax* is distinguished on pp. 177 and 225.

Uca (Minuca) [galapagensis] pugnax virens Salmon & Atsaides, 1968

(Gulf of Mexico, from Ocean Springs west and south to Coatzacoalcos, Mexico)

MORPHOLOGY

With the characteristics of the species.

In both sexes *pugnax virens* is distinguished from *p. pugnax* as well as from two sympatric *Minuca* along the northern Gulf coast, *minax* and *rapax longisignalis*, by never having the eyebrows strongly inclined; from *rapax longisignalis*, *p. virens* differs clearly, in both sexes, by the absence of pile on the lower margins of the ambulatories.

Examination of additional material has shown that several characteristics given in the type description, although sometimes diagnostically helpful, are unreliable, because of variation in *p. virens* or in the other two sympatric *Minuca* in the north. While the tuberculate ridge inside the major carpus is always interrupted, being divided into two parts, as described, in *p. virens*, a similar interruption sometimes occurs in *r. longisignalis* and in *minax*. Again, while in most specimens of *p. virens* the front appears somewhat truncate, at least in an antero-dorsal view, this characteristic is occasionally ambiguous or absent. The tuberculation of the palm also proves to be too variable to provide trustworthy characters.

Finally, the gonopod, as in related forms in this subgenus, shows no characteristics of diagnostic reliability, its form and extent of variation appearing closely similar to those of *p. pugnax* and about equally similar to both subspecies of *rapax*. Examination of the type material showed that the spines around the pore, referred to in the type description and illustrated in its accompanying figure, are a combination of stiff bristles (often very deceptive in these species) with the narrow tips of the inner process and both flanges, all of which project slightly beyond the pore. Although the anterior flange usually projects distinctively this projection forms at best a supplementary character; sometimes, too, its narrowness is apparent only, since it is then joined to the pore by a slanting, transparent section. In one individual the tip of the anterior flange appeared broken, perhaps during copulation.

In Tampico, Mexico, and doubtless slightly farther south, *p. virens* is sympatric with *vocator*; in the male the absence of pile on the ventral surfaces of the ambulatories in *p. virens* distinguish it easily, as do the strong development of the oblique ridge inside the palm and the very different form of the gonopod; in the female the flat edge of the gonopore contrasts with the uneven, tuberculate margin found in *vocator*.

REFERENCES AND SYNONYMY

Uca (Minuca) pugnax (Smith, 1870)

Note. U. pugnax is one of the species of *Uca* that have been used extensively in laboratory investigations. These have concerned principally endogenous rhythms of various kinds and the hormonal control of chromatophore activity. Key references appear on p. 448.

TYPE DESCRIPTION. See under *U. (M.) pugnax pugnax*, below.

Uca (Minuca) pugnax pugnax (Smith, 1870)

Gelasimus vocans (not *Cancer vocans* Linnaeus)

Gould, 1841: 325 (part). U.S.A.: Massachusetts. Taxonomy, habitat, and habits (digging), all treated briefly.

Gelasimus vocans, var. A

De Kay, 1844: 14; Pl. 6, Fig. 10. U.S.A.: New York. Taxonomy; habitat.

"A Fiddler-crab with Two Large Hands"

Smith, 1869.2: 557. Preliminary note on the type-specimen now extant.

Gelasimus pugnax

TYPE DESCRIPTION. Smith, 1870: 131; Pl. 2, Fig. 1; Pl. 4, Figs. 2-2d. Type-locality (= of specimens figured): New Haven; other localities on northeast coast, U.S.A. (Yale, 1 male !)

Verrill, 1873: 367, 377, 466, 468, 545. U.S.A.: Massachusetts. General (not local) distribution; basic ecology and habits; no display or combat.

Verrill & Smith, 1874: 73, 83, 172, 174, 251. A reissue of Verrill, 1873.

Kingsley, 1878: 321. U.S.A.: Georgia. Local habitats.

Uca pugnax

Rathbun, 1900.2: 585. Key.

Rathbun, 1905. U.S.A.: northeastern states. Local distribution.

Fowler, 1912: 454. U.S.A.: New Jersey. Taxonomy; habits.

Schwartz & Safir, 1915. U.S.A.: New York. Ecology; habits.

Hay & Shore, 1918: 451. U.S.A.: North Carolina. Taxonomy; color; ecology; habits, including breeding season.

Rathbun, 1918.1: 395; Pl. 139 (photos). East coast of U.S.A. from Provincetown, Massachusetts, to Myrtle Bush Creek, South Carolina. Not the specimen from New Orleans, Louisiana. Taxonomy. (Part USNM !)

Pearse, 1928: 231ff. U.S.A.: North Carolina. Ecology; tests on toleration of desiccation and fresh water.

Boone, 1930: 220; Pl. 74. U.S.A.: New York. Taxonomy.

Crane, 1943.3: 217. U.S.A.: northeastern states. Taxonomy; color; waving display.

Crane, 1957. U.S.A.: northeastern states. Preliminary classification of waving display.

Teal, 1958. Illus. U.S.A.: Georgia. Local distribution; ecology; experiments on tolerances for different foods, temperatures, and salinities, as well as on substrate preferences; comparisons with sympatric species.

Teal, 1959. Illus. U.S.A.: Georgia. Respiration and its relation to ecology.

Miller, 1961. U.S.A.: North Carolina. Feeding mechanisms and their adaptations.

Miller, 1965. U.S.A.: Morphology, physiology, and ecology in relation to distribution.

Salmon, 1965. Illus. U.S.A.: North Carolina. Waving display and sound production in courtship.

Salmon, 1967: 450; Illus. U.S.A. Distribution in Florida. Waving display; sound production.

Salmon & Atsaides, 1968.2. Illus. U.S.A. Included in their general paper on visual and acoustical signaling in *Uca*.

Uca (Minuca) pugnax virens
Salmon & Atsaides, 1968

Uca virens

TYPE DESCRIPTION. Salmon & Atsaides, 1968.1: 281; Text Figs. 2, 3, 5, 6, 7. U.S.A.: Mississippi: Ocean Springs; Texas: Port Aransas. Color; waving display; sound production. (USNM !)

Salmon & Atsaides, 1968.2. Illus. U.S.A. No new material. Included in their general paper on visual and acoustical signaling in *Uca*.

35. *UCA (MINUCA) ZACAE* CRANE, 1941

(Tropical eastern Pacific)

PLATE 28 *A-D*. MAP 14.
FIGURES 68 *F*; 100. TABLES 9, 10, 19, 20.

INTRODUCTION

Uca zacae at the moment appears to be a somewhat unimpressive member of its subgenus. The crabs are almost the smallest in the group, dark in color, with only a single jerk in their waving display, infrequent drumming, minimal combats, and a known range covering a short strip of coast. Nevertheless a few populations observed briefly in life were all easily accessible, crowded, lively, and showing hints of aggressive behavior that nowadays would demand analysis. These fiddlers will certainly repay some concentrated attention.

MORPHOLOGY

Diagnosis

Front wide; orbits strongly but not extremely oblique, with antero-lateral margins short but definite. Outer major pollex with a shallow, triangular depression supraventrally at its base; palm without an oblique, tuberculate ridge which is represented if at all by a few minute tubercles close to carpal cavity; center palm entirely smooth; a few granules present subdorsally; proximal ridge at dactyl base not diverging from distal; manus not tumid; no backwardly directed heel. Gonopod with anterior flange projecting scarcely or not at all beyond posterior flange; inner process narrow, flat, tapering. Gonopore crescentic.

Description

With the characteristics of the subgenus *Minuca* (p. 154).

MALE

Carapace. Front only moderately wide for the subgenus, contained about 4 times in width of carapace between antero-lateral angles, where it is widest; angles project scarcely or not at all, being about rectilinear; orbits strongly but not extremely oblique; antero-lateral margins distinct, short, straight or slightly convex, slightly converging, broadly angled posteriorly; dorso-lateral margins moderately converging. Upper postero-lateral stria long; lower short, weak, or absent. Suborbital region and adjacent floor of orbit not densely setose; crenellations moderately developed externally, weak to absent internally. H-form depression with traces of pile. 3rd to 6th abdominal segments partly fused.

Major Cheliped. Ventral margin of merus with two rows of fine serrations proximally, changing distally, especially in anterior row, to small tubercles. Manus of normal thickness and not unusually broad in comparison to its length, being at least 1.5 times longer than broad; fingers clearly longer than palm in specimens with a carapace length approaching 6 mm. Lower part of outer manus with a row of tubercles supramarginally, small and beaded proximally but large and separated in distal half and stopping at pollex base; a shallow, triangular depression, distinctly bounded, at base of pollex. Large tubercles covering less than upper half of outer manus. Palm without tuberculate, oblique ridge; central part of palm smooth, not granulate or tuberculate; beaded inner edge of dorsal margin curving down around carpal cavity, there being no connection from carpal cavity to depression of upper palm; this depression is weakly tuberculate; proximal ridge at base of dactyl not diverging from distal ridge. Pollex and dactyl both normally shaped for the subgenus, not short, deep, and thick, the dactyl curving down beyond pollex as usual. Prehensile edge of pollex with inner row of tubercles obsolescent to absent and outer row weak, although median row is strong; just before the tip a large tubercle stands close to outer edge, followed distally by two small ones; these three apparently belong to the middle, rather than outer, row and represent in rough form the small, outer, distal crest found in the superspecies *galapagensis*. Proximal half of dactyl's prehensile edge usually with several, rounded tubercles, enlarged and about equally spaced.

Minor Cheliped. Palm normal, not broad and thick. Gape narrow, but broader than usual in the subgenus.

Ambulatories. Meri slender, their dorsal edges practically straight. Legs without pile, except for sparse patches sometimes present dorsally on carpus and proximal manus of each, but easily abraded and perhaps variable.

Gonopod. Flanges broad, of similar breadth and length, scarcely concave. Inner process narrow, tapering, thin enough to be translucent, closely applied to posterior flange, the anterior being completely uncovered as is the path of the canal throughout the length of the flanges except at their extreme base; pore opening posteriorly, not distally, the flanges appearing fused in front of canal. Thumb well developed, but arising far below flanges and ending below their base.

FEMALE

Gonopore. Crescentic; outer lip thickened but without tubercle.

Measurements (in mm)

	Length	Breadth	Propodus	Dactyl
Holotype male	6.9	11.0	15.1	9.5
Largest female	6.1	9.1	–	–

Morphological Comparison and Comment

Although not as extremely paedomorphic in form as *pygmaea*, *zacae*'s most obvious specializations are along the same lines, the orbits being markedly oblique, claw not strongly leptochelous, and with tuberculate, oblique ridge of palm absent. In comparison with more characteristic members of the subgenus *Minuca* its front is relatively narrow, the gape in the small chela somewhat wide, the legs unusually narrow and, finally, the gonopod more similar to those of certain *Celuca*. All of these characters, in fact, have much in common with the less specialized members of that subgenus, although it shares a majority of characters with *Minuca*, including the jerking form of its waving display. Undoubtedly it should be considered an intermediate species.

Adults of *zacae* are easily confused with young *galapagensis*, from which they differ as follows: Orbits more oblique in *zacae*, fingers longer than palm, merus of each ambulatory slender, relatively so even in females. Finally, the male gonopods do not have the anterior flange projecting distinctly beyond the posterior, as is always true in *g. herradurensis*, while the inner process is relatively inconspicuous, unlike its form in other *Minuca*, both adults and young.

Two males and a female of the subgenus *Minuca*, all now in bad condition, were taken on Cocos Island, well off the coast of Costa Rica. The larger male, measuring about 9 mm long with a propodus of 22.5 mm, is much larger than other known specimens and its peculiarities may be due only to size. Its characteristics include less oblique orbits than are found in *zacae* as now understood, and there are traces of an oblique ridge on the palm. The other specimens are small and in even less good condition.

Certain identification must await more material. Because of the geographical interest combined with the questionable identity, the large male is listed on p. 327.

Color

Displaying males: Display whitening absent. Carapace dark, marbled with gold color and black; major cheliped with outer manus reddish brown to pinkish orange; fingers white.

SOCIAL BEHAVIOR

Waving Display

Wave lateral straight, the cheliped unflexing obliquely upward, usually with one jerk during elevation; sometimes, instead, the elevation is unbroken while a jerk occurs on the descent. During elevation one or both of the middle ambulatories on each side are raised high off ground, held there at wave's peak, then brought down along with cheliped with apparent force. At higher intensities cheliped is not brought all the way to ground between waves. Body held fairly high during waving and a few steps usually are taken during each wave. Duration of wave about 4 to 5 seconds. (Components 4, 6, 11, 12; timing elements in Table 19, p. 656.)

Acoustic Behavior

Major manus drumming (component 9) observed but not filmed at burrow mouth during courtship. Its use, and that of other possible acoustic components, not observed clearly enough during agonistic behavior for description.

RANGE

Tropical eastern Pacific from Los Blancos, El Salvador, to Golfito, Costa Rica; questionably from Cocos Island.

BIOTOPES

Mud and muddy sand flats close to mouth of streams, usually near mangroves, sometimes on the stream banks themselves when flat or gently sloping; sometimes on flats behind the fringing mangroves; occasionally in almost or wholly fresh water slightly upstream from mangrove limits. (Biotope nos. 9, 12, 14, 15.)

SYMPATRIC ASSOCIATES

As in *U. galapagensis* (p. 186).

MATERIAL RESULTING FROM FIELD WORK

(The complete list of specimens examined is given in Appendix A, p. 607.)

Observations and Collections. Nicaragua: Corinto; San Juan del Sur. Costa Rica: Golfito.

Films. Golfito.

TYPE MATERIAL AND NOMENCLATURE

Uca zacae Crane, 1941

HOLOTYPE. In Smithsonian Institution, National Museum of Natural History, Washington. Male, cat. no. 137426 (formerly New York Zoological Society cat. no. 381,112); Golfito, Costa Rica; length 6.9 mm (see p. 207). Also deposited: 27 paratypes, males and females, cat. no. 137427 (formerly New York Zoological Society cat. no. 381,113): Golfito, Costa Rica. (!)

REFERENCES AND SYNONYMY

Uca (Minuca) zacae Crane, 1941

Uca zacae

TYPE DESCRIPTION. Crane, 1941.1: 175; Text Figs. 4c, 5; Pl. 2, Fig. 5. Nicaragua; Costa Rica. (USNM !)

Holthuis, 1954.1: 41. El Salvador. Habitat.

Holthuis, 1954.2: 162. Reference to 1954.1, above.

Barnwell, 1968.1. Illus. Costa Rica. Chromatophoric responses to light and temperature.

Uca macrodactyla glabromana

Bott, 1954: 168 (part). El Salvador. Some of paratypes should be referred to *zacae*. (Frankfurt !)

36. *UCA (MINUCA) SUBCYLINDRICA* (STIMPSON, 1859)

(Western north Atlantic: Gulf of Mexico)

PLATE 28 *E-H*.
FIGURES 67; 100.
MAP 11.
TABLE 9.

INTRODUCTION

A small crab with an almost cylindrical body, *Uca subcylindrica* is a challenging species of which we know almost nothing. So far we have no information whatever on the living fiddler, and only a handful of preserved specimens, all from southern Texas and northern Mexico.

In addition to an unusually rounded body, this fiddler has a gonopod of striking form. The proportions of the parts are so distorted from their customary arrangements that it is not at all certain that the species should be included in the subgenus *Minuca*. Once we have learned something of its display and other characteristics, a change in its suggested systematic position will perhaps be indicated.

MORPHOLOGY

Diagnosis

Front broad; no oblique tuberculate ridge on major palm; body subcylindrical; gonopod with distal part of inner process divided, appearing fringed.

Description

With the characteristics of the subgenus (p. 154).

MALE

Carapace. Highly arched, hepatobranchial and branchial regions being the most tumid, boundaries between regions indistinct to absent. Front contained about 3 times in carapace breadth; widest part of carapace is behind antero-lateral angles which are almost rectilinear, not produced; antero-lateral margins slightly diverging, sharply angular posteriorly at least on one side; dorso-lateral margins weak to absent, sometimes present only on major side, sometimes not connecting with antero-lateral margin; upper postero-lateral stria moderately long, but weak, sometimes absent; lower stria very weak to absent. Suborbital crenellations small but well formed throughout, similar, scarcely enlarged around orbit's outer angle; suborbital region with long, curved setae but no pile. Abdomen unusually broad.

Major Cheliped. Ventral margin of merus with two rows of tubercles, sometimes irregular, sometimes partially joined by transverse rugosities. Manus with outer surface entirely covered with small tubercles of similar size throughout, scarcely or not at all larger near upper margin; slightly bent over; subdistal groove absent but a strong row of close-set tubercles present along most dorsal part of manus, ending distally in a direct line with upper border of proximal, subdorsal groove on dactyl; ventral margin with a tuberculate keel and adjacent groove above it, the tubercles close set, almost beaded, throughout, largest distally near pollex base where it ends abruptly; lower distal part of manus also with a tuberculate ridge that continues to run along middle of pollex where the tubercles decrease in size and die out, although the ridge itself may continue almost to pollex tip; no marked groove above this ridge; pollex base above ridge depressed, the depression continuing partway along pollex as a shallow groove but not as well marked as in Indo-Pacific *Deltuca*; no enlarged heel on manus. Palm with lower triangle smooth; oblique, tuberculate ridge absent but central area and region of ridge covered with small tubercles; beaded, upper edge of carpal cavity scarcely curving downward, so that cavity connects by a trench, scarcely interrupted by granules, with the predactyl area; both predactyl ridges strongly tuberculate, with proximal ridge parallel to adjacent groove. Prehensile edge of pollex with tubercles of outer and inner rows weak to obsolescent, but of inner row strong; a small, distal, tuberculate crest, similar to that found in the superspecies *galapagensis*, but with its tubercles usually not individually distinct. Dactyl little longer than manus with a strong, subdorsal groove, a faint suggestion of a lateral groove beneath it, running at least half the outer dactyl's length, and a row of large tubercles proximally on prehensile edge.

Minor Cheliped. Fingers both broad, gape almost absent.

Ambulatories. Moderately slender, the dorsal margins of meri being usually slightly convex; meral armature extremely weak, and virtually absent on ventral margins; these margins, as well as the usual areas more distally, have exceptionally long, curved, spaced, moderately stiff setae.

Gonopod. Anterior flange narrow and short, projecting little or not at all beyond pore; posterior flange very broad, its base apparently traceable far down shaft, its outer half (farthest from pore) greatly produced; this portion of the flange alone is covered by the minutely divided distal half of the inner process, the most anterior "fringes" being short while the rest are progressively longer, the most posterior extending slightly beyond distal projection of posterior flange. Pore large, its opening distal, between the flanges. Thumb vestigial, represented by a broad, slight shelf, displaced by the tip's torsion from its usual relation to the other elements.

FEMALE

Gonopore extremely large, oval, with the long axis directed antero-posteriorly, the margin slightly raised except antero-internally. A small tubercle, variable in size, at the rim's antero-external boundary; in one of the three cotypes this tubercle engages closely with a similar tubercle on the concealed, anterior part of a projecting spermatophore. The spermatophores in all three examples are partly soft, smooth, and slightly arched over the sternum. The appearances and large sizes of gonopore and spermatophore are unique in the genus.

Measurements (in mm)

	Length	Breadth	Propodus	Dactyl
Largest male (Cotype)	11.5	18.5	21.5	15.5
Largest male (USNM)	11.0	(damaged)	25.0	16.5
Moderate male	10.0	16.0	18.0	11.0
Largest female (Cotype)	12.0	20.0	–	–

RANGE

Known only from the United States, near the southwest shore of Texas, and from the adjacent part of Mexico, on the Rio Grande.

TYPE MATERIAL AND NOMENCLATURE

Gelasimus subcylindricus Stimpson, 1859

COTYPES. In Museum of Comparative Zoology, Harvard University, Cambridge, Massachusetts: cat. no. 1327; 1 male, 3 females, type-locality: Matamoros, Mexico, on the Rio Grande; collected by M. Berlandier; deposited by Lieut. Couch, U.S. Army. (!) Measurements: see above.

In the Muséum d'Histoire Naturelle, Paris, are 2 male *subcylindrica* collected by Stimpson in the type-locality. They are listed by the museum as questionable "types non spécifiés." Rathbun (1918.1: 419) concurs: "Perhaps these also are cotypes." (!)

Since no confusion exists, no lectotype has been designated.

REFERENCES AND SYNONYMY

Uca (Minuca) subcylindrica
(Stimpson, 1859)

Gelasimus subcylindricus

TYPE DESCRIPTION. Stimpson, 1859: 63. Matamoros, Mexico. (MCZ !)

Smith, 1870: 137; Pl. 4, Figs. 6-6b. Types discussed and illustrated.

Uca subcylindrica

Rathbun, 1900.2: 585. Key.
Rathbun, 1918.1: 419; Pl. 155; Pl. 160, Fig. 5. Corpus Christi and near Santa Rosa, Cameron Co., Texas, U.S.A. (USNM !)

IX. *CELUCA* SUBGEN. NOV.

Typus: *Uca deichmanni* Rathbun, 1935. (East and west coasts of America; Indo-Pacific)

PLATES 29-43.
GONOPOD DRAWINGS: FIGURES 68 (part), 69-71.
DENDROGRAMS: FIGURES 96, 101.
MAPS 7, 15, 16, 17, 21.
TABLES 1, 8, 9, 10, 12, 14, 19, 20.

MORPHOLOGY

Diagnosis

Very small to moderate-sized *Uca* with front moderate to wide; antero-lateral margins varying from short to long but always posteriorly angled, not curving. Carapace profile strongly arched (except in *inaequalis* and allies) and sometimes fully semi-cylindrical. Eyestalks of moderate length and diameter, not nearly filling orbits; suborbital marginal crenellations, although rarely absent, usually definite and often large throughout, never obscured by projecting setae. Special armature sometimes present on lower proximal palm and anterior surface of 1st major ambulatory; carpal cavity often with beaded ridge on upper margin; cavity sometimes extended distally; major pollex sometimes with ventral carina; tip of major dactyl sometimes with a minute crest; major manus with all tubercles small except sometimes near upper margin which is then strongly bent over, often flattened. Small cheliped various but with gape sometimes very wide with serrations vestigial to absent. Gonopod with flanges present or absent, the tip often a projecting tube; thumb often vestigial, represented by a nubbin or shelf far from gonopod's tip. Female gonopore never with a tubercle except in one subspecies of *speciosa*. Segments of male abdomen sometimes partly fused.

Description

With the characteristics of the genus (p. 15).

MALE

Carapace. Front moderately wide, its breadth between eyestalks always more than twice basal breadth of erected eyestalk; not spatulate, the narrowest part not lying between eyestalk bases, but tapering gradually to a broadly rounded tip; frontal breadth measured between inner eyebrow margins contained from about 6 to about 3 times in carapace width between tips of antero-lateral angles. Orbits usually slightly oblique; rarely moderately to strongly oblique (*argillicola, helleri, triangularis*) or virtually straight (*crenulata, latimanus*); always moderately sinuous, because of projecting inner eyebrow and, sometimes, antero-lateral angle. Antero-lateral angles usually rectangular or nearly so, uncommonly acute and produced antero-laterally. Antero-lateral margins distinct from dorso-laterals except in *triangularis*; short except, rarely, as long as is characteristic of the subgenus *Minuca*; edge straight or slightly concave; direction ranging from slight convergence through straight to divergence; continuing into dorso-lateral margins either with an angular or rounded turn. Dorso-lateral margins ending about mid-cardiac region; well developed except in several species; never absent. Postero-lateral striae present except in *triangularis*, although variable within species, sometimes weak, and in some individuals absent; usually one pair only of raised, obliquely transverse lines, curved or straight, roughly parallel to end of dorso-lateral margin, about midway between end of that margin and postero-lateral angle of carapace; rarely two pairs of striae. Vertical lateral margin strong and continuing to angle between antero-lateral and dorso-lateral margins, except for weakness in the dorsal half in *beebei, stenodactylus, triangularis,* and *lactea,* and its total obsolescence in *musica.* Armature of the above dorsal and lateral margins and raised lines is confined to minute beading, often extremely fine and difficult to distinguish even under high power and with a fine needle; sometimes it appears altogether lacking; there is individual variation within species; when present in one locality it is usually also visible in the others; serrations, tubercles, and spinules are altogether absent; beading of vertical lateral margin, when present, usually continues posteriorly along ventral margin of carapace above bases of 3rd and 4th ambulatories, ending at the postero-lateral angle. Lateral carapace margins little converging except in *triangularis.* Carapace profile in lateral view strongly arched, except in *inaequalis* and several related forms; it is virtually semi-cylindrical in *beebei, stenodactylus,* and *latimanus;* hepatic and branchial regions in the most convex species usually tumid and confluent; branchial region sometimes partly divided by a longitudinal furrow; furrows between regions deeply marked only in *oerstedi*. Dorsal surface smooth, except for minute granules antero-laterally in several species, and for patches of pile in *speciosa*, *oerstedi*, *inaequalis*, and *tomentosa*.

Suborbital region flat or slightly concave; setae confined, at least in preservative, almost always to a submarginal row. Pterygostomian region moderately convex, strongly so in all species having the carapace nearly semi-cylindrical; slightly rugose or tuberculate, especially antero-laterally; always covered entirely with setae, sparse to moderately plentiful, which thin out laterally.

Frontal margin slightly raised, usually continuous, rarely weak or absent distally; entire margin always completely smooth with no trace of beading. Eyebrow always distinct, slightly or moderately inclined, the lower margin curved, broadest toward end of inner quarter or third, tapering externally; inner boundary a definite, oblique, raised line, well beyond outer base of eyestalk, confluent with base of frontal margin; outer boundary except in *saltitanta*, *lactea*, and *deichmanni* always extending beyond level of base of eye when eyestalk is depressed. Maximum width ranging from less than one-quarter the diameter of adjacent part of depressed stalk to slightly more than its breadth; both extremes are rare. Upper and lower raised margins always well developed, the lower slightly the stronger except in *lactea*; fine beading almost always distinguishable on both and traceable externally beyond eyebrow, on upper orbital margin, as far as base of antero-lateral angle; this beading, however, is sometimes so little developed that it is distinguishable chiefly by contrast with the smoothness of the adjacent front and antero-lateral angles. Like the margin itself the beading is almost always stronger on lower than upper edge.

Floor of orbit smooth, without mound, ridge, or tubercles, except in some *triangularis*; a single submarginal row of short setae. Suborbital margin projecting but scarcely or not at all rolled outward, although the marginal crenellations are set obliquely, being directed in general forward and upward; sinuosity of margin slight to moderate. Crenellations always present even internally except in some *inaequalis* and its allies, although often fused except distally; inner smaller than outer, often much smaller; extent from close to inner corner of orbit at least to antero-external angle and sometimes, as isolated, often slightly smaller tubercles, continuing completely around external margin of orbit; details, both of this outer extent and of the individual forms of crenellations and tubercles, often notably variable within species, including the two sides of the same individual. A narrow, channel-like groove, free of tubercles, always present on outer, posterior margin of orbit, beneath base of antero-lateral angle. Lower side of antero-lateral angle sometimes rounded but usually

marked by a definite edge, blunt or sharp; probably always in intact specimens with a row or cluster of setae in its basal part, rarely with one or two minute basal tubercles or spinules; distal portion of edge, whether sharp or blunt, always unarmed (except in some *triangularis* and *lactea*), extending to apex of antero-lateral angle. Eyestalks relatively short and thick (except in *lactea*), although still not fitting orbits closely.

Second Maxilliped. Spoon-tipped hairs moderate to numerous, attaining the largest numbers in the genus.

Third Maxilliped. Ischium with central portion ranging from almost flat to strongly convex, the longitudinal grooves weakly marked, medially absent in convex forms. Merus very short. Gill small to vestigial, without distinct books.

Major Cheliped. Ischium with or without a few minute tubercles, variable, on some part of distal margin.

Merus without high crests or large tubercles, but always with rugosities on parts of dorsal and posterior surfaces, and with small tubercles, serrations, rugosities, or a combination of these armatures along the margins.

Postero-dorsal margin developed only proximally, except in *triangularis*; the corresponding region distally consists of the unangled, convex merging of dorsal and posterior surfaces. In its developed, proximal portion this margin is usually marked by rugosities which are longest and strongest toward end of margin, before middle of merus; here the latter projects posteriorly in a more or less pronounced heel-like convexity; often the margin is a low, blunt ridge, unarmed except for a few rugosities at its end on the convexity; rarely the margin is simply the faintly angled boundary between dorsal and posterior surfaces.

Dorsal surface proximally always almost or completely flat; smooth, except, sometimes, for a few setae; this area is of variable length and breadth, usually about one-fourth to slightly more than one-half length of merus and roughly trapezoidal in shape, the distal margin being oblique. The remaining, more distal part of dorsal surface, which projects beyond carapace, is bent at a slight angle to the proximal part, and is slightly convex or flat but never concave; it is abruptly rugose, the rugosities all definitely tuberculate, strong or weak, and always weakest in postero-distal part of surface.

Antero-dorsal margin convex at least distally, sometimes strongly arched throughout; always distinct for entire length of merus, though sometimes blunt proximally. Tubercles, serrations, rugosities, or some combination of these forms present at least distally, where they are always largest, often as the culmination of a graduated series. Extreme, cuff-like end of merus with or without minute distal tubercles dorsally and antero-dorsally.

Anterior surface vertical, flat, usually smooth except for a small cluster of tubercles or rugosities distally just below end of antero-dorsal margin; rarely this cluster is lacking; rarely the entire anterior surface is finely granulous.

Ventral marginal armature ranging from an irregular band of small tubercles, extending slightly submarginally on each side, to a single row of large, strong serrations; armature always strongest distally.

Posterior surface ranging from moderately rugose to, rarely, practically smooth. The roughest portion is always that of the posterior protuberance, while the rugosities are always weaker and less numerous there than on highest parts of dorsal surface.

Details of the armature of all parts of the merus are unusually variable even within populations. On the dorsal and posterior surfaces, as well as proximally on the postero-dorsal margin, the rugosities vary in spacing, length of the striae, and size of the surmounting tubercles. The armatures of the antero-dorsal and ventral margins are even more variable. Sometimes a particular series is best described in one individual as a continuous band of small tubercles, in a second as a row of serrations, in a third as a margin of short rugosities, and in a fourth as a combination of these structures. Often, also, the proximal part of any margin may be wholly unarmed in some specimens, minutely tubercled or rugose in others. Because of the prevailing variation, only unusual characteristics of diagnostic value, or of potential functional interest, will be mentioned in the descriptions of species.

Carpus: Postero-dorsal surface usually rugose, at least in dorso-distal half; rarely tuberculate or smooth. Anterior margin usually sharp, bent over, with a few small tubercles proximally; distal portion often concealed by pile; rarely poorly marked and unarmed. Anterior surface usually with an indistinct, low, blunt, smooth, oblique ridge that is rarely surmounted by one or more tubercles.

Manus: Outer surface usually appears smooth in comparison with species in *Deltuca* and *Afruca*; small to very minute tubercles, however, always cover the entire area, except close to pollex and dactyl. Usually the tubercles are slightly enlarged on upper manus, but again smaller or altogether absent near the dorsal margin, or the surface is often roughened by various devices; adjacent to the subdorsal groove the tubercles are often minute or altogether absent; tubercles smallest near ventral margin, absent from neighborhood of pollex base, and almost absent from cuff bounding base of dactyl. Rarely a longitudinal band of slightly enlarged tubercles covers a secondarily convex area near base of lower dactyl and of gape. Sometimes part of upper manus appears eroded because of irregular linear arrangements of the tubercles which are sometimes raised on rugosities and sometimes enclose smooth cells.

Upper part of outer manus usually characterized by definitely bending toward the inside and by broadening, flattening, and conspicuous roughening of the dorsal margin which, because of the bending, proximally overhangs the carpal cavity. These specializations are achieved by a variety of means, largely species-specific and, in comparison with merus armatures, only moderately variable within species. The general range is as follows: About at level of cuff, upper third or less of manus begins to bend gradually inward to varying extents, so that dorsal margin coincides proximally with that of upper part of carpal cavity, which it overhangs; more distally, the margin broadens, usually to form a narrowly triangular area that continues to base of dorsal part of dactyl. With a few exceptions, the triangle is more or less flat and partly or wholly tuberculate; usually the tubercles differ from those of nearby outer surface in various characteristics; often these tubercles are smaller, sometimes pointed and slanted distally, sometimes arranged in oblique rows directed distally and internally. A submarginal groove usually bounds most of area externally, starting proximally and continuing almost to dactyl base. Above and contiguous with the groove is the margin's best-defined portion, usually consisting of a beaded edge along uppermost part of carpal cavity and usually continuing distally, after cavity edge has curved downward, as a row of more or less regular tubercles. Inner boundary of triangle, however, is usually indistinct, differentiated from adjacent upper palm chiefly by the beginning of the palm's verticality, and often by the absence of tubercles on adjacent area inside palm. Occasionally there is a definite inner edge, characterized by special tubercles; sometimes neither outer nor inner edge of dorsal margin is defined in distal portion.

Cuff outside base of dactyl almost always smooth except for a vertical row of several large tubercles marking start of outer subgape row of tubercles on pollex. Rarely a few minute tubercles are scattered on ventral portion; very rarely the cuff is absent, there being no groove bounding the region at base of dactyl from rest of manus.

Lower distal outer manus at pollex base always flattened about to level of adjacent pollex, almost always smooth and sometimes with a large, indistinct, very shallow depression (exceptions are *argillicola* and *triangularis*, each having the depression small, definite, and relatively deep). Along lower edge of this flattened area a keel sometimes starts which continues along the outer side of pollex and is sometimes tuberculate. Ventral margin of manus ranging from merely angled, through a regular row of tubercles or beading, to strongly keeled; keeled margins are usually surmounted by tubercles; the margin, whatever the type of armament, is often flanked externally by an adjacent groove. These ventral specializations always start slightly distal to proximal end of manus, which is always roughened, and end either at pollex base or continue throughout part or all of pollex. Tubercles and groove are sometimes strongest proximally, rarely distally, under pollex base, and sometimes they are equally developed throughout. Heel always well rounded but never strongly projecting posterolaterally.

Palm with proximal lower part forming a characteristic triangular area bounded by carpus, lower carpal cavity, the steeply sloping oblique ridge, and ventral margin; it is always covered at least in lower part by minute or extremely minute tubercles; usually they extend over entire surface and almost always are larger proximally and near ventral margin than elsewhere; sometimes they show a faintly vertical or reticulate formation, the spaces between being smooth; sometimes there are groups of specially enlarged tubercles in lower, proximal part only, and rarely the area is crossed obliquely by a series of parallel striae.

Oblique ridge inside palm ranging from high to rudimentary; rarely it lacks tubercles. Apex almost always near horizontal level of base of gape, rather than lower; apex high or low in height, sometimes lower than portion of ridge midway to pollex base. Tubercles always small at distal end of ridge near pollex base, usually largest on apex, but sometimes below level of apex when that part of ridge is higher; tubercles sometimes stopping short at apex, sometimes continuing partway or entirely to level of predactyl area below dorsal margin, rarely turning at apex and following lower edge of carpal cavity, which in these cases is unusually sharp and thin, for a short distance proximally. Arrangements of tubercles on oblique ridge and around cavity various, but usually in a moderately or wholly regular row of single tubercles, contiguous or slightly separated, and often graduated between pollex base and apex, but more or less irregular, compound and often interrupted when continued toward dorsal margin.

Central area of palm flat, usually smooth, merging gradually with the slight depression at pollex base, the latter depression almost always smooth; rarely a definite depression in upper part of palm, below predactyl depression.

Carpal cavity showing a wide range of size and depth, lower margin thick, thin, or part of carpal attachment; distal part either gradually or steeply sloping; upper part sometimes with a long extension into predactyl area, its width and depth diminishing distally; in other species the cavity is more or less sharply cut off from this area by the downcurving, beaded, inner edge of dorsal margin, by other tubercles, and by a predactyl ridge or plateau.

Predactyl area, in species lacking a long extension from carpal cavity, usually with a small cavity, more or less surrounded by small tubercles and set off from the adjacent midpalm area by ridging, or merging with it without barriers; predactyl area sometimes rugose, or almost smooth, instead of tuberculate.

Proximal and distal ridges inside base of dactyl usually present and tuberculate, the proximal almost always strongly developed and parallel to adjacent groove and dactyl base, longer than distal ridge, the tubercles larger and the more ventral tubercles largest; tubercles of proximal ridge rarely subequal. Tubercles of distal ridge (cuff) in ventral portion only, small, few, sometimes vestigial or absent. In *lactea* only the distal row, instead of the proximal, is well developed. Groove between ridges almost always well developed.

Pollex and dactyl: Almost all species of *Celuca* are clearly leptodactylous, brachydactylism being rare, and there being only a slight range within species, in contrast with the prevalence of wide variation within species of Indo-Pacific subgenera. Pollex and dactyl always tapering and moderately compressed; almost always clearly longer than manus, slender, the dactyl being clearly convex along dorsal margin in about distal half or less, curving downward just beyond tip of the usually straighter pollex, ending slightly below it; viewed from above, the dactyl also curves slightly inward, its distal portion, when closely apposed to that of pollex, almost always lying against a slight depression on inner surface of pollex tip. Pollex usually deeper than dactyl; sometimes it is very deep proximally, giving it a triangular shape, the proximal end of dactyl being then strongly oblique and upper part of manus correspondingly obliquely truncate. Gape width near middle usually slightly greater than depth of adjacent part of pollex, but rarely either narrower or wider. Long external grooves always absent, although a broader depression is rarely present. Outer, proximal, subdorsal groove on dactyl present or absent, but proximal dorsal tubercles always present, varying in size and extent. Sides of both pollex and dactyl, externally and internally, either smooth or covered, partly or wholly, by tubercles which are either minute or, rarely, small. All three rows of gape tubercles represented to some degree in both pollex and dactyl; distally all marginal rows are usually weak but sometimes form keels, with or without tuberculation. Pollex with inner marginal row almost always continuing proximal row of tubercles at base of dactyl, while the outer marginal row sometimes similarly continues tuberculation of outer cuff: the inner marginal row proximally is usually far separated from median row, the intervening space being smooth except, sometimes, for profuse pile; sometimes the space is achieved through the proximal absence of the median row; uncommonly, the outer marginal row is separated from the median. Armature of median row in both pollex and dactyl various and variable but usually without either striking enlargements of individual tubercles or large teeth with tuberculate edges; an arrangement often occurring in individuals throughout the subgenus consists of one tubercle slightly enlarged near middle of pollex and two tubercles on dactyl, well separated from both dactyl ends and from each other. Curved distal part of dactyl sometimes with median row forming a keel surmounted by a graduated series of minute, contiguous tubercles, always sharp, sometimes serrate; in this area enlarged tubercles or tuberculate teeth never occur in any species, whether or not the specialized keel is present. Pile on proximal pollex at base of gape present or absent, sometimes long and dense.

Minor Cheliped. Merus with dorsal margin weakly armed with spinulous tubercles, or with continuations of the more or less vertical rugosities of posterodorsal surface; antero-ventral margin smooth and rounded; postero-ventral margin always with a distinct row of tubercles, usually spinulous; posterodorsal surface with vertical or oblique rugosities, usually tuberculate, often few and feeble and concentrated dorsally and distally; in *triangularis* only this surface is markedly tuberculate; anterior and ventral surfaces smooth; dorso-distal margin either with slight irregularities or a series of minute tubercles; predistal, dorsal ridge, similarly armed, well developed. Carpus with subdorsal ridge ranging from distinct with minute tubercles to absent. Manus rarely conspicuously slender or deep. Mano-pollex ridge almost always well developed, starting at or near middle of manus and continuing throughout most of pollex. Dactyl ridges and intervening groove weak to absent. Dactyl almost always clearly longer than manus; dactyl and pollex tips usually normal but in several species neither distally expanded nor meeting perfectly, being slender, tapering, and not making contact at the tips. Gape various, ranging from narrow, with regular serrations, except proximally and distally, on prehensile edges to extremely wide, both pollex and dactyl being arched and without armature. Setae always sparse, except for the usual brush on inner distal margins of dactyl and pollex; no setae in close-set row or tuft on upper part of palm.

Except for the almost constant mano-pollex ridge and for the strong serrations on the prehensile edges in some species, the armature of the minor cheliped in this subgenus is comparatively weak throughout. Its general trend is from stronger armature in rough-

ly the first half of the species in the order listed to conspicuous weakness in the variously specialized species placed in the second half. That is, rugosities, ridges, and serrations are weakest and sometimes absent in the latter forms. Throughout the subgenus all these characters show considerable individual variability in the degree of weakness. With few exceptions, only the width of the gape and the form of the armature on the prehensile edges are useful in practical taxonomy.

Ambulatories. Usually slender in comparison with those of other subgenera, the meri of even the 2nd and 3rd legs moderate to narrow, their dorsal margins slightly convex to practically straight; in only several species are these meri notably wide with strongly convex dorsal margins. Armature of margins weak. Posterior surfaces of meri with finely tuberculate vertical striae present, vestigial or absent. Anterior surfaces of part or all of carpus, manus, and distal part of merus with 1st ambulatory, usually of major side only, often with ridges or tubercles in species having corresponding specializations on manus of major cheliped. Pile rarely present, then confined to intersegmental areas, ventral surface of manus or dorsal parts of carpus and manus.

The weakness and variability of the armature of the margins and posterior surfaces in this subgenus make them impractical characters for taxonomic use, although definite species differences do exist and would yield, as in the case of minor characteristics on the major cheliped, to detailed analysis (see p. 462).

Abdomen. Middle segments sometimes more or less fused.

Gonopod. Various, ranging from forms with flanges wide to absent, the pore from small, below lip of confluent flanges to large and funnel-like at extreme end of the protruding, narrow corneous shaft. The only characteristic in common is the relation of pore to thumb, the latter always ending well below the former, often far below; in some of the latter examples the thumb is vestigial. Inner process always flat, closely applied to shaft or flange, inconspicuous, often translucent; usually either narrow and spinous or triangular, the apex sharp. Torsion absent or slight.

FEMALE

With the characteristics of females in the genus (p. 17), when the following annotations are taken into account. Otherwise as in male, as described above.

Because of the superficial feminization of the carapace in males through its arching and armature, the carapace in females differs little or not at all from that of males in breadth, degree of arching, shallowness of grooving, and length of antero-lateral margins, although in almost all species the orbits are slightly less oblique, while the eyebrow is wider and less inclined, as characteristic of females in other subgenera. The armature, including ridges, beading, tuberculate striae, and serrations on carapace and cheliped, is in most species slightly stronger than in males, as elsewhere in *Uca*; only the ambulatories, however, are always clearly much more strongly armed than in males, their meri as usual somewhat broader. The suborbital crenellations, which in some other subgenera are larger and more extensive than in males, in *Celuca* are little if at all larger and then only toward the inner end; in the more specialized species, arranged in this study in about the second half of the subgenus, the crenellations of both sexes are markedly large and extensive and, in several speies, are even stronger in males than in females.

A detail of interest in the armature of the female cheliped concerns the variable presence of minute tubercles or granules on the antero-ventral margin of the merus. This area in males is apparently always rounded and unarmed, save for the usual short proximal and distal ridges. In females, one side is almost always armed with minute tubercles surmounting the ridges, with or without non-linear granules between; there is considerable individual variation, and sometimes the roughening is present on both sides; in only one species, *festae*, does it take the constant form of a well-marked row of small tubercles throughout the margin's length.

As elsewhere in the genus, fine granulation over part or all of the carapace sometimes occurs, although absent in males except in one species (*latimanus*). When pile occurs on the carapace in males in a species-specific pattern, it is found similarly in females; in several species characteristic patches are present in females only.

Gonopore usually without characteristics of taxonomic convenience. Tubercles almost always absent, as are sculpturing of the rim and other external complexities.

Abdomen with all segments always distinct.

Size

Small.

Color

A high degree of display lightening occurs in at least some populations of 15 among the 26 species of *Celuca* recognized in this study. Of these 15, 9 attain a striking state of polished white; in most of the group completely white individuals are sometimes seen, but the extreme phase is usual only in *saltitanta*

and *l. lactea*; in *saltitanta* it is confined to males; in *l. lactea* it seems to be the prevailing color of both sexes, probably even in the non-breeding seasons. A pre-white yellow phase has not been found in this subgenus and general bright yellow is known only in some populations of *triangularis* where the color is apparently not at all connected with display periods. Occasional yellow dots in both sexes often form part of the pattern in the marbled and mosaic effects of the dark carapaces of *inaequalis* and its allies. Eyestalks are sometimes greenish yellow. Spectrum orange is unknown, the nearest being the orange yellow claws of 2 subspecies of *lactea*. Reds occur frequently but usually only on the major cheliped and occasionally on the ambulatories; in both areas the hue ranges from orange red or spectrum red to purple red, and is most vivid in displaying populations; sometimes, as is of frequent occurrence in *lactea annulipes* and *stenodactylus*, the intensity of the red increases as the carapace lightens. A similar correlation also takes place in the orange yellow claws of *lactea perplexa*, particularly in the most eastern populations. Except for uncommon adult females of *lactea annulipes* which are entirely rose red without obvious relation to their state of receptivity, red does not occur on carapaces. The nearest approach to these hues is a purple or purple red spot on some *pugilator*. Blue and green are uncommon and usually appear on the carapace, as in *uruguayensis*, *cumulanta*, *beebei*, *stenodactylus*, and *deichmanni*; in several other species the eyestalks show these colors; sometimes the small chelae are blue, especially in adult females, erratically distributed in a population. An exception to these localized blues and greens is *oerstedi*, in which not only the carapace, both anterodorsally and anteriorly, but the frontal aspects of the 1st ambulatories are an intense blue green. Another exception is *lactea*, where blue often combines variously with a darker or lighter transverse pattern across the carapace. In crabs that remain generally dark even during display, somewhat lighter tones, often orange brown as in *inaequalis*, distinguish the lower major manus and sometimes the merus, at least on its inner side. The vivid phases summarized above are highly fugitive, while the young throughout the subgenus are dull.

SOCIAL BEHAVIOR

The variety of components as shown by single species of *Celuca* in waving display, threat and combat is high, the only other subgenus with forms attaining comparable diversity being the related *Minuca*. Special components confined to high intensity courtship are less rare than usual, and attain the maximum complexity found in the genus. Sound components and their ritualized versions are often incorporated into waving display. Lateral waving in a number of species is at the fastest rate attained in the genus; because of the small amplitude of vertical waves in, for example, *Deltuca*, the rates of vertical and lateral waves are not considered comparable. Again, more time is devoted to waving display than in other subgenera by at least a number of species, both as individuals and in the duration of the intertidal display period. Territoriality is strongly developed. In construction chimneys are absent but turrets and hoods, with two exceptions in *Minuca*, are confined to *Celuca*. Displacement activities are prevalent. Females are normally attracted below ground, matings or attempted matings on the surface being rare.

RELATIONSHIPS

Comparisons with other groups will be found on p. 19 in connection with the discussion of the subgeneric dendrogram.

The 26 species recognized in the subgenus *Celuca* defy tidy subdivisions. Although some of them are of course obviously more closely related to one another than to the rest of the group, practically no morphological and behavioral characters are held in common only by members of one of these subgroups, since the character in question also crops up in one or more species in one or more other parts of the subgenus. Most of the species are confined to the short stretch of coast extending from El Salvador to northern Peru; therefore allopatric speciation has been limited.

Nevertheless the subgenus does divide fairly naturally into two major groups, with a number of anomalous species showing less clear-cut relationships. When I first started serious work on the phylogeny of this subgenus, after becoming aware of the importance of gonopod structure in other parts of the genus, I was confused for a long time; a very obvious division in *Celuca*, in accordance with strongly tubular tips and those with large flanges, did not agree at all with divisions concerned with other basic characters, including form of the small chelae and overall pattern of armature. For example, the group of species eventually included in the superspecies *crenulata* and its allies showed contrasting forms of gonopods, although their relationships in a number of other details seemed very close. The answer, of course, as in some members of the subgenera *Deltuca* and *Uca*, proved to be that the form of the gonopod is on occasion plastic, apparently in conformity to the needs of sympatry.

The other early difficulty in seeing relationships within the subgenus was that the definition of a *rap*—termed *drumming* in this study—broke down. It is now clear, first, that drummings on the ground occur in other subgenera, and, second, that not all apparent drummings, even within small groups of related

species, touch the ground. In some the cheliped vibrates in the air and in some the drumming is against the suborbital regions, while some species perform all versions of this general vibration of the major cheliped. Therefore one of the old distinctions between crabs of Groups 4 and 5 (Crane, 1941.1) has proved to be invalid.

In spite of the deficiencies of both gonopod form and drumming techniques in dividing large groups of species, in *Celuca* a basic division into two informal categories remains useful; they correspond in general to the old Groups 4 and 5, and are shown as the two largest assemblies on the dendrogram, on the left and right, respectively. In addition the subgenus includes a number of associated species, shown as separate branches arising more or less closely to the bases of the large groups.

In an overall view, the large group on the left includes *crenulata* and its many allies; in general these species are less specialized for a semi-terrestrial life than are those of the second group, of which *deichmanni* is a characteristic example. In addition, the members of the first group are less active socially, although their morphological specializations for threat and combat are not only comparable in complexity to those of the second group but most of the components appear sporadically in both; usually however they are more prevalent in one group than in the other. On a subbranch placed about midway between the two groups are *beebei* and *stenodactylus*. Three species, *argillicola*, *pugilator*, and the Indo-Pacific *triangularis*, appear to have no close relations although their morphological characteristics place them firmly in *Celuca*. The relationships will now be reviewed in more detail.

The first and largest group of species is characterized by small chelipeds with the narrow gapes and fine serrations that occur throughout the rest of the genus; most of them live in mud or sandy mud and are little subject to desiccation; the time devoted to display is—with the notable exception of *saltitanta*—moderate but not excessive. The arching of the carapace accordingly does not—except in *saltitanta*—reach the extremes found elsewhere in the subgenus. Many of the species show unique specializations on the outer, upper manus; here sharp bending is sometimes accompanied by conspicuous flattening of the area and a patterning of the tubercles. The pollex is often proximally broad with a triangular projection, but the tips of the major cheliped are simple, lacking in particular a crest on the dactyl's tip.

This large group is viewed as composed of 15 species. Five of the species have ranges that are geographically distinct; these forms are therefore treated as a superspecies. They are: *uruguayensis* in the south Atlantic; *cumulanta* in the tropical Atlantic; *speciosa* (with two subspecies) also in the Atlantic but ranging from the north tropical to the warm temperate zone; *crenulata* (two subspecies) in the subtropical and warm temperate Pacific; and *batuenta* in the tropical Pacific. Display whitening is apparent in four of the five and is strong in two; palm-ambulatory stridulating apparatus occurs in three; specializations of the upper manus are notable in all five. Both the pollex and dactyl are normally elongate, without anomalies except in *batuenta* which shows in moderate form structures that are exaggerated to various degrees in other related species in the Pacific. On the whole the superspecies members are less specialized morphologically than their closest relations and, except for *batuenta*, their waving displays are not elaborate; only *cumulanta* is known to erect structures, and these are turrets or poorly formed hoods, always of rare occurrence.

One member of the superspecies, *batuenta*, is grouped with three other species in the tropical Pacific to form a local alliance; by definition the four forms are closely sympatric in the strictest sense. In a certain order these species are *inaequalis*, *oerstedi*, *batuenta*, and *saltitanta*. The arrangement is from the least to the most convex, from the least to the highest degree of display lightening, and from the least to the most active in waving display, although in any other subgenus each would be considered extremely active. Each of them has a unique and striking display; in each except *oerstedi* drumming components, sometimes ritualized, are incorporated into the waving display; palm-ambulatory apparatus is developed in only one of the four, *inaequalis*; there it is used not only in threat but at least sometimes as an integral part of combat. Although no structures are built by members of the alliance, in these other particulars they appear definitely to be more specialized than do the members of the superspecies.

Three other members of the group, *tenuipedis*, *tomentosa*, and *tallanica*, are obviously closely related to the local alliance, both by morphology and in waving display; the display characteristics are known through the recent work of von Hagen in Peru (1968.2). Since we do not yet know whether they occur in close sympatry with the members of the alliance, I have omitted them on the dendrogram from inclusion in the unshaded bracelet marking the local alliance. In general they have most in common with *inaequalis*, including a deep pollex, patterned pile on the carapace, and major-manus-drumming incorporated in display; display white is not known to develop.

The remaining four species in the group are *leptochela*, *helleri*, *festae*, and *dorotheae*. The first two are so little known that no constructive remarks can be made concerning them here, except that the form of their gonopods indicates clearly that both belong in this general section of the subgenus, while the rela-

tively wide gape in the small chela of at least *helleri* probably points toward the second group; they are thus given satellite positions. A species clearly less closely related to the superspecies *crenulata* and its allies is *festae*; here the elongation of the chela is more extreme than in any other *Uca*; this length affects the armature of the appendage; major-manus-drumming is not incorporated in its waving display, and the display is relatively simple; in general the species is probably best viewed as belonging on the fringes of the large group, with its affinities at least equally close to the stock that culminates in *beebei* and *stenodactylus*. The last species in the group of 15 is *dorotheae*, also of an intermediate position and therefore placed closest on the dendrogram to the *beebei-stenodactylus* branch. The form of its predactyl ridges and of its major cheliped place it with the large group, its small chela tends toward the second large group (that includes *deichmanni*), while the form of the ambulatory apparatus, its usual color, and its construction of turrets are characteristics having more in common with those of *beebei*.

The elongate branch that divides distally for *beebei* and *stenodactylus* is designed to indicate both their intermediate position and the presence of unique specializations. The form of the small chela, with narrow gape and serrations, is characteristic of the left-hand group on the dendrogram; the serrations are stronger, however, than anywhere else in the genus except in (*Australuca*) *seismella*. The carapace shape is strongly arched, although not quite to the extent found in the extreme forms centering around *deichmanni*; these fiddlers spend prolonged periods waving, and the display is relatively unspecialized though rapid; one of them, *beebei*, attains the fastest circular wave in the genus. The armature of the claw shows in full measure the specializations of both large groups, with strong palm-ambulatory equipment, upper manus peculiarities, and a tuberculate crest near the dactyl's tip. In *stenodactylus* the herding of females is more strongly developed than in any other *Uca*, being incorporated into courtship display. Finally, *beebei* constructs turrets.

The right-hand group on the dendrogram is composed of seven species: *lactea*, *leptodactyla*, *limicola*, *deichmanni*, *musica* (two subspecies), *latimanus*, and the aberrant *triangularis*. In the light of recent work, it now seems likely that the Indo-Pacific species *lactea* should be diagrammed as a more widely separated form; its distinctions from the rest of the group consist chiefly of the less advanced structure of the oblique ridge and carpal cavity on the palm, the arrangement somewhat resembling those more characteristic of non-American subgenera; in addition waving periods in *lactea* are more restricted than usual in *Celuca* and more firmly tied to short portions of the lunar cycle; these characteristics, it seems likely, are indications of the early branching of the ancestral American stock of *Celuca* (p. 436). In spite of these dissimilarities, *lactea* seems related most closely to *deichmanni* and its relations, as shown by its suitability for semi-terrestrial life and the complex similarity of all forms of its social behavior; the resemblances seem to have developed too many details for them to be the result of parallelism. Similar remarks apply to the other Indo-Pacific species, *triangularis*, which does not appear to be closely related to any species of *Celuca*, and which is shown at the base of the second group.

Of the five remaining species, *leptodactyla* occurs in the Atlantic and the other four in the eastern Pacific. These are found only in the tropics except for one subspecies, *m. musica*. The Atlantic *leptodactyla*, in morphology and relatively simple waving display, appears to have much in common with *limicola*, and the two are considered to form a superspecies. The two subspecies, *m. musica* and *m. terpsichores*, are exceedingly similar except for size and for details of the palm-ambulatory equipment; its palmar component attains a degree of specialization unknown elsewhere in the subgenus and is shared in any form in the right-hand group only by *deichmanni*. Three of the five species sometimes live in close sympatry, and have much in common morphologically. These three, consisting of *limicola*, *deichmanni*, and *m. terpsichores*, are therefore treated as forming the third alliance shown on the dendrogram. The last species, *latimanus*, appears aberrant because of the juvenile form of its major cheliped; in all other features it is an advanced member of its group. Several of these species, in fact, represent at least morphologically the pinnacle not only of the subgenus *Celuca* but also of the genus, showing extreme adaptations to a semi-terrestrial life and specializations for threat and combat; it is here that semicylindrical carapaces combine with widely gaping small chelae fitted for picking up sandy substrate. Here too are remarkable elaborations of the cheliped's armature, in particular the stridulating ridge on the palm in *musica*, and the minute precision of distal crests. No other species, except for *saltitanta*, equal these fiddlers in the vigor, prevalence, and duration of waving display. Finally, in three of the species—*leptodactyla*, *m. terpsichores*, and *latimanus*—the peak of construction capability is reached, the hoods achieving a perfection of form, relative size, and smooth finish found in no other species.

NAME

Celuca: From the Latin adjective *celer*, "swift," in recognition of the rapid motions of these highly active *Uca*. The name, therefore, in the selected meaning of the root, signifies "fast-moving fiddler."

37. *UCA (CELUCA) ARGILLICOLA* CRANE, 1941

(Tropical eastern Pacific)

PLATE 29 *A-D*.   MAP 15.
FIGURES 68 *J*; 101.   TABLES 9, 10.

INTRODUCTION

Living close to fresh water, *U. argillicola* has striking paedomorphic characteristics of shape and armature along with a persistently pale color. On my second visit to the type locality in Pacific Costa Rica, it was clear that the name *argillicola* is partially unsuitable, since the species is not confined to clay substrates.

As yet we have learned nothing about its social behavior. Even the general waving pattern is unknown, in spite of observations which supplemented a week's sporadic observation of field populations with two months' close checking of a small group brought to the Trinidad crabberies. At the time of these observations (1961) I had not yet become aware of the importance of threat stridulation and combat behavior in relation to morphology, and took no note of any such activities that may well have occurred. The impression in both field and crabbery was of inactive individuals, their burrows grouped in close proximity, as in the field; although they fed normally, they spent long periods without motion, even though tides and weather were optimal for other species, which in parts of the field and in the crabberies, waved and were otherwise active with peripheral intermingling. Inevitably, the relative social lethargy of *argillicola* recalled that of certain species in the subgenus *Deltuca* and, among *Celuca, triangularis*.

The peculiar characteristics of *argillicola*'s shape and weak armature make further observations highly desirable. In addition, an adequate collection must still be made.

MORPHOLOGY

Diagnosis

Front moderately broad; orbits strongly oblique; antero-lateral margins practically absent; palm without oblique, tuberculate ridge; heel of manus without backwardly directed projection; fingers clearly shorter than palm.

Description

With the characteristics of the subgenus *Celuca* (p. 211).

MALE

Carapace. Frontal breadth contained about 5 times in that of carapace. Orbits moderately oblique. Antero-lateral angles broadly acute, slightly produced. Antero-lateral margins variably very short, always moderately convergent and rounding into dorso-lateral margins. Two pairs of postero-lateral striae; on each side the upper stria is very long and bent sharply, the angle located about midway between end of dorso-lateral margin and postero-lateral angle of carapace; anterior arm extending far forward, paralleling dorso-lateral margin while the 2nd arm is short, paralleling posterior border of carapace; lower stria very short, lying just above postero-lateral angle of carapace. Breadth of eyebrow more than half diameter of adjacent part of depressed eyestalk. Suborbital margin with crenellations near antero-external angle little enlarged, continuing as separated teeth completely around outer margin of orbit, ending on posterior side of channel. Edge of lower side of antero-lateral angle blunt to moderately sharp.

Major Cheliped. Merus: Antero-dorsal margin straight, sharp and unarmed except at extreme distal end, which is arched and serrate. Ventral margin with a single row of large, regular serrations, largest in distal third. *Carpus*: Anterior margin poorly marked, scarcely bent over, with only several proximal irregularities.

Manus: Bending of outer, upper surface minimal. Outer tubercles graduated in size from minute ventrally to moderate subdorsally. Dorsal outer groove present. Dorsal margin distal to carpal cavity flat, its edges diverging distally, each definite and tuberculate, tubercles of inner edge serrate; dorsal surface between these edges smooth except for sparse tubercles. Cuff outside dactyl base rough with tubercles except dorsally; no vertical row of tubercles starting here and continuing on subgape region of pollex; groove weak. Flat area outside pollex base with a distinct, narrowly triangular depression adjacent to middle of pollex side. Ventral margin with a row of large, graduated tubercles, largest distally and ending on pollex about a third of way to tip. Ventral, outer groove shallow.

Palm with lower triangle smooth except for a few

widely scattered minute pits and, near the rudimentary oblique ridge, well-spaced tubercles, small and low. Oblique tuberculate ridge almost absent, traceable as a low, broad blunt ridge stopping short on the very low apex at carpal cavity; tubercles small, low, non-linear, covering the ridge and continuing with the slightly smaller ones adjacent on mid-palm and the very small ones below on triangle, where they die out. Center palm with moderate-sized tubercles starting below end of circumcarpal beading; tubercles smallest close to oblique ridge. Carpal cavity with lower margin very short, occupied by carpal attachment; distal part of cavity shallow, sloping gently to merge with center palm. Beaded, inner edge of proximal dorsal margin curving disto-ventrally around upper cavity, stopping abruptly about level of middle of dactyl. Predactyl area completely smooth, with a slight, elongate, median hollow, cut off from dactyl base by a few tubercles.

Pollex and *dactyl*: Clearly shorter than palm; both broader than usual, but pollex not strikingly broad proximally; dactyl wider and longer; gape toward middle narrower than breadth of adjacent pollex. Pollex with depression outside pollex base continued slightly distally, and with corresponding depression inside continued farther, diminishing to a point. Dactyl with proximal subdorsal groove traceable throughout outer proximal half of dactyl, diminishing in size; upper proximal dactyl very rough with pointed tubercles that slant distally; they are arranged in several rows, the uppermost being weakest but longest, and traceable almost to dactyl tip. Gape teeth, especially proximally, large, low, rounded, close-set. Pollex with inner marginal row of gape tubercles in proximal half far separated from and lower than median row, the intervening area being broad, concave and smooth, slanting inward, dying out distally; proximal tubercles of inner row large; subdistal tubercles of median row large, low and rounded. Pollex tip trifid, each point being the terminal tubercle of a row. Dactyl with tubercles of median row few and large, one near proximal end much enlarged, several subdistal less so, being lower and rounded. Gape pile present.

Minor Cheliped. Gape at base narrower than width of pollex, decreasing distally. Serrations strong, not in contact.

Ambulatories. All meri unusually broad for a *Celuca*; dorsal margins moderately to strongly convex. First carpus on both sides anteriorly with a regular, longitudinal row of minute, close-set tubercles along middle of segment throughout almost its entire length.

Gonopod. Corneous tip alone shows torsion in orientation. Flanges well developed, about equal in breadth, confluent and forming the gonopod's distal, truncate edge just beyond pore. Pore narrow; proximal to the pore the shaft projects slightly from flange level. Inner process short, flat, triangular, the apex pointed, arising and ending near free edge of the posterior flange, not overlying shaft. Thumb well developed, moderately short, not nearly reaching corneous base of specialized tip.

FEMALE

Gonopore without tubercles; outer rim clearly higher than inner.

Measurements (in mm)

	Length	Breadth	Propodus	Dactyl
Male holotype	7.8	12.2	13.5	8.0
Female paratype	7.3	10.8	–	–
Female paratype (immature)	5.6	8.5	–	–

Morphological Comparison and Comment

U. argillicola is the only *Celuca* in the eastern Pacific without a tuberculate oblique ridge on palm but with the orbits strongly oblique. Its relatively narrow front distinguishes it from all the members of the subgenus *Minuca*—*panamensis*, *vocator*, *pygmaea*, and *zacae*—that also lack the oblique ridge on palm. In addition, *argillicola* is apparently a larger species than *pygmaea* and *zacae*, while its orbits are less strongly oblique; in contrast, it is much smaller than *panamensis* and *vocator*, with orbits more oblique. The gonopods of all the species are distinctive from one another, while other characters in one or more of the species, such as pile on carapace, projecting heel on manus, and form of tuberculation of pollex tip, are minor aids in identification. Fortunately, two males from the American Musuem of Natural History were available to compare with the holotype.

Color

Carapace and ambulatories clear, creamy white; this color is not fugitive and is characteristic of both sexes in young as well as adults. Both chelipeds of males and females with dorsal parts of segments apricot buff. In the type locality several populations occurred on yellowish white clay, with which the crabs blended well; on darker substrates the observed individuals were always conspicuously pale, closely similar to the others.

RANGE

Tropical eastern Pacific from the Gulf of Dulce in Costa Rica to Buenaventura, Colombia.

BIOTOPES

The largest groups were found on the steep banks of streams little affected by tides, the water probably almost fresh, close to or slightly beyond the upstream limit of mangroves. Several populations were in vertical banks of white clay or brown loam. Two others were on high, flat ground, under tall ferns, behind the mangroves that fringed a stream; the soil here was either sandy or clayey. Scattered individuals or very small groups occurred further downstream in muddy or sandy soil, rarely among mangrove shoots, sometimes with the ground covered with rotting coconut husks, among rank ferns and monocotelidons, or between buttresses of swamp palms and other trees. All these areas were largely shaded, the sun reaching them briefly and in patches. (Biotopes 14, 15.)

SYMPATRIC ASSOCIATES

The largest aggregations of *argillicola* were not mixed with other *Uca*; they occurred farther upstream than any eastern Pacific fiddler except *brevifrons*. The small groups and scattered individuals found closer to and within the mangrove limits were associated with three species of the subgenus *Minuca*—*vocator ecuadoriensis, galapagensis herradurensis,* and *zacae.*

MATERIAL RESULTING FROM FIELD WORK

(The complete list of specimens examined is given in Appendix A, p. 607.)

Observations and Collections. Golfito, Costa Rica. Several specimens also sent alive to the crabberies in Trinidad from Buenaventura, Colombia; they were positively identified by me before being placed, with other live examples from Golfito, in the crabberies; unfortunately they were not secured as specimens when they died.

TYPE MATERIAL AND NOMENCLATURE

Uca argillicola Crane, 1941

HOLOTYPE. In Smithsonian Institution, National Museum of Natural History, Washington. Cat. no. 137400 (formerly New York Zoological Society cat. no. 381,134); 2 paratype females, cat. no. 137401 (formerly New York Zoological Society cat. no. 381,135). Measurements: p. 221. Type-locality: Golfito, Costa Rica. (!)

REFERENCES

Uca (Celuca) argillicola Crane, 1941

TYPE DESCRIPTION. Crane, 1941.1: 183; Text Figs. 4j; Pl. 1, Fig. 3; Pl. 2, Fig. 6. Costa Rica: Golfito. (USNM !)

Garth, 1948: 61. Panama: Perlas Is., on San José I., banks of lower Rio Marina. Colombia: Humboldt Bay. (!)

38. *UCA (CELUCA) PUGILATOR* (BOSC, 1802)

(Subtropical and temperate western north Atlantic)

PLATE 29 *E-H*. MAP 16.
FIGURES 37 *K*; 69 *F*; 101. TABLES 9, 10, 12, 14, 19, 20.

INTRODUCTION

Probably more biologists know the sand fiddler, *U. pugilator*, by name than any other *Uca* except its familiar associate, *pugnax*. For many years the two species have provided much of the material for experimental work on crustacean hormones and biological rhythms. During the first half of this century *pugilator* was already suspected of signaling to females by drumming on the ground, and throughout the 1960s the species was a chief subject of pioneer experiments on acoustical signaling in the genus. Recently we have learned that this species, as well as the eastern Atlantic *tangeri*, both use a sun compass in a simple form of homing. Key contributions on these varied topics are listed under the annotated references for *pugilator* (p. 227) and are noted as well in appropriate chapters of Part Two.

Louis Bosc, a keen naturalist, watched the crabs in Carolina and, when he named the species in 1802, gave us the first detailed account of living fiddlers. He called the crabs *pugilator* because "as soon as a man or an animal appeared in their midst they bent their great claw, holding it in front of them as though it were a challenge to combat, and then escaped, running sideways and keeping it always in the same position." Apparently he never had the pleasure of watching a combat between males, or observed either waving or digging. He was especially impressed by the vast numbers of the crabs on the seashore and on the banks of tidal streams, ". . . thousands and even millions. . . ." He continued, "They have a great number of enemies among otters, bears, birds, turtles, alligators, etc. but their abundance is such that the devastation accomplished by these animals is not discernible."

In similar places, all through the days of the young coastal settlements, small boys and grown-ups, whether Indians or Europeans, doubtless used the crabs for bait. Nowadays these fiddlers, already reduced by pollution and the destruction of their habitats, are rarely plentiful enough to supply the sheds that offer "LIVE BAIT—FIDDLERS." Instead, in northeastern states summer fishermen learn that the crabs come from Florida, shipped by the crateful by collectors who also supply laboratories. How long the species can stand the several threats forms the usual regrettable question.

A number of *pugilator*'s characteristics are unique or rare in the genus. It lacks close relations. It is one of only five species with ranges primarily in the temperate zone. With them it is adapted to subfreezing temperatures, hibernating regularly where cold is encountered. Apparently in adaptation to short northern summers, *pugilator* actively and regularly courts both by day and at night; in the dark it uses acoustical signals instead of waving display in attracting the attention of females. In the more southern parts of its range the rare characteristic of droving is highly developed. Its combat behavior, still incompletely known, appears to be unusually complex, both in form of components and in the relation of ritualization to forceful fighting; it is one of the few species in the genus that lacks an oblique ridge inside the palm, and the only one where the relation of this lack to combat technique has so far been learned. Finally, although there seems to be no reliable morphological basis for distinguishing subspecies, slight differences in waving display are shown between New England and southern populations, while droving seems to occur only in the south; as usual, a morphological comparison of adequate population samples from numerous localities has not been undertaken. In a study still incomplete, the peripheral population on Cape Cod, the species' northern boundary, appears to be relatively inactive socially for undetermined reasons. In view of all these distinctions, the survival of numerous populations—large, vigorous, well distributed and in the midst of normal habitats—must forcefully be encouraged: without any question, this species counts as one of biology's prized assets.

MORPHOLOGY

Diagnosis

Front wide; no oblique, tuberculate ridge on palm; antero-lateral margins slanting strongly outwards; meri of ambulatory legs in male slender, with mid-dorsal margin of 3rd merus straight in both sexes: carapace in female granulate.

Description

With the characteristics of the subgenus *Celuca* (p. 211).

MALE

Carapace. Front very wide, similar to those in the subgenus *Minuca*, contained less than 3 times in carapace breadth. Antero-lateral margins also as in most *Minuca*, being long, slightly diverging, and rounding into dorso-lateral margins. Breadth of eyebrow about half diameter of adjacent part of depressed eyestalk. Suborbital margin with crenellations near antero-external angle little enlarged, continuing as separated teeth along outer orbital margin almost to channel. Edge of lower side of antero-lateral angle blunt.

Major Cheliped. Merus: Antero-dorsal margin straight except for arching near distal end; blunt and rugose throughout. All rugosities strong for the subgenus, continuing across upper part of anterior surface; lower part of this surface finely granulate. *Carpus:* Dorso-posterior surface covered with close-set, rounded tubercles; oblique ridge of inner surface with indistinct tubercles.

Manus: Bending of outer, upper surface distinct. Outer tubercles large for the subgenus and mingled in two sizes throughout most of lateral surface; tubercles near ventral margin abruptly smaller; subdorsally largest. Dorsal outer groove present. Dorsal margin distal to carpal cavity broad, slightly convex, strongly tuberculate throughout but without specially armed edges. Cuff outside dactyl base rough with tubercles except dorsally. Flat area outside pollex base not smooth but covered with minute tubercles; lower edge of area with a weak, tuberculate keel that starts more proximally, at about middle of manus, well above ventral margin, and continues along most of pollex in its ventral half; keel often poorly developed and lacking tubercles distally or altogether. Ventral margin with a row of large and similar tubercles that die out in proximal third of pollex. Ventral, outer groove shallow, developed near pollex base only.

Palm with lower triangle covered with minute tubercles, somewhat rugose proximally, merging with larger tubercles of center palm on and near area of oblique ridge. This ridge is practically absent, represented only by an obtuse edge marking the usual distinction between lower triangle and center palm; apex and linear tubercles lacking, the area being covered with a continuation of the minute tubercles of lower triangle. Center palm distinctly convex, covered with tubercles of varying sizes ranging from small to moderate, all much larger than those of lower triangle, largest near base of gape. Pollex cavity moderate for the subgenus, not slight. Carpal cavity with lower margin thin; distal part of cavity steeply sloping except dorsally, turning at an angle

into center palm. No downward turn of inner edge of dorsal margin, the upper part of carpal cavity extending distally in a narrow, shallow, smooth projection that almost fills predactyl area. Tubercles of proximal ridge at dactyl base sometimes in an irregular, angled, or multiple row; those of distal row always multiple.

Pollex and *dactyl*: Pollex with tip turned slightly upward; dactyl with entire distal half clearly convex; gape near middle much wider than adjacent part of pollex. Outer pollex with a faint, long crease above the adjacent low keel that starts on lower manus. Dactyl with subdorsal, external, proximal furrow moderate. Both pollex and dactyl completely covered externally with minute tubercles; those of upper, proximal dactyl larger; slightly serrate on dorsal margin where they are traceable, diminishing regularly in size, throughout most of dactyl's length. No gape tubercles conspicuously enlarged; locations of minor enlargements variable. Pollex tip distinctly trifid. Gape pile present or absent.

Minor Cheliped. Gape at base narrower than width of pollex, decreasing distally. Serrations few, strong, in middle portion only, not in contact.

Ambulatories. 2nd and 3rd meri slender, their dorsal margins scarcely convex. 1st ambulatory on major side with merus, carpus, and manus roughened anteriorly with tubercles on lower anterior surface.

Gonopod. Flanges well developed, about equally broad; each a little produced beyond the subdermal pore. Inner process narrow, blunt, covering canal, ending just below pore. Thumb long, slender, not nearly reaching base of flange.

FEMALE

Entire dorsal and postero-lateral surface of carapace finely granulate, except for posterior margin. Suborbital armature stronger than in male. Gonopore roughly triangular, with outer edge slightly concave externally and inner edge medially angled.

Measurements (in mm)

	Length	Breadth	Propodus	Dactyl
Large male (Florida)	14.0	21.0	35.0	27.0
Large male (New York)	13.0	20.0	34.0	26.0
Moderate male (New York)	11.0	17.0	25.0	20.0
Large female (New York)	14.0	19.0	–	–
Moderate female (New York)	11.0	16.0	–	–

Morphological Comparison and Comment

Throughout almost all its range, *pugilator* may be distinguished from other *Uca* by the absence of an oblique tuberculate ridge inside palm; the only exception, *subcylindrica*, which is also ridgeless, is known only from close to the Rio Grande and differs strikingly from *pugilator* in its cylindrical form, small size, short fingers, and unique gonopod. In *pugilator* the strong, tuberculate ridge along the side of the major pollex is another distinctive character among United States fiddlers.

Along the Atlantic coast north of Florida, the only other species of *Uca* are *minax* and *pugnax*. Diagnostic differences between *pugilator* and *minax* are summarized on p. 177. From *pugnax*, with which it often mingles marginally, *pugilator* differs not only in the absence or presence, respectively, of the palm and pollex ridges mentioned above, but also in its more arched, less uneven carapace which is smooth in the male but granulate in the female. Finally, in characteristic biotopes populations of *pugilator* include larger adults than do those of *pugnax*.

The affinities of *pugilator* remain puzzling. The morphology of both carapace and, in part, the major cheliped has a number of features in common with most *Minuca*; yet in details of its armature and, particularly, in its waving display *pugilator* seems obviously closer to *Celuca*. Yet it shows no outstanding similarities to any of the subgroups in either of the subgenera. As usual, its anomalous position adds to the interest of the species.

In the future, work with large population samples may make desirable a division of *pugilator* into species, or, more probably, subspecies. Already we have a growing knowledge of regional differences in display, acoustic behavior, and physiology. Yet the necessary morphological background is missing and any present attempt at subdivision seems to be unwarranted.

Color

Display whitening conspicuously present and fugitive in both sexes, but the hue is usually yellowish white and the shiny, high reflectance found in some tropical forms is never attained. The other distinctive characteristic of displaying males is a fugitive, purplish violet patch occupying the cardiac region that often disappears with further whitening: in more southern populations including non-displaying groups, it is sometimes pink rather than purplish and confined to the gastric region. In non-display phases the carapace is often variously marked in semi-reticular patterns of brown, or is completely brown with small gold-colored or light brown spots. Major cheliped in displaying males buff to yellowish white; base of chela often pale orange. Eyestalks buff or grayish white; never green. Third maxilliped white, standing out from the duller underparts. Minor cheliped with manus and chela white. Ambulatories attain white, but anterior sides of 1st ambulatory meri sometimes show persistent dark purplish brown or reddish purple.

SOCIAL BEHAVIOR

Waving Display

Wave lateral-circular, but in the reverse direction characteristic of other species. Specifically, in *pugilator* the cheliped is raised at once in front, still partly flexed, then is unflexed outward, brought down and folded inward. The usual pattern in the genus, in the numerous species having characteristic circular waves, is for the cheliped to be unflexed first outward, close to ground level, then raised and finally brought down to the flexed, resting position. In *pugilator* there is a short pause at the wave's highest point. In northern populations, including those of Cape Cod, Connecticut, and Long Island, no jerking of the wave has been detected, either by observation or in film analyses. In contrast, in the most southern populations, particularly those near Miami, the cheliped during elevation makes a single jerk midway to its highest reach; sometimes this irregularity is detectable only in film analysis and usually occurs only in displays of moderate intensity. Carapace raised and lowered during each wave. Minor cheliped makes a small roughly corresponding motion. Ambulatories not raised or kicked outward, except for a rare elevation of a leg during a high stretch. At high intensity 4 or 5 displays form a series, the cheliped is not lowered quite to the ground between waves, and the body does not sink completely to rest position. In a series there is no pause between waves, so that the accent comes on the brief hesitation at high point of wave. Chela remains practically closed during display, the dactyl not being raised. The crab often remains in one spot during waving but sometimes moves several steps to the side. Manus-drumming sometimes occurs in conjunction with waving at mouth of male's burrow, during high intensity courtship, after the male has attracted a female's attention and just before the male descends. (Waving components 4, 7, 9, 10, plus the occasional single jerk, component 6, described above; timing elements in Table 19, p. 656.)

Precopulatory Behavior

Female follows a male, after his major-manus-drumming, down his burrow, at least during diurnal courtship (p. 502).

Acoustic Behavior

Salmon and his associates (Salmon & Atsaides, with refs., 1968, 1969; Horch & Salmon, 1969) have accumulated extensive data on the acoustical characteristics of major-manus-drumming in *pugilator*, its apparent means of perception by the female, and the employment of this component during courtship; the work will be reviewed under the general discussion of sound production, p. 480. Burkenroad observed (1947) that the major manus, when vibrating vertically at the mouth of the burrow, did not touch the ground in the individuals he was watching, and he therefore argued that the motions called "rapping" (now calling "drumming") that Dembowski (1926) and Crane (1943) believed might well produce sounds, must have been similar vibrations in the air. Salmon's work has since confirmed the existence of actual drumming on the ground. But analysis of films I have made on Long Island and in Florida, along with subsequent field observations, have shown clearly that Burkenroad was also correct in his own conclusions that the claw did not touch the ground: not only is major-manus-drumming common in the species but also major-merus-drumming (against the suborbital region and, probably, the side of the carapace) occurs frequently, especially in agonistic situations; I do not yet know if it is ever used in courtship. Burkenroad additionally did not detect, through observation, this contact of the cheliped with the carapace; it is probable that in some instances the motion is confined as he believed to vibration in the air (p. 483). The list of acoustic components in *pugilator* that are now known to me are as follows: major-merus-rub (component no. 1); ?palm-leg-rub, inferred from morphology (4); leg-wagging (5); major-manus-drum (9); leg-stamp (11).

Combat

Combat in this species was closely observed only on the southern shore of Cape Cod, in a marginal population. In general it was notable for its rarity; when present, high ritualization was shown between males; however most combats took place when a burrowless crab, returning from feeding at lowest tide levels to the burrow area, forcefully expelled a smaller crab of either sex from a hole. The expelled individual then promptly seized a burrow belonging to a still smaller crab. Eventually the smallest dispossessed individual, ending a series of 4 or 5 evictions, simply crept under a bit of shell as the tide washed in. This type of behavior was observed day after day during June and July of 1968. The rare encounters between adult males included the following components: manus-rub with taps, pollex-rub, pollex-under-and-over-slide, pollex-base-rub, dactyl-slide, upper-and-lower-manus-rub, dactyl-submanus-slide, interlace, pregape-rub, and supraheel-rub. In the absence of an oblique ridge on the palm, the pregape-rub and supra-heel-rub apparently take the place of the single component, the heel-and-ridge, that is so common in a number of other fiddlers; in the pregape-rub the pollex longitudinally rubs the strongly tuberculate, cen-

tral palm, while in the supraheel-rub the dactyl moves vertically up and down on the lower, proximal part of the outer manus; meanwhile the pollex, held away from the palm, remains inactive; these two components are described on p. 490. (Component nos. 1, 2, 3, 5, 6, 7, 8, 9, 10, 13.)

RANGE

Eastern and Gulf coasts of the United States of America. Rare on northern shore of Cape Cod (Massachusetts), occurring at least to the Little Pamet R., near Truro; locally common on the Cape's south shore. From there south locally abundant in suitable habitats as far as Key West, Florida. Present also on the west coast of Florida and the Gulf of Mexico west to Corpus Christi, Texas. Single records: Old Providence I., Bahamas and Santo Domingo.

BIOTOPES

Sheltered shores with sandy to muddy sand substrate, often with scattered shells and stones. Where the burrows are in sand, there is usually a layer of mud not far below the surface, or at least mud forms the substrate near low tide levels, where the crabs then often feed. (Biotopes 5, 6.)

SYMPATRIC ASSOCIATES

The only other *Uca* that occur in the more northern temperate part of *pugilator*'s range are *pugnax* and *minax*. Both live typically in mud or sandy mud, rather than in sandier habitats, yet individuals of all the species sometimes mingle, especially near habitat margins. Since both *pugnax* and *minax* are members of the subgenus *Minuca*, they are not at all close sympatric associates.

In southern Florida, (*Celuca*) *s. speciosa* and *pugilator*, closer sympatrics, sometimes occur in adjacent populations; their waving patterns and tempo are markedly distinct. In Florida populations of *pugilator* also sometimes mingle marginally with (*Minuca*) *rapax*.

MATERIAL RESULTING FROM FIELD WORK

(The complete list of specimens examined is given in Appendix A, p. 607.)

Observations and Collections. United States: Cape Cod, near Truro and Cotuit; Connecticut: Greenwich; New York: Stony Brook, Port Jefferson, Huntington, Cold Spring Harbor, Pelham Bay; New Jersey: Brigantine; Florida: Fort Lauderdale, Key Biscayne.

Films. Massachusetts, New York, Florida.

Videotapes. Massachusetts.

Type Material and Nomenclature

Uca pugilator (Bosc, 1802)

Bosc's type of *Ocypoda pugilator* is not extant.

Type-locality: Caroline. Since no taxonomic confusion exists at present, the designation of a neotype does not seem to be indicated.

References and Synonymy

Uca (Celuca) pugilator (Bosc, 1802)

Note. U. pugilator is one of the species of *Uca* that have been used extensively in laboratory investigations. These have concerned principally endogenous rhythms of various kinds and the hormonal control of chromatophore activity. Key references appear on p. 448.

Ocypoda pugilator

TYPE DESCRIPTION. Bosc, 1802: 197. "Caroline." Color; habitat; habits (p. 187).

Olivier, 1811: Vol. 8: 418. "Caroline." No new material.

Bosc, edited by Desmarest, 1830: 241-44, 250. "Caroline." Morphology; confused synonymy; habits, paraphrased from the type description.

Ocypode pugilator

Latreille, 1802-1803: 47. America. No new material.

Say, 1817-1818: 70, 443. America. Taxonomy; general habitat and habits (digging).

Gelasima pugillator

Latreille, 1817.2: 519. America. No new material.

Gelasimus vocans (not *Cancer vocans* Linnaeus)

Gould, 1841: 325 (part). U.S.A.: Massachusetts. Taxonomy; habitat; habits (digging); all comments brief.

De Kay, 1844: 14 (part); Pl. 6, Fig. 9. U.S.A.: New York. Taxonomy.

Gelasimus pugilator

Desmarest, 1825: 123. Apparently no new material. Brief taxonomy.

Milne-Edwards, 1852: 149; Pl. 4, Fig. 14. Taxonomy.

Stimpson, 1859: 62. U.S.A.: south side of Cape Cod in Massachusetts to Texas. Record from Mexico questionable.

Smith, 1870: 136; Pl. 4, Fig. 7. U.S.A.: Massachusetts to Gulf States. Taxonomy.

Verrill, 1873: 336, 364, 462, 464, 467, 545. U.S.A.: Massachusetts. General distribution (no local limits); basic ecology and habits; no display or combat.

Verrill & Smith, 1874: 42, 70, 168, 170, 173, 251. A reissue of Verrill, 1873.

Bolau, 1878: 149. Probably the first record of specimens of *Uca* kept in captivity (Hamburg Aquarium).

Kingsley, 1878: 321. U.S.A.: Georgia. Local habitat. Spelling: *pugillator*.

Kingsley, 1880.1: 150 (part). Only those specimens recorded from the east coast of the United States from Nantucket in Massachusetts to Florida. Taxonomy.

Kingsley, 1880.2: 399. U.S.A.: North Carolina: Beaufort; Florida: Sarasota Bay.

Young, 1900: 274 (part). Only those specimens recorded from the United States; no detailed localities. Taxonomy.

Yerkes, 1901: 417. Morphological variation; right- and left-handedness.

Uca pugilator

Rathbun, 1900.2: 585; Text Figs. 3, 4. Key.

Rathbun, 1905: 1. U.S.A.: northeastern states. Local distribution.

Sendler, 1912: 191. U.S.A.: New York. Taxonomy.

Fowler, 1912: 446. U.S.A.: New Jersey. Taxonomy; habits.

Hay & Shore, 1918: 452. U.S.A.: North Carolina. Taxonomy; color; ecology; habits.

Rathbun, 1918.1: 400; Pl. 141; Pl. 160, Fig. 2. U.S.A.: Numerous localities throughout the species' range from Corn Hill and Truro (near Boston), Massachusetts to Key West, Florida and, along the Gulf of Mexico, to Galveston, Texas. Taxonomy. Brief habits. (USNM, part !)

Hyman, 1920. Illus. U.S.A.: North Carolina. Post-embryological development.

Stephenson, 1921. Illus. U.S.A.: Biology. [Reference not translated.]

Hyman, 1922. Illus. Habits.

Dembowski, 1925. U.S.A.: Massachusetts. Habits, including possible sound production.

Dembowski, 1926. U.S.A.: Massachusetts. Habits.

Pearse, 1928: 231ff. U.S.A.: North Carolina. Ecology; tests on toleration of desiccation and fresh water.

Maccagno, 1928: 44; Text Fig. 28 (claw). U.S.A.: Eastern coast, including North Carolina at Beaufort. Taxonomy.

Crane, 1943.3: 220. U.S.A.: Connecticut; New York. Taxonomy; waving display; "rapping."

Coventry, 1944: 542. Bahamas: Old Providence I. Predation by birds.

Burkenroad, 1947. U.S.A. Courtship; experiments on sexual discrimination; discussion of possible methods of sound production; nocturnal activity.

Fingerman, 1957. U.S.A.: Mississippi. Tidal rhythms in relation to burrow position.

Crane, 1957. U.S.A.: Connecticut; New York; Florida. Preliminary classification of waving display.

Teal, 1958. U.S.A.: Georgia. Local distribution; ecology; experiments on tolerances for different foods, temperatures, and salinities, as well as on substrate preferences; comparisons with sympatric species.

Teal, 1959. U.S.A.: Georgia. Respiration and its relation to ecology.

Salmon & Stout, 1962. U.S.A.: North Carolina. Sexual discrimination and sound production in courtship.

Salmon, 1965. U.S.A.: North Carolina. Waving display and sound production.

Miller, 1965. U.S.A. Morphology, physiology, and ecology in relation to distribution.

Vernberg & Costlow, 1966. Development of right- or left-handedness.

Herrnkind, 1966. U.S.A.: Florida. Orientation by polarized light. (Abstract.)

Knopf, 1966. U.S.A.: Massachusetts. General ecology; behavior, all non-social except for displacement feeding.

Salmon, 1967. U.S.A.: Distribution in Florida. Sound production.

Herrnkind, 1967. U.S.A.: Florida. Development of celestial orientation. (Abstract.)

Herrnkind, 1968.1. U.S.A.: Florida. Illus. Visual orientation.

Herrnkind, 1968.2. U.S.A.: Florida. 4 text figs. U.S.A.: Florida. Breeding in captivity and mass-rearing of progeny.

Herrnkind, 1968.3. U.S.A.: Florida. 11 text figs. Adaptive, visually-directed orientation.

Miller & Vernberg, 1968. U.S.A. Thermal requirements in relation to geographic distribution.

Salmon & Atsaides, 1968.2. U.S.A. Included in their general paper on visual and acoustical signaling in *Uca*.

Salmon & Atsaides, 1969. U.S.A.: Florida. Experiments on sensitivity to substrate vibration.

Horch & Salmon, 1969. U.S.A.: North Carolina. Experiments on the production, perception, and reception of acoustic stimuli.

39. *UCA (CELUCA) [CRENULATA] URUGUAYENSIS* NOBILI, 1901

(Subtropical western south Atlantic)

PLATE 30 *A-D*.
FIGURES 68 *I*; 101.
MAP 15.
TABLES 9, 10, 12, 20.

INTRODUCTION

This species, along with the west African *tangeri*, extends farther south than any other Atlantic members of the genus. *U. uruguayensis*, a distinct form known only from southern Brazil and Uruguay, belongs to a group of five allopatric forms and five other closely related species that occur in the tropics of both the Atlantic and Pacific shores of America.

Some share similar structures, apparent acoustical equipment, on the lower part of the major palm and on the 1st ambulatory, and all have characteristic combat armature on the claw.

MORPHOLOGY

Diagnosis

Front moderately wide. 1st ambulatory in male on each side with a ridge on lower, anterior side of manus. Major cheliped with outer manus strongly tuberculate dorsally, its subdorsal ridge and associated groove well developed, as is the proximal, subdorsal furrow on dactyl; no inner keel on pollex tip. Female gonopore with outer margin almost straight.

Description

With the characteristics of the subgenus *Celuca* (p. 211).

MALE

Carapace. Frontal breadth contained about 5 times in that of carapace. Antero-lateral margins straight or slightly concave, scarcely converging. The single pair of postero-lateral striae located close behind ends of dorso-lateral margins, rather than about midway to postero-lateral carapace angle. Carapace profile practically semi-cylindrical, but hepatic and branchial areas not specially expanded above rest of carapace. Breadth of eyebrow about equal to diameter of adjacent part of depressed eyestalk. Suborbital margin with crenellations near antero-external angle enlarged, continuing as separated teeth along outer orbital margin almost to channel. Lower side of antero-lateral angle blunt.

Major Cheliped. Merus: Rugosities weak except on posterior convexity. Antero-dorsal margin proximally straight, blunt, and practically unarmed, arched and armed only distally. Anterior surface finely tuberculate.

Manus: Bending of outer, upper surface distinct and, beside dorsal margin, flattened. Outer tubercles all small, increasing slightly in size upward, becoming suddenly much larger near dorsal margin. Dorsal outer groove present. Dorsal margin distal to carpal cavity with outer edge distinct and tuberculate; inner edge indistinct; intervening surface tilted inward, scarcely flattened, armed with irregularly scattered, sharp tubercles. Ventral margin with a row of minute and similar tubercles that die out close beyond pollex base. Ventral, outer groove absent. Dorsal pile usually conspicuous.

Palm with lower triangle covered with minute tubercles arranged, at least in center of area, in irregularly vertical rows, not raised on rugosities. Oblique tuberculate ridge high, its central part slightly higher than apex, and its tubercles slightly larger than on apex; they continue a short way up distal margin of carpal cavity. Center palm with small tubercles in upper part, which show tendency to arrangement in vertical rows. Carpal cavity with lower margin moderately thick; distal part of cavity steeply sloping except dorsally, turning at an angle into center palm. Beaded inner edge of proximal dorsal margin curving disto-ventrally for a short distance only, stopping above beginning of distal extension of carpal cavity; this extension is narrow, shallow, and smooth, almost filling predactyl area.

Pollex and *dactyl*: Dactyl with subdorsal, external, proximal furrow moderate. Both pollex and dactyl completely covered externally with extremely minute tubercles; those of upper, proximal dactyl much larger and on dorsal margin traceable, diminishing in size, along more than half length of dactyl. No gape tubercles conspicuously enlarged; a series of several slightly enlarged tubercles usually present toward middle of pollex and a similar series proximally on dactyl, other minor enlargements being present or absent. Pollex tip weakly bifid. Gape pile present.

Minor Cheliped. Gape at base narrower than width of pollex, narrowing distally. Serrations few, strong, in middle portion only, not in contact.

Ambulatories. 2nd and 3rd meri moderately broad, their dorsal margins slightly convex to almost straight. A distinct ridge, the crest entire, on lower, anterior manus of 1st ambulatory on both sides; extending along about the distal two-thirds of the segment near ventral margin, it is apparently unique in the genus; in position it corresponds roughly to the anterior, ventral margin of the manus that is usually weak or absent in *Uca*.

Gonopod. Flanges large, of similar width, the anterior slightly produced. Pore practically terminal although flanges are continuous in front of it. Inner process triangular, with a broad base overlying canal and posterior flange, the pointed tip reaching gonopore. Thumb short, slender, not nearly reaching base of flange.

Abdomen. 4th to 6th abdominal segments in male fused, although boundaries between segments are sometimes faintly indicated.

FEMALE

Suborbital armature scarcely or not at all stronger than in male, but sometimes extending slightly farther laterally on outer orbital margin and even onto lower margin of orbital angle. Gonopore with outer margin thin, almost straight; inner margin strongly curved.

Measurements (in mm)

	Length	Breadth	Propodus	Dactyl
Largest male (Maldonado)	10.5	16.5	30.0	19.0
Moderate male (Maldonado)	8.0	13.5	19.1	11.5
Largest male (Rio de Janeiro)	7.0	12.0	22.0	15.0
Moderate male (Rio de Janeiro)	6.5	10.0	14.5	9.5
Largest female (Maldonado)	10.5	16.5	–	–
Moderate female (Maldonado)	9.0	14.0	–	–

Morphological Comparison and Comment

The only other small *Uca* with ranges coinciding in part with that of *uruguayensis* are *leptodactyla* and, very rarely, *cumulanta*. The diagnosis should differentiate males of the species satisfactorily. The females are superficially similar, differing principally in proportions. The carapace of *uruguayensis* shows an intermediate degree of arching between that of *cumulanta* and *leptodactyla* while the legs are of

similarly moderate length and breadth; the form of the gonopore is subject to distortion, apparently by spermatophores, and can also be an unsatisfactory character, although more useful than the others. There is, however, no substitute, in attaining certain identifications, for familiarity with the females of all three species. At present the known range of *uruguayensis* coincides with those of the two northern species only near Rio de Janeiro; *cumulanta* is known to occur there only from several specimens in the Smithsonian Institution.

Color

Displaying males: Display white strongly developed, preceded by a green carapace, the major cheliped ranging from dark red to orange red. Sometimes the cheliped remains red while the rest of the crab whitens; sometimes it too becomes polished white.

SOCIAL BEHAVIOR

Waving Display

Lateral, strongly circular. (Observation casual only.)

RANGE

From Rio de Janeiro (Guanabara Bay) in Brazil south to Buenos Aires, Argentina.

BIOTOPES

Sandy mud to mud, in full sunshine near mangroves, or partially in their shade. (Biotope nos. 9, 12.)

SYMPATRIC ASSOCIATES

The only other *Celuca* with a range coinciding at all with that of *uruguayensis* is *leptodactyla*. The populations appear to be separated, at least in the Rio de Janeiro area, by differences in habitat, since *uruguayensis* congregates on muddy, not relatively sandy, substrates. On Ilya Pinheiro, populations of both (*Minuca*) *burgersi* and *rapax* were active within a few meters.

MATERIAL RESULTING FROM FIELD WORK

(The complete list of specimens examined is given in Appendix A, p. 608.)

Observations and Collections. Rio de Janeiro, Brazil.

TYPE MATERIAL AND NOMENCLATURE

Uca uruguayensis Nobili, 1901

HOLOTYPE. In Regio Museo Zoologico di Torino. Label follows name with "Tipo. Dr. F. Silvestri.

1900." Measurements in mm: length 9.5; propodus 27. (!) Type-locality: La Sierra, Uruguay.

Type Material of *Uca olympioi* Oliveira, 1939. In Instituto Oswaldo Cruz, Rio de Janeiro. Types not seen by me, but Dr. Oliveira showed me the living crabs, which I collected, on Ilya Pinheiro, the type-locality in Guanabara Bay, Rio de Janeiro. Now that the characteristics of gonopods and palm-ambulatory stridulating mechanisms are known in the group, it is clear that *olympioi* should be considered a synonym of *uruguayensis*. Oliveira's 4 specimens (1939.1: 130) that he referred to the latter species were all unusually large for the island; the smaller ones, on which he based *olympioi*, were from characteristic, displaying populations. The major chelipeds are in general more leptochelous and attain adult proportions at a smaller size, in the Rio de Janeiro region than in the specimens I have seen from Uruguay; nevertheless, since Oliveira did secure large, brachychelous examples on Pinheiro, a division into subspecies appears at present to be unwarranted. The superficial resemblance of leptochelous *uruguayensis* to *leptodactyla* led him to believe that *olympioi* was most closely related to that species, rather than to *uruguayensis*.

REFERENCES AND SYNONYMY

Uca (Celuca) uruguayensis Nobili, 1901

Uca uruguayensis

TYPE DESCRIPTION. Nobili, 1901.2: 14. Uruguay: La Sierra. (Torino !)

Rathbun, 1918.1: 413; Pl. 150. Brazil: Rio de Janeiro and, in Santos, East Piassaguera. Uruguay: Maldonado and elsewhere. Taxonomy.

Luederwaldt, 1919.1: 384, 400. Brazil: Santos. Color.

Luederwaldt, 1919.2: 435. Brazil: São Sebastião, Santos.

Maccagno, 1928: 38; Text Fig. 23 (claw). No new material. Taxonomy of type.

Luederwaldt, 1929: 54. Brazil: São Sebastião I.

Oliveira, 1939.1: 130; Pl. 8, Figs. 45, 46. Brazil: Rio de Janeiro. Taxonomy.

Gerlach, 1958.1: 672. Brazil: Estado São Paulo (near Cananéia). Ecology.

Uca olympioi

Oliveira, 1939.1: 128; Pl. 8, Figs. 41, 42, 48; Pl. 14, Figs. 63, 64. Brazil: Rio de Janeiro. Type description; color (p. 140).

40. *UCA (CELUCA) [CRENULATA] CRENULATA* (LOCKINGTON, 1877)

(Subtropical and warm temperate eastern north Pacific)

PLATE 30 *E-I*.   MAP 15.
FIGURES 70 *D, G*; 101.   TABLES 9, 10, 20.

INTRODUCTION

Although the form here designated *U. crenulata crenulata* still occurs in southern California on the Pacific coast of America, it has become scarce. Certainly it is the only *Uca* now found in the eastern Pacific north of the Mexican border. The few biological observations were made before the middle of this century (p. 233). We know nothing of the biology of the form here regarded as a subspecies, *c. coloradensis*, and restricted to the Gulf of California.

MORPHOLOGY

Diagnosis

Front moderately wide. Major cheliped with upper part of manus bent over moderately to sharply, its top more or less flattened, with a distinct marginal ridge and groove; surfaces of pollex and manus finely tuberculate or granulate; oblique, tuberculate, ridge inside palm strong; Gonopod with thumb well developed. Female carapace sometimes with two patches of pile posteriorly.

Description

With the characteristics of the subgenus *Celuca* (p. 211).

MALE

Carapace. Frontal breadth contained less than 4 times in that of carapace. Orbits practically straight. Antero-lateral angles either slightly acute and produced or rectangular. Antero-lateral margins straight, diverging, turning at an angle into dorso-lateral margins which are either strong throughout or weak posteriorly. Often 2 pairs of postero-lateral striae, but upper pair, close behind ends of dorso-lateral margins, sometimes weak or absent. Vertical lateral margins slightly weak dorsally. Eyebrow more than half as wide as diameter of adjacent part of depressed eyestalk. Suborbital margin with strong crenellations, differing little in size, not continuing around outer orbital margin. Lower side of antero-lateral angle rounded.

Major Cheliped. Merus: Antero-dorsal margin moderately arched throughout; blunt but tuberculate.

Anterior surface granulate. *Carpus:* Oblique ridge of inner surface granulate.

Manus: Bending of outer, upper surface moderate to strong, notably flattened. Outer tubercles minute in lower three-quarters, abruptly large in upper, outer, bent-over quarter and, proximally, near articulation with carpus; tubercles forming variable, irregular rugosities near dorsal margin in some individuals and, stopping, leaving a nearly smooth band, alongside the strong outer dorsal groove. Dorsal margin distal to carpal cavity with outer edge distinct and tuberculate, inner edge indistinct, intervening surface relatively narrow, little flattened, covered closely with sharp tubercles in irregular, obliquely transverse rows, directed intero-distally. Ventral margin with a row of minute and similar tubercles that die out at pollex base. Ventral, outer groove absent.

Palm with lower triangle covered with minute tubercles either completely or only in about proximal half. Oblique tuberculate ridge moderate, apex and ridge of about uniform height or apex slightly higher; tubercles close set and of similar size throughout, continued far upward around carpal cavity's low distal margin to its distal extension. Carpal cavity with ventral margin low, thick, convex; distal part of cavity steeply sloping to tubercles on upward extension of oblique ridge. Beaded inner edge of proximal dorsal margin curving disto-ventrally for a short distance only, stopping above beginning of distal extension of carpal cavity; this extension is short, diminishing gradually in depth and width, bounded above and below by tubercles as described above and incompletely separated from predactyl area by scattered tubercles. Predactyl area centrally smooth, marked by a small depression and bounded ventrally by a low, blunt ridge with one or more coarse tubercles.

Pollex and *dactyl:* Similar to *uruguayensis*, but with minute tubercles or granules differently distributed and sometimes present on inner as well as outer surfaces; tubercles on proximal, dorsal part of dactyl smaller and less extensive. Gape pile present or absent.

Minor Cheliped. Gape much narrower than pollex. Serrations in general well developed but variable, not in contact.

Ambulatories. 2nd and 3rd meri moderately broad, their dorsal margins little convex to almost straight. 1st carpus on major side, minor side, or both anteriorly with a longitudinal row of minute tubercles or granules, often vestigial, in proximal half of segment.

Gonopod. Differs from that of *cumulanta* (p. 241) in lacking a projection on inner edge of pore lip, in sometimes having the external distal edge of tube produced, in having a longer inner process, and in sometimes having an exceptionally long thumb.

FEMALE

A pair of large patches of pile on posterior part of carapace, one above base of each 4th ambulatory, present or absent. Suborbital crenellations markedly more deeply excavate than in male, with less fusion at outer anterior angle of orbit. Gonopore either straight or curved.

Measurements (in mm)

	Length	Breadth	Propodus	Dactyl
c. coloradensis (All from near mouth of Colorado R.)				
Largest male (holotype)	12.5	20.0	33.0	29.0
Moderate male	10.5	16.0	22.0	16.5
Largest female	12.0	17.5	–	–
Moderate female	9.5	12.5	–	–
c. crenulata (All from west coast of Lower California)				
Largest male	13.0	19.0	36.0	25.5
Moderate male	11.0	17.0	29.0	21.0
Largest female	13.0	18.5	–	–
Moderate female	12.0	17.5	–	–

Morphological Comparison and Comment

U. crenulata should be easily distinguished as follows from the few other moderate- to broad-fronted *Uca* with ranges known to be partially coincident, or that may prove to be so. *U. c. coloradensis* differs from *m. musica* in its less arched carapace, wider front, flattened dorsal manus, and lack of parallel striae on triangular area of major palm. *U. c. crenulata* differs from *stenodactylus* in its wider front, marginal ridge with groove on upper manus, and large thumb on gonopod; it differs from two *Minuca*, *brevifrons* and *galapagensis*, in its narrower front and, especially, in its straight antero-lateral margins with sharp posterior angles.

The forms here considered subspecies differ so slightly, when compared with other members of the superspecies, that their reduction from specific rank seems justified. This step is taken in spite of, or rather partly because of, a collection of specimens in the Museum of Comparative Zoology, Harvard, that contains representatives of both forms from Guaymas. When, however, the material is carefully compared with examples from the head of the Gulf of California and from the west coast of Lower California— that is, from characteristic parts of the range of each form—anomalies are apparent in some specimens. For example, the degrees of bending and flattening of the dorsal part of the major manus show more variation than usual, appearing intermediate in several individuals. The Guaymas series (listed under both subspecies under Material Examined, p. 608), are insufficient for a comparative study which, it is hoped, someone will soon undertake in this distributionally interesting locality. Specimens from San Felipe, on the Gulf's northwest coast, although less ambiguous than some of those from Guaymas, show traces of *coloradensis* characteristics; they are referred here, however, to *c. crenulata*.

A comparison of all the members, including *crenulata*, of the superspecies *crenulata* and their allies will be found on p. 217.

Color

(I made the following observations on September 30, 1944, in bright sunshine during a tidal period that reached low in mid-afternoon, at Newport Beach, Los Angeles. Only several males displayed; they were of small size and presumably belonged to the year's brood, although they may have been of normal size for their latitude.) Fugitive display whitening present, starting development an hour or more before low. Lightest males, including the two displaying individuals: carapace grayish white. Major cheliped with merus bright crimson red outside; inner merus, outer manus, and outer chela brilliant white; palm and inner chela creamy to white. Eyestalks ranging from yellow green to pale turquoise. Minor chela pinkish. Ambulatories with anterior merus of 1st crimson red; otherwise all segments marked with brown and cream.

SOCIAL BEHAVIOR

Waving Display

Each of the two displaying individuals gave a single series of lateral, straight waves, each wave lasting less than .75 second; jerks absent; several waves in series with a small pause between each 2 waves. No signs of courtship were observed. (Observed component: 4.)

RANGE

Western shores of Mexico and United States of America, from Tenacatita Bay in Jalisco and the Gulf of California north to near Los Angeles. Distribution details under subspecies. More southern records, from Costa Rica (Boone, 1930: 221) and from Salvador (Bott, 1954: 171), refer to other species.

BIOTOPE

At Newport Beach the crabs lived on the edges of salt lagoons well sheltered from the sea. The burrows were in sloping banks, the lower portions of which were muddy changing to sandy at upper levels. (Biotope no. 16.)

MATERIAL RESULTING FROM FIELD WORK

(The complete list of specimens examined is given in Appendix A, p. 608.)

Observations and Collections. Los Angeles (Newport Beach).

TYPE MATERIAL AND NOMENCLATURE

Uca crenulata (Lockington, 1877)

HOLOTYPE of *Gelasimus crenulatus*. Not extant. Type-locality: Todos Santos Bay, Lower California. A male labeled "tipo" from the type-locality is deposited in the museum of the University of Torino (cat. no. 872). (!)

Type Material of *Gelasimus gracilis* Rathbun, 1893. In Smithsonian Institution, National Museum of Natural History, Washington. Cat. no. 4622. 1 male specially labeled "type" plus 33 additional males and 6 females from La Paz, Lower California. (!)

Uca crenulata coloradensis (Rathbun, 1893)

HOLOTYPE of *Gelasimus coloradensis*. In Smithsonian Institution, National Museum of Natural History, Washington. Male holotype cat. no. 17459. Measurements on p. 233. Type-locality: Horseshoe Bend, Colorado R. (!)

Uca (Celuca) [crenulata] crenulata coloradensis (Rathbun, 1893)

(Head of Gulf of California; Guaymas)

MORPHOLOGY

With the characteristics of the species. *Carapace*: Antero-lateral angles practically rectangular. Dorso-

lateral margins weak posteriorly. Upper pair of postero-lateral striae sometimes present. Female with a pair of large patches of pile on posterior part of carapace. *Major Cheliped*: Upper, outer part of manus sharply bent over and strongly flattened. Palm with apex and ridge of oblique ridge of about uniform height, the tubercles of similar size throughout. Pollex and dactyl with external granules confined to surfaces close to gape. Gape pile present or absent. *Gonopod*: External distal edge of tube produced; thumb not exceptionally long. *Female Gonopore*: Anterior, outer edge very slightly raised, thin, curved.

Uca (Celuca) [crenulata] crenulata crenulata (Lockington, 1877)

(From Tenacatita Bay in Jalisco north on the Gulf of California's east shore to Guaymas; on its west shore from La Paz north to San Felipe; on the west coast of Baja California and California from Todos Santos Bay to Newport Beach, near Los Angeles)

MORPHOLOGY

With the characteristics of the species. *Carapace*: Antero-lateral angles slightly acute and produced. Dorso-lateral margins well marked throughout. Upper pair of postero-lateral striae always absent. Female without patches of pile on posterior part of carapace. *Major Cheliped*: Upper, outer part of manus, while moderately bent over, is less sharply so than in *c. coloradensis*, and less flattened. Palm with apex slightly higher than in *c. coloradensis*, with its tubercles a little larger than those on remainder of oblique ridge. Fine tuberculation of pollex and dactyl not confined to surfaces close to gape, as in *c. coloradensis*, but generally distributed externally, as in *uruguayensis*, and also extending over inner surfaces as well, except along middle of inner side of pollex; there the slight depression originating on palm at pollex base extends more distally than usual. Gape pile absent. *Gonopod*: External distal edge of tube less produced, resembling instead the rounded edge found in *cumulanta*. Thumb much longer than in any related form, almost reaching corneous base of specialized tip instead of ending far below it. *Female gonopore*: Slit-like, its axis anterior-posterior.

REFERENCES AND SYNONYMY

Uca (Celuca) crenulata (Lockington, 1877)

TYPE DESCRIPTION. See under *U. (C.) crenulata crenulata*, below.

Uca (Celuca) crenulata crenulata (Lockington, 1877)

Gelasimus crenulatus

TYPE DESCRIPTION. Lockington, 1877: 149. Mexico: Todos Santos Bay on west coast of Lower California.

Gelasimus gracilus

Rathbun, 1893: 244. Mexico: La Paz, at south end of Lower California. Type description.

Uca gracilis

Rathbun, 1898.2: 603. Mexico: Gulf of California (Pichilinque Bay). Taxonomy.
Rathbun, 1900.2: 586. Key.

Uca crenulata

Holmes, 1900: 75; Pl. 1, Figs. 7-9. No new record. Taxonomy; reported reexamination of Lockington's material.
Rathbun, 1918.1: 409; Pl. 146. U.S.A.: San Diego. Mexico: Lower California from Todos Santos

Bay to San Felipe: ? Guaymas; Mazatlan. (USNM !)
Schmitt, 1921: 279; Text Fig. 164. U.S.A. California. No new record. Taxonomy.
Rathbun, 1923.1: 632. Mexico: Lower California.
Rathbun, 1924.4: 377. Mexico: Gulf of California: Balandra Bay, Carmen I.
Maccagno, 1928: 43; Text Fig. 27 (claw). Mexico: Todos Santos Bay. Taxonomy.
Crane, 1941.1: 198. No new records. Taxonomy.
Garth, 1960: 110. Mexico: west coast Baja California. Distribution.

Uca (Celuca) crenulata coloradensis (Rathbun, 1893)

Gelasimus coloradensis

TYPE DESCRIPTION. Rathbun, 1893: 246. Mexico: Horseshoe Bend, Colorado R. (USNM !)

Uca coloradensis

Holmes, 1900: 76. Mexico: Guaymas. Taxonomy.
Rathbun, 1900.2: 586. Key.
Rathbun, 1918.1: 210; Pl. 147. Mexico: Colorado R., opposite mouth of "Hardy's Colorado" R., Sonora; Guaymas. Taxonomy. (USNM !)
Garth, 1960: 113. Mexico: mouth of Colorado R. Distribution.

41. *UCA (CELUCA) [CRENULATA] SPECIOSA* (IVES, 1891)

(Subtropical and warm temperate western Atlantic)

PLATE 31.   MAP 15.
FIGURES 68 *G*, *K*; 101.   TABLES 9, 10, 12, 19, 20.

INTRODUCTION

Two forms hitherto considered separate species, *Uca speciosa* and *U. spinicarpa*, are here classed as subspecies, since the chief specific characters have proved to be unreliable. Both are known principally from the subtropical and warm temperate shores of the United States of America, although one extends south to Yucatan. At least in Florida *speciosa* uses sound production in nocturnal courtship, a behavior pattern that is proving to be common among species from higher latitudes (pp. 238, 501).

MORPHOLOGY

Diagnosis

Front broad; antero-lateral margins straight, sharply angled posteriorly. Major cheliped with oblique, tuberculate ridge present on palm and continued strongly upward around carpal cavity to subdorsal area; tuberculate ridge on outer side of pollex absent; upper, outer manus smooth to casual inspection up to level of dorsal groove and ridge. Female either with plentiful pile, a tubercle beside gonopore, and a normally rounded front, or with practically no pile, no tubercle beside gonopore, and a front with posterior edge of distal marginal border truncate.

Description

With the characteristics of the subgenus *Celuca* (p. 211).

MALE

Carapace. Front wide, contained about 3 to 3.5 times in carapace breadth. Antero-lateral angles broadly acute, slightly produced. Antero-lateral margins long, straight or slightly concave, scarcely converging, angling sharply into dorso-lateral margins which are sometimes weak posteriorly. One or two pairs of postero-lateral striae, strong or weak, the short, lower striae being sometimes absent. Pile on dorsal part of carapace very variably present in patches on hepatic and branchial regions and in H-form depression.

Major Cheliped. Merus: Antero-dorsal margin moderately to strongly arched throughout, the distinct edge covered with minute tubercles. *Carpus:* With or without an enlarged tubercle at upper end of tuberculate, oblique ridge of inner surface; it may be present or absent in both subspecies and varies in size.

Manus: Bending of outer, upper surface distinct, moderately to strongly flattened near dorsal margin. Outer tubercles minute throughout, scarcely to distinctly enlarged dorsally, sometimes very slightly larger near base of gape. Dorsal outer groove shallow to pronounced. Dorsal margin distal to carpal cavity with outer edge distinct, slightly raised, and tuberculate; inner edge indistinct; intervening surface scarcely flattened, narrow, armed with tubercles larger than any on outer surface and arranged in irregular, oblique rows directed interno-distally. Ventral margin blunt, scarcely angled proximally, definitely rounded distally, the minute tubercles of outer manus continuing uninterrupted onto palm except for a regular row of very minute tubercles which do not have an outer groove adjoining them.

Palm with lower triangle covered with minute tubercles. Oblique tuberculate ridge moderate below apex which is moderately to strikingly high; tubercles largest on apex and continued far upward around carpal cavity's low, distal margin to the region of its distal extension, if any, into the predactyl area of upper palm. Center palm with a scattering of small to moderate tubercles in predactyl area only. Carpal cavity and predactyl area about as in *crenulata*, but the cavity is sometimes confluent with predactyl area instead of always set off by a low ridge.

Pollex and *dactyl*: Differ from both *uruguayensis* and *crenulata* as follows: Dactyl in one subspecies less convex, the gape being narrower, sometimes barely wider near middle than adjacent pollex. Dactyl always with outer, proximal, subdorsal groove absent or rudimentary, its dorsal and subdorsal tuberculation being small and virtually confined to proximal end; extremely minute tubercles present marginally and submarginally on both outer and inner sides of pollex and dactyl, but absent in central areas except on outer side of pollex, which is entirely covered. Gape pile absent.

Minor Cheliped. As in *crenulata*.

Ambulatories. 2nd and 3rd meri moderately slender, their dorsal margins slightly convex to almost straight. No anterior modification of 1st ambulatory.

Gonopod. Flanges well developed, continuous distally in a truncate or slightly convex edge beyond the subdistal pore on inner surface. Flanges almost equal in width. Inner process triangular, with or without a broad base, overlying canal and posterior flange, not reaching distal margin of flange. Thumb moderately short and slender, its tip not reaching base of flange.

Pile. Very variably present in patches on posterior hepatic regions, in H-form depression, postero-dorsally on major merus and carpus, dorsally on major manus, and, on each of first 3 ambulatories, on dorsal merus and on dorsal and posterior parts of carpus and manus. Through abrasion, variability, subspecific differences, or all three, pile is often practically absent.

FEMALE

Carapace and legs sometimes with very abundant pile, sometimes with practically none, depending on abrasion and the subspecies. Gonopore in one subspecies with a well-developed tubercle on antero-lateral part of margin, unique in the subgenus *Celuca*.

Measurements (in mm)

	Length	Breadth	Propodus	Dactyl
U. speciosa speciosa				
Largest males				
(Tortugas, Florida)	10.0	15.0	31.0	24.0
(Sarasota, Florida)	10.0	15.0	27.0	21.0
Moderate male				
(Miami, Florida)	8.5	12.0	17.5	11.5
Largest female				
(Sarasota, Florida)	10.0	14.5	–	–
Ovigerous female				
(Miami, Florida)	8.5	11.5	–	–
U. speciosa spinicarpa				
Largest males				
(Mississippi)	12.0	19.0	32.0	24.5
(Texas)	11.0	17.0	25.5	19.0
Moderate male				
(Alabama)	8.0	11.5	17.0	11.5
Largest female				
(Mississippi)	13.0	18.5	–	–
Moderate female				
(Alabama)	7.0	9.5	–	–

Morphological Comparison and Comment

The diagnosis should easily distinguish *speciosa* from *pugilator*, the only other *Celuca* in its range. The distinctively rotund *(Minuca) subcylindrica*, which, like *pugilator*, lacks an oblique ridge on the major palm, is in its shape and gonopod completely different. The diagnosis distinguishes *speciosa* as well from *burgersi*, *rapax*, and *pugnax*, all of the subgenus *Minuca*, which are the only other small *Uca* in the area, all of which may be expected to occur with it more or less sympatrically.

No series of either *s. speciosa* or, in particular, *s. spinicarpa* appear to have been collected giving numbers sufficient for satisfactory comparisons of the subspecies. It is interesting that all of the taxonomically reliable characters distinguished on the major cheliped combine to provide stronger armature in *s. spinicarpa* than in *s. speciosa*; the associated specializations extend even to the broader gape in *s. spinicarpa* which, as usual, accompanies an outstandingly high apex. All of these subspecific distinctions on the cheliped are functionally concerned, in species where the combat behavior is known, in the component here termed the heel-and-ridge (p. 490).

A comparison of all the members, including *speciosa*, of the superspecies *crenulata* and their allies will be found on p. 217.

Color

Displaying males near Miami: Carapace in February dark brown, except anteriorly where there was a wash of yellowish; in August some of the displaying individuals lightened to dull yellowish white. Major cheliped (both seasons): brown except for lower outer manus and entire chela, which were brilliant white, outside and in. Eyestalks greenish. Ambulatories, especially meri, often greenish, possibly due to algae. Females marbled with light and dark.

SOCIAL BEHAVIOR

Waving Display (Known in s. speciosa only)

Wave usually a straight lateral, but sometimes circular; cheliped stretched only slightly up, the movement being confined almost altogether to an outward and inward sweep of manus and chela; jerks absent; a brief pause between waves of a series, never a peak; fingers held constantly noticeably separated. Timing at greatest speed 2 waves per second, usually slower. Minor cheliped makes a corresponding motion. Body raised only slightly at beginning of a series and held there throughout; during peak display periods more than 12 waves sometimes occur in a single series. Ambulatories not raised as a regular part of display, although occasionally one leg elevates on minor side, apparently for balance only. (Components 4, 5, 9, 11; timing elements: Table 19, p. 656.)

Precopulatory Behavior
(Known in s. speciosa only)

Female follows male down his burrow, preceded by rapid waving display and by major-manus-drumming at the burrow's mouth, just before his descent. This observation agrees with that of Salmon (1967), who also observed and recorded sounds of courtship at night (see below and p. 501). On February 9, 1944, a warm day, a small group in Miami was displaying strongly; 1 ovigerous female was collected.

Acoustic Behavior (Known in *s. speciosa* only)

In connection with nocturnal courtship Salmon (1967) found that sound production replaced waving altogether, and concluded that the signals were of only one kind.

Observations made diurnally in the course of the present study showed that major-manus-drumming (component 9) occurs not only in high-intensity courtship but also before combats and between combat rounds. Major-merus-rubbing apparently occurs, but the filmed evidence is indistinct.

Combat (Known in *s. speciosa* only)

The drumming under these conditions of combat also occurs occasionally in alternation with waving; furthermore, single waves with drumming were sometimes performed in turn by each opponent before combat and between rounds in a fashion analogous to behavior since observed in *lactea perplexa* in Fiji and New Guinea (p. 491). At the time of the work on *speciosa* the complexities of combat were unknown, details of components were not noted, and unfortunately no films of the patterns were obtained.

RANGE

Southern United States of America, eastern Mexico, and western West Indies, as follows. *U. s. speciosa*: Southeastern Florida from Vero Beach south to the Florida Keys; western Florida north to Piney Island, near the state's western border; Peninsula of Yucatan (holotype only); Cuba. *U. s. spinicarpa*: The limited material has been collected in Alabama, Mississippi, and Texas in the United States, and from northeastern Mexico.

The specimens previously recorded from the island of Curaçao are all representatives of the allopatric species, *cumulanta*, as is the single record from Jamaica.

BIOTOPES

In southern Florida, *s. speciosa* occurs above mid-tide level on sheltered shores near mangroves, the muddy substrate being either bare or covered with mangrove debris (biotopes 6 and 9). The only information on the biology of *s. spinicarpa* is given by Rathbun (1918.1, p. 411) as follows: "Mr. J. D. Mitchell says of the habitat of this species: 'Drain runs half a mile into the land from Matagorda Bay; in the head of this drain there are several shallow wells of fresh water for cattle, walled on three sides. These crabs were all around these wells.' "

SYMPATRIC ASSOCIATES

U. s. speciosa is separated by habitat preference from *pugilator*, the only other *Celuca* known from Florida. *U. rapax* occupies the zone above it and *thayeri* below.

MATERIAL RESULTING FROM FIELD WORK

(The complete list of specimens examined is given in Appendix A, p. 608.)

Observations and Collections. Key Biscayne, Miami, Florida.

Films. Key Biscayne.

TYPE MATERIAL AND NOMENCLATURE

Uca speciosa (Ives, 1891)

HOLOTYPE. *Gelasimus speciosus*: Not now in Academy of Natural Sciences, Philadelphia, although it was deposited there according to Rathbun (1918.1). Type-locality: Port of Silam, Yucatan.

Uca spinicarpa Rathbun, 1900

TYPE. In Smithsonian Institution, National Museum of Natural History, Washington; cat. no. 22183. Type-locality: Galveston, Texas. Measurements in mm: length 9.5; breadth 13.8 (Rathbun). Holotype not designated in lot of 2 males, 1 female; formerly dried, now in poor condition; the major claws of both males, however, are intact. (!)

Uca (Celuca) [crenulata] speciosa speciosa (Ives, 1891)

(Florida; Cuba; Yucatan)

MORPHOLOGY

Carapace: Front slightly broader than in *s. spinicarpa*, the posterior edge of its distal border slightly rounded; carapace finely granulate on anterior part of dorsal surface; posterior part of dorso-lateral margins and upper pair of postero-lateral striae weak. *Major Cheliped*: Tubercles of outer manus slightly larger proximal to base of gape than subdorsally, where the tubercles are remarkably small for the area, which is only moderately bent over and flattened with the subdorsal groove shallow; apex of oblique ridge on palm only moderately high; tubercles and ridge proximal to predistal area of palm strong but somewhat irregular; distal tubercles of oblique ridge in irregular, multiple rows; gape slightly narrower than in *s. spinicarpa*; subdorsal, proximal groove on dactyl absent. *Gonopod*: Inner process with base very broad, tip reaching pore; thumb arising far down in shaft and ending far below base of flange. *Female gonopore*: Tubercle present. *Pile*: When not abraded, plentiful in both sexes but especially on female, where its distribution is as follows. Carapace covered dorsally with dense, short pile except, apparently, near front and orbital margins; ambulatories with dorsal and posterior surfaces similarly covered on merus, carpus and manus of at least 3rd and 4th legs and, to a lesser extent, of the more anterior appendages including minor cheliped.

Uca (Celuca) [crenulata] speciosa spinicarpa Rathbun, 1900

(Alabama, Mississippi, Texas; northern Mexico)

MORPHOLOGY

Carapace: Front slightly narrower than in *s. speciosa*, the posterior edge of its distal border markedly truncate; carapace smooth on anterior part of dorsal surface; dorso-lateral margins and upper pair of postero-lateral striae strong. *Major Cheliped*: Tubercles of outer manus distinctly larger subdorsally than elsewhere, this area flatter than in *s. speciosa*; dorsal submarginal groove well developed; apex of oblique ridge on palm strikingly high; tubercles of this ridge both in upper part beside carpal cavity and in distal part near ventral margin regular, with little or no development of multiple rows; gape broader than in *s. speciosa*; subdorsal, proximal groove on dactyl present, although rudimentary. *Gonopod*: Inner process with base relatively narrow, its tip not nearly reaching pore; thumb moderately short, arising closer to base of flange than in *s. speciosa*, and with its tip almost reaching flange. *Female Gonopore*: Tubercle absent. *Pile*: Apparently far less plentiful in both sexes, even when not abraded, than in *s. speciosa*. A number of males lack it except for traces in H-form depression and on legs, while it is sometimes plentiful on the flattened, dorsal part of outer major manus. In the several females examined it is virtually absent.

In the several specimens from Tampico, the antero-lateral margins are relatively rounded, not strongly angled.

REFERENCES AND SYNONYMY

Uca (Celuca) speciosa (Ives, 1891)

TYPE DESCRIPTION. See under *U. (M.) speciosa speciosa*, below.

Uca (Celuca) speciosa speciosa (Ives, 1891)

Gelasimus speciosus

TYPE DESCRIPTION. Ives, 1891: 179; Pl. 5, Figs. 5, 6. Mexico: Yucatan: Port of Silam (= "possibly Dzilam de Bravo"—Chace & Hobbs, 1969).

Uca speciosa

Rathbun, 1918.1: 408; Pl. 145. U.S.A., from south and west Florida. (Not specimens listed from Curaçao, now transferred to *U. cumulanta*.) Taxonomy. (USNM !)

Crane, 1957. U.S.A.: southern Florida. Preliminary classification of waving display.

Salmon, 1967: 450; Figs. 1, 2, 3, 4, 6, 7. Distribution in Florida; waving display; sound production.

Miller, 1965. U.S.A. Morphology, physiology, and ecology in relation to distribution.

Chace & Hobbs, 1969: 215; Text Figs. 73c, d. Distribution to date, following their reidentification of Rathbun's Curaçao specimens as *U. cumulanta* and her Jamaican specimen identified as *U. spinicarpa* now referred to *U. speciosa*. New record from Cuba. Taxonomy.

von Hagen, 1970.1: 227. Distribution in West Indies (no new records). Taxonomy.

Uca (Celuca) speciosa spinicarpa Rathbun, 1900

Uca spinicarpa

TYPE DESCRIPTION. Rathbun, 1900.2: 586. U.S.A.: Texas: Galveston. Description confined to key and designation of type. (USNM !)

Rathbun, 1918.1: 411. U.S.A.: Alabama: near Mobile; Mississippi: Biloxi Bay; Texas: drain near Matagorda Bay. Mexico: Maron: Laguna Madre; Tampico. (Not specimen from Kingston, in Jamaica [USNM 22313 = *cumulanta*] or from Brazil.) Taxonomy. (USNM !)

42. *UCA (CELUCA)* *[CRENULATA] CUMULANTA* CRANE, 1943

(Tropical western Atlantic)

PLATES 32 *A-D*; 47 *B*.
FIGURES 37 *N*; 56 *G*; 60 *J*, *K*; 70 *L*; 101.

MAP 15.
TABLES 9, 10, 12, 14, 19, 20.

INTRODUCTION

Among tropical Atlantic fiddlers, displaying males of *cumulanta* and *leptodactyla* alone are known to build hoods. As in the other hood-building forms, only some members of a few known populations construct them.

Within its range, *Uca cumulanta* is always the smallest fiddler in a mixed population, the other *Celuca*, *leptodactyla*, preferring sandier habitats. Found on muddy banks near low-tide levels, *cumulanta* is usually recognizable at once in the field through the bright blue green carapaces of most displaying males.

MORPHOLOGY

Diagnosis

Crab size small, front moderately broad; antero-lateral margins straight, sharply angled posteriorly. In male, no abdominal segments fused; 1st ambulatory with a row of minute tubercles on anterior side of major carpus. Ambulatories in both sexes without pile on carpus or manus. Gonopod tip appearing broadly tubular, there being no projecting flanges; a high tubercle on antero-inner edge of the wide, terminal pore. Female carapace sometimes with two patches of pile posteriorly.

Description

With the characteristics of the subgenus *Celuca* (p. 211).

MALE

Carapace. Frontal breadth contained about 4 times in that of carapace. Antero-lateral margins straight or slightly concave, very slightly diverging or converging; angling sharply into dorso-lateral margins. Each of the single pair of short postero-lateral striae curves inward to parallel the posterior margin of carapace. Breadth of eyebrow about half diameter of adjacent part of depressed eyestalk. Suborbital margin with strong crenellations, differing little in size, not continuing around outer orbital margin. Lower side of antero-lateral angle rounded.

Major Cheliped. Merus: Antero-dorsal margin moderately arched throughout; proximally forming a thick blunt ridge which is wholly unarmed; distally tuberculate and rugose. Surface rugosities weak except on posterior convexity. *Carpus*: Practically smooth except for a few distal rugosities. Oblique ridge of inner surface indistinct and tuberculate.

Manus: Bending of outer, upper surface distinct, flattened near dorsal margin. Outer tubercles minute; slightly larger in outer part of bent-over portion, but minute or absent alongside the strong, dorsal, outer groove. Dorsal margin distal to carpal cavity with outer edge distinct but with tubercles few or weak; inner edge indistinct; intervening surface an elevated, broad, somewhat convex lip, sometimes almost smooth, sometimes with traces of irregular, oblique rows of low, elongate tubercles directed intero-distally; dorsal margin close to dactyl base with a variable cluster of low tubercles and pits. Cuff outside dactyl base with a scattering of minute tubercles, as on nearby outer surface of manus, in addition to the usual row of tubercles starting here and continuing on subgape region of pollex. Ventral margin a rudimentary keel, low and blunt, with tubercles scarcely indicated or absent, variable, stopping at pollex base. Ventral outer groove absent.

Palm with lower triangle covered with extremely minute tubercles. Oblique ridge high throughout, apex highest; tubercles tending to irregularity, usually in single row between pollex base and apex, where they are largest, sometimes bicuspid; continued, still large and both irregular and variable in form and arrangement, practically or entirely reaching down-curving inner edge of dorsal margin. Carpal cavity with ventral margin low, thick, convex; distal part of cavity sloping moderately, its upper region without a distal extension, being set off from the spacious predactyl area by the two sets of tubercles just described. Predactyl area smooth or slightly lumpy, with a small depression, not set off from center palm by either ridge or tubercles.

Pollex and *dactyl*: Differ from *c. coloradensis* as follows. Dactyl with outer, proximal subdorsal groove shorter; dactyl's dorsal and subdorsal tubercles few, low, rounded, and traceable only through about proximal third of segment. Gape pile absent.

Minor Cheliped. As in *crenulata.*

Ambulatories. 2nd and 3rd meri moderately broad, their dorsal margins slightly convex. 1st carpus on major side with a short, longitudinal row of minute tubercles along middle of anterior surface.

Gonopod. No projecting flanges; pore very wide, terminal, with a high tubercle on its antero-inner edge. Inner process narrow, scarcely tapering, overlying canal; tip rounded, almost reaching pore. Thumb short, not nearly reaching base of projecting corneous section.

Pile. Two small patches of pubescence present or absent in H-form depression.

FEMALE

Dorsal and, behind vertical lateral margin, lateral parts of carapace finely granulate. A pair of patches of pile on posterior part of carapace, present or absent, large or small, one above base of each 4th ambulatory, as well as pile in H-form depression. Suborbital crenellations definitely stronger than in male.

Measurements (in mm)

	Length	Breadth	Propodus	Dactyl
Largest male (Trinidad)	10.5	15.5	28.0	21.5
Moderate male (Trinidad)	8.0	13.0	22.0	16.0
Smallest displaying male (Turiamo)	4.9	6.8	9.0	5.8
Largest female (ovigerous) (Turiamo)	8.5	13.0	–	–
Moderate female (Turiamo)	6.0	8.5	–	–

Morphological Comparison and Comment

The diagnosis aims in particular to distinguish *cumulanta* from *leptodactyla,* the only other *Celuca* in its usual range, as well as from small adults of *Minuca.*

All of the specimens at hand of female *cumulanta* with pile posteriorly on the carapace come from Turiamo, Venezuela, and from Georgetown, Guyana. It is interesting that two patches of pile similarly restricted are known only in females of *crenulata coloradensis,* from the head of the Gulf of California. The small patches usually found in the H-form depression of *cumulanta* are often absent in males from Trinidad and Turiamo but regularly present elsewhere.

A comparison of all the members, including *cumulanta,* of the superspecies *crenulata* and their allies will be found on p. 217.

Color

Displaying males: No display whitening. Carapace iridescent or shining greenish or bluish, at least anteriorly and sometimes completely, the color varying among populations from bottle green to pale turquoise; remainder of carapace, if any, mottled, or marbled with dark and light. Major cheliped: merus outside brown, inside reddish to reddish brown; outer manus yellowish brown; palm bluish or yellowish; fingers sometimes pinkish at base and tips, otherwise white or yellowish outside and in; ambulatories reddish or reddish brown, sometimes banded with shades of gray. Females dull.

SOCIAL BEHAVIOR

Waving Display

Wave lateral, either straight or circular; jerks absent; no pause at peak, which is higher than in Florida *speciosa;* minor cheliped makes corresponding motion; body moderately elevated with each wave; ambulatories remain on the ground. Display at rate of about 1 wave per second, the waves often separated within the irregular series by 2 seconds or more except during high intensity courtship. (Component nos. 4, 5, 9, 10; timing elements: Table 19, p. 656.) Many series, whether in threat, ambivalence, or definite courtship, end with major-manus-drums against the substrate close to the burrow mouth. Major-merus-drums against the carapace that sometimes occur between and after waves appear to be confined to agonistic situations between males. Often a small, natural elevation close to the burrow is climbed and used as the display site. Only one individual was ever seen to display from the top of his hood, which he did repeatedly.

Precopulatory Behavior

The male sometimes strokes the female with his ambulatory dactyls at the surface after high intensity waving and before drumming on the substrate at the mouth of his burrow. No attempt to copulate at the surface has ever been seen. The usual pattern is for the female to follow the male below ground, at the end of high intensity waving and final drumming. Several males, each of which was not promptly followed under these conditions by an attentive female, reemerged, mounted a small, natural mound nearby, and displayed vigorously with drumming. In each of the three examples the female suddenly dashed down his burrow, in advance of the male, which promptly followed. One of these three crabs had a well-constructed hood, but did not display from its summit.

Acoustic Behavior

In addition to the major-merus-drums and major-manus-drums already described, leg-wagging occurs as usual in both sexes. (Component nos. 7, 9, 5 respectively.) Major-manus-drumming has been seen clearly to occur, well inside the burrow mouth, in response the threatened entry of an intruder as well as of an attracted female; oscillograms from recordings made in each case are included in Pl. 47.

Combat

The following components are known to be included in *cumulanta*'s repertory: Manus-rub, dactyl-slide, interlace, and heel-and-ridge, with and without taps (component nos. 1, 6, 9, 12). Burrow-holders close to the entrance to their burrows often use either the major-merus-drum against the carapace or the major-manus-drum against the substrate, with or without waving, just before a combat. Alternating displays, each composed of both drumming and waving, by the two opponents have not been noted, unlike the observation already described for the closely related *speciosa*.

Hood and Chimney Construction

Only three populations have been found in which some displaying males build hoods, the activity which suggested the name for the species. These places were the type-locality at Pedernales, Venezuela, Turiamo, also in Venezuela, and on the island of Curaçao. No trace of hoods was ever found in Trinidad, where *cumulanta* was often observed, both in the field and in crabberies, nor were they found in any of a number of populations checked in Guyana, Surinam, and Brazil. The hoods observed ranged from rough and poorly formed to examples as symmetrical as those of any fiddler except *musica terpsichores*. A typical hood at the type-locality measured about 38 mm in height and slightly more in width, outside dimensions. The single observation of a hood's top being used as a display site was mentioned above, under the heading Waving Display.

In Turiamo and Curaçao a few examples were seen of small chimneys that apparently belonged to the females of this species. They were being reinforced from the inside. The individuals were not captured, and their identity remains to be checked.

RANGE

Known from the Caribbean coast of Panama at Colon; from Venezuela, Guyana, Surinam, and Brazil, and from the islands of Jamaica (one record), Curaçao, and Trinidad; apparently absent from any other islands in the Caribbean.

BIOTOPES

Sheltered banks and shores of mud or sandy mud close to mangroves but not in their shade (biotope nos. 11, 12). Populations are usually concentrated somewhat below mid-tide levels.

SYMPATRIC ASSOCIATES

Since *cumulanta* does not share habitats with *leptodactyla*, it has no close sympatric associate. In typical associations elsewhere, it shares the lower shore with *maracoani*, *thayeri*, or both, while *rapax* lives in the upper zone, the two species mingling at the edges. It will be noted that all four belong to different subgenera. Other members of the genus wtih coincident ranges usually occur in less saline habitats, except for the uncommon *major* which lives in habitats similar to that of *cumulanta*.

MATERIAL RESULTING FROM FIELD WORK

(The complete list of specimens examined is given in Appendix A, p. 608.)

Observations and Collections. Venezuela: Pedernales (Delta Amacuro); Turiamo (Aragua). Trinidad. Curaçao: observations but no collections.

Films. Pedernales; Trinidad.

Sound Recordings. Trinidad.

TYPE MATERIAL AND NOMENCLATURE

Uca cumulanta Crane, 1943

HOLOTYPE. In Smithsonian Institution, National Museum of Natural History, Washington. Male, cat. no. 137402 (formerly New York Zoological Society cat. no. 42423). Type-locality: Pedernales, Delta Amacuro, Venezuela. Measurements in mm: length 8.4; breadth 13.9; propodus 24.5. (!)

Additional type material. Deposited in the same institution, from the type-locality: 7 male paratypes, length 6.3 to 7.8 mm; 6 female paratypes, length 6.9 to 8.4 mm: cat. no. 137403 (formerly New York Zoological Society cat. no. 42423a. (!)

References and Synonymy

Uca (Celuca) cumulanta Crane, 1943

Uca speciosa (not *Gelasimus speciosus* Ives)

Rathbun, 1918.1: 408 (part). Specimens listed from Curaçao. (USNM !)

Rathbun, 1924.3: 19. Curaçao. Local distribution. (USNM !)

Uca spinicarpa (not of Rathbun)

Rathbun, 1918.1: 412 (part). Specimen listed from Jamaica. (USNM !)

Uca cumulanta

TYPE DESCRIPTION. Crane, 1943.2: 42; Text Figs. 1g, h, i; Pl. 1, Figs. 4-6. Venezuela. Color; waving display; hoods; habitat. (USNM !)

Crane, 1957. Venezuela; Trinidad. Preliminary classification of waving display.

Holthuis, 1959.3: 274; Text Fig. 68a; Pl. xiv, Fig. 3; Pl. xv, Fig. Suriname. Taxonomy; habitat.

von Hagen, 1967.2. Trinidad. Tape-recordings secured. (Preliminary statement.)

Chace & Hobbs, 1969: 211; Text Figs. 71e, f. Taxonomy. Reidentification of Rathbun's (1918.1: 409; 1924: 19) specimens of *speciosa* from Curaçao as *cumulanta*.

von Hagen, 1970.1: 226. Colon, Panama (= record of material in Leiden museum). Relationships; distribution.

von Hagen, 1970.4. Illus. Trinidad. Adaptations to a particular intertidal level.

43. *UCA (CELUCA) [CRENULATA] BATUENTA* CRANE, 1941

(Tropical eastern Pacific)

PLATE 32 *E-H*.   MAP 15.
FIGURES 70 *A*; 93; 101.   TABLES 9, 10, 12, 19, 20.

INTRODUCTION

A very small, partly white fiddler, *Uca batuenta* has a waving display of special interest. The conspicuous drumming motions that end most individual waves rarely or never actually touch either the ground or the carapace. These vibrations should, it seems, be considered a ritualization of a common acoustic mechanism, the major-manus-drum. This component occurs in proven functional form in two other members of the same superspecies, *speciosa* and *cumulanta*, in both of which tape recordings have been secured.

During each wave, *batuenta* holds its cheliped for an instant at the highest point attained, a habit that usually distinguishes it easily in the field even from *saltitanta*, its otherwise similar, close sympatric. The only other small fiddler in the eastern Pacific known to have a wave similarly accented is *deichmanni*, which usually does not share habitats with these mud-loving species and moreover shows no display whitening on the carapace.

MORPHOLOGY

Diagnosis

Size very small. Front moderately broad; eyebrow less than half width of eyestalk; antero-lateral angles rectangular, not produced. Suborbital crenellations obsolescent except externally where, however, there is no large isolated tooth. Major pollex at base little or no deeper than base of dactyl, merging with manus dorsally in the usual concave line; no large, triangular teeth on dactyl but pollex usually with a single, very large, tuberculate projection in distal half; no supraventral keel on outer pollex. Minor cheliped with palm not unusually broad and thick. Gonopod with thumb represented by a shelf; pore without a projection on lip. Segments of male abdomen incompletely fused.

Description

With the characteristics of the subgenus *Celuca* (p. 211).

MALE

Carapace. Differs from *cumulanta* as follows: Postero-lateral striae shorter, sometimes indistinct. Vertical lateral margin weak. Carapace profile more arched, the grooves less well marked. Frontal margin distally weak. Breadth of eyebrow slightly narrower than in *cumulanta*. Suborbital crenellations near antero-external angle fused into a single elongate crest.

Major Cheliped. Merus: Antero-dorsal margin straight proximally, convex distally; proximally with a low ridge which is either practically unarmed or slightly rugose. Anterior surface almost or entirely unarmed dorso-distally. *Carpus*: Practically smooth.

Manus: Bending of outer, upper surface distinct, moderately flattened near dorsal margin. Outer tubercles extremely minute throughout, with enlargement barely indicated in outer part of bent-over portion, but minute or absent alongside the weak, dorsal outer groove. Dorsal margin distal to carpal cavity with outer edge raised and thickened, with its top weakly tuberculate; inner edge indistinct; intervening surface unusually narrow, smooth except for a few small pits and for weak, non-tuberculate, obliquely transverse rugosities directed intero-distally. Ventral margin a rudimentary keel, marked only proximally by small tubercles and dying out before pollex base. Ventral, outer groove rudimentary, indicated only along proximal tubercles.

Palm without tubercles on lower triangle; instead, its proximal portion is slightly lumpy, with minute pits; median area with small, shallow furrows, chiefly longitudinal. Distal half of oblique ridge low, blunt, tubercles small to practically absent; apex abruptly high with large tubercles; smaller tubercles continuing irregularly upward as far as the slight distal extension of carpal cavity. Center palm with small tubercles that merge proximally with large ones around carpal cavity; palm proximal to gape with small pits. Carpal cavity with ventral margin low, forming only a broad angle where the two surfaces meet; distal part of cavity sloping moderately, its upper region, although without a distal extension, not separated from predactyl area by a downward continuation of its dorsal margin or other tubercles, but only by the flat, higher plane of predactyl region. Latter spacious, without a depression, smooth except for slight pitting, not set off from center palm.

Pollex and *dactyl*: Pollex unusual in shape, its upper margin proximally concave and, in distal third, a large triangular tooth, its distal slope, ending at pollex tip, concave, and both edges serrate; lower margin of pollex slightly sinuous. Gape in proximal third slightly wider than adjacent part of pollex but narrow more distally because of tooth. Inner side of pollex with depression from pollex base continued beyond middle. Inner side of dactyl also with a faint, median concavity. Attachment of dactyl to manus slightly oblique. Dactyl with outer, proximal, subdorsal groove absent; dorsal and subdorsal tubercles also absent. Both pollex and dactyl wholly smooth on outer and inner surfaces. Inner marginal row of gape tubercles on pollex practically or wholly lacking; tubercles of outer marginal rows very small, forming proximal edge of triangular tooth by swerving medially; median row absent proximally but apparently giving rise to crest of triangular tooth and to the tubercles along its distal concave edge. Pollex tip simple. Dactyl with inner marginal row practically lacking; tubercles of external marginal row minute; median row complete with a few enlarged tubercles proximally. Gape pile present or absent.

Minor Cheliped. As in *crenulata*.

Ambulatories. 2nd and 3rd meri moderately wide, their dorsal margins slightly convex. No anterior modifications of 1st ambulatory.

Gonopod. Flanges narrow, but distinct; pore small, subterminal. Inner process narrow, tapering, blunt. Thumb vestigial, represented by an oblique shelf far below base of flange.

FEMALE

Manus of cheliped not especially broad and thick; some pile present on ambulatories, at least in a small, dorso-distal patch on posterior side of 1st and 2nd carpus.

Measurements (in mm)

	Length	Breadth	Propodus	Dactyl
Largest male (holotoype) Canal Zone	4.8	7.6	11.8	8.2
Moderate male (paratype) Canal Zone	4.5	7.0	11.0	7.5
Large ovigerous female (paratype) Costa Rica	4.1	6.5	–	–
Moderate ovigerous female (paratype) Costa Rica	3.5	5.0	–	–

Morphological Comparison and Comment

The male's slender major pollex combined with its predistal protuberance is the most obvious, single diagnostic character for *batuenta*, setting it off conveniently from the superficially similar *saltitanta*, as well as from related species with a protuberance also, but with the pollex broad; as usual this distinction must be used with care because it can be ambiguous in leptochelous individuals. The young of *batuenta* are closely similar to adults of *tenuipedis*; young males of *batuenta* may be distinguished at once by the presence of an oblique ridge inside major palm, even when the chela is still so short that it resembles that of the other species; the greater width of the ambulatory meri of *batuenta* is the other major specific character readily discernible in the young.

A comparison of all the members, including *batuenta*, of the superspecies *crenulata* and their allies will be found on p. 217.

Color

Displaying males: Display whitening only partly developed. Carapace pale brown or yellowish, speckled and marbled variably with white. Major cheliped with merus and carpus, both outer and inner surfaces, pinkish brown; outer manus and palm, bluish white; chela polished white. Buccal and pterygostomian regions about like dorsal part of carapace. Ambulatories with meri anteriorly pinkish brown; rest of ambulatories and entire minor cheliped brown, speckled and banded with white. Eyestalks sometimes green. Females similar to males.

SOCIAL BEHAVIOR

Waving Display

Wave lateral; irregularly circular at least at high intensity; a brief pause at highest point reached by the chela. The irregularity of the circular motion occurs because the claw is brought down to a flexed position in front of the starting point; from here it usually returns to rest position in front of the buccal area through a series of bouncing motions, consisting of about 3 to 7 vibrations; some of these may hit the ground and so qualify as major-manus-drums, but the majority, as shown clearly in film analyses, do not touch the substrate at all; often the vibrations start in mid-air, before the cheliped is more than half lowered, so that they could be labeled minute jerks if their imminent development into drumming motions were not understood. True jerks, as the word is used in this contribution (p. 496), are altogether absent from the display. Major chela opens and closes during each wave. Minor cheliped sketches in a synchronous motion, without vibration, only occasionally. Body raised and lowered during each wave. Ambulatories remain on ground. The crab often moves a few steps while the cheliped is raised. The entire display, including drumming motions, lasts about a second. (Component nos. 5, 9, 10; timing elements in Table 19, p. 656.)

Acoustic Behavior

It seems likely, because of its incorporation into waving display and its partial or complete ritualization into visually conspicuous motion, that the major-manus-drumming behavior described above is not in this species an acoustic component (no. 9); observations and films so far obtained are inadequate bases for a decision, and no recordings of any of the components have been attempted. The other acoustic components observed, all detected in film analyses, are minor-merus-rubs, leg-wagging, and major-merus-drums (component nos. 2, 5, 7). Each occurred in agonistic situations, closely associated with waving display.

This species is one of several *Celuca* in which displacement-cleaning of the major cheliped by the minor is often prevalent during display. Film analysis has raised the suspicion that stridulation will prove to take place during these motions, but the stroking of the claw by the minor chela occurs in so many areas, reached by a variety of routes, that the designation of any of the motions as a component would be premature (see general discussions of cleaning and of acoustic behavior, pp. 472 and 480).

RANGE

Pacific coast of El Salvador to northern Peru.

BIOTOPES

Mud, among unshaded mangrove shoots; sometimes on open mud flats with mangroves nearby. (Biotope nos. 8, 11, 12.)

SYMPATRIC ASSOCIATES

In rich habitats *batuenta* is sometimes found within inches of any of three species closely related to it, *saltitanta*, *oerstedi*, and *inaequalis*, usually along with a prodigal quantity of other *Uca*, as is characteristic of good habitats in the tropical eastern Pacific.

MATERIAL RESULTING FROM FIELD WORK

(The complete list of specimens examined is given in Appendix A, p. 609.)

Observations and Collections. Costa Rica: Puntarenas; Ballenas Bay. Panama: near Old Panama. Canal Zone: Balboa. Colombia: Buenaventura (specimens sent from here to Trinidad crabbery, where observations were made but specimens not secured for preservation). Ecuador: Puerto Bolivar: observed but not collected.

Films. Near Old Panama.

TYPE MATERIAL AND NOMENCLATURE

Uca batuenta Crane, 1941

HOLOTYPE. In Smithsonian Institution, National Museum of Natural History, Washington. Male, cat. no. 137405 (formerly New York Zoological Society cat. no. 4121). Type-locality: La Boca, Balboa, Canal Zone. Measurements on p. 245. (!)

Additional type material. Deposited in same institution: paratypes, male and female, from the type-locality, from Puntarenas, Costa Rica, and from Ballenas Bay, Costa Rica, cat. nos. 137406, 79399, 137404 (formerly New York Zoological Society nos. 4122, 381136, 381137). (!)

REFERENCES AND SYNONYMY

Uca (Celuca) batuenta Crane, 1941

Uca batuenta

TYPE DESCRIPTION. Crane, 1941.1: 187; Text Figs. 4n, 5, 8; Pl. 6, Fig. 26. Costa Rica and Panama. (USNM !)
    Crane, 1957. Panama; Ecuador. Preliminary classification of waving display.
    von Hagen, 1968.2: 411. Peru: Puerto Pizarro. Waving display résumé.

Altevogt & Altevogt, 1967. E 1292. Film of waving display.

Uca saltitanta batuenta

Bott, 1954: 178. El Salvador: Puerto el Triunfo. (Frankfurt !)
    Peters, 1955: 433; Text Fig. 4. El Salvador: Puerto el Triunfo. Waving display.

44. *UCA (CELUCA) SALTITANTA* CRANE, 1941

(Tropical eastern Pacific)

PLATE 33 *A-D*.
FIGURES 39 *I*; 46 *O*; 70 *B*; 81 *M*; 93; 101.

MAP 15.
TABLES 9, 10, 12, 19, 20.

INTRODUCTION

A strikingly small species, *U. saltitanta* shows in several particulars the highest degree of development in its general group. This group comprises both the members of its tropical eastern Pacific alliance and of the superspecies *crenulata*. The outstanding specializations of *saltitanta* include its strongly semi-cylindrical carapace, which doubtless assists in moisture conservation during long periods of vigorous display on dark mudflats in blazing sunshine; a uniquely shaped pollex, yet to be observed during combat; the highest degree of display whitening; the fastest single waves in the group and among the fastest in the genus; and the greatest number of waves in unbroken series of any species of *Uca*.

MORPHOLOGY

Diagnosis

Size very small. Front moderately broad; eyebrow less than half width of eyestalk; antero-lateral angles acute and slightly produced. Suborbital crenellations obsolescent, except externally where there is a strong series including, in male, a large, isolated tooth on orbit's outer margin. Major pollex at base broad, merging with manus dorsally in a straight line continuous with its prehensile edge; chela usually with two large, triangular teeth on dactyl but never with a large projection in distal half of pollex. Minor cheliped with palm broad and thick, especially in female. Gonopod with thumb represented by a shelf. Segments of male abdomen incompletely fused.

Description

With the characteristics of the subgenus *Celuca* (p. 211).

MALE

Carapace. Frontal breadth contained about 4 to 5 times in that of carapace. Orbits almost straight. Antero-lateral angles ranging from the usual rectangular to broadly acute and slightly produced. Antero-lateral margins concave and strongly diverging. Vertical lateral margin complete but weak. Cara-

pace profile practically semi-cylindrical, the grooves scarcely marked, but hepatic and branchial areas not specially expanded above rest of carapace. Eyebrow shorter than usual and very narrow, less than one third diameter of adjacent part of depressed eyestalk. Suborbital margin with crenellations very small throughout except, near antero-external angle, for two broad tubercles, the outer being isolated. Lower side of antero-lateral angle with a short edge basally produced into a minute tubercle or spinule.

Major Cheliped. Merus: Rugosities all very weak except on posterior convexity; sometimes almost absent even there. Postero-dorsal margin practically absent, being smooth and rounded even at base. Antero-dorsal margin very strongly convex, more definitely a crest than in any other *Celuca*, edged throughout its length with a single row of minute serrations except at extreme distal end; here there is a small, deep cavity filled with pilous setae which project beyond it; similar setae in a patch on adjacent cuff. Anterior surface unarmed dorso-distally. *Carpus*: Practically smooth; anterior margin concealed in pile.

Manus: Breadth almost or altogether equal to length; upper distal boundary, giving rise to dactyl, very oblique, leaving the dorsal margin unusually short while pollex base is relatively wide, giving rise to the characteristic, triangular pollex. Area of dorsal margin unusually unspecialized, except for its shortness. Bending of outer, upper surface of manus slight, with little flattening near dorsal margin. Outer tubercles large for the subgenus over the entire lateral surface, largest in middle of side; near ventral margin they are abruptly minute, as well as adjacent to dorsal margin and to cuff at base of dactyl. Dorsal outer groove absent. Dorsal margin distal to carpal cavity with both outer and inner edges indistinct, the narrow intervening surface virtually smooth or slightly pitted. Cuff outside dactyl base oblique, chiefly smooth or slightly granulate, with clustered small tubercles ventro-distally replacing the row of tubercles that usually continues on subgape region of pollex. Area outside pollex base sometimes with a definite, large, but very shallow depression, instead of the usual flattening; lower edge of area with a distinct keel that continues distally along most of pollex in its ventral half. Ventral margin with a row of minute tubercles throughout its length; proximally they are slightly elevated on a low keel, with a faint, ventral, outer groove alongside; keel and groove both absent more distally, the tubercles becoming increasingly weak and dying out on proximal part of pollex.

Palm with lower triangle smooth except for a few small pits and, distally near oblique ridge, minute tubercles. Oblique ridge low to moderate, apex lower than middle of ridge; tubercles in distal half weak to almost absent, sometimes in multituberculate row; proximally always stopping short at apex. Center palm with a few scattered tubercles proximally and, sometimes, minute tubercles distally. Carpal cavity shallow, with ventral margin very short, low, thick, blunt; distal part of cavity sloping gradually into center palm; inner edge of proximal, dorsal margin not curving downward, the cavity having a distal extension that is smooth, shallow, and narrow, almost reaching dactyl base. Surrounding predactyl area also smooth, not set off from center palm. Distal ridge at base of dactyl with tubercles obsolescent.

Pollex and *dactyl*: Pollex notably triangular, being proximally both broad and thick; ventral margin almost straight; dorsal margin with a large subdistal tooth, triangular or convex, its margins tuberculate. Dactyl unusually long, curving far down below pollex tip, its attachment to manus very oblique; ventral margin with two well-separated triangular teeth, the edges tuberculate and each apex tubercle somewhat enlarged. Gape much narrower than adjacent pollex. Outer carina on lower pollex traceable almost to pollex tip. Dactyl with outer, proximal subdorsal groove absent; proximal upper inner (not median) part of dorsal margin with an irregular row of minute tubercles, multituberculate proximally, traceable more than half length of dactyl. Dactyl distally with a faint internal groove in downcurving part of tip; the same region, on both inner and outer sides, covered with very minute tubercles and, sometimes, pits, the roughening usually extending farther dorsally on inner side than outer. Pollex and dactyl otherwise laterally smooth. Gape with all three rows of tubercles well developed on pollex, except that outer row ends abruptly farther from tip than usual and that inner row is sometimes obsolescent in middle; outer (not inner) row and middle row widely separated in pollex' proximal half; dactyl with inner row missing but median and outer rows complete; triangular teeth described above in both pollex and dactyl arising entirely from middle rows. Pollex tip simple. Gape pile absent.

Minor Cheliped. Manus deep. Pollex and dactyl shorter or scarcely longer than manus. Gape very narrow, absent in middle where the strong, uneven teeth of prehensile edges articulate.

Ambulatories. 2nd and 3rd meri moderately slender, their dorsal margins slightly convex. No anterior modification of 1st ambulatory. Pile as in female, but less persistent, more variable, or both.

Gonopod. Flanges distinct, the inner slightly produced; pore subterminal. Inner process narrow, tapering. Thumb vestigial, represented by a shelf-like projection, arising far below base of flange.

Abdomen. 3rd to 6th abdominal segments incompletely fused.

FEMALE

Suborbital armature weaker than in male and manus of cheliped broader and thicker than in male's minor cheliped, and with the serrations in its gape larger. Ambulatory pile, as in *batuenta*, in a small patch on dorso-distal part of posterior surface of carpus and on dorsal and proximal part of manus, at least on 1st and 2nd legs; most persistent on 2nd.

Measurements (in mm)

	Length	Breadth	Propodus	Dactyl
(All from Balboa, Canal Zone)				
Largest male (holotoype)	6.0	8.8	12.2	10.4
Large male	5.5	7.0	12.0	10.0
Largest female	5.3	6.8	–	–
Ovigerous female	4.6	6.5	–	–

Morphological Comparison and Comment

The strikingly triangular form of the major pollex, combined with the strongly oblique upper distal margin of the manus, make the appearance of *saltitanta*'s cheliped unique in the genus; the two widely separate enlarged teeth on the dactyl are almost always well enough developed to be a convenient diagnostic aid.

A comparison of all the members, including *saltitanta*, of the superspecies *crenulata* and their allies will be found on p. 217.

Color

Displaying males: During peak periods, polished display whitening completely envelops most individuals, although occasionally the carapace remains grayish or yellowish, or the white is sparsely marked with brown. Females dark.

SOCIAL BEHAVIOR

Waving Display

Wave lateral, either straight or with a slightly irregular circularity that is less than in *batuenta*, the cheliped on return to the ground being closer to the buccal area from which it started; the shape of the waves in both species and the inclusion of drumming motions at the end of each wave are otherwise similar. Unlike *batuenta*'s wave, however, *saltitanta*'s shows no pause at the peak. No films that have yet been secured show clearly whether or not the drumming motions actually touch the ground or whether, as in *batuenta*, they are largely or wholly vibrations in the air; again, tape recordings have not been made. In the field 3 or 4 vibrations were counted after each wave. Jerks absent. With every wave, the major chela opens and shuts, the minor cheliped unflexes partway and flexes again, the body is raised and lowered, and the ambulatories are kept on the ground, except in making the several steps to one side or the other that often accompany a wave. After an instant's pause in rest position, the display is repeated. At moderate, territorial intensity the rate is less than half a second per wave, and upwards of 100 waves occur in a single series. (Component nos. 4, 5 with irregularity, 9, 10; timing elements in Table 19, p. 656.)

Whenever a small eminence exists even as much as several inches from the burrow, the male climbs to its summit and displays only from this position except in the last stages of courtship. (No observations that include male intruders have yet been made.)

Precopulatory Behavior

At the approach of a wandering female the male's display becomes even faster before he races down his small hill and resumes display after bracing the legs of his minor side inside the burrow mouth. Usually the female circles him at a distance of several inches. Just before he descends he stretches the ambulatories of his major side straight out, clear of the ground, the distal segments rigid as the meri vibrate in apparent contact. Almost certainly they are stridulating and both position and motion are identical with those typical of unreceptive females throughout the genus. The crab remains in this position for a second or less, then suddenly descends into the burrow. The female has been seen to follow a number of times. This is the only example I know from personal experience where leg-wagging assuredly is used in a non-ambivalent, courtship situation (cf. Salmon, 1967, quoted in the present study on pp. 194 and 195, who recorded leg-wagging in (*Minuca*) *rapax* and *pugnax*).

Acoustic Behavior

In addition to its use in high-intensity courtship, leg-wagging (component no. 5) has been seen and photographed during mildly agonistic situations between neighboring males. As also described, major-manus-drumming (9) may or may not be acoustically functional.

RANGE

Pacific coast of El Salvador to Buenaventura, Colombia.

BIOTOPES

U. saltitanta occurs in greatest numbers on open flats formed of deep and sticky mud near river mouths; fringing mangroves usually grow within about a hundred meters of the populations, although in very rich areas the species both feeds and displays even farther out. A few individuals sometimes live among pioneer mangrove shoots, but never in even partial shade. (Biotope no. 8.)

SYMPATRIC ASSOCIATES

When this small fiddler occurs among these pioneer shoots it sometimes mingles with *batuenta*, a species closely related and even smaller. The only usual associate of *saltitanta*, however, is (*Uca*) *ornata*, the giant fiddler that contrasts with it in practically every particular.

MATERIAL RESULTING FROM FIELD WORK

(The complete list of specimens examined is given in Appendix A, p. 609.)

Observations and Collections. Costa Rica: Puntarenas; Panama and Canal Zone: near Old Panama, Balboa; Colombia: Buenaventura; material sent from there examined and then observed in Trinidad crabbery but not secured and preserved.

Films. Old Panama.

TYPE MATERIAL AND NOMENCLATURE

Uca saltitanta Crane, 1941

HOLOTYPE. In Smithsonian Institution, National Museum of Natural History, Washington. Male, cat. no. 137407 (formerly New York Zoological Society cat. no. 4123). Type-locality: La Boca, Balboa, Canal Zone. Measurements on p. 248. (!)

Additional type material. In same institution: paratype males and females from type-locality (cat. nos. 79403 and 137408, both formerly New York Zoological Society cat. no. 4124). (!)

REFERENCES AND SYNONYMY

Uca (Celuca) saltitanta Crane, 1941

Uca saltitanta

TYPE DESCRIPTION. Crane, 1941.1: 189; Text Figs. 4o, 5; Pl. 2, Figs. 10, 11; Pl. 3, Fig. 14; Pl. 6, Fig. 25. Costa Rica and Canal Zone. Color; waving display; habitat. (USNM !)

? Peters, 1955. Illus. El Salvador: Puerto el Triunfo. Morphology; waving display; ecology.

Crane, 1957. Panama. Preliminary classification of waving display.

Uca saltitanta saltitanta

Bott, 1954: 177; Text Fig. 22; Pl. 19, Figs. 21a, b. El Salvador: Puerto el Triunfo. Taxonomy. (Frankfurt !)

45. *UCA (CELUCA) OERSTEDI* RATHBUN, 1904

(Tropical eastern Pacific)

PLATE 33 *E-H*.   MAP 15.
FIGURES 70 *I*; 93; 101.   TABLES 9, 10, 12, 19, 20.

INTRODUCTION

Small, waving fiddlers showing greenish blue when seen from the front should turn out to be *Uca oerstedi*. At high intensity the display of this species includes a component apparently unique in the genus, the vibration of the first ambulatories; their meri appear to human eyes to be particularly vivid, approaching peacock blue. This exhibit of intense hue through a motion of display makes a good example of the untouched opportunities for experimental work on the possible functions of color in social behavior.

This species is the third member of the local alliance.

MORPHOLOGY

Diagnosis

Front moderately broad; antero-lateral margins long, strongly diverging and posteriorly angled; eyebrow less than half width of eyestalk. Carapace in male unusually areolate, with paired patches of pile, all easily abraded, on gastric, cardiac, and branchial regions; each branchial region divided by a longitudinal furrow along outer boundary of pile. Major pollex with a median projection, long and low, on prehensile edge and a long, low, supraventral keel. Ambulatories in both sexes without pile on postero-dorsal surfaces of carpi and mani. Gonopod with thumb represented by a shelf; pore with a projection on internal (anterior) lip.

Description

With the characteristics of the subgenus *Celuca* (p. 211).

MALE

Carapace. Frontal breadth contained about 5 to 6 times in that of carapace. Orbits almost straight. Antero-lateral margins long, straight, divergent, angling sharply into dorso-lateral margins. Each member of the single pair of short, postero-lateral striae is strong and curves inward to parallel the posterior margin of carapace, close behind end of dorso-lateral margin. Grooves between carapace regions deep for this subgenus, especially outside gastric and cardiac regions, where they are continuous with postorbital furrow. Inner half of each branchial region depressed, covered with pile, bounded externally by a longitudinal groove, better developed than in the several related species in which it is usually traceable; pile also sometimes present on each side of gastric and cardiac regions, giving 4 additional patches; all pile is easily abraded. Breadth of eyebrow more than half diameter of adjacent part of depressed eyestalk. Suborbital margin almost straight, with crenellations represented internally by minute irregularities; proceeding outwards these increase very slightly in size; near antero-external angle are 3 to 6 true crenellations, still small, usually truncate and usually separated; they extend little or not at all around outer orbital margin. Lower side of antero-lateral angle blunt.

Major Cheliped. Merus: Antero-dorsal margin proximally a low ridge, straight and unarmed, flaring only in subdistal part of distal third into a short, low, convex, serrate crest. *Carpus*: Oblique ridge of inner surface with several proximal tubercles.

Manus: Bending of outer, upper surface moderate, flattening near dorsal margin slight. Outer tubercles large for the subgenus except near heel and ventral margin, where they are somewhat smaller; tubercles near dorsal margin of similar size to those below, but a few are usually raised on irregular, low rugosities. Dorsal outer groove strong, but present only as far as level of distal edge of carpal cavity. Dorsal margin above groove high, erect, its outer side with small tubercles; distal to groove and carpal cavity both outer and inner edges of dorsal margin indistinct; intervening surface convex, sometimes forming a blunt ridge, always smooth except for a few tubercles or rugosities. Area outside pollex base with a large, very shallow depression and with a low tuberculate keel along lower edge; keel traceable more proximally as a row of small tubercles and continued distally, but with tubercles minute or absent, along most of pollex, close to lower margin (mentioned under pollex below, as the supraventral keel). Ventral margin of manus with low tuberculate ridge, proximally slightly enlarged and serrate, the tubercles dying out at pollex base. Ventral, outer groove present proximally only.

Palm with lower triangle smooth except for slight pitting proximally and extremely minute tubercles distally. Oblique ridge moderate, the apex higher with its tubercles abruptly larger; small tubercles continue upward around carpal cavity, the row being almost regular or slightly multituberculate, as far as predactyl area. Carpal cavity with ventral margin very low, broad, blunt; distal part of cavity sloping moderately; inner edge of proximal dorsal margin not curving downward, the cavity having a shallow, short, broad, distal extension that slopes gradually into the smooth surface, flat or convex, of the short predactyl area.

Pollex and *dactyl*: Upper margin of pollex with a tuberculate convexity, or a bluntly triangular tuberculate tooth, near middle. Dactyl unusually long, curving distally far down below tip of pollex. Gape much narrower than adjacent pollex. Pollex externally with a low, broad, supraventral keel continuing its entire length, less outstanding than in Rathbun's (1918.1, Pl. 152) photograph of the large holotype. Dactyl with a faint, external concavity, wide proximally and diminishing distally to end about middle of segment; outer, proximal subdorsal groove only faintly indicated but sometimes traceable almost half length of dactyl, bounded above by a row of widely spaced small tubercles and, below it proximally, by a few more. Outer and inner surfaces of both pollex and dactyl entirely smooth. The usual three rows of gape tubercles all present in both pollex and dactyl, all the marginal rows ending well before tip; tubercles edging convexity of pollex in gape arising completely from median row. In addition to convexity of pollex, a few tubercles are sometimes enlarged proximally in dactyl and a single tubercle about middle of dactyl appears to occur regularly. Pollex tip simple. Gape pile plentiful.

Minor Cheliped. Gape at base narrower than width of pollex, decreasing distally. Middle section with a few, long, strong teeth almost in contact.

Ambulatories. 2nd and 3rd meri unusually broad for a *Celuca*, their dorsal margins moderately convex. No anterior modification of 1st ambulatory.

Gonopod. Flanges vestigial; pore large, its internal (anterior) lip lower than the external, with a projection. Inner process small, short, flat, not reaching pore. Thumb represented by an oblique shelf far proximal to the corneous base of the specialized tip.

FEMALE

Carapace finely granulate, including sides behind vertical lateral margin; the only smooth portions are the cardiac and intestinal regions and the area near

posterior margin. Pile characteristic of male carapace usually rudimentary or absent. Suborbital armature not stronger than in male.

Measurements (in mm)

	Length	Breadth	Propodus	Dactyl
(All specimens from Old Panama)				
Largest male	8.0	11.8	16.5	8.0
Moderate male	7.0	10.2	12.2	8.5
Largest female (ovigerous)	7.0	10.0	–	–

Morphological Comparison and Comment

A variable amount of pile is present near center of carapace, usually in two large patches, in all specimens in this collection. The larger male paratype examined at the Smithsonian Institution, however, has no pile, and Rathbun mentions none in her description. In her diagnosis (1918.1: 414) she includes "oblique and distal ridges on inner side of palm meeting." When this area is examined from the point of view of our currently more extensive knowledge of palm armature in *Celuca*, the statement can be misleading, since the oblique ridge, although continued well up around the carpal cavity, actually merges to a variable degree with a group of predistal tubercles in this narrow region, and they, in turn, are quite distinct from the well-formed, unexceptional, proximal, predistal ridge.

The most convenient single character distinguishing *oerstedi* promptly from other *Celuca* in the eastern Pacific is the strong divergence of the long, straight, antero-lateral margins which posteriorly are sharply angled. A comparison of all the members, including *oerstedi*, of the superspecies *crenulata* and their allies will be found on p. 217.

Color

Displaying males: No display whitening. Carapace, 3rd maxillipeds, dorsal and posterior surfaces of all segments of both chelipeds except distally, and all segments of ambulatories posteriorly are usually dull purple or dull purplish blue. Lower outer manus of major cheliped white; dactyls of both chelipeds white, inside and out. All remaining inner surfaces of chelipeds, the pterygostomian areas, subhepatic regions, and the anterior surfaces of the ambulatories are turquoise blue, with the meri, especially of the 1st legs, more intensely colored than the other segments.

SOCIAL BEHAVIOR

Waving Display

Wave lateral, straight to circular, raised upward comparatively little. Jerks absent, although the major cheliped is returned to rest position more rapidly than it is unflexed, giving the superficial effect of a single jerk. Minor cheliped synchronously unflexed and flexed, sometimes vibrated, perhaps stridulating. Body usually raised and lowered with each wave but sometimes held in that position throughout the waves of a single series. During waving the ambulatories remain on the ground except during the usual steps to one side or the other of the burrow, or around it, chiefly during display of moderate intensity. At high intensity the 1st ambulatories vibrate in the midst of each wave, when both chelipeds unflex maximally, thus exposing fully the intense green blue of the quivering meri. The social conditions eliciting this behavior remain unknown. The wave is rather slow for the group, requiring more than a second for completion; there is a short pause between waves within a series. (Component nos. 4, 5, 9, 11, plus the unique ambulatory vibration described; timing elements in Table 19, p. 656.)

Acoustic Behavior

The observed and filmed examples consist with certainty only of leg-wags (component no. 5) and major-merus-drums (7). One short scene shows good, but not conclusive, evidence of a palm-leg rub (4).

RANGE

Gulf of Fonseca, El Salvador, to Panama City, R.P.

BIOTOPES

Beaches of sandy mud, close to the open water of large bays, but always with some protection from the sea such as a small stand of mangroves close to low tide levels or a nearby jetty. Also found on the edge of mud flats near mouths of streams, close to mangroves and often among their shoots. (Biotope nos. 6, 12.)

SYMPATRIC ASSOCIATES

The closely related species, *batuenta*, sometimes occurs among mangrove shoots with *oerstedi*, but the latter usually lives closer to open water as described, where it often occurs with two other *Celuca*, *beebei* and *deichmanni*, as well as with *heteropleura* of the subgenus *Uca*.

MATERIAL RESULTING FROM FIELD WORK

(The complete list of specimens examined is given in Appendix A, p. 609.)

Observations and Collections. Panama: Old Panama. Canal Zone: La Boca, Balboa.

Films. Old Panama.

TYPE MATERIAL AND NOMENCLATURE

Uca oerstedi Rathbun, 1904

HOLOTYPE. In Universitetets Zoologiske Museum, Copenhagen. Male. Measurements in mm: length 12; breadth 16.8 (Rathbun). Type-locality: Punta Arenas (= Puntarenas?) Costa Rica.

Additional type material. Female in Copenhagen. Male paratype (cat. no. 31506) in Smithsonian Institution, National Museum of Natural History, Washington. Length in mm: 11; propodus broken. (!)

REFERENCES

Uca (Celuca) oerstedi Rathbun, 1904

Uca oerstedi

TYPE DESCRIPTION. Rathbun, 1904: 161. Costa Rica: Punta Arenas. (part = USNM !)
  Rathbun, 1918.1: 414; Pl. 152, Figs. 1, 2. Taxonomy of type material.

Crane, 1941.1: 184; Text Figs. 2, 3, 4k. Panama and Canal Zone. Taxonomy; color; waving display; habitat. (USNM !)
  Bott, 1954: 178; Text Fig. 18; Pl. 19, Fig. 23a, b. El Salvador: Puerto el Triunfo. Taxonomy.
  Crane, 1957. Panama. Preliminary classification of waving display.

46. *UCA (CELUCA) INAEQUALIS* RATHBUN, 1935

(Tropical eastern Pacific)

PLATE 34 *A-D*.
FIGURES 26 *G*; 31 *F*; 70 *K*; 101.

MAP 15.
TABLES 9, 10, 12, 14, 20.

INTRODUCTION

U. inaequalis, the fourth of the species in the local alliance, is small and largely brown, the distinguishing tufts of pile on the carapace often hidden in life by clinging mud.

According to present knowledge, *inaequalis* appears more prone to combat than its close relations; before encounters and between rounds the opponents sometimes wave in alternation, in a fashion similar to that of *lactea perplexa* in New Guinea and Fiji; at these times palm-leg rubbing is also prevalent, *inaequalis* being provided with armature that is moderately well developed but not as striking as in *musica*.

For a long time I believed that *inaequalis* uses typical major-manus-drumming (formerly called "rapping") as a characteristic end to many high intensity waves, just as does, for example, the Atlantic species, *cumulanta*. Film analysis has now shown that there is probably no touching of the substrate whatever, just as there is little or none in *batuenta* and *saltitanta*. Instead, *inaequalis* definitely vibrates the major merus against the outer suborbital part of the carapace, a motion which apparently does not occur as part of waving display in either *batuenta* or *saltitanta*. The motions in all four species, superficially similar to the human eye, are of such special evolutionary interest that they would repay detailed comparison.

MORPHOLOGY

Diagnosis

Front moderately broad; male carapace with 8 conspicuous pilous elevations across carapace, plus other lower similar structures; all present but weaker in female. 1st ambulatory on major side with a short row of small tubercles on both carpus and distal merus. Major cheliped with pollex proximally very broad with a predistal, triangular, tuberculate tooth; no distinct, pilous depression outside pollex base. Gonopod with a conspicuous projection on edge of its large, terminal pore; thumb represented by an oblique shelf.

Description

With the characteristics of the subgenus *Celuca* (p. 211).

MALE

Carapace. Frontal breadth contained about 3.5 times in that of carapace. Orbits moderately oblique. Antero-lateral margins straight or slightly concave, their direction almost straight, but varying from slightly concave to slightly convex, angling into the dorso-lateral margins either bluntly or sharply. Carapace with 12 or more small elevations covered with pile, the latter easily detached and often partly or wholly missing in preserved specimens. Frontal margin weak. Eyebrow breadth half or less than half diameter of adjacent part of depressed eyestalk. Suborbital margin entire, except, starting at antero-external angle, for one or more enlarged crenellations or tubercles; their size, shape and number are variable; sometimes there is a single broad-based sharp tubercle, sometimes several wide crenellations that extend partway around outer orbital margin. Lower side of antero-lateral angle blunt.

Major Cheliped. Merus: Postero-dorsal margin smooth and practically absent; posterior convexity, however, strongly developed. Antero-dorsal margin strongly convex throughout, although most arched in distal third; edge a distinct low crest, proximally unarmed, distally serrate. *Carpus:* Oblique ridge of inner surface with or without granules.

Manus: Bending of outer, upper surface slight, with slight flattening beside dorsal margin. Outer tubercles extremely minute except for slight enlargement and traces of rugosities near dorsal margin; tubercles smallest near ventral margin. Dorsal outer groove present. Dorsal margin distal to carpal cavity with both outer and inner edges distinct and tuberculate; intervening surface flattened, tilted toward the palm, almost smooth except for a few distal pits and tubercles or sometimes rugosities directed intero-distally. Lower edge of flattened area outside pollex base with traces, usually faint, of a broad, blunt, supraventral keel that continues distally along lower part of pollex; individual variation, as in *oerstedi*, is considerable. Ventral margin with a row of small, similar tubercles dying out on pollex, midway to tip. Ventral, outer groove absent.

Palm with lower triangle covered by moderate tubercles, largest proximally. Oblique ridge low, the apex little or not at all higher than median portion; tubercles small and similar throughout including distally, but stopping on or before apex. Center palm with a broad band of small tubercles proximally. Carpal cavity shallow, with both ventral and distal margins broad, blunt, and indistinct. Beaded edge of proximal dorsal margin slanting slightly downward to bound on its dorsal side the distal extension of cavity; extension is shallow, broad, pointed, and almost reaches dactyl base, the small predactyl area being convex and smooth, not set off from center palm. Ridges at base of dactyl both reduced, the proximal with only several ventral tubercles and the distal with none; groove between ridges scarcely indicated.

Pollex and *dactyl*: Pollex broad; tip curving upward; dactyl not much longer than manus; its attachment to manus very oblique. Gape narrow, more so even than width of dactyl. Pollex externally with variable traces of a supraventral keel, as stated above, in description of manus. Inner surfaces of pollex and dactyl slightly concave throughout. Upper part of dactyl proximally very thick, flattened, a narrow groove on external side of inner part of dorsal margin (representing the usual outer, proximal, subdorsal groove); groove flanked internally by a row of minute tubercles dying out beyond groove toward middle of dactyl; extremely minute tubercles also present in vicinity of groove. Pollex and dactyl with outer and inner surfaces wholly smooth. Pollex with all three rows of gape tubercles well developed, but with inner row ending far short of tip; outer (not inner) and middle rows widely separated in proximal half on both pollex and dactyl. Dactyl with inner row represented only by a few proximal tubercles. Tuberculate triangular teeth in gape weak or absent, but one tubercle of middle row usually moderately enlarged toward middle of both pollex and dactyl, the enlargement on pollex sometimes tending toward the triangular tuberculate form characteristic of related species. Pollex tip simple. Gape pile plentiful.

Minor Cheliped. Manus and fingers slender. Gape at base narrower than width of pollex, decreasing distally. Serrations strong, even, in contact beyond middle.

Ambulatories. 2nd and 3rd meri moderately broad, their dorsal margins slightly convex. 1st carpus on major side anteriorly with a row, sometimes irregular, of 7 to 9 minute tubercles; in addition, the ad-

jacent 1st merus sometimes has 3 or 4 granules antero-distally, forming an angle or curve.

Gonopod. Flanges absent; pore very large, terminal, its internal (anterior) lip with a large projection extending beyond rest of pore margin. Inner process reaching almost to pore margin; broad, flat. Thumb represented by an oblique shelf far proximal to corneous part of specialized tip.

FEMALE

Carapace finely granulate, especially posteriorly and laterally, including sides behind vertical lateral margin. Suborbital armature about as in male. Pilous elevations less distinct than in male.

Measurements (in mm)

	Length	Breadth	Propodus	Dactyl
(All specimens from Guayaquil)				
Largest male	6.2	10.0	13.5	9.2
Moderate male	5.5	7.5	9.0	6.5
Largest female (ovigerous)	7.0	9.0	–	–
Moderate ovigerous female	5.5	7.2	–	–

Note. Rathbun's (1935) type measurements: length 8.0; breadth 11.2; (propodus 16.0; dactyl 11.0– J.C.). Von Hagen's (1968.2) largest specimen: length 7.2; breadth 10.5.

Morphological Comparison and Comment

Although the dorsal patches of pile on the carapace are more persistent in this species than in most others where they are importantly diagnostic, they are still subject to abrasion. Nevertheless, even when pile is practically absent the 2 transverse rows of small elevations across the middle of the carapace are distinctive, each consisting as it does of 4 similar mounds. These 8 are almost always larger than the 4 or more additional elevations, with or without pile, which are also usual, especially, it seems, in more northern populations. No quantitative work on their size and occurrence has yet been undertaken.

A comparison of all the members, including *inaequalis*, of the superspecies *crenulata* and their allies will be found on p. 217.

Color

Displaying males: Display whitening absent. Carapace dark brown marbled with white; buccal and pterygostomian regions brown. Major cheliped dark brown except as follows: upper surfaces of merus, carpus, manus, and base of pollex rufous (reddish orange brown); lower half of manus also rufous, brighter than the other areas; distal half of dactyl and entire pollex white. Anterior sides of meri of at least 1st and 2nd ambulatories sometimes purple to purplish red; ambulatories otherwise dark brown marbled with white, like carapace. Females dark.

SOCIAL BEHAVIOR

Waving Display

Wave lateral, either straight or at high intensity irregularly circular, the cheliped then being brought to the ground in front of rest position and vibrated back toward the buccal area, as in *batuenta* and *saltitanta*. In the few examples where the activity can be clearly seen, all in films, the crab at these times is lifting the cheliped's merus against an adjacent part of the carapace, in a major-merus-suborbital component that is usually a drumming but sometimes a rubbing. The motion occurs frequently enough during the display of *inaequalis* to be considered an integral part of it. In the clearest examples the backward tilting of the upper part of the major merus toward the carapace is distinct during the vibrations. A single series of these motions that end a wave number between 10 and 14—many more than have been counted in either *batuenta* or *saltitanta*. Jerks, actual or apparent, absent, although the cheliped is lowered very abruptly. Minor cheliped synchronously unflexed and flexed. Body raised and lowered during each wave. Ambulatories remain on the ground except for the usual few steps. Wave slow, lasting more than a second, including vibrations, with at least an equal period between waves. (Component nos. 4, 5 with irregularity, 9, 10.)

Acoustic Behavior

The following motions have been observed and filmed. During moderate intensity waving display, as described above: major-merus-rubs (component no. 1), major-merus-drums (7), and, rarely and questionably, major-manus-drums (9). During agonistic encounters, including both threat and actual combat: minor-merus-rubs (2), palm-leg-rubs (4) and leg-wags (5).

Combat

A. D. Blest of the University of London filmed an excellent sequence of a long and forceful combat in Panama that gives our only clear evidence of the occurrence of two elements of special interest in the combat of this species. First is the alternation of waving by the opponents, mentioned in the Introduction to *inaequalis*. Second is an associated, alternating use of palm-leg-rubs. The filmed combat is heterochelous and mutual; it consists of several rounds, all ending forcefully with the push-off, fling, or actual upset of the same individual; as often happens in *lactea* combats in the South Pacific, the overthrown crab repeatedly returned for another round. Except for manus-rubs and one set of manus-taps or pushes (component no. 1), the actual combat components were too incomplete and irregular to isolate; potential interlaces (9) were incomplete and the eventual grips irregular or indistinct. In each of three rounds, however, a type of forceful push-off, not observed before, was delivered when the actor unflexed his cheliped and struck his palm against the palm or outer manus of his opponent. Also included in this same combat were clear examples of tripping and of down-pointing (agonistic component 2), both noted also in the field, and both occurring before formal combat in the wholly unrelated *tangeri*, *maracoani*, and *ornata*.

RANGE

Pacific coast of El Salvador to northern Peru.

BIOTOPES

Mud, often stony mud, often shaded by mangroves, the area sometimes separated from the lower part of a stream by fringing mangroves; rarely on unshaded, open mud flats. (Biotope nos. 9, 11, 12, 14.)

SYMPATRIC ASSOCIATES

Depending on the biotope, *inaequalis* is sometimes associated with one or another of the other members of the alliance—*batuenta*, *beebei*, or *oerstedi*. Not characteristically mingling with any particular species of the numerous *Uca* living in various combinations of the same habitats; perhaps most often found with *beebei*, although the optimum habitat of the latter species occurs in somewhat more saline localities.

MATERIAL RESULTING FROM FIELD WORK

(The complete list of specimens examined is given in Appendix A, p. 609.)

Observations and Collections. Nicaragua: Corinto. Costa Rica: Puntarenas, Ballenas Bay, Golfito. Canal Zone: Balboa. Panama: near Old Panama. Ecuador: Puerto Bolivar.

Films. Old Panama.

TYPE MATERIAL AND NOMENCLATURE

Uca inaequalis Rathbun, 1935

HOLOTYPE. In Smithsonian Institution, National Museum of Natural History, Washington. Male, cat. no. 70833 (labeled "TYPE"); in same jar under same number are 6 additional males and 3 females. Type-locality: Salada (= El Salado), Guayaquil, Ecuador. (!)

REFERENCES

Uca (Celuca) inaequalis Rathbun, 1935

Uca inaequalis

TYPE DESCRIPTION. Rathbun, 1935.1: 51. Ecuador: Guayaquil. (USNM !)

Crane, 1941.1: 185; Text Fig. 4L, Pl. 2, Figs. 8, 9; Pl. 3, Fig. 12. Nicaragua to Panama. Taxonomy; color; waving display; habitat. (USNM !)

Bott, 1954: 176; Text Fig. 21; Pl. 18, Fig. 20a, b. El Salvador: Puerto el Triunfo. Taxonomy.

Peters, 1955: 435; Text Figs. 4, 5. El Salvador: Puerto el Triunfo. Waving display.

Crane, 1957. Panama, Ecuador. Preliminary classification of waving display.

von Hagen, 1968.2: 410; Text Fig. 4a. Peru: Puerto Pizarro. Taxonomy; color; waving display.

Altevogt, 1970. Illus. Peru. Form and function of vibration signal.

47. *UCA (CELUCA) TENUIPEDIS* CRANE, 1941

(Tropical eastern Pacific)

PLATE 34 *E-H*.   MAP 15.
FIGURES 68 *H*; 101.   TABLES 9, 10, 20.

INTRODUCTION

Notable for slender legs and small size, *Uca tenuipedis* was unknown in life until von Hagen (1968.2) found it in Peru. Now, thanks to his work on its waving display, we have behavioral evidence that it should be considered a member of the group composed of *crenulata* and its allies, in accordance with its morphological characteristics of form and armature.

MORPHOLOGY

Diagnosis

Front moderately broad. Ambulatories unusually slender in both sexes; antero-lateral margins long, slightly diverging, posteriorly angled. Carapace without pile. No oblique ridge on major palm; pollex very broad at base. Gonopod with large flanges but thumb represented by an oblique shelf.

Description

With the characteristics of the subgenus *Celuca* (p. 211).

MALE

Carapace. Frontal breadth contained almost 4 times in that of carapace. Antero-lateral margins long, straight or slightly concave, slightly diverging, angling into dorso-lateral margins. A variable amount of pile present on lower, posterior sides of carapace. Frontal margin distally weak. Eyebrows narrow, about one-fourth diameter of adjacent part of depressed eyestalk. Suborbital margin with internal crenellations minute, those in outer half much larger, but variable; the most external, around anterior part of outer orbital margin, are very large and separated. Lower side of antero-lateral angle blunt.

Major Cheliped. Merus: Postero-dorsal margin practically absent, although area is rugose as usual and posterior convexity is strongly developed. Antero-dorsal margin proximally a low ridge, straight and unarmed, flaring only in distal third into a low, convex crest which is irregularly tuberculate. *Carpus:* Apparently smooth except for pile, and apparently lacking tubercles on anterior margin.

Manus: Wide, being only about one-fifth longer than broad. Bending of outer, upper surface slight, with slight flattening beside dorsal surface. Outer tubercles extremely minute except for slight enlargement near dorsal margin; there they are in short, irregular rows, most of them vertical, some horizontal, forming a roughly reticulate pattern; the tubercles, unlike tuberculate striae, not set on ridges; intervening spaces smooth. Dorsal outer groove shallow, dying out subdistally. Dorsal margin with outer edge above groove beaded, confluent with margin above carpal cavity only proximally, at extreme end; indistinct distal to groove; inner edge indistinct; intervening surface broad, flattened, covered with sharp, crowded tubercles midway in size between those of upper and lower side of manus. Lower edge of flattened area outside pollex base with a keel that continues distally along most of pollex in its ventral half. Ventral margin with a row of large tubercles extending its entire length; just before pollex base they become smaller and irregularly double, dying out about halfway to tip. Ventral, outer groove absent.

Palm with lower triangle smooth to faintly reticulated in upper portion; tubercles wholly lacking. Oblique ridge, including apex, rudimentary, being low, broad, blunt, and reaching carpal cavity unusually far below gape level, even when the great breadth of pollex is considered; tubercles lacking throughout. Center palm smooth except for a band of sparse, very minute tubercles in upper part adjoining predactyl area; in largest specimen slight depressions give a reticulated aspect. Carpal cavity small and shallow, ventral margin very short, moderately broad; distal margin indistinct, the cavity merging with center palm. Edge of proximal dorsal margin turning slightly downward distally, but not bounding distal extension of carpal cavity; the extension is short, proximally broad, distally pointed, and merges gradually with predactyl area. Predactyl area spacious, convex, and covered closely with small, sharp tubercles that continue unbroken from adjacent dorsal margin.

Pollex and *dactyl:* Differ from *inaequalis* as follows: Dorsal half of dactyl covered with sharp tubercles, except for a well developed, smooth, outer groove that extends about half length of dactyl. Gape with inner row of tubercles on pollex extending almost to tip; median enlargement on pollex gape definitely a tuberculate tooth, the margin distal to its apex being concave; no gape teeth clearly enlarged on dactyl.

Minor Cheliped. Gape very narrow. Strong serrations practically in contact.

Ambulatories. All segments very slender. Dorsal margins even of 2nd and 3rd meri almost straight. 3rd merus extending more than one-quarter of its length beyond antero-lateral angle when laid forward. No anterior modification of 1st ambulatory. Inconspicuous pile present or absent, then perhaps abraded, dorsally on carpus, particularly of 2nd and 3rd legs.

Gonopod. Tip very similar to those of *uruguayensis* and *speciosa*. Flanges large, the anterior the larger; pore subterminal, the flanges continuing beyond it, their common distal margin narrowly rounded. Inner process triangular with a broad base, overlying canal, the pointed tip almost reaching gonopore. Thumb absent, represented only by a few bristles arising from a shelf; the latter slants obliquely toward base of appendage, far below base of flange.

FEMALE

Ambulatories remarkably slender, including the meri, differing very slightly in this character from those of the male, and consequently forming a serviceable character for easily distinguishing females of the species.

Measurements (in mm)

	Length	Breadth	Propodus	Dactyl
Largest male (von Hagen, Peru)	5.3	7.1	9.7	
Holotype male (Costa Rica)	5.0	6.5	8.1	5.8
Largest female (Costa Rica)	5.0	6.6	–	–
Largest female (von Hagen, Peru)	5.0	6.8	–	–
Smallest ovigerous female (von Hagen, Peru)	4.4	5.7	–	–

Morphological Comparison and Comment

The combination of a wide manus, short fingers, and absence of an oblique ridge on the palm gives a paedomorphic appearance to the major cheliped. In contrast to such species as *pygmaea*, however, the form of the carapace suggests no paedomorphism, while the slenderness of the legs is unusual in adults of any subgenus and is never in itself a juvenile character.

A comparison of all the members, including *tenuipedis*, of the superspecies *crenulata* and their allies will be found on p. 217.

Color

(After von Hagen, 1968.2.) Display whitening absent. Males and females similar. Carapace dull brown, marbled with gray. Eyestalks and minor chelipeds gray brown. Ambulatories somewhat lighter pale gray to light horn color, on the posterior side with dark gray or brown bands. Anterior side of major cheliped of males: palm brown above, gray white below. Dactyl brown above, pollex gray white.

SOCIAL BEHAVIOR

Waving Display

(After von Hagen, 1968.2.) Wave apparently lateral-straight, with an irregular circularity occurring because the claw is brought down to a flexed position in front of starting point, as in *batuenta*. A very brief pause at highest point reached by chela. Claw brought down to a flexed position abruptly, thus making the first stroke of a major-manus-drumming against the substrate; it is brought back into the original position through the series of bouncing motions that compose the drumming, consisting of 1 to 7 vibrations (usually 4 to 5). During waving minor cheliped participates. Body raised and lowered during each wave. Ambulatories remain on ground. Duration of entire display, including drumming, about 0.5 to 0.75 second. (Component nos. 4, 5, 9, 10.)

RANGE

Costa Rica and Peru.

BIOTOPE

Mudbanks on shores of mangrove estuaries. (Biotope no. 12.)

TYPE MATERIAL AND NOMENCLATURE

Uca tenuipedis Crane, 1941

HOLOTYPE. In Smithsonian Institution, National Museum of Natural History, Washington. Male, cat. no. 137409 (formerly New York Zoological Society cat. no. 381,143). Type-locality: Ballenas Bay, Costa Rica. Measurements on this page. (!)

Additional Type Material. 9 male and 2 female paratypes from the type-locality in the same institution, cat. nos. 79404 and 137410 (formerly New York Zoological Society cat. no. 381144). (!)

References

Uca (Celuca) tenuipedis Crane, 1941

Uca tenuipedis

TYPE DESCRIPTION. Crane, 1941.1: 186; Text Fig. 4m; Pl. 2, Fig. 7; Pl. 3, Fig. 13. Costa Rica: Ballenas Bay. (USNM !)

von Hagen, 1968.2: 410. Peru: Puerto Pizarro. Taxonomy (= measurements and key characters); color; waving display.

48. *UCA (CELUCA) TOMENTOSA* CRANE, 1941

(Tropical eastern Pacific)

PLATE 35 *A-D*.   MAP 15.
FIGURES 70 *E*; 101.   TABLES 9, 10, 20.

INTRODUCTION

Uca tomentosa is one more small fiddler that seems clearly, if not closely, related to the superspecies *crenulata*. This one has no outstanding morphological peculiarities, is dull in color, and has confusingly various amounts of pile on its carapace. The unique male holotype of *U. mertensi* Bott, 1954, and the series from Peru referred to *mertensi* by von Hagen, 1968.2, are examples of *tomentosa*. Von Hagen's field work has provided basic data on *tomentosa*'s color and waving display.

MORPHOLOGY

Diagnosis

Front moderately wide. Major pollex not unusually broad proximally and without a strong projection on prehensile edge. 1st ambulatory on major side without special armature on anterior surface; merus of each ambulatory, especially of 2nd and 3rd, broad. Carapace in both sexes with patches of pile, but patterns various, easily abraded. Gonopod with pore large and terminal, its edge somewhat uneven but with no trace of a projection; thumb well developed.

Description

With the characteristics of the subgenus *Celuca* (p. 211).

MALE

Carapace. Frontal breadth contained about 5 times in that of carapace. Antero-lateral margins straight, scarcely convergent, angling bluntly into dorso-lateral margins. Dorsal surface of carapace with short pile on branchial regions in a reticulated pattern of varying extent and partly covering also mesogastric, cardiac, and hepatic regions; one patch in holotype on branchial region immediately interior to the longitudinal groove that, as in *oerstedi*, *tallanica*, and other related forms, partly subdivides branchial region. Breadth of eyebrow more than one-half diameter of adjacent part of depressed eyestalk. Suborbital margin with crenellations only slightly enlarged near antero-external angle; they then extend, slightly reduced in size, around outer orbital margin as far as

channel. Lower side of antero-lateral angle moderately sharp.

Major Cheliped. Merus: Antero-dorsal margin proximally a low ridge, straight and unarmed, flaring only subdistally into a low crest, scarcely convex, and irregularly tuberculate. *Carpus*: Anterior margin practically erect, instead of bent over; oblique ridge of inner surface poorly marked but with a row of 7 distinct tubercles.

Manus: Bending of outer, upper surface moderate but practically without flattening. Outer tubercles extremely minute in lower half, increasing gradually to moderate size toward upper margin; tubercles absent adjacent to the distinct dorsal outer groove. Dorsal margin with outer edge confluent with that of carpal cavity only at extreme proximal end, distinct and tuberculate throughout entire length of margin; inner edge beyond cavity indistinct; intervening surface flattened, smooth proximally, roughened distally by an irregular longitudinal row of small, separated tubercles. Lower margin weakly keeled, surmounted by coarse, low, blunt beading that dies out near pollex base. Faint, ventral, outer groove present proximally.

Palm with lower triangle evenly covered with extremely minute tubercles. Oblique ridge high, the apex slightly higher than median portion but the tubercles there only slightly larger than elsewhere; tubercles continued beyond apex upward around carpal cavity to predactyl area, but few, small, and irregularly spaced. Carpal cavity with ventral margin thick, blunt; distal edge moderately sloping; no distal extension, the beaded inner edge of proximal dorsal margin curving downward to meet tubercles continued upward from oblique ridge, and hence bounding upper distal part of carpal cavity. Predactyl area spacious, practically flat, smooth, rounding into the tuberculate dorsal margin not set off from center palm.

Pollex and *dactyl*: Pollex slightly less deep than dactyl. Dactyl little longer than manus. Gape narrow, less than depth of pollex. Pollex with a faint inner depression extending through about half of length. Dactyl with outer, proximal, subdorsal groove well developed, surrounded by strong tubercles that dorsally extend beyond it through about half dactyl's length. Both pollex and dactyl with all lateral surfaces smooth. Gape in both pollex and dactyl with outer and median rows of tubercles well developed but inner row absent on pollex except proximally and weak on both pollex and dactyl. Pollex tip simple. In gape near pollex tip a small, tuberculate convexity which is absent in the larger holotype of *mertensi*, instead of enlarged into the triangular tooth characteristic of most related species; a single enlarged tubercle near middle of dactyl. Gape pile present.

Minor Cheliped. Slender; gape at base narrower than width of pollex, decreasing distally. Middle section with serrations strong, distally in contact.

Ambulatories. All meri broad for a *Celuca*: 2nd and 3rd with dorsal margins strongly convex. No anterior modification of 1st ambulatory.

Gonopod. Flanges absent; pore large, with its internal (anterior) tip lower than the external but with no trace of a projection; external (posterior) lip broadly angled. Inner process narrowly triangular, flat; its rounded tip extending slightly beyond adjacent lip of pore. Thumb well developed but short, extending slightly more than its own length proximal to corneous base of specialized tip.

FEMALE

Carapace pile about as in male.

Measurements (in mm)

	Length	Breadth	Propodus	Dactyl
Largest male (von Hagen; Peru)	11.0	16.8	30.6	
Male (holotype of *mertensi*; Bott; Salvador)	9.5	15.0		
Holotype male (Costa Rica)	6.6	10.2	14.5	9.0
Largest female (von Hagen; Peru)	11.4	17.9	–	–
Largest female paratype (Costa Rica)	7.3	11.9	–	–

Morphological Comparison and Comment

A comparison of all the members, including *tomentosa*, of the superspecies *crenulata* and their allies will be found on p. 217.

Color

(After von Hagen.) No display whitening observed. Male: Carapace with mosaic pattern of dark green, light green and yellow. Eyestalks emerald green. Ambulatories anteriorly light gray flecked with darker; posteriorly marbled brown and yellow. Minor cheliped light gray. Outer side of major cheliped:

manus and dactyl in juveniles light brown; in older individuals manus gray, sprinkled above with yellow; whitish below, as well as on pollex. Females: Carapace with mosaic pattern of dark brown, green, and yellow, infrequently red and olive; ambulatories anteriorly light gray or pale rose. Underside of ischium rose. Otherwise as in male.

SOCIAL BEHAVIOR

Waving Display

(After von Hagen, 1968.2.) Wave lateral-circular. A distinct pause at higher intensities at highest point reached by chela. Claw lowered into a flexed position abruptly, giving impression of jerks as it strikes the ground in front of the rest position and is then moved back into its usual place, apparently in a fashion similar to that found in *tenuipedis* and *batuenta*; unlike those species, however, *tomentosa* makes only a single stroke against the ground, instead of a series of vibrations; hence it is not a fully developed major-manus-drumming. During waving minor cheliped participates. Body tilted strongly during each wave, doubtless through bending of the more posterior ambulatories and extension of those in front. Some ambulatories sometimes raised during display. Duration of entire display, including the single stroke of major cheliped against ground, about 1.25 to 1.5 seconds. (Component nos. 5, 9, 10, 12.)

RANGE

From Los Blancos, El Salvador, to Puerto Pizarro, Peru.

BIOTOPES

Clay-like mud flats to muddy sand; smooth, superficially encrusted sand (von Hagen) (= over mud). (Approximate biotope nos. 5, 6, 8.)

TYPE MATERIAL AND NOMENCLATURE

Uca tomentosa Crane, 1941

HOLOTYPE. In Smithsonian Institution, National Museum of Natural History, Washington. Male, cat. no. 137411 (formerly New York Zoological Society cat. no. 381,132). Type-locality: Puntarenas, Costa Rica. Measurements on this page. (!)

Additional Type Material. 4 female paratypes from the type-locality in the same institution. Cat. no. 137412 (formerly New York Zoological Society cat. no. 381,133). (!)

Type Material of *Uca mertensi* Bott, 1954 (= holotype male). In Forschungsinstitut Senckenberg,

Frankfurt a.M. Cat. no. 1863; length 9.5 mm. (!)
Type-locality: Los Blancos, El Salvador.

The small size of the holotype of *tomentosa*, in comparison with both the holotype of *mertensi* and of the Peruvian material in von Hagen's collection,

may be one factor in the differences in amount of pile. Except for less pile and the expectably longer major chela in the holotype of *mertensi*, the specimens are closely similar, including details of armature and gonopod.

References and Synonymy

Uca (Celuca) tomentosa Crane, 1941

Uca tomentosa

TYPE DESCRIPTION. Crane, 1941.1: 179; Text Figs. 4h, 6. Costa Rica: Puntarenas. (USNM !)

Uca mertensi

Bott, 1954: 169; Text Fig. 11; Pl. 16, Fig. 11a, b.

El Salvador: Los Blancos. Type description. (Frankfurt !)

Altevogt & Altevogt, 1967. E 1290. Film of waving display.

von Hagen, 1968.2: 425; Text Fig. 4b, c. Peru: Puerto Pizarro. Taxonomy; color; waving display; habitat (Text Fig. 20).

49. *UCA (CELUCA) TALLANICA* VON HAGEN, 1968

(Tropical eastern Pacific)

PLATE 35 *E-H*.　　MAP 15.
FIGURE 70 *J*.　　TABLES 9, 10.

INTRODUCTION

Von Hagen's publication of the type description of *U. tallanica* adds one more species, notably small and remarkably interesting, to this group of closely related forms. For years it seemed unwise to base a new species on the few imperfect male specimens I collected in Ecuador. Now the name *tallanica* is available, based on a series from Peru. In the following description I have combined von Hagen's data with that from the Ecuadorian specimens; the details of the armature and gonopod, however, are derived entirely from the latter.

MORPHOLOGY

Diagnosis

Front moderately wide. Major cheliped with a small, pilous depression with definite margins at base of outer pollex; center of palm tuberculate; both ridges at base of dactyl well developed, tuberculate; a strong, tuberculate projection subdistally on pollex. 1st ambulatory on major side with a long row of minute tubercles on anterior surface of carpus; merus of each ambulatory strikingly broad, including that of 4th. Male carapace with usually 6 small, curved patches of pile; female usually with 2. Gonopod with pore large and terminal but without a projection on its edge; thumb represented by an oblique shelf.

Description

With the characteristics of the subgenus *Celuca* (p. 211).

MALE

Carapace. Frontal breadth contained about 3.3 times in that of carapace. Orbits moderately oblique. Antero-lateral margins short, straight or slightly concave, converging, turning further inward posteriorly at a broadly obtuse angle. Each member of the upper pair of postero-lateral striae is long and strong and, at extreme inner end, curves sharply inward, paralleling posterior border; lower striae sometimes a single pair, sometimes represented by several, at least on one side, always short, always far lateral. Carapace with small, paired patches of pile, the larger curved

and comma-shaped, usually numbering 3 pairs, as follows: on each side of the mid-line one patch on upper and one on lower longitudinal groove of the H-formation, plus a smaller patch, sometimes straight, on branchial region immediately interior to the longitudinal groove that, as in *U. oerstedi*, partly subdivides branchial region. Breadth of eyebrow less than half diameter of adjacent part of depressed eyestalk. Suborbital margin with crenellations distinct throughout, although small internally; they increase gradually to moderate size in neighborhood of outer angle and continue, somewhat separated, almost to outer orbital channel. Lower side of antero-lateral angle blunt on major side, moderately sharp on minor.

Major Cheliped. Merus: Postero-dorsal margin smooth and practically absent; posterior convexity strongly developed. Antero-dorsal margin practically straight; distinct and sharp throughout but not high enough to be termed a crest; unarmed in proximal half but finely serrate, the serrations increasing progressively in size, distally. *Carpus*: Oblique ridge of inner surface sometimes with several very small tubercles.

Manus: Bending of outer, upper surface moderate, flattening near dorsal margin slight, although a broad area below the subdorsal ridge and groove is practically smooth. Outer tubercles in lower half small but then increasing regularly until, by the time they stop abruptly along lower edge of subdorsal, smooth area, they are large for the subgenus. A few enlarged tubercles, comparable in size to the largest ones subdorsally, in an irregularly vertical row behind base of gape. Dorsal outer groove strong, continuing entire length of margin. Dorsal margin above groove high throughout, armed except proximally with a regular row of similar, rounded tubercles, so nearly contiguous that they almost constitute beading. Dorsal margin itself almost flat with variable, scattered, sharp tubercles; inner edge distal to carpal cavity indistinct. Area outside pollex base with a small, triangular depression filled with pile, the base of the triangle, paralleling ventral margin, being much longer than its altitude; the triangle is margined ventrally by small tubercles that continue a short way proximally beyond the depression, in a weak version of the supraventral keel found in *oerstedi* and others; however the tubercles do not continue at all distally, onto the pollex. Ventral margin of manus with a row of strong tubercles, adjoined externally by a groove, running entire length but stopping abruptly beneath pilous triangle at pollex base. The tubercles are graduated in size, from minute proximally to large before and beneath triangle; pile is conspicuous between the large tubercles and in the adjoining groove.

Palm with lower triangle finely granulate. Oblique ridge low, broad, poorly marked by a row of tubercles scarcely larger than those of adjoining center palm area; apex scarcely evident, slightly lower even than adjoining ridge; tubercles not continuing upward around carpal cavity, except insofar as they are barely distinguishable from the center palm armature, which extends to the cavity's distal edge. Carpal cavity's ventral margin very low, broad, blunt; distal part of cavity sloping very gently; inner edge of proximal dorsal margin finely tuberculate, almost beaded, curving only slightly downward and so permitting a distal extension of the cavity, short and narrow; this extension slopes up into the smooth surface of the short predactyl area. Predactyl ridges both with well-developed tubercles.

Pollex and *dactyl*: Pollex not markedly triangular, the base, although broad, being little or not at all wider than corresponding part of dactyl; prehensile edge of pollex subdistally only with a triangular, tuberculate tooth, its distal edge being longer. Dactyl long, its strongest curvature being only distal, as it continues below tip of pollex; latter is practically straight ventrally, not curved upward. Gape moderately narrow proximally, practically absent distally in region of triangular tooth of pollex. Outer, proximal, subdorsal groove on dactyl broad, extending as much as about a third length of dactyl, flattened, smooth, flanked above and below by well-developed tubercles continuing dorsally beyond groove. Pollex and dactyl with outer and inner surfaces wholly smooth. Pollex and dactyl with outer and inner rows of gape tubercles proximally weak, distally vestigial, or represented by ridges only. Inner and middle rows widely separated proximally on pollex only; tubercles edging convexity of pollex in gape arising completely from median row; convexity is confined to about distal third, with a short proximal edge, a triangular apex, and a long, almost straight distal edge fully armed with small, sharp tubercles, similar except sometimes for one slightly enlarged near middle of the edge; no projections or enlarged tubercles elsewhere in gape, except occasionally for one tubercle in distal third of dactyl; both median rows proximally composed of very low, practically contiguous tubercles, either broadly rounded or almost flat, especially near base where gape pile is plentiful.

Minor Cheliped. Gape at base narrower than width of pollex, decreasing distally. Middle section with serrations strong, even, in contact beyond middle.

Ambulatories. All ambulatories, including 4th, strikingly broad for a *Celuca*, more so even than in *oerstedi*; the only strongly convex dorsal margin, however, is that of 3rd leg. A long row of minute tubercles on carpus of 1st ambulatory on major side; in addition, the nearby antero-ventral margin of the merus of the same ambulatory is distally finely tuberculate.

Gonopod. Flanges absent; pore very large, terminal, without a projection extending beyond rest of pore margin. Inner process ending slightly proximal to pore, moderately broad, flat, and closely appressed to tube, except at distal end where its spinous tip projects freely, as the end of the tube curves slightly away from it, immediately before the pore. Thumb represented by an oblique shelf far proximal to corneous part of specialized tip.

This gonopod closely resembles that of the sympatric *inaequalis* except for the apparently narrower inner process and the distinct lack of a projection on margin of pore. It is even more closely similar to that of *helleri*, from the Galapagos, but does not have even the rudimentary indication of a pore projection and, again, the inner process is less extensive.

FEMALE

(From von Hagen's type description and key, plus examination of 2 immature paratypes.) Usually with only 2 comma-shaped patches of pile, but sometimes with more (4 to 6 patches in 5 out of 20 females). Lower suborbital margin with crenellations laterally separate and gradually diminishing in size, ending at channel.

Measurements (in mm)

	Length	Breadth	Propodus	Dactyl
Largest male (von Hagen; Peru)	7.5	11.5	17.3	
Moderate male (present collection; Ecuador)	7.1	10.3	14.5	9.5
Holotype (von Hagen; Peru)	6.9	10.3	13.3	9.0 (J.C.)
Largest female (von Hagen; Peru)	8.0	11.6	–	–
Smallest ovigerous female (von Hagen; Peru)	5.8	8.4	–	–

Morphological Comparison and Comment

A comparison of all the members, including *tallanica*, of the superspecies *crenulata* and their allies will be found on p. 217.

Color

(After von Hagen.) Male: No display whitening observed. Carapace dark brown patterned with pale yellow, sometimes with a violet sheen; posterior quarter sometimes light yellow; eyestalks gray to gray green. Ambulatories anteriorly violet, elsewhere like carapace; minor cheliped gray violet. Outside of major cheliped: outer manus like cheliped, gray below; pollex white; dactyl proximally and above gray brown. Female: Yellow pattern of the dark brown carapace darker and not as conspicuous as in male. Chelae whitish, speckled with gray. Otherwise as in male.

Social Behavior

Waving Display

(After von Hagen.) Only lowest intensity observed, accompanied by feeding motions. Similar to low intensity of *batuenta*, the major cheliped making a weak circular motion.

Range

Known only from Puerto Bolivar, Ecuador, and Puerto Pizarro, Peru.

Biotopes

(From von Hagen, Text Fig. 20.) Upper levels of mudbanks of estuaries, very close to mangroves, including in their shade. The single specimen taken by Crane in Puerto Bolivar was in a similar, but flatter, habitat, and not in shade. (Biotope nos. 8, 9.)

Type Material and Nomenclature

Uca tallanica von Hagen, 1968

HOLOTYPE. In Rijksmuseum Natuurlijk Historie, Leiden. Male, cat. no. D 23046; measurements on p. 265. Type-locality: Puerto Pizarro, Peru. (!)

Additional Type Material. In same institution: 5 male and 3 female paratypes (cat. no. 23047). (!)

Reference

Uca (Celuca) tallanica von Hagen, 1968

Uca tallanica

TYPE DESCRIPTION. Von Hagen, 1968.2: 412; Text Figs. 5b, c, 6. Peru. Color; waving display. (Leiden !)

50. *UCA (CELUCA) FESTAE* NOBILI, 1902

(Tropical eastern Pacific)

PLATE 36 *A-D*. MAP 16.
FIGURES 71 *C, D*; 101. TABLES 9, 10, 12, 19, 20.

INTRODUCTION

The fingers of the major cheliped in *Uca festae* attain record lengths in relation to the size of the crab, although the armature shows no uncommon peculiarities. Notes made many years ago in Ecuador indicate that fighting was more prevalent, during those particular low tides, than in any other species I have encountered. Our present knowledge of combat components leads to the expectation that the long fingers will prove to be concerned with some combat characteristic, as well as with any physical and psychological advantages of large size and high visibility (p. 487). A possibly related point in the crab's evolution is that display color is dull except for the dazzling white fingers; in motion their flashing shows up well against the dark background of river shore or salt flat.

MORPHOLOGY

Diagnosis

Front moderately broad. Major cheliped with fingers strikingly long and slender, straight except for dactyl's tip; outer manus with a small group of large tubercles close to base of gape. Gonopod with a short thumb and without a projection on pore margin. Female with pile postero-laterally on dorsal part of carapace, and on posterior part of its sides.

Description

With the characteristics of the subgenus *Celuca* (p. 211).

MALE

Carapace. Frontal breadth contained about 4.5 times in that of carapace. Antero-lateral margins straight, parallel, or slightly diverging, angling into dorso-lateral margins. Hepatic and branchial regions fused and arched moderately above level of surrounding carapace. Eyebrow breadth more than half diameter of adjacent part of depressed eyestalk. Suborbital crenellations in outer half large and well separated, continuing slightly around outer orbital margin, but stopping well short of channel. Lower side of antero-lateral angle moderately sharp with a minute basal tubercle.

Major Cheliped. Merus: All rugosities sparse and weak; tubercles of antero-lateral and ventral margins small.

Manus: Bending of outer, upper surface strong and markedly flattened. Outer tubercles extremely minute near lower and upper margins; in middle portion of palm they become abruptly much larger, especially distally, near base of lower part of dactyl; this entire middle area of larger tubercles is separately convex, projecting beyond the usual convexity of outer surface; upper flattened portion with a broad, longitudinal depression or series of pits extending through distal half. Dorsal, outer groove distinct, curving outward distally and ending just before dactyl base. Dorsal margin with outer edge distal to carpal cavity weakly beaded or tuberculate; inner edge indistinct; intervening surface narrow, flat, marked by a few indistinct tubercles that are sometimes confluent with several blunt, distal rugosities on adjacent palm. Lower margin with edge blunt and without special tubercles, the only armature being a continuation of the minute tubercles of lower, outer surface. Ventral, outer groove absent.

Palm with lower triangle covered with extremely minute tubercles arranged in irregular, chiefly vertical rows, the tubercles largest proximally. Oblique ridge moderate, the apex high; tubercles increasing regularly in size to apex and just above it, where they are largest; above apex slightly smaller tubercles continue upward along carpal cavity, the row varying from regular to irregular, then turn distally below predactyl area where they die out. Center palm sloping steeply upward to tubercles bordering carpal cavity; a slight depression in upper portion, below predactyl area. Carpal cavity with ventral margin steep-sided, being only moderately thick, but the edge itself is broad and blunt; distal part of cavity sloping moderately to the almost straight (vertical) distal boundary surmounted by the tubercles of upward extension of oblique ridge; no true distal extension into predactyl area, from which the cavity is incompletely set off by the area's flatter plane and, usually, lumpiness; inner edge of proximal dorsal margin not curving downward. Predactyl area small, narrow, flat, or with a faint depression; slightly lumpy or, sometimes, smooth; bounded ventrally, as noted above, by end of upward extending tubercles from oblique ridge. Distal ridge at base of dactyl better developed than proximal, the tubercles being larger and extending farther dorsally.

Pollex and *dactyl*: Exceedingly long, the longest relative to length of manus (and probably to length of crab) of any species in the genus; for example, while the propodus attains a length of about 4 times the length of the carapace, the dactyl measures around 7 times that of manus. Both fingers very slender; dactyl slightly broader than pollex, notably convex throughout, curving unusually far beyond and below pollex. Gape moderate at middle, clearly wider than adjacent part of pollex. Dactyl with a slight, outer, proximal subdorsal groove or concavity, sometimes minutely tuberculate like surrounding area; both area and groove sometimes smooth in large specimens. Pollex and dactyl laterally smooth except near ventral and dorsal margins respectively, and near gape, where they are covered with extremely minute tubercles, inside and out, especially in smaller specimens. On gape, outer and inner rows of both pollex and dactyl inconspicuous and become indistinct before middle of length; tubercles of median row only slightly larger and, while distinct beyond middle, are also diminished more distally and merged among the general covering of minute tubercles. Pollex tip simple. Pollex gape without enlarged tubercles; dactyl sometimes with a single tubercle slightly enlarged at about end of proximal third.

Minor Cheliped. Gape very narrow. Strong serrations practically in contact.

Ambulatories. 2nd and 3rd meri moderately slender, their dorsal margins scarcely convex to almost straight. No anterior modification of 1st ambulatory.

Gonopod. Flanges absent; pore very large, its internal (anterior) lip lower than the external, without a projection. Inner process broad, flat, triangular, almost or quite reaching pore. Thumb small but distinct, varying in length even within populations, ending far proximal to the corneous base of the specialized tip.

FEMALE

Carapace regions definitely less marked than in male, the difference being unusually strong for a *Celuca*. Strong granulation dorsally and laterally on carapace; in postero-lateral part of dorsal region as well as on sides behind the vertical lateral margins the armature takes the form of true tubercles rather than granules. Pile plentiful above and below postero-lateral margins and in region of posterior striae. Cheliped on one side with antero-ventral margin armed more strongly than in any other species, there being a fairly regular row of distinct tubercles throughout its length. Suborbital margin with crenellations near outer anterior orbital angle more closely

set than in male but not larger. On 3rd and 4th ambulatories, pile present posteriorly on merus, carpus, and manus.

Measurements (in mm)

	Length	Breadth	Propodus	Dactyl
From El Salvador (Bott's material, 1954)				
Male, identified by Bott as *U. festae* (cat. no. 2102)	10.0	16.0		
Male, holotype of *U. orthomana*; juvenile	7.0	11.0		
From Ecuador (present collection)				
Largest male	13.0	21.5	53.0	46.0
Moderate male	9.0	14.5	30.0	28.0
Largest female (ovigerous)	13.0	19.0	–	–
Smallest ovigerous female	8.0	12.5	–	–

Morphological Comparison and Comment

The differences between the poorly known northern form of *festae* and the southern representatives seem too slight even to warrant the erection of subspecies. The distinctions at present appear confined to the gonopod, which has a slightly thicker thumb and thicker tip in Bott's material from El Salvador. Possibly also the gape of the major cheliped and thickness of the fingers remain of juvenile proportions when the crabs are longer than in the known populations of Ecuador.

Several juveniles from Panama in the present collection are even smaller, measuring about 5 mm in length, than are Bott's examples upon which the species *U. orthomana* and part of *U. leptochela* were erected; furthermore the Panama material is in poor condition. Nevertheless all of their specific characters are distinct and all are well developed except the gonopod thumb; this structure is represented only by an oblique shelf, as in some related species. Juveniles from Ecuador show a similarly late development of this structure. Finally, a male measuring 7.5 mm long and of mature proportions, from Contreras Island, off Panama, also has the thumb represented only by a shelf.

This species appears to have affinities slightly closer to members of the superspecies *crenulata* and its allies (p. 217) than to *beebei* and *stenodactylus*. It is essentially an intermediate form, however, unique only in the striking elongation of the major cheliped.

Color

Displaying males: Display whitening absent. Carapace at palest appears brownish gray; buccal regions and 3rd maxillipeds clear white. Major cheliped: Merus, carpus and manus dull greenish brown inside and out, except for narrow orange ends of connecting segments; chela shining white, as in minor chela and its manus; merus and carpus of minor cheliped and of all ambulatories purplish, at least anteriorly; elsewhere these appendages are dull monochrome, sometimes darker than the carapace. Eyestalks dull yellow or yellow orange in both sexes. Females otherwise dark.

SOCIAL BEHAVIOR

Waving Display

Wave lateral, strongly circular at high intensity, with the carpus then kept elevated, so that the tips of the fingers slant downward at the wave's low point but do not touch the ground during a series; a brief pause at peak; the chela then swings downward abruptly and another brief pause precedes the next wave; fingers usually slightly apart, parallel, during raising of cheliped, closed on lowering. (It seems probable that during the descent of the chela in a high-intensity wave a major-merus-rub or -drum takes place; the requisite motions were described at Guayaquil, but do not appear in the films made of waving at moderate intensity in the Trinidad crabbery.) Body raised high and the position held, at least anteriorly, throughout a series, the ambulatories remaining on the ground except when one or two are raised, apparently as an aid to balance. Waving was at a maximum rate of slightly less than one to a second in the field at Guayaquil; in the crabbery the lone male waved at much slower rates. In both field and crabbery more than 50 waves sometimes occurred in single series. (Component nos. 4, 5, 9, 11; timing elements in Table 19, p. 000.)

Precopulatory Behavior

During the observations at Guayaquil, males often chased females with the major cheliped extended, probably in a form of herding (component no. 14). When a female appeared attentive, a displaying male without any chasing approached her and stroked her carapace. No attempts at copulation were observed, and although excited males descended their own burrows as usual in *Celuca*, no females were seen to follow.

Acoustic Behavior

As already described under waving display, major-merus-rubs (component 1) or major-merus-drums (7) probably occur during some waves of high in-

tensity. Below, in the comments on combat, the frequent use of probable minor-merus-rubs is mentioned (2). Leg-wags (5) were seen both during waving display and before combat. Although I was watching for it particularly, no drumming against the substrate, either real or apparent, took place. Although a single individual sent us from Buenaventura lived and waved for months in the Trinidad crabberies, no stridulatory motions were ever observed.

Combat

The numerous combats seen at Guayaquil often included three and even four opponents; after a time the third individual would leave or be ousted, while a group of four would split into conventional pairs. In species I have since come to know well, most combats occur close to the time of low tide, and have almost or completely ceased long before the rising water approaches their burrows. Contrastingly, the *festae* fights continued to be plentiful even as the water washed over the legs of the struggling males. Combats were often preceded by rapid motions of the minor cheliped which almost certainly will prove to be minor-merus-rubs. Displacement cleaning of the major cheliped by the minor between rounds was so prevalent that it should probably be considered a part of combat; it also followed numerous encounters; in this species, particularly, stridulation during these motions should be suspected.

The several low-tide periods of observation at Guayaquil started April 20, before and during new moon, which fell on April 22, with low water each day occurring around mid-afternoon.

RANGE

U. festae is known from El Salvador, Panama (near Panama City), Colombia (Buenaventura), and Ecuador.

BIOTOPES

At Guayaquil, the type-locality, *U. festae* was present in largest numbers on flat banks of the river in the town itself; it also lived in the frequently dry salt flats nearby known as "El Salado." (Biotope nos. 14, 15); in Panama the only known specimens were found in mangrove mud near a stream mouth.

SYMPATRIC ASSOCIATES

At the Guayaquil level of the Guayas River *festae* was the only *Uca*; at El Salado, *vocator ecuadoriensis* and *g. galapagensis* also occurred in numbers, both members of the subgenus *Minuca*.

MATERIAL RESULTING FROM FIELD WORK

(The complete list of specimens examined is given in Appendix A, p. 610.)

Observations and Collections. Guayaquil, Ecuador.

Films. Photographed in Trinidad crabberies from specimens sent from Buenaventura, Colombia.

TYPE MATERIAL AND NOMENCLATURE

Uca festae Nobili, 1901

HOLOTYPE. In Museo di Zoologia della Università di Torino. Male, cat. no. 1438. Type-locality: Rio Daule, Ecuador. Measurements in mm: length 12; breadth 18 (Rathbun). (!)

Type Material of *Uca guayaquilensis* Rathbun, 1935. In Smithsonian Institution, National Museum of Natural History, Washington. Male, cat. no. 70831. Type-locality: Salada (= El Salado), Guayaquil, Ecuador. Measurements in mm: length 9.1; breadth 14.3 (Rathbun); under same number in same vial male and female. (!)

Type Material of *Uca orthomanus* Bott, 1954. In Forschungsinstitut Senckenberg, Frankfurt a.M. Male, cat. no. 1873. Type-locality: Puerto el Triunfo, El Salvador. Measurements on p. 268. 2 paratype males, cat. no. 2074. (!)

Type Material (part) of *Uca leptochela* Bott, 1954. In Forschungsinstitut Senckenberg, Frankfurt a.M. At least 2 of the 5 male paratypes are juvenile examples of *U. festae* (cat. 2073 part). (!)

REFERENCES AND SYNONYMY

Uca (Celuca) festae Nobili, 1902

Uca festae

TYPE DESCRIPTION. Nobili, 1901.4: 51. Ecuador: Rio Daule Inferiore. (Torino !)

Rathbun, 1918.1: 420. Taxonomy. No new material.

Maccagno, 1928: 32; Text Fig. 18 (claw). Ecuador: Guayaquil. Taxonomy, including that of type material. (Part !)

Bott, 1954: 171; Text Fig. 12; Pl. 17, Figs. 12a, b. El Salvador: El Zunzal. (!)

Crane, 1957. Ecuador. Preliminary classification of waving display.

Uca guayaquilensis

Rathbun, 1935.1: 50. Ecuador: Guayaquil. Type description. (USNM !)

Uca orthomana

Bott, 1954: 175; Text Fig. 19; Pl. 18, Figs. 18a, b. El Salvador: Puerto el Triunfo. Type description. (Frankfurt !)

Uca leptochela

Bott, 1954: 176 (part: 2 paratypes). El Salvador: Puerto el Triunfo. (Frankfurt !)

51. *UCA (CELUCA) HELLERI* RATHBUN, 1902

(Tropical eastern Pacific: Galapagos Islands)

PLATE 36 *E-H*.   MAP 15.
FIGURES 70 *H*; 101.   TABLE 9.

INTRODUCTION

Uca helleri appears to be endemic in the Galapagos, and remains the only species in the genus with a distribution confined to a small group of islands. Few specimens have been collected, and information on its biology appears to be lacking. Films of waving fiddlers, lent by visitors, do not show the display characters clearly enough to describe them.

MORPHOLOGY

Diagnosis

Front moderately broad. Orbits oblique; antero-lateral margins short, not at all divergent. Major cheliped with the minute tubercles on outer manus and triangle on proximal palm in a reticulate pattern; oblique tuberculate ridge strong but apex low; pollex not unusually wide, its prehensile edge without either a triangular protuberance or small distal crest. Minor cheliped with gape moderate, serrations weak. Gonopod without flanges, with the thumb represented by an oblique shelf, and with a small protuberance on pore's margin. Female without pile on carapace but with persistent pile posteriorly and ventrally on 3rd and 4th ambulatories.

Description

With the characteristics of the subgenus *Celuca* (p. 211).

MALE

Carapace. Frontal breadth contained more than 3 times in that of carapace. Antero-lateral margins straight, slightly converging, rounding gradually into the weak dorso-laterals. Vertical lateral margins weak. Frontal margin distally weak. Eyebrow breadth more than half diameter of adjacent part of depressed eyestalk. Suborbital crenellations of moderate size in outer half, continued around outer orbital margin as far as channel. Lower side of antero-lateral angle blunt.

Major Cheliped. Merus: Rugosities weak. Postero-dorsal margin represented only by a blunt angle, un-armed, between dorsal and posterior surfaces. Antero-dorsal margin convex throughout, proximally blunt-edged and unarmed; distally with small rugosities. Ventral margin proximally notably blunt and unarmed. *Carpus*: Practically smooth, the anterior margin scarcely indicated and unarmed.

Manus: Bending of outer, upper surface strong, dorsal portion broad and almost flat. Outer tubercles extremely minute, very slightly larger on flattened, upper surface and everywhere with a reticulate pattern; in lower third the intervening cells are smooth; in upper, the tubercles tend to be in transverse, vertical rows. Dorsal, outer groove practically absent. Dorsal margin distal to carpal cavity simple, little thickened, blunt, not flattened, with irregular, low tubercles continuous with those of upper outer manus. Cuff smooth, without the usual vertical row of tubercles. Area outside pollex base with a large, slight, flat depression, definitely triangular, continuing onto proximal part of pollex, and bounded ventrally by a blunt, low, indistinct keel. Lower margin with edge scarcely angled, set off only by a row of minute tubercles scarcely larger than those of adjacent surfaces, and dying out at about base of pollex. Ventral, outer groove absent.

Palm with lower triangle covered with extremely minute tubercles arranged in a reticulate pattern that is more pronounced than on outer manus, the intervening cells being smooth. Oblique ridge moderate, scarcely if at all higher at apex, where the tubercles are slightly enlarged; tubercles start well above ventral margin and, beyond apex, continue around carpal cavity practically to dorsal margin; the most dorsal tubercles, proximal to predactyl area, in an irregular group. Center palm with low, irregular lumpiness or rugosities in upper part. Carpal cavity with ventral margin low, thick, blunt; distal margin sloping moderately; inner edge of proximal, dorsal margin sloping slightly downward distally, but not blocking carpal cavity from predactyl area; no distal extension of cavity into predactyl area, from which it is incompletely separated by the clustered tubercles of upward continuation of oblique ridge. Predactyl area smooth, its boundaries indefinite, with a slight depression. Distal ridge at base of dactyl weak, tubercles practically absent; groove shallow.

Pollex and *dactyl*: Gape wider than adjacent pollex. Dactyl with outer, proximal, subdorsal groove represented by a small, shallow depression; tuberculation in the area consisting only of a few minute tubercles. Both pollex and dactyl smooth inside and out, except for some extremely minute tubercles externally around gape. On gape, pollex with outer row of tubercles complete, inner strong proximally and traceable throughout; median row absent proximally, the space between outer and inner rows broad and flat; dactyl with outer row complete, inner row absent proximally, median almost absent in middle portion. Pollex tip bifid. All gape tubercles small, several slightly enlarged toward middle of pollex and proximally in dactyl. Gape pile almost absent.

Minor Cheliped. Gape slightly wider throughout than width of pollex. Serrations small, few irregular, sometimes almost lacking.

Ambulatories. Merus of first 3 moderately broad, that of 4th slender, its dorsal margin practically straight. Pile absent.

Gonopod. Very similar to that of *inaequalis*, but projecting margin of lip not so much produced. Flanges absent; pore very large, terminal, its internal (anterior) lip with a slight, obtuse projection extending beyond rest of pore margin; inner process large, triangular, reaching pore margin or slightly beyond; thumb represented by an oblique shelf far proximal to corneous part of specialized tip.

FEMALE

Carapace slightly granulate posteriorly and laterally. Cheliped sometimes, as in type, with the small serrations larger and more numerous than in male. Ovigerous type female (the only female in good condition examined): merus of first 3 ambulatories scarcely wider than in male, but notably thicker, as in that of 4th; latter also is slightly wider. 3rd ambulatory with abundant, persistent pile as follows: on postero-ventral surface except proximally and distally, and extending onto a short, median section of ventral surface; carpus with a small, subdistal patch near postero-distal margin; manus with posterior and ventral surface entirely covered except at extreme proximal end. Pile also present on 4th leg on merus and manus near middle of ventral surface.

Measurements (in mm)

	Length	Breadth	Propodus	Dactyl
Male (holotype)	5.6	8.1	10.5	7.0
	(Rathbun)	(Rathbun)	(J.C.)	(J.C.)
Female type				
(ovigerous)	7.0	8.3	–	–
	(J.C.)	(J.C.)		
Male, large				
(Garth, 1946)	8.2	12.3	21.4	15.5
Female (Garth, 1946)	8.0	11.0	–	–

Morphological Comparison and Comment

Although the form of the gonopod is closely similar to that of *oerstedi, inaequalis,* and *leptochela,* the other characteristics of *helleri* combine in such a way that it appears to have no very close relations. The scarcity and fragility of material are major handicaps in understanding its position. On the major cheliped the reticulated appearance both of the outer manus and of the palm's proximal triangle is variable, never conspicuous, and sometimes almost lacking.

RANGE

Known only from the Galapagos, from the following islands: Narborough, Albemarle, Tower, and Indefatigable.

BIOTOPE

"Sandy mud beneath mangrove roots." (Garth, 1946.)

SYMPATRIC ASSOCIATES

"*U. helleri* and *U. galapagensis* Rathbun are found in separate colonies and on separate islands except at Academy Bay, where they occur in adjacent coves." (Garth, 1946.)

TYPE MATERIAL AND NOMENCLATURE

Uca helleri Rathbun, 1902

HOLOTYPE. In Smithsonian Institution, National Museum of Natural History, Washington. Male, cat. no. 24829. Distinguished from 2 smaller males in same jar under same number by Rathbun's measurements.

Additional type material, same jar and number: 1 ovigerous female. Type-locality: Mangrove Point, Narborough I., Galapagos. Measurements above. (!)

REFERENCES AND SYNONYMY

Uca (Celuca) helleri Rathbun, 1902

Uca helleri

TYPE DESCRIPTION. Rathbun, 1902.3: 277; Pl. 12, Figs. 3, 4. Galapagos Is. (USNM !)

Rathbun, 1918.1: 415; Text Fig. 170; Pl. 151. Galapagos Is.: Narborough and Tower Is. Taxonomy.

Boone, 1927: 278; Text Fig. 98. Galapagos Is.: Tower I.

Crane, 1941.1: 198; Text Fig. 4r. No new material. Taxonomy.

Garth, 1946: 517; Pl. 87, Figs. 5, 6. Galapagos Is.: Indefatigable I. (Academy Bay); Albemarle I. (Black Bight); Tower I. (Darwin Bay); Narborough I. (Mangrove Point).

Uca leptochela leptochela (not *Uca leptochela* Bott, 1954)

Bott, 1958: 210. Galapagos Is.: Indefatigable I. (Frankfurt !)

52. *UCA (CELUCA) LEPTOCHELA* BOTT, 1954

(Tropical eastern Pacific: El Salvador; Galapagos Islands)

FIGURE 71 *E.* TABLE 9.
MAP 15.

INTRODUCTION

This little-known species, *Uca leptochela*, has been recorded from El Salvador. It appears to be closely related to *Uca dorotheae*. A similar form, established as a subspecies, was collected in the Galapagos.

MORPHOLOGY

Diagnosis

Apparently about as in *dorotheae*, except for a narrower gape on the minor cheliped and, on the gonopod, a wider distal pore that lacks a marginal protuberance (data from Bott, 1954, Text Fig. 20 and Bott, 1958, Text Fig. 2). Armature, if any, on anterior aspect of 1st ambulatory unknown.

SOCIAL BEHAVIOR

Waving Display

(Data from Peters, 1955: 441.) About as in *dorotheae*, with a distinct pause at the peak; the tempo, however, is much slower, the wave itself lasting up to 1.5 seconds exclusive of pause at peak, which Peters timed at about 0.7 second.

TYPE MATERIAL AND NOMENCLATURE

Uca leptochela Bott, 1954

HOLOTYPE. In Forschungsinstitut Senckenberg, Frankfurt a.M. Male, cat. no. 2072. Type-locality: Puerto el Triunfo, El Salvador. Measurements in mm: length 6; breadth 9 (Bott). (!; incomplete examination only.)

Additional Type Material. Of 5 paratype males from the type-locality in the same institution (cat. no. 2073) at least 2 are juvenile examples of *U. festae*; each was distinguished in particular by the presence of a developed thumb on the gonopod and a protuberance on the margin of its pore. (!)

Uca leptochela eibl Bott, 1958

HOLOTYPE AND PARATYPES. In Forschungsinstitut Senckenberg, Frankfurt a.M. Male, cat. no. 2535. Type-locality: Tower I., Galapagos. Measurements in mm: length 8, breadth 11. The distinctness of this specimen remains at present in question; accompanying the type description (1958: 210) are 2 drawings, Text Fig. 1 showing the outside of a major propodus and dactyl and Text Fig. 2 depicting the tip of a gonopod. Neither one is stated to be drawn from the holotype, and the claw appears certainly to be that of one of the paratypes, at least 2 of which are unquestionably juvenile examples of *U. galapagensis*. The figure of the gonopod shows the oblique shelf representing the thumb, characteristic of both the mainland holotype of *leptochela* and the Galapagos species, *helleri*, as well as other mainland species; unlike *helleri* it does not show a marginal protuberance.

Additional Type Material. In same institution: paratypes (cat. no. 2541): "many males and females" (Bott) from the type-locality. See reidentification of 2 males, (!), received on loan, above.

REFERENCES AND SYNONYMY

Uca (Celuca) leptochela Bott, 1954

Uca leptochela

TYPE DESCRIPTION. Bott, 1954: 176 (part); Text Fig. 20; Pl. 18, Figs. 19a, b. El Salvador. (Not 2 of paratypes; see under type material.)
   Peters, 1955: 441; Text Fig. 10. El Salvador. Waving display. Frontal view in illustration, a photo, is taxonomically helpful, showing inside of major cheliped, etc.

Uca leptochela eibl

TYPE DESCRIPTION. Bott, 1958: 210 (? part); ? Text Fig. 2; not Text Fig. 1. Galapagos. (Not 2 of paratypes; see under type material.)

53. *UCA (CELUCA) DOROTHEAE* VON HAGEN, 1968

(Tropical eastern Pacific)

PLATE 37 *A-D*.   MAP 15.
FIGURE 71 *F*.   TABLES 9, 10, 20.

INTRODUCTION

Uca dorotheae appears to hold an interesting position in the subgenus, intermediate between members of the group that includes the superspecies *crenulata* with its allies and that composed of *beebei* and *stenodactylus*. Although in the most specialized parts of its armature, on the upper manus and anterior ambulatory, *dorotheae* resembles *beebei*, its more basic structures, as well as the form and tempo of its waving display, have more in common with, for example, *cumulanta* or *festae*. Its closest relation may well prove to be the Salvadorean form of *leptochela*, at present poorly known.

Like *beebei* and several other *Celuca*, displaying males of *dorotheae* sometimes build pillars or hoods.

MORPHOLOGY

Diagnosis

Front moderately wide, its distal margin strong. Small cheliped in both sexes with the serrations minute and similar to absent and with gape of moderate but variable width. Major manus outside bent sharply and flattened subdorsally, but without a separate, longitudinal convexity adjoining the narrow, flattened area; palm with the usual 2 predactyl ridges; gape wider than pollex base. 1st ambulatory on major side with minute tubercles on anterior surface of distal merus, carpus, and proximal manus; middle ambulatory meri moderately long and slender. Gonopod with thumb represented by an oblique shelf; a low protuberance on margin of pore. Female without pile on sides of carapace posteriorly.

Description

With the characteristics of the subgenus *Celuca* (p. 211).

MALE

Carapace. Frontal breadth contained about 3.5 times in that of carapace. Antero-lateral margins poorly marked, straight or slightly concave, angling into dorso-lateral margins which, immediately behind angle, become obsolescent to absent. One or both pairs of postero-lateral striae variable on the two sides, weak, or absent. Vertical lateral margin weak ventrally, obsolescent to absent in dorsal part. Carapace profile practically semi-cylindrical, grooves not well marked, no external division between hepatic and branchial regions, which are little arched above level of surrounding carapace. Frontal margin distally usually strongly marked (only moderately so in holotype). Eyebrow broad, about equal to diameter of adjacent part of depressed eyestalk. Suborbital margin with crenellations small, low, and close set, although not contiguous, internally, but changing rather abruptly to much broader and higher near antero-external angle; they continue to be close set (unlike for example in *beebei* and *festae*), completely around outer orbital margin as far as channel. Lower side of antero-lateral angle rounded, with or without a tubercle.

Major Cheliped. Merus: Rugosities weak. Antero-dorsal margin strongly arched and moderately thin throughout; proximal portion unarmed, distal finely serrate. *Carpus* with about 5 to 8 tubercles on proximal part of anterior margin.

Manus: Bending of outer, upper surface strong and markedly flattened, but the bending commences more dorsally than usual and is abrupt; no secondary longitudinal convexity in area of the bend. Outer tubercles small, largest dorsally, a few sometimes enlarged also at base of gape; not absent adjacent to dorsal margin, although they show an irregularly rugose arrangement on the entire flattened area. Dorsal, outer groove absent. Dorsal margin proximally, where it coincides with upper edge of carpal cavity, distinct, smooth; distal to carpal cavity indefinite, merging with upper palm, and marked only by an irregular band of small tubercles. Area outside pollex base with a shallow, indistinct depression, without definite boundaries; no beaded keel below it or continuing along side of pollex. Lower margin with a beaded edge and faint adjacent outer groove, the groove ending at pollex base, and the beaded edge either ending with it or continuing slightly beyond.

Palm with lower triangle slightly concave in center, especially distally, almost smooth in upper half but with lower covered with minute tubercles in irregular, vertical bands. Oblique ridge, including apex, low but strongly tuberculate, in a multiple row, irregular and variable, extending up around distal margin of carpal cavity. Carpal cavity with ventral margin low, moderately thick, the sides not steep; distal margin of cavity low, the slope gradual; inner, beaded edge of proximal dorsal margin not curving downward, but cavity is cut off from predactyl area by the continuing tubercles of upper section of oblique ridge and by the tubercles that are occasionally present on upper palm, the two sets being in this upper region then wholly merged. Predactyl area mostly smooth, as is lower part of center palm, the two regions being separated by the convex upper center palm. Both predactyl ridges well developed.

Pollex and *dactyl*: Dactyl slightly arched throughout; pollex varying somewhat in ventral profile from straight to slightly convex. Gape slightly to much wider than adjacent pollex. Pollex, although it lacks an outer beaded keel on lower, outer side, is marked by a broad, blunt, submarginal ridge that merges in distal portion with the ventral margin and throughout accentuates a very shallow, broad depression— it is not distinct enough to term a groove—that runs practically entire length of middle of outer surface. Dactyl with outer, proximal, subdorsal groove extremely short, almost absent, surrounded by irregular tubercles that continue for a short distance beyond it near margin. Surfaces of pollex and dactyl very finely granulate proximally outside, smooth distally and throughout inner surfaces. On gape, outer and inner rows are similar on both pollex and dactyl, being composed of very minute tubercles spaced widely proximally but close set distally; tubercles of median rows, although much larger, are still small, the row complete in both pollex and dactyl, the form of the tubercles rounded and similar in proximal half, less so in distal, with a slightly enlarged tubercle at about middle of pollex and, sometimes, a series of separated, similar tubercles in its distal half; dactyl usually with at least one or two slightly enlarged tubercles toward middle; pollex tip simple; dactyl without a subdistal crest. Gape pile plentiful.

Minor Cheliped. Notable for the variable width of its gape, which in other species is a reliable taxonomic character. In *dorotheae*, the variation is not extreme, however, but only from moderately narrow to moderately broad. Mano-pollex ridge strong. Prehensile edges with serrations minute to practically absent; when present they are similar in size, none ever enlarged, and the gape never narrow enough for them to be in contact. Tips of chela not extremely slender, articulating well.

Ambulatories. Moderately long and slender, being intermediate in proportions between those of *beebei* and *stenodactylus*. 1st merus, carpus, and manus anteriorly on major side with bands of minute, irregular tubercles. On the merus the band occurs distally; on carpus the multiple row of tubercles extends throughout its length; on manus it is confined to a small patch proximally, although some roughness may extend among the setae which arise in the usual row more distally.

Gonopod. Flanges absent, pore large, its internal (anterior) lip lower than the external and without a definite, low projection. Inner process flat, triangular, about reaching pore. Thumb absent, represented only by a shelf slanting obliquely toward base of appendage.

Abdomen. 3rd to 6th abdominal segments with traces only of partial fusion.

FEMALE

Suborbital crenellations not stronger than in male. No pile on sides of carapace posteriorly. Since the general characteristics of front, small chelipeds, suborbital crenellations and leg proportions apply to both sexes, they serve adequately for diagnosis.

Measurements (in mm)

	Length	Breadth	Propodus	Dactyl
Largest male (Puerto Bolivar)	8.2	13.5	24.0	17.5
Moderate male (Puerto Bolivar)	6.5	11.0	17.5	12.0
Largest female (von Hagen; Peru)	7.3	11.7	–	–
Smallest ovigerous female (von Hagen; Peru)	6.3	9.8	–	–

Morphological Comparison and Comment

Although its specific characters are of course unique in combination, *dorotheae* lacks unusual structures of diagnostic help. The nearest approach is the presence anteriorly on the 1st major ambulatory of tubercles on three, not only on the one or two segments where they occur in other species with similar armature. From the superficially similar *beebei* it can be most easily distinguished by the strong, distal border on the front combined with the lack of enlarged tubercles in the small cheliped's gape.

Color

Displaying males: Display whitening moderately developed in some individuals in Puerto Bolivar, Ecuador, in April (personal observation), but without definite display lightening in Puerto Pizarro, Peru (von Hagen). In both localities the color was otherwise similar, the carapace being in general olive green speckled with yellow; eyestalks olive green; major cheliped with brownish red to wine red on lower parts of all segments of major cheliped inside and out, except, as usual, for most of the white pollex; red also on the anterior parts of major meri on 1st, 2nd, and 3rd legs. Upper parts of major cheliped, except for white dactyl, apparently usually lightening only to yellowish brown. Von Hagen noted one or two brown spots on distal part of posterior sides of all ambulatories.

SOCIAL BEHAVIOR

Waving Display

Wave lateral, straight, or, at high intensity, circular. Jerks absent; pause at peak strongly present; there is no drumming against either carapace or ground and the cheliped does not vibrate in the air. Minor cheliped usually makes a corresponding motion. Body raised and lowered with every wave except at highest intensity. In response to a wandering female a displaying crab sometimes runs rapidly from side to side, or approaches her in zigzags, his cheliped held motionless in the air, somewhat as in *stenodactylus*. Duration of single wave: slightly more than 1 second. (Component nos. 4, 5, 9, 10, 12.)

Pillar Construction

In Peru, von Hagen found that many waving males constructed pillars similar to those of *beebei* (1968.2: 431; Text Fig. 18).

RANGE

Tropical eastern Pacific; known from the following localities: Canal Zone: Balboa. Ecuador: Puerto Bolivar. Peru: Puerto Pizarro.

BIOTOPES

Sheltered flats and the sloping banks of lagoons, on sandy mud to mud. (Biotope no. 11.)

MATERIAL RESULTING FROM FIELD WORK

(The complete list of specimens examined is given in Appendix A, p. 610.)

Observations and Collections. Canal Zone (2 specimens; not seen in life); Puerto Bolivar, Ecuador.

TYPE MATERIAL AND NOMENCLATURE

Uca dorotheae von Hagen, 1968

HOLOTYPE. In Rijksmuseum van Natuurlijk Historie, Leiden. Male, cat. no. D 23054. Measurements in mm: length 7.1; breadth 11.7; propodus 19.7 (von Hagen). Type-locality: Puerto Pizarro, Peru. (!)

Additional Type Material. In same institution: 25 male and female paratypes, cat. no. D 23055; same locality. (!)

REFERENCE

Uca (Celuca) dorotheae von Hagen, 1968

Uca dorotheae
TYPE DESCRIPTION. von Hagen, 1968.2: 429; Text Figs. 11, 12a, 18. Peru. Color; waving display; pillars. (Leiden !)

54. *UCA (CELUCA) BEEBEI* CRANE, 1941

(Tropical eastern Pacific)

PLATES 37 *E-H*; 50 *B*. MAP 16.
FIGURES 40 *C, D*; 49 *C, D*; 71 *A*; 93; 101. TABLES 9, 10, 12, 19, 20.

INTRODUCTION

Adapted to a number of common habitats, *Uca bee-bei* is probably the most abundant of all the fiddlers within its range, even though its local populations are often rather small. As with *lactea* in the Indo-Pacific, several dozen individuals sometimes thrive on bits of suitable shore only a few meters long.

In most places, green on the carapace and purplish brown on the lower cheliped distinguish displaying males at a glance among any variety of species. The extremely rapid, simple, "beckoning" wave makes an even better recognition mark, reliable even in populations in which display white is prevalent. Once it has become familiar, this rhythm will identify the crab as far as it can be seen.

MORPHOLOGY

Diagnosis

Front moderately wide, its distal margin obsolescent. Small cheliped in both sexes with large, uneven serrations in its narrow gape. Major manus outside with a subdorsal, longitudinal convexity adjoining its flattened area; palm with only one predactyl ridge. 1st ambulatory on major side with minute tubercles on anterior surface of carpus and manus but not on distal merus; middle ambulatory meri moderately broad. Gonopod with thumb represented by an oblique shelf; no protuberance on margin of pore. Female without pile on posterior sides of carapace.

Description

With the characteristics of the subgenus *Celuca* (p. 211).

MALE

Carapace. Frontal breadth contained 4 times or less in that of carapace. Antero-lateral margins slightly concave, divergent, angling into dorso-lateral margins, which are weak or obsolescent and end before middle of cardiac region. Each of the single pair of postero-lateral striae is either very weak or obsolescent. Vertical lateral margin obsolescent or absent in dorsal portion. Carapace profile practically semi-cylindrical, grooves not well marked, with no external division between hepatic and branchial regions;

these regions are, however, little arched above level of rest of carapace. Frontal margin distally obsolescent or absent. Eyebrow broad, about equal to diameter of adjacent part of depressed eyestalk. Suborbital margin with crenellations small, low, and close set internally, changing gradually to broader, higher, and separated near antero-external angle; they continue, widely separated, around outer orbital margin as far as channel. Lower side of antero-lateral angle rounded.

Major Cheliped. Merus: Rugosities weak. Antero-dorsal margin strongly arched, only the distal portion showing little curvature; proximal portion a low, blunt ridge, unarmed; distally less a distinct ridge and variably minutely serrate or tuberculate. *Carpus*: About 5 or 6 distinct tubercles on proximal part of anterior margin.

Manus: Bending of outer, upper surface strong and markedly flattened, but with a secondary longitudinal convexity that starts, variably, near level of middle of dorsal margin above carpal cavity, runs midway between beginning of bend and margin, and ends at dactyl base. Outer tubercles small, showing little difference in size, slightly larger along secondary convexity and at base of gape; absent adjacent to dorsal margin. Dorsal, outer groove distinct proximally, weak to absent distally. Dorsal margin proximally, where it coincides with upper edge of carpal cavity, thick and convex, covered with small tubercles; distal to carpal cavity indefinite, convex, irregularly tuberculate, usually not set off clearly from either outer manus or upper palm. Area outside pollex base with a large, distinct, triangular depression, continuing onto proximal part of pollex, covered with very minute tubercles and bounded ventrally by a beaded keel traceable throughout pollex. Lower margin with a row of small, close-set tubercles traceable throughout length of pollex. Ventral, outer groove slight, present proximally only.

Palm with lower triangle covered with extremely minute tubercles, very slightly larger near ventral margin, with some in highly variable arrangements of short bands, leaving smooth areas between, either as vertical spaces or minute cells. Oblique ridge high throughout, the apex relatively moderate, being a little lower than middle part of ridge; tubercles largest on this middle part; they continue upward from apex around the much lower edge of carpal cavity, always small and highly variable in arrangement and extent. Carpal cavity with ventral margin only moderately thick, the sides being steep; edge itself, however, is blunt; distal margin of cavity low, the slope being gradual; inner, beaded edge of proximal dorsal margin curving downward, partly setting off carpal cavity from predactyl area; cavity, which wholly lacks a distal extension, is also bounded by a broad, tuberculate ridge across proximal end of predactyl area. Predactyl area with a slight, smooth depression, set off from the smooth center palm only by a convexity. Distal predactyl ridge and tubercles wholly absent; proximal well developed, continuing as usual outer prehensile row of pollex although sometimes with a short, tubercle-free space between its lower end and prehensile row.

Pollex and *dactyl*: Dactyl notably straight, not arched, before distal end, which as usual curves down slightly beyond pollex. Gape slightly wider than adjacent pollex. Outer keel that starts below depression at pollex base continued along most of lower part of pollex. Dactyl with outer, proximal, subdorsal groove short, surrounded by several rows of minute tubercles which converge and continue near margin for a short distance. Surfaces of pollex and dactyl both covered with extremely minute tubercles except along middle of inner sides; irregular rows of tubercles curve around outside proximal end of gape. On gape, outer row of tubercles complete in both pollex and dactyl; inner row on both in distal part is a virtually non-tuberculate keel; median row on pollex absent; median row on dactyl distally set on a minute crest, marked outside by a crease at base of the minute, sharp, regularly diminishing tubercles. Pollex tip simple. Gape pile plentiful.

Minor Cheliped. Notable for weakness of all ridges and armature, except for serrations and teeth of prehensile edges; even mano-pollex ridge is weak. Manus and fingers broad; fingers little longer than palm. Gape at base narrower than pollex, decreasing distally. Prehensile edges serrate and, in middle portion, coarsely toothed, although rarely (including in holotype) the serrations are low and appear worn; sizes and arrangement variable; usually, however, the region has three or four large teeth in middle of each prehensile edge, flanked by fine serrations which articulate almost or completely.

Ambulatories. Short, with moderately broad meri, in comparison with the ambulatories of *stenodactylus*. 3rd merus on minor side extending about a quarter of its length beyond antero-lateral angle when laid forward, its breadth equal to about two-fifths of its length. 2nd and 3rd meri with dorsal margins slightly convex to almost straight. 1st merus, carpus, and manus anteriorly on major side with a band of minute, irregular tubercles; the band occurs on median distal part of merus and, on carpus and manus, on lower half of each segment; at most, each row is about 6 tubercles wide; considerable variation is shown in this armature, the tubercles being sometimes absent on the manus and, especially in northern populations, may consist of single rows on the carpus.

Gonopod. Flanges absent, pore large, its internal (anterior) lip lower than the external, without a projection. Inner process flat, triangular, about reaching pore. Thumb absent, represented only by a row of bristles, set far proximal to base of corneous tip and arising from a shelf slanting obliquely toward base of appendage.

Abdomen. 3rd to 6th abdominal segments incompletely fused.

FEMALE

Suborbital crenellations not stronger than in male. No pile on sides of carapace posteriorly. Carapace dorsally very finely granulate.

Measurements (in mm)

	Length	Breadth	Propodus	Dactyl
Largest male (Ecuador)	8.0	13.0	22.5	17.5
Holotype male (Canal Zone)	7.4	10.4	19.2	
Moderate male (Canal Zone)	5.5	9.0	17.0	11.0
Largest female (Canal Zone) (paratype)	7.9		–	–
Largest ovigerous female (Canal Zone) (paratype)	5.2		–	–
Smallest ovigerous female (von Hagen; Peru)	4.2	6.5	–	–

Morphological Comparison and Comment

Comparisons with *beebei*'s two closest relations, *dorotheae* and *stenodactylus*, appear on pp. 275 and 283. The nearer affinity of *beebei* seems unquestionably to be with *stenodactylus*.

Color

Displaying males: Display whitening usually absent but fully developed at least at Puerto Bolivar, Ecuador. The usual, dark form is as follows. Carapace anteriorly brilliant iridescent green (turquoise blue in Puerto Bolivar), posteriorly gray. Major cheliped: outer lower merus, carpus, and manus, as well as lower palm, purple to purplish brown; upper outer manus and upper palm varying from ochraceous yellow to rose pink; anterior merus and mid-palm, gray; fingers sometimes yellow proximally, otherwise entirely white. Ambulatories gray except as follows: on major side, coxa and proximal merus usually purple both anteriorly and posteriorly; corresponding parts on minor side sometimes greenish. Females with dark mottlings or spots on a somewhat lighter ground.

SOCIAL BEHAVIOR

Waving Display

Wave lateral, straight, or, at high intensity, strongly circular, the chela tip reaching well above the eyes. Cheliped brought down abruptly, but jerks are absent; also absent is a pause at the peak; there is no drumming against either carapace or ground and the cheliped does not vibrate in the air. Minor cheliped usually makes a synchronous motion. Body normally held fairly high and not further raised during display, although the carapace is often slightly raised in front and depressed posteriorly during high intensity display, and by individuals living in soft mud. In courtship the crab sometimes revolves before the female during waving, thus from time to time presenting a view of his carapace and posterior ambulatories, often brightly colored to human eyes, toward the female. This action occurs so frequently that it should probably be considered an integral part of the species' display. During many waves one or more ambulatories are kicked upward, their meri not in contact. Display appears exceptionally vigorous and the motion is faster than that of any other species with a broadly circular wave, the fastest rate being at more than 2 to the second and the waves continuing without pause in series that sometimes attain 20 or more. (Component nos. 4, 5, 9, 11 modified, 12; timing elements in Table 19, p. 656.)

Altogether absent in *beebei*'s display is the racing to and fro with cheliped held aloft, as well as any herding of females; both of these activities are characteristic of *stenodactylus*.

Precopulatory Behavior

As usual in American fiddlers, the female characteristically follows the male down his burrow. As in a number of other species, however, atypical surface mating has been observed. Twice a male and female were seen in the copulatory position close to the female's burrow. In each case the male was a mature burrow-holder, not an aggressive wanderer, and the female did not have her legs stiffened in the non-receptive position.

Hood and Pillar Construction

Of six species of *Celuca* known to build pillars or hoods, those of *beebei* are among the most variable and the least well formed. As in the other species,

they are constructed only by males in display condition, only by some displaying males in a population on a given day, and are known to occur in only a few of the numerous populations that have been observed. In *beebei* I have seen them only in and near Panama City, and in part of a population in Golfito, Costa Rica. In both localities pillars were formed almost altogether by individuals living on muddy sand, not mud, the consistency of the mud being unsuitable during the hour or so before low tide, when most construction is undertaken; during that period the moisture has not drained fully away. In *beebei* the shapes sometimes attain the form of definite hoods arching over the hole; more often they are pillars, partly smoothed and with only a slight hollow on the side facing the burrow mouth; sometimes they are merely low cones, scarcely higher than the builder, with the individual packets of which they are made showing plainly on their surface.

Acoustic Behavior

The following components have been observed in the field and recur in films: minor-merus-rubs (component no. 2) and leg-wagging (5), with contact between the meri clearly visible. Minor-merus-drums (8) probably occur, but the filmed record is indistinct. Since appropriate armature is well developed on both major palm and 1st ambulatory, the inclusion of palm-leg-rubs (4) in the repertory is inferred on morphological grounds.

RANGE

Pacific coast of El Salvador to northern Peru.

BIOTOPES

This species occurs in a variety of habitats. It is most abundant on muddy sand that is often mixed with small stones, near mangroves, on sheltered flats around and within the mouths of streams, and usually below mid-tide levels; also common in similar surroundings on substrates of sandy mud or mud, often mixed with stones; occasionally on more open mud flats close to mangroves; rarely among mangrove shoots. (Biotope nos. 6, 8, 9, 11, 12, 14.)

SYMPATRIC ASSOCIATES

Uca beebei often mingles marginally with its close relation, *stenodactylus*, although most of any such population of *beebei* occurs nearer to low-tide levels. The species associates more or less closely in individual localities with one or more of all the other *Celuca* in the tropical Eastern Pacific except those living well up tidal streams. These occasional associates include *batuenta*, *saltitanta*, *oerstedi*, *inaequalis*, *limicola*, *latimanus*, and, rarely, *musica terpsichores*. Among other subgenera, *(Uca) heteropleura* and *stylifera* often share muddy sand beaches with *beebei*, while young *ornata* occur with it in the mud. Species of *Minuca* are almost always represented only adventitiously, or in immature stages, in populations of *beebei*.

MATERIAL RESULTING FROM FIELD WORK

(The complete list of specimens examined is given in Appendix A, p. 610.)

Observations and Collections. Nicaragua: Corinto. Costa Rica: Puntarenas; Golfito. Panama and Canal Zone: Bahia Honda; Panama City; Old Panama; Balboa. Ecuador: Puerto Bolivar.

TYPE MATERIAL AND NOMENCLATURE

Uca beebei Crane, 1941

HOLOTYPE. In Smithsonian Institution, National Museum of Natural History, Washington. Male, cat. no. 137413 (formerly New York Zoological Society cat. no. 4129). Type-locality: La Boca, Balboa, Canal Zone. (!)

Additional Type Material. 16 male and female paratypes in same institution, cat. no. 137414 (formerly New York Zoological Society cat. no. 4130). (!)

REFERENCES AND SYNONYMY

Uca (Celuca) beebei Crane, 1941

Uca stenodactylus (not *Gelasimus stenodactylus* Milne-Edwards & Lucas, 1843)

Rathbun, 1918.1: 417 (part); Pl. 152, Fig. 3; Pl. 153. Smithsonian Institution, National Museum of Natural History, Washington. Cat. no. 32322. Costa Rica: Puntarenas. Taxonomy. (USNM !)

Uca beebei

TYPE DESCRIPTION. Crane, 1941.1: 192; Text Fig. 4p; Pl. 4, Fig. 16; Pl. 5, Fig. 20; Pl. 6, Fig. 27. Corinto, Nicaragua to Old Panama, Panama. Color; waving display; pillar construction; general habits. (USNM !)

Crane, 1944: Color change.
Holthuis, 1954.1: 41. El Salvador.
Holthuis, 1954.2: 162. El Salvador.
Altevogt & Altevogt, 1967. E 1289. Film of waving display.
von Hagen, 1968.2: 432; Text Fig. 12b. Peru: Puerto Pizarro.

Uca stenodactyla beebei

Bott, 1954: 175; Text Fig. 17; Pl. 18, Figs. 17a, b. El Salvador: Puerto el Triunfo; Rio Zunzal. Taxonomy. (Frankfurt !)
Peters, 1955: 432 ff.; Text Figs. 2, 20, 50, 51. El Salvador. Morphology controlling claw-bending; waving display; pillar construction; ecology.

55. *UCA (CELUCA) STENODACTYLUS* (MILNE-EDWARDS & LUCAS, 1843)

(Tropical, and perhaps subtropical, eastern Pacific)

PLATES 38 *A-D*; 50 *B*.
FIGURES 24 *K-M*; 25 *D-F*; 27 *K, L*; 30 *C, D*; 32 *D-F*;
34 *D*; 35 *G, H*; 36 *E*; 37 *O*; 45 *S-UU*; 46 *P*; 47 *E, F*;
48 *E, F*; 53 *D*; 71 *B*; 81 *L*; 101.

MAP 16.
TABLES 9, 10, 12, 14, 19, 20.

INTRODUCTION

A striking sight when he is in full display, a male *Uca stenodactylus* often races to and fro, holding a great pink claw high overhead or stretching it to the side as he chivies a small female toward his burrow.

One of the most interesting aspects of the species is the coincidence of much of its range with that of its closest relation, *beebei*. Often, as in the alliance that includes *batuenta* and its relations, the species mingle at least marginally in the same biotope. Understandably they have been confused in collections, or not given full specific status because of their close morphological similarities. Yet, once distinguished, there is no question of their separateness, even when the contrasts of their appearance in life and of their displays have not been encountered.

MORPHOLOGY

Diagnosis

As in *beebei* excepts as follows. Major manus with subdorsal, longitudinal convexity less pronounced and without a smooth area above it; dactyl slightly arched throughout, instead of curving downward only distally. 1st ambulatory on major side with minute tubercles of anterior surface obsolescent to absent; present if at all on carpus only. Middle ambulatory meri slender, their dorsal edges almost straight in male; all ambulatories in both sexes notably longer and more slender than in *beebei*. Small chelae in both sexes with finger tips more slender, pointed, not meeting exactly. Gonopod with thumb represented by a transverse shelf or a short stump, not by an oblique shelf.

Description

With the characteristics of the subgenus *Celuca* (p. 211).

MALE

Carapace. Differs from *beebei* as follows. Orbits almost straight; antero-lateral margins often not dis-

tinct from dorso-laterals but rounding smoothly into them. Carapace profile even more strongly convex, with the fused hepatic and branchial regions notably arched above level of surrounding carapace; the latter extends slightly farther laterally than in *beebei*, so that its sides are vertical. Frontal margin distally weak but always complete. Eyebrow broader than in *beebei*, in most individuals distinctly more than equal to diameter of adjacent part of depressed eyestalk. Lower side of antero-lateral angle usually a distinct, though blunt, edge; sometimes microscopically granulate.

Major Cheliped. Merus: As in *beebei*, but rugosities except in smaller specimens are even weaker. *Carpus*: Smooth except for fine granules distally on dorso-posterior surface; anterior margin with 10 or more tubercles which are proximally large and of irregular sizes, and distally very small and even.

Manus: Very similar to that of *beebei*, except that there is no smooth area close to dorsal margin, while the ventral, outer groove is stronger and continues to base of pollex.

Palm differs from that of *beebei* as follows: first, the tubercles of lower triangle, although slightly enlarged all along a narrow area above ventral margin as in *beebei*, are even larger in its distal portion, close to distal end of oblique ridge. Second, proximal ridge at base of dactyl, although lacking tubercles as in *beebei*, is better developed, as is the groove between it and distal ridge.

Pollex and *dactyl*: Differ from *beebei* as follows. Dactyl slightly arched throughout, instead of being curved downward only distally. Gape accordingly is usually slightly wider. Dactyl with outer, proximal, subdorsal groove absent.

Minor Cheliped. As in *beebei*, except that manus is less deep, the fingers are almost half as long again as manus and more slender, while the finger tips are not dilated but tapering, and not in contact (as in almost all other species of *Uca*) but with the tip of the dactyl falling inside that of the pollex, as is characteristic of major chelipeds. Again as in *beebei*, the enlarged teeth, and more rarely the serrations as well, occasionally appear worn or damaged, especially in large specimens.

Ambulatories. Differ from those of *beebei* in having the legs longer and the meri more slender; 3rd merus on minor side reaches more than half its length beyond antero-lateral angle when laid forward, its breadth about a fourth to a third of its length. 1st carpus and 1st manus anteriorly on major side with tubercles obsolescent, represented if at all by a few irregular tubercles on carpus only.

Gonopod. Differs from *beebei* in having the thumb represented by a horizontal shelf, which occasionally is produced as a short stump, especially in larger specimens; the shelf is never oblique.

Abdomen. 3rd to 6th abdominal segments incompletely fused.

FEMALE

In *stenodactylus* the females are most readily distinguished from those of *beebei* by their long, slender ambulatories and, on the chelae, by the narrow, pointed tips that do not come into contact, as described above under *male*.

Measurements (in mm)

	Length	Breadth	Propodus	Dactyl
Lectotype male (Chile)	10.5	16.0	33.0	24.0
Largest claw (body not secured) (Nicaragua)	–	–	30.0	22.0
Large male (Nicaragua)	8.0	13.0	21.5	16.0
Moderate male (Nicaragua)	7.0	11.0	19.0	14.0
Largest ovigerous female (von Hagen; Peru)	7.2	11.3	–	–
Smallest ovigerous female (Nicaragua)	5.1		–	–

Morphological Comparison and Comment

Although they are superficially similar and obviously closely related, *stenodactylus* and *beebei* appear wholly distinct even in preservative, once several points of difference have become familiar. The most easily and quickly observed are, first, the contrast between the general form of the ambulatories, which are contrastingly long and slender in *stenodactylus*; in this comparison the breadth of the last ambulatory merus, often helpful in such comparisons, should be disregarded, since it is relatively slender in both species, the dorsal margin being almost straight. Second, the tips of the small chelae are so reliably different that in the field their examination scarcely calls for a hand lens, the slender, crossed tips of the closed chela showing clearly in *stenodactylus*. In the laboratory the gonopods' slight difference, consisting wholly

of the shape of the vestige that represents the thumb, proves trustworthy. (See also p. 218.)

Color

Displaying males: Display whitening apparently never fully developed dorsally although present on regions below the eyes and on parts of the chelipeds. Dorsal part of carapace gray blue to intense violet blue anteriorly, pale pink to white posteriorly. Major cheliped: Outer and inner merus and carpus yellowish white to white. Outer lower manus and outer pollex violet pink to pink; outer upper manus and outer dactyl pinkish white to white; palm pale orange pink; inner pollex and inner dactyl scarlet orange to orangish white. Eyestalks lemon yellow. 3rd maxillipeds, suborbital area and pterygostomian regions greenish or yellowish white. Minor cheliped with merus and carpus yellowish, manus and chela pink or pale orange. Ambulatories anteriorly flame scarlet, the merus brightest; posteriorly pale rose pink, the 3rd and 4th duller. For course of color development before display, cf. p. 467. Females dark, mottled or spotted; chelipeds with manus and chela white to violet; eyestalks often yellow.

SOCIAL BEHAVIOR

Waving Display

Wave lateral, wholly or almost straight rather than circular. Starting from a flexed position well above the ground, the cheliped unflexes straight outward and only slightly upward, so that at its highest reach during moderate intensity waving the chela is only slightly above the eyes. No pause occurs at peak and no jerks, drumming, or aerial vibrations, the cheliped being flexed into rest position practically in the same plane as the one in which it was extended. Both major and minor chelae held slightly open. Minor cheliped makes a similar, synchronous motion. Body held raised throughout a single series. Ambulatories not raised during waves, except for the few rapid steps to one side or the other that are usually taken. Wave roughly only half as fast as in *beebei*, lasting more than a second except during highest intensity courtship at burrow mouth.

During the motion of an attentive female, particularly of a wandering individual, the form of the male's resultant display is unique in the genus. Ceasing to complete his waves, he races back and forth holding his large cheliped high overhead. Since a wandering female stimulates a number of neighboring males at once, a flash of quivering motion may sweep far across a group, more striking than in any other species I know. Sometimes one or more males lower the cheliped to a straight lateral position, chasing the dodging females in the usual form of herding (p. 496). Sometimes a male partly surrounds a pursued individual with his cheliped and seems actively to urge her toward his burrow; touching apparently does not occur and the observed females always soon escaped, either descending their own burrows nearby or continuing to wander through the population. Whether or not the pursued females are immature in this species, as in both *vocans* and *lactea*, has not been determined. (Component nos. 4, 9, 11, 14; timing elements in Table 19, p. 656.)

Precopulatory Behavior

Although, as in *beebei*, two pairs were seen in mating position at the surface, the usual pattern conforms to that characteristic of American crabs. Having attracted the attention of a female, the male speeds up his rate of waving at the mouth of his burrow and descends, the female following.

Once an apparent copulation, preceded by mutual stroking, took place on the top of a female's chimney.

Chimney Construction

Chimney-building has been observed in only one group of adult, non-ovigerous females during a single low-tide period in the Canal Zone. As mentioned above, one copulation was observed on top of a chimney during the same observational period. The identity of the builders, and of the copulating couple, was checked. No chimney building is known to occur elsewhere in *Celuca*, with the apparent exception of sporadic female *cumulanta* (p. 242).

Acoustic Behavior

As in *beebei* (p. 280).

Combat

The single filmed example shows a homoclawed, low-intensity, mutual combat composed of forceful manus-pushes alternating with manus-rubs (component 1). Both components were formed unusually, since during pushing the lower part of one manus came in contact with the upper part of the other, while during the ritualized rubbing the actor of the moment rubbed vertically (up and down), rather than longitudinally. During neither pushing nor rubbing were the chelae in contact. At the end, both chelae, one after the other, made rough **down-points**, an uncommon threat motion observed additionally only in *tangeri*, *maracoani*, *ornata*, and *inaequalis*. Pushing was resumed during which one crab performed a leg-wag. Promptly thereafter his opponent backed away.

RANGE

From the Pacific coast of El Salvador at least to northern Peru. The type was reported to have been taken at Valparaiso, and Porter (1913: 316) wrote of eight more recent specimens: "We collected them on the beaches and bays of Valparaiso, Quintero and Algarrobo." Since then, however, the species has apparently not been found in Chile.

BIOTOPES

Characteristically found on the upper half of sheltered beaches of muddy sand or sandy mud, or on salt flats more or less cut off from open water by mangroves and sometimes covered only by spring tides; *stenodactylus* shows, in fact, none of *beebei's* acceptance of more muddy substrates. (Biotope nos. 5, 6, 11.)

SYMPATRIC ASSOCIATES

Occasionally small populations mingle marginally with *beebei*, which occurs in the zone below it on gently sloping shores. On salt flats no close associates are known.

MATERIAL RESULTING FROM FIELD WORK

(The complete list of specimens examined is given in Appendix A, p. 610.)

Observations and Collections. Nicaragua: Corinto. Costa Rica: Port Parker; Golfito. Panama and Canal Zone: Panama City; Balboa. Ecuador: Puerto Bolivar.

Films. Panama City.

TYPE MATERIAL AND NOMENCLATURE

Gelasimus stenodactylus Milne-Edwards & Lucas, 1843

TYPE-SPECIMEN. Muséum National d'Histoire Naturelle, Paris. Male, listed as a type by the museum. Label: "*Gelasimus stenodactylus* Edw. & Luc. M. d' Orbigny. Valparaiso." Specimen dried but in good condition except for 2 missing ambulatories, the 3rd on the left side and the 4th on the right. When the gonopod was relaxed its characteristics agreed well with those of *stenodactylus* as described in this study, the thumb being represented by a minute nubbin, not by a slanting shelf. The entire specimen agrees well with the illustration of the type (1847, Atlas Pl. 11, Fig. 2). The only disagreement is in the larger size reported by the authors (length 13 mm; breadth 18 mm; measurements of this specimen, 10.5 mm long, given above, p. 283). Except for a female, perhaps belonging to another species, in the box with the male just described, no other type material, or any other similar material reported to be from Valparaiso, was found. (!) Apparently these 2 specimens are the cotypes examined by Rathbun (1918.1: 417).

Type Material of *Gelasimus gibbosus* Smith, 1870. In Museum of Comparative Zoology, Harvard University, Cambridge, Massachusetts. Consists of holotype male only, cat. no. 5911. Type-locality: Gulf of Fonseca. Specimen in alcohol, in good condition, although most appendages are detached with some ambulatories missing. All of the specific characters are clear and agree excellently with those of *stenodactylus* as described above. Gonopod with thumb a distinct, short stump. Measurements in mm: length 9; propodus 24; dactyl 18. (!)

References and Synonymy

Uca (Celuca) stenodactylus (Milne-Edwards & Lucas, 1843)

Gelasimus stenodactylus

TYPE DESCRIPTION. Milne-Edwards & Lucas, 1843: 26; Pl. 11, Fig. 2, 2a. Valparaiso. (Paris !)
    Nicolet, 1849: 165. Chile. Taxonomy; color.
    Milne-Edwards, 1852: 149. Taxonomy of type.
    Smith, 1870: 139. No new record. Listing of types.
    Cano, 1889: 234. Ecuador.

Gelasimus gibbosus

    Smith, 1870: 140. Central America: Gulf of Fonseca. (MCZ !)

Uca stenodactyla

    Porter, 1913: 315. Chile: Valparaiso: Quintero and Algarrobo.
    Crane, 1941.1: 195; Text Fig. 4q; Pl. 4, Fig. 15; Pl. 5, Fig. 21; Pl. 6, Fig. 28; Pl. 9, Figs. 41, 42. Nicaragua; Costa Rica; Panama; Canal Zone. (USNM !) Taxonomy; color; waving display; mating behavior; habitat.
    Holthuis, 1954.2: 163. El Salvador.
    Garth, 1957: 107. Listing of Chilean records.

Uca stenodactylus

    Rathbun, 1918.1: 416 (part); not Pl. 152, Fig. 3, or Pl. 153. Costa Rica; Panama. (USNM !)
    von Hagen, 1968.2: 433; Text Fig. 12c. Puerto Pizarro, Peru.

Uca stenodactyla stenodactyla

    Bott, 1954: 173; Pl. 17, Fig. 16a; Pl. 18, Fig. 16b. Not Text Fig. 16. El Salvador. (Frankfurt !)
    Peters, 1955: 442ff.; Text Figs. 11, 12, 25, 30, 50, 51. El Salvador. Morphology controlling claw bending; waving display; ecology.

56. *UCA (CELUCA) TRIANGULARIS* (A. MILNE-EDWARDS, 1873)

(Tropical Indo-Pacific)

PLATE 38 *E-L*.
FIGURES 24 *I, J*; 32 *N, O*; 50; 51; 59 *A-C*; 68 *C-E*; 101.

MAP 7.
TABLES 8, 10, 12, 20.

INTRODUCTION

Among the species of the subgenus *Celuca* recognized in this study, only *triangularis* and *lactea* occur in the Indo-Pacific. The species contrast with each other in several ways, forming a classic example of differences that occur between related forms sharing broadly coincident ranges.

While *lactea* thrives in a variety of biotopes flooded regularly by tides of high salinity, *triangularis* in any locality lives only in one or two secluded habitats that are normally cut off from marine tides and often watered only by scarcely brackish streams. Nevertheless populations of both species sometimes intermingle marginally and most are subject to mixing after the sudden local changes in terrain that result from floods, typhoons, and earthquakes. It is therefore of special interest that in morphology they differ strongly only in the tips of the gonopods, which show contrasting shapes and, it seems, must present a functional barrier to interbreeding.

To an observer in the field, however, the most striking difference between the two lies in their levels of social activity. While *lactea* is highly active at suitable seasons during at least one part of every lunar cycle, *triangularis* can only be called lethargic. It is clear from Altevogt's account (1957.1) that in southeast India the local subspecies has a season of waving display well before the rains; Feest (1969) reports that in the same region two principal mating seasons exist, coinciding with the two monsoons. Although I checked a New Guinea population almost daily for eight weeks during a period when the other *Uca* displayed regularly, in *triangularis* waving by one and two individuals, respectively, was observed on only two days. Nocturnal social activity was not evident and even diurnal feeding and moving about were usually minimal. Although my acquaintance with other populations—in Penang, in several parts of the Philippines, and in New Caledonia—was confined to a few days in each locality, inactivity of the crabs was always the rule: half a dozen displaying individuals in Joló and a single waving male near Nouméa made the highlights of all the field work done on those particular days. In short, these fiddlers have proved to be so inactive under a wide variety of seasons, tidal rhythms, lunar phases, times of day and night, habitats, and geographic localities that the more lethargic species of *Deltuca* seem almost lively in comparison. Keys to their obscure rhythms remain to be found, perhaps by someone lucky enough to have a window on their biotope and binoculars on his sill.

MORPHOLOGY

Diagnosis

Front moderately broad. Merus of minor cheliped posteriorly flattened with a row of tubercles running along its lower surface, close above ventral margin; distally the row curves abruptly upward; in females the structures occur on both chelipeds. Orbits strongly slanting. Gonopod ending in a prolonged, slender, tapering tube, its flanges almost lacking and invisible without detailed examination.

Description

With the characteristics of the subgenus *Celuca* (p. 211).

MALE

Carapace. Frontal breadth contained about 4.5 times in that of front. Orbits strongly oblique. Antero-lateral angles strongly acute, antero-laterally produced. Antero-lateral margins absent, the dorso-laterals proceeding directly postero-internally in an unbroken line from the antero-lateral angle, strongly converging, ending opposite anterior edge of cardiac. Postero-lateral striae absent or, rarely, a single small one is detectable, chiefly by touch. Vertical lateral margin dorsally absent. Sides of carapace strongly convergent. Eyebrow narrow, about one-third diameter of adjacent part of depressed eyestalk. Floor of orbit with or without a tuberculate ridge, stronger on minor side. Suborbital crenellations low and contiguous internally, high and partly separated near antero-external angle, continuing around outer orbital margin but stopping before channel; largest on minor side. Lower side of antero-lateral angle moderately sharp with a row of spinules; because the antero-lateral angle is produced, this ventral margin forms a long edge.

Major Cheliped. Merus: Postero-dorsal margin extending entire length of segment as an armed, distinct ridge; thick rugosities on proximal half or more; these change abruptly about at the crest of the usual convexity to a single row of small, sharp tubercles or serrations; the convexity is directed more dorsally than posteriorly and consists wholly of the strongly arched profile of the margin. Antero-dorsal margin scarcely arched. Posterior surface with minute tubercles rather than rugosities, which, however, tend where they are sparse to form linear arrangements; they are most numerous proximally and near both postero-dorsal and ventral margins.

Manus: Bending of outer, upper surface slight, without flattening. Outer tubercles on lower half of manus extremely minute, arranged in irregular, low, indistinct rugosities, some around shallow pits; a few of the pits give rise to single setae; tubercles close to dorsal, outer groove minute to absent; remaining tubercles of upper, outer half slightly larger than elsewhere, especially proximally, and not arranged in patterns. Dorsal, outer groove distinct, curving outward distally and ending just before dactyl base. Dorsal margin with outer edge confluent with that of carpal cavity only at extreme proximal end, distinct throughout entire margin, slightly thickened, surmounted by tubercles which, beyond carpal cavity, may vary in size and may form irregularly double or triple rows; inner edge absent; region usually occupied by intervening surface tuberculate, almost vertical and virtually indistinguishable from palm. Cuff at base of dactyl smooth except for a few variable granules; crease traceable ventrally only or altogether absent. Area outside upper part of pollex base with a small, shallow, irregular hollow, its edges indefinite. Ventral margin with a distinct, blunt keel, usually beaded, the beads largest and curving slightly above outer side of margin in their distal region, before stopping suddenly near pollex base. Ventral, outer groove along beaded edge present, being widest and deepest distally.

Palm with lower triangle covered with extremely minute tubercles, not as closely set as in *beebei*, for example. Oblique ridge moderate, the apex located opposite pollex, not gape, and scarcely or not at all higher than middle of ridge; tubercles large, largest either on apex or more distally; at apex the row turns proximally, not dorsally, two or three tubercles occupying entirely the short ventral margin of carpal cavity. Center palm with entire upper part covered with moderate tubercles extending slightly over distal margin of carpal cavity onto the sloping wall of cavity itself, and continuing upward to dorsal margin. Carpal cavity with distal margin near apex low and smooth, above that higher, tuberculate as just described; no downward extension of inner edge of dorsal margin; a very short, broad, distal extension of

carpal cavity into predactyl area, its distal end tuberculate. Predactyl area spacious, with a slight depression distally; entire area covered with tubercles which are continuous with those of upper palm and dorsal margin. Proximal ridge at base of dactyl broad and short with a few large tubercles that in rare individuals are lacking; distal ridge represented only by a scattering of extremely minute tubercles; groove absent.

Pollex and *dactyl*: Both moderately slender but relatively wide in comparison with those of *lactea* and long-fingered neotropical species. Pollex arched throughout, its ventral margin notably convex, its gape margin concave; dactyl also arched, but less so than pollex, usually slightly wider than corresponding part of pollex, except proximally and distally. Outer surface of dactyl proximally notably convex; surfaces of distal dactyl and all pollex somewhat less so. Gape with median portion equal to or slightly wider than adjacent part of pollex. Dactyl outside with two broad, shallow grooves, one just above gape, the other in the usual proximal, subdorsal position; although less developed than in *Deltuca*, the upper is longer and deeper than usual in *Celuca*, although highly variable, while the lower is traceable almost to the segment's tip; both grooves in some individuals are so weak, however, that they might easily be overlooked. Proximal part of area near upper groove rough with tubercles, largest in a dorsal marginal band and on dorsal flattened base of dactyl; the marginal band continues, marked by minute tubercles, throughout dactyl's length. Surfaces of pollex and dactyl otherwise virtually smooth except for minute pits on outer pollex, continuing those of outer manus. On gape, in both pollex and manus, tubercles of outer row are complete and usually include in distal part a conspicuously enlarged tubercle, the base of which extends across most of the thickness of the gape; tubercles of inner row smaller than in outer, absent proximally; median row weak or absent in middle and, sometimes, distal portion, with at least one notably enlarged tooth proximally. Pollex tip simple. Gape pile scanty.

Minor Cheliped. Merus posteriorly flattened, with distinct tubercles, not rugosities, on postero-dorsal surface. They are arranged in roughly vertical rows, except for a strong, longitudinal row above postero-ventral margin; the latter row curves abruptly upward at its distal end, a short distance before end of segment; this structure alone is completely diagnostic for the species. Gape at base narrower than width of pollex, decreasing distally. A few strong serrations in distal half, almost in contact.

Ambulatories. Legs short for a *Celuca*, the 3rd merus on minor side about reaching antero-lateral angle when laid forward, or at most extending barely beyond it. 2nd and 3rd meri moderately broad, their dorsal margins scarcely convex to almost straight, depending on the subspecies. On minor side, antero-dorsal surfaces, especially proximally, slanting forward, so that they lie at an angle both to the dorsal margin and to the rest of the anterior surface, which, of course, is vertical as usual; the modified portion is strongly rugose; the effect is best developed on the 1st and 2nd legs and almost lacking on the 4th; it is much weaker or practically absent on all legs of major side. No other anterior modifications of 1st ambulatory on either side.

Gonopod. Superficial appearance of tip is that of an elongate, slender tube. In reality extremely narrow flanges extend throughout the length of the corneous distal part of the shaft. Beyond the subterminal pore the flanges are confluent, the distal edge either rounded, uneven, or, sometimes, slightly expanded and almost truncate. Inner process narrowly triangular, flat, almost reaching pore. Thumb well developed, closely applied to corneous shaft, long, its length varying with the subspecies but always ending close to base of inner process, at beginning of distal corneous portion of gonopod.

FEMALE

Suborbital armature slightly stronger than in male; ventral margin of antero-lateral angle minutely tuberculate. Cheliped merus as in minor cheliped of male, with diagnostic flattening and tuberculation. Modification of antero-dorsal surfaces of ambulatory meri as in its maximum development in male, but present on all legs on both sides. Abdomen of adult female unusually narrow. Gonopore with a low tubercle on antero-external margin.

Measurements (in mm)

	Length	Breadth	Propodus	Dactyl
U. triangularis triangularis				
Largest male (Zamboanga)	10.0	16.5	21.5	14.0
Largest claw (detached) (Zamboanga)	–	–	23.0	15.0
Moderate male (Zamboanga)	7.5	13.0	17.0	11.0
Largest female (Madang)	7.0	12.5	–	–
Largest ovigerous female (Madang)	6.0	9.0	–	–
U. triangularis bengali				
Holotype male (Penang)	8.0	14.0	20.5	13.0
Largest male (Penang)	9.0	15.0	22.0	13.0
Largest female (Penang)	8.0	13.5	–	–
Largest ovigerous female (Penang)	8.0	12.0	–	–

Morphological Comparison and Comment

The differences between the two subspecies to be described are suitably minor, being confined to the occurrence of tubercles on orbital floor, degree of tuberculation on center of major palm, strength of its predactyl ridge, slight differences in shape of ambulatory meri, length of gonopod thumb, and form of gonopore.

Remarks on the relationships of *triangularis* will be found on p. 218 and in the introduction to the species (p. 286).

Color

Both subspecies include some non-displaying populations with at least the carapace wholly orange, yellow, or white, either throughout the day and night or only during periods of bright sunshine. These pale groups have been observed in southeast India by Altevogt and in Penang and New Guinea by me. On dull days and during the falling tide in Penang, the yellow phase was preceded by a barred pattern that was the only one observed in the southern Philippines and in New Caledonia. The markings consist of 2 to 5 transverse pale bands, usually blue and often broken, on a dark ground and are closely similar to those characteristic of the local *lactea*. Major cheliped: In the western subspecies, brightening to uniform orange; in New Guinea, sometimes orange to white; in New Caledonia, with yellowish manus; in the Philippines, dull except for pale lower manus; in all eastern groups a speckling of fine brown spots at least on manus is diagnostic. Fingers white. Ambulatories in some individuals of populations of both subspecies sometimes ranging from orange to yellow or white; otherwise dark. Carapace and appendages of females similar to those of local males.

SOCIAL BEHAVIOR

Waving Display

Wave with vertical, semi-unflexed and lateral-circular components, the circular wave having been observed in both subspecies. In both Joló and Nouméa only the circular display was seen and it appeared close to the corresponding form in *lactea*, although no curtsy was included; single waves were at a rate of slightly more than 1 to the second. The only examples of vertical and semi-unflexed waves were found in New Guinea, where a single male several times raised and lowered his major cheliped in a low, fast motion with the hint of a curtsy, as he faced an apparently attentive female 2 inches away; in the same population a week earlier, 2 males gave several brief, semi-unflexed waves; no lateral displays were ever seen there. Altevogt (1957.1), in describing the display in a population in southeast India, also remarked on the lateral form of the wave and gave similar figures for its duration. (Component nos. 1, 3, 5, ?13.)

Acoustic Behavior

Unknown. Especially to be watched for is stridulation using the unique armature of the minor cheliped's posterior merus. The unusual tuberculation of the antero-dorsal parts of the ambulatory meri will probably prove to be part of the mechanism. See Figs. 50 and 51.

RANGE

From eastern India and Burma to New Caledonia; north to the Nansei (Ryukyu) Islands.

BIOTOPES

As indicated in the Introduction to this species, *triangularis* usually occurs on brackish flats and on the banks of tidal streams, always sheltered from open water as well as from prolonged exposure to sun. Large populations have been found in Penang and Zamboanga among the shoots of mangroves and in the mud sheltered by the overhanging branches of the parent trees. In Joló a group in a small lagoon was separated from the sea only by a narrow bar of soil and sand. Often crowded populations occur on the upper levels of steep banks slightly upstream from the limits of the mangrove. (Biotope nos. 9, 13, 14, 15.)

SYMPATRIC ASSOCIATES

U. triangularis mingles marginally and rarely with its close relation, *lactea*. In Penang it is associated with *rosea*, which was more abundant closer to low-tide levels of a secluded mangrove cove, while the burrows of *triangularis* peppered the banks on its upper slopes. In the Philippines and New Guinea, both within the mangrove zone and upstream from it, *triangularis* burrows in similar positions above those of *coarctata* and, less often, *dussumieri*, all the species being established up tidal creeks in the same small coves.

MATERIAL RESULTING FROM FIELD WORK

(The complete list of specimens examined is given in Appendix A, p. 610.)

Observations and Collections. Malaya: Penang. Indonesia: near Surabaja. Philippines: Mindanao, in neighborhood of Davao and Zamboanga; Sulu on

Joló and Tawi Tawi; Palawan. Territory of Papua & New Guinea: near Madang. New Caledonia: near Nouméa.

Films. Penang; near Davao; near Madang.

Sound Recordings. Near Nouméa.

TYPE MATERIAL AND NOMENCLATURE

Gelasimus triangularis

TYPE. In Muséum d'Histoire Naturelle, Paris. Male, listed by museum as a "type nonspecifié." Label in box: "*Gelasimus triangularis* A. Edw. Collect. A.M. Edwards 1903. Nlle. Caledonie." On back of box is the same wording plus "auct. det." Measurements in mm: length 8; breadth 12; propodus 17; dactyl 11. Photo of claw: present contribution, Pl. 38 *I, J*.

The specimen is in exceedingly poor condition, but its specific characteristics remain unmistakable. There seems to be no need at present to designate one of the specimens since collected (1965) by me in New Caledonia as a neotype. When I examined the type (1959), the specimen, dried and varnished, was glued directly on a soft wood base. After the base had been dissolved off through placing it, with no prior manipulation of the specimen, in relaxing solution, the abdomen, left gonopod, and the entire tip of the right gonopod proved to be missing. On the right gonopod a bit of subdistal tissue probably is the thumb. The specimen lacks tubercles on orbital floor, while the distal predactyl ridge has the tubercles poorly developed; these characters are now known to be within the range of variability of *t. triangularis* males, as they occur in the type-locality and elsewhere. (!)

Apparent Type Material of *Gelasimus triangularis* var. *variabilis* de Man, 1891. In Rijksmuseum van Natuurlijk Historie, Leiden. 1 male, cat. no. 1247, is labeled "*Uca triangularis* (AMEd) var. *variabilis* de Man Ludeking 1863. Amboina," the wording being the same on 2 labels, one being in a vial that holds a smaller vial and the crab, with the 2nd label. Both are written in de Man's hand (*fide* Holthuis). Length approx. 7.5 mm; propodus 21; thumb on gonopod even longer than in Fig. 59*B* of present contribution. A group of males and females, cat. no. 1537, are labeled as follows: "*Uca triangularis* (A. M. Edw): vari. variabilis (de Man) syntypes Amboina E.W.A. Ludeking, 1863." (!)

Uca triangularis bengali subsp. nov.

HOLOTYPE. In Smithsonian Institution, National Museum of Natural History, Washington. Male, cat. no. 137674. Type-locality Penang, Malaya. Measurements on p. 288. (!)

Uca (Celuca) triangularis triangularis (A. Milne-Edwards, 1873)

(Labuan, Indonesia, Philippines, Nansei [Ryukyu] Islands, Palau Islands, New Guinea, and New Caledonia. ?Caroline Islands)

MORPHOLOGY

With the characteristics of the species.

Tubercles on floor of orbit always present at least in females and in all or practically all males at least on minor side. The tubercles are best developed in Philippines populations, but even here they are highly variable; the maximum number totals around 20, with the addition of grouped spinules near or beyond inner end of row; in some males the tubercles are reduced to less than 10 and show irregularities of size and arrangement, or are represented only by a few spinules. On minor side in male and on both sides in female the row runs well behind suborbital margin and is often convex posteriorly in the middle; in contrast, on the male's major side the row parallels and lies very close to the margin. The range of variation indicated occurs within at least some populations, including New Caledonia, New Guinea, and the south Philippines.

As indicated in the description of *t. bengali*, the palm in *t. triangularis* is often smoother, with the proximal predactyl ridge relatively short and the tubercles fewer or, rarely, absent; meri of ambulatories in general narrower, the upper edge of the 3rd being definitely straight; gonopod thumb longer, being more or less equal to exposed part of tube; gonopore larger, with a definite apex directed antero-internally.

Color

No pale phase was encountered in the Philippines; entirely white or yellow white individuals were the rule in New Guinea, with orange males appearing occasionally; in New Caledonia the coloring was as in the Philippines, except that the manus was tinged with yellow. The major cheliped of all individuals, even when they are otherwise white, is speckled above with minute dark round spots ranging from brown to purplish brown; at most they cover the dorsal and outer parts of merus, carpus, manus, and lower palm; at the least they are scattered on the upper outer manus.

Uca (Celuca) triangularis bengali subsp. nov.

(Eastern India, Burma, and western Malaya)

MORPHOLOGY

With the characteristics of the species.

Tubercles on floor of orbit altogether lacking in both sexes. Palm often more tuberculate than in *t. triangularis*, with proximal ridge at dactyl base longer; meri of ambulatories at least on minor side broader than in most *t. triangularis*, the upper edge of the third being definitely convex; gonopod thumb shorter, about one-half length of exposed part of tube; gonopore with a small, blunt, inconspicuous tubercle.

Color

Pale phases ranging from orange to white appear to be usual, while brown speckles on the major cheliped have not been reported.

REFERENCES AND SYNONYMY

(When I have neither seen the specimens listed nor made collections from near the same localities, the references are placed in accordance with the known geographical distribution of the subspecies. No examples of coincident ranges or apparent hybrids are yet known.)

Uca (Celuca) triangularis (A. Milne-Edwards, 1873)

TYPE DESCRIPTION. See under *U. (C.) triangularis triangularis*, below.

Uca (Celuca) triangularis triangularis (A. Milne-Edwards, 1873)

Gelasimus triangularis

TYPE DESCRIPTION. A. Milne-Edwards, 1873: 275. New Caledonia. (Paris !)

Kingsley, 1880.1: 150. No new record. Taxonomy of type.

de Man, 1892: 307. Sumatra.

de Man, 1895: 577 (part). East Indies (not Penang). Taxonomy.

Gordon, 1934: 11. Dutch East Indies.

Gelasimus triangularis var. *variabilis*

de Man, 1891: 47. Amboina. Type description. (Leiden !)

Tweedie, 1937: 14. Sumatra: Simalur (= Simeuloe) I.

Uca variabilis

Ortmann, 1897: 353. No new record. Taxonomy.

Ward, 1941: 3. Philippines: west coast, Gulf of Davao.

Uca triangularis

Nobili, 1899.2: 274. East Indies: Andai. Brief taxonomy.

Estampador, 1937: 543. Records from the Philippine Is.

Estampador, 1959: 100. Records from the Philippine Is.

Uca novaeguineae (not of Rathbun, 1913)

Sakai, 1936: 171; text fig. Pelew (= Palau) Is. Taxonomy; color. (Yokohama !)

Miyake, 1938: 110. Micronesia: Ngardok; Babelthaob; Palau Is. (Part may be *U. chlorophthalmus crassipes, q.v.*)

? Miyake, 1939: 223, 242. Micronesia: Palao Is.; Caroline Is. (Kusaie). (Part or all may be *U. chlorophthalmus crassipes, q.v.*)

Uca triangularis variabilis

Tweedie, 1950.1: 357. Labuan. (Raffles !)

Uca (Celuca) triangularis bengali subsp. nov.

Gelasimus perplexus (not of Milne-Edwards, 1837)

Heller, 1865: 38; Pl. 5, Fig. 4. Ceylon. Southeast India: Madras. Taxonomy.

Gelasimus triangularis

de Man, 1887.1: 119. Mergui Archipelago: Kisserain I. Taxonomy. (Leiden !)

? de Man, 1892: 307. Sumatra: eastern coast at Batu Bahra. (May be *U. t. triangularis.*)

Henderson, 1893: 388. Southeast India: Madras; Ennore. Taxonomy; habitat. (BM !)

de Man, 1895: 577 (part). Malaya: Penang. (Not records from East Indies.) Taxonomy.

? Nobili, 1899.3: 518. Sumatra: Siboga. Taxonomy. (May be *U. t. triangularis.*)

Alcock, 1900: 356. No new record. Taxonomy.

Gelasimus variabilis (not *G. triangularis* var. *variabilis* de Man)

? Ortmann, 1894.2: 758. Singapore. Taxonomy. (This is the only record of the occurrence of *triangularis* in Singapore; may be *U. t. triangularis.*)

Uca triangularis

Nobili, 1903.2: 20. Southeast India: Pondichéry.

Maccagno, 1928: 31. No new record. Taxonomy.

Chopra & Das, 1937: 421; text fig. Mergui Archipelago: Bockachaung. In fresh water.

Panikkar & Aiyar, 1937: 295. Statement that these authors, unlike Henderson, 1893, did not find *triangularis* in the Madras area.

Altevogt, 1957.1. Illus. Southeast India (from near Madras to Cape Cormorin). Waving display; ecology; general habits.

Feest, 1969. Illus. Southeast India: Porto Novo. Breeding biology; ecology; post-embryological development.

57. *UCA (CELUCA) LACTEA* (DE HAAN, 1835)

(Indo-Pacific)

PLATES 39; 40; 45 *A*; 47 *C, D*; 50 *A*.
FIGURES 17; 18; 19; 20; 24 *N, O*; 26 *D*; 27 *I, J*; 29 *D*;
31 *E*; 32 *L, M*; 37 *L*; 41 *A, B*; 46 *L*; 54 *H-KK*;
69 *B-E*; 74 *L-N*; 81 *O*; 83 *B*; 91 *A-D*; 92; 101.

MAP 21.
TABLES 5, 6, 8, 10, 12, 14, 19, 20, 22.

INTRODUCTION

One temptation in this study is to write about *Uca lactea* in superlatives. Among fiddlers these small crabs have one of the four widest ranges on earth. They form perhaps the most abundant species. They vary the most widely in often bright and complex color patterns. Above all they present us with more challenges, as they unfold the refinements of their social behavior, than any other fiddler I know.

A report on these behavior patterns, particularly those of agonistic behavior, will be published separately; it is too detailed for appropriate inclusion here and, moreover, includes the results of recent field work (1969-1970) which has not been fully analyzed. One result, perhaps of the most general interest, is that the frequency of combat components, although not their form, differs in different parts of the range. Less striking differences are found in the components of waving display, especially in those of high intensity courtship, one or more of them being absent in one part of the range, or occurring only in low intensity display in another area.

In years to come, when the difficulties of fiddler-rearing have been reduced to laboratory trivia, this species may prove to be well suited for genetic studies of behavior. Meanwhile, the question of the basic value of combat now appears to be complicated rather than clarified by this recent work. In brief, one can say with confidence that *lactea* will continue to repay study, both as an entire, lively animal in its own habitat and in suitable crabberies; in addition it should make an excellent subject for correlated physiological research.

This species is one of those showing a particularly strong lunar or semi-lunar rhythm, and it is therefore important to plan field work accordingly. During some seasons and in some places, *lactea* proved to be socially active only around new moon, only around full moon, or around both; in South Africa and in Japan it does not display during the local winters, and in Japan, at least, hibernates during severe weather. In general when planning short trips it is probably safest to arrange to be in a given locality around new moon rather than full; the exact selection

of dates must depend on whether the investigator's chief interest lies in waving display and the associated courtship or in combat behavior. Waving display and courtship reach their height between about one-half and two hours after low tide and are especially prevalent, and sometimes only to be observed, when low tide occurs in the late afternoon and in the late morning. Combat, on the other hand, rarely furnishes much observation material except between about one-half hour before low tide and an hour or so afterwards, and only when the time of low tide is between about 8:30 and 11 A.M. Where tides are very unusual, active combat behavior still seems to await the eventual occurrence of mid-morning lows, no matter how small the tidal levels at the time. For research purposes useful amounts of combat, waving, and courtship may be found during longer periods and at other times of the day, but it seems clear that one is most likely to find populations in a state of high social activity at these times, when only a few days are available for work. If, however, a traveling biologist with, for instance, five days for work confidently aims for a row of morning lows near a particular new moon, he may well find that his selected population is ruled by the full moon instead. Therefore, before following any advice in this paragraph, *caveat lector*.

Nocturnal activity, except for the usual acoustic responses to intruders, seems to be poorly developed in *lactea*, judging by my repeated checks of populations in various localities. Possibly inconspicuous nocturnal courtship behavior accompanying acoustic components occurs, but it has not yet been investigated. A foundation was laid by Altevogt (1957.1), who found that some *lactea* remain active at twilight.

After de Haan first described the species from Japan in 1835, similar forms became familiar under other names, in particular *annulipes* and *perplexa*, which were collected in other parts of the Indo-Pacific. In the present study, because of strong evidence of interbreeding as well as close similarities in morphology and behavior, only a single species, *lactea*, is recognized; except in the interesting areas where two forms coincide, four subspecies can be readily distinguished—*lactea, annulipes, perplexa,* and *mjobergi.*

MORPHOLOGY

Diagnosis

Broad-fronted Indo-Pacific *Uca*; major cheliped with oblique tuberculate ridge inside palm present, proximal predactyl ridge not diverging dorsally from dactyl base, no hook-like predistal tooth on dactyl, and no small, sharply bounded depression outside pollex base; posterior merus of minor cheliped without vertical row of tubercles; merus of last ambulatory slender, its dorsal margin practically straight; one postero-lateral stria on each side; gonopod with large flanges, the tube not projecting; gonopore without tubercle.

Description

With the characteristics of the subgenus *Celuca* (p. 211).

MALE

Carapace. Frontal breadth contained about 3 to 4 times in that of carapace. Antero-lateral angles acute, antero-laterally produced. Antero-lateral margins approximately straight, converging, rounding into dorso-lateral margins. Each of the single pair of postero-lateral striae is directed almost antero-posteriorly, rather than in the usual nearly transverse position. Vertical lateral margin highly variable, being sometimes weak or absent in dorsal fifth. Carapace profile almost semi-cylindrical; hepatic and branchial regions not fused, only moderately arched. Tip of front little rounded, appearing almost truncate in dorsal view. Eyebrow shorter than usual, and narrow, measuring half or less than half the diameter of adjacent part of depressed eyestalk, which itself is unusually slender in comparison with that usual in the subgenus; lower margin of eyebrow weaker than upper. Suborbital crenellations minute in inner half, increasingly large near antero-external angle, continuing around outer orbital margin, separated, almost to channel. Lower side of antero-lateral angle moderately sharp, usually minutely serrate.

Major Cheliped. Merus: Antero-dorsal margin proximally scarcely arched and not ridged, extremely weak, consisting only of the slightly angular meeting of dorsal and anterior surfaces, usually completely unarmed, occasionally with a few irregular granules; distal portion more distinct, marked principally by a cluster of tubercles which are sometimes large and numerous. *Carpus:* Dorso-posterior surface smooth.

Manus: Bending of upper, outer surface apparent near dorsal margin only, where both bending and moderate flattening are apparent. Outer tubercles extremely minute, although variable among individuals; slightly larger and more widely spaced only in a narrow band adjacent to dorsal margin. Dorsal outer groove absent. Dorsal margin above carpal cavity strong, formed of large, sharp tubercles; distally it is weak, marked only by small tubercles and lacking both inner edge and intervening space to set it off from palm. Area outside pollex base not smooth, the minute tubercles of outer manus continuing across it and onto pollex; in southwestern populations of *l. annulipes,* lower part sometimes with indications of a row of slightly larger tubercles, not elevated on a keel, that continues onto pollex. Ventral margin or supramargin of manus with a row of tubercles, small and faint in proximal half, slightly larger distally, and dying out at pollex base or along its proximal part. Ventral, supramarginal outer groove present in distal half of manus only, where it is strong; it often continues, faintly or strongly, throughout most of pollex.

Palm with lower triangle covered relatively sparsely with extremely minute tubercles. Oblique ridge usually high and thin, the highest point usually slightly higher than apex and distal to it; tubercles largest on highest point of ridge; they either stop at apex or, in some individuals of all subspecies, turn proximally for a short way along ventral margin of carpal cavity. Center palm definitely convex, with variable tuberculation ranging from minute and sparse to moderate in both tubercle size and distribution. Carpal cavity with ventral margin always low, thin, and sharp, whether or not surmounted by tubercles continued from apex; distal margin of cavity sloping gently, merging gradually with center palm; inner edge of dorsal margin not curving downward; upper part of carpal cavity with a distal extension occupying almost entire predactyl area. Edges of latter area indefinite, either smooth or minutely tuberculate.

Pollex and *dactyl:* Highly variable within and between subspecies, leptodactylous and brachydactylous forms being well developed. Both pollex and dactyl always long and compressed, at least one usually having the gape edge slightly concave, the dactyl's always being more or less convex dorsally and always wider than the pollex, sometimes strikingly so. Gape wider, sometimes much wider, than adjacent pollex. Outer pollex sometimes with a low supramarginal keel with or without tubercles, and a shallow groove contiguous to its upper side; keel not continuous with the row of tubercles on ventral margin of manus. Along lower side of pollex, above position of pollex keel and groove, whether or not they are present, a straight row of small, close-set tubercles sometimes occurs. Dactyl with outer, proximal, subdorsal groove absent; tubercles of the area well developed. Pollex and dactyl covered externally with minute tubercles, often larger on pollex; both fingers internally smooth or with extremely minute tubercles. Gape with outer marginal rows in both pollex and dactyl sometimes clearly traceable only about half-way to tip of segment; inner rows sometimes obsolescent, with only the most proximal tubercles clearly referable to this row and not to the general tuberculation near gape; tubercles of median row ranging from absent to moderate with up to several enlarged tubercles or tuberculate, triangular teeth. Tip of pollex simple. A predistal, triangular tooth, sometimes much enlarged and its distal margin concave, often present. Gape pile sparse.

Minor Cheliped. Gape throughout about as wide as pollex. Serrations weak, widely and unevenly spaced, not in contact, occurring more proximally on pollex than on dactyl; sometimes absent.

Ambulatories. 2nd and 3rd meri moderately slender to moderately broad, their dorsal margins scarcely convex to almost straight. 4th merus always strikingly slender, its dorsal margin practically straight. No anterior modifications of 1st ambulatory.

Gonopod. Flanges strongly developed, the anterior or posterior being the larger, depending on position of the canal, which varies with the subspecies; one or both flanges may have its round distal edge produced slightly or far beyond the small subterminal gonopore. Inner process narrow, tapering to a blunt end which overlies canal but may or may not reach gonopore. Thumb short but well developed; very variable in position, it may stop far below base of anterior flange or overlap it.

FEMALE

Carapace regions definitely less well marked than in male, the difference being unusually strong for a *Celuca*. Suborbital crenellations better developed than in male on inner half, but about equal to male's in outer portion; some individuals, especially in some populations of *l. annulipes*, with a row of tubercles on orbital floor, sometimes clearly supported by a ridge; their occurrence, sizes, and numbers are however exceedingly variable, usually even within populations. Gonopore with outer margin arched strongly toward midline, so that the pore itself appears crescentic and the margin a rounded lip; this lip slants up into the gonopore depression, the degree of slant and its direction depending on the subspecies; in all except one subspecies (*annulipes*), it is rimmed conspicuously with corneous brown, which sometimes also occurs on the surrounding tissue. These minor gonopore differences are unfortunately the only reliable

distinctions among the subspecies that have yet become apparent.

Size

Moderately small. Measurements in descriptions of subspecies.

Morphological Comparison and Comment

U. lactea differs conspicuously as follows from the other species of Indo-Pacific fiddlers having moderately broad fronts. It differs from both subspecies of *inversa* in the presence of an oblique tuberculate ridge on palm, in the small, slender propodus of the small cheliped in both sexes, and in having only one pair of postero-lateral striae on each side; it differs additionally from *i. inversa* in East Africa in the absence of a terminal hook on major dactyl. It differs from *chlorophthalmus* in having the proximal predactyl ridge parallel to, not diverging from, the major dactyl's base; in lacking a small, distinctly bounded depression outside base of major pollex; and in having slender meri in both sexes on the last ambulatories. Finally, it differs from *triangularis* in lacking, in both sexes, a vertical row of tubercles predistally on posterior minor merus, and in not having a projecting terminal tube on gonopod.

The species divides satisfactorily into four subspecies, based principally on the shapes of the gonopod flanges, gonopore tubercle, and, less reliably, major chela. Two of the four subspecies have not been found to show any zones of coincidence or apparent interbreeding. Their ranges are small, being restricted to northwest Australia (*l. mjobergi*) and from Hong Kong to Japan (*l. lactea*). The other two subspecies, *l. annulipes* and *l. perplexa*, cover between them the entire tropical Indo-Pacific, from eastern Africa to Samoa, excluding only *mjobergi*'s limited range in Australia.

No sharp geographical boundary divides *annulipes* from *perplexa*. Populations showing coincidence and signs of hybridizing occur from southeast India to the central Philippines. These mixed populations and the incidence of individuals of the atypical subspecies are rare in the west and in the northeast. In the intermediate areas of Indonesia and the southern Philippines, on the other hand, they are common. The subspecies *annulipes* apparently occurs alone in Africa, West Pakistan, western India, and Ceylon, while *perplexa* appears unmixed in the north Philippines, New Guinea, eastern Australia, and the islands of the tropical Pacific, reaching the Ryukyu Islands in the north and Samoa in the east.

As in the apparently hybridizing subspecies of *vocans*, the male gonopods show distinct characteristics of either one subspecies or the other. Only in one or two individuals from the north-central Philippines is the shape definitely ambiguous. Female gonopores, on the other hand, while usually characteristic of either *annulipes* or *perplexa*, sometimes show intermediate states of the tilting of the tubercle tip and its orientation.

In the subspecific characters of the major cheliped, on the other hand, the characters are independently assorted, showing every degree of mixing throughout the broad region where both types of gonopods occur in the same population. Dactyls that are to varying degrees broad, as in *perplexa*, occur on individuals with gonopods of *annulipes* shape; these broad-dactyled crabs may or may not have a triangular tooth on the pollex that approaches or equals the size of this structure characteristic of *perplexa* in the east. Similarly, slender claws turn up in the zones of coincidence on individuals with gonopods wholly typical of *annulipes*.

Since the gonopod form appears practically always to be entirely characteristic of one subspecies, while the other subspecific characters vary considerably even in apparently unmixed populations, specimens have been arbitrarily assigned to one subspecies or the other in accordance with the form of their gonopods alone. This course has seemed preferable to indicating apparent crosses, even when other characters in an individual and its population show clear signs of interbreeding; it makes possible a preliminary estimate, however unreliable, of the amount of mixing that occurs in various localities. Until genetic studies are undertaken and the dominance relations of particular traits established, obviously no definite results can be obtained. At present the only two conclusions to be drawn seem to be, first, that *annulipes* and *perplexa* give every evidence of frequent hybridization within the area of occasional coincidence and, second, that the heart of the area of coincidence, as in *vocans* and certain *Deltuca*, lies in the region of the Sunda Shelf.

Beyond the area of coincidence where both types of gonopods occur, claws are sometimes found that are clearly atypical of the region. For example, an unusually wide dactyl or a large tooth on the pollex infrequently turns up in Africa, although the proportions are typical in the east. Similarly, although a tuberculate ridge on the major pollex is strong and prevalent only from Zanzibar to Mozambique, the character appears, although rarely, on claws as far away as Fiji; almost always in these individuals it is a weak structure and probably scarcely functional. These characters would lend themselves well to statistical analysis.

The relationship of *lactea* to other members of the subgenus is discussed on p. 218.

Color

Displaying males: Polished white fully developed over entire crab, in some individuals of some populations in three of the four subspecies. In *mjobergi* alone such a development of white was not found. The population of *l. lactea* in northwest Taiwan in May was the only one in which this high development of white occurred in almost all displaying males and, incidentally, on the females too. More characteristic of all other populations were the following general ranges of color, which are summarized in tabular form in Table 5.

Carapace. Sometimes attaining a homogeneous pale color in all subspecies, particularly in Philippine populations of *perplexa*; the most prevalent in *perplexa*, as well as in East African populations of *annulipes*, is fine marbling, the component colors being black with white or blue. In *mjobergi* the marbling is brown with buff or white. In *l. annulipes* from the non-African part of its range the pattern is characterized instead by strong transverse markings, exceedingly variable within populations; often the ground color is black to blue with three or four paler bands, either solid or divided into spots, crossing the entire width of the carapace and ranging in hue from light blue to white; at a later stage of display lightening the pattern appears instead as black or blue markings on a blue or white background. In some pure-culture populations of *l. perplexa* (that is, without admixture of *l. annulipes*) marbled carapaces sometimes have 1 or 2 dark, *annulipes*-like bands, but posteriorly only.

Major Cheliped. Polished white chelipeds occur in single populations of each subspecies except *mjobergi*, even though the carapace in these populations does not whiten strongly. Except for these three, and representatives of other populations attaining complete white, the cheliped in all subspecies shows some tint or shade of yellow or red, not only in every displaying male, but almost always in all non-displaying individuals also, although the hues are then usually dulled and, even in displaying males, often confined to the lower, outer manus. In all subspecies except *annulipes* the color is basically yellow, ranging from pale buff to orange yellow of high saturation. The most intense yellows were found in *perplexa* in Fiji and in *mjobergi*, in both Darwin and Broome. The color is usually brightest on the entire outer manus and proximal part of chela, but the inner merus is often similarly colored in spite of strong display whitening of the rest of the crab. In contrast, yellow does not occur in *annulipes* in populations showing no coincidence with those of *perplexa* (see below). Instead, red is found in a wide range from pale pink to highly saturated spectrum red or, sometimes, scarlet red; it attains its most vivid form in Ceylon (where there is no coincidence with *perplexa*) and in Singapore (where the two subspecies coincide and apparently interbreed); in these two localities, especially, the inner side of the merus often remains bright red even though the outer manus has paled to pink during carapace lightening. In all subspecies the fingers are white at least distally, and the palms almost always pale.

Eyestalks usually gray, sometimes yellowish or greenish. Anterior aspects of orbits, pterygostomian regions, and buccal area usually lighten to about same degree as carapace, although sometimes they attain a higher degree of white, often being clear white in New Guinea, for example, even when the carapace remains in its usual marbled phase; in New Guinea as well as in the Philippines and in Fiji these anterior aspects are sometimes pale yellow, partly or wholly. Minor cheliped is most often white but rarely marked with red, yellow, or blue, these unusual phases cropping out in both *perplexa* and *annulipes* in various localities.

The ambulatories attain whiteness in all populations where this degree of display lightening occurs. More often red, particularly dark red but sometimes (in *mjobergi*) scarlet, occurs on the anterior sides of all the ambulatories in all subspecies; sometimes it is confined to the legs on the minor side; often it occurs only on the merus, or merus and carpus, of both sides, a distribution that is prevalent in *annulipes*. The non-red aspects are confined to the occasional occurrence of yellow instead of red in some populations of *perplexa*, and, on all parts of the ambulatories not colored yellow or red or lightened to white, to the light and dark banding which gave *annulipes* its name. Often however the legs even of strongly displaying males are completely dark, including the anterior meri, with even the bands not apparent.

Where populations of *annulipes* and *perplexa* coincide and are suspected strongly of interbreeding (p. 294), the characteristic red-marked cheliped and banded carapace of *annulipes* appear usually to be paired, while the solidly pale or marbled carapace and yellow-marked cheliped of *perplexa* occur together at least most of the time. Whether the colors match the gonopods, however, is at present unknown.

Females often attain about the same degree of lightening as males. Rare individuals in *perplexa* and *annulipes* throughout their ranges are dark red or rose red, sometimes including the appendages. These reddish females are sometimes but not always among the few females in any population being actively courted on a given low-tide period. As in many other species, small dark females with no lightening at all sometimes wander the most actively, eliciting high intensity waving from numerous males.

SOCIAL BEHAVIOR

Waving Display

Wave always lateral at medium or high intensities, ranging through degrees of lateral-straight to wholly lateral-circular; circular waving occurs in fully developed territorial waving, in many displays toward males, and in all fully developed courtships. Diminishing circular waves present or absent. Vertical waves sometimes occur at lowest intensity in a threat situation or in young individuals. Jerks always absent. Minor cheliped usually makes corresponding motion. Body raised or not on extended ambulatories at both low and high intensities, often during high intensity held raised throughout a series. Curtsy in some form present at highest intensity in all subspecies, sometimes confined to courtship; the cheliped is then extended frontally or, usually, laterally, the circular wave is suppressed, and instead the cheliped makes a downward motion so that sometimes the lower edge of the propodus strikes the ground. Sometimes in threat situations the wave is similarly eliminated, but the cheliped is then held flexed, not extended, and alternately lowered and raised, either actually touching the ground or not, the ambulatories also sometimes bending simultaneously in a curtsy. During circular waving of moderate or high intensity one or more ambulatories are raised in leg-waves; in threat situations, however, leg-waving is replaced by sound-producing leg-wagging. During display the crab often moves a few steps while cheliped is raised or laterally extended. Seriality ranges from pronounced to absent. Non-forceful herding sometimes present. Synchronous waving noted rarely. Display rate very variable, depending both on type and intensity of wave and general level of waving activity in the population. The fastest circulars photographed were at rate of about 3 per second, the slowest at 2.3 seconds per wave. (Component nos. 1, 4, 5, 9, 10, 11, 12, 13, 14, plus the downstrokes of extended cheliped described above. Timing elements in Table 19, p. 656.)

Precopulatory Behavior

Male attracts female down his own burrow. No attempts at copulation seen at surface, except by rare wandering males and in captivity in crabberies. Altevogt (1955) reported surface copulations to be rare in western India. Feest (1969), however, found them plentiful in the southeastern part of the same country, on a semi-lunar schedule.

Hood Construction

Fully formed hoods were constructed by displaying males of *l. annulipes* in West Pakistan in June and by a similarly active group of *l. lactea* in Taiwan in May; rudimentary hoods, poorly formed and less than 5 mm high, were detected beside the burrows of rare individuals in 2 other populations of displaying males, in *l. annulipes* in Ceylon in May and in *l. perplexa* in northeast New Guinea in June and July.

Acoustic Behavior

The following components have been filmed: major-merus-rub (component no. 1), leg-wagging (5), major-merus-drum (7), major-manus-drum (9), minor-chela-tap (10) and, questionably, leg-stamp (11). Whirls have been recorded on tape when an individual of either sex was introduced into a burrow inhabited by a male; these recordings were secured in the Red Sea, southwestern India, New Caledonia, and the Fiji Islands (Pl. 47 *C, D*). Videotapes of diurnal leg-wagging in agonistic situations were made in northeastern New Guinea.

Combat

The emerging results of combat in *lactea* have been mentioned in the introduction to the species (p. 292) and in the general account of combat (p. 516). The following components have been identified: manus-rub (component 1); manus-and-chela-rub (1a); pollex-rub (2); chela-rub (2a); pollex-under-and-over-slide (3); subdactyl-and-subpollex-slide (4); dactyl-slide (6); chela-tips-slide (6a); upper-and-lower-manus-rub (7); interlace (9); heel-and-hollow (11); heel-and-ridge (12); as in heel-and-ridge, but ridge only rubbed (12a).

RANGE

Tropical and subtropical Indo-Pacific, from eastern Africa to Samoa; from near Massawa in the Red Sea, Karachi in Pakistan, and Fukuoka in southern Japan south to the Umngazana River, Cape Province, in Africa, Broome in Western Australia, Gladstone in Queensland, Australia, and Tonga, South Pacific.

The four subspecies here recognized range roughly as follows: *annulipes* from eastern Africa to the Philippines, being rare east of Borneo and coincident with *perplexa*, with which it apparently hybridizes, in southeast India (Pondichéry), Malaysia, Indonesia, and the Philippines; *perplexa* from Java, the Philippines, New Guinea, and northeast Australia to Samoa; rare west of Java, occurring in the regions as listed above, coinciding with *annulipes*; *lactea* from southern Japan southwest to Hong Kong; *mjobergi* is known only from northwest Australia and northwest New Guinea.

BIOTOPES

Sheltered shores near large bays or the open sea, sometimes protected only by reefs or offshore mud flats; substrate ranging from sandy mud to muddy sand; less often populations occur in mud without noticeable sandy admixture, as well as in sand with very little silt, especially close to mangroves occurring in such a substrate. Often adjoining mangroves, the burrows being sometimes among their rhizophores; socially active parts of populations, however, are never in partial shade; frequently along the shores of estuaries; when close to river mouths the crabs usually live on one or both sides of the mouth itself, rather than slightly upriver; when populations do occur within the mouth and are distributed farther upstream, the individuals become progressively smaller as the water becomes more brackish, although social behavior, apparently normal, continues. Burrows rarely found in steep banks. (Biotope nos. 6, 9, 11, 12, 14.)

SYMPATRIC ASSOCIATES

Throughout most of its range, *lactea* occurs characteristically with (*Thalassuca*) *vocans*; although the burrows often mingle, *lactea* occupies higher levels on the shore. In some parts of East Africa, as in a population in Pemba, the species occurred not with *vocans* but with (*Amphiuca*) *chlorophthalmus* and (*Deltuca*) *urvillei*, which occupied, respectively, successively lower levels on the shore, *vocans* being absent from that particular community. In the Red Sea area (Massawa), Zanzibar, and Karachi, *lactea* occurred close to and even mingled with (*Amphiuca*) *inversa*, although *inversa* characteristically but not always occurred on higher, drier, more inland areas.

Farther east, populations of *lactea* often occur close to those of (*Celuca*) *triangularis*, but always in a more exposed position, and, in my experience, without their burrows intermingling; in many localities, however, the barriers between are fragile, temporary, or both, as when a group of *lactea* on the shore of a small, largely enclosed bay is separated only by a barricade of sand, less than a meter high and sometimes scarcely more across the top, from the tiny lagoon inhabited by *triangularis*. As in several such instances known in the Philippines, the lagoon may be cut off for months from the sea; yet the barriers are subject to swift destruction by storms or unusual tides. In an equally frequent example, populations of the two species occur on the opposite sides of natural estuaries, tidal marshes, or even artificial fishponds—*lactea* on suitable strips of shore close to the outlet to the sea, and *triangularis* near the inland boundary of the habitat, often slightly inside the mouths of small streams, with its burrows, in such cases, on the steep banks. (See also pp. 286, 289, and 297.)

MATERIAL RESULTING FROM FIELD WORK

(The complete list of specimens examined is given in Appendix A, p. 611.)

Observations and Collections. Ethiopia: Massawa. Pemba and Zanzibar. Mozambique: Inhaca I. Aden. Pakistan: Karachi. India: near Bombay; Ernakulam. Ceylon: Negombo. Malaya: Penang. Singapore. Sarawak: Santubong. Java: near Surabaja. Northwest Australia: Darwin; Broome. New Guinea: near Madang. New Caledonia: near Nouméa. Fiji Is.: Viti Levu (various localities). Samoa: near Upolu. Philippines: Tawi Tawi; Joló; Zamboanga; Gulf of Davao (various localities); near Manila. Hong Kong. Northwest Taiwan: Tamsui. Japan: Kyushu, near Fukuoka.

Films. Massawa, Pemba, Zanzibar, Inhaca, Karachi, Negombo, Singapore, Santubong, Darwin, Broome, Madang, Nouméa, Fiji, Zamboanga, Gulf of Davao, Manila, Hong Kong, and Tamsui.

Sound Recordings. Massawa, Ernakulam, Negombo, Madang, Nouméa, and Fiji.

Videotape recordings. Negombo and Madang.

TYPE MATERIAL AND NOMENCLATURE

Uca lactea (de Haan, 1835)

LECTOTYPE of *Ocypode (Gelasimus) lacteus* de Haan. In Rijksmuseum van Natuurlijk Historie, Leiden. 1 male of a group under cat. no. 254, all with the label "*Uca lactea* (de Haan) Burger Type Japan." Selected by J. Crane. Material includes 3 other adult males, all in excellent condition, in alcohol, as well as several smaller males and females. This group, along with a jar of other specimens from Japan (cat. no. 1575), show an excellent range of claw diversity; gonopods as described in present study for *U. l. lactea*. (!)

In Paris 2 males and 1 female bear the locality label "Japon" and are listed as possible cotypes, having been received from Leiden. (!) (Pl. 40 A-B.)

Uca lactea annulipes (Milne-Edwards, 1837)

LECTOTYPE of *Gelasimus annulipes* Milne-Edwards. In Muséum National d'Histoire Naturelle, Paris. One male from a group of 3 in a single box with the label "*Gelasimus annulipes* Edw. Mer. des Indes M. Reynaud." Selected by J. Crane. They are listed by the museum as "types non spécifiés." Condition excellent, although dried, and showing the now well-known variation in claw form. All 3 specimens were relaxed for gonopod examination; all gonopods are typical of those specimens referred in the present study to *U. lactea annulipes*. Measurements in mm: length 10.5; breadth 18.5; propodus 31 (= lectotype). (!)

Uca lactea perplexa (Milne-Edwards, 1852)

TYPE MATERIAL. In Muséum National d'Histoire Naturelle, Paris. 2 males in same box with the label "*Gelasimus perplexus* M. Besukuj Javae." They are listed as "types non spécifiés." Condition very poor, the specimens having been dried, wired, and somewhat crushed. It seems undesirable to designate a lectotype or neotype at present. The larger specimen, about 9 mm long with the propodus 27 mm, was relaxed; the gonopod is clearly of the form referred in the present study to the subspecies, *U. lactea perplexa*; the claws are also of characteristic shape. (!)

Uca lactea mjobergi Rathbun, 1924

HOLOTYPE of *Uca mjobergi*. In Naturhistoriska Riksmuseet (sektionen för evertebrat zoologi), Stockholm: Male; measurements in mm: length 8.1; breadth 13.3; propodus 22 (Rathbun). Type-locality: Broome, Australia.

Additional Type Material. In Smithsonian Institution, National Museum of Natural History, Washington: 2 male paratypes, cat. no. 56418. (!)

Type Material of *Gelasimus forceps* Milne-Edwards, 1837. The type description of this species is from parts of 2 specimens, belonging to different subgenera, in the Muséum National d'Histoire Naturelle, Paris, and joined by wire. The claw is that of a specimen of *lactea*; it is the one illustrated by Milne-Edwards, 1852, Pl. 3, Fig. 11a. The situation is described in the present contribution on p. 323. (!)

Type Material of *Gelasimus porcellanus* Adams & White, 1848. In British Museum (Natural History), London. Label: "*Gelasimus porcellanus* White 1847. Lectotype (selected 1911, W. T. Calman) Borneo. 901. 444.106." Also a similarly labeled paratype. Both are clear examples of *U. lactea annulipes*. (!)

Type Material of *Gelasimus annulipes* var. *albimana* Kossmann, 1877. Not located. Red Sea.

Type of *Uca annulipes* var. *orientalis* Nobili, 1901. In Regio Museo Zoologico, Torino, cat. no. 1521. Samarinda, Borneo, 1 male. (!) Here synonymized with *U. lactea perplexa*.

Material labeled *Uca consobrinus*, a name given in MS only, by de Man. 7 males with this name on the label are deposited in the Zoological Museum,

Amsterdam, but not designated as types; deposited in a single jar with 3 labels, as follows: (1) "*Gelasimus consobrinus* de Man Batavia Collection de Man." (2) "g" [stands for position on shelf] "Zoolog Museum. Amsterdam. Colleckie de Man." (3) "*Gelasimus consobrinus* de Man Dr. J. Verwey Batavia 1928." Lengths 8 to 11 mm. These specimens all proved to be characteristic brachychelous examples of *lactea annulipes*. In answer to a question from me, Dr. Verwey wrote as follows in a letter (11 July 1959): ". . . you will have seen from my note on p. 185 of my paper on mangrove crabs that De Man, who identified my animals, gave the name *consobrinus* to the species in question, because he had become convinced that it was different from *annulipes* (Latr.) H. Milne-Edwards. I suppose that the description of his *consobrinus* was not published because he died a short time after having identified my animals, but I am not quite sure that his paper was not published after his death without my knowing it." It seems now to be definite that de Man did not publish the paper, and that Verwey, in his 1930 contribution, is the only author to have used the name. (!)

Uca (Celuca) lactea annulipes
(Milne-Edwards, 1837)

(East Africa to Singapore; occasionally to central Philippines).

MORPHOLOGY

With the characteristics of the species.

Gonopod practically without torsion; anterior flange longer than posterior and clearly wider, the pore being located in a narrow notch near posterior margin; thumb ending distinctly below flange base. Gonopore with marginal lip strongly tilted, its tip far above the sternal surface (and therefore appearing deeply depressed in the usual ventral view of the observer); no corneous brown pigment on rim of lip. Major cheliped: Outer pollex in African populations from Zanzibar and farther south almost always with a straight row of small tubercles along lower side; this structure is present only very rarely elsewhere and then weakly; supramarginal keel and groove present or absent, the keel tuberculate or smooth; predistal triangular tooth characteristically small or absent, except in eastern zones of mingling with *perplexa*, where it sometimes attains proportions typical of that eastern subspecies. Dactyl with upper margin convex throughout, notably flattened, central portion not wider than adjacent part of gape.

Measurements (in mm)

	Length	Breadth	Propodus	Dactyl
Largest males				
(Java: Leiden 2012 part)	13.5		43.0	
(Mozambique)	11.5	19.0	39.0	28.5
Moderate male				
(Mozambique)	9.0	15.0	26.0	19.0
Smallest displaying male				
(Ethiopia)	4.0	7.0	9.0	5.5
Largest female				
(ovigerous)(Singapore)	10.0	14.0	–	–
Smallest ovigerous female				
(Mozambique)	7.0	11.0	–	–
Largest female				
(Mozambique)	9.0	15.0	–	–

SOCIAL BEHAVIOR

Waving Display

Diminishing waves absent except in Massawa, Ethiopia, at medium low intensities. Curtsy, confined to courtship, starts earlier than in *l. perplexa* or *l. lactea*, immediately after pause, as cheliped begins outward sweep of next wave. Otherwise about as in *perplexa*, including downstrokes. Body held high on extended ambulatories during an entire series of circular waves (without curtsies). Vertical waves observed only in eastern populations and then only in the young, the motion being prevalent in Sarawak.

Uca (Celuca) lactea mjobergi
Rathbun, 1924

(Northwest Australia and northwest New Guinea)

MORPHOLOGY

With the characteristics of the species.

Gonopod with slight torsion; anterior flange longer than posterior but scarcely wider, the pore being located in a broad and shallow notch, thumb ending distinctly below flange base. Gonopore with marginal lip only slightly tilted, scarcely projecting up into pore's cavity, much less than in *l. annulipes*; rim of lip corneous brown. Major cheliped: Oblique tuberculate ridge inside palm unusually low for the species. Outer manus, pollex, and dactyl markedly smooth with little variation, individuals with relatively rough claws, such as occur in all other subspecies, not having been found in *mjobergi*. Outer pollex without a row of small tubercles along lower side; marginal keel and supramarginal groove absent; predistal triangular tooth very small to absent. Dactyl with upper margin convex throughout; breadth greater than that of adjacent part of gape except in large specimens, especially leptochelous individuals, where the dactyl usually is clearly narrower than gape.

Measurements (in mm)

	Length	Breadth	Propodus	Dactyl
Largest male (Broome)	10.0	15.5	25.0	15.5
Moderate male (Broome)	7.0	11.5	15.5	9.5
Largest female (Darwin)	7.0	11.0	–	–

SOCIAL BEHAVIOR

Waving Display

Diminishing waves pronounced during moderate to high intensity display, each series starting with a regular, circular wave and followed by two to four circles of decreasing amplitude. Body raised and lowered only during the first wave. Curtsies, confined to courtship, start at beginning of a series, the cheliped usually being held extended, with the waves interrupted throughout a series of curtsies, but with the cheliped not making downstrokes. Vertical waves not seen (no observations made during onset of a display period, especially in the morning, when they are most likely to occur). Synchronous waving observed briefly, when neighboring males were courting a single, wandering female.

Uca (Celuca) lactea lactea (de Haan, 1835)

(Hong Kong to Japan)

MORPHOLOGY

With the characteristics of the species.

Gonopod with slight torsion; posterior flange longer and wider than anterior, the pore located in a notch both wide and deep, each flange usually extending well beyond it, although, as usual in the species, there is variation; thumb longer and larger than in any other subspecies, reaching well beyond flange base. Gonopore with a marginal lip moderately tilted up into pore's cavity; rim of lip corneous brown. Major cheliped: Oblique tuberculate ridge on palm sometimes continued slightly along lower margin of carpal cavity. Outer pollex without a straight row of small tubercles along lower side; supramarginal keel and groove present or absent, the keel minutely tuberculate or not, the tubercles when present sometimes non-linear, continued ventrally around margin; groove frequently represented by a broad, smooth depression, often present even in individuals lacking supramarginal keel; predistal, triangular tooth usual-

ly present but always poorly marked, being low, blunt and with a long base. Dactyl dorsally convex throughout, not notably flattened, not wider in central portion than adjacent part of gape except in the small specimens having claws otherwise of adult form.

Measurements (in mm)

	Length	Breadth	Propodus	Dactyl
Largest male (Kyushu, Japan)	13.0	21.0	33.0	26.0
Largest female (Kowloon, Hong Kong)	8.0	14.0	–	–
Largest ovigerous female (Tamsui, Taiwan)	8.0	12.0	–	–
Smallest ovigerous female (Kowloon, Hong Kong)	7.5	11.0	–	–

SOCIAL BEHAVIOR

Waving Display

Diminishing waves absent. Curtsy starts with cheliped extended, usually laterally, most waves then being omitted while a number of curtsies are performed in succession, with or without downstrokes of cheliped. Curtsy occurs during both threat and courtship.

Uca (Celuca) lactea perplexa (Milne-Edwards, 1852)

(Tropical West Pacific; occasionally to eastern India)

MORPHOLOGY

With the characteristics of the species.

Gonopod with strong torsion, posterior flange slightly longer than anterior and clearly wider, the pore being set in a rather broad and very shallow notch; both flanges relatively short and broad compared with those of *l. annulipes*; thumb not as long or thick as in *l. lactea*, but similar and reaching beyond flange base. Gonopore with marginal lip only slightly tilted, scarcely projecting up into pore's cavity, much less than in *l. annulipes*; rim of lip corneous brown; axis of lip directed obliquely forward, not toward midline as in the other subspecies. Major cheliped: Outer pollex very rarely with an indistinct row of minute tubercles, discernible among a usually general distribution of tubercles, making the outer pollex distinctly rough in most individuals of this subspecies; supramarginal keel and groove absent; predistal tooth moderate to large except in areas of mingling with *annulipes*, where it is sometimes small. Dactyl arched only in about distal fifth, starting at beginning of the strong, downward curve; more proximally its upper margin is virtually straight and almost parallel to edge of gape; the dactyl's breadth beyond proximal end is therefore greater in most individuals than usual in other subspecies, and in many individuals the dactyl is widest slightly distal to the middle; gape correspondingly narrowed; accordingly, the dactyl's width is greater than that of adjacent part of gape; outer surface of dactyl always smooth and somewhat flatter than usual in other subspecies.

Measurements (in mm)

	Length	Breadth	Propodus	Dactyl
Largest male (Ryukyu Is.)	11.5	19.5	37.5	24.0
Largest male (Joló, Philippines)	10.0	17.0	34.0	23.0
Moderate male (Joló, Philippines)	7.0	12.0	21.0	14.0
Largest female (Singapore)	10.0	16.0	–	–
Largest ovigerous female (Singapore)	10.0	14.0	–	–
Smallest ovigerous female (Zamboanga, Philippines)	6.0	10.0	–	–

SOCIAL BEHAVIOR

Waving Display

Diminishing waves absent. Curtsy about as in *l. lactea*, starting with cheliped already extended laterally, waves being often omitted during a series of curtsies, and the cheliped often making downstrokes instead. In this subspecies curtsies with the cheliped extended are confined to courtship. In addition, similar motions of the cheliped, toward or touching the ground, are made with the appendage flexed, apparently only during high intensity threat. At these times a regular curtsy sometimes accompanies the flexed downstroke. During circular waving of moderate intensity, the claw moves forward and slightly outward at lowest point of wave before coming to rest. Low intensity verticals present at least in some populations (New Caledonia and New Guinea). Alternating, pre-combat waves prevalent between confronting pairs of males. Synchronous waving of a group of males observed several times; in each case a single wandering female was the focus of the waves, which shifted to the curtsy sequence as she approached any of the displaying crabs.

REFERENCES AND SYNONYMY

Uca (Celuca) lactea (de Haan, 1835)

The references to this species are not distributed among the subspecies because of apparent hybridization of the forms in parts of the range (see p. 294).

Ocypode (Gelasimus) lacteus

TYPE DESCRIPTION. de Haan, 1835: 54. Japan. (Leiden !)

Gelasimus forceps (part; see p. 323)

Milne-Edwards, 1837: 52 (part).
Milne-Edwards, 1852: 148 (part); Pl. 3, Fig. 11a but not Fig. 11. (Paris !)

Gelasimus marionis (not of Desmarest, 1825)

Milne-Edwards, 1837: 53. Locality not given.

Gelasimus annulipes

TYPE DESCRIPTION. Milne-Edwards, 1837: 55; Pl. 18, Figs. 10-13. "La mer des Indes." (= Type description of subspecies *U. lactea annulipes*, as used in present contribution.) (Paris !)

White, 1847: 36. India: Pondichéry.
Dana, 1852: 317. Singapore. Taxonomy.
Milne-Edwards, 1852: 149, Pl. 4, Figs. 15, 15a, 15b. "Mers d'Asie." Taxonomy.
Heller, 1865: 38. Ceylon; Nicobars; Madras.
Hilgendorf, 1869: 85. Zanzibar.

Hoffmann, 1874: 18. Madagascar: Nossy Faly; Nossy-Bé. Taxonomy.
Kossmann, 1877: 55. Red Sea.
Hilgendorf: 1879: 803. Mozambique: Inhambane. Taxonomy.
Miers, 1879.2: 488. Rodriguez I. Taxonomy.
Kingsley, 1880.1: 148; Pl. 10, Fig. 22. Australia; Singapore; Zanzibar.
Richters, 1880: 155. Mauritius: Fouquets. [Not seen.]
de Man, 1880: 69. East Indies; Madagascar. Taxonomy.
Miers, 1880: 310. Moluccas: Batjan (BM !). Malaysian region. Taxonomy.
Lenz & Richters, 1881: 423. Madagascar. Taxonomy.
Miers, 1886: 244. Philippines. Fiji Is.: Matuku. Taxonomy.
de Man, 1887.1: 118. Mergui Arch. Taxonomy.
de Man, 1887.2: 353. East Indies: Amboina (Leiden !); Insel Noordwachter. Taxonomy.
Pfeffer, 1889: 29. Zanzibar. Tanganyika: Bagamoyo.
de Man, 1891: 39. Mergui Arch. Taxonomy.
de Man, 1892: 307. Celebes. Taxonomy.
Alcock, 1892.2: 415. India. Habits.
Henderson, 1893: 388. India: Madras and neighborhood. Taxonomy; habitat.
Alcock & Anderson, 1894: 202. India. Record.

Ortmann, 1894.1: 59; 67. Zanzibar. Tanganyika: Lindi. Color; habits; habitat.

Ortmann, 1894.2: 758. Samoa: Upolu. Taxonomy.

Zehnter, 1894: 178. Amboina. Taxonomy.

de Man, 1895: 577. East Indies: Malakka; Atjeh; Pontianak; West Borneo. Malaya: Penang. Taxonomy. (All specimens except from Atjeh: Leiden !)

Alcock, 1900: 353. Coasts of India and adjacent shores from Karachi to Mergui. Taxonomy.

de Man, 1902: 483. Ternate. Taxonomy.

Lanchester, 1902: 549. Malaya: Trenggano. "16 males right-handed, 16 left-handed."

Lenz, 1905: 365. Zanzibar: Kokotoni. Taxonomy.

Lenz, 1910: 558. East Africa: Witu I. (Patta). Taxonomy.

Sewell, 1913: 339, 344. Southeast Burma, including Tavoy I. Color; habits.

Bouvier, 1915: 301; fig. Mauritius: Grand Port. Taxonomy.

Kemp, 1915: 221. India: Chilka Lake, Orissa. Taxonomy.

Kemp, 1918: 227. Siam. Color.

Raj, 1927: 148. Ceylon: Gulf of Manaar (Krusadai Is.). Taxonomy; color.

Gordon, 1934: 10. East Indies: Bali; S. Manoembaai. Taxonomy.

Tweedie, 1937: 141; Text Fig. 1a.

Sakai, 1940: 32. Japan. Geographic distribution.

Chapgar, 1957: 508; Pl. 13. Western India: Bombay; Karwar; Okha; Kolak; Umarsadi.

Gelasimus lacteus

Krauss, 1843: 14, 39. South Africa. Brief color, morphology, habitat, habits.

Milne-Edwards, 1852: 150; not Pl. 4, Fig. 16: see p. 323. China (Macao). Taxonomy.

Stimpson, 1858: 100. China; Macao. Habitat.

Miers, 1879: 36. Korean and Japanese Seas. Taxonomy.

Kingsley, 1880.1: 149; Pl. 10, Fig. 28. Japan; Pondichéry. Taxonomy.

Cano, 1889: 92; 234. China: Amoy. Taxonomy.

Ortmann, 1894.2: 759. New Guinea: Kaiser Wilhelmsland. South Seas. Samoa: Upola.

Alcock, 1900: 355. Andamans; Karachi. Taxonomy.

Stimpson, 1907: 108. Résumé in English of Stimpson, 1858. "Cum-sing-moon and Macao, China." Brief morphology; color in life chalk white; habitat.

Sakai, 1940: 32. Japan. Geographic distribution.

Shen, 1940: 231. Hong Kong: Kowloon and neighborhood.

Lin, 1949: 26. Taiwan.

Ono, 1959. Japan: Fukuoka, in estuary of Tuatera River. Ecology.

Gelasimus porcellanus

White, 1847: 36. No description.

Adams & White, 1848: 50. Borneo. Type description. (BM !)

White, 1848: 86. No new material.

Milne-Edwards, 1852: 151. No new material. Taxonomy.

Kingsley, 1880.1: 155. No new material. Type description quoted.

Gelasimus annulipes var. albimana

TYPE DESCRIPTION. Milne-Edwards, 1852: 150; Pl. 4, Figs. 18, 18a. Java. (!) (= Type description of subspecies *U. lactea perplexa*, as used in present contribution.) Author lists his reference (1837, above) to *G. marionis* under heading of *G. perplexus*.

A. Milne-Edwards, 1873: 274. New Caledonia.

Gelasimus perplexus

Kossmann, 1877: 53. Red Sea. Type description.

Kossmann, 1878: 258. Reference to type description.

Uca annulipes

Ortmann, 1897: 354. No new record.

Nobili, 1899.2: 274. Australo-Malaysia; Andai.

Nobili, 1899.3: 518. Sumatra: Siboga.

Doflein, 1899: 193. Indian Ocean.

Lanchester, 1900.1: 754. Singapore. Heterogony; color; habits.

Lanchester, 1900.2: 258. Malaysia. Taxonomy.

Schenkel, 1902: 580. Celebes: Kema. Taxonomy.

Nobili, 1903.2: 20. Western India. Taxonomy.

Nobili, 1906.2: 150. Persian Gulf. [No further information on locality given.] Taxonomy; color. [The collector, M. Tramiet, was not a member of the expedition.]

Nobili, 1906.3: 312. Red Sea.

Borradaile, 1907: 66. Seychelles. Habitat.

Borradaile, 1910: 408. Aldabra I.

Rathbun, 1910: 322. Siam. Habitat.

Pearse, 1912.2: 113; text fig. Philippines. Habits.

Kemp, 1915: 221. India: Chilka Lake, Orissa. Taxonomy.

Laurie, 1915: 416. Red Sea.

Roux, 1917: 614. New Guinea: Siari.

Stebbing, 1917: 16. Natal. Taxonomy; color.

Symons, 1920: 309; text figs. Ceylon. Habits.

Balss, 1924: 15. Red Sea.

Boyce, 1924: 250. South Africa: Durban Bay. Color; courtship; habits.

Maccagno, 1928: 35; Text Fig. 20 (claw). Gulf of Aden; Italian Somaliland; Persian Gulf; western and southeast India; Singapore; East Indies (Samarinda and Amboina). Includes some material previously published by Nobili. Taxonomy.

Sakai, 1936: 170. Palao Is.

Miyake, 1936: 511. Ryukyu Is.: Miyara mangrove swamp.

Estampador, 1937: 542. Philippines. Local distribution.

Panikkar & Aiyar, 1937: 295, 301. India: Madras area. Local distribution.

Suvatti, 1938: 74. Siam: Lem Ngob; Koh Chang.

Miyake, 1938: 109 (probably part only). Micronesia: Ngardok; Babelthaoh; Palau Is.; Caroline Is.; Saipan; Mariana Is.

Miyake, 1939: 190, 222, 241; Pl. 16, Fig. 2. Micronesia: Palau Is.

Sakai, 1939: 616. Japan: Misaki (*fide* Parisi); Loo Choo (= Ryukyu = Nansei) Is.: Yaenama.

Sakai, 1940: 28. General distribution.

Chace, 1942: 202. Tanganyika Territory: Lindi.

Vatova, 1943: 24. Italian Somaliland.

Stephensen, 1946: 189, 210. Iranian Gulf. No new material. Lists Nobili, 1906.3 as giving the only record of occurrence of *Uca* in Iranian [Persian] Gulf.

Buitendijk, 1947: 280. Malaya: Port Dickson.

Barnard, 1950: 97; text fig. South Africa. Taxonomy.

Tweedie, 1950.1: 356. Borneo: Labuan.

Fourmanoir, 1953: 89. Madagascar, near Canal de Mozambique. Color; ecology.

Altevogt, 1955.1: 702; text figs. India: near Bombay. Morphology; habits.

Altevogt, 1955.2: 501; text figs. India: near Bombay. Morphology; behavior.

Day & Morgans, 1956: 277, 305. South Africa: Durban Bay. Ecology.

Gordon, 1958: 238. Mozambique: Inhaca I. Synchronous claw-waving.

Macnae & Kalk, 1958: 39, 67, 125. Mozambique: Inhaca I. Color; general behavior; ecology.

Estampador, 1959: 100. Philippines. Local distribution.

Sankarankutty, 1961: 113. Bay of Bengal: Andaman/Nicobar Is.

Forest & Guinot, 1961: 141. New Caledonia. Taxonomy.

Macnae, 1963: 23. Mozambique (Inhaca I.) to Cape Province, South Africa (mouth of Umnagazana R.). Local distribution; ecology.

Feest, 1969: 159; text figs. Southeast India. Breeding biology; post-embryological development.

Gelasimus annulipes var. *lacteus*

Ortmann, 1894.2: 759. South Seas; Samoa; New Guinea (Kaiser Wilhelmsland). Taxonomy.

Uca lactea

Ortmann, 1897: 355. No new record. Includes *G. a.* var. *orientalis*. Taxonomy.

Schenkel, 1902: 580. Celebes: Kema. Taxonomy.

Stebbing, 1910: 327. Annotated references.

Pesta, 1913: 57. Samoa: Upolu. Taxonomy. [Spelling of species name: *lactaea*.]

Stebbing, 1917: 16; figs. South Africa: Durban Bay. Taxonomy.

Parisi, 1918: 92; fig. Naviagori Is.; Formosa; Chichijimi; Bonin Is. Taxonomy.

Tesch, 1918: 39. East Indies.

Gee, 1925: 165. China: Cum-Sing-Moon; Amoy. Macao. Hong Kong.

Maccagno, 1928: 29; Text Fig. 15 (claw). Borneo: Labuan.

Kellogg, 1928: 356. China: Amoy.

Gordon, 1931: 528. Hong Kong.

Boone, 1934: 199; Pl. 103. New Caledonia.

Sakai, 1934: 320. Japan: Nagasaki.

Kamita, 1935: 61, 69 (= English résumé). Yellow Sea: western Korea.

Takahasi, 1935: 78ff. Formosa. Habits.

Sakai, 1936: 171. Palao Is.

Miyake, 1936: 511. Ryukyu Is.: Miyara mangrove swamp.

Miyake, 1938: Palau Is.: Babelthaob; Ngardak.

Balss, 1938. Fiji Is.

Sakai, 1939: 618. Japan: Ise Bay; Tosa Bay; coast of Miyazaki-ken.

Miyake, 1939: 222: 242. Micronesia: Palau Is.

Buitendijk, 1947: 280. Malaya: Port Dickson.

Barnard, 1950: 96. South Africa. Taxonomy.

Ono, 1962. Japan: Fukuoka, in estuary of Tatara River. Ecology, with special reference to substrate; interspecific behavior with other arthropods.

Ono, 1965. Japan: Fukuoka, in estuary of Tatara-Umi River. Ecology; feeding in relation to morphology of mouthparts.

Macnae, 1966: 77, 79, 80. Australia: Queensland (Thursday I. to Port Curtis). Color; ecology.

Uca annulipes var. *orientalis*

Nobili, 1901.1: 13. Borneo: Sarawak. Type description.

Nobili, 1903.1: 21. Borneo: Samarinda.

Maccagno, 1928: 36; Text Fig. 21 (claw). Borneo: Buntal. Taxonomy.

Uca mjobergi

TYPE DESCRIPTION. Rathbun, 1924.2: 9. Western Australia: Broome. (= Type description of subspecies *U. lactea mjobergi*, as used in present contribution.)

Uca consobrinus (= MS name of de Man. Material in Amsterdam; !; see p. 298).

Verwey, 1930: 172ff. Java. Ecology.

58. *UCA (CELUCA) [LEPTODACTYLA] LEPTODACTYLA* RATHBUN, 1898

(Tropical western Atlantic)

PLATE 41 *A-D*.   MAP 17.
FIGURES 37 *M*; 56 *F*; 60 *N, O*; 69 *K, L*; 101.   TABLES 9, 10, 12, 19, 20.

INTRODUCTION

Uca leptodactyla, uruguayensis, and *major* are the only Atlantic species that attain the dazzling white shown by a number of forms in other parts of the tropics. Also, *leptodactyla* and two other *Uca*, *minax* and *cumulanta*, are the only builders of hoods in the Atlantic. As a final distinction, *leptodactyla* shares with the north temperate *pugilator* a preference for sandier, saltier habitats than those frequented by other *Uca* on the east coast of America.

MORPHOLOGY

Diagnosis

Front moderately wide; carapace strongly arched. Anterior side of 1st ambulatory in male without a row of tubercles or a ridge on any segment. Major cheliped with triangular section of lower, proximal palm, practically smooth, the granules very minute; no subdorsal ridge and associated groove on upper, outer manus; pollex tip with a minute, inner keel. Small chela in both sexes with gape moderately wide with serrations obsolescent to absent. Gonopod with large flanges, both of which project clearly and unevenly beyond pore. Female gonopore with outer margin strongly curved, the horns of the crescent directed outward.

Description

With the characteristics of the subgenus *Celuca* (p. 211).

MALE

(Cf. *U. deichmanni*, p. 311)

Carapace. Differs from *deichmanni* as follows. Front slightly wider; antero-lateral margin less concave; 1 pair of postero-lateral striae present, though often weak. Carapace profile slightly less arched, although still almost semi-cylindrical. Suborbital crenellations very small to almost absent internally, but near outer orbital margin approaching strength and form found in *deichmanni*; they continue around outer orbital margin, being there even more numerous than the corresponding series in *deichmanni*; the most posterior is on the far side of the channel directly under the base of the antero-lateral angle, and hence approaching in position *deichmanni*'s single tubercle on underside of antero-lateral angle; number and exact arrangement of crenellations and separated tubercles variable.

Major Cheliped. Merus: Unusually long and slender. Antero-dorsal margin moderately arched, weak, not ridged except for a blunt, low thickening in large specimens; unarmed proximally or with a few granules or small tubercles variously distributed; distally with the edge more distinct, armed principally with small rugosities.

Manus: Bending of outer, upper surface distinct and moderately flattened, especially proximally. Outer tubercles minute except on bent-over region, where they are larger and in an irregularly reticulate pattern, or partly in transverse rows. Dorsal, outer groove indistinct. Dorsal margin distal to carpal cavity with outer and inner edges indistinct; intervening surface flattened, smooth except for a few tubercles, some of them in obliquely transverse rows directed antero-internally. Cuff at base of dactyl sometimes with a few granules in lower part in addition to tubercles in the usual row starting here and continuing along pollex below gape. Ventral margin without a carina but regularly beaded as far as pollex base, the beads being small and similar throughout except for slight enlargement distally. Ventral, outer groove absent.

Palm with lower triangle covered with extremely minute tubercles, slightly larger distally. Oblique ridge moderate, highest and the sides steepest in middle of ridge rather than at apex; tubercles largest on highest part of ridge; beyond apex they continue slightly upward around carpal cavity but die out in small tubercles continuous with those of center palm. Center palm convex, with extremely minute tubercles or none; depression at pollex base moderate, opposite upper part of pollex rather than lower. Carpal cavity with ventral margin broad and blunt; distal margin above upward extension of oblique ridge very low, the cavity merging gradually with center palm; upper part of cavity completely set off from predactyl area by beaded inner edge of dorsal margin which turns downward, sometimes with interruptions or irregularities in this portion, and by a proximal ridge bounding predactyl area. Latter area slightly depressed, smooth, bounded dorsally also by a ridge; ventrally its margin slopes more gradually to center palm convexity. Distal ridge at dactyl base weak to absent.

Pollex and *dactyl*: Dactyl curved throughout most of length, slightly deeper than pollex except distally. Gape in middle section wider than that of adjacent pollex. Dactyl with proximal, outer, subdorsal groove short to vestigial; entire proximal dorsal area with minute tubercles that, much diminished in size, continue along dorsal margin beyond middle of segment and sometimes almost to tip; submarginal tubercles, always very minute, sometimes present along gape in both pollex and dactyl. Gape with outer and inner rows of tubercles sometimes weak, the inner row sometimes wholly absent on pollex in middle section and, distally, represented by a keel ending in a sharp angle close to tip; inner row on dactyl variably weak; middle row in both pollex and dactyl highly variable, never strongly marked, portions often irregular or minutely multituberculate. Pollex tip weakly bifid, sometimes almost trifid, where a slightly enlarged median tubercle approaches tip. Gape pile sparse.

Minor Cheliped. Gape as wide or nearly as wide as pollex throughout most of its length. Serrations weak and not in contact, or absent.

Ambulatories. 2nd and 3rd meri moderately slender, their dorsal margins scarcely convex to almost straight. No anterior modification of 1st ambulatory.

Gonopod. Differs from the similar gonopod of *deichmanni* (p. 312) in having the pore terminal, not subterminal; posterior flange wider than anterior; edges of flanges variably truncate or produced, not rounded in an unbroken edge; inner process extending over posterior flange, covering canal partly or not at all. Thumb larger, almost reaching base of flange.

Abdomen. 3rd to 6th abdominal segments incompletely fused.

FEMALE

Suborbital crenellations better developed than in male on inner half, but about equal to male's in outer portion. Gonopore strongly crescent-shaped, the horns directed outward.

Measurements (in mm)

	Length	Breadth	Propodus	Dactyl
Largest male (São Salvador)	6.5	11.0	23.0	18.0
Large male (Turiamo)	6.0	10.5	19.5	14.0
Moderate male (São Salvador)	5.0	8.5	14.0	11.5
Largest female (Turiamo)	6.0	10.0	–	–

Morphological Comparison and Comment

This species is the smallest *Uca* in the Atlantic. It is there distinguished from *cumulanta* and *uruguayensis*, the only other *Uca* of similar size, by the characters given in the diagnosis. In particular, it differs from *cumulanta*, its occasional northern sympatric, as follows. Middle abdominal segments fused; 1st major ambulatory without tubercles anteriorly on carpus; gonopod tip not tubular but with large, projecting flanges; gonopore crescentic. From *uruguayensis*, with which it coincides near Rio de Janeiro, it is distinguished by having no antero-ventral ridge on manus of 1st major ambulatory; by the leptomorphic meri of major cheliped and ambulatories, which are relatively long, slender, and almost straight; and by having on the gonopod both flanges projecting beyond level of pore; for remarks on female identification see p. 230, under *uruguayensis*.

The closest relation of *leptodactyla* appears to be *limicola* in the Pacific. They share all major similarities, including both the moderate convexity of the carapace and the general form and strength of their armature. A minute subdistal crest on pollex is confined to *leptodactyla*.

In the morphological description of *leptodactyla* another close relation, *deichmanni*, was selected for special comparison. This course was followed because, in contrast to *limicola*, *deichmanni* is known from material that is both relatively plentiful and not delicate. See also p. 219.

Leptochelous examples with the fingers strikingly attenuated appear to be more prevalent in Brazilian populations than in the north. In such specimens the manus is often smoother. Whether or not minor variations in the gonopod flanges are correlated with geographic distribution is not known. Adequate material may eventually make a subdivision of *leptodactyla* desirable.

Color

Displaying males: Display whitening usually fully developed on carapace, often extending over all appendages. The white phase is sometimes preceded by a yellow stage that ranges from orange to pale lemon, the carapace usually whitening in advance of the cheliped and legs. When not white, the major cheliped shows varying degrees of yellow, orange, or red; any of these hues may envelop the entire appendage except for the white finger tips, or it may be confined to one part of a single segment. Ambulatories and minor cheliped usually ranging, like the major cheliped, through a stage of red, orange, or yellow before attaining white. An interesting point is that members of populations even living in partial shade often whiten maximally.

Females also whiten fully.

Social Behavior

Waving Display

Wave always lateral, straight to regularly circular except for a slight irregularity during high intensity. At these times the fingers are sometimes brought down slightly in front of the rest position and jerked abruptly into place, somewhat as in *batuenta* and its allies, but without vibrations. An additional motion, apparently confined to courtship, consists of a single small lunge of the cheliped toward the female at the end of each wave. One or both of these motions may prove to involve stridulation, but observations, none of which were made after the start of acoustic studies, were inadequate to determine details and the motions do not show in the films. Minor cheliped may or may not make a corresponding motion during each wave; at times it certainly hangs motionless. Carapace is raised at least anteriorly with every wave, but no raising of the ambulatories occurs except in taking a few steps from side to side. At moderate intensities the waves range in length from about 1 to 2 seconds, each with a slight pause between but none at the peak position; at high intensity the rate is close to 2 waves per second. (Component nos. 4, 5, 9, 10, plus the special motions mentioned; timing elements in Table 19, p. 656.)

Hood Construction

As usual in species with structure-making patterns, the hoods of *leptodactyla* are built sporadically by only part of the displaying males in few populations. In the present species I found them only in Recife and Fortaleza, while none were seen in Rio de Janeiro, Salvador, Venezuela, or Tobago.

Acoustic Behavior

Von Hagen (1967.2) reports successful recording of acoustic signals from *leptodactyla* in Trinidad.

Range

West Indies, from Bahamas and Puerto Rico south; Mexico (see p. 307); South America from Turiamo, State of Aragua, Venezuela to Florianapolis, State of Santa Catarina, Brazil. The species has not been reported from Florida in recent years, and has never been recorded between Trinidad and Recife.

Biotopes

This species usually occurs only on shores washed by tides of fully marine salinity. The preferred substrates are relatively sandy for a *Uca*, with little or no admixture of mud, although sand then overlays a foundation of coral and mud on beaches protected chiefly by offshore reefs. In such areas *leptodactyla* often feeds far from its burrows close to low-tide levels, moving in droves along the water's edge where food is more plentiful, as does *pugilator* in equivalent temperate habitats. Sometimes it occurs at higher levels, among marine grasses in sandy soil that is covered only by spring tides; here, during dry spells, the fiddlers aestivate. Some populations occupy more muddy or clayey habitats in the partial shade of mangroves, but almost always in pioneering stands, on the edges of large bays or of islands exposed to waters of the open sea. (Biotope nos. 4, 5, 6, 9.)

Sympatric Associates

None in the sandier localities; among mangroves sometimes seaward of *rapax* or, unusually, *burgersi*.

Material Resulting from Field Work

(The complete list of specimens examined is given in Appendix A, p. 613.)

Observations and Collections. Puerto Rico; Turiamo, Venezuela; Tobago; Trinidad; Recife and São Salvador, Brazil.

Films. Recife.

Type Material and Nomenclature

Uca leptodactyla Rathbun, 1898

TYPES. In Smithsonian Institution, National Museum of Natural History, Washington. One male and one female, cat. no. 22315. Type-locality: near Fort Montague, Nassau, New Providence, Bahamas. Length in mm: male 5, female 7. (!)

I agree with Chace & Hobbs (1969: 212), who write as follows concerning the spelling of *leptodactyla*: "It seems obvious that the selection of the specific name of this species was intended as adoption of Guérin's manuscript name, a noun in apposition to the generic name [*Gelasimus*], which therefore should have been spelled '*leptodactylus.*' There is no absolute proof from the original description, however, that this was the intention, and we have therefore followed the advice of L. B. Holthuis to use the original spelling of the name. This decision was influenced further by the fact that Rathbun used this spelling in 1918, even though another species in the same genus was spelled '*stenodactylus*' in that publication."

In the Academy of Natural Sciences at Philadelphia three males are labelled as the types of Guérin's species, described, in manuscript only, as *Gelasimus leptodactylus*; their present catalogue number is 9-2965. (!) They are doubtless the specimens mentioned by Rathbun (1918.1: 420, 421, and references). The only locality given is Mexico, which constitutes the only record for either Mexico or Central America.

REFERENCES AND SYNONYMY

Uca (*Celuca*) *leptodactyla* Rathbun, 1898

Gelasimus leptodactylus

Guérin, in MS. [Not seen.]

Uca leptodactyla

TYPE DESCRIPTION. Rathbun, 1898.1: 227. [In Rankin, 1898.] (USNM !.) Bahamas.

Rathbun, 1900.3: 136. Brazil: Rio Parahyba do Norte.

Rathbun, 1902.1: 7. Puerto Rico. Taxonomy.

Rathbun, 1918.1: 420; Pl. 156. Bahamas; Cuba; Jamaica; Puerto Rico; Mexico; Brazil south to Santos. ? U.S.A.: western Florida. Taxonomy. (USNM ! most; Philadelphia !: Guerin's specimens from Mexico.)

Luederwaldt, 1919.1: 384, 400. Brazil: Santos.

Luederwaldt, 1919.2: 435. Brazil: São Sebastião; Santos.

Rathbun, 1924.3: 19. Curaçao. Local distribution.

Maccagno, 1928: 41 (part). (Not Text Fig. 25.) Brazil: Bahia. Nobili's specimens, referred by him (1899.1) to *U. gibbosa*, reidentified. Also specimens from São Sebastião (Torino !).

Luederwaldt, 1929: 54. Brazil: São Sebastião I.

Matthews, 1930; illus. Brazil. Color; hood construction; general habits.

Oliveira, 1939.1: 126; Pl. 5, Figs. 25-28; Pl. 6, Fig. 29; Pl. 8, Fig. 47; Pl. 13, Figs. 61, 62. Brazil: Rio de Janeiro. Taxonomy; color (p. 140).

Oliveira, 1939.2: 496. Brazil: Rio de Janeiro. Ecology.

Oliveira, 1939.3: 523. Brazil: Rio de Janeiro. Ecology: analysis of substrate.

Crane, 1957. Trinidad; Brazil. Preliminary classification of waving display.

Gerlach, 1958.2. Brazil: Estado São Paulo at Cananéia. Social behavior including that of females during courtship.

von Hagen, 1967.2. Trinidad. Tape recordings secured. (Preliminary statement.)

Chace & Hobbs, 1969: 212; Text Figs. 71g, h. Taxonomy. Comments on spelling of *leptodactyla* (see also present study, p. 306).

von Hagen, 1970.1: 227. Caribbean distribution; taxonomy.

Uca gibbosa (not *Gelasimus gibbosus* Smith)

Nobili, 1899.1: 5. Brazil: Bahia. Taxonomy. (See, under *Uca leptodactyla* above, Maccagno, 1928.)

59. *UCA (CELUCA) [LEPTODACTYLA] LIMICOLA* CRANE, 1941

(Tropical eastern Pacific)

PLATE 41 *E-H*.   MAP 17.
FIGURES 70 *F*; 93; 101.   TABLES 9, 10, 12, 19, 20.

INTRODUCTION

Uca limicola is known only from a few specimens, a brief observation in Panama, and two filmed sequences. On this evidence it lacks outstanding peculiarities; morphologically its close relationship to *leptodactyla* in the Atlantic seems clear.

MORPHOLOGY

Diagnosis

Front moderately wide; carapace strongly arched. Major cheliped with gape moderate, both fingers slender and tapering, the few enlarged teeth scarcely larger than the rest; no parallel ridges on triangle of lower palm. In both sexes outer suborbital crenellations gradually, not abruptly, enlarged, with spaces between them; largest beside channel. Gape of small chelae moderate with serrations very small to absent. Gonopod with a long thumb; no protruberance on pore margin but inner process projects slightly beyond it.

Description

With the characteristics of the subgenus *Celuca* (p. 211).

MALE

Carapace. Frontal breadth contained slightly more than 4 times in that of carapace. Antero-lateral margins slightly concave, slightly converging, angling with variable sharpness into dorso-lateral margins. Breadth of eyebrow about half diameter of adjacent part of depressed eyestalk. Suborbital crenellations moderately well developed throughout, with little size increase near antero-external angle; they continue, with wider separations, around outer orbital margin to channel. Lower side of antero-lateral angle moderately sharp.

Major Cheliped. Merus: Antero-dorsal margin almost straight proximally, definite but without a ridge, armed with minute tubercles; distally strongly arched, thick, armed with small tubercles and rugosities.

Manus: Bending of outer, upper surface definite, beginning lower than usual, making the resultant dorsal area exceptionally broad proximally; flattening slight. Outer tubercles small, slightly larger near dorsal margin, where they are also more widely spaced. Dorsal, outer groove indicated only proximally where the tuberculated dorsal margin coincides with dorsal margin of carpal cavity. Dorsal margin distal to cavity poorly defined, without definite outer or inner edges, but marked only by a narrow band of low, sparse tubercles; these are slightly larger than those of the adjacent outer manus, and a few are set in roughly transverse rows. Ventral margin with an indistinct row of multiple tubercles only slightly larger than those on adjacent surfaces, dying out at pollex base. Ventral, outer groove absent.

Palm with lower triangle covered with extremely minute tubercles, slightly larger in ventral third, especially distally; middle portion of triangle with a slight tendency to arrangement in narrow, vertical, or oblique bands, with smooth spaces between. Oblique ridge moderate, highest at apex. Tubercles in a regular row, graduated upward, largest on apex; beyond apex they continue, slightly reduced in size and regularity, up around carpal cavity, meeting downturning beaded inner edge of dorsal margin at level of predactyl area. Carpal cavity with ventral margin low, thick, blunt; distal side of cavity sloping steeply to upward continuation of tubercles beyond apex; upper part of cavity completely set off from predactyl area by beaded inner edge of dorsal margin which turns downward, and by the tubercles continued from apex. Predactyl area spacious, without a depression, covered with longitudinal bands of sharp tubercles that are virtually continuous with those of dorsal margin.

Pollex and *dactyl*: Pollex with lower margin slightly convex. Gape wider than adjacent part of pollex. Dactyl with outer, proximal, subdorsal groove very short; surrounding area covered with small, low tubercles that die out toward middle of length. Pollex and dactyl with extremely minute tubercles covering outer surfaces entirely and inner ones marginally; small pits unusually numerous. Gape rows all developed. Pollex tip bifid. Enlargement of several or more tubercles on pollex and dactyl minimal and their location on gape variable. Gape pile almost absent.

Minor Cheliped. Gape slightly narrower than width of pollex, but width maintained throughout to the normally expanded tips of pollex and dactyl. Serrations very small to vestigial or absent.

Ambulatories. 2nd and 3rd meri moderately slender, their dorsal margins scarcely convex to almost straight. No anterior modifications of 1st ambulatory.

Gonopod. Flanges vestigial; pore large, with its internal (anterior) lip lower than the external but with no trace of a projection; external (posterior) lip broadly angled. Inner process broadly triangular, flat, extending slightly beyond adjacent lip of pore. Thumb well developed, long, both arising and ending far proximal to corneous base of specialized tip.

Abdomen. 3rd to 6th abdominal segments incompletely fused.

FEMALE

With the obvious exceptions, as in male. For taxonomic purposes the following characters in combination are probably the most useful: raised frontal margin; suborbital crenellations as in diagnosis; cheliped gape moderately wide with vestigial serrations; merus of 4th ambulatory slender.

Measurements (in mm)

	Length	Breadth	Propodus	Dactyl
Largest male (paratype)	6.6	10.0	16.0	11.0
Holotype male	5.8	9.2	15.8	9.8
Largest female (paratype)	6.4	9.8	–	–

Morphological Comparison and Comment

Although *limicola* is superficially similar to *dorotheae*, the gonopod is different in every component, most obviously in its long thumb. The major cheliped of *limicola* differs clearly in having the tubercles of the oblique ridge in a strong, regular row of single, not multiple, tubercles, while the suborbital crenellations are externally separate, not close set.

The nearest relation of *limicola* is probably *leptodactyla* in the Atlantic, which it resembles closely in most characteristics (p. 305).

Color

Displaying males: Whitening absent, but individuals observed alive only once. Carapace and appendages brown except for the paler major manus and white fingers.

SOCIAL BEHAVIOR

Waving Display

Wave lateral, strongly circular. No pause at wave's peak, but a definite pause in some waves before the peak, giving the effect of 2 jerks which are apparently unique during circular waves. No drumming. Minor cheliped usually does not make a synchronous motion. Body slightly raised with each display during which legs are not raised except in running from side to side, and during leg-wagging. Duration of each wave about 1.5 to more than 2 seconds, with short or long pauses between; series not well marked. (Component nos. 5, sometimes with pause before peak; 9 occasional; 10; timing elements in Table 19, p. 656.)

Acoustic Behavior

Leg-wagging (component no. 5) occurred both during and between waves.

RANGE

Known from El Salvador, Costa Rica (Golfito), and Panama (near Old Panama).

BIOTOPES

Muddy flat inside mouth of a small stream, close to mangroves but not in shade; also slightly farther upstream on banks in partial shade. (Biotope nos. 12, 13, 14.)

SYMPATRIC ASSOCIATES

During the field observations in Panama a number of *Celuca* displayed in association with *limicola*. Intermingled was *beebei*; established several meters away were populations of *batuenta* and *saltitanta* and, within 50 meters more, *oerstedi* and *deichmanni* were also active. The specimens collected in Costa Rica from farther upstream were not observed in life.

MATERIAL RESULTING FROM FIELD WORK

(The complete list of specimens examined is given in Appendix A, p. 613.)

Observations and Collections. Costa Rica: Golfito; type series collected. Panama: near mouth of Rio Abajo, close to Old Panama. Three filmed specimens observed, collected, and brought alive to Trinidad. Positive identification was made before releasing them in the crabberies; unfortunately they were not recovered.

Films. Rio Abajo, Panama.

TYPE MATERIAL AND NOMENCLATURE

Uca limicola Crane, 1941

HOLOTYPE. In Smithsonian Institution, National Museum of Natural History, Washington. Male cat. no. 137415 (formerly New York Zoological Society cat. no. 381,152). Type-locality: Golfito, Costa Rica. Measurements on p. 309. (!)

Additional type material. One male and one female paratype from same locality in same institution, cat. no. 79401 (formerly New York Zoological Society cat. no. 381,153 part). (!) 17 additional male and female paratypes, now in very poor condition because of an accident in New York; from same locality in same institution, cat. no. 137416 (formerly cat. no. 381,153 part, New York Zoological Society. (!)

REFERENCES AND SYNONYMY

Uca (Celuca) limicola Crane, 1941

Uca limicola

TYPE DESCRIPTION. Crane, 1941.1: 198; Text Fig. 4t; Pl. 4, Fig. 17; Pl. 5, Fig. 22; Pl. 6. Fig. 29. Costa Rica: Golfito. (USNM !)

Holthuis, 1954.1: 41. El Salvador. Size record: Length 8.2 mm.

Holthuis, 1954.2: 163. Record of preceding reference.

Crane, 1957. Panama. Preliminary classification of waving display.

Uca coloradensis (not *Gelasimus coloradensis* Rathbun)

Bott, 1954: 171. El Salvador: El Zunzal. (Frankfurt !)

60. *UCA (CELUCA) DEICHMANNI* RATHBUN, 1935

(Tropical eastern Pacific)

PLATE 42 *A-D*.　　　MAP 17.
FIGURES 69 *A*; 93; 101.　　TABLES 9, 10, 12, 14, 19, 20.

INTRODUCTION

Usually found on somewhat sandier, more open shores than most *Uca*, *deichmanni* sometimes lives in close association with two other common *Celuca*, *oerstedi* and *beebei*. The waving displays of this trio can often be compared without even shifting focus, giving in capsule form an example of the obvious differences in display rhythm among the wealth of sympatric species that occur in the eastern Pacific. The pause at *deichmanni*'s point of highest reach, the lack of pause in *beebei* combined with its more horizontal wave, and *oerstedi*'s slower, widely circling motion of the cheliped make each species unmistakable.

MORPHOLOGY

Diagnosis

Front moderately broad; carapace strongly arched; posterior part of branchial region in male divided longitudinally by a shallow furrow. Small cheliped with a wide gape, the obsolescent serrations scarcely visible. Major cheliped with upper, outer manus sometimes appearing eroded because of linear arrangements of the tubercles; pollex always with a subdistal, tuberculate tooth or crest. Gonopod with flanges well developed and with a moderate thumb. Female ambulatories slender, with dorsal margin of 4th merus straight.

Description

With the characteristics of the subgenus *Celuca* (p. 211).

MALE

Carapace. Frontal breadth contained about 4.5 times in that of carapace. Antero-lateral margins slightly concave, slightly converging or diverging, angling bluntly into dorso-lateral margins. One or both pairs of postero-lateral striae may be present, but always obsolescent; sometimes both pairs are absent. Carapace profile semi-cylindrical; hepatic and branchial regions fused, strongly arched above surrounding carapace; posterior branchial region divided longi-

tudinally by a shallow groove. Eyebrow shorter than usual, its breadth slightly less than diameter of adjacent part of depressed eyestalk. Suborbital crenellations notably strong throughout, although small internally; most are long and separated, and continue around outer orbital margin, more widely separated, to channel. Lower side of antero-lateral angle variably sharp, with a basal sharp tubercle similar to the crenellations adjacent on the other side of channel.

Major Cheliped. Merus: Antero-dorsal margin moderately arched proximally, blunt, without a ridge, and unarmed; distally margin is practically straight, with small tubercles, serrations, or rugosities. Proximal part of dorsal surface, when intact, covered with exceptionally long, soft setae. *Carpus* practically smooth.

Manus: Bending of outer, extreme upper surface definite; flattening apparent, especially proximally. Outer tubercles minute, a little larger in a slightly convex area behind upper base of pollex, as well as on bent-over dorsal portion where they are also more widely separated. Erosions mentioned in type description inconspicuous or absent; when present, they are due to effect of arrangement of some tubercles in upper half in vertical or oblique rows with small, smooth spaces, or smaller tubercles, between. Dorsal, outer groove absent. Dorsal margin distal to carpal cavity marked only by cessation of tubercles of bent-over surface at edge of the vertical palm. Area outside pollex base slightly granulate, as is adjacent pollex, rather than completely smooth; bounded on lower edge by a weak, tuberculate keel that starts more proximally in middle of palm and continues throughout length of pollex. Ventral margin throughout covered with small tubercles, most of which continue the tuberculation of adjacent parts of the manus; the outermost tubercles, however, form a regular row, and are slightly enlarged and of similar size throughout; tubercles end at pollex base. Ventral, outer groove strong.

Palm with lower triangle covered with extremely minute tubercles, slightly larger near ventral margin and, especially, in distal part of triangle. Oblique ridge moderate, highest in middle of ridge rather than at apex. Tubercles largest on highest part of ridge; beyond apex a few tubercles continue upward a short distance around carpal cavity, dying out in irregularities. Carpal cavity with ventral margin very low, thick, blunt; distal side of cavity sloping gently, its margin scarcely marked beyond ending of tubercles above apex; inner edge of dorsal margin not turning downward to bound upper distal part of cavity; no distal extension of cavity, the boundary being indicated by a definite edge marking proximal end of the flat predactyl area. The latter area is smooth or slightly bumpy, with a small, smooth, median depression and, usually, with a few minute tubercles near its proximal and ventral boundaries.

Pollex and *dactyl*: Pollex with ventral margin slightly convex. Gape toward middle equal to or wider than adjacent part of pollex. Dactyl with outer, proximal, subdorsal groove absent; entire area covered with small tubercles which continue, diminishing in size, along dorsal margin. Pollex and dactyl with minute tubercles covering outer surfaces entirely and inner ones marginally. Gape rows all developed, the inner rows both very weak in middle sections and ending distally in a keel, either smooth or very minutely tuberculate. Pollex tip simple. Gape subdistally on pollex with a large, convex, tuberculate tooth or crest, appearing to arise partly from median and partly from outer row; extreme distal part of dactyl's median row faintly keeled with minute, sharp tubercles, very close set and diminishing regularly in size distally; in addition, a slightly enlarged tubercle sometimes present near dactyl's middle. Gape pile almost or entirely absent.

Minor Cheliped. Gape wider than pollex throughout, the slender fingers tapering regularly, not expanded distally. Feeble serrations barely distinguishable.

Ambulatories. 2nd and 3rd meri slender, their dorsal margins practically straight. 1st carpus on major side roughened anteriorly by at most several minute tubercles.

Gonopod. Flanges well developed, continuous distally in a smoothly rounded edge beyond the subdistal gonopore on inner surface. Anterior flange wider than posterior. Inner process overlying distal part of tube, ending at gonopore, its tip blunt. Thumb moderately short and slender, its tip not nearly reaching base of flange.

Abdomen. 3rd to 6th abdominal segments with slight traces of fusion.

FEMALE

Suborbital armature not stronger than in male. Even 2nd and 3rd ambulatories almost as slender as in male, the dorsal margins of their meri only slightly convex; merus of the extremely slender 4th leg is dorsally completely straight.

Measurements (in mm)

	Length	Breadth	Propodus	Dactyl
Largest male (holotype) (Rathbun)	7.9	12.0		
Large male	6.4	10.0	18.0	12.0
Moderate male	5.5	9.5	14.5	10.5
Largest female	5.5	9.5	–	–
Ovigerous female	5.0	9.0	–	–

Morphological Comparison and Comment

Although this species bears a general resemblance to *limicola* and *leptodactyla*, the armature of its major cheliped is more specialized and its carapace breadth relatively even greater, probably increasing the capacity of its branchial chambers in adaptation to its somewhat more exposed and active life. The other species, such as *festae*, that show a longitudinal groove on the branchial regions are equipped with pile in the area, which is not true of *deichmanni*.

Color

Displaying males: Display whitening absent except on major manus and chela. Carapace greenish brown to pale gray. Major cheliped: Posterior merus and carpus brown to pale bluish green; outer manus and entire chela polished white; anterior merus, carpus, and palm purple red, pink, or pale yellow to white. Orbits, 3rd maxillipeds, and the suborbital and pterygostomian regions pale blue green to violet blue; subhepatic region green to blue green. Minor cheliped pale orange pink to pink. At least 1st, 2nd, and 3rd ambulatories with anterior meri pale orange pink to pale yellow; posterior aspects and remaining segments dark.

SOCIAL BEHAVIOR

Waving Display

Wave lateral-straight. Major cheliped extended maximally obliquely upward, the chela tip reaching high above eyes; a definite pause occurs in this position while there is no pause with the cheliped flexed between the waves of a series. During each wave the dactyl is raised slightly to parallel pollex, minor cheliped does not make a synchronous motion, body is held raised throughout a series, and the ambulatories are not raised except to take a few steps to one side or the other. At both moderate and high intensities, display sometimes attains the rate of about 2 waves per second and 12 to 15 waves usually occur in a series. (Component nos. 4, 10, plus pause at peak: timing elements in Table 19, p. 656.)

Often closely associated with waving is major-mero-suborbital drumming; as described below it superficially resembles major-manus-drumming against the ground, which forms an integral part of waving display in *cumulanta* and other species.

Acoustic Behavior

It was formerly thought that in all species which vibrated the major cheliped the manus was always rapping against the ground, in the component here termed the major-manus-drum. In *deichmanni* film analyses show that in every distinct example of the vibration the manus does not touch the ground; instead the merus vibrates against an adjacent part of the carapace in a typical major-merus-drum. Synchronous drumming during combat is described in the next section. Unlike *batuenta*, for example, where some of the vibrations proved to be aerial only, in *deichmanni* contact with the carapace seems always to be made. Because of the indistinctness of some of the filmed examples, the occurrence of the major-manus-drum is included here as a questionable component to be further checked. Additional components observed in the field and in films are minor-claw-rubs, leg-wags, minor-chela-taps, and palm-leg-rubs. (Component nos. 1, 3, 4, 5, 7, ?9, 10.)

During displacement cleaning, which is prevalent in this species, stridulation is strongly suspected.

Combat

The only known ritualized component is the dactyl slide (no. 6), which occurred between forceful pushing and a forceful grip with incomplete fling. One of 2 such filmed combats was interrupted between pushes by a series of mero-suborbital drummings given synchronously by both opponents; the 5 separate contacts of the merus with the carapace were divided by each crab into 2 groups of 3 fast and 2 slow vibrations, and the timing even of these separate vibrations coincided.

RANGE

Port Parker, Costa Rica, to Old Panama, Panama City, R.P.

BIOTOPES

Bay shores open to marine tides, often on substrates of muddy sand, frequently mixed with small stones, or in pure sand close to its marginal merging with a mud flat; frequently near large, scattered stones. (Biotope nos. 4, 5, 6.)

SYMPATRIC ASSOCIATES

Groups of *deichmanni* usually live adjacent to or mingling with one or more of the following *Celuca*, depending on the substrate and on any muddier area with which it shares a border: *oerstedi, beebei, stenodactylus, latimanus,* and *m. terpsichores.* In addition, *deichmanni* often occurs along with *heteropleura* and *stylifera,* both members of the subgenus *Uca.*

MATERIAL RESULTING FROM FIELD WORK

(The complete list of specimens examined is given in Appendix A, p. 613.)

Observations and Collections. Costa Rica: Port Parker, Piedra Blanca, Uvita Bay, and Golfito; Panama and Canal Zone: Bahia Honda, Panama City, Old Panama and nearby Balboa.

Films. Near Old Panama.

TYPE MATERIAL AND NOMENCLATURE

Uca deichmanni Rathbun, 1935

HOLOTYPE. In Smithsonian Institution, National Museum of Natural History. Washington. Male, cat. no. 70832. Panama. Measurements on p. 312. (!)

REFERENCES

Uca (Celuca) deichmanni Rathbun, 1935

Uca deichmanni

TYPE DESCRIPTION. Rathbun, 1935.1: 51. Panama.
Crane, 1941.1: 199; Text Fig. 4u; Pl. 4, Fig. 18; Pl. 5, Fig. 23; Pl. 6, Fig. 30. Costa Rica and Panama. Taxonomy; color; waving display. (USNM !)
Crane, 1957. Panama. Preliminary classification of waving display.

61. *UCA (CELUCA) MUSICA* RATHBUN, 1914

(Tropical and subtropical eastern Pacific)

PLATES 42 *E-H*; 43 *A-D*; 49.
FIGURES 26 *H*; 31 *G*; 39 *E, F*; 46 *M*; 49 *A, B*; 69 *G, H*;
101.

MAP 17.
TABLES 9, 10, 12, 20.

INTRODUCTION

The outstanding structural characteristic of both sub-species of *Uca musica* is a distinctive stridulating mechanism. This consists of parallel striations on the manus and a row of tubercles on the 1st ambulatory, their positions being characteristic of less remarkable armature in a number of other *Celuca*. Although *m. terpsichores* has been observed in life, the functioning of this apparatus, presumably in the component termed the palm-leg-rub, has not yet been observed or recorded. *U. m. terpsichores* is also remarkable for fashioning the most perfectly shaped hoods known in the genus. These structures are also the largest in comparison with the size of the crab, and no populations have yet been found in which at least some of the displaying males did not make hoods on any given day.

Since color, display, and biotope are all unknown in *m. musica*, they are described only under the heading of *m. terpsichores*.

MORPHOLOGY

Diagnosis

Front moderately wide, its anterior margin obsolescent; eyebrows very wide. Body cylindrical. Major palm with an oblique stridulating ridge on lower, proximal part, composed of parallel striae. A row of tubercles anteriorly on merus and carpus of 1st major ambulatory. Small chelae in both sexes with gape extremely wide and serrations absent. Female carapace microscopically granulate.

Description

With the characteristics of the subgenus *Celuca* (p. 211).

MALE

Carapace. Frontal breadth contained about 3.5 to 4.5 times in that of carapace, depending on measurement method, since the eyebrow's inner margin in this species is somewhat ambiguous. Antero-lateral angles acute, produced laterally. Antero-lateral margins long, straight, diverging, rounding into dorso-laterals. Two pairs of postero-lateral striae, short and weak, sometimes obsolescent. Vertical lateral margin very weak. Carapace profile semi-cylindrical; hepatic and branchial regions fused, arched strongly above surrounding carapace; posterior, vertical, branchial furrow indicated. Frontal margin distally absent. Upper margin of eyebrow very weak; eyebrow wide, about equal to diameter of adjacent part of depressed eyestalk. Suborbital crenellations small but distinct internally, very much larger near antero-external angle; they continue, separated and of equal height, around outer orbital margin almost to channel. Lower side of antero-lateral angle with moderately sharp edge.

Major Cheliped. Merus: Antero-dorsal margin strongly arched throughout, with a rudimentary proximal ridge armed with a single row of minute, separated tubercles; distally the margin is not ridged, but armed with small rugosities and serrations. *Carpus* with sharp rugosities postero-ventrally; otherwise dorso-posterior surface is closely covered with minute sharp tubercles. Antero-dorsal margin sharp, sometimes finely tuberculate.

Manus: Bending of outer, upper surface only near dorsal margin, moderate; flattening slight. Outer tubercles minute over entire surface, except for slight enlargement on bent-over portion; there they are variably arranged in transverse or oblique rows, with occasional smooth cells among them, especially proximally; middle, outer palm sometimes (*m. terpsichores*) with the tubercles in irregular vertical rows. Dorsal, outer groove absent. Dorsal margin distal to cavity not at all defined, rounding gradually into upper palm, and covered with tubercles similar to those of outer manus, not arranged in rows. Cuff at base of dactyl smooth except ventrally, where a few minute tubercles occur in addition to the usual row that starts here and continues along pollex below gape. Area outside pollex base with a slight depression, the boundaries indefinite; area either practically smooth or covered with minute tubercles. Ventral margin with a low, beaded keel, the beads of similar size throughout. Ventral outer groove present but shallow and narrow. Keel, beading, and groove all end at pollex base, although pollex is ventrally granulate.

Palm's lower triangle with minute tubercles either covering it completely or absent in upper part. In addition, the triangle is marked by a wide ridge running obliquely from carpal attachment to ventral margin; this structure is formed of about 8 to 16 longitudinal, parallel striae; area above and distal to striae slightly concave. Oblique, tuberculate ridge high, highest slightly distal to apex. Tubercles very large throughout, except at ventral and dorsal ends; the largest, on highest part of ridge, are unusually outstanding, not contiguous, rather conical; beyond apex tubercles continue upward in a regular row along edge of carpal cavity as far as level of predactyl area, where they almost touch the downcurving inner edge of dorsal margin. The usual slight depression at pollex base is either finely tuberculate or smooth. Carpal cavity deep, its ventral edge definite, thick but moderately sharp; distal slope steep; upper part sharply and completely set off from predactyl area, without a distal extension, by the tubercles of the downward curve of the weakly beaded, inner edge of dorsal margin, almost meeting tubercles of upper end of row extending dorsally from apex. Predactyl area covered by small tubercles which in ventral portion are set on a slight ridge dividing the area from center palm; a small, central depression present or absent. Proximal predactyl ridge with tubercles unusually large and regular, but distal ridge and tubercles obsolescent to absent.

Pollex and *dactyl*: Fingers notably slender; dactyl arching throughout its length. Gape very wide, being, throughout middle portion, about twice or more width of adjacent pollex. Dactyl with outer, proximal, subdorsal groove present or absent. Pollex and dactyl with minute tubercles covering outer and inner surfaces at least marginally. Gape in both pollex and dactyl with external rows indistinct, except proximally on pollex; elsewhere the tubercles usually merge with those generally distributed; inner rows with tubercles distinct proximally only on pollex and usually not even proximally on dactyl; represented on both distally by a low keel which, on dactyl only, is very minutely tuberculate; median row present throughout on pollex; distal part of median row on dactyl keeled, surmounted by small tubercles, somewhat serrate, very close set, and diminishing regularly in size distally. Middle section of dactyl's middle row very weak or absent. Pollex tip simple. A few slightly enlarged teeth present or absent on pollex and, proximally only, on dactyl. Gape pile almost or entirely absent.

Minor Cheliped. Mano-pollex ridge extremely weak, almost absent. Manus wide, the pollex arising wholly from its lower, distal portion, so that the gape's base is notably wide. Pollex and dactyl much longer than manus, slender and arched, so that the gape is extremely wide throughout, being twice or more width

of pollex at all points. Serrations entirely absent. Fingers tapering to sharp points that meet imperfectly, the dactyl falling either inside or outside the pollex when apposed (in preserved specimens).

Ambulatories. 2nd and 3rd meri moderately slender, their dorsal margins scarcely convex to almost straight. 1st carpus and distal end of 1st merus both with single rows of tubercles anteriorly on major side. The row on merus is oblique, running from near distal end of dorsal margin to distal end of segment, above ventral margin. The row on carpus extends along almost full length of segment, about midway between dorsal and ventral margins.

Gonopod. Anterior flange well developed, strongly curved, its distal margin sometimes extending slightly farther than that of pore; posterior flange absent. Pore moderately small, terminal, its inner (anterior) lip slightly lower than outer. Inner process broad, flat, overlapping part of pore and extending very slightly beyond it. Thumb well developed but short, ending more than its own length below corneous base of specialized tip.

Abdomen. 3rd to 6th abdominal segments almost completely fused.

FEMALE

Carapace very finely granulate, especially anterolaterally. Suborbital armature not stronger than in male.

Measurements (in mm)

	Length	Breadth	Propodus	Dactyl
m. terpsichores				
Largest male (holotype)	6.3	10.4	16.0	
Moderate male (Panama)	4.5	7.0	10.5	8.0
Largest female (paratype)	6.4	11.0	–	–
m. musica				
Largest male (Gulf of California)	8.0	14.5	22.0	17.5
Moderate male (Isla Marguerita)	6.0	10.5	16.0	12.5
Female: Tepoca Bay (Lower California)	7.0	11.0	–	–

Morphological Comparison and Comment

The two forms here considered subspecies differ in a number of minor details tending in each form in a single direction. In *m. musica* the combat armature of the claw is very slightly stronger while the elements—striae and tubercles—of the palm-ambulatory stridulatory apparatus are more numerous. In size *m. musica* is somewhat larger than is *m. terpsichores*.

Females and clawless males can be distinguished readily from those of *latimanus* by the obsolescence of the anterior marginal line of the front, and from those of *deichmanni*, *limicola*, and *crenulata* by the strikingly broad eyebrows.

RANGE

From Baja California and the Gulf of California to Puerto Pizarro, Peru. It seems that records, substantiated only by photographs, from Vancouver and Seattle (Rathbun, 1918.1) must be incorrect. Perhaps formerly present near San Diego, California.

MATERIAL RESULTING FROM FIELD WORK

(The complete list of specimens examined is given in Appendix A, p. 614.)

Observations and Collections. Nicaragua: Corinto. Costa Rica: Port Parker; Golfito. Panama and Canal Zone: Old Panama; Balboa. Ecuador: Puerto Bolivar.

TYPE MATERIAL AND NOMENCLATURE

Uca musica Rathbun, 1914

HOLOTYPE. In Smithsonian Institution, National Museum of Natural History, Washington. Male, cat. no. 22081. Measurements in mm: length 8; breadth 12.9 (Rathbun). Type-locality: Pichilinque Bay, Gulf of California, Lower California. (!)

Uca terpsichores Crane, 1941

HOLOTYPE. In Smithsonian Institution, National Museum of Natural History, Washington. Male, cat. no. 137417 (formerly New York Zoological Society cat. no. 4144); measurements on p. 315. Type-locality: La Boca, Balboa, Canal Zone. (!)

Additional type material. 1 male and 3 female paratypes in the same institution, cat. nos. 79405 and 137418 (formerly New York Zoological Society cat. no. 4145) from the same locality. (!)

Uca (Celuca) musica terpsichores Crane, 1941

(Corinto, Nicaragua, to Puerto Pizarro, Peru; questionably from El Salvador)

MORPHOLOGY

With the characteristics of the species.

Major Cheliped: Tubercles of outer manus smaller than in *m. musica* and, in middle region, tending to be arranged in irregular, vertical rows; area outside pollex base practically smooth, instead of covered with minute tubercles. Palm with tubercles absent from upper part of lower triangle; striae of triangle's oblique ridge number about 8 to 12, the ridge area being wider than in *m. musica*; depression inside pollex base smooth. Dactyl slightly less arched than in *m. musica*, resulting in a slightly narrower gape; dactyl's outer, proximal, subdorsal groove absent. Pollex and dactyl with outer, as well as inner, minute tubercles confined to marginal surfaces. Gape with distal, median, file-like row on dactyl not as strongly serrate as in *m. musica*. *Minor Cheliped*: Gape slightly narrower than in *m. musica*, with the fingers correspondingly less arched. *Ambulatories*: Tubercles on anterior merus and carpus of major side slightly less numerous than in *m. musica*, those on carpus being more widely spaced. *Gonopod*: Inner process slightly broader than in *m. musica* and thumb arising somewhat more proximally.

Color

Displaying males: Display whitening often completely attained except for major cheliped. Carapace when not polished white has dull yellow markings dorsally. Major cheliped sometimes polished white except for bright pink on lower outer manus, base of pollex, and anterior merus; more often, when carapace is marked with yellow the entire merus is also dull yellow, instead of white and pink. Minor cheliped and ambulatories entirely white. Females gray, not spotted.

SOCIAL BEHAVIOR

Waving Display

From observations only; films not secured. Wave lateral, circular. Cheliped starts from an unusual position with the chela extended straight in front of crab, touching or nearly touching ground; after the claw has reached its low peak it is brought down abruptly and without pause into the same position. During each wave the minor cheliped makes a weak, synchronous motion. Body held high off ground, the legs being stretched up throughout a series; ambulatories at this time are not raised except in making occasional steps from side to side. At moderate intensity waving is at the rate of about 2 waves per second and at least 10 waves, separated by brief pauses, may occur in a single series. (Component nos. 5, 9, 11, and, as described below, possibly 12.)

During high intensity courtship the cheliped sometimes stretches stiffly to the side and is either held motionless or vibrated, probably against the carapace in a mero-suborbital-rub or drum. Meanwhile various ambulatories are raised and lowered, apparently at random, but whether as stridulating leg-wags or as purely ritualized leg-waves is unknown. In the absence of recent observations and of any films, these suspected acoustic components are omitted from Table 12.

Hood Construction

General comments on these remarkable structures are given in the introduction to this species. Examples measured up to 30 mm high and 20 mm wide, inside dimensions, while their thickness attained 5 mm; in contrast, the height of the crab, which measures less than 7 mm long, is always strikingly less than that of the hood, while the burrow's diameter is less than half the hood's breadth. All the examples seen were perfectly formed, smoothed on the inside, and with their edges overhanging the burrow's mouth.

BIOTOPES

Bay shores with a substrate of more or less muddy sand, usually with the mouth of a stream nearby; also found on an island, Jambeli, at the mouth of the large Guayas River, Ecuador. (Biotope no. 6.)

SYMPATRIC ASSOCIATES

The populations of *m. terpsichores* known to me were usually somewhat apart from other *Celuca* on the same beach. Nearby were often one or more of the following species: *stenodactylus, beebei,* or *deichmanni*; in the subgenus *Uca, stylifera* was an occasional associate and, in *Minuca, panamensis* sometimes frequented stonier areas at the end of the beach.

Uca (Celuca) musica musica
Rathbun, 1914

(Mexico: Baja California, west coast, north at least to Magdalena Bay; Gulf of California north at least to San Felipe and Guaymas; south of Gulf to San Blas in Nayarit)

MORPHOLOGY

With the characteristics of the species.

Major Cheliped: Tubercles of outer manus slightly larger than in *m. terpsichores*, not in vertical rows on middle region; area outside pollex base covered with tubercles slightly smaller than those of adjacent outer manus and continuing on proximal part of pollex. Palm with lower triangle covered with minute tubercles; in upper part they are relatively wide-spaced and, near ventral margin, closer, larger, and longitudinally elongate; striae of triangle's oblique ridge number about 14 to 16, the ridge area being narrower than in *m. terpsichores*; depression inside pollex base covered by minute tubercles continuing onto pollex. Dactyl slightly more arched than in *m. terpsichores*, resulting in a slightly wider gape; dactyl's outer, proximal, subdorsal groove well developed; surrounding area covered with small, sharp tubercles that die out dorsally among the generally distributed smaller ones that extend toward dactyl tip. Pollex and dactyl with minute tubercles covering outer surfaces entirely and inner ones marginally. Gape with distal, median, file-like row on dactyl more strongly serrate than in *m. terpsichores*. *Minor Cheliped*: Gape slightly wider than in *m. terpsichores*, the fingers being more arched. *Ambulatories*: Tubercles on anterior merus and carpus of major side more numerous than in *m. terpsichores* and on carpus set closer together. *Gonopod*: Inner process slightly narrower than in *m. terpsichores* and thumb arising slightly more distally.

REFERENCES AND SYNONYMY

Uca (Celuca) musica Rathbun, 1914

TYPE DESCRIPTION. See under *U. (C.) musica musica*, below.

Uca (Celuca) musica musica Rathbun, 1914

Gelasimus gibbosus (not of Smith)
    Streets, 1877: 113. Mexico: Lower California: La Paz.
    Lockington, 1877: 150. Mexico: Lower California.

Uca stenodactylus (not *Gelasimus stenodactylus* Milne-Edwards & Lucas)
    Rathbun, 1898.2: 603. Mexico: Gulf of California (Pichilinque Bay).

Uca musica
TYPE DESCRIPTION. Rathbun, 1914.2: 127; Text Fig. 5; Pl. 10. Mexico: Gulf of California (Pichilinque Bay). (USNM !)
    Rathbun, 1918.1: 417; Pl. 154. Mexico: Gulf of California; Guaymas Bay; La Paz; Mazatlan. Photo-

graphic records, almost certainly erroneous, from U.S.A., labeled Seattle, and from Canada, labeled Vancouver I. Taxonomy. (USNM !)

Balss, 1921: 698. Figure of stridulating organ. No new record.

Schmitt, 1921: 280; Text Fig. 165. No new records. Taxonomy.

Rathbun, 1924.4: 376. Mexico: Gulf of California (Tepoca Bay).

Uca (Celuca) musica terpsichores
Crane, 1941

Uca terpsichores

TYPE DESCRIPTION. Crane, 1941.1: 202; Text Fig. 4w; Pl. 4, Fig. 19; Pl. 5, Fig. 24; Pl. 6, Fig. 31; Pl. 7, Fig. 37. Corinto, Nicaragua to Panama. Color; waving display; hood construction. (USNM !)

? Bott, 1954: 173 (? part). (Not Text Fig. 15; not Pl. 17, Figs. 15a, b.) El Salvador: La Union. Taxonomy.

Crane, 1957. Panama; Ecuador. Preliminary classification of waving display.

Altevogt & Altevogt, 1967: E 1293. Film of waving display.

von Hagen, 1968.2: 428. Peru: Puerto Pizarro. Taxonomy; summary of waving display.

62. *UCA (CELUCA) LATIMANUS* (RATHBUN, 1893)

(Tropical eastern Pacific)

PLATES 43 *E-H*; 48.   MAP 17.
FIGURES 45 *P-RR*; 46 *N*; 53 *C*; 70 *C*; 101.   TABLES 9, 10, 12, 19, 20.

INTRODUCTION

The major cheliped of adult *Uca latimanus* closely resembles the claws of most half-grown fiddler crabs. The entire appendage is short, its manus broad and thick; the palm practically lacks an oblique ridge and the stumpy fingers are not only straight but far shorter than the palm, while the gape is a narrow slit edged with feeble tubercles. In all other characteristics of both morphology and behavior, this species is one of the most advanced in the genus. Everyone with an interest in the relation of structure to function can only await with impatience some knowledge of combat techniques in this crab.

MORPHOLOGY

Diagnosis

Front moderately wide; carapace strongly arched. Small chela with gape extremely wide, serrations absent. Major cheliped with manus short and broad; fingers shorter than manus; gape extremely narrow; oblique, tuberculate ridge on palm practically absent. One pair of strong postero-lateral striae. Gonopod with well-developed flanges and a moderate thumb.

Description

With the characteristics of the subgenus *Celuca* (p. 211).

MALE

Carapace. Frontal breadth contained about 4 times in that of carapace. Orbits straight. Antero-lateral margins long, sometimes concave, slightly diverging, angling into dorso-lateral margins. Each of the single pair of postero-lateral striae is always strong, but varies in length and curvature, if any; sometimes, on each side, two other striae, both very faint, occur near postero-external angle of carapace. Carapace profile semi-cylindrical; hepatic and branchial regions fused, arched strongly above surrounding carapace; antero-lateral part of hepatic region granulated. Eyebrow broad, at least equal to diameter of adjacent part of depressed eyestalk. Suborbital crenellations strong throughout, continuing, separated, around outer orbital margin almost to channel. Lower side of antero-lateral angle with sharp edge.

Major Cheliped. Merus: Antero-dorsal margin almost straight proximally, not ridged, armed with a single row of minute tubercles or wide-spaced, small rugosities; distally strongly arched, with larger rugosities and serrations. *Carpus*: Small tubercles of anterior margin usually largest distally.

Manus: Bending of upper, outer surface practically absent. Outer tubercles small and similar over entire surface, being enlarged slightly near dorsal margin, where many are set on broad, low rugosities, the spaces between being smooth; most rugosities are vertical or oblique ridges; a few are mound-like. Dorsal, outer groove faintly indicated, with interruptions, or absent. Dorsal margin bent over; relatively broad throughout but, as usual, broadest distally; slightly convex, not flat; except for the finely beaded dorsal edge of carpal cavity, no longitudinal rows of tubercles mark outer or inner edge; entire surface, from narrow portion adjoining carpus to broad distal end covered with sharp tubercles slanted in a distal direction, smaller than those on outer palm and more closely set. Cuff at base of dactyl with lower half to two-thirds with tubercles like those on adjacent manus. Area outside pollex base not smooth, the tubercles of adjacent manus extending over entire area and continuing onto pollex. Ventral margin a blunt edge, the tubercles of outer manus continuing over it and merging with the smaller tubercles of adjacent palm. Ventral outer groove absent.

Palm with lower triangle covered closely with minute tubercles, largest in upper part of triangle, smallest along entire length of ventral margin, where they are most closely set; this ventral band of smallest tubercles is widest proximally; in brief, the usual size relationships of tubercles on the triangle are reversed. Oblique ridge almost absent throughout, represented only by a very broad, low, blunt, angling of the meeting surfaces; apex very low. Tubercles in usual area of oblique ridge altogether absent except for a slight extension of the minute tubercles of lower triangle over the distal part only. Carpal cavity with lower margin very broad, low, blunt; distal margin very low, upper portion sloping gradually to predactyl area, from which it is partly cut off by an irregular row of non-contiguous, small to minute tubercles curving downward, continuing inner edge of dorsal margin; predactyl area tuberculate with an elongate smooth depression; tubercles between depression and center palm extremely small and sparse, so that the area is not clearly separated from center palm.

Pollex and *dactyl*: Both always much shorter than manus, straight, wide, thick, the dactyl slightly wider than pollex, proximally broad, the dactyl not curving at all beyond pollex, the two tips instead meeting distally; gape much narrower throughout than either pollex or dactyl. Dactyl with a short, outer, proximal, lateral groove; the usual outer, proximal, subdorsal groove represented by a crease only, the strong tubercles of area practically uninterrupted and continued to tip of segment. Outer pollex and dactyl covered completely with moderate-sized tubercles continued from outer manus, except in lateral dactyl groove; inner side of pollex also tuberculate except for the distal extension of the usual slight depression at pollex base; inner surface of dactyl smooth except near margins. Gape in both pollex and dactyl with outer row of tubercles distinct throughout and inner row indistinct distally; median row strong and complete in both, on pollex far separated proximally from both marginal rows, the tubercles contiguous throughout, rounded, largest proximally, diminishing in size distally but remaining of similar shape. Tip of pollex notably wide and thick, not pointed, the small distal teeth of median row continuing regularly to its end. Gape pile plentiful.

Minor Cheliped. Gape wider throughout than width of pollex. Serrations entirely absent. Fingers tapering, but to rounded, not pointed, ends; when apposed, the tips meet.

Ambulatories. 2nd and 3rd meri wide for a *Celuca*, their dorsal margins moderately convex. 1st carpus anteriorly on major side with a broad, low ridge, the crest entire, running longitudinally along middle of proximal two-thirds of segment.

Gonopod. Flanges well developed, the anterior the broader; gonopod practically terminal; inner process broad for the subgenus, covering most of posterior flange and canal region; tip rounded, extending beyond proximal part of lip of gonopore; thumb short, not nearly reaching base of flange.

Abdomen. 3rd to 6th abdominal segments almost completely fused.

FEMALE

Carapace with antero-lateral regions finely granulate as in male. Suborbital armature not stronger than in male. With the obvious exceptions, the diagnosis applies equally well to both sexes.

Measurements (in mm)

	Length	Breadth	Propodus	Dactyl
Largest male (Ecuador)	9.5	15.3	16.5	8.2
Moderate male (Panama)	8.0	13.4	15.5	7.9
Largest female (Panama)	8.6	13.5	–	–
Moderate female (Panama)	7.5	11.0	–	–
Holotype male (Mexico)	(Rathbun)	(Rathbun)	(J.C.)	(J.C.)
	6.3	10.0	10.0	5.0

Morphological Comparison and Comment

As with most of the *Uca* that show some paedomorphic characteristics, the closest relations of *latimanus* are not evident. Judging by the form of its carapace, it belongs among the *Celuca* which have specialized for an active life through guarding against desiccation; the minor cheliped is similar to those of its presumed relations (p. 219). See also introductory remarks, p. 319.

Color

Displaying males: Full display whitening developed on carapace only. Carapace dorsally when not polished white is sparingly marked with gray blue or brown. Major cheliped with entire merus, carpus, and manus orange brown in varying shades, the outer manus always brightest; chela always entirely white. 3rd maxillipeds, pterygostomian areas, subhepatic regions, and minor cheliped white. Ambulatories with anterior surfaces of meri of 1st, 2nd, and 3rd purplish red; remainder of legs anteriorly gray blue, posteriorly white, marked with gray blue or brown. Females olive brown, speckled with gold-colored spots.

SOCIAL BEHAVIOR

Waving Display

Wave lateral, straight at low intensities, fully circular at high. The straight wave sometimes is broken by 2 jerks during the unflexing out and up, and there may or may not be a pause at the peak of highest reach. In the circular wave of high intensity, jerks are almost or wholly absent, but a distinct pause at the peak is usual. Dactyls of both chelae held slightly raised throughout. Body kept moderately high during a series. Ambulatories are not raised during display and steps are not usually taken except in high intensity courtship, when the male turns as needed to keep facing the female. Wave widely various in duration. Slight chasing of females in a poorly developed form of herding has been observed several times. (Component nos. 4, 5, 6, 11, 14, plus the irregular pause at peak; timing elements in Table 19, p. 656.)

During high intensity waving, including courtship, the minor cheliped makes circling mero-suborbital rubs, apparently as an integral display component. Ordinary unflexings of the minor cheliped in a motion synchronous with display have not been observed.

Hood Construction

While not attaining the relative size and symmetry of those constructed by *musica terpsichores*, hoods are well formed and can be found in some part of most displaying populations. Their construction appears sometimes to be dependent on semi-lunar rhythms which differ in two levels of the same population (Crane, 1941.1).

Acoustic Behavior

The following components occur repeatedly in the several filmed sequences at hand: major-merus-rub, minor-merus-rub as described above, and leg-wag. An additional probable component, not shown with complete clarity, is a minor-claw-rub in which the minor chela reaches across the crab and rubs or plucks the dorsal part of the major merus. (Component nos. 1, 2, 3, 5.)

RANGE

La Paz, Lower California, to Puerto Bolivar, Ecuador.

BIOTOPES

Probably never exposed to large bodies of water or to fully marine salinities, but occurring in a variety of sheltered habitats. The substrate ranges from muddy sand through sandy mud to mud on the flats and banks of small lagoons and tidal streams, usually close to mangroves; the crabs sometimes live among their shoots. While occasional small groups have burrows in partial shade the most active part of any population always occurs in full sunlight. (Biotope nos. 9, 11, 12, 14.)

SYMPATRIC ASSOCIATES

The most frequent associates of *latimanus* are probably *beebei* and *inaequalis*. As with most hood-builders, individual *latimanus* usually group closely together without a mixture of other species, although a number of others may be only inches away from the small colony.

MATERIAL RESULTING FROM FIELD WORK

(The complete list of specimens examined is given in Appendix A, p. 614.)

Observations and Collections. Mexico: Tenacatita Bay; Nicaragua: Corinto and San Juan del Sur; Costa Rica: Port Parker and Golfito. Panama and Canal Zone: Panama City and Balboa. Ecuador: Puerto Bolivar.

Films. Golfito.

TYPE MATERIAL AND NOMENCLATURE

Gelasimus latimanus Rathbun, 1893

HOLOTYPE. In Smithsonian Institution, National Museum of Natural History, Washington. Male, cat. no. 17500. Type-locality: La Paz, Lower California. Measurements on p. 320. (!)

REFERENCES AND SYNONYMY

Uca (Celuca) latimanus (Rathbun, 1893)

Gelasimus latimanus

TYPE DESCRIPTION. Rathbun, 1893: 245. (USNM !)

Uca latimana

Nobili, 1901.3: 52. Colombia: Tumaco.

Maccagno, 1928: 42; Text Fig. 26 (claw). No new record. Taxonomy.

Bott, 1954: 172; Text Fig. 14; Pl. 17, Fig. 14a, b. El Salvador: Puerto el Triunfo. Taxonomy.

Peters, 1955: 438ff.; Text Figs. 8, 26, 31, 32. El Salvador. Waving display; general behavior.

Uca latimanus

Rathbun, 1918.1: 422; Pl. 157 (= holotype). Costa Rica: Gulf of Dulce: Santo Domingo. Taxonomy. (USNM !)

Crane, 1941.1: 201; Text Figs. 2, 3, 4v; Pl. 6, Fig. 33; Pl. 7, Fig. 36; Pl. 8, Figs. 38, 39, 40. Mexico to Panama. Taxonomy; color; waving display; hoods; habitat. (USNM !)

Crane, 1944. Panama; Ecuador. Note on color changes in field.

Garth, 1948: 61. Colombia: Humboldt Bay.

Crane, 1957. Panama; Ecuador. Preliminary classification of waving display.

Systematic Uncertainties

SPECIES NAMES, GEOGRAPHICAL RECORDS, AND UNDETERMINED SPECIMENS

This section has three divisions. The first gives an annotated list of species names which cannot be placed satisfactorily, whether as valid species or as synonyms; it includes comments on geographical records where they are concerned in the discussion of the unacceptable names. The second section lists questionable geographic records alone, each under the name of a species here regarded as valid. In each of these two sections the forms are listed alphabetically under the name of the species, with a cross-reference, when pertinent, to other comments in the text. The last division lists several specimens of *Uca* of special interest that are not given specific names in this contribution.

Current locations of the specimens discussed are preceded by the word *in*, followed by the name of the city. The full name of the institution appears in the list of abbreviations, Appendix E, p. 678.

1. SPECIES NAMES

The following names are not treated taxonomically in the systematic section:

GELASIMUS AFFINIS Guérin, 1829: Pl. 1, Fig. 3. (Not *Gelasimus affinis* Streets; see p. 171, in the account of *Uca* (*Minuca*) *burgersi*, p. 168.) Guérin's species name rests on his illustration of a major claw, showing in outline form only the outer view, and its caption, reading in full as follows: "Fig. 3. Serre du Gélasime semblable (Gelasimus affinis Guér.)" The figure is placed next to Guérin's dorsal view of *Gelasimus duperreyi* (1829: Pl. 1, Fig. 2), which Guérin himself (1838: 10) rightly synonymized with *Gelasimus tetragonon* (= *Uca* (*Thalassuca*) *tetragonon* in the present study; see p. 81). He also included in his synonymy of 1838 his other species, *affinis*. He implies, but does not state, that his specimens of *affinis* also were taken on Bora-Bora. The claw depicted in the reference of 1829: Pl. 1, Fig. 3 certainly does not represent that of a specimen of *tetragonon*, although it could reasonably be referred either to *chlorophthalmus* or to *lactea*. While *chlorophthalmus* occurs on Bora-Bora, *lactea* does not The material is not in Paris, nor does the name of the species appear on lists of the collections. In addition, it is not deposited in Philadelphia. Presumably this material is no longer extant. It seems clear that the name should be discarded on the bases of the absence of a specimen, the lack of descriptive material, and the ambiguity, inadequacy, and incompleteness of the illustration.

GELASIMUS AMAZONENSIS Doflein, 1899: 193. In Torino (formerly in Munich). Clearly labeled "Tipo. Munich Teffè-Amazzon" and so considered by Maccagno, 1929: 27, Text Fig. 14. This specimen proved to be an unmistakable example of *U. c. chlorophthalmus*, with a range restricted to the east coast of Africa. (!) The locality given, Teffè, spelled "Tefe" on modern maps, is far upriver from Manaos. I have found no trace of any other specimen so labeled, and since the locality given in the description is an incredible one for the genus as known, some mistake obviously occurred. In the present contribution *amazonensis* will be found in the synonymy of *chlorophthalmus* (pp. 101, 103). (Note added in proof: von Hagen has meanwhile found in the Munich museum, the original depository, type-specimens of *G. amazonensis* which he concluded independently are examples of the species here referred to as *U. chlorophthalmus* [personal communication]. His report is in preparation.)

GELASIMUS BASIPES. In Paris. Apparently a manuscript name. Source unknown. The following label, in ink, was the only one legible in a bottle of mixed Indo-Pacific narrow-fronted *Uca*: "Gelasimus basipes. Museum Paris. Ajuthia. Bocourt. 1862." The only Ajuthia I have traced was the former capital of Thailand, located well upriver even before the delta was as extensive as at present. (!)

GELASIMUS CUNNINGHAMI Bate, 1868: 447; fig. Rio de Janeiro, Brazil. *Trichodactylus* not *Uca* according to Göldi, 1885, 1886.

GELASIMUS DEXIALIS. In Paris. Apparently a manuscript name of Latreille. Label: "Gelasimus dexialis, Latr. MM. Peron et Le Sueur. Australie." The single specimen is small, dried, and broken. (!)

ACANTHOPLAX EXCELLENS Gerstaecker, 1856: 138. Rathbun (1918.1: 385) referred this specimen from Veragua, western Mexico, to *U. insignis* after examining it in the Berlin Museum. Since some of Rathbun's material has proved to belong to the species *U. ornata* (Smith, 1870) and some to *U. maracoani insignis* (Milne-Edwards, 1852), the identity of *excellens* remains in question. The type was apparently lost during the Second World War; if it is found, and proves to be an example of the species here termed *ornata*, the name *excellens* will have priority. (Pp. 146, 153.)

GELASIMUS FORCEPS Milne-Edwards, 1837: 52; 1852: 148; Pl. 3, Figs. 11, 11a. In the type description, "Australasie" is given as the locality; in 1852, "Australie." In 1959 in Paris I examined the dried contents of a box with the following label: "Gelasimus forceps Latr. Type. Australie." The material superficially resembled a normal male *Uca* that had been dried, varnished, and glued to the bottom of the box, a procedure often followed with small crustaceans toward the middle of the nineteenth century. Further inspection showed that while the carapace was in poor condition, having been crushed and glued together, the anterior region was intact. This area was unquestionably typical of Indo-Pacific, narrow-fronted subgenera, while the major cheliped was clearly that of an example of (*Celuca*) *lactea*. When the specimen was relaxed for gonopod study the claw proved to be attached to the body by a splinter or toothpick that had been run carefully through both parts. Furthermore, the body turned out to have a female's abdomen curving around the remains of a mass of eggs.

Comparison of both carapace and claw with Milne-Edwards' illustrations (1852) showed that the drawings are excellent representations of the composite "specimen" under examination. Therefore the accident that led to the erroneous assembly of parts must have occurred before the drawings were made for the 1852 paper, but not necessarily before the type description of 1837. I could not certainly identify the female, the carapace of which measures about 11 mm in length.

In the same museum case, adjacent to the box holding the material labeled "*forceps*" as described above, were two boxes labeled *lactea*, received from Japan through the museum in Leiden. One of them contained another monster, the major claw of a *Deltuca* male, perhaps an example of *coarctata flammula*; the propodus, 32 mm long, had been glued to the body of a *lactea*, the carapace of which measured 11 mm. The claw appears in Milne-Edwards' paper of 1852 as Pl. 4, Fig. 16. The second box contained good representatives, two males and one female, of

l. lactea. No body was found that could belong with the *Deltuca* claw; the claw in the *forceps* box probably was that of the specimen of *lactea* to which the *Deltuca* claw had been attached.

It will be seen that at least three specimens were involved in the mix-up: an ovigerous female probably of the subgenus *Deltuca*, a male of the same subgenus, and a male (*Celuca*) *lactea*. There seems to be no doubt but that the name *forceps* should be discarded.

This name has appeared several times in the older literature (White, 1847: 36; Hess, 1865: 146; Kingsley, 1880: 144; Ortmann, 1897: 350); Ortmann synonymized *bellator* and *signatus* with *forceps*. In Paris specimens of *lactea* from Australia were labeled by A. Milne-Edwards in 1903 as "? *forceps*," showing that he was not aware of the ancient confusion.

GELASSIMUS HUTTONI Filhol, 1886: 386. Type-locality: Campbell I., New Zealand; *Uca huttoni*, Chilton & Bennett, 1929: 761. No new material. "Type—Paris Museum." I was unable to find the type, apparently the unique specimen, of this species in Paris, or to hear news of it elsewhere. Dr. R. K. Dell, of the Dominion Museum, Wellington, reports that all of Filhol's types went to France (personal communication). He and his colleagues believe it to be almost completely certain that *Uca* does not occur in New Zealand. "I think we now know the littoral crab fauna fairly well (not that the systematics are completely worked out but we do know what is here) and nothing approaching *Uca* has been seen. In this latter case it would be most surprising if the genus were to occur at Campbell Island where the crab fauna is particularly sparse and is obviously Subantarctic in derivation and relationship. . . . From our point of view we do not consider [*U. huttoni* and *U. thomsoni*] part of the New Zealand fauna." Filhol's description is insufficient for determining the specimen's affinities, and he published no illustration.

UCA LEPTOSTYLA Nutting, 1919: 182. No description. Antigua. Apparently a *lapsus calami* for some name already published, since the author refers to Rathbun's figures (1918.1).

UCA MEARNSI Rathbun, 1913: 616. Known only from the female holotype from the Philippines. It has already been placed as a questionable synonym of *U. (Deltuca) coarctata* (pp. 53, 55, 57). Enough uncertainty remains to make it desirable to keep the name in mind by listing it here.

GELASIMUS MINOR Owen, 1839: 79. Type-locality: Oahu, Sandwich Is.

Type-specimens originally deposited in the collections of the Royal College of Surgeons in London.

Dr. Jessie Dobson, curator of the College's Anatomy Museum, reports that these collections were destroyed during the bombing of the College in May 1941 (personal communication through Dr. Isabel Gordon of the British Museum). There is no record of their having been transferred earlier to the British Museum or elsewhere. Although the type-specimen of *G. minor* is not mentioned by name in Owen's catalogue, which is still at the college, two specimens are described under the heading "Gelasimus duperreyi?" with a reference to the type illustration of *duperreyi* (Guérin, 1826, Pl. 1). Owen may well have later decided to describe the two males as new; the locality given is "Oahu, Sandwich Island," which was given by G. Tradescent Lay, the collector of these and other specimens on the "Blossom" in 1826 and 1827. Owen's catalogue mentions both that the claw was orange yellow and the body "liver-coloured"; these terms would apply well to a *vocans* or a *lactea perplexa*, rather than to a *tetragonon* (= *duperreyi*) or *chlorophthalmus crassipes*, the only two species reaching the longitude of Hawaii. The few morphological details given could apply to examples of any of the four species, although the drawing seems rather clearly to depict a *chlorophthalmus* or *lactea*. However, no *Uca* material that appears certainly to have been collected in Hawaii seems to be extant. Edmondson (1933: 270, 1946: 311, and in personal conversation, 1956), was never able to find any trace of *Uca* on any of the islands he visited in the group, although he did extensive field work over many years and was alert to the possibility of the crabs' presence. To sum up, it seems probable that an error was made on the original locality label and that specimens, as indicated by the color notes and drawing, should have been referred to *lactea*. The name *lactea* de Haan, 1835, has priority over *minor*. In any case, the name *minor* should be unavailable because of its use by Linnaeus, 1758, in the trinomial *Cancer vocans minor*, which he certainly applied to a member of the modern genus *Uca* (see below, p. 326).

GONEPLAX NITIDA Desmarest, 1817. The references to this fossil crab are as follows:

Goneplace luisant, *Goneplax nitida*
  Desmarest, 1817: 505. Type description. Specimen in "collection du Muséum d'Histoire naturelle de Paris." Type-locality unknown.

Gélasime luisante—*Gelasima nitida*
  Desmarest, 1822: 106; Pl. 8, Figs. 7, 8. Further description with detailed illustrations.

Gelasimus
  Desmarest, 1825: 124. Reference, at end of his account of living species in the genus, to the fossil described in the references above.

Gélasime luisante—*G. nitidus*
  Milne-Edwards, 1837: 55. Cites above references; states that this fossil appears to be close to *G. maracoani* but has minor differences; and appends the abbreviation "(C.M.)," indicating that the specimen was still in the collection of the museum.

I can find no trace that the specimen was ever again discussed. In his 1852 account of the genus Milne-Edwards omitted a record of the fossil, but mentions *Gelasimus nitidus* Dana, 1851, from the Fiji Is. It is Dana's species to which subsequent authors have referred, usually as a synonym of the species treated in this study as *Uca (Thalassuca) vocans* (pp. 89, 94).

The fossil specimen is no longer extant, according to the curators concerned in Paris, who undertook in 1972 a thorough search of the collections of fossil crustaceans. The clear description and illustrations provided by Desmarest still cannot make the generic identification certain, since, as Desmarest points out, only the proximal part of one, very large cheliped is preserved, the second cheliped being either destroyed or wholly hidden. The surviving merus appears long and smooth, as occurs often in living species of *Uca*, while the carpus is large, apparently more coarsely tuberculate than in living forms, and ending in a large crest that is unique in both size and orientation. The manus and chela are unfortunately altogether missing. The carapace and the partially visible ambulatories agree well with those of some living species of *Uca* in the subgenera *Deltuca*, *Thalassuca*, and *Uca*. In view of the rarity of fossil specimens of *Uca*, even the data on this vanished specimen may one day prove of help in working out the evolution of armature in the genus *Uca* or related groups.

GELASIMUS PALUSTRIS. Milne-Edwards (1852: 148; Pl. 4, Fig. 13) heads his confused synonymy of *G. palustris* with the pre-Linnean reference "*Cancer palustris*, Sloane, Hist. of Jamaica, vol. II, p. 269 (1725)." It seems, however, that Milne-Edwards is technically the author of the name, because Sloane uses the phrase *Cancer palustris* only in the descriptive heading introducing his account, as follows: "*Cancer palustris cuniculos sub terra agens*. Maracoani. Margr. p. 184, ed. 1648," in contrast to his preceding heading, "*Cancer terrestris* . . ." Sloane continues, "This crab agrees in everything to the Description of Marcgrave . . ." As already mentioned in the present study (p. 146) there is no later record of *Uca (Uca) maracoani* in Jamaica. Furthermore, Sloane's comments suggest that he was observing the smaller, more active *rapax* or *burgersi* rather than *maracoani*: "It is frequent in all marish Grounds and among the Mangroves by *Passage-Fort*. It makes itself Burrows and runs into them without any choice of this or that, but into every one large enough to receive it as our Coneys often do."

Milne-Edwards gives "Antilles" as the habitat. He mentions no material; even in 1837: 54, under "*Gelasimus vocans*, Desmarest," which he lists among his synonyms in the 1852 contribution, he does not include his frequently used note, "coll. du Museum" or "(C.M.)."

In Paris, however, the collection includes two boxes of "types non spécifiés," each labeled "Gelasimus palustris; Guadeloupe; M. Duchassang." (!) The first contains a single male clearly referrable to the species known in the present publication as *Uca (Boboruca) thayeri*, carapace length 18 mm, propodus 59 mm; the second contains a younger example of similar length but smaller propodus (35 mm); this one was relaxed and proved to have a gonopod altogether characteristic of *thayeri*. A third box also contained dried crabs labeled *palustris* but not regarded as types: three male *rapax* from "Sto. Domingue," and one male *thayeri* from Martinique. A fourth box contained two *rapax* and two *burgersi* from Marie Galante, and a vial of small examples of *rapax* from "Sto. Domingue." Other specimens, all dried, labeled *palustris* included examples of *rapax* from Rio de Janeiro, of *thayeri* from Brazil, and of *minax* from South Carolina. The front illustrated by Milne-Edwards was not that of one of the *thayeri* considered to be "types non spécifiés" but rather probably represented *rapax*; the figure of the claw, on the other hand, was not drawn from any of the specimens in the group and its identity remains uncertain. None of the material was labeled simply "Antilles," although Dr. Guinot suggested in conversation that Milne-Edwards could well have used that general term in his paper, rather than the name of a particular island, and that he did not necessarily base his description on one specimen, or on examples from a single locality. From the short published description the salient characters cannot be determined.

Rathbun (1918.1: 397) placed *palustris*, preceded by a question mark, in the synonymy of *pugnax rapax*. Acceptance of the name *palustris*, along with the designation of a lectotype, would of course place in synonymy any of the younger but well-established names cited above. The obvious choice would be *thayeri*, on the basis of the specimens termed "types non spécifiés"; the illustrations, the non-correspondence between the localities on labels and in the type description, and the associated additional specimens argue against that course, as does the poor condition of the specimens. In view of all these circumstances, it seems that the name *palustris* should be discarded.

GELASIMUS POEYI. A name given by Guérin in manuscript to specimens from tropical America. Placed

in the synonymy of *stenodactylus* by Kingsley (1880.1: 154), along with Atlantic specimens now designated *leptodactyla*. I did not find the specimens at the Academy of Natural Sciences in Philadelphia.

GELASIMUS RECTILATUS Lockington, 1877: 148; *Uca* rectilata, Holmes, 1900: 76; Pl. 1, Figs. 10-14; redescription without new material; *Uca rectilatus*, Rathbun, 1918.1: 405; text fig. after Holmes. Type-locality: west coast of Lower California. The two type-specimens, male and female, were destroyed in the San Francisco fire. Nothing similar to the illustrations has since been reported from Lower California. The description is inadequate for present use. The drawings show the oblique orbits and simple armature of a juvenile, but several species occur in the eastern Pacific retaining similar characters as adults.

GELASIMUS ROBUSTUS White, 1847: 35. Name only. Not found in BM.

GELASIMUS RUBRIPES Jacquinot, 1842-1853: Pl. 6, Figs. 2, 2B; Jacquinot & Lucas, 1853: 66. The type description states that the type-locality was unknown and that the specimen had not been given to the museum in Paris. In that institution a box now contains two dried male *Uca*, listed by the museum as "types non spécifiés," labeled as follows: "Gelasimus rubripes Hombr. & Jacq. M. Jacquinot. Raffles Baie." (!) The only Raffles Bay I have located is in northwest New Zealand and was a port of call on the "Astrolabe" expedition, on which the collections described were made. Difficulties at once arise. Not only has *Uca* almost certainly been absent from New Zealand at least throughout historical times (see *Gelasimus huttoni*, above), but the type-locality was unknown to the authors when they described the species. In addition, the two presumed type-specimens in Paris prove to be examples of two western Atlantic species—*rapax* and either *mordax* or *burgersi*. (*U. burgersi* had not been distinguished from *mordax* when I examined the material in 1959.) Furthermore, the illustration definitely shows a narrow-fronted Indo-Pacific form, although the species cannot be determined from it. Finally, the narrative of the expedition relates that the only American stops on the trip were around Tierra del Fuego and at Concepción, Chile. To add to the confusion, the ship had sailing orders for Rio de Janeiro—a locality within the range of *rapax* and *burgersi*. Perhaps because of those sailing orders, a chapter is headed "Teneriffe a Rio Janeiro" [*sic*], although it contains no reference to Rio whatever; possibly the title was inserted during the final editing after d'Urville's death. The conclusion is that labels were obviously mixed, but whether the accident happened on the voyage or later at the museum will probably remain a mystery. Since then the name *rubripes* has been applied to a number of Indo-Pacific species. In view of the various confusions described above, the name should, it seems clear, be discarded.

GELASIMUS TENUIMANUS White, 1847: 35. Name only. Swan R., Australia. Not found in BM.

GELASIMUS THOMSONI Kirk, 1880: 236; text fig. of major claw. Habitat given as Wellington, New Zealand; *Uca thomsoni*, Chilton & Bennett, 1929: 731. No new material. Location of type material unknown; Dr. Dell reports (personal communication) that it has not been in the museum at Wellington for some time, although a number of Kirk's other types are still there. There is no record of the specimens in the British Museum. Information is insufficient for certain identification, but the form of the illustrated chela makes it almost sure that *thomsoni* is a synonym of *coarctata*, which is unknown south of Queensland. Kirk wrote: "Two males and one female of this singular and pugnacious looking crab were brought to me some time ago by one of the local fishermen." These circumstances make it exceedingly likely that the specimens were actually captured within their usual range far to the north, and reached the Wellington fisherman at second or third hand. If the specimens had been reported from within the range of *coarctata*, I would have included *thomsoni* without hesitation among the synonyms of *coarctata*. For Dr. Dell's comments on the unlikelihood of the occurrence of a *Uca* in New Zealand, see p. 323.

GELASIMUS VARIEGATUS Heller, 1862: 21. Type-locality: Madras. I have been unable to locate any material referred to this species. In particular the type-specimen was not found in the museum in Vienna when I asked for it in 1963. According to the short type description in Latin, this species is close to *annulipes* but includes a denticulate crest on the major merus. Almost certainly *variegatus* and *inversa* are synonymous. However, *inversa* has not been recorded from eastern India, while *variegatus* was described from Madras. The specimens were collected on the round-the-world expedition of the "Novara." According to oral reports, mistakes on the labels have been encountered in other material resulting from that trip, and it is not unlikely that the type-material of *variegatus* was collected farther west.

GELASIMUS VOCANS, as figured by Milne-Edwards, ?1836, Pl. 18, Fig. 1, with caption. As stated on p. 20, this single illustration stands for the type-species of the genus *Gelasimus*. Although important features of the figure do not depict with any accuracy the appearance of members of the species *Uca vocans*, by the standards of the day the drawing can be considered an acceptable representation.

Figs. 1a and 1b on the same plate certainly represent other species. The first, designated as "Gelasime platydactyle," is a realistic view of the orbital area, with eye and style, of a species later (1852) described by Milne-Edwards as *G. styliferus* (p. 142). Fig. 1b, again attributed to *G. vocans*, shows part of an anterior view, including a very broad front found only in members of the subgenus *Minuca*. 1c-g, representing maxillipeds, a sternum, and abdomens of both sexes, are specifically indeterminate.

CANCER VOCANS MINOR Linnaeus, 1758: no. 14; Herbst, 1781: 81; Pl. 1, Fig. 10. Indeterminate. Specimen apparently not extant.

"A small *Luzone crab* with a large fight Claw, cat. 189. These burrow in holes on the Shore." Petiver, 1767: Pl. 78, Fig. 5. The caption, cited in full above, is the only text and appears in the section entitled "Gazophylacii Naturae & Artis, *Decas Septima & Octava*." This plate is listed in the synonymy of *Cancer vocans* by Fabricius, 1775, under *Ocypode heterochelos* var. *minor* by Lamarck, 1801, and in the synonymy of *bellator* by Kingsley 1880.1.

It seems likely that the illustration represents *Uca coarctata*, a species common near Manila. The orbits are shown as suitably oblique and sinuous, while the major chela has a large, subterminal projection on the dactyl, the gape of both dactyl and pollex being otherwise unarmed.

According to his admirer, Sir Hans Sloane (1725), James Petiver did not usually take good care of his specimens, which he received from all over the world and drew with indefatigable zeal. It is therefore not surprising that the specimen of his "small *Luzone crab*" has apparently not survived, except as a figure in this long-posthumous edition.

2. QUESTIONABLE GEOGRAPHIC RECORDS

This section consists of a checklist of dubious locality records that hold special interest. They range from the inherently impossible to the somewhat unlikely. All of the specimens concerned have been examined by me, with three indicated exceptions.

The forms are listed alphabetically, using the name of the species under which the material is recorded in the Systematic Section and in Appendix A. Details will be found on the pages listed below in parentheses, following each locality. For geographic confusions in association with species names omitted from the Systematic Section, see the list starting on p. 322.

> *Uca bellator.* Nancouri, Nicobar Is.; Sydney. (Pp. 66, 595.)
> *U. burgersi.* West Africa. (Pp. 327, 604).

U. coarctata c. Adriatic; Odessa; Tahiti. (Pp. 594, 595.)

U. demani typhoni. Tuamotu Arch. (Pp. 41, 593.)

U. d. dussumieri. Tuamotu Arch. (Pp. 41, 593.)

U. dussumieri spinata. Madagascar. (Pp. 35, 593.)

U. forcipata. Zamboanga, at 10 fath.; Tuamotu Arch. (Pp. 41, 594.)

U. maracoani maracoani. Jamaica. (Pp. 148, 324; material not found.)

U. maracoani insignis. Valparaiso, Chile. (Pp. 146, 438.)

U. musica. Seattle; Vancouver. (P. 316; material not found.)

U. pugilator. Santo Domingo. (Pp. 226, 607.)

U. tangeri. Bahia. (P. 123; material not found.)

U. tetagonon. Hawaii. ("Sandwich Is."). (Pp. 324, 597.)

U. urvillei. Vanikoro. (Pp. 60, 595.)

3. Undetermined Specimens

The following specimens I have not been able to determine; each may well prove to represent a new species of *Uca*. For lack of adequate material it seems unwise to describe and name them. All except no. 4 are deposited in the Smithsonian Institution, National Museum of Natural History, Washington. No. 4 is deposited in the collections of the Allan Hancock Foundation, University of Southern California, Los Angeles, California.

1. USNM cat. no. 43439. Labeled *Uca tetagonon*. Reef off Cebu, Philippine Is. This single male should undoubtedly be referred to the subgenus *Australuca*. The only species of this subgenus so far known from the Philippines is *bellator*, from which this form is clearly distinct. It also differs from the remaining species in the subgenus. Carapace length *ca.* 8 mm. Since specimens of *Uca* do not occur literally on reefs, a confusion in locality may exist, at least on the local level.

2. NYZS cat. no. 41500. La Boca, Balboa, Canal Zone; Jan. 1941. Collected by J. Crane. Omitted from Crane, 1941.1. *Uca.* sp. One male in poor condition. Interesting, intrasubgeneric characters.

3. NYZS cat. nos. 44200, 44201. Puerto Bolivar and La Salada, Guayaquil, Ecuador, respectively; April 1944. Collected by J. Crane. Two males of the subgenus *Celuca*, near members of the superspecies *crenulata* and their allies. Characterized most conspicuously by antero-lateral patches of pile on carapace.

4. Hancock Foundation. Cocos I. off Costa Rica. One male, tentatively identified as *Uca zacae*; its large size, with a carapace length of 9mm, far exceeds that of recorded specimens. (See p. 207.)

In an ambiguous category are five museum specimens with labels stating that they were collected in western Africa. They consist, first, of a molting female from Liberia (USNM 21847) and, second, of three males and one female, all in very poor condition, from the Cameroons (MCZ 8962); each of the males lacks the major cheliped. They are certainly members of the American subgenus *Minuca*, and probably examples of *burgersi*. At present it must remain uncertain whether the locality records are erroneous, whether the specimens are transatlantic strays, or whether they are members of an undescribed species resident in Africa.

Figures

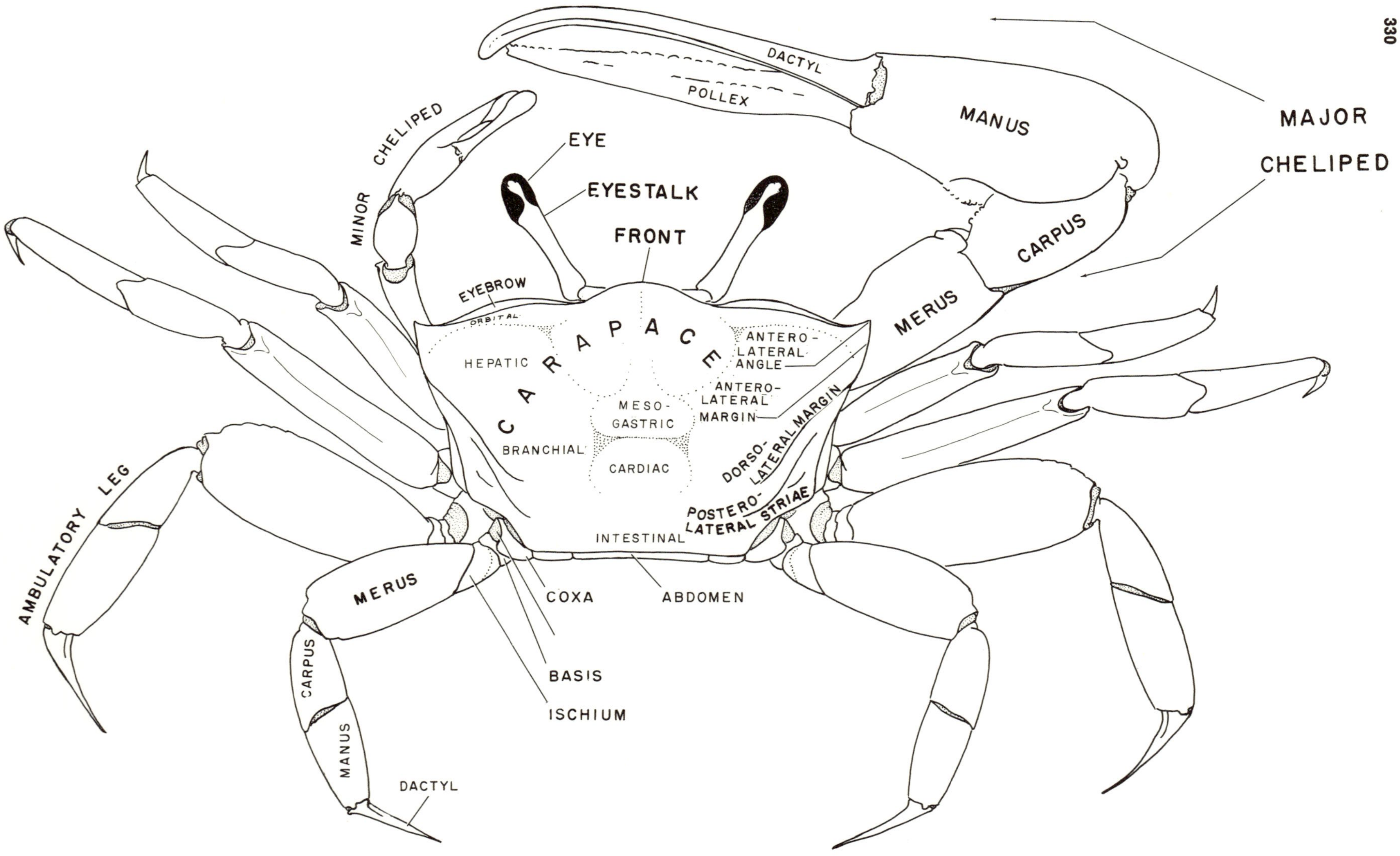

Fig. 1. Diagrammatic dorsal view of a fiddler crab (genus *Uca*), showing some of the anatomical terms used in this contribution. See also Figs. 2, 3, 4, 42, 43, 44, and pp. 448ff.

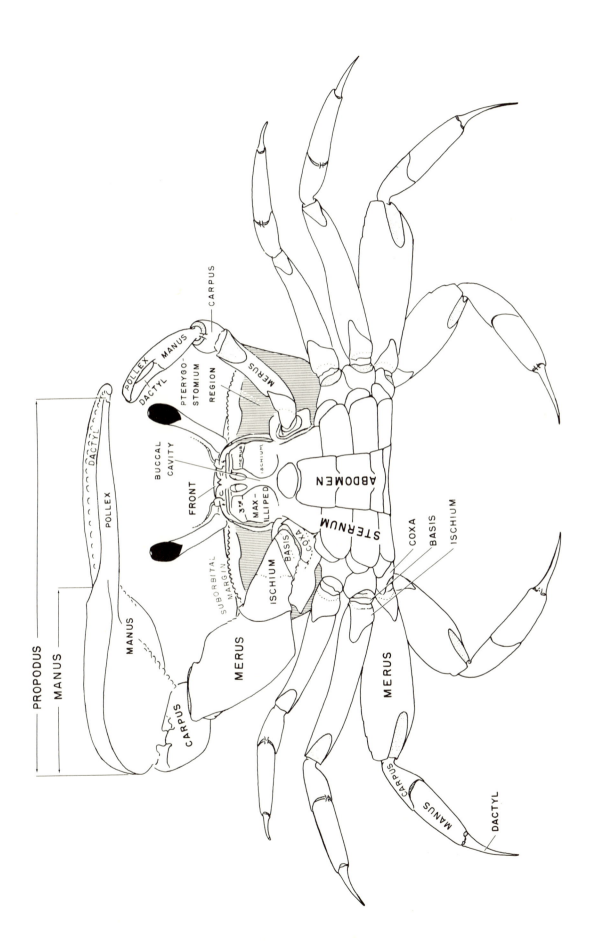

Fig. 2. Diagrammatic ventral view of a fiddler crab (genus *Uca*). Other information as in Fig. 1.

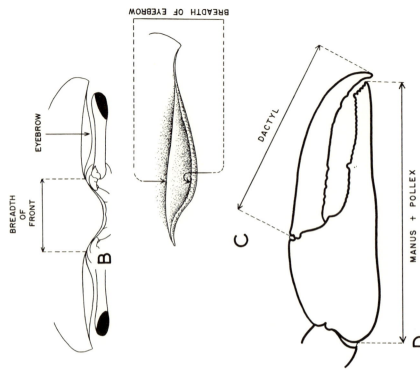

Fig. 4. Methods of measurement. *A*, carapace; in this contribution the breadth is always measured between the tips of the antero-lateral angles, which is usually, but as shown above, not always the widest part of the carapace. *B*, front; *C*, eyebrow; *D*, major claw. (Pp. 449, 456.)

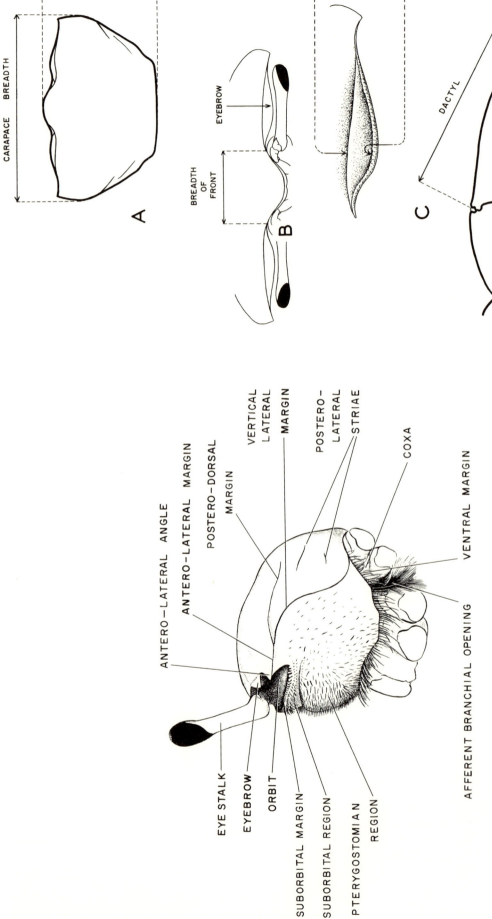

Fig. 3. Diagrammatic lateral view of a fiddler crab (genus *Uca*). Other information as in Fig. 1.

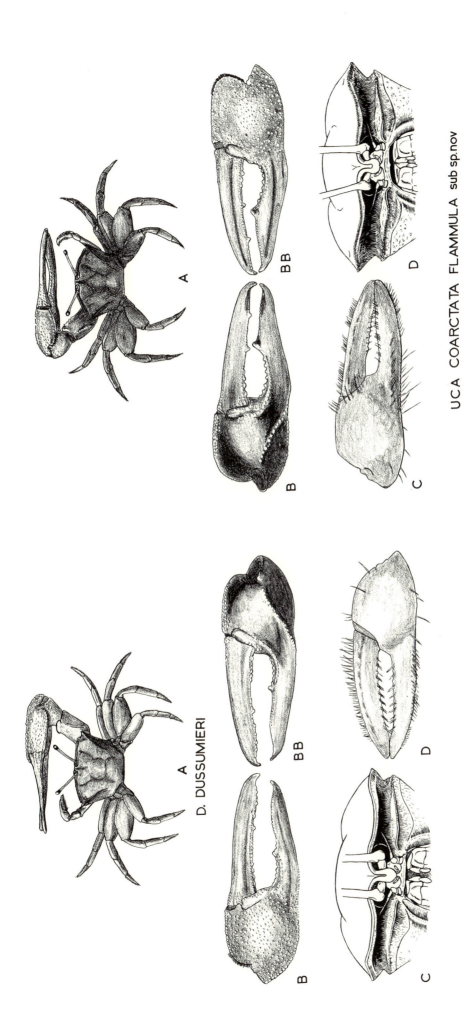

UCA COARCTATA FLAMMULA sub sp.nov

FIG. 6. *Uca* (*Deltuca*) *coarctata flammula* subsp. nov.: male; Port Darwin, Australia; carapace length, 16 mm. *A*, dorsal view; *B*, major claw, inner view; *BB*, same, outer view; *C*, minor claw, outer view; *D*, carapace, anterior view. (P. 56.)

D. DUSSUMIERI

FIG. 5. *Uca* (*Deltuca*) *dussumieri dussumieri* (Milne-Edwards): example from Zamboanga, Philippine Is.; carapace length, 13 mm. *A*, dorsal view; *B*, major claw, outer view; *BB*, same, inner view; *C*, carapace, anterior view; *D*, minor claw, outer view. (P. 37.)

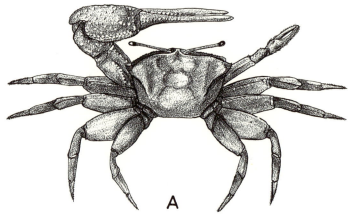

A

URVILLEI

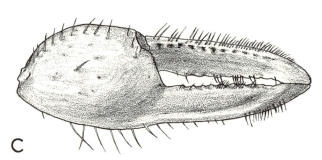

B

BB

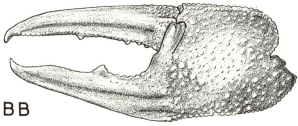
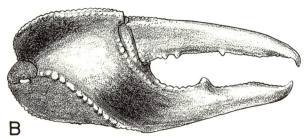

C

Fig. 7. *Uca* (*Deltuca*) *urvillei* (Milne-Edwards): example from Pemba I., Zanzibar, Tanzania; carapace length, 16 mm. *A*, dorsal view; *B*, major claw, inner view; *BB*, same, outer view; *C*, minor claw, outer view. (P. 58.)

Fig. 8. Type-specimens of *Gelasimus* in the Muséum National d'Histoire Naturelle, Paris. All described by Milne-Edwards; 1852. Localities are those given by Milne-Edwards and by the labels on the specimens. See text under name in parentheses. Anterior views of carapaces. *A, Gelasimus dussumieri*; Samarang (= *Uca dussumieri dussumieri*, lectotype). *B, Gelasimus dussumieri*; Malabar (= *Uca urvillei*). *C, Gelasimus brevipes*; Chine (= *Uca arcuata*). *D, Gelasimus urvillei*; Vanikoro (= *Uca urvillei*, holotype). (Pp. 35, 47.)

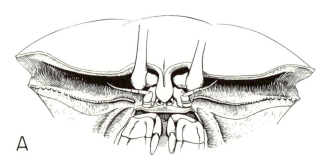
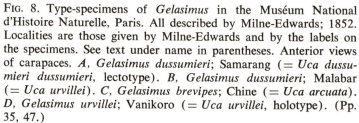

A

B

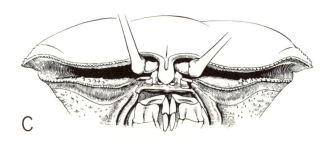
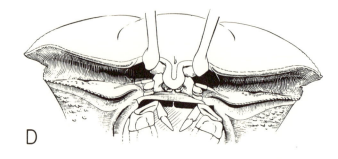

C

D

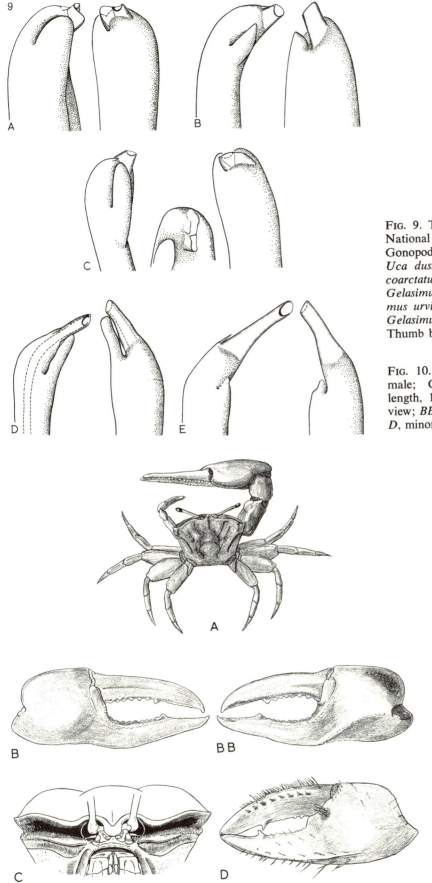

9

Fig. 9. Type-specimens of *Gelasimus* in the Muséum National d'Histoire Naturelle, Paris. Data as in Fig. 8. Gonopods. *A, Gelasimus dussumieri*; Samarang (= *Uca dussumieri dussumieri*, lectotype). *B, Gelasimus coarctatus*; Odessa (= *Uca coarctata*, lectotype). *C, Gelasimus brevipes*; Chine (= *Uca arcuata*); *D, Gelasimus urvillei*; Vanikoro (= *Uca urvillei*, holotype). *E, Gelasimus dussumieri*; Malabar (= *Uca urvillei*). Thumb broken off. (Pp. 35, 47.)

Fig. 10. *Uca (Australuca) polita* sp. nov.: holotype male; Gladstone, Queensland, Australia; carapace length, 15 mm. *A*, dorsal view; *B*, major claw, outer view; *BB*, same, inner view; *C*, carapace, anterior view; *D*, minor claw, outer view. (P. 72.)

UCA POLITA sp.nov.

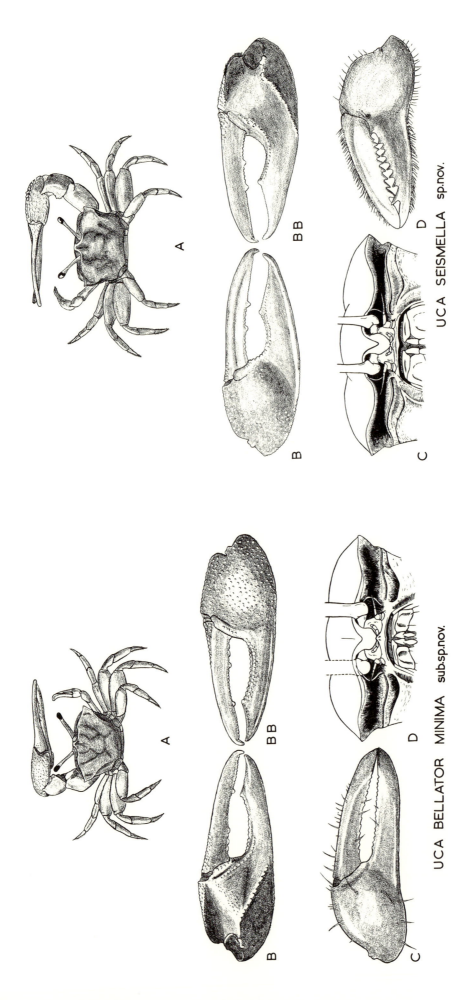

UCA SEISMELLA sp.nov.

UCA BELLATOR MINIMA subsp.nov.

Fig. 12. *Uca* (*Australuca*) *seismella* sp. nov.: holotype male; Port Darwin, Australia; carapace length, 8 mm. *A*, dorsal view; *B*, major claw, outer view; *BB*, same, inner view; *C*, carapace, frontal view; *D*, minor claw, outer view. (P. 70.)

Fig. 11. *Uca* (*Australuca*) *bellator minima* subsp. nov.: holotype male; Port Darwin, Australia; carapace length, 6.8 mm. *A*, dorsal view; *B*, major claw, inner view; *BB*, same, outer view; *C*, minor claw, outer view; *D*, carapace, anterior view. (P. 68.)

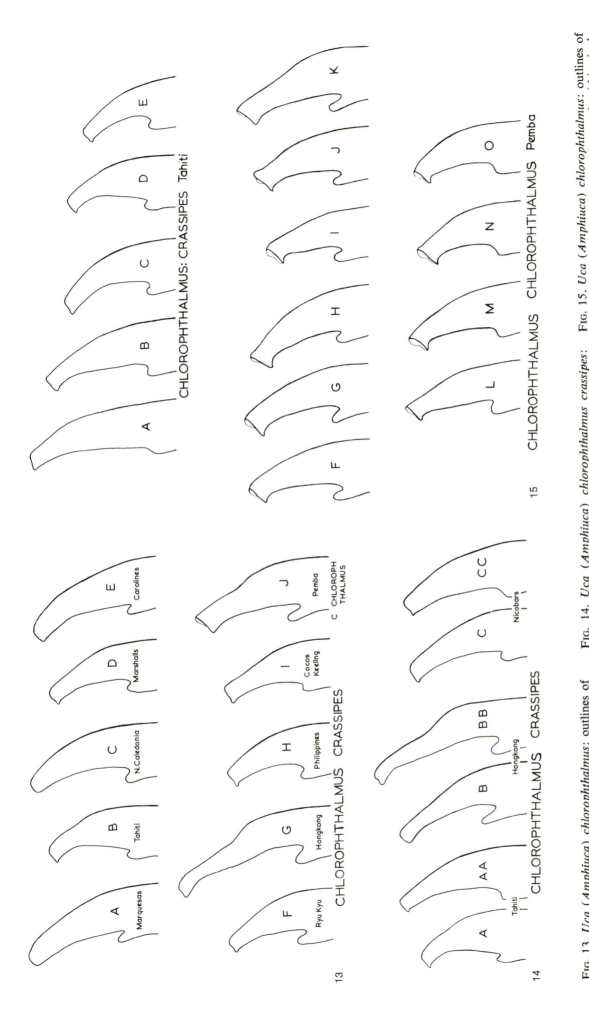

Fig. 13. *Uca (Amphiuca) chlorophthalmus*: outlines of gonopod tips from various parts of range. *A-I, chlorophthalmus crassipes*; *J, chlorophthalmus chlorophthalmus*. All examples from mature specimens. (Pp. 98ff.)

Fig. 14. *Uca (Amphiuca) chlorophthalmus crassipes*: outlines of gonopod tips from three localities, showing variation within populations. *A-CC* as labeled. All examples from mature specimens. (Pp. 98ff.)

Fig. 15. *Uca (Amphiuca) chlorophthalmus*: outlines of gonopod tips to show changes with growth within single populations. *A-O* as labeled. Carapace lengths in mm: *A*, 4.2; *B*, 6.2; *C*, 8.3; *D*, 12.0; *E*, 13.0; *F*, 5.4; *G*, 5.6; *H*, 8.5; *I*, 10.0; *J*, 11.0; *K*, 12.0; *L*, 5.6; *M*, 6.2; *N*, 8.9; *O*, 11.0 (Pp. 98ff.)

338

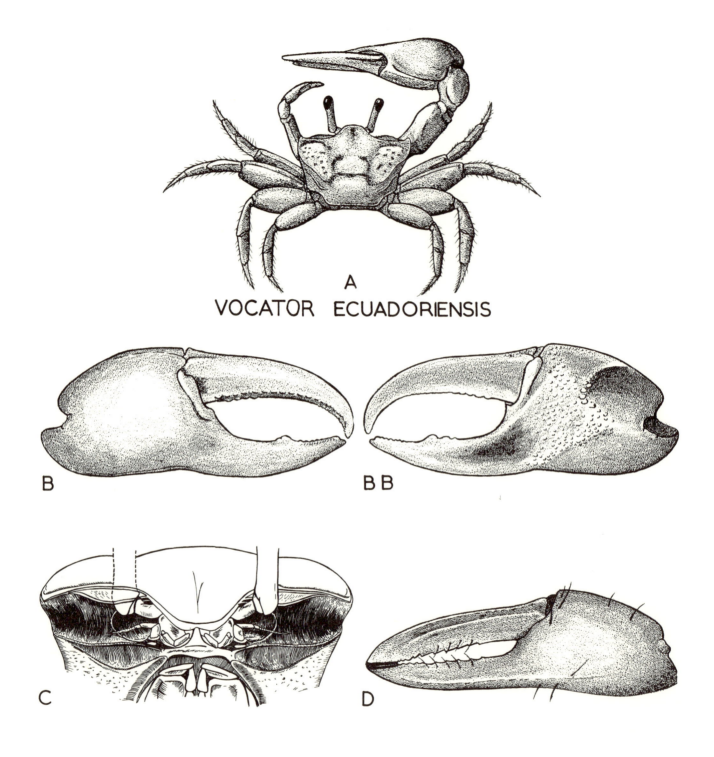

A

VOCATOR ECUADORIENSIS

B

BB

C

D

Fig. 16. *Uca (Minuca) vocator ecuadoriensis* Nobili:
example from Guayaquil, Ecuador. Carapace length: 11
mm. *A*, dorsal view; *B*, major claw, outer view; *BB*,
same, inner view; *C*, carapace, anterior view; *D*, minor
claw, outer view. (P. 166.)

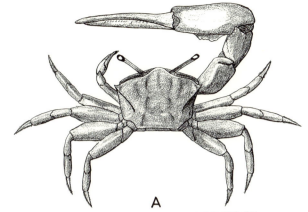

A

UCA LACTEA MJOBERGI

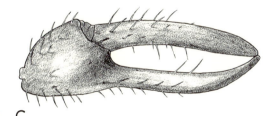

B

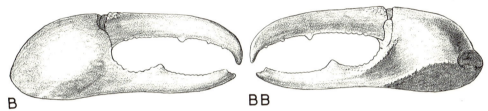

BB

FIG. 17. *Uca (Celuca) lactea mjobergi* Rathbun: from Broome, Northwest Territory, Australia. *A*, dorsal view; *B*, major claw, outer view; *BB*, same, inner view; *C*, minor claw, outer view. (P. 299.)

FIG. 18. *Uca (Celuca) lactea*: outlines of major claws, showing variation in shape in adults of two subspecies. *A-F*, as labeled. All examples from mature specimens. (Pp. 293ff.)

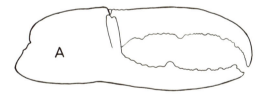

17 C

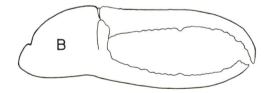

A

D

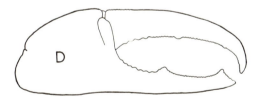

B

E

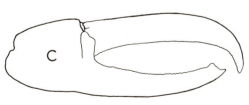

C

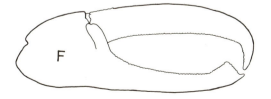

F

18   LACTEA ANNULIPES   LACTEA PERPLEXA

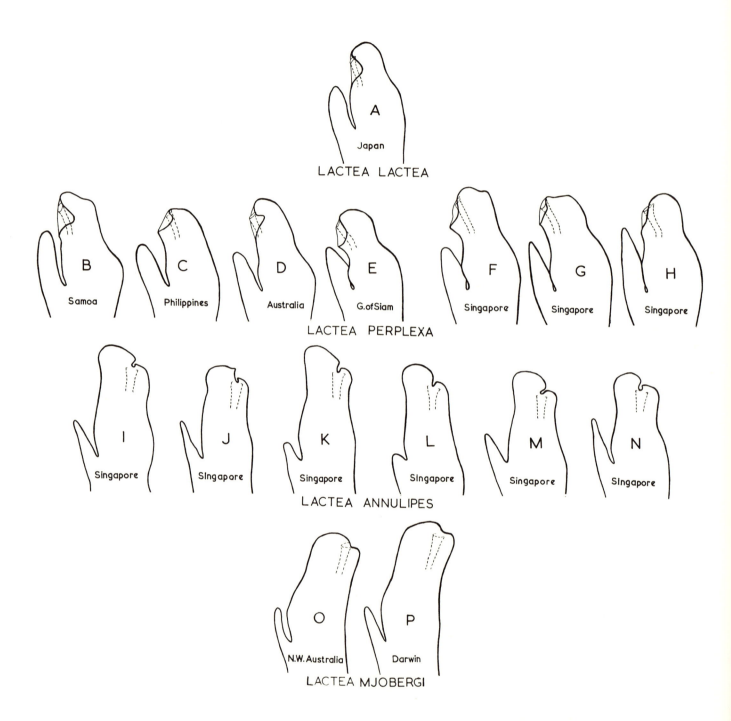

Fig. 19. *Uca* (*Celuca*) *lactea*: outlines of gonopod tips of all four subspecies, showing differences between forms and variations within the subspecies. *A-P*, as labeled. All examples from mature specimens. (Pp. 294ff.)

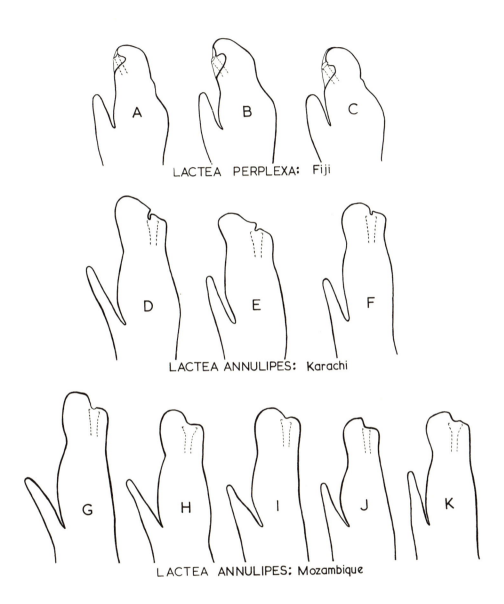

LACTEA PERPLEXA: Fiji

LACTEA ANNULIPES: Karachi

LACTEA ANNULIPES: Mozambique

FIG. 20. *Uca* (*Celuca*) *lactea*: outlines of gonopod tips in two subspecies to show changes with growth within single populations. *A-K*, as labeled. Carapace lengths in mm: *A*, 6.0; *B*, 8.0; *C*, 10.0; *D*, 9.3; *E*, 10.0; *F*, 11.0; *G*, 4.0; *H*, 7.7; *I*, 10.0; *J*, 11.0; *K*, 12.0. (Pp. 294ff.)

342

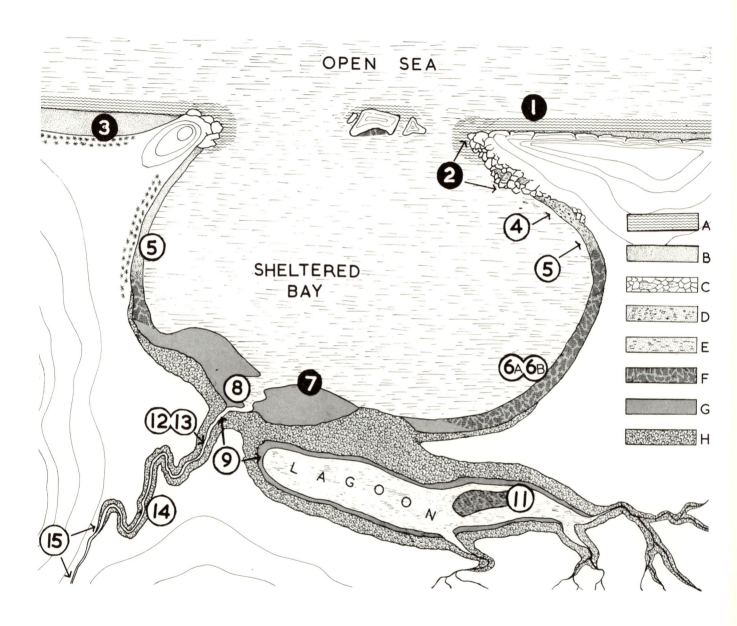

Fig. 21. Diagrammatic plan of principal biotopes supporting *Uca*. For key see Table 10; discussion on pp. 444ff.

22

23

FIG. 22. Same as Fig. 21, but semi-realistic view.

FIG. 23. Biotopes, as diagrammed in Figs. 21 and 22, visible from near mouth of river.

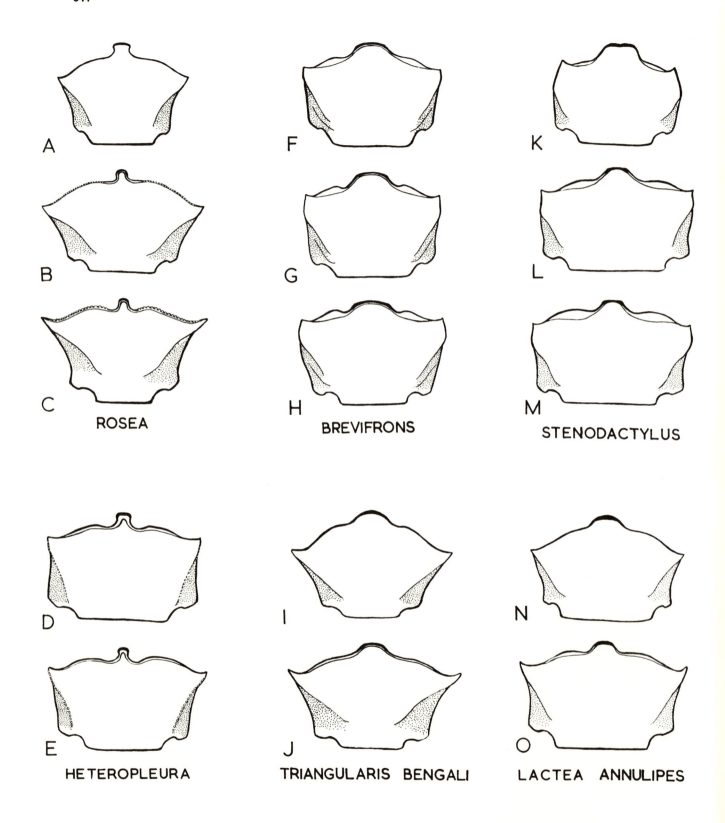

FIG. 24. Outlines of carapaces in six species of *Uca*, to show differences among species and changes with growth in each species. Carapace lengths in mm are as follows. (*Deltuca*) *rosea*: *A*, 1.7; *B*, 2.9; *C*, 13.0. (*Uca*) *heteropleura*: *D*, 4.0; *E*, 14.0. (*Minuca*) *brevifrons*: *F*, 3.4; *G*, 4.7; *H*, 18.0. (*Celuca*) *triangularis*: *I*, 2.5; *J*, 7.6. (*Celuca*) *stenodactylus*: *K*, 1.4; *L*, 2.0; *M*, 9.0. (*Celuca*) *lactea annulipes*: *N*, 4.0; *O*, 10.0. (Pp. 449ff.)

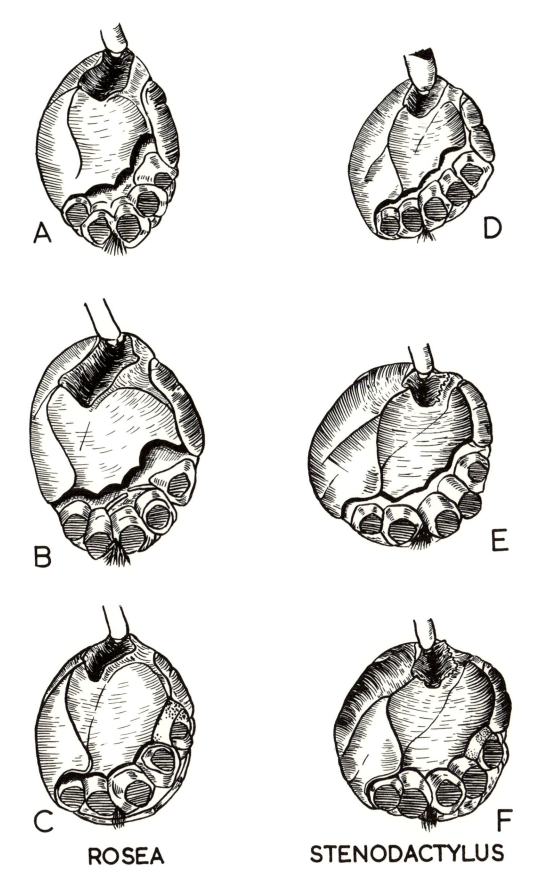

FIG. 25. Lateral views of carapaces in *Uca* to show changes with growth. Carapace lengths in mm are as follows. (*Deltuca*) *rosea*: A, 1.7; B, 2.9; C, 13.0. (*Celuca*) *stenodactylus*: D, 1.4; E, 2.0; F, 9.0. (Pp. 449ff.)

ROSEA　　　STENODACTYLUS

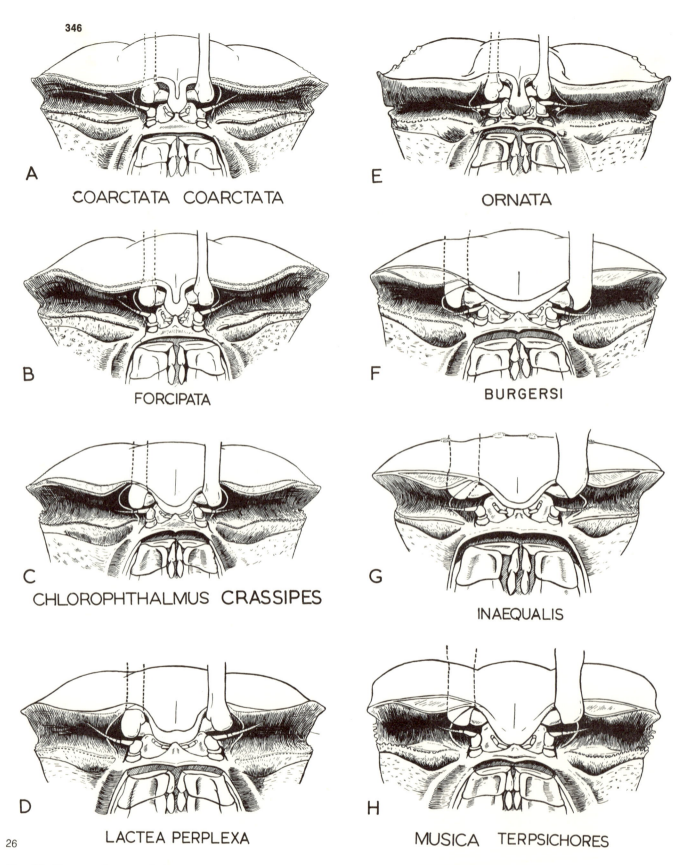

346

A COARCTATA COARCTATA

B FORCIPATA

C CHLOROPHTHALMUS CRASSIPES

D LACTEA PERPLEXA

E ORNATA

F BURGERSI

G INAEQUALIS

H MUSICA TERPSICHORES

26

FIG. 26. Comparative morphology of anterior aspect of carapaces in *Uca*, with special reference to the front, eyebrows, armature of orbits, and dorsal profile. Subgenera: *A, B, Deltuca; C, Amphiuca; D, G, H, Celuca; E, Uca; F, Minuca.* Drawn from mature males. (Pp. 452-454.)

FIG. 27. Comparative morphology of anterior aspect of carapaces in *Uca*, to show changes with growth; with special reference to the front, eyebrows, armature of orbits, and dorsal profile. Carapace lengths in mm are as follows. (*Deltuca*) *dussumieri spinata*: *A*, 4.0; *B*, 7.6; *C*, 20.0. (*Afruca*) *tangeri*: *D*, 3.0; *E*, 6.0; *F*, 24.0. (*Deltuca*) *urvillei*: *G*, 5.0; *H*, 18.0. (*Celuca*) *lactea annulipes*: *I*, 4.0; *J*, 10.5. (*Celuca*) *stenodactylus*: *K*, 2.0; *L*, 9.0. (Pp. 452-454.)

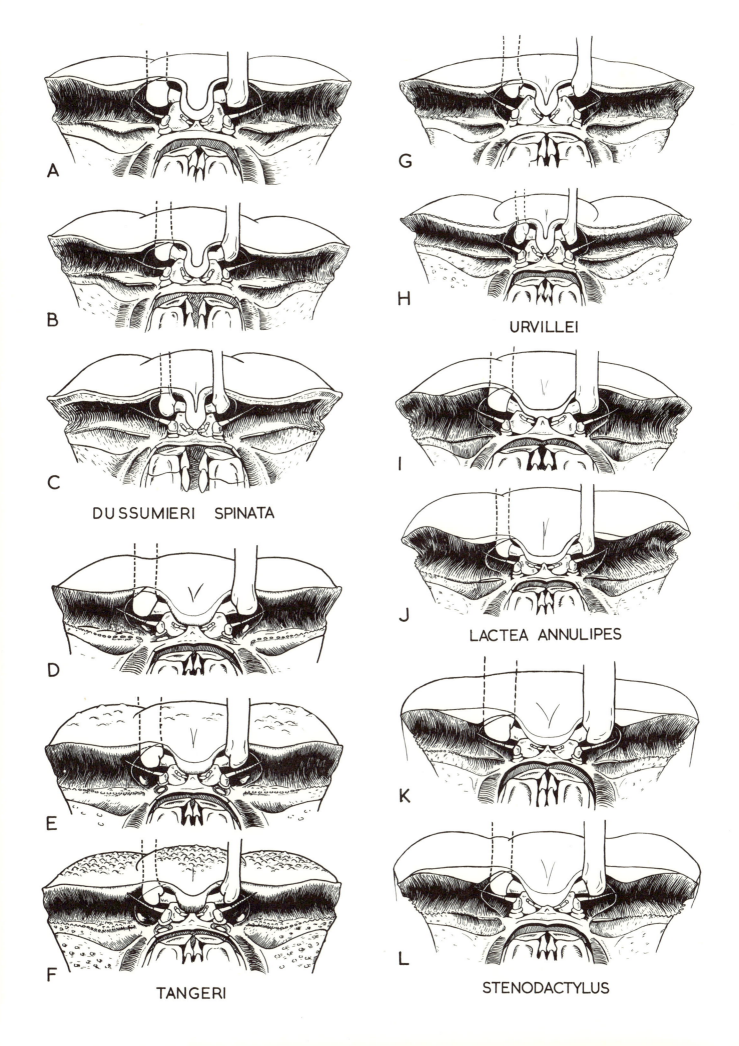

A

B

C

DUSSUMIERI   SPINATA

D

E

F

TANGERI

G

H

URVILLEI

I

J

LACTEA ANNULIPES

K

L

STENODACTYLUS

348

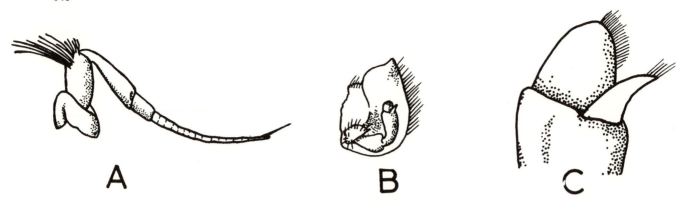

FIG. 28. *Uca* (*Uca*) *maracoani*: *A*, antenna. *B*, *C*, antennule. (P. 454.)

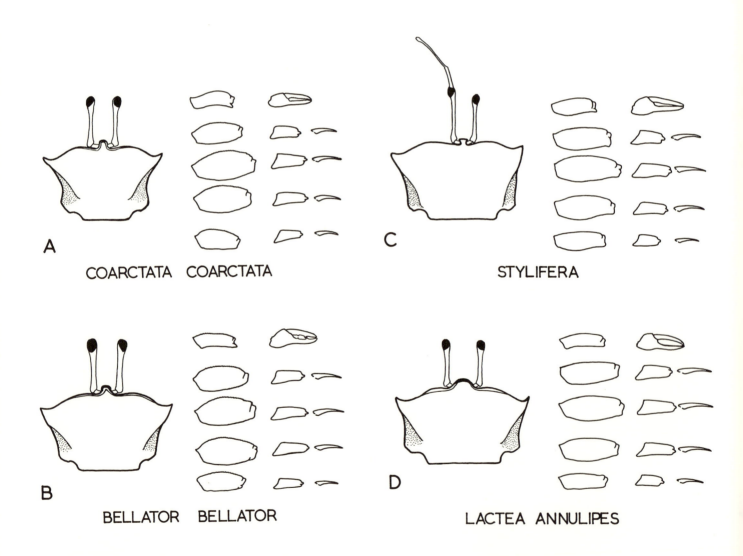

COARCTATA   COARCTATA

STYLIFERA

BELLATOR   BELLATOR

LACTEA ANNULIPES

FIG. 29. Dorsal views in examples of four subgenera of *Uca*, showing general similarities of proportions. Subgenera: *A*, *Deltuca*; *B*, *Australuca*; *C*, *Uca*; *D*, *Celuca*. Segments of minor chelae illustrated: merus, propodus, and dactyl; of ambulatories: merus, manus, and dactyl. Drawn from mature males. (P. 461.)

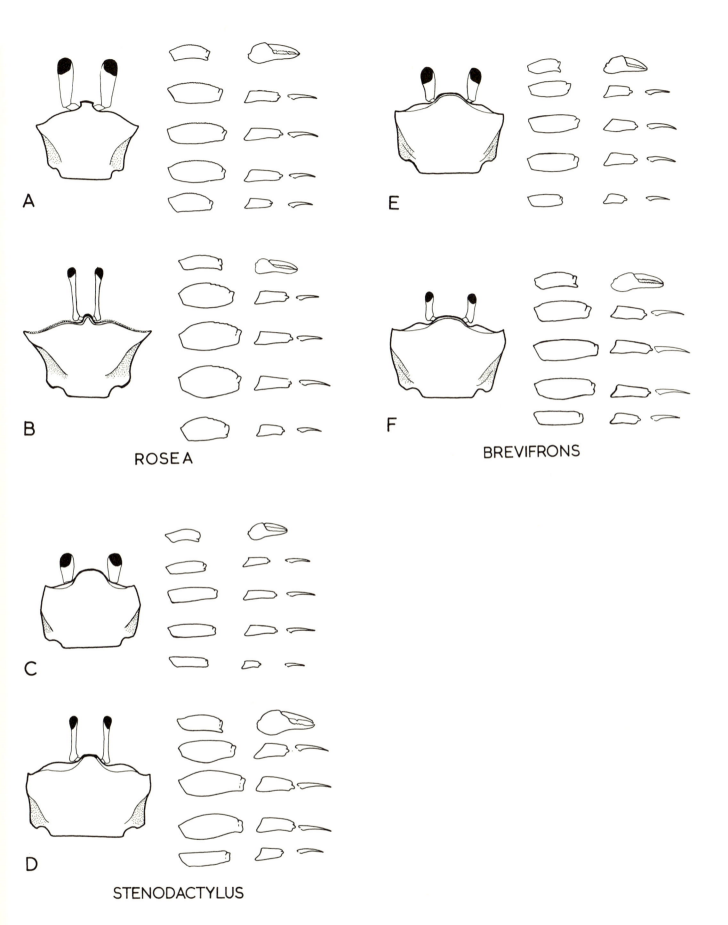

A

B

ROSEA

C

D

STENODACTYLUS

E

F

BREVIFRONS

Fɪɢ. 30. Changes in proportions occurring with growth, in examples of three subgenera of *Uca*. In each of the three species the upper figure is a young male, the lower a mature male. Carapace lengths in mm are as follows. (*Deltuca*) *rosea*: *A*, 1.7; *B*, 13.0. (*Celuca*) *stenodactylus*: *C*, 1.4; *D*, 9.0. (*Minuca*) *brevifrons*: *E*, 3.4; *F*, 18.0. (Pp. 451, 462.)

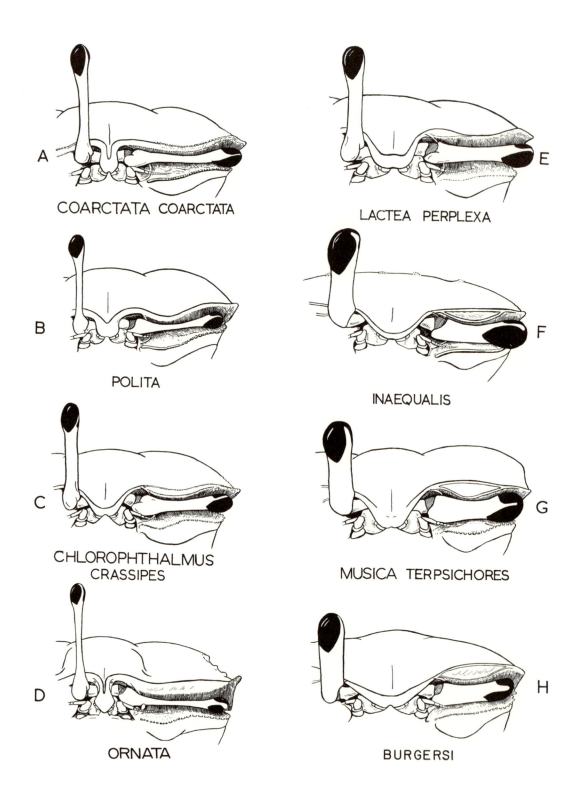

A COARCTATA COARCTATA

B POLITA

C CHLOROPHTHALMUS
CRASSIPES

D ORNATA

E LACTEA PERPLEXA

F INAEQUALIS

G MUSICA TERPSICHORES

H BURGERSI

FIG. 31. Frontal views in mature males of eight species of *Uca*, showing range of differences in shape and proportions of eyes and eyestalks. Subgenera: *A, Deltuca; B, Australuca; C, Amphiuca; D, Uca; E, F, G, Celuca; H, Minuca.* (P. 454.)

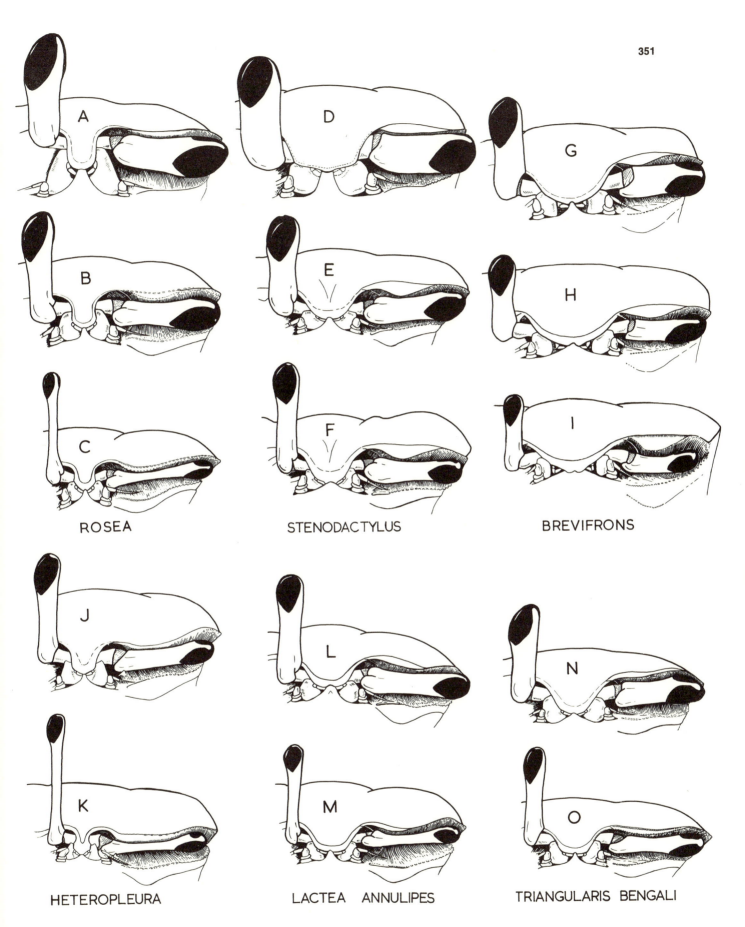

Fig. 32. Changes in proportions of eyestalks occurring with growth, in examples of six species of *Uca*. Drawn from males. Carapace lengths in mm are as follows. (*Deltuca*) *rosea*: *A*, 1.7; *B*, 2.9; *C*, 13.0. (*Celuca*) *stenodactylus*: *D*, 1.5; *E*, 2.7; *F*, 9.0. (*Minuca*) *brevifrons*: *G*, 3.4; *H*, 4.7; *I*, 18.0. (*Uca*) *heteropleura*: *J*, 4.0; *K*, 14.0. (*Celuca*) *lactea annulipes*: *L*, 4.0; *M*, 9.0. (*Celuca*) *triangularis bengali*: *N*, 2.5; *O*, 7.6. (P. 455.)

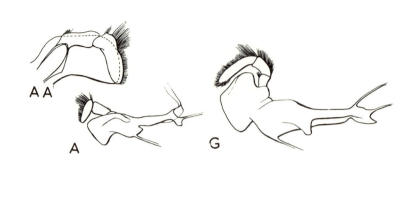

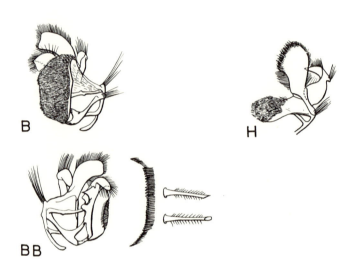

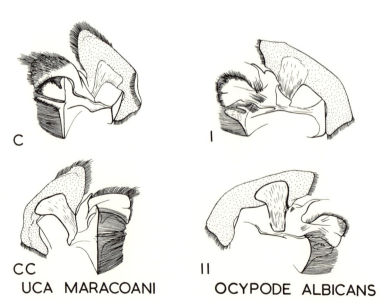

FIG. 33. Mouthparts in adult males of *Uca maracoani* and *Ocypode albicans*. *A-FF, Uca maracoani; G-LL, Ocypode albicans*. Mandible: *A, AA, G*. First maxilla: *B, BB, H*. Second maxilla: *C, CC, I, II*. First maxilliped: *D, DD, J*. Second maxilliped: *E, K*. Third maxilliped: *F, L*. Gill of third maxilliped, enlarged, *FF, LL*. (P. 455.)

UCA MARACOANI

OCYPODE ALBICANS

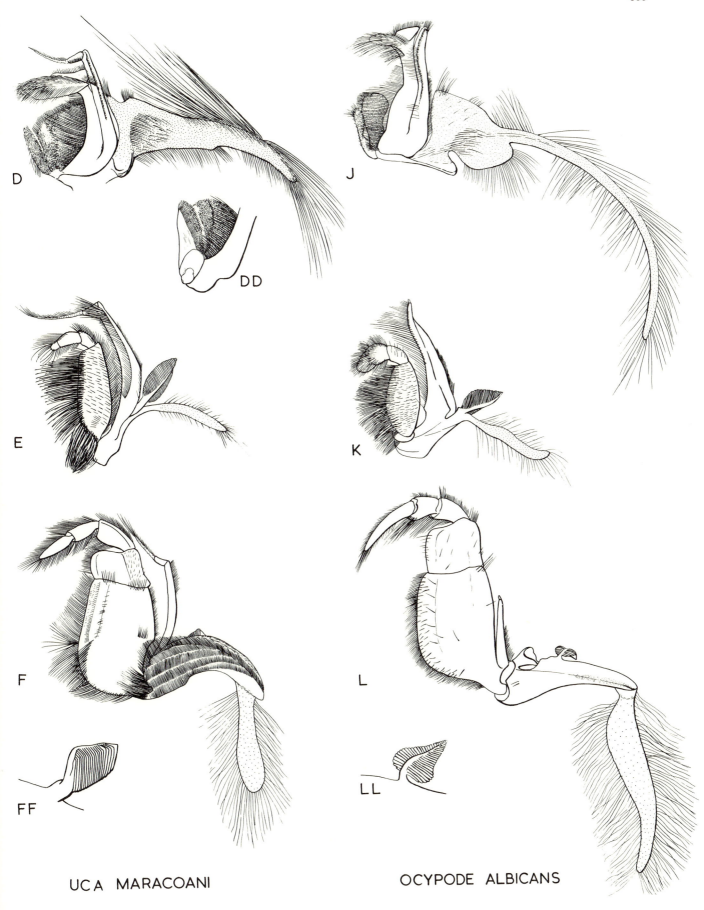

D

DD

E

J

K

F

FF

L

LL

UCA MARACOANI

OCYPODE ALBICANS

354

34

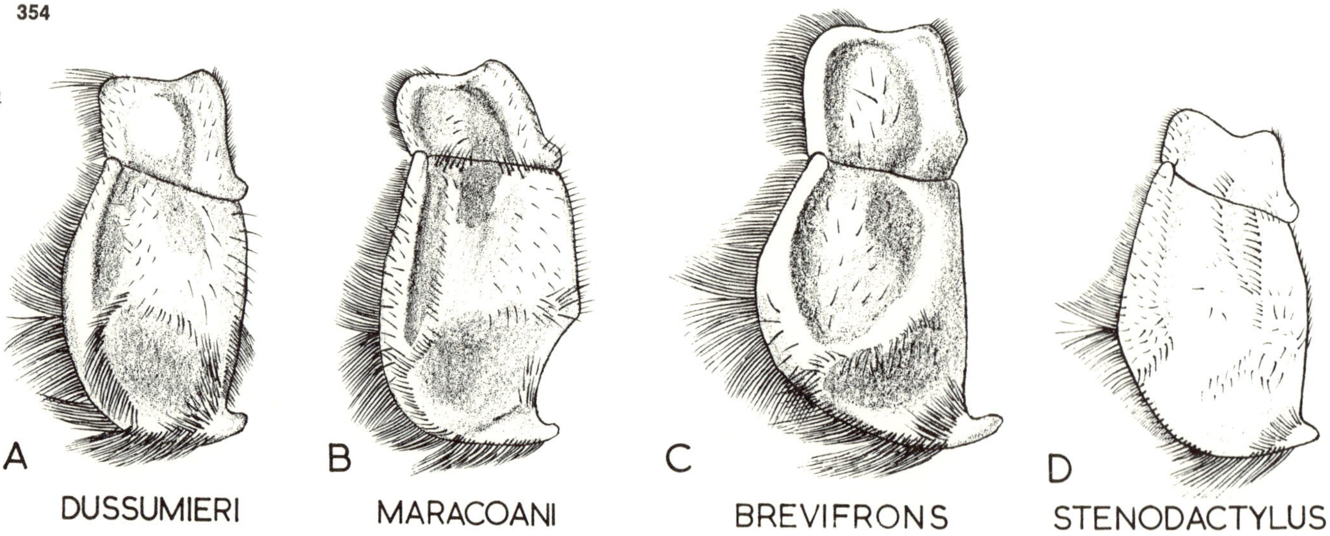

A
DUSSUMIERI

B
MARACOANI

C
BREVIFRONS

D
STENODACTYLUS

35

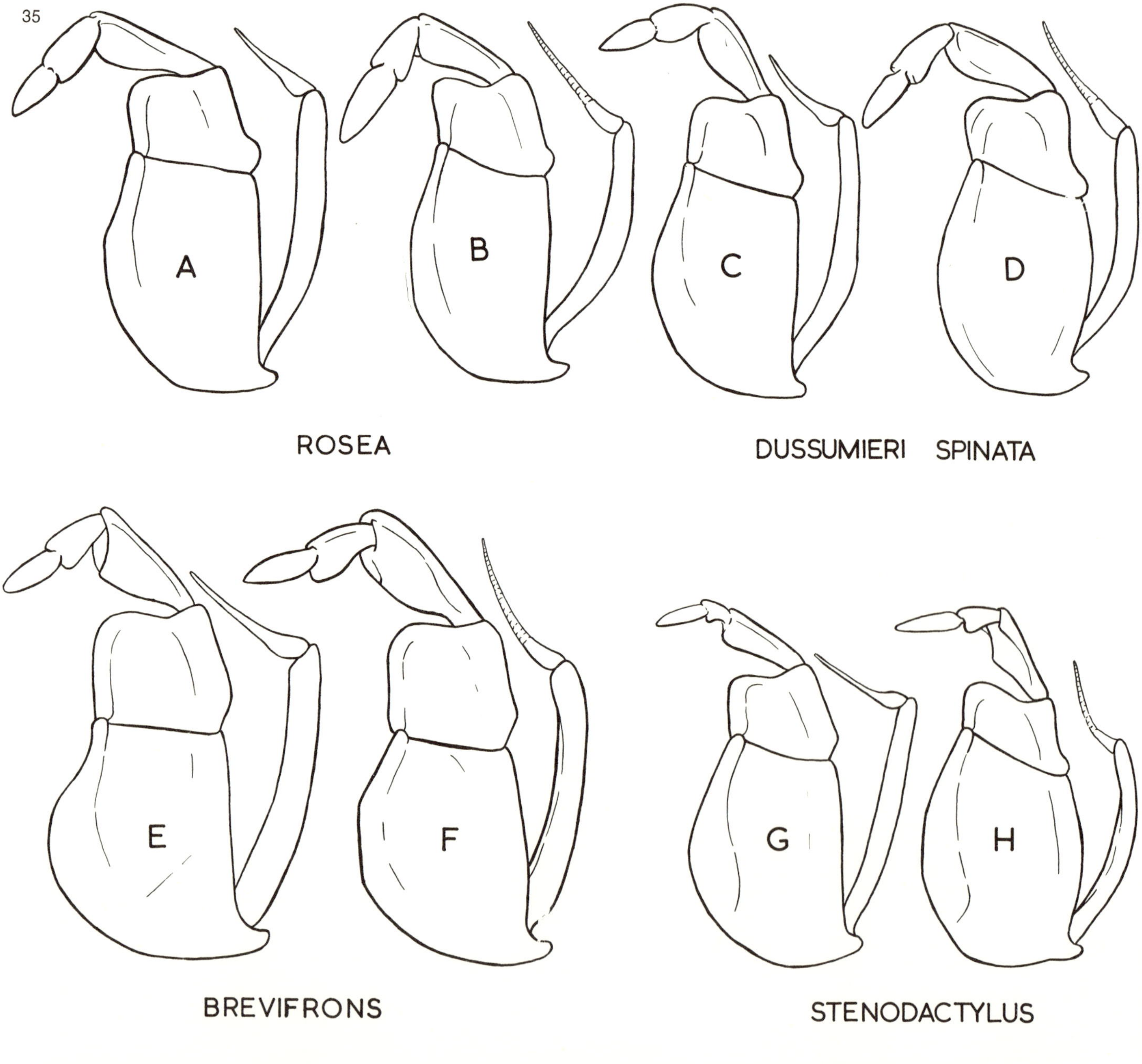

A

B

C

D

ROSEA

DUSSUMIERI   SPINATA

E

F

G

H

BREVIFRONS

STENODACTYLUS

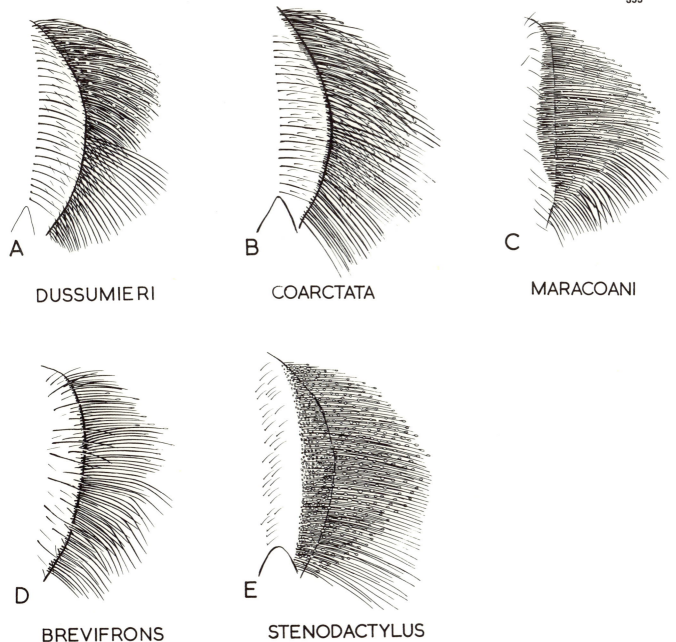

DUSSUMIERI

COARCTATA

MARACOANI

BREVIFRONS

STENODACTYLUS

FIG. 34. Ischium and merus of left third maxilliped in examples of four subgenera in *Uca*. *A*, (*Deltuca*) *dussumieri*; *B*, (*Uca*) *maracoani*; *C*, (*Minuca*) *brevifrons*; *D*, (*Celuca*) *stenodactylus*. Throughout, the length of the ischium is used as the constant. Drawn from mature males. (P. 455.)

FIG. 35. Ischium and merus of left third maxilliped in *Uca* to show comparison in young and mature crabs. Other data as in Fig. 34. Carapace lengths in mm are as follows. (*Deltuca*) *rosea*: *A*, 1.7; *B*, 13.0. (*Deltuca*) *dussumieri*: *C*, 4.0; *D*, 20.0. (*Minuca*) *brevifrons*: *E*, 3.4; *F*, 18.0. (*Celuca*) *stenodactylus*: *G*, 1.6; *H*, 9.0. (P. 455.)

FIG. 36. Merus of second maxilliped in five species of *Uca*, to show relative abundance of spoon-tipped setae. Subgenera: *A*, *B*, *Deltuca*; *C*, *Uca*; *D*, *Minuca*; *E*, *Celuca*. (P. 455.)

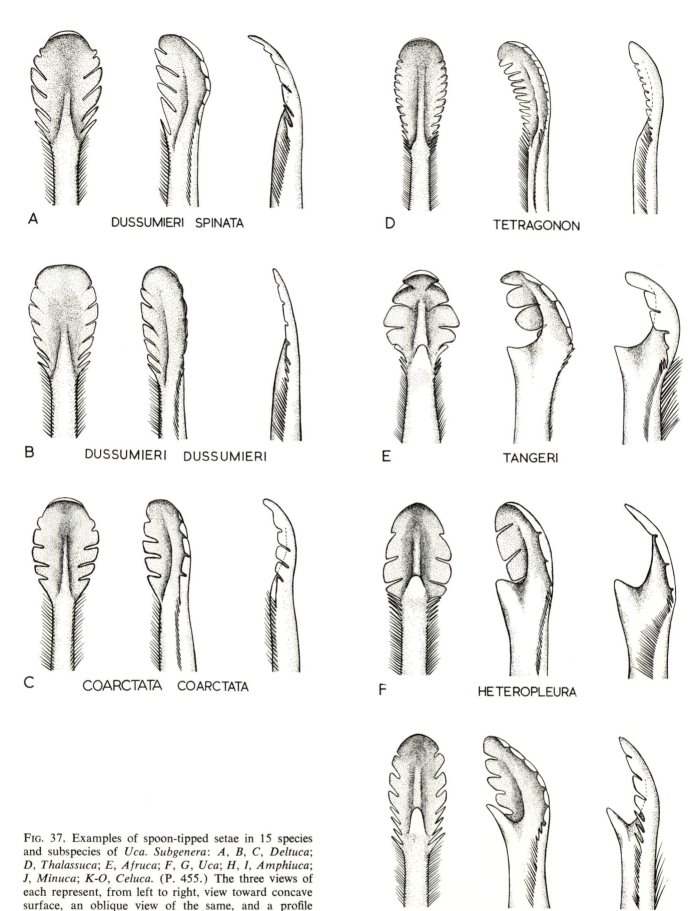

A   DUSSUMIERI SPINATA

B   DUSSUMIERI DUSSUMIERI

C   COARCTATA COARCTATA

D   TETRAGONON

E   TANGERI

F   HETEROPLEURA

G   MARACOANI MARACOANI

FIG. 37. Examples of spoon-tipped setae in 15 species and subspecies of *Uca. Subgenera: A, B, C, Deltuca; D, Thalassuca; E, Afruca; F, G, Uca; H, I, Amphiuca; J, Minuca; K-O, Celuca.* (P. 455.) The three views of each represent, from left to right, view toward concave surface, an oblique view of the same, and a profile view showing amount of distal curvature.

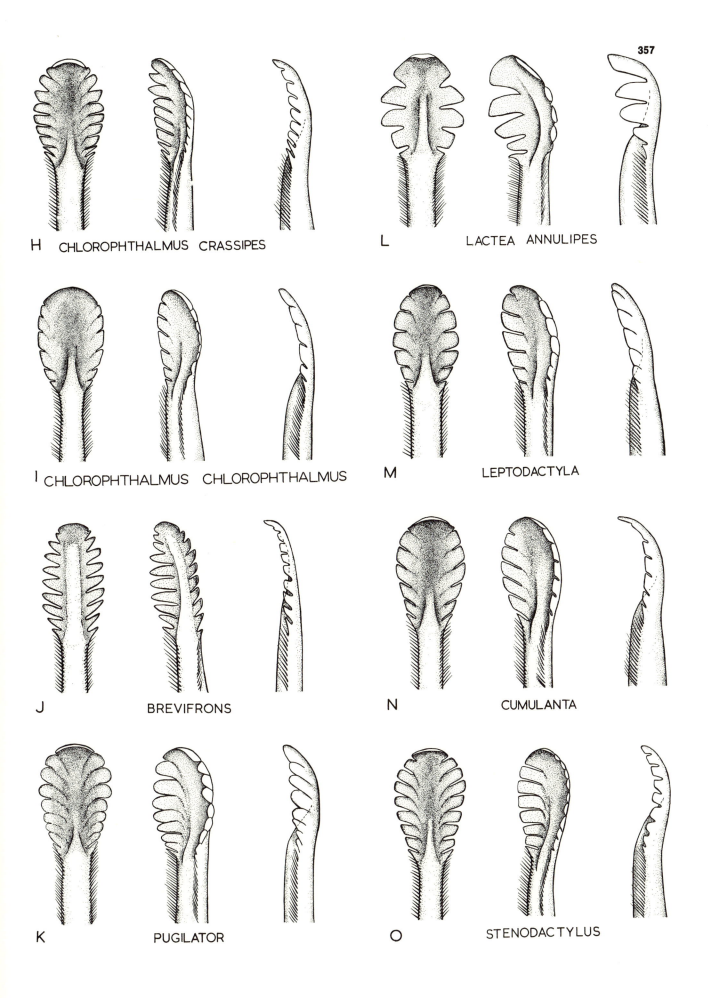

H CHLOROPHTHALMUS CRASSIPES

L LACTEA ANNULIPES

I CHLOROPHTHALMUS CHLOROPHTHALMUS

M LEPTODACTYLA

J BREVIFRONS

N CUMULANTA

K PUGILATOR

O STENODACTYLUS

358

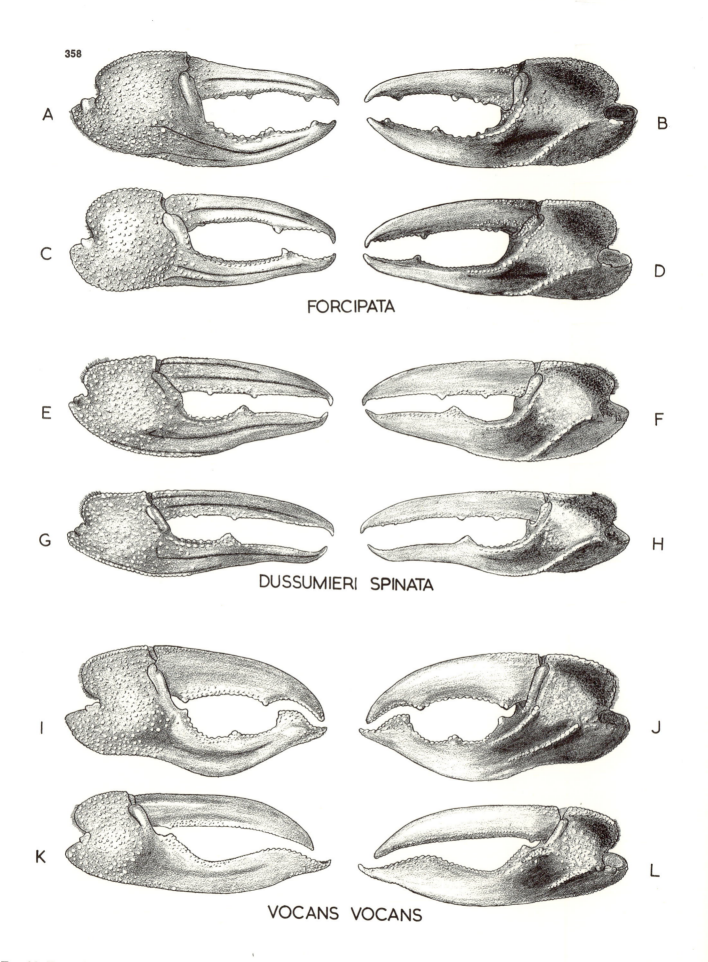

FORCIPATA

DUSSUMIERI SPINATA

VOCANS VOCANS

FIG. 38. Examples of dimorphism in claw of major cheliped in the subgenera *Deltuca* (*A-H* and *M-X*) and *Thalassuca* (*I-L*). In each species, outer views are on the left, inner on the right; brachychelous form above, leptochelous below. (P. 459.)

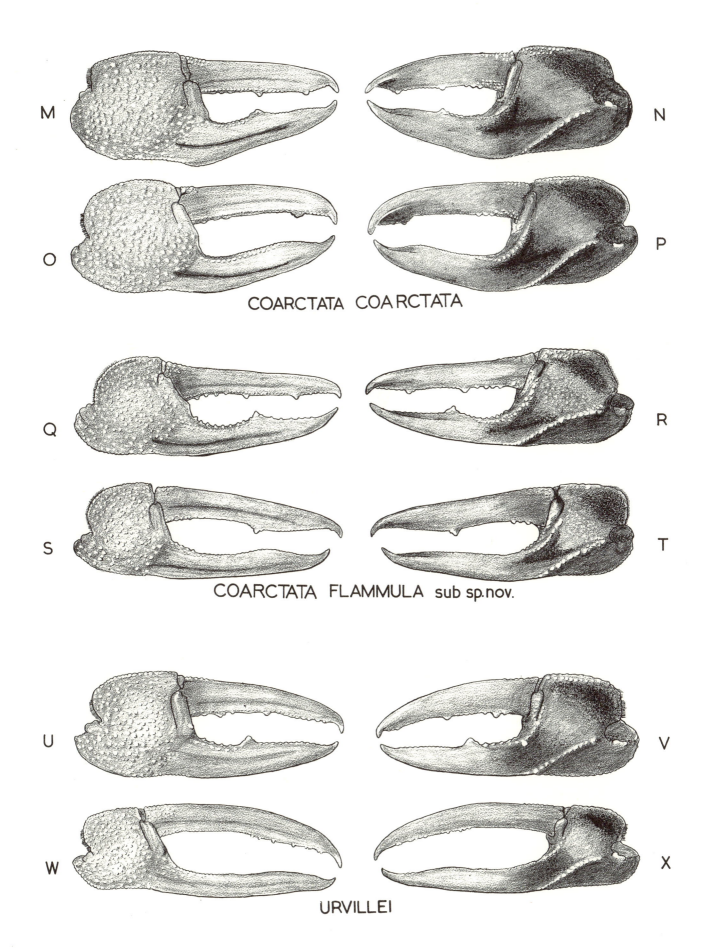

M

N

O

P

COARCTATA COARCTATA

Q

R

S

T

COARCTATA FLAMMULA sub sp.nov.

U

V

W

X

URVILLEI

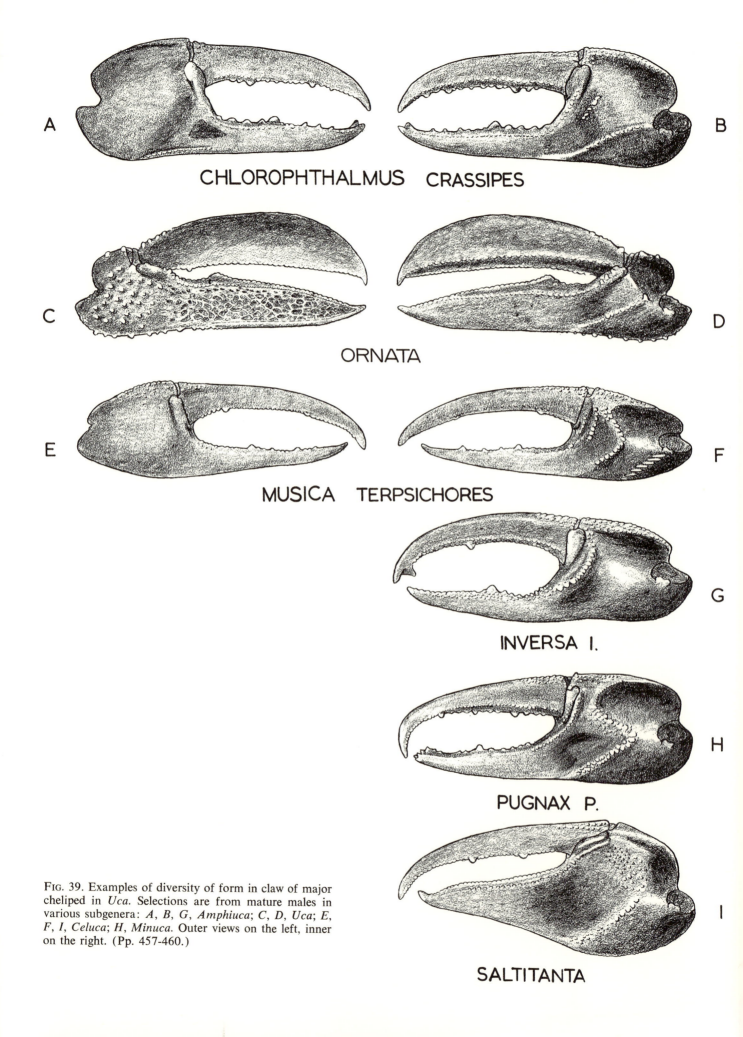

A

CHLOROPHTHALMUS CRASSIPES

B

C

ORNATA

D

E

MUSICA   TERPSICHORES

F

G

INVERSA I.

H

PUGNAX P.

I

SALTITANTA

Fig. 39. Examples of diversity of form in claw of major cheliped in *Uca*. Selections are from mature males in various subgenera: *A, B, G, Amphiuca*; *C, D, Uca*; *E, F, I, Celuca*; *H, Minuca*. Outer views on the left, inner on the right. (Pp. 457-460.)

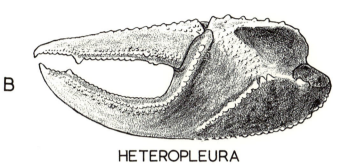

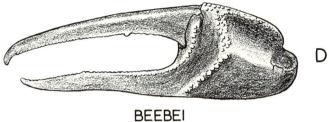

HETEROPLEURA

BEEBEI

FIG. 40. Examples of dimorphism in claw of major cheliped in the subgenera *Uca* (*A*, *B*) and *Celuca* (*C*, *D*), for comparison with Fig. 38. (P. 459.)

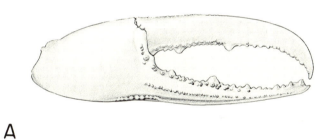

A

B

FIG. 41. Outer side of major claw in two subspecies of *U.* (*Celuca*) *lactea*: *A*, *l. annulipes*, from Inhaca I., Mozambique; pollex ridge here often reaches maximum development. *B*, *l. perplexa* from the Fiji Is.; dactyl breadth and triangular structure at pollex tip are usually more strongly developed in this eastern part of the species range than elsewhere. (Pp. 295, 639.)

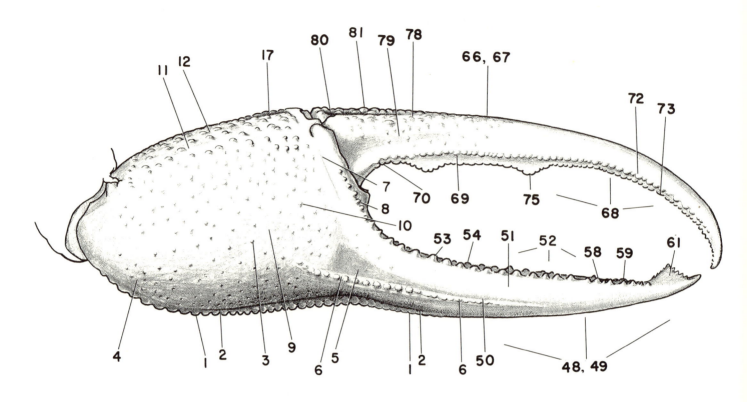

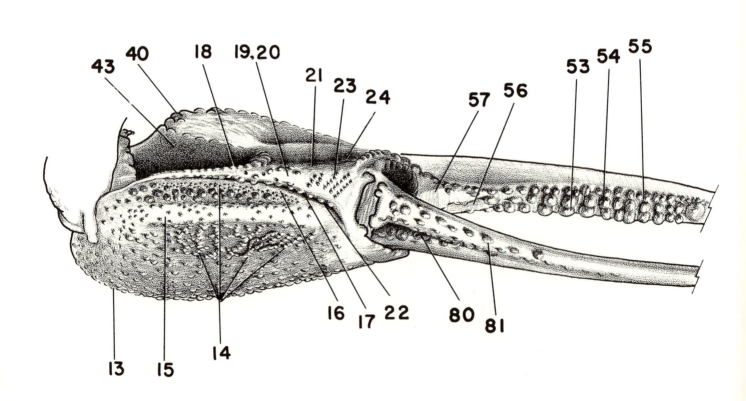

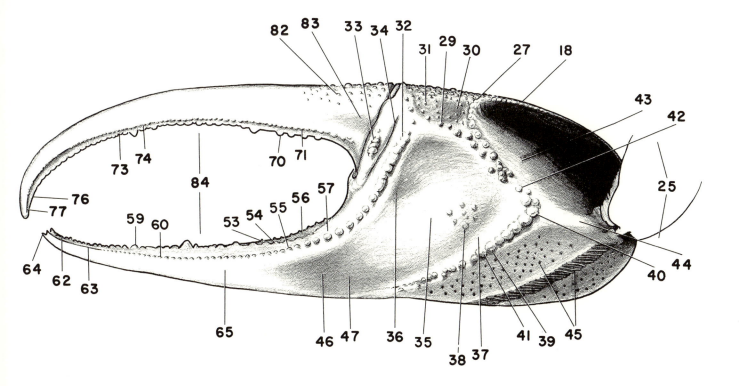

Fig. 42. Semi-diagrammatic view of outer claw in *Uca*, to show location and general appearance of contact areas functioning during intermale combat. The organization of the numbered regions into combat components is shown in Table 13. See also Table 14 and text, pp. 457ff and 487ff. The drawing is composite with characteristics drawn from a number of species.

Fig. 43. As in Fig. 42, except that the view is obliquely dorsal, showing parts of both outer and inner surfaces of claw. In this composite other examples have been incorporated.

Fig. 44. As in Figs. 42 and 43, except that the view is of the inner (palmar) surface. Again, other examples have been used in constructing the composite; for example, the armature of the gape in this drawing does not correspond to that shown in Fig. 42.

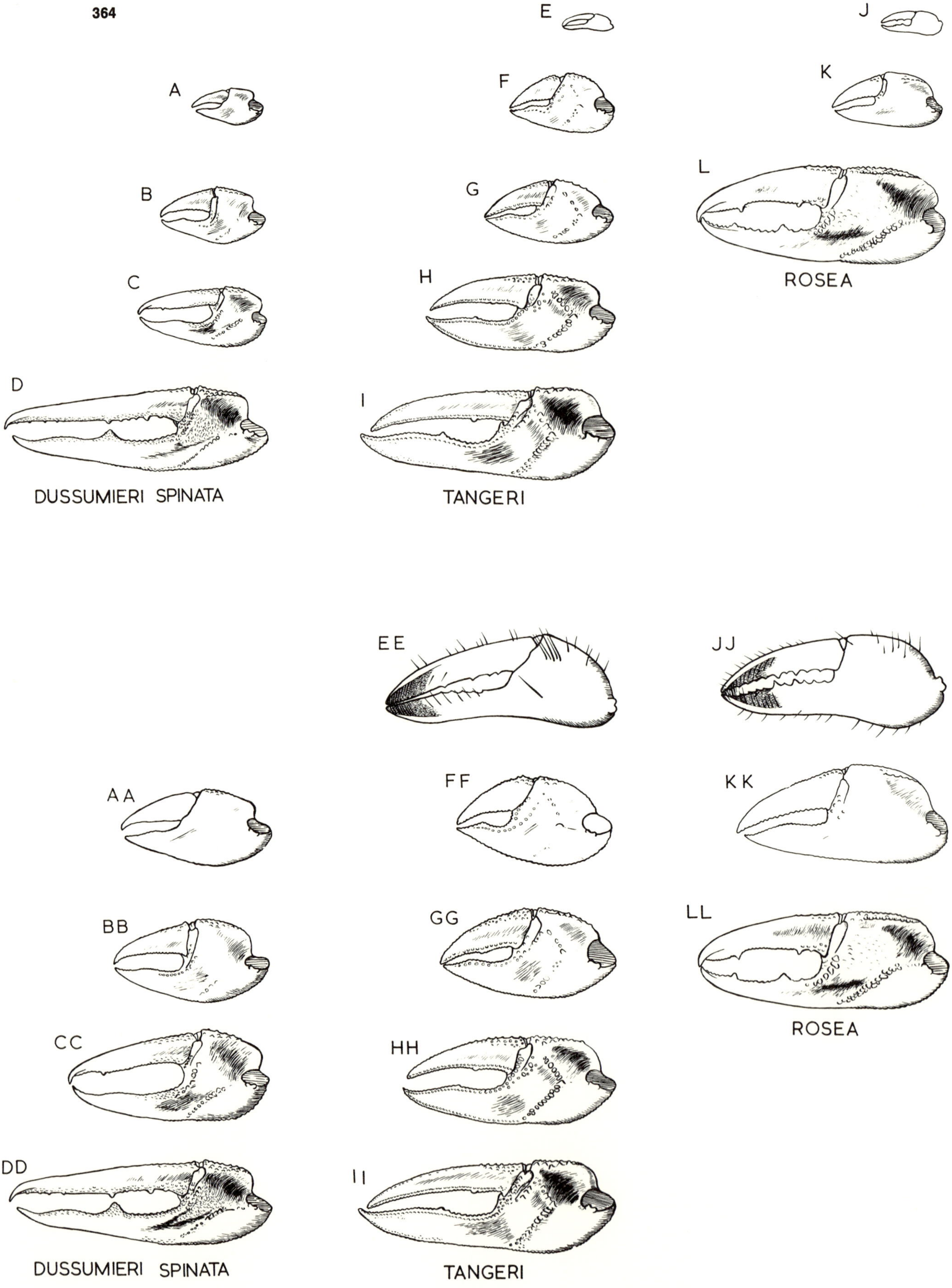

A

B

C

D

DUSSUMIERI SPINATA

E

F

G

H

I

TANGERI

J

K

L

ROSEA

EE

JJ

AA

FF

KK

BB

GG

LL

CC

HH

ROSEA

DD

II

DUSSUMIERI SPINATA

TANGERI

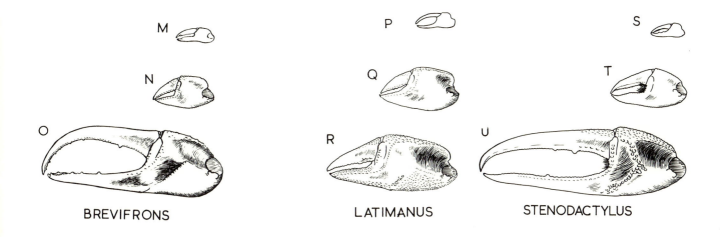

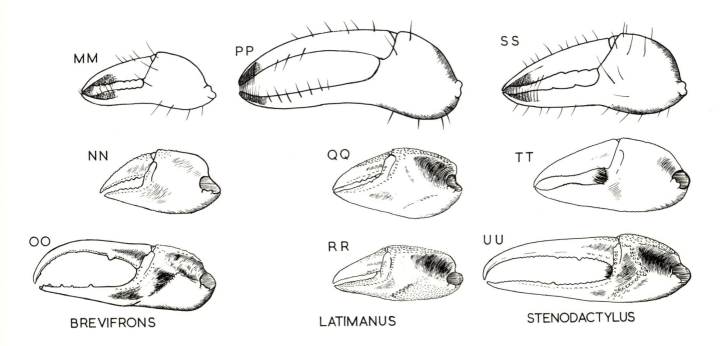

FIG. 45. Growth of major claw in six species of *Uca*. *A-U*, each claw drawn in proportion to length of that specimen's carapace; *AA-UU*, drawn with the length of manus constant throughout, to show changes in lengths of dactyl and pollex relative to manus. Carapace lengths in mm are as follows. (*Deltuca*) *dussumieri spinata*: *A, AA*, 4.0; *B, BB*, 7.6; *C, CC*, 12.0; *D, DD*, 20.0. (*Afruca*) *tangeri*: *E, EE*, 3.0; *F, FF*, 4.7; *G, GG*, 6.0; *H, HH*, 14.0; *I, II*, 24.0. (*Deltuca*) *rosea*: *J, JJ*, 1.7; *K, KK*, 3.4; *L, LL*, 13.0. (*Minuca*) *brevifrons*: *M, MM*, 18.0; *N, NN*, 3.4; *O, OO*, 4.7. (*Celuca*) *latimanus*: *P, PP*, 3.0; *Q, QQ*, 4.6; *R, RR*, 8.1. (*Celuca*) *stenodactylus*: *S, SS*, 1.6; *T, TT*, 20.0; *U, UU*, 9.0. (P. 459.)

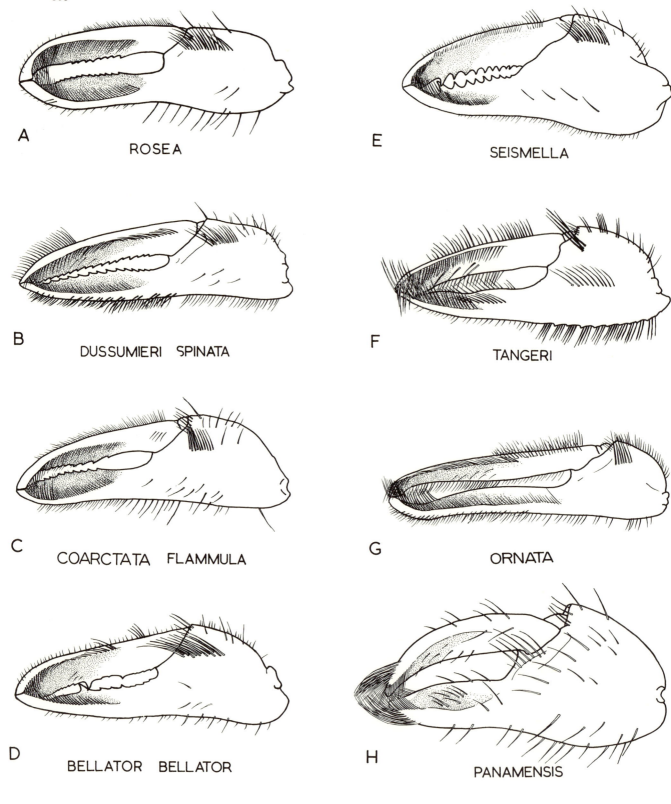

A ROSEA

B DUSSUMIERI SPINATA

C COARCTATA FLAMMULA

D BELLATOR BELLATOR

E SEISMELLA

F TANGERI

G ORNATA

H PANAMENSIS

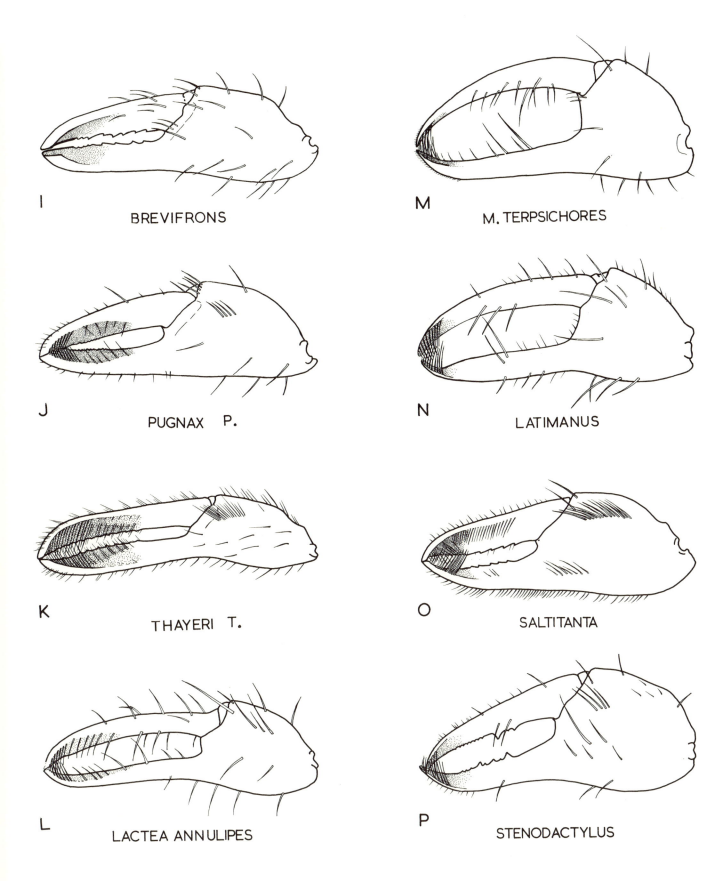

I BREVIFRONS

M M. TERPSICHORES

J PUGNAX P.

N LATIMANUS

K THAYERI T.

O SALTITANTA

L LACTEA ANNULIPES

P STENODACTYLUS

FIG. 46. Examples of diversity of form in claw of minor cheliped. Seven subgenera in *Uca* are illustrated by one or more examples, all drawn from mature males. Subgenera: *A-C, Deltuca; D-E, Australuca; F, Afruca; G, Uca; H-J, Minuca; K, Boboruca; L-P, Celuca.* (P. 460.)

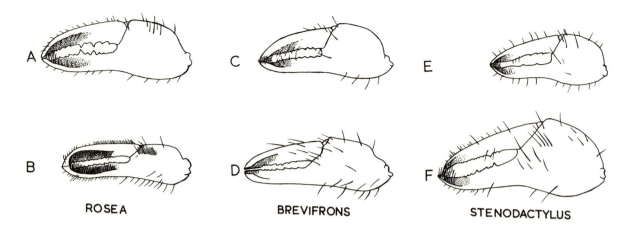

FIG. 47. Growth of minor claw in three species of *Uca*, each claw drawn in proportion to length of that specimen's carapace. Principal setae included. Carapace lengths in mm are as follows. (*Deltuca*) *rosea*: *A*, 1.7; *B*, 13.0. (*Minuca*) *brevifrons*: *C*, 3.4; *D*, 18.0. (*Celuca*) *stenodactylus*: *E*, 1.4; *F*, 9.0. (P. 461.)

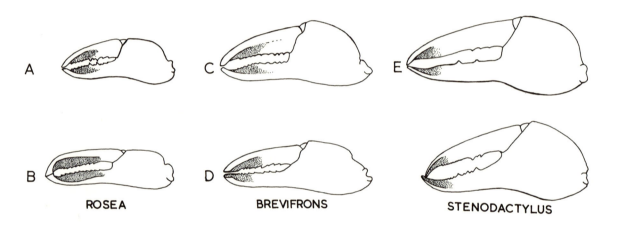

FIG. 48. Growth of minor claw in three species of *Uca*, drawn with the length of the manus constant throughout, to show changes in lengths of pollex and dactyl relative to manus. Setae omitted. Carapace lengths in mm are as follows. (*Deltuca*) *A*, 1.7; *B*, 13.0. (*Minuca*) *brevifrons*: *C*, 3.4; *D*, 18.0. (*Celuca*) *stenodactylus*: *E*, 1.4; *F*, 9.0. (P. 461.)

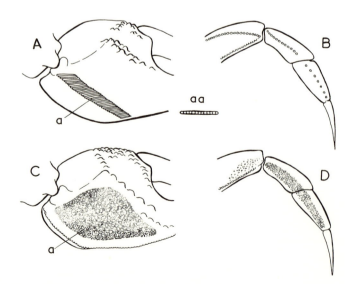

FIG. 49. Armature of proximal triangle on palm and of first ambulatory in two species of the subgenus *Celuca*. *A, B, musica musica*: *C, D, beebei*; *a*, striae and tubercles; *aa*, enlargement of single stria. (Pp. 278, 315, 458.)

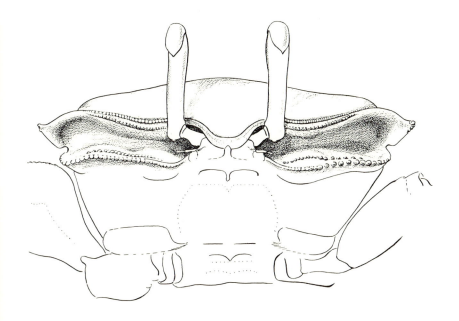

FIG. 50. Armature of orbits in (*Celuca*) *triangularis triangularis*. Note differences in position and spacing of tubercles, forming a row on floor of each orbit, on major and minor sides. (Pp. 286, 290.)

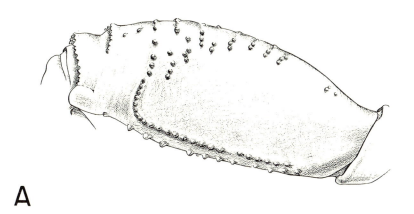

A

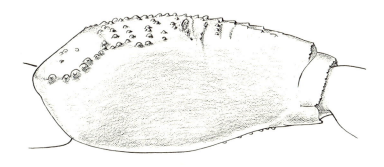

B

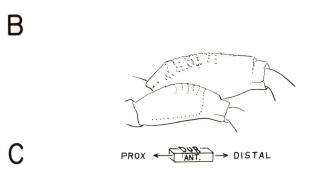

C

PROX ← DOR / ANT. → DISTAL

FIG. 51. Details of certain elements of armature in (*Celuca*) *triangularis triangularis*: *A*, merus of minor cheliped, posterior view; *B*, first ambulatory merus, anterior view tilted slightly forward to show tubercles otherwise hidden; *C*, meri of minor and first ambulatory, anterior view, to show juxtaposition. The stippling illustrates positions of tubercles on posterior side. (Pp. 287, 290.)

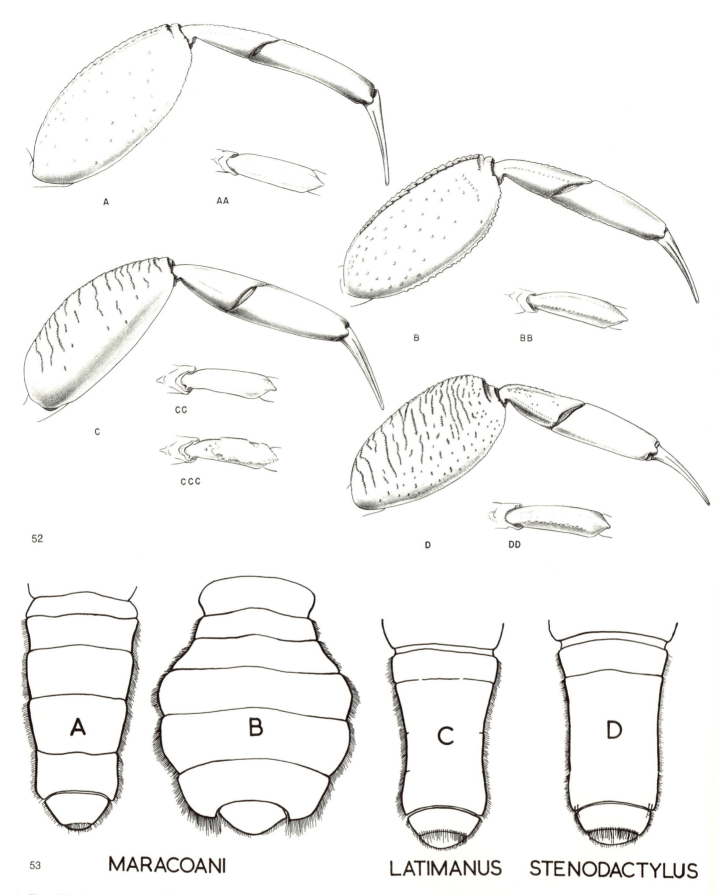

52

53

MARACOANI

LATIMANUS   STENODACTYLUS

FIG. 52. Armature on third ambulatory in two subgenera of *Uca*. *A*, posterior view, *AA*, dorsal view of
carpus in male (*Deltuca*) *dussumieri*; *B*, *BB*, same in
female. *C*, *CC*, corresponding structures in male
(*Minuca*) *rapax*, the pile having been removed from
CC; *CCC*, pile in place; *D*, *DD*, same in female (pile on
carpus always absent). (Pp. 16, 17, 462.)

FIG. 53. Form of abdomen in *Uca*, dorsal (external)
views. *A*, usual form in male; *B*, female. *C*, male showing partial fusion of some segments characteristic of a
few species of *Celuca*; *D*, male showing maximum fusion. (P. 463.)

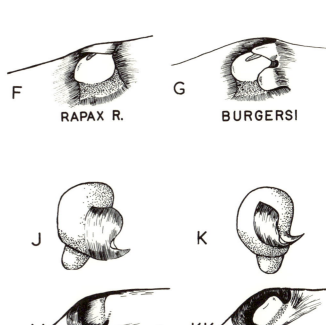

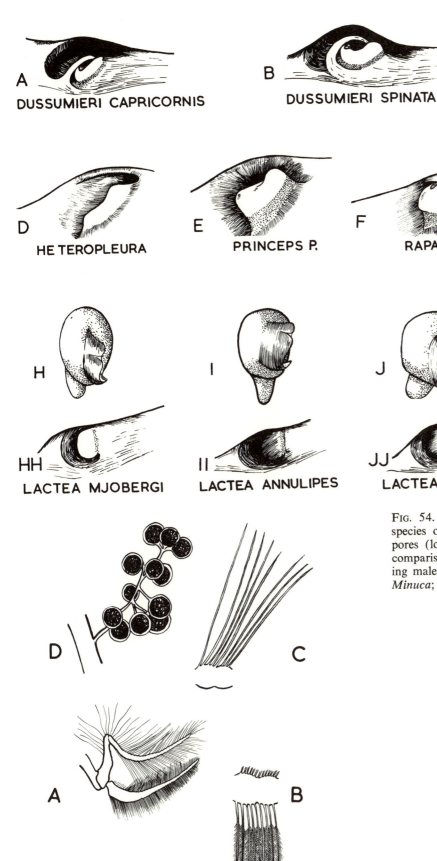

DUSSUMIERI CAPRICORNIS

DUSSUMIERI SPINATA

DUSSUMIERI DUSSUMIERI

HE TEROPLEURA

PRINCEPS P.

RAPAX R.

BURGERSI

LACTEA MJOBERGI

LACTEA ANNULIPES

LACTEA LACTEA

LACTEA PERPLEXA

FIG. 54. Gonopores in examples of species and subspecies of *Uca*. *A-G*, gonopores alone; *H-KK*, gonopores (lower row) of subspecies of (*Celuca*) *lactea* in comparison with distal view of gonopod in corresponding male. Subgenera: *A-C, Deltuca; D, E, Uca; F, G, Minuca; H-KK, Celuca.* (P. 465.)

FIG. 55. Eggs and pleopods in *Uca (Uca) maracoani*. *A*, pleopod; *B*, setae on exopodite; *C*, setae on endopodite; *D*, attachment of eggs to setae on endopodite. (P. 465.)

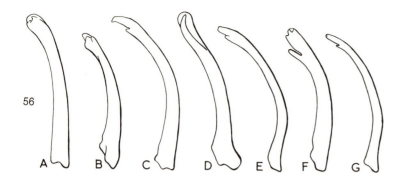

56

FIG. 56. Examples of gonopods in *Uca*, to show general range of shape. Outer views of right gonopods. *A*, (*Deltuca*) *dussumieri spinata*; *B*, (*Thalassuca*) *vocans vocans*; *C*, (*Amphiuca*) *chlorophthalmus crassipes*; *D*, (*Uca*) *maracoani maracoani*; *E*, (*Boboruca*) *thayeri thayeri*; *F*, (*Celuca*) *leptodactyla*; *G*, (*Celuca*) *cumulanta*. (P. 463.)

FIG. 57. Examples of gonopod tips in six genera of the family Ocypodidae. (P. 463.)

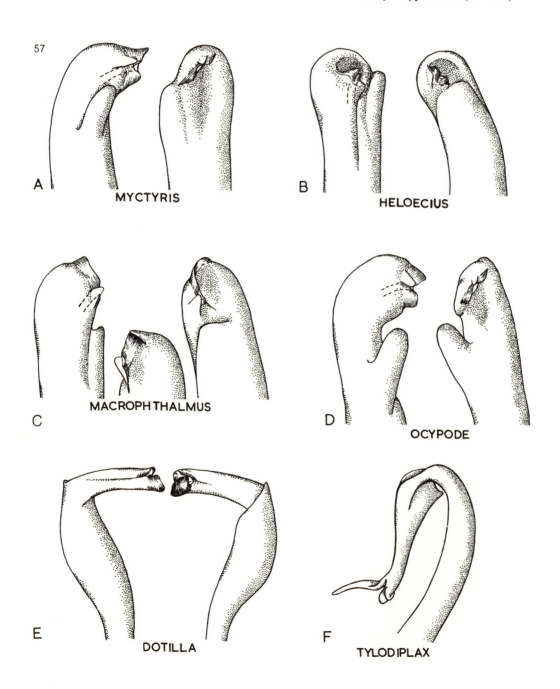

57

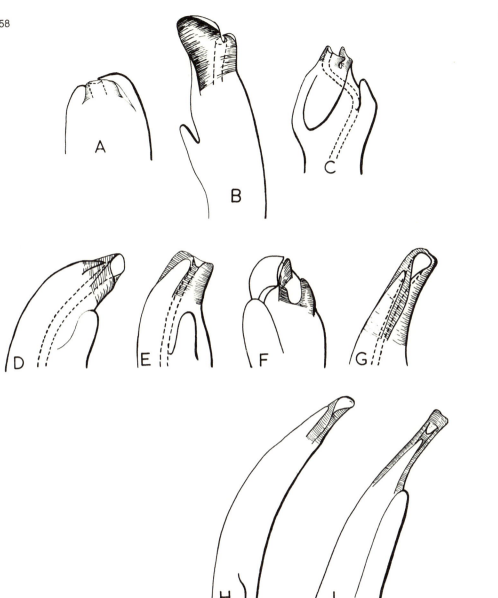

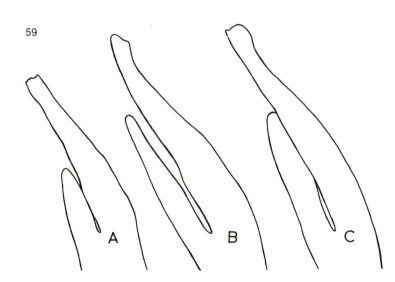

FIG. 58. Examples of characteristics of gonopod tips in *Uca. A*, simple flanges. *B*, large flange; thumb ending well below flange base. *C*, strong torsion. *D*, inner process spiny; genital pore large. *E*, inner process flat; genital pore small. *F*, inner process tumid; one flange deeply excavate; thumb large, reaching beyond flange base. *G*, inner process transparent; distal shaft of canal tubular, corneous, projecting; flanges inconspicuous to absent. *H*, edges of distal tubular shaft of canal clearly overlapping; thumb absent. *I*, canal appears as a simple, narrow tube fastened against a single flange that projects beyond it; thumb very long. (P. 463.)

FIG. 59. *Uca (Celuca) triangularis*: variation in tips of gonopod. *A, triangularis bengali*, Penang, Malaya; *B, triangularis triangularis*, Magnanud R., Victorias, Negros Occidentalis, Philippine Is.; *C, triangularis triangularis*, Madaum, Mindanao, Philippine Is. Approximate carapace length in each, 8 mm. (Pp. 286, 463.)

374

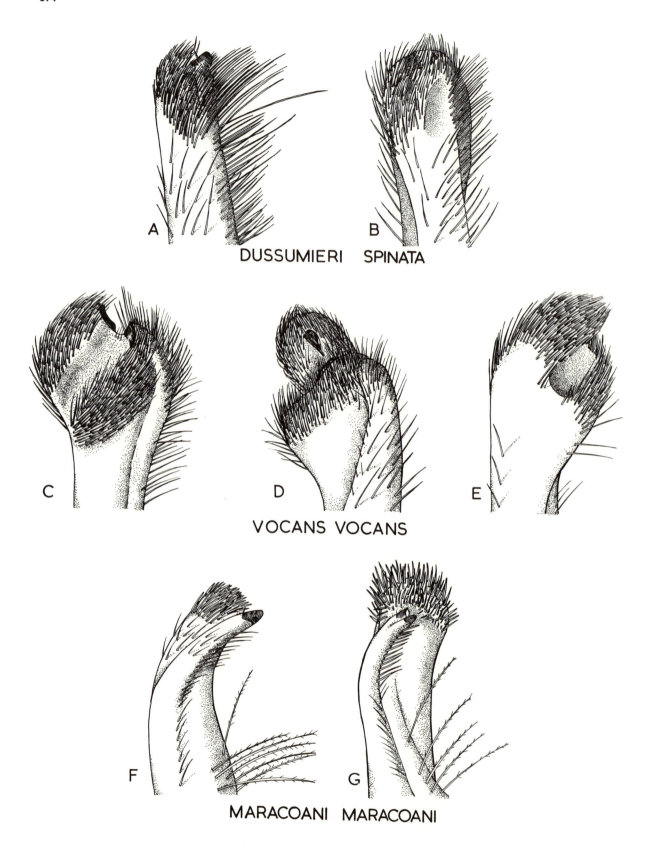

A

B

DUSSUMIERI SPINATA

C

D

E

VOCANS VOCANS

F

G

MARACOANI MARACOANI

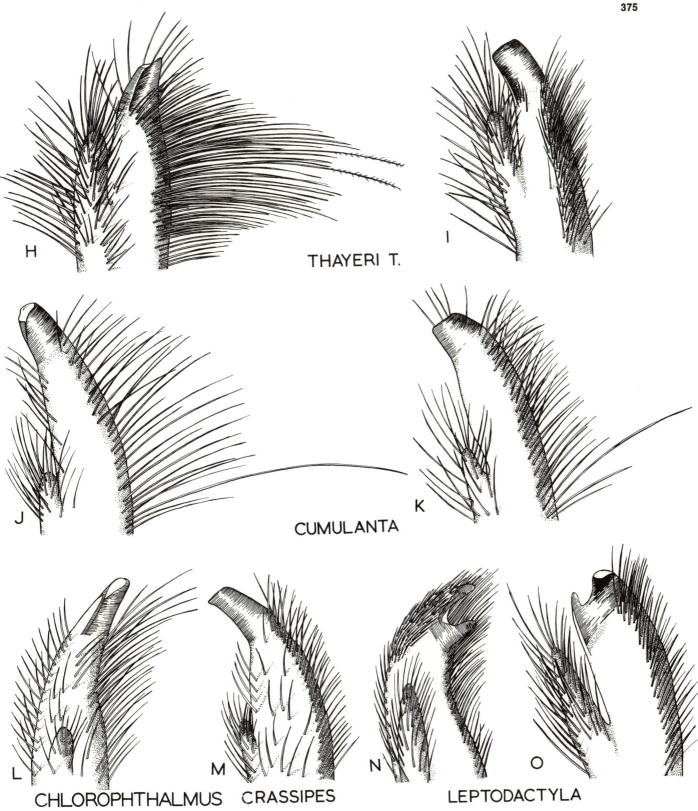

THAYERI T.

CUMULANTA

CHLOROPHTHALMUS CRASSIPES LEPTODACTYLA

FIG. 60. Examples of gonopod tips in seven species of *Uca*, showing setae intact. Subgenera: *A-B, Deltuca*; *C-E, Thalassuca*; *F-G, Uca*; *H-I, Boboruca*; *J-K, N-O, Celuca*; *L-M, Amphiuca*. (Setae have been removed in all other text-figures of gonopods.) (P. 463.)

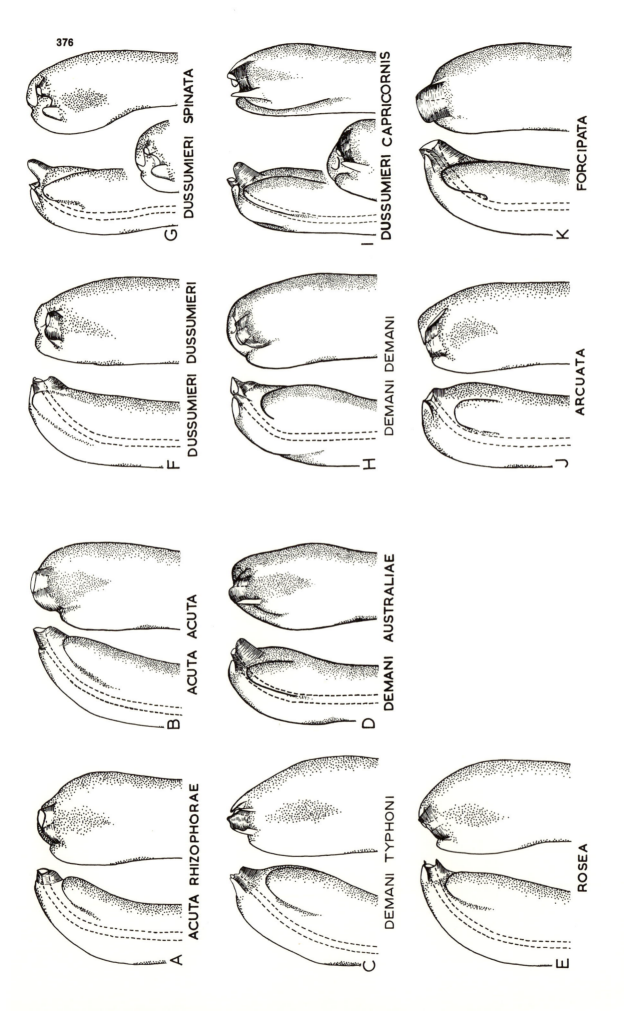

A ACUTA RHIZOPHORAE

B ACUTA ACUTA

C DEMANI TYPHONI

D DEMANI AUSTRALIAE

E ROSEA

F DUSSUMIERI DUSSUMIERI

G DUSSUMIERI SPINATA

H DEMANI DEMANI

I DUSSUMIERI CAPRICORNIS

J ARCUATA

K FORCIPATA

FIG. 61. Gonopod tips in the subgenus *Deltuca* (part). In these and all subsequent drawings through Fig. 73, the right gonopod is depicted. Unless otherwise noted, the left drawing in each pair represents approximately an antero-lateral view, while the right is approximately the lateral view from the outer side. Minor differences in orientation, not specially noted, were occasionally required because of torsion, in order to facilitate interspecific comparisons. (Pp. 22ff, 463.)

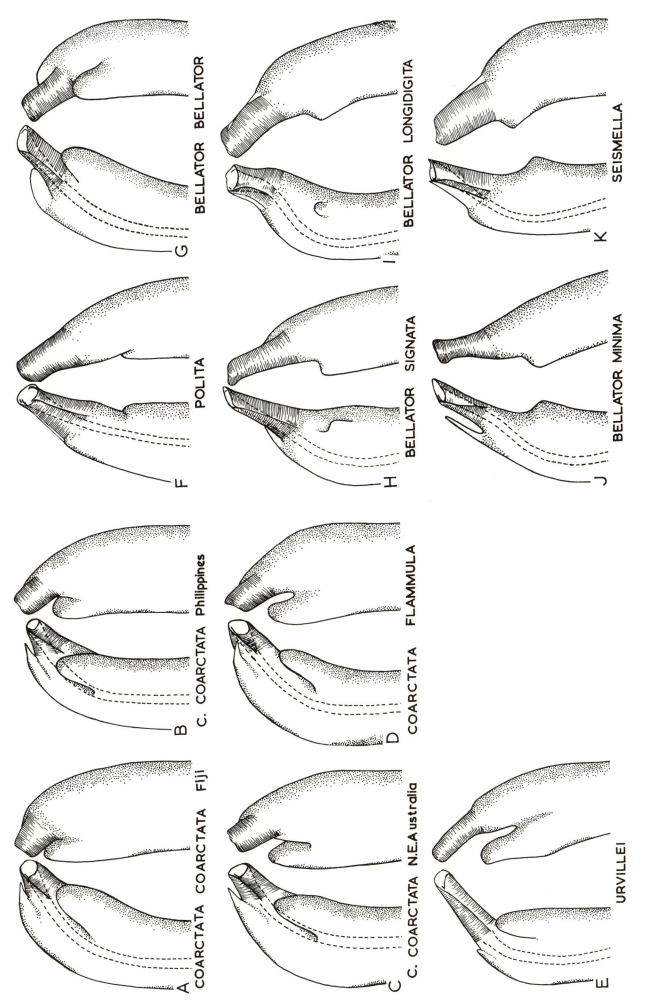

FIG. 62. *A-E*, gonopod tips in the subgenus *Deltuca* (concluded). *F-K*, gonopod tips in the subgenus *Australuca* (complete). Orientation data as in Fig. 61. (Pp. 22, 63, 463.)

A COARCTATA COARCTATA Fiji

B C. COARCTATA Philippines

C C. COARCTATA N.E.Australia

D COARCTATA FLAMMULA

E URVILLEI

F POLITA

G BELLATOR BELLATOR

H BELLATOR SIGNATA

I BELLATOR LONGIDIGITA

J BELLATOR MINIMA

K SEISMELLA

FIG. 63. *A-C*, gonopod tips in the subgenus *Thalassuca* (part). *D*, gonopod tip in (*Afruca*) *tangeri*, the unique member of the subgenus. Middle drawing in *A*, distal view. Middle drawing in *D*, posterior view. Orientation data otherwise as in Fig. 61. (Pp. 75, 117, 463.)

FIG. 64. *A-F*, gonopod tips in the subgenus *Thalassuca* (concluded). *AA-FF*, female gonopores in corresponding subspecies of *vocans*. Orientation data on gonopods as in Fig. 61. The small central drawings show relative sizes of flanges, otherwise partly concealed, from various aspects. Left gonopores depicted, all from an antero-ventral view. (Pp. 75, 463.)

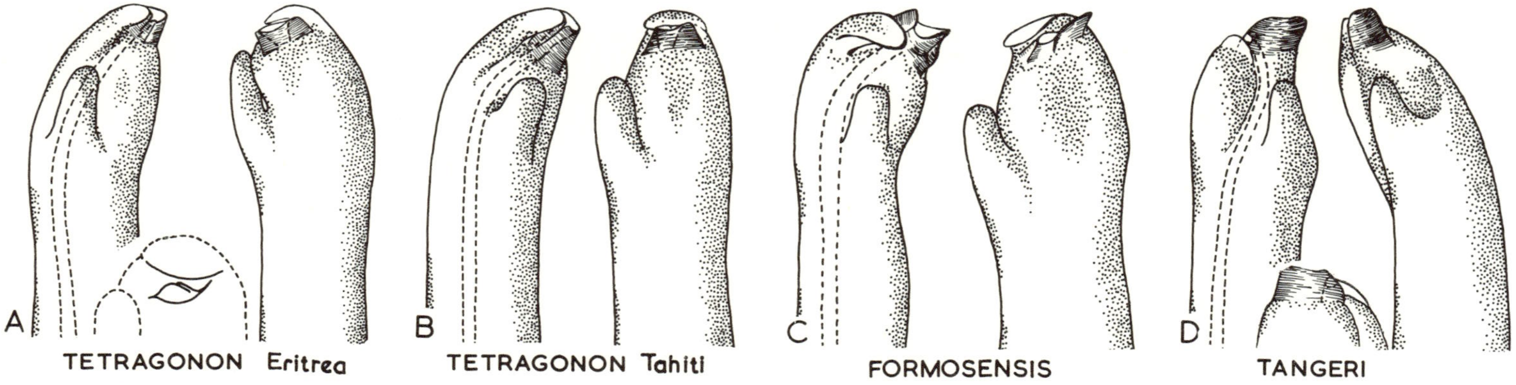

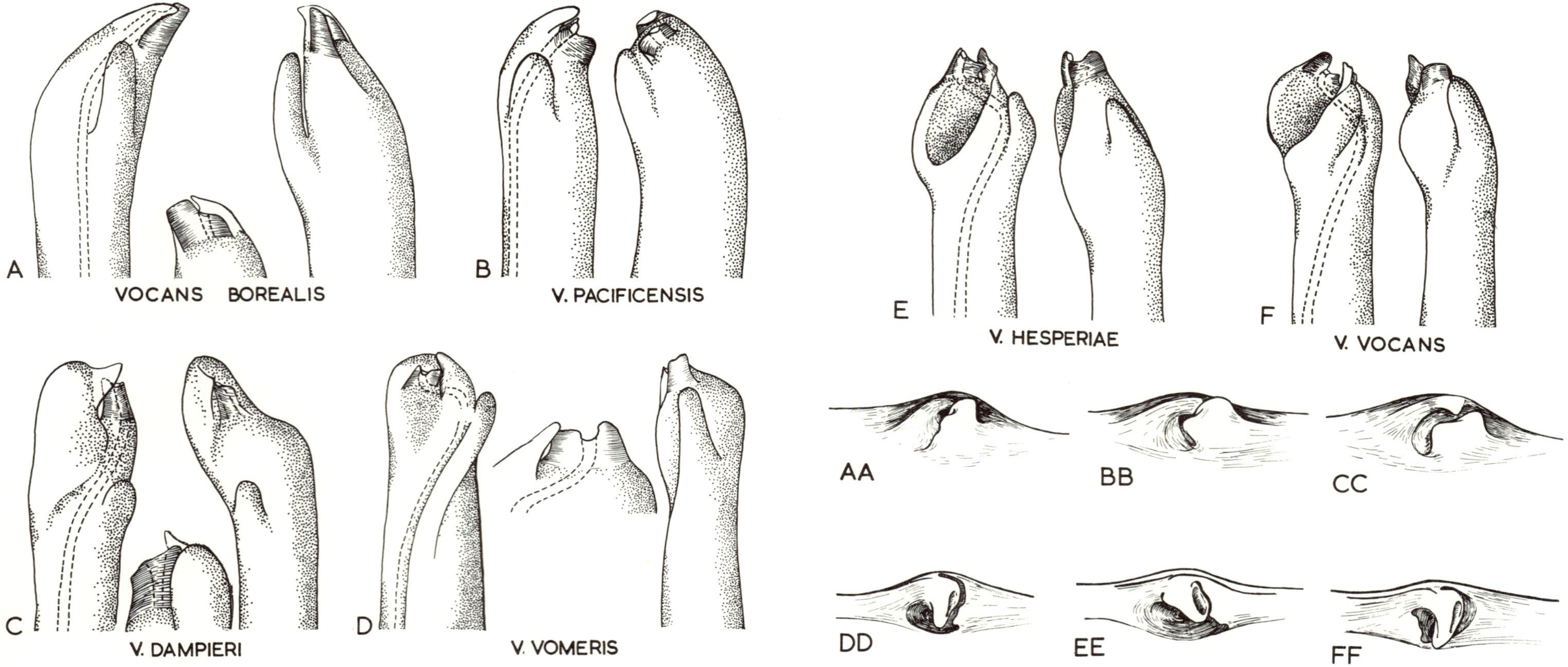

TETRAGONON Eritrea

TETRAGONON Tahiti

FORMOSENSIS

TANGERI

A   B   C   D

VOCANS BOREALIS

V. PACIFICENSIS

V. HESPERIAE

V. VOCANS

A   B   E   F

V. DAMPIERI

V. VOMERIS

C   D

AA   BB   CC

DD   EE   FF

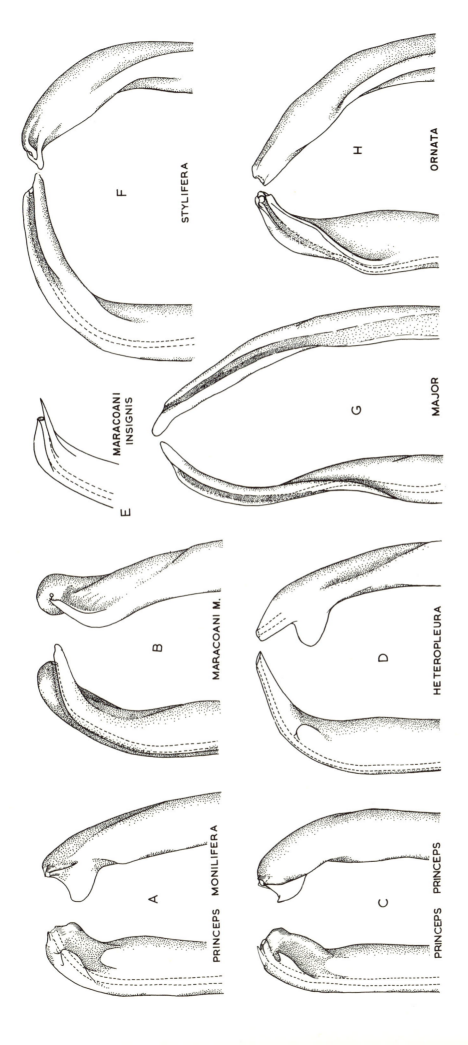

FIG. 65. Gonopod tips in the subgenus *Uca* (complete). Orientation data roughly as in Fig. 61, but most of these species show strong torsion. (Pp. 126, 463.)

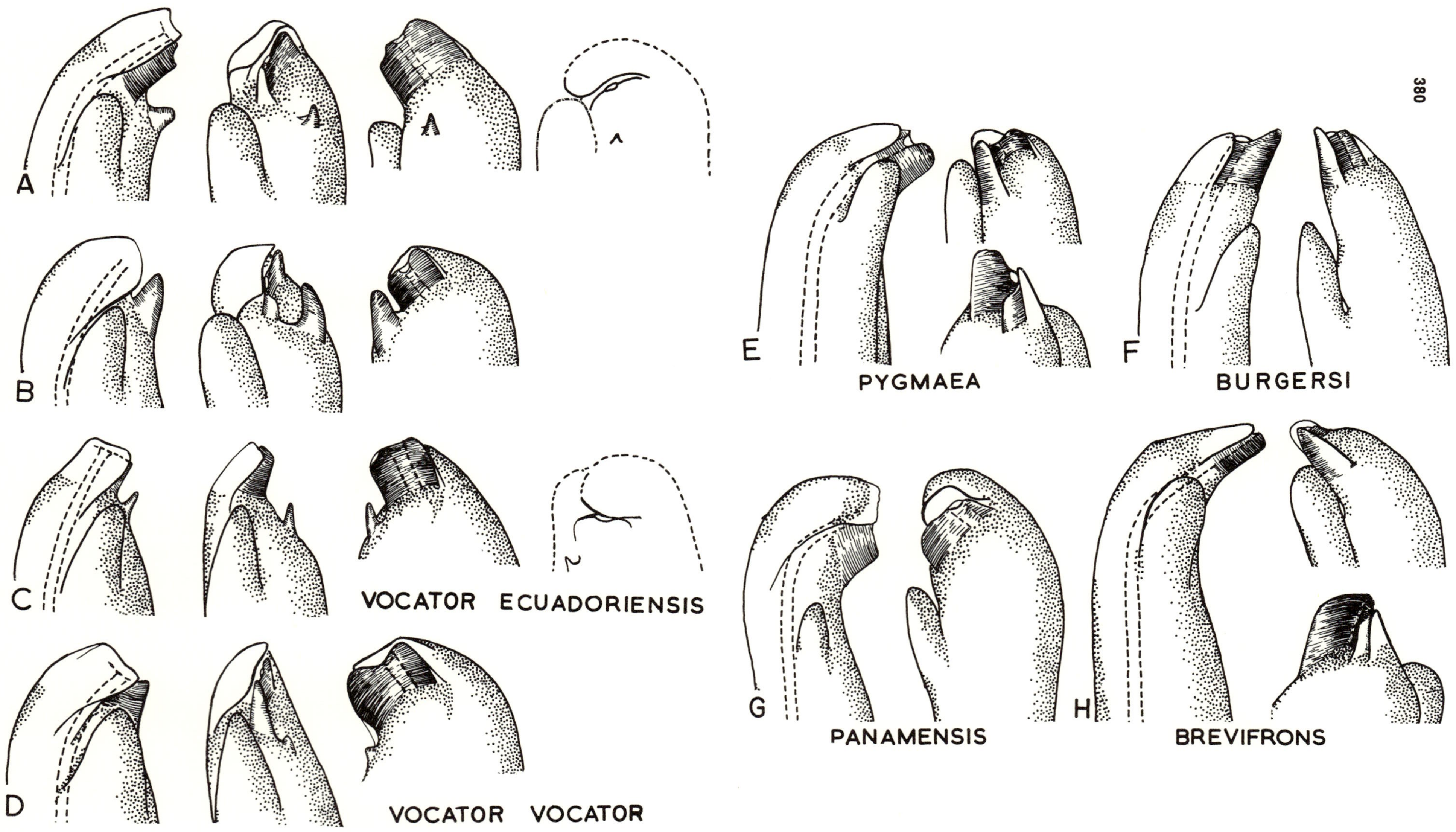

BURGERSI

F

PYGMAEA

E

BREVIFRONS

H

PANAMENSIS

G

VOCATOR ECUADORIENSIS

VOCATOR VOCATOR

A

B

C

D

Fig. 66. Gonopod tips in the subgenus *Minuca* (part). Orientation data as in Fig. 61, except that third, or third and fourth, drawings when present illustrate posterior and partly distal views. (Pp. 155, 463.)

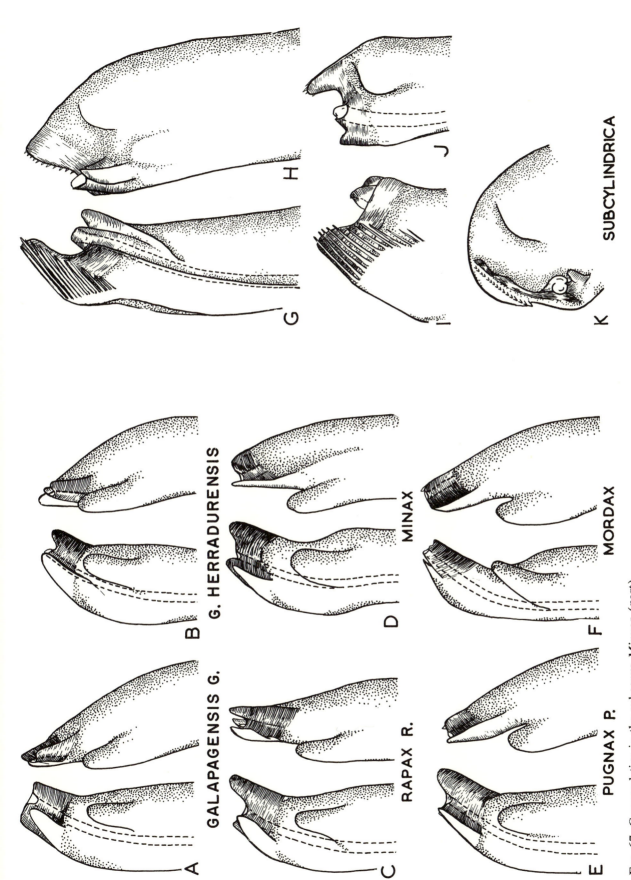

Fig. 67. Gonopod tips in the subgenus *Minuca* (part). *A-H,* orientation data about as in Fig. 61; *I, J, K,* approximately inner-posterior, posterior, and distal views, respectively. (Pp. 155, 463.)

GALAPAGENSIS G.

G. HERRADURENSIS

RAPAX R.

MINAX

PUGNAX P.

MORDAX

SUBCYLINDRICA

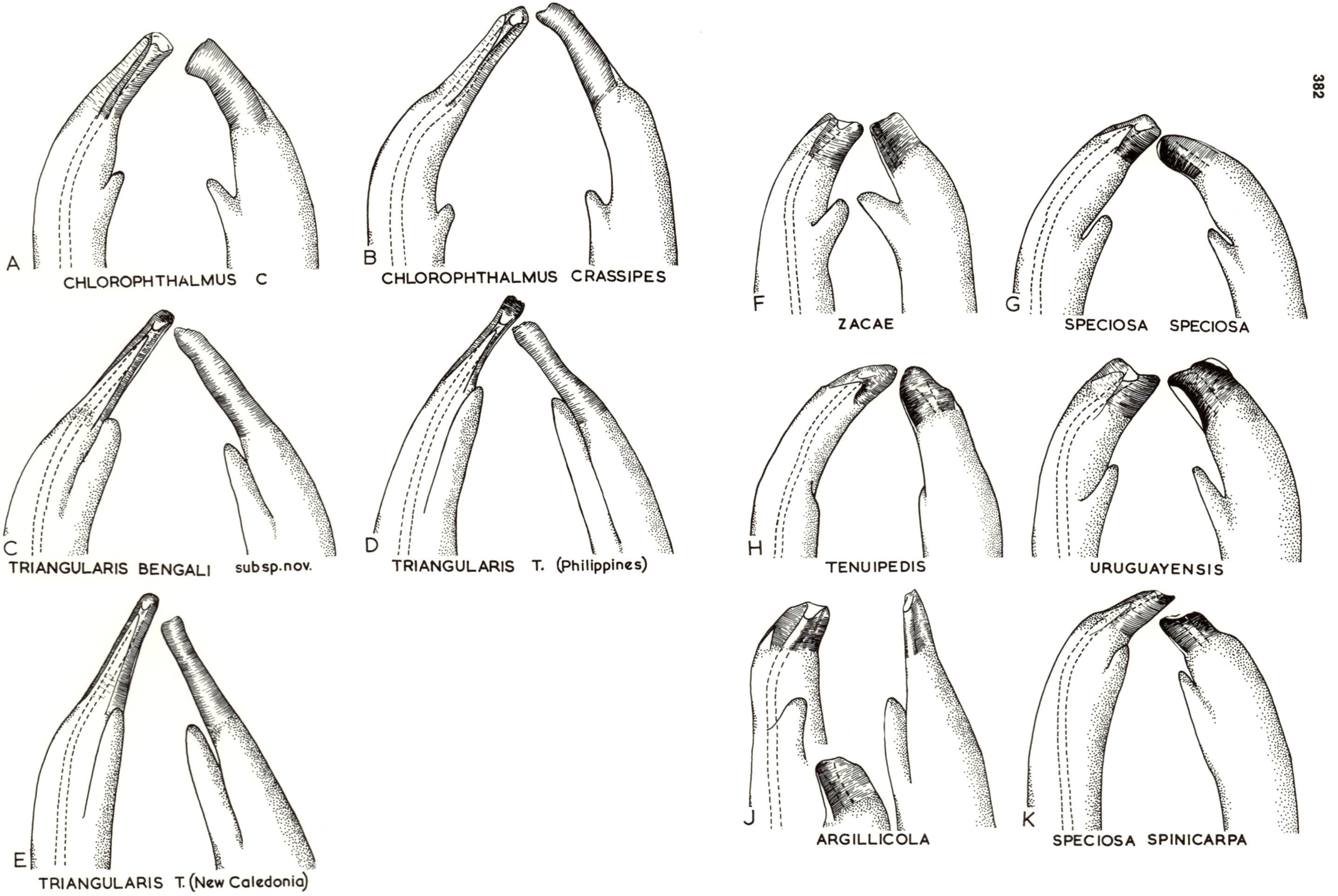

FIG. 68. *A*, *B*, gonopod tips in the subgenus *Amphiuca* (part). *C-E* and *G-K*, same, *Celuca* (part). *F*, same, *Minuca* (concluded). Orientation data as in Fig. 61. (Pp. 97, 155, 215, 463.)

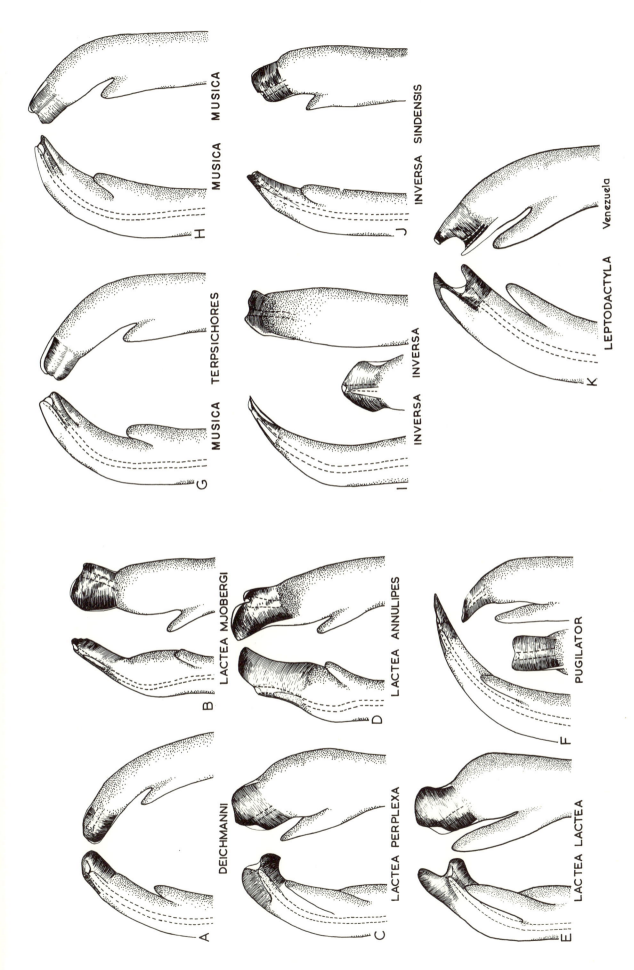

Fig. 69. *A-H* and *K*, gonopod tips in the subgenus *Celuca* (part). *I-J*, same, *Amphiuca* (concluded). Orientation data as in Fig. 61. (Pp. 155, 215, 463.)

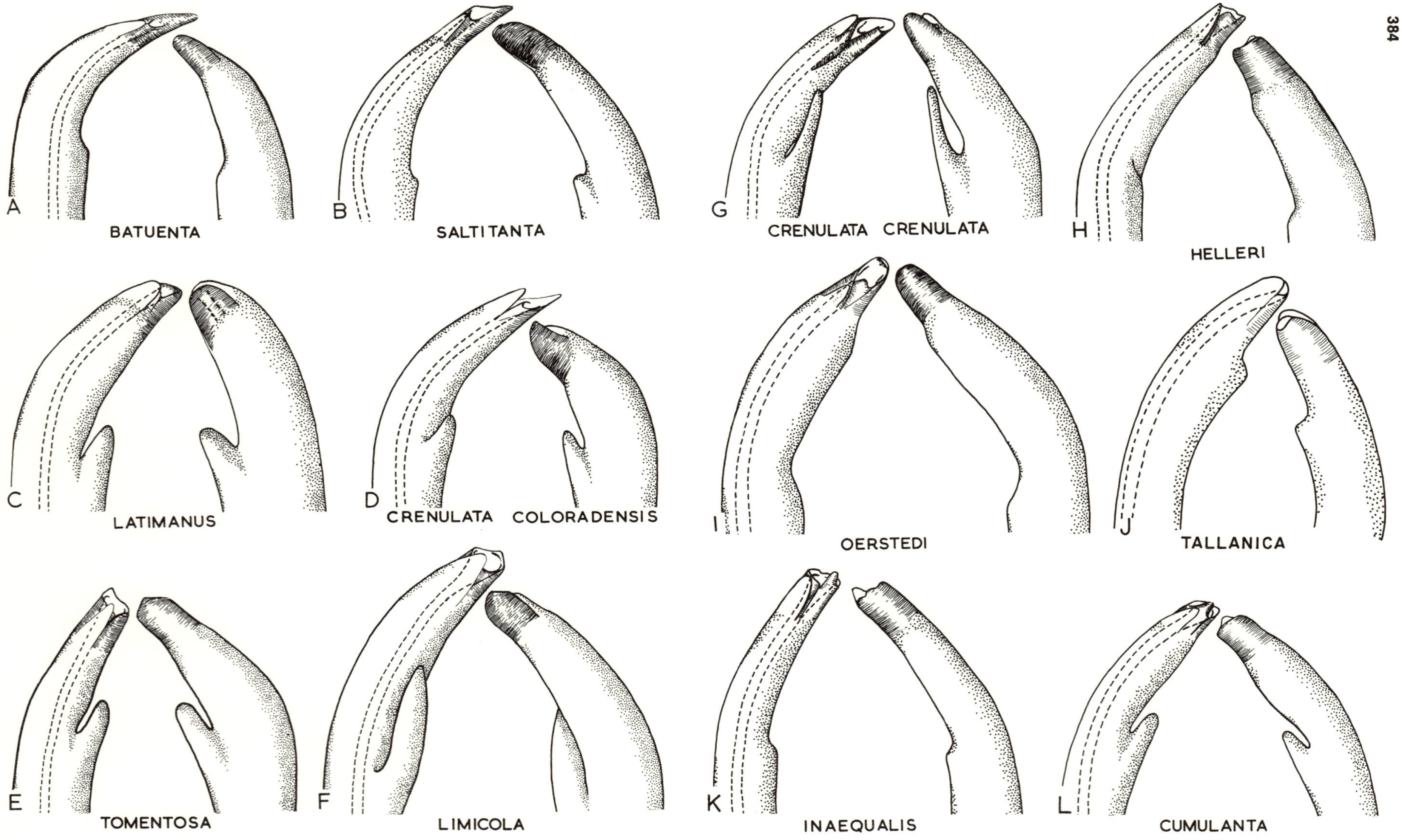

Fig. 70. Gonopod tips in the subgenus *Celuca* (part). Orientation data as in Fig. 61. (Pp. 215, 463.)

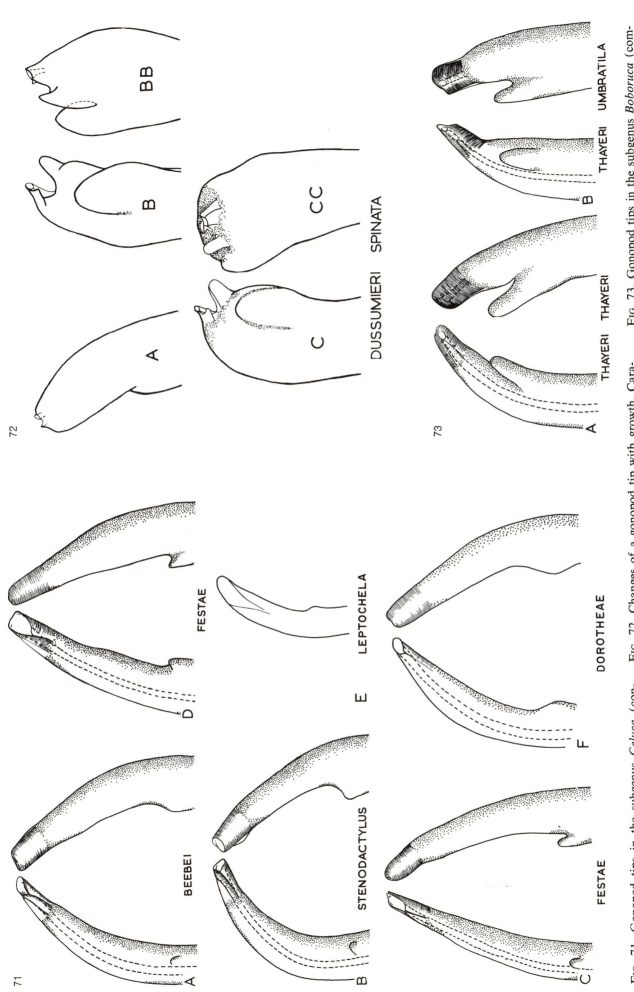

71

FESTAE

BEEBEI

STENODACTYLUS

E LEPTOCHELA

DOROTHEAE

FESTAE

72

BB

B

A

DUSSUMIERI   SPINATA

CC

C

73

THAYERI   THAYERI   THAYERI   UMBRATILA

A   B

FIG. 71. Gonopod tips in the subgenus *Celuca* (concluded). Orientation data as in Fig. 61. (Pp. 215, 463.)

FIG. 72. Changes of a gonopod tip with growth. Carapace lengths in mm: *A*, 4.4; *B*, anterior view, 7.4; *BB*, same, lateral view; *C*, anterior view, 15.0; *CC*, same, lateral view, tilted slightly to show distal aspect. Subgenus; *Deltuca*. (P. 464.)

FIG. 73. Gonopod tips in the subgenus *Boboruca* (complete). Orientation data as in Fig. 61. (Pp. 110, 463.)

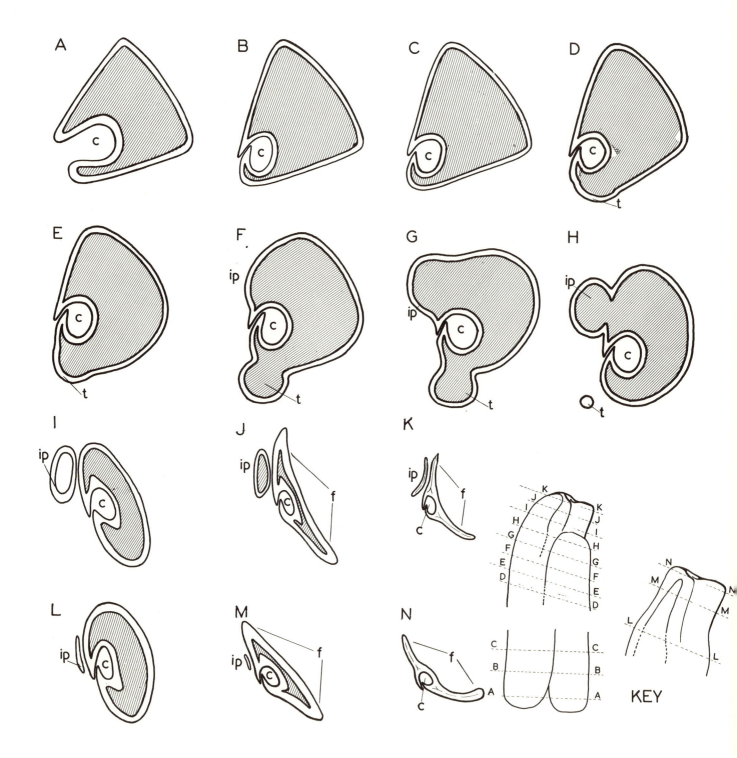

Fig. 74. Sections through a gonopod of two species in *Uca* with well-developed flanges. Middle section of gonopod omitted, as shown in Key. *A-K, (Deltuca) dussumieri dussumieri; L-N, (Celuca) lactea perplexa,* tip only. Abbreviations: c, canal; t, thumb; ip, inner process; f, flanges. (P. 463.)

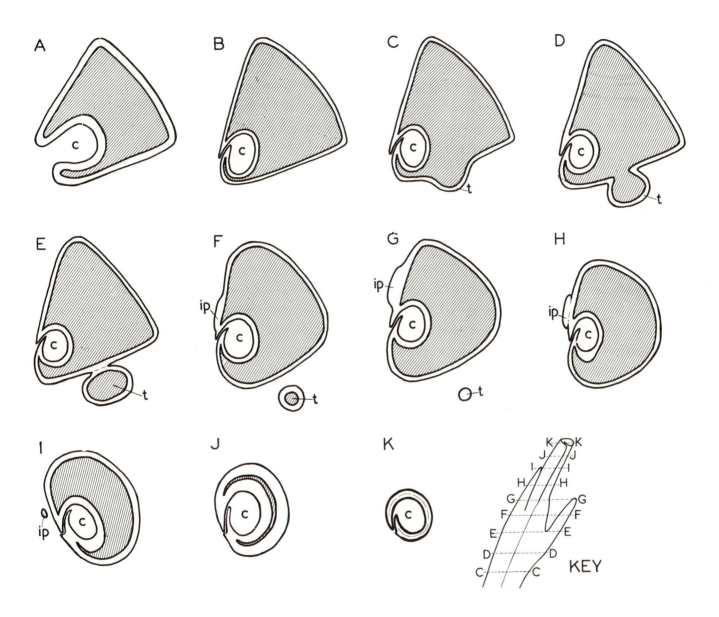

Fig. 75. Sections through gonopod of (*Deltuca*) *urvillei*, a flangeless species with the tip tubular and projecting. In Section *G* the tip of the thumb, t, is shown closer to the shaft than in the Key, to facilitate alignment of drawings. Abbreviations as in Fig. 74. (P. 463.)

388

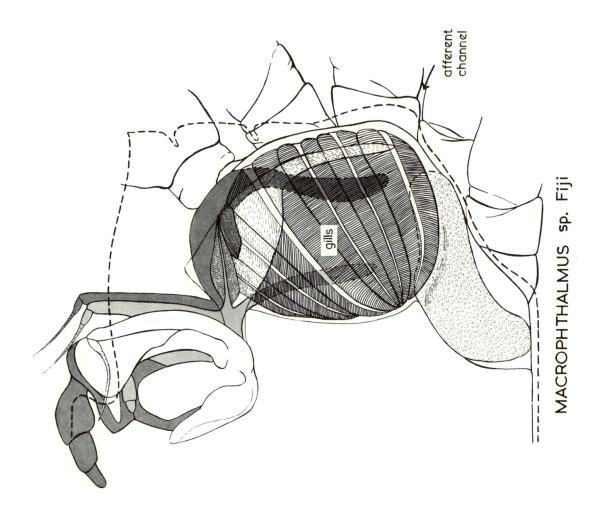

afferent
channel

gills

MACROPHTHALMUS sp. Fiji

Fig. 76. Semi-diagrammatic drawing of gill system and maxillipeds, dorsal view in *Macrophthalmus* sp. Setae omitted. Key to shading as in Fig. 78. (Pp. 455, 469.)

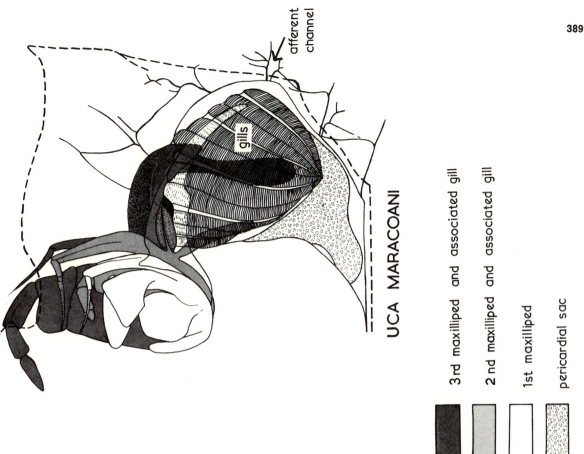

UCA MARACOANI

3 rd maxilliped and associated gill

2 nd maxilliped and associated gill

1st maxilliped

pericardial sac

Fig. 78. *Uca (Uca) maracoani maracoani*. Data as in Fig. 76. Ischium of third maxilliped drawn to length equal to that of *Macrophthalmus*.

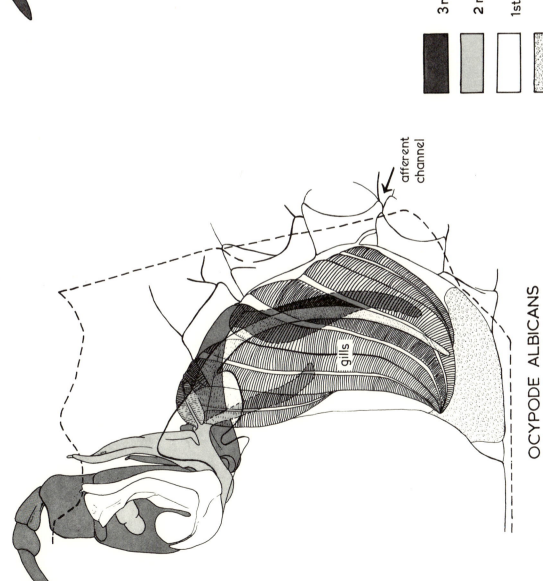

OCYPODE ALBICANS

Fig. 77. *Ocypode albicans*. Data as in Fig. 76. Ischium of third maxilliped drawn to length equal to that of *Macrophthalmus*.

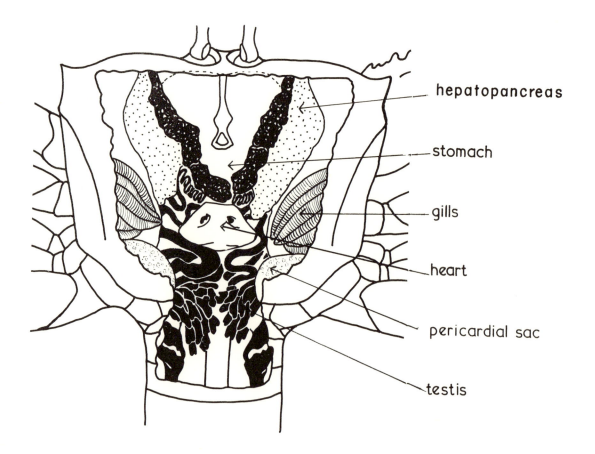

hepatopancreas

stomach

gills

heart

pericardial sac

testis

FIG. 79. Diagram of internal organs of male *Uca (Uca) maracoani maracoani*. Gonads partly developed. Kidneys omitted. Preliminary comparative work shows them to be far more extensive in some other subgenera, for instance in *Minuca*, than in *(Uca) maracoani*. (P. 469.)

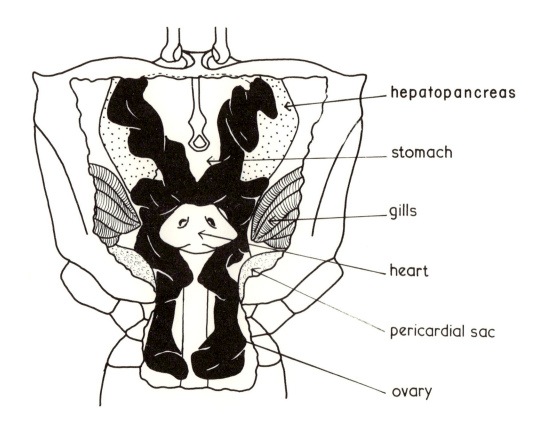

hepatopancreas

stomach

gills

heart

pericardial sac

ovary

FIG. 80. Diagram of internal organs of female *Uca (Uca) maracoani maracoani*. Other data as in Fig. 79.

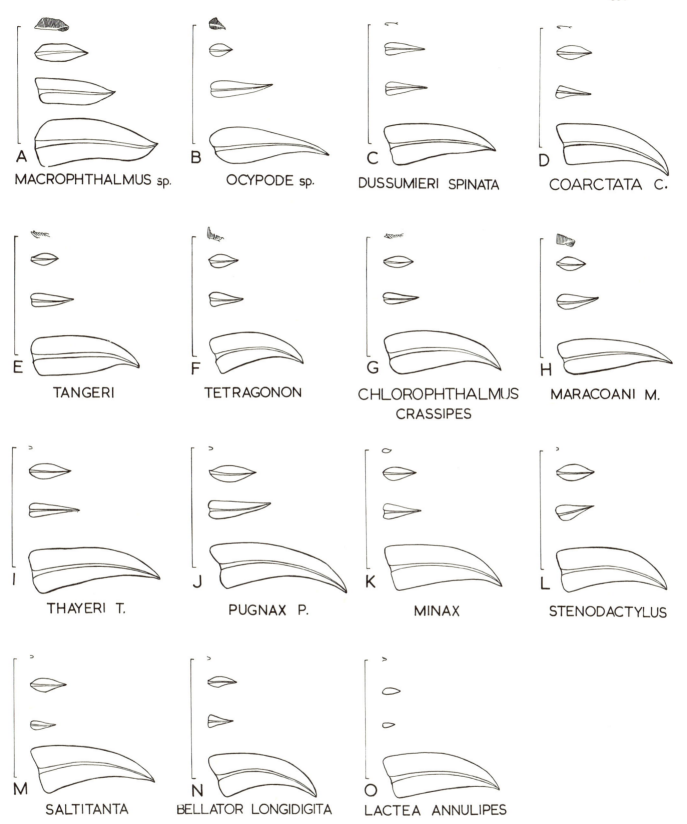

391

A. MACROPHTHALMUS sp.

B. OCYPODE sp.

C. DUSSUMIERI SPINATA

D. COARCTATA C.

E. TANGERI

F. TETRAGONON

G. CHLOROPHTHALMUS CRASSIPES

H. MARACOANI M.

I. THAYERI T.

J. PUGNAX P.

K. MINAX

L. STENODACTYLUS

M. SALTITANTA

N. BELLATOR LONGIDIGITA

O. LACTEA ANNULIPES

Fig. 81. Comparative sizes and shapes of gills in representative species of *Macrophthalmus*, *Ocypode*, and *Uca*. In each example the gills are drawn in proportion to carapace length, indicated by the bar on the left of each series, of a standard length throughout. Top drawing: gill on third maxilliped; below it, gill on second maxilliped; below that, first regular gill; lowest drawing, largest (third) gill. Subgenera of *Uca*: C-D, *Deltuca*; E, *Afruca*; F, *Thalassuca*; G, *Amphiuca*; H, *Uca*; I, *Boboruca*; J-K, *Minuca*; L, M, O, *Celuca*; N, *Australuca*. (P. 469.)

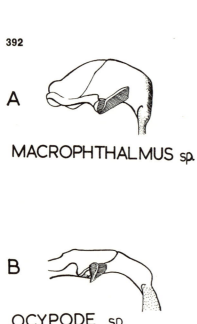

A MACROPHTHALMUS sp.

B OCYPODE sp.

C MARACOANI M.

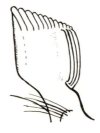

D HETEROPLEURA

E TETRAGONON

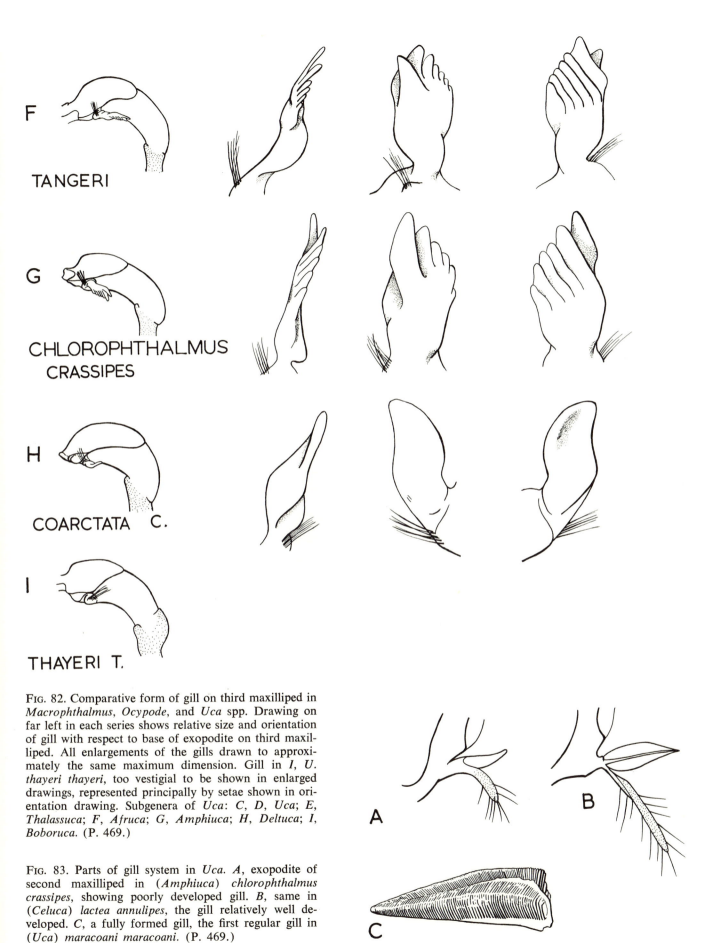

F **TANGERI**

G **CHLOROPHTHALMUS CRASSIPES**

H **COARCTATA** C.

I **THAYERI** T.

FIG. 82. Comparative form of gill on third maxilliped in *Macrophthalmus, Ocypode,* and *Uca* spp. Drawing on far left in each series shows relative size and orientation of gill with respect to base of exopodite on third maxilliped. All enlargements of the gills drawn to approximately the same maximum dimension. Gill in *I, U. thayeri thayeri,* too vestigial to be shown in enlarged drawings, represented principally by setae shown in orientation drawing. Subgenera of *Uca: C, D, Uca; E, Thalassuca; F, Afruca; G, Amphiuca; H, Deltuca; I, Boboruca.* (P. 469.)

FIG. 83. Parts of gill system in *Uca. A,* exopodite of second maxilliped in *(Amphiuca) chlorophthalmus crassipes,* showing poorly developed gill. *B,* same in *(Celuca) lactea annulipes,* the gill relatively well developed. *C,* a fully formed gill, the first regular gill in *(Uca) maracoani maracoani.* (P. 469.)

394

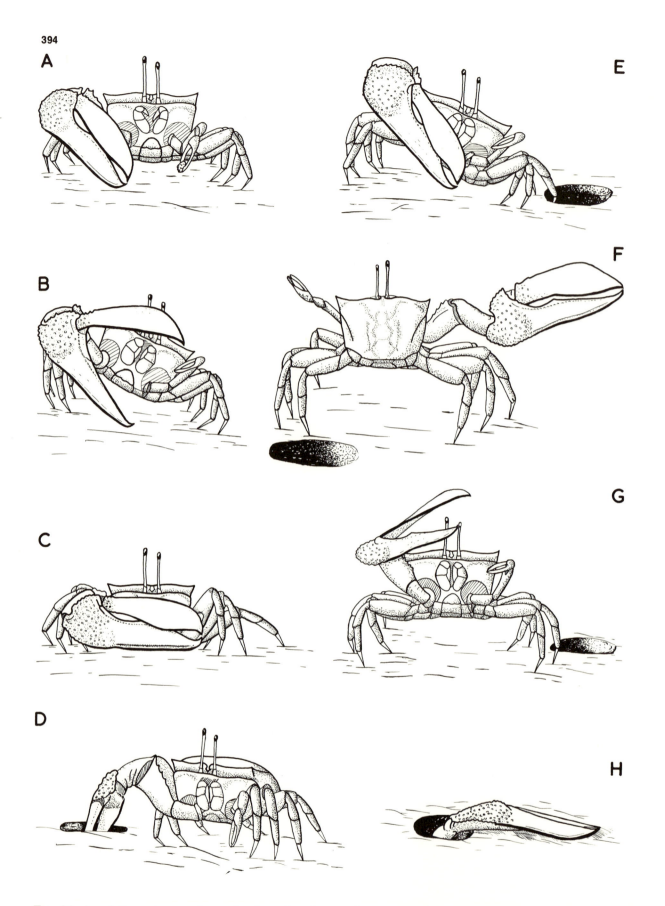

Fig. 84. Agonistic postures and associated motions in *Uca* (*Uca*) *maracoani maracoani*. *A, B, E,* the raised-carpus, showing several intensities (agonistic component no. 1 in text, p. 479). *C,* the creep (no. 11). *D,* aggressive wanderer reaching down burrow of a burrow-holder. *F,* the lateral-stretch (no. 10). *G,* low intensity waving display directed toward a potential intruder into a burrow-holder's display territory. *H,* the flat-claw (no. 8). (Pp. 478ff.)

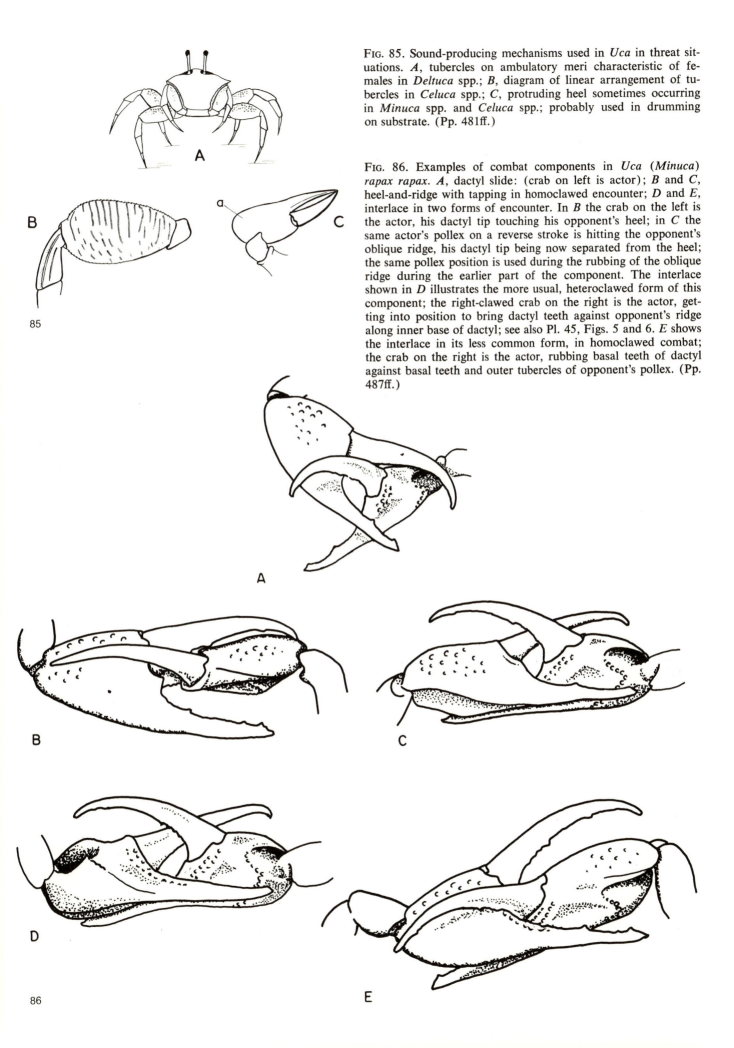

FIG. 85. Sound-producing mechanisms used in *Uca* in threat situations. *A*, tubercles on ambulatory meri characteristic of females in *Deltuca* spp.; *B*, diagram of linear arrangement of tubercles in *Celuca* spp.; *C*, protruding heel sometimes occurring in *Minuca* spp. and *Celuca* spp.; probably used in drumming on substrate. (Pp. 481ff.)

FIG. 86. Examples of combat components in *Uca* (*Minuca*) *rapax rapax*. *A*, dactyl slide: (crab on left is actor); *B* and *C*, heel-and-ridge with tapping in homoclawed encounter; *D* and *E*, interlace in two forms of encounter. In *B* the crab on the left is the actor, his dactyl tip touching his opponent's heel; in *C* the same actor's pollex on a reverse stroke is hitting the opponent's oblique ridge, his dactyl tip being now separated from the heel; the same pollex position is used during the rubbing of the oblique ridge during the earlier part of the component. The interlace shown in *D* illustrates the more usual, heteroclawed form of this component; the right-clawed crab on the right is the actor, getting into position to bring dactyl teeth against opponent's ridge along inner base of dactyl; see also Pl. 45, Figs. 5 and 6. *E* shows the interlace in its less common form, in homoclawed combat; the crab on the right is the actor, rubbing basal teeth of dactyl against basal teeth and outer tubercles of opponent's pollex. (Pp. 487ff.)

396

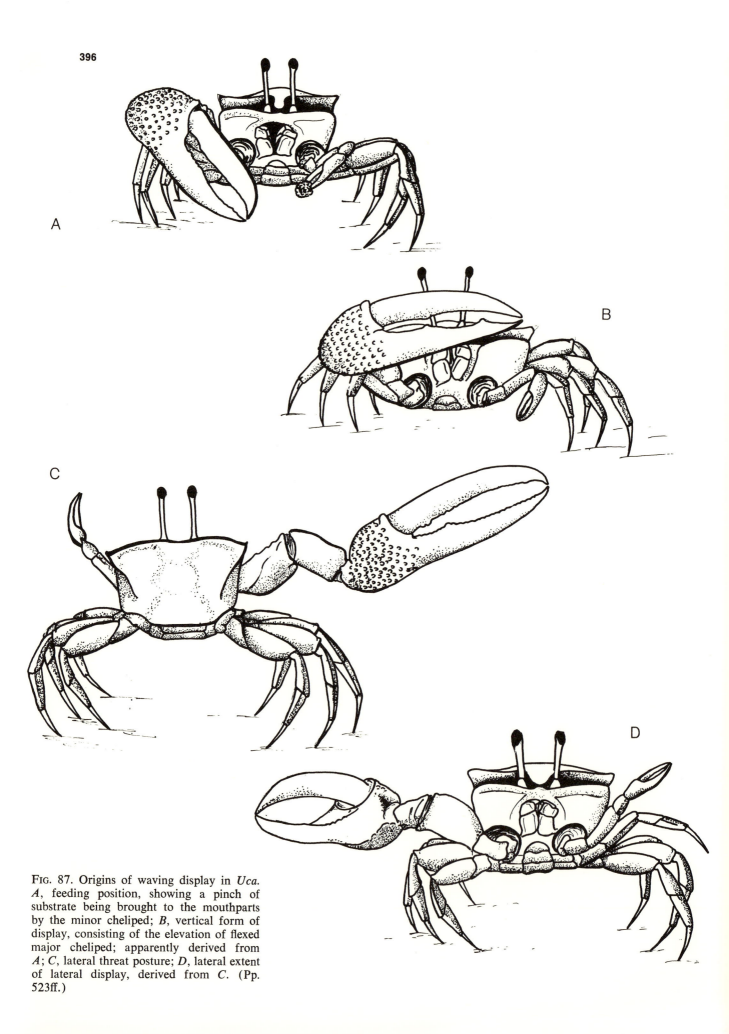

FIG. 87. Origins of waving display in *Uca*. *A*, feeding position, showing a pinch of substrate being brought to the mouthparts by the minor cheliped; *B*, vertical form of display, consisting of the elevation of flexed major cheliped; apparently derived from *A*; *C*, lateral threat posture; *D*, lateral extent of lateral display, derived from *C*. (Pp. 523ff.)

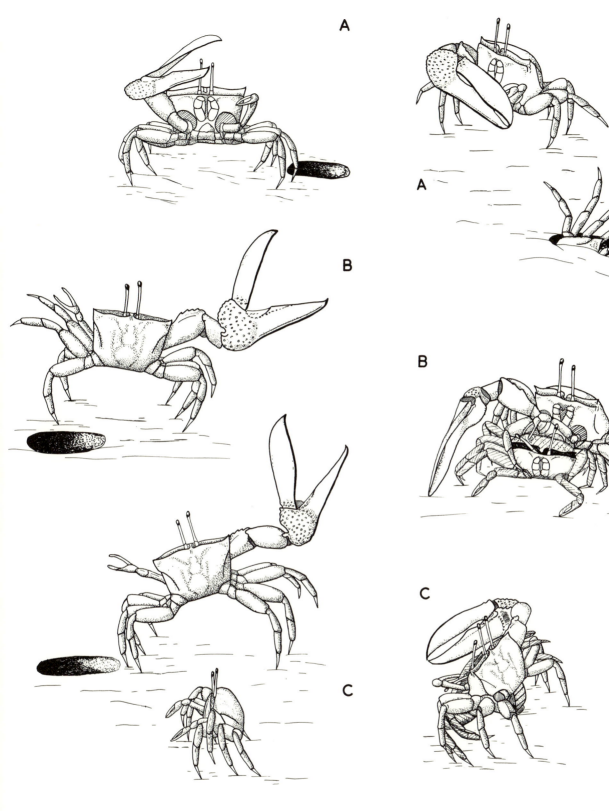

FIG. 88. Waving display in *Uca (Uca) maracoani maracoani*. *A*, male waving at low intensity beside burrow. *B*, waving, high intensity, as female approaches behind male. *C*, female nearing burrow and about to descend it ahead of male, as waving display continues at highest intensity. Increased intensity shown by upward stretching of ambulatories, higher upward reach of major cheliped, the chela held open and describing a circular motion, and the outward kicking of one or more ambulatories. (Pp. 501ff.)

FIG. 89. Precopulatory behavior and surface copulation in *Uca (Uca) maracoani maracoani*. *A*, one of postures assumed at approach of male by an unreceptive female (agonistic component no. 14, p. 480). *B*, plucking motions by male with minor chela of female's carapace. *C*, copulating position. (Pp. 503ff.)

398

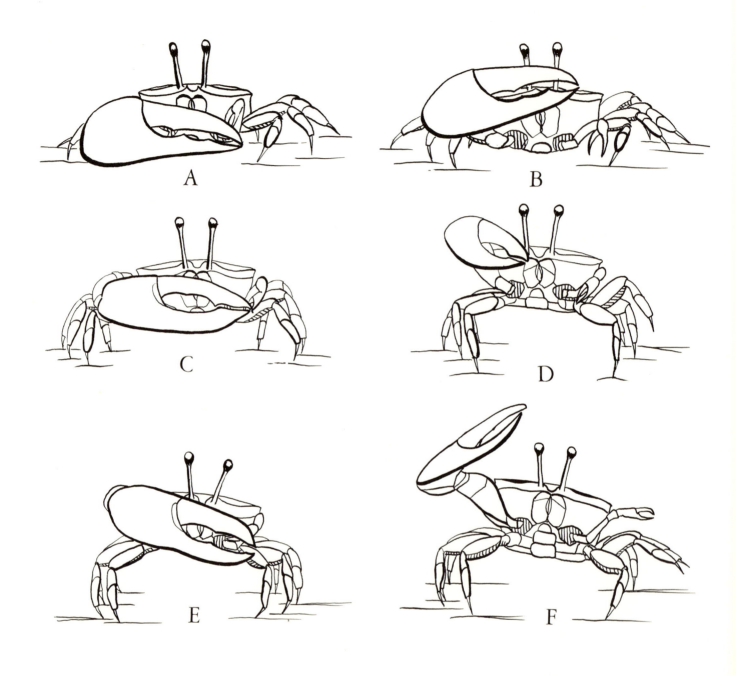

Fig. 90. Vertical display in *Uca*. All drawings made from selected frames of motion picture films; examples on left show rest positions between waves, those on right the maximum elevation of the cheliped, which is raised and lowered in a single plane. Note in the series, reading from the top down, the progressively higher reach of the cheliped and greater elevation of the carapace. *A, B, Deltuca acuta rhizophorae* (photographed in Singapore); *C, D, (Australuca) bellator bellator* (Philippine Is.); *E, F, (Deltuca) demani demani* (Philippine Is.). (Pp. 496ff.)

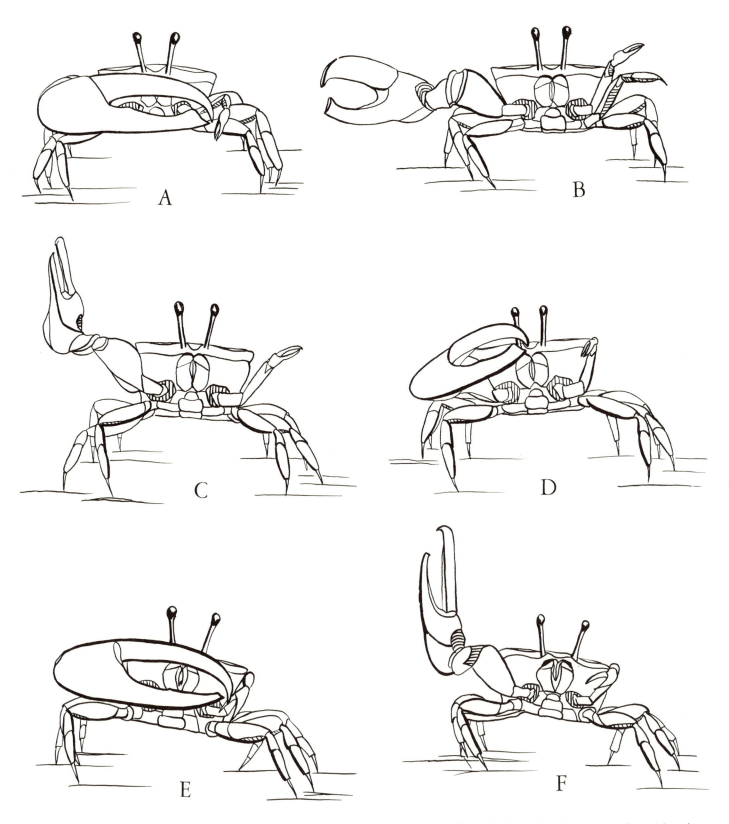

FIG. 91. Lateral display in *Uca*. All drawings made from selected frames of motion picture films. *A-D, (Celuca) lactea perplexa* (photographed in the Fiji Is.) showing maximum development of the lateral-circular wave, in which the cheliped starting from the flexed position (*A*) is unflexed outward (*B*), then raised (*C*), and finally returned (*D*) to the starting point. This wave is best developed in displays of moderate intensity; at low intensity or at high intensity during advanced display, the wave may be of a vertical or lateral single plane type. *E, F, (Minuca) rapax rapax* (Venezuela). Rest position and maximum cheliped reach of jerking-oblique wave, characteristic of moderate intensity display. Cheliped is unflexed outward, raised and lowered in a series of jerks. (Pp. 496ff.)

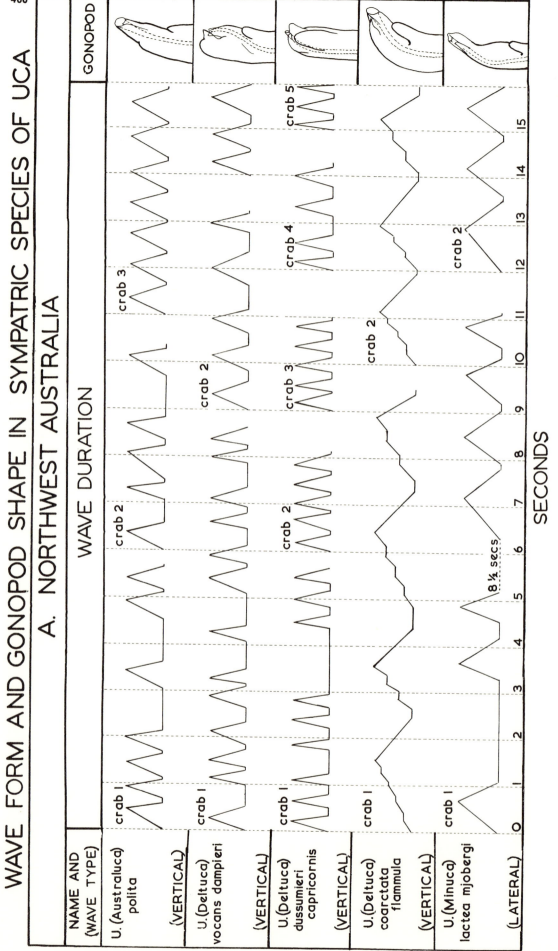

Fig. 92. Wave form and gonopod shape in sympatric species of *Uca*. From films made in Broome, northwest Australia. Base line in each represents period between waves; apex of each wave shown by peak. It will be noted that only two of these species, *(Deltuca) dussumieri capricornis* and *(Deltuca) coarctata flammula*, are closely related and hence sympatric in the narrow sense of the term. (Pp. 528ff.)

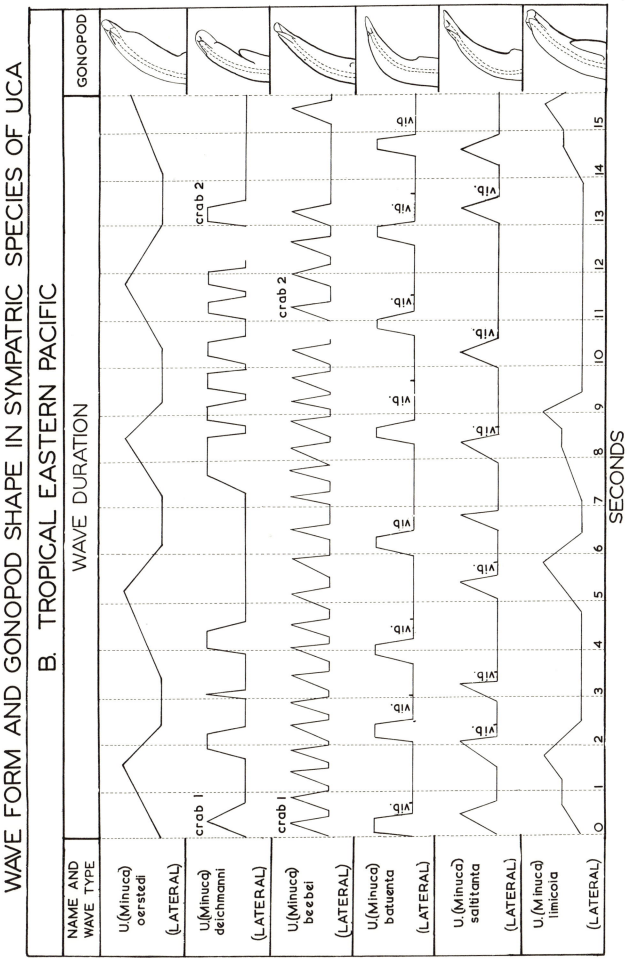

WAVE FORM AND GONOPOD SHAPE IN SYMPATRIC SPECIES OF UCA

B. TROPICAL EASTERN PACIFIC

Fig. 93. Wave form and gonopod shape in sympatric species of *Uca*. From films made in the tropical eastern Pacific, near Old Panama, R.P. Since all these species are members of the subgenus *Celuca*, they are rather closely related. Only three, *oerstedi*, *batuenta*, and *saltitanta*, share so many characteristics that they merit the term "closely sympatric." When the cheliped is held in place briefly at a wave's peak, the pause is indicated in the diagram by a plateau. "VIB": vibration, the strokes being too rapid to show individually on motion picture film, exposed at speed of 1/48 sec. (24 frames per sec.). (Pp. 528ff.)

Fig. 94. Activity phases of *Uca* (*Uca*) *maracoani maracoani* in an outdoor crabbery. Phase sequences in behavior of six adult males. The position of each dot indicates the highest types of activity attained by an individual on a particular date, when these types are arranged in a series from least social to most social. The complete natural sequence appears to range from uninterrupted inactivity underground, through simple maintenance (feeding and digging) activities, wandering, aggressive wandering, territoriality, and, finally, display. Gaps in the diagrams represent days when observations were missing or inadequate. The graph of each individual's activity ends with the day before its death. These specimens were selected for illustration because of their longevity. The shorter records of fifteen other individuals showed similar charactefistics. (Pp. 505ff.)

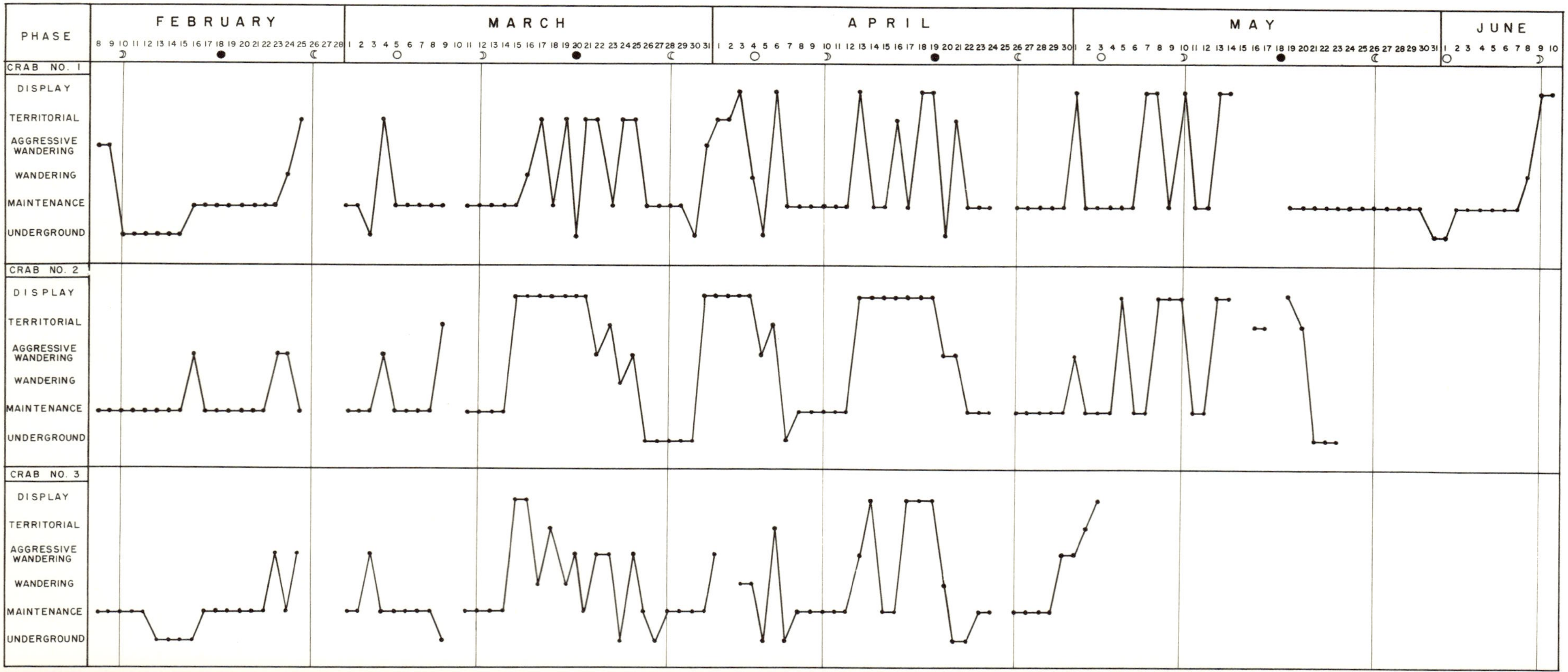

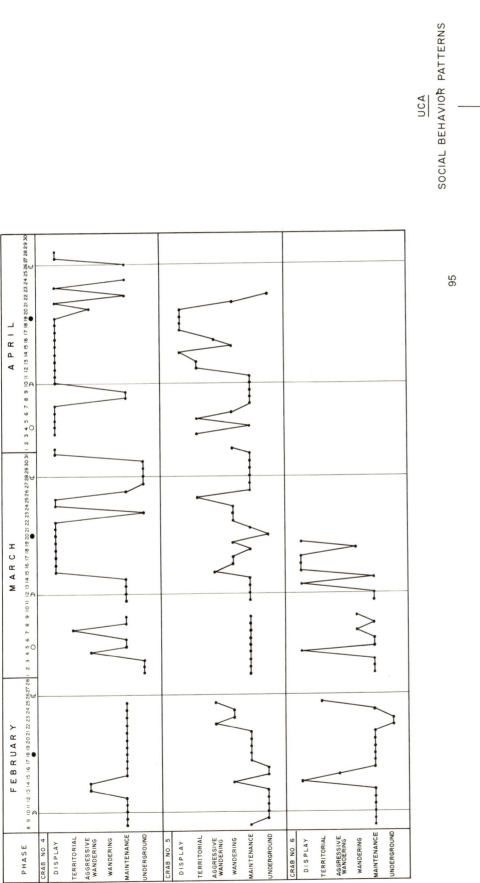

95

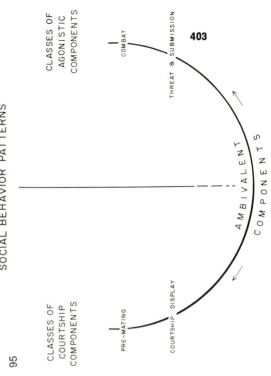

SOCIAL BEHAVIOR PATTERNS

UCA

Fig. 95. Diagram indicating the extent of ambivalence in social behavior patterns in *Uca*. (Pp. 517ff.)

404

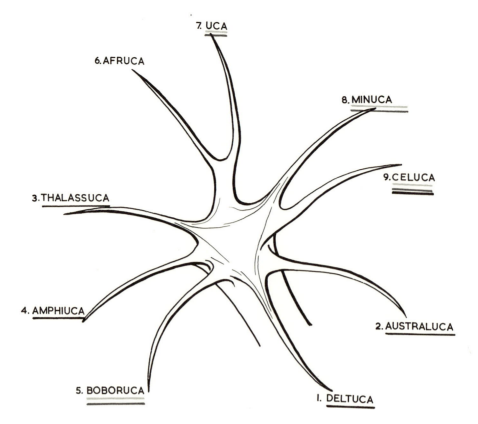

FIG. 96. Dendrogram: subgenera of *Uca*. Key to under-scoring of names: black, occurs in the Indo-Pacific; pale gray, eastern Pacific; dark gray, western Atlantic. (Pp. 18, 531.)

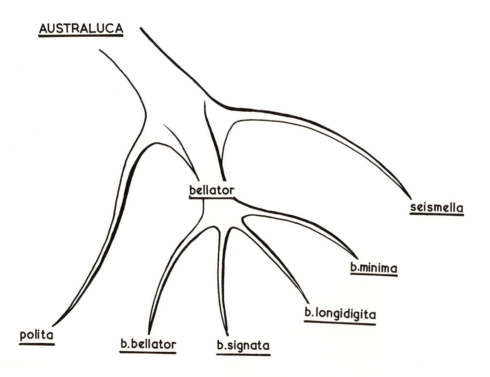

FIG. 97. Dendrogram: subgenus *Australuca*. Key to underscoring as in Fig. 96. (Pp. 63, 531.)

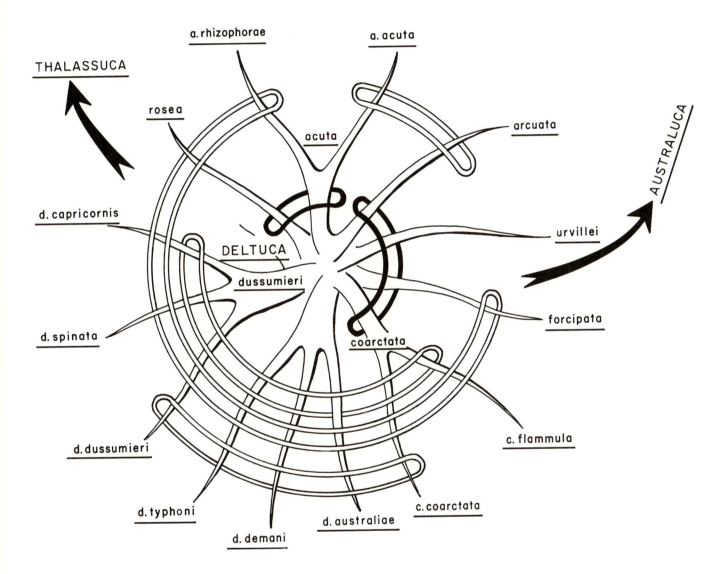

THALASSUCA

AUSTRALUCA

a. rhizophorae

a. acuta

rosea

acuta

arcuata

d. capricornis

DELTUCA

dussumieri

urvillei

d. spinata

coarctata

forcipata

d. dussumieri

c. flammula

d. typhoni

c. coarctata

d. demani

d. australiae

Fig. 98. Dendrogram: subgenus *Deltuca*. Black bands, superspecies. Hollow bands, sympatric forms; exception: *demani typhoni* and *demani demani*, which replace each other allopatrically. Key to underscoring as in Fig. 96. (Pp. 24, 531.)

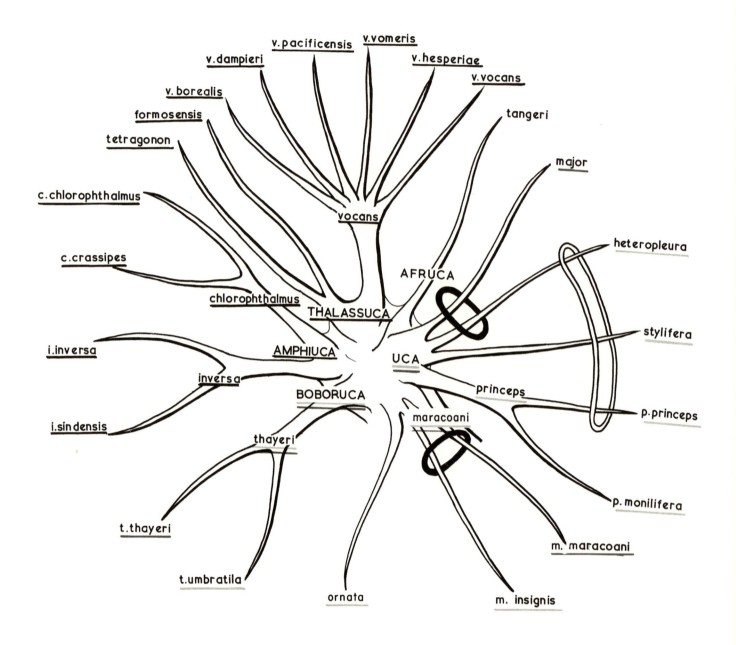

Fig. 99. Dendrogram: subgenera *Thalassuca, Amphi-uca, Boboruca, Afruca, Uca.* Key to underscoring as in Fig. 96. Black bands, superspecies. Hollow bands, sympatric forms. Key to underscoring as in Fig. 96. (Pp. 76, 97, 111, 117, 127, 532.)

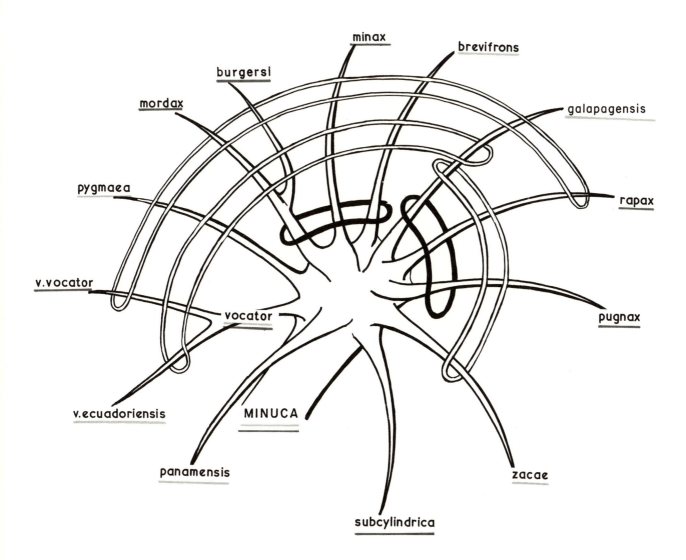

FIG. 100. Dendrogram: subgenus *Minuca*. Black bands, superspecies. Hollow bands, sympatric forms. Key to underscoring as in Fig. 96. (Pp. 156, 533.)

408

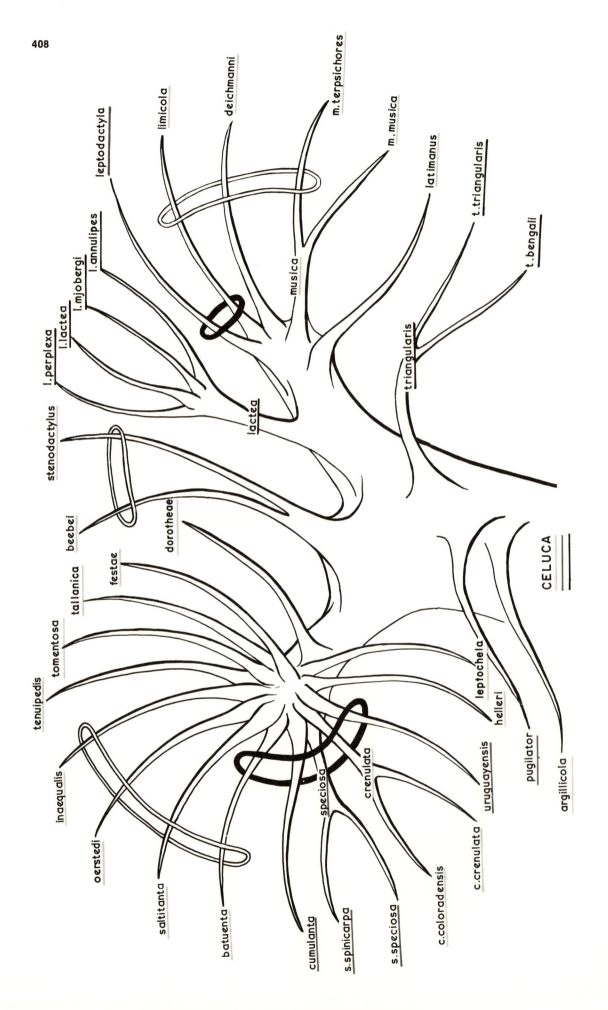

Fig. 101. Dendrogram: subgenus *Celuca*. Black bands, superspecies. Hollow bands, sympatric forms. Key to underscoring as in Fig. 96. (Pp. 217, 533.)

Maps

INTRODUCTION

The 21 maps that follow consist of 17 in black-and-white and four (nos. 18-21) partly in color.

The first illustrates the distribution of the genus *Uca*, along with degrees of concentration of its species.

Numbers 2-17 show the distribution of most of the species, usually arranged in the sequence of subgenera employed in this contribution. Because of the prevalence of sympatry, however, and the exigencies of photoreproduction, many species are not presented in taxonomic sequence. In particular, the last maps, nos. 18-21, follow the American series, but show, with the aid of color, the complex distribution of certain Indo-Pacific forms that are taxonomically scattered.

To aid rapid comparison of ranges, an alphabetical list of species with corresponding map numbers follows this introduction.

Screens. Throughout the series the extent of each screen represents my conclusion on the probable distribution of the species at the present time, based on information that appears to me to be reliable. This information consists first of published records which seem to me trustworthy, either because the contribution is the work of an experienced specialist, or the record of a species in a part of its range where it is otherwise known to occur, or because I have examined the material in the museum where it is deposited. Second, the ranges are based on additional specimens in museums which have never been recorded in print, or for which references to the published accounts were not certainly determined, if at all. The third source of information is provided by specimens I collected in the field (Table 24).

Records that seem to me wholly erroneous, such as the occurrence of a species in Odessa or Yugoslavia, have been omitted from the maps, although they are discussed under the species concerned and listed in the section on questionable geographic records (p. 326). Where boundaries are uncertain because of the absence of recent records, as in Chile, the screen ends with an irregular edge; where records are wholly lacking throughout a wide area between populations, the sections of screen are rounded off.

Symbols. For all the Indo-Pacific species dots and other symbols mark the localities where specimens that I have personally examined were collected. On maps where a number of such localities are very close together, as on the northwest coast of the Gulf of Davao, a single dot represents two or more sites. In some older museum specimens the label gives only a general locality, such as "Japan" or "Madagascar." If I did not find material to examine from particular localities in the same area, I used an open diamond symbol on the map; otherwise the more general name is not represented. An arrow indicates an extension of a range beyond the confines of the map.

Of special interest on Maps 18-21 are the colored symbols falling outside the usual boundaries of the species or subspecies, within the range of an allopatric form. These distributions are discussed under the species headings in the systematic section, in Chapter 1 and in Chapter 7; see also Tables 3, 6, and 22.

No symbols appear on any American map, since they proved in some cases to be impractical because of extensive sympatry and in all to be of doubtful value. These numerous American forms are usually characterized by short or narrow ranges in uncomplicated patterns, while their taxonomic histories are, with a few famous exceptions, less tortuous than those of most Indo-Pacific species. It seemed, therefore, that on American shores the precise localization on a crowded map of the origins of examined specimens would not be very helpful, either in clarifying allopatric situations or in evaluating evidence for particular distributions.

All of the records of material examined, both precisely and imprecisely localized, are listed in Appendix A. Ranges of the species and subspecies appear under appropriate headings in the systematic section. The chapter dealing with zoogeography starts on p. 431.

Excluded from the maps and from Appendix A is museum material I examined early in the course of the study before I had attained sufficient knowledge of the forms involved to refer them with confidence to species or subspecies. This limitation applies particularly to the collection of Australian material in Sydney. The indicated ranges of some species occurring in Australia are, therefore, almost certainly already in need of extension.

Alphabetical List of Species of the Genus *Uca*, with Number of Map on Which Each Appears

Species Name	Number of Map	Species Name	Number of Map	Species Name	Number of Map	Species Name	Number of Map
acuta	2	*festae*	16	*mordax*	12	*stenodactylus*	16
arcuata	19	*forcipata*	19	*musica*	17	*stylifera*	10
argillicola	15	*formosensis*	4	*oerstedi*	15	*subcylindrica*	11
batuenta	15	*galapagensis*	14	*ornata*	10	*tallanica*	15
beebei	16	*helleri*	15	*panamensis*	12	*tangeri*	8
bellator	3	*heteropleura*	10	*polita*	4	*tenuipedis*	15
brevifrons	13	*inaequalis*	15	*princeps*	9	*tetragonon*	4
burgersi	12	*inversa*	7	*pugilator*	16	*thayeri*	11
chlorophthalmus	5, 6	*lactea*	21	*pugnax*	10, 14	*tomentosa*	15
coarctata	19	*latimanus*	17	*pygmaea*	11	*triangularis*	7
crenulata	15	*leptochela*	15	*rapax*	14	*uruguayensis*	15
cumulanta	15	*leptodactyla*	17	*rosea*	2	*urvillei*	19
deichmanni	17	*limicola*	17	*saltitanta*	15	*vocans*	20
demani	2	*major*	10	*seismella*	4	*vocator*	13
dorotheae	15	*maracoani*	9	*speciosa*	15	*zacae*	14
dussumieri	18	*minax*	12				

411

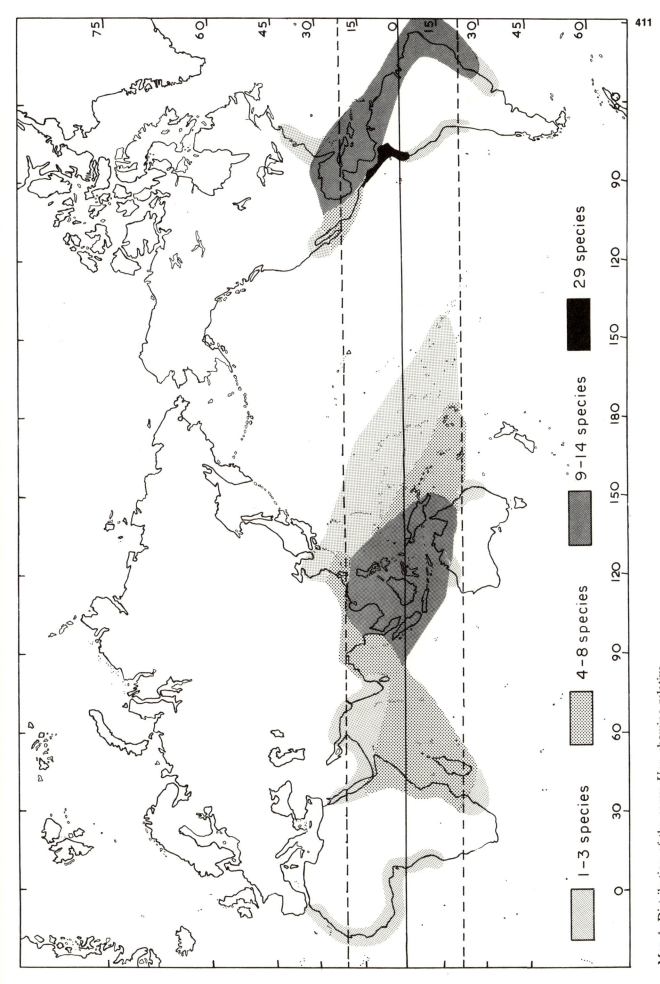

MAP 1. Distribution of the genus *Uca*, showing relative concentrations of species in different parts of the range. See p. 409ff.

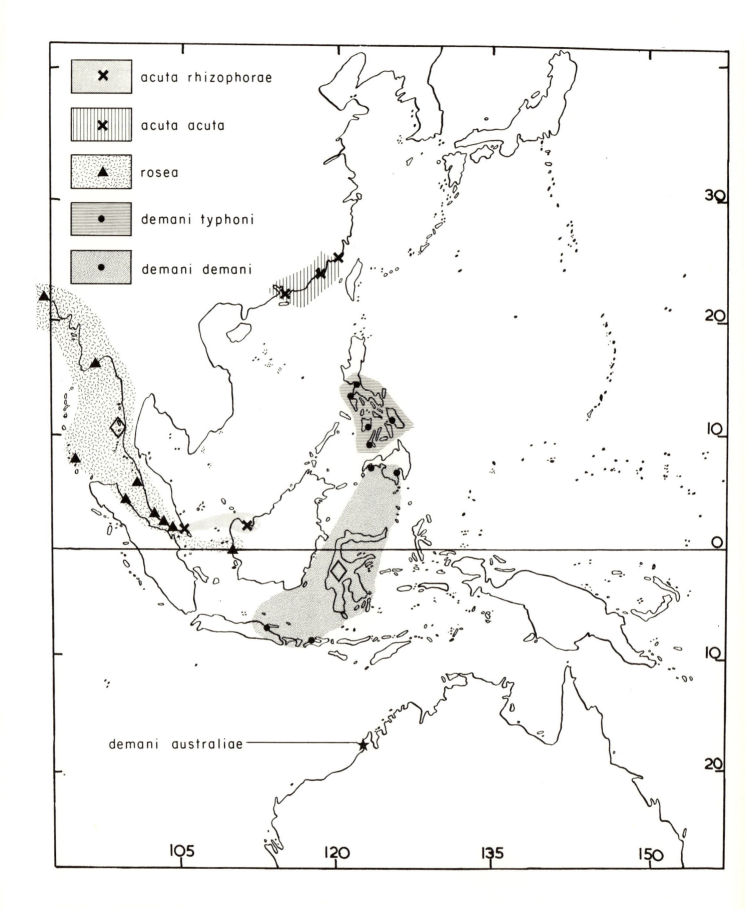

MAP 2. Distribution of the subgenus *Deltuca* (part).
(General explanation: p. 409.)

Legend:
- ✕ acuta rhizophorae
- ✕ acuta acuta
- ▲ rosea
- ● demani typhoni
- ● demani demani
- ★ demani australiae

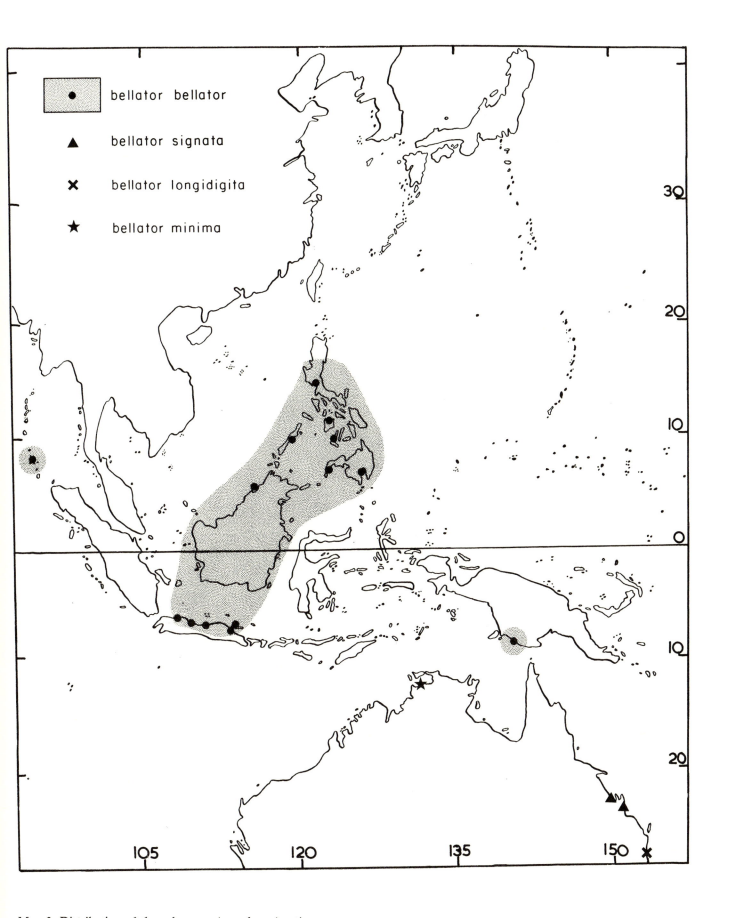

MAP 3. Distribution of the subgenus *Australuca* (part).
(General explanation: p. 409.)

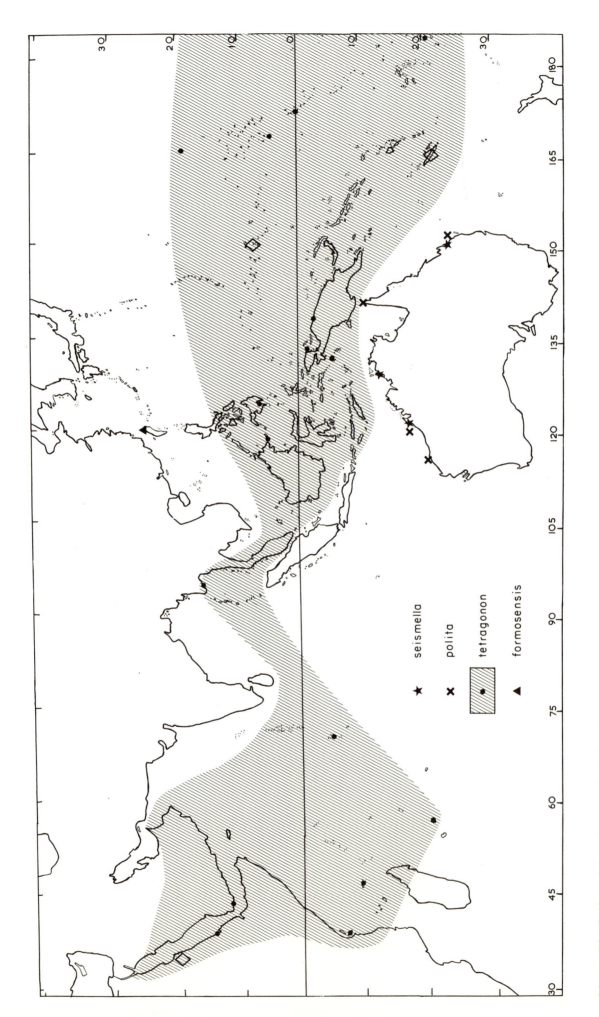

MAP 4. Distribution of the subgenera *Australuca* (concluded) and *Thalassuca* (part). (General explanation: p. 409.)

seismella ★

polita ✕

tetragonon ▨ ●

formosensis ▲

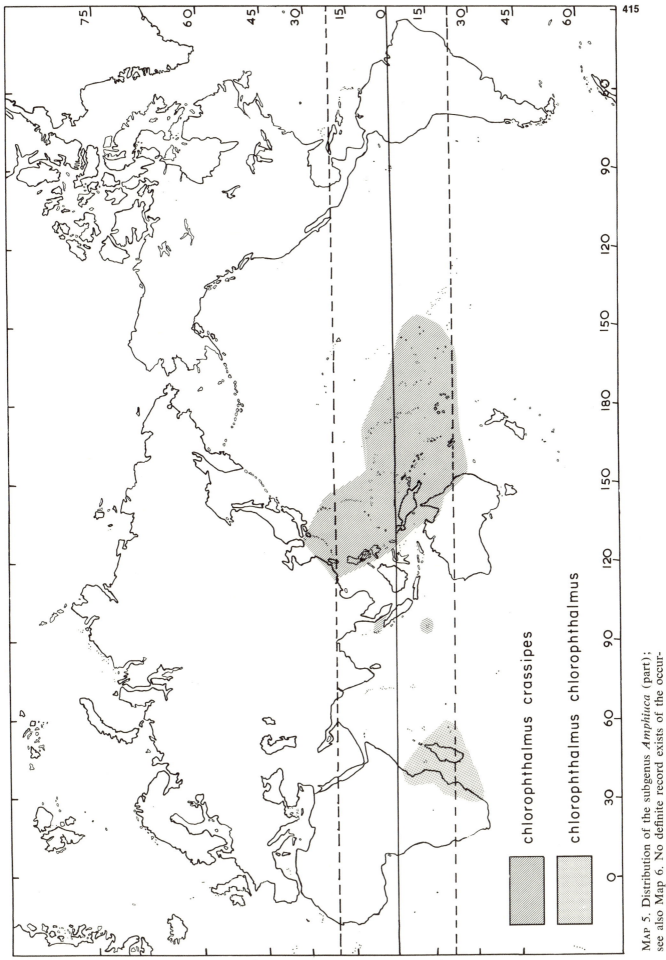

MAP 5. Distribution of the subgenus *Amphiuca* (part); see also Map 6. No definite record exists of the occurrence of *chlorophthalmus* in Australia. (General explanation: p. 409.)

chlorophthalmus crassipes

chlorophthalmus chlorophthalmus

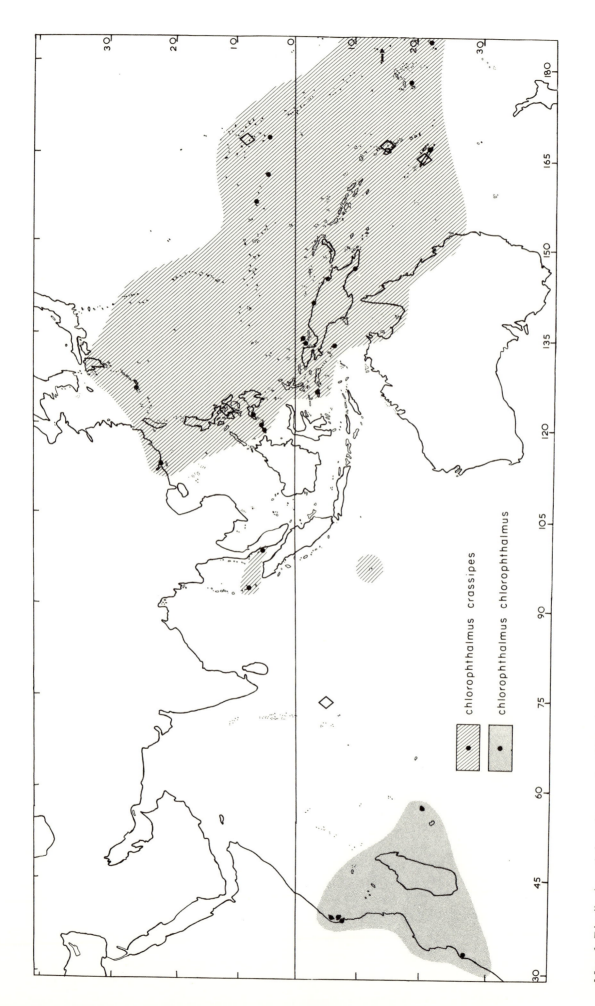

MAP 6. Distribution of the subgenus *Amphiuca* (part); see also Map 5. No definite record exists of the occurrence of *chlorophthalmus* in Australia. (General explanation: p. 409.)

417

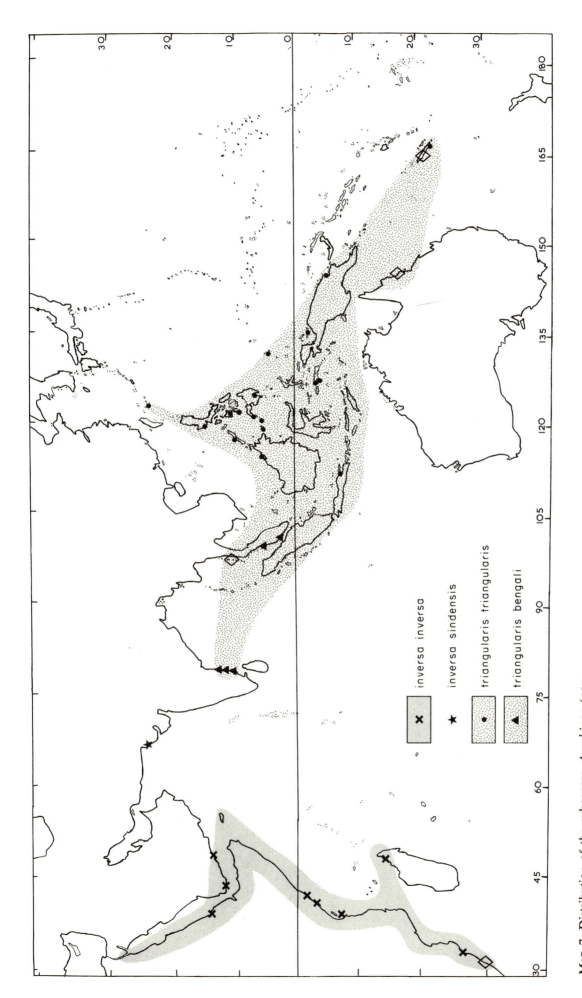

MAP 7. Distribution of the subgenera *Amphiuca* (concluded) and *Celuca* (part). Only one screen is used showing the distribution of (*Celuca*) *triangularis* because the normal boundaries between subspecies, particularly in Sumatra, are not yet known. (General explanation: p. 409.)

418

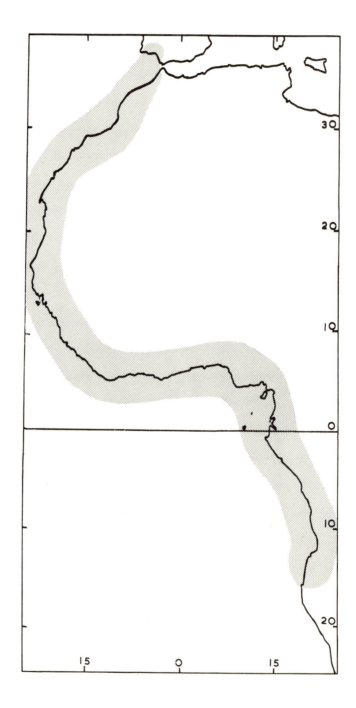

MAP 8. Distribution of the subgenus *Afruca* (complete),
represented by the species *tangeri*. (General explanation: p. 409.)

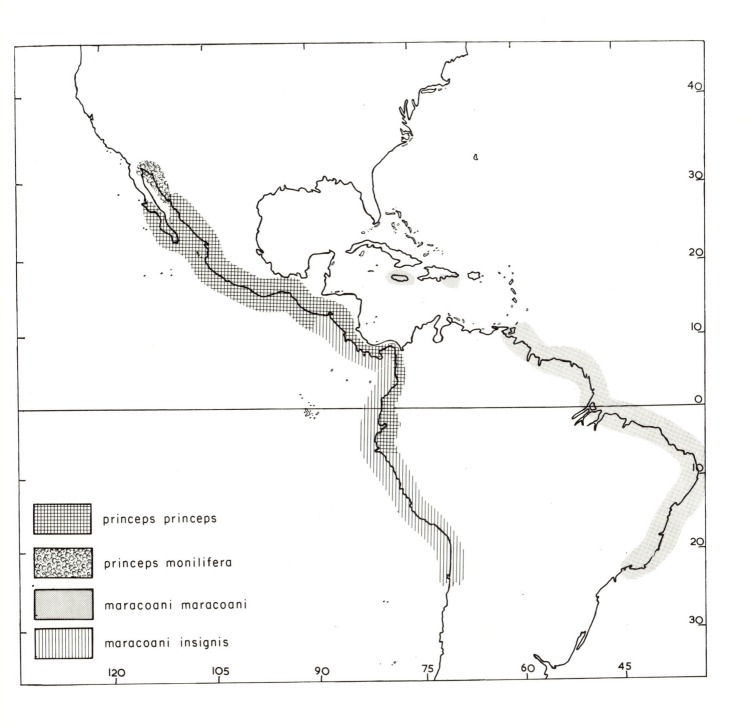

MAP 9. Distribution of the subgenus *Uca* (part). (General explanation: p. 409.)

420

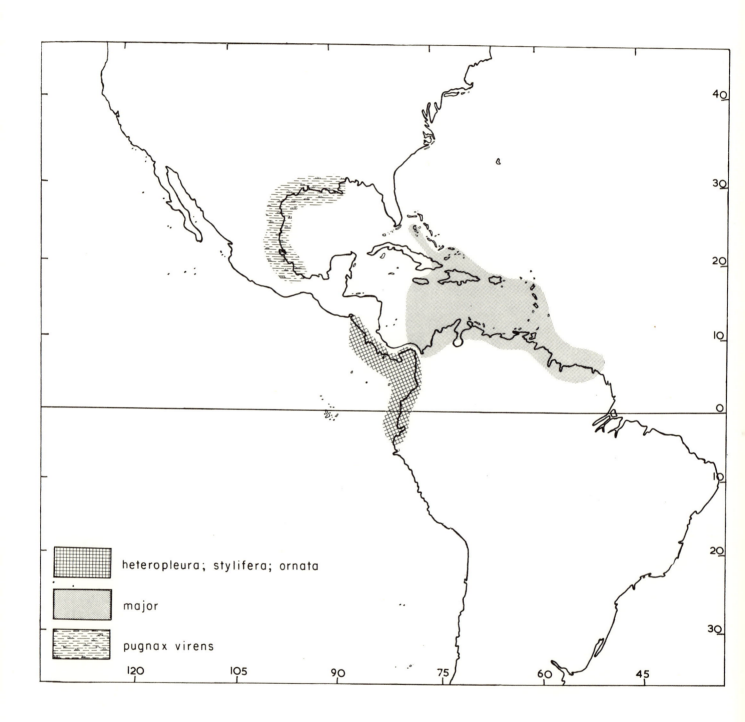

MAP 10. Distribution of the subgenera *Uca* (concluded) and *Minuca* (part); for the distribution of (*Minuca*) *pugnax pugnax*, see Map 14. (General explanation: p. 409.)

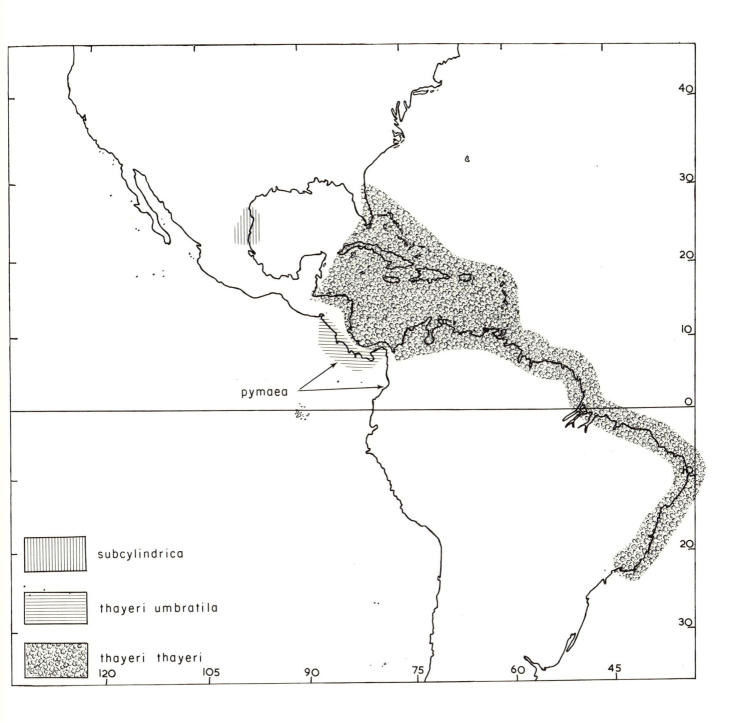

MAP 11. Distribution of the subgenera *Minuca* (part) and *Boboruca* (complete). (General explanation: p. 409.)

422

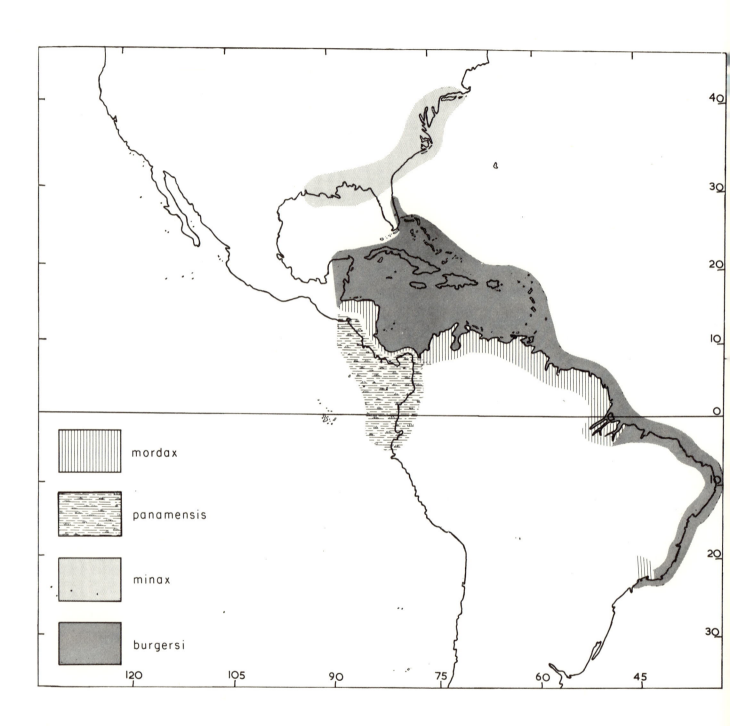

MAP 12. Distribution of the subgenus *Minuca* (part).
(General explanation: p. 409.)

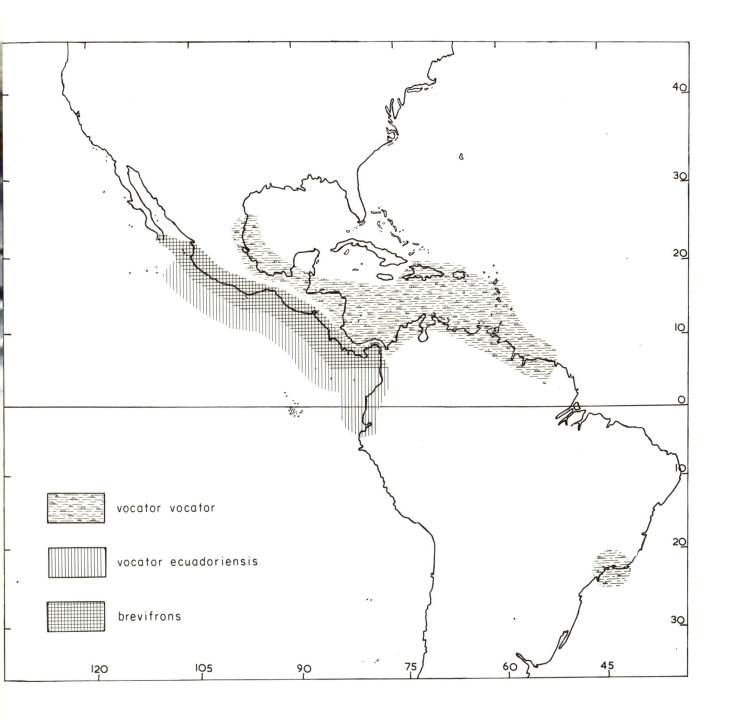

MAP 13. Distribution of the subgenus *Minuca* (part).
(General explanation: p. 409.)

424

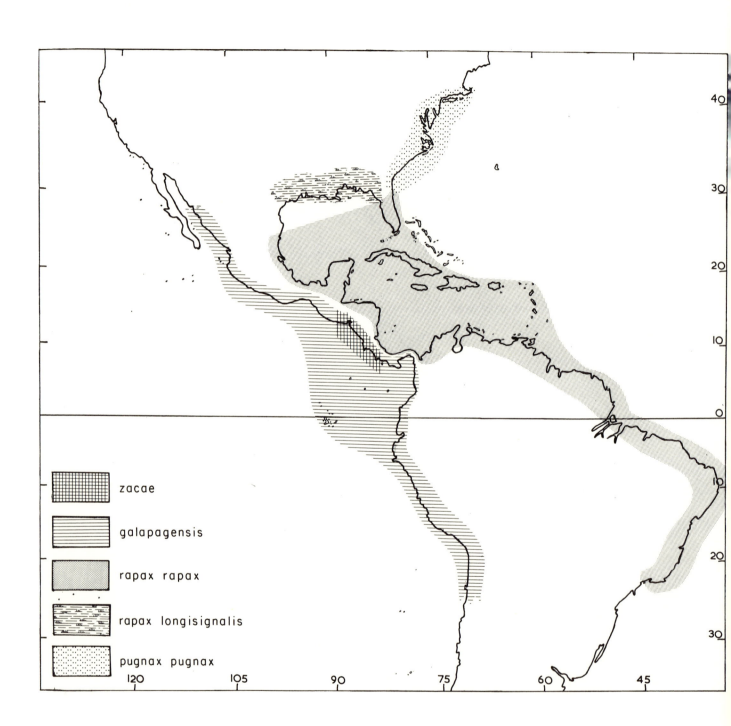

zacae

galapagensis

rapax rapax

rapax longisignalis

pugnax pugnax

MAP 14. Distribution of the subgenus *Minuca* (concluded); for the distribution of (*Minuca*) *pugnax virens*, see Map 10. (General explanation: p. 409.)

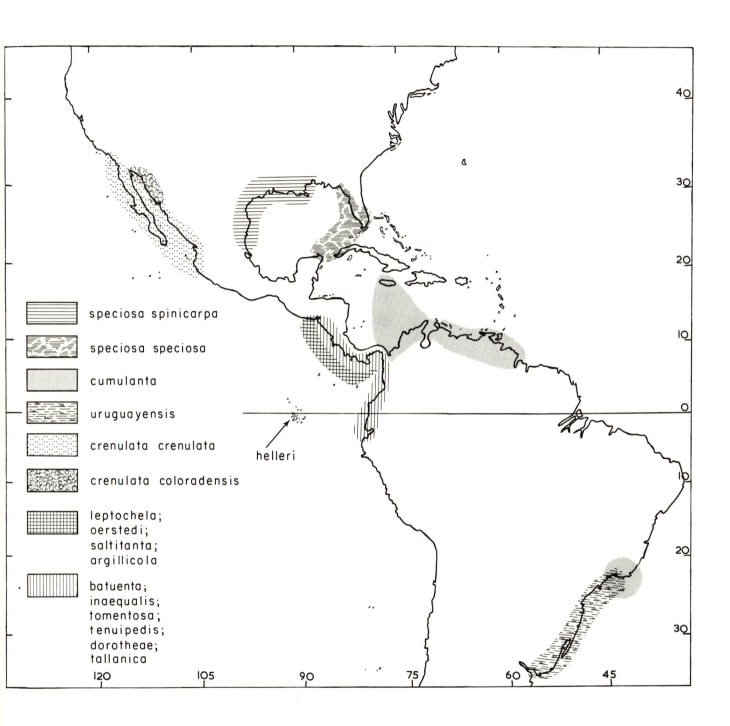

speciosa spinicarpa

speciosa speciosa

cumulanta

uruguayensis

crenulata crenulata

crenulata coloradensis

leptochela;
oerstedi;
saltitanta;
argillicola

batuenta;
inaequalis;
tomentosa;
tenuipedis;
dorotheae;
tallanica

helleri

Map 15. Distribution of the subgenus *Celuca* (part).
(General explanation: p. 409.)

426

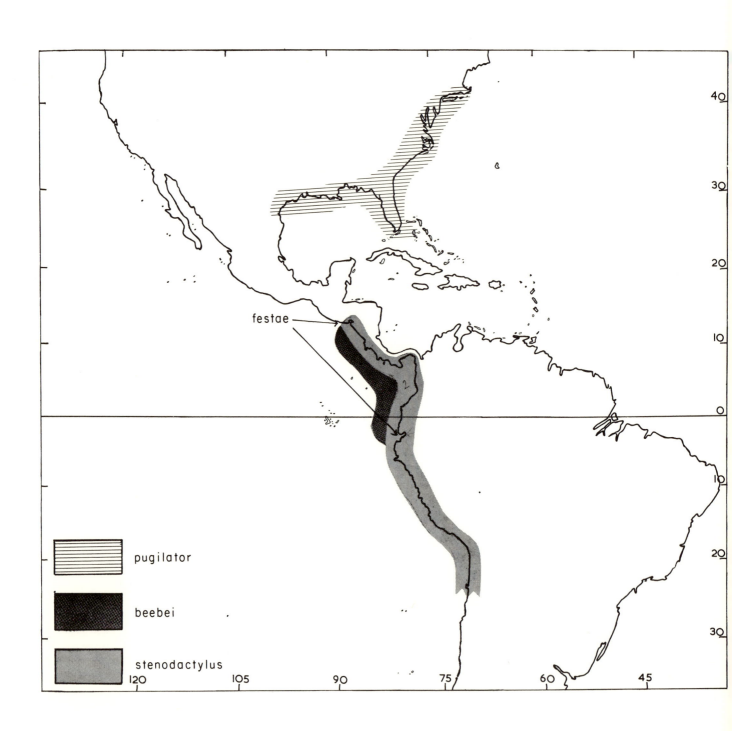

festae

pugilator

beebei

stenodactylus

120   105   90   75   60   45

MAP 16. Distribution of the subgenus *Celuca* (part).
(General explanation: p. 409.)

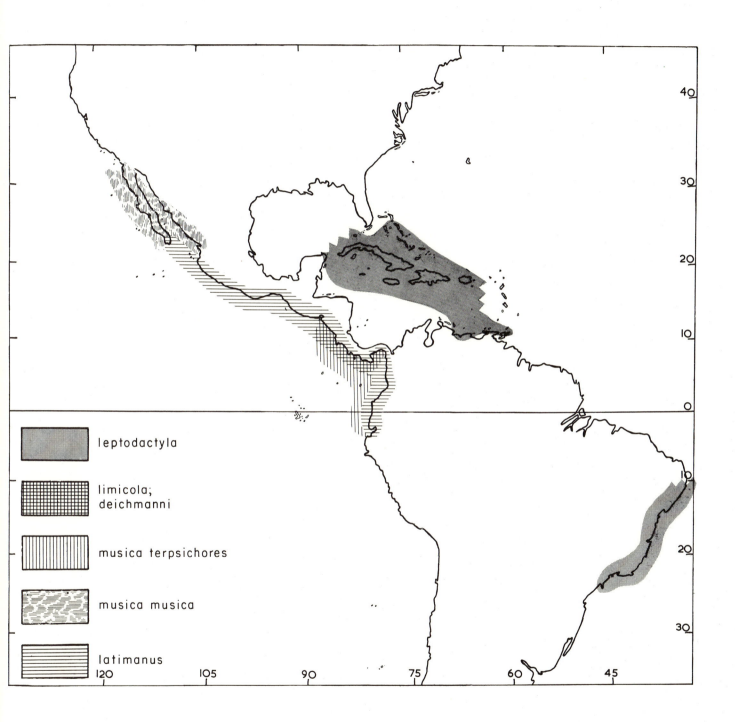

MAP 17. Distribution of the subgenus *Celuca* (part).
(General explanation: p. 409.)

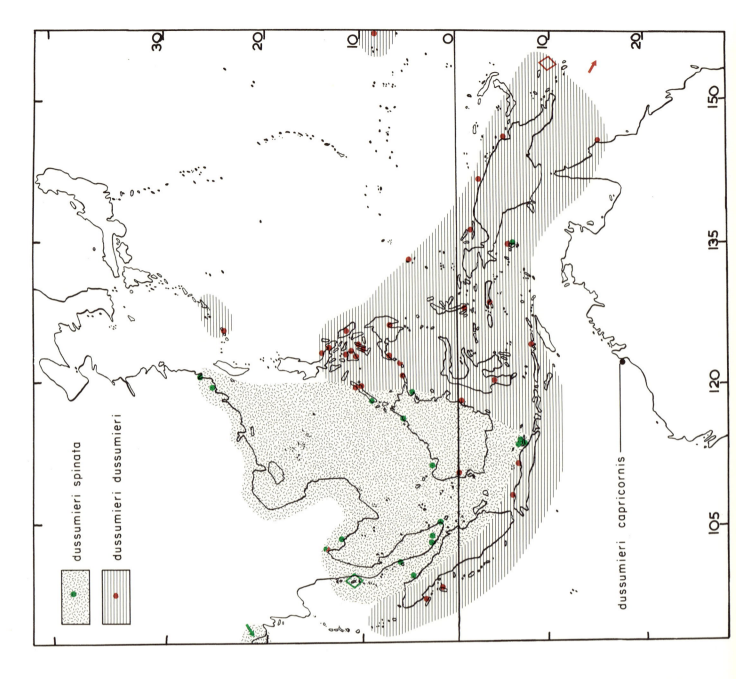

dussumieri spinata

dussumieri dussumieri

dussumieri capricornis

MAP 18. Distribution of the subgenus *Deltuca* (part). *U. dussumieri capricornis* is also known farther south, from the Monte Bello Is., through a single young male (p. 592). The known range of *d. dussumieri* has recently been extended south to Moreton Bay, near Brisbane (p. 38). (General explanation: p. 409.)

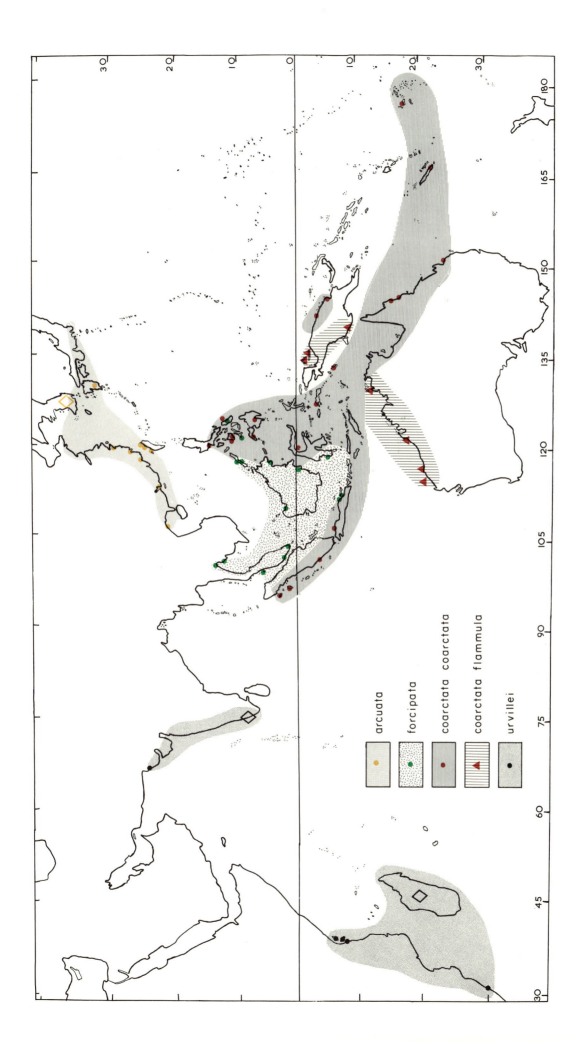

Map 19. Distribution of the subgenus *Deltuca* (concluded). The known range of *c. coarctata* in Australia has recently been extended south to Moreton Bay, near Brisbane (p. 57). (General explanation: p. 409.)

arcuata

forcipata

coarctata coarctata

coarctata flammula

urvillei

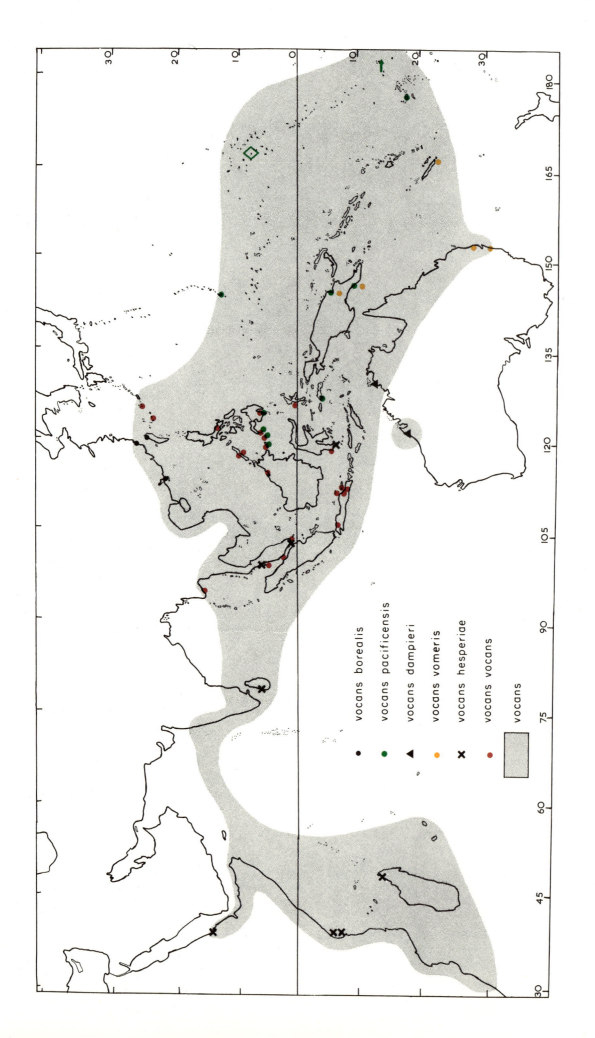

MAP 20. Distribution of the subgenus *Thalassuca* (concluded). Only a single screen is used because of uncertain boundaries among the subspecies (p. 409). *U. vocans* has been reported once in recent years near Sydney (p. 94). (General explanation: p. 409.)

vocans borealis

vocans pacificensis

vocans dampieri

vocans vomeris

vocans hesperiae

vocans vocans

vocans

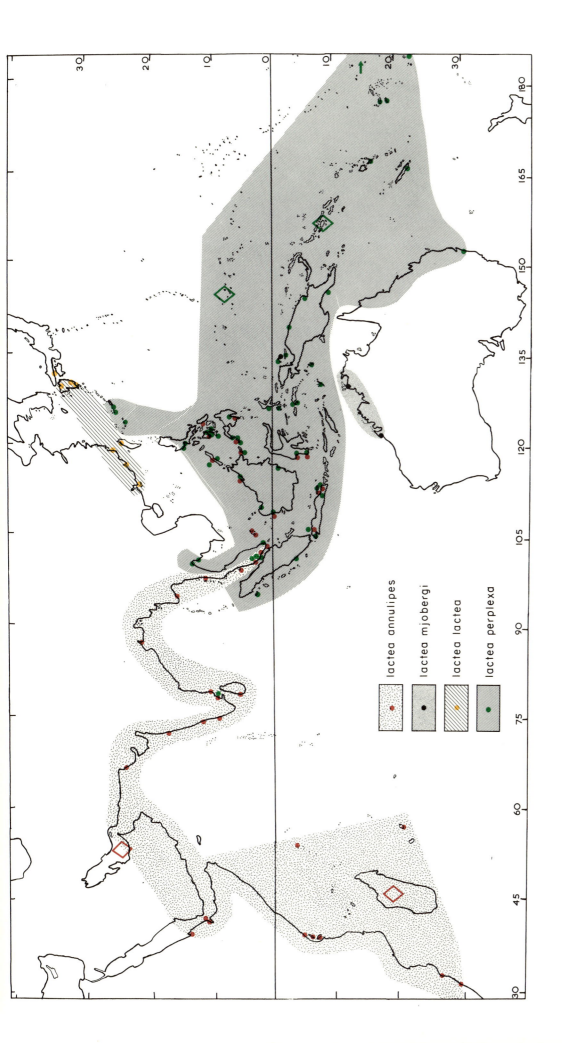

lactea annulipes

lactea mjobergi

lactea lactea

lactea perplexa

MAP 21. Distribution of the subgenus *Celuca* (concluded). (General explanation: p. 409.)

Part Two · Toward an Evolutionary Synthesis

Chapter 1. Zoogeography

OUTLINE

A. INTRODUCTION

Fiddler crabs, comprising the homogeneous genus *Uca*, are widely distributed throughout the tropics and subtropics of the world. Nine subgenera, totaling 62 species, are recognized in this contribution. More than two-fifths of these species are found only along the coast of the tropical eastern Pacific. Less than one-third occur in the entire Indo-Pacific region from East Africa to the Marquesas. On most coasts from one to three species extend into temperate regions.

The puzzles of the origins of *Uca* distribution yield both clues and permanent uncertainties in abundance. The clues, as usual, encourage hypotheses, while the uncertainties promptly reduce most reasoning to the level of speculation. The virtual lack of fossils remains a familiar obstacle.

The following survey first presents factual data on the modern distribution of the genus. A subsequent section considers the historical aspects of its zoogeography and presents several alternative explanations of the origins of the fauna; it is concluded that

an Indo-Pacific center is most likely. The final sections in the chapter cover the distribution of the species among the subregions, with special reference to evolutionary aspects.

Contributions which, along with their references, will provide background material for the topics discussed include Durham (1950); Durham & Allison (1960); Ekman (1953); Garth (1960, 1965, 1966); Hedgpeth (1953, 1957.2); Hubbs (1960); Kummel (1961); Mayr (1963); MacGinitie (1958); Sverdrup, Johnson, & Fleming (1942); Trewartha (1954); Troll & Paffen (1963); Tweedie (1954); Umbgrove (1949). See also Hallam, ed. (1973).

B. SURVEY OF CURRENT DISTRIBUTION OF *Uca*

1. *Regional Distribution of Subgenera.* The distribution of the nine subgenera is shown in Table 7. It will be seen that four are confined to the Indo-Pacific,

three to the coasts of America, and one to the eastern Atlantic. The ninth, and largest, is primarily American, but has two representatives in the Indo-Pacific. Representatives of each of the subgenera found in America occur on both the Atlantic and Pacific sides of the Isthmus.

2. *Regional Distribution of Species.* The general distribution of the 62 species recognized in this contribution is shown in Map 1 and Tables 8 and 9. Eighteen occur in the Indo-Pacific, 43 on the coasts of America, and one in the eastern Atlantic. Fifteen of these American species are found in the Atlantic and 31 in the Pacific; the American total is limited to 46 because of the fact that three species are represented by forms so closely related on the two coasts that they are here considered subspecies, in spite of their current complete isolation by the Isthmus of Panama.

The richest area by far, with 30 species, is a small region of the tropical eastern Pacific, centering along the shores of the Gulf of Panama. Twenty-two of these 30 species are endemic (see below).

In the eastern Atlantic the genus is represented by one species, *tangeri*, that ranges the full length of the coast from the warm temperate climate of south Portugal to the subtropics of Angola. This species is related to American species of the subgenus *Uca*; it is, however, here given a subgenus of its own (*Afruca*). A member of the American subgenus *Minuca* perhaps occurs in West Africa (p. 327).

3. *Subregional Distribution of Species: Preliminary Survey.* The richest region of the Indo-Pacific is the area embracing the shores of northern Borneo, northern Sumatra, and the Bay of Bengal; here 11 species occur. Two other regions, the Philippines–East Indies Axis and northeastern Australia, are almost as rich, with 10 species each.

To the west, north, and east, numbers decrease, as follows: northwest Australia has eight species, East Africa–Western India, six; China–West Taiwan–Japan, seven; and the Pacific Islands, five.

On American coasts the following distribution prevails. From southern California to the southern border of Mexico, five species; from the Gulf of Fonseca to Peru, 30 species; from Chile, perhaps as many as three species. From the temperate eastern United States (excluding northeast Florida), three species; from the Gulf states, six; subtropical Florida, five or six; tropical western Atlantic, from Mexico and the Caribbean to Rio de Janeiro, Brazil, 12; southern Brazil, five or six; Uruguay and Argentina, one.

These subregional distributions will be considered in more detail, in relation to the evolution of the species, in the sections starting on p. 435.

4. *Allopatric Distribution in the Indo-Pacific Region.* Among the 18 Indo-Pacific species, one (*tetragonon*) ranges, without regional variations making useful the erection of subspecies, throughout the tropics from the Red Sea to the Tuamotus. Three species, divided here into two to six subspecies each, show an almost equally wide range (*vocans, chlorophthalmus, lactea*). Four species, each comprising two to four subspecies, have ranges roughly confined to the area bounded by eastern India, the Philippines, and the tropical coasts of Australia (*dussumieri, demani, bellator, triangularis*); a partial exception, *dussumieri*, alone occurs north of the tropics in China and the Nansei (Ryukyu) Islands.

Two groups of species are regarded as superspecies (*acuta* and *coarctata*). One of them, *coarctata*, is composed of four species and ranges throughout the Indo-Pacific. The second, *acuta*, is found from Fukien in northern China to Indonesia and from the Ganges delta to Java; it includes two species of which one subspeciates.

5. *Allopatric Distribution in the American Region.* All of the 12 species in the tropical western Atlantic have close relations across the Isthmus of Panama. In three of them, *maracoani, thayeri*, and *vocator*, the allopatric representatives are here given only subspecific rank. A fourth, *rapax*, is morphologically so close to *galapagensis* that the specific standing accorded them may be illogical (p. 183). The remaining eight species in the western Atlantic appear sufficiently distinct from their Pacific counterparts to be counted without hesitation as full species.

Along both the Pacific and Atlantic coasts the erection of subspecies occasionally appears both justified and desirable. So far there is little or no evidence of the formation of clines; adequate meristic data, however, remain virtually absent.

6. *Endemism.* In the entire Indo-Pacific region only three species have narrow ranges. *U. polita* and *seismella*, both in the subgenus *Australuca*, are confined to Australia. *U. (Amphiuca) inversa* is known only in eastern Africa and in Pakistan, at Karachi, with a distinct subspecies in each of the two areas.

The contrast with the situation in America is striking. It has already been mentioned that 22 species are confined to a small tropical area on the Pacific coast of America. This strip extends only from about lat. 14° N. to about lat. 3°30′ S., or a straight distance of about 1,100 miles. The region includes part of the Pacific coast of El Salvador, and extends to the extreme northern edge of Peru, on the southern margin of the Gulf of Guayaquil.

Endemism is also apparent near the latitudinal extremes of the genus on three of the four American coasts. One species is endemic on the west coast of California and in northwest Mexico (*crenulata*); northern subspecies of two others (*m. musica* and *princeps monilifera*) are virtually confined to the

Gulf of California; two endemic species are found on the south and east coasts of the United States (*minax*, *pugilator*); one occurs only from subtropical Brazil to Buenos Aires, Argentina (*uruguayensis*).

In the Indo-Pacific, on the contrary, the few species ranging into temperate climates are, with two exceptions, merely widespread species from the tropics. The exceptions are *arcuata* and *formosensis*. Three other species extending into Indo-Pacific areas of winter chill are represented by subspecies of narrow range (*acuta*, *vocans*, *lactea*). Finally, a tropical subspecies of *dussumieri* ranges without obvious morphological changes north to near Fu-chou.

The present extremes of distribution in the genus are summarized in Table 7.

7. *Summary of Current Distribution of Species.* By far the richest area in number of species is the west coast of tropical America (eastern Pacific), with 30 species. The tropical American east coast (western Atlantic), the Indo-Malayan region (tropical west Pacific) and the northeast Australian (Solanderian) region are comparable to one another, having 10 to 12 species, one-third to two-fifths of the number found in the tropical East Pacific. Only six species occur in eastern Africa, while the fauna is poorest near the northern and southern limits of the genus range (one to two species), the eastern boundary of the Indo-Pacific (two) and, especially, the eastern Atlantic (one or, possibly, two).

C. HISTORICAL ZOOGEOGRAPHY

The question of the genus origin may now be discussed. Later the background of speciation in the geographical subregions will be considered.

Among the references listed on p. 431, those of the most direct aid in preparing this section have been Durham (1950), Ekman (1953), Kummel (1960), MacGinitie (1958) and, for recent views on several dates, Hallam, ed. (1973).

1. *Factors Concerned in Postulating the Geographical Origin of* UCA. As in all such questions, five major lines of evidence must be examined: first, the geological and climatic history of the world; second, the present distribution of the group; third, its fossil history; fourth, evidence from the group's systematics, especially from comparative morphology and social behavior; fifth, the group's ecological requirements. Although the present discussion is concerned primarily with the first two, the other aspects, discussed in more detail elsewhere, will be mentioned from time to time.

As mentioned in the beginning of this chapter, the fossil record is almost nonexistent. Only three specimens referable to *Uca* appear to exist, two from Pliocene beds, one from Pleistocene, and all in

America; all are similar to contemporary American species (pp. 127, 157). A fourth fossil, attributed by Desmarest (1822) to the genus *Gelasimus* (= *Uca*) has disappeared (p. 324). In addition, an ocypodid from the Pliocene of Jamaica resembles *Macrophthalmus*, an Indo-Pacific genus (Rathbun, 1919).

Three major areas seem to be possible centers of early *Uca* evolution and dispersal. The genus may have evolved in the circumtropical area during the greatest expansion of the Sea of Tethys, or, somewhat later, in either tropical America or in the Eastern Pacific. A temperate or polar origin seems unlikely because of the warm-climate distribution of almost the entire highly various family. In the present section each of these major possibilities is considered in turn.

The following annotated list assembles a number of geological, climatic and ecological factors which were probably, as in the evolution of so many other marine animals, of primary importance in the history of the group. This attempt to understand *Uca*'s particular characteristics requires their review in juxtaposition, in spite of the familiar nature of some of the material.

1. A warm global climate existed from the Cretaceous through the Eocene, with the tropics extending far north along the Pacific and Atlantic coasts.

2. The climate gradually cooled from the Oligocene through the Pleistocene.

3. During these periods the Atlantic Ocean cooled drastically. This was particularly true of the eastern Atlantic, in contrast to the eastern Pacific and Indo-Pacific regions where cooling was, respectively, slight and absent. There was a consequent striking impoverishment of Atlantic fauna, which has been demonstrated in many groups of animals. In other marine and littoral groups, however, including echinoderms and certain crabs, the Caribbean is now richer in species than is the tropical eastern Pacific.

4. The vast Tethyan seaway gave access at its maximum extent from the eastern Pacific all the way eastward to China. After the beginning of the Oligocene it dwindled, and from the Miocene onward it was impassable as a route between the Atlantic and Indian Oceans.

5. The last possible exchange of marine faunas between the tropical Atlantic and Pacific took place near the end of the Pliocene, perhaps 5,000,000 years ago, ceasing when, about that time, the Isthmus of Panama became completely emergent.

6. The major islands of the Philippines and East Indies, at present so rich in species in many groups

of animals, were apparently rarely above water before the Eocene.

7. The Sunda Shelf was emergent for much of the period from the Pliocene to the end of the Pleistocene. This area extended the coast of southeast Asia to include Borneo, Java, Sumatra, and many lesser islands, acting as an effective barrier between the China Sea and Philippines area on one hand and the Bay of Bengal and western regions on the other. After the Pleistocene, seas covered this shelf region, the East Indies assumed much of their present form, and free dispersal of organisms once more was possible between the Indian Ocean and the north tropical Pacific. This submersion of the Sunda Shelf probably took place not more than 10,000 years ago.

8. A similar barrier, the Sahuli Shelf, probably connected northern Australia and New Guinea from time to time over long periods, separating the littoral fauna of western and eastern Australia. It seems likely that the last submersion of that barrier also occurred soon after the Pleistocene's end.

9. Intense orogenic activity took place in the Philippines–East Indies area from the Pliocene onward.

10. In contrast, the shores of the Indian Ocean, both east and west, were then quiet.

11. The East Pacific Barrier is as potent a force in this group of animals as in practically all other groups. There is no evidence that any modern Indo-Pacific species of *Uca* has ever reached America or vice versa, much less become established there. Nevertheless, two considerations make caution in this area of importance. First, the timing of metamorphosis of some pelagic larvae into littoral crustaceans is partly controlled by characteristics of the water, the molts to postpelagic stages apparently being triggered by chemical factors occurring near land (Garth, 1966, and refs.). Second, the Indo-Pacific ocypodid, *Ocypode ceratophthalmus*, occurs on Clipperton Island along with 15 other species of brachyuran crabs that are of Indo-Pacific origin and only 18 others that are American (Garth, 1965). The distance from Clipperton to the nearest point on the Mexican coast is about 600 nautical miles.

12. The mid-Atlantic, at least in the period from the late Tertiary to the present, has been an almost equally potent barrier.

13. *Uca* is essentially a tropical genus, always reaching its greatest development within the true tropics. A few species, however, have adapted to freezing winter temperatures, hibernating for months at a time. The genus is absent from cold temperate and boreal climates.

14. As in other groups of littoral animals, regions with high tides support higher numbers of species, other factors being equal.

2. Uca *as a Possible Member of the Old Mesozoic Circumtropical Fauna.* According to this view, *Uca* would have formed part of the fauna evolving on the shores of the Tethys, when the vast sea reached its greatest extent. At this time, between the late Mesozoic and the Eocene, no barriers of land or climate stood from the eastern Pacific eastward to China. If the genus evolved under these conditions, no Tethyan neighborhood is more likely than another to have been the center of *Uca* evolution. Similarly, no early differentiation of American and Indo-Pacific forms would be expected.

Ekman (1953) has been the strongest recent advocate of descent of circumtropical organisms from marine animals of this old warm water shelf. This view is certainly the correct explanation of the present-day distribution of a number of groups, notably corals, echinoderms, and some fishes. Ancestral forms no doubt circulated freely throughout the Tethyan realm and on across the Pacific where they were stopped—as today—by the East Pacific Barrier. Ekman does not discuss the possibility of migration of warm water shelf species across the Behring bridge in the days when temperatures were higher.

As listed briefly under points 3 to 5, p. 433, three great geological and climatic events broke up the circumtropical fauna, starting roughly in the Oligocene, culminating in the Pleistocene, and ending with that epoch's close.

First, the Tethys became reduced, resulting in the ultimate blocking off of the Atlantic from the Indian Ocean by emergent Asia Minor, Iran, and Suez. The last narrow passage closed sometime in the Miocene; the western end of the Mediterranean had also come to have such a narrow transit to the Atlantic that as a funnel for dispersal it had long been impractical.

Second, the last Central American passage between the Atlantic and Pacific was closed by land emerging near the end of the Pliocene, about five million years ago. Until then, via several seaways from Tehuantepec in the north to one across present-day Colombia in the south, passages of various widths had been freely open. During much of the time, wide seaways existed.

Third, the climate of the Atlantic deteriorated progressively from the Oligocene on. The eastern Atlantic was most affected. The West Indies were also chilled from time to time, although not so drastically; nevertheless, it was enough to kill off many formerly prominent elements of the fauna. A notable example is coral of the genus *Pocillopora*, which now remains abundant and ecologically important both in the eastern Pacific and from mid-Pacific to the Red Sea. At intervals only the South American shelf, the region of Amazonica, was left as a dubious refuge for creatures of tropical shores. In contrast, the tropical eastern Pacific was little chilled, and the Indo-Pacific region not at all.

These events, then, are the well-established background explaining the poverty of some groups of the Atlantic littoral today. They also indicate the generous amounts of time that have allowed Indo-Pacific and American faunas to diverge.

The weaknesses of the view of a continuous circumtropical fauna, in connection with *Uca* origins, are three. First, the mid-Atlantic barrier is almost as formidable for littoral animals with short pelagic larval stages as is the eastern Pacific barrier. Second, the consensus of thought now holds that no convenient land bridges extended even around the northern part of the Atlantic over which littoral animals could migrate. Granted that the northern seas and shores were much warmer then than now, it still seems unlikely that the Tethyan route was used importantly; although they were not subjected to known cooling, the species of East Africa and India are few in number. Finally, it seems again unlikely that only two genera and only advanced species of all this large family should be found in America, while all other genera and all the species showing most primitive characters are found only in the Indo-Pacific.

3. *The Panamic Region as a Possible Center of Dispersal.* If only the species numbers of present distribution were considered, the center of dispersal could logically be the crowded Panamic region. Even with no regard for geological history, however, this area is virtually ruled out as the center of origin on grounds of morphology and behavior. The vast majority of species in this richest area show characters which are specialized in comparison with those shown by 16 out of 18 species in the Indo-Pacific. These characters include carapace shape, armature of carapace and appendages, and social behavior patterns.

Again, out of 14 genera referred on good grounds to the Family Ocypodidae, only two—*Uca* and *Ocypode*—are found in the Americas. All others occur only in the Indo-Pacific, an area centered chiefly in the Indo-West-Pacific region.

Therefore, if the Americas, on the basis of sheer numbers of present-day species, are considered to be the center of distribution, these formidable objections exist: (1) They are all specialized forms, in comparison with the majority of Indo-Pacific forms. (2) All of the numerous related genera, with one cosmopolitan exception (*Ocypode*), are confined to the Indo-Pacific.

4. *The Possible Colonization of America via the Behring Strait.* This explanation of the present distribution of the genus seems most likely. The multi-

tude of forms in America can be explained by the crossing and successful establishment of even a single Indo-Pacific species. The affinities of the migrants, of however many species, would almost certainly be with the root stock of the present subgenus *Thalassuca* (including *tetragonon* and *vocans*). Possible further affinities with *Australuca* (including *bellator*) are indicated if a second stock is involved. In the absence of a helpful fossil record, further speculation is inappropriate.

In the Eocene and during the early Oligocene, subtropical seas extended at least to the 55th parallel and a warm temperate climate far up into Alaska. Fiddler crabs are notably adaptable to cold weather, even within a single species and without external signs of subspeciation (*tangeri*, *seismella*, *polita*). It therefore seems the necessary crossing could have been made up the warm coast of east Japan, with time to breed along the bridge from summer to summer. For millions of years winters in that region may not have been as severe as in present-day Connecticut and southern Massachusetts, where three species now occur. All of these forms have close relations in the tropics.

If this hypothesis is correct, the migrants, once down the west coast of America, found an abundance of suitable habitats. Similar ecological conditions are found in tropical swamps, estuaries, and protected bays from Fiji to Zanzibar and from West Africa to Costa Rica. Yet nowhere in America, it seems likely, did many ocypodid members of the mangrove community then exist. The *Macrophthalmus*-like fossil of the Jamaican Pliocene (Rathbun, 1919) certainly has no present-day descendants. Conditions, for whatever reason—the absence of competitors or the presence only of those giving inferior competition— must have been ripe for a burst of speciation.

The second possible means of a migration from Asia is via island-hopping and the crossing of the East Pacific Barrier. A few years ago this route appeared discredited but, as discussed by Garth (1966), in some pelagic larvae the duration of instars can be prolonged, with obvious possible effects on distribution.

On the whole, then, the hypothesis is favored in this contribution that the origin of *Uca* was in the Indo-Pacific region and that one or two of the groups there gave rise, after migration across the Behring Bridge, or possibly the mid-Pacific, to the American species.

5. *Evolutionary Aspects of* UCA *Zoogeography.* Two American subgenera of *Uca* are morphologically closest to the Indo-Pacific stock. These are the subgenus *Uca*, including *maracoani* and the rest of the American "narrow-fronts," and the subgenus *Minuca*, erected for the specially broad-fronted spe-

cies. They, West Africa's *tangeri*, and the Indo-Pacific subgenus *Thalassuca* (including the ubiquitous *vocans* and *tetragonon*) clearly had common ancestors. In America, however, social behavior developed well beyond the *Thalassuca* activity patterns, along lines parallel to those found in any other waves of migrants, presumably of *Australuca* (*bellator*) stock (p. 435). Examples of such parallel social development even intergenerically are found throughout the family; the Indo-Pacific genus *Ilyoplax* is an excellent example (p. 494). From these advanced behavioral characters, as well as from strictly morphological traits, it is held here that *tangeri* is derived from the stock of the American subgenus *Uca* (*maracoani*, etc.), rather than directly from the Indo-Pacific *Thalassuca*.

Uca thayeri, for which the subgenus *Boboruca* has been erected, is of great interest from two points of view. As already mentioned, it is one of the few American forms in the genus which have changed so slowly in the past five million years that the forms on both sides of the Isthmus cannot logically be regarded as more than subspecies on the basis of the usual criteria. Second, this species shows combinations of characters, both morphological and behavioral, suggesting definite affinities with the Indo-Pacific narrow-fronts of the subgenus *Deltuca*.

The other species represented on each side of the Isthmus by forms considered of subspecific rank are (*Uca*) *maracoani* and (*Minuca*) *vocator*.

The remaining two groups of American *Uca*, placed here as the sole representatives of Bott's subgenus *Minuca*, include the most advanced species in the genus, both morphologically and in social behavior. Whether they developed from the same root stock as the subgenus *Minuca* (*rapax*, etc.), as seems most likely, or whether they represent a separate crossing, with their nearest relations among the ancestors of the subgenus *Australuca*, it is impossible to say.

Celuca has speciated most profusely in the tropical east Pacific, and it is from this group that the two Indo-Pacific species from American stock are derived (*triangularis* and *lactea*). Apparently they are descended from one or two successful westward migrations. This view seems more likely than the alternative, which would demand an Indo-Pacific origin for *Celuca*, making all crossings then from west to east. The closer affinities of *Celuca* are undoubtedly with the other American groups rather than with root stocks near the subgenus *Australuca*, the only candidates with relations among present-day Indo-Pacific crabs. The first hypothesis, then, demanding one or two east–west crossings, seems the more likely. The matter will be further considered in the last chapter.

Unless a mid-Pacific route was used, the crossings, like those made by the original migrants, would have had to be made before the onset of really cold weather. Only three of the contemporary forms in these groups extend into regions regularly subjected to temperate or upper subtropical winters. These are *pugilator* in the northeastern United States, *crenulata* in California and at the mouth of the Colorado River, and *lactea* in Japan, Pakistan, and southeast Africa.

6. *Conclusions Concerning the Origin and Distribution of* Uca. The hypothesis favored in this contribution is that *Uca* originated in the Indo-Pacific region, that ancestors of one or two of the Indo-Pacific subgenera there gave rise, after migration across the Behring Bridge, to four new subgenera in America; that representatives of one of these groups migrated westward via the same route and gave the Indo-Pacific fauna two living representatives of these advanced American subgenera. The chilling of the Atlantic can in part explain the paucity of Atlantic species compared with their abundance in the eastern tropical Pacific. The large tide ranges in the Pacific area may explain the wealth of recent species in that restricted region (see Chapter 2, p. 443).

D. Subdivisions of the Indo-Pacific Region in Relation to *Uca* Distribution (Table 8)

No particular order of listing is very satisfactory, since the regions are related in different ways, depending on the species groups examined. Still, very definite conclusions can be drawn on the subject. A study of the list of Indo-Pacific species and their allopatric affinities to one another leads to the following statements. The most specialized forms, with no exceptions, are found along the Philippines–Java axis; the least specialized are peripheral, particularly in northwest Australia, the China–northwest Taiwan–Japan region, and, to a lesser extent, the mid-Pacific region. This is in contrast to the frequently existing condition in birds and other animals, where specializations occur in peripheral isolation or other conditions of partial isolation. The situation will now be examined in more detail.

Uca in the Indo-Pacific, as shown by the distribution of species and subspecies, indicate the division of the vast stretches of the Indo-Pacific into seven major subdivisions, as follows.

1. Northwest Australia (Dampierian Region).
2. China; northwest Taiwan; Japan.
3. Pacific Islands.
4. Northeast Australia and New Caledonia (Solanderian Region).
5. Philippines–East Indies Axis; New Guinea.
6. Northern Borneo; northern Sumatra; Malaya to eastern India.
7. Western India and eastern Africa.

Table 8 lists the species by regions and indicates allopatric relationships. Characteristics of their respective *Uca* populations are described below.

1. *Northwest Australia (Dampierian Region).* This area has two remarkable characteristics. First, of the eight species recorded, at least two are allopatric forms showing less advanced waving display characters than do their representatives in other regions. Second, except for *demani* and the three members of the typically Australian subgenus, *Australuca*, the affinities are definitely with the *Uca* fauna of the other coasts of the Indian Ocean, rather than with the Pacific faunas to the north and east.

2. *China; Northwest Taiwan; Japan.* The six species concerned (*acuta, dussumieri, arcuata, formosensis, vocans,* and *lactea*) are, with the exception of *dussumieri*, distinct forms with more conservative morphological and, where known, behavioral characteristics than have the neighboring forms in the Philippines–East Indies area. *U. dussumieri* alone is a surprise. Although it reaches seasonally cold Fu-chien Province, it belongs, according to all discernible morphological evidence, to the subspecies *spinata*. This form is found typically in the tropics, in Singapore, northwest Borneo, and the west coast of Malaya. At the present time only *arcuata* and diminishing populations of *lactea* occur in Japan. *Uca formosensis*, known only from northwestern Taiwan, is possibly a cold-water allopatric representative of the far-ranging species, *tetragonon*. More probably it is a conservative member of its subgenus (*Thalassuca*), with resemblances to *Deltuca*.

Species reported from the Nansei (Ryukyu) Islands have, with the exception of *arcuata*, proved to be identical with the tropical forms of the Philippines–East Indies axis; *v. vocans* and *lactea perplexa* are of usual occurrence; *d. dussumieri* is apparently an occasional stray, where it was probably reported as *dubia* by Stimpson (1858); specimens agreeing with his original description proved upon examination to be examples of *d. dussumieri*.

3. *Pacific Islands.* These are expectably characterized by a progressive impoverishment eastward. Only two species, *tetragonon* and *chlorophthalmus*, are found east to the Society Islands and beyond; *lactea* and *vocans* reach Samoa; *coarctata* filters out in Fiji. One of them, *vocans*, is a relative conservative among the six subspecies; *coarctata* and *lactea* cannot be usefully distinguished taxonomically from the advanced forms found in the Philippines–East Indies axis.

4. *Northeast Australia and New Caledonia (Solanderian Region).* This rich area is chiefly a southward extension across New Guinea of the Philippines–East Indies axis. It differs in three ways: first, some northern species are not present; second, typically Australian members of *Australuca* are strongly repre-

sented; third, two species, *vocans* and *bellator*, form separate subspecies here.

5. *Philippines–East Indies Axis; New Guinea.* This area, very rich in number of species, is the habitat of the most specialized subspecies, the specialization being shown particularly in genital morphology and display behavior. The region extends in general to include part of Sumatra in the west and New Guinea in the east. The *Uca* extending northeast into the Nansei (Ryukyu) Islands have all proved to belong to this fauna rather than to the section typical of the cold-current shores to the west (2, above). Several of the forms in New Guinea have affinities with, respectively, northwestern Australia (*coarctata flammula* in parts of West Irian), northeastern Australia (*vocans vomeris*), and the Pacific Islands (*vocans pacificensis*); both of the latter subspecies were found in eastern New Guinea and show evidence of hybridization. The other species all belong to subspecies occurring in Philippines and, in most cases, south and east of New Guinea as well.

6. *Northern Borneo; Northern Sumatra; Malaya to Eastern India.* This region was separated from the Philippines–East Indies region throughout the Pleistocene and much of the earlier Cenozoic by the Sunda Shelf. It forms an area slightly richer in species of *Uca* and other ocypodids than any other in the Indo-Pacific, but its subspecies are usually identical with those of eastern Africa. In other words eastern Africa is an impoverished extension of the shores of the Bay of Bengal. To the east the subdivision extends around the tip of the Malay Peninsula and up the coast to Thailand. The north coast of Sumatra, and northwestern Borneo are typically included.

7. *Western India and Eastern Africa.* This fauna is poor in species compared with the wealth of forms farther to the east. Only one species, *inversa*, does not have allopatric representatives elsewhere, or, like *tetragonon*, range beyond the region. The allopatric forms are all moderately to extremely advanced members of their respective groups either in genital morphology, waving display, or both.

8. *Areas of Coincidence.* One of the most interesting aspects of this entire study concerns the areas of coincidence in distribution. These occur in the general region extending from Ceylon and Madras around the tip of the Malay Peninsula (south of Penang), and in Sumatra, Java, Borneo, Celebes, and the southern Philippines. In these areas there is clear juxtaposition and in some cases evidence of hybridization in adjacent western and eastern populations. The groups involved are the subspecies of *dussumieri, vocans,* and *lactea,* each in a different subgenus, along with the full species *forcipata* and *coarctata* of the superspecies *coarctata* (subgenus *Deltuca*). All of this mingling might normally be ex-

pected through the classical lifting of a barrier. In this case the barrier, the Sunda Shelf, was removed by its flooding, probably most recently some 10,000 years ago. The situation is discussed under the species involved as well as in the chapter on speciation, p. 530.

9. *Conclusions.* It seems that the broad paleozoogeographical picture for the genus may be as follows. In the days of the maximum development of the Tethys, during the Cretaceous and early Tertiary, an access to its usually placid shores was available from the Australian shield all the way to China, India (then a great island), and a very small Africa.

The land of west Australia has been perhaps least disturbed, and its fauna therefore least subject to evolution since that time; particularly, it lacked the stimulus of true isolation. Also, its simple coastline provided relatively few habitats. The fauna of both west Australia and the China–northwest-Taiwan–Japan axis had eventually to adapt to moderately cold winter weather. This too, through decrease in numbers of generations, could slow evolutionary rates in comparison with tropical *Uca*, most of which breed through the year.

All the other regions, with the exception of the Java–Philippines axis, show various geological histories but are relatively lacking in striking events.

In contrast, the Java–Philippines axis starting in the Tertiary has undergone violent volcanic activity which continues to the present day. Coastlines have been changed repeatedly; islands have risen and subsided; shores are convoluted; a multitude of niches has appeared and vanished. All of this apparently proved highly stimulating to evolution in a genus of *Uca*'s requirements. The subspecies all show the most highly developed displays and the most complex gonopods in their species.

E. Subdivisions of the American Region in Relation to *Uca* Distribution (Table 9)

The dates of migration to and from North America must of course remain problematical. Because of lowered temperatures the movements could scarcely have occurred later than the Oligocene in either direction, in any use of the Behring route. If the first migrations eastward occurred in the Paleocene, and the last during the middle of the Oligocene, a period of some 30 to 55 million years was available for the development of American groups. When the changes are considered that have occurred in some allopatric species within perhaps the last five million years, after the closing of the American isthmus, it seems likely that the long, warm stretch of the early Tertiary was altogether adequate for the evolution of the new subgenera.

1. *Eastern Pacific.* In the eastern Pacific the distribution of *Uca* divides clearly into three zones. The first extends from the southern part of California and the mouth of the Colorado River to the southern part of the State of Oaxaca, in south Mexico; the second, from El Salvador and the Gulf of Fonseca to the southern edge of the Gulf of Guayaquil; and the third, from Peru south of the Gulf to northern Chile. In contrast to the supremely rich tropical subdivision, the number of species in the northern zone is mediocre. Garth (1960) has examined in detail the composition of the brachyuran fauna in the region, including *Uca*. The southern zone, down the almost straight west coast of Peru and northern Chile, is one of the most impoverished coasts in the world, from the point of view of support for populations of *Uca*. Not only is this third subdivision almost devoid of suitable habitats, but the coast is relatively cool for its latitude because of the Humboldt Current.

In the northern zone only one species, *crenulata*, is endemic. Two others are considered in this contribution to be subspecies of species occurring also in the south. None of the two or three species counted as questionably ranging south to Chile appears to differ from northern forms; recent material, however, is non-existent.

2. *Western Atlantic.* On the west coast of the Atlantic, four zones are counted. The first covers the temperate shore of the United States; the second those of the states bordering the northern part of the Gulf of Mexico; the third, tropical North and South America, including the Caribbean islands, as far south as Guanabara Bay in Brazil; the fourth, the extreme southern part of Brazil, adjacent Uruguay, and, from a single record, the northern shore of Argentina near the mouth of the Rio de la Plata.

In the northern zone three species occur farther north than anywhere else in the world and are regularly subjected to freezing winter temperatures. Their boundary at present is Cape Cod, although there is some evidence that one of the species once reached the southern part of Maine. The zone's southern boundary is northern Florida, where the three species of the north are supplemented during clement periods by two stragglers from the tropics, their presence favored by the Gulf Stream. The shifting and precarious footholds held by these southern forms correspond to the fluctuating distribution of several species in the Nansei (Ryukyu) Islands, which are similarly affected by the Japanese Warm Current.

The second subdivision is a rather anomalous zone, just beginning to be understood (Salmon & Atsaides, 1968.1, 1968.2). Here along the northern part of the Gulf Coast of the United States and northwest Mexico, three of six forms of *Uca* are somewhat different from their allopatric representatives, both temperate and tropical, on the Atlantic coast and on the southern part of the Gulf. Another interesting occurrence in the region is an endemic species, *subcylindrica*, from Texas and the adjoining coast of Mexico.

The five species now found in the southern half of Florida are of varied geographical affiliations in keeping with the intermediate and fluctuating climate of the area; consequently the region is not viewed as a subdivision. Three of the species are tropical and here, at the northern edge of their extensive ranges, only one seems well established. A fourth species is confined to the subtropics and the Gulf coast, while the fifth, although it thrives in south Florida, is characteristic also of the temperate coast to the north. A sixth species, wholly tropical, has not been reported for many years.

The third zone in the Atlantic covers the tropics more completely than in the Pacific, extending from the West Indies and the tropical part of Mexico to Rio de Janeiro, Brazil, at the Tropic of Capricorn. Even with 12 species, the subdivision supports only two-fifths as many as the more restricted tropical zone in the eastern Pacific.

South of the tropics, in the fourth zone, a single endemic species occurs. While distinct, with a short range from Rio de Janeiro to Buenos Aires, *uruguayensis* is clearly the allopatric representative of a widespread group of northern species. Five or six of the tropical species filter out not far south of Rio.

With the exception of *subcylindrica*, mentioned above, every form occurring in the western Atlantic has at least one close relation in the eastern Pacific.

3. Conclusions. A relative impoverishment of the Atlantic fauna in comparison with that of the Pacific shows strikingly when the numbers of species of *Uca* are compared on the two sides of the Isthmus. Two additional factors, besides the presumed influence of the Ice Age climate, should be kept in mind as possible contributory causes to the disparity in species. If, as seems most likely, the ancestors of American *Uca* migrated from Asia, then their first differentiations

were likely to take place on the nearest Pacific shores, even if the Atlantic happened then, and often subsequently, to be fully accessible. Again, although a mangrove swamp in Venezuela or Brazil now appears as richly suitable for *Uca* as its equivalent in Costa Rica or Ecuador, differences in such complex biotopes, rather than climate, may have controlled in part the rate of speciation in the neighboring oceans.

F. SUMMARY

The 62 species of *Uca* recognized in this contribution occur throughout the warmer parts of the world. The majority are confined to the tropical eastern Pacific, while on all continents the warm temperate regions are impoverished, as is the entire coast of west Africa. It is considered most likely that the genus arose in the Indo-Pacific, and that the American species are descendants of migrants moving along the Behring Bridge during the early Tertiary. It is further suggested that one or two later migrations in the opposite direction gave rise to Indo-Pacific members of a subgenus otherwise American. In the Indo-Pacific region seven subzones are recognized, of which the richest are those included in the tropical areas from India to the Philippines and northeastern Australia. The development of allopatric species and of subspecies was clearly influenced strongly by the repeated emergences of the Sunda Shelf. The Sahuli Shelf also probably played a role, as has the associated orogenic activity that has been characteristic of the entire region. The most conservative area, judging by the morphology of the local *Uca*, has been the northwest coast of Australia, which has had a contrastingly quiet history during the Cenozoic, the presumed period of the group's evolution. Several zones are apparent in both oceans bordering the American continents, each of the series running north and south in conformity with the continental coasts. Aside from climatic factors limiting distribution, the most important events were the sporadic emergences of land bridges that partly or wholly divided the faunas of the Atlantic and Pacific; the current bridge, the Isthmus of Panama, has served as a complete barrier for perhaps five million years.

Chapter 2. Ecology

OUTLINE

A. INTRODUCTION

Almost all adult fiddler crabs live in the intertidal zones of sheltered bays and estuaries, digging and feeding in sandy mud and muddy sand throughout the warmer parts of the world. These habits show basic demands for tidal action, for a soft and nutritious substrate, and for warmth.

As usual in *Uca* biology, each of these needs turns out to have wide variability, and all illustrate well the capacity of the genus to adapt to a wide range of conditions.

In this chapter the chief physical and biological factors in their environments will be examined. Because of the overlapping and interdependence of categories, none of the usual groupings of ecological factors will be made. Instead, the categories presented range roughly from the most physical to the most biological in nature.

The ecology of *Uca* may be considered from two distinct points of view. The first of these aspects is the distribution of the species and the second the activity of the populations under various conditions. These divisions will be considered under each heading.

Since very little is known of the ecological requirements of the pelagic young, these stages will be mentioned below only incidentally.

General accounts of climatic factors, the intertidal zone, estuarine environments, mangrove associations, and chemical aspects of ecology will be found in appropriate textbooks and isolated papers. Particularly useful in providing background are Allee, Emerson, Park, Park, & Schmidt (1949); Hesse, Allee, & Schmidt (1951); Ekman (1953); Trewartha (1954); Day & Morgans (1956); Macnae (1956, 1957, 1968); Macnae & Kalk, eds. (1958); Moore (1958); Teal (1958, 1959); Edney (1960, 1961, 1962); Macnae & Kalk (1962); Bliss (1968); Bliss & Mantel, eds. (1968); Miller & Vernberg (1968); and Newell (1970).

Hedgpeth (1957.1, 1957.2) makes sensible comments on ecological terminology. Since, as he points out, there is still no close agreement, particularly internationally, on the use of new terms, it would be inadvisable to bar the use of such inexact but useful words as "environment," "habitat," and "community." He also suggests that when more technical terms are used they be defined at the time by each author.

Since even now ecologists do not seem all to be in perfect agreement on appropriate uses even of such established terms as "biotope" and "biocoenosis," less widely accepted words will be discarded here in favor of the simplest terminology consistent with lack of confusion. For example, "Intertidal Zone" will be used throughout to indicate the general habitat of all *Uca*. The zone will be divided into "upper," "middle," and "lower" portions, rather than into more precise subdivisions in which various prefixes are placed before the general word "littoral." Although the more technical terms are in other studies often preferable, their use here would be inadvisable, for two different reasons. First, we are still too ignorant of exact ecological needs in the genus. Second, many species have widely variable ecological distributions.

It is hoped that this use of a majority of non-technical terms will be acceptable even to professional ecologists, since the present treatment does not aim to present a technical account of the ecology of the genus. Desirable as such a contribution would be, too little work has been done by anyone to make it possible. As in many aspects of *Uca* biology, only a résumé can be given of the beginnings, along with key references, to serve as a basis for future work.

Finally, it is cheering to note that Hedgpeth's pleasant comment of some years ago (1957.2: 48) has suddenly gone out of date. He said: "There are many still unreconstructed people to whom ecology is an unnecessary synonym for natural history." Perhaps our technical knowledge of the ecology of fiddlers will swiftly grow because of our present realization of the overwhelming importance of the ecology of the planet. And the survival of many now endangered *Uca* populations may incidentally be prolonged.

B. TEMPERATURE

Fiddler crabs are typically adapted to warm climates. In the rainy tropics they are active all the year around, and breeding individuals are found during every month. In the subtropics reproduction in some species is restricted by the dry season, rather than by temperature. The few species found in the warm temperate zone, however, are strongly controlled by temperature, both in the limits of their distribution and in their seasons of activity. In these forms reproduction is confined to the warmest months, while during cold weather the crabs hibernate in burrows.

1. *Geographical Ecology and Temperature.* The world climate map given by Ekman in the English edition (1953) serves admirably for discussion of *Uca*'s distribution in relation to temperature. The simple zonation is in keeping wtih the distribution of this genus, as with that of other littoral forms. Three of its main divisions cover *Uca*'s range—tropics, subtropics, and the warm temperate zone. The complex subdivisions required in terrestrial climatology are inapplicable to this intertidal group.

2. *Low Temperatures.* The lowest levels at which temperate and subtropical populations display vary widely. For example, in the latitude of New York City (42° N) in mid-June, *minax* fed and moved about above ground during periods of low tide. Yet the crabs did not display on days when the weather bureau temperature during the preceding night fell below 21° C (70° F); this behavior occurred in spite of the fact that daytime temperatures regularly reached above 29° C (84° F). The readings were made on sunny days with the instrument held barely clear of the ground and shaded.

During 1968, a series of temperature readings was taken on the southern coast of Cape Cod in the northeastern United States, a peninsula marking the northern limit of *pugilator*. The nightly minima were read

from a point just above high-tide level, with the base of the frame of the maximum-minimum thermometer in contact with the ground, and the instrument sheltered by a post on one side and a piece of wood above it. The diurnal readings, obtained as in *minax*, were all made in the midst of the observation area. During June the nightly low ranged from 7° C (44° F) to a single reading of 15° C (59° F), while the daytime highs, regardless of sunshine, lay between 9° C (49° F) and 29° C (84° F). On June 1, a sunny day, several individuals waved at low intensities when the temperature stood at 19° C (66° F), the low of the previous night having been 10° C (50° F). Although similar records were made several times later in the season, they mean little from a functional viewpoint, since the waves were not fully developed and no other trace of social behavior of any kind was seen. The season's maximum amount of social activity, including presumably actual breeding, occurred in the middle weeks of June, the population being notably less active in July in spite of higher temperatures. The July lows ranged from 7° C (44° F) on July 30 to 21° C (70° F) on July 18; the highs extended from 22° C (72° F) to 31.5° C (89° F). At no time during the season did the level of social activity in the species appear equivalent to that found in more southern populations.

In comparison, on the west coast of Australia, at Broome (17° S), four of the five local species were active socially in early May, the Australian autumn, although the temperature nightly fell to the neighborhood of 15° C (59° F); the days, however, were tropically warm.

It is probable that in subtropical localities near the temperate zone, such as Florida, Rio de Janeiro, and the north Philippines, populations of truly tropical species are usually killed off by the occasional freezing temperatures. In these cases the populations are doubtless quickly replaced through the arrival of pelagic larvae, carried by currents from more tropical regions.

Edney (1962) found in the laboratory that temperatures of 10° C (50° F) were lethal for all the South African specimens with which he worked, the material having been taken on Inhaca Island, Mozambique. In July and August the island's temperature in the mangrove water sometimes falls to 17° C (62.6° F) (Macnae & Kalk, 1962). Although, as Macnae (1957) says, the species of *Uca* on Inhaca are all tropical, still these populations are clearly tolerant of mildly subtropical temperatures.

Physiological studies by a number of investigators (Miller & Vernberg, 1968, and refs.) show that low temperatures are directly important in limiting the distribution into temperate latitudes of this essentially tropical genus. For example, the tropical crab, *r. rapax*, acclimated to 18° C (64° F) in the laboratory, soon succumbed to a temperature of 10° C (50° F), although this temperature is readily tolerated by its close northern relation, *p. pugnax*, with which, in the northern half of Florida, it is sometimes sympatric. Unusually low temperatures also interfere with growth, molting, and other processes in both larvae and adults.

Several species—*crenulata* in the eastern Pacific, *rapax* and *thayeri* in the western Atlantic, and *tangeri* in southern Europe and West Africa—range from the deep tropics through the subtropics and well over the border line of the warm temperate zone without apparent morphological changes.

Still, behavior differences related to temperature are notable in these species ranging widely through climate zones. In the tropics, *rapax* is socially active perennially in all except the driest periods; in the subtropics seasonally; and in the warm temperate zone activity is controlled largely by temperature.

Temperature directly controls the speed of display in subtropical populations, as in *tangeri* (Altevogt, 1959).

3. *High Temperatures.* According to my observations (unpublished), it seems highly probable that the four forms of *Uca* that remain active at the highest temperatures are Red Sea populations of *tetragonon*, *vocans hesperiae*, *inversa inversa*, and *lactea annulipes*. Of these, work at Massawa showed that *tetragonon* proved most resistant to heat, continuing to wave in full sunlight when the mercury reached 41° C (106° F) in the sun at the surface, providing there was an intermittent breeze. With or without a breeze, *tetragonon* always descended its burrow at about 42° C (108° F). At 41° C (106° F) *lactea* was inactive, in its burrow. The other species descended at about 40° C (104° F), although display and courtship stopped at 39° C (102° F). Feeding and threat stopped earlier, at about 38.5° C (101° F). In contrast, most *Dotilla* and all hermit crabs descended by 38.5° C (101° F.). In Singapore, where maximum ambient temperatures are lower than in the Red Sea, the temperature tolerance of *U. lactea* was lower by 5.5° C (almost 10° F).

Species of the subgenus *Celuca*, the group best adapted for land, best withstand the highest temperatures, whether in the Indo-Pacific, eastern Pacific, or western Atlantic. The dazzling display whitening assumed by a number of these species may help keep the body temperature down, but whitening is by no means found in all hot locations or in all the expectable species. The extreme of whitening also occurs sporadically in *Thalassuca* and *Uca* and does not seem to be related to their endurance of heat during display, although it well may prove to be a factor.

It must be remembered that two temperatures are important to all—that on the inside of the burrow and that on the surface outside the mouth. The inside temperature is of course less extreme in each direction. At Massawa in July, when surface temperatures were about 42° C (108° F), four inches down the burrow the reading was only 40° C (104° F). Extreme readings on a breezeless protected flat were 43.5° C (112° F) on the surface, and 40.5° C (105° F) four inches down in the burrow of an *inversa*.

Macnae has shown that because of transpiration body temperatures in *Uca* in the open sunlight are 1° C or more lower than the air temperatures (1957).

The same author, always working on Inhaca Island in Mozambique, found that the upper limits of tolerance varied with the species, the individual, and even the season. In the Mozambique summer (January) toleration was roughly 2° C higher than during the winter. Nevertheless, 45° C (113° F) was lethal for all fiddlers. Of these *inversa* was notable for more tolerance than the rest; *lactea annulipes* followed; next came *vocans* and *chlorophthalmus*; *urvillei* was least tolerant of heat. The order is expectable in view of the habitats, *inversa* occurring usually on baking salt flats protected from sea breezes, *urvillei* often in the shelter of partial mangrove shade, and the others variously between. The habitat of *vocans*, near low-tide levels in very damp ground, would reduce heat exposure.

Verwey (1930), Orr (1955), and Teal (1958) found somewhat similar lethal upper temperatures.

The most direct climatic comparisons in acclimatization are those of Vernberg and Tashian (1959). They compared temperate and tropical American species, finding that at 42° C (107.6° F), or 44° C (111.2° F) tropical species survived longer than temperate forms. If the tropical forms were kept first for seven days at 15° C (59° F), the difference in survival time disappeared. Temperate zone species, as mentioned in the preceding section, were more cold resistant than tropical forms.

A biological clock mechanism seems to be concerned in the relation of seasonal temperature to display. Specimens of *U. arcuata* brought alive to the William Beebe Tropical Research Station in Trinidad from south Japan in May lived about 18 months. Individuals displayed well through the first and second summers, but did not display at all during the intervening winter, even though temperatures remained tropical, differences in day length were only very minor, and no dry season conditions were in evidence in the out-of-door crabbery. Unlike the winter behavior of the crabs in Japan, however, whether they hibernated partially or altogether, they fed and moved about normally in Trinidad and showed some territorial threat behavior. As said above, they did not wave and no courtship or mating seems to have occurred. Observation was not constant and if waving occurred very rarely it might, of course, have passed unnoticed; it could not however have taken place frequently.

C. WATER

1. *Tides.* As in many marine animals, in *Uca* a direct relation exists between the number of species and the breadth of the tidal range. This relationship exists only of course where climate and substrate are suitable and geographical and historical factors appropriate. Even though *Uca* rarely takes advantage of the lowest tidal levels, wide intertidal bands encourage so many forms of life that these filter-feeding crabs, living on the middle and upper levels, undoubtedly benefit from the biological richness of the general environment. When more is understood both of their food requirements and of the actions of pheromones, the interdependence of factors will be better understood.

Whatever the detailed causes, more species live in the neighborhood of the Gulf of Panama than anywhere else in the world (Map 1). This region is notable for extremely wide tidal ranges. In the region of Tahiti the usual 6-inch tides can only be partly blamed for the presence of only two species, since this area is on the extreme edge of the East Pacific Barrier; few kinds of pelagic larvae from the west have been able to cross the long stretches between islands. In southeast Asia and the Philippines another major factor is doubtless responsible for the relatively few *Uca*, in addition to mediocre tides. This factor is the use of niches by other genera in the family not found in America (*Ilyoplax, Scopimera, Dotilla,* and others). In the tropics of the western Atlantic, which would normally make a good comparison with conditions across the Isthmus, the effects of the glacial epochs are no doubt still being felt (p. 433).

The tide, whatever its range, has a powerful effect on the activity of *Uca.* Usually, of course, the crab spends the entire period of high tide underground, emerging only after the water has receded from the level of the burrow's mouth. With the exception of unusual local populations of *burgersi* (p. 168) and *pugnax* (unpubl.), I know of no record of *Uca* performing waving displays, or engaging in combats or courtships underwater. Occasionally fiddlers feed when beyond the edge of the tide, or in shallow runoffs coursing through flats or lagoons; usually their eyes project above the surface. Often they escape threats by running a short distance into the water when they are at a distance from their burrows. Otherwise, they habitually stay away from the sea and descend their burrows as the tide advances.

Endogenous tidal rhythms play important parts in the activities of fiddlers. The rhythm is apparently fixed for an individual in accordance with the tidal schedule in effect at the locality where the postmegalops comes ashore, although, as usual in biological rhythms, the rhythm can be altered within a few days.

2. *Rainfall.* Rainfall is of special importance in bringing about high activity, including display, in fiddlers living near high-tide levels. These populations occur on the banks of tidal streams and especially on tidal flats reached only by spring tides. In periods of drought crabs remain in their burrows for days, weeks, or months at a time. The first showers bring them out to feed actively; within a week or less most of them start display and courtship. In the dry season even populations of the same species, such as *rapax* in the West Indies and northern South America, which live close to the shore and are exposed to regular semi-diurnal tides, do not display or breed during the dry season, although they feed regularly and are otherwise active. This seasonal activity in well-watered populations indicates the existence of still another biological clock.

Other good examples of this behavior occur in the same subgenus, *Minuca,* on both sides of the Isthmus of Panama; in both regions the dry season is severe and only the highest tides reach the flats. In the rainy season these localities show high activity daily, regardless of the tidal level.

In temperate zones warm temperatures and the rains usually coincide, as on the east coast of America and in southern Japan. A major reason for the poor *Uca* representation in California, where only *crenulata* occurs, is perhaps the fact that here the rainy season comes during the colder weather.

D. SALINITY

Fiddler crabs, both as species and as individuals, are highly tolerant of changes in salinity; a review of the complex means by which osmoregulation is accomplished in *Uca* and other semi-terrestrial decapods is given by Bliss (1968). This adaptability is unsurprising, since so many species live under estuarine conditions which entail an impressive range. The extremes extend from salinity near or equaling that of the open ocean, characteristic of the dry season, to water which is practically fresh, in delta country during the height of the rains.

On the other hand, in many localities rich in *Uca,* testing shows that even during the height of the rains the water at high tide is almost equivalent in salinity to that of the nearby sea. Such a locality is found near Port of Spain, Trinidad. Another stabilizing factor of great importance is the frequently poor drainage in the burrows; a result of this condition is that the salts are held in the substrate, the burrow water showing a notably higher salinity than the tidal waters flooding the area. Hence, when crabs descend into their burrows to replenish the moisture supply in their gills, they do so from the standing water of high salinity, and so conserve their own supply of salts.

The tolerance to changes in salinity does not mean that salinity preference is not a species-specific characteristic, regardless of the breadth of the range. All *Thalassuca* and most *Celuca* live close to the open sea, where waters at high tide at all seasons approach in salinity those of the adjacent ocean. In contrast, *Deltuca* and *Minuca* in their respective hemispheres are tolerant of extremely low salinities, descending to about one part per thousand.

Abnormally high salinity appears to be intolerable. When the straits between the mainland and the island of Singapore were blocked by a causeway built soon after 1945, the adjacent shores became grossly impoverished. In contrast, in the experience of Dr. M.W.F. Tweedie, they had formerly been a rich habitat for many ocypodids. It seemed to both of us, when Dr. Tweedie showed me the locality in 1955, that the heightened salinity must have been the key factor in the decline of *Uca*. At that time it was about 40/00; the area did not appear to Dr. Tweedie to have changed very much in other ways, the human population had not notably risen in the neighborhood, and no pollution was detectable.

One of the areas of highest salinity inhabited by fiddler crabs is the Red Sea. At the surface of the open sea the salinity ranges from 37/00 to 40/00 (Ekman, 1953). Since at Massawa, where I worked, there are no backwaters where *Uca* are found, these salinities are probably fairly typical of the shore.

E. Light

Fiddler crabs are primarily diurnal animals. Most of their activities, including feeding, digging, fighting, display, and mating, usually take place in daylight. This rhythm is particularly strong in tropical species, in which nocturnal activity seems to be confined chiefly to a little feeding by moonlight, to warning sounds made in periods of alarm within the burrows, and to wandering by crabs in the appropriate phases (p. 504). Nevertheless, as found by von Hagen (1970.3), nocturnal activity even in the tropics is clearly more plentiful than I formerly believed.

In the temperate zone, however, considerable courtship activity is nocturnal in both species investigated in detail. These forms are *tangeri* in southern Spain and north Africa, studied by Altevogt (1959, 1962) and von Hagen (1961, 1962), and, in the eastern United States, *pugilator* as shown by Burkenroad (1947), Salmon & Stout (1962), and Salmon (1965). It may well be that the shorter breeding season available in the north has resulted in selection toward nocturnal activity. This subject will be continued in a later chapter (p. 502).

It is certainly true that all populations of *Uca* show more waving activity in sunshine than in cloudy weather, providing only that the temperature is not too high (p. 442). Regardless of adequately warm temperatures and low tides, intensive feeding does not start in tropical populations until about two hours or more after sunrise. Display starts even later. At the end of the day, however, there is a higher tolerance for lower light levels than during the early morning. In an active population, both display and feeding are sometimes continued late in the afternoon, when low tide falls at that time near new and full moons. Sometimes waving activity then continues until sunset or even slightly later, as found in *vocans* near Bombay by Altevogt (1957) and in *vocans* and *lactea* in Fiji by Crane (unpublished).

F. The Intertidal Habitat

The general habitat most characteristic of *Uca* lies in the tropics, somewhere near the mouth of a silt-laden stream. The stream flows into a sheltered bay partly fringed by mangroves. Beyond the mud flats near the river mouth, along the shores of the bay, curve wide beaches of sandy mud and muddy sand. Tidal lagoons often fill depressions behind the shore. Sometimes the locality consists only of a narrow stream as it flows from nearby hills into a small bay on an island's lee. Sometimes it embraces the vast delta of one of the great rivers of the tropical world.

As shown in Figs. 21-23, almost all species of *Uca* live in some part of this general habitat. Only a few forms occur either in more exposed localities or far upstream in altogether fresh water.

In particular, no *Uca* live on surf-beaten shores, pure sandy beaches, rocky shores, or in tide pools. Two species are borderline exceptions. One is *tetragonon*, which in some localities is found only on shores where a thin layer of sandy mud covers a firm substrate of coral and shelly conglomerate. The second is *panamensis*, found only among stones at the ends of sandy mud beaches.

No *Uca* live in the middle of broad open flats of soft, deep mud, probably because the deficiency of oxygen causes a food shortage. Only *maracoani* and a related form, *ornata*, extend far out in such mud. Even they do not inhabit its middle portions. Finally, no *Uca* live in the deep shade of thickly growing mangroves.

In brief, fiddler crabs may be viewed as animals living in environmental borderlands. Just as many birds and butterflies occur typically on the edges of fields and forests, so *Uca* is adapted not only to the major boundary between land and sea, but within that tidal zone to the edges of rivers, mud, and mangroves.

G. Biotopes

The term *biotope* will be used to denote the subdivisions of the general *Uca* habitat described in the preceding section. This is characterized by Hesse, Allee, & Schmidt (2nd edition, 1951, p. 165) as "an area showing uniformity in the principal habitat conditions."

In a discussion of *Uca* at the present time there seems to be no advantage in subdividing the biotopes listed below or in grouping them into larger units. This conclusion has been reached simply because we do not yet know enough about factors that limit particular species of fiddler crabs to special sections of the shore. For example, food requirements are for many species undoubtedly an important limiting factor of which we now know practically nothing.

Because of these uncertainties, the 16 biotopes proposed for use are listed with no suggestion of hierarchies. It is certain, however, that all are not of equal importance and that they are not sufficiently subdivided. Nevertheless the list should prove of distinct pragmatic value, since it is based on wide personal experience in observing the ecological distribution of species in the field. Figures 21-23 and Table 10, used in conjunction, give an overall view of the occurrence of *Uca* as it is now known. Further details will be found under the species headings in the systematic section.

The list below gives the outstanding superficial characteristics of the biotopes. They are numbered in accordance with the above-mentioned figures and table. The same numbers are used in the species descriptions in the systematic section under the heading Biotopes.

List of Intertidal Biotopes

1. (Surf-battered rocky shore: no *Uca*.)
2. (Rocks and tide pools: no *Uca*, except for *panamensis*, which rarely occurs in tide pools.)
3. (Unprotected sandy beach: no *Uca*.)
4. Stony sand with underlying mud, ranging sometimes to a hard substrate of coral or shelly conglomerate, with mud in interstices and overlying the surface in a thin layer.
5. A thin layer of sand overlying mud.
6. Muddy sand to sandy mud.
7. (Open flats of soft, deep mud, far from shore, or middle portion of a broad, protected expanse of similar mud: no *Uca*.)
8. Open mud flats near mangroves.
9. Muddy sand to mud, partly shaded by edge of mangroves.
10. (Mud in full shade of dense mangroves: no *Uca*.)
11. Flats of muddy sand, sandy mud, or mud in or close to mangrove-fringed lagoon; some of these flats are flooded only by occasional spring tides.
12. Muddy stream banks, flats, near mouth; close to mangroves.
13. Same as no. 12, but with steep banks.
14. Same as nos. 12 and 13, but water brackish.
15. River banks, usually steep, upstream from mangroves; substrate muddy, clayey, or partly sandy; water ranging from almost fresh to fresh.
16. Subtropical and temperate marshes (mangroves absent).

Mud may be defined according to Macnae (1957) in the definition which seems best to fit the current situation as "mineral alluvium mixed with organic matter." Sand, on the other hand, is almost pure mineral matter, composed of particles larger than in mud, and to which consequently far less edible matter sticks. The mouth-parts of *Uca* are well adapted to these differences. The normal habitat of a given species, whether sandier or muddier, can be guessed merely by examining the setae on the endites of the second maxilliped; the more spoon-tips, the more nearly sandy the environment will prove to be (morphology, p. 455; behavior, p. 456).

Pheromones, secretions deposited by animals externally, may form a valuable part of the *Uca* environment. Their roles are not yet known in this group.

H. Communities

Animals and plants associated with *Uca* would make a formidable study all by themselves. Macnae has contributed most to the subject in his reviews of South African fauna from ecological viewpoints (1956, 1957) and in his general account of mangrove flora and fauna of the Indo-west-Pacific region (1968). The present review will be confined to general comments. In the systematic section remarks are included under most species headings on their associates within the genus *Uca*, in both the broad and narrow uses of the word "sympatric."

1. *Associated Vegetation.* Mangroves are the most typical plant formation associated with *Uca*. When one approaches an unfamiliar tropical coast by land, sea, or air, the best way of locating *Uca* as rapidly as possible is simply to watch for the patch of mangroves lying nearest to the town and proceed there promptly at the next low tide. Mangroves, of course, consist of a variety of superficially similar trees and shrubs living in association, both the genera and species varying in different parts of the world. Two of the most common genera are *Rhizophora* and *Avicennia*. The stilt-rooted plants are found throughout the tropics and subtropics of the world, except on coasts so arid that there is neither a reliable rainy

season nor streams running into the sea. They are also absent from mid-Pacific islands. Their latitudinal range is similar to that of corals, extending from about lat. 32° S in southeast Africa to Bermuda (32° N), an exceptional northern record. Although *Uca* do not rely directly on mangroves for either food or shelter, they are of great indirect importance to the genus. The rotting leaves and associated animal life enrich the surrounding mud and muddy sand in which they grow and from which the fiddler crabs strain their food.

When the water becomes too fresh upriver for mangroves to grow, their place is taken in fiddler ecology by assorted plants, of which a good example in the American tropics is the arum *Montrichardia arborescens* ("mucka mucka"; "boroboro"). *Pandanus* serves similarly in many parts of the Indo-Pacific, such as in the great rivers of Borneo.

In temperate climates, too far north for mangroves, grasses are the principal substitutes. *Spartina* and *Phragmites* are good examples in the northeastern United States.

2. *Animal Associations.* (*a*) PREDATORS. The principal enemies of fiddler crabs among mammals are the crab-eating raccoons of the neotropics; their footprints are unmistakable. In the Georgia salt marshes of the United States, Teal (1958) found that raccoons and clapper rails both fed regularly on fiddlers, while aquatic predators of possible importance included fish and large crabs. As a whole, birds do not seem to take much advantage of the vast numbers of fiddlers that are sometimes available, although on any suitable shore occasional birds, ranging from herons to crows, attempt to seize a few members of the genus. Usually such hunters are so infrequent that they are negligible nuisances during observation of fiddler behavior.

In January 1971 near Bathurst, Gambia, John Endler listed the following predators of *tangeri*, along with their methods of hunting (personal communication): reef heron (eats adults in water); lemon black-backed gull (lies in wait for adults); ringed plover (stalks adults and young); whimbrel (stalks adults and, especially, young); crested lark (pounces on young in short flights); hooded vulture (lies in wait for adults). In addition he found that lizards (*Agama*) stalk the juveniles. This is the most extensive list I have ever seen that was compiled in one locality; comparison material from other areas would be of great interest.

Raut (1943) also observed whimbrels catching fiddlers, in this instance on an island near Bombay. The whimbrels' method of capturing the crabs has already been described (p. 4).

Kingfishers use still another technique. Unlike a whimbrel, which seizes a fiddler by the large claw, a kingfisher grasps the body. In both New Guinea and Ceylon I watched such a capture. The bird, crab in beak, flew with it to a perch, shook the fiddler vigorously until the major cheliped autotomized, and then swallowed the crab; in each case I picked up the claw and checked that the prey was a good-sized adult of *vocans*. One year at our mountain station in Trinidad the population of our most isolated crabbery suddenly declined; we found that a local kingfisher had learned to wait during our artificial low tides, and pounce.

In Trinidad and Venezuela I have also seen scarlet ibis seize fiddlers, but was not close enough to observe details of their technique. For these birds as well as for other species, the crabs seem usually to descend their burrows too quickly for capture. The successes for birds, perhaps, are usually insufficient to encourage the necessary effort. Possibly the crabs are distasteful, but we have no evidence whatever that this is so. Only among certain Indo-Pacific species are the crabs so vividly colored at all ages that aposematism may be suspected (p. 469).

Von Hagen (1969) reports that in Trinidad a grackle and a poecilid fish seized eggs from females of *vocator*.

The fish *Periophthalmus* has not been seen to attack *Uca*, although I have watched them often in association. The two animals frequently share a mud flat under crowded conditions; they appear to disregard one another entirely.

In southern Spain and Portugal, man is a direct predator of fiddlers, since *tangeri* claws are "harvested," the crab being thrown back to grow another (p. 118). Furthermore, in southern Japan a limited supply of canned *arcuata* is seasonally available (p. 44). Elsewhere, fiddlers do not seem at present to be eaten regularly by man.

(*b*) ASSOCIATED INVERTEBRATES. The most typical completely sympatric associates in the Indo-Pacific are grapsid crabs and other genera of ocypodids. Particularly notable are *Ilyoplax*, *Scopimera*, *Dotilla*, and *Macrophthalmus*. Outside the grapsoids are a few xanthids, hermit crabs, and small mollusks and worms, all of which no doubt contribute importantly to the biological richness of the substrate. On American shores other ocypodids are absent except for *Ocypode*, which overlaps with *Uca* only marginally.

(*c*) EFFECT OF MAN. Human activities are undoubtedly and strongly affecting the distribution and existence of *Uca*. A surprising result is that some of these effects are in favor of the fiddler crabs.

For example, a dock-side cove beside the Pacific entrance to the Panama Canal may be contrasted with a river mouth on the west coast of Palawan—one of the most undisturbed coasts in the world. The Panama spot, arched around the outlet to a large sewer, supported a cluster of straggling mangroves, received a wealth of garbage tossed there from time to time—and nourished more kinds of fiddlers than any other spot I ever found. Sixteen species lived in an area of less than four hundred square yards. The place has long since been ruined by oil and eclipsed by construction, yet the principle it illustrates remains unchanged. In general, the richest site for fiddlers will not be a shore devoid of humanity, but some spot close to a city, and even closer to a thriving village—preferably with stilt-legged huts and no plumbing.

In contrast, northwest Palawan was the worst tropical place of the kind for fiddlers in my experience, although it seemed ecologically ideal. The coast was wild, warm, humid, and rich with streams, mud, sandy mud, and mangroves. The whole shore was backed up by dense forest, and there was no trace of any source of pollution.

The negative side of man's activities is altogether expectable. Pesticides, roads, harbors, airports, swamp drainage, factory pollution, oil dumped from ships, overcut mangroves, and sheer numbers of people are all having drastic effects on fiddlers. These changes are so rapid that they can be detected with ease when one revisits a site only a short time later. The Panamanian cove mentioned above is an example. Another is a neighborhood on Singapore Island, formerly known as Pandan Swamp, which, during 1955, was a wonderfully rich locality for numerous kinds of ocypodids. The day I left, in September, workmen began to cut its mangroves to make still another useful fish pond. I hear it is now a thriving suburb.

In 1962 I could find no trace of *dussumieri* in Semarang, Java, which was its type-locality, while fiddlers of all kinds were, if still present, inaccessible anywhere near the great city of Djakarta, which was well known (as Batavia) in Dutch records of *Uca* during the past century. These days military installations screen almost all the city's shores and the burgeoning population has overrun the rest.

Although the southwest coast of India, near Cochin, has a topography and climate ideal for *Uca*, only two species occur there. In 1965 very small populations were found. The pressure of humanity has probably been severe here not only for a few years but for centuries.

Near New York City, on the New Jersey meadows, fiddlers seem to be extinct; fortunately populations at present remain in relatively accessible spots on Long Island and in Connecticut.

Still more fortunately, rivers still flow in the tropics—in Malaya and Brazil, in Borneo, Ecuador, and Costa Rica—where *Uca* still thrive. William Beebe once called a Venezuelan mangrove delta "the Land of a Single Tree." Thanks to the mangroves' inhospitality to man, each delta is also the land of a million fiddlers.

I. SUMMARY

Most *Uca* live in warm climates along the sandy mud shores of protected bays, the waterways of mangrove swamps, sheltered flats near river mouths, and the muddy banks of tidal streams. On the other hand fiddlers are never found on sandy beaches, rocky coasts, coral reefs, or any shore regularly exposed to strong surf. Aside from these limitations, their adaptability is worthy of respect. Both as species and individuals they can adjust to wide ranges of temperature, moisture, and salinity. In New England and Japan several species hibernate. In excessive heat fiddlers everywhere descend their burrows, whether in the middle of a tropical day or, in a state of aestivation, for many weeks in response to heat and drought. Nevertheless species-specific differences are increasingly apparent, not only in physiological tolerances but in the temperatures at which normal feeding and social activities occur. Tides are another important factor in *Uca* ecology; wide amplitudes encourage the development of a flourishing intertidal community, including *Uca*, when other conditions are sufficiently favorable. The activities of fiddler crabs are closely meshed with tidal rhythms, the crabs remaining in their burrows throughout the periods of high tide. After a drought rainfall stimulates social activity, although fiddlers go underground during a hard rain. All species are most active in bright sunlight, especially socially; at night waving display is always absent but sound production sometimes increases. Predation on *Uca* is limited, considering the abundance of both crabs and shore birds. Man, in progressively polluting and destroying habitats, is as usual the chief enemy. Except near large cities, however, fiddlers in the tropics seem to be holding their own.

Chapter 3. Structures and their Functions

OUTLINE

A. Introduction

In live fiddler crabs structure and function are challenging subjects that continue to provide biologists with a series of surprises. This chapter concerns selected morphological characteristics and their relations to certain ecological requirements and behavior patterns.

Most of the characters discussed are external and include all of those most typical of the genus. Where our knowledge is severely limited, as with antennae and antennules, the structures are briefly mentioned to underline our ignorance. The survey includes the range of variation, post-larval development, apparent adaptive values, and the functional relation, in social behavior, of the parts to one another. Descriptions of characters of taxonomic importance are given in the systematic section, beginning on p. 8.

While gills and mouthparts are superficially reviewed, internal anatomy, its physiology, endocrinology, and the bases of biological rhythms are omitted as outside the scope of this book. Key contributions, with references, include Barnwell (1968.1), Bliss & Mantel, eds. (1968), Démeusy (1957), Hoffmann (1971), Passano (1960), Sandeen (1950), Teal (1959.2), Vernberg & Tashian (1959), Waterman, ed. (1960, 1961), and Newell (1970).

B. The Crab as a Whole

Male fiddler crabs stand out among all other animals in their general appearance, and a human observer can recognize them instantly, even when they are sitting motionless on a tropical mud flat among a bewildering throng of other small crustaceans. Their outstanding characteristic is the contrast between the two chelipeds, one being strikingly large while the other is minute; no other group shows a comparable disparity.

In other respects, the shapes of all *Uca* resemble those of many other shore crabs. The carapace is usually smooth and always convex, with roughly four to six sides. Its divisions are not deeply set off from one another, although sometimes the branchial regions project well above the dorsal surface. The eyes are on stalks. The eight walking legs in an overall view are not remarkable.

When any *Uca* is examined closely, the prevalence of tubercles, serrations, ridges, and furrows is striking. They occur especially on the sides of the carapace, around the orbits, on both chelipeds, and on the ambulatories. They divide functionally into two groups. Those on the minor cheliped and on the more proximal segments of the major cheliped and ambulatories rub against the carapace or against one another during agonistic situations. In contrast, the manus and chela of the major cheliped and at least the mani and dactyls of the ambulatories are used primarily in ritualized combat, engaging with correlated structures on the opposing crab. The continuing discovery of unsuspected uses of armature details remains one of the great surprises in the study of fiddler crabs.

C. Size and Allometric Growth

Size is of special interest in fiddler crabs because parts of their anatomy grow at very different rates. The large claw, especially, is far longer in large individuals than in small adults of the same species. As a result its development has often attracted the attention of observers, who have ranged from curious naturalists to biologists studying growth in a variety of animals. The general term allometric growth is currently applied to such an inequality of rate. Gould (1966: 629) defines allometry in part as ". . . the study of proportion changes correlated with variation in size of either the total organism or the part under consideration . . ."

No detailed contribution to either size or allometry in *Uca* has been attempted in the present study. The following paragraphs give only a summary of the range of sizes in the genus and of their distribution among the subgenera, along with a brief introduction to their growth ratios. All measurements in this section are in millimeters.

The fiddler crab with the largest dimensions known to me is a specimen of (*Afruca*) *tangeri* from West Africa in the collections of the American Museum of Natural History; its carapace measures 33 long and 47 broad, while the propodus attains 109. The next largest is a (*Uca*) *maracoani insignis* I took in Ecuador, measuring 31.5 long with a propodus of 94. Probably individual *maracoani* in life surpass in weight *tangeri* of equal carapace lengths because of the greater breadth of each finger of the claw.

The smallest *Uca* of mature proportions is (*Celuca*) *batuenta*; the largest individual known, the holotype, measures 4.8 long and 7.6 wide, with a propodus of 11.8. Two other species, (*Minuca*) *pygmaea* and (*Celuca*) *tenuipedis*, are both known only from specimens less than 6 long; all three are from the tropical eastern Pacific, the region characterized by the small size of most of its numerous species of fiddler crabs (pp. 533, 534).

Between these groups of extremes small species are far more numerous than large, as shown by the chart below. It is furthermore true that the largest sizes within a species are rare, the most numerous and active members of a breeding population usually being considerably shorter. Because longer individuals not only have much longer claws than short crabs but are proportionally broader and thicker as well, a slightly longer crab is of much greater bulk than is indicated by the length dimensions of the carapace and appears much larger in the field.

These differences can be of great importance to an investigator when he is selecting species for various kinds of research, ranging from neurophysiology to behavior observations in indoor crabberies; further comments on these practical aspects will be found in Appendix D, p. 670.

In addition to size range the following chart shows the longest claws attained within each group.

Distribution of Size in the Genus *Uca* (mm)

(As Indicated by Lengths of the Largest Known Individuals in Each Species)

Size-group	Carapace lengths	Propodus lengths	No. of Species*	Subgenera represented
Gigantic	30-33	87-109	3	*Afruca, Uca*
Very large	25-29	66-80	3	*Deltuca, Uca*
Large	20-24	50-71	10	*Deltuca, Uca, Minuca*
Moderate	15-19	31-58	11	All except *Afruca* and *Celuca*
Small	10-14	22-53	22	All except *Boboruca, Afruca,* and *Uca*
Very small	5-9	10-24	22	*Australuca, Amphiuca, Minuca,* and *Celuca*

*Sometimes represented, through subspecies, in two size-groups.

In the systematic section a few sample measurements follow the morphological description of each species. These are intended merely as guidelines to future investigators, showing the largest specimens that have been on hand, or those of unquestioned identity appearing in past publications; these figures are the bases of the chart above. Also included in the systematic section are measurements of moderate-sized males that are more typical of most specimens to be encountered in the field and in collections. These measurements, in combination, of the largest and intermediate males, also give a preliminary indication of the allometric curve when the ratio of carapace length to propodus length is compared at these two points.

In his classic paper on allometry in *Uca p. pugnax* and in *minax*, Huxley (1924) remarked that in groups characterized by such continuing allometric growth the usual bases for morphological species designations were useless since the individual had no fixed form. He suggested that in such cases the k values, which indicate the rates of proportional growth in various pairs of characters, might prove to be important specific distinctions.

This ratio, whether based on measurements or, as in Huxley's study, on weight, proves to be of little or no value for taxonomic purposes in *Uca*. First, there are so many species of similar size and proportions that their differential growth ratios in many instances coincide. Second, and far more interesting, adult proportions are attained at different lengths in different populations of the same species. Just one example, that of a group of (*Minuca*) *rapax* populations, is given here (Table 11). I have found similar differences in two populations of (*Celuca*) *lactea annulipes* taken in different biotopes on the island of Zanzibar, as well as of *lactea perplexa* on the northwest side of Viti Levu, Fiji. One of the groups of *lactea* lived farther upstream than usual, while the other occupied a normal habitat near mangroves. Again, because of the different proportions of leptochelous and brachychelous individuals, k values can differ in accordance with the prevailing claw form of the invariably few individuals of large size. The difference can be striking in Indo-Pacific subgenera where extremes of the two forms occur in almost every population. Samples of adequate size would, of course, take care of the difficulty. Finally, when comparisons of claw size to crab size are made by weight, variations exist in degree of tissue shrinkage, size of air spaces, and amount of liquid; these differences depend only partly on whether freshly dead or preserved crabs are selected for the work; it seems preferable, therefore, to use carapace length in relation to length of propodus (manus and pollex) rather than weight of crab in relation to weight of claw.

In every species of *Uca* the propodus attains a length of at least twice the length of the carapace. Both fingers of the chela—the pollex portion of the propodus and the opposing dactyl—also elongate allometrically with respect to the manus. The extreme of both forms of elongation is reached in (*Celuca*) *festae*; in one individual the longest propodus measures 4 times the carapace length, while the dactyl is almost 7 times the length of the manus.

Early in the study a practical rule was needed to determine the size at which an individual male should be considered mature, whether alive or preserved. The indication used has been the dactyl's length; when it is clearly longer than the manus the crab is considered mature; except in a very few species, such as *latimanus*, all characterized by paedomorphic claws, the gauge seems to work. By the time the dactyl has surpassed the manus the specific characteristics of armature and gonopods are well developed, and individuals with shorter fingers only rarely display and court.

During growth the major merus also increases allometrically in direct relation to the similar growth of the propodus and dactyl, all of which of course are functionally closely related. In addition, the breadth of the carapace especially anteriorly increases with size, the degree differing with the species. Further changes in proportion are mentioned later in this chapter in connection with individual structures. It seems likely that the proportions of the rare giants in each species illustrate size limitations imposed by sheer weight.

Altevogt (1955.1, 1955.2) recorded in India the total weights and weights of the major chelipeds in a series of *vocans*. His largest specimen, weighing 7.40 gms, carried a cheliped of 3.56 gms—more than 48 percent of its total weight. In a series of *maracoani* we weighed in Trinidad, the heaviest claw proved to be slightly more than 40 percent of the crab's total weight, its live weight being 8.09 gms and the weight of the autotomized cheliped 3.32 gms. Although in both species these proportions are rare exceptions, maximum weights of about one third of the total weight are common in species both large and small distributed widely throughout the genus. Males within the normal size range of socially active adults perhaps usually carry chelipeds weighing a quarter or more of their total weights. Comparative series of weights, made on representative species in the various subgenera and using samples of adequate size, would prove of interest.

Female *Uca* are almost always smaller in their carapace dimensions than the largest known conspecific males, and as a rule smaller also than many of the displaying males in the actively breeding part of a population. Females less than two-thirds the maximum female size known in a species sometimes carry eggs, the abdomen being apparently always of adult breadth before they do so. Small, immature females are occasionally courted vigorously by mature males, usually with herding (p. 503), and small males with major chelipeds of juvenile proportions sometimes try to mate, apparently without success (p. 121).

Biologists wishing to inaugurate studies on fiddler allometry will find useful discussions and references in Huxley, 1924; Huxley & Callow, 1933; Gray & Newcombe, 1938.1, 1938.2; MacKay, 1943; Gould, 1966; Herrnkind, 1968.2; and Mayr, 1969. In the present study the following figures illustrate allometric growth in various structures: 24, 25, 27, 30, 35, 37, 45, 47, and 48.

The subject of regeneration properly belongs in

the realm of physiology and so will be excluded here except for two statements that still appear to be needed, judging by continuing verbal enquiries and recent non-technical publications. First, when a major cheliped is autotomized the regenerated major cheliped always grows on the same side of the crab on which it was previously located; second, when differentiation begins, at or soon after the second post-megalopal stage, it does not result from the loss of a cheliped. Incidentally minor chelipeds in males and either cheliped in females are only rarely lost at any age. These points have been established by Vernberg & Costlow (1966), by Feest (1969) and by me (unpublished).

Large claws occur about equally often on the right and left. As examples, in populations of *tangeri*, *rapax*, and *lactea* 50 males of each were checked.

D. GENERAL SHAPE

When viewed from above, the carapace ranges from almost quadrilateral to clearly hexagonal. The differences depend on the width of the front, the degree of the backward slant of the orbits, and the convergence of the sides as they approach the posterior border. In most species the orbits of adults are so slightly oblique and the sides so little convergent that the crabs seem almost rectangular; the breadth is always a little greater than the length. Females usually have both the orbits and sides slightly less slanting than do males.

The orbits of young crabs always slant more than do those of adults, sometimes strikingly so. Nevertheless, species having more horizontal orbits as adults show this tendency, in comparison with slanted species, from the early post-megalopal stages. (Figs. 1, 24.)

The paedomorphic carapace shape is marked in only a few species. In the Indo-Pacific these include *rosea* and *triangularis*, and, in the Americas, *pygmaea* and *zacae*. The shape is not apparently associated with paedomorphism in other characters, such as the size, proportions, or armament of the major cheliped.

Slanting orbits are associated in both young and old with dorso-lateral carapace margins that converge strongly posteriorly. To a lesser extent the vertical lateral margins also converge most in forms with slanting orbits. Another feature of the carapace associated with scarcely oblique orbits and sides is the presence of distinct, antero-lateral margins, converging little or not at all, or even diverging, before they turn to continue as the dorso-lateral margins. Sometimes, as in the subgenus *Minuca*, the antero-laterals are extensive. In other examples, as in (*Deltuca*) *dussumieri*, they are rudimentary and individually variable. Throughout the genus they vary even on the two sides of the same individual, especially in males, where they tend to be longer on the side of the major

cheliped. In most species they are less developed in the young than in adults; an exception is (*Celuca*) *stenodactylus* (Fig. 24 *K, L, M*).

In adults, throughout the genus, the carapace is wider than long, the length being roughly 60 percent of the width. In young crabs the breadth is always relatively less than in adults.

The regions of the carapace are poorly marked, especially in a few species that are periodically subject to desiccation. Here the hepatic and branchial regions are often altogether fused and project well above the general level of the carapace, which in the same species is also exceptionally arched (Fig. 25). The difference between *rosea*, one of the least arched, and *stenodactylus*, one of the most curving, is considerable even in the smallest crab stages.

The same figure shows the difference in orbital slant, and the correlated ventral curvature of the sternum and abdomen.

The aspects of the carapace so far considered are degree of orbital slant, presence of antero-lateral margins, degree of lateral convergence, relation of breadth to length, projection of hepatic and branchial regions, and degree of arching of the carapace as a whole.

All of these characteristics increase the crab's volume, particularly that of the empty space covered by the hepato-branchial regions. They seem obviously to be associated with a single ecological adaptation, the avoidance of drying out. Not one of the traits varies oppositely from the others. For example, a crab with strongly slanting orbits and strongly converging lateral margins never has long antero-lateral margins, a high-arched carapace, or projecting hepato-branchial regions.

These occasional needs to conserve moisture fit in well with the variations in carapace characteristics both among the species and in post-megalopal development. For example, all young crabs, being both small and equipped with slanting orbits, tend to live closer to low-tide levels than do adults, and accordingly are less subject to desiccation. Similarly, all species having strongly slanting orbits as adults live in habitats normally moist and partly shady.

In contrast, whenever the shape of a species results in an unusually large volume, its members always prove to be exposed at times to risks of drying out. The hazards may occur in at least one of four ways. First, during a tropical dry season any burrows above normal tidal levels may be dry for weeks, except, no doubt, for periodic subsurface moistening by spring tides; during these weeks or even months the crabs do not emerge at all. Second, some crabs burrow on flats always dried out during the neap tides, and are only active during the fortnightly spring periods; how much risk of desiccation they run at the bottom of their burrows is still unknown; the fact is that these species always show a large volume for their size.

Third, several species hibernate in the frozen mud of nearly freshwater swamps and river mouths. Finally, a number of very small, tropical species display for long periods in blazing sun, without moistening their gills in the water at the bottoms of their burrows.

Good examples of highly arched species that aestivate seasonally, during neap tide periods, or both are some populations of *galapagensis* in the eastern Pacific and *inversa* around the shores of the Indian Ocean. Both *arcuata* in Japan and *minax* in the northeastern United States regularly hibernate in those parts of their ranges and are notably arched. Since the larger specimens always live in almost fresh water, their burrows are probably often frozen solidly in winter. It is notable that *pugnax*, a species broadly sympatric with *minax*, is not shaped in such a way that it attains maximum volume for its size, and it is always smaller than *minax*; expectably, it lives in more saline marshes where tides probably reach the bottoms of hibernation burrows even during prolonged periods of surface freezing. A comparative study of the ecology and physiology of aestivation and hibernation in these species would be most interesting.

U. saltitanta, a very small species in the eastern Pacific, will serve as a good example of a form that displays with prolonged vigor and shows all the features resulting in a large volume. Its range is strictly tropical and its burrows always close to low-tide levels, so that neither seasonal nor semi-lunar desiccation could ever be threats. Yet its preferred substrate is always dark mud in open sunlight, while its peak display periods extend through the noonday heat, all characteristics promoting maximum temperatures at the time of their greatest activity. Yet *saltitanta* displays as vigorously as any known species, and for astonishingly long sessions. One individual continued for 20 minutes. Its endurance must be aided not only by its shape but by its brilliant white display color.

Several other American species, *stenodactylus*, *leptodactyla*, and *latimanus*, not only are very small and yet display in baking sun for many minutes at a time, but also often have burrows subject fortnightly to desiccation during neap tides. Their large volumes must, it seems, be doubly useful.

E. Front

The front of the carapace is not associated with the adaptation to avoid desiccation. This tongue extends ventrally between the eyes (Figs. 1, 26) and varies in width from less than a fifteenth to about two-fifths the width of the carapace. The front's breadth and shape are similar within subgenera, but there is often marked individual variation, and females are as a rule slightly wider than males. Both the narrowest and broadest examples (subgenera *Uca* and *Minuca*) are American. A shallow, central furrow usually is evident, as is a low rim continuous with the orbit's dorsal margin. This rim is almost always broad and smooth, but very rarely is minutely beaded on one or both edges.

In all species except those of *Minuca*, the relative width of the front decreases during growth. In *Minuca*, however, characterized by broad fronts, the width increases relatively as well as actually. (Fig. 32.)

The adaptive values both of broad fronts and the associated shortened eyestalks remain puzzling. A narrow front obviously permits longer eyestalks and therefore makes visible more distant objects. Long stalks also serve as periscopes when a crab is feeding in shallow water (von Hagen, 1970.2). Yet some species of *Minuca*, for example *burgersi* and *rapax*, also often feed in shallow water, even though they have the shortest eyestalks in the genus, and continue to feed even when wholly submerged. Perhaps efficiency of vision has been sacrificed in *Minuca* to advantages in the morphology of the underlying nervous centers. As noted above, armature, except for the unlikely possibility of the rim, is absent; so far no appendage has been observed to rub against it.

F. Dorsal and Lateral Armature of Carapace

In addition to the equipment forming part of the orbits themselves (p. 453), the carapace bears assorted organized roughnesses that interrupt to various degrees the smoothness of its surfaces and margins. All are localized and variable, especially in males. Always, too, they are stronger and less variable in females. Only one species, the West African *tangeri*, has the dorsal carapace generally covered by large tubercles. Many species, however, are more or less finely tuberculate on the hepatic and branchial regions, especially near the antero-lateral margins. Sometimes the roughness is achieved partly or wholly by weak rugosities.

On the dorso-lateral, lateral, and posterior parts of the carapace lie a group of striae which rank with the most variable characteristics of the genus. The variability extends through all groups, among the subgenera within species, and between sexes. All of the structures, when present, are stronger in females and sometimes are only developed in that sex; all of them are sometimes altogether absent. All are formed of striae which rarely develop to the point where they may be termed ridges; they range from smooth to beaded and, rarely, strongly tuberculate. The group consists of the antero-lateral margins, postero-dorsal margins, postero-lateral striae, vertical lateral margins, and posterior margin (Figs. 1, 3).

The antero-lateral margins alone are sometimes finely serrate; usually they are minutely beaded or smooth; as described earlier (p. 451), they range from absent to moderately long. Continuing inwards and posteriorly, the dorso-lateral margins occur most typically in males as smooth or feebly beaded striae which show stronger beading in females. These striae are weak to absent and especially variable in Indo-Pacific *Deltuca* and related subgenera; contrastingly, one American species, (*Uca*) *ornata*, has them strongly tuberculate in both sexes. When present, the two postero-lateral striae lie one above the other behind and below the dorso-lateral margins. Always absent in *Deltuca*, they range through varying degrees of predictability and strength to the American *Minuca* where the upper stria is always long and strong, curves forward and laterally, and is sometimes beaded; in addition the short lower stria in *Minuca* is sometimes supplemented by similar structures, always short, which are irregularly arranged on the posterior sides. The vertical lateral margin in *Deltuca* is always weak in males, being absent at least in its upper portion. In American crabs, on the contrary, it is distinct, except uncommonly in the subgenus *Uca*, and often beaded. Behind it, continuing around the lower edge of the carapace laterally and posteriorly, usually lies a broad, smooth, raised rim similar to that of the front. It sometimes has its upper edge finely beaded, at least laterally, in continuation with the vertical lateral margin.

In spite of the variation of all these structures throughout the genus, striae, viewed as a group, are clearly weakest in Indo-Pacific narrow-fronts, particularly in *Deltuca*, and strongest in the American *Uca*, *Minuca*, and *Celuca*, where the males sometimes show as much strength in the structures as do the females. All develop gradually during growth.

Not one of these structures has yet been clearly and certainly observed in action. The nearest approach to data are short motion picture scenes of several *Celuca*. In these, the meri of the middle ambulatories appear to be rubbing against either the vertical lateral margin, or the postero-lateral striae, or both; these obscure motions are in association with the definite rubbing of the ambulatories against one another in threat situations (pp. 482, 483).

G. ORBITAL REGION

1. *Introduction.* (Figs. 3, 26.) The orbits form two large sockets in the carapace opening anteriorly; within them the depressed eyestalks and eyes fit loosely. The stalk is often much more slender than the diameter of the socket, although the eye itself fits more snugly, especially in *Minuca* and *Celuca*. Throughout the genus the orbit not only furnishes obvious protection to the eye when the crab is under-

ground, but is equipped in various fashions to drain off the muddy or sandy water that accumulates in the socket, particularly while the crab is digging. All species have a major drain at the inner, posterior corner of the orbit, where a channel passes between the base of the antennule and the end of the suborbital ridges. A second channel is always free of tubercles in the orbit's extreme outer, posterior corner. The spaces between the crenellations of the suborbital margin also provide a series of drains.

In all species, through the varied ridges, tubercles, and crenellations of its margins, the orbit forms an apparently important part of the signaling equipment of the carapace. At present we are only beginning to learn how it is used. In the following account of the orbit's divisions, this equipment will be described in more detail.

2. *Eyebrow.* The eyebrow, a curving, double-bordered section of the dorsal edge of the orbit, lies immediately above the central part of the depressed eyestalk. It ranges from almost non-existent in Indo-Pacific narrow-fronts to wider than the stalk's thickness in some *Celuca* (Fig. 26). It varies similarly in its degree of verticality; sometimes it is completely visible in a dorsal view of the crab and sometimes it can scarcely be seen except from the front. Its proximal and distal borders are uncommonly marked by smooth striae, but usually by finely beaded edges, of which one or the other is often much the stronger throughout related groups of species. The eyebrow and its beading are well developed in young crabs. Its function at any age is unknown.

3. *Antero-lateral Angle.* This fixed point on the carapace ranges from bluntly obtuse to acutely produced; its shape on two sides of the same male and even female are often somewhat different; in males the more produced angle is on the side of the major cheliped. The angle's sides and the adjacent anterior and lateral margins of the carapace are sometimes minutely serrate or tuberculate as is, rarely, the ventral margin of the angle. There is little change with growth, except that in wide-angled species the angle is more acute in small crabs, while in all species any inequality on the two sides increases with age. In at least several species the angle is rubbed by at least the major merus in threat situations; details are not yet known.

4. *Suborbital Margin.* In most Indo-Pacific *Uca*, the antero-ventral margin of the eye is almost entire, with low crenellations confined to the outer half of the lower margin of the orbit. Toward the middle or inner part of the orbit's floor, these crabs often have either a distinct ridge or mound, or else there is a row of tubercles, which varies considerably within species and is sometimes found only in females.

These tubercles occur in males in some *Deltuca*, in *Australuca*, and in the west African *tangeri* (*Afruca*); very rarely they occur in female *Celuca*. In all species with weak crenellations but with prominences, tubercles, or granules on the orbit's floor, the general aspect of this part of the orbit is rolled out.

In contrast, species of *Thalassuca* and *Amphiuca* in the Indo-Pacific, the West African *tangeri*, and almost all American forms have well-developed crenellations extending the full width of the lower orbit and sometimes continuing along its outer margin. In every species the crenellations are accentuated close to the margin's antero-external angle. Often they are simply larger than elsewhere; sometimes they are distally truncate and directed strongly outward; sometimes a series is fused into a strong, projecting ridge; often the spaces between them are wider and more deeply incised. Throughout the genus, weak crenellations are often associated with profuse setae and, sometimes, pile, both on the floor of the orbit and on the suborbital and pterygostomian regions. Often only the crenellations of the outer angle are unobscured.

Without exception, species that live on sandier rather than muddier shores have stronger crenellations, fewer setae, and less pile. Except in many *Celuca*, which are often notable for strong crenellations in both sexes, this marginal equipment is usually stronger in females than in males and they sometimes have less pile around the crenellations.

Our knowledge of the functions of the crenellations is rudimentary. It is certain, both from field observations and from motion pictures, that at least the crenellations of the outer angles are sometimes rubbed by the dorsal and postero-dorsal armature of the major merus in males; at least in some *Celuca*, the action of the merus is probably best described as a vibratory tapping, rather than as a rub. It occurs in threat situations and, perhaps, in some unmixed courtship displays. In addition we know that the minor cheliped in males and both chelipeds in females rub against the crenellations. In no single case are the exact juxtapositions of segments and their parts clearly established (p. 482).

5. *Suborbital and Pterygostomian Regions.* These areas, immediately below the orbits and each other, may for convenience be briefly considered along with the orbits. Functionally they have perhaps more in common with the third maxillipeds which lie between them, since their pile and setae, profuse or sparse, apparently play a role in the aeration of water recirculated through the gills, the associated cooling of the body (Altevogt, 1969.2), and on occasion of the eggs (p. 472).

In all species the region immediately below the lower orbital margin is somewhat flattened and bordered posteriorly by a sinuous ridge; in general it is concave anteriorly and stops short of the orbit's inner corner; it apparently forms part of the system draining muddy water from the orbits. The surface of the suborbital region is usually more nearly naked than is the pterygostomian area, which is always well provided at least with short, sparse setae; these setae extend laterally to the vertical lateral margin, where they abruptly cease. In general the suborbital and pterygostomian regions are similarly equipped in any given species, being in some setose and relatively flat, and in others nearly naked and strongly convex; often the pterygostomian region is tuberculate. Since the roughness is frequently completely concealed by setae, it does not seem likely that the tubercles can then function in any form of sound production; our ignorance of the subject, however, remains complete. As with the dorsal part of the carapace and the third maxillipeds, mud-dwellers show less convexity than do species living in sandier habitats.

6. *Growth of Orbital Structures.* In young *Uca* the orbits are relatively deeper dorso-ventrally than in adults, in correlation with the wider eyestalks. Eyebrows are narrow or absent in the youngest crabs. Prominences, granules, or tubercles are discernible on the orbit's floor, when characteristic of a species, in the youngest post-megalopal individuals; sometimes, as in *dussumieri* (*Deltuca*), prominences are better developed in young crabs than in adults, although this is not true of granules and tubercles. Marginal crenellations are always weak or undeveloped in the smallest crabs. The characteristics of the suborbital and pterygostomian regions develop gradually. (Fig. 27.)

H. ANTERIOR APPENDAGES

1. *Antennae and Antennules.* These organs are small and inactive in *Uca*, in comparison with their size and importance in aquatic forms. Comparative counts of the segments of the antennal flagellum have not been undertaken, except for a token examination of three widely separated species, *dussumieri*, *maracoani*, and *minax*. In each of these forms, the segments in three males numbered between 30 and 40.

In common with other crabs, *Uca* has no statocysts and there is at present no evidence that the antennules play any part in the detection of vibrations. At least when a fiddler crab is on the surface, they are kept folded in their sockets below and behind the front (Figs. 26, 28); they never show overt motion when in the presence of possible airborne chemical stimuli, as during courtship. In short, at present any comment on the functions of these organs simply underscores ignorance.

2. *Eyes and Eyestalks.* As will be seen from Figs. 31 and 32, the length and relative thickness of the eyestalk and the size of the terminal eyes are exceedingly

variable throughout the genus. The species with the longest stalks are members of the Indo-Pacific sub-genus *Deltuca* and the American subgenus *Uca*; all of the longest stalks are also the most slender, lying loosely in the orbits. In broad-fronted crabs the stalks are of course shorter, since in these species the cara-paces are not correspondingly wider. In short-stalked forms the stalks are also relatively thicker. Crabs with slender stalks usually have the diameter of the eye greater than that of the stalk; especially in the superspecies *coarctata*, the eye is notably bulbous (Fig. 31 *A*). In contrast, in very broad-fronted crabs (Fig. 31 *H*) the eye projects scarcely at all beyond the circumference of the stalk. In a few species (Fig. 31 *E*) the depressed eye projects distinctly beyond the presumably protective antero-lateral margin of the orbit; these crabs do not live in the mud.

In several species of *Uca* the eyestalk and orbit on the major side are considerably longer than those on the minor, growing allometrically. The best exam-ples, both in the subgenus *Uca*, are *maracoani* and *ornata*.

In two related species, *heteropleura* and *stylifera*, the eye on the major side in adult males is sometimes equipped with a terminal style of variable length, similar to those found in some species of *Ocypode*. In almost all species of the same subgenus, *Uca*, the style occurs in some young males but not in adults. The structure is never strikingly colored to human eyes, even when the crab has attained the polished white of display coloration; it seems to be too slender to add to the crab's conspicuousness. It plays no part in waving display or advanced courtship. Work on possible endocrine involvement has not been under-taken. In short, the style's function remains unknown. Von Hagen (1970.2) has recently reviewed the problem.

The young of all species, whether Indo-Pacific or American, have the eyestalk relatively thicker, with the eye larger, than do adults (Fig. 32). As the front is proportionately wider in young than adults in all subgenera except *Minuca*, so the eyestalks are shorter in the young with the same exception. In other words, almost all species in *Minuca*, both in the shape of the front and length of the eyestalks, accentuate the post-megalopal characters of the genus, in a distinct form of paedomorphism. In very young *Uca* the eyes in the rest position tend to project beyond the antero-lateral angles, as is usual in megalopa.

3. *Maxillipeds and Other Mouthparts.* As in other crabs, the maxillipeds are concerned with the inges-tion of food and involved in the respiratory system. In addition, the flagellum of the third maxilliped is one of the appendages used in cleaning the eye. Fig. 33 shows the structure of the three pairs of maxil-lipeds, in comparison with those of *Ocypode*, the other genus considered to be in the same subfamily;

see also Figs. 76, 77 and 78. Comparison may also be made with the mouthparts of *Cancer*, figured in Pearson's monograph (1907-1908). The most nota-ble specializations in *Uca* are the development of spoon-tipped setae on the second maxilliped and of the gill on the third. The gill will be discussed along with the gill system (p. 469).

Both the sculpturing and patterns of external setae on the third maxillipeds are basically similar through-out the group and in detail they have not proved very helpful as taxonomic characters. Channeling, prob-ably as part of the external drainage system, is nota-ble in species of *Minuca*, most of which live in mud or sandy mud rather than in muddy sand. In species of *Celuca* with strongly arched carapaces and sterna, the ischium and merus of this outermost maxilliped are also unusually convex. In *Minuca*, the merus of the third maxilliped is relatively large in respect to the length of the ischium. This can be viewed as a minor paedomorphic character, since the merus throughout the genus is slightly reduced in relative size during growth.

On the second maxilliped the inner edge of the an-terior half of the merus, along with the tip of the palp, usually has some setae ending in concave, pecti-nate expansions; these are here termed "spoon-tipped setae." In some species they are almost lack-ing, while in others they may total several hundred individuals, arranged in numerous rows and covering more than half the inner surface of the entire length of the maxilliped. Examination of a majority of spe-cies in the genus has shown that those species living primarily on muddy substrates have few spoon-tipped setae, while those in environments most nearly approaching sand have the most. Examples of the various forms occurring are given in Figs. 36 and 37. The question of ranges of variation among different populations living on somewhat different substrates has not been investigated. It is not yet known whether the number of scallops on the spoon vary in accord-ance with habitat. One characteristic of probable phylogenetic importance is the proximal, spine-like protuberance found only in *tangeri* from West Africa and in all six members of the American subgenus *Uca.* (Fig. 37, *E, F, G.*) Spoon-tipped setae appear early in the life of the crab, but do not reach their final numbers until the individual approaches matu-rity.

Uca's method of feeding appears simple. To casual observation the crabs seem literally to be eating mud. At low tide a fiddler scrapes up pinch after pinch of substrate with a small cheliped and thrusts the pellet into the buccal region. Here, however, the inner max-illipeds function together in a complex straining pro-cedure that separates edible matter from most of the inorganic particles. The organic portion, still slightly mixed with substrate, then passes into the stomach, while mineral matter is rejected. Because of the activities of the mouthparts fiddler crabs are usually called filter feeders.

The first account of *Uca* which attempted analysis of both food and method was that of Verwey (1930). Crane (1941) described the spoon-tipped setae and suggested their association with muddy or sandy habitats. Altevogt (1955.1, 1955.2, 1956.1) carried analysis and description much farther, including an account of the process by which edible particles are floated away from the heavy inorganic matter. Finally, Miller (1961) described the functions of the many appendages and their parts involved in the complex process, contributing new facts and probabilities. He also presented additional ecological evidence for the differences observed among the mouthparts of three species. More recently, Ono (1965) provided data on *Uca lactea* in Japan.

The following summary of the role of the inner maxillipeds and other normally concealed mouthparts is taken essentially from Miller. A general account of feeding, as a major behavior pattern, starts on p. 472.

The rapid and complex motions in the buccal cavity are unfortunately wholly invisible under natural conditions. A good idea of them may, however, be obtained by removing the third maxillipeds, placing a bit of substrate on the exposed inner mouthparts, and observing the subsequent activity through a dissecting microscope. The ingested particles consist principally of light-weight edible matter and coarser inorganic particles. All are trapped in the long setae of the first maxillipeds. Against these the matted setae of the second maxillipeds now press. The area is simultaneously flooded by water from the branchial chambers. The second maxillipeds in vibrating, lateral sweeps free the coarser particles from the first maxilliped, catching them between specialized setae, including those with spooned tips; during the sweeping motion these and other setae scour the particles free of filmy food material, which is left caught in the setae of the first maxillipeds.

From there the comb-like setae of the vibrating maxillae pass the material backward, probably removing further inorganic matter. Finally it is drawn, still somewhat mixed with mineral substrate, between the mandibles.

Meanwhile, when the second maxillipeds reach the lateral ends of their rapid sweeps, they break away from their closely appressed, scouring position against the first maxillipeds. At this instant the heavy inorganic particles are washed down and back to the base of the buccal cavity.

Here, when an intact crab feeds normally, a wet gobbet appears and grows rapidly in size. Usually a small cheliped wipes it off, after every few clawsful of substrate, and drops it on the ground as a pellet; sometimes it falls off of its own accumulating weight.

I. Major Cheliped

1. *Introduction.* Nothing approaching the exaggeration of this appendage is found in other crustaceans, even among lobsters. For comparisons in the animal kingdom one must hunt far; deep-sea Gulper Eels (*Eurypharynx*) with their elongate jaws are the best examples that come to mind. Antler-bearing mammals, including even moose and Irish elk, show in contrast minor specializations. In fiddler crabs the claw alone, at its maximum relative weight, reaches almost half the total weight of the crab. In length, the same part of the fiddler sometimes measures more than three times the length of the carapace. At its extreme, the claw attains four times the carapace length and, in the same species, the major dactyl seven times the length of the manus.

The segments divide functionally into three groups (Figs. 1 and 2). First and most proximal are the basis and ischium; the merus and carpus form the second pair; finally, the segments of the claw itself form the third group, composed of the propodus (manus and pollex) and the dactyl. As in other decapods, the pollex is a fixed extension of the lower distal part of the manus; attached to the latter's upper distal end is the dactyl or movable finger. The dactyl and pollex together are termed the chela, with prehensile edges formed by the upper margin of the pollex and the lower of the dactyl.

The basis and ischium, while important in the movement of the merus and hence of the entire appendage, are small, relatively invariable, and unarmed, except for occasional granules or spinules which are always minute and variable and have never been observed in use. These two proximal segments will not be considered further.

In addition to their muscular functions during display and combat, the merus and carpus both are equipped with a variety of structures apparently used only in threat, or possibly in courtship situations, and never in combat. At these times, roughnesses on the merus are rubbed or rapidly tapped against correlated structures on the anterior and antero-lateral parts of the crab's own carapace, or against the merus of an ambulatory. Sound is undoubtedly usually produced by these motions, but the recorded evidence is still scanty.

Finally, the claw itself is used principally in waving display, where it is rhythmically moved, and in combat with other males. Additional uses occur in sound production, when it sometimes vibrates against the ground in both threat and courtship. In a few species, certain tubercles apparently also signal by rubbing against an ambulatory. When viewed as a whole, the appendage shows a striking functional dichotomy. The armature of the merus is self-directed, chiefly against correlated structures on the carapace. That of the claw, with rare exceptions, is directed toward particular structures on the claw of another individual. Yet, regardless of additional minor uses in the acoustics of courtship, the entire complex armature of both merus and claw forms a vast system reserved for intermale behavior.

2. *Merus and Carpus.* These segments are larger or smaller in conformity with the allometric growth of the manus, pollex, and dactyl. The result is that the tip of the chela always, in the folded rest position, lies either directly in front of the buccal region or somewhat beyond it, on the side of the minor cheliped.

With a few exceptions, the merus is a stout segment. It is always convex and partly roughened posteriorly and dorso-distally. Dorso-proximally and anteriorly it is flat and almost always smooth. Details of the range of its armature are given in the taxonomic description of the genus (p. 15). Table 12 gives the known instances of the armature's use.

This armature is almost always strongest along the antero-dorsal margin, particularly at its distal end where adjacent tubercles also often occur in the upper distal end of the anterior surface. In young crabs this distal complex can come into rubbing or vibratory contact with the crenellations of the outer suborbital margin. In larger individuals, however, allometric growth puts the segment's end altogether out of reach of the carapace. Perhaps any stridulatory or vibratory function this area may have is important only in juveniles, since in mature males waving display may supplant certain acoustical threats.

The postero-dorsal margin in most subgenera is strongly angled only in its proximal part and throughout is scarcely armed; distally it tends to obsolescence and is sometimes absent. On each side of this distal marginal area, the dorsal and posterior surfaces are more or less roughened. Even in the many species where these regions are nearly smooth, some rugosities persist dorso-posteriorly near the segment's distal end. This area, at least, seems sometimes in young crabs to vibrate against the suborbital region, the antero-lateral angle, or both. As with the antero-distal tubercles of the dorsal margin, allometric growth carries these rugosities beyond the possibility of such contact. More proximal and posterior rugosities apparently serve the larger crabs, and even the postero-distal armature is for them within reach of at least the first ambulatory.

The smooth, proximal section of the dorsal aspect of the merus comes most easily into contact with the adjacent outer pterygostomian region; such contact would seem, from a signaling point of view, to be ineffective, since the region is always setose. The fact that this part of the carapace, like the entire antero-dorsal and lateral areas, covers chiefly empty space

should, however, be kept in mind; it may well prove true that some of the vibratory motions of the merus consist only of tapping this smooth portion against the pterygostomian region; preliminary observations and films give some support to this notion. Perhaps the underlying airspace provides a drum-like resonating chamber. This structure might also reinforce the more conventional forms of stridulation suspected to result from the frication of some of the meral armature against the suborbital, antero-lateral, and lateral structures of the carapace.

Rubbing also certainly takes place against the dorsal meral margins of at least the more anterior ambulatories. Which merus in the latter contacts does the rubbing is not yet clear, but it seems that both the cheliped and the leg may at times be active.

Distally and subdistally the dorsal part of the merus shows two parallel ridges which are usually armed with minute tubercles or serrations. Their use is unknown.

The third margin, the ventral, is always sharply angled and more or less regularly equipped with small tubercles or serrations. Its function, again, is unknown.

The carpus, usually almost as broad as long, like the merus is anteriorly flat and essentially smooth, while it is rounded and usually rough posteriorly. Dorsal flattening is characteristic, in conformity to the dorsal part of the carpal cavity of the manus, into which it fits when the cheliped is flexed at rest. The dorsal and posterior armature ranges from rugose to tuberculate and is sometimes almost lacking. The dorsal margin, the ventral, and a low, oblique ridge on the anterior surface usually have a few tubercles, always very small except for rare enlargement on the anterior ridge. None of them has been observed in use; they do not seem suitably located for stridulation in conjunction with any part of the adjacent inner manus (palm).

3. *Manus, Pollex, and Dactyl.* This section, the claw, forms a functional whole which may be viewed in several ways. It attracts an observer's attention first in waving display; here its large size, rhythmic motions, and often striking colors all make it conspicuous. It also acts as a grasping organ, whether the object seized by its prehensile edges is another claw or a human finger. Finally, it is one of the most highly and variously specialized organs known to zoology, and certainly is unsurpassed in the number of adaptations for ritualized combat. The emphasis in this section will be on the last aspect.

The variety of armature on the claw appears at first bewildering, even when the basic functions of the structures have become apparent. A count of distinct structures on the claw of a single subgenus, *Celuca*, totals 84 (Figs. 42, 43, 44; Table 13); each of these may be justly considered a unit, it seems, since each on occasion varies independently of its neighbors. The table also indicates the range of variation within each unit. Of these 84 structures, about 60 are known to be used in at least one combat component in at least one species of *Celuca* or the related genus, *Minuca*. Each of the two species in which the forms of combat are now most familiar employ more than 40 of these structures in their combat repertories (Table 14). The particular structures used in a given combat depend not only on the forms of the combat components employed, but also on whether the combat is between homoclawed or heteroclawed individuals, on notable differences in size, and even on whether the first actor approaches his opponent from the outer side of the latter's manus or from the palm side.

The structures as a whole are arranged in orderly groups, regardless of their wide range of variations. They divide in general into ridges, mounds, grooves, pits, depressions, and associations of tubercles; almost all are highly localized and usually of limited variability within species.

Only the outer side of the manus is relatively homogeneous in contour and armature, being in general convex and usually covered with tubercles. Even here the tubercles are of at least different sizes on different parts of the manus, and are often differently arranged in the upper and lower portions; often they are further specialized near a depression at the pollex base. Often the upper third or less is bent over toward the inside and more or less flattened, functionally forming a part of the dorsal margin.

In contrast, the palm has some wholly smooth areas contrasting with characteristic arrangements of tubercles, while cavities and grooves are juxtaposed with mounds and ridges. In an overall view, seven general regions are characteristic of the genus. Most proximal is the carpal cavity, a deep depression into which fits the carpus when the cheliped is folded, and the edges of which vary greatly among species in height, shape, and armature. The second characteristic structure is the oblique ridge, usually tuberculate, running approximately from the carpal cavity to the ventral margin near the pollex base. Often sharp, with a high apex beside the carpal cavity and regularly tuberculate, the ridge is sometimes low, blunt, and nearly smooth. The third major area consists of one or two roughly vertical ridges at the dactyl base, usually tuberculate; the proximal continues ventrally, rounding into the inner prehensile row of the pollex. The fourth area is the depression at the pollex base, ranging from slight to deep and smooth to deeply grooved or tuberculate. The fifth region, while less conspicuous than the preceding four and at present less well known functionally, must be of great importance, judging by the variety and complexity of

its specializations. This zone is that of the dorsal margin and the adjacent submarginal region of the palm, which functionally appear closely related; a similarly close connection doubtless exists, as indicated in the preceding paragraph, between the dorsal margin and the submarginal area of the outer manus. On the upper palm the marginal and submarginal areas are equipped with various combinations of an impressive assortment of tubercles arranged in reticular and linear formations and of grooves and depressions, and often show as well notable flattenings or convexities. The sixth area, that of the center palm, varies from a distinct trough connecting the carpal cavity with the pollex depression to a convex area almost continuous with the dorsal side of the oblique ridge; finally, a proximal triangular area is bounded above by the lower edge of the carpal cavity and by the oblique ridge; it slopes from there proximo-ventrally to the ventral margin. In *Celuca* this triangular region shows a number of specializations known in one species to be used in threat, in correlation with tubercles on the first ambulatory; it is suspected of being concerned in other species in both threat and combat; these are the only structures on the claw in *Uca* known to be used in rubbing against another appendage on the same crab (p. 482). It is strongly suspected, however, that some cleaning motions made against the claw by the minor cheliped are actually stridulating activities (p. 461).

The lower margin of the manus shows a set of structures, ranging as usual independently of one another, and consisting maximally of keel, linear tubercles and, externally, an adjacent groove; sometimes they continue unbroken to the pollex tip.

Finally, the upper margins of the dactyl and the inner and outer surfaces of pollex and dactyl range from smooth to tuberculate and are sometimes grooved. The proximal outer end of the dactyl is usually tuberculate, with a submarginal groove; the tubercles often extend beyond the groove well along the dorsal margin of the dactyl. Except for this roughness, proximal tubercles usually present on the pollex, and sometimes minute tubercles near the prehensile edges, the outer surfaces of both pollex and dactyl are usually smooth; the inner surfaces are almost always so, again with the exception that minute tubercles sometimes occur near the prehensile edges.

The prehensile edges of both pollex and dactyl consistently show three rows of tubercles on each, of which the median is usually best developed and, except distally, the location of the enlarged tubercles or the rare tuberculate teeth. Any of the rows may be absent in any part of its length. Proximally the rows vary independently of the median and distal parts, and often only at their extreme proximal and distal ends are one or more rows irregular.

In general shape the chelae vary widely, as already

mentioned, in length, slimness, and degree of gape. Fig. 39 gives the range of the extremes. It should be mentioned here that while *maracoani* and the related *ornata* have the broadest and most compressed chelae, other members of the subgenus *Uca*, as well as *(Thalassuca) vocans*, *(Afruca) tangeri*, and, in the pollex form only, *(Australuca) seismella* and *(Celuca) saltitanta*, are also notably deep and compressed. *U. saltitanta* is an end form, in which the beginning of pollex expansion and compression is suggested by others in the series, notably *inaequalis* and *oerstedi*.

The morphology of the claw is further complicated by the fact that each species throughout the genus tends to dimorphism of the cheliped, although the forms intergrade. In American species, the differences are much less pronounced than in the Indo-Pacific *Deltuca* and *Thalassuca*. In *vocans*, particularly, the forms are so distinct that they received variety names (*forma typica* and *var. nitida*), even though it was known at the time that both varieties occur in single populations.

Fig. 38 shows pairs of these dimorphisms in six Indo-Pacific species, and Fig. 40 in two American forms. In each case, the upper figures illustrate the heavy, high, short-fingered form with highly developed prehensile tubercles, while the lower figures are examples of individuals in the same species with the manus short and low while the fingers are slender with reduced tubercles. They will be referred to as brachychelous and leptochelous forms. Because a leptochelous chela weighs far less than an equally long brachychelous example, and because of the allometric growth of the claw, it is not surprising that the largest examples in any population often are strikingly leptochelous. It seems probable, in fact, that in the largest species, such as *urvillei*, *dussumieri*, *tangeri*, and even in American *ornata* and *maracoani*, markedly brachychelous examples eventually molt into a size making reproduction and, in time survival, unlikely, since display would be ineffective and feeding inadequate. There is evidence for this hypothesis in the subgenera *Afruca* and *Uca* (p. 512).

It should be emphasized that both leptochelous and brachychelous claws are found even among quite young crabs, on the same biotope; it seems that the differences cannot be due to differences in nourishment. Nevertheless, when chelipeds are regenerated in crabs already well grown, the result seems usually to be along leptochelous lines. Irregularities and abnormalities in tuberculation always give clues even in dubious cases to the presence of partial or complete regeneration.

The leptochelous tendency of most of the American groups is very noteworthy, and differences between the two forms in any population are relatively slight.

In development (Fig. 45), the direction of growth is from a small claw indistinguishable from the other member of the pair and resembling in all essentials the adult form of the minor cheliped. Even in the first post-megalopal instar, however, the two claws are microscopically distinct. At this stage the crab feeds with both claws, as do females throughout life, and the fingers are long in comparison with the manus (Fig. 45 *EE, JJ, MM, PP, SS*). Thereafter, with differentiation between the two chelipeds the manus and chela both grow allometrically in respect to the length of the carapace (Fig. 45 *A-D, F-I, K-L, N-O, Q-R, T-U*) but, in all except *latimanus*, the chela (composed of pollex and dactyl) grows at a greater rate than the manus. *U. latimanus*, therefore, in this character is strikingly paedomorphic in comparison with other species.

The armature of the pollex and dactyl develops gradually, beginning probably in the second instar, and certainly shortly after the onset of allometric growth. The proximal tuberculate ridge at the base of the dactyl and the more proximal tubercles in the gape seem always to appear very early. Young crabs, with the chela still notably short, in some species have the ventral margin of the manus tuberculate only at this time; later this armature becomes obsolescent or absent. The last major series of structures to be developed are those connected with the oblique ridge and with the dorsal margin. In large *Minuca* the tubercles and ridges of the palm tend to be reduced, apparently not through wear; in these individuals the palm may in fact lack most of its specific characters. In contrast, the prehensile edges of the chela sometimes show distinct signs of being worn down in combat. No detailed comparative study of the development of the armature has yet been undertaken; in association with a study of the development of combat, it would be sure to yield rewarding results. (See also p. 515.)

When ritualized combats between fiddlers are observed with understanding, the numerous structures involved divide into two groups. The first provides the active crab with a selection of structures used in rubbing or tapping other structures included in the second group, and located on the motionless claw of his opponent. The members of these two functional groups will be termed instruments and contact areas.

It is now clear that each area equipped with one or more of these specializations is a point of contact during the forceful pushing and grasping motion of some form of unritualized fighting. It is similarly apparent that every known component of ritualized encounters is based firmly on these varied grips. The subject will be amplified in succeeding chapters (pp. 487 and 516).

A study of Figs. 42, 43, and 44 and of Tables 13 and 14 will give a better idea of the functional relationships of combat structures than would many paragraphs of text. In ritualized combat, all of the structures on the palm are always contact areas, never instruments. The remaining zones consist of the outer manus, outer pollex, outer dactyl, lower pollex, upper dactyl, and both prehensile edges; they are all used on occasion either as instruments or as contact areas. The prehensile edges usually serve as the only instruments in high-intensity combat, while the outer parts of the manus and chela become instruments only in low intensity components, such as the manus rub.

Even in *Minuca* and *Celuca* the forms of combat are still known too incompletely to provide correlations in more than elementary or suggestive form. The following examples, based on Table 14, show the kinds of relationships that are emerging between structure and function. The components of combat mentioned are described in the glossary and in the section on patterns of combat behavior (p. 489f.).

All species in which the heel-and-ridge is a known component have in common the following three pairs of characteristics. A strongly curved dactyl is associated with a well-rounded manus heel. A moderate to wide gape is associated with a moderate to high oblique ridge on the palm. An absence of greatly enlarged teeth in either prehensile edge is associated with the contact in combat of a fairly long portion of dactyl or pollex respectively with outer manus or palm.

All species known to use the interlace have at least the proximal predactyl ridge well developed and one of several specializations resulting in a broad, smooth, median space on the proximal, prehensile edge of the pollex.

Finally, certain structures used only as areas of contact are related to other zones of similar function. The best example so far known is the outer, upper, proximal groove on the dactyl in certain *Minuca* and *Celuca*. When it is well developed, a similar groove almost always runs the full length of the outer manus just below its upper margin; the two grooves, exactly at the same level, are separated only by the junction of dactyl and manus. As far as is now known, this two-segment groove is used only as a guide for the tip of the opponent's dactyl in attaining the position for heel-and-ridging.

J. MINOR OR SMALL CHELIPED

In comparison with the corresponding appendages of most other ocypodids, the minor chelipeds are almost as specialized as the majors in general shape and size, although in the opposite direction of reduction. In every species they are far smaller than in other members of the family and, in fact, all crabs with the exception of certain spider crabs. Even in the latter examples, the entire appendage is elongated and it is only the propodus and dactyl that are strongly reduced, while in *Uca* all segments are proportionately small. Few similarities of trend can be traced between the major and minor chelipeds in any species, and often their characteristics are contrasting. For example, the minor chela in *ornata* is extremely slender, yet the major pollex and dactyl are both very deep.

The armature of the minor cheliped, while far simpler than that of the major, still varies considerably among the species. Except for the equipment of the prehensile edges, comparative study of the tubercles, ridges, grooves, and setae has not been undertaken in detail. These structures will doubtless prove to be of importance when their functions in social behavior are better known.

Like the major merus, that of the minor is typically rough on the antero-dorsal and ventral margins and on the rounded surfaces of the dorsal and posterior regions. The armature includes various combinations of serrations, tubercles, spinules, and rugosities. The anterior surface is almost always flat and smooth. The carpus often is flat above with a curved, outer, subdorsal longitudinal ridge. The lower half of the outer manus always has a longitudinal ridge at least distally that continues along the pollex, sometimes for most of its length. The dactyl often shows a faint longitudinal groove on its upper half between two low ridges; sometimes only a single ridge is present and no groove.

Minor chelipeds (Fig. 46) are notably similar in general characteristics within species groups, particularly in the width of the gape and in the relative size and character of the armature on the prehensile edges. The narrowest gapes have a series of fine serrations and are characteristic of many mud-livers, such as in Indo-Pacific *Deltuca* and American *Minuca*. Some female *Deltuca* and *Australuca* have a single enlarged tooth on both pollex and dactyl in at least one cheliped; no such enlargement is found in the male minor cheliped, except in *seismella* (*Australuca*).

At the other extreme from the narrowly gaping, serrate, minor chelipeds are the forms with wide gapes in which tubercles are absent. These are confined to species of *Celuca*, most live in muddy sand or, for *leptodactyla* and *musica terpsichores*, the substrate surface may be almost sandy, with little mud or clay admixture.

In the mud-living forms from the Indo-Pacific, plentiful setae are arranged in series both along the inside of the middle of the dactyl and pollex and marginally. In these and the species immediately following, the chelae are deeply excavate distally, forming setae-fringed spoons. In *tangeri*, as well as in *ornata* and other members of the subgenus *Uca* and in *thayeri*, two rows of subdistal setae on both pollex and dactyl form a complex "basket." This basket is highly evolved in the aberrant species *panamensis* (*Minuca*), forming a prominent tuft of stiff brown bristles.

An exceptional shape is shown by (*Celuca*) *stenodactylus* which has a well-formed basket along with chela tips that are pointed and crossed.

Finally, in the species characterized by slender chelae and wide gapes, there are few setae except for a distal basket, and the distal excavations are relatively slight.

In females, both chelipeds resemble those of the males; sometimes the appendages on the two sides differ in small details.

When post-megalopal forms are investigated, the minor chelipeds are found to be similar to those of adults except that they are relatively larger in comparison with the carapace in *rosea* and smaller in *brevifrons* and *stenodactylus* (Fig. 47). The chelae of all young crabs are less setose than those of adults, and with the onset of allometry the characteristics of the adult are rapidly acquired. The fingers show allometric growth with respect to the manus in all three of the above species, but most so in *rosea*, among the several unrelated species investigated (Fig. 48).

Although the small cheliped is used in a number of ways, its primary function is undoubtedly that of carrying food to the mouthparts. Obvious adaptations to this role are the perfectly meeting, spooned tips of the chela and the serrations of the prehensile edges, especially in mud-living crabs that must lift wet and sticky substrate. Pellets composed chiefly of sand, on the other hand, are lifted in the basket of curved setae at the tip of a widely gaping chela. *U. panamensis* scrapes algae from stones with thick tufts of stiff bristles. *U. tangeri* has a similar habit, but also lives and feeds, depending on its physiological phase, in a wide range of habitats from the wettest of low-tide mud to high on beaches of almost pure sand. Its small chela, equipped with both serrations and abundant bristles, well reflects these varied needs. After nourishment has been extracted from several pellets by the mouthparts, the small cheliped wipes away a gobbet of material that has accumulated below the posterior part of the third maxillipeds.

The minor cheliped also plucks, pats, and strokes the carapaces of courted females in all species in which precopulatory behavior has been filmed or observed in daylight above ground (p. 503).

In addition, the appendage usually makes small motions during display that roughly correspond to the movements of the major cheliped and, except in a few dubious cases, that do not at all come in contact with the carapace or other appendages.

It now appears certain that the small cheliped also acts as a sound-producing mechanism of importance. Salmon (1965) reported the rapping or knocking by a single (*Celuca*) *pugilator* female against the substrate. In addition, two other species, (*Celuca*) *deichmanni* and (*Uca*) *ornata*, were filmed during the present study; here the minor chela of a male briefly taps the ground. The behavior is shown only in a single example of each species.

More usual is the vibration or rubbing of the merus of the minor cheliped, apparently against an adjacent part of the carapace. Successful recordings have not yet been made, but films show the motions clearly in seven species, where the area of contact appears variously to be the crenellate suborbital margin, the outer orbital margin, or the antero-lateral angle. The species are (*Deltuca*) *demani*, (*Australuca*) *seismella*, (*Minuca*) *galapagensis*, and, in *Celuca*, *batuenta*, *inaequalis*, *beebei*, and *stenodactylus*. Rubbing between the minor cheliped merus and that of the first ambulatory is strongly suspected in films of (*Uca*) *heteropleura* and, in *Celuca*, of *inaequalis* and *limicola*.

In addition, it is highly probable that during some cleaning motions made by the minor, which moves toward their distal ends along the major pollex and dactyl, the crab is actually producing sound. Well suited for this are the ridges on the carpus, outer manus, and chela. Sometimes, too, the crab rubs its prehensile edges along the corresponding portions of the major. Heretofore, I have considered these motions pure displacement cleaning, when the crab was displaying with an already polished cheliped, and with evidence of a conflict situation. As in the other forms of apparent stridulation by the minor cheliped, recorded evidence of sound production has not yet been secured. Filmed examples total 23 species, distributed through every subgenus; 11 of the species are members of *Celuca*.

True cleaning by the small cheliped of another appendage can occasionally be seen, when the chela seizes the partly depressed eye itself between its prehensile edges and rubs along it, from the tip to near the base of the stalk.

K. Ambulatories

The walking legs are similar in form and proportions in all parts of the genus, in comparison with wide-ranging characters such as the front, eyes, chelipeds, and gonopods (Figs. 29, 30). The order of length is always the same, extending from the long second through the third and first to the short but fully functional fourth. Corresponding legs among the species differ somewhat in length but never extremely. More notable differences occur in the relative breadth of the meri and the degree of arching of its dorsal margin. In females the legs are regularly shorter and the meri wider and more arched than in males of the same species.

The armature of all legs varies more in amount and strength than in type or distribution. Most of it occurs on the merus and carpus. The merus shows a single dorsal marginal ridge, two ventral marginal ridges, and, on its posterior surface, scattered tubercles or vertical rows of tuberculate striae. These tubercles and striae are irregularly placed but are always more numerous in the segment's dorsal half than more ventrally; sometimes they occur only near the dorsal margin. The scattered tubercles occur in the Indo-Pacific subgenera *Deltuca*, *Australuca*, and *Thalassuca*; some female *Amphiuca* have both scattered tubercles and tuberculate striae; in the other subgenera all or almost all the tubercles surmount definite striae (Fig. 52). Any or all of the marginal ridges may be partly or wholly spinulous, or tuberculate, whether strongly or weakly. Often certain margins are wholly smooth and rounded, a weakness that occurs most frequently on the fourth leg; in contrast, the second and third are always more strongly equipped than the others. The posterior ventral margin always has stronger armature than the anterior, and this margin is usually stronger distally than proximally, while the opposite is true of the weaker, anterior margin. Dorsally, at the distal end of the merus, two parallel ridges, subdorsal and distal, are often equipped with minute tubercles or spinules. Variation often is found even on the two sides of the same individual.

The carpus is also importantly and variably armed. Two dorsal ridges, with or without some form of roughening, are usual on the second and third legs; the first leg usually has only a posterior ridge and the fourth only an anterior; a third longitudinal ridge sometimes occurs on the posterior surface of one or more of the first three legs, where tubercles or striae are also usual, similar to those on the merus. Like the armature of the merus, that of the carpus varies within species, and on two sides of the same individual.

Anteriorly on the first ambulatory on the major side, some male *Celuca* have minute tubercles, usually in a longitudinal row and sometimes also on the distal part of the adjacent merus and the proximal end of the manus. These special structures are correlated with the development of other equipment on the major cheliped, found on the triangular area of the lower, proximal part of the manus.

On both merus and carpus, the armature is always notably stronger in females than in males of the same species (Fig. 52).

Armature of the ambulatory mani is weak or absent in both sexes, although dorsal, lateral, and ventral ridges occur; the segment is always thinner, in the antero-posterior dimension, than the merus and carpus. The tapering dactyl is six-sided, the divisions separated by weak ridges. Both manus and dactyl are well provided with setae, often in the form of long, stiff bristles. On the more proximal segments, setae are instead usually sparse.

Pile is often importantly present on the ambulatories (p. 465).

Reference to Fig. 30 shows that the ambulatories change only in minor ways with growth. The degree of change is less than in *Ocypode* (Crane, 1937) but in the same direction of increased length. The change in length compared with carapace length is particularly evident in the growth of the merus. Again as in *Ocypode*, both merus and manus become relatively thicker. The armament develops by degrees, the legs of early instars being almost smooth.

In *Uca* the legs contrast both with those of fast-running *Ocypode* and with those of many sedentary xanthids, such as *Xanthodius sternberghii* that spends the low-tide period under stones. Although fiddler crabs are active, and sometimes exceedingly so, as in displaying *stenodactylus*, they are never racers and the moderately efficient ambulatories reflect their way of life. The most active species usually live on firmer substrates, have the longest legs, and, at least in males, more slender meri. Mud-livers tend to have shorter legs and broader meri. It is possible that slightly shorter and broader legs reduce the tendency to sink in wet mud during feeding; a wide spread of long legs would seem, however, to be more useful. If short, broad legs are useful in the mud, their frequent presence in females only may be explained. When females are carrying eggs and hence extra weight, they usually feed on soft substrate closer to low-tide levels than do males. The location would seem, nevertheless, to be more important as a means of keeping the eggs from drying out.

The following suggestions seem more likely. The shorter legs may represent the unspecialized form of the appendage; in a few male *Celuca* they have become adapted for faster motion on firmer substrates as well as for the attainment of conspicuous height in visual display. The wide meri, on the other hand, may be viewed as another adaptation in females for a more efficient expanse of stridulating surface.

The ambulatories perform an assortment of functions. All digging and the erection of chimneys, pillars, and hoods beside the burrow are performed by them, sometimes with some aid from the small cheliped. In carrying the excavated pellet, the stiff bristles of manus and dactylus form basket-like supports, regardless of any sensory functions they may prove also to have. The ambulatories often play parts in visual display, perhaps partly due to the need for balance, and also adding to the male's apparent size. Unreceptive females, in warding off males, sometimes rear high up on the tips of the middle dactyls; sometimes they proceed partly down their own burrows, leaving the ambulatories of one side extending stiffly up in the air (Fig. 89). In several species in which a curtsy forms part of display, some of the turned-under dactyls stamp on the substrate at the curtsy's low point. Salmon (1965) secured a sound record of this behavior in *pugilator*. Stamping also occurs in agonistic situations between males, especially in the subgenus *Uca*. The ambulatories also tap and stroke females in precopulatory behavior, usually in conjunction with similar motions of the minor cheliped.

The role of ambulatory armature in social behavior is certainly of great importance. In both sexes it appears to be confined to agonistic situations. Rubbing of the meri against one another occurs throughout the genus and is one of the most prevalent of social activities. It is seen even more often in females than in males; it is doubtless pertinent that their armature is always stronger than in males of the same species. The meri also certainly sometimes rub against dorso-lateral and lateral structures on the carapace, although the film evidence is still scanty. The carpi also often rub against each other and against the meri of adjacent legs.

Film analyses have shown no rubbing patterns characteristic of species or species groups. In all species photographed at some length, the meri of all four legs are shown making rubbing motions and coming clearly into contact with adjacent legs, although only rarely in the same motion picture sequence. Very often two alternate legs on the same side are used as motionless supports, while the meri of the other two are tilted so that the dorsal serrations rub against the roughness of the posterior surface of the adjacent legs. Sometimes adjacent legs, especially the second and third, are simultaneously raised, bent at the mero-carpal joint, and rubbed against each other. Sometimes the carpal ridges rub against the meri or, perhaps, each other. Often, but not always, meral rubbing occurs first on the side of the crab nearest the apparent stimulus. In all, variation and lack of stereotypy are notable in the behavior of individuals.

A quite different use of the ambulatories corresponds to ritualized combat in males. Here two females line up parallel to each other and mutually rub the extended mani and dactyls of their adjacent sides with such speed that the motion appears to be a prolonged quivering. It may be that some or all of the setae are sensory and reacting with one another. Sometimes the dactyls reach the meri of partners, and so could well be rubbing with their ridges against the marginal or posterior armature. No good photographic evidence is yet available. The females may face the same way or oppositely. Similar behavior has been observed once or twice between males and, briefly, as part of courtship. Details are still unknown.

L. Abdomen

The abdomen in both sexes of *Uca* has seven segments (Fig. 53 *A, B*). In some *Celuca* characterized by strongly arched carapaces and sterna, several of the segments are partly or wholly fused (Fig. 53, *C, D*).

M. Gonopod (First Pleopod) in Male

1. *Introduction.* Although the gonopods in *Uca* are usually a primary aid in the taxonomic determination of species, they must be used with care as keys to relationships and evolutionary trends. Parallelism occurs widely. Rarely the opposite tendency forms a pitfall, where a variety of shapes occurs in a single subgenus (*Uca*). Again, the gonopod of one or two subspecies in a clearly allopatric group may be aberrant. Within subspecies, gonopod characteristics are constant with minor exceptions.

A gonopod consists basically of a shaft enclosing a tubular canal and ending, in its distal one-fifth to one-sixth, in a modified tip (Fig. 56). The shaft is three-sided, its anterior, flat surface curving forward in conformity with the sternum's convexity.

The modified tip (Fig. 58) consists of the genital opening, here termed the pore, in association with a maximum of four specialized structures. Each of the four ranges independently of the others, in various parts of the genus, from absent or vestigial to extremes of hypertrophy. In this contribution, these structures are termed flanges, inner process, thumb, and setae. An additional characteristic in certain groups is a distal torsion of the shaft, along with some or all of its parts.

The basic structure of two principal forms of gonopods is shown in Figs. 74 and 75.

2. *Flanges.* These calcified wings, when present, occur on both sides of the terminal part of the canal (Fig. 58 *A*), or may extend above the pore (*B*), the canal appearing to be fastened against the postero-internal surface of the large, apparently single flange (*E, I*). The arrangement of simple wings on each side of the terminal pore occurs in other ocypodids, namely in *Macrophthalmus, Myctiris,* and *Ocypode* (Fig. 57 *C, A, D*). In *Uca* it is found in parts of the subgenera *Deltuca, Thalassuca, Uca,* and *Minuca.*

Expanded flanges, in which the end of the canal appears to be fastened directly to their continuous, convex, postero-internal side, are characteristic of (*Boboruca*) *thayeri* and (*Celuca*) *lactea.*

In the superspecies *coarctata*, flanges are progressively reduced, while the end of the tube throughout shows the antero-external edge of the tube's tip overlapping the postero-internal edge. This overlapping, with concomitant absence of flanges, occurs also in

Australuca, in (*Amphiuca*) *chlorophthalmus*, and in some species of *Celuca*. In these flangeless species, the calcified portion of the tip is often elongate (Fig. 58 *H*).

Reinforcing struts, often strongly calcified, sometimes support the flanges, and sometimes are even partially separated into spines (*dussumieri spinata*, *vocator*).

3. *Inner Process.* This structure arises at the base of the exposed terminal portion of the tube or flanges, on its inner side, toward the midline of the crab. It may take various forms, all clearly homologous; it is occasionally difficult to see, in some species because of its small size, in others because of the surrounding flesh and setae, and in still others because of its thinness and transparency. All species have it in some form, and it is clearly represented also in *Myctiris*, *Heloecius*, *Macrophthalmus*, and *Ocypode*; its occurrence in other genera apparently has not been investigated. Its three principal forms are as follows:

(*a*) A sharp, slender spine of varying extent, sometimes curved or twisted, at base of the exposed end of the tube. This spine is characteristic of the superspecies *coarctata*, the subgenus *Australuca*, and part of the subgenus *Uca*. (Fig. 58 *D*.)

(*b*) A fleshy structure, setigerous, sometimes greatly tumid and sometimes twisted anteriorly to rest against the tube and flanges at an angle. This form is found throughout *Deltuca*, except in the superspecies *coarctata*, in *Thalassuca*, in *Afruca* (*tangeri*), and in some species of *Uca* and *Minuca*. (Fig. 58 *E, F*.)

(*c*) A flat plate or shelf, in shape triangular, an elongate oval, or narrow throughout and distally pointed. Sometimes this process is practically fused with the tube, transparent, naked, and discernible with difficulty. This type occurs in some *Minuca*, in *Amphiuca*, and in *beebei* and related species in *Celuca*. (Fig. 58 *H*.)

4. *Thumb.* This structure is usually present, and typically almost terminal (Fig. 58 *E, F*). Sometimes it is strongly reduced and arises far down the shaft (Fig. 58 *B*). Rarely, it is represented only by a small shelf (Fig. 58 *H*) as in *beebei* and *stenodactylus* (*Celuca*), or is indistinguishable, as in some members of the subgenus *Uca*. Except in cases of torsion, to be described, it arises on the antero-external side of the gonopod. It is present in other genera in the family. In species where it arises well below the tip, the thumb may be highly variable even within single species in the same population (Fig. 59).

5. *Setae.* As in other structures, these vary widely in form and arrangement throughout the genus, but are distally present in profusion on every species. Because of their frequent concealment of the underlying structures, they have been omitted altogether from the drawings, except for a general view of the types in Fig. 61; here extremes of variation are shown, with examples taken from the various subgenera. It will be seen that the pattern usually includes a "mane" of rather long setae, often with certain ones isolated and elongate. In species with little or no torsion, this series is postero-external. One or more subterminal longitudinal series of setae may occur on other facets of the gonopod. Very short setae, thickly set, often spine-like and some of them curved, are found distally in most species of the subgenera *Deltuca*, *Thalasucca*, and *Uca*; they often occur on the inner process. A naked area characteristically is found antero-externally. The thumb is usually well covered with setae whenever it approaches the gonopod's tip.

6. *Torsion.* This characteristic is as notable in some groups of *Uca* as any of the structures just described.

In the untwisted gonopod, the canal proceeds up the shaft, near its antero-internal angle, until, in the terminal region, it curves externally to the pore. The specialized tip of the shaft, in conformity, is tilted externally, the inner process curving with it; one flange, termed the posterior, is directed postero-internally, the other, the anterior, extends antero-externally; the thumb arises from the anterior side of the shaft, slightly external to the path of the canal. (Fig. 58 *E*.)

In species showing torsion, on the other hand, the tip of the gonopod is twisted toward the outside (Fig. 58 *C*). Sometimes the entire tip appears slightly twisted below the base of the thumb, so that the terminal parts maintain their usual relationships toward one another, although their orientations are changed. Sometimes the torsion starts distal to the thumb, as in some species of *Minuca*, where the usual relation of thumb to inner process and flanges is changed. Torsion is most extreme in certain *Thalassuca* and *Uca*. Here the thumb may even arise postero-laterally and, continuing the direction of torsion itself, end as a strictly posterior instead of anterior structure (*vocans vocans*). In three out of six *vocans* subspecies, the canal, after proceeding in the direction of antero-external torsion, doubles back on itself terminally, to exit in something like the original position between two flanges which have themselves undergone somewhat independent torsion. In the subgenus *Uca*, torsion, along with reduction and hypertrophy of various structures, reaches such lengths that the general form of the gonopods, in five out of eight species and subspecies, is almost useless in tracing relationships.

7. *Development.* Differentiation of the gonopods starts early in post-megalopal stages; species charac-

ters, when diagnostic, are usually reliable by the time the ridging (although not the tuberculation) is well developed on the inside of the major cheliped—probably in the second or third post-megalopal instar.

8. *Function.* The invariable function of the gonopod is to pass spermatophores into the genital openings of the female. These openings, the gonopores, are described below. In no *Uca* does it seem that more than the extreme tip of the organ is inserted, although perhaps further insertion may occur in some species with elongate tips; it is apparent from comparisons that in most species it is necessary and indeed possible only for the two pores to come into contact. The fantastic variations and exaggerations found in the genus serve at most to orient the gonopod tip to the gonopore and hold it in position during spermatophore transfer; there is no trace of a complex "lock-and-key" arrangement. Gonopores are usually simple; if a tubercle is present in one position or another on its margin, manipulation shows it to be probable that the inner process on the gonopod, or a flange, or the thumb is oriented to the tubercle in a certain way. Among the subspecies of *vocans* and *lactea*, where modifications of the gonopore might be expected to occur in close correspondence to the exaggerated torsions of the gonopod, only slight adjustments have evolved, in relative sizes of tubercles and ridges; no torsion exists in the female organ. (Fig. 54.)

9. *Spermatophores.* No comparative work has been undertaken. Spermatophores usually appear in the female to be carried largely or altogether internally. In a few *Minuca*, particularly in *vocator*, they have a conspicuous external section that makes the gonopore appear more tuberculate or sculptured than is in fact the case, and can easily lead to taxonomic misidentifications. It seems likely, at least in these cases, that the protuding spermatophore prevents multiple matings. The entire subject is obviously in need of study from several points of view.

N. Gonopore in Female

These structures are a pair of orifices on the third abdominal segments of females. Specialization ranges from a simple marginal tubercle to varying types of semi-circular ridges (Figs. 54, 64 *AA-FF*). As mentioned above, they do not show obvious torsion associated with that of their conspecific or consubspecific males. Any sculpturing does not become complete until the abdomen reaches its apparent maximum breadth with respect to its coverage of the sternum; series of measurements have not, however, been undertaken.

O. Second Pleopod in Male

The second pleopod in *Uca* appears unspecialized and varies little. It may serve to support the gonopod during copulation.

P. Pleopods in Female

As in other genera, fiddler crab eggs are attached to the pleopods, as shown in Fig. 55.

Q. Pile

Many species of *Uca* have characteristic areas of close-set, short, setae superficially resembling fur. The terms "pile," "pubescence," and "tomentum" are here considered synonymous; "pile" is perhaps preferable because of brevity. In a given species of *Uca* pile may occur on any part of the carapace, on the major cheliped, on some or all of the ambulatories, or on all of these regions. Usually it is more abundant in females than in males of the same species; an interesting exception occurs in (*Minuca*) *vocator vocator*, where pile is always present on certain lower parts of the ambulatories in males but absent in females.

Sometimes the form and distribution of pile show little variation within a species and serve as a reliable taxonomic character. Examples are eight persistent tufts on the dorsal part of the carapace in (*Celuca*) *inaequalis*, and pubescence, even more firmly attached, on the lower parts of the ambulatory carpi and mani in (*Minuca*) *mordax* of both sexes.

In most species, however, pile is both highly variable and extremely subject to abrasion. Even more irritating than this unreliability as a taxonomic character is our ignorance of its functions.

Its most obvious use is in courtship. In most or all species of *Uca*, during the final phases the male strokes, taps, or plucks at the female's carapace with the minor cheliped. Expectably the pilous areas hold mud longer than does the smooth carapace and so facilitate, and perhaps stimulate, the usually ritualized feeding motions of the male; it seems likely that the sensation of these plucking motions may be sexually stimulating at least to the female. In groups where courtship is least developed (*Deltuca* and *Thalassuca*) this behavior always precedes the aboveground mating. In groups with advanced courtship, where the crabs typically mate underground, the pattern usually takes place when the female stiffens her legs, the genus-wide signal of non-receptiveness. This occurs when the male is not fully in the mating phase or when, in unexplained instances, complete copulations take place on the surface (pp. 502-504).

In most subgenera pile occurs in at least a few

species on the carapace of the female and sometimes on that of the male as well. In (*Celuca*) *cumulanta* distinctive patches are found in the females of certain populations only; in (*Celuca*) *crenulata* patches are found in one subspecies but not in the other. In two of three subspecies of (*Deltuca*) *dussumieri*, in one of two (*Deltuca*) *acuta*, and in (*Uca*) *major* the females are distinguished by a border of pile near the ventral margin of the last ambulatory merus; its function remains totally unknown.

A frequent location of pile, particularly in some *Celuca*, lies at the base of the gape in the major cheliped. As a speculation without factual support, it may be suggested that pile serves as a kind of buffer during the interlace component of combat; perhaps the pile masks noise or tactile sensations that might interfere with stimuli resulting from the rubbing of the actor's gape tubercles along the predactyl ridges of his opponent (p. 490). In general the pubescence on the major cheliped in any species is so variable and fragile that for the most part its occurrence is not specially mentioned in the systematic descriptions.

In some species, including (*Afruca*) *tangeri* and (*Uca*) *maracoani*, it seems that algae are associated with pile, especially on the propodus between the tubercles and within the pits.

In conclusion, there is no question but that pile merits close comparative study when broods of *Uca* eventually come to be routinely reared.

R. COLOR

The colors of fiddler crabs are so varied and often vivid that the only surprise is that they have not attracted more attention. Their study by zoologists has taken two main directions, which have until now remained separate. When the approaches merge we shall begin to understand the subject's intricacies, and to perceive the relations of color and color changes to ecology, physiological conditions, and behavior.

The first approach is physiological; it is concerned largely with the endogenous rhythms that control paling and darkening, and with the hormones that activate these basic color changes; the work has necessarily been confined almost entirely to the laboratory and depends on experimental procedures. The second deals with the coloration of *Uca* in the field and with their changes in association with waving display; so far this study has remained observational; the subject now invites, and urgently needs, experimental results to determine the functions of color in the genus.

Almost all color changes in *Uca* are caused by the expansion and contraction of monochromatic chromatophores, most of which are located in the epi-

dermis. Each has permanent branching processes, so that the pigment may be either concentrated at the center, and so practically excluded from the general pattern of the crab, or dispersed into the processes. Four pigments are known in brachyurans—black or dark brown, red, yellow, and white; a fifth, blue, is found outside the chromatophores. The five apparently all occur in *Uca*; their chemistry still awaits investigation.

Fiddler crabs were among the first animals used in the study of biological clocks. During the early decades of this century the work progressed on several fronts. Observers saw that *Uca* on the northeast coast of the United States appeared darker during the day than at night. Physiologists soon found that even when they kept the crabs in total darkness the diurnal rhythm of change persisted for a long time. The investigators then discovered a similar, persistent darkening–paling rhythm that depends for its timing on the tidal schedules of the locality in which the crabs are taken. This tidal rhythm, superimposed on the diurnal rhythm, results in a semi-lunar effect, caused by the coincidence of the dark-color peaks of each rhythm and, during the same short period, the coincidence of their pale-color peaks, the effect occurring about every two weeks. At the same time that these investigations were going forward, other research showed that expansion and contraction of the chromatophores are controlled by hormones located in the sinus glands of the eyestalks and in the central nervous organs.

The contributions of Kleinholz (1942) and Brown (1944) give comprehensive surveys of the early work on the endogenous rhythms of *Uca* and their control, along with full references. For recent reviews and developments see Brown (1961) and Barnwell (1968.1, 1968.2).

The remainder of this survey of color in the genus will be devoted to the second approach to the subject—the observation of color and color changes in the field. The number of species in which color in life is known is 59 of the total of 62 species recognized in this contribution. In some our knowledge is extensive and in others scanty; only in (*Thalassuca*) *formosensis*, (*Celuca*) *helleri*, and (*Celuca*) *leptochela* have the colors never been recorded. Wherever, in the summaries given below, percentages are included, all refer only to the 59 species on which we have information.

All *Uca* change color at least to the extent, as described above, of becoming darker during the day and somewhat paler at night; the tidal rhythm induces darkening at the time of low tide and paling during high water, when the crabs are normally in their burrows. These palings are not at all similar to the temporary assumption of polished white, presumably through pigment dispersal, by some species during the display phase; instead they result principally or altogether from the simple concentration of dark pigment into the centers of the chromatophores.

All species also darken under two other conditions. When a fiddler shows display coloration as described below, it will slowly lose its polished white and any intense colors it has assumed whenever it descends into its damp burrow and remains there for more than a few minutes. No darkening effect occurs after brief stays below ground, such as are caused by alarms, short periods of digging, or descents for gill-moistening.

Fiddlers in display coloration also darken when an observer seizes and holds them. This change, in contrast to those previously described, is rapid. A released crab, even if returned promptly to its own burrow, often does not resume display coloration on the same day.

Many *Uca* change color little or not at all when in display condition. Some parts of the males, particularly of the major cheliped, are often more or less distinctively colored at all times: but the color is somewhat more intense in many displaying individuals. Other species during display show polished white to varying degrees on the carapace alone at least in some populations, while the appendages are then more or less vividly colored with yellow, orange, red, or purple. Intense blue and green are sometimes display colors, but not on the major cheliped. Finally some individuals or populations attain a polished display white that suffuses the entire surface of carapace and appendages. The following paragraphs survey the genus in regard to both general color and to color change in connection with display.

In 57 species the tips, at least, of the major chela are white or, in several forms, very pale pink or yellowish. The two with dark tips are (*Uca*) *maracoani* and (*Uca*) *ornata*.

All have some tint or shade of yellow, orange, red, or intermediate hues on the lower outer manus. This color is usually persistent except that in display phases it often becomes more intense or, unusually, suffused with polished white. Often the lower, outer manus is the only area of definite color on the crab. Where other parts of the major cheliped are similarly hued, the color in this area is the most intense. Sometimes it is so characteristic, as in (*Thalassuca*) *vocans*, that it makes a good field character.

In 12 species, roughly 20 percent, a complete color change to polished display white occurs in some individuals of some populations. Only in (*Celuca*) *l. lactea* does it seem to be the usual color of the crabs, regardless of season or display phase. In all other forms of that species, as well as in other species, its occurrence is subject to geographical differences as well as to age, sex, phase, and season. Often many individuals in a population on a particular day dis-

play and court vigorously and even successfully, attracting the attention of females and engaging in combat with other males, and yet do not assume display coloration; I have not found the converse to be true.

In 24 species, about 40 percent of the total, at least the carapace sometimes changes during the display phase to polished white (Pl. 48). Even on the carapace alone polished white has never been observed in any species of *Deltuca*, or in the two monospecific subgenera, *Afruca* and *Boboruca*. In *Australuca*, *Thalassuca*, and *Amphiuca* whitening takes place in only one species each and then rarely and usually incompletely. In the remaining subgenera—*Uca*, *Minuca*, and *Celuca*—are distributed the majority of species sometimes attaining a carapace of polished white; all these subgenera are confined to America except for two species of *Celuca*. In all subgenera, except for (*Celuca*) *l. lactea*, the full assumption of polished white even on the carapace alone is confined to fully tropical forms.

Among species in which the carapace does not become polished white, but only pales or brightens slightly if at all, several characteristics stand out.

First, only in the Indo-Pacific are found bright red and bright blue, often with white or bluish spots; here these vivid hues cover most or all of the carapace and sometimes the ambulatories as well. In American subgenera blues and greens are uncommon and confined to display coloration on the anterior part of the crab only, while red crabs in America are represented only by some populations of a single subspecies, (*Minuca*) *vocator ecuadoriensis*; their pervading hue, not connected with display, is dull rose red. In the Indo-Pacific, in contrast, certain members of *Deltuca*, *Thalassuca*, and *Amphiuca* are distinguished by the prevalence of intense reds and blues; these hues are characteristic of both sexes and persistent; they are brightest in young crabs and dullest in adult males.

Second, in a number of species of *Celuca* a marbled or mottled carapace with copper- or gold-color combined with paler or darker tones is frequent.

Third, in (*Amphiuca*) *i. inversa* and in many populations of (*Celuca*) *lactea*, particularly of *l. annulipes*, a striking pattern of transverse pale bands or marblings on a dark background is common; when display coloration is assumed, the proportions change, so that the crab eventually appears as polished white with narrow bands of dark blue or black.

Fourth, many species of *Minuca* and *Celuca* lighten somewhat when in a display phase from their usual grays, browns, and buffs. Familiar examples in the western Atlantic temperate zone are (*Minuca*) *pugnax* and (*Celuca*) *pugilator*.

Fifth, certain individuals in populations with strong display whitening go through a brief, bright stage of orange or chrome yellow that sometimes suffuses the entire crab and sometimes the carapace alone; this phase is usually succeeded by a lightening to yellowish white before polished white is assumed. I have observed these intermediate phases in certain populations of (*Thalassuca*) *vocans*, (*Uca*) *princeps*, and (*Minuca*) *galapagensis*; they probably occur in some populations of (*Celuca*) *triangularis bengali*. In each of the species the yellow phase may be the terminal one for individuals not in display phase; perhaps in *triangularis bengali* it is the terminal hue of entire populations, whether displaying or not. This suffusion of yellow occurs additionally, as far as known, only in the early crab stages of (*Deltuca*) *rosea*.

Purplish or dark red sometimes colors the ambulatories in a number of species, whether or not the carapaces whiten. Outstanding are (*Deltuca*) *demani*, (*Afruca*) *tangeri*, (*Uca*) *major*, (*Uca*) *stylifera*, and (*Minuca*) *vocator ecuadoriensis*. Bright red orange, orange red, or crimson are usual in (*Thalassuca*) *tetragonon*, (*Amphiuca*) *chlorophthalmus*, and (*Celuca*) *lactea annnulipes*.

Only one or two general statements can be made about the colors of females. In most species members of this sex are duller and often darker than the males, while individual breeding females usually assume polished display white, if they ever do so, less often than males of the same populations. Exceptions are the females of (*Thalassuca*) *vocans* and, in the subgenus *Uca*, of *major* and *stylifera*, all of which show strong display whitening in some populations. In addition, females of (*Celuca*) *l. lactea* as a whole always attain an equivalent white to that of their males. An uncommon, rose red phase occurs in individual females of *lactea annulipes* and *l. perplexa*, whether or not they are attracting courtship behavior and responding to waving males; it is also found in occasional ovigerous females of the same species; the color seems to be confined to adults.

Changes to display coloration take place gradually, usually during the hour or so before diurnal low tides. Usually they proceed to their fullest development on warm, bright days. The time needed for the change varies in different individuals within the same population, and even in the same individual on different days. The fastest such change I have seen was by individuals of (*Uca*) *stylifera*; in these the change was fully accomplished in 15 minutes or less; one of slowest was *latimanus*, which sometimes spent two hours or more effecting the change. It may be that sunlight, high temperatures, and dryness are all important. Further speculations on this topic and on the possible roles of the several pigments in effecting changes to display coloration were given in Crane (1944); none was based on experimental or analytical work and no such research seems yet to have been undertaken.

The first record of color change in *Uca* appeared as a footnote in Müller's "Facts and Arguments for Darwin," originally published in 1864. In the English translation (1869: 36) the note reads: "This smaller *Gelasimus* is also remarkable because the chamelion-like change of colour exhibited by many crabs occurs very strikingly in it. The carapace of a male which I have now before me shone with a dazzling white in its hinder parts five minutes ago when I captured it; at present it shows a dull grey tint at the same place."

A few years later Darwin continued the account: "I am informed by Fritz Müller, that in the female of a Brazilian species of Gelasimus, the whole body is of a nearly uniform greyish-brown. In the male the posterior part of the cephalo-thorax is pure white, with the anterior part of a rich green, shading into dark brown; and it is remarkable that these colours are liable to change in the course of a few minutes—the white becoming dirty grey or even black, the green 'losing much of its brilliancy.' It deserves especial notice that the males do not acquire their bright colours until they become mature. From these various considerations it seems probable that the male in this species has become gaily ornamented in order to attract or excite the female." (*Descent of Man*, ed. of 1874: 275.) Müller's own detailed account appeared in 1881 (p. 373).

Still later Alcock (1892, 1902) was of the opinion that both color and waving played roles in courtship, serving as recognition devices to the females, which also appreciated the display's color and motion. After that observers became sharply divided on the subject, as reviewed in the general account of waving display (p. 494). To this day the value of display coloration in relation to behavior remains unsettled, since even elementary experiments, save for those of von Hagen (1970.3), have not been undertaken. Its merely sporadic appearance in many species where it attains a high development is one of its most puzzling characteristics.

For experiments on the functions of color the following species would, it seems, prove to be the most rewarding, since the changes occur in many or most populations and are confined to displaying males and, sometimes, adult females: on American shores, in the subgenus *Uca*, *major* in the Atlantic and *stylifera* in the eastern Pacific; among *Celuca*, *leptodactyla* in the Atlantic and, in the eastern Pacific, *saltitanta* and *musica terpsichores*. In the Indo-Pacific no species achieves display white in most populations; in occasional places in some seasons it reaches full development in (*Thalassuca*) *vocans* and (*Celuca*) *lactea*; less often it occurs in (*Amphiuca*) *i. inversa*. Additional species that also show strong display coloration, but only sporadically, are, in the subgenus *Minuca*, in the eastern Pacific *g. galapagensis*; in *Celuca*, in the eastern Pacific, *beebei* and *latimanus*;

in the western Atlantic, *uruguayensis*; in the Indo-Pacific, *lactea annulipes* and *l. perplexa*.

Among the colored areas that appear especially promising for experimental work on their functions are the following: the lower, outer manus on the major cheliped, prominent in some threat postures throughout the genus and in vertical waves; coloration of the remainder of the cheliped, particularly when the inside differs from the outside, the two areas contrasting strikingly during lateral waving; the occurrence of contrasting colors on the anterior sides of at least the first ambulatories; further observations, in addition to experiments, on *beebei* and other species that revolve during display and that have both anterior and posterior sides of the ambulatory meri colored; the blues and greens sometimes found on the anterior parts of carapaces and adjacent areas, as in *cumulanta* and *beebei*; the large, white or blue white spots on the posterior ambulatory meri in some *Deltuca*, where they are conspicuous when raised in threat display toward the rear.

Also in need of investigation is the possibility that some of the persistently vivid colors in Indo-Pacific *Deltuca* spp., (*Thalassuca*) *tetragonon*, and (*Amphiuca*) *chlorophthalmus* may be aposematic. Similarly, while general observation shows that *Uca*, like many other crustaceans, alter their prevailing hues when not in display coloration to tone in with a variety of substrates and so achieve a kind of cryptic coloration, no precise work on the problem seems to have been effected. Finally, in spite of an occasional supposition that diurnal darkening has a cryptic value, the question has not received the benefits of experimental attention.

The course of color development in young fiddler crabs varies greatly among the species. Only among a few *Deltuca* are the very young strikingly colored; usually they appear at least as dull as non-displaying crabs in species with strong developments of display coloration. A few species have not only the tips of the two chelae but each manus as well shining white in the young of both sexes. Tones of orange or red usually appear on the major manus in otherwise dull adolescents at earlier stages than do the paler or more vivid colors of other regions of the body.

S. GILLS

The only internal structures which will be specially considered here are the gills, with emphasis on those of the third maxilliped. These gills alone show major differences and apparently have phylogenetic implications. (Figs. 76, 77, 78, 81, 82, 83.)

In Fig. 81, gills from representative species of *Macrophthalmus*, *Ocypode*, and *Uca* are drawn, using the length of the carapace as a standard. *Macrophthalmus*, which leads a life by far the least ter-restrially adapted, has the best developed gills. In *Uca*, however, reduction of the anterior gills is not in conformity with the ecological requirements of the species. In any case, throughout the group these small gills at best could probably be of little use. When the gill on the third maxilliped is only a nubbin, it cannot, of course, function at all.

In no *Uca* does the third maxilliped gill approach its development in *Macrophthalmus*, or the perfection of shape in *Ocypode* (Fig. 82). *U. maracoani*, with only a few books missing from one side and orientation unchanged, is best developed; members of the subgenera *Thalassuca*, *Afruca*, and *Amphiuca* all have clearly formed books with much in common. In other subgenera, these gills show all stages of reduction. In the few species in which a long series has been examined, the character is subject to variation within populations, although some members of each will show the maximum development of the gill characteristic of the species. The species best known in this respect are *vocans*, *chlorophthalmus*, and *tangeri*.

The gill on the second maxilliped and the first regular gill, also reduced, vary among the species; either one may be the larger, with little respect to relationships shown by other characters. The species with the greatest reduction of all three small gills is *lactea annulipes*.

T. INTERNAL ORGANS

A semi-diagrammatic view of the arrangement of the internal organs is provided in Figs. 79 and 80. It is designed for only two purposes. First, for workers unfamiliar with the internal organs of crabs, it will serve as a guide in locating the gonads and determining their readiness for breeding. Second, it shows the relative positions of branchial chambers, gills, afferent branchial orifices, and pericardial sacs; some of these play not only the usual role in basic ecological adaptations but apparently also in sound production. The latter aspects will be considered in the chapter on social behavior.

A comparative study of the respiratory, digestive, and reproductive systems of representative species of *Uca*, still to be undertaken, would assuredly furnish illuminating results. Key studies based on other crustaceans including crabs are Pearson (1907-1908), Waterman, ed. (1960, 1961), and Bliss & Mantel, eds. (1968).

U. SUMMARY

The morphology of *Uca* is described, with the principal emphasis on structures that play or probably play direct roles in behavior or that are of taxonomic importance. Larval stages are omitted. The uses, known and hypothesized, of the individual structures are considered, in preparation for a discussion in later chapters of their functioning as groups in behavior patterns. Viewed as a whole, the major and minor morphological features of fiddler crabs divide into two classes—those concerned primarily with the maintenance of the animal in its environment, and those connected with social behavior. In the first group are the crab's basic shape and the equipment for respiration, feeding, and digging. In the second are the numerous specializations concerned primarily or altogether with social behavior. They include the varieties of hypertrophy in the major cheliped, the complex armature of carapace and appendages, and the diversity of gonopods. On the claw alone, comprising the manus and chela, 84 different specializations are counted that appear to function only in intermale combat. In fact, the great majority of morphological specializations for social behavior in *Uca* apparently are used only in agonistic behavior, chiefly between males, rather than in courtship. In the chapter the post-larval growth of each principal structure is briefly described and the range of variation indicated. While notable allometric growth is a characteristic of most morphological features, including its extreme expression in the major cheliped, it is not of practical use in the determination of species. Particularly in Indo-Pacific species, dimorphism of the claws is strongly apparent, although intermediate forms always occur. Color in some species varies along geographic lines and in others on the general tone of the substrate; in individuals it also depends on circadian rhythms, on breeding condition, and on whether a male is in display phase; in some species daily color change in displaying crabs is striking; even in dull species parts of the major cheliped contrast somewhat with the substrate and carapace. In practical taxonomy the most useful structures for species determination in males are unquestionably the gonopods, in combination with the shape and breadth of the front and the sculpturing of the major cheliped. Usually of secondary usefulness are the dentition and gape width of the minor cheliped, the armature of the orbits, and the shape and armature of both carapace and ambulatory meri. Of rare but important helpfulness are special armatures on the first ambulatory and the presence or distribution of pile on carapace or ambulatories. Females are distinguished from males chiefly in having two small and equal claws, similar to or identical with the minor cheliped of the male; the armature of carapace and ambulatories is invariably somewhat stronger than in the male, apparently in connection with sound production and combat, the signaling devices being more restricted than in the male through lack of a major cheliped. Taxonomically females are often difficult to determine in preserved collections from areas in which the local communities are unknown. The best key characters usually include any or all of the following: gonopore sculpturing, resemblance of front to that of male, location of pile, dentition and gape width of chela, and, used with great caution, the shape of the carapace and leg segments, all in comparison with those of their male conspecifics. In females the orbits are never more oblique than in males, the antero-lateral margins shorter, the carapace flatter, or the width of the ambulatory meri less; each of these characters always differs if at all from the expression in the male in the opposite direction, the characters together increasing the volume of the carapace and hence helping avoid desiccation.

Chapter 4. Non-Social Activities

OUTLINE

A. INTRODUCTION

The behavior of fiddler crabs may be roughly divided into maintenance activities, defenses against predators, and social patterns. As usual in such classifications, and particularly in any subdivision of *Uca*, the dividing line is movable and a number of activities are multifunctional. Burrows provide an example. Seldom dug by the current occupant, they are nevertheless kept clear and, if necessary, enlarged by that individual. The burrows probably originated as an underwater defense against fish and other predators and continued as protection also against desiccation, heat, and terrestrial enemies during low tide, as the ancestors of fiddler crabs assumed an increasingly active intertidal life. Today, at least, burrows also play fundamental roles in social behavior.

The present chapter will be confined to brief descriptions of certain non-social activities, selected principally as a foundation for the consideration of social behavior. Categories will be listed in rough order, from the most generalized activities, such as walking, to those most characteristic of the genus. Comments on the more physiological forms of behavior, such as respiration, will be confined to aspects particularly associated with other *Uca* activities, or with its ecology. Throughout, for amplification of such basic patterns see Waterman, ed. (1960, 1961), Marler & Hamilton (1966), and contributions to the *Symposium on Terrestrial Adaptations in Crustacea* (Bliss & Mantel, eds., 1968).

The various ecological factors and endogenous rhythms concerned in the activities of fiddler crabs are discussed in the chapter on ecology (pp. 440ff.) and in the section on rhythms in social behavior (p. 504).

B. RESPIRATION

As in all ocypodid crabs, when *Uca* are wholly or partially underwater, water is drawn into the gill chambers on each side through the afferent orifice between the second and third pairs of ambulatories. From here it circulates around the gills and passes out through the efferent channel at the antero-external angle of the merus of the third maxilliped. During low tide the crab descends its burrow the bottom of which is either damp or under water, at irregular intervals.

A supplementary system of maintaining moisture in the branchial chambers during low tide is provided by the crab's habit of suddenly lowering the body against the ground. In this position the brushes of setae fringing the afferent orifice are brought into contact with the damp substrate and requisite moisture drawn up into the pericardial sacs.

In addition to its primary functions of bringing oxygen to the gills and preventing desiccation, the water circulating in the branchial chambers is concerned in two other activities. One of these is feeding, since the water is led through the rhythmic action of the mouthparts into the buccal cavity, where it helps separate coarse, inedible material from finer organic matter. This process has been described on p. 456.

Finally, respiratory water apparently is sometimes of use in making acoustic signals (p. 484).

Uca and *Ocypode* are the two most perfectly amphibious crustaceans. Of the two, *Uca* is the less terrestrial. Although tested species can survive indefinitely when completely submerged, the same forms die even in humid air after a few hours. Nevertheless, strongly displaying species and those that aestivate and hibernate must often be taxed, and doubtless then the crab relies on the air spaces in the epibranchial regions which are lined with blood vessels, as well as on moisture stored in the pericardial sacs.

The epibranchial regions probably also serve as resonating devices when appendages rub against the carapace during sound production (p. 483).

In the respiration category may be mentioned the

aeration of eggs by ovigerous females. These individuals may be seen now and then, either crouching low or standing high, while masses of bubbles emerge from the branchial orifices at the third maxillipeds and pass backwards over the egg mass. Similar behavior, here termed *bubbling*, has been noted in both sexes kept in small containers when the shallow water becomes foul, and also in large terraria, where Schöne (1968) found it occurred regularly, apparently as a cleaning mechanism. Altevogt (1969.2) has presented evidence that bubbling is a temperature regulating mechanism. Finally, in the burrow bubbling sometimes functions as an acoustic signal (Crane, 1966 and p. 484).

C. LOCOMOTION

Since it is a typical crab, *Uca* walks sideways. Either the large claw or the minor may lead, depending on the social situation. Also, depending on the crab's phase, the body may be held very high, or slung low. A fiddler can run rapidly, in spite of its moderately short legs, but only for a few seconds at most. In many species the crabs are adept at changing direction and proceeding in brief dashes. Such speed and maneuvering are used only during emergencies associated with escape, combat, and courtship. Manton's pioneer studies on crustacean locomotion (especially those of 1952 and 1959) would make an excellent basis for a comparative study among the species of *Uca*.

D. FEEDING

To any human being who watches a fiddler crab eat, the feeding process looks simple. At low tide the crab sits on the shore with the buccal area almost vertical, the most posterior part being also the lowest. In this position the fiddler repeatedly scrapes up a bit of substrate in a small cheliped and places it between the inner edges of the third and second maxillipeds, held ajar during feeding. Males use the single minor cheliped, apparently at a faster rate than the two similar chelipeds of females, which use them alternately. Sometimes single scrapes are made, sometimes four or more at a time, before the material is carried to the mouth. There do not seem to be specific differences in this behavior; the number of scrapes is probably related to the character of the terrain. In general small crabs feed faster than large ones and, expectably, heat speeds the tempo. Displacement feeding motions and those combined with low intensity waving are predictably slower than the highly motivated gathering of substrate following a crab's emergence after high tide. Miller (1961) in his detailed examination of the feeding process, gives rates of routine feeding in several species. The course of the pinch of substrate within the buccal area,

based principally on Miller's work, has been reviewed in this contribution starting on p. 456.

The spoor of active *Uca* in the Indo-Pacific may briefly be confused with that of other genera but can always be identified by the lines of pellets radiating from a burrow. The crab's habit is to move slightly ahead as it feeds, whether it is at the time a burrow resident or a wanderer.

The only detailed analyses of the pellets discarded in feeding and of fecal pellets have been published by Verwey, Altevogt, and Miller (all *loc. cit.*, p. 456). Verwey's work was in Java; Altevogt's analyses were wholly on *vocans*, near Bombay; Miller worked on species in the eastern United States. There is general agreement that microscopic algae and protozoans living in the substrate, as well as organic matter brought in by the tide, form the main items of diet. Diatoms, although often ingested, seem usually to be excreted.

The zoeae and megalopa are efficient predators of planktonic animals (Herrnkind, 1968.2, and refs.).

In addition, I often observed both *panamensis* (in Costa Rica and Panama) and *tangeri* (in Angola) supplement filter-feeding with the plucking of algae growing on stones. Furthermore, in the females of a number of species, such as *vocator* and *maracoani*, the pile sometimes appears to give support to living algae. In both these species and many others a plucking motion with the minor cheliped from the female's carapace plays a definite role in courtship (p. 503).

In Costa Rica I once saw *brevifrons* eating mammalian feces, although *Uca* certainly is not normally a scavenger.

When a male loses the minor cheliped, he attempts to use the major in feeding. If the degree of hypertrophy is not too great, he may successfully support himself for weeks or months, until regeneration is accomplished, as has been observed in the crabberies. Crabs losing both chelipeds bring the buccal region in contact with the substrate. No data are at hand to show the long-term success of this procedure.

E. DEFECATION

The anus is at the end of the abdomen; the small fecal pellets are deposited by a partial lowering of the abdomen. These pellets are smaller than the cylinders of substrate discarded at the posterior end of the third maxilliped and, once recognized, the two kinds of discarded matter cannot be confused. Large gobbets found near fiddler burrows are merely loads brought up during digging operations.

F. CLEANING ACTIVITIES

The major cleaning pattern is simply submersion. A crab muddied from digging will often emerge dripping and free of mud, after a brief descent into his burrow in which water is still standing, just before starting a display period. Cleanliness is definitely associated with the display phase (p. 687); crabs in other phases, particularly when in the wandering or early aggressive periods, sometimes do not remove mud except from the eyes and eyestalks.

Special cleaning motions, moderately well stereotyped, play roles both in actually removing mud clinging to parts of the carapace and appendages and in social behavior (p. 461). The eyes and their stalks are cleaned primarily by depressing them alternately into freshly wet sockets. Secondary agents are the distal, setose segments of the third maxillipeds and the serrations and setae of the minor chelipeds. The direction of the wiping motion, in contrast to that used in cleaning other appendages, is from the distal to the proximal end, doubtless as a means of avoiding the addition of mud from the stalk to the functional part of the appendage. Since this direction is contrary to the usual pattern in arthropod cleaning motions with which I am familiar, the very practical adaptation is of special interest. Mantids, for example, in cleaning their forelegs with their mouthparts, work from the proximal to distal ends of the segments.

The minor cheliped cleans the major, always spending most time and moving most often over the outer surface of the manus and chela. Sometimes it reaches within the gape to clean part or all of the outer pollex as well as, less often, the tubercles of the gape. Motions along the inner side of manus and chela are less frequent, and those against the carpus and merus unusual. The outer sides of the major merus and carpus, especially proximally, as well as of the minor cheliped and ambulatories, are cleaned principally by rubbing adjacent meri against one another. Among the ambulatories the other segments are used similarly, particularly the ambulatory dactyl.

In cleaning no specific differences are apparent. No usual sequence has been observed. All are well developed in the early crab stages. All are prevalent forms of displacement behavior (p. 520).

G. BASIC DEFENSE

This section concerns only defense against predators, which, in *Uca*, are restricted chiefly to birds, crab-eating raccoons, dogs, and human beings. Protection against ecological hazards are discussed starting on p. 451, and in the following section on burrow-digging (p. 474). Intraspecific agonistic behavior is treated in later chapters.

Against predators, actual or potential, the primary defense is a simple escape down a burrow, preferably the hole which the crab has inhabited for a longer or shorter period but, failing that, any one that can be reached by a brief run.

If a fiddler escapes into a burrow that is already occupied, the intruder will be tolerated by the resident, when acoustic signals fail to drive it out, until the intruder senses that it can emerge and go elsewhere. This behavior is further described under territoriality, p. 511. Repeated quick forays are made to the brink of the burrow, so that the eyes can briefly overlook the surface. Although none of the necessary experimental work has been done, part of the emergence signal is undoubtedly visual, the positive stimulus apparently being a lack of unaccustomed objects in motion. If the strange object, such as an observing human, remains motionless long enough, at a highly variable distance of toleration, the crab emerges in spite of the addition to the surroundings. It may also stay below in response to vibration signals, such as steps near the burrow. Populations of the same species and subspecies vary widely in their toleration distances. Groups that are frequently disturbed, such as populations beside roads or on village shores, are often expectably more tolerant. Populations in more isolated places are unpredictable; the reasons for the local variations could form a most rewarding study.

When a fiddler crab is pursued while feeding or wandering near the low-tide level at a distance from all burrows, it has available several lines of defense. First it may try to sink into the mud with a rotating motion. If the substrate is too firm for the crab to corkscrew downward but soft enough for easy excavation, the fiddler can quite quickly dig a burrow with the ambulatories. Obviously, there is rarely time for this activity in the presence of a true predator. I have, however, watched it occasionally when I have been the cause of the alarm. Again, such crabs sometimes escape threats by running into the water far enough to be wholly submerged. The final defense is the universal crab gesture of rearing back by folding the posterior ambulatories, simultaneously straightening the front legs to raise the front of the carapace, and spreading both chelipeds out, with the chelae open. If the crab is chivied when in this position with beak, paws, stick, or finger, the major chela moves to seize the threatening object.

A crab in its burrow sometimes seizes a similar thrusting threat in the same way; usually, however, it burrows downward as fast as possible, and grabs the intruder only when further digging is blocked.

Although some components of acoustic behavior may prove to be elicited only by potential predators, no example is yet known. On the other hand, sound production is a usual response of a burrow resident to small intruders, in particular to conspecific *Uca* (p. 485).

H. BURROWING

Although burrowing motions appear identical throughout the genus, it used to seem that good specific differences would emerge in the forms of the burrows. These characters doubtless do exist, but they have proved to be so dependent on ecological and seasonal conditions, and so much comparative work would have to be done in order to draw valid conclusions, that this kind of species-specific behavior will not be described here in more than general terms.

Takahasi, for example (1935), described what appeared to be typical burrows of *arcuata* and *vocans borealis* in northwest Taiwan. Investigations of the same locality showed in the spring of 1963 (Crane, unpublished) that, although the habitat investigated by Takahasi was optimal for *vocans*, it was not favorable for *arcuata*. Since the latter species lives typically in crowded populations in upriver mud, the burrows there are quite different from those of the sparse, atypical population living among *vocans* in muddy sand on the somewhat stony shore. Takahasi's method of taking plaster casts of numerous individual burrows seems to be the only one that would be altogether valid for detailed studies. Casts would, of course, have to be made of many groups of examples, from various levels on the shore and in various ecological and geographical parts of the range of each species.

In view of these requirements, it will only be said that most *Uca* appear to have simple burrows, proceeding diagonally downward. The deepest burrows, which may end more than three feet underground, are made by the larger individuals, in river banks and on shores near high-tide level, as well as by some aestivating or hibernating crabs. The shallowest, sometimes only two or three inches deep, are produced by very small species or by individuals in moist habitats. Enlarged niches are sometimes found in the shafts, as are, I think rarely, forks and more than one entrance.

Regardless of the direction or depth of the digging, the method is always the same. The crab digs only with the legs of the minor side, which precede the crab into the hole and curve around the excavated mass, forming a basket with the setae. All material from burrows is carried at least a short distance from the mouth and either dropped on the spot, or more rarely tossed away. The dropped loads are always much larger than the feeding pellet, but are often well compressed into balls.

A single exception occurs to the rule that males dig only with the legs of the minor side. In attempting to force another individual from a burrow, the intruder sometimes briefly thrusts his major chela below ground, making with it prying or scooping motions that rarely bring out a small amount of substrate or even successfully flip out a smaller crab. Usually the motions are ineffectual; it seems probable that the descending cheliped continues to act as a threat sym-

bol, doubtless sometimes engaging the claw of the other male (p. 515).

Very young crabs do not make their own burrows, but when pursued escape into those of larger individuals, or simply sink into soaking wet terrain at the water's edge. Their first burrows are very shallow, and close to the low-tide level. The larger displaying males are typically found nearest to high-tide marks. The subject will be further discussed in the following chapters.

The burrow is often, but by no means always, stopped up with a plug of material from either the outside or inside before it is covered by the tide or in the midday heat. It has been suggested that a column of air is kept by this means in the burrow. This does not seem possible, since apparently nothing is done to line the burrow with waterproof material, save for a few notable exceptions; *lactea*, for example, in its sandier habitats, lines the shaft with wet mud from below. The releaser for the plugging behavior is wholly unknown.

I. ORIENTATION

The critical study of orientation in *Uca* has begun very recently. Altevogt (1963.1) showed how *tangeri* find their way back to a particular area of the beach, but not necessarily to their former burrows, by polarized light. A more general paper on orientation in *tangeri* followed (Altevogt & von Hagen, 1963). Herrnkind (1966, 1967, 1968.1, 1968.3) described similar behavior in *pugilator*. In both species the crabs use, as appropriate, polarized sky light, sun navigation, and landmarks for orientation at a distance. At close range, however, *tangeri* appear to find their burrows with the help of kinesthesia rather than vision (Altevogt & von Hagen, 1963).

J. LEARNING

Altevogt (1963.2) performed the first modern experiments on learning in *Uca* during his investigation of visual discrimination ability in *tangeri*. Using artificial burrows as rewards, he succeeded in training between four and six individuals in a total of 23 to select a particular direction, and to discriminate between lights of different intensities. He was unable to train any of the group to distinguish between colored lights of equal intensity or between lights of different planes of polarization.

Langdon (1971), after finding that shape discrimination is well developed in *pugilator*, concluded: "Training studies suggest that the preference for vertical vegetation-like stimuli is a learned one and is subject to modification to adapt to changing characteristics of the environment. However, these experiments do not rule out the possibility that innate components exist and cause a predisposition to prefer vertical visual stimuli. . . ."

K. SUMMARY

The non-social behavior included in this chapter consists of maintenance activities and defenses against predators. Most of the selected patterns clearly form the foundations on which are based patterns of social behavior. For example, territoriality depends on the habit of digging shelters in the intertidal substrate. The overt facets of respiration are reviewed, as are procedures in walking, feeding, defecation, cleaning, avoiding predators, and digging. Recent studies of orientation and learning are noted.

Chapter 5. Components of Social Behavior

OUTLINE

A. INTRODUCTION

In fiddler crabs, social behavior consists almost altogether of agonistic activities, their ritualized derivatives, and courtship. Most but not all aggressive behavior, whether ritualized or not, is also clearly connected with reproduction. With two exceptions—droves and synchronous waving—none of the patterns shows collaboration between more than two crabs in a joint activity. Fiddler crabs neither dig communal burrows, for example, nor threaten predators in unison. There is no trace of the group territoriality found especially in some mammals, where an area is defended by its inhabitants from trespass by another group of conspecifics. Furthermore, since the larvae are released at hatching, there are no mutual, behavioral responses between parent and mobile offspring, even to the extent found briefly in scorpions.

Yet, as described in the chapter on structures and functions, the specializations for threat, combat, and courtship are unsurpassed. Although fiddler crabs cannot be termed socially cooperative in any usual sense, they must, it seems, be vitally dependent on one another. The forms of this dependence remain unclear. As will be shown, males do not have to fight to obtain empty burrows. There is no visible struggle for food. There seems to be ample space in every population for courtship. Successfully breeding males do not seem to be congregated in the middle of leks, although quantitative work in this field has not been done. Combats with other males are not prerequisites enabling them to mate. Females are not attentive to encounters between males, nor do they select the brightest or most actively waving individual. In several subgenera, waving is not even always a part of successful courtship, although it always occurs dur-

ing the breeding season in every population. Successful mating, even in subgenera in which courtship is elaborate, will take place in a finger bowl, with no preliminaries, no food, and no tide. Consequently, we seem to be left with the vague conclusion that both intermale behavior and waving courtship display serve under natural conditions as mechanisms for stimulating and synchronizing reproductive behavior in the population as a whole. They doubtless also discourage copulation by males not in optimal breeding condition. The single certainty is that we know almost nothing of the physiological mechanisms underlying these values.

As usual in social animals, some of the components in agonistic behavior and courtship are ambivalent. The most striking example is "waving," the motion of the large cheliped in males which, in most species, is an indispensable part of courtship and, in all species, is performed most rapidly and vigorously during courtship. Waving is also an important part of threat behavior which, in fact, almost certainly underlies much of the evolution of the wave. Recent investigations are bringing to light increasing numbers of examples where the same means of sound production is used in both aggression and courtship. The structures sometimes built outside fiddler burrows are probably similarly ambivalent.

In the following pages, the general repertory of social components will first be enumerated. The great majority are concerned with agonistic behavior, not with courtship. They will be divided, for convenience rather than from evolutionary logic, into seven groups: first, droves; second, agonistic postures and associated motions; third, sound components, both known and probable, in threat and courtship; fourth, combat components; fifth, the components of waving display; sixth, constructions beside burrows; seventh, copulation. Subsequently, rhythms and phases of social activities will be considered, along with examples of unusual behavior.

For general accounts, with bibliographies, of the bases of social behavior in animals, see Marler and Hamilton, 1966, Eibl-Eibesfeldt, 1970, and Hinde, 1970; for reviews of social behavior in Crustacea, see Schöne, 1961 and 1968.

B. DROVES

In the lowest rank of fiddler sociality belong the kinds of behavior grouped here under *droves*, a term occasionally applied to moving aggregations of *pugilator* in the southern United States. No driving or herding of the crabs is implied. Like numerous other moving groups in the animal kingdom, including schools of fish and traveling caterpillars, the individuals in fiddler droves take no apparent notice of any particular individuals. Although their formation changes rather freely, they do move as a crowd, in a roughly similar but easily altered direction, and may stay together throughout almost an entire period of low tide. All of this behavior is in contrast to the types of social behavior which will be described later.

The droves of *pugilator* have been familiar to naturalists, children, and, unfortunately for conservation, commercial collectors for a long time. Thanks to Herrnkind's recent work (1968.3 and refs.) on visual orientation in the species, this part of *Uca* behavior is at last receiving modern professional attention.

Droving appears to be of at least two distinct kinds, representatives of both of which were noticed by Herrnkind in the course of his study. In the first, feeding aggregations of both sexes move down from the upper beach to low-tide levels. Second, "small groups of either sex occasionally migrate along the shore for considerable distances, up to 100 m, and apparently establish residence in new localities."

The first class occurs, as Herrnkind points out, in populations of the species holding burrows high up on the beach, rather than among those living and feeding on flatter shores. As in all other species of *Uca* with which I am familiar, individuals with burrows on the higher levels, where the food supply is less rich, often come down to feed at the water's edge. Droving becomes conspicuous in many *pugilator* populations because they are adapted to relatively sandy habitats, the upper levels of which are poor in microorganisms compared with lower levels, so that great numbers of crabs, "from a few dozen to thousands of individuals" (Hyman, 1922) sometimes move down to feed.

The second class is of greater interest from a social point of view. These crabs, consisting only of males in the examples I have observed, move more as an associated group. They may or may not spend much of their time feeding. In *pugilator*, the only droves I saw during brief observations near Miami, all consisted of males behaving as do single tropical males in the non-aggressive wandering phase, except that they moved almost as a unit. The members almost touched one another, feeding below the levels of the burrows as in the mixed groups of the first class but along a strip starting a few feet farther along the beach than the nearest holes. No droves formed of both sexes, or of females, occurred in this population. When they were startled by me the males stayed together, moving obliquely away toward the water, then surging back a few moments later.

I have watched large droves composed wholly of males in four other ocypodids, in *Uca tangeri* in Angola, *Dotilla* spp. in Zanzibar and western Australia, and *Myctiris* (soldier crabs) in Brisbane. In each, the individuals numbered between 500 and 1,500. In *tangeri* and in both observations on *Dotilla* the droves appeared only in conjunction with spring tides, which of course uncovered more potential feeding area than was usually available to them. Two points in this connection are interesting. First, these tides occurred in the morning, in correlation with a natural peak of *Uca* social activity (p. 505). Second, except in the case of Zanzibar *Dotilla*, the crabs did not proceed beyond the usual limits uncovered, although they did not avail themselves of this area on other days. Fortunately Cameron (1966) has provided a season-long study of drove behavior in Australian *Myctiris*, such as is greatly needed in *Uca*. In fiddlers the corresponding pattern is so little known that no further reference to it will be made in later sections.

It is worth noting here that I did not see any shore bird, the most usual predator on fiddlers, pay apparent attention to the large and conspicuous droves of the second class described above, much less seize a crab.

C. AGONISTIC POSTURES AND ASSOCIATED MOTIONS

Certain postures are prevalent in *Uca* during territorial, aggressive, and submissive behavior. Most of them will probably prove to be characteristic of the entire genus; others appear to be confined to several subgenera. These postures all may and usually do continue as equally typical motions. They do not follow one another in fixed sequence, although their occurrences and order are sometimes predictable in a given social situation.

The postures and their associated motions, described below, will be given names suggestive of their form.

1. *Raised-carpus*. Male. A low intensity posture. The mano-carpal joint of the flexed, major cheliped is raised, with the fingers ajar and pointing obliquely down. Sometimes the carapace on the major side is also slightly raised. The raised joint is usually pointed toward a nearby male that is either approaching from a distance or encroaching on the feeding area of the posturing individual. More rarely, it is aimed towards a crowding female or a member of another species, especially when all are feeding in a closely spaced population. The posture is sometimes accompanied by the vibration of the major merus against an adjacent part of the carapace (p. 482). (Fig. 84 *A*, *B*, *E*.)

2. *Down-point*. Male. High intensity threat often leading to combat. Apparently a direct derivative of the raised-carpus, which, however, does not necessarily precede it. It has been observed in only four species, *tangeri* (subgenus *Afruca*), *maracoani* and *ornata* (both in the subgenus *Uca*), and *pugilator* (*Celuca*). The posture is characterized by the extreme vertical position of the chela. The finger tips usually touch the ground immediately in front of the crab while the mano-carpal joint is maximally raised. Frequently, two potential opponents approach each other closely before both, almost simultaneously, assume the position. High intensity combat with linked chelipeds sometimes follows immediately.

3. *Frontal-arc*. Male. Low intensity. Similar to the raised-carpus, except that the flexed cheliped, fingers slightly to widely apart, is entirely parallel to the ground and almost touches it. Although the posture is sometimes held for moments at a time, the claw usually sweeps slowly forward and back in a small arc. It occurs frequently in a crowded feeding situation and the actor often does not interrupt the rhythm of his minor cheliped.

4. *Forward-point*. Male. Moderate intensity. Uncommon. The major cheliped, held low and level with the ground, is half unflexed, so that the wide-open chela points forward, toward an approaching opponent.

5. *Lunge*. Male. High intensity. Often follows a raised-carpus or forward-point. With the major cheliped partly flexed, the crab makes a feint toward a potential opponent, the body and cheliped being thrust suddenly in the direction of the other crab while the ambulatories shift position little or not at all.

6. *After-lunge*. This motion appears identical with the lunge and is separated from it here only because of its close association with combat, and because the name was used in an earlier contribution (Crane, 1967). The after-lunge often follows an encounter so closely that it may prove desirable in the future to term it a full component of combat.

7. *Carpus-out*. Male. Usually follows a raised-carpus. The crab, at the continued approach of a potential opponent, sometimes partly withdraws into his burrow, leaving the still-flexed cheliped protruding more or less above the ground, the mano-carpal joint still highest.

8. *Flat-claw*. Male. A partially withdrawing crab frequently leaves the entire manus and chela perfectly flat on the surface. (Fig. 84 *H*.)

9. *Chela-out*. Male. A third posture occasionally assumed in partial withdrawal occurs when only the extreme tips of the fingers remain above ground, ajar and pointing straight up.

10. *Lateral-stretch*. Male. This position is typical of burrow-holders, is almost certainly ubiquitous in the genus, and occurs with great frequency. In earlier contributions (Crane 1957, 1966) it was called the "lateral threat posture." The major cheliped is widely outstretched, either far to the side or obliquely forward. The minor cheliped may or may not be similarly extended. The stance is reserved in social behavior for threat toward a potential opponent passing either to the rear or obliquely to the side; it is sometimes accompanied by vibration or rubbing of the major merus against the side of the carapace (p. 482). This posture corresponds to the classic threat position found in many kinds of crabs and perhaps occurs in all brachyurans; Schöne (1968) gives examples of this and similar postures in a variety of forms. In the field I have observed it in the portunids *Portunus* and *Callinectes*, in many xanthids such as *Carpilius*, *Xanthodius*, and *Menippe*; in the grapsid genera *Grapsus*, *Pachygrapsus*, and *Sesarma*; in the potamonid *Pseudothelphusa*; in the gecarcinids *Gecarcinus* and *Cardisoma*; and, finally, in various ocypodids other than *Uca*. All of these crabs, as well as *Uca*, when cornered or overtaken by predators or humans, may give this response. (Fig. 84 *F*.)

11. *Creep*. Male. A crab in neither an aggressive nor a display phase often walks past displaying males with the body held rather low; occasionally, but not always, the minor side leads. The same behavior is sometimes accentuated when the crab has been physically overturned in a forceful fight, or has had his burrow taken from him. At these times the sternum barely clears the ground. (Fig. 84 *C*.)

12. *Prance*. Male. Several species have been observed walking stiffly with their dactyls turned under. Some-

times the behavior occurs in an obvious intermale threat situation; sometimes no potential opponent is in sight. It seems likely that the odd method of walking may produce sounds through stamping; the stimulating crab may well be underground.

13. *High-rise.* Male and female. Both sexes sometimes raise their bodies high on their extended ambulatories, some of which often do not touch the ground. The posture in males is often associated with waving, in both threat and courtship. In females the stance indicates unreceptiveness and is usually accompanied by ambulatory stridulation (p. 482). Her stiffened legs and increased height make it difficult or impossible for the male to grasp and turn her into the copulatory position.

14. *Legs-out.* Female. A frequent form of unreceptive behavior. The female partly descends her burrow, leaving the legs of one side projecting stiffly upright in the air or obliquely out. This action effectively blocks the attempts of the male to dislodge her or, when her burrow is large enough, to follow her below. (Fig. 89 *A.*)

D. Sound Production

I. *Introduction and Historical Review*

Reviews of mechanisms for sound production in crabs have been contributed by Guinot-Dumortier & Dumortier (1960) and by Dumortier (1963). They present a wide variety of mechanisms, almost all stridulatory in nature, in nine families of brachyurans. Sound has been actually detected in few of these species, but the specializations are so apparent that there seems to be little question but that sounds can be elicited and recorded as soon as investigators give attention to the problem. Most of the mechanisms consist of a series of close-set parallel ridges or slightly raised striae which are readily rubbed by or against a row or cluster of tubercles; the latter structure is often less specialized than the parallel ridges. Most of the examples given concern the following pairs of opposable parts of brachyuran anatomy: (1) suborbital region against merus or carpus of cheliped; (2) pterygostomian region against manus or dactyl of cheliped; (3) third maxilliped against cheliped's dactyl; (4) one segment of the cheliped against another; (5) cheliped manus against part of an ambulatory; (6) part of one ambulatory against part of another.

The production of sounds by ocypodids was first reported by Alcock (1892) and Anderson (1894) after they heard stridulation in India from burrows of two species of *Ocypode*. Suspected in *Dotilla* by Aurivillius (1893), sound production was confirmed more than sixty years later by Altevogt (1957.2). It was suspected in *Ilyoplax* and *Macrophthalmus* by Tweedie (1954). Stridulatory motions have been observed and photographed, but not tape-recorded, in both these genera by Crane (unpublished). Finally, *Ocypode gaudichaudii* has been heard producing sounds when in its burrow (Crane, 1941.2).

Dembowski (1925) first reported sounds from *Uca* when he worked on *pugilator* in the eastern United States. Rathbun (1914.2, 1918.1) described juxtaposable ridges and granules, which she presumed must be stridulatory, on the major manus and first ambulatory in *musica*.

Crane (1941.1, 1943.3, 1957) mentioned the possibilities of sound production in courtship, threat display, or both in the genus. The mechanisms suggested were stridulation and "rapping" on the ground with the major cheliped. Three other species, *inaequalis, beebei* and *terpsichores*, were shown in the 1941 contribution to have specializations similar to those of *musica*. In the same three references, rappings on the surface of the ground beside the burrow were reported as visible elements in certain species-specific courtship displays. These species were *inaequalis, cumulanta, batuenta, saltitanta,* and *pugilator*.

Meanwhile, Burkenroad (1947) reported on the use of sounds by *pugilator* in nocturnal courtship. He found that during darkness drumming or knocking underground replaced waving as a stimulus to the female. Burkenroad did not then believe that the ground was actually hit at the mouth of the burrow by the cheliped during visual display during the daylight.

In 1959 Altevogt reported work on sounds made by *tangeri*, as part of a detailed study of the ethology and ecology of that species in southern Spain. The sounds were considered to be primarily part of nocturnal courtship, and showed a correspondence in rhythm with that of diurnal waving. They were of two kinds, a long whorl and a short. When drumming was made on the surface of the ground by a crab above the burrow of another crab, it served as "an irresistible signal for the occupant to come out."

So far none of the observers had yet obtained tape recordings of sound production by any species, whether actually already heard by human ears, or suspected through morphological and behavioral clues.

Recently the subject has been actively pursued with rewarding results. Salmon and Stout (1962) published an analysis of tape recordings of *pugilator* in their burrows, and once more reported vibration against the burrow's rim. In the same year Altevogt and von Hagen published analyses of records of the populations of *tangeri* they were continuing to study in Spain.

Since those first studies, additions to the list of species shown by tape-recordings definitely to produce sounds have increased yearly, through the work of

Salmon (1965, 1967) and Salmon & Atsaides (1968.1, 1968.2) in the United States, Crane (preliminary report, 1966) in Trinidad and the Indo-Pacific, and von Hagen in Trinidad (preliminary report, 1967.2). At present, the number of species so recorded is about 20, with every subgenus represented except *Amphiuca*. Morphological and observational data give less complete evidence for other forms. There remains, in short, no question whatever but that sound plays an important role in social behavior throughout the genus.

The preceding sentence is now the only general statement on sound production in *Uca* that can be made without hedging. The numerous means by which sounds are produced, the range of variation of their acoustical characteristics, the stimuli that elicit them, and their functions are just beginning to be understood. The sensory receptors are still unknown. There is even disagreement on the medium carrying the stimuli, since all stimuli may be perceived through substrate vibrations, as seems likely to von Hagen (1962), or some may be airborne, as considered highly probable by Salmon (1965, 1967). Investigation of the subject continues (Horch & Salmon, 1969).

Rapping or drumming, as reported in contributions through 1962, was apparently done by vibrating the flexed cheliped against the ground; this took place at the burrow entrance and, it was suggested, when the crab was underground. In a preliminary report in 1966, Crane listed tape-recordings of sounds made not only by the usual drumming but also by rubbing the ambulatory meri together, by vibration of membranes at the base of the minor cheliped, and by emission of bubbles from the efferent branchial channels; examples of these were played back at the 1965 Ritualization Discussion; Pl. 47 of the current contribution is the first published record of the oscillographs. Meanwhile, Salmon (1967) reported ambulatory vibration as distinct from stridulation by Florida species of *Minuca*, and published representative oscillographs; he was then of the opinion that this behavior was subgenerically typical, while drumming of the cheliped seemed restricted to *Celuca* (*pugilator*, *speciosa*). Von Hagen's 1967 contribution, a preliminary report, lists the species he recorded in Trinidad, but does not comment on the methods of sound production or include oscillographs. Finally, Crane (1966, 1967) reported that the sounds of ritualized combats between males are sometimes audible to man; not one of them has yet been recorded.

All the investigators agree that *Uca* sounds so far recorded are of low frequency and that the tempo increases with a rise in temperature. The contributions of Altevogt and von Hagen, and of Salmon and his co-workers, agree that sound production replaces waving as a part of courtship, both under certain conditions when the crab is underground and at night. In addition, von Hagen (1962), enlarging on Altevogt's original observation, described for the first time an aggressive element in the drumming of *tangeri*. When a burrow-holder emerged at night in response to drumming on the surface by a passing male, two results might follow. If the emerging crab was a female, copulation might ensue, but if another male, aggressive behavior could follow. All investigators agree that spontaneous sounds at night are usual in several species now acoustically familiar. Altevogt (1964, 1970) showed that in (*Afruca*) *tangeri* and (*Celuca*) *inaequalis* antiphonal drumming may be elicited. This behavior resulted when he tapped a finger on the surface of the ground close to the mouths of burrows occupied by male. Crane (1966) remarked that sounds can be most easily elicited by an observer by arranging for a conspecific male to enter a burrow occupied by another male. Salmon (1967) and Salmon & Atsaides (1968.1) also report successful use of this technique.

Recently, my reexamination of morphological specializations and of motion picture films, as well as further field observations, and video-tape-recordings have all led me to believe that sound production in general functions primarily in intermale behavior and in behavior by unreceptive females, while its use in courtship, at least on the surface, is more restricted. This viewpoint will be amplified in later sections.

II. *Components of Sound Production in* Uca

The means of sound production, known and suspected, in fiddler crabs divide conveniently into a number of categories and subdivisions, containing, in all, a minimum of 16 components. Several of these components are clearly compound, but in the present state of our knowledge it is impractical to fracture them further. In the present section, each component is provided with a short name, for ease of use in later sections, and briefly characterized. Descriptions of morphological and operational characteristics are given in more detail throughout Chapter 3 on Structures and Their Functions. A relevant page number is given at the end of each component description. Table 12 gives a utilitarian—not a phylogenetic—classification of these components and records their known occurrence in the genus, along with the kind of evidence at hand for including them.

III. *Sound Production by a Single Individual*

(*a*) STRIDULATION

The term stridulation will be used in its narrow sense. This will confine it to sounds heard, tape-recorded, or suspected that result from the juxtaposition, with rubbing, of two anatomical parts of the same individual. Components in which one part is tapped or vibrated ("drummed") against another part are placed in the next category (*b*). Guinot-Dumortier & Dumortier (1960) applied to Brachyura in general a nomenclature essentially the same as that long used in Orthoptera, where each kind of apparatus is divided into a *pars stridens* and a *plectrum*. The *pars stridens* is typically a series of striae or other elevations moved against an opposable ridge or sharp edge forming the *plectrum*. In *Uca* the distinction is often unclear or even wholly inapplicable; hence these otherwise useful terms will not be used in this contribution.

1. *Major-merus-rub.* The antero-dorsal or postero-dorsal armature on the merus of the major cheliped rubs against some part of the adjacent armature of the carapace. The suborbital crenellations, antero-lateral angle, antero-lateral margin, and perhaps the pterygostomian and subbranchial regions are the most probable structures involved. The rubbing appears typically to be a back-and-forth motion, but the details have not been worked out; doubtless the direction of rubbing depends on the alignment of the particular underlying structure concerned. In a few instances, a distinctly circular direction to the rubbing motion shows clearly in film sequences. Whether rubbing by the antero-dorsal or postero-dorsal surface of the merus takes place seems usually to depend on the position of the crab toward which the threat is directed. For example, when a potential intruder is passing behind a burrow-holder, the latter often extends the major cheliped in a lateral-stretch (p. 479), and may simultaneously perform a major-merus-rub; here his postero-dorsal armature rubs against the nearest part of the carapace, which in this instance will probably be the antero-lateral margin. Sometimes this component is performed when the burrow-holder is part-way down a hole, with the flexed cheliped and adjacent anterior carapace projecting.

In a single film sequence, showing two *deichmanni* facing each other, the crabs each made a major-merus-rub, one after the other, repeating the alternation several times without pause. This is apparently the only example yet at hand of antiphonal behavior during rubbing components.

The strongest part of the antero-dorsal meral armature is always distal. This portion of the merus, however, can come into contact with the carapace only in young crabs, because of the later effect of allometric growth. Any other function of the region in mature crabs is not yet known.

The component is so far known only in threat situations. Observed frequently in the field; filmed. Pp. 452, 457.

2. *Minor-merus-rub.* Similar to the major-merus-rub in form and use. Details unknown. Filmed only. Pp. 454, 461.

3. *Minor-claw-rub.* The manus and chela, or the chela alone, of the minor cheliped rubs against the suborbital crenellations. Details unknown. Filmed in one species only (*deichmanni*), but observed fairly often in the field. P. 461.

4. *Palm-leg-rub.* Structures on the anterior side of the first ambulatory adjacent to the major cheliped rub against, or are rubbed by, structures on the lower proximal triangle of the major palm. Known only in the subgenus *Celuca*, almost entirely by inferences drawn from the morphological specializations. These are best developed in six species. A film made by David Blest of *inaequalis* forms the only record to date of the structures in use. Here, between stages of a combat between two males, the first leg of one crab reaches forward and rubs against the proximal lower palm of his own cheliped. Pp. 458, 462.

5. *Leg-wag.* Structures, principally on the meri of ambulatories, rub against one another (Fig. 85 *A*, *B*). Rarely, the first ambulatory merus also rubs against, or is rubbed by, the merus of the minor cheliped. Both sexes behave similarly, the action being extremely common throughout the genus and usually occurring close to the mouth of the stridulating crab's burrow. It forms the most widely used threat pattern of females, in which this armature is always notably stronger than in their conspecific males. Often it forms part of the rejection pattern of unreceptive females.

Salmon (1967) and Salmon & Atsaides (1968.1, 1968.2) have recorded sound produced by similar leg motions in males of several species of *Minuca* in Florida. It appeared to these investigators, however, that the vibrating legs did not come into contact with one another, the sound apparently being made by their passage through the air as they vibrated in the dorso-ventral plane. The frequencies recorded also proved lower than would be expected from stridulatory mechanisms. The vibration occurred between contacts of the legs with the ground, which produced sounds of higher frequency (see the leg-stamp component, p. 484). During diurnal courtship the vibratory motions apparently sometimes followed the curtsy component of waving display (p. 496), just after the male had attracted a female close to the burrow mouth, curtsied with leg-stamps, and descended. At these times sounds were made that were similar to those recorded from leg vibrations made on the surface at night; because of the crab's position underground, however, diurnal tape-recordings were indistinct. Nocturnal sound production occurred both spontaneously and by experimental elicitation through tactile stimuli of various kinds, as well as by scratching the surface of the ground near the burrow with a twig. The investigators also found that in *rapax* a male sometimes increased his rate of sound production when another male was artificially stimulated. Differences among the species occurred in the duration of sound, intersound intervals, and number of pulses per second.

In the films made in connection with my own observations, leg-wagging appears in more than 20 species, but always inadvertently, since its daytime occurrence is sporadic, sudden, and almost always of brief duration. It has been tape-recorded in daylight on the surface in (*Thalassuca*) *vocans* and (*Celuca*) *lactea* (Crane, 1966); both of these examples were males in threat situations. (Pl. 47). Finally, during 1969 in New Guinea (Crane, unpublished), video-tape-recordings were made of the leg-wagging of burrow-holding *lactea* when approached by aggressive wanderers; for the first time we have definite proof of the synchrony of leg-wagging motions with sound production, along with evidence of the social situation in which the behavior occurs.

In fact, according to all my observations of leg-wagging, its function in daylight, above ground, appears always to be a part of agonistic behavior; often it appears in conflict situations. Judging by the similarities of sound, it almost certainly occurs as one of the warning signals underground in the presence of an intruder. Stridulations of this kind apparently rarely if ever play a role in normal courtship during daylight. However, since similar movements of the legs are often a part of waving display, as used in both threat and courtship, it is sometimes difficult or impossible to decide, even from a well-exposed film, whether the surfaces of adjacent legs actually come into contact, or whether the motion is ritualized, as described on p. 524, into a purely visual signal. The nocturnal vibrations described by Salmon & Atsaides (*loc. cit.*) are very much faster than the slow movements of ritualized wagging. P. 463.

6. *Leg-side-rub.* The meri of the more posterior ambulatories apparently rub against the antero-lateral, dorso-lateral, or vertical lateral margins, or the postero-lateral striae, of the carapace. Filmed only, the juxtaposition showing with certainty in only one species. P. 463.

(*b*) VIBRATION AGAINST CARAPACE

7. *Major-merus-drum.* As in the major-merus-rub, noted under 1 above, except that the motion is a series of rapid blows instead of a rub. Details remain unclear, but the smooth proximo-anterior surface of the merus strikes rapidly and repeatedly against the carapace, apparently on the outer pterygostomian or subbranchial region. The sound may well be amplified by the overlying airspace. All of the species so far known to make this motion are among those equipped with the deep bodies and expanded branchial chambers adapted for prolonged exposure to air. Known only in threat situations. Easy to confuse, during field observation and casual inspection of film, with the major-manus-drum (component 9, below). Filmed and tape-recorded. Pp. 453, 457.

8. *Minor-merus-drum.* Similar to preceding component, but performed by merus of minor cheliped. So far observed only in males. Filmed. Pp. 460, 461.

(*c*) VIBRATION AGAINST SUBSTRATE

9. *Major-manus-drum.* The proximo-ventral part of the manus of the major cheliped strikes repeatedly against the substrate. This is the first recognized and most familiar of the sound-producing components. It has been called by several terms: *trommelwirbel* and *wirbel* by Altevogt and by von Hagen; *vibration* and *drumming* by Burkenroad; *rapping* by Crane and by Salmon, Salmon & Atsaides, and Salmon & Stout (all references *loc. cit.*). All the terms refer to striking the ground rapidly, usually several or more times in a series, with the lower proximal part of the major manus or sometimes with the entire lower edge of the propodus. The term *drumming* will be used in this contribution, as the most descriptive term in English.

In drumming, the major cheliped is usually strongly flexed. The only discernible morphological specialization is an occasional enlargement of the proximal, ventral part of the manus (Fig. 85 *C*). Such an enlargement is not present in all species now known to drum, and on the other hand it is strongly developed in some of those forms in which drumming has not yet been detected (Pl. 22 *C*).

Drumming is known to be used in both threat and courtship, both by day and by night, on the surface beside the burrows at night as well as partway inside. It has been variously described, filmed, and tape-recorded by the investigators concerned in the studies cited on pp. 480-485, as well as in the course of the present work. During the latter I have watched and photographed drumming on the surface in *urvillei*, *vocans*, and *lactea* in the Indo-Pacific, as well as in *tangeri* in Portugal and Angola, and *pugilator*, *speciosa*, and *cumulanta* in the western Atlantic. In addition, I now have tape-recordings which appear to be the result of underground drumming in *dussumieri*, *coarctata*, *tetragonon*, three subspecies of *vocans*, and two subspecies of *lactea* in the Indo-Pacific. M. Flinn and I, working together, have also secured recordings of underground drumming in *vocator* and *cumulanta* in Trinidad; in each species recorded, the crab was drumming so close to the mouth of the burrow on one or two occasions that the vibrating cheliped was fully visible during the recordings; in these cases, therefore, there is no question that another means of sound production might have been responsible for the result.

The major-manus-drum is easy to confuse, as remarked above, with the major-merus-drum (component 7). P. 458.

10. *Minor-chela-tap.* The tips of the pollex and dactyl of the minor cheliped strike rapidly, several times in succession, against the substrate. Seen on film only, apparently in threat situations. P. 461.

11. *Leg-stamp.* In threat situations, the dactyls of two or more ambulatories are turned under and the crab in walking raises them higher than usual and brings them forcefully down; known only in film, but in all probability sounds with a signal value are produced. Similarly, in the threat movements termed lunges, after-lunges, and prances (pp. 479-480), acoustic signaling may well be involved. Finally, in high intensity threat and courtship, when some species lower the body in the component termed the curtsy, the legs during some curtsies make a stamp similar to that described above. Salmon (1967) and Salmon & Atsaides (1968.1) secured tape-recordings of the resultant sound. They also found that some of the multiple pulses obtained from leg-wagging were preceded by single pulses marking the momentary touching of the ground by the legs otherwise raised during vibration. P. 463.

(*d*) SOUND CONNECTED WITH RESPIRATION

12. *Bubbling.* This method of sound production seems to be unquestionably under the control of the crab, and to serve as a warning signal to intruders. The sound is produced by bubbles of air, enclosed by water from the gill, that appear rhythmically from the efferent branchial orifices at the distal outer corners of the meri of the third maxillipeds. The bubbles are not nearly as profuse and extensive as those occasionally produced by ovigerous females, perhaps for cleaning and aerating the eggs, and by both sexes in a number of grapsoids, used sometimes as a cleansing agent (Schöne & Schöne, 1963) and sometimes as a regulator of body temperature (Altevogt, 1969.2).

At present bubbling in the current sense is known to occur only in *maracoani* (subgenus *Uca*). Judging by the similarities of sound produced in the burrow under similar conditions, it seems probable that it occurs also in *Minuca*. Work on this signal is in a very preliminary stage. It may prove to be the most generally distributed of all warning signals, since no specialization at all, except for the control of the bubble formation, seems to be needed. Tape-recorded in males. Not yet tested in females. P. 472. A similar mechanism has been reported by von Hagen (1968.1) in *Ocypode* and in two species of *Sesarma* (Grapsi-

dae). The resultant hissing (*zischende*) serves as a threat signal.

13. *Membrane-vibration.* A very different signal is also made partly by means of the respiratory apparatus. Here water, or air and water, from inside the epibranchial chambers is vibrated against the membrane at the base of one or both chelipeds. Striations in the membranes, clearly visible only in life, appear to be species-specific, as are the sounds produced. These have been observed and recorded outside the burrow, when the crab was held under a dissecting microscope; identical sounds in response to intruders were recorded from within the burrows. So far, this vibration of the membranes has been observed with certainty and recorded only in males of *vocator* and *rapax*, both members of the subgenus *Minuca*. P. 165.

IV. *Sound Produced by Contact between Two Crabs*

14. *Claw-rub.* Always by the major chelipeds of two males during combat. The term given here is broken down into distinct components and described in the section on combat, starting on p. 488.

15. *Claw-tap.* The remarks concerning component 14 apply.

16. *Interdigitated-leg-wag.* Two males or two females sometimes line up beside each other, with at least the more distal of their ambulatory segments in contact, and, sometimes, vibrating. Threat situations only. Observed in field but neither filmed nor tape-recorded.

V. *Characteristics of Sounds Produced by* Uca

It seems clear from the preceding section that acoustic signaling devices in *Uca* are more varied than in any invertebrate group so far investigated. Because this study is still in its active infancy, remarks on the characteristics of the sounds will be only illustrative and accordingly brief.

The sounds so far recorded on tape in studies of the genus all consist of rhythmic pulses, or groups of pulses, of low frequency. Examples of resultant oscillograms appear in Pl. 47. Detailed analytical work on sounds produced by other members of the genus has been done by Altevogt and von Hagen, using populations of (*Afruca*) *tangeri* in Spain, and by Salmon and his associates on a number of species in the United States. These investigators, in references cited below, agree that almost all the characteristics of each of the sounds vary widely. The variations appear in response both to temperature and to the social or experimental situation prevailing at the time of the recording. The figures that follow were extracted from the detailed data given in their contributions; they are presented in order to show the general nature of the characteristics involved and of their ranges of variation.

Altevogt (1962) and von Hagen (1962), when they examined the two kinds of whirls (major-manus-drums) made by *tangeri*, found the following characteristics. The short whirl consisted of 1 to 4 beats (that is, contacts of the manus with the ground), while the long whirl ranged from 7 to 12 beats. Each short whirl lasted about 0.5 second, while the long whirl extended over a period of 0.84 to about 2 seconds. The intervals between short whirls ranged from 0.8 to 1.4 second with the intervals between single beats ranging only from 0.8 to 0.10 second—a notably constant character. The intervals between short whirls were found to be equal to those between single waves during daylight under comparable conditions. In long whirls, single beats ranged from 0.12 to 0.16 second; long whirls were not repeated at regular intervals.

Salmon & Stout (1962) and Salmon (1965) found that in the corresponding component in (*Celuca*) *pugilator* the rapping (major-manus-drums) consisted of only one type, not a long and a short whirl. The number of pulses (beats) in each sound (rapping; drumming) ranged from 2 to 11, intervals between sounds from about 0.3 to more than 5 seconds, and the rate of production from 3 to 32 sounds per minute.

Recently (1971), Salmon showed that these manus-drum sounds made by *pugilator* contained maximal energies between 600 and 2,400 Hz. In contrast, sounds of "honking" made by uncertain means (present contribution, below and p. 195), which he recorded from (*Minuca*) *rapax*, contained maxima between 300 and 600 Hz. He also reported that *rapax* proved significantly more sensitive to vibrations between 480 and 1,000 Hz than was *pugilator*. Additional related contributions are those of Salmon (1967), Salmon & Atsaides (1968.1, 1968.2, 1969), and Horch & Salmon (1969).

To human ears, the sounds produced by *Uca* fall very roughly into four categories: first, those of a rasping nature, probably resulting from conventional types of stridulation where one surface of a crab is rubbed against another; second, sounds resulting from the various types of vibration; third, the rhythmic, hollow noises that appear to originate from mechanisms based on the respiratory system; and, finally, distinct series, always from underground, of unknown origin. The most characteristic of these unknown noises, called *honking* in the description of *rapax* behavior (p. 195), is not included in the table; the mechanism may well prove to be one already listed.

To me, it seems highly probable that male *Uca* of at least most species can make at least three different signals, using wholly different mechanisms and excluding any acoustic signals resulting from combat. In some *Celuca*, the number is doubtless higher. The repertory of *deichmanni*, for example, includes five apparent acoustic signaling motions, all of which appear on film, although they have not been tape-recorded. These are the major-merus-rub, minor-claw-rub, leg-wag, major-manus-drum, and minor-chela-tap (components 1, 3, 5, 9, and 10). It seems likely that one or more components, perhaps including mechanisms not yet suspected, are usually restricted to sound production underground.

Although some of our tape-recordings of *cumulanta* in Trinidad resulted from the introduction of a conspecific male as stimulus, others reproduced a resident male's drumming at the burrow mouth during courtship. As in the cited work of Altevogt, von Hagen, Salmon, and co-workers concerning other species, these surface drummings were closely similar to subsurface sounds elicited by male intruders. In the case of *tangeri*, it will be recalled, identical drummings on the surface by a passing male could elicit either aggressive or courtship behavior depending on the sex of the emerging burrow-holder, according to von Hagen (1962). In a film of the same species I made in Angola, an aggressive wanderer induced a burrow-holding male to emerge and give up his burrow merely by drumming on the surface; no combat ensued; instead the dispossessed crab promptly departed in the creep position (p. 121).

These examples indicate that social sound production, as with other branches of *Uca* behavior, is undoubtedly ambivalent.

E. Components of Combat and Their Organization

I. *Introduction*

Preliminary Note. Much of this section is taken directly from an earlier paper on the subject (Crane, 1967). An exception, aside from minor alterations, is the description of additional components of combat, resulting from field work done between 1968 and 1970. Further laboratory work has also led to a better understanding of the relations between individual structures on the claw and their uses in combat. This aspect has already been reviewed in the section on the morphology of the major cheliped, beginning on p. 456. Tables 13 and 14 show the distributions of structures and components as they are known today. The locations of the structures are shown in Figs. 42-44. A contribution on results of the recent field work is in preparation and will be published separately.

Male fiddler crabs sometimes seize each other's large claws at the climax of a fight. Physical damage practically never occurs, although the stronger sometimes flings the weaker inches away or flips him altogether upside down. Almost all combats, however, stop short of a violent finish.

The work described below shows that most combats are so fully ritualized that the observer can detect no element of force. In these encounters the end activity is the rubbing or tapping of a correlated structure by a different part of the opponent's claw. Morphological specializations include ridges, tubercles, and other structures.

At times the effective meeting of parts seems to be achieved through cooperative movements of the less active crab. In the most elaborate combats the components are performed by each individual in turn, while his partner holds still.

Preliminary observations on ritualized combat in *Uca* were made in the Indo-Pacific (Crane, 1966). In some species the claws, partly engaged, vibrated back and forth with a clicking sound audible to the observer. Detailed descriptions were not secured in the field, although motion pictures recorded the pattern. In other forms, pits and tubercles apparently served as deterrents to forceful linkage of the chelipeds. These observations, made incidentally during a study of waving display, all showed the need for a concentrated study of combat.

Such field work was accordingly undertaken in Trinidad during 1966. A socially advanced neotropical species, *rapax*, was selected as the principal subject; comparative observations on other species have begun. The results give some basic information on the occurrence and organization of combat. In this section are also included data on the results of a group of individual combats. Discussions of the functions and origins of combat are reserved, as for other social components described in this chapter, for later consideration. Since then field work on combat has continued in the northeastern United States, in New Guinea, and in Ceylon.

II. *Historical Review*

The fighting proclivities of male fiddler crabs have long been familiar to naturalists strolling on suitable shores. Compared with other conspicuous fiddler activities, however, combats are so uncommon, short, fast, and superficially similar that it is not entirely surprising that their patterned complexities have been overlooked. The infrequent reports were only roughly descriptive and wholly unanalytical. Examples include Pearse (1912.2); Dembowski (1925); Verwey (1930); Crane (1941.1, 1958, 1966).

Intermale combat in other brachyurans occasionally has been reported. Observations have been made on ocypodids, in addition to *Uca*, as follows: in *Dotilla* by Tweedie (1950.3) and Altevogt (1957.2); in *Heloecius* by Tweedie (1954) and Griffin (1968); in *Ocypode quadrata* by Schöne (1968). I have seen combats, but not hitherto reported them, in *Dotilla* and *Heloecius*, as well as in *Scopimera*, *Macrophthalmus*, and *Ilyplax*. Schöne's contribution consists of an illustrated survey of agonistic and sexual display in a wide variety of brachyuran crabs, both aquatic and semi-terrestrial; his list of references is comprehensive. His application of the adjectives "wild" and "irregular" to fighting without apparently formal patterns corresponds to part of the usage of "forceful fighting" in the present contribution; he reports its occurrence in one or more species in the families Cancridae, Portunidae, Xanthidae, Majidae, Parthenopidae, and Grapsidae, as well as in Ocypodidae.

He reports mutual pushing with the chelipeds in a more formal pattern in *Ocypode quadrata*, a member of the genus most closely related to *Uca*. After an account of its threat display Schöne continues, "The most common type of fight is a formalized interaction. The threatening crabs raise themselves higher, lift the first walking legs, close up, touch the fronts of the chelae, and push. After the push, which can be very short, they part, and one or both walk away. Wild attacks have rarely been observed."

Schöne observed similar behavior in two spider crabs, *Maja verrucosa* and *Euryneme aspera*. Of the first species Schöne writes as follows:

In the irregular fight of *Maja verrucosa*, the opponents thrust and strike with chelae and first walking legs. Often one seizes the other crab's legs and twists. This sometimes leads to loss of a leg in the smaller animal. A formalized type of fight has been observed between males, especially when the male defends the female during the mating period. The male turns against the approaching crab, raises the chelipeds to low-intensity threat, and sometimes strikes. Both crabs close to touching distance and extend the chelipeds to the sides. Each one tries to adjust the chelipeds so that they touch those of the opponent over the whole length. This may last for several seconds. Often the pincers hold each other. Then the crabs press and shift the chelipeds forwards-upwards. The fight continues in the irregular manner as described above.

Additionally, Griffin (1968) has described the combat of the grapsoid *Hemiplax latifrons* and Warner (1970) of the grapsids *Aratus pisoni* and *Goniopsis cruentata*.

None of these examples of relatively regular pushing procedures, although they are formal in comparison with irregular and wild fighting, attain either the ceremonial character or the complexity of high-intensity "ritualized combat" in the sense used in the following pages.

Although the study of acoustical behavior in *Uca* is now progressing (ref.: pp. 480ff), the sounds of combat have yet to be recorded; occasionally they are audible to man. A review by Guinot-Dumortier & Dumortier (1960) on the morphology of stridulation in crabs gives no examples where one crab is presumed to stridulate against another part of another individual.

References to the concept of ritualization in animals are given in the next chapter (p. 519). Ritualized combat is specifically discussed by Lorenz (1964, 1966.1, 1966.2).

Temporary physiological conditions termed phases (Crane, 1958), important in combat, are described on pages 505 and 687.

III. *Organization of Combat in* Uca

(a) RELEVANT CHARACTERISTICS OF OPPONENTS IN *U. rapax*

The pattern of a combat between *rapax* males depends largely on three factors: the phase of each individual; the relative size of the two crabs; and the location—whether on the right or left side—of the large claw of each.

Phase. Males engage in combat only when they are in either the aggressive wandering phase or the waving display phase. In the latter condition they are always burrow-holders. Combats take place only between an aggressive wanderer and a burrow-holder or between two burrow-holders; encounters between two aggressive wanderers are unknown. When a wanderer elicits combat from a burrow-holder irregular components are frequent and a ritualized encounter sometimes ends as a forceful fight. In contrast, combats between neighboring males are usually composed of highly regular, ritualized components and practically never end in force. These categories and their results will be described beginning on p. 492.

Relative Size. The larger crab in all combats usually has the advantage. It is interesting, therefore, that, in the majority of combats including an aggressive wanderer, the latter instigates combat with a burrow-holder larger than he.

Claw of Same or Opposite Side Enlarged. In *rapax* as usual in *Uca* (p. 451) right- and left-clawed individuals occur in equal numbers. In the total count given in Table 15, no tendency is shown in combat for instigators to approach opponents having the same claw enlarged as their own rather than the claw on the opposite side. Therefore, homoclawed and heteroclawed combats are about equally likely to take place. Table 16 shows clearly that the frequency of several of the components described below is affected by this characteristic.

(b) COMPONENTS OF COMBAT

In the earlier account of combat five separate ritualized components were recorded, only one of which, the heel-and-hollow, was not observed in Trinidad *rapax*. This number has now been increased to 15, through field observations on *pugnax* (subgenus *Minuca*) and, in the subgenus *Celuca*, *pugilator* and *lactea*.

In fiddler crabs many combats can be clearly divided between those that are forceful and those in which pushing and gripping are absent. Other combats obviously include both forceful and non-forceful components. In still others, particularly when the outer surfaces of the two mani are in contact, it is often impossible to determine whether or not force is involved.

The components first described below will be the basic, forceful elements of unritualized fights. Their distribution appears to be genus-wide. In addition to their use of obviously strong pressures of various kinds, they are distinguished by variations in the form of the shoves and seizures as well as by the unpredictability of their sequences. They are not included in Table 14.

The second group includes the 15 types of action now considered to be distinct, non-forceful components of ritualized combat. Since the earlier paper another component, there termed "taps," has been included as a part of each of the several other components in which it sometimes forms a climax. Separate paragraphs will, in addition, be devoted to it at the end of the descriptions of the regular components.

In any ritualized combat the components are very rarely or never performed simultaneously by both crabs. The temporarily active crab is here termed the "actor" and his inactive opponent the "inactor." Parts of the claw of the actor serve as the "instrument" for the performance of each component, while correlated parts of the inactor's claw form the "friction area." As shown in Table 14, each component involves the direct use of from three to twelve or more structures, almost always different, by both actor and inactor.

(c) FORCEFUL COMPONENTS

1. *Manus-push.* The chelipeds are held flexed, the chelae partly open through slight lifting of the dactyls. Meanwhile the lower, smoother halves of the mani are pushed against each other. It is often impossible to decide whether or not a manus-push includes a rub component or vice versa. (Pl. 45 *A*.)

2. *Grips, Flings, and Upsets.* The occasional forceful end of a combat is composed of irregular elements that appear to be largely or wholly unritualized. They may be grouped under grips, flings, and upsets—descriptive terms for actions that merge into one another. (Pl. 45 *B*.)

A grip occasionally follows an unsuccessful attempt by an aggressive wanderer to get into the heel-and-ridge position; the fingers then slip beyond the normal position and firmly seize the base of the manus with one finger hooked into the carpal cavity; even the carpus itself may be grasped. Sometimes grips occur when two crabs are grossly mismatched in size; then the larger may seize the entire manus of the smaller crosswise between his fingers. More often the forceful component consists only of a longitudinal grip, with perhaps an undetected push, the actor then opening his fingers. Almost always the crabs then separate.

The term "fling" here includes those actions, always starting with a grip, that result in a skid or partial upset of the opponent. The momentum of the actor's pushing grip carries the released crab sliding backward, or he is thrust off balance with some of his ambulatories off the ground. "Upset" is confined to actions resulting in the complete overturn of the opponent onto his carapace, with all of his ambulatories in the air.

Both the occurrence and the progress of these forceful endings have so far proved to be unpredictable. The following figures, gathered in the study on *rapax*, are relevant. In 180 combats in which the endings were adequately observed, 15 (8.3 percent) ended forcefully. In 14 fights the opponents were an aggressive wanderer and a burrow-holder; the wanderer was larger than the burrow-holder in 7 combats, smaller in 6, while his identity was uncertain in 1; homoclawed and heteroclawed combats were equally divided. In 6 fights the forceful component consisted of a grip only, 5 others ended in flings, and only 3 in total upsets. The burrow-holder was the actor both during the preceding ritualized component and in the grip and subsequent action in 7 combats, the aggressive wander in 2, 1 of them resulting in the eviction of the burrow-holder; in 2 fights both ritualized and force components were mutual, 1 of them resulting in the second observed eviction of a burrow-holder; in 1 fight the action consisted wholly in the eviction of an aggressive wanderer that had slipped in while the burrow-holder was engaged with a neighbor. The ritualized components immediately preceding forceful endings were interlaces in 7 fights and well-developed heel-and-ridges with taps in 3; antecedents to the grip in the remaining examples were irregular or improperly observed. Finally, 5 fights were followed by reduced aggressiveness of the wan-

derer and 2 by the dispossession of a burrow-holder by a wanderer.

The single forceful ending to a combat between burrow-holders was very short; the slightly smaller instigator lightly seized the opponent's entire palm, then released it and went home; both crabs promptly resumed waving.

In brief, about 1 in 12 combats had a forceful ending, usually consisting only of a brief grasp by one crab of the other's manus; 14 out of 15 fights were between an aggressive wanderer and a burrow-holder; in most the burrow-holder was the active crab at the end of the fight, seizing the wanderer and administering the final push, fling, or upset. Forceful components preceded reductions in the wanderer's aggressiveness in 5 combats and preceded the taking over of his opponent's burrow in 2. These changes in behavior will be further discussed beginning on p. 493 and are included in Table 18.

(*d*) RITUALIZED COMPONENTS

In the first two components below, only the outer surfaces of the claw come into contact; they are viewed as constituting the low-intensity components of combat behavior. In the remaining components, a part of each claw enters the gape of the opponent and they are all considered to be of high intensity.

1. *Manus-rub* (Fig. 39). The outer sides of the opposing mani are rubbed back and forth, longitudinally, against each other. In most species the lower halves are nearly smooth and it is there that the surfaces usually come into contact. It is possible that the rubbing over the smooth surface facilitates the attainment of high-intensity components. At these times the manus-rub continues past the bases of the gapes until the chelae are in contact externally and are free to proceed to one or more of the following components. Tapping very rarely follows the rubbing; when it does so it is as distinct as in the components where it often occurs.

2. *Pollex-rub.* An extension of the manus-rub, in which the outer sides of the pollices rub longitudinally. On the outer propodus *U. pugilator* has a low keel with an adjacent groove beside its upper edge. Starting on the lower distal part of the manus, the keel and groove continue out along the lower side of the pollex, well above the ventral margin. During both the manus-rub and the pollex-rub *pugilator* tilts the lower half of the propodus outward. In this way, the smoothest parts of the opposing mani and the full lengths of the keels are clearly concerned in the rubbing.

3. *Pollex-under-and-over-slide.* The actor rubs the prehensile edge of his pollex along the ventral mar-

gin of his opponent's pollex, from near its distal end to a point ranging from near the pollex base to slightly beyond it, on the lower edge of the manus. The actor then momentarily disengages. Without a pause he promptly slides the ventral edge of his own pollex along the prehensile edge of that of his opponent.

4. *Subdactyl-and-subpollex-slide.* The actor slides the dorsal edge of his dactyl along the prehensile edge of his opponent's dactyl while, simultaneously, he also slides the prehensile edge of his pollex along the ventral margin of the opponent's pollex. Single points on the proximal half of the actor's dactyl and pollex form the instruments, but the slide may be along the full length of the opponent's chela.

5. *Pollex-base-rub.* (In each of the several examples seen of this component a burrow-holder was partly down his burrow, with his flexed major cheliped still visible, the mano-carpal joint highest, but the pollex and dactyl still exposed, their tips touching the ground. The encounters were all heteroclawed.) The actor approaches the burrow-holder, the cheliped being slightly flexed and the upper part of the manus and the dactyl being tilted toward the inside. In this position the actor's pollex is at a convenient height to reach through the burrow-holder's gape, from its inner side. The pollex then, with the tubercles of the middle region of its prehensile edge, rubs vertically up and down. The contact area is the flat area at the base of the burrow-holder's pollex. The motion in the examples seen was repeated in several short series of two or three up-and-down motions each.

6. *Dactyl-slide* (Fig. 86 *A*). With the chelae of both crabs partly open, the dactyl of one moves on top of that of the opponent at about the middle of its length, more or less at right angles, while the pollex passes within the gape. The approach may be from either the inner or the outer side of the claw. Both chelae are by then widely opened and the pollex does not touch the opponent's fingers. Gentle maneuvering for this position may continue for several seconds. Once it is achieved, the rounded teeth of the actor's dactyl slide longitudinally back and forth along the middle portion of the upper edge of the opponent's dactyl. No force seems to be used by either crab and no attempt is made to use the claws as pincers, all four tips being at all times free in the air. Except for the sliding motion both crabs remain almost motionless. An infrequent climax is the vibratory tapping of the uppermost dactyl against the one held quietly beneath it. In *rapax*, at least, the vibration is performed only by the crab with the dactyl on top. Afterward the two crabs break suddenly apart. When tapping does not occur, the encounter either breaks off after at most several seconds of slide or, infrequently,

passes into component 9 or 12 below. Dactyl slides occur more often in heteroclawed than in homoclawed combat.

7. *Upper-and-lower-manus-rub.* In a commonly occurring forceful grip, the actor seizes the opponent's entire manus between the prehensile edges of his dactyl and pollex. In this fully ritualized version, designated component 7, the initial position is identical, but no pressure whatever appears to be used. Instead, the actor rubs the prehensile edges of his claw longitudinally back and forth at least along the dorsal and ventral margins of the manus and sometimes also continuing distally onto the corresponding margins of dactyl and pollex. The timing is relatively slow and tapping has not been seen. In the earlier account of combat (Crane, 1967), this ritualized component was not differentiated from its forceful version, which was included without special comment among the grips, flings, and upsets. Since I became aware of this highly ritualized form through observaitons of *pugnax*, *pugilator*, and *lactea*, a review of the early notes and films of *rapax* has shown that it occurs in that species too. At the time I considered it only one more example of indeterminate, irregular behavior by aggressive wanderers.

8. *Dactyl-submanus-slide.* The burrow-holder is partway down his burrow, with the flexed cheliped obliquely raised, about as in component 5, the ventral edge of the manus being clear of the ground. Only three encounters are known, all homoclawed. The opponent-to-be, the actor, approaches with his cheliped flexed and the mano-carpal joint similarly raised, as in threat component 1. The actor then gently slides the prehensile edge of his dactyl along the ventral margin of the manus of the burrow-holder, the longitudinal rub being repeated once or twice. The actor then stops. The burrow-holder promptly emerges and, reversing roles with his opponent, repeats the motions. The former actor has meanwhile tilted his manus so that its lower edge is more accessible to the dactyl of the burrow-holder. With both crabs now on the surface, the necessary postures appeared awkward for both, but they did not hesitate in any of the observed encounters. In each the burrow-holder made a number of rather slow slides. Then the original actor went away. The terms "instigator" and "aggressive wanderer" are not used, since in the northern species in which the component was observed the social situations were not clearly evident.

9. *Interlace* (Figs. 86 *D, E*; Pl. 45 *E, F*). In this component the fingers of each claw overlap the opponent's manus, so that the bases of the gapes are almost or wholly in contact. The sequence is charac-

teristic of heteroclawed combats although not confined to them. Typically the position is assumed by the crab that has its dactyl against the inner side of the opponent's manus rather than against the outer side, which is the position for a heel-and-ridge (component 12). In a fully developed interlace the chela of each crab is wide open through high elevation of the dactyl; the tips of both pollex and dactyl are wholly free from contact with any part of the opponent. In this position the most proximal teeth of the pollex come into contact with one or both of the tuberculated ridges paralleling the base of the opponent's dactyl and rub up and down along their course. At highest intensity the rub follows along the longer, less variable subdistal ridge that continues from the dactyl base down around the base of the gape and out along the proximal, upper, inner portion of the pollex.

The climax usually consists of frication by the pollex teeth as just described, whereupon the encounter ends. This component may culminate in serial taps similar to those following a normal heel-and-ridge component but made, instead, by the basal gape teeth against the subdistal ridge of the manus in the interlace position. While a tapping finale is the normal end of a fully developed heel-and-ridge component, it is uncommon after an interlace and, as when it follows a dactyl slide, cannot be regarded as typical (p. 491).

An interlace is usually preceded by a manus-rub and less often by a dactyl-slide. It also occurs in mutual heteroclawed encounters when one opponent, with the dactyl against the inner side of the other's manus, performs the interlace. Usually it is the temporarily more active crab, in passing from a low-intensity manus-rub to high-intensity, that assumes the heel-and-ridge position, with the dactyl outside the manus; hence the second crab arrives automatically in a position appropriate for the interlace.

10. *Pregape-rub.* This component occurs more often in homoclawed than in heteroclawed encounters, as does the heel-and-ridge (component 12). The positions are similar in both, the chelae being partly linked with the dactyl of the actor in contact with the outer manus of his opponent, while his own pollex lies against the opponent's palm. In the pregape-rub, however, both friction areas are near the base of the gape, while the movement is longitudinal, not vertical or oblique. The tips of the chela seem chiefly to be used in the rub, which has been observed clearly only three times, in *pugilator* only. Its unritualized counterpart is a grip, with the point of seizure in the same position, that occurs frequently in the genus.

11. *Heel-and-hollow.* This component is known only in Indo-Pacific species. The degree of ritualization it attains apparently varies with the subgenus. Force is sometimes clearly involved, yet at least in *lactea* (subgenus *Celuca*) highly ritualized tapping sometimes forms the entire component. In the component, the tip of the pollex is inserted in or does not pass beyond the hollow near the pollex base. The dactyl holds a position similar to that in the pregape rub or as in *lactea*, the heel-and-ridge (components 10 and 12), since it lies against the outer manus.

12. *Heel-and-ridge* (Figs. 86 *B*, *C*; Pl. 45 *C*, *D*). In *rapax* and many other species the dactyl, longer than the pollex, curves downward beyond it. This characteristic proves to be of definite use in heel-and-ridging, where the dactyl arches around the curving heel outside the manus of the opponent. On its way toward the heel, the dactyl tip appears to feel its way, using as a guide the outer crease along the base of the dorsal marginal ridge. Afterward, upon reaching the proximal part of the manus, the dactyl tip does not touch the heel except during the climax to be described. Meanwhile the pollex, shorter than the dactyl and virtually straight, comes into contact with the oblique tuberculated ridge of the inner manus. The pollex teeth, in the examples where an adequate view or film was secured, rub up and down along the oblique ridge. At the climax, however, the actor taps the ridge rapidly three or four times with his pollex; on opposite strokes, when the pollex is away from the ridge, the teeth near the dactyl tip come into contact with the manus heel. At highest intensity the tapping is faster and of smaller amplitude; the effect is vibratory (p. 491). It is always the actor of the moment—not necessarily by now the original instigator—that performs the tapping. No attempt at seizing and gripping has ever been detected when the claws are in the heel-and-ridge position. In *rapax* a heel-and-ridge may be preceded by a manus-rub or infrequently by a dactyl-slide. Occasionally in mutual, heteroclawed combat a heel-and-ridge is followed by or alternates with an interlace.

13. *Supraheel-rub.* The actor places his dactyl around the outside of the manus, the pollex against the palm, about as in the heel-and-ridge (component 12). The position of the dactyl is however higher, against the upper, proximal part of the other's manus. Here the distal part of the actor's dactyl makes rapid, up-and-down rubs, different in direction as well as position from those occurring in a heel-and-ridge. In each of the few examples seen the position and motion of the pollex was invisible. It is possible that the positions assumed by the dactyl on the outer manus in the pregape-rub and supraheel-rub (components 10 and 13) are associated with the vestigial development of the oblique ridge on the palm in *pugilator*, along with its characteristic tuberculation, which may well be involved in the positions of the dactyl as well as of the more directly concerned pollex.

14. *Dactyl-along-pollex-groove.* The actor places the distal part of the prehensile edge of his dactyl against the groove in the proximal part of the inactor's outer pollex, and rubs back and forth along the groove; sometimes the rubbing continues into the hollow at the distal end of the manus. This component, only rarely observed, occurs in both homoclawed and heteroclawed combat. The position is obviously more easily attained in heteroclawed combats; when homoclawed opponents perform it, the actor tilts his claw appropriately.

15. *Subdactyl-and-suprapollex-saw.* With the median and distal part of his dactyl's prehensile edge, the actor rubs at right angles across the distal half of the prehensile edge of the inactor's pollex, between the two large projections that often occur on the pollex in *vocans*, the only species yet known to perform this component.

(*e*) TAPPING

In threat and courtship behavior a number of species make drumming motions of chelipeds or legs against the ground, their own carapaces, or both. These motions all appear to form separate components. In combat, on the other hand, a similar motion now and then climaxes five different components, and is always preceded by one of them, the tapping occurring without a change of position immediately after the characteristic form of rubbing. It seems appropriate, therefore, not to give tapping the rank of a component. The five components it is known to follow are the manus-rub, dactyl-slide, interlace, heel-and-hollow, and heel-and-ridge (numbers 1, 6, 9, 11 and 12).

An exceptional form of tapping occurs in recently studied populations of *lactea*. Here, tapping of the outer surfaces of the mani does not necessarily follow closely on any component, and often in fact inaugurates the encounter, or forms the only part of it. It is, in fact, elaborated and intensified to such a degree that it must be viewed as a separate component. In a subsequent contribution, now in preparation, it will be termed *clacking*.

The following details of the more usual form of tapping were all assembled during the study of *rapax* in Trinidad. In this species tapping occurs most frequently at the end of a heel-and-ridge component. Sometimes it is preceded by an interlace or dactyl-slide; rarely by a manus-rub. Always it consists of the rapid tapping of the dactyl or, in most heel-and-ridge sequences, of the dactyl and pollex alternately, against a particular part of the opponent's claw. Rather slow tapping of wide amplitude is performed both by aggressive wanderers and by burrow-holders. Rapid taps of narrow amplitude are performed only by a burrow-holder, usually at the end of an encoun-

ter beside his own hole. The several examples recorded on motion picture film appear as blurs on frames exposed at 1/48 second. After tapping, a combat often breaks off abruptly.

(*f*) ACTIVITIES ASSOCIATED WITH COMBAT

1. *Withdrawals.* Certain associated activities of combat may be termed withdrawals. In all, a burrow-holder descends partly or wholly underground in the presence of an opponent, whether potential or actual. He may either refuse to respond at all to the initial threat of an approaching male, or he may withdraw in the midst of a combat, even when he has the advantage of size. The reasons for these withdrawals of larger males are worthy of a separate investigation. Partial withdrawals leaving only part of the chela itself on the surface are exceedingly common throughout the genus, and almost always occur before the trespassing crab comes within reach of the burrow-holder. The various degrees of projection of the chela have already been included among the groups of threat components (p. 479). An aggressive wanderer may or may not make prying motions at the claw with his own chela (Fig. 84 *D*) or ambulatories, or may even stamp on a flat claw before passing on. The flat-clawed partial withdrawals appear very rarely if at all to occur in encounters between neighboring burrow-holders, and seem most often to be prevalent in crowded populations with plentiful aggressive wanderers. Counts and distance measurements have still to be begun. Withdrawal wholly underground almost always occurs when the crab's opponent, whether an aggressive wanderer or a trespassing male from another burrow, is larger than he. Very rarely an opponent tries to dig out the burrow-holder that has withdrawn from an actual combat.

A second class of withdrawals exists which perhaps should be differentiated and termed false-withdrawals, although they sometimes are followed by total withdrawal. A false-withdrawal is always partial and only apparent, since the burrow-holder clearly uses the upper part of his burrow as a firm foothold for the prosecution of a combat. Three recently identified components, the pollex base-rub, dactyl-submanus-slide, and supraheel-rub (numbers 5, 8, and 13) in the examples seen have almost always at least begun when a burrow-holder has partly descended, leaving most of the flexed cheliped thrusting vertically into the air. In addition, I have seen these half-withdrawn burrow-holders engage most effectively in all classes of combat, both forceful and ritualized.

2. *Down-pushes.* This forceful form of behavior has been seen in a number of species, but was observed in detail only four times, all ending combats in *rapax*. In each of these a crab actively pushed his opponent

down the latter's own burrow. Three of the combats were between burrow-holders, the actor using the low-intensity manus-push. In one of these the instigator was the smaller crab; the larger pushed him back to his own burrow, alternating manus-pushes with manus-rubs. The fourth down-push marked the end of a meeting between an aggressive wanderer and a burrow-holder; the down-push was the only noteworthy part of the brief combat, which included a manus-rub and an irregular interlace. In this fight the down-push was delivered by a final grip.

3. *After-lunges.* The after-lunge is included among the threat components (p. 479) since it appears to be identical with the usual lunge of pure threat behavior. It often follows combat, apparently throughout the genus; in *rapax* after-lunges were counted in almost half the combats in which its presence or absence was noted. Sometimes it follows a combat so closely that perhaps it should then be considered a component. Such distinctions between after-lunges, however, would be too blurred to be practical.

4. *Removal of Burrow Intruders.* This behavior has frequently been seen in a number of species. It usually occurs when a smaller crab slips unnoticed down the hole while the burrow-holder is engaged in combat. The returning crab then digs up the intruder, or, when the latter is small enough, thrusts his major cheliped down the hole and flips the little one out. The motions used are also found under certain conditions among crabs not in display phase but seeking burrows for shelter from the tide (p. 511).

(g) CATEGORIES

This section is used practically unchanged from the previous contribution on combat (*loc. cit.*, pp. 62-63). Although the figures included are all derived from the detailed study of *U. rapax* in Trinidad, the general statements appear to be widely applicable to the organization of combats in a number of other species.

The preceding accounts of behavior components have mentioned two general classes of combats, those between an aggressive wanderer and a burrow-holder and those between two burrow-holders. The characteristics of each will now be considered as a basis for a review of combat results. Mutual combat occurs in both categories but will be examined more closely under subheading 3 below.

1. *Combats between an Aggressive Wanderer and a Burrow-Holder.* These combats are notable for their irregularity, for the prevalence of forceful grips, and for the frequent withdrawal of a vigorous burrow-holder from an encounter. In first determining the combat repertory of any species of *Uca*, in fact, it is misleading to concentrate on combats involving aggressive wanderers. Such descriptions might easily be as inaccurate as reports on the nest-building behavior of birds drawn from observations of individuals that have not fully reached breeding condition.

Unfortunately for the observer, the combat activities of fiddler crabs in this phase are often more numerous and certainly more easily foreseen than are encounters between burrow-holders. It is temptingly convenient to select for attention an active aggressive wanderer and watch him on his progress through the population. Such a course is usually conveniently marked by the threats of burrow-holders on his route. If his aggressive phase is well established, a number of combats may be thus observed. Once the repertory of a species is partially known, of course, the combats instigated by these wanderers form a rich source of information.

The combats may be composed of one or most of the components previously described. Often, however, the high-intensity components—dactyl-slides, heel-and-ridges, and interlaces—are imperfectly attained or the motions are atypical; tapping, if any, is of wide amplitude.

For example, several aggressive wanderers, after an uneventful manus-rub, inserted both pollex and dactyl through the gape of the burrow-holder's chela, instead of inserting only the pollex. No regular combat development by the wanderer seemed possible from this position. One such contest ended with the burrow-holder's opening his chela wide and freeing himself. In another the crabs broke apart after awkward shaking and shoving. After each of these encounters the wanderer departed.

Similarly, an apparently clumsy attempt is sometimes made by the wanderer to attain a dorsal-slide position. This the opponent thwarts by raising his own dactyl high. The maneuver does not happen in fully ritualized fights between burrow-holders, when the opponent usually appears wholly unresisting and—one is tempted to say—cooperative. When a slide position is obtained, the wanderer sometimes saws back and forth transversely across a single spot on the opponent's upper dactyl rather than in the normal longitudinal direction.

The behavior of a burrow-holder approached by an aggressive wanderer is often atypical of his normally aggressive display phase; this is true even when, as is usually the case, the wanderer is the smaller crab. As described under Withdrawals (p. 491), the burrow-holder sometimes goes partly or completely underground either when approached or later in the course of the encounter. When the wanderer has passed, the burrow-holder emerges promptly and resumes display.

In a series of 180 combats in *rapax*, all except one of the 15 forceful endings took place at the close of a burrow-holder's fight with an aggressive wanderer.

2. *Combats between Two Burrow-Holders.* These encounters usually occur at the mouth of the burrow of one of the participants. In contrast to the usual procedure in combats started by an aggressive wanderer, the instigator in this category is usually the larger crab. A minority of encounters between burrow-holders take place on or close to the boundary between two territories.

The same two neighboring crabs occasionally proceed through highly ritualized combats a number of times during a single low tide. Sometimes a burrow-holder seems to be attracted to his neighbor's vicinity by a combat between the neighbor and another crab, usually an aggressive wanderer. Irregularities in the performance of the components are rare, combats brief, and forceful endings practically absent.

3. *Mutual Combats.* About one-third of all combats in the *rapax* series were strongly mutual in the sense that each crab performs at least one of the ritualized components. In the 154 combats analyzed in Tables 15 and 16, mutual elements were detected in 38 percent of the combats between an aggressive wanderer and a burrow-holder, and in 31 percent of those between two burrow-holders. In 87 percent of the mutual combats including an aggressive wanderer, the latter was smaller than the burrow-holder he approached; this figure is higher than the proportion (67 percent) of smaller wanderers in the entire combat sample.

In a few additional combats, which are included in Tables 15 and 18, an aggressive wanderer's only activity appeared to be the initial approach to the burrow-holder. After that the burrow-holder seemed to be the sole actor and the wanderer eventually departed. Similar cases were observed twice in combats between burrow-holders. No doubt many more such examples were seen than are cited in the tables; they are not included since the instigator's inactivity, whether he was a wanderer or a neighbor, could be only suspected because the angle of observation was unsatisfactory.

The heart of the problem of instigator inactivity lies in the observation of low-intensity components, the manus-rubs. The actor or actors are especially difficult to detect, even in motion picture close-ups, because manus-rubbing is the only component in which mutual action can be performed simultaneously by both crabs; this of course is because the necessary juxtaposition of parts—outer manus to outer manus—is much less precise than in the subsequent components of high-intensity combat. Even in the analyses of highly mutual combats, the counts for mutual manus-rubs are doubtless low; when uncertain about the mutuality of a rub, I counted it as absent.

In ritualized components of high intensity—dactyl-slides, heel-and-ridges, and interlaces—the actors either perform the slide in turn or exchange roles, the former actor holding still while the second crab performs the next component. Any necessary shift in position is made, including those demanded by a different size of claw. Sometimes the temporarily inactive crab is not only quiescent during and after the shift but even appears to take a cooperative part in it. For example, after a slide a retiring actor brought his dactyl below that of his opponent, into the latter's gape, and then stopped moving. This left the other crab in the actor's position for the slide that promptly followed.

In mutual heteroclawed combats the heel-and-ridge is performed by the individual's having his dactyl outside the manus, the interlace by the crab with the dactyl against the inner surface, as described on p. 490. When the second crab becomes the actor, he needs only to engage the claws farther, to the gape base, to attain fully the interlace position.

A total of 11 heteroclawed combats were observed in the *rapax* series which included this alternate performance of heel-and-ridging with interlacing; one record was secured on motion picture film; all except one of these examples took place between an aggressive wanderer and a burrow-holder.

Although some mutual encounters were among the best examples of ritualized combat, with no discernible trace of force, in others aggressive elements were scarcely disguised. In these combats the shift from one actor to another was clearly accompanied by irregular pushing or by abruptly jerking motions of a claw in the midst of a component.

(h) DURATION

The duration of a group of combats has been investigated only in *rapax*. Here 104 combats, of all classes, each observed from its beginning, were approximately timed. The great majority turned out to be very short encounters, most of them lasting between about 3 and 8 seconds, and all but 9 lasting less than 20 seconds. Each of these 9 continued for more than a minute and included high-intensity components. Proportionately more of these long combats occurred between right- and left-clawed crabs, more had mutual components and more ended forcefully than did short combats.

IV. *Post-Combat Behavior*

After most encounters between *rapax* males, the opponents promptly resumed their pre-combat activity. Aggressive wanderers passed on through the population, instigating new combats and engaging in other activities (pp. 487, 492, 505, and 687). Burrow-holders, returning with equal completeness to all their former activities, first resumed waving. Similar sequels to combat have now been observed in a number of other species.

Almost one quarter of all encounters in *rapax* which were sufficiently observed, however, were followed by detectable changes in behavior. In all the combats of the series, both opponents were watched long enough in 148 examples to form a suitable basis for an examination of such changes and of combat composition when subsequent changes did not occur. These alterations in behavior were of two kinds: either the aggressiveness of an aggressive wanderer was reduced or there was an appreciable delay in the resumption of waving by a burrow-holder. Reduction of aggressiveness in a wanderer and delayed waving by the burrow-holder never followed the same combat, nor was waving ever delayed by both opposing burrow-holders.

Table 18 breaks down the 148 combats where subsequent behavior was observed into a number of potentially relevant subdivisions. The first column, headed Result, divides the group into those with behavior unchanged, waving delayed, or aggression reduced. In the second column, Combat Class, the opponents' phases and relative size are indicated, as in previous tables, as well as, where necessary, the instigator and the site of the combat. Under General Combat Composition selected characteristics are isolated; these show the relative prevalence of low- and high-intensity combats, forceful components, tapping, and mutual components. It seemed that one or more of these aspects of combat might be correlated with behavior changes or the lack of them. However, no clear-cut correlation emerges. For example, neither tapping nor mutual components preclude either a delay in resumption of waving or reduced aggression; similarly, forceful endings are not necessarily followed by subsequent behavior changes. Nevertheless, certain trends are indicated.

Five points emerge that seem noteworthy in spite of the small samples. First, in combats followed by the reduced aggressiveness of a wanderer, forceful endings were more numerous than in combats either not followed by behavior changes or with a subsequent delay in resumption of waving. Second, long combats were most numerous in the class followed by delayed waving, less so among those resulting in reduced aggression, and rare among encounters with no detectable results. Third, mutual components were relatively fewer in combats followed by changes in behavior. Fourth, tapping was usually absent from encounters with forceful endings; this absence is probably correlated with the frequently prompt cessation of combat after tapping. When tapping did occur in the course of a fight ending forcefully, subsequent behavior was changed. Finally, after combat any changes in behavior were usually shown by the smaller crab.

Since the study of combat in *rapax* was made entirely in the field among unmarked crabs, few hints of summation were observed. When this important aspect is suitably investigated, summation will almost certainly prove to play a part in the effects of combat.

A study comparable to the investigation of *rapax* is in preparation on combat in (*Celuca*) *lactea*. The most striking difference between the two species is the relative incidence of force. Whereas in the *rapax* series only 9 percent of high-intensity combats ended forcefully, in *lactea* the percentage ranged from 48 to 65. Again, after the rare forceful ending in *rapax* combats the final inactor seemed never to return for another engagement; in *lactea*, on the other hand, the same individual often picked himself up after an upset and came back promptly for one or more additional rounds. Each of the subsequent fights often ended in a fashion similar to the first. As in *rapax*, occupancy of a burrow was not an immediate aim. Finally, in *lactea*, the equivalent of manus-rubs or -taps was a strongly forceful component that was almost always a major element (p. 491). Yet both the positions and motions were so stereotyped and of such an exaggerated nature that they appeared as ritualized as any of the complex components known in other species. Thus force itself has been ritualized, and my early definition of a ritualized combat or component in *Uca* as one in which force plays no apparent part breaks down. To cover the changed situation, I am now substituting a more satisfactory phrase. For the purposes of this contribution, ritualized combats or components lack the ingredient of irregular force.

F. WAVING DISPLAY

I. *Introduction*

As the exaggerated size of one cheliped is the most typical morphological characteristic in fiddler crabs, so the "waving" of this appendage in visual display is unique among appendages in the variety and complexity of its patterns of movement.

In this contribution, the term *waving display* is used in a general sense to include not only the motions of the major cheliped, but all other movements associated with the waving category of apparently visual display. Equivocal displays that may include sound components, although they are mentioned below, will be discussed in more detail in the next chapter.

II. *Historical Review*

Rhythmic motions of both chelipeds have been reported in a number of other ocypodids. Those of *Dotilla* were described by Tweedie (1950.3) and Altevogt (1957.2); *Macrophthalmus* and *Ilyoplax* by

Tweedie (1954); and *Heloecius* by Ward (1928), Tweedie (1954), and Griffin (1968). I have also observed waving display in all these genera as well as in *Scopimera* (Crane, unpublished). In some of these forms the carapace is raised and lowered, adding to the conspicuousness of the behavior. In all, the "waving" consists of a relatively simple elevation and extension of the chelipeds, usually up and out, with a return to the rest position. In none are the entire body and a number of appendages also conspicuously involved, nor are the patterns of display as different interspecifically as among the various forms of *Uca*. Roughly similar motions are also known in a number of other grapsoids; pertinent contributions include Tweedie (1954), Schöne (1961, 1968), Schöne & Schöne (1963), Reese (1964), Salmon & Atsaides (1968.2), Wright (1968) and Warner (1969, 1970), along with their respective references.

The first author to mention waving behavior in *Uca* was Rumphius, whose account of East Indian fiddlers was published in 1705. He wrote, ". . . during ebb tide one sees it strenuously waving the largest claw continuously, as if it wanted to call people, and when one approaches it, it hides in the sand. Its name in Latin is *Cancer Vocans* and in Malayan, Cattam Pangel, that is, the Caller. . . ." Although Linnaeus (1758) adopted the Latin form as an official binomial, and the common names "fiddler-crab" and "winkerkrabbe" came into use (e.g. Smith, 1870.2, Müller, 1881), waving itself seems to have been ignored in print for decades.

Finally came the observations of Alcock (1892, 1902) and Pearse (1912.2, 1914.1, 1914.2), the principal early contributors, who regarded waving, along with colors, as a means of attracting the female. Pearse's 1912 paper, however, noted that fiddlers in the Philippines sometimes waved when no female was present, and in the non-breeding season at that. Starting about the same time, brief remarks on waving, also with the emphasis on the courting character of the display, were contributed by Schwartz & Safir (1915), Symons (1920), Johnson & Snook (1927), Beebe (1928), and Matthews (1930); all of these references except the first noted that males waved more rapidly when a female appeared, as did the more detailed contributions of Crane (1941.1ff.), Burkenroad (1947), von Hagen (1961, 1962), and Salmon & Stout (1962). Altevogt (1955.1, 1955.2, 1957.1, 1959) also favored the courtship interpretation of waving, while the recent work of Salmon (1965, 1967) and of Salmon & Atsaides (1968.2), in the sections where it concerns waving, has concentrated on its relation to courtship. Meanwhile, several other observers decided that the primary, and probably the only, function of waving was to delimit territory; this view was held by Verwey (1930), Hediger (1933.2, 1934), and Gray (1942). To Crane (1941.1, 1957, 1966) and to Schöne & Schöne (1963) waving appeared ambivalent; a similar view will be expanded later (pp. 501, 517).

The history of the descriptions of particular kinds of waving display is shorter. Crane (1941.1, 1943.2, 1943.3) reported specific distinctions in the movements, and attempted descriptions indicating phylogenetic relationships among species in the eastern Pacific, along with a few from the western Atlantic. Since the early accounts of waving, descriptions have become more precise, with the relation of the waving to sound production receiving increasing attention. Burkenroad (1947) discussed correlations between waving and sound production in *pugilator*. Altevogt (1955.1, 1955.2, 1957.2, 1962), Peters (1955), von Hagen (1962, 1968.2), Salmon & Stout (1962), Salmon (1965, 1967), and Salmon & Atsaides (1968.1, 1968.2) have all described visual display in many species in detail, securing data on timing and its variation and on directions of movement through analyses of films.

Crane (1957, 1966), also aided by films, reported a rough division of types of visual display into two categories, termed vertical and lateral waves, each of which was associated with a particular form of daytime courtship. Among species characterized by vertical waves, the male pursues the female toward her burrow, with or without waving; copulation takes place on the surface of the ground; this behavior is characteristic of most species in the Indo-Pacific. Among lateral wavers, on the other hand, the male in normal diurnal courtship attracts the female down his own burrow, which he enters first; this sequence Crane reported to be most characteristic of American species. Intermediate types of display, to be further discussed below, were stated to occur, a circumstance also reported by Salmon (1967), Salmon & Atsaides (1968.2), and von Hagen (1968.2). Recent work, as will be seen, has uncovered further complexities.

Wright (1968), in a comparative study of visual display in semi-terrestrial crabs, observed that waving displays in *Uca*, in spite of their variety, all fit into his general category of Lateral Merus Displays, but that the subdivisions he has distinguished in a number of semi-terrestrial groups are not present. Described here as the Forward-Point (p. 479), Wright's other principal category, the Chela Forward Display, occurs in several species of *Uca*, but only as a threat posture.

The present section is confined to a description of the components, the organization, and the timing of waving display, including its associated motions. In addition, indications are included on the extent of the variations of these displays, both within and among populations.

III. *Components of Waving Display*

(*a*) MOVEMENTS OF MAJOR CHELIPED

1. *Vertical-wave.* The flexed cheliped moves up and down in a single plane in front of the body. In some cases it may strike the ground, then becoming, of course, an example of sound component no. 9 (p. 483). In some Indo-Pacific species the movement forms the entire waving display. (Fig. 90.)

2. *Jerking-vertical-wave.* As in the vertical wave, above, but with definite pauses that appear as jerks. These may be confined to several in the rising period, but usually one or more jerks also occur as the claw is returned to rest position.

3. *Semi-unflexed-wave.* Cheliped partly unflexed during wave, pushing outward and sometimes up to an acute angle with the front. A development of 1 or 2, above, with the jerks largely smoothed out.

4. *Lateral-straight-wave.* The cheliped is completely unflexed, either straight outward or obliquely upward, and returned to the rest position as in 2, in a single plane.

5. *Lateral-circular-wave.* The cheliped is again completely unflexed outward, or obliquely upward, as in 4. Instead of returning to rest position in the same plane, however, the claw is raised, then flexed and brought down to rest position from above the eyes and buccal region. (Fig. 91 *A-D.*) For the only known exception, see *pugilator* (p. 225).

6. *Jerking-oblique-wave.* As in 4 or 5 but with jerks, the initial direction being always obliquely upward, not sometimes straight outward. The jerks range in number from one or two during the unflexing to more than 20; the number on the return descent usually does not exceed several and may be absent. (Fig. 91 *E, F.*)

7. *Reversed-circular-wave.* As in 5, but cheliped is first brought up in front of buccal region, somewhat like the beginning of a vertical wave. After an unflexed high reach, it returns to rest position by being lowered laterally and flexed as it passes parallel to the ground. So far noted only in *pugilator.*

8. *Overhead-circling.* The cheliped does not return to the ground during a series of waves. A development of component 5, it is characteristic of large, heavy crabs in the subgenus *Uca.* Here the motion, found only in high intensity display, is probably an aid to balance. Sometimes occurs in other species when a male is partway down a burrow, with the projecting cheliped continuing to wave.

(*b*) MOVEMENTS OF MINOR CHELIPED

9. *Minor-wave.* The minor cheliped makes a motion similar to that of the major. The small appendage is sometimes almost certainly involved in sound production during some of its motions, but often not at all.

(*c*) MOVEMENTS OF AMBULATORIES

10. *Leg-stretch.* The crab raises its body with each wave, through straightening the leg joints, the crab being supported on the tips of the dactyls. At highest intensity only the two pairs of middle legs, always the longest, touch the ground. The high point of the wave coincides with the greatest elevation of the body. (Fig. 90 *C, D.*)

11. *Prolonged-leg-stretch.* As in 8, except that the position is held throughout a series of waves. (Fig. 90 *E, F.*)

12. *Leg-wave.* This position is closely similar to the acoustical leg-wag (p. 482), two or more legs being raised on a side and moved up and down. Unlike the leg-wag, however, in leg-waving the meri do not touch at all, and the component is purely visual. The raised legs usually reach their highest point at the peak of the major cheliped's elevation, which also coincides with maximum elevation of the body. (Fig. 88 *B, C.*)

13. *Curtsy.* The ambulatories are momentarily depressed, the crab bobbing repeatedly down and up. Depending both on species and circumstances, the curtsy precedes, coincides with, or follows the peak of the wave, or at highest intensity may replace waving.

14. *Herding.* A male stops display, extends the major cheliped laterally, or flexes it above the front, and approaches an attentive female. He then maneuvers her toward his burrow, darting and zigzagging with great agility, while the female just as actively seeks to dodge away from him. Usually he does not touch her until he has chivied her almost to the burrow's edge; if she does not escape, he then sometimes literally pushes her down, following her at once. Sometimes several neighboring males simultaneously attempt to herd a single female.

(*d*) TIMING

15. *Duration of Single Waves.* An important, but highly variable component. Discussed below.

16. *Pause at Peak of Wave.* When present, forms a notable characteristic of the display.

17. *Pause Duration between Waves of a Single Series.* Forms a fairly reliable characteristic only in relation to the tempo of the particular wave series being analyzed.

18. *Number of Waves in Series.* Sometimes a definite and fairly useful characteristic, but not sufficiently well known to include in Table 19.

(e) MOTIONS ASSOCIATED WITH WAVING DISPLAY

The preceding list includes only motions or elements of timing that can be expected to add to the visual effect of the waving display; in general they increase the crab's apparent size, make him more conspicuous, or contribute to the distinctiveness of the waving rhythms. There are in addition two general kinds of motions which perhaps also contribute to the visual value of the display; since they belong to other areas of behavior, they are not included either in the annotated list above or in Table 20. These are, first, the motions, sometimes very conspicuous, associated with sound production; a good example is the drumming found in some forms of high intensity courtship. Second are the various kinds of displacement behavior, at times taking visually prominent forms. An example is incomplete drumming motions, where the cheliped touches neither the ground nor, through the merus, the carapace. Some small motions of the minor cheliped, mentioned under visual display component no. 9, above, probably also belong in this anomalous group.

IV. *Organization of Components in Waving Display*

Table 20 gives an idea of the known distribution of components throughout the genus, while Table 19 assembles current data on the timing of the waving part of display. The paragraphs below will comment on this material. Details for each species are given in the systematic section. General discussion is reserved for following chapters.

The clearest examples of relatively simple displays occur in *Deltuca* and *Thalassuca*; here the waves are very low verticals, usually in series, probably with an acoustic drumming component sometimes associated with the downbeat of the cheliped manus. The cheliped is never raised high or stretched out, and during high intensity there are no special kicks, stretches upward, or curtsies. Sometimes the cheliped makes several brief pauses during the raising of the claw, each giving the effect of a jerk; these jerking displays are usually not in series. Sometimes both simple and jerking waves are found in the same display. In one subspecies of *coarctata*, the display of very young crabs consists of jerks, while older individuals give simple waves. Copulation is not always preceded by waving; it occurs at the surface with the male following the female and patting, plucking, or stroking her carapace.

Other Indo-Pacific species, particularly in *Australuca, Thalassuca,* and *Amphiuca,* tend to raise the carapace on the stretched legs, slightly to considerably, either with each wave or during a series of waves. While all of them give a low, chiefly vertical wave, the cheliped is sometimes swung obliquely outward to a semi-lateral position during its elevation; this occurs when the object of the display, whether potential threat or potential mate, is passing to the side or rear. None of these shows any trace of a circular lateral motion of the cheliped, nor do special display movements occur before copulation. Except in individual *bellator* (*Australuca*), copulation occurs at the surface, as described above.

In the remaining subgenera, the display is definitely lateral and, in its extreme form, includes high circular motions. Circularity is present in at least some part of the display in all American crabs, except in *Boboruca*, in which the jerking display unfolds the cheliped obliquely upward, somewhat as in an exaggerated Indo-Pacific vertical wave, and returns it in the same plane to rest position. In addition, lateral-circular waves occur in *triangularis* and *lactea* in the Indo-Pacific, these two forms belonging to the subgenus *Celuca*, which is otherwise American.

One point of particular interest is the occasional occurrence of the vertical type of wave, characteristic of socially less advanced forms in the Indo-Pacific, in the displays of highly specialized species which at moderately high intensities make lateral waves. Every lateral-waving species known to me by more than the briefest observation proves to have these subsidiary displays, always exceedingly similar to those of Indo-Pacific vertical wavers. Always they occur either during low intensity display, or among juveniles, or in both these situations. Often the atypical vertical display is combined with feeding. When moderate to high intensity phases of either threat or courtship take place, the wave form changes promptly to the lateral motions more characteristic of the species.

In contrast, lateral displays never occur among the most typical vertical-wavers, either in young crabs or during periods of low intensity. On the contrary, whenever weakly lateral tendencies occur, they are elicited at rather high intensity only, when a potential opponent or possible mate is passing to the side or rear; it is always a poorly developed, oblique, lateral-straight wave, tending toward component 4, and never a lateral-circular-wave.

A prevalent variation with lateral-waving species is the frequent change with intensity from lateral-straight waves to lateral-circulars. Sometimes the circularity is so slight that it becomes apparent only in detailed film analysis. This is especially true of some of the jerking *Minuca*. In *speciosa* (subgenus *Celuca*), Salmon (1967) noted the tendency for the wave to return to rest position in the same plane in a threat situation (here called a lateral-straight-wave), and to be circular in courtship.

In lateral-wavers, the female is attracted close to the burrow of the male. After intensification and often specialization of his display, he descends. Sometimes she follows him below, where in completed courtships copulation occurs. The preliminary specializations of visual display include special steps, stamps, or kicks of the ambulatories, rhythmic lowering of the carapace in curtsies, and vibrations of the cheliped, which are often visually conspicuous as well as acoustically efficient.

Herding (visual display component no. 14) is a variety of courtship behavior which appears to occur only sporadically in the genus, and is known among both vertical- and lateral-wavers. When it does occur successfully, it must replace the precopulatory behavior characteristic of the species. It apparently occurs only in certain populations, or at least is more frequent in some than in others. First observed in *stenodactylus* of the subgenus *Celuca* (Crane, 1941.1), it is now known also in *Minuca*, represented by Florida populations of *burgersi*, *rapax*, and *pugnax* as described by Salmon (1967), as well as in seven additional species, as shown in Table 20 (Crane, unpublished). Completion of the pattern, where the male succeeded in bringing about the descent of the female ahead of him into his burrow, has so far been observed only in *stenodactylus* and in *bellator*; in the latter species the entire puzzling sequence was filmed.

The timing of display, as suggested above, is so variable that it is disappointing as a refined taxonomic character, while its evolutionary aspects remain obscure. Few general statements can be made, in spite of the modest but well-distributed amount of sampling that has been done. It is true that the fastest waves are non-jerking verticals not raised far off the ground. Lateral waves are, in general, slower than verticals, but in highly evolved species the tempo is sometimes rapid, as in *beebei*, where most waves are completed in less than a half-second. Slowest of all are the multiple jerks of *rapax*, which at most, for an entire wave, clock at about 6 seconds, and at slowest at the rate of only one complete wave every 18 seconds.

Waving in general, particularly in socially advanced species, is speeded up not only by higher temperatures but especially by the heightened intensity of advanced courtship. Pre-combat threat waving sometimes approaches but never exceeds the speed of display in the last moments before a male descends his burrow, after he has attracted the near approach of a female.

Under natural conditions waving occurs only by day. A correspondence between its time components and the rate of sound production occurs in *tangeri*, as already noted on p. 480.

When motion picture films are analyzed, the variations in timing even of parts of waves are striking. This remains true when the films have been made of waves of approximately the same intensity, in similar social situations, and photographed at similar temperatures. When different intensities are compared, even greater ranges expectably occur. The parts of waves, such as upstrokes, temporal divisions by jerks, duration of pauses at peak of wave, and duration of the downstrokes also vary extensively, as do pauses between waves. When populations are compared that are well separated geographically, the range of temporal characteristics is likely to be further extended.

In some species wave counts and timings show a wider range in the systematic treatment than under the corresponding listing in Table 19. This discrepancy is due to the fact that the description of waving display includes both field notes and the results of film analyses, while the table is based only on film analyses.

In order to obtain adequate statistical material, unrealistic amounts of film would need to be exposed with the most exacting attention to factors of ecology, breeding season, distance of population from limit of range, semi-lunar rhythms, tidal conditions, sunlight, temperature, time of day, and social situation. Subsequent analysis of a caliber to uncover all but the grossest differences would have to employ the usual, time-consuming, single-frame techniques. It is difficult to see how the most generous and intelligent use of computer time could curtail the necessary preliminaries sufficiently to make feasible a study of adequate scale to give reliable data for comparative ethology. In short, my current conclusion is that time samplings, as presented in the recent papers of Salmon and his co-workers, by von Hagen, and in the present contribution should be continued at present—and only temporarily—on a very small scale, and that it be considered of minor importance even as a taxonomic tool, particularly in the study of allopatric populations. This course should provide extra time needed more immediately for basic experimental work. When this research has disclosed the essential factors in the timing of displays, it will then certainly be feasible to collect and process adequate data for rewarding results.

This program would parallel the course which seems currently needed in considering display color in *Uca* (p. 468). Contrary to earlier expectations, color has turned out to be of little use in taxonomy and a variable attribute of questionable importance in social behavior. As in display timing, both the physiological investigations and the experimental work on behavior remain to be done.

During earlier work on fiddler crabs, I was convinced that waving display not only is species-specific, but that the characteristics most distinctive to the human observer hold throughout the species range. This conclusion undoubtedly holds true for many species, including several with an extensive geographical range. One such form is *tangeri*, found from

Spain to Angola. In split-frame motion pictures, showing waving individuals from Portugal and Angola side by side, waves of similar intensity match each other almost precisely. This stretch of coastline covers some 46 degrees of latitude. Here, with the possible exception of extremely rare accidentals, *tangeri* is the only *Uca*.

On the other side of the Atlantic, *rapax* provides a contrasting example. This species ranges from north-central Florida to southern Brazil, a distance approaching the latitudinal distribution of *tangeri*. Throughout the tropical part of its range, some local populations of *rapax* mingle with those of *burgersi*, although the latter species is found characteristically in biotopes that are slightly more sheltered and less saline. The favored biotopes of both species do show wide zones of merging, although as usual in such cases the species for which the habitat is the more suitable predominates. In other examples, contiguous populations of the two species are often in full sight of each other, without visual obstructions. Similar situations occur more rarely in overlapping or contiguous populations of *rapax* and *mordax*.

Throughout this western Atlantic range, the number of jerks in *rapax* is extremely high, while the jerks in both *burgersi* and *mordax* number only up to four or five, during both unflexing and flexing. In comparison, *rapax* shows at least six jerks while unflexing, and the more usual number approaches 20. (Crane, unpublished; see p. 193.)

In eastern Florida, one limit of the range of *r. rapax*, distribution coincides for a short distance with that of *p. pugnax*, which is its allopatric representative in the temperate zone (Tashian, 1958). Here *pugnax* shows little or no trace of jerks, while in the north its jerks are definite (Tashian, 1958; Crane, 1943.3). These examples show how local differences in wave characteristics help to differentiate the visual displays of otherwise similar sympatric species. They give excellent illustrations of the principle of emphasized distinctions in zones of coincidence.

The question of display intensity in *Uca* is both important and difficult. Differences result in many display variables, of both timing and components, all of which often shift gradually into one another (Fig. 88 *A, B*). In spite of efforts, it has proved altogether impracticable to set up, for the species descriptions, criteria of "typical intensity." Yet, as any observer will agree, the species of *Uca* on a given stretch of shore can all be readily distinguished as far as they can be seen, providing only that the individuals are waving at moderately high intensity.

There is no doubt but that eventually species descriptions will precisely and reliably present the essential features of waving display. That time, however, will not come until electrophysiological work

unites with experimental ethology in determining the important releasers among sympatric species. At present we know from experiment that the sight of the major cheliped under certain conditions releases aggressive behavior in males which are in the requisite physiological phase. We also know that any crab within a size-range appropriate to the species, without a large cheliped and passing or approaching a male in suitable condition, sometimes elicits courtship behavior (Altevogt, 1957; Salmon & Stout, 1962; Crane, unpublished). Beyond these beginnings the field is clear.

In conclusion, waving display should undoubtedly be regarded as species-specific, providing only that typological definitions are avoided and three circumstances kept in mind. First, the observed samples of display in a given population must include high intensity courtships. Second, the waving displays of a number of populations must be compared through as wide a geographical range as possible, just as in the case of morphological characteristics. Third, the displays of the other species of *Uca*, living within sight of a given population, must be taken into account; this caution is, of course, even more important in the case of closely sympatric forms.

G. Construction Activities Beside Burrows

I. *Introduction*

A few species of *Uca*, along with many other littoral and terrestrial crustaceans, sometimes pile up some kind of structure close to their burrows. Examples are known, for instance, in gecarcinids (Silas & Sankarankutty, 1960), while the well-formed "castles" of certain Indo-Pacific and West African *Ocypode* sometimes make striking additions to a sandy shore. Linsenmair (1967) found that in *Ocypode* the structures played an important role in reproductive behavior, having signal functions both between males and in courtship.

In *Uca* the structures take the form of walls completely surrounding the mouth of the burrow, here termed a *chimney*, of a rough little *pillar* beside the burrow, or, in a further development of the pillar, of a symmetrical and smoothly arched *hood*; the latter is always concave on the side closest to the burrow.

These structures have three features in common. With two known exceptions, they are erected only by adults, whether females or males, and at least in the males only by individuals in breeding condition. Second, construction of chimneys is sporadic throughout the genus. Third, pillars and hoods, although almost entirely confined to certain species of the subgenus *Celuca*, are absent over whole areas of the ranges of the species, apparently without regard to ecological differences. They are also absent in part of a population or within populations in a majority of displaying individuals. Fourth, all of the structures, except for some material in chimneys in at least *arcuata*, *forcipata*, and *urvillei*, are made by carrying substrate with the ambulatories of the minor side from beyond the burrow to its edge, not from below ground.

II. *Chimneys*

In the subgenus *Deltuca* chimneys are characteristic of at least breeding females in *arcuata*, *forcipata*, and *urvillei*, and in some populations of *coarctata*; all these species are allopatric forms. The only examples known of constructions by young *Uca* are found among them: In Hong Kong I saw two very small *arcuata* females building chimneys. Similarly, in Fiji only a few young *c. coarctata* of both sexes built chimneys; no structures were found around large burrows. The subspecies *coarctata flammula*, in northwest Australia, apparently makes none at all. Chimneys are also sometimes built by (*Amphiuca*) *chlorophthalmus*. In America the structures seem to be confined to the Pacific-Atlantic subspecies, (*Boboruca*) *thayeri umbratila* and *t. thayeri*, and to (*Minuca v. vocator*). The building behavior of *vocator ecuadoriensis* in the Pacific, if any exists, is not known.

III. *Pillars and Hoods* (Pls. 48 B, 49)

These two categories represent varying degrees of quality in the same type of structure. They are fashioned only by certain displaying males in the subgenus *Celuca* and, in *Minuca*, only in *minax* and, from one observation, *pugnax* (p. 202).

The distribution of the structures in *lactea* is of particular interest. Many displaying males of *l. lactea* were making large, well-formed hoods in northwest Taiwan in May and of *l. annulipes* in West Pakistan in June, although not in September. In north-central New Guinea several structures, more like pillars than hoods and not as high as their builders, were discovered during June and July in a population of *l. perplexa*; this particular group of fiddlers was subjected to prolonged, daily observation during the period and it would have been impossible for me to miss any additional construction activity. In spite of careful search from Fiji to Zanzibar, wherever displaying populations of *lactea* were encountered, no other examples have been found.

In the tropical Atlantic the structures occur in *leptodactyla* and *cumulanta*, and in the eastern Pacific, in *beebei*, *latimanus*, and *musica terpsichores*, always in the usual scattered distribution.

H. PRECOPULATORY BEHAVIOR AND COPULATION

Under this heading are listed, first, the species on which observations have been made. There follows a consideration of behavioral components that are included in display—whether visual, acoustic, or both—in which the male appears to be directing them toward a particular female, and which may therefore be termed courtship components. Mating sites are then discussed along with the associated behavior of courted females. The section closes with a description of the assumption of the position for copulation—apparently the same throughout the genus.

High intensity courtship behavior has now been observed in 25 of the 62 species recognized in this contribution. Copulation itself was first described by Pearse (1914.1), who observed *pugilator* in the laboratory. Pairings on the surface of the ground that showed every evidence of being complete are now known from field observations on 20 species as follows: in the subgenus *Deltuca*: all eight species (present study); *Australuca*: *bellator* (present study); *Thalassuca*: *vocans* (Altevogt, 1955.1, 1955.2, 1957.1; present study); *Amphiuca*: *chlorophthalmus* (present study); *Boboruca*: *thayeri* (present study); *Afruca*: *tangeri* (Altevogt, 1959; von Hagen, 1961, 1962; present study); *Uca*: *stylifera* (Crane, 1941); *maracoani* (Crane, 1958; present study); *Celuca*: *pugilator* (Burkenroad, 1947; Salmon, 1965); *beebei* (Crane, 1941); *stenodactylus* (Crane, 1941); *lactea* (Altevogt, 1955; Feest, 1969); *Minuca*: *vocator* (von Hagen, 1970.3). For a general account of mating in the Brachyura, see Hartnoll, 1969.

Caution is always needed in attributing to normal courtship any display component or other activity. In the displaying portion of a thriving population the individuals are constantly reacting toward neighbors and nearby wanderers of both sexes. Under these conditions both conflict and displacement behavior are sometimes prevalent. Consequently difficulties arise in interpretation, particularly but not exclusively when the species or subspecies is poorly known and observation time is limited.

Special caution is also needed under crowded conditions when it is not known whether or not the male in that particular taxon sometimes or usually courts with the dorsal part of his carapace directed toward the female. When he does so, the carapace is tilted more or less vertically, exhibiting a conspicuously large area. I have seen this posterior orientation used by males at least rarely in one or more species of every subgenus, and its use may be a general component characteristic of the genus; its prevailing use seems to lie in first attracting the attention of a passing female, especially when the male is already engaged in a threat display toward another male in front of him. In such cases the male sometimes then swings around and proceeds with characteristic frontal courtship behavior. In other examples the rear view is presented to a passing male. In a few species, however, notably (*Thalassuca*) *vocans* and (*Celuca*) *beebei* the male vigorously displays so often with the female behind him that it must be considered a regular part of the species-specific courtship pattern. In these exceptional cases display toward the rear has been observed in the field when no other individuals of either sex were nearby except for the courted female—a situation giving, of course, the most reliable evidence of a component used in courtship. I have also seen it in the New York crabberies when (*Minuca*) *minax* displayed in this fashion a number of times when no other crabs shared the tank with the single pair under observation.

Most display components as mentioned earlier are ambivalent (p. 495; Fig. 95). In the great majority of species a burrow-holding male increases the tempo of his display whether he is threatening another male or directing his activities toward a particular female. Not only does he wave faster, but his display motions are usually accentuated, the cheliped reaches higher into the air, and the wave when of a lateral type becomes more spacious, the form of the change depending on the shape of the wave at low intensities. The change often proceeds from a straight lateral motion, as in *speciosa*, in which the claw returns to the flexed position through the same arc at which it was extended, to become at high intensity definitely circular. If the wave is already somewhat circular, with the tip of the chela sketching ovals in the air, the narrow diameters become broader while pauses between waves are shortened; in (*Uca*) *maracoani* and its allies the ovals at high intensity become almost fully circular and continuous, with the chela tip directed almost straight upward (Fig. 88 *C*). Some specific characteristics of low-intensity waves are accented, either through prolongation of a pause at the wave's apex or increased speed in a vibratory component. Sometimes, as in *lactea*, the wave is elided altogether, being replaced by other motions. In a number of *Celuca* major-manus-drumming or its ritualized counterpart, then not acoustically functional, also increases in prevalence. All of these signs of high-intensity display are changed little or not at all when unequivocal courtship is under way.

In contrast, only several components of waving display appear to be restricted to courtship. The only widespread example is the curtsy, as it occurs in (*Afruca*) *tangeri* and in several *Celuca*. In *Minuca* this component is also present in the repertory of several species, but in each it is ambivalent. Another courtship component of waving display is the quivering of the outstretched first ambulatories in *oerstedi*; their anterior surfaces, along with contiguous parts

of the carapace and buccal regions, are bright blue to human eyes and particularly conspicuous. A third courtship component is the vibration of the stiffened ambulatories on the major side in *saltitanta*, just before a male precedes an attracted female down his burrow; whether or not the motion produces sound is unknown. Finally, a special development of display toward the rear seems, at least in *beebei*, to be characteristic of courtship alone; in this species the male revolves in front of the female; only from the rear is the vivid green on the carapace visible, along with the purple aspects of the legs. Unfortunately, this juxtaposition of color and motion remains only a challenge for future work; experiments on the possible role of color in the social behavior of *Uca* have not yet been undertaken.

The well-substantiated cases where sound plays a definite part in the final stages of courtship are confined to (*Afruca*) *tangeri*, (*Minuca*) *pugnax*, (*Celuca*) *pugilator*, (*C.*) *speciosa*, and (*C.*) *cumulanta*. In these examples sound tapes have been secured during well-established instances of courtship. The work with *tangeri* was reported by Altevogt (1962) and von Hagen (1962), with *rapax*, *pugilator*, and *speciosa* by Salmon & Stout (1962) and by Salmon (1965, 1967), and with *cumulanta* in the present contribution. The results in *tangeri* and *pugilator* are discussed by Salmon & Atsaides (1968.2), as well as in the systematic section of the present contribution under each of the species concerned. Altevogt, von Hagen, and Salmon and his co-workers consider that the sounds they recorded were wholly or partly concerned with courtship, particularly at night. The examples of sound production with which I am familiar, whether recorded or inferred through stridulatory or drumming behavior of the individuals, were for the most part made during agonistic or, at most, ambivalent situations. The matter is further discussed under sound production (p. 481); see also under *tangeri* in the systematic division (p. 121).

The methods of proven sound production that are certainly used at times in courtship consist of major-manus-drumming in *tangeri*, *pugilator*, and *cumulanta*, and of leg-wagging in *pugnax*. Proved use of sound in the final stages of courtship is to be expected in many other species, but to date the evidence is observational only; sound tapes secured from activities of other species were made either during potentially ambivalent situations on the surface or when the crabs were underground, with no evidence available as to the particular stimulus situation, if any, that elicited the sound. It should also be pointed out, as stated under the descriptions of display behavior in a number of species of *Celuca*, that a number of my early descriptions of drumming ("rapping") with the major cheliped in courtship situations were based on faulty observation. When the films were analyzed, the motions were found to consist wholly or partly of ritualized drumming, with the cheliped not coming in contact with the ground. In other cases mero-suborbital contacts, not drumming on the ground, were products of the observed vibrations (pp. 226, 244, 245, 483).

In three of the nine subgenera copulation appears always to take place on the surface near the female's burrow. These groups, all confined to the Indo-Pacific, consist of *Deltuca*, *Thalassuca*, and *Amphiuca*. In the fourth Indo-Pacific subgenus, *Australuca*, copulation seems characteristically to occur at the surface, but several times a male was seen to edge a female more or less forcibly toward his burrow and push her down ahead of him.

In the remaining subgenera—*Boboruca*, *Afruca*, *Uca*, *Minuca*, and *Celuca*—circumstantial evidence indicates that the crabs habitually copulate underground in the burrow of the male, after the female has been attracted by waving display and, sometimes, by sound production. Copulations that appear complete occur rarely at the surface in each of these groups. (Notably, in [*Minuca*] *vocator* von Hagen [1970.3] found surface copulations to be prevalent in certain populations that inhabited areas where dense vegetation restricted vision.) The majority of copulatory positions attained, however, are imperfectly oriented or the assumption of position is incomplete; in particular the abdominal flaps of one or both partners fail to be extended away from the sternum. These pairings appear to break off prematurely, and most often represent attempts at copulation by aggressive wanderers (see below). Incomplete surface copulations apparently occur less rarely in the subgenera *Afruca* and *Uca* than in *Minuca* and *Celuca*.

Nocturnal copulation at the surface appears to be usual in eastern United States populations of *pugilator* and *pugnax* and in Spanish populations of *tangeri* (Burkenroad, 1947; Salmon & Stout, 1962; Salmon, 1965, 1967; Salmon & Atsaides, 1968.2; von Hagen, 1961, 1962); in these species nocturnal sound production is prevalent, replacing waving.

One continuing need in the study of *Uca*'s courtship behavior and copulation is the use of infrared instruments during observations at night. There is no question, thanks to the contributions just cited, but that nocturnal courtship does occur in at least temperate populations of the listed species. These authors have all also reported actual copulation at the surface. Males of all these species, however, during the day attract females underground, matings at the surface being certainly rare, except in von Hagen's report of Spanish *tangeri* (see present contribution, p. 120). It seems possible that at least some of the reported pairings above ground at night are incomplete copulations performed by wandering males, or with females not wholly receptive, as occurs also at the surface during daylight. Use of adequate infrared illumination, now available but apparently not yet used by observers of *Uca*, should settle the relative numbers of complete and incomplete copulations in darkness.

The point is, of course, of particular interest in relation to the apparent advantages of underground mating; with temperate species the short breeding season may be offset by the addition of nocturnal reproductive activity (see Salmon, 1965, and present contribution, p. 176). If so, it is difficult to understand why sound could not wholly replace waving, attracting females underground instead of only to the vicinity of the males' burrows; undoubtedly sound in some species of *Minuca* and *Celuca*, both temperate and tropical, almost or wholly replaces waving in daytime in the final moments before the male descends and the female follows. The final selective step, it seems, could well have been taken long ago, with mating underground following acoustic components of courtship.

The behavioral role played by females in courtship remains largely unknown. The most prevalent and conspicuous form of female courtship behavior is the wandering stage of receptive individuals. This activity does not seem to exist in Indo-Pacific subgenera, in which the male characteristically does not attract the female to his own burrow but instead approaches and mates with her on the surface. In *Afruca*, *Uca*, and, especially, *Minuca* and *Celuca*, on the other hand, the great majority of courtship activity is directed toward wandering females (Pl. 50). The situation in *Boboruca* varies and needs further observation. In exact contrast to wandering males it is the wandering females that are always, and, in some species at least, uniquely responsive to the male's courtship display and that eventually sometimes follow him into his burrow underground. In species where the pattern is best known and, perhaps, best developed an experienced human observer can detect a receptive female at a distance without binoculars, not merely by the increased intensity of the display of the males among which she passes (p. 495) but by her own movements. Her progress is characteristically in short spurts, often somewhat jerky, with her body held low on the ground and her ambulatories little extended. This is the opposite of her non-receptive posture where the legs are spread stiffly and the body raised high; it will also be noted that this attitude is closely similar to that of a male in submissive posture, as when a non-aggressive wanderer passes close to a waving burrow-holder (Fig. 84 *C*).

A wandering female often progresses for many yards through displaying populations, sometimes not pausing at all, but eventually stopping to feed and perhaps take over an empty burrow far from where she was first observed. Usually her course zigzags, as she dodges among groups of males excited by her passing. Sometimes she is briefly attracted by an individual male, as shown by her pausing and perhaps approaching closer to his burrow. Less often she approaches closely enough to elicit any high-intensity courtship component in his repertoire before he descends his own burrow; usually the female then moves on; rarely she follows him below. If she does so she usually remains only a few seconds; this stay is presumably not enough for successful copulation, judging by the longer periods observed at the surface. Very rarely the male plugs up the burrow after the female has descended and neither crab emerges during the remainder of that period of low tide.

In the subgenera confined to the Indo-Pacific the females are far less active and wandering seems to be minimal or absent. Sometimes individual pairs maintain adjacent burrows for several days or more; in a population of *chlorophthalmus* in New Guinea the record for a single pair was four weeks—from one new moon to the next. In that population, as well as in others of the same species, of *dussumieri*, and of *coarctata*, courtships were observed repeatedly between members of the same pairs of neighboring residents. This behavior occurred on two or three successive days during favorable semi-lunar periods.

In all the subgenera a male is often stimulated to intensive display by a nearby female that suddenly emerges from a burrow and moves, even when normally feeding, either briefly toward a male or past him. Under these conditions the male's burst of activity is usually short.

We still know almost nothing of the requirements for receptive behavior in the female. Feest (1969) published a study on the ontogeny and sexual biology of *triangularis* and *lactea annulipes* in southern India. Information on the physiology of reproduction is not included, but clear data are presented showing peaks of copulation and egg-bearing around new and full moon. In *triangularis* maxima occur in July and October, during each of the monsoons; *l. annulipes*, she found, breeds throughout the year. Data are given on numbers of surface copulations observed in the field between March and May in *l. annulipes*, in relation to the moon, hour, tides, and weather, but, as the author states, the table serves to indicate days on which strongly increased sexual activity was observed in the males. The question of completeness of these surface copulations and a description of any special activity of the females are not included in the discussion. Both early and recent contributions on other species, principally those of the temperate zone, only indicate that breeding seasons reach peaks in the earlier parts of the northern summer; no precise work has yet been carried out (p. 441). All observers agree that in *Uca* females

mate when the carapaces are hard; this is in contrast to some other brachyurans (Hartnoll, 1969 and refs.).

Ryan (1966) recorded for the first, and apparently only, time the occurrence of a pheromone in decapods, a swimming crab in Hawaii. He concludes "these experiments indicate that a pheromone in the form of a sex attractant permits males to detect the premolt condition of *P. sanguinolentus* females. This does not eliminate an important role of submissiveness or other behavior on the part of the female in mating. These experiments also indicate that the pheromone is released through the excretory pores. Origin of the pheromone, its chemical nature, and the way it is detected by males remain to be determined."

On the basis of my observations both in the Trinidad crabberies and in the field it seems likely that a similar mechanism may operate throughout the genus *Uca*. Two facts are suggestive. First, in the crabberies individual adult females of *maracoani*, *chlorophthalmus*, and *dussumieri*, after spending several days underground, emerged from marked burrows slightly larger than before but with hard carapaces. It was these individuals that caused intense courtship activity by conspecific males which were also well established in the crabberies and individually known through carapace markings. The second suggestive fact is that in both *vocans* in Fiji and *lactea* in New Guinea, males sometimes herded females (p. 496); when I captured these females, all proved to be slightly immature; the examples examined totaled about seven in each species. In New Guinea I kept additional, herded, immature *lactea* in our small field crabberies for more than a week (until our departure terminated the work) in the hope that these individuals would molt; although they did not do so, this circumstance is without value because of inadequate conditions in the crabbery.

Judging by the examples seen of copulation at the surface, the position assumed appears identical throughout the genus. Once the two sexes have come into contact, the subsequent pattern consists of two stages. First, the male climbs the carapace of the female from the rear. The female, when receptive, remains in a more or less normal rest position with her carapace horizontal; her legs are bent and held close to the body, which almost or quite touches the ground. The male advances until his front is at least at the level of her mesogastric region. With his minor cheliped he usually taps or strokes the anterior part of her carapace and often plucks at it. In females having pile in the area at least some of the motions are directed toward this pubescence; when pile has been abraded, or when it is not characteristic of the species, the male often makes precisely similar motions; observations have been insufficient to determine the patterns more definitely; it is nevertheless certain that whether or not pile is present and plucked at, or whether the female's carapace is clean or muddy, the male often carries his minor cheliped toward his buccal region, making feeding gestures that are sometimes complete and sometimes show the incompleteness often found in ritualized patterns. The male's ambulatories, either singly or in various combinations, meanwhile almost always make stroking or tapping motions. (Fig. 89 *B, C.*)

In the three eastern Pacific species—*stylifera, beebei,* and *stenodactylus*—in which I have seen copulation at the surface, the female's ambulatories also rubbed against those of the male. I have not observed these species during copulation since originally describing it (1941.1), and it seems likely that the female's motions in each case were actually part of her non-receptive pattern; perhaps they included agonistic stridulation (p. 484) against the ambulatories of the male.

After some seconds or minutes of stroking, plucking, or tapping with the minor chela, the male turns the now quiescent female upside down and holds her at an angle to the ground while the abdominal flaps of both individuals project from the sterna. In all the examples I have seen the male both turns and holds the female with his ambulatories alone, the major cheliped being held flexed, in rest position. Altevogt (1955.1, 1955.2), in contrast reported that male *vocans* used the major cheliped to seize the female and hold her in place; later both Altevogt (1959) and von Hagen (1962) saw the claw similarly used in *tangeri* while Darwin (1871) speculated that some such use would be found for the major cheliped. It seems likely that the instances observed by Altevogt and von Hagen were examples of the unusual behavior that crops up so frequently among aggressive wanderers (pp. 507, 687).

Only the tips of the gonopods are inserted in the gonopore and, in the species where the thumbs are reduced and arise far down the shaft, it seems that these elements cannot be functional; it is worth noting that variation is usually characteristic of thumbs in these positions.

In surface matings the crabs remain in contact for periods ranging up to an hour or more (von Hagen, 1962 and present study). Sometimes, for whatever reason, the copulating position is held for only a minute or so. Pairs in which the gonopods are apparently fully engaged can sometimes with care be approached and picked up.

The form that copulation takes underground is unknown, since the necessary arrangements for observation in crabberies have not been performed.

I have seen ovigerous females copulating, particularly in *vocans,* and spermatophores are often found in place in preserved ovigerous specimens throughout the genus.

I. RHYTHMS OF SOCIAL BEHAVIOR

The effects of temperature, tides, rainfall, and light on activity in *Uca* were discussed in an earlier chapter (p. 440). Here their social aspects will be reviewed. Social activity throughout the genus is largely confined during the daytime to several hours before and after low tide. As the water begins to ebb, almost the only activity of all members of the population is feeding, with, sometimes, minor excavations of the burrows from which they emerged. Although some individuals continue waving display until the water almost touches them, most again feed and deepen the burrows during the last hour or two of rising tide. Some species attain the peak of their waving activity at least an hour before low water; others are most active shortly after the tide has started in.

Social activity at night never includes waving display, but at least some species substitute acoustic components during nocturnal courtship (refs. on p. 504). The hours before midnight are, so far as known, those during which most acoustic activity takes place.

Acoustic responses from within the burrow to an intruder have been elicited experimentally by day as well as by night in a variety of species (p. 481).

Intermale combat of all kinds seems to peak during the day, about an hour earlier than waving display for the species; in any case it virtually vanishes within an hour after low tide. Neither combat nor threat behavior has been directly reported at night, although von Hagen (1962) reports the emergence of male *tangeri* at night in response to surface drumming by another male.

Recent laboratory work on endogenous circadian activity rhythms (Barnwell, 1968.2, and refs.) shows that a maximum is apt to come in the morning and a second, minor peak around sunset, with minimal activity in the early hours of the afternoon. This result agrees with my field observations (Crane, 1958, and later, unpublished); these indicate that when low tide occurs between about 1300 and 1600 social behavior is less than on other days, even when all other factors, such as temperature, are favorable.

It also agrees with observations of Altevogt, of von Hagen, and of Salmon and his associates on the beginning of nocturnal acoustical activity at dusk, although this can also be interpreted as a response to decreased light. Experimentally these investigators have induced the diurnal substitution of sounds for waving merely by covering a visually displaying male with a box (Salmon & Atsaides, 1968.2, and refs.).

A few field and crabbery observations show in addition that a pre-sundown period of increased waving sometimes occurs when the tide is low at that time. Both Altevogt (1957.1) and I have noted it in *vocans*, in India and Fiji, respectively. Pre-sundown waving is particularly evident in at least Trinidad populations of *thayeri*, where I have checked it sufficiently often to be certain of its regular occurrence. I have also noted it sporadically in a number of other species, in several subgenera. *U. thayeri* is one of the forms in which the peak of waving activity occurs, when the tide is favorable, as early as 0700 to 0800.

In general, the period of maximum social activity takes place fortnightly when low tide falls during the full daylight hours between about 0800 and 1200. On many shores, where the tides are those characteristic of many, rather even, continental coastlines, this period begins a day or two after new and full moons. Because of the simultaneous occurrence of both a response to the ebbing tide and to change of light, it now seems feasible to eliminate the idea of a third, semi-lunar factor to explain the high degree of social activity when low tide falls in the morning; mutual reinforcement of circadian and tidal endogenous rhythms appears a sufficient explanation.

It is worth noting that representatives of more than 20 species maintained a fortnightly rhythm when they were kept for weeks on an unnatural tidal schedule. The phenomenon was observed in crabberies out-of-doors in Trinidad. Although the water level was lowered only once a day, during the morning, and raised in the afternoon, the peak of waving and other social activities still occurred on days close to full and new moon. The conditions and observations were not sufficiently continuous and complete to warrant formal presentation here. The effect, however, was so reliable that it was used to schedule in advance concentrated periods for observation of waving by individuals flown to Trinidad, over a period of five years, from overseas. The crabs came from the eastern Pacific, Japan, and the Philippines, all, by chance, from shores where spring tides, following new and full moons, occurred in the morning. All the fiddlers accommodated within a few days, as usual, to both the local light cycle and to the single low-water period provided every morning. For weeks, and, in a few cases for almost a year, they continued to show clear traces of the fortnightly rhythm in their waving display, courtship behavior, and construction of towers or pillars on the edges of the holes. This effect corresponded to that already reported (Crane, 1958) in the same Trinidad crabberies when *maracoani* and other local species were provided with normal tidal cycles.

Investigation of rhythms in populations from shores with single, diurnal tides and from those with minimal tidal effects are now beginning to be made (Barnwell, 1968.2 and refs.).

In the tropics social behavior, including reproduction, continues all the year around, unless it is locally impeded by monsoon floods, as in southeast Asia, or by drought, as for *Minuca* in some parts of the neotropics. Here, when conditions are favorable, the crabs emerge and feed, but they often do not wave at all for weeks at a time, while combat and other forms of social behavior are minimal or absent. In the subtropics such as in the southern half of Florida, social behavior is strongly reduced or absent only during the winter. In at least some subtropical populations, combat and occasional waving appear seasonally before courtship, as was shown in *inversa* when a population was observed in Mozambique in late August, at the species' southern limit (Crane, unpublished).

In temperate populations the actual breeding season is apparently restricted to one or two months, although waving and incomplete courtships extend throughout warm weather. Adequate data are still not available even for the northeastern United States. During the summer of 1968 I observed the onset and development of social behavior in *pugilator*, on the northern boundary of its range. Waving appeared in some individuals at or soon after on their first appearance following hibernation, at the end of May; yet throughout the breeding season, ending in early August, combat behavior was almost lacking, except for a small amount of burrow seizing from small crabs by larger individuals; the minimal, ritualized combat between males peaked in mid-June, as did the infrequent waving displays and high intensity courtships. Non-social activities, however, including feeding, burrowing, and quickness of response to moving objects appeared altogether normal.

J. Phases of Social Behavior

In addition to the rhythms described above, all tropical fiddler crabs sufficiently studied prove to have individual, endogenous phases which pass from below-ground inactivity through one or more days of unusually long periods of feeding and passive sitting to a wandering phase. This is succeeded by at least two social phases, that of aggressive wandering and that of waving display and mating in association with a particular burrow. Each has been already characterized in detail in preceding sections on combat, waving displays, and copulation. An intermediate phase, marking the onset of territoriality but without waving, is present or absent. When present, it may be demonstrated only during part of a single period of low tide. These elements are known in moderate detail only for *maracoani* (Crane, 1958 and Figs. 92, 93), but all of the elements have been exhibited, both in the field and in crabberies, by examples of all the subgenera. The phases are recurrent, and of such brief duration, at least in the tropics, that they cannot be due simply to the seasonal state of the gonads. It may well be that in the temperate zone, where the breeding season is short, no such differentiation exists. The point obviously needs study.

Under crabbery conditions, dominance hierarchies are held at most for a few days. Crabs at the top of the day's hierarchy are almost always in full display phase, in which the crab waves and defends a territory. Only when in this phase does a male copulate, although males in the phase of aggressive wandering often try unsuccessfully to do so. The burrow held by a waving crab is then often the center of combat between the two; occasionally the wanderer seizes the burrow, but if so, he soon abandons it. Males at the peak of a display phase sometimes continue to wave vigorously at times of the day when other members of their species are inactive.

Crabs in the non-social phases are unaggressive, move about with the body held close to the ground, and sometimes remain secluded for days in their burrows. These periods underground are only rarely associated with molting.

Accumulating examples indicate that capturing and handling active crabs, whether they are aggressive wanderers or displaying individuals, sometimes alters their subsequent behavior, changing their phase to one less active socially.

Droving, rare in the genus as a whole, shows intermediate characteristics. The participants are obviously in a phase at once non-aggressive, non-territorial, and non-waving; yet the movements, somewhat synchronized as they are, are unmistakably a form of social behavior.

A final phase variation needing comment has been only briefly observed in the Indo-Pacific. Rare individuals, showing every other sign of being aggressive wanderers, wave as they walk even when nowhere near a burrow they have occupied. Altevogt first reported such behavior in *vocans* in India (1955.1, 1955.2). I have also seen it in *vocans* in Zanzibar, as well as in *dussumieri* in the Philippines and in *urvillei* in Zanzibar. The two subgenera concerned are *Thalassuca* and *Deltuca*.

K. UNUSUAL BEHAVIOR

The term *unusual behavior* may perhaps serve less inaptly than some other heading for this section, covering the range of examples better than *abnormal activities*, for instance, or *irregular patterns*. Still, the choice is really a catchall to cover our ignorance. Once we understand the circumstances under which an activity occurs, even if the function remains obscure, the bit of pattern usually fits into the ethological scheme. Also, once we have learned something about it, we may meet it often, as with a new word mastered and adopted with enthusiasm.

A good example is the "posing" of fiddler crabs. Observers now and then come upon a *Uca* that spends many minutes without moving and is unusually slow to take flight. The individuals most likely to attract attention are large males as they stand beside their burrows, bodies raised, major chelipeds extended up and out, and their carapaces tilted toward the sun. Almost any population visited several times will yield at least one conspicuous example. A number of writers (Pearse, 1912.2, 1914.2; Crane, 1943.3; Altevogt, 1957.1; von Hagen, 1962) commented on the habit. Pearse's suggestions that it might be a threat posture or part of courtship proved unsatisfactory and the behavior remained unexplained. Jansen (1970) has now found that posing, at least in temperate zone populations of *U. tangeri*, occurs before the molt, and summarizes his findings in part as follows (p. 58):

> The concentration of the hemolymph is correlated with external factors (salinity, temperature) and the moulting processes. Changes of environmental salinity lead to adaptive hemolymph concentration by ional shifts, but there is no shift of the hemolymph water content. . . . There is an adaptation to the seasonal temperatures in *Uca tangeri* leading to a rise of upper thermal death limits during the summer. *Uca tangeri* regulates its body temperature by transpiration, extrusion of buccal foam, and sheltering in places of favourable microclimate. Posing is due to a deficiency of water- and ion-balance. It is only during the premoult stages that a critical hemolymph concentration is met with in *Uca tangeri*, and only during this time the crab is especially sensitive to environmental concentration.

The seasonal changes described cannot be responsible for the prevalence of posing also in the tropics, but Jansen's work certainly provides the key to the behavior as it occurs throughout the genus. Posing can no longer properly be listed as "unusual behavior," although before Jansen's explanation I probably would have included it among the examples below.

Most forms of behavior that both occur infrequently and remain unexplained are probably activities of males in the aggressive wandering phase (p. 687). A striking example is given by Pearse (1912.2), in his account of fiddler crab behavior in Manila:

> Some of the activities of the fiddlers were like those displayed by higher animals while at play. The crabs frequently darted about apparently

without a serious purpose, and were sometimes downright mischievous. On one occasion a male was half-heartedly pursuing a female. She went to her burrow, secured a plug near by, and shut herself in. The male then came directly to the burrow, seized the plug, and cast it to one side. The female was just emerging from the burrow when the writer ended the episode by frightening the participants by a sudden movement. Another time, two males (an *Uca marionis nitida* and *U. forcipata*) of medium size were seen running about for perhaps half an hour over an area about 12 meters in diameter. They kept close together and acted like two mischievous sailors ashore. The tide was coming in rapidly, and in their rambles the pair came to a place where a large slow-moving *U. forcipata* was carrying a plug to close his burrow. They waited until the plug had been pulled down over the owner, then the *U. forcipata* went to the hole and removed it; and, as the outraged owner emerged, the plug remover and his mate scuttled off toward the former's burrow some 4.5 meters away. He soon closed his own burrow, for the advancing water threatened to inundate it, and his companion hurried away down the *estero*. The writer watched him until he had gone more than 11 meters and was lost to view at the edge of the advancing water. To all appearances activities such as these just described were carried out in a spirit of sport.

The account is more revealing than Pearse's later (1914.2) shortened version of the same episode, since the quoted paragraph gives the distances traversed by the fiddlers. This progress through the population, the "half-hearted" pursuit of the female, and the generally destructive character of the males' activities are frequent characteristics of aggressive wanderers. I have never seen two such males moving about close together as Pearse described, but any one who spends enough time watching fiddlers will not be greatly surprised by the account. Such an observer may also sometimes decide not to delete the anthropomorphic comparisons that turn up in his field notes; much of the lively variety of activities in a group of fiddlers is peculiarly difficult to convey in undefiled scientific terms.

A second example I would also now unhesitatingly attribute to the crab's being in an aggressive wandering phase. The episode took place in Panama and the male fiddler concerned, *U. musica terpsichores*, was individually known to me from observations made on several previous days on the same spot. In this account (Crane, 1941.1: 160), the word *shelter* was used for structures beside the burrow now termed *pillars* and *hoods* (present study, p. 500).

One of the most individualistic, inexplicable performances I saw was that of a moderate-sized but apparently adult male *terpsichores*. His display coloration was not well developed on the day in question, his usually white carapace being heavily streaked with dull yellow and his cheliped scarcely pink. He did not build a hood or display, but enlarged his burrow and fed energetically. Then, suddenly, he went straight over to the newly erected shelter of a neighbor fully eighteen inches away. Without any provocation or preliminaries he undermined the shelter from the rear and pushed it down on top of its owner; the two crabs then spent 15 minutes fighting, in the course of which both darkened rapidly, losing all trace of display coloration, and the shelter owner lost the tip of his pollex. Finally, the aggressor let the owner go, then went directly to the next hood, six inches from the first, and repeated the episode exactly. In this case, too, the owner was powerless and was constantly thrust down his own hole, although he put up a good fight. At last, after another 25 minutes of uninterrupted struggle, the aggressor released this crab also, and returned, without any hesitation, to his first victim, who by now was cleaning himself up and had regained most of his display coloration. At the approach of his former antagonist, the victim tried to flee down his hole, but was seized from behind. Another duel, lasting no more than several minutes this time, followed, and ended as on the first two occasions by the aggressor's abruptly releasing his victim. This time the former returned slowly but directly to his own hole, cleaned himself, and began to feed. Neither of the two victims rebuilt their shelters on that day, although the tide was only slightly past dead low at the time.

On another visit to Panama in 1957 I observed similar behavior in a male *beebei*, which pulled down and stamped on the pillar of a neighbor; as in the male *m. terpsichores* described above, the *beebei* neither displayed nor built a pillar on the day concerned.

As suggested in the section on precopulatory behavior, it seems likely that the seizure of females in the large cheliped observed by Altevogt (1957.1) and von Hagen (1962) may also have been performed by aggressive wanderers.

In an early paper (Verrill, 1873) appears an account of fiddlers' taking bits of vegetation down their burrows, presumably for use as food. I have not been able to check this behavior in *Uca*, although it has also been reported in several popular accounts and I have seen examples of it in *Ocypode gaudichaudii* in the eastern Pacific. In Tahiti a small land crab (Gecarcinidae) appeared by daylight at the mouth of his burrow as soon as I dropped a red pen beside it; this he seized in his claw and dragged underground.

As a final example the following observation made in Fiji in 1965 (unpublished) will end this section, which could continue at great length. It concerns courtship behavior that became irregular, in *c. coarctata*. A male and female, with burrows about 18 inches apart, performed a sequence I have not seen in any other *Uca*. For about ten minutes they had been in copulatory position beside the female's burrow, the normal location in this species. Possibly disturbed by me, the female suddenly went underground and the male returned to his own burrow and also descended. Five minutes later the female emerged, went straight to the male's hole, and reached down it with the ambulatories of one side. She then returned to her own burrow, moving fast. Meanwhile the male surfaced promptly, followed her as he waved, caught up with her close to her burrow's mouth, and they started mating again.

L. SUMMARY

The categories of social behavior distinguished in *Uca* consist of droving, agonistic postures and motions, sound production, combat components, waving display, construction activities, and precopulatory behavior. Several of these classes include many different components; 14 agonistic postures and motions have been distinguished, 13 components of ritualized combat, and 14 of waving display exclusive of timing elements. In addition, 16 methods of sound production are enumerated, counting both those known through tape recordings and those presumed to occur from combined morphological and behavioral evidence. When to all these figures are added the several components in each of the remaining categories, the number of known components in *Uca*'s social repertory totals more than 70; at least 50 of them certainly occur in each of a number of species. Many of the components are widely distributed throughout the genus; some are ubiquitous; a few are apparently restricted to several species. Almost all social patterns are connected at least indirectly with reproduction. The majority are concerned with aggressive behavior rather than with courtship; others are ambivalent, including most components of waving display. Only a few are confined to courtship. Social behavior is controlled by a number of rhythms, all more or less endogenous. They include responses to tidal, circadian and seasonal factors; in addition there is possibly a separate semi-lunar rhythm. At least in the males of many tropical species, individual rhythms are responsible for short behavioral phases, mating behavior being restricted to a period termed the display phase. Many examples of unusual behavior are probably due to a male's being in the midst of an aggressive wandering phase.

Chapter 6. Territoriality, Functions of Combat and Display, and the Origins of Social Patterns

OUTLINE

A. INTRODUCTION

The last chapter examined aspects of social behavior which could be objectively described. In colloquial brief, it dealt with the *whats* and *whens* of these patterns. This chapter, in contrast, deals with the *what fors* and the *how comes*, which, together, constitute the *whys*. Both subjects obviously demand speculation, which in many contributions can be quarantined at the end. In this discussion such a course would be illogical and confusing, since the heart of the subject consists of speculation. Therefore I have tried simply to base most non-factual statements directly on objective knowledge which is either reviewed in the same section or else supported by references given to other pages; more general speculation is, I hope, labeled with clarity.

The chapter, then, will attempt to make some sort of functional and evolutionary sense from the wealth of variety in the social life of fiddler crabs. Looked at as a whole, this variety is astonishing. The components reviewed in the preceding pages listed seven general categories—droves, threat postures and associated movements, sound components, combat, waving display, constructions beside burrows, and copulation. Fourteen different threat postures and motions emerged; 16 methods of sound production were counted, both proved and assumed; 15 components of combat were described, along with 18 components of waving display.

In combat alone, 84 different morphological specializations were designated on the claws; sometimes more than 40 of these structures appear in a single species. Some of the combat components should in time certainly be subdivided, as they become better known, and the 84 structures on propodus and dactyl could even now, by viewing them differently, be increased with propriety, without further knowledge of their functions.

At the end of the chapter a review of the rhythms influencing the timing of social behavior added more complexities to the subject. Occurrence of the various activities is controlled by endogenous rhythms responsive at the least to tidal levels and to light. In addition, a separate factor of temporary, recurrent, physiological phases is individual and apparently wholly under internal control. We have also noted that social behavior is not always closely connected with the state of an individual's gonads, although practically all of it is confined to populations in their breeding season.

Because of its association with all forms of social behavior in *Uca*, territoriality becomes the subject of the first section of the present chapter. The second division discusses the functions, known and suspected, of the social patterns, while the final section considers the role of ritualization and other factors in their evolution.

B. TERRITORIALITY

Although one useful definition describes a territory as "any defended area," the phrase is somewhat inadequate for use in discussing fiddler crabs. A review of various aspects of territoriality in the genus must, in fact, make use of most current concepts of the subject, as well as of related topics.

One of these related aspects is covered by the term "home range." It applies very well to the basic condition of entire local populations of a species, where in any one low-tide period most individuals are neither waving nor fighting. Instead, they pass the greater part of their time in feeding close to the burrow in which they spent the preceding period of high tide, and at most wander, feeding, nearer the water's edge. This minor wandering depends both on the individual's phase and on the richness of the substrate in providing food close to the burrow. During an aggressive wandering phase, the range extension is among displaying males of the local population, usually higher on the shore, rather than down near the low-tide level. Thus the home range of an individual may change from day to day, and may usefully be viewed as almost the same as that of the local population. In these terms it normally extends throughout the local area that is ecologically suitable for the population. Where such a biotope is extensive, as on a long strip of homogeneous shore, it is possible that no single individual ever travels throughout the area, even during several wandering phases. The necessary work with marked crabs, started by Altevogt (1954), remains to be done. When it is, care must be taken either to avoid handling during marking, or to allow several days for the released crabs to recover before relying on any data collected. This procedure is essential because of the effects of handling on the crabs; the indications are strong that their individual phases are affected.

Animal territories in the more restricted sense may be divided into three general groups, each of which helps fulfill several related biological needs. First, a territory often supplies protection from environmental stress and from predators. Second, it frequently promotes successful reproduction in a variety of ways; it may serve as a center to attract and stimulate sexual partners and to deal with male intruders, threatened or actual; the resulting displays and combats almost certainly are mutually stimulating, regardless of their outcomes, and in many animals ensure fertilization of most females by the stronger or more energetic males; in some animals territories also serve importantly as centers for rearing the young. Third, a territory may aid in the distribution of individuals in a population in accordance with the food supply.

In the coming discussion, the view is presented that in *Uca* only the first two categories are chiefly concerned. All three will now be described as they occur in fiddler crabs.

I. *Basic Territoriality: The Burrow as a Shelter*

For these crabs as for numerous other intertidal animals, the burrow is primarily a protection. During high tide it provides shelter against aquatic predators, and during low water against the triple threat of desiccation, heat, and terrestrial enemies. Burrows serve these functions whether or not a crab defends a particular hole.

Most individuals during most low-tide periods remain unchallenged holders of the particular burrow from which they emerged when the water receded; they feed in the neighborhood, excavate the burrow a bit, and retire into it as the tide rises. A break in this pattern usually comes only if the resident enters a wandering phase, or if it is dispossessed.

A crab that for any reason lacks a burrow generally has little difficulty in finding a new one. Biotopes normally have more burrows than crabs when the tide is out, since some wanderers always move out of their holes as the water level falls. A burrow that gives even a fair fit will be taken over and remodeled as needed. A corollary is that although any crab except the smallest can, capably and in minutes, dig a new one, a burrowless crab, unless he is in an aggressive phase, with few exceptions hunts first for an empty hole. This habit has perhaps adaptive value through conservation of time and energy, when all activities must be accomplished during short periods.

Very young crabs do not use burrows at all. When pursued or when overtaken by the tide, they simply sink into the damp terrain wherever they happen to be. Herrnkind (1968.2) gives data on the instars at which laboratory-reared *pugilator* start to dig burrows.

When feeding far out on the moist flats, alarmed crabs sometimes sink into the substrate in a fashion used by the very young. When such a crab returns to the burrow area ahead of the rising tide, it does not necessarily approach its former burrow, but seeks out an empty hole, as described above. Sometimes, as repeatedly noted in a northern population of *pugilator*, the burrowless crab quickly ousts a smaller crab by prying it out with the major cheliped or digging it up with the ambulatories. In the examples observed, the dispossessed individual, always in a non-display phase, either sought an empty burrow nearby or, often, ousted a still smaller crab. In each case, there was no sign of protest by the evicted individual. The final, smallest crab in such a series several times sought shelter under a shell. In each example the dispossessions took place low on the beach, close to the level of the rising water. Females as well as males were dug up, but I never saw a female oust a male; sometimes they traveled many feet before finding an empty burrow of suitable size.

When startled or pursued in the burrow area but at a distance from his own hole, a fiddler often shelters temporarily in one already occupied. When an intruding *Uca* enters such a burrow, the resident makes acoustic signals. If the alarm persists, the signal stops and the two crabs temporarily tolerate the double occupancy. Usually the intruder emerges and moves away soon after the emergency has passed. When an intruder is suddenly and somewhat forcefully inserted by an investigator, sounds of scuffling often follow and normal acoustic signals are lacking.

This acoustic warning system is clearly genus-wide. Specific differences in the recorded signals have not yet been investigated, nor is it known how both signals and tolerance levels vary with respect to size, sex, and physiological phase.

Since this kind of burrow defense takes place only below ground, the territory is obviously confined solely to the burrow itself.

II. *Display Territoriality: The Burrow as the Center of a Defended Area*

When a male crab becomes ready for courtship and mating, the mouth of a particular burrow quite suddenly becomes for him the center of a small territory which he vigorously defends on the surface, primarily through visual signals of threat and of waving display. Sound production at the surface, as well as underground, sometimes also plays a part in this second territorial pattern. Intermale combats occur in any part of the territory, but are usually close to the burrow's mouth; only in encounters between neighbors do they occasionally take place on the borders. Finally, in the subgenera which are socially most advanced, receptive females are attracted to the defended area where, in the ultimate stages of court-

ship, they are induced to descend the burrow by intensive waving and other motions, sound production, herding, or a combination of these activities. When the burrow serves, following successful courtships, as a mating site, it offers obvious protection.

In these advanced forms, too, the burrow mouth itself perhaps acts as a signal to the female. Perhaps it shows conspicuously as a dark object in the paler terrain, and so may direct or even attract her after the displaying male has vanished into it.

Adult females, especially when carrying eggs, uncommonly also defend burrows on the surface by means of threat postures, stridulation, and, sometimes, the interlacing of ambulatories in apparently forceful combat with other females. Almost nothing precise is yet known of the nature of territorial behavior in this sex.

The size of the defended ground around the burrow mouth varies widely depending on the species, the size of the individual and the degree of crowding on the biotope. The display energy of the crab on a particular day is probably also a factor. Usually a fiddler defends a small circle with a radius equaling several of his own widths, with the crab measured in the maximum stance of full display. In species of *Deltuca*, where the male pursues the female and display is vertical, the territory seems to be smallest, scarcely more than standing room with the cheliped extended in lateral threat position. In species with advanced display requiring more space, as in *stenodactylus*, the defended area is relatively much larger; an excited individual races back and forth, its elongate cheliped thrust out and up, and the crab probably covering more than six times its maximum dimension in an ordinary beat. When females defend a burrow on the surface, they confine their activity to standing-room beside the mouth.

In flourishing populations, boundaries of displaying males often adjoin and the territories are smaller. Females and young, passing or feeding, are tolerated close to or even within territorial limits; members of other species are similarly accepted, providing that a male is not closely similar in size and shape to the burrow-holder. Often, however, some of these tolerated individuals are warned off by low intensity threat motions when the resident male is feeding.

During the onset of display territoriality, fiddler crabs tend to go to the upper levels of their intertidal habitat or to pick some slight eminence for their burrow. This behavior probably has the double effect of giving them more time daily for display and acts, in combination with upward stretches of cheliped and body, to make them more conspicuous. Finally, the drier substrate doubtless provides a better resonator for acoustic drumming (von Hagen, 1962).

In many populations the very largest individuals

are found farther from the water, often in quite un-suitable localities. In these cases the tendency to go higher seems to be an example of hypertrophied be-havior, analogous to the impractically large chelipeds found in senescent males. If the territory is long held the crabs may be definitely undernourished, although this result has not been proved. Unexplained deaths, perhaps due to senescence in some sense, occurred almost daily during a week I spent observing *tangeri* in Angola. All the dead crabs were among the larg-est males, all occupying a row of the highest burrows, along the upper edge of the intertidal area; each day a few of these large crabs were in the display phase, but no females were ever seen to approach the zone. All of the nine dead crabs examined had atrophied digestive glands (livers). A similar state was found in a number of large *maracoani* kept for months in the Trinidad crabberies; these individuals, also, oc-cupied burrows on the highest available substrate, and occasionally displayed one day and died the next.

Another such instance of the hypertrophy of this part of territorial behavior was well shown by a sin-gle large *demani* male near Davao in the Philippines. He waved, almost without rest periods, all during the single low-tide period of observation. His burrow mouth surmounted a nearly desiccated elevation. No other member of his species was closer than 50 feet, and all were lower down on damp ground, suitable for feeding. Presumably the lone male's display area was covered by water at times of spring tides; at present however there was no trace even of other burrows.

The frequent concentration of displaying males in a restricted section of rather high ground superficially resembles a lek formation. It must be emphasized, however, that these gatherings have not yet been found to be characteristic of all observed populations in even a single species. Very often the displaying males are scattered, although in clear sight of one another, throughout the greater part of a population. The relative breeding success of individual males with burrows in different locations has not yet been investigated.

Females in breeding condition in some species, particularly in the socially less specialized subgenera of the Indo-Pacific where mating takes place on the surface, occupy burrows close to those of waving males. In the advanced groups, receptive females move actively through the displaying part of the population. After mating, while the eggs are develop-ing, they usually inhabit burrows closer to the low-tide levels than those of any other section of the population.

Early reports (Schwartz & Safir, 1915 and ref.) state that single males and females occupy the same burrow for indefinite periods. The instances I have found of pairs in burrows are of three kinds. First, a female, after an apparently successful courtship, often or usually stays in the burrow with the male throughout the following high tide; this statement is based on about eight reliable field observations of five species in assorted subgenera, as well as on at least that many others in the Trinidad crabberies. Second, I once collected a series of about 50 adult *ornata* in Panama; three of the burrows were occu-pied by adult pairs; none held two crabs of the same sex; the three pairs had perhaps been startled by my approach or had just mated. Third, as described ear-lier, when fiddlers are startled while feeding away from their burrows, such as when a human being approaches, they take refuge in any nearby burrow of convenient size, regardless of the presence or sex of its occupant; I have often collected such individ-uals. It seems that future investigators of the dura-tion of double occupancy by pairs should take special care to exclude startled crabs from the data.

A display territory may be defended for only part of one low tide, or held for many weeks. Often terri-tory-holding, as in *Scopimera* and *Dotilla*, is so tem-porary that a new burrow is dug at every low tide. Then the crab sinks and twists in a characteristic fashion wherever it happens to be when nearly over-taken by the rising water. Burrows in these genera, as in *Uca*, are defended at the surface only by crabs in display phase.

Species of the *Uca* subgenera *Deltuca* and *Thalas-suca* do not usually keep the same burrow more than a few days. Often in *vocans* and other species living close to low-tide levels, burrows seem to be changed freely, even between low and high tides of the same day, and display territories are sometimes held only for part of a single low-tide, as I found definitely to be true in a few individuals observed in Fiji, during 1965. Long before, Altevogt reported (1954) simi-lar mobility in the same species in western India, and, unfamiliar at the time with many other species, con-cluded that the crabs should not properly be called territorial. It is now clear, however, that true terri-torial behavior does occur, even in *vocans*, and that the duration of defense should be left out of any definition of territorial behavior.

A contrasting example was afforded by two *dussu-mieri* from the Philippines, a male and female. These crabs were kept for months in a Trinidad crabbery, along with a large and changing population of other species. The female moved the mouth of her burrow from time to time, laid eggs, and eventually died. The male, which came into the display phase repeat-edly and maintained threat and combat relations with neotropical males of comparable size, kept the mouth of his burrow within an area of about 6 inches square for more than 6 months. The slight shifts in the position of the mouth usually occurred only after a fresh layer of mud was added to the substrate. This crab also went through brief phases of wandering and aggressive wandering from time to time, but returned to the burrow daily.

Additional examples of territories definitely held for more than one week include the following observations, each on only one or two individuals: *stylifera* in Panama (Crane, 1941); *chlorophthalmus* in New Guinea; and *arcuata, chlorophthalmus, maracoani, thayeri, rapax,* and *festae,* all in the Trinidad crabberies.

Numerous other observations both in the field and in the crabberies indicate that display territories in *Uca* are usually held for several days at most, throughout the period of one display phase. As in related areas of fiddler study, only extensive field work with marked crabs will give an accurate picture of the situation.

III. *Territory and Dispersal: The Relation of Burrow Defense to Food Supply*

In many animals territoriality is instrumental in distributing members of the population in accordance with the food supply, a view amply demonstrated in many of Wynne-Edwards' examples (1962).

In *Uca* such a spacing-out function may be performed by basic territoriality, where the burrow serves as a shelter. As described above, occupancy of a burrow by more than one crab is discouraged by acoustic signals. It is not yet clear why this is so. In at least one related genus (*Potamocypoda,* Tweedie, 1954; Crane, unpublished), large burrows are at least sometimes altogether communal. In these groups the individuals are all smaller than are most fiddler crabs, and all live in very wet, muddy environments, so that food supplies are probably more abundant than in the habitats of many *Uca.* In the latter genus, the adaptive value of single occupancy very likely involves efficient distribution and undisturbed feeding over an adequate-sized patch of terrain.

It does not seem possible, however, that either basic or display territoriality in *Uca* can ensure the maintenance, through successive generations, of a stable population consonant with the food supply on a particular biotope. As in most crabs, the larvae are pelagic, and there seems to be no mechanism for ensuring that surviving young are cast up on the parental shore. In addition, the population of a given strip of ecologically suitable shore must be determined primarily by the food supply available to the youngest crabs, near the water's edge.

Very young crabs do not fight and hold no terri-

tories. As the tide approaches, partly grown males at most sometimes take over burrows low on the shore from smaller crabs, which make no apparent effort to stay in possession, but simply move on to an empty burrow, or oust some still smaller individual of either sex.

In contrast, mature crabs in full display phase and defending territories high up on the shore often do little feeding and, since both growing algae and tidal jetsam are less abundant, have less food than do those farther down. These individuals sometimes leave their burrows and descend to feed among the non-territorials. The center of the breeding individuals usually lies in the middle parts of a habitat. Here, among the displaying males, non-breeding as well as receptive females are tolerated along with young of assorted sizes. The very young move about a lot, and females may or may not do so, depending on their phase.

These most actively breeding parts of a population do not usually contain the largest males, which more often are found on the highest usable terrain, only briefly moistened by the tide. Gigantic males, apparently senescent and perhaps undernourished but displaying characteristically and defending their boundaries from similar neighbors, sometimes are the only occupants of this unfavorable zone.

In short, display territoriality in *Uca* apparently does not at all promote an optimum distribution of a population. Basic territoriality on the other hand, in discouraging occupation of a burrow by more than one individual, may aid in distributing individuals over the biotope. It seems that neither form of territoriality, however, can play any part in keeping the population stable.

C. FUNCTIONS OF COMBAT AND DISPLAY

I. *Introduction*

The preceding section described the display territory as a center of social behavior, including visual threat display, sound production, intermale combat, and waving display. In this section suggestions will be made concerning the functions of all these activities and their relations both to territoriality and to one another. The basic question of the role of agonistic behavior in relation to courtship and reproduction will also be examined.

The selection of a logical order of comment is difficult. Combats rarely take place unless preceded by threat postures or by waving, and usually by both. Yet, as will be shown later, threat postures appear to be derived from combat while some waving components just as clearly are based on threat. Sound components crop up throughout the categories. On the whole, combat can be viewed as the most basic of the activities concerned, even though it occurs more rarely than any of the patterns, and, in its wealth of modern complexities, is certainly the most specialized of all.

The functions of intermale combat and its ritualizations, therefore, will be considered in the first two sections. Parts of both are taken with little alteration from a previous contribution (Crane, 1967), since they still express my present viewpoint. Several results of subsequent field studies are also briefly considered in these and later sections, because they affect some of the conclusions in the published paper; a formal contribution is in preparation.

II. *Functions of Intermale Combat*

Possible explanations of fighting remain far from clear. It is unsurprising that each crab coming into territorial and display phases acquires a burrow in a spot appropriate for display and mating. But there appears to be no need to obtain such a burrow through combat. As described earlier, there is never a shortage of empty burrows during a period of low tide. Furthermore, few areas are so crowded that there does not seem to be ample space, empty and undefended, among displaying males where new burrows could be dug.

Yet an aggressive wanderer, unless his aggressiveness has been reduced by a fight with a burrow-holder, either pays no attention to these empty burrows, or pokes into them with cheliped or ambulatories briefly and superficially; then he moves on.

Although occasionally an aggressive wanderer stops slightly longer at a burrow occupied by a non-displaying crab, any attempt to dig out such a crab is rare and little effort is expended. I have never knowingly seen a combat between a crab that is not in display phase and a wanderer, or between two wanderers.

In fiddler crabs no harems are maintained and single females seem never to be direct causes of intermale combat. Occasionally a male even abandons an advanced courtship attracted by a combat between two other males.

The immediate goal of an instigator does not in fact seem to be the taking over of a suitable burrow as a center for display or a direct competition for females. Rather, the apparent aim is a combat with a displaying male.

The combat itself is characteristically partly or fully ritualized; only in *lactea* are components known that are at once forceful and stereotyped (p. 494); in most cases the combat results in no detectable change in the subsequent behavior of either crab. One or the other withdraws his claw from contact with that of his opponent; the aggressive wanderer

resumes his progress through the population, threatening and entering into new combats; the burrow-holder promptly resumes display, its intensity undiminished. In one such *rapax* combat in nine, however, a wanderer's aggressiveness was reduced; in one in 45, the burrow-holder was dispossessed and the wanderer took over. Less intensive observation of other advanced species have yielded corroborative observations: the wanderer's behavior is similar and only rarely is there a detectable result. In those instances where a wanderer actually takes over a burrow he sometimes assumes the display phase at once; more often, he does not wave, but shortly abandons the burrow and moves on, his aggressiveness maintained and his territorial drive still in abeyance.

With these figures in mind it seems likely that combat may sometimes either advance or retard the assumption of territorial and waving phases by the wanderer. Summation, as suggested on p. 494, may well play a part here that the field techniques in use could only suggest. Combat, then, may serve as a mechanism for ensuring that suitable burrows for display are not taken over by males in subbreeding condition; nevertheless, the availability of empty burrows, noted above, forms an obvious argument against this view.

The function of combat will now be examined from the point of view of the burrow-holder. If this displaying crab is not vigorous enough or sufficiently motivated to fend off an aggressive wanderer, he may be in an inadequate condition for breeding and should not, from the point of view of selection, be left in a position to attract receptive females. Yet many vigorous burrow-holders, in other species as in *rapax*, withdraw partly or wholly from an incipient combat, even in the frequent instances where the approaching wanderer is the smaller crab; then the burrow-holders resume waving and courting promptly and strongly when the wanderer has departed. The role of this withdrawal behavior in the pattern of combat remains puzzling.

After combat, however, one *rapax* in six delays waving, while one in 45 loses his burrow, with a consequent postponement of resumed display. These relative numbers agree well with impressions received in numerous more casual observations of combat in other species.

In examining the possible selective values of fighting and its ritualization, a distinction should be kept in mind between these two visible results—namely reduced aggressiveness and delayed waving. Since all burrow-holders are in the display phase and, as part of that phase, in a threatening and fighting mood toward both aggressive wanderers and trespassing neighbors, it seems that a post-combat reduction in aggressiveness by a wanderer normally would result only if he were not ready for territorial-display-mating behavior. In that case the "loss" of a combat would be a selective advantage. On the other hand, a reduction of display time for a burrow-holder through prolonged combat would be a disadvantage.

A reasonable suggestion, therefore, appears to be that the ultimate value of combat, regardless of the role of ritualization, lies in preventing suboptimal males from wasting the breeding time of the population by attracting receptive females. This explanation does not, however, account for all the facts. It takes no account of combats, largely or fully ritualized and resulting very rarely in waving delays, between vigorous burrow-holders. Again, the function of down-pushes remains unexplained. Here one burrow-holder, far from endeavoring to take over the burrow of his neighbor or at least to dig the occupant out and engage him in combat, simply thrusts him forcefully underground before returning to his own burrow and resuming display.

When viewed as a whole it seems that the function of combat may lie primarily in stimulating and synchronizing mating behavior. As in so many other groups of animals where such an effect is suspected, proof awaits work in endocrinology and neurophysiology.

Similarly in need of the attention of physiologists are two strong impressions that recur during field work on *Uca*. One is that combat may serve to release tension in the actively courting section of the population. The other impression, particularly compelling when one is watching ritualized mutual encounters, is that combat appears often to be in progress for its own sake. The attention of a third crab is sometimes drawn to a nearby combat; he may then either interrupt or engage one of the participants after the end of the first encounter. Even more suggestive are the sequences of high ritualization discussed below. It is noteworthy that recent experimental work indicates the existence of an "appetance for aggression" in two species of fish and in squirrel monkeys (Thompson, 1963, 1964; Azrin, Hutchinson, & McLaughlin, 1965; Rasa, 1971); the subject is reviewed by Eibl-Eibesfeldt (1970: 326).

III. *Questions of Adaptive Values in Combat Ritualization*

As shown in previous sections, a large majority of combats in *Uca rapax* show no detectable element of force and hence may be termed fully ritualized. More casual observations on other species indicate that ritualization is similarly prevalent throughout the genus. Finally, in one component prevalent in *lactea* combats, force itself appears to be ritualized. In searching for the selective advantages of ritualization, the immediate effects of individual combats have proved unilluminating. As is well known, even

the most violent fights in *Uca* practically never result in physical damage; no injury at all was ever seen in *rapax*. The exceptions are healed puncture wounds on a manus that are apparently inflicted by an opponent's chela tips. As soon as one learns their appearance, the small pits or discolorations can be found on occasional males in almost any collection of preserved fiddlers. I now suspect that the great majority are not received in combats on the surface. Instead, they probably result from the engagement of two claws in a burrow shaft, when a male reaches down and tries to pull another fiddler out. At least in a northern population of *pugilator*, I often saw young crabs dispossess still smaller crabs in this way when the tide was approaching (p. 511). Particularly if the burrow occupant has recently molted, an intruding claw might easily pierce the soft integument. Possibly a deep puncture would leave visible traces after subsequent molts (Pl. 46).

It seems, therefore, that a protective function, which has been considered obvious in the ritualized encounters of many well-armed animals, is not now of importance in fiddler crabs.

Again, the data on *rapax* give no evidence that ritualized encounters are any more likely than the uncommon forceful fights either to promote or to prevent behavior changes in an opponent. This is true in general both of reductions in the aggressiveness of a wanderer and of delays in resumption of waving by a burrow-holder.

In *rapax* the only apparent advantage of ritualization seems, rather, to lie in the shortening of combats. The counts so far made indicate clearly that ritualized encounters are not only far more numerous than those including components of force but also that they are shorter, most lasting less than 10 seconds. In contrast, forceful fights continuing more than one minute are usual. This is true whether or not a forceful combat results in subsequent visible behavior changes for either crab. While this difference in duration appears to have no obvious importance for an aggressive wanderer, the shortening of combats through ritualization may well be a selective advantage through its effects on burrow-holders.

This suggestion is based on both the ecology and the mating behavior of *Uca*. Since they court only during low tide, and are usually further restricted by other requirements, both meteorological and physiological, their periods for courtship and mating are limited. Combat and courtship cannot proceed simultaneously and, in *Uca*, the combats of males seem to hold no attraction whatever for females. Therefore it seems clear that, by shortening combats, ritualization ensures that courtship opportunities are minimally reduced.

It may be that an important factor in waving display lies in its stimulating effect on other males or in the synchronizing of breeding activities. Here, too, a shortening of each combat would advantageously shorten the time during which one or two wavers did not contribute to the communal effect.

An unresolved objection to the suggested advantages of shortened combats has arisen in my recent work (unpublished) on two species of the subgenus *Celuca*, *pugilator* and *lactea*. In these forms almost all combats take place close to the hour of low tide, while most waving displays and most courtships occur later. In two thriving populations of *lactea*, in fact, the last combat of the day and the first courtship were often separated by more than an hour.

One characteristic of ritualized combat becomes increasingly apparent with continued observation. This consists in the leisurely, formalized, and wholly unforceful cooperation sometimes apparent between the two opponents. A highly ritualized encounter in *rapax* may run about as follows. An instigator, whether wanderer or neighbor, approaches a burrow-holder. A rub by one or both crabs, outer manus against outer manus, usually follows. Next, the instigator sometimes holds perfectly still while his opponent slowly eases his chela into the actor's slide position; the two crabs may then reverse the role, the shift being accomplished slowly, without fumbling, and with the apparent cooperation of the crabs. In a few moments they may progress to a similar alternation of heel-and-ridging or, in hetero-clawed encounters, to an alternation of heel-and-ridging with interlaces. In other examples a single opponent may be the actor throughout, the second crab holding himself quietly. When the actor breaks off, both crabs move apart and resume their pre-encounter activities.

Observation of these encounters gives a strong impression that they provide one or both crabs with satisfactions that are not concerned in direct goals, such as taking over a burrow or evicting a trespasser; the activity itself seems to serve as the goal. Current work on *lactea* (p. 494) suggests that forceful combats also provide their own rewards. We know nothing at all yet about the means of conferring satisfaction—whether through the performance of the motions, or through the reception of associated sensory stimuli.

If ritualization does indeed sometimes operate selectively through shortening combats and thus providing more time for courtship, then an obvious pressure would be toward even shorter ritualized encounters. Ultimately the action might be reduced to a token touch of mani or single rubs of ridges by briefly overlapping chelae.

This trend is not apparent. According to our present knowledge, the socially advanced species have the largest repertoire of combat actions and the most extensive structural specializations for high-intensity encounters. If ritualization shortens combats, then further elaboration could nullify the effect. Occasional prolonged encounters in *rapax*, fully ritualized and elaborately mutual, suggest that this process may prove to be a factor in the continuing evolution of the species.

IV. *Functions of Threat Postures, Threat Motions, and Sound Production*

When the functions of threat postures, motions, and sound production are considered in relation to the above conclusions on combat, an obvious question arises. If ritualized combat is such a short, efficient, and pleasant way of stimulating, synchronizing, or otherwise promoting reproduction, why does threat behavior persist in the genus? After all, these postures and motions seem most effectively to discourage the realization of many potential combats. The apparent answer has a number of parts, based on both direct and indirect evidence and, ultimately, on speculation.

First, basic forms of at least visual threat display are certainly far older than *Uca*'s components of ritualized combat, which necessarily evolved in association with the specialized major cheliped. Schöne (1968) describes and illustrates the occurrence of basic agonistic postures in a wide variety of brachyuran crabs, including some fully aquatic forms; its prevalence as a behavior pattern throughout much of the animal kingdom does not need comment. It seems that such a basic form of behavior would be unlikely wholly to be eliminated from the genetic constitution of a small group.

Second, basic threat postures are used by fiddlers, as by other crabs, not only in intermale situations but also as a pre-final defense against predators (p. 473). Since actual seizing of an active threat—whether bird, crab-eating raccoon, or human finger—may well result in the loss of a slow-growing appendage, even if the crab escapes, the use of a strongly deterrent threat display appears to be a strong advantage. This fact alone would encourage the retention of at least the basic postures of lateral and frontal threat.

Third, as described on p. 479, threat signals are freely used in *Uca* not only between aggressive wanderers and burrow-holders, or between two burrow-holders, but also, at low intensities, in gently warding off an encroaching female or a young crab when a male is feeding. Sometimes both young and females use threat signals in similar fashion, as do members of two species sharing the biotope. Threat signals are also the primary means used by unreceptive females in warding off males. The use of auditory signals by burrow-holders underground, apparently to ward off intruders, was described on p. 481. Antiphonal sound production by male *tangeri* was reported by Altevogt (1964.1).

Fourth, successful threat displays obviate all risk of time-consuming forceful combats. Such is not the case with ritualized encounters, which neither always replace prolonged fights, nor always prevent a forceful ending to a combat with early ritualized components (p. 488).

Fifth, most threat signals are less time-consuming even than most full ritualized combats (p. 493), making more time available for courtship.

The sixth and last function of threat to be considered is wholly speculative. In common with other conspicuous or audible elements of social behavior, threat postures, motions, and sounds may well be part of the pattern of general social stimulation—to the actor, to his potential antagonist, and even to his neighbors. It will be remembered (p. 492) that at least one threat action, the after-lunge, is made by a burrow-holder after a combat; it is always directed toward the receding figure of his former opponent. Comparable behavior is of course exceeding common in a variety of animals. I have seen it a number of times even in rhinoceros beetles, in a pattern described by Beebe (1946). Always the component followed an intermale combat in which one beetle was, as usual in the species, taken between the "horns" of his antagonist and flung, upside down, to the ground. The successful beetle usually then moved his own body rapidly up and down several times, by stretching and bending his legs, somewhat as in fiddler curtsy components; all the while the beetle's head faced his upset opponent. As far as I know, the physiological explanations for such patterns have not yet been provided, and the components perhaps should not be included under threat behavior. Even if a mere release of excess energy, not then needed for further combat, is involved, the effect of further, stereotyped motion is also perhaps self-stimulating to the animal.

In summary, then, the above position on the functions of threat display in *Uca*, including postures, associated motions, and sound components, may be stated as follows. The basic postures of threat display, appearing widely among brachyuran crabs, provide defenses against predators through reducing the risk of injurious or fatal contact. Some visual and acoustic threat signals are of social use in fiddlers of both sexes and different ages. In agonistic relations between adult males, threat signals have two advantages over fully ritualized combats: when effective, they altogether preclude a time-consuming forceful combat, and they are usually shorter even than brief, ritualized encounters. Finally, it also seems likely that threat behavior, in common with other forms of social display, provides both to the displaying crab

and to conspecifics stimulation that somehow promotes reproduction.

V. *Functions of Waving Display*

In the categories of social behavior so far discussed—combat and threat displays—there has been no question of ambiguity. All the components are directly associated with agonistic behavior, not with courtship. In the next category, that of waving display, the functions of most of the components are clearly ambivalent, since they are employed both in wholly intermale situations and in pure courtship (Fig. 95). As described elsewhere (p. 501), a few components occur only in courtship; they are then confined to periods of high intensity. On the other hand courtship in many species has no components differing from those of waving display directed toward males. Waving itself can and does take place out of sight of other displaying, or even apparently attentive individuals, and rarely in the Indo-Pacific even occurs without a display territory. A low-intensity, vertical form of waving uncommonly is seen even in immature individuals (p. 497). The functions of waving display cannot, therefore, be resolved in any simple fashion. According to our present understanding, its probable uses are as follows.

First, waving is an advertising display almost always indicating the presence of an adult male close to breeding condition and displaying close to a particular burrow at the center of a small defended area. There he will threaten or enter into combat with intruding males in appropriate physiological condition and, through increased intensity and often elaboration of the waving pattern, follow or attract potential mates. Waving serves, then, most obviously as an identification mark. This mark, just as in birds and many other animals, is treated quite differently depending on the sex and physiological state of the viewing conspecific.

Second, it seems unarguable in this particular category that the display is stimulating to females and, in appropriate species, perhaps the distinction should be made that it is also directive. With few exceptions, waving immediately precedes normal attempts to copulate. Again, wandering females often clearly change course and approach a crab that suddenly begins intensive display (p. 503).

Whatever future work discloses about lek-like characteristics in groups of displaying males, it seems probable that they will share with many highly developed lek patterns the advantage of arousing a female gradually to mating readiness as she wanders through the display area. So far there is no evidence that females are attracted most to males that are larger, brighter, whiter, more vigorous in their displays or with burrows near the center of the group.

Third, waving display often or even principally can be viewed primarily as a particularly efficient component of threat behavior that is confined to intermale situations. As listed in the fourth function of threat display (p. 517), waving probably reduces the frequency of actual combats, which do not at all attract females; unlike strictly threat displays, however, waving displays in highly economical fashion serve simultaneously to repel males and attract females. Thus, not only is time saved for courtship through the avoidance of long combats, but even the waste of momentary threat motions is usually eliminated. A short period of observation of any healthy population in actively breeding condition will strongly advocate this view: the intensity of waving is often or usually clearly increased by a burrow-holder at the approach of another male, even when the potential intruder's major cheliped is in full view of the waving crab. (Thus there is no question in these examples that the burrow-holder is temporarily mistaking the male for a female, as sometimes happens, stimulating the burrow-holder briefly to increase his waving toward the tempo of intensive courtship.) Very often the approaching male, whether a wanderer, aggressive wanderer, or neighboring burrow-holder, passes promptly on or withdraws, without stimulating the displaying crab to change his ambivalent wave to monovalent threat.

Fourth, whether waving is also stimulating to males is more debatable, since, as in combat and threat, evidence is lacking. Suggesting the probability of stimulation are populations in which waving display of moderate intensity is the only social behavior apparent for several hours at a time. Especially in socially advanced species of *Minuca* and *Celuca*, waving over entire display areas appears frequently to be uninterrupted by the approach of other individuals of either sex, combat and threat behavior often being virtually confined to earlier hours before the onset of waving. An excellent example is *lactea* (p. 292). At these times waving display should perhaps be regarded as another example of hypertrophied behavior, analogous to the ascent of old males to suboptimal levels of the shore, rather than as a mechanism for stimulation. The concentrated waving occurs when the tide is already on the way in and does not seem ever to be followed by bouts of agonistic or courting activity at atypically late periods during the same low tide; any stimulating effect would have to carry over until the next retreat of the water. This protracted effect may well occur, however, the results of waving perhaps being cumulative over a period. The several examples of synchronous waving (p. 300; Gordon, 1958) I have seen have all been in prosperous populations in which aggressive wanderers were at the time scarce or absent and in which neighboring males stayed close to their own burrows. While the synchrony may have an effect only on females in the display area, it seems more likely that the males themselves are stimulated.

Fifth, even if there would be no interbreeding among sympatric species in the absence of waving, interspecific differences in waving display must be an important time-saver in the intertidal hours available for courtship. Females with only the rarest exceptions are not attracted toward males of other species and then only briefly (Crane, 1941.1 and unpublished).

Sixth and finally, waving displays may serve directly as one barrier to interbreeding among intermingled allopatric forms. These last two suggested functions will be considered further in the next chapter.

In summary, waving is an advertising display, almost always centered on a territory, and characteristic of a male ready to court females and to behave aggressively toward certain other males. Such display appears to be unquestionably important in directing the attention of receptive females and in stimulating them to cooperate in copulation. It also serves as a form of threat display toward other males, reducing the frequency of time-consuming threat displays and combats while simultaneously serving as a courting signal. It also seems to be a reasonable speculation that waving display, as well as threat and combat behavior, is stimulating to other males in or near breeding condition. Again, the existence of interspecific differences in waving apparently aid the efficiency of simultaneous breeding seasons among sympatric forms by avoiding incompatible courtships. Finally, these differences may serve as an important barrier in maintaining distinctions between closely associated populations of allopatric forms.

VI. *Functions of Display Territories*

The need remains to return to the role of territory in social behavior, this time in relation to the functions considered above of combat, threat, and waving display. Display territories may act, it seems, rather as artifacts concerned with reproduction. As such they have much in common with hypertrophied claws and complex display motions. All are devices which help bring males and females into a state of readiness for mating, ensure that potential partners are physiologically in condition for copulation, and, finally, bring the members of a pair together.

To go a step further, territories may be viewed as the ritualization of an artifact, a burrow, which most of the time is purely a genus-wide defense against predators and desiccation. Only during certain physiological phases of an individual is it turned into a status symbol. At these times, as already remarked, a combat, not a burrow, seems to be of prior importance to the crab.

It was also suggested that in socially advanced subgenera, in which the female follows the male below ground, the burrow mouth itself may serve as a sign stimulus to the female, perhaps merely as a dark object seen from crab height. In the ultimate specialization in this connection, it seems possible that vertical structures erected beside the burrow by some displaying males (p. 499) may be explained as ritualized burrow mouths, serving as supernormal sign stimuli. This explanation would not, of course, preclude their simultaneous usefulness as acoustic amplifiers, as suggested by Salmon & Atsaides (1968.2).

VII. *Conclusion*

When the complex of social patterns is regarded as a whole, it is difficult to avoid the conclusion that the varied activities have a stimulating effect on the population. The attractions of these views persist, in spite of our regrettable dearth of physiological data. I think that no one has expressed this attitude as well as did Fraser Darling, in a short paper entitled "Social Behavior and Survival" (1952), which in part is based on his earlier contribution on avian sociality (1938). In commenting on the latter paper Darling states:

It held the dual thesis of the reality of social stimulation to reproductive condition in such birds as are social or colonial at some state of, or throughout, the breeding cycle; and the existence of a threshold of numbers in some colonial species, which might be critical as to whether the birds bred or not.

Naturally, the extent to which the social factor enters into the life of birds varies greatly. In some it appears to be sporadic, in others seasonal, and in others it constitutes the whole way of life. Whereas the benefits of sociality in the lower animals as studied by Allee (1931, 1938) and others appear to be physiological in origin, operation, and result, the basic element of *stimulation* in avian sociality seems to be psychological and psycho-physiological. . . .

The aggressive quality of bird song has, I think, been overemphasized. Proclamation, yes; *apparently* aggressive, yes; no more combative than a military tournament of befrogged dragoons, but probably even more stimulating. So-called fighting, and singing, are in my opinion often a form of social stimulation and have indirect survival value as aids to development of reproductive con-

dition. I should think the term "aggressive behavior" could be dropped for a great deal of true display.

It seems that the above remarks apply as well to fiddler crabs as to birds and mammals.

D. RITUALIZATION AND THE ORIGINS OF THE COMPONENTS OF SOCIAL BEHAVIOR

I. *Introduction*

Anatomists usually trace with ease the descent of even the most unusual anatomical structures. Nobody argues with the conclusion that the "fishing-rods" on the snouts of angler fish, in spite of their worm-like bait and luminous bulbs, evolved from the first spine of the dorsal fin, which migrated forward and burst into specialization. The line of descent of the major cheliped in fiddler crabs is far more obvious. Although bizarre and hypertrophied, it undoubtedly developed directly from a homologous appendage found in numerous other animals. Segments have been neither added nor lost and there is no question, in spite of altered shapes, of their identities. Even the allometric growth of this appendage, so exaggerated in *Uca*, occurs to some degree in many other crabs.

Unlike the derivation of morphological characteristics, the evolution of behavior patterns is often compound in nature, and, in the absence of fossils, conclusions can never be proved to the satisfaction of everyone. The simpler movements involved in basic activities, such as locomotion, can usually be satisfactorily traced far back in an animal's ancestry. As soon as social patterns are considered, however, complexities multiply. Particularly in reproduction the sequences often turn out clearly to be formed largely from pieces of feeding, cleaning, fighting, and other behavior patterns resulting in curious hodge-podges that somehow work.

In the course of evolution these parts of patterns have with changed functions become simplified, exaggerated, or both. Hence they are termed *ritualized* in the senses used by Huxley (1914), Lorenz (1941), Baerends (1950), Tinbergen (1952, 1953), Huxley *et al.* (1966), and Eibl-Eibesfeldt (1970). Often, too, associated morphological structures are enhanced by increased size, altered shape, or changed color. Through these modifications they are made more conspicuous in visual display or otherwise contribute toward the production of an unambiguous signal in communication. These changes from the original form may be startling. Nevertheless, when a number of related species, in various stages of display evolution, are available for observation, the behavior in question can frequently be satisfactorily traced. And the evidence is often formidable. On such evidence the bases of ritualization appear most frequently to lie in intention motions and displacement behavior. These in turn typically result from situations originally involving inadequate motivation, conflict, or frustration.

In classic examples of the derivation of displays from such activities, Lorenz (1941) showed the development of certain duck displays from preening and Baerends of stickleback courtship from combinations of escape, fighting, and nest-building motions (Baerends, 1950; Baerends & Baerends-van Roon, 1950). Similarly Tinbergen (1952) traced the courtship of herring gulls through ritualization of postures showing conflict between tendencies to fight and to submit. In gulls food-gathering motions are also incorporated into courtship display.

In *Uca* striking parallelisms are apparent with such developments in vertebrates. Without the concept of ritualization, in fact, both the motions of the large claw and the subtler movements of other appendages would be unintelligible from the point of view of evolutionary biology.

II. *Displacement Activities*

Displacement behavior is not in itself a regular part of any social category, although it is often closely associated with threat, combat, and waving display. Because it plays, according to the view held in this contribution, such an important role in the origins, through ritualization, of many social components in *Uca*, an account of its characteristics has been deferred to this section.

This kind of behavior was first reported in the genus by Gordon (1955), who described incomplete feeding motions in *vocans* in South Africa. Displacement activities also undoubtedly occur throughout most if not all of the genus. I have watched them often in all species with advanced visual display that are well known to me, as well as in the intermediate *vocans*. They are of questionable occurrence only in *Deltuca* in the Indo-Pacific, the subgenus in which visual display is least developed. I have observed good examples also in the related genus *Ilyoplax*, both in the field in Japan, the Philippines, and Java, and in the crabberies in Trinidad.

Displacement behavior in the sense used here is defined as the release of energy, accumulated through the frustration of one or more drives, by activity characteristic of another drive. For instance, a bird often meets a situation where its urge to fight is in conflict with its tendency to flee. Under these conditions, it may stand still and preen its feathers or even go to sleep.

Fiddler crabs, depending on species and circumstances, often combine feeding or cleaning motions with waving display. They also sometimes stridulate or drum when the occasion does not normally elicit this response. Often these actions are incomplete or ineffective. In a typical example, a crab's minor cheliped, in a travesty of feeding motions, pinches air near the ground and then raises the claw toward the buccal area, where the third maxillipeds are appropriately ajar; the claw then is lowered, sometimes without even touching the mouthparts. Sometimes this feeding motion is even more sketchy, consisting of a brief, vertical arc of motion, all in mid-air. Much rarer examples emerged from a recent review of old motion picture sequences. These showed that several instances of drummings by the major cheliped were incomplete; the appendage, though vibrated, barely failed to touch either the ground with the manus or the carapace with the merus.

Displacement activities are usually clearly distinct from intention motions which typically are rather easily recognized as low-intensity, preliminary versions of actions that may or may not follow. They are also usually, but not always, distinct from the simultaneous performance of two different categories of behavior, most commonly true feeding and waving. In these situations, one pattern or the other usually proceeds at least at moderate intensity, while the second, although complete and efficient, is at low intensity. In the unequivocal examples of waving-cum-feeding, there is no apparent stimulus to waving display, in the form of either a threatening or approaching male or an approaching or attentive female. The waving continues at a low level, but functional feeding proceeds, normal pellets being rhythmically discarded.

In recent years a useful distinction has often been made between displacement and redirected activities. Examples of the latter term's use concern the substitutions of an inappropriate object as the focus of aggressive or courtship behavior which is frustrated in its natural expression. In fiddler crabs the displacement or redirection is often, in single individuals and within a few seconds, so rapidly changed and so often only partly ritualized, that it seems best to keep terminology at a minimum.

Such, then, is the type of material on which, I believe, are based many of the social components which have evolved in *Uca*. Even when in a given species a displacement motion has most clearly been ritualized into a characteristic part of some kind of display, its additional use as an occasional displacement activity is sometimes very apparent. When unfamiliar species are observed, a useful clue to the history of a puzzling action may sometimes be gained from color accents. If, for example, the minor cheliped is strikingly whitened, its use during visual display may well indicate ritualized motions rather than continuing displacement activity.

Sometimes an effective way of observing displacement behavior in the field is simply to watch or photograph the crabs at exceedingly close range. The procedure works reliably, it is probably needless to say, only with small, socially advanced species at the height of a vigorous display period. Their conflict, then, is between the urge to continue waving and the urge to retreat down their burrows. Under these conditions display sometimes becomes complicated by a variety of extra motions. When observation is based on a sufficient knowledge of normal behavior in the species and of other members of the genus, the atypical grouping of motions can be enlightening from an evolutionary point of view (p. 520).

III. *Derivations of Social Components*

The apparent origins of the components in three major categories of social behavior will now be considered. Combat, agonistic postures with their associated motions, and waving display will be reviewed in that order. Sound components both proved and, through morphological evidence, apparent will be apportioned among the several categories, since sound appears sometimes to have been the basis and sometimes a result of the development of a component belonging to another category.

Precopulatory and copulatory behavior are not included in the discussion because nothing is known of their form in underground matings. Furthermore, since we know nothing as yet of the part played by

chemical factors in behavior, their possible role in the derivation of components will obviously have to be omitted. Once an adequate foundation of further research has been laid, future workers will doubtless find open to them another attractive source of reasoned speculation.

For ease of reference all the components are relisted together in Table 21, numbered in accordance with their descriptions in the chapter starting on page 476. The table also gives page numbers covering descriptions in each category along with a reference to the earlier table showing their known occurrence in the genus. No table was supplied for the category erected for agonistic postures and motions because most of them are found widely or ubiquitously in the genus. When the posture is rare the known examples are mentioned in the discussion. For further ease of reference, each component name will be followed by its number prefixed by a letter: C for combat component numbers, T for threat, W for waving, and S for sound. The accompanying diagram shows the general pattern of apparent derivations.

(a) COMBAT

In the descriptions the first two ritualized components, manus-rubs and pollex-rubs (C1, C2), are regarded as low-intensity combat in distinction to the remaining high-intensity components.

It seems likely that the low- and high-intensity ritualized components have been derived from different sources. These are, respectively, from low-intensity, forceful manus-pushes and from high-intensity, forceful fighting with linked chelipeds. If, as seems certain, the usual direction in combat evolution has been toward the reduction of forceful fighting through ritualization, one logical point for the application of a deterrent would be immediately prior to the actual grip. The high-intensity components (C3-C13) all appear to have originated directly from unritualized forceful grips. All take place with the two claws partly or wholly in a position for grasping each other; when, rarely, a ritualized encounter proceeds to a grip, little or no change in basic claw position is made. As in social sequences in many other animals, the specializations appear to have been added one in front of another, the interlace (C9) being perhaps the closest now known to the original fighting grip.

Threat gestures in many animals are themselves certainly to be understood as ritualizations of fighting, where a weapon, impressive in size or other potential advantage, is effectively exhibited. By this criterion the threat gestures of crabs including *Uca* all qualify as ritualized fighting.

We therefore emerge with the following view. From forceful fighting in ancestral *Uca* were derived two apparently distinct classes of ritualization. The

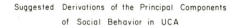

Suggested Derivations of the Principal Components of Social Behavior in UCA

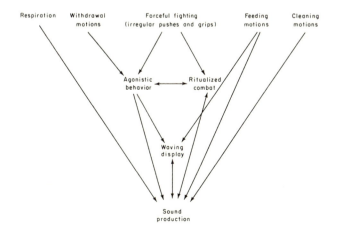

first consists of threat components, most of which, as will be suggested below, are derived largely from intention motions of fighting, including exhibition of the potential weapon. The origins of the second class, composed of the components of ritualized combat described above, are less clear. Ritualized they certainly are, in the sense that the original motions have been changed and the associated structures transformed. Yet no component of ritualized combat appears to be built on a pattern of frustration or on a conflict of drives, on intention motions, or on displacement behavior. Rather, it seems most likely that the principal roots lay in states of low motivation, the components evolving through the minute, cumulative advantages of brief, less-than-violent encounters, with actual contacts nevertheless at the usual points and along the usual edges. In short, when the doubtless more venerable threat components were not quite strong enough to control contact fighting, the forceful grips themselves became transformed into additional effective shorteners of the fray. But there is no likelihood that the two classes form a single sequence of evolution.

In the low-intensity combats formed by manus-pushes and manus-rubs (C1) a similar course of origin may be postulated. Here, again, the forceful push is considered basic; I have observed it, although casually and imperfectly, in several other ocypodids. In *Uca* combat its usual ritualized form appears to be the manus-rub (C1).

Three sound components, namely claw-rubs, claw-taps, and interdigitated leg-wags (S14-S16), are all associated with combat. The leg-wag is characteristic of females. Doubtless all developed in the closest association with the diminishing of force during the ritualization of combat. More specifically, it seems likely that the occasional occurrence of tubercles on areas of frequent contact during combat, and the consequent production of adventitious sound, led through natural selection to the variety of structures and behavior patterns now characteristic of ritualized combat. Since we know nothing yet of the role, if any, played by tactile stimulation, or indeed by sound production, in combat, further comment would be gratuitous.

(b) AGONISTIC POSTURES, ASSOCIATED MOTIONS, AND THEIR SOUND COMPONENTS

The first ten postures and motions (A1-A10) all have in common the prominent exhibition of the major cheliped and are hence confined to males. The first five of these appear to have as their bases rather simple motions of intention to fight with linked chelipeds, or to push away by force a potential opponent approaching, or being approached, from the front. The raised-carpus (A1) sometimes precedes high-intensity combat but more often leads to more in-

tense frontal threat behavior (A2-A5), or to the forceful manus-push or to the ritualized manus-rub (C1); usually it subsides without sequel. In short, it serves as a non-explicit threat signal of very low intensity, with roots that are perhaps similarly plural.

The derivation of the down-point (A-2) if viewed from the point of view of *Uca* alone, appears to be much more directly concerned with high-intensity combat than does the raised-carpus, since it apparently always precedes high-intensity linkage of chelipeds in the four species where it has been observed. Thus it seems most likely that the component is derived rather simply from a motion of fighting; the linkage, when it takes place, proceeds directly from the posture, the four claw-tips coming into interdigitated contact as they leave the ground. An interesting point is that this posture, rare in *Uca*, is widely distributed in other Brachyura (Wright, 1968). Since actual combat is little or not at all known in these other examples, it is premature to comment further. The known examples in *Uca* are *maracoani* and *ornata* (subgenus *Uca*), *tangeri* (*Afruca*), and *inaequalis* (*Celuca*).

The frontal-arc (A3), with its to-and-fro warding-off motion, appears in contrast to the down-point (A2) to be derived from manus-pushing, rather than from high-intensity fights.

The forward-point and lunge (A4, A5), on the other hand, appear again to be intention motions of fighting. The after-lunge (A6) differs little or not at all from the lunge in posture and motion, yet its occurrence almost always follows an actual combat and is directed toward a retiring opponent. Once more, our ignorance of some of the factors makes additional discussion without value.

The carpus-out, flat-claw, and chela-out (A7-A9) each represents a progressive degree of withdrawal into the burrow by a burrow-holder when a potential opponent is nearby. Each of these components may be viewed most simply as a signal of readiness, of minimal intention, to fight; in form each is virtually only one of the preceding front threat postures, standing on end and reduced to minimum intensity. Sometimes only the extreme tip of the chela projects above ground. Consisting as it does solely of this motionless display of part of the potential weapon, each can also be considered an ultimate expression of a ritualized fight.

The lateral-stretch (A10), on the other hand, nearly ubiquitous in brachyurans, differs little or not at all in the social behavior of *Uca* from the threat posture sometimes assumed in the face of a predator. Here the threat is of injury to another species, but its origin in an intention posture of attack and display of the weapon appears similar.

Motions corresponding to the creep (A11) are so widespread in the animal kingdom that its occurrence in fiddlers scarcely needs further comment. It forms one more instance of the signal value of apparent size-reduction and withdrawal of weapon-display in situations requiring or inviting withdrawal from or appeasement of another's aggressive behavior. As in the high-rise (A13), below, its origins must lie in the practical importance for many animals of large physical size.

Although a true creep does not seem to occur in females, a low posture in this sex, as in so many other groups of animals, is a signal of receptivity.

The prance (A12), now noted with moderate care only in males of *maracoani, ornata*, and *pugilator*, will almost certainly prove to be of wide occurrence in *Uca*. At least occasionally it almost certainly is associated with the leg-stamp (S11), which sometimes, in the absence of any apparent surface stimulus, appears to be directed toward an individual underground. The same posture used in threat situations between males within sight of one another sometimes show the turned-under dactyls not quite touching the ground, and therefore incapable of producing sound; at these times the prance may be viewed as the ritualization of leg-stamping.

The high-rise (A13) is characteristic of both sexes, unlike the preceding components of agonistic behavior. It represents the opposite of the creep, exaggerating as it does the size of the threatening crab. It reflects, once more, the importance of bigness as a signal in social behavior. In *Uca* males, as in other animals, the fact upon which it is based is that larger crabs in combat induce their opponents to go rapidly away more often than do smaller crabs. The high-rise is often combined with threat motions of the major cheliped. Unreceptive females performing a high-rise when approached by a male are often effective in deterring him without contact from an attempt to mate. Both sexes, but especially females, sometimes combine the high-rise with the acoustical signals of leg-wagging (S5).

It seems likely that the conflict during high-rises between tendencies to stay on the surface, feed, wave, or enter into combat, depending on circumstances, may be closely concerned in the origins of leg-wagging (S5). The motions are closely similar to those involved in true leg-cleaning. Displacement leg-cleaning is of frequent occurrence in such a situation, and the eventual development of acoustical equipment through ritualization of leg-cleaning appears to be a logical sequel.

The female's legs-out posture (A14) appears to be a special case of the high-rise (A13), indicating a stronger degree of rejection. Sometimes it, too, is associated with leg-wagging (S5).

All of the sound components now known appear to be primarily threatening in use, whether or not they are directly associated with an agonistic posture or motion, as often is true of the leg-wag (S5), as mentioned above, or with the threatening use of waving display. Only one of the group seems also to be a characteristic part of intensive courtship in a number of species; this is the major-manus-drum (S9); it is of course to be expected that others of the group are used in courtship underground or at night; Salmon & Atsaides (1968.2) considered leg-wagging in the evening at the surface to be a courtship signal in two species of *Minuca*.

Most of the components also crop up in ambiguous situations in the midst of waving display, when the displaying crab appears to have his attention divided between a female and either another male or an outside threat, such as an observer or camera close by. In these examples, the motions normally associated with sound components appear usually to be incomplete, and are therefore then to be classified as displacement activities. Feasible bases seem to be confined to cleaning and feeding motions. The major-merus-rub, minor-merus-rub, minor-claw-rub, palm-leg-rub, leg-wag, and leg-side-rub (S1-S6) seem to be more allied to cleaning, while the drumming of major and minor meri against the carapace (S7, S8) are possibly derived from feeding, and the minor-chela-tap (S10) more probably from the same source. No origin for the leg-stamp (S11) suggests itself. Bubbling (S12) and membrane vibration (S13) are both derived directly from respiratory processes by way, perhaps, of displacement behavior.

Special mention may be made of the minor-merus-drum (S8), which seems to be a good example of both apparent double ritualization and of ambivalent behavior. Whether or not the component was originally derived from feeding, the merus sometimes makes apparent drumming contact against some part of the suborbital region, although the sound has not yet been recorded; this occurs in threat situations only. In high-intensity waving display, however, *festae* and *lactea* sometimes shake the flexed minor cheliped in the air well in front of the suborbital region. This of course could be merely displacement minor-merus-drumming. At its most conspicuous, however, the minor cheliped is constrastingly lightened, and the motions seem to be an incorporated (though occasional) part of the high-intensity waving display; the shaking occurs in both courtship and threat, when no conflict seems to be involved. Under these conditions, it is a good illustration of the concept of ritualization.

(c) WAVING DISPLAY AND ASSOCIATED SOUND COMPONENTS

The first eight of the 14 motion components distinguished in waving display are characterized by movements of the major cheliped. Of these, the vertical-wave (W1) and the jerking-vertical-wave (W2) seem more likely to have evolved through ritualized displacement feeding than from any threat component, none of which is characterized by the strongly flexed cheliped shown in typical displays of *Deltuca*, in which social behavior is least complex (Fig. 87 *A, B*). It also seems likely that the vertical wave arose in at least some species through an intermediate acoustic element, the major-manus-drum (S9). In the wave-form shown in certain *Deltuca*, especially in *acuta*, the likelihood is clear. Additional suggestive examples are given in a few motion-picture close-ups, made with the camera on or near the ground. Here the major manus, although vibrating, does not reach the ground; nevertheless actual drumming is an integral part of waving display in the two examples, *saltitanta* and *deichmanni*, both of which belong to the socially advanced subgenus, *Celuca*. The occurrence of incomplete drumming in other species, when it is not a part of regular display, doubtless represents displacement drumming.

In contrast to vertical-waves, all waves with distinctly lateral, or obliquely lateral, characteristics seem clearly to have evolved from the basic threat posture of the lateral-stretch (A10). This derivation applies to the semi-unflexed-wave (W3), lateral-straight-wave (W4), lateral-circular-wave (W5), jerking-oblique-wave (W6), reversed-circular-wave (W7), and overhead-circling (W8). (Fig. 87 *C, D*.)

In many of the displays of many species, the classification of a wave-form as simply vertical or lateral would be inaccurate, even in unambiguous displays of high intensity, since the cheliped motions show characteristics of each, pushing slightly outward, obliquely upward, or both. Nevertheless the basic derivations even of such complex examples as jerking *Minuca* and the overhead circling of *maracoani* or *ornata*, when films are carefully analyzed, seem to have more in common with typical lateral wavers, their motions appearing more likely to be developments of warding-off threat gestures, than of feeding motions.

All jerking waves (W2, W6) appear unmistakably to be compound waves developed from simple waves, either vertical or obliquely lateral; each step in the ascent following the pause starts at the level where the preceding wave reached its peak, without or almost without descent.

The leg-stretch and prolonged leg-stretch (W9, W10) are both versions of the high-rise (A13) incorporated into waving display, without change in its function of increasing apparent size. The leg-wave (W12) is a clear example of a double ritualization, having come, it seems evident, straight through ritualization of the leg-wag (S5), which in turn seems equally obviously to have been derived from cleaning motions. In the field all stages of the sequence, including functional displacement motions

in appropriate social situations, can sometimes be readily observed. The curtsy (W13), characteristic of high-intensity courtship in a number of species, apparently derives directly from an intention movement, with display and burrow descent tendencies in conflict. Salmon & Atsaides (1968.2) found it sometimes to be preceded in *Minuca* by the leg-stamp (S11).

The occasional incorporation of the minor-merus-drum (S8) into high-intensity waving display was described at the end of the preceding section.

(d) EVOLUTION OF CHIMNEYS, PILLARS, AND HOODS

The chimney, a wall closely surrounding the burrow, may have a double origin. The simplest derivation would be from bringing spoil from below, during excavation, and not tossing it to a distance. Such a heap, particularly after development into an encircling wall, would certainly serve to aid in keeping the burrow moist and cool, of special importance to ovigerous females. *U. urvillei*'s chimney is partly made in this manner.

Most chimneys, however, as mentioned earlier, are built through scraping mud from a distance of some inches, even though excavated mud is still carried or flung by the same crab to a similar distance.

The origin of this behavior must, it seems, be through a quite different pattern from that of excavation. The only likely activity seems to be the genus-wide habit of plugging the burrows with a stopper brought from some inches away. Since a similar function, in the same species and even individuals, is often performed by pushing mud up from below, it may well be that the origin of the wall is also double. It seems most unlikely that chimneys, prevalent in the less-specialized subgenera, have evolved from the highly specialized, rarely occurring, pillars and hoods, limited almost wholly to *Celuca*. The subgenus, it will be recalled, is the most specialized subgenus of all, both in adaptation to littoral life and in social behavior.

The sporadically occurring pillars and hoods are probably derived through phylogenetic simplification of chimneys. Each pillar or hood is made by bringing all material from beyond the burrow mouth; all differ strongly from chimneys in being built only by males during their display phases.

Salmon & Atsaides (1968.2) suggest plausibly that the structures may serve as amplifiers of acoustic signals. Still, if hoods are amplifiers they apparently do not serve for the amplification of sounds at night; there is no evidence that they are constructed during nocturnal low tides.

It also seems likely that pillars and hoods have arisen through ritualizations of chimneys, signaling to males the presence of a display territory and to females a potential mate. Possibly each structure represents, larger than life, the displaying male himself, visible outside the burrow to aggressive males and passing females even when he is not stretched to display height and even when he is underground.

Still farther out on the speculative limb is a suggestion which, if it proves experimentally to be valid, could be the most interesting explanation of all. In each of the species concerned, the female, when sufficiently stimulated, follows the male below ground. After the male vanishes, the only visual guide or stimulus for the female in the last several inches is the sight of the burrow—a dark object in the terrain. This is, of course, in addition to any acoustic signal the male may be making, or any stimulating effect of a previous visual image the female may retain.

It seems possible that if such a dark object is a releasing or directing mechanism, then a *vertical* dark object may be a better mechanism for attracting a female than one that is flat and foreshortened. In the highest development of hood-making, the structure fashioned by *musica terpsichores* alone is usually found on light-colored muddy sand; the shading resulting from the concavity would be an effective darkener.

Throughout the pillar-and-hood-making group, however, the principle appears the same. Setting a facsimile of the stimulus on edge makes it effective long before it would otherwise be clearly visible. Such a development would parallel the rising of a crab to dactyl-tip to wave. If the structure does stand for the hole, then it can be regarded as both a supernormal stimulus and a ritualized burrow mouth. A special sense for these terms is, then, needed, since the original stimulus, the hole, still exists.

Perhaps, finally, the structure functions in all three ways, amplifying sound, increasing the apparent size of the crab, and, ultimately, representing the burrow.

Unquestionably certain populations of pillar-makers are physically unable to fashion the structures because of inappropriate substrates. Nevertheless, the erratic appearance of pillar-building among and within populations remains mysterious. Possibly this activity is a case of genetically controlled behavior in polymorphic forms.

E. Summary

The view is presented that in *Uca* the burrow has two chief functions—to provide shelter and, during the display phase of a male, to serve as the center of a defended area. These uses are termed *basic territoriality* and *display territoriality*. In this genus territoriality seems scarcely concerned with aiding the distribution of individuals in accordance with the food supply or of stabilizing their numbers. The principal function of combat may lie primarily in stimulating and synchronizing mating behavior, as well as

in releasing tension during the reproductive period; combats often seem to progress for their own sake, particularly in highly ritualized mutual encounters. While in forceful fighting physical injury virtually never occurs, these combats are often prolonged. The behavior changes that sometimes follow combat are similar, whether the combat has been long and forceful or short and ritualized. Therefore the main advantage of ritualization in some species appears to be the shortening of combats, allowing more time both for waving display and for actual courtship. This interpretation is inadequate to explain the situation in other species, now under investigation; in these forms force itself is ritualized, forming a regular part of many combats; furthermore, during the course of a single period of low tide, combat and courtship are largely confined to different hours. In *Uca* basic threat postures provide defenses against predators, as in numerous other crabs; in addition the postures, motions, and acoustic signals have various signaling functions in both males and females; in intermale relations they often discourage contact of any kind, thus saving even more time than ritualized combat; they are commonly used in females to ward off unacceptable males, as well as in occasional interfemale encounters. Threat postures and acoustic signals also may well prove to have stimulating value, both to the actor and to conspecifics. In waving display most of the components are employed both in intermale situations and in pure courtship; a very few occur only in high intensity courtship. It seems certain that waving display serves as an identification mark, treated variously by conspecifics, depending on sex, physiological stage, and age; and that it is stimulating at least to females. Probably it also stimulates males; by repelling most other males it can in addition avoid the time needed even for threat postures. Interspecific differences in waving patterns discourage incompatible courtships among closely related species. The burrow of a waving crab, as well as any structure beside it, may be viewed as an artifact of display. Displacement behavior is discussed. A speculative review follows of the likely derivations of the principal components of combat, threat, waving display, and their associated sound components through the ritualization of fighting, withdrawal, feeding, cleaning, and respiratory activities. The derivation appears often to be through the medium of displacement activities, as well as, in waving display, from threat postures. The more specialized structures beside the burrows of some displaying crabs are probably derived from moisture-conserving chimneys that occur rather widely in the genus; their functions during display may amplify sound, increase the apparent size of the crab and possibly represent the burrow. The need for physiological research is stressed in several sections.

Chapter 7. Speciation, Phylogeny, and Directions of Evolution

OUTLINE

A. INTRODUCTION

In *Uca* the evidence on evolutionary trends remains indirect. Fossils are confined to three specimens, one of them a dactyl, while in related groups they are similarly rare. Genetics is untapped. Even the indirect evidence is restricted, since our knowledge of comparative development, basic physiology, sense organs, and neurophysiology is rudimentary or wanting. Experimental work on behavior is starting, but has scarcely begun to extend to the comparative approach; only several releasers have been studied experimentally.

On the other hand, both morphology and behavior have yielded information of aid in working out the probabilities of fiddler crab descent. Wide distribution, diverse habitats, numerous forms, and social complexities all invite comparative studies. Many of the results now at hand combine to give persuasive evidence on the group's development.

On the basis of present knowledge, then, the following pages will discuss roots of speciation and directions of evolution in *Uca*. With material from earlier chapters, connections among the subjects will be stressed. The chief of these relationships are outstanding features of morphology and behavior to each other, as well as to certain aspects of distribution and ecology. The principal viewpoints are the extent of variability, the role of allopatry, and the adaptations related to sympatry. Stimuli to evolution and rates of change are considered. A phylogeny of the group is proposed. Finally a brief discussion comments on directions of evolution within the group.

References are largely omitted in this chapter, since it is based on material presented in the systematic section (Part One), discussed in the preceding chapters of Part Two, and included in the indexes. General references to zoogeography, ecology, and behavior appear in the introductory sections and subsections of the pertinent chapters. Documented discussions of allopatry, sympatry, and related topics are given, in particular, by Grant (1963) and by Mayr (1963, 1970).

B. PLASTICITY

The chief characteristics of *Uca* can usefully be viewed in two groups, one including the conservative essentials of maintenance and the other the diverse features connected with reproduction.

Adaptations in the first group deal with the individual's basic needs as an amphibious, filter-feeding crab that lives its entire adult life in the intertidal zone. They control respiration under fluctuating conditions, help retain moisture, extract nourishment from the substrate, promote efficient movement out of water, and, through vision, vibration receptors, and burrowing behavior, provide means of escape from predators. All of these functions take place in a complex of rhythms controlled at the least by light, tide, and seasons. The extremes of ecology to which fiddler crabs are exposed are nonetheless limited and repetitious; their related adaptations are therefore as a whole conservative. For this reason the structure of the gills and mouthparts, for example, have limited usefulness in tracing the evolution of the group, although some of the details are clearly responsive to ecological needs. Similarly, the general shapes of the carapace differ little, all species being roughly four-sided and slightly broader than long, as are numerous grapsoids; the eyes are always erectile on definite stalks; the ambulatories show only modest ranges in shape and length. Behavior patterns associated with maintenance vary even less; walking, digging, and feeding motions are virtually identical in all fiddler crabs.

Most physiological characteristics in *Uca* have not

yet been examined from a comparative point of view. Some topics are beginning to furnish rewarding data in relation to ecology; these include species-specific tolerances to temperature, salinity, and drought, as well as differing requirements in food. All of these factors, vital as they are to an understanding of evolution, depend on characters which for the practicing taxonomist are now largely cryptic and for the evolutionary biologist still too little known to be related to his information from other disciplines.

The second group, ultimately concerned with reproduction, includes the most striking characteristics of appearance and behavior in fiddler crabs. Often they prove useful in showing relationships among subgenera, species, or both. In morphology the principal characters are connected with the shape and armature of the major cheliped, the armature of the orbits and ambulatories, and the shape of the gonopod tip. Since parallelism and convergence are extremely common in gonopods, these appendages must be used with particular caution as phylogenetic clues. In behavior the chief variables are the waving patterns of the major cheliped along with associated movements of other appendages, a few components of combat and threat, the two basic forms of the behavior of females during courtship, and the shapes of any structures fashioned beside the burrows. In different categories of social behavior are the relative amounts of time devoted to waving display, the peak hours of social activity, and the prevalence of displacement activities. All these forms of social behavior are under the control of at least two internal rhythms, which are in turn partly associated with the ecologic rhythms listed above. One of these internal rhythms is connected with the state of the gonads; the other is controlled by short, physiological phases. Both groups of rhythms show some differences within the genus, although little information is as yet available.

One characteristic that should be specially mentioned is the sporadic occurrence in *Uca* of a species or subspecies that differs strikingly from its close relations in gonopod morphology or waving display. As will be shown in the section on sympatry, the evolutionary pressure for the peculiarities seems clearly to be the maintenance of barriers among forms that are sometimes sympatric. The point to emphasize here is that the tendencies toward startling change are in themselves definite attributes of a subgenus or a species.

Two morphological characters related to each other definitely differ among some subgenera, and are hence of basic taxonomic usefulness and potential evolutionary interest. These are the breadth of the front and the lengths of the eyestalks. Unfortunately functional reasons for these differences remain puzzling.

C. ALLOPATRY

As in other advanced groups of animals with many surviving species, forms of *Uca* with different ranges often seem clearly to have developed from common ancestors. Sometimes their differences from one another warrant specific status, their relations being indicated in this contribution by their grouping into the informal category of superspecies. Sometimes the differences appear to require only subspecific status. Again, certain populations show very minor and variable differences, which either are too poorly known to justify the proposal of a subspecific name, or else the atypical traits appear sporadically in other parts of the species range.

The strength of these tendencies to differ geographically varies with the terrain. Along the unbroken coasts of Africa and the warmer parts of America, the allopatric splitting appears to have been less than on the fragmented shores of southeast Asia and the adjacent seas.

For example, no overt differences exist between populations of *Uca* in the Red Sea and those in Mozambique that make desirable the proposal of subspecies. Although some individuals in the more southern populations of *lactea annulipes* show minute differences in claw proportions, these distinctions are minimal in comparison with geographic distinctions in this species that center in the Sunda region to the east. Similarly, along the entire coast of west Africa none of the forms into which *tangeri* has been occasionally divided appears to be justified.

In the western Atlantic, other examples give more scope for discussion. Two species range from Florida or the Bahamas to southern Brazil, while five more occur from the West Indies to Rio de Janeiro. None of them seems to me to give valid grounds for the erection of subspecies. Yet the direction of the equatorial current in the northeast part of the bulge of Brazil is westward, the counter-equatorial current chancy, and the drought-ridden dunes of the region inhospitable to fiddler crabs. It may be of course that differentiating traits within species on continuous warm coasts are chiefly physiological, as we are finding is true of temperate forms in the United States (p. 441). If so, these cryptic characters contrast with the morphological and ethological traits which are conspicuous whenever allopatric forms are given taxonomic recognition and which, in the considered forms, suit a variety of habitats.

More probably, the widest-ranging species of the Americas and Africa do owe their homogeneity, as first suggested, to the simple fact of the continuities of the coasts and to their relative stability. Even along the barren shores of northeast Brazil a few havens offer refuge, breaks in the dunes to streams behind. Spring tides and rare rains carry megalopa shoreward and infrequently bring hatchlings to the open sea.

In the heart of the Indo-Pacific region, in contrast, well-marked subspecies are the rule. This condition may be due to the lower hardiness of Indo-Pacific larvae under adverse conditions or to shorter development periods. Yet these suggestions seem unlikely. Rather, the explanation must lie partly in the sheer prevalence of islands and of broken coastlines, as well as in the distances between landfalls. Finally, beyond these possible aids to subdivision, the key stimuli have certainly been the temporary isolations of a continuing volcanic past.

The characteristics of marginal populations in wide-ranging species of *Uca* have not yet been compared in detail with those of populations of the same species that are more centrally located. Nevertheless it is already clear that the outpost groups show no striking tendencies to differentiation. Apparent reasons for their continuing similarities to the central stocks differ in species that range widely from west to east from those with distributions extending far to the north.

Longitudinally, the only examples of marginal populations are *tetragonon* and *chlorophthalmus*. These two species alone reach mid-Pacific. In this tropical climate, on the western edge of the Eastern Pacific Barrier, and with the major currents directed back westward, these populations might be expected to show at least tendencies to subspeciation. Such is not the case. The survival efficiency of their larvae is probably formidable and may be the key factor in keeping open the gene pool of these most oceanic of *Uca*. It also seems likely that at least the westward counter-equatorial current may be a thoroughfare.

In populations of *Uca* at the latitudinal extremes of species ranges, accommodation to cold is the usual problem, and there is continuity with populations from warmer waters. Evolution here may well be lethargic because of the few generations possible, compared with the nearly continuous breeding found in wet tropical regions.

The topic of mingled populations of allopatric forms is crucial to any survey of evolution in the genus. In a few cases these overlapping boundaries between allopatric species and between subspecies are strikingly apparent. Viewed pragmatically, the populations are of course sympatric in the most restricted sense of the word. They will therefore be considered in the next section.

D. SYMPATRY

(Figs. 92, 93)

Twenty years ago the genus *Uca* appeared to be crowded with forms confusingly similar yet occurring in the same places and showing no evidence of interbreeding. Accordingly, they seemed to meet the criterion of species status, "sympatric coexistence without interbreeding," although their numbers appeared excessive even for a group as diversified as fiddler crabs.

Field work, reviews of the literature, and the study of museum collections have now somewhat reduced the problem. Potentially sympatric species of *Uca* turned out to be of four kinds. Some, recorded from a single locality, proved to occur as adults in distinct habitats. Many other records indicating sympatric occurrences were based on taxonomic confusions. In a third group, juveniles were described or identified as different species from the adults. Finally, many forms reported from the same place do in fact coexist in perfect sympatry, two or more species of great similarity mingling on the same strip of shore. The members of this final group are the subjects of this section.

The geographic barriers which led eventually to instances of sympatry are occasionally apparent. Examples are the repeated topographical changes in southeast Asia and emergences of Panama. In particular, the common ancestors of the superspecies *acuta* and of *dussumieri* in the Tethys area may well have been temporarily separated merely by an earlier version of the Sunda Shelf. After its submergence, the species held distinct. Still later they subdivided, through reemergence of the land, to varying degrees that still persist. Nevertheless in most cases of sympatry among very closely related forms the origins of the situation are so uncertain that speculation is unrewarding.

As Mayr illustrates in the course of his discussion of sympatry (1963: 449ff.), when similar species in any group become better known, biologically and morphologically, differences between them turn out to be numerous. In *Uca* no instances appear of the infrequent, anomalous sort of which an example is the *pipiens* group of *Culex*. Always, when we have enough information, these morphologically similar fiddler crabs prove to be distinct in assorted structural and behavioral characters. Usually the preferred microhabitats also differ.

As in other animals a number of conditions appear to have been prerequisites for sympatry. These are, first, a period of geographic isolation, during which differentiation of two or more forms became genetically fixed; second, removal of the barriers; third, migration into areas of competition; fourth, additional adaptations to food or substrate as needed to permit geographic coexistence; fifth, adjustment of reproductive morphology, physiology, and mating behavior in such ways that productive matings are promoted, wasted germ cells avoided, and so, incidentally, specific barriers maintained.

In *Uca* the disadvantage of wasting time is probably still another factor in the evolution of distinctive mating patterns. Because of the limited intertidal hours available for courtship, behavior that discourages the attraction of unsuitable partners should have positive value.

The morphological characters that usually aid most dependably in distinguishing species that are closely sympatric are the tip of the gonopod and, sometimes, the form of the female's gonopore. Behaviorly two such species almost always are strikingly distinct in their waving displays. Sound probably will prove also to differ significantly among sympatric forms. Almost certainly chemical factors provide the remaining principal barriers, but they have not been explored.

Differences in habitat preferences, while often pronounced, are at best inefficient dividers between *Uca*. Even species normally found in distinct biotopes often intermingle both in areas rich in nourishment and along adjacent margins of their preferred niches. Differences in breeding seasons, which in many other animals form important barriers, are almost nonexistent in *Uca*.

A final point of interest concerns the increasingly apparent tendency of species to differentiate locally in waving display, gonopod form, or both, in such a way that their characteristics become distinct in relation to the species most closely related to them with which they are sometimes sympatric. The principle is known in certain birds and other animals when allopatric forms occur together in boundary zones. Here their differentiating characteristics are often more pronounced than in parts of the ranges where the two forms are not in contact.

Nevertheless, the similar phenomenon in *Uca* appears more complex, since characters which would be viewed ordinarily as of specific or at least subspecific importance if a restricted part of the range were examined prove themselves to be highly plastic. It was only when the differing sympatric associates of certain allopatric populations were considered that the sometimes abrupt differences began to be understandable, as the following examples will suggest.

The occurrence of jerking in the waving displays of the superspecies coarctata *and the species* dussumieri. In the Indo-Pacific the most widely ranging allopatric *Uca* is the informal superspecies *coarctata*. Found from east Africa to Fiji and Japan, it is divided formally in this study into four species, one of which is subdivided into two subspecies. Table 23 shows these divisions, their ranges, and the waving characteristics under consideration, along with those of their closest sympatric relations. Figure 90 presents the displays of two of the subspecies in diagrammatic form, for comparison with those of the

species sympatric with them. With two exceptions the members of the superspecies are the only jerkers in the observed sympatric assemblies. In one exception both *forcipata* and *rosea* proved to be jerkers, although their rhythm, as far as could be determined from insufficient observation on *rosea*, is quite different. The second exception concerns populations of *c. coarctata* that are regularly intermingled in the Philippines with *d. dussumieri*—a close relation of similar size and appearance. Here both *coarctata* and, in a complete change of its own usual pattern, *dussumieri* jerk vigorously. The display of *coarctata*, however, is fully distinguished by a series of forceful diminishing waves that follows each jerking primary wave; the diminishing waves are of greater amplitude, and hence more conspicuous, than in populations in Fiji, where *dussumieri* does not occur, or in northwest Australia, where *dussumieri* does not jerk. In all of these pairs of Indo-Pacific forms the gonopods are distinct.

In Indo-Pacific narrow-fronts, it will be recalled, mating is sometimes not prefaced by waving display, although the male is always in a waving phase during the low-tide period concerned. Also, the female is approached close to her own burrow by the male, instead of attracted to his; when unreceptive, she either assumes one of the appropriate threat postures or simply goes underground down a shaft, which is usually too small for the male to use. Thus, the importance of waving display in courtship and as a species barrier is more questionable than in the socially advanced subgenera. Yet the example of sympatric differences just given indicates a value to the species of waving display that is more directly concerned with particular potential copulations than with any of the other functions of this behavior that have been postulated (p. 517).

Local display differences among forms with similar gonopods. In *Minuca* of the eastern United States an evolutionary situation of great interest is becoming apparent, through the work of Salmon (1967) and of Salmon & Atsaides (1968.1, 1968.2). The species most concerned are *rapax*, its northern allopatric representative, *pugnax*, and two related forms on the Gulf of Mexico, described by Salmon & Atsaides as species and viewed here as subspecies (p. 190). Of major interest is the observation that Atlantic coast populations of *pugnax* differ in the distinctness of the jerks in their waving display. Once more the explanation seems to lie in their sympatric associations. Where the ranges of *rapax* and *pugnax* coincide in eastern Florida, the jerking of *pugnax* is absent, while that of *rapax* is as pronounced as usual. Farther north, *pugnax* jerks are distinct, indicating that this very common component of waving in many *Minuca* has been weakened in the coincident area under the pressure of sympatry. Differences of similar interest, including both speed of waving and acoustic characters, exist among the forms occurring sympatrically along the Gulf coast.

It is noteworthy, in view of these behavioral differences, that in *Minuca* the gonopods of groups of closely related species, even where these regularly occur sympatrically, are so similar that they are often of little taxonomic use. This is particularly true since individuals even within populations show considerable variation in the one or two details, such as the degree of flange projection, that might be helpful. We may expect more cases of clear behavioral distinctions to come to light in this group where gonopod differences appear to be so slight.

Regional hypertrophy of gonopod characteristics. Throughout practically all the rest of the genus the gonopods are specifically distinct and, on the taxonomic level, thoroughly reliable. Intraspecific variation is minimal and confined to such non-functional parts as vestigial thumbs. Gonopods are also one of the best indicators of phylogenetic relationships within the group.

Yet in an Indo-Pacific species, *vocans*, allopatric populations show such exaggerated and contorted variations of their parts that in any other instance these differences would be considered excellent taxonomic characters for defining full species. All other morphological and display characteristics, including the rather conservative gonopores of the females, show at most characters of subspecific value, while in colors and ecological niches all the allopatric forms are closely similar. Finally, apparent hybrids occur in New Guinea and the Philippines (Table 3). To consider these forms full species would certainly be unwarranted.

The explanation for the development of these gonopod distinctions seems once more to lie in the sympatric associates of *vocans*. The species shares virtually all of its range with *tetragonon*, a member of the same subgenus with a more conservative gonopod; *tetragonon* is usually found in more exposed locations and shows no definite regional variations from Africa to the Tuamotus. Occasionally the two species are locally closely sympatric. *U. vocans*, however, appears regularly to extend its range farther north than does *tetragonon* and the gonopod of this northern subspecies approaches the simplicity found in *tetragonon*.

The sympatric crowding in the Philippines appears responsible in *vocans* for the opposite effect. Here the local subspecies often associates with from one to five species of the neighboring subgenus *Deltuca*, each of which has a gonopod of basic design similar to that of *tetragonon* and *vocans*. On these shores the gonopod of *vocans* attains its most contorted form.

There seems to be no question but that in the even more crowded parts of the eastern Pacific similar explanations will be found for some of the abrupt shifts in gonopod form characteristic of the subgenus *Celuca*, where the change from flanged to tubular gonopods and perhaps back again has apparently occurred several times.

In a wholly American group, the subgenus *Uca*, occurs an example of gonopod differences with so few species involved that the evolutionary pressures appear clear. These species are clearly related to the Indo-Pacific *vocans*. Several of the group of six have gonopods recognizably similar to the old Indo-Pacific pattern, as well as to *tangeri* in the eastern Atlantic and, of course, to one another. In the remainder of the subgenus, including their allopatric members on opposite sides of the Panamanian isthmus, the gonopods when considered alone appear to belong to unrelated forms. Included is an Atlantic species, *major*, in which these organs are flangeless, thumbless, and tubular, while their Pacific counterparts, as well as *major*'s occasional sympatric, *maracoani*, have opposite characteristics of a wholly conservative nature. Again, the likely explanation for the striking difference appears to be the sympatric association.

E. STIMULI TO SPECIATION AND RATES OF CHANGE

Probably the greatest single stimulus to evolution in *Uca* was the presence of suitable habitats in America that were incompletely occupied by animals of similar needs, or by animals that succumbed to new competition. Here the postulated migrants across the Behring Bridge during the early Tertiary could flourish in the necessary isolation. Under such conditions a consequent burst of speciation would not be surprising. A somewhat similar stimulus was the blocking off of the Indian Ocean from the Pacific by the Sunda and, probably, Sahuli Shelves. These events, repeated a number of times to various degrees of completeness, gave recurrent spells of isolation. A third stimulus, comparable to the second, was the emergence of the Isthmus of Panama, most recently about five million years ago. A final impetus occurred more recently in the Philippines and East Indies; during the orogenic activity of the late Cenozoic, species probably evolved rapidly with the appearance of temporary barriers and rich new land.

These four events give some idea of evolutionary rates within the genus.

If the hypothesis of Behring migrations in the early Tertiary is accepted, along with the corollary that two Indo-Pacific *Uca* came from American stock, then four American subgenera, totaling 45 species, developed during some 30 to 55 million years. These totals comprise almost half the subgenera and more than two-thirds of the known species, including all those that are most specialized for littoral life morphologically and most advanced in the development of social behavior. The specializations and developments unique to the American subgenera (and to *tangeri*, its one eastern Atlantic derivative), include the following: enlargement of branchial chambers; maxilliped adaptations for dealing with food particles sifted in partly dry, muddy sand rather than in wet mud or sandy mud; development of the lateral-circular-wave; development of male-fashioned pillars and hoods; development of the female's following of the male, leading to copulation beneath the surface; and development of special acoustic signaling and combat components that do not apparently occur in Indo-Pacific subgenera.

The last closing of a seaway through Central America near the end of the Pliocene gives us our most exactly known date for the beginning of a period of complete isolation in *Uca*. This land barrier last became entire perhaps five million years ago. There is no indication whatever that larvae, much less adults, have migrated through the Panama Canal.

During these past five million years five pairs of Pacific-Atlantic allopatric species have evolved; the 10 species total one-quarter of all American forms considered in this publication to hold specific rank. In three other species the allopatric populations are so similar that, for consistency's sake, the members of both pairs have been designated subspecies. Most of the east–west allopatrics however differ in major characteristics of gonopods, gonopores, cheliped ridges, and pattern of waving display, the distinctions being clearly on the specific level.

Unfortunately, the duration of the Sunda barrier is uncertain. Similarly uncertain is the length of time during which took place the major orogenic activity in the islands to the east and south of Sunda. Nevertheless, it is fairly well agreed that the old Sunda barrier broke down very recently, perhaps within the last 10,000 years. Since then there has been freedom of mingling, so that it is possible to determine how fixed had become the reproductive barriers among the allopatric forms fostered by the presence of the Sunda Shelf.

Without exception, only minor morphological and behavioral differences evolved during the short period of isolation. For example, structural changes show best in the size of gonopod flanges and degree of torsion; in the shape and modeling of the front; minor differences in the suborbital system of tubercles and ridges; ill-fixed, small changes in the tuberculation of the prehensile edges of the claws. In behavior the differences, confined chiefly as usual to waving display, are even slighter, comprising distinctions in the prevalence of jerking, or of seriality in the waves; there are also sometimes stronger or weaker tendencies to build chimneys or pillars. In four species these differences are considered to be on the subspecific level, particularly in view of evidence of interbreeding where populations coincide; in two superspecies the differences are here regarded as of specific importance. As usual in such cases, an argument could almost as easily be made for considering even these superspecies as species composed of particularly distinct subspecies, in which no evidence of interbreeding has been found. (See Tables 2, 3, 6, 22, and 23, along with the associated portions of the text.)

F. PHYLOGENY

(Figs. 96-101)

The distinctness of *Uca* as a genus has been mentioned several times in these pages. *Ocypode* is generally agreed to be its closest relation. Yet the basic differences are numerous and no intermediate forms remain.

Uca itself, in contrast, is composed of a wealth of radiating species, many of them closely related and some of these intermediate between groups. Nevertheless, in spite of difficulties, the genus does divide into categories which are very distinct, when end forms are compared. For evolutionary studies it is fortunate that corresponding ecological niches exist in the Indo-Pacific and in America. By comparing adaptations in the two hemispheres, it is possible to separate similarities indicating relationships from those which probably represent ecological parallelisms and convergences. By this means hypothetical ancestors and evolutionary trends can be suggested.

With this background in mind, then, what were the probable characteristics of ancestral *Uca*?

It seems that this crab must have been only moderately specialized for an amphibious life or for a particular substrate. Its waving display was a simple vertical raising and lowering of both flexed chelipeds, as in various grapsids and ocypodids, as well as in the *Uca* subgenus *Deltuca*. Both visual and acoustical displays were used, but only in basic territoriality and intermale relationships, and little or not at all in courtship. Forceful combat certainly existed, since it is prevalent in decapods, and probably also, as in some other crabs, simple forms of ritualized encounters. The gill system was well developed. The front was moderately narrow. The gonopod was of the basic type, with well-developed flanges and with the broad inner process found in various ocypodid genera as in most members of *Deltuca*, *Thalassuca*, and *Minuca*.

The *Uca* living today that are most similar to this hypothetical ancestor comprise the Indo-Pacific subgenera *Deltuca* and *Australuca*. Their patterns and armature of social behavior are less highly developed than in the other groups. In general they specialized for mud-living through a number of devices that seem to facilitate the drainage of liquid mud from carapace and appendages. The lower margins of the orbits are rolled out, while the suborbital crenellations, always easily clogged, are low and the series short. The eyestalks fit loosely in wide sockets. Before the use of cheliped armature was observed in the combats of American crabs, it seemed to me that the function of chela grooves in Indo-Pacific forms must also be to aid in drainage; that potential use now seems more dubious but still undetermined. The small chelae on mud-dwellers are all well toothed, preventing moist mud from sieving through the gape on the way to the mouth. Ambulatories are usually broadened, perhaps originally only as a hindrance to sinking in deep mud, but at least at present thus affording an increased area for stridulatory armature. In females throughout the genus the ambulatories are more frequently broadened and to a greater degree, perhaps in adaptation both to the female's carrying heavy eggs and to her spending much of her ovigerous period in very wet areas, near the tide's edge; the broadening also provides space for additional stridulatory armature, also better developed in females as elsewhere in the genus, her threat and combat components being largely restricted to activities of the ambulatories. Display whitening does not occur in *Deltuca* and *Australuca*. Certain colors, particularly in the young, are often striking; carapace spots of blue and red or white in adults have given some evidence of use in threat postures.

In social behavior species of *Australuca* are slightly more advanced than those of *Deltuca*; they spend more time in waving display, the wave is of somewhat greater amplitude, and courtship shows the beginning of complexities. In particular, a male sometimes herds a female toward his own burrow, pushes her down it, and then follows.

In contrast to the foregoing mud-livers, another Indo-Pacific subgenus, *Thalassuca*, is adapted to relatively exposed marine habitats—muddy sand near low-tide levels, shell-encrusted rocks and dead coral in even more exposed localities, and pelagic islands. Specializations have perhaps included a longer larval period, or at least a period capable of prolongation. These crabs lack most of the apparent adaptations for mud-living. The orbits are not rolled outward; the crenellations are well formed and extensive, appearing equally suitable for water drainage and for stridulation. Grooves on the major chela are reduced or absent. The small chela gapes more and has smaller teeth, since the gobbets of food-bearing substrate, large but sandy, apparently are thus brought more efficiently to the mouth in the setigerous tips of the chelae. Unlike the situation in *Deltuca*, copulation is always preceded by waving display, which,

however, remains almost purely vertical and without complexities. *Thalassuca* spends far more time in waving display even than *Australuca*.

From close to the base of *Thalassuca* probably came *Amphiuca* (*chlorophthalmus* and *inversa*). The two groups show affinities in their third maxilliped gills and in the presence of an accentuated depression, almost certainly of importance in combat, on the outer pollex base. They differ markedly in the gonopods which, as in aberrant *Deltuca*, have lost or modified the flanges, reduced the inner process, and, in *chlorophthalmus*, elongated the tip into a projecting tube; similar changes in gonopod structure happen sporadically in the subgenus *Celuca*. They differ also in the wider front, smoothness of the manus, and in more distinct traces of an incipient lateral wave than in any other Indo-Pacific subgenus.

Except for the broad front and the wave, *chlorophthalmus* and *inversa* show clearly their less-specialized origins. The wave is still altogether in a single plane and without circularity, in spite of its partial laterality. Again, the male still approaches the female and mates on the surface. It is interesting, in view of the hypothetical origin of American forms developed below, that *chlorophthalmus* females in breeding condition build chimneys, as do both some *Deltuca* in the Indo-Pacific and some *Minuca* in America. *Thalassuca*, on the other hand, build no structures whatever. Like *Thalassuca*, *chlorophthalmus* is an adaptable colonizer, and in mid-Pacific is found on poorly protected shores; *inversa*, in contrast, builds no structures and is confined to protected flats in the extreme western part of the Indo-Pacific.

It seems likely that American groups developed from one or more stocks ancestral at least to *Thalassuca* and probably to *Amphiuca*. From such a base the present-day American subgenera could all logically have developed, including *Uca*, *Minuca*, *Boboruca*, and *Celuca*.

All of these, along with *Afruca* (*tangeri*), show such similar elaborations of social behavior that it seems they must have a common basis. These elaborations consist of the following patterns: the attraction of the female to the burrow of the male, with mating underground; the development of a circular lateral wave widely used in high-intensity display; the frequent use of the mero-carpal joint, lower side of the manus, and pollex of the major cheliped in drumming on the ground; and the sporadic use of curtsies in the final stages of courtship. With the possible exception of ground-drumming, none of these specializations occurs in *Deltuca*, *Australuca*, *Thalassuca*, or *Amphiuca*, all of which are restricted to the Indo-Pacific.

Beyond these similarities, however, the American subgenera differ in outstanding particulars when intermediate forms are disregarded. *Uca* and its relation, *Afruca*, are closest to the original stock of *Thalassuca*. Their basic similarities comprise gonopod structure, including the tendency of those appendages to striking exaggerations and differences among related forms; carapace and cheliped shape; gill structure of the third maxilliped; and the basic form of display with the lateral circular element usually weak. When in fact a *heteropleura* (subgenus *Uca*) is waving on a shore in Panama he bears a striking resemblance, even in color, to a *vocans* (*Thalassuca*) on the other side of the Pacific. Nevertheless his is strictly an American design, having, in addition to the *Thalassuca* resemblances, all the characteristics listed in the preceding paragraph.

As a whole, the closely related members of *Minuca* are mud-dwellers and correspond ecologically to *Deltuca* in the Indo-Pacific. Their gonopods are of the conservative ocypodid type found in most *Deltuca* and all *Thalassuca*, with flanges and inner process well developed, unlike *Australuca* and *Amphiuca* in the Indo-Pacific. Chimney-making, a pattern often found in Indo-Pacific mud-dwellers far back on protected shores, is prevalent in *Minuca* too; it may prove to be chiefly a superficial similarity, resulting from equivalent ecologies. Its drainage arrangements apparently depend on short sockets, smoother appendages, and slender chelae, rather than on rolled-out orbits, and, perhaps, the channeling aid of grooves on the claw. The broadening of the front reaches its maximum in *Minuca* and remains unexplained. It may be nothing more than an incidental effect of orbit shortening, resulting in faster cleaning. Yet it seems unlikely that this adaptation would have evolved along with necessarily shortened eyestalks, reducing the field of vision. The slender stalks found in Indo-Pacific groups and accentuated in the subgenus *Uca* would, it seems, have proved a less expensive cleaning system. Perhaps a broad front, simply but importantly, adds space for neural connections. Finally, *Minuca* is especially notable for adapting better and more frequently than any other group to extremes of drought and cold, through aestivation and hibernation.

The subgenus *Boboruca*, erected for the Atlantic and Pacific allopatric forms of *thayeri*, in some ways resembles *Deltuca* more closely than do any of the other American subgenera. The similarities lie in the shape of front and carapace and in its social behavior; yet its armature is characteristic of American forms. As a whole it appears to be merely a conservative descendent of one of the postulated migrants, rather than a more direct representative of an Indo-Pacific group. The fact that the Atlantic and Pacific populations have changed so little since the isthmus last emerged further supports this interpretation.

Except for mud-living adaptations, most speciali-zations reach their height in *Celuca*. Waving display is elaborate, with a great increase in waving speed and in the time devoted to it; morphological specializations for ritualized combat are most numerous; structures beside the burrow, built only by displaying males, appear sporadically among the species. In *Celuca* also occurs the highest toleration of exposure to air; the equipment to extract nourishment from a relatively sandy substrate; the greatest elongation of walking legs, providing relative speed in movement on the ground; the greatest elongation of the claw, and (along with several members of the subgenera *Uca* and *Thalassuca*) the prevalence in end-forms of highly developed display-whitening. Finally, one characteristic making possible the explosive evolution of *Celuca* in the eastern Pacific has certainly been their small size.

This is the only American subgenus to occur also in the Indo-Pacific, where it is represented by two species. One of them, *lactea*, is not only the widest-ranging *Celuca* but one of the most successful species in the entire genus.

G. Directions of Evolution

Fiddler crabs have specialized in three principal directions. Almost all of their outstanding peculiarities now seem rather clearly to indicate these trends. The first leads toward a more terrestrial environment, the second toward the sharing of habitats by closely related forms, and the third toward sociality. These trends appear not only when the genus is viewed as a whole but within the larger subgenera.

Adaptations toward a more terrestrial environment. Modern *Uca* are already well adapted to intertidal life. All can withstand exposure to the air, all can move efficiently in that thin medium, and all feed during low tide by sifting the substrate. Furthermore, all use burrows efficiently as protection from desiccation and predators, while all depend on vision and substrate vibration as warning systems.

Within the group only the adaptations to life out of water and to sifting substrate show notable differences. Repeatedly the phylogenetic trend has been to move to higher levels of the littoral that are uncovered for longer periods, or to colonize flats that are flooded only intermittently. In both these situations even the bottoms of the burrows are sometimes dry; moisture conservation then becomes of prime importance. The most obvious structural adaptation of these species is the increased volume of the carapace, attained through arching in all dimensions. A similar response is found in the several hibernating species.

The second obvious adaptation of many species living higher on the shore is the development of specialized setae on the mouthparts. These setae, found in species on somewhat sandy, sloping biotopes, are shaped for rubbing off food particles clinging to the grains of sand. Simpler setae strain out organic matter from the finer silt of more muddy levels.

Adaptations to sympatry. Although coastlines are impressive in mileage, their area is negligible compared with those of other habitats on sea and land. And of all the stretches of climatically suitable shore, only a few offer support to *Uca* and fewer still provide abundant food. It is in such favored places that *Uca* congregate. This sharing of rich and restricted biotopes by related forms requires both that food supplies remain adequate and that species barriers be maintained. Fiddler crabs have dealt with the first requirement partly by miniaturization. This solution shows supremely well in the sympatric associations of *Celuca* in the tropical eastern Pacific and, less strikingly, of *Deltuca* in the Philippines. Adjacent or coincident microniches with differing foods will doubtless also prove to be of major importance for sympatry. In these favored localities genetic barriers are reinforced principally by differences in waving display, form of gonopod, or both; color and acoustic behavior will probably be found to be of secondary importance.

Adaptations to increased sociality. The trend toward sociality appears closely related to the first trend mentioned—that toward a more terrestrial environment. Throughout the genus the species living closest to low-tide levels devote less time to waving display during each daytime low; an obvious explanation is that there is not enough time both to feed sufficiently and to display at length. Although this conclusion is probably sound, it does not cover the general principle that in the less social subgenera, in particular most *Deltuca*, waving display is largely confined to a few days every two weeks; in socially advanced groups the restriction is far less rigid. Since the upper levels are nearly always less rich in food than the lower, the upward migrations of breeding males in socially advanced species solve the time problem at the expense of convenient food; it will be remembered that these males feed relatively little during their display phases, apparently making up for their fasting during other parts of their cycle.

In all these advanced subgenera the males not only seek out higher, more open, or conspicuous territories and wave for longer periods but share additional social characters. Their attachment to particular territories is more sustained; their high-intensity waving display is faster, more complex, or both; display color change in three subgenera is often striking; mating is normally underground instead of on the surface, with the female attracted to the male; the

morphological specializations for stridulation and ritualized combat reach their apex; displacement behavior is prevalent. All of these developments become most elaborate in end-forms of *Celuca*, although in some *Uca* and *Minuca* similar complexities are achieved. In short, the trend is to make social behavior more active, more time-consuming, and more complex.

The development of species-specific social behavior makes possible the full use of coincident and adjacent niches by closely similar forms. Behavioral diversity makes the crowding possible and probably contributes, in a literal feedback mechanism, to the richness and attractiveness of the environment. The results of crowding by individuals and species are not only probable enrichment, for *Uca*, of the substrate through glandular and other contributions. They also provide the social stimulus of active animals.

The high value of social behavior to *Uca* is strongly indicated by the eastern Atlantic species, *tangeri.* With the possible exception of rare strays from unrelated subgenera, this fiddler crab shares the coast with no other *Uca.* Yet, in spite of a total lack of sympatric associations, it maintains intact a large and highly evolved repertoire of social components.

We do not know with any certainty the advantages to *Uca* that these varied components provide. Only a few are fairly apparent when direct courtship behavior is observed; here the activities are comparable to those of courtship in many other animals.

On the other hand the emphasis on combat—its prevalence, patterns, and equipment—remains a puzzle. It is all very well to toss off the whole astonishing production as a mechanism for avoiding damage, or an activity useful in stimulating and coordinating reproduction, or a device for saving time to spend, instead, on the immediacies of courtship. Doubtless all these notions are true some of the time, to some degree, and in some species. Yet, even allowing for evolutionary inefficiency and for the importance of minute selective advantage, these suggestions do not seem to be enough. They simply do not explain why the greatest variety of morphological details in fiddlers are used only in intermale combat.

So, the ultimate reasons for *Uca*'s trend toward sociality remain mysterious. Vast numbers of marine invertebrates, from corals upward, have lived successfully under crowded conditions with no notable development of social behavior at all. *Uca*, in contrast, has developed its wealth of social intricacies. Threatening, fighting, rubbing, and tapping one another, stridulating, drumming, building, chasing, waving, curtsying, droving—these and all their further complexities have evolved from simple beginnings traceable quite distinctly in living species. Limitations in the basic amphibious plan of fiddler existence will probably discourage much further evolution. Meanwhile all over the world their habitats diminish.

Yet right now these lively crabs are flourishing and await the further study which is their due. Our understanding is small and their achievement great. Opportunities for a practicing biologist are just about unlimited.

H. SUMMARY

As in many other animals, the characters in *Uca* that are most helpful in deciphering relationships and phylogenetic history are those connected ultimately with reproduction. In contrast more conservative structures and behavior patterns maintain the individual. The traits showing the greatest diversity are the armature of appendages and orbits, form of gonopod, and patterns of waving display. Allopatric divisions are most prevalent in the heart of the Indo-Pacific, where orogenic action subjected the fauna to temporary partitions; marginal populations, on the other hand, in this genus are conservative; intermediate are forms living along the relatively unbroken coasts of Africa and America.

Differences in waving display and gonopod shape are apparently important means through which forms living in close sympatry avoid wasted courtship time and ineffective matings. In several species either the display or the gonopod or both show changes from their expectable patterns when the range coincides in part with that of a close relation. Since in sympatric communities breeding periods are usually synchronous and habitats overlapping, seasons and niches are largely ineffective in maintaining specific barriers.

The strongest stimuli to speciation in *Uca* were probably the presence of available biotopes in America and the intermittent barriers of both the Panamanian isthmus and the Sunda Shelf. The subgenera showing the highest specializations in both structure and behavior appear to have arisen in America. The degrees of differences are described that developed during the known periods of isolation. A phylogeny is proposed in which the Indo-Pacific subgenus *Deltuca* is held to be closest to the ancestral stock, while the American subgenus *Celuca* shows the greatest number of species and extremes of specializations. The genus as a whole has evolved in three principal directions. The first of these has been toward a more terrestrial environment, through devices for avoiding desiccation and for utilizing food from the upper littoral; the second, toward sympatric sharing of habitats, is aided by miniaturization as well as by structural and behavioral differences; the third and last trend has been toward increasing complexity of social organization.

Plates

INTRODUCTION

In every plate in which four aspects of a specimen or specimens are illustrated that are members of one species or subspecies, *A* and *E* are dorsal views; *B* and *F*, anterior views; *C* and *G*, outer views of the major claw (propodus and dactyl); *D* and *H*, inner views of the major claw. For nomenclature of the parts, see Figs. 1, 2, 3, 42, 43, and 44, as well as Table 13. In plates where four aspects are not illustrated, the views are individually identified; the word "claw" always refers to the major claw.

In the captions the carapace length of the specimen photographed is stated, in the belief that it will give a better idea of the size of the specimen than would the degree of magnification or the dimensions of the front or claw.

Abbreviations of institution names, given in full on p. 678, indicate the sources of photographed specimens and include catalogue numbers where these are available. All NYZS specimens are now on deposit in the USNM (see p. 591).

Pl. 1. *A-D. Uca (Deltuca) acuta rhizophorae.* Sarawak: Santobong. Carapace lgth. 10 mm. NYZS. (P. 27.)

E-H. Uca (Deltuca) acuta acuta. China. Carapace lgth. 13 mm. USNM 57033. (P. 28.)

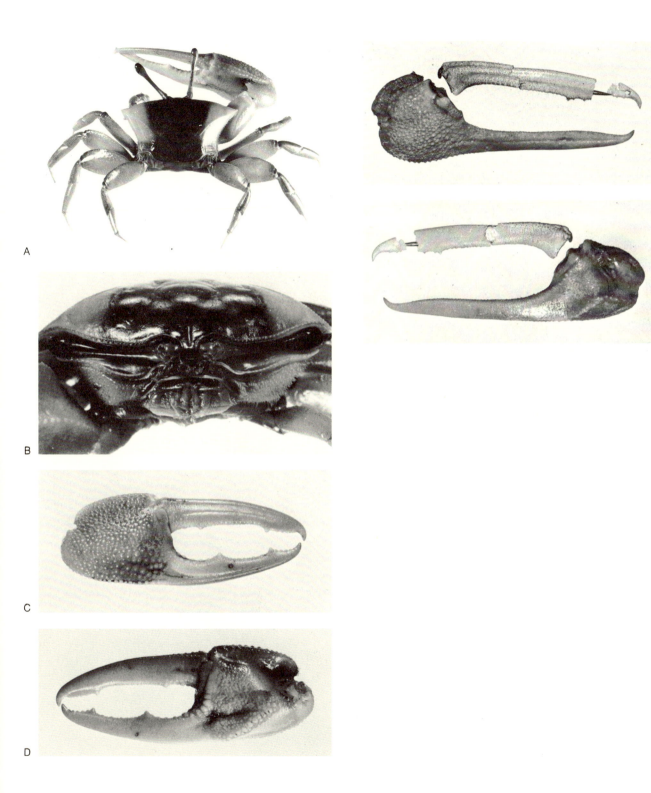

PL. 2. *A-D. Uca (Deltuca) rosea.* Malaya: Penang. Carapace lgth. 12 mm. NYZS. (P. 29.)

E-F. Uca (Deltuca) dussumieri dussumieri. Java: Semarang. Carapace lgth. 14 mm. Leptochelous individual. Paris: lectotype of *Gelasimus dussumieri*, claw. (P. 35; Figs. 8, 9.)

Pl. 3. *A-D. Uca* (*Deltuca*) *dussumieri dussumieri.*
Philippine Is.: Zamboanga. Carapace
lgth. 13 mm. Leptochelous individual.
NYZS. (P. 37.)

E-H. Uca (*Deltuca*) *dussumieri spinata.* Singapore. Carapace lgth. 20 mm. Brachychelous individual. NYZS. (P. 36.)

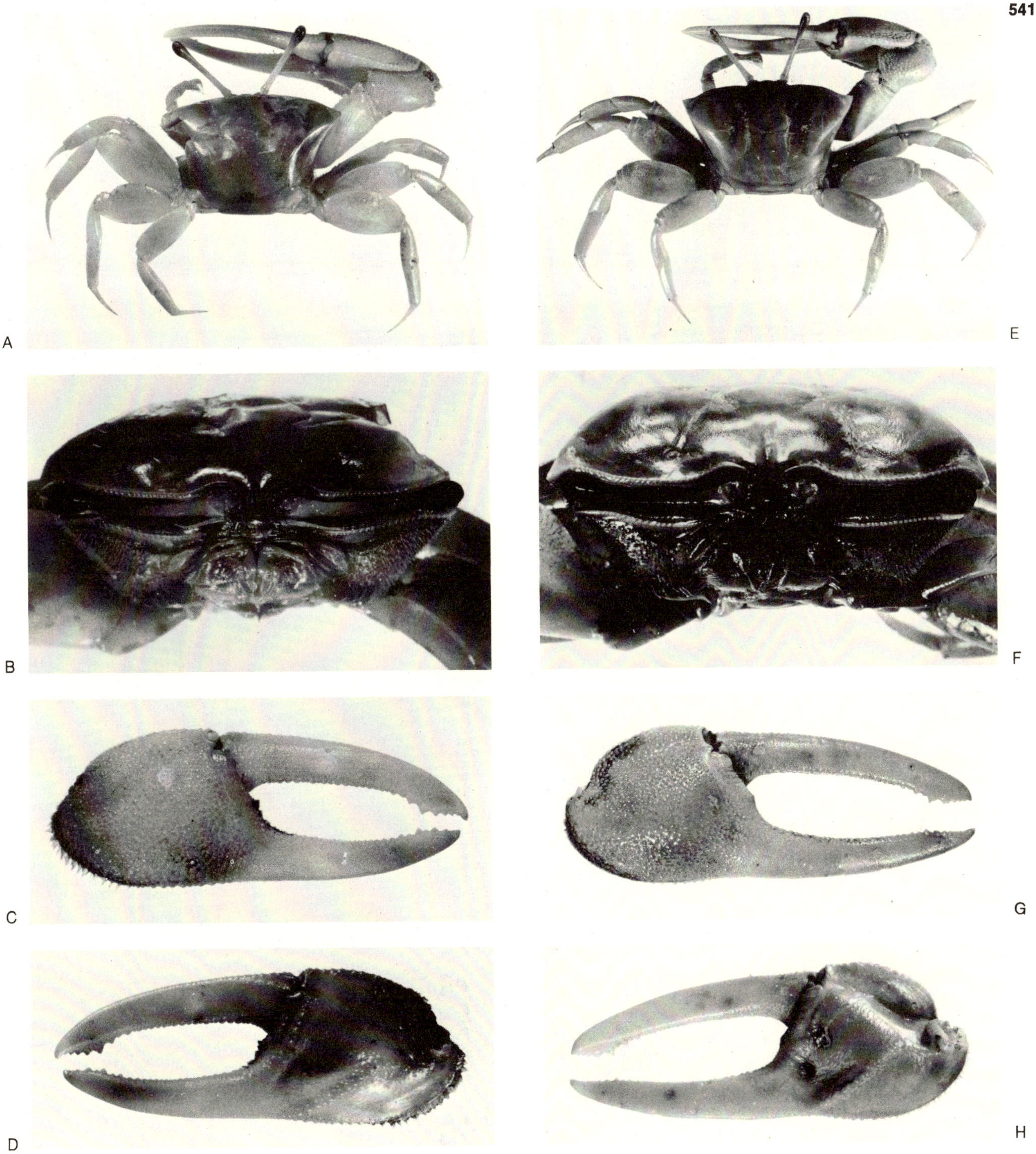

PL. 4. *A-D*. *Uca* (*Deltuca*) *demani australiae*. Australia: Broome. Carapace lgth. 14.5 mm. USNM 64250. (P. 41.)

E-H. *Uca* (*Deltuca*) *demani typhoni*. Philippine Is.: Iloilo. Carapace lgth. 18 mm. USNM 73201. (P. 41.)

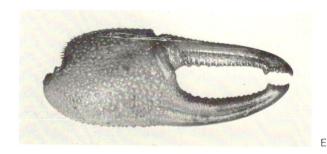

A

B

C

D

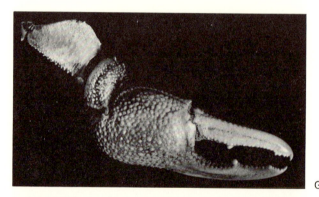

E

F

G

H

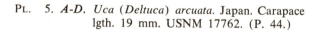

PL. 5. *A-D.* Uca (*Deltuca*) *arcuata.* Japan. Carapace lgth. 19 mm. USNM 17762. (P. 44.)

E-F. Uca (*Deltuca*) *arcuata.* China. Carapace lgth. 15 mm. Paris: from type material of *Gelasimus brevipes*; claw. (P. 47; Figs. 8, 9.)

G-H. Uca (*Deltuca*) *forcipata.* Malaya. Carapace lgth. 10.5 mm. BM reg. no. 44.106: holotype of *Gelasimus forcipatus*; claw. (P. 51.)

PL. 6. *A-D. Uca (Deltuca) forcipata.* Singapore. Carapace lgth. 16 mm. NYZS. (P. 48.)

E-H. Uca (Deltuca) coarctata coarctata. Fiji Is. Carapace lgth. 15 mm. NYZS. (P. 55.)

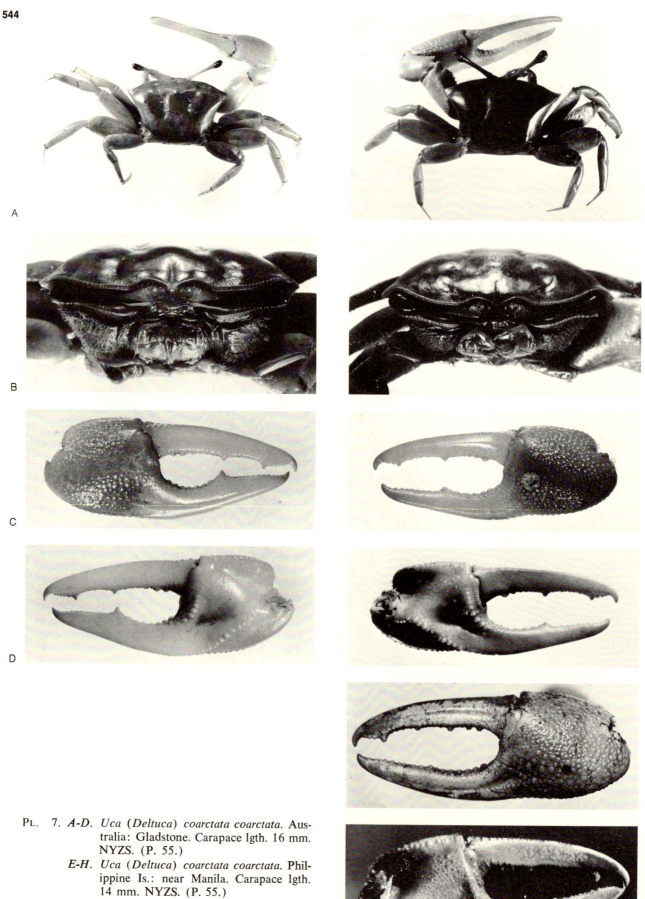

PL. 7. **A-D.** *Uca* (*Deltuca*) *coarctata coarctata.* Australia: Gladstone. Carapace lgth. 16 mm. NYZS. (P. 55.)

E-H. *Uca* (*Deltuca*) *coarctata coarctata.* Philippine Is.: near Manila. Carapace lgth. 14 mm. NYZS. (P. 55.)

I-J. *Uca* (*Deltuca*) *coarctata coarctata.* "Odessa." Carapace lgth. 11 mm. Paris: from type material (not lectotype) of *Gelasimus coarctatus*; claw. (P. 55.)

A

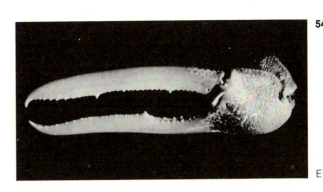

545

E

B

C

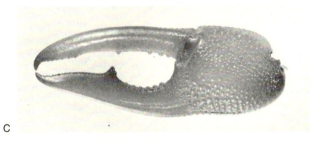

D

PL. 8. *A-D. Uca (Deltuca) coarctata flammula* subsp.
nov. Australia: Port Darwin. Carapace
lgth. 17.5 mm. NYZS. (P. 56.)

E. Uca (Deltuca) coarctata flammula. Australia: Cossack (lat. 20°40′ S., long.
117°5′ E.). Propodus lgth. 75 mm. Collection of Wm. Macnae; outer claw.

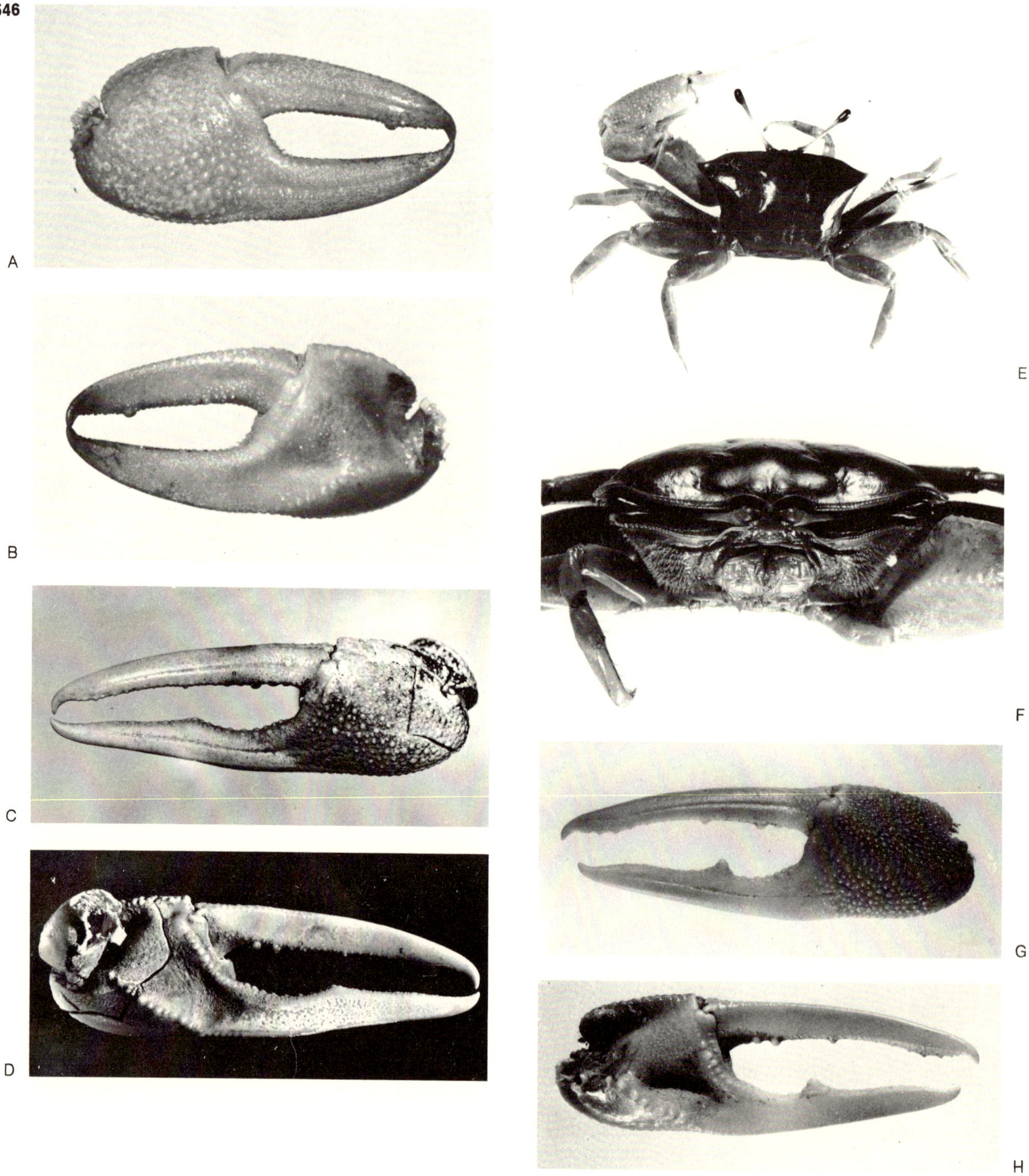

PL. 9. *A-B.* Uca (*Deltuca*) *urvillei*. "Vanikoro." Carapace lgth. 11 mm. Paris: Type of *Gelasimus urvillei*; claw. (P. 60; Figs. 8, 9.)

C-D. Uca (*Deltuca*) *urvillei*. India: Malabar Coast. Carapace lgth. 17 mm. Paris: from type material of *Gelasimus dussumieri*. (P. 60; Figs. 8, 9.)

E-H. Uca (*Deltuca*) *urvillei*. Tanzania: Pemba I. Carapace lgth. 18.5 mm. NYZS. (P. 58.)

PL. 10. *A-D. Uca (Australuca) bellator bellator.* Philippine Is.: near Manila. Carapace lgth. 10 mm. NYZS. (P. 66.)

E-F. Uca (Australuca) bellator bellator. Philippine Is. Carapace lgth. 12.5 mm. BM Reg. No. 43.6: holotype of *Gelasimus bellator*; claw. (P. 66.)

G-J. Uca (Australuca) bellator signata. Australia: Gladstone. Carapace lgth. 9.5 mm. NYZS. (P. 67.)

PL. 11. *A-D. Uca (Australuca) bellator longidigita.* Australia: Brisbane. Carapace lgth. 10.5 mm. NYZS. (P. 68.)

E-H. Uca (Australuca) bellator minima subsp. nov. Australia: Port Darwin. Carapace lgth. 5 mm. NYZS. (P. 68.)

PL. 12. *A-D. Uca (Australuca) seismella* sp. nov. Australia: Port Darwin. Carapace lgth. 7 mm. NYZS. (P. 70.)

E-H. Uca (Australuca) polita sp. nov. Australia: Gladstone. Carapace lgth. 9.5 mm. NYZS. (P. 72.)

Pl. 13. *A-D*. *Uca* (*Thalassuca*) *tetragonon*. Ethiopia: Massawa. Carapace lgth. 16 mm. NYZS. (P. 77.)

E-H. *Uca* (*Thalassuca*) *tetragonon*. Society Is.: Tahiti. Carapace lgth. 22 mm. NYZS. (P. 77.)

PL. 14. *A-D*. *Uca* (*Thalassuca*) *formosensis*. Taiwan: Taihoku. Carapace lgth. 15 mm. USNM 55386. (P. 83.)

E-H. *Uca* (*Thalassuca*) *vocans vocans*. Philippine Is.: Madaum. Carapace lgth. 12.5 mm. (P. 92.)

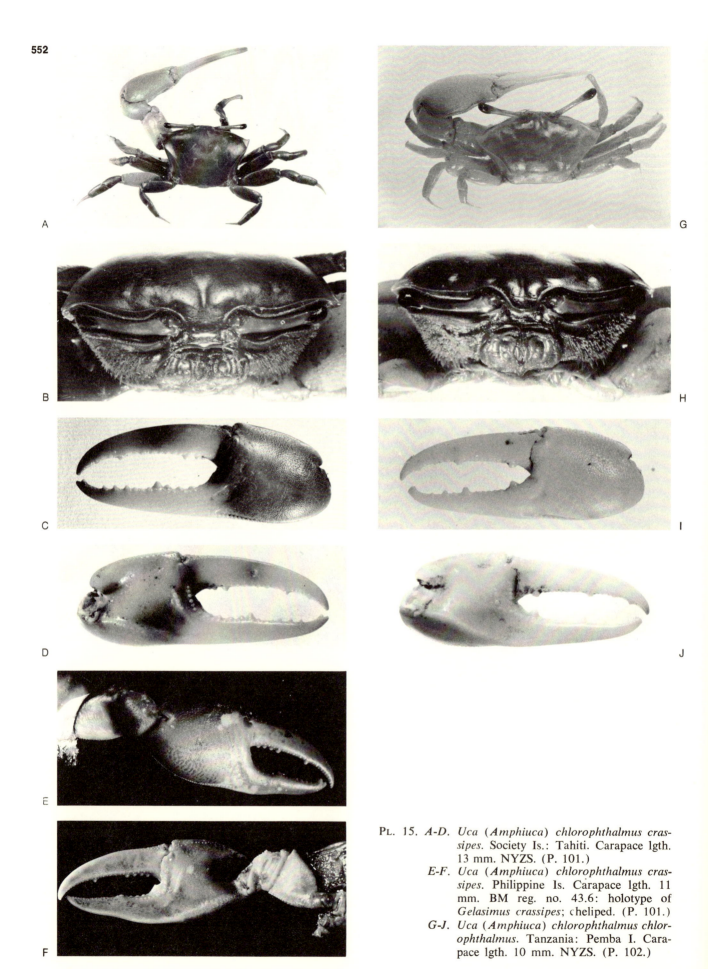

PL. 15. *A-D. Uca (Amphiuca) chlorophthalmus crassipes.* Society Is.: Tahiti. Carapace lgth.
13 mm. NYZS. (P. 101.)

*E-F. Uca (Amphiuca) chlorophthalmus cras-
sipes.* Philippine Is. Carapace lgth. 11
mm. BM reg. no. 43.6: holotype of
Gelasimus crassipes; cheliped. (P. 101.)

*G-J. Uca (Amphiuca) chlorophthalmus chlor-
ophthalmus.* Tanzania: Pemba I. Carapace lgth. 10 mm. NYZS. (P. 102.)

PL. 16. *A-D. Uca (Amphiuca) inversa inversa.* Tanzania: Zanzibar. Carapace lgth. 11 mm.
NYZS. (P. 107.)

E-H. Uca (Amphiuca) inversa sindensis. Pakistan: Karachi. Carapace lgth. 7 mm.
NYZS. (P. 108.)

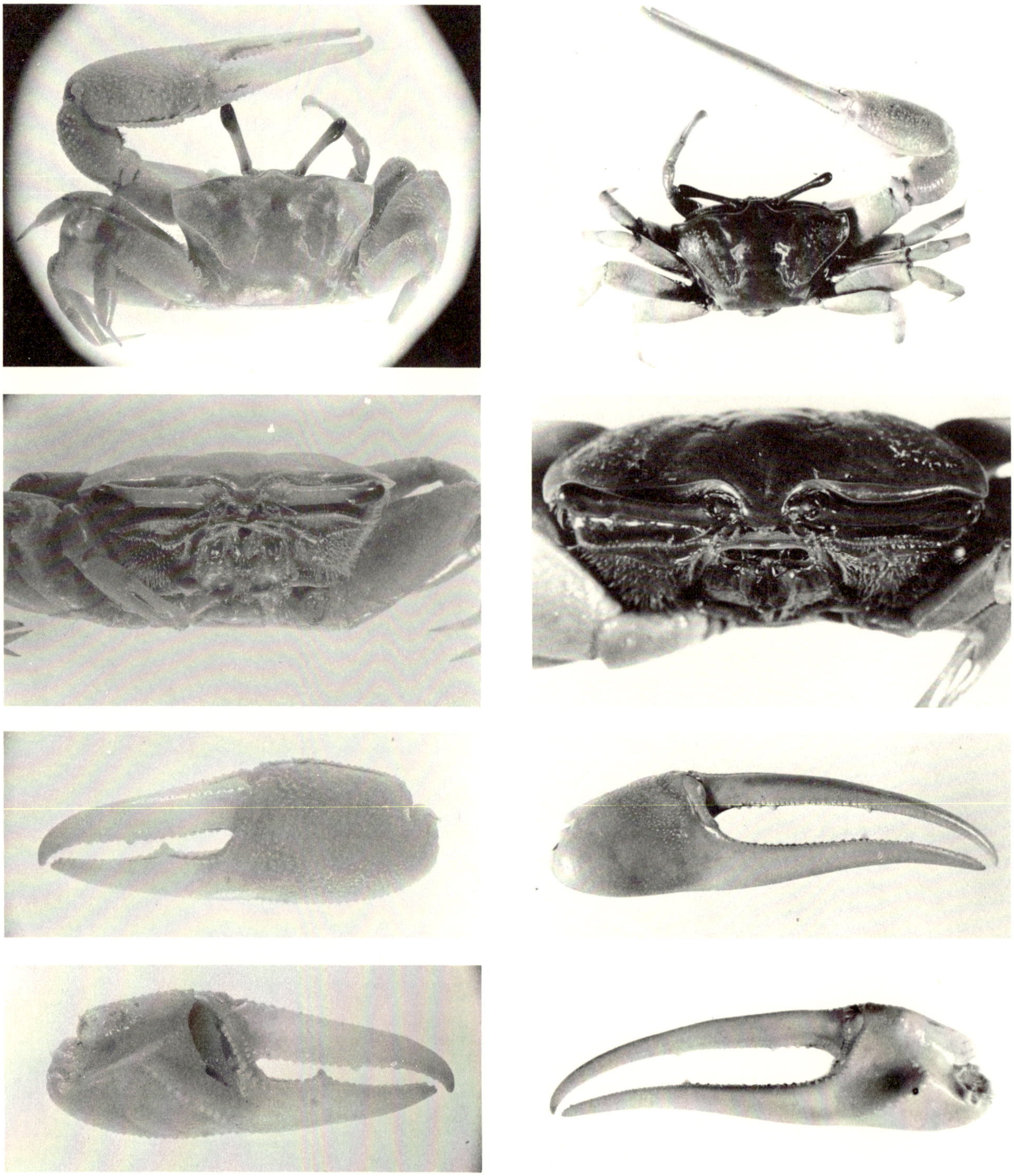

PL. 17. *A-D. Uca (Boboruca) thayeri umbratila.* Eastern Pacific: Costa Rica (Golfito). Cara-
pace lgth. 16 mm. NYZS. (P. 113.)

E-H. Uca (Boboruca) thayeri thayeri. West
Indies: Trinidad. Carapace lgth. 17.5
mm. NYZS. (P. 114.)

A

E

B

F

C

G

D

H

Pl. 18. *A-D.* Uca (*Afruca*) *tangeri*. Angola: Luanda. Carapace lgth. 25 mm. NYZS. (P. 118.)

E-H. Uca (*Uca*) *princeps monilifera*. Mexico, in Gulf of California: San Felipe. Carapace lgth. 29.5 mm. USNM 67735. (P. 131.)

PL. 19. *A-D. Uca (Uca) princeps princeps.* Eastern Pacific: Costa Rica (Golfito). Carapace lgth. 23 mm. NYZS. (P. 131.)

E-H. Uca (Uca) heteropleura. Eastern Pacific: Panama (Panama City). Carapace lgth. 12 mm. NYZS. (P. 133.)

PL. 20. *A-D. Uca (Uca) stylifera.* Eastern Pacific: Nicaragua (Corinto). Carapace lgth. 18 mm. NYZS. (P. 140.)

E-H. Uca (Uca) major. Venezuela: Turiamo. Carapace lgth. 14 mm. NYZS. (P. 136.)

PL. 21. *A-D. Uca (Uca) maracoani maracoani.* Guyana: Georgetown. Carapace lgth. 25 mm. NYZS. (P. 147.)

E-H. Uca (Uca) ornata. Eastern Pacific: Panama (Panama City). Carapace lgth. 29 mm. NYZS. (P. 150.)

PL. 22. *A-D. Uca (Minuca) panamensis.* Eastern Pacific: Colombia (Humboldt Bay). Carapace lgth. 12 mm. AMNH 7605. (P. 158.)

E-H. Uca (Minuca) pygmaea. Eastern Pacific: Colombia (Buenaventura). Carapace lgth. 6 mm. NYZS. (P. 161.)

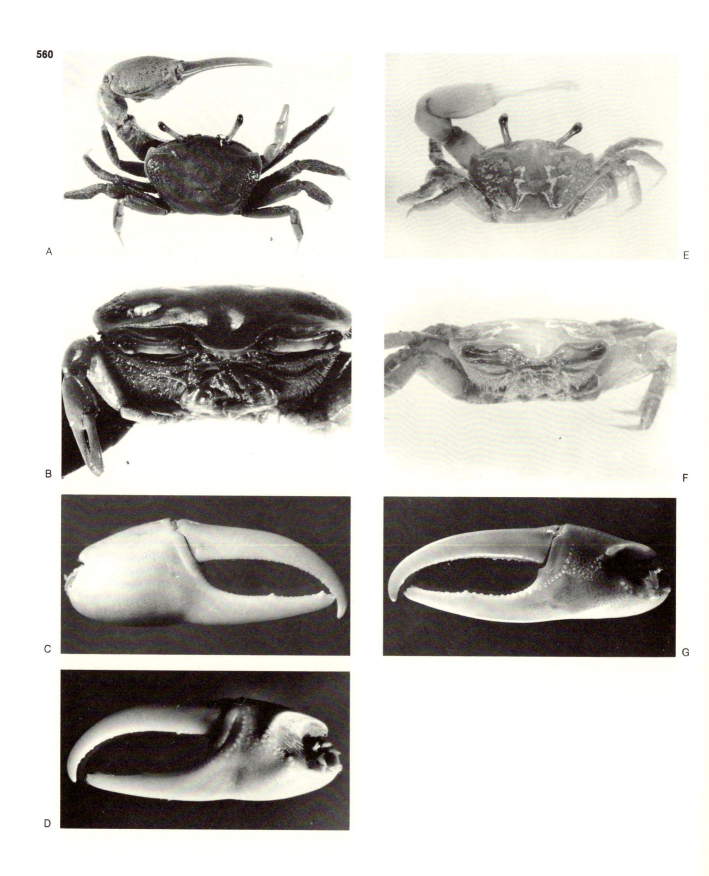

PL. 23. *A-D. Uca (Minuca) vocator ecuadoriensis.* Eastern Pacific: Ecuador (Guayaquil). Carapace lgth. 13.5 mm. NYZS. (P. 166.)

E-G. Uca (Minuca) vocator vocator. Venezuela: Maracaibo. Carapace lgth. 14.5 mm. *G* = inner side of claw. NYZS. (P. 166.)

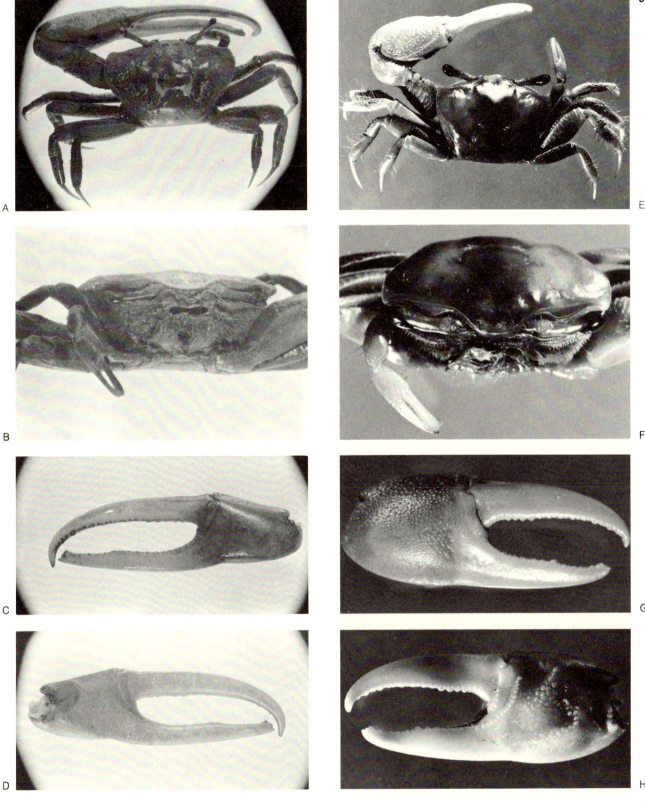

A

B

C

D

E

F

G

H

PL. 24. *A-D. Uca (Minuca) vocator vocator*. West Indies: Trinidad (Caroni Swamp). Cara-
pace lgth. 17.5 mm. Leptochelous form.
NYZS. (P.166.)

E-H. Uca (Minuca) burgersi. West Indies:
Tobago. Carapace lgths.: *E, F*, 9 mm;
G, H, 10.5 mm. NYZS. (P. 168.)

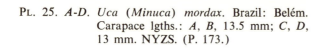

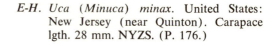

PL. 25. *A-D. Uca* (*Minuca*) *mordax*. Brazil: Belém. Carapace lgths.: *A, B*, 13.5 mm; *C, D*, 13 mm. NYZS. (P. 173.)

E-H. Uca (*Minuca*) *minax*. United States: New Jersey (near Quinton). Carapace lgth. 28 mm. NYZS. (P. 176.)

PL. 26. *A-D. Uca (Minuca) brevifrons.* Eastern Pacific: Panama (Piñas R.). Carapace lgth. 14 mm. AMNH 10702. (P. 180.)

E-H. Uca (Minuca) galapagensis galapagensis. Eastern Pacific: Ecuador (Guayaquil). Carapace lgth. 14.5 mm. NYZS. (P. 187.)

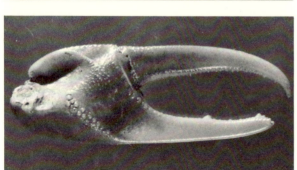

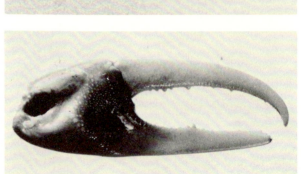

PL. 27. *A-D. Uca (Minuca) rapax rapax*. West Indies: Trinidad (Cocorite Swamp). Carapace lgth. 12 mm. NYZS. (P. 196.)

E-H. Uca (Minuca) pugnax pugnax. United States: New York (Oyster Bay). Carapace lgth. 12 mm. NYZS. (P. 203.)

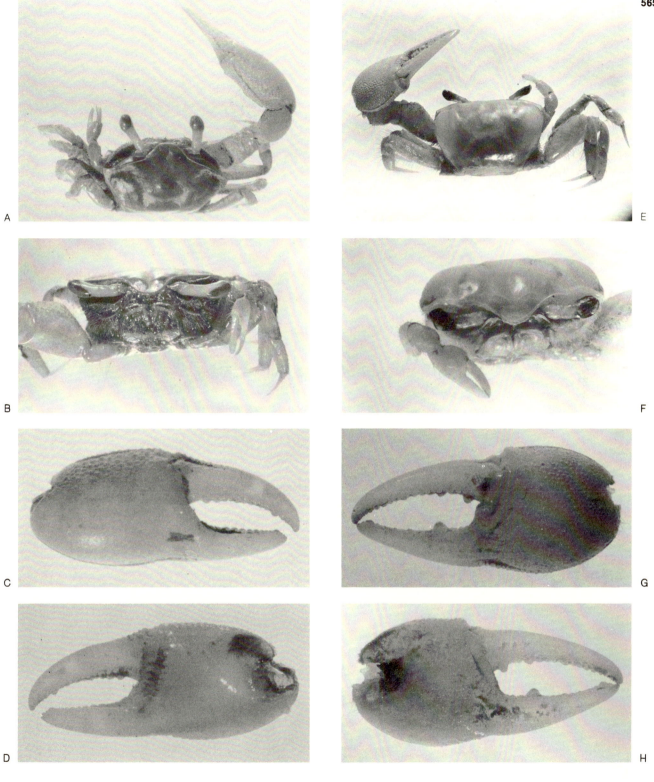

PL. 28. *A-D*. *Uca* (*Minuca*) *zacae*. Eastern Pacific:
Costa Rica (Golfito). Carapace lgth. 6
mm. NYZS. (P. 206.)

E-H. *Uca* (*Minuca*) *subcylindrica*. United
States: Texas, Carapace lgth. 9 mm.
USNM 99826. (P. 209.)

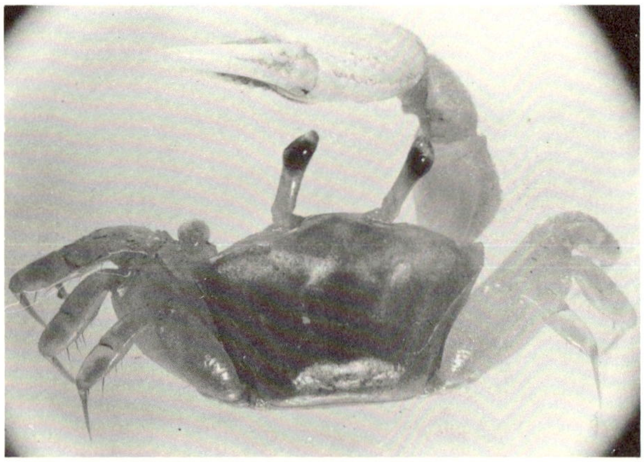
A

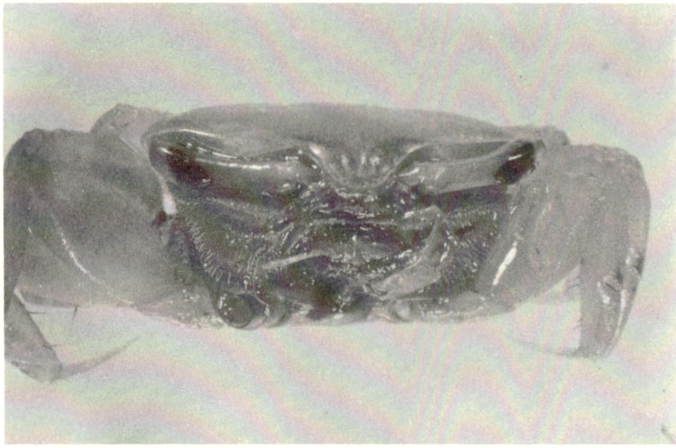
B

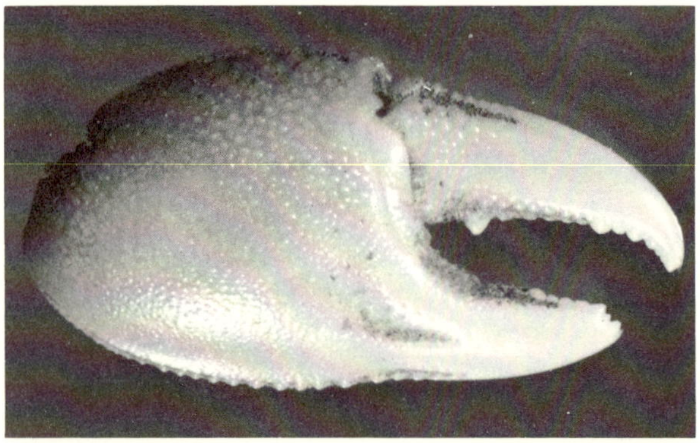
C

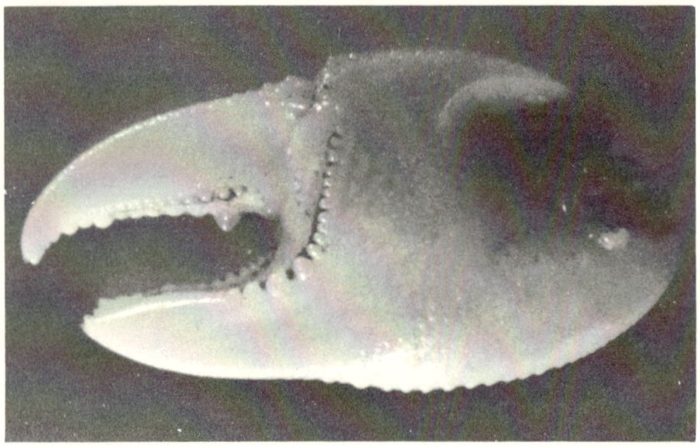
D

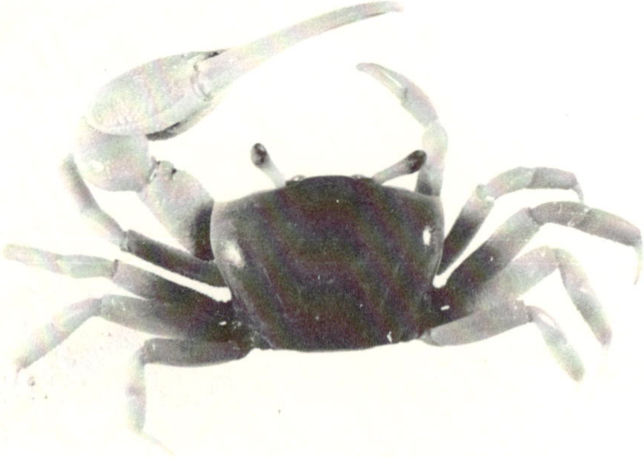
E

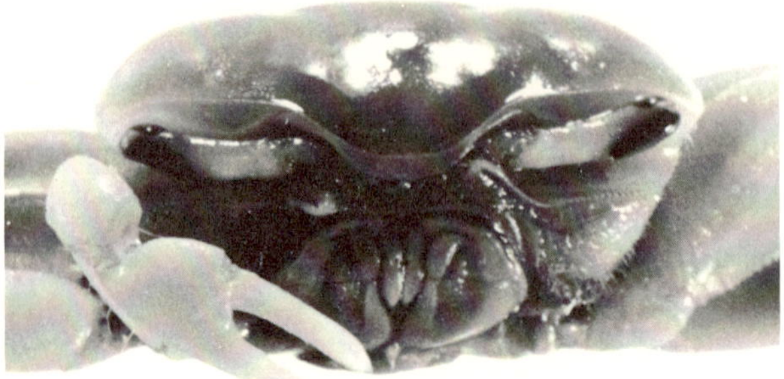
F

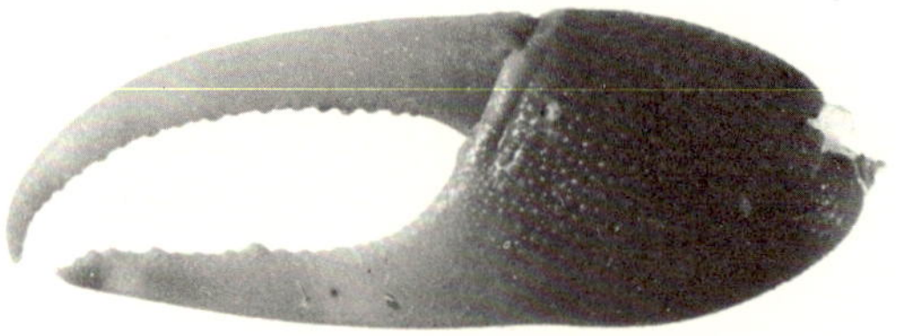
G

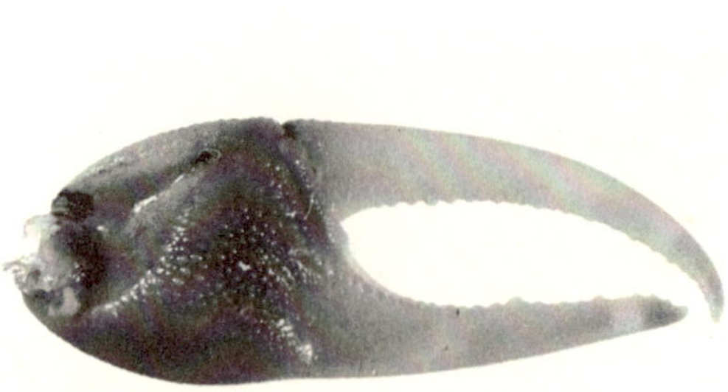
H

PL. 29. *A-D*. *Uca* (*Celuca*) *argillicola*. Eastern Pacific: Panama (Pearl Is.). Carapace lgth. 9 mm. (P. 220.)

E-H. *Uca* (*Celuca*) *pugilator*. United States: Florida (near Daytona Beach). Carapace lgth. 13 mm. NYZS. (P. 223.)

PL. 30. *A-D. Uca (Celuca) uruguayensis.* Uruguay: Maldonado. Carapace lgth. 8 mm. MCZ 5926. (P. 229.)

E-H. Uca (Celuca) crenulata crenulata. Mexico: Lower California, west coast (Todos Santos Bay). Carapace lgths.: *E, F*, 12 mm.; *G, H*, 11 mm. USNM 19033. (P. 234.)

I. Upper: *Uca (Celuca) crenulata coloradensis.* Mexico: Gulf of California, near mouth of Colorado R. Carapace lgth. 11 mm. USNM 18292. (P. 234.)
Lower: *Uca (Celuca) crenulata crenulata.* Mexico: Lower California, west coast (Todos Santos Bay). Carapace lgth. 10.5 mm. USNM 19033. (P. 234.)
Obliquely dorsal view of major manus and dactyl in each subspecies.

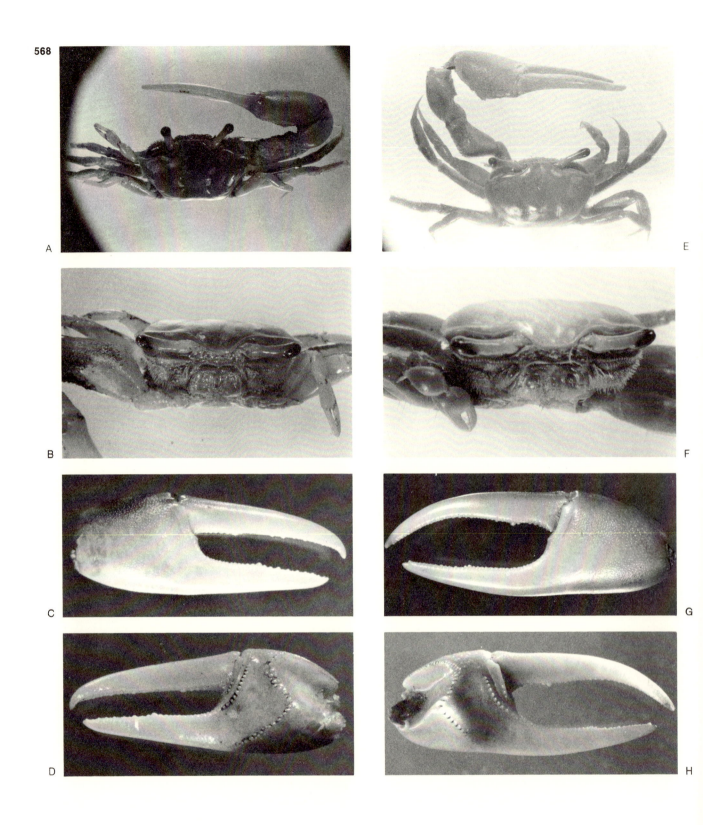

PL. 31. *A-D. Uca (Celuca) speciosa speciosa.* United States: Florida (Miami). Carapace lgth. 8 mm. NYZS. (P. 238.)

E-H. Uca (Celuca) speciosa spinicarpa. United States: Mississippi. Carapace lgth. 11 mm. USNM 90305. (P. 239.)

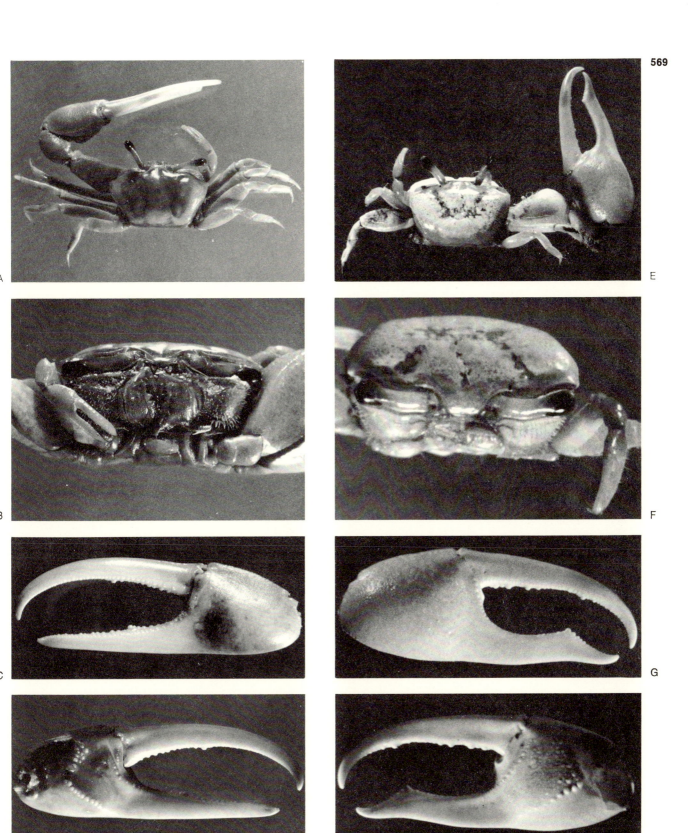

A

E

B

F

C

G

D

H

Pl. 32. *A-D*. *Uca* (*Celuca*) *cumulanta*. West Indies: Trinidad (Diego Martin). Carapace lgths.: *A*, *B*, 8.5. mm; *C*, *D*, 9 mm. NYZS. (P. 240.)

E-H. *Uca* (*Celuca*) *batuenta*. Eastern Pacific: Panama Canal Zone (Balboa). Carapace lgth. ca. 4.5 mm. NYZS 4122 = USNM 137406; paratype. (P. 244.)

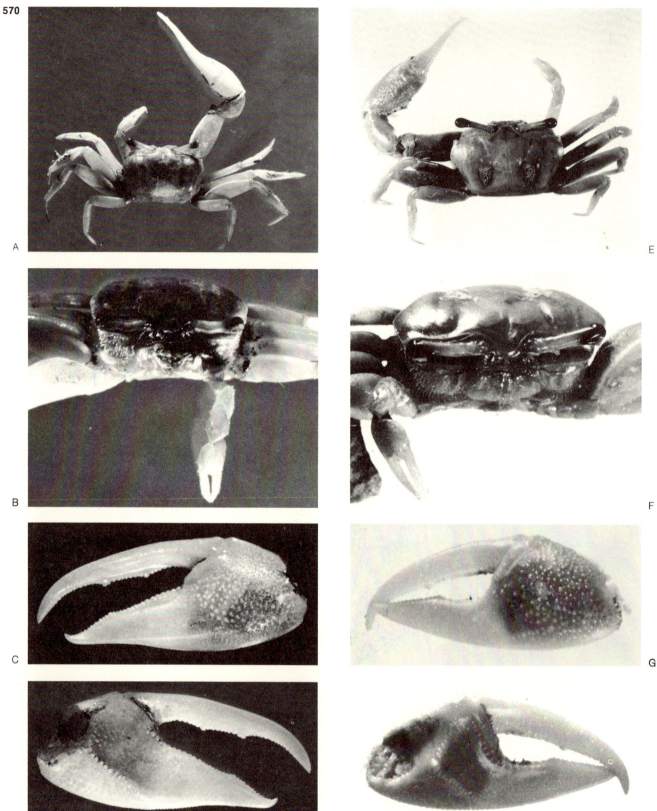

PL. 33. *A-D. Uca (Celuca) saltitanta.* Eastern Pacific: Panama Canal Zone (Balboa). Carapace lgths.: *A, B,* 5.7 mm; *C, D,* 5.5 mm. NYZS. (P. 247.)

E-H. Uca (Celuca) oerstedi. Eastern Pacific: Panama (Panama City). Carapace lgth. 6.5 mm. NYZS. (P. 251.)

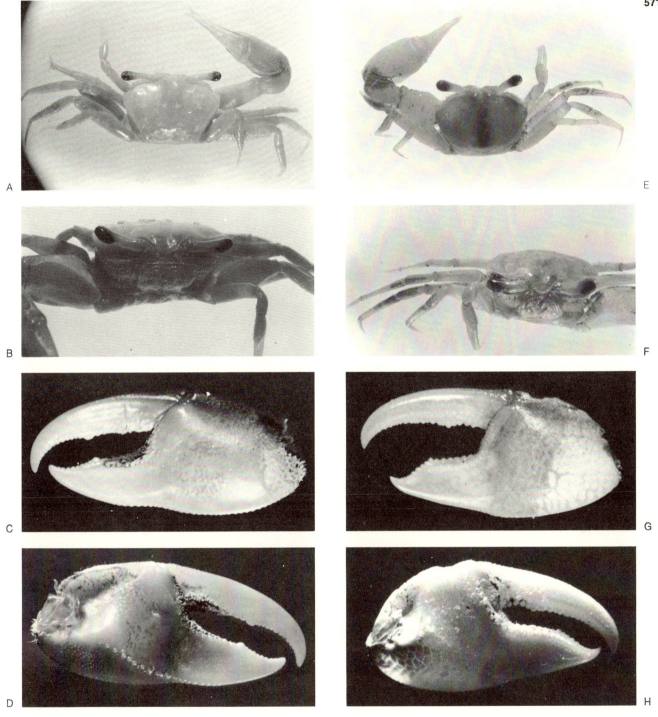

PL. 34. *A-D*. *Uca (Celuca) inaequalis*. Eastern Pacific: Ecuador (Guayaquil). Carapace lgths.: *A*, *B*, 6 mm; *C*, *D*, 6.2 mm. NYZS. (P. 254.)

E-H. *Uca (Celuca) tenuipedis*. Eastern Pacific: Costa Rica (Ballenas Bay). Carapace lgth. 5 mm. NYZS 381,144 = USNM 137410; paratype. (P. 258.)

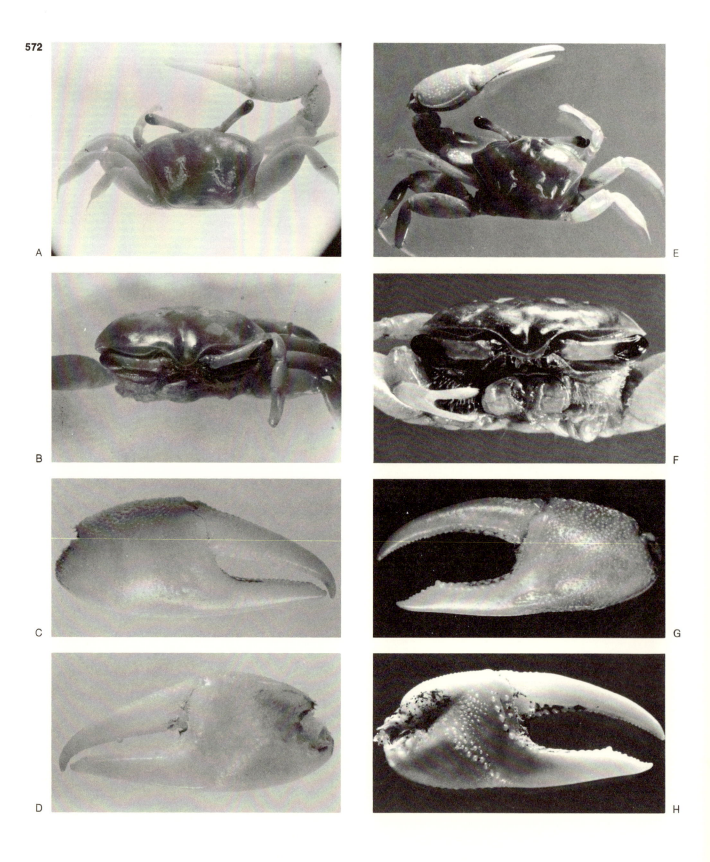

PL. 35. *A-D*. *Uca* (*Celuca*) *tomentosa*. Eastern Pacific: Costa Rica (Puntarenas). Carapace lgth. 6.6 mm. NYZS 381,132 = USNM 137411; holotype. (P. 261.)

E-H. *Uca* (*Celuca*) *tallanica*. Eastern Pacific: Ecuador (Puerto Bolivar). Carapace lgth. 7 mm. NYZS. (P. 264.)

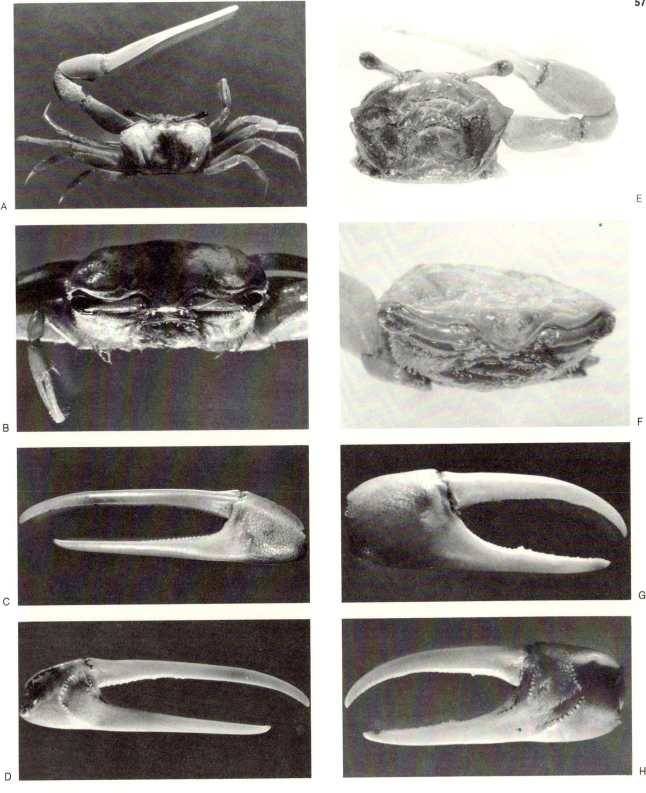

PL. 36. *A-D*. *Uca* (*Celuca*) *festae*. Eastern Pacific: Ecuador (Guayaquil). Carapace lgths. *A*, *B*, 7.2 mm; *C*, *D*, 8 mm. NYZS. (P. 267.)

E-H. *Uca* (*Celuca*) *helleri*. Eastern Pacific: Galapagos Is. Carapace lgth. 6.8 mm. USNM 25666. (P. 271.)

PL. 37. *A-D. Uca (Celuca) dorotheae.* Eastern Pacific: Ecuador (Puerto Bolivar). Carapace lgths.: *A, B,* 7 mm; *C, D,* 8 mm. NYZS. (P. 275.)

E-H. Uca (Celuca) beebei. Eastern Pacific: Panama (Panama City). Carapace lgth. 5.5 mm. NYZS. (P. 278.)

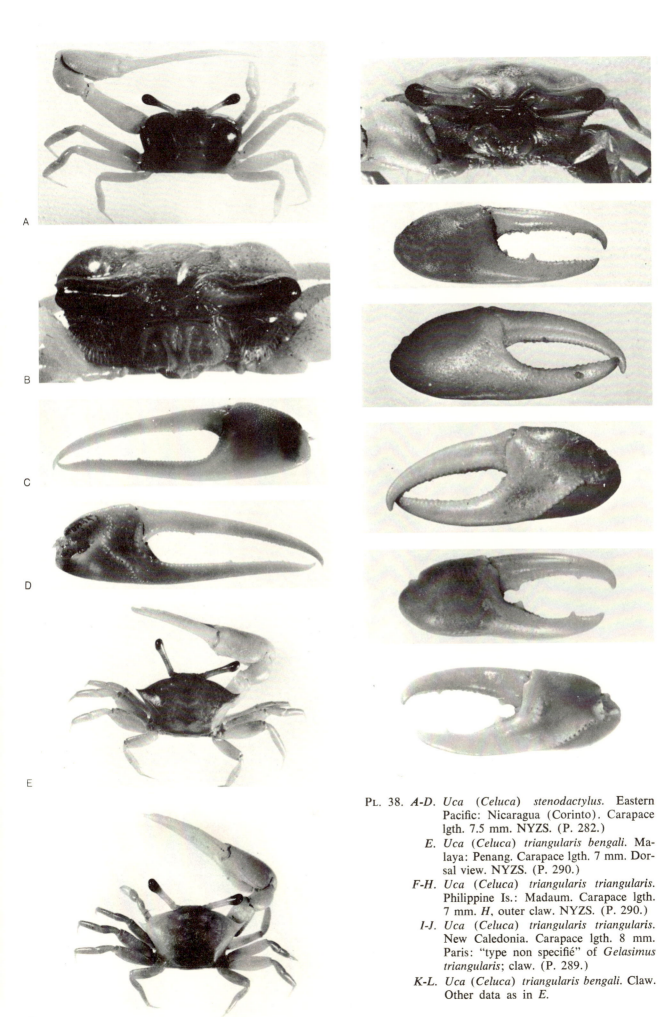

PL. 38. *A-D. Uca* (*Celuca*) *stenodactylus.* Eastern
Pacific: Nicaragua (Corinto). Carapace
lgth. 7.5 mm. NYZS. (P. 282.)

E. Uca (*Celuca*) *triangularis bengali.* Malaya: Penang. Carapace lgth. 7 mm. Dorsal view. NYZS. (P. 290.)

F-H. Uca (*Celuca*) *triangularis triangularis.*
Philippine Is.: Madaum. Carapace lgth.
7 mm. *H*, outer claw. NYZS. (P. 290.)

I-J. Uca (*Celuca*) *triangularis triangularis.*
New Caledonia. Carapace lgth. 8 mm.
Paris: "type non spécifié" of *Gelasimus
triangularis*; claw. (P. 289.)

K-L. Uca (*Celuca*) *triangularis bengali.* Claw.
Other data as in *E.*

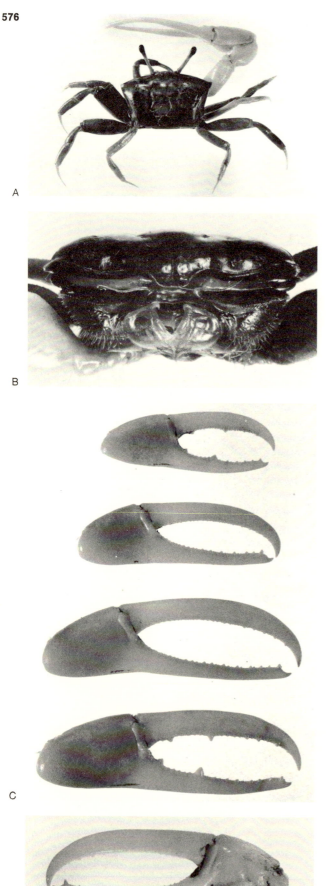

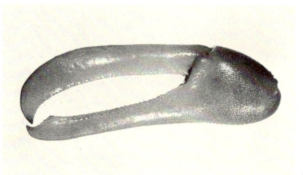

A

B

C

D

E

F

PL. 39. *A-D. Uca (Celuca) lactea annulipes.* Mozambique: Inhaca I. Carapace lgth. of specimen providing all views except outer views of claws below the uppermost in *C*, 10.5 mm. Additional claws from specimens of similar size in same population. NYZS. (P. 299.)

E-F. Uca (Celuca) lactea perplexa. Fiji Is. Carapace lgth. 9 mm. *E*, dorsal view; *F*, outer claw, showing extreme example of pollex profile characteristic of this subspecies. NYZS. (P. 300.)

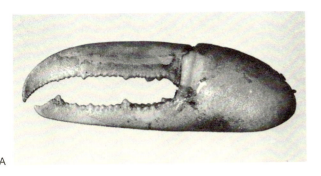

A

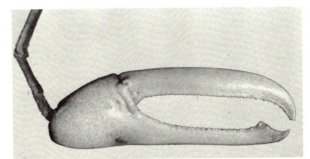

E

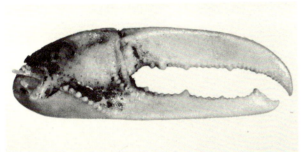

B

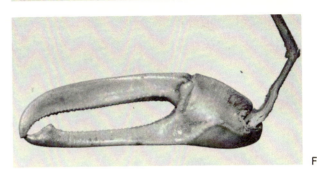

F

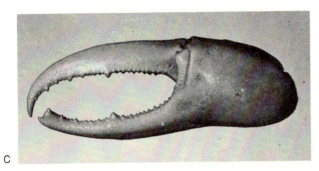

C

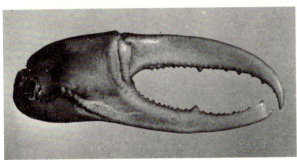

D

PL. 40. *A-B.* *Uca* (*Celuca*) *lactea lactea.* Japan. Carapace lgth. 11.5 mm. Paris, from Leiden; perhaps a cotype of *Gelasimus lacteus*; claw. (P. 298.)

C-D. *Uca* (*Celuca*) *lactea annulipes.* "Mer des Indes." Carapace lgth. 10.5 mm. Paris: lectotype of *Gelasimus annulipes*; claw. (P. 298.)

E-F. *Uca* (*Celuca*) *lactea.* Australia. Propodus lgth. 32 mm. Paris: claw of composite type of *Gelasimus forceps.* (P. 298.)

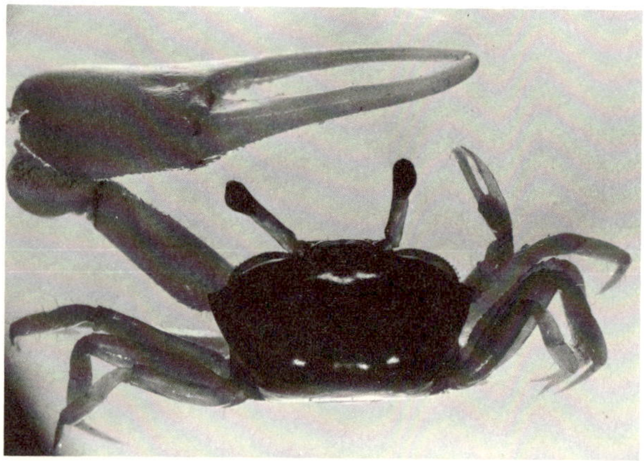
A

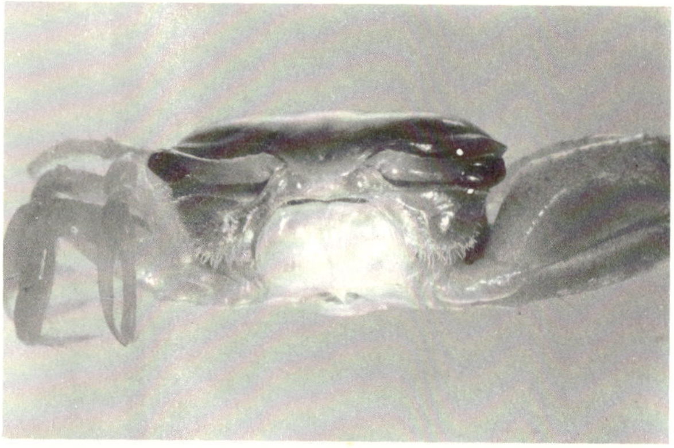
B

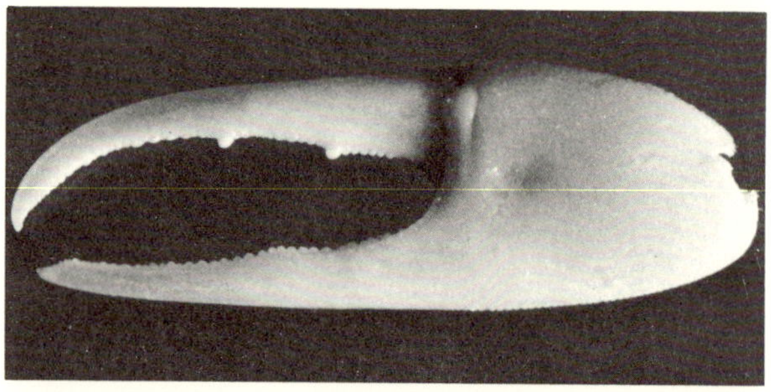
C

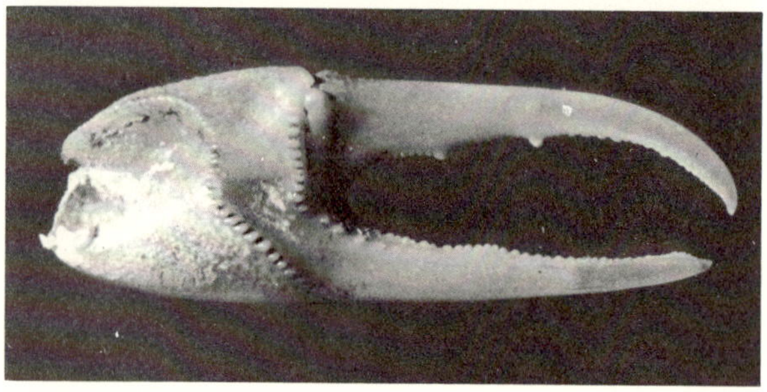
D

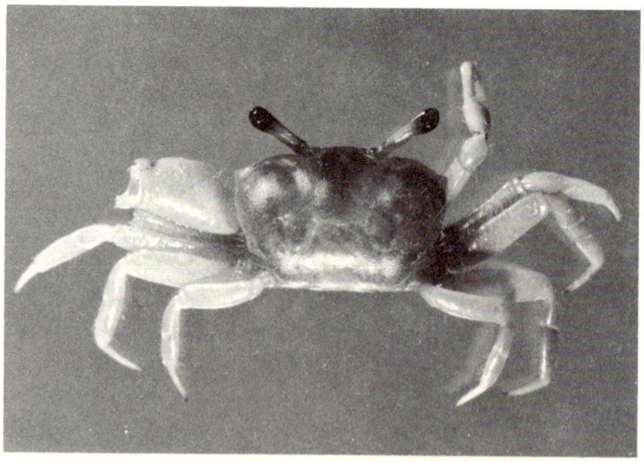
E

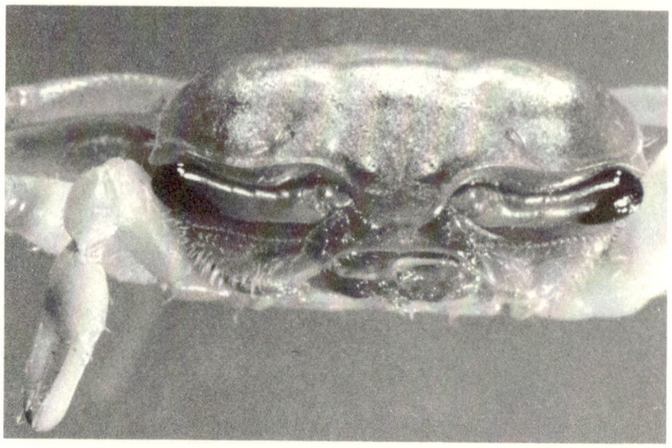
F

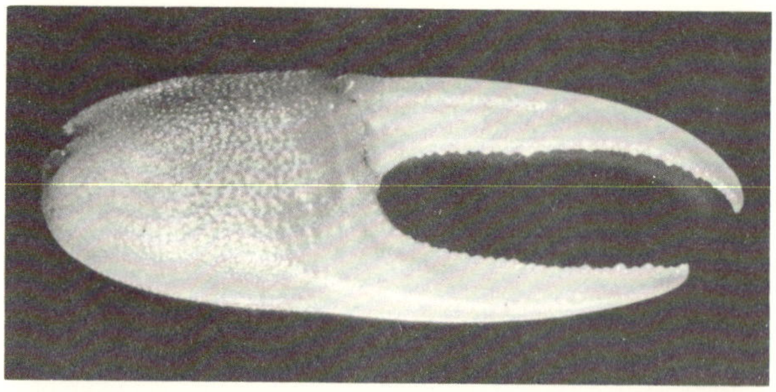
G

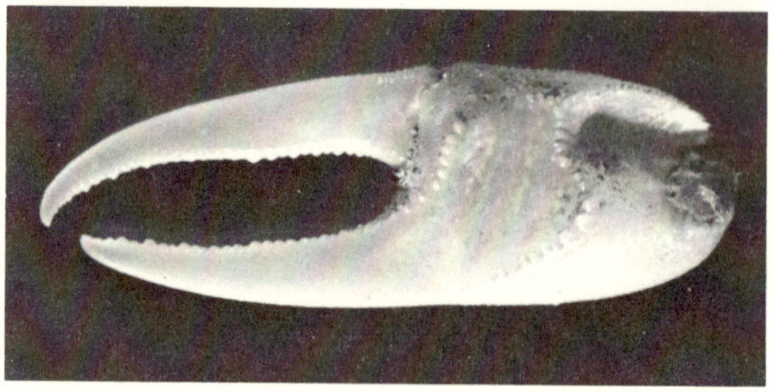
H

PL. 41. *A-D*. *Uca* (*Celuca*) *leptodactyla*. Venezuela: Turiamo. Carapace lgths.: *A*, *B*, ca. 5.2 mm; *C*, *D*, 5.5 mm. NYZS. (P. 304.)

E-H. *Uca* (*Celuca*) *limicola*. Eastern Pacific: Costa Rica (Golfito). Carapace lgths.: *E*, *F*, 6.2 mm; *G*, *H*, 7 mm. NYZS. 381,-153 = USNM 79401; paratypes. (P. 308.)

PL. 42. *A-D. Uca (Celuca) deichmanni.* Eastern Pacific: Costa Rica (Golfito). Carapace lgths.: *A*, *B*, 6 mm; *C*, *D*, 6.2 mm. NYZS. (P. 311.)

E-H. Uca (Celuca) musica terpsichores. Eastern Pacific: Costa Rica (Golfito). Carapace lgths.: *E*, *F*, 5.5 mm; *G*, *H*, 6 mm. NYZS. (P. 316.)

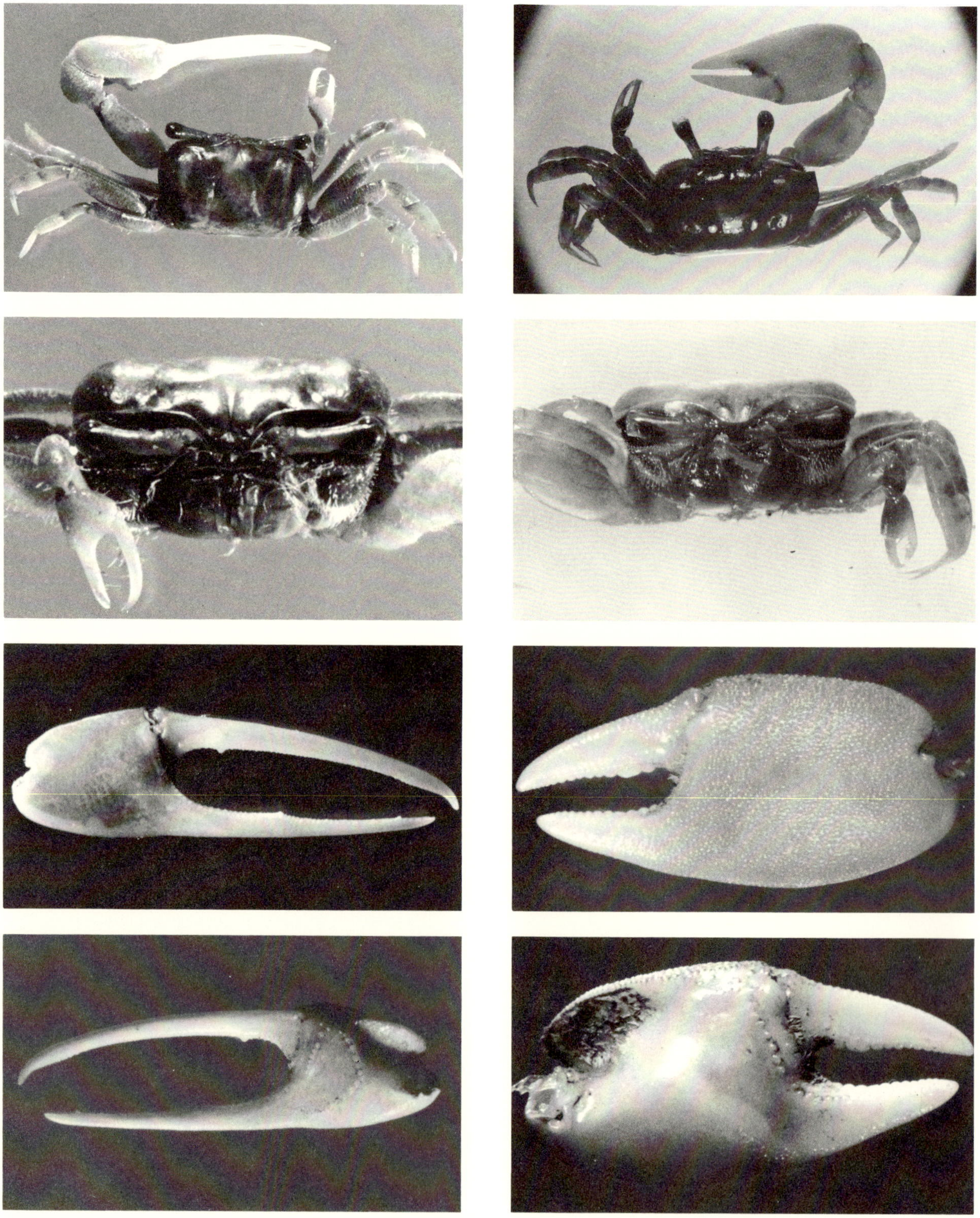

580

PL. 43. *A-D. Uca (Celuca) musica musica.* Mexico: Gulf of California, east coast (San Blas). Carapace lgths.: *A, B,* 6 mm; *C, D,* 5.8 mm. USNM 99755. (P. 317.)

E-H. Uca (Celuca) latimanus. Eastern Pacific: Panama Canal Zone (Balboa). Carapace lgth. 8 mm. NYZS. (P. 319.)

A

B

PL. 44. *A*. One excellent habitat for fiddler crabs: a
sheltered bay in the tropics, with a wide
tidal range, a beach of sandy mud, and
mangroves standing close to the mouth
of a stream. Locality: the Pacific coast
of Costa Rica, near Port Parker.

B. A living male fiddler crab, *Uca* (*Uca*)
maracoani maracoani, in Trinidad. (P.
147.)

A

C

B

D

E

F

PL. 45. Combat in *Uca*. (P. 485.) All photographs made from 16 mm color motion picture frames.

A. Low intensity: the manus-push in (*Celuca*) *lactea annulipes* in a combat between two burrow-holders. Singapore.

B. An irregular, forceful fight between a burrow-holder and an aggressive wanderer in (*Uca*) *maracoani maracoani*. Trinidad.

C. High-intensity ritualized combat in (*Minuca*) *rapax rapax*, in Trinidad. Tap following a heel-and-ridge in homoclawed combat. The actor is the crab on the right. His dactyl is striking the heel of his opponent's manus while his pollex is free.

D. Same combat as in *C*. Alternate stroke showing the actor's pollex against his opponent's invisible oblique ridge, on inner side of manus.

E. A heteroclawed combat occurring in the same population, showing the interlace component. The actor is on the right. The teeth near his dactyl's base are starting to rub downward against the ridges of his opponent's inner manus, which parallel the dactyl's base.

F. Same combat as in *E*, near end of the downward stroke.

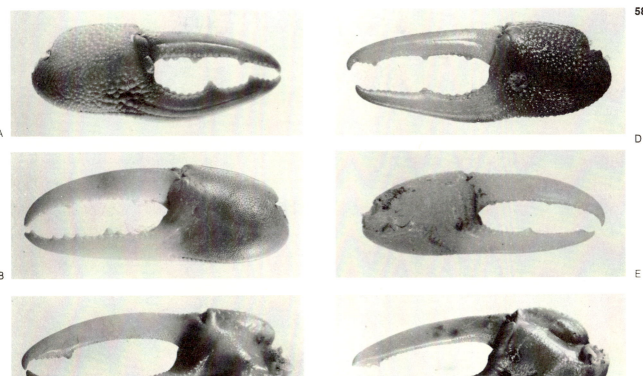

PL. 46. Major claws in *Uca*. *A-C*, depressed areas at base of pollex that apparently facilitate seizure by an opponent during forceful combat. *D-F*, healed puncture wounds, apparently received in combat. (P. 515.)

A. (*Deltuca*) *forcipata*; outer claw. Data as in Pl. 6 *A-D*.

B. (*Amphiuca*) *chlorophthalmus crassipes*; outer claw. Data as in Pl. 15 *A-D*.

C. (*Deltuca*) *coarctata coarctata*; inner claw. Data as in Pl. 6 *E-H*.

D. (*Deltuca*) *coarctata coarctata*; outer claw. Data as in Pl. 7 *E-H*.

E. (*Australuca*) *seismella*; inner claw. Data as in Pl. 12 *A-D*.

F. (*Deltuca*) *demani typhoni*; data same as in Pl. 4 *E-H*.

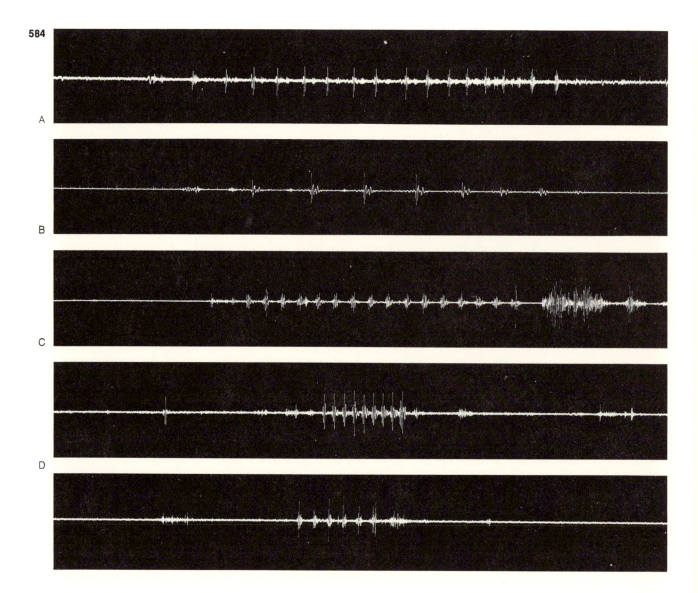

PL. 47.   Examples of sound production in *Uca*.
            All the sounds were made by burrow-
            holding males following the withdrawal
            of an intruding conspecific male. (P.
            484.)
       A. (*Thalassuca*) *vocans pacificensis*. Fiji Is.
            Time scale: 1 sec. = 12.5 cm.
       B. (*Celuca*) *cumulanta*. Trinidad. Time
            scale: 1 sec. = 12.5 cm.
       C. (*Celuca*) *lactea annulipes*. Red Sea.
            Time scale: 1 sec. = 12.5 cm.
       D. (*Celuca*) *lactea perplexa*. New Cale-
            donia. Time scale: 1 sec. = 12.5 cm. The
            lower strip is a continuation of the upper.

PL. 48.   Activities by a single male, performed in   *A.* Feeding.
sequence around his burrow. (*Celuca*)   *B.* Building hood.
latimanus. Panama Canal Zone: Balboa.   *C.* Waving display. The crab has now as-
(P. 319.)   sumed display white.

PL. 49.   An example of hood construction at the peak of its development, both in relative size and in symmetry. Its builder is (*Celuca*) *musica terpsichores*; carapace lgth. ca. 7 mm. Panama Canal Zone: Balboa. (P. 500.)

A

B

PL. 50. Courtship in *Uca*. (P. 500.)

  A. A group of males, chelipeds in rest position, face a wandering female as she crouches momentarily motionless. As soon as she started to walk again, the males resumed waving, turning as necessary to keep their anterior aspect directed toward her. *(Celuca) lactea annulipes*. Singapore.

  B. Two males displaying toward a conspecific female, *(Celuca) stenodactylus*. The third male, facing to the left and not waving, is a member of a closely related species, *(Celuca) beebei*. Panama Canal Zone: Balboa.

Appendixes

Appendix A. Material Examined

It would be misleading, it seems, to list the total number of specimens of each species examined in the course of this study: the significance of the word *examined* changes from one species to the next. Sometimes, a great deal of time and effort was spent on a few specimens; in other cases, large numbers of individuals were examined, as in *lactea*, but the detailed work on most of these extended only to the gonopods and general claw shape, in order to establish subspecies and determine their ranges; only in series from selected localities did detailed study and comparison extend to many characters.

Comparison of the space consumed in the listing gives an idea of the totals, and in any individual case the numbers can be added up; general remarks on abundance or scarcity of a species in nature are usually included, when either attribute is noteworthy, in the introduction to each species.

Again, in most species the numbers of individuals observed in life far exceeded those I collected for later study or examined in other institutions. Even more than with the preserved material, of course, the detail with which living specimens were observed varied from species to species. An indication of my familiarity with each is given in the species treatments under the heading Field Material.

Probably the most useful function of this appendix is to show the present location of the material. The full names and addresses of the institutions abbreviated (USNM, BM, etc.) in the lists below appear on p. 678.

All of the specimens collected in the course of this study and formerly in the collections of the New York Zoological Society (NYZS) are now in the Smithsonian Institution, National Museum of Natural History, Washington (USNM). Type specimens included in that collection have already been assigned numbers by the museum and these are included here. Numbers for the rest of the specimens, indicated below simply by "(NYZS)," can be obtained directly from the museum: enough data are given in the listing to facilitate their selection there for future examinations; these numbers range approximately from 138043 to 138877. The great majority of the total number of preserved *Uca* in the world are deposited in the Smithsonian; even before the addition of the New York Zoological Society material the collection was by far the largest in existence; the total collection probably now numbers about 16,000 specimens, of which more than 14,000 have resulted from the present study.

Collections were also examined or material received on loan through the generous cooperation of other appropriate institutions in New York, Philadelphia, New Haven, Cambridge (Massachusetts), Los Angeles, London, Paris, Leiden, Amsterdam, Copenhagen, Göttingen, Frankfurt, Torino, Singapore, Port Moresby, Sydney, Yokohama, Fukuoka, and Honolulu. The Australian Museum in Sydney has a fine collection of *Uca* from Australia and neighboring shores. Unfortunately, when I stopped there briefly in 1956 I had not done sufficient taxonomic work on the group to profit by the material; it will reward future attention.

The subgenera, species, and subspecies are listed in the order given in the table of contents and the systematic section.

The authorities and references for all synonymized names included in these lists will be found in the systematic section, under the headings of Type Material and Nomenclature and of References and Synonymy. They also appear in the Index to Scientific Names (p. 719).

The order in which geographical names are listed is not always similar from one species to the next. In these exceptions the arrangements were selected because they seemed to be more helpful than strict consistency in making certain distributional comparisons among, for example, allopatric or sympatric forms. In no case was any attempt made to suggest palaeogeographic dispersal.

Geographical names are spelled in accordance with the form used on recent maps in English. In cases where another name or spelling is much more familiar than the currently accepted form, the older name is placed in brackets.

The abbreviations *G.* and *U.*, preceding species names in the listing of type-specimens examined, refer respectively to the generic names *Gelasimus* and *Uca.*

SUBGENUS *DELTUCA*

Uca (*Deltuca*) *acuta* (Stimpson, 1858). (P. 25.)

U. (*D.*) *acuta rhizophorae* Tweedie, 1950

MALAYSIA. *Sarawak*: Santobong (near mouth of river, below Kuching): HOLOTYPE of *U. rhizophorae*, m. (BM); 12 m., 2 f. (NYZS).

SINGAPORE. Geylang R. (near mouth): 4 m., 2 f. (NYZS). Jurong R. (near mouth): 1 m. (NYZS).

U. (*D.*) *acuta acuta* (Stimpson, 1858)

HONG KONG. Kowloon (Castle Peak area): NEOTYPE of *U. acuta*, m. (NYZS = USNM 137665); 37 m., 6 f. (NYZS).

CHINA. *Fukien*: Amoy: 4 m., 2 yg. f. (USNM 57033); 1 m. (USNM 125702); 2 m. (USNM 64953); 3 m. (USNM 64954); 1 yg. m., 1 yg. f. (USNM 74906); 1 m. (BM 2463 part). Tsimei: 2 m. (USNM 57812); 1 m. (USNM 57826). Chin B.: 1 m. (USNM 57055). Foochow: 2 m. (USNM 59096); 1 m. (USNM 59097). Guantoo (near Foochow): 1 m. (USNM 61845). Santu: 1 m. (USNM 61844); Kiaohsien [Kiaochow]: (BM; at least 2 m. = *a. acuta*). Liuwutien: 3 yg. m., 3 yg. f., ? 1 very yg. m., ? 1 very yg. f. (USNM 61843).

Uca (*Deltuca*) *rosea* (Tweedie, 1937). (P. 29.)

INDIA. Port Canning: 2 m., 1 f. (USNM 63673); 1 m., 1 f., + 7 yg. (USNM 63674). *Nicobar Is.*: Xancovri: 2 m., 1 f. (USNM 19711).

BURMA. Rangoon: 1 m. (MCZ 5787). *Mergui Archipelago*: 2 m. (Leiden reg. no. 4); 4 yg. (BM 86.52); 3 m. + yg. (Amsterdam: de Man collection).

MALAYSIA. *Malaya*: *Penang*: Penang: 95 m., 18 f. (NYZS). *Selangor*: Port Swettenham: HOLOTYPE of *G. roseus*, m. (BM); 2 m., 1 f. (NYZS; gift of M.W.F. Tweedie). *Negri Sembilan* (near Sungei Dua ferry): 29 m., 8 f., + 20 yg. (NYZS). *Malacca*: Malacca: 1 m. (USNM 19552).

INDONESIA. *Kalimantan* [*Borneo*]. Pontianak: 3 m., 1 f. (USNM 39162); 2 m. (Leiden 1535). *Sumatera* [*Sumatra*]. Atjeh: 1 yg. m., 1 yg. f. (USNM 39161).

Uca (*Deltuca*) *dussumieri* (Milne-Edwards, 1852). (P. 32.)

U. (*D.*) *dussumieri capricornis* subsp. nov.

AUSTRALIA. *Western Australia*: Broome: HOLOTYPE of *U. d. capricornis*, m. (NYZS = USNM 137675); 43 m., 30 f. (NYZS); Monte Bello Is.: 1 yg. m. (BM 1938.6. 13-1-2, part: lgth. 12 mm).

U. (*D.*) *dussumieri spinata* subsp. nov.

INDIA. Mouth of Hooghly R. (near Calcutta): 2 m. (BM 60.15).

BURMA. *Mergui Archipelago*: 1 m., 1 f. (Leiden 5); 1 yg. m. (Amsterdam); 1 m. (BM).

MALAYSIA. *Malaya*: *Penang*: 1 m. (Amsterdam). *Negri Sembilan*: near Sungei Dua: 1 m., 5 f. (NYZS). *Malacca*: Malacca: 1 m. (USNM 39174); 2 m. (Amsterdam). "Borneo." 1 m. (BM: Genard). *Sarawak*: Santobong (near mouth of river below Kuching): 4 m., 2 f. (NYZS; ex-Raffles, gift of M.W.F. Tweedie). *Labuan*: 2 m., 1 f. (Raffles). *Sabah* [N. Borneo]: Sebatic I.: 35 m., 35 f. (USNM 43428).

SINGAPORE. HOLOTYPE of *U. d. spinata*, m. (NYZS = USNM 137677); 24 m., 12 f. (NYZS); 3 m., 2 f. (Raffles).

INDONESIA. *Sumatera* [*Sumatra*]. Atjeh: 1 m. (USNM 39173). *Djawa* [*Java*]. 1 m. checked among 5 (Leiden 1242). Pasuruan (near Surabaja): 14 m., 1 f. (NYZS). Madoera: including 4 m. (Amsterdam); 2 m. (Leiden 2015). Zuidk (near Madoera): 1 m. (Leiden 2013). *Aru* [*Aroe, Arrou*] *Is.*: S. Manumbai: 4 m. (BM).

PHILIPPINE IS. *Palawan*: Nokoka Bay: 5 m. (USNM 43429).

THAILAND. Bangpoo (near Bangkok): 1 m. (USNM 94418); 1 f. (USNM 125708). Ajuthia: 2 m. (Paris). Lem Ngob: 1 f. (USNM 43296). Paknam: 2 m., 1 f. (USNM 63665); 2 m., 1 f. (USNM 63667.

CHINA. *Fukien*: Amoy: 1 m. (USNM 64959); 3 m. (Leiden 1244). Foochow: (Minhow): 2 m., 2 f. (USNM 61840); 1 m., 1 f. (USNM 58740); 2 m. (USNM 59091); 1 m. (USNM 59092). Liuwutien: 2 yg. m. (USNM 61841).

Locality Record Questionable. MADAGASCAR. Nossy Faly (on northwest coast): 1 m. (Leiden 1243 part). This is the specimen mentioned by de Man (1891: 26) as one of those figured by Hoffmann (1874: Pl. 3, Fig. 22); it is the only individual reported from the island according to Crosnier (1965). See page 35.

U. (D.) dussumieri dussumieri (Milne-Edwards, 1852)

INDONESIA. *Sumatera [Sumatra]*: Simeulue [Simaloer]: Sinabang [Sinaboung]: 10 m. checked in series (Leiden 2146). Nias: 3 m. (Amsterdam). Sibigo [Siboga]: 1 m. of series checked (Torino—see Nobili 1899). *Djawa [Java]*: 2 m. (BM 74.57). Samarang: LECTOTYPE of *G. dussumieri*, m. (Paris). Djakarta [Batavia]: 1 m. (Paris: Blecker); 1 m. (Paris: Blecker); 2 m. (Paris: Blecker). *Sulawesi [Celebes]*: Parepare: 2 m. (Amsterdam). *Adonara*: Sagoe: 2 yg. m. (Amsterdam). *Molucca Is.*: Batjan: 1 m. (BM). Ambon [Amboina]: 2 m. (Leiden 1245). *Aru [Aroe, Arrou] Is.*: 3 m. (Paris).

NEW GUINEA. *West Irian*: Japen: Seroei: Laut: 1 m. (Leiden). Mimika R.: 1 m. (BM). Hollandia Haven: 3 m. (Leiden). "Om geving Base," C. Hollandia: 2 m., 1 f. (Leiden: Holthuis collection 143). Territory of Papua and New Guinea: near Madang: 5 m., 4 f. (NYZS).

SOLOMON IS. 1 m. (Paris 1386).

AUSTRALIA. *Queensland*: Cooktown: 1 m. (Paris 4234); 1 m. (Paris 4239).

NEW CALEDONIA. 2 m. (Paris); 1 m. (Paris: A. le Comte). Near Nouméa: 5 m., 4 f. (NYZS).

CAROLINE IS. *Ponape*: 2 m. (Leiden 265). *Palau Islands*: 2 m., 2 f. (NYZS; collected by Takahasi, 1935; gift from T. Sakai, U. of Yokohama).

PHILIPPINE IS. *Sulu*: Tawi Tawi: 3 m., 1 f. (NYZS). Joló: 13 m., 18 f. (NYZS). *Mindanao*: Zamboanga: 3 m., 1 f. (USNM 43430); 113 m., 29 f. (NYZS). Gulf of Davao: Padada Beach: 1 f. (AMNH 8318). Davao (Agnow Swamp): 1 yg. f. (AMNH 8555 part). In and near mouth Padada R.: 1 yg. m., 2 yg. f. (AMNH 8393 part). Sasa: 1 f. (NYZS). Iling R., N. of Sasa: 10 m. (NYZS). Madaum: 1 yg. m., 3 yg. f., 1 mating pair (NYZS). *Negros Occ.*: 1 f. (USNM 25225). Magnanod R., Victorias, Bangi-Bangi: 1 m. (USNM 64981); 1 m. (USNM 64982); 2 f. (USNM 73273). *Lubang*: Looc: 2 m., 1 f. (USNM 43431). *Cebu*: Maston I.: 1 m. (USNM 43276). *Panay*: Jaro R.: 2 m., 2 f. (USNM 73184). Iloilo: 4 m., 1 f. (USNM 73201 part); 1 m. (USNM

125710); 1 f. (USNM 73221); 1 m. (USNM 73224); 1 f. (USNM 125703). "Shore around Iloilo R.": 1 m. (USNM 43433). *Guimares*: Jordan R.: 6 m., 2 f. (USNM 43432). *Samar*: Catbalogan: 1 m. (NYZS). *Palawan*: Puerto Princesa: 7 m., 7 f. (NYZS). Northwest coast, near mouth of Baheli R.: 6 m., 2 f. (NYZS). *Luzon*: Lamao: 1 m. (NYZS). Ragay Gulf: 7 m., 7 f. (USNM 43434).

NANSEI [RYUKYU or LOO CHOO] IS. *Ishiyaki I.*, mouth of Miyara R.: 1 m. (NYZS; collected by Ohishima, gift from T. Sakai, U. of Yokohama).

Additional Material with Incomplete or Questionable Locality Records.

SOUTHERN PACIFIC. (USNM 17764.)

TUAMOTU ARCH.: Toan Atoll: 1 m. (USNM 93283 part). *Locality doubtless in error; see p. 41.*

Uca (Deltuca) demani Ortmann, 1897. (P. 39.)

U. (D.) demani australiae subsp. nov.

AUSTRALIA. *Western Australia.* Broome: HOLOTYPE of *U. d. australiae*, m. (USNM 64250).

U. (D.) demani typhoni subsp. nov.

PHILIPPINE IS. *Negros Occ.*: 1 m. (USNM 25223); 1 m. (USNM 65012). *Panay*: Iloilo 1 m., 1 f. (USNM 73163 part); 16 m., 6 f. (USNM 73201 part); 3 yg. m., 2 yg. f. (USNM 73211 part); 4 m., 1 yg. m. identity?, 2 yg. f. (USNM 73221 part). Jaro R.: 1 m. (USNM 73222 part). *Lubang*: Looc: 1 m., 2 yg. f. (USNM 43437 part). Basus R.: 1 yg. f. (USNM 43435). *Samar*: Catbalogan: 1 m., 1 f. (NYZS). *Luzon*: Manila: HOLOTYPE of *U. d. typhoni*, m. (USNM 43041).

TUAMOTU ARCH.: Toan Atoll: 2 m., 1 f. (USNM 93283 part). *Locality doubtless in error; see p. 41.*

U. (D.) demani demani Ortmann, 1897

INDONESIA. *Djawa [Java].* Pasuruan (near Surabaja): 1 m. (Amsterdam, one examined of several labeled *demani*). *Sumbawa*: Baii van Bima: 1 m. (Amsterdam; de Man's 2nd specimen, referred to *forcipata*, 1892.1, included in Ortmann's proposal of the new name, 1897). *Sulawesi [Celebes]*: LECTOTYPE of *G. demani*, m. (Leiden 1257; de Man's first specimen, 1891.3:32).

PHILIPPINE IS. *Mindanao*: Zamboanga: 6 m. (USNM 43307: Type series, including type, of *Uca zambo-*

angana); 53 m., 6 f. (NYZS). Gulf of Davao: Padada Beach: 16 m., 14 f., most of them yg. (AMNH 8304, 8306 part, 8307, 8308 part, 8310, 8313, 8330, 8331, 8332, 8380 part). In and near mouth Padada R.: 8 m., 15 f., most of them yg. (AMNH 8342, 8343, 8386, 8387, 8388, 8390, 8391 part, 8392, 8393 part). Malalag: 3 m. (NYZS). Davao Beach: 3 m., 2 f. (AMNH 8421, 8422, 8554). Sasa: 1 m. (NYZS). Iling R., N. of Sasa: 2 m. (NYZS).

Uca (Deltuca) arcuata
(de Haan, 1835). (P. 44.)

NORTH VIETNAM. *Tonkin*: Pointe du Scorpion ("brought by a fisherman"). 1 m. (Paris 1028).

HONG KONG. 1 m., 2 f. (BM 1935.3.19); Kowloon (Castle Peak area): 5 m. (NYZS).

CHINA. "Chine": Type material of *G. brevipes* (Paris). *Fukien*: Amoy: 1 m. (BM 2463); 2 m., 1 f. (USNM 57813). Foochow: 1 f. (USNM 57613); 1 f. (USNM 59093); 1 m. (USNM 59094); 1 m. (USNM 59095); 1 m. (USNM 61847); 1 m., 1 f. (USNM 61849); 3 m., 1 f. (USNM 63709); 1 m., 1 f. (USNM 75441). Guantoo (near Foochow): 1 m. (USNM 61848). Muiwha (near Foochow): 1 m., 3 f. (USNM 61842). Santuao 2 m., 2 f. (USNM 61850). *Chekiang*: Wenchow: 2 m., 4 f. (USNM 74604); Liuwutien: 1 yg. f. (USNM 61846).

TAIWAN [FORMOSA]. Takao: Shenshoshi: 3 m. (USNM 55424). Ampin: 1 m. (USNM 57492). Taichu: Soroku: 1 f. (USNM 55398). Shinchika: 1 f. (USNM 55400). Toutti: 2 m., 1 f. (BM 84.10). Tamsui: 158 m., 4 yg. m., 1 yg. f. (NYZS). Taihoku: 1 m. (USNM 55399).

KOREA [CHOSEN]. 2 m., 1 f. (USNM 19708); yg. (BM 92-12-15-1-7).

JAPAN. LECTOTYPE of *G. arcuatus*, m., + 3 m., 1 f., labeled "*Gelasimus arcuatus* de Hn. Types" (Leiden 243) + series (Leiden 244, 245); 1 m. (USNM 17762). *Kyushu*: Ariake Bay: 1 f. (NYZS).

Uca (Deltuca) forcipata
(Adams & White, 1848). (P. 48.)

MALAYSIA. *Malaya. Penang*: 10 m., 1 f. (NYZS); 1 f. (USNM 39163); 1 yg. m., 1 yg. f. (USNM 43306); 1 m. (Paris). *Negri Sembilan*: Sungei Dua: 11 m., 13 f. (NYZS). "Borneo." HOLOTYPE of *G. forcipatus*, m. (BM "Borneo 44.106; 4476"). *Sarawak*: Kuching: 1 m. (Torino). *Sabah* [N. Borneo]:

Sebatic I.: 1 m. (USNM 125705); 1 m., 3 f. (USNM 43438; note inside vial says "m. compared by Calman with type of *forcipata*").

SINGAPORE. 4 m. (NYZS): 3 m. (NYZS; ex-Raffles, gift of M.W.F. Tweedie).

INDONESIA. *Kalimantan* [*Borneo*]. Samarinda: 1 m. (Torino). *Sumatera* [*Sumatra*]. Buttikofer: 2 m. (Leiden 2033). *Djawa* [*Java*]: near Surabaja: swamp at Trengglunga: 79 m., 2 f. (NYZS). *Near Surabaja*: swamp at Paseruan: 12 m. (NYZS). *Sulawesi* [*Celebes*]. Makassar: 2 m., 1 f. (MCZ 7255).

PHILIPPINE IS. *Mindanao*: Zamboanga: 1 m. (BM 84.31. HMS *Challenger*. Locality record questionable, since designated depth ["10 faths."] impossible. Name on label: *rubripes*. *Samar*: Silaga R.: 2 m., 2 f. (USNM 43436). *Negros Occ.*: Magnanod R., Victorias: 1 m. (USNM 73269). Bangi-Bangi: 2 m. (USNM 64991); 1 m. (USNM 73271); 3 yg. m. (USNM 73221 part). West coast of *Palawan*: Baheli R. 9 m., 4 f. (NYZS). *Palawan*: Puerto Princesa: 4 yg. m., 1 yg. f. (NYZS).

THAILAND. Bangpoo (near Bangkok): 3 m., 1 f. (USNM 94419). Paknam: 2 yg. m., 3 yg. f. (USNM 125706); 1 m. (USNM 63666). Lem Ngob: 6 m., 2 f. from type material of *U. manii* Rathbun (Copenhagen); 4 m. from same series, including one compared by Calman with type of *G. forcipatus* (USNM 39714).

TUAMOTU ARCH. Toan Atoll: 1 m., 1 f. (USNM 93283 part). *Locality doubtless in error*; see p. 41.

Uca (Deltuca) coarctata
(Milne-Edwards, 1852). (P. 52.)

U. (D.) coarctata coarctata
(Milne-Edwards, 1852)

[RUSSIA.] "Odessa": LECTOTYPE of *G. coarctatus*, m. (Paris). *Locality name unquestionably erroneous*; ? 1 f. same locality, separate box, poor condition.

INDONESIA. *Sumatera* [*Sumatra*]: Simeulue [Simaloe] I.: 3 m., 3 f. (NYZS; ex-Raffles, gift of M.W.F. Tweedie). Nias: series (Amsterdam). Benkulen: 6 m., 2 f. (USNM 75865). *Djawa* [*Java*]: 1 m. (BM 74.57). Djakarta: (Paris). *Sulawesi* [*Celebes*]: Togian I.: 1 m. (USNM 19556). *Seram* [*W. Ceram*]: Piroe: yg. (Amsterdam). *Aru* [*Aroe, Arrou*] *Is.*: Manumbai: 3 m. (BM 1935.1.28.1-3).

NEW GUINEA. *West Irian*: Noord [Lorentz] River: 2 yg. m. (Amsterdam). Mimika R. near Kokonau: 1 m. (BM 1911.8.1.25). Near Hollandia: 1 m. (Leiden). Territory of Papua and New Guinea: 3 m., 4 f. (NYZS).

AUSTRALIA. *Queensland*: Cooktown 1 m. (Paris 2089). Cairns: 12 m., 3 f. (MCZ 5786); 1 m. (USNM 22198). Gladstone: 5 m. (NYZS); 1 m. (Paris 1981).

NEW CALEDONIA. 2 m. (Paris, M. Balansa); series, poor condition (A. Milne-Edwards). Near Nouméa: 9 m., 3 f. (NYZS).

FIJI IS. *Viti Levu*: Suva: paratype of *U. ischnodactylus*, 1 m. (USNM 99261). 5 m., 4 f. (NYZS). Near mouth of Wainibokasi R., near Rewa: 5 m. (NYZS). Raki-Raki: 9 m., 3 f. (NYZS). Tavua: 53 m., 13 f. (NYZS). Lautoka: 5 m., 3 f. (NYZS).

? TAHITI. 1 m. (USNM 19662). *Locality error probable.*

PHILIPPINE IS. *Mindanao*: Zamboanga: 24 m., 3 f. (NYZS). Gulf of Davao: Padada Beach: 1 yg. f. (AMNH 8306 part); 1 f. (AMNH 8338); 1 m. (AMNH 8339); 3 f. (AMNH 8377); 1 yg. m., 1 yg. f. (AMNH 8378); 2 yg. m. (AMNH 8379); 7 m., 3 f. (AMNH 8380 part). In and near mouth of Padada River: 1 yg. f. (AMNH 8344); 1 m. (AMNH 8394). Davao Beach: 2 yg. m. (AMNH 8308 part). Davao: Agdow Swamp: 3 yg. m. (AMNH 8555 part); ? 1 f. (USNM 43383; type of *U. mearnsi*). Madaum: 1 m., 1 yg. f. (NYZS). Iling R. (N. of Sasa): 2 m. (NYZS). *Panay*: Jaro R.: 1 m. (USNM 125707); 1 m. (USNM 73225 part). Iloilo: 1 m. (USNM 125709). *Samar*: Catbalogan: 1 m. (NYZS). *Luzon*: Bataan (Samal): 1 m. + drawn gonopod (body missing) (NYZS). Manila: 2 m. (BM 72.7). Dagadagatan R.: Navotas (near Manila): 1 m. (NYZS). Manila: 2 m. (USNM 43040; labeled *U. rathbunae* Pearse; collected and identified by him; see p. 55).

U. (D.) coarctata flammula subsp. nov.

NEW GUINEA. *West Irian*: Boshek near Biak: series (Leiden). Japen I.: Seroei: 1 yg. (Leiden). Merauke: series (Amsterdam). Merauke (near Kampong Sawa): 1 yg. m. (Leiden).

AUSTRALIA. *Northern Territory*: Darwin: HOLOTYPE of *U. c. flammula*, m. (NYZS = USNM 137676); + 20 m., 4 f. (NYZS). *Western Australia*: Broome: 2 m., 2 f. (USNM 56419); 2 m., 1 f. (USNM 56420); 25 m., 17 f. (NYZS). Port Walcott (near Dampier Archipelago): 1 m. (BM 89.5.12.2). Nicol B.: 2 m. (BM 69.38). Monte Bello Is.: 1 m. (USNM 46348); 1 m. (BM 1938.6.13.1.2). Mermaid Strait: 1 m. (USNM 19550).

Uca (Deltuca) urvillei (Milne-Edwards, 1852). (P. 58.)

[NEW HEBRIDES.] "*Vanikoro*": LECTOTYPE of *G. urvillei*, m. (Paris). *Locality name unquestionably erroneous.*

INDIA. *Malabar*: 1 m. (Paris). This specimen is one of 2 m. labeled "Type" of *G. dussumieri* Milne-Edwards.

PAKISTAN. Karachi: 1 m. (BM 97.9.12.1); 1 m. (BM 82.3).

EAST AFRICA. *Tanzania. Pemba Island*: Kiburunzi (near Chake Chake): 36 m., 19 f. (NYZS). *Zanzibar*: 3 lots of dried specimens (Paris). *Zanzibar* (near Muongoni): 3 m., 1 f. (NYZS). *Tanga*: Dar es Salaam (near Mtoni): 2 m. (NYZS).

MADAGASCAR. 2 m. (Leiden 1243); (Paris, several series, including 266, 339, and 130.96).

SOUTH AFRICA. Durban: 1 m. (Amsterdam 1894).

SUBGENUS *AUSTRALUCA*

Uca (Australuca) bellator (Adams & White, 1848). (P. 64.)

Uca (A.) bellator bellator (Adams & White, 1848)

INDIA. *Nicobar Is.*: Nankauri: 1 m. (USNM 22286).

MALAYSIA. *Labuan*: 2 m. (NYZS; gift of Raffles); 6 m., 4 (1 ovig.) f. (NYZS; gift of Raffles).

INDONESIA. *Djawa* [*Java*]: "Java": small series (BM 74.51). Djakarta [Batavia]: small series (Leiden 1255); 1 m., which is apparently part of type material of *G. signatus* var. *angustifrons* (Leiden 1536); series (Leiden 2016). Near Tilliwong (northwest coast): 1 m. (Leiden 2017). Bahduheira: small series (Leiden 2662). Semarang: 83 m. with claws, 102 m. bodies, many detached claws, 72 (25 ovig.) f. (NYZS). Madoera: 1 m. (Leiden 2037). Near Surabaja (at Trengglunga): 1 m. (NYZS).

NEW GUINEA. *West Irian*: Merauke: 1 m. (Torino); near Merauke: 3 m. (Leiden).

PHILIPPINE IS. "*Philippines*": HOLOTYPE of *G. bellator*, m. (BM). *Mindanao*: Zamboanga: 1 m. (NYZS). Gulf of Davao: Iling R. (north of Sasa): 1 m. (NYZS); Madaum: 4 m. (NYZS). *Negros Occ.*: Magnanod R. (near Victorias at Bangi-Bangi): 1 f. (USNM 65009). *Panay*: Jaro R.: 1 m. (ex-USNM 73225). *Palawan*: Puerto Princesa: 1 f. (NYZS). *Luzon*: Dagadagatan R. (near Manila, above Navotas): 11 m., 4 f. (NYZS).

U. (A.) bellator signata (Hess, 1865)

AUSTRALIA. "*Australia*": TYPE-SPECIMEN of *G. signatus*, m. (Göttingen 3665, in 53b). "*East Coast Australia*": 2 m. (Leiden 1261). *Queensland*: Port Curtis: 14 m. & f. (BM 8131). Gladstone: 6 m., 4 f. (NYZS). Boyne R. (above Gladstone): 4 m., 2 f. (NYZS).

U. (A.) bellator longidigita (Kingsley, 1880)

AUSTRALIA. *Queensland*: seaward from Brisbane: S. bank of Brisbane R., near mouth, Sandgate, and Shorncliffe: 22 m., 6 (5 ovig.) f. (NYZS).

U. (A.) bellator minima subsp. nov.

AUSTRALIA. *Northern Territory*: near Darwin: Ludmilla Creek, near junction with Golf Club R.: HOLOTYPE m. (NYZS = USNM 137668) + 5 m., 7 f. (NYZS).

Uca (Australuca) seismella sp. nov. (P. 70.)

AUSTRALIA. *Western Australia*: Broome: 2 yg. m. (NYZS). *Northern Territory*: near Darwin: Golf Club R.: HOLOTYPE m. (NYZS = USNM 137666) + 29 m., 2 ovig. f. (NYZS). *Queensland*: Gladstone: 7 m., 4 (1 ovig.) f. (NYZS).

Uca (Australuca) polita sp. nov. (P. 72.)

AUSTRALIA. *Western Australia*: Nicol B.: 4 specimens (BM 69-38). Broome: 13 m., 11 f. (NYZS). *Queensland*: Gladstone: HOLOTYPE m. (NYZS = USNM 137667) + 7 m., 4 (ovig.) f. (NYZS). *Thursday I.*: 2 m. (NYZS: gift; donor unknown).

SUBGENUS *THALASSUCA*

Uca (Thalassuca) tetragonon (Herbst, 1790). (P. 77.)

EGYPT. "*EGYPTE*": NEOTYPE of *U. tetragonon*, m. (Paris) + 1 f. (Paris: both marked "Type" by A. Milne-Edwards).

ETHIOPIA. *Eritrea*: 1 m. (Torino 1763). Near Bosso: 1 m. (Torino 2276). Isole Key: 1 m. (Torino 2072). Massawa: 2 m., 1 f. (AMNH 4323 part); 23 m., 3 f. (NYZS); 2 m. (Torino 665 & 2098). Near Massawa: Green I.: 39 m., 24 f. (NYZS).

ADEN. 1 m. (USNM 19324).

EAST AFRICA. *Tanzania*: *Zanzibar*: between Chakwani and Mbeni, Kombeni B. and Chwaka: 52 m., 10 f. (NYZS).

INDIAN OCEAN. *Aldabra I.*: 1 m. (USNM 17749). *Mauritius*: 2 m., 1 f. (USNM 19654); 1 m., 3 f. (Philadelphia 2932: labeled *G. desjardinii* Guérin). *Chagos Is.*: Diego Garcia: 1 m. (BM 1910.4.30.18).

BURMA. Rangoon: 1 m. (MCZ 5779 part).

PHILIPPINE IS. *Sulu*: Tawi Tawi: 2 m., 2 f. (NYZS).

Mindanao: Gulf of Davao: Padada Beach: 1 yg. m. (AMNH 8367).

NEW GUINEA. 1 m. (Torino). *West Irian*: Japen I., near Seroei: small series (Leiden: Holthuis collection, 25 Feb. 1955). Noemfoor: 2 m. (Leiden: Leg. Hendriksen, 1953). New Hollandia: 1 m. (Leiden: Leg. Van Hout, 1955). Reef at Marine Kazerne: 2 + ? 1 yg. m. (Leiden 683).

AUSTRALIA. Sydney: Type material of *G. variatus*, 2 m. (Göttingen 52a; *exact locality questionable*; see p. 80). "*Australien*": Type material of *G. variatus*, 1 m. (Göttingen 52b).

NEW CALEDONIA. 1 m., 1 f. (Paris: both marked "Type" of *G. tetragonon* by A. Milne-Edwards).

PACIFIC OCEAN. *Caroline Is.*: small series (BM 98.11.1.1-5 part). *Wake I.*: Series (Bishop). *Marshall Is.*: Jaluit, Marchalsoerne: 1 m., 1 f. (USNM 19709). *Gilbert Is.*: Apiang: 2 m., 2 (1 ovig.) f. (USNM 93282). *Ki Is.*: 1 m. (BM 1910.3.29.18). *Samoa*: Savaii I.: 1 m. (BM 74.54). Apia: 9 m., 10 f. (NYZS). *Friendly Is.*: Tonga: Tongatabu: 5 m. (USNM 33164). *Line Is.*: Fanning I.: Inner lagoon:

1 f. (USNM 46682). Penrhyn I.: 8 m. (AMNH 7454); 1 f. (AMNH 7476); 1 m. (AMNH 7477). *Society Is.*: Bora-Bora: 21 m., 18 f. (NYZS). Raiatea: 2 m., 1 f. (NYZS). Tahiti: 1 f. (MCZ 5806); (at Faaa): 16 m., 14 f. (NYZS). *Tuamotu Is.*: Raroia Atoll (S. end of Oneroa I.): 31 m. and f. (USNM 94558).

? *Hawaii.* "Isles Sandwich": 1 m., 1 f. (Paris: both marked "Type" of *G. tetragonon* by A. Milne-Edwards). *Locality questionable*; "Iles Sandwich" perhaps an error for Bora Bora; see pp. 80 and 324.

Not Examined. About 50 additional specimens at USNM from localities from which other representatives are included in the material listed above.

Uca (Thalassuca) formosensis Rathbun, 1921. (P. 83.)

TAIWAN [FORMOSA]. *Taichu*: Rokko: TYPE SERIES of *U. formosensis*, 2 m. (USNM 54472); 1 f. (USNM 55385). Tamsui: 2 m., 1 f. (NYZS; gift of T. Sakai, U. of Yokohama). Giran (northeast coast): 1 m. (USNM 55386).

Uca (Thalassuca) vocans (Linnaeus, 1758). (P. 85.)

For additional specimens examined, but not referred to subspecies, see note at end of listing for the species, below (p. 598).

U. (T.) vocans borealis subsp. nov.

HONG KONG. HOLOTYPE of *U. vocans borealis*, m., (NYZS = USNM 137669) + 80 m., 84 f. (NYZS); 1 m. (USNM 44358).

CHINA. *Fukien*: Chin Bey (near Amoy): 1 ovig. f. (USNM 57051). Tsimei: 2 ovig. f. (USNM 57825); 3 m., 2 (1 ovig.) f. (USNM 57826).

TAIWAN [FORMOSA]. Tamsui: 3 m. (NYZS).

U. (T.) vocans pacificensis subsp. nov.

INDONESIA. *Ambon [Amboina]*: 2 m. (Leiden D 275).

PHILIPPINE IS. (probably hybridizing with *U. vocans vocans*). *Sulu*: Tawi Tawi: 26 m., 11 f. (NYZS). Joló: 1 m., 2 f. (NYZS); Princess Tarhata Kiram's Beach): 3 m., 2 ovig. f. (NYZS). *Mindanao*: Zamboanga: 16 m., 8 f. (NYZS). Gulf of Davao: Padada Beach: 1 m. (AMNH 8362); 1 m. (AMNH 8401 part). Madaum: 1 m. (NYZS).

NEW GUINEA. *Territory of Papua and New Guinea* (probably hybridizing with *U. vocans vomeris*): Near Madang on north central coast: Maiwara: 4 m., 1 f. (NYZS). Port Moresby: Taurama Beach and Bootless B.: 2 lots: 3 m. (UPNG; collected by M. Cahill).

PACIFIC OCEAN. *Marshall Is.*: 3 m. (MCZ 5773). *Guam*: 12 m., 2 f. (Bishop). *Fiji Is.*: *Viti Levu*: Suva: HOLOTYPE of *U. vocans pacificensis*, m. (NYZS = USNM 137670) + 28 m., 14 f. (NYZS). Rewa (flats at mouth of Wainibokasi R.): 18 m., 22 f. (NYZS). *Samoa Is.*: 15 m. (NYZS); 1 m., 1 f. (Leiden 2665).

U. (T.) vocans dampieri subsp. nov.

AUSTRALIA. *Northern Territory*: Darwin: 28 m., 19 f. (NYZS). *Western Australia*: Broome: HOLOTYPE of *U. vocans dampieri*, m. (NYZS = USNM 137671) + 33 m., 40 f. (NYZS).

U. (T.) vocans vomeris McNeill, 1920

NEW GUINEA. *Territory of Papua and New Guinea*: near Madang on north central coast: Maiwara: 13 m., 16 f. (NYZS). Port Moresby: Fisherman's I.: 1 m. (UPNG; collected by M. Cahill).

NEW CALEDONIA. Near Nouméa: mouth of Dumbea R.: 6 m. (NYZS).

AUSTRALIA. *Queensland*: Shorncliffe (near mouth of Brisbane R.): 1 m. (NYZS). Sandgate (near mouth of Brisbane R.): 2 m. (AMNH 4884). *New South Wales*: Trial B.: 2 m., 2 f. (now in poor condition; Bishop 601); 3 m., 2 f. (Bishop 2482).

U. (T.) vocans hesperiae subsp. nov.

ETHIOPIA. *Eritrea*: Massawa: 17 m., 6 f. (NYZS).
EAST AFRICA. TANZANIA. *Pemba I.*: Chake Chake: 2 m. (NYZS). Zanzibar: Chakwani Fish Ponds:

HOLOTYPE of *U. v. hesperiae*, m. (NYZS = USNM 137672) + 4 m., 10 f. (NYZS): between Chakwani and Mbeni: 17 m., 6 f. (NYZS).

MADAGASCAR. Eau Saumâtre: 1 f. (Paris). Nossy Bé: 9 m. (Leiden D 274, labeled *marionis excisa*; see p. 89).

CEYLON. *Negombo* (near Colombo): 1 m. (NYZS).

MALAYSIA. *Malaya*: Penang (probably hybridizing with *U. vocans vocans*): 1 m. (NYZS).

SINGAPORE. (Probably hybridizing with *U. vocans vocans*): 1 m. (NYZS; gift of M.W.F. Tweedie).

INDONESIA. *Sulawesi [Celebes]*: Padang. 1 m. (Leiden 2751).

U. (T.) vocans vocans (Linnaeus, 1758)

BURMA. Rangoon: 9 f. (MCZ 5779 part).

MALAYSIA. *Malaya*: Penang: 1 m. (NYZS). Port Dickson: 2 m., 1 f. (Leiden 5269). *Borneo*: *Sabah* [*N. Borneo*]: Labuan: 1 m., 1 f. (NYZS; gift of M.W.F. Tweedie).

SINGAPORE. 6 m. (NYZS; gift of M.W.F. Tweedie). Geylang R. (near mouth): Kallong: 21 m., 29 f., + 1 copulating pair (NYZS).

INDONESIA. *Bawean*: Sangapura: 30 m., 3 f. (NYZS; gift of H. Oesman). *Djawa* [*Java*]: Tanjong Priok (near Djakarta): 1 m. (Leiden 1491). North coast: 4 m. (Leiden 2048). Trengglunga (near Surabaja): 9 m., 3 f. (NYZS). Pasuruan (near Surabaja): 8 m., 18 f. (NYZS); Besoeki: 6 m. (Leiden 277, 1674); Madoera; 4 m. (Leiden 2045). *Sulawesi* [*Celebes*]: Makassar: 1 m. (Leiden 5268). *Halmahera*: Ternate: 1 m. (Leiden 1533).

PHILIPPINE IS. *Sulu*: Tawi Tawi: 5 m., 3 ovig. f. (NYZS). Joló: 2 m., 2 f. (NYZS); Princess Tarhata Kiram's Beach: 3 m. (NYZS). *Mirdanao*: Zamboanga: 117 m., 38 f. (NYZS). Gulf of Davao: Padada Beach: 1 f. (AMNH 8325); near mouth of Padada R.: 3 m. (AMNH 8401 part); flats about 2 miles south of Padada: 2 m. (AMNH 8556). Sasa: 1 f. (NYZS). Madaum: NEOTYPE of *U. vocans*, m. (NYZS = USNM 137673) + 33 m., 16 f. (NYZS). *Palawan*: Puerto Princesa: 36 m., 21 f. (NYZS). Northwest coast, near mouth of Baheli R.: 8 m., 2 f. (NYZS). *Luzon*: Lamao: 1 m. (NYZS).

NANSEI [RYUKYU or LOO CHOO] IS. 1 m. (MCZ 1333). Ishiyaki I. (mouth of Miyara R.): 3 m. in 2 lots. Kume-jima: 3 m. (NYZS; collected by Minei, gift of T. Sakai, U. of Yokohama).

Additional Material of Uca vocans Examined Sufficiently for Species Identification, but not Referred to Subspecies. At the time of visits to several European institutions made during the first years of this taxonomic revision, I had not yet distinguished the characteristics upon which the subspecies have since been based. Although for the most part, as with any African specimens, the subspecies can be safely presumed from the locality, individuals from Indonesia, New Guinea, and the Philippines may be from mixed populations. They are all therefore listed here only under the name of the species, *U. vocans*.

RED SEA. 1 m. (Leiden 1493).

MADAGASCAR. Nossy Bé. Short series (Leiden 275; labeled, as are the 9 m. already listed under *U. v. hesperiae*, Leiden 274, "*U. marionis excisa*," see p. 89).

NEW GUINEA. *West Irian*: A long series from Kotabaru [Hollandia], Japen and their vicinity (Leiden, several lots, collected by Holthuis, 1954ff.; specimens not examined individually).

PHILIPPINE IS. Holotype of *G. cultrimanus*, 1 m. (BM 43.6 Cuming collection).

CAROLINE IS. 1 m. British Museum.

Uca vocans: *Specimens not Examined.* More than 1,000 specimens at USNM from a variety of localities throughout the range of the species. The longest series, totaling 727 m. + 140 f., is from Iloilo, Panay I., in the Philippines and would form an excellent basis for future work on variations and the possible hybridization of subspecies (USNM 73164). Other outstanding series are 42 m. from Benkulen, Sumatra (USNM 75873) and 33 m. + 5 f. from Apia, Samoa (USNM 43300). This collection is currently labeled and catalogued under the following names: *U. cultrimana, marionis, marionis* var. *nitidus,* and *marionis vomeris.*

SUBGENUS *AMPHIUCA*

Uca (Amphiuca) chlorophthalmus (Milne-Edwards, 1837). (P. 98.)

U. (A.) chlorophthalmus crassipes (Adams & White, 1848)

INDIA. Car Nicobar: 3 m., 1 f. (NYZS; gift of Raffles Museum, Singapore).

INDIAN OCEAN. Cocos or Keeling Is.: 6 m., 3 f. (NYZS; gift of Raffles, Singapore).

MALAYSIA. *Malaya*: *Penang*: 1 f. (USNM 43370).

HONG KONG. 2 m. (USNM 43849); Kowloon (Castle Peak area, east of mouth of Deep Bay): 2 m. (NYZS).

JAPAN. *Tokaito coast*: Oho Sima: 2 f. (USNM 22288).

NANSEI [RYUKYU or LOO CHOO] IS. Kume-jima: 1 m. (USNM 73258); carapace only (USNM 73259).

PHILIPPINE IS. "Sinquejor": LECTOTYPE of *G. crassipes* (BM 902 43b). *Sulu*: Tawi Tawi: 3 m. (NYZS); Tapiantana I.: 1 m. (USNM 43451). *Mindanao*: Zamboanga: 1 f. (USNM 43371); 72 m., 47 (18 ovig.) f. (NYZS).

INDONESIA. *Molucca Is.*: Bourou I.: 1 m. (Phil. Acad.). *Aru [Aroe, Arrou] Is.*: 8 m., 1 f. (Paris).

NEW GUINEA. "New Guinea": type series of *U. novaeguineae*, 2 m. (USNM 6374). *West Irian*: in and near Kotabaru [Hollandia], Noemboor, and Biak: numerous m. & f. (Leiden 635 [part], 714, 723, 743 [part], and unnumbered; collected by Holthuis, 1954-1955; samples only examined in detail by Crane). *Territory Papua and New Guinea*: near Madang on north central coast: Maiwara: 6 m., 4 (1 ovig.) f. (NYZS). Port Moresby: 10 m., 2 f. (UPNG; collected by M. Cahill); Bootless B.: 4 m., 2 f. (UPNG; collected by M. Cahill).

PACIFIC OCEAN. *Caroline Is.*: Kusaie: 1 m. (USNM 33167); 4 m., 1 f. (MCZ 8845); Strong I.: 7 m., 3 f. (MCZ 5792); Ponape [Ascension]: 1 f. (MCZ 5967). *Marshall Is.*: 2 m., 2 f. (USNM 81379); 15 m., 18 f. (MCZ 5790); Ebon: 6 m., 1 f. (MCZ 5791). *New Hebrides*: Port Resolution: 2 m. (Paris, from Australian Museum). *New Caledonia*: short series (Paris). Balansa: 1 m., 1 f. (USNM 20299). Nouméa: 1 m. (NYZS). *Fiji Is. [Viti]*: 1. (Paris). *Viti Levu*: Suva: 2 (1 ovig.) f. (NYZS). *Samoa Is.*: Upolu: Apia: 3 m. (Torino); 12 m., 4 (3 ovig.) f. (NYZS). *Friendly Is.*: Tongatabu: "type non specifié" of *G. gaimardi,* m. (Paris). *Society Is.*: Bora-Bora: "type non specifié" of *G. latreillei*, 1 m. (Paris); 22 m., 9 f. (NYZS). Raiatea: 12 m. (AMNH 2482); 4 m. (NYZS). Tahiti: small series (Paris); Faaa: 79 m., 69 (5 ovig.) f. (NYZS). *Marquesas Is.*: 1 m. (USNM 81415 part). ["Iles Sandwich": 2 m. (Paris). Perhaps an error for Bora Bora; see p. 324].

Note. The material in Paris was originally labeled variously *chlorophthalmus, gaimardi, latreillei*, and *pulchellus*; in BM *crassipes*; in USNM *gaimardi* and, from Hong Kong, *splendidus*; in Torino *gaimardi*.

Not Examined. More than 200 additional specimens in USNM, listed as *U. gaimardi*, from Samoa and the Society Is.

U. (A.) chlorophthalmus chlorophthalmus (Milne-Edwards, 1837)

EAST AFRICA. TANZANIA. *Pemba Island*: Kiburunzi (near Chake Chake): 30 m., 22 f. (NYZS). *Zanzibar*: (Muongoni): 13 m., 1 ovig. f. (NYZS). *Tanga*: Dar es Salaam near Kondeni (first stream north of city): 1 m. (NYZS).

MOZAMBIQUE. *Inhaca Island* (east lagoon): 9 m., 5 f. (NYZS).

INDIAN OCEAN. "Indian Ocean": 1 m. (Paris). *Mauritius*: Ile de France: "TYPE NON SPECIFIÉ" of *G. chlorophthalmus*, m. (Paris). Ile Maurice: small series of small m. & f. (Paris).

BRAZIL. Teffe on the Amazon: Type of *U. amazonensis* (Torino), m. *Locality certainly erroneous*; see p. 322.

Uca (Amphiuca) inversa (Hoffmann, 1877). (P. 105.)

U. (A.) inversa inversa (Hoffmann, 1877)

ETHIOPIA. Eritrea: 1 m. (Torino). Massawa: 16 m. (NYZS).

ARABIA. *Hadramaut*: Aden: 1 m. (BM part of series, gift of Yerberg). Mukalla: 1 m. (BM 88A).

SOMALILAND. Laguna di Honio: 1 m. (Torino).

KENYA. Near Lamu: 1 m. (BM 93.11.9.6).

TANZANIA. *Zanzibar*: Chakwani Fish Ponds: 84 m., 37 f. (NYZS).

MOZAMBIQUE. *Inhaca I.* (east lagoon): 4 m. (NYZS).

MADAGASCAR. Nossy Fally: LECTOTYPE of *G. inversus*, m. (Leiden 251; see comment on additional type material, p. 106).

NATAL. Perhaps from type material of *G. smithii*: 1 m. (Philadelphia 3003; labeled *inversa*).

Not Examined. A collection of more than 300 m. and f. from East Africa, near Mombasa, at Changawe and its strait (USNM 43465, 43466, 43467).

U. (A.) inversa sindensis (Alcock, 1900)

PAKISTAN. Mud between Karachi & Clifton (beyond refugee camp): 15 m. (NYZS).

SUBGENUS *BOBORUCA*

Uca (Boboruca) thayeri Rathbun, 1900. (P. 112.)

U. (B.) thayeri umbratila Crane, 1941

EL SALVADOR. La Herradura & La Union: type material of *U. thayeri zilchi* Bott, 1954 (Frankfurt: Senck. 2058, 2070). El Triunfo: 1 m. listed by Bott, 1954, as *U. t. thayeri*; has regenerated claw (Frankfurt: Senck. 2057).

COSTA RICA. Puntarenas: HOLOTYPE of *U. umbratila*, m. (NYZS 381,129 = USNM 138132); PARATYPE f. (NYZS 381,130 = USNM 138133). Golfito: 3 m., 5 f. (NYZS 38,509a; inadvertently omitted from Crane, 1941.1). Ballenas B.: 1 f. (NYZS 381,131).

PANAMA. Near Panama City: east of Old Panama, near mouth of Rio Abajo, by water pump: 2 m. (NYZS).

CANAL ZONE. Balboa at La Boca: PARATYPES of *U. umbratila*: 1 m., 1 f. (USNM 79407); 18 yg. m., 12 yg. f., 2 post-megalopal carapaces (NYZS 4118 = USNM 138134).

U. (B.) thayeri thayeri Rathbun, 1900

UNITED STATES OF AMERICA. *Florida*: St. Augustine: 3 m., 2 f. (NYZS). Miami: 1 m. (NYZS). Marco-Coxambas Cut (near Coon Key): 1 m., 1 f. (USNM 74493). Sanibel I.: 1 f. (MCZ 10183).

GUATEMALA. Puerto Barrios: 1 m. (NYZS).

GUADELOUPE. 2 ovig. f. (NYZS).

TRINIDAD AND TOBAGO. *Tobago*: near Pigeon Point: 3 m. (NYZS). *Trinidad*: near Port of Spain (Cocorite, Laventille & Caroni Swamps): 85 m., 72 (37 ovig.) f. (NYZS).

VENEZUELA. *Aragua*: Turiamo: 2 m. (NYZS). *Delta Amacuro*: Pedernales: 3 f. (NYZS).

BRAZIL. *Parahybo do Norte*: Parahybo R. at Cabedello. TYPES of *U. thayeri*: 7 m., 1 f. (USNM 23753). *Pernambuco*: Recife (swamp close to Olinda Causeway): 9 m., 3 (2 ovig.) f. (NYZS). *Bahia*: São Salvador [Bahia] (on Itaparica I.): 10 m., 3 f. (NYZS). Plataforma: 1 m., 1 f. (USNM 40619). *Rio de Janeiro*: east shore Ilya Pinheiro (lagoon): 2 m., 1 ovig. f. (NYZS).

Not Examined. About 50 additional specimens at USNM from various localities throughout the range of the subspecies.

SUBGENUS *AFRUCA*

Uca (Afruca) tangeri (Eydoux, 1835). (P. 118.)

PORTUGAL. *El Algarve*: Faro: 213 m., 75 (9 ovig.) f. (NYZS). Vila Réal de São Antonio: 4 m. (NYZS).

TANGIER. TYPE MATERIAL of *G. tangeri*, 1 m., 1 f. (Paris 3478-1, 3478-2; "types non specifiés"); 1 m., 1 f. (Paris 3477); 1 m. (Phil. Acad. 9-3028, from Guérin; "type").

"GUINEA COAST," probably including GHANA. "Boubry," "Acre d'Elmina," and "Kust van Guinea": type material of *G. perlatus* (Leiden 259, 260, 261, 262).

NIGERIA. Lagos (at Tarkwa B.): 29 m., 17 f. (NYZS).

CONGO. Banana: 44 m., 20 f. (AMNH 3071 part); 2 m., 2 f. (AMNH 3073); 5 m., 6 f. (AMNH 3075 part).

ANGOLA. Santo Antonio do Zaire: 1 m., 1 f. (AMNH 3072). Near Loanda: Samba: 68 m., 34 f. (NYZS): Samba Pequeña: 33 post-megalopal yg. (NYZS); Ilya de Cabo: 94 yg. m., 98 yg. f. (NYZS). Lobito: 6 m., 4 f. (AMNH 5916); 3 m., 2 f. (AMNH 5917 part).

WEST AFRICA. 2 m. (AMNH 5922); 14 m., 4 f. (AMNH 5933 part); 39 m., 17 f. (AMNH 5934 part); 10 m., 8 f. (AMNH 5935 part); 14 m., 3 f. (AMNH 5936 part).

Not Examined. More than 50 specimens at USNM from various localities in west Africa, between Dakar and Loanda.

SUBGENUS *UCA*

Uca (Uca) princeps
(Smith, 1870). (P. 128.)

U. (U.) princeps princeps (Smith, 1870)

MEXICO. *Lower California*: *West Coast*: Santo Domingo: 1 m. (USNM 51107). Abreojos Point: 5 m., 3 f. (USNM 20689). *Gulf of California (South District)*: La Paz: 1 m., 2 f., 1 yg. (USNM 14826). *Sonora*: Empalme: 2 m. (USNM 57066). *Nayarit [Tepic Terr.]*: San Blas: 1 yg. m., 1 yg. f. (USNM 99750); 6 m. (USNM 99751); 4 m. (USNM 99752); 1 m., 3 f. (USNM 20654).

NICARAGUA. Corinto: COTYPES of *G. princeps*, 9 m. + 3 m. (MCZ 5813, 5814).

COSTA RICA. Golfito: 2 m., 1 f. (NYZS 38588). Ballenas B.: 1 m. (NYZS 38355). Santo Domingo, Gulf of Dolce: 1 yg. m. (USNM 19440).

CANAL ZONE. Balboa (La Boca): 3 disintegrated specimens (NYZS 4135).

ECUADOR. Puerto Bolivar (near mouth of Guayas R.): 11 m., 3 f. (NYZS).

PERU. B. of Sechura: salt marshes back of Chulliyache: 3 m. (USNM 40467). Rio Zarumilla: salt flats at Puerto Grande: 1 m. (USNM 40468).

U. (U.) princeps monilifera Rathbun, 1914

MEXICO. *Gulf of California (North District)*: San Felipe: 1 m., 1 f. (USNM 67735). Montague I. (mouth of Colorado River): 1 m. (USNM 48829). *Sonora*: Guaymas: HOLOTYPE of *U. monilifera*, m. (MCZ 1578) + 7 m. "out of type lot" (MCZ 8272); PARATYPE, m. (USNM 22180, received from MCZ, from 1578).

Uca (Uca) heteropleura
(Smith, 1870). (P. 133.)

EL SALVADOR. Gulf of Fonseca: HOLOTYPE of *G. heteropleurus*, m. (MCZ 5819).

COSTA RICA. Punta Arenas: 1 yg. m. (USNM 39099). Golfito: 2 m., 2 f. (NYZS 38589).

PANAMA. Bahia Honda: 7 m., 5 f. (NYZS 38698). Panama City: 1 m. (NYZS 4138); Bellavista: 43 m., 14 f. (NYZS).

CANAL ZONE. Balboa: 2 m. (NYZS 4137).

ECUADOR. Puerto Bolivar (near mouth of Guayas R.): 8 m. (NYZS).

Uca (Uca) major
(Herbst, 1782). (P. 136.)

WEST INDIES. *Bahamas*: Bimini: 1 m., 1 f. (USNM 91740). *San Salvador* or *Watling's Island*: 2 m., 1 f. (USNM 11375); 1 m. (MCZ 6366). *Cuba*: Baracoa: 6 m., 1 f. (USNM 25549). *Jamaica*: 1 m. (USNM 42918). *Guadeloupe*: Type of *G. grangeri* ("type non specifié"), 1 f. (Paris). *Trinidad*: 1 m. (NYZS; gift of H. O. von Hagen). *Curaçao*: Carmabi: 1 m. (NYZS; gift of A.C.J. Burgers).

CANAL ZONE. Colon: 1 yg. m. (USNM 57744).

VENEZUELA. *Aragua*: Turiamo: 3 m. (NYZS).

CAYENNE. Type of *G. platydactylus* ("type non specifié"), 1 m. (Paris).

Uca (Uca) stylifera
(Milne-Edwards, 1852). (P. 140.)

NICARAGUA. Gulf of Fonseca: Cotypes of *G. heterophthalmus*, 4 m. (MCZ 5818). West coast: 2 m. (USNM 74428). Corinto: 4 m., 1 f. (NYZS 3813); 2 f. (MCZ 5815 part; published by Smith, 1870: 126, as *U. ? princeps*; third specimen is missing).

COSTA RICA. Puntarenas: Pacific-Estero side: 2 m. (USNM 32325). Punta Arenas: 1 m. (USNM 61592). Golfito: 8 m., 9 (1 ovig.) f. (NYZS 38589).

CANAL ZONE. Balboa (La Boca): 2 m. (NYZS 4136).

ECUADOR. Guayaquil: LECTOTYPE of *G. styliferus* ("type non specifié"), 1 m. (Paris). Puerto Bolivar (near mouth of Guayas R.): 1 m. (NYZS).

Uca (Uca) maracoani
(Latreille, 1802-1803). (P. 143.)

U. (U.) maracoani maracoani (Latreille, 1802-1803)

WEST INDIES. *Santo Domingo*: Sanchez: 1 f. (AMNH 2466). *Trinidad*: Port of Spain (Cocorite): 16 m., 16 (1 ovig.) f. (NYZS).

VENEZUELA. *Delta Amacuro*: Pedernales: 14 m., 3 f. (NYZS 42414).

GUYANA. Near Georgetown: 1 m. (AMNH 4637). Georgetown: 2 m., 1 f. (AMNH 6645); mouth of Demerara R. and Kitty Village: 3 m., 2 f. (NYZS); Kitty Village: 26 m., 18 (3 ovig.) f. (NYZS).

BRAZIL. *Maranhão*: 13 m., 5 f. (USNM 17657 part); 2 m., 2 f. (USNM 22190). Tutoya Creek: 4 m., 1 ovig. f. (AMNH 4657). *Rio Grande do Norte*: Natal: 6 m., 1 f. (USNM 25697). *Bahia*: São Salvador [Bahia]: Bahia de Todos Santos: Ilya Itaparica: 1 m., 1 f. (NYZS). Porto Seguro: 2 m., 1 f. (USNM 22191). Plataforma: 4 m., 3 f. (USNM 40613); 7 m., 2 f. (USNM 40614). *Parana*: Paranagua: 1 m. (USNM 71177).

U. (U.) maracoani insignis
(Milne-Edwards, 1852)

NICARAGUA. Gulf of Fonseca: Type of *G. armatus*, 1 m. (MCZ 5816).

ECUADOR. Puerto Bolivar (mouth of Guayas R.): 5 m., 1 f. (NYZS).

PERU. Piura (mud flats at mouth of Rio Tumbez): 1 m. (USNM 71321); Chulliyache (B. of Sechura; salt marshes): 1 m. (USNM 40489).

"CHILI." HOLOTYPE of *Acanthoplax insignis* ("type non specifié"), f. (Paris).

Uca (Uca) ornata
(Smith, 1870). (P. 150.)

WEST COAST OF CENTRAL AMERICA. HOLOTYPE of *G. ornatus*, f. (MCZ 5817).

COSTA RICA. Puntarenas: 1 m. (USNM 61036).

PANAMA. Mouth of Rio Abajo (near Old Panama): 4 m., 1 ovig. f. (NYZS; gift of M. D. Burkenroad); 25 m., 31 (3 ovig.) f. (NYZS). Panama City: (Bellavista): 1 m. (NYZS 4140).

CANAL ZONE. Balboa (La Boca): 2 m., 1 yg. f. (NYZS 4139).

PERU. Puerto Pizarro: holotype of *U. pizarri*, m. (Leiden D 23061) + paratype, 1 m. (Leiden D 23062).

SUBGENUS *MINUCA*

Uca (Minuca) panamensis
(Stimpson, 1859). (P. 158.)

NICARAGUA. Gulf of Fonseca, near Potosi R.: 2 m., 1 yg. f. (NYZS 37701). Corinto: Cardon: 9 m., 5 f. (NYZS 37735).

COSTA RICA. Port Parker: 7 + 1 yg. m., 8 + 2 yg. f. (NYZS 3861 & 3892). Culebra B.: 1 m. (NYZS 38119). Piedra Blanca: 5 m., 4 f. (NYZS 38166); 3 m., 3 f. (NYZS 38197). Ballenas B.: 1 f. (NYZS 38345). Uvita B.: 5 m., 3 yg. f. (NYZS 38456). Golfito: 7 m., 5 f. (NYZS 38517).

PANAMA. Bahia Honda: 3 m., 1 f. (NYZS 38706). Patillo Point: 9 m., 8 f. (AMNH 5974); 4 m. (AMNH 6642); 2 f. (AMNH 6663). Panama B.: north shore Tabogilla I.: 2 m., 6 f. (AMNH 5978); 1 yg. f. (AMNH 5985). Pearl Is.: Pacheca I.: 2 m., 1 f. (AMNH 5983). Panama City: 1 yg. m., 2 yg. f. (poor condition) (NYZS 4149).

COLOMBIA. Gorgona Island: Gorgonilla: 4 m., 4 (2 ovig.) f. (NYZS 38855); 13 m., 9 f. (NYZS 38878). Humboldt B.: 9 m., 2 f. (AMNH 10703); 1 m. (AMNH 10704).

PERU. Payta: 4 m., 1 f. (MCZ 5908).

Not Examined. About 15 additional specimens at USNM from within the range of the species.

Uca (Minuca) pygmaea
Crane, 1941. (P. 161.)

COSTA RICA. Golfito: HOLOTYPE of *U. pygmaea*, m. (NYZS 381,110 = USNM 137419); PARATYPES, 10 m. (NYZS 381,111, part = USNM 137420), 2 m. (NYZS 381,111, part = USNM 79402).

COLOMBIA. Buenaventura: 1 m. (NYZS; gift of T. Collett, U. of London).

Uca (Minuca) vocator
(Herbst, 1804.) (P. 163.)

U. (M.) vocator vocator (Herbst, 1804)

MEXICO. *Vera Cruz*: Tampico: 1 m. (USNM 43353 part). Arroyo de la Renal, tributary of Tonchochapa R.: 1 m. (NYZS: gift of G.A.W.).

BRITISH HONDURAS. Near Belize: 1 m. (USNM 21373 part).

GUATEMALA. Puerto Barrios: 3 m., 2 f. (NYZS).

NICARAGUA. Greytown: 1 m. (USNM 18434 part).

WEST INDIES. *Santo Domingo*: Sanchez: 1 f. (MCZ 9860 part); 1 m., 2 f. (AMNH 2441 part). *Puerto Rico*: San Juan: 2 m., 1 f. (NYZS), 1 m. (AMNH 2690). *Guadeloupe*: 1 m. (NYZS). *Trinidad*: Blanchisseuse (north coast): 1 m. (NYZS). Cocorite Swamp (near Port of Spain): 2 yg. m. (NYZS). Behind pump station, Caroni Swamp: 5 m. (NYZS). Caroni Swamp: 4 m. (NYZS). Lavantille Swamp: 4 m. (NYZS). Near stream mouth, Manzanillo (east coast): 2 m. (NYZS).

VENEZUELA. *Zulia*: Maracaibo: 14 m., 1 f. (NYZS 42419; published as *U. murifecenta* in *Zoologica* 28 (7): 38; 1943); 2 f. (NYZS 42421 part; published as *U. pugnax rapax, loc. cit.*). Lagunillas: 4 m., 4 f. (NYZS). *Sucre & Monagas*: near mouth of San Juan R.: holotype of *U. murifecenta*, m. (NYZS 42167 = USNM 137424); paratypes of *U. murifecenta*, 5 m., 6 f. (NYZS 42417 = USNM 137425); 25 m., 20 (3 ovig.) f. (NYZS 42418; published as *U. murifecenta, loc. cit.*). *Delta Amacuro*: Pedernales: specimens wholly desiccated, pulverized; discarded (NYZS 42416; published as *U. murifecenta, loc. cit.*).

GUYANA. Georgetown: Kitty Village (flats near seawall): 10 m., 5 (1 ovig.) f. (NYZS); foot of Kitty Village: 1 f. (NYZS).

BRAZIL. *Estado Rio de Janeiro*: Serra de Masaché: 2 m. (USNM 50690 part).

SURINAME. Shore of Suriname R., near Leonsburg, north of Paramaribo: NEOTYPE of *U. vocator*, m. (Leiden 1208).

U. (M.) vocator ecuadoriensis
Maccagno, 1929

MEXICO. *Nayarit* [*Tepic Terr.*]: San Blas: holotype of *U. schmitti*, m. (USNM 80451 out of 22306, formerly labeled *U. mordax*) + paratypes, 3 m. (USNM 22306, also formerly labeled *U. mordax*); 1 m. (USNM 99756). *Guerrero*: Acapulco: paratype of *U. schmitti*, 1 m. (MCZ 5892, part, formerly labeled *U. mordax*).

NICARAGUA. San Juan del Sur: paratype of *U. schmitti*, m. (NYZS 381,116 = USNM 137421) published first as *U. mordax* in *Zoologica* 26 (19): 176; 1941.1); 1 yg. m. (NYZS 381,120; published first 1941.1, *loc. cit.*, as *U. brevifrons*).

COSTA RICA. Negritos I.: paratype of *U. schmitti*, m. (NYZS 381,117 = USNM 137423) published first as *U. mordax, loc. cit.* Golfito: paratype of *U. schmitti*, m. (NYZS 381,118 = USNM 137422) published first as *U. mordax, loc. cit.*); 2 m. (NYZS).

PANAMA. Bahia Honda: 5 m., 1 f. (NYZS 38681; omitted from *Zoologica* 26 (19); 1941).

COLOMBIA. Buenaventura: 2 m., 2 f. (USNM 158,353, part); 1 m. (NYZS: gift of T. Collett, U. of London); 2 m., 2 f. (NYZS; gift of Brother Iñez, Bogotá, Colombia).

ECUADOR. Esmeraldas: TYPE MATERIAL of *U. ecuadoriensis*, 2 yg. m. (Torino); 1 m. (USNM 98041). Puerto Bolivar: 6 m., 6 f. (NYZS). Guayaquil: 1 m. (USNM 70867); El Salado: 1 m. (USNM 98039); 96 m., 20 (8 ovig.) f. (NYZS). Near Guayaquil: 3 m. (USNM 98038).

PERU. Puerto Pizarro: holotype of *U. lanigera*, m. (Leiden D 23049); paratypes, 10 m., 5 f. (Leiden D 23050).

Uca (Minuca) burgersi
Holthuis, 1967. (P. 168.)

UNITED STATES. *Florida*: Ft. Lauderdale (Dania Beach, mangroves to south): 3 m. (NYZS).

MEXICO. Yucatan: Progreso: 20 m., 10 f. (MCZ 8628); 2 m., 2 f. (USNM 81381); near Telchac—P. Motul: m. + ? yg. (USNM 95526 part).

BRITISH HONDURAS. Belize: Yarborough: Loyola Park: 1 m. (NYZS; gift of collectors C. Fairweather and J. Lopez). Near Belize: 2 f. (USNM 21373 part); 8 m. (USNM 22604 part).

GUATEMALA. Puerto Barrios: 12 m., 2 f. (NYZS).

CANAL ZONE. Toro Point: 2 m. (USNM 59307).

WEST INDIES. *Swan I.*: 6 m. (MCZ 8392). *Cuba*: Santa Clara: Damuge R.: 11 m., 4 f. (AMNH 3166); Isla Turiguano: 10 m., 1 yg. f. (NYZS; gift of W.R.W. Riggs). *Bahamas*: Abaco I.: near Water Cay: 1 m. (MCZ 8632). Andros I.: 9 m., 3 f. (AMNH 2527); Mangrove Cay: 1 f. (MCZ 8643). East side Long I.: Grays: 65 m., 9 f. (MCZ 9440). Great Inagua I.: Mathew Town: 23 m., 10 f. (MCZ 10333); Savannah 3 mi. N. of Mathew Town: 11 m., 10 (3 ovig.) f. (MCZ 10361). *Haiti*: Manneville: 12 m., 6 (1 ovig.) f. (MCZ 8690). Near Port-au-Prince: Etang Saumâtre: 1 m. (NYZS; collected by W. Beebe). *Santo Domingo*: Sanchez: 28 m., 20 (5 ovig.) f. (AMNH 2444). Monte Cristi: 1 m. (MCZ 9880 part). *Puerto Rico*: San Juan: 13 m.,

4 f. (AMNH 2983). Enseñada: 1 m., 1 f. (AMNH 2984 part); 2 m. (AMNH 3011 part). *Virgin Islands*: St. Croix: Buck Island Swamp: 3 m., 1 f. (NYZS; gift of H. Beatty). St. Thomas: 13 m. (NYZS); Red Hook: 81 m., 78 f. (NYZS). *St. Martin*: Crab Hole Cistern: 2 m., 2 ovig. f. (NYZS; gift of P. W. Hummelinck); Little B. Pond, east: 4 m., 4 ovig. f. (NYZS; gift of Dr. Hummelinck); pond of Point Blanche: 2 m. (NYZS; gift of Dr. Hummelinck). *Antigua*: 2 m., 3 f. (AMNH 2505). *Barbados*: Payne's B. (St. James Sector): 45 m., 18 (2 ovig.) f. (NYZS); 40 m., 9 f. (NYZS). *Tobago*: Pigeon Point Swamp (across road from beach): 28 m., 6 (1 ovig.) f. (NYZS). *Trinidad*: Blanchisseuse (north coast): 6 m., 2 f. (NYZS); sandy ground near Nariva Swamp Bridge: 1 m. (NYZS; gift of H.-O. von Hagen). *Aruba*: Balashi, by Spaans Lagoen (mangrove swamp): 1 f. (NYZS; gift of P. W. Hummelinck). *Curaçao*: HOLOTYPE of *U. burgersi*, m. (Leiden D 23012). *Jamaica*: 1 m. (USNM 18533).

VENEZUELA. *Aragua*: Turiamo: 7 m. (NYZS). *Miranda*: Laguna de Tacarigua: 1 m. (NYZS; gift of W. H. Phelps, Sr.).

BRAZIL. *Céara*: Fortaleza (semi-closed swamp beside Avenida Getulio Vargas): 30 m., 8 f. (NYZS). *Bahia*: São Salvador [Bahia]: Bahia de Todos Santos: Ilya Itaparica: 5 m., 1 f. (NYZS). *Rio de Janeiro*: 2 m., 2 f. (USNM 19971); Ilya Pinheiro: 4 m. (NYZS), Niterói [Nictheroy]: 1 m. (USNM 74436). *São Paulo*: Santos: São Sebastião: 4 m. (NYZS; gift of F. J. Vernberg); Ilya Bela: 1 m. (USNM 74437).

Locality and identification questionable. AFRICA: CAMEROONS: Metet: 3 yg. m., major chelipeds missing; 1 f. (MCZ 8962). LIBERIA: 1 f. (USNM 21847). See p. 327.

Not Examined. Numerous additional specimens at USNM from the West Indies, labeled *Uca mordax*. Casual inspection of sample jars shows these specimens to be *U. burgersi*.

Uca (Minuca) mordax
(Smith, 1870). (P. 173.)

GUATEMALA. Puerto Barrios: 3 m., 2 f. (NYZS).

NICARAGUA. Greytown: 8 m., 2 f. (USNM 18433).

COLOMBIA. Turbo: 2 m. (MCZ 5891).

VENEZUELA. *Zulia*: Lagunillas: 1 m. (NYZS). *Monagas*: Caripito: 8 m., 1 f. (NYZS 4252); 22 m., 18 (4 ovig.) f. (NYZS). *Sucre*: Guanoco: 35 m., 24 f. (NYZS 4254). *Sucre & Monagas*: near mouth of

San Juan R.: 26 m., 17 f. (NYZS 42166). *Delta Amacuro*: Tapure, near Pedernales: 74 m., 47 (4 ovig.) f. (NYZS 42415); 2 yg. m., 1 yg. f. (NYZS 42420 part). *Curiapo*: 1 m., 1 f. (USNM 95992).

TRINIDAD. Blanchisseuse, north coast of Trinidad: 1 m., 1 f. (NYZS); near mouth of Oropouche R., eastern Trinidad: 3 m., 1 f. (NYZS).

GUYANA. Georgetown: 2 m. (NYZS collected by W. Beebe); foot of Kitty Village: 2 m., 2 f. (NYZS).

SURINAME. Near Paramaribo: 14 m., 13 (2 ovig.) f. (NYZS); 18 m., 9 f. (NYZS). Mouth of Marawyne R. (West Indian Coconut Plantation): 41 m., 12 f. (NYZS).

BRAZIL: Maranhão: Pará: Belém (canals in city): COTYPES of *G. mordax*, 7 m., 5 f. (MCZ 5882); 50 m., 10 f. (NYZS). Itabapuana: 4 m., 4 f. (USNM 22186). *Pernambuco*: Recife [Pernambuco]: 1 m. (USNM 25698). *Rio de Janeiro*: "Terra de Masahe" [Serra de Masaché]: 2 m. (USNM 47834).

Uca (Minuca) minax
(LeConte, 1855). (P. 176.)

UNITED STATES. *Massachusetts*: Wareham: 1 f. (AMNH 1028). *Connecticut*: Old Lyme: 10 m., 5 f. (NYZS). Clinton: 1 m. (AMNH 6490). *New Jersey*: Dennis Creek: apparent TYPE MATERIAL of *G. minax*, 2 m., 1 f. (Philadelphia Academy 3581). Lower Alloway Creek near Quinton: 11 m., 15 f. (NYZS). *Florida*: Jacksonville: 8 m., 9 f. (AMNH 3986); 11 m. (AMNH 3981). Key West: 2 m. (MCZ 5900). Pensacola: 1 m. (MCZ 5901). *Alabama*: Mobile B.: 1 m. (USNM 21681). *Mississippi*: west end of Back B. Bridge, east Biloxi: 1 m., 1 f. (USNM 90302). Ocean Springs on Biloxi B.: 4 m., 2 f. (USNM 101102). *Louisiana*: Pont Chartrain: Point Platte: 1 f. (USNM 98144). Cameron: 1 m. (USNM 30570). Chauvin: 2 m. (AMNH 9800); 10 m., 33 f. (AMNH 9814). Gulf of Mexico, south of Morgan City: 5 f. (AMNH 9831).

Uca (Minuca) brevifrons
(Stimpson, 1860). (P. 180.)

MEXICO. *Lower California (South District)*: west coast (near Cape St. Lucas): lagoon at Todos Santos: from TYPE LOT, 1 f. (MCZ 1332). *Guerrero*: Acapulco: 3 m. (MCZ 5910); 1 claw (MCZ 5892 part). Puerto Angeles: 1 claw (NYZS 381,119).

COSTA RICA. Port Porker: 5 m., including yg. + claws, 4 f. (NYZS 381122). Negritos I.: 3 m. + growth series, including 23 yg. m. and f. (NYZS

381123). Cocos I.: 1 f. (USNM 63152). Uvita B.: 3 m., 2 f. (NYZS 381124). Golfito: 2 m., 1 f. (Torino); 1 + 2 yg. f. (NYZS 381125). Quebrada Chavarria: 3 m., 4 f. (USNM 19435). Boca del Rio Jésus Maria: 1 m. (USNM 32323); 1 m., 1 f. (USNM 32324). Pigres: 9 m., 1 f. (USNM 43352). Parida I.: 2 m., 2 f. (NYZS 381126).

PANAMA. "Panama": 6 m., 2 (1 ovig.) f. (MCZ 5909); 1 m. (USNM 22185). Yarisa: 2 m. (AMNH 6641). Banks of Rio Piñas: Piñas B.: 8 m., 1 f. (AMNH 10702). Darien: Rio Lara: holotype of *U. brevifrons* var. *delicata*, m. (Torino). Marraganti: 1 m. (USNM 48277). Rio Calabre: 4 m. (USNM 43988).

CANAL ZONE. La Capitana: 1 m. (USNM 43848); 3 m. (USNM 44320); 2 large claws (USNM 49097).

Uca (Minuca) galapagensis Rathbun, 1902. (P. 183.)

U. (M.) galapagensis herradurensis Bott, 1954

MEXICO. *Sonora*: Guaymas: ? 1 m. (MCZ 5913). *Jalisco*: Tenacatita B.: ? 1; very yg. in poor condition, discarded (NYZS 381,127; published as *U. macrodactyla* in *Zoologica*: 26 (19): 178; 1941).

EL SALVADOR. La Herradura: HOLOTYPE of *U. herradurensis*, m. (Frankfurt 1865). Los Blancos: holotype of *U. macrodactyla glabromana*, yg. m. (Frankfurt 1842).

NICARAGUA. Corinto: 2 m., 1 f. (MCZ 5912); 12 m., 1 f. (NYZS 381,128; published as *U. macrodactyla*, *loc. cit.*).

COSTA RICA. Boca del Rio Jésus Maria: 1 m., 1 f. (USNM 32320).

PANAMA. Near Panama City: 1 m. (AMNH 5979); 2 yg. m. (AMNH 5984). Between Patillo Point and Old Panama: 1 m. (AMNH 5992). Rio Abajo (swamp near mouth; near Old Panama): 89 m., 17 (1 ovig.) f. (NYZS).

U. (M.) galapagensis galapagensis Rathbun, 1902

COLOMBIA. Tumaco: 1 yg. m., subspecies questionable (Torino); Tumaco (Puntilla Sta. Elena): 1 m. (Torino).

ECUADOR. *Galapagos Is.*: Indefatigable I.: HOLOTYPE of *U. galapagensis*, m. (USNM 22319) + 5 additional m. under same number; 1 m. (NYZS 2367). James I.: 3 m., 1 f. (USNM 57743); 1 m. (NYZS 2624 part). Eden I.: 1 m. (NYZS 2042; see p. 188). Guayaquil: 16 m., 10 (1 ovig.) f. (NYZS). Puerto Bolivar (mouth of Guayas R.): 17 m., 18 f. (NYZS).

PERU. Puerto Grande on Zarumilla R. (near Capon salt flats): 1 m. (MCZ 12179).

[CHILE. Valparaiso: ? type material of *G. macrodactylus* (Paris; see p. 186).]

Not Examined. A few additional specimens at USNM from within the ranges of both subspecies and catalogued under the names of *U. galapagensis* and *U. macrodactylus.*

Uca (Minuca) rapax (Smith, 1870). (P. 190.)

U. (M.) rapax rapax (Smith, 1870)

UNITED STATES. *Florida*: Ft. Lauderdale: Dania Beach (mangroves to south): 3 m. (NYZS); Miami: Key Biscayne (St. V mudbank drainage): 2 m. (NYZS; gift of U. of Miami, Oceanographic Lab. Collector: H. K. Voris); causeway to Tahiti Beach, south Miami: 13 m., 8 f. (NYZS); 18 m., 8 f. (NYZS). Everglades: 2 m. (AMNH 2523).

MEXICO. *Tamaulipas*: Laguna de la Madre Austral: 3 miles south of south point of rocks: 1 m., 2 (1 ovig.) f. (USNM 96475).

GUATEMALA. Puerto Barrios: 22 m., 4 f. (NYZS).

BRITISH HONDURAS. *Belize*: 2 m., 2 f. (USNM 22604 part); 1 m., 1 f. (USNM 50950). Loyola Park, Yarborough: 23 m., 6 f. (NYZS; gift of C. Fairweather & J. Lopez).

CANAL ZONE. Fort Randolph, Galeta Pt.: 6 m. (USNM 89572).

WEST INDIES. *Bahamas*: Andros I.: 30 m. (AMNH 2449). *Haiti*: Bizoton: 8 m. (NYZS 27273; collected by W. Beebe). *Santo Domingo*: Sanchez: 5 m., 1 f. (MCZ 9860 part); 23 m., 14 (1 ovig.) f. (AMNH 2441 part). Monte Cristi: 6 m. (MCZ 9880 part). *Puerto Rico*: Enseñada: 3 m., 4 f. (AMNH 3011 part); 1 m. (AMNH 2984 part). Ponce: 4 m. (AMNH 3012); 22 m., 20 f. (NYZS). San Juan: 5 m., 1 f. (AMNH 3013); 69 m., 14 (8 ovig.) f. (NYZS). *Virgin Is.*: St. Thomas: 100 m., 17 f. (NYZS); Red Hook: 1 m. (NYZS; gift of Dr. Hummelinck). *Tobago*, near Pigeon Point: 3 m., 3 (1 ovig.) f. (NYZS); 4 m. (NYZS); 43 m., 5 (1 ovig.) f. (NYZS). *Trinidad*: Blanchisseuse (north coast): 3 m. (NYZS). Chacachachari I.: 1 m. (AMNH 2452). Port of Spain and vicinity: 2 m. (AMNH 2530); 7 yg. f. (NYZS); 4 f. (NYZS); Diego Martin R.: 127 m. (NYZS); Cocorite Swamp: 11 yg. m. (NYZS); 88 m. (NYZS): Laventille and Cocorite Swamps: 46 m. (NYZS). Monkey Point, near San Fernando: 4 m. (NYZS). *Curaçao*: Carmabi: 3 m., 1 f. (NYZS; gift of A.C.J. Burgers). *Aruba*: Spaans Lagoen, Balashi: 3 m., 1 f. (NYZS: gift of P. W. Hummelinck). *Jamaica*: 1 m., 2 f. (USNM 18553); 1 m. (USNM 22307).

COLOMBIA. Cartagena: 14 m., 9 f. (NYZS).

VENEZUELA. *Zulia*: Maracaibo: 425 m., 73 f. (NYZS 42421, 42422, 42426, 42427, parts; 5 collecting stations from Yacht Club and vicinity extending 4 miles north. Note: Maracaibo collection made during World War II and necessarily preserved in rum in the field. Many specimens in poor condition but saved because of potential future use in studies of growth and variation). Lagunillas: 1 m. (NYZS). *Yaracuy*: mouth of Yaracuy R.: 1 m. (NYZS). *Carabobo*: Puerto Cabello: 54 m., 26 f. (NYZS). *Aragua*: Turiamo: 31 m., 11 (1 ovig.) f. (NYZS). Islas Los Roques: 4 m., 3 f. (NYZS; gift of W. H. Phelps, Sr.); 33 m., 7 f. (NYZS: gift of W. H. Phelps & A. L. Haight). *Miranda*: Laguna de Tacarigua: 11 m. (NYZS: gift of W. H. Phelps, Sr.). *Anzoa-Tegui*: Puerto la Cruz: 2 m. (NYZS). *Sucre & Monagas*: near mouth of San Juan R.: 8 m., 9 (2 ovig.) f. (NYZS 42168). *Delta Amacuro*: Pedernales: 73 m., 32 (1 ovig.) f. (NYZS 42420).

GUYANA. Georgetown (foot of Kitty Village): 24 m., 6 f. (NYZS); 194 m. (including 66 with claws attached), 16 f. (NYZS).

SURINAME. Near Paramaribo (near river mouth): 13 m., 2 f. (NYZS).

BRAZIL. *Maranhão*: São Luiz (swamp paralleling Avenida Getulio Vargas, behind mental hospital): 36 m., 2 f. (NYZS). *Céara*: Fortaleza: (semi-closed swamp beside Avenida Getulio Vargas): 4 m. (NYZS). *Pernambuco*: Recife [Pernambuco] (swamp before beginning of Olinda Causeway): 11 m., 5 (1 ovig.) f. (NYZS). *Bahia*: São Salvador [Bahia]: Bahia de Todos Santos: Ilya Itaparica: 4 m., 1 f. (NYZS). *Rio de Janeiro*: Ilya Pinheiro: 83 m., including 59 with claws attached), 29 (1 ovig.) f. (NYZS); Serra de Masaché: 1 m. (USNM 50690).

Not Examined. Numerous additional specimens at USNM from many localities within the range of the subspecies; they are catalogued under the names *U. pugnax rapax* and *U. rapax*; also a few small specimens at AMNH, similarly catalogued.

U. (M.) rapax longisignalis
Salmon & Atsaides, 1968

UNITED STATES. *Florida*: Yankeetown: 14 m., 3 f. (NYZS; gift of M. Salmon). *Mississippi*: coast of Mississippi: 1 m. (USNM 74902). Ocean Springs:

HOLOTYPE of *U. longisignalis*, m. (USNM 121599), PARATYPES, 5 m. (USNM 122204); 2 m. (NYZS; gift of M. Salmon); 4 m., 2 f. (USNM 215461); Biloxi B.: 1 m. (USNM 21845). *Louisiana*: near New Orleans: Grand Isle: 3 m. (USNM 2259). *Texas*: Galveston: Offat's Bayou: 1 m. (USNM 82110 part). Matagorda B.: 1 m. (MCZ 12178); 13 m., 1 f. (USNM 33035).

Uca (Minuca) pugnax
(Smith, 1870). (P. 200.)

U. (M.) pugnax pugnax (Smith, 1870)

UNITED STATES. *Massachusetts*: Amamesse I.: Hadley Harbor: 8 m., 4 f. (AMNH 1031). Cape Cod: near Provincetown: SW of Telegraph Hill: 3 m. (NYZS, July, 1972); mouth of Little Pamet R.: 1 m. (NYZS, July, 1972). Harwich: 12 m., 1 f. (NYZS; gift of G. Cuyler); Woods Hole: 6 m., 2 f. (AMNH 97); North Falmouth: 36 m. (AMNH 1030); 1 m. (AMNH 1027 part). Martha's Vineyard: Katona B.: 5 m., 1 f. (AMNH 2435). Wareham: 7 m., 8 f. (AMNH 1026). *Connecticut*: New Haven: TYPE MATERIAL of *G. pugnax*, 1 m. (Yale 1060). *New York*: Oyster B.: 36 m., 15 f. (NYZS). Fire Island: 2 m. (AMNH 9358). Sheepshead B.: 1 m. (AMNH 2454). Staten Island: Princes B.: 29 m., 9 (mostly ovig.) f. (AMNH 2447). *New Jersey*: Bergen Beach: 6 m. (AMNH 2438); 3 m. including 1 yg. (AMNH 2440); 1 m. (AMNH 2610); 11 f. (AMNH 6119). West Keansburg: 1 f. (AMNH 2611). Near Brigantine: 11 m. including 3 yg., 10 f. including 1 yg. (NYZS). *Florida*: St. Augustine: 29 m., 8 f. (NYZS).

Not Examined. Numerous additional specimens at USNM from many localities throughout the range of the subspecies.

U. (M.) pugnax virens
Salmon & Atsaides, 1968

UNITED STATES. *Louisiana*: Cameron: 1 m. (USNM 30570). *Texas*: Matagorda B.: 1 m. (USNM 33031). Port Aransas: HOLOTYPE of *U. virens*, m. (USNM 121598); PARATYPES, 5 m. (USNM 122205); 3 m. (NYZS; gift of M. Salmon); 2 m.,

1 f. (NYZS; gift of M. Salmon). Texas coast: 4 m., 1 ovig. f. (USNM 72132); "Gulf coast, Texas": 1 m., 1 f. (USNM 72131). Corpus Christi: 7 m. (NYZS).

MEXICO. *Tamaulipas*: Tampico: 1 m. (USNM 18689); 4 m. + ? 1 juv. m. (USNM 43353 part). *Vera Cruz*: Arroyo de la Renal, tributary of Tonchochapa R. (Tonala drainage): 10 m., 6 f. (NYZS; gift of G.A.W.). Arroyo Amate of Uxpanapa R. of Coatzacoalcos (drainage ca. 8 km east of Minatitlan): 2 m. (NYZS; gift of G.A.W.); 8 km from Coatzacoalcos on new road to Minatitlan: 2 f. (NYZS; gift of G.A.W.).

Uca (Minuca) zacae
Crane, 1941. (P. 206.)

NICARAGUA. Corinto: 2 m., 2 f. (NYZS 381,114 = USNM 79408). San Juan del Sur: 2 m. (NYZS 381,115).

COSTA RICA. Golfito: HOLOTYPE of *U. zacae*, m. (USNM 137426, formerly NYZS 381,112); PARATYPES, 16 m., 11 f. (USNM 137427, formerly NYZS 381,113); 1 m. (NYZS). Cocos I: ? 2 m., ? 1 f. (U. of Southern California, Allan Hancock Foundation; see p. 207).

Uca (Minuca) subcylindrica
(Stimpson, 1859). (P. 209.)

UNITED STATES. *Texas*: Corpus Christi: 1 m. (USNM 23655). 3 miles south of Armstrong: Kenedy Co.: 4 m. (USNM 99826). Cameron Co.: near Santa Rosa: 1 m. (USNM 17807).

MEXICO. Matamoros (on the Rio Grande): COTYPES of *G. subcylindricus*, 1 m., 3 f. (MCZ 1327). "Mexico": 2 m. (Paris. Perhaps also COTYPES; see p. 210).

SUBGENUS *CELUCA*

Uca (Celuca) argillicola
Crane, 1941. (P. 220.)

COSTA RICA. Golfito: HOLOTYPE of *U. argillicola*, m. (NYZS 381,134 = USNM 137400); PARATYPES, 2 f. (NYZS 381,135 = USNM 137401).

PANAMA. *Pearl Is.*: San José (banks lower Marina R.): 2 m. (AMNH 9873).

COLOMBIA. Buenaventura: several specimens identified with certainty but not preserved (NYZS).

Uca (Celuca) pugilator
(Bosc, 1801-1802). (P. 223.)

UNITED STATES. *Massachusetts*: Cape Cod: Truro or Corn Hill: 1 m. (USNM 43355); mouth of Little Pamet R.; 1 m. (NYZS, July, 1972); North Falmouth: 85 m., 12 f. (AMNH 1027 part). *New York*: Long Island: Easthampton: 7 m., 2 f. (AMNH 2405); near Oyster B.: 15 m., 8 f. (NYZS). New York Harbor: 2 m. (AMNH 55). *New Jersey*: Bergen Beach: Growth series of yg. (AMNH 2613). West Keansburg: 1 yg. (AMNH 2560). *Florida*: 5 m. (AMNH 2387). Ft. Lauderdale (Dania Beach, mangroves): 2 m., 2 f. (NYZS). Miami: 5 m., 2 f. (AMNH 3014). Key West: 2 m. (USNM 18552). Punta Gorda: 1 m. (AMNH

2600); 3 m., 1 f., 1 yg. (AMNH 2631); growth series of yg. (AMNH 2886). Seven Oaks: 2 m. (AMNH 2389). *Texas*: Galveston: Offat's Bayou: 1 m. (ex-USNM 82110). Corpus Christi: 3 m., 1 f. (NYZS).

WEST INDIES. *Bahamas*: Andros I.: 1 m., 1 f. (AMNH 2406).

? SANTO DOMINGO. 2 yg. males. (AMNH 2555; *locality questionable*; see p. 226).

Not Examined. The USNM has a large collection of additional specimens from many localities on the eastern and southern coasts of the United States. All of these records fall within the boundaries of the examples listed above. Most of the catalogue numbers apply to single specimens or to short series numbering fewer than 10 of both sexes. Exceptionally large series include the following: *Massachusetts*: Woods Hole: 85+ specimens (USNM 3212). *Virginia*: Smiths Island: 34 m., 41 f. (USNM 74452, 74453). *North Carolina*: Beaufort: 25+ yg. specimens (USNM 54471); Gallant Point: Beaufort Harbor: 30+ specimens (USNM 63329). *South Carolina*: "South Carolina": 65 m., 62 f. (USNM 2061); Charleston: 30 m., 9 f. (USNM 3148). *Florida*: Pine Key: 79 m., 7 f. (USNM 6440); Big Gasparilla Pass: 26 m., 23 f. (USNM 15253).

Uca (*Celuca*) *uruguayensis*
Nobili, 1901. (P. 229.)

BRAZIL. *Rio de Janeiro*: 2 m., 1 f. (USNM 40624);
Guanabara B.: Ilya Pinheiro: 11 m., 2 f. (NYZS);
Ilya Pinheiro (south side by wharf): 14 m., 7 f.
(NYZS); Paqueta: 21 m. + 2 claws (USNM 71181);
4 m. (USNM 71182 part); 1 m. (USNM 71183
part). *São Paulo*: Santos: E. Piassaguera: 1 m.
(USNM 47870); Santos Estuary, between canals 4
and 5: 15 m. (USNM 71187 part). *St. Catarina*:
São Francisco do Sol: 1 yg. m. (USNM 71180 part).
Florianopolis: 11 m., 7 f. (USNM 71188 part); end
of Praia da Fora: 1 m., 1 f., 1 yg. (USNM 73445).

URUGUAY. Maldonado (near Montevideo): 19 m., 5
f. (MCZ 5926). La Sierra: HOLOTYPE of *U. uruguayensis*, m. (Torino). *Canelones*: Santa Lucia R.:
2 m. (USNM 72320).

ARGENTINA. Buenos Aires: Lavalle (Ajo): 6 m.
(USNM 54716).

Uca (*Celuca*) *crenulata*
(Lockington, 1877). (P. 232.)

U. (*C.*) *crenulata coloradensis*
(Rathbun, 1893)

MEXICO. *Gulf of California*: *Sonora*: near mouth of
Colorado R.: Horseshoe Bend: HOLOTYPE of *G.
coloradensis*, m. (USNM 17459). Opposite mouth
of "Hardy's Colorado R.": 31 m., 12 f. (USNM
18292); 3 m., 1 f. (MCZ 4263). Guaymas: 7 m.,
4 f. (MCZ 5916).

U. (*C.*) *crenulata crenulata*
(Lockington, 1877)

UNITED STATES. *California*: Newport Beach: 2 m.
(NYZS). San Diego (Mission B.): 11 m., 1 f.
(USNM 55220).

MEXICO. *Lower California* (*South District*): west
coast: Magdalena B. (Mangrove I.): 2 m. (AMNH
5477). Todos Santos B.: 35 m. + 1 extra claw, 9 f.
(USNM 19033). *Gulf of California* (*South District*): La Paz: type material of *G. gracilis*, 34 m.,
6 f. (USNM 4622). Pichilinque B.: 22 m., 7 f.
(USNM 22080). Head of Concepción B.: 7 m.
(AMNH 5516). (*North District*): San Luis Gon-
zaga B.: 4 m. (USNM 17458). San Felipe: 11 m.,
2 f. (USNM 67725). *Sonora*: Guaymas: 13 m., 3 f.
(MCZ 1594). *South of Gulf of California*: *Jalisco*:
Tenacatita B.: 2 m., 1 f. (NYZS 381,151).

Uca (*Celuca*) *speciosa*
(Ives, 1891). (P. 236.)

U. (*C.*) *speciosa speciosa* (Ives, 1891)

UNITED STATES. *Florida*: Miami: Key Biscayne: 6
m., 6 f. (NYZS); ditch by road to Tahiti Beach: 5
m., 4 (1 ovig.) f. (NYZS). Card Sound: 4 m., 2 f.
(USNM 15256). Coon Key: 1 m. (USNM 73418).
Key West: 3 m. (USNM 71290). Tortugas: 20 m.,
7 f. (USNM 65942). Wakulla Beach: 1 m. (USNM
90606). Duck Rock (southwest coast): 1 yg. m.
(AMNH 9622); 1 yg. f. (AMNH 9623). Sanibel I.:
Heller's Cove or Clam Bayou: 2 m. (MCZ 10182).
Sarasota B.: 1 m., 2 f. (USNM 42617); 6 m., 2 f.
(USNM 73417). Manatee Co.: Sugarhouse Creek
at Travertine Quarry: 4 m., 1 f. (USNM 71289).

WEST INDIES. *Cuba*: *Matanzas*: Hicacos Peninsula:
Laguna Chaco, Xanadu: 8 m., 4 f. (USNM 99966).
Laguna de Paso Malo Varadero: 3 m. (USNM
96456).

U. (*C.*) *speciosa spinicarpa* Rathbun, 1900

UNITED STATES. *Alabama*: near Mobile: 3 m., 1 f.
(USNM 22312). Mobile: E. shore Cedar Pt.: 6 m.,
1 f. (NYZS; gift of D. A. Archer, U. of Alabama,
1948). *Mississippi*: Biloxi B.: Ocean Springs: 3 m.,
1 f. (USNM 10103). Biloxi: B. shore: 1 m. (USNM
21684). Biloxi: Ocean Springs Bridge: 3 m. (USNM
73419). Biloxi: 1 f. (USNM 54335). E. Biloxi below Back B. Bridge: 1 m., 1 f. (USNM 90305).
Texas: Galveston: TYPES of *U. spinicarpa*, 2 m., 1 f.
(USNM 22183); Offatt's Bayou: 1 m. (USNM
82110 part). Drain near Matagorda B. 1 m. (USNM
25034).

MEXICO. "Maron, Lagoon Madre, Mex. from oysters,
saltwater—E. Palmer by purchase": 1 yg. f. (USNM
43364). Tampico: 1 m. (USNM 22311); 2 m.
(USNM 43353 part).

Uca (*Celuca*) *cumulanta*
Crane, 1943. (P. 240.)

WEST INDIES. *Jamaica*: Kingston Harbor: 1 m.
(USNM 22313). *Curaçao*: 7 m., 1 f. (USNM
22310). Caracas B.: 1 m. (USNM 56909). *Trini-
dad*: near Port of Spain: Diego Martin: 53 m.
(NYZS); Cocorite: 4 m. (NYZS); Lavantille: 5 m.
(NYZS).

VENEZUELA. *Aragua*: Turiamo: 34 m., 12 f. (NYZS).
Delta Amacuro: Pedernales: HOLOTYPE of *U. cumu-*

lanta, m. (NYZS 42423 = USNM 137402); PARA-TYPES, 7 m., 6 f. (NYZS 42423a = USNM 137403); 30 m., 20 f. (NYZS 42425 part; others listed by Crane 1943, now missing).

GUYANA. Georgetown: 13 m., 8 f. (NYZS).

BRAZIL. "Mamonguape stone reef": 1 m. (USNM 25700). *Rio de Janeiro*: Ilya Governador: 1 m. (USNM 73446). Paqueta: 1 m. (USNM 71182 part); 2 m., 1 f. (USNM 71171).

Uca (Celuca) batuenta
Crane, 1941. (P. 244.)

COSTA RICA. Puntarenas: PARATYPES of *U. batuenta*, 2 m., 1 ovig. f. (NYZS 381,136 = USNM 79399); Ballenas B.: Additional PARATYPES, 4 m., 2 ovig. f., 1 extra claw (NYZS 381,137 = USNM 137404).

CANAL ZONE. Balboa (La Boca): HOLOTYPE of *U. batuenta*, m. (NYZS 4121 = USNM 137405); additional PARATYPES, 3 m. (NYZS 4122 = USNM 137406).

PANAMA. Near Old Panama: Abajo R. (near mouth by water pump): 1 m. (NYZS).

ECUADOR. Puerto Bolivar: 3 m. (NYZS).

Uca (Celuca) saltitanta
Crane, 1941. (P. 247.)

COSTA RICA. Puntarenas: 3 m., 1 f. (NYZS 381,138).

CANAL ZONE. Balboa (La Boca): HOLOTYPE of *U. saltitanta*, m. (NYZS 4123 = USNM 137407); PARATYPES, 7 m., 7 f. (NYZS 4124 = USNM 137408); TOPOTYPES, 4 m., 4 f. (NYZS 4125 part = USNM 79403); 80 m., 50 (most ovig.) f. (NYZS 4125 part; others listed by Crane, 1941 now missing).

COLOMBIA. Buenaventura: several specimens identified with certainty but not preserved (Crane).

Uca (Celuca) oerstedi
Rathbun, 1904. (P. 251.)

COSTA RICA. Punta Arenas: PARATYPE of *U. oerstedi*, 1 m. (USNM 31506).

CANAL ZONE. Balboa (La Boca): 9 m., 5 f. (NYZS 4119).

PANAMA. Old Panama: 6 m., 1 f. (NYZS 4120); west of Police Station: 44 m., 4 (3 ovig.) f. (NYZS).

Uca (Celuca) inaequalis
Rathbun, 1935. (P. 254.)

NICARAGUA. Corinto: 2 m. (NYZS 381,139).

COSTA RICA. Puntarenas: 6 m., 3 f. (NYZS 381,140). Ballenas B.: 3 m., 1 f. (NYZS 381,141). Golfito: 10 m., 3 f. (NYZS 381,142).

CANAL ZONE. Balboa (La Boca): 10 m., 11 f. (NYZS 4126).

ECUADOR. Puerto Bolivar: 6 m., 3 f. (NYZS). Guayaquil: El Salado: TYPE of *U. inaequalis*, m. + 6 m., 3 f. (USNM 70833; locality spelled on records "Salada"); 1 m. (USNM 70900, "Salada"); 22 m., 10 f. (NYZS).

Uca (Celuca) tenuipedis
Crane, 1941. (P. 258.)

COSTA RICA. Ballenas B.: HOLOTYPE of *U. tenuipedis*, m. (NYZS 381,143 = USNM 137409); PARATYPES, 8 m., 1 f. (NYZS 381,144, part = USNM 137410); PARATYPES, 1 f. (NYZS 381,144, part = USNM 79404).

Uca (Celuca) tomentosa
Crane, 1941. (P. 261.)

EL SALVADOR. Los Blancos: holotype of *U. mertensi*, m. (Frankfurt, 1863).

COSTA RICA. Puntarenas: HOLOTYPE of *U. tomentosa*, m. (NYZS 381,132 = USNM 137411); PARATYPES, 3 f. (NYZS 381,133, part = USNM 137412); PARATYPE, 1 f. (NYZS 381,133, part = USNM 79406). Golfito: 1 m. (NYZS).

PANAMA. Near Old Panama: Abajo R. (near mouth by water pump): 3 yg. m., 3 yg. f. (NYZS).

Uca (Celuca) tallanica
von Hagen, 1968. (P. 264.)

ECUADOR. Puerto Bolivar: 7 m. (NYZS).

PERU. Puerto Pizarro: HOLOTYPE of *U. tallanica*, m. (Leiden D 23046); PARATYPES, 5 m., 3 f. (Leiden D 23047).

Uca (Celuca) festae
Nobili, 1902. (P. 267.)

EL SALVADOR. Puerto El Triunfo: holotype m. and paratypes, 2 m., of *Uca orthomana* (Frankfurt 1873 and 2074); 2 of paratypes of *Uca leptochela* (Frankfurt 2073 part). La Union: 3 m. (Frankfurt 2102).

PANAMA. Contreras I. (tide pool in jungle): 1 m. (AMNH 11562). Near Old Panama: Abajo R. (near mouth by water pump): 3 yg. m., 1 f. (NYZS).

ECUADOR. Guayaquil: El Salado: type material of *U. guayaquilensis*, 2 m., 1 f. (USNM 70831); 2 m., 1 f. (USNM 70901); 1 m., 2 ovig. f. (NYZS); El Rio: 48 m., 42 (8 ovig.) f. (NYZS). Rio Daule Inferiore: HOLOTYPE of *U. festae*, m. (Torino 1438). *El Oro*: Gualtaco (SW Ecuador): 4 m. (USNM 97908); 3 f. (USNM 97926).

Uca (Celuca) helleri
Rathbun, 1902. (P. 271.)

GALAPAGOS IS. 1 yg. m., 1 yg. f. (USNM 63154). Albemarle I.: Black Bight: 1 m. (USNM 25666). Narborough I.: Mangrove Point: TYPE MATERIAL of *U. helleri*, 3 m., 1 f. (USNM 24829).

Uca (Celuca) leptochela
Bott, 1954, 1958. (See p. 274.)

Uca (Celuca) dorotheae
von Hagen, 1968. (P. 275.)

CANAL ZONE. Balboa (La Boca): 1 m., 1 f. (NYZS).

ECUADOR. Puerto Bolivar: 9 m., 3 f. (NYZS).

PERU. Puerto Pizarro: HOLOTYPE of *U. dorotheae*, m. (Leiden D 23054); PARATYPES, 25 m. & f. (Leiden D 23055).

Uca (Celuca) beebei
Crane, 1941. (P. 278.)

NICARAGUA. Corinto: 2 m. (NYZS 381,148).

COSTA RICA. Puntarenas: 3 m. (USNM 32322; 46263); 2 m., 2 f. (NYZS 381,149). Boca del Rio Jesús Maria: 1 m. (USNM 32321).

PANAMA. Bahia Honda: 1 m. (NYZS 381,150). Panama City (Bellavista): 16 m., 6 f. (NYZS 4133). Old Panama: 7 m., 3 f. (NYZS 4134).

CANAL ZONE. Balboa (La Boca): HOLOTYPE of *U. beebei*, m. (NYZS 4129 = USNM 137413); PARATYPES, 8 m., 8 f. (NYZS 4130 = USNM 137414); 30 m., 19 f. (NYZS 4131 part); 2 m., 2 f. (NYZS 4131 part = USNM 79400); 5 yg. (NYZS 4132).

ECUADOR. Puerto Bolivar: 7 m., 1 f. (NYZS).

Uca (Celuca) stenodactylus
(Milne-Edwards & Lucas, 1843). (P. 282.)

EL SALVADOR: *G. of Fonseca*: holotype of *G. gibbosus*, m. (MCZ 5911).

NICARAGUA. Corinto: 24 m., 7 f. + post-megalopal yg. (NYZS 381,145).

COSTA RICA. Port Parker: 1 m. (NYZS 381,146). Puntarenas: 3 m. (USNM 39098); 1 m. (USNM 46263). Golfito: 1 m. (NYZS 381,147). San Lucas: 2 m. (USNM 76133); 2 yg. m. (USNM 76140).

PANAMA. Agallero B.: Chitre: 1 m. (NYZS; gift of M. D. Burkenroad). Panama City: (Bellavista): 13 m. (NYZS 4128); 2 f. (NYZS).

CANAL ZONE. Balboa (La Boca): 1 m. (NYZS 4127).

ECUADOR. Puerto Bolivar: 3 m., 1 f. (NYZS).

CHILE. Valparaiso: TYPE-SPECIMEN of *G. stenodactylus*, m. (Paris).

Uca (Celuca) triangularis
(A. Milne-Edwards, 1873). (P. 286.)

U. (C.) triangularis triangularis
(A. Milne-Edwards, 1873)

MALAYSIA. *Labuan [North Borneo]*: 2 m. (NYZS; gift of Raffles).

INDONESIA. *Djawa [Java]*: Near Surabaja: Pasuruan: 10 m., 1 f. (NYZS); Trengglunga: 2 m. (NYZS). *Moluccas*: Amboina: 1 m. (Amsterdam); 1 m. (Leiden 1247); syntypes of *U. t. variabilis*, 3 m. (Leiden 1537). West Ceram: 1 m., 1 claw (Amsterdam).

NEW GUINEA. *West Irian*: Near Biak: Seroei-Japen: 5 m. (Leiden 724, part). *Territory of Papua and New Guinea*: near Madang on north-central coast: Maiwara: 6 m., 4 (1 ovig.) f. (NYZS).

? AUSTRALIA. "Great Barrier Reef Expedition": 5 yg. (BM 1950.12.1. 17-21; only record found that may be from Australia).

NEW CALEDONIA. "Nlle. Caledonie": TYPE of *G. triangularis*, 1 m. (Paris). Near Nouméa: 5 m., 5 f. (NYZS).

PHILIPPINES. *Sulu*: Tawi Tawi: 1 m. (NYZS). Joló (Princess Tarhata Kiram's Beach): 3 m. (NYZS). *Mindanao*: Zamboanga: 1 m. (USNM 43376); 1 m. (USNM 43454); north fishponds shorewards: 32 m. (NYZS). Gulf of Davao: Padada Beach: 2 m., 1 f. (AMNH 8305); 1 f. (AMNH 8306 part); 1 m. (AMNH 8312); 1 m. (AMNH 8314); 1 m. (AMNH 8315); 1 m. (AMNH 8316); 1 m., 1 f. (AMNH 8317); 1 m., 1 f. (AMNH 8336); Padada R. mouth: 1 m. (AMNH 8341); in and near Padada R. mouth: 2 m., 3 f. (AMNH 8383); 1 m., 1 f. (AMNH 8384); 2 m., 2 f. (AMNH 8385); Davao Beach: 9 m. (AMNH 8552). Mangrove swamp and adjacent muddy beach: 1 m. (AMNH 8553). Iling R., north of Sasa: 2 m. (NYZS). Madaum: 9 m., 1 f. (NYZS). *Negros Occ.*: Magnanod R., Victorias: 4 m. (USNM 73270). *Panay*: Jaro R.: 2 m. (USNM 73182). *Palawan*: Baheli R.: 4 m. (NYZS). *Luzon*: Mariveles: 1 m. (USNM 46652).

PALAU IS. 1 m. (Yokohama).

NANSEI [RYUKYU or LOO CHOO] IS. *Iriomote I.*: 1 m. (NYZS; gift of T. Sakai, U. of Yokohama).

U. (C.) *triangularis bengali* subsp. nov.

INDIA. Madras: 10 m. & f. + series of yg. (BM 92.7.15. 209-18). Karikal Marais: 1 m. (Paris 2-1946, part). Pondichéry: 1 m. (Torino 961).

BURMA. Mergui Archipelago: 2 m. (Amsterdam); 1 m. (Leiden 6).

MALAYSIA. *Malaya*: *Penang*: HOLOTYPE of *U. t. bengali*, m. (NYZS = USNM 137674); 49 m., 13 (1 ovig.) f. (NYZS); 1 m. (NYZS; gift of D. Tam); 2 m. (Leiden 1250); 2 m. (USNM 39172). Selangor: Port Swettenham: 1 m., 1 f. (NYZS; gift of M.W.F. Tweedie, Raffles).

Uca (Celuca) lactea
(de Haan, 1835). (P. 292.)

U. (C.) *lactea annulipes*
(Milne-Edwards, 1837)

ETHIOPIA. *Eritrea*: Massawa: 49 m., 18 f. (NYZS); 1 m. (AMNH 4323 part). Green I. (near Massawa): 4 m. (NYZS). Perim I.: 1 m. (Paris).

SOMALILAND. Obock: 2 m. (Paris).

KENYA. Near Mombasa: Gazia: Small series, including 3 m. checked (Paris).

"BRITISH EAST AFRICA": strait at Changamwe: 7 m., 4 f. (USNM 43379).

TANZANIA. *Pemba*: Near Chake Chake: Kiburunzi: 5 m. (NYZS). *Zanzibar*: 1 m. (USNM 19558); 5 m., 3 f. (USNM 22197). Chwaka: 2 m. (NYZS). Muongoni: 20 m. (NYZS). Chakwani Fish Ponds: 146 m., 22 f. (NYZS). Between Mbeni and Chakwani Fish Ponds: 3 m. (NYZS).

MOZAMBIQUE. *Inhaca I.*: east lagoon: 393 m., 90 (1 ovig.) f. (NYZS).

MADAGASCAR. 1 m. (Paris); northwest Madagascar: 1 m. (USNM 19554).

SOUTH AFRICA. *Natal*: Durban: 1 m. (AMNH 1153).

INDIAN OCEAN. "Mer des Indes": LECTOTYPE of *G. annulipes*, m., + 2 m., totaling 3 m. listed as "types non specifiés" by museum (Paris). *Mauritius*: 2 m. (BM); 1 m. (Paris). *Seychelles*: Mahé I.: 7 m. (BM 196).

PERSIAN GULF. 1 m. (Paris).

PAKISTAN. Kurrachee [? = Karachi]: 4 m. (USNM 13877). Karachi: 86 m., 19 (1 ovig.) f. (NYZS). Between Karachi and Clifton: 6 m., 1 f. (NYZS).

INDIA. Bombay: 2 m. (NYZS: identified in field; not saved). Mahé: 3 specimens (Torino). Ernakulam: 5 m., 2 f. (Marine Fisheries Lab., Ernakulam). Pondichéry: small series (Paris). Tuticorin: 5 m. (BM 90.10.20. 11-14). Karikal Marais: 1 m. (Paris). Madras: 2 m. (USNM 19710). Near Madras: Pamban: 32 m. examined in long series (BM Harding). "Calcutta": 1 m. (MCZ 5799).

CEYLON. 1 m. (BM 1.16.158). Negombo: 54 m. (NYZS).

BURMA. Rangoon: 3 m. (MCZ 5800). Mergui: 1 m., 1 f. (Amsterdam).

MALAYSIA. *Malaya*: *Penang*: 3 m. (NYZS); 1 m. (NYZS, gift of D. Tam, Bureau Fisheries). *Malacca*: Malacca: 3 m. (Leiden 1742): 3 m. (Amsterdam); 2 m. (USNM 39175 part). Port Dickson: 1 m. (Leiden 5240, part). Malacca Straits: 2 m. (Leiden 4891). *Labuan* [*North Borneo*]: 1 m. (NYZS; gift of Raffles through M.W.F. Tweedie).

SINGAPORE. 4 m. (NYZS; gift of Raffles through M.W.F. Tweedie). Gaylang R. by Kallong Airport: 49 m., 41 f. (NYZS).

"BORNEO." lectotype of *G. porcellanus*, m. and paratype m. (BM 901.444.106).

INDONESIA. *South China Sea*: Anambas Is.: Pulo Siantan: 1 m. (USNM 23365); Pulo Lankawi: 4 m. (USNM 23875). *Kalimantan* [*Borneo*]: Pontianak:

1 m. (Amsterdam). *Djawa* [*Java*]: Djakarta [Batavia] at Tanjong Priok: 5 m. (Leiden 2010, part); 7 m. (Leiden 2012); 7 m., not designated types, but labeled "*Gelasimus consobrinus* de Man," as stated on p. 298 (Amsterdam). North coast: 3 m. (Leiden 2031). Madura [Madoera]: 4 m. examined (Leiden 2035, part). Besuki [Besoeki]: 1 m. (Leiden 238, part). Bantam Pruput: 1 m. (USNM 43378). *Celebes*: Makasar [Makassar]: 2 m. (Leiden 2608); 1 m. (Leiden 5239); 29 m., 3 f. (MCZ 7247). Parepare: 10 m. (Amsterdam, part).

PHILIPPINE IS. *Sulu*: Tawi Tawi: 28 m. (NYZS). Tapiantana I.: 1 m. (USNM 43449). *Mindanao*: Sasa: Gulf of Davao: 6 m., 1 f. (NYZS). Davao: Malalag: 5 m. (NYZS). *Panay*: Iloilo: 8 m. (USNM 73198 part). *Samar*: Catbalogan: 3 m. (NYZS; gift of Sohrab Boya, Araneta Institute of Agriculture). *Palawan*: Ulugan B.: 1 m.; gonopod has some *l. perplexa* similarities (USNM 43450).

U. (C.) lactea mjobergi Rathbun, 1924

AUSTRALIA. *Northern Territory*: Darwin ("Buffalo Creek" and Dinah B.): 16 m., 28 f. (NYZS); Golf Club Creek: 23 m., 10 f. (NYZS). *Western Australia*: Broome: PARATYPE of *U. mjobergi*, m. (USNM 56418); 245 m., 5 f. (NYZS). Near Broome: Cape Boileau: 1 m., 1 f. (BM 1932.11.30. 166-168).

NEW GUINEA. *West Irian*: Bosnek: Mokmer, near Biak: 1 m. (Leiden); west of Bosnek: 3 m. (Leiden).

U. (C.) lactea lactea (de Haan, 1835)

HONG KONG. Kowloon (Castle Peak Area): 39 m., 10 (1 ovig.) f. (NYZS).

CHINA. "Chine": 2 m. (Paris). *Fukien*: Amoy: 2 m., 1 f. (BM 1935.3.19.7); 1 m. (Leiden 252); 3 m., 3 ovig. f. (USNM 61836). Tsimei: 4 m. (USNM 57827). Chin Bey: 2 m. (USNM 57045). Jau Ab: 1 m. (USNM 59166). Liuwutien: 3 m., 1 f. (USNM 61839).

TAIWAN [FORMOSA]. Tamsui: 45 m., 2 (1 ovig.) f. (NYZS). Shinchiku: 1 m. (USNM 55384).

JAPAN. LECTOTYPE of *G. lacteus*, m. (Leiden 254 part), + additional TYPE MATERIAL (Leiden 254 part); additional specimens (Leiden 1575). *Kyushu*: Kagoshima: 1 m. (USNM 48475). Miyazaki Prefecture: 1 m. (NYZS; gift of Kyushu U.). Fukuoka: Tatara R.: 6 m. (NYZS; gift of T. Sakai, U. of Yokohama). *Honshu*: Hiroshima Prefecture: Onomichi: 4 m. (USNM 43377).

U. (C.) lactea perplexa (Milne-Edwards, 1837)

INDIA. Pondichéry: small series (Paris).

MALAYSIA. *Malaya*: Negri Sembilan (near Sungei Dua ferry): 1 m., 2 f. (NYZS). *Malacca*: Port Dickson: 1 yg. m. (Leiden 5240); 2 m. (Leiden 5270 part); Malacca: 2 m. (USNM 39175 part). *Sarawak*: Santubong: 1 m. (BM 1900.12.1.22). *Labuan* [*North Borneo*]: 3 m. (Torino).

[See also under INDONESIA for other localities on Borneo.]

SINGAPORE. 1 m. (NYZS; gift of Raffles through M.W.F. Tweedie). Gaylang R. by Kallong Airport: 21 m., 19 f. (NYZS).

THAILAND. *G. of Siam*: Laem Ngop [Lem Ngob]: 2 m. (USNM 39713). *Chon Buri*: Bong Saen: 4 m. (USNM account no. 214063-2 parts). Bangpoo (near Bangkok): 2 m. (USNM 94420; lost by Crane after examination).

INDONESIA. *Kalimantan* [*Borneo*]: Pontianak: 2 m. (Amsterdam part). Samarinda: type of *U. annulipes* var. *orientalis*, 1 m. (Torino 1521). *Sumatera* [*Sumatra*]: Simeuleu [Simaloer]: 4 m. (Leiden 2084). Benkulen: 10 m. (USNM 75867). *Djawa* [*Java*]: "Javae": 2 m., listed by museum as "TYPES NON SPECIFIÉS" of *G. perplexus* (Paris). West coast: 1 m. (Leiden 2030). Djakarta [Batavia], at Tandjong Priok: 1 m. (Leiden 2010 part). Bantam Pruput: 1 m. (USNM 43378). Madura [Madoera]: 23 m. (Leiden 2032). Near Surabaja: Trengglunga: 1 m. (Amsterdam); 4 m., 1 f. (NYZS); Pasuruan: 1 m. (Amsterdam); 50 m., 23 f. (NYZS); Besuki [Besoeki]: 2 m. (Leiden 238 part). *Sulawesi* [*Celebes*]: Makasar [Makassar]: 4 m., 1 f. (USNM 39475); 1 m. (MCZ 7247 part). Parepare: 5 m. (Amsterdam, part). *Molucca Is.*: Batjan (off Halmahera): 1 m. (BM). Tawang I. (near Pahaji B., western Halmahera): 19 m., 1 f. (MCZ 11184). *Tenimber* [*Timorlaut*] *Is.*: 1 m. (BM). *Aru* [*Aroe, Arrou*] *Is.*: 1 m. (Paris). *Seram* [*W. Ceram*]: Piru [Piroe]: 2 m. (Amsterdam); 3 m., 1 f. (Amsterdam). Ambon [Amboina]: 1 m. (Amsterdam).

PHILIPPINE IS. *Sulu*: Tawi Tawi: 10 m. (NYZS). Joló: (Princess Tarhata Kiram's Beach): 10 m. (NYZS). Tapiantana I.: 1 m. (USNM 43449). *Mindanao*: Zamboanga (northeast fishponds, seaward of dikes): 161 m., 31 (7 ovig.) f. (NYZS). Gulf of Davao: Sasa: 7 m. (NYZS). Iling R. (north of Sasa): 3 m. (NYZS). Madaum: 3 m. (NYZS). Malalag: 3 m. (NYZS). Mataling R., Malabang: 25 m., 3 yg. f., + 12 extra claws and several bodies in poor condition (USNM 43382). *Cebu Channel*: Waiming: 2 m. (USNM 26201). *Negros Occ.*:

Bangi Bangi, Magnanod R.: 8 m. (USNM 64979); 1 m. (USNM 64986); 3 f. (USNM 64989); 3 f. (USNM 64990); 2 f. (USNM 65008). *Panay*: shore about Iloilo R.: 1 m. (USNM 43452). Iloilo: 20 m., 4 f. (USNM 73198 part). Jaro R.: 11 m. (USNM 73191); 2 m. (USNM 73199). *Guimares*: 1 m. (USNM 26202); Jordan R.: 6 m. (USNM 43453). *Palawan*: Puerto Princesa: 6 m., 3 (1 ovig.) f. (NYZS). Ulugan B.: 1 m. (USNM 43450). *Luzon*: Novatas, near Manila (mouth of Dagadagatan R.): 1 m. (NYZS). Bataan: Lamao: 23 m., 2 f. (NYZS; gift of Miss Beatrice Haygood, Araneta Institute of Agriculture). Mariveles: 25 m. (USNM 46649). Casaguran: 4 m. (USNM 43381).

NANSEI [RYUKYU or LOO CHOO] IS. "Loo Choo": 2 m. (BM 87.5). Ishigaki I.: mouth of Miyara R.: 4 m., 1 f. (NYZS; gift of T. Sakai, U. of Yokohama). Kume-jima (Loo Choo Is.): 2 major chelipeds (USNM 73260). Okinawa: Baten: 1 m. (NYZS; gift of T. Sakai, U. of Yokohama).

JAPAN. Tokaito Coast: Oho Sima: 1 m. (USNM 22287).

NEW GUINEA. *West Irian*: Kotabaru [Hollandia], Japen I., and Biak: series, m. & f. (Leiden, collections by H. B. Holthuis, years 1952-1955). *Territory Papua & New Guinea*: near Madang on north-central coast: Maiwara: 21 m., 3 f. (NYZS). Port Moresby: Taurama Beach: 1 m. (U. of Papua & New Guinea).

AUSTRALIA. *New South Wales*: Trial B.: 1 m. (Bishop 2484); 2 m. (USNM 64607).

NEW CALEDONIA. 2 m. (USNM 20300). Near Nouméa: 13 m., 6 f. (NYZS).

PACIFIC OCEAN. *Caroline Is.*: 1 m. (BM 98.11.1.1-5). Lawi-Kondo: 5 m. (Bishop 4435). *Solomon Is.*: 2 m. (Paris). *New Hebrides*: Malle Kula: 2 m. (Paris). *Fiji Is.*: Viti Levu: Rewa (near mouth Wainibokasi R.): 9 m., 3 f. (NYZS). Suva: (Queen Elizabeth Drive and flat between town and seaplane base): 31 m., 10 f. (NYZS). Tavua: 19 m., 17 f. (NYZS). Kandavu Island: Tavuki: 12 m. (MCZ 5801). Bau I.: 1 f. (MCZ 5803). Lasema: 2 m. (MCZ 9080). Nukulau I.: 1 m. (MCZ 5802). *Friendly Is.* [*Tonga*]: Vavau: 5 m. (BM 1911.9.18. 27-29). *Samoan Is.*: 2 yg. m. (USNM 72522). Pago Pago: 1 m. (USNM 6587). Apia: 9 m., 1 ovig. f. (USNM 43380).

Uca (Celuca) leptodactyla
Rathbun, 1898. (P. 304.)

MEXICO. 3 m., the basis for Guérin's manuscript name, *Gelasimus leptodactylus* (Phil. Acad. 9-2965).

WEST INDIES. *Bahamas*: New Providence: Nassau (near Fort Montague): TYPE MATERIAL of *U. leptodactyla*, 1 m., 1 f. (USNM 22315). Bimini: north of Lerner Marine Lab.: 3 m., 3 f. (USNM 91738). Turks I.: 1 m. (AMNH 2296). *Puerto Rico*. San Juan: San Antonio Bridge: 13 m., 9 f. (USNM 24546). *Tobago*: near Pigeon Point: 11 m., 3 f. (NYZS). Lagoon off Buccoo Reef: 7 m. (NYZS).

VENEZUELA. *Aragua*: Turiamo: 14 m., 9 f. (NYZS).

BRAZIL. *Pernambuco*: Recife [Pernambuco]: 14 m., 3 f. (USNM 40617); swamp before beginning of Olinda Causeway: 5 m., 5 f. (NYZS); ? Pernambuco: 12 m., 7 f. (USNM 40618); *Bahia*: São Salvador [Bahia]: Bahia de Todos Santos: 3 m., 3 f. (USNM 48297); Ilya Itaparica: 29 m. (NYZS); 80 m., 24 f. (NYZS). Porto Securo: 3 m., 1 f. (USNM 22189). Plataforma: 10 m., 3 f. (USNM 40616). *Rio de Janeiro*: 21 m., 10 f. (MCZ 5923); Guanabara B.: Paqueta I.: 10 m. (USNM 71183 part); 4 m., 1 f. (USNM 74484); 2 claws, 7 f. ? 2 yg. (USNM 71182); Ilya Governador: 1 m. (USNM 71185); Niterói [Nictheroy]: Conto do Rio: 1 m., 1 f. (USNM 74438). São Francisco: 1 f. (USNM 73490). *Rio de Janeiro* and *São Paulo*: Parahyba R.: 1 m. (USNM 25701). *São Paulo*: Santos: between canals 4 and 5 of estuary: 2 m., 2 f. (USNM 71187). São Sebastião: 1 m. (USNM 47850). *Parana*: Paranagua: 5 chelipeds (USNM 71184). *St. Catarina*: São Francisco: 3 m., 9 f. (USNM 71186); 13 m., 7 f. (USNM 71180); 1 f. (USNM 73490); 8 m., 11 f. (USNM 71179). Florianopolis: 1 m. (USNM 71188 part). Also from Brazil: Paiuny: Amarracão: 1 m. (USNM 48894). Maruim: 4 m. (USNM 22318).

Not Examined. Additional specimens at USNM from throughout the West Indies.

Uca (Celuca) limicola
Crane, 1941. (P. 308.)

EL SALVADOR. El Zunzal: 2 m. (Frankfurt 1862, 1864).

COSTA RICA. Golfito: HOLOTYPE of *U. limicola*, m. (NYZS 381,152 = USNM 137415); PARATYPES, 17 m. & f. (NYZS 381,153 part = USNM 137416); 1 m., 1 f. (NYZS 381,153 part = USNM 79401); 1 m. (NYZS).

Uca (Celuca) deichmanni
Rathbun, 1935. (P. 311.)

COSTA RICA. Port Parker: 9 m., 7 f. (NYZS 381,154). Piedra Blanca: 1 m. (NYZS 381,155). Uvita B.: 7

m., 5 f. (NYZS 381,156). Golfito: 10 m. (NYZS 381,157).

PANAMA. Bahia Honda: 1 m. (381,158). Shore of Panama: HOLOTYPE of *U. deichmanni*, m. (USNM 70832). Panama City: 7 m., 6 f. (NYZS 4142); (Bellavista): 1 m. (NYZS). Old Panama: 2 m., 2 f. (NYZS 4143); below Police Station: 2 m. (NYZS).

CANAL ZONE. Balboa (La Boca): 5 m. (NYZS 4141).

Uca (Celuca) musica
Rathbun, 1914. (P. 314.)

U. (C.) musica terpsichores Crane, 1941

NICARAGUA. Corinto: 1 f. (NYZS 381,159).

COSTA RICA. Port Parker: 1 m., 3 f. (NYZS 381,160). Golfito: 4 m., 1 f. (NYZS 381,161).

PANAMA. Old Panama: 2 m. (NYZS 4146).

CANAL ZONE. Balboa (La Boca): HOLOTYPE of *U. terpsichores*, m. (NYZS 4144 = USNM 137417); PARATYPES, 1 m., 1 f. (NYZS 4145 part = USNM 79405); PARATYPES, 2 f. (NYZS 4145 part = USNM 137418).

ECUADOR. Jambeli I., near mouth of Guayas R.: 3 m. (NYZS).

U. (C.) musica musica Rathbun, 1914

MEXICO. *Lower California (South District)*: West Coast: Magdalena Bay: 1 m. (USNM 50632). Isla Margarita: 2 m. (USNM 95527). *Gulf of Califor-*

nia: 42 m., 3 f. (AMNH 2424). *(South District)*: La Paz: 6 m. (USNM 2294). Pichilinque B.: HOLOTYPE of *U. musica*, m. (USNM 22081). *Near Puertocito*: 10 m., 1 f. (USNM 106176). *(North District)*: San Felipe: 1 m., 1 f. (USNM 67733); 2 m., 2 f. (MCZ 9334). Tepoca B.: 1 f. (USNM 58112). *Sonora*: shore of Guaymas B.: 1 m. (USNM 31512). *Sinaloa*: Mazatlan: 2 m. (USNM 5054). *South of Gulf of California*: *Nayarit [Tepic Terr.]*: San Blas: 63 m., 7 f. (USNM 99755).

Uca (Celuca) latimanus
(Rathbun, 1893.) (P. 319.)

MEXICO. *Gulf of California (South District)*: La Paz: TYPE of *G. latimanus*, m. (USNM 17500). *Sinaloa*: Escuinapa: 1 f. (USNM 60230). *Nayarit [Tepic Terr.]*: San Blas: 10 m. (USNM 99754). *Jalisco*: Tenacatita B.: 6 m. & f. (NYZS; specimens lost).

NICARAGUA. Corinto: 1 m., 1 f. (NYZS 381,163). San Juan del Sur: 1 m., 1 f. (NYZS 381,164).

COSTA RICA. Port Parker: 1 m., 1 f. (NYZS 381,165). Culebra B.: 1 m. (NYZS 381,166). Gulf of Dolce: Santo Domingo: 1 m. (USNM 19442). Golfito: 4 m., 4 f. (NYZS 381,167); 3 m. (NYZS).

PANAMA. Panama City: 3 m., 2 f. (NYZS 4148). Contreras I.: 1 m. (AMNH 11563).

CANAL ZONE. Balboa (La Boca): 24 m., 8 f. (NYZS 4147).

ECUADOR. Puerto Bolivar: 2 m. (NYZS).

Appendix B. Keys

INTRODUCTION

The following keys are regional and artificial, not phylogenetic. It would no doubt have been possible to make one mammoth key to the entire genus, but the phylogenetic aspects of this genus have, it is hoped, been covered in text, tables, and diagrams. On the other hand, a means of practical identification of species is essential in this confusing group; a single key, it seems, would have been only a tour de force of little pragmatic value. Therefore any phylogenetic order that appears in the keys is adventitious.

Because of the fragility of specimens, the frequent absence in collections of some important appendages, and the fact that half of every population is female, effort has been made to include several characters for each species. Nevertheless in most species the most conspicuous characters are on the major cheliped only, while the most reliable, with few exceptions, show on the gonopod. The characters on the latter appendage are omitted in the keys whenever feasible because of the inconvenience and sometimes difficulty of adequate examination; when they are used, an attempt is made to refer only to the more macroscopic characters. Most field workers are in the habit of carrying at least a x10 lens; this should be adequate to settle many dubious cases even on a mud flat. It goes without saying, however, that the keys are only a preliminary guide to be used in close association with figures, plates, diagnoses, and descriptions.

One difficulty in the construction of the keys has been the allometric character of many attributes—particularly the proportions of fingers to manus in the major cheliped. The characteristics affected by growth as well as secondary sexual characters should be kept in mind during use of the keys. The more obvious attributes in these two groups are as follows. In the young the major fingers are much shorter than in adults, while the entire major cheliped is proportionately short; the carapace is narrower in the young and has the orbits more oblique. In females the orbits are usually less oblique than in males, the ambulatory meri broader, the suborbital crenellations larger, and the tubercles on the ambulatories both larger and more numerous. Any reference in the keys to slender ambulatory meri or extremely oblique orbits, or to other primarily non-sexual characters, apply to both sexes unless otherwise specified; that is, there may still be slight differences between the sexes in the species concerned, but the character indicated is clearly applicable to both sexes in comparison with its expression in the alternate choice in the key.

When the words *palm*, *pollex*, *dactyl*, and *fingers* appear, those of the major cheliped are to be understood, unless the term *minor cheliped* or *small cheliped* is used; *minor cheliped* is reserved for the small cheliped of the male; *small* applies to both sexes or to the female.

Species recorded from highly questionable localities are not included in the regional keys concerned. An example is *tangeri*, which was recorded once from Brazil. Others are listed on pp. 326-27.

In first planning these keys I had hoped to include behavioral and ecological characters. Because of the unexpected extent of both variation and convergence that has emerged, the idea proved impractical. On the other hand, any worker staying for a time in one subregion can easily construct a serviceable and biologically enlightening key, including unmistakable attributes of habitat, waving display, and sound production. When such keys are available and biologists using *Uca* in various ways come to know their material in the field, the overlong alternates on some of the following pages will no longer, we may hope, seem necessary.

The Guide to Characters that follows gives references to illustrations of the principal characters employed in the keys.

Guide to Characters

(For location of characters on the crab, see Figs. 1-3; 42-44 and Table 13.)

Character	Examples	
	Figures	*Plates*

CARAPACE (DORSAL PART)

Front

Narrow	26 *A, E*	13 *F,* 19 *B*
Wider	26 *C, F*	15 *H,* 17 *F,* 25 *F*
Base moderately constricted	26 *A*	13 *F*
Base strongly constricted	26 *E*	19 *B*
Anterior margin distinct	26 *G*	–
Anterior margin obsolescent	26 *H*	–

Orbital Margin

Scarcely oblique	24 *E, H*	3 *E,* 27 *E*
Strongly oblique	24 *J*	7 *A,* 22 *E,* 28 *A*

Antero-lateral Angles

Acute and produced	24 *C*	7 *E,* 22 *A*
Rectangular	24 *H*	23 *A*

Antero-lateral Margins

Short to absent	24 *C, J*	9 *E;* 22 *E;* 38 *E*
Long	24 *H, M*	5 *A;* 27 *A*
Curving gradually into dorso-laterals	–	23 *A*
Turning at an angle into dorso-laterals	–	34 *A;* 36 *A*

Dorso-lateral Margins

Weak to absent	–	18 *E*
Well-marked	–	19 *A*
Tuberculate	–	21 *E, F*

Carapace Convexity

Slight	25 *C*	–
Almost semi-cylindrical	25 *F*	–

ORBITAL REGION

Eyebrows

Lower margin absent	26 *A, E*	–
Lower margin present	26 *C, G*	–
Narrow	26 *C*	–
Broad	26 *G*	–

Suborbital Margin

Rolled out	26 *A*	–
Erect	26 *D*	–
Crenellations small to absent (except usually externally)	26 *G*	–
Crenellations large	26 *H*	–

Floor of Orbit

With tubercles	26 *A*	–
With a ridge or mound	26 *B*	–
Without elevations	26 *C*	–

Eye

With a style or stylet	29 *C*	19 *E, F;* 20 *A, B*
Eye diameter greater than that of erected stalk	31 *A*	–
Eye diameter similar to that of erected stalk	31 *H*	–

Character	Examples	
	Figures	*Plates*
MAJOR CHELIPED		
Merus (Antero-lateral Margin)		
With distal tooth or teeth	–	14 *E*; 21 *E*
With row of tubercles	7 *A*	3 *E*; 21 *A*
With crest	10 *A*	18 *E*
Without large structures	5 *A*	22 *A*
Outer Manus		
Tubercles large	38 *A-W*; 39 *C*	5 *C*; 21 *G*
Tubercles small	39 *A, E*	12´*G*; 30 *C*
Upper part bent over, with ridge and adjacent groove	43 (nos. 17, 16)	30 *I*
Tubercles largest near dorsal margin	–	15 *C*
Tubercles largest near pollex base	–	14 *G*
With proximal projection	85 *Ca*	22 *C*
Palm		
Oblique tuberculate ridge strong	44 (no. 41); 39 *B*; 40 *C*	26 *H*
Oblique tuberculate ridge weak	39 *I*	23 *D*
Oblique tuberculate ridge obsolescent to absent	39 *G*	22 *D*
Proximal Ridge at Dactyl Base		
Parallel to distal ridge	44 (no. 32)	27 *D*; 30 *D*; 38 *H*
Diverging from distal ridge	–	24 *H*; 25 *D*
Beaded Edge above Carpal Cavity		
Present	44 (no. 27); 39 *H*	–
Short and weak to absent	39 *I*	–
Lower Proximal Palm		
With enlarged or patterned tubercles	49 *Ca*	–
With parallel ridges ("stridulating ridges")	49 *Aa*	–
Dactyl (Outer)		
With 2 long grooves	38 *E, G*	–
With 1 long groove	38 *A, C*	–
With no long grooves	38 *I, K*	–
Broad and flat	39 *C*	–
Pollex (Outer)		
Base with definite small depression	39 *A*	15 *I*
Base with broad, shallow depression	38 *I*	14 *G*
Base without a definite depression	38 *A*	16 *C*
Shape somewhat triangular, the base deep	–	33 *C, G*
Shape normal, the base not wider than adjacent part of dactyl	–	37 *C*
Supraventral keel present	42 (no. 6)	37 *C*
Conspicuous pits present	–	21 *C*
Gape		
Dactyl with subdistal hook-like projection	39 *G*	16 *D*
Pollex and dactyl with forceps-like tip	38 *A, B*	–
Pollex with triangular projection	42 (no. 61); 12 *B, BB*	10 *C, D*; 32 *G, H*; 33 *G, H*
Pollex with a subdistal outer crest	39 *E*	–
Pollex with a subdistal tooth in middle row	–	26 *H*; 27 *C, D*
MINOR OR SMALL CHELIPED		
Merus		
Posteriorly flattened, with special tubercles	51 *A*	–

Character	*Examples*	
	Figures	*Plates*

Gape

Narrow	46 *A, I*	–
Broad	46 *L, M*	–
Serrations present	46 *B, K*	–
Serrations vestigial or absent	46 *G, L, N*	–
Enlarged teeth present	46 *E, P*	–
Enlarged teeth absent	46 *B, J*	–

Chela with Thick Distal Brushes	46 *H*	–

AMBULATORIES

Merus

With simple tubercles not raised on striae	52 *A, B*	–
Tubercles in rows raised on vertical striae	52 *C, D*	–
Segment slender	–	3 *E;* 24 *E;* 39 *A*
Segment broad or enlarged on 4th leg	–	2 *A;* 6 *A*
Dorsal margin almost straight on 4th leg	–	39 *A*
Dorsal margin convex on 4th leg	–	2 *A*

1st Ambulatory Anteriorly with Tubercles or a Ridge	49 *B, D*	–

Pile on Ventral Surface of Some Segments	–	25 *A*

GONOPOD

Tip thick and blunt	56 *A;* 58 *A*	–
Tip a projecting tube	58 *H*	–
Tip with short, spinous projection	61 *G, I*	–
Flanges moderate to wide	58 *A, B, E*	–
Flanges vestigial to absent	58 *G, H, I*	–
Thumb large, subdistal	58 *D, E, F*	–
Thumb long	56 *F;* 58 *I*	–
Thumb arising far from tip	58 *B*	–
Thumb a nubbin	58 *H*	–
Thumb a straight shelf	71 *B*	–
Thumb an oblique shelf	71 *A*	–
Inner process large, truncate	58 *E;* 66 *B*	–
Inner process thin, tapering	58 *D, G, H, I*	–
Inner process tumid	58 *A,* 65 *B*	–

GONOPORE *(Female)*

With tubercle or protuberance	54 *B, E, KK*	–
Without tubercle	54 *D, II*	–

ABDOMEN *(Male)*

Segments all distinct	53 *A*	–
Some segments more or less fused	53 *C, D*	–

MAXILLIPEDS

Spoon-tipped Setae on 2nd

Each with a pointed basal process	37 *E, F, G*	–
None with a pointed basal process	37 except *E, F, G*	–

Gill on 3rd

Large, with distinct books	82 *C, D, E, F, G*	–
Small to vestigial, without books	82 *H, I*	–

KEY TO THE GENERA OF OCYPODINAE

Eyes elongate, occupying most of ventral surface of eyestalks. Chelipeds very unequal in both sexes. Antennular flagella rudimentary, completely hidden beneath front. .... *Ocypode*

Eyes round, terminal on slender eyestalks. Male with 1 cheliped enormously enlarged; the other, and both chelipeds in female, are minute and similar to one another. Antennular flagella small, not hidden beneath front. .... *Uca* (p. 15)

KEY TO THE SUBGENERA OF *Uca*
(Guide to Characters: p. 616)

1. Front narrow: narrowest between eyestalk bases, its minimum breadth subequal to, rarely, 1.5 times, basal breadth of erected eyestalk. 2

   Front wider: narrowest below eyestalk bases, its breadth between them twice or more basal breadth of eyestalk. 5

2(1). Suborbital margin rolled out, its crenellations very small to absent, except sometimes near outer angle; floor of orbit often with tubercles, a ridge or a mound; gill on 3rd maxilliped small to vestigial, without books. 3

   Suborbital margin erect, its crenellations large; floor of orbit without any elevations; gill on 3rd maxilliped large, with distinct books in at least most individuals of a population. 4

3(2). Major merus without large structures on antero-dorsal margin; tubercles on outer manus large, usually largest near pollex base; pollex never with a ventral carina; ambulatory meri always with simple tubercles that are never raised on vertical striae; gonopod tip usually thick and blunt, rarely a projecting tube. .... *Deltuca* (Indo-Pacific; p. 21)

   Major merus with an antero-dorsal crest; tubercles on outer manus moderate to small, always largest in dorsal half; pollex sometimes with a ventral carina; ambulatory meri sometimes with some tubercles on vertical striae; gonopod tip always a projecting tube. .... *Australuca* (Indo-Pacific; p. 62)

4(2). Lower margin of eyebrow present; base of front between eyestalks moderately constricted, its breadth more than equal to diameter of the erected stalk; tubercles of outer major manus enlarged only at pollex base; pollex without a ventral carina; spoon-tipped setae on 2nd maxilliped never with a pointed basal process; eye never with a distal style. .... *Thalassuca* (Indo-Pacific; p. 75)

   Lower margin of eyebrow absent; base of front between eyes strongly constricted, its breadth much less than diameter of erected stalk; tubercles of outer manus enlarged subdorsally as well as at pollex base; pollex with a ventral carina; spoon-tipped setae always with a pointed basal process; eye on major side sometimes with a distal style, especially in young. .... *Uca* (America; p. 125)

5(1). Carapace largely covered with prominent tubercles; orbital floor with a large, spinous tubercle near inner corner; lower edge of eyebrow absent; spoon-tipped setae of 2nd maxilliped with a pointed basal process .... *Afruca* (Eastern Atlantic)*

   Carapace practically smooth; orbital floor without elevations; lower edge of eyebrow present; spoon-tipped hairs without a pointed basal process. 6

6(5). Gill on 3rd maxilliped in most individuals large with many books; a small depression outside pollex base with definite boundaries; no beaded ridge on carpal cavity's upper margin; predactyl ridges diverging. .... *Amphiuca* (Indo-Pacific; p. 96)

   Gill on 3rd maxilliped small to vestigial, never with books; any depression outside pollex base without definite boundaries; beaded ridge usually present along carpal cavity's upper margin. 7

7(6). Front moderately narrow, its breadth between eyestalk bases only about twice diameter of erected eyestalk at its base; eye diameter distinctly greater than that of adjacent stalk; carapace convexity slight. .... *Boboruca* (America; p. 109)

_____

* Recorded once from Brazil. A single species, *tangeri*.

Front broader, usually much broader, its breadth between eyestalk bases always more than twice basal breadth of erected eyestalk; eye diameter similar to that of adjacent stalk; carapace convexity moderate to semi-cylindrical. 8

8(7). Antero-lateral margins long (excepting in *pygmaea*), curving into dorso-laterals; pollex never with a ventral carina; 1st ambulatory on major side never with special armature anteriorly; gonopod never ending in a projecting tube; thumb always large, subdistal. .... *Minuca*
(America; p. 154)

Antero-lateral margins short, or, if long, angling sharply into dorso-laterals; pollex sometimes with ventral carina; 1st ambulatory on major side often with special armature anteriorly; gonopod often ending in a projecting tube; thumb usually arising far from gonopod's corneous tip and often vestigial. .... *Celuca*
(America; 2 spp. Indo-Pacific; p. 211)

REGIONAL KEYS

1. KEY TO THE SPECIES OF *Uca* IN EAST AFRICA AND THE RED SEA
(Guide to Characters: p. 616)

Note on distribution: All of the species except *U. tetragonon* occur in suitable habitats from Somaliland to Natal, including offshore islands; *tetragonon* has apparently not been reported in Africa south of Zanzibar. In addition, the ranges of *tetragonon*, *vocans*, *inversa* and *lactea* extend northward to various points on both coasts of the Red Sea. For Madagascar see Key 2.

1. Front narrow: narrowest between eyestalk bases, its minimum breadth subequal to, rarely 1.5 times, basal breadth of erected eyestalk. 2

Front wider: narrowest below eyestalk bases, its breadth between them twice or more basal breadth of erected eyestalk. 4

2(1). A long, distinct furrow outside major dactyl; floor of orbit with a row of tubercles; suborbital crenellations small and indistinct; gonopod ending in a slender tube; female gonopore without a tubercle. .... *urvillei*
(p. 58)

No long, distinct furrow outside major dactyl; no tubercles on floor of orbit; suborbital crenellations large & distinct; gonopod tip thick & blunt; female gonopore with or without a tubercle. 3

3(2). Major dactyl strikingly broad & flat; oblique ridge on palm high with large tubercles; small cheliped in both sexes with fingers much longer than manus. Female: no

pile posteriorly on carapace sides; a large protuberance beside gonopore. .... *vocans*
(p. 85)

Major dactyl of conventional shape; oblique ridge on palm low, the tubercles small & irregular; small cheliped in both sexes with fingers shorter than manus which is notably broad. Female: a patch of pile posteriorly on carapace sides; no protuberance beside gonopore. .... *tetragonon*
(p. 77)

4(1). Major dactyl with a large distal tooth; major merus with a crest; gonopod with large flanges; in both sexes small cheliped unusually large & ambulatory meri without tuberculate striae. Female: gonopore neither depressed nor triangular, but with its postero-outer margin slightly raised. .... *inversa*
(p. 105)

No large distal tooth on major dactyl; no crest on major merus; small cheliped of usual proportions; ambulatory meri always with some tuberculate striae. Female gonopore either depressed or triangular. 5

5(4). A triangular depression outside base of major pollex; oblique tuberculate ridge on palm moderate to low, predactyl ridges diverging; gonopod ending in a tube; ambulatory meri broad in both sexes. Female: middle ambulatories with meri having tubercules, not tuberculate striae, on lower, posterior surfaces which are bent forward; gonopore triangular. .... *chlorophthalmus*
(p. 98)

No triangular depression outside base of major pollex; oblique tuberculate ridge on palm high; predactyl ridges parallel; gonopod ending in wide flanges; ambulatory meri narrow in both sexes. Female: tuberculation of middle ambulatory meri on vertical striae, the segments not bent; gonopore appearing deeply depressed, the marginal lip strongly tilted. .... *lactea*
(p. 292)

2. KEY TO THE SPECIES OF *Uca* IN MADAGASCAR, PAKISTAN, INDIA, CEYLON, BURMA, AND WESTERN MALAYA

(Guide to Characters: p. 616)

1. Front narrow: narrowest between eyestalk bases, its minimum breadth subequal to, rarely 1.5 times, basal breadth of erected eyestalk. 2

   Front wider: narrowest below eyestalk bases, its breadth between them twice or more basal breadth of erected eyestalk. 7

2(1). No long, distinct furrow outside major dactyl; suborbital crenellations large and distinct; no tubercles on floor of orbit; female chelae never with enlarged teeth in gape. 3

   1 or 2 long distinct furrows outside major dactyl; suborbital crenellations small and indistinct; tubercles sometimes present on floor of orbit; female chelae usually with 1 or 2 enlarged teeth on at least 1 side. 4

3(2). Major dactyl strikingly broad & flat; oblique ridge on palm high with large tubercles; small cheliped in both sexes with fingers much longer than manus. Female: no pile posteriorly on carapace sides; a large protuberance beside gonopore. *Occurs throughout this key's area.* .... *vocans*
(p. 85)

   Major dactyl of conventional shape; oblique ridge on palm low, the tubercles small & irregular; small cheliped in both sexes with fingers shorter than manus, which is notably broad. Female: a patch of pile posteriorly on carapace sides; no protuberance beside gonopore. *Occurs throughout this key's area.* .... *tetragonon*
(p. 77)

4(2). No row of tubercles on floor of orbit; 2 long furrows on major dactyl; gonopod with distinct flanges adjoining distal tube. Female: sometimes with antero-lateral margins practically absent but if so gonopore not triangular with highest point at postero-inner angle; 4th ambulatory merus sometimes edged posteriorly with pile. 5

   One long furrow on major dactyl; gonopod without apparent flanges, the pore at end of a more or less projecting tube. Female: gonopore either triangular with highest point at postero-inner angle or a simple depression; 4th ambulatory merus never edged with pile. 6

5(4). Antero-lateral margins absent, strongly converging from angles; no isolated, subdistal tooth on major merus; several subdistal teeth on dactyl and pollex forming forceps-like tip; 4th ambulatory merus broad, its dorsal margin convex; gonopod tip without spinous projection. Female: no postero-ventral pile on 4th ambulatory merus. *Bay of Bengal.* .... *rosea*
(p. 29)

   Antero-lateral margins present, not strongly converging; an isolated, subdistal tooth on major merus, distinct from the small crest; merus of last ambulatory slender; gonopod tip with a short, spinous projection. Female: 4th ambulatory merus edged postero-ventrally with pile. *Bay of Bengal; ? Madagascar.* .... *dussumieri*
(p. 32)

6(4). A row of tubercles on floor of orbit; distal tube of gonopod elongate; tip of chela without forceps-like, enlarged teeth; female gonopore a slight, simple depression. *Madagascar; Pakistan; western India.* .... *urvillei*
(p. 58)

   No row of tubercles on floor of orbit; distal tube of gonopod not elongate; tip of chela forceps-like; female gonopore partly rimmed, with highest point at postero-inner angle. *Bay of Bengal.* .... *forcipata*
(p. 48)

7(1). Small chelipeds in both sexes with merus posteriorly flattened, armed with tubercles arranged in a supraventral row that curves up distally; orbits very oblique; gonopod tip tubular, greatly produced. *Bay of Bengal.* .... *triangularis*
(p. 286)

   Small chelipeds with merus not posteriorly flattened & not with a row of tubercles; orbits not strongly oblique; gonopod tip with or without flanges. 8

8(7). No oblique tuberculate ridge on palm; a crest on major merus; small chelae exceptionally large; postero-lateral striae absent; ambulatory meri without tuberculate striae. *Madagascar; Karachi, Pakistan.* .... *inversa*
(p. 105)

Oblique tuberculate ridge on palm present; no crest on major merus; small chelae of usual size; ambulatory meri always with some tuberculate striae. 9

9(8). A small depression with definite margins outside base of major pollex; oblique tuberculate ridge on palm moderate to low; predactyl ridges diverging. Gonopod ending in a tube. Ambulatory meri broad in both sexes. Female: middle ambulatories with meri having tubercles, not tuberculate striae, on lower, posterior surfaces which are bent forward. *Indian Ocean; western Malaya.* .... *chlorophthalmus*
(p. 98)

No definite depression outside base of major pollex; oblique tuberculate ridge on palm high; predactyl ridges parallel. Gonopod ending in wide flanges. Ambulatory meri narrow in both sexes. Female: tuberculation of middle ambulatory meri on vertical striae, the segments not bent. *Occurs throughout this key's area.* .... *lactea*
(p. 292)

3. Key to the Species of *Uca* in Malaysia, Indonesia, Thailand, the Philippines, New Guinea, and the Oceanic Islands of the Pacific
(Guide to Characters: p. 616)

1. Front narrow: narrowest between eyestalk bases, its minimum breadth subequal to, rarely 1.5 times, basal breadth of erected eyestalk. 2

Front wider: narrowest below eyestalk bases, its breadth between them twice or more basal breadth of erected eyestalk. 10

2(1). No long, distinct furrow outside major dactyl; suborbital crenellations large & distinct; no tubercles on floor of orbit; female chelae never with enlarged teeth in gape. 3

1 or 2 long furrows outside major dactyl (sometimes faint); suborbital crenellations small & indistinct; tubercles sometimes present on orbit's floor; female chelae usually with at least 1 enlarged tooth on at least 1 side. 4

3(2). Major dactyl strikingly broad & flat; oblique ridge on palm high with large tubercles; small cheliped in both sexes with fingers much longer than manus. Female: no pile posteriorly on carapace sides; a large protuberance beside gonopore. *Range as in key title, east to Samoa.* .... *vocans* subspp.
(p. 85; subspp. Table 3)

Major dactyl of conventional shape; oblique ridge on palm low, the tubercles small & irregular; small cheliped in both sexes with fingers shorter than manus, which is notably broad. Female: a patch of pile posteriorly on carapace sides; no protuberance beside gonopore. *Range as in key title, east to Tahiti.* .... *tetragonon*
(p. 77)

4(2). Males & females with a row of distinct tubercles on orbital floor (excluding scattered, minute granules or a mound, found in *acuta*, below); a long lateral furrow on major dactyl present, but sometimes faint. 5

Males never, females in rare individuals only, with a row of tubercles on orbital floor; major dactyl with 1 or 2 long, distinct lateral furrows. 7

5(4). Row of orbital tubercles very long (up to 22; very rarely vestigial); lateral furrows on pollex & major dactyl often faint; no single, large tooth on either finger but a series of subdistal tubercles on each usually slightly enlarged; flanges present on gonopod, the anterior curving strongly & decreasing in height away from pore; female gonopore trapezoidal with a posterior tubercle and a low rim. *Philippines & Indonesia.* .... *demani*
(p. 39)

Row of orbital tubercles shorter; 1 long, strong, lateral furrow at least on major dactyl; no flanges on gonopod; female gonopore with a postero-outer tubercle. 6

6(5). Major manus with outer tubercles very large, largest at pollex base; dactyl with a large, predistal, hook-like structure; pollex with tip slender & no triangular structure near its middle; gonopod tip only moderately produced; female cheliped with dactyl's dorsal margin nearly naked. *Sumatra to Fiji Is.; New Guinea; Philippines.* .... *coarctata* subspp.
(p. 52)

Major manus with outer tubercles small, largest dorsally; pollex with a triangular projection beyond middle; no special structure on dactyl; tip of gonopod produced in a slender tube; female cheliped with dactyl's dorsal margin setose. *Nicobar Is.; Indonesia; western New Guinea; Philippines.* .... *bellator*
(p. 64)

7(4). 2 long furrows on major dactyl; gonopod with distinct flanges. 8

1 long furrow on major dactyl; gonopod with or without distinct flanges. 9

8(7). Antero-lateral margins absent, strongly converging from angles; no isolated, subdistal tooth on major merus; several subdistal teeth on major dactyl & pollex forming forceps-like tip; 4th ambulatory merus enlarged; gonopod tip always without spinous projection. Female: no postero-ventral pile on 4th ambulatory merus. *Western Malaya.* .... *rosea*
(p. 29)

Antero-lateral margins present, not strongly converging; an isolated, subdistal tooth on major merus distinct from the small crest; no forceps-like tip on chela; 4th ambulatory merus slender, its dorsal margin practically straight; gonopod tip with or without a short, spinous projection. Female: 4th ambulatory merus edged postero-ventrally with pile. *Range as in key title, east to New Caledonia.* .... *dussumieri* subspp.
(p. 32)

9(7). Outer major manus with large tubercles; several subdistal teeth on major dactyl & pollex forming a strong forceps-like tip; no mound or minute granules on orbital floor; gonopod flanges vestigial, represented by struts each narrower distally than diameter of pore; female gonopore partly rimmed, with highest point at postero-inner angle. *Malaysia; Thailand; Indonesia; Philippines.* .... *forcipata*
(p. 48)

Outer major manus with large tubercles only near pollex base; forceps-like tip on chela weak; a distinct mound on orbital floor near inner corner, sometimes with a scattering of granules nearby; gonopod flanges large; female gonopore with an outer tubercle. *Singapore; Sarawak.* .... *acuta rhizophorae*
(p. 27)

10(1). Small chelipeds in both sexes with merus posteriorly flattened, armed with tubercles arranged in a supraventral row that curves up distally; orbits very oblique; gonopod tip tubular, greatly produced. *Range as in key title, east to New Caledonia.* .... *triangularis* subspp.
(p. 286)

Small chelipeds with merus not posteriorly flattened and not with a row of tubercles; orbits not strongly oblique; gonopod tip with or without flanges. 11

11(10). A shallow, triangular depression outside pollex base; oblique, tuberculate ridge on palm moderate to low; 4th ambulatory merus broad, its dorsal margin convex; gonopod tip without flanges, tubular. Female: middle ambulatories with lower margins of meri bent forward. *Range as in key title, east to Marquesas* .... *chlorophthalmus crassipes*
(p. 101)

No triangular depression outside pollex base; oblique, tuberculate ridge on palm strong; 4th ambulatory merus slender, its dorsal margin practically straight; gonopod tip with large flanges. Female: middle ambulatories not with bent lower margins of meri. *Range as in key title, east to Samoa.* .... *lactea* subspp.
(p. 292; subspp. Table 6)

4. KEY TO THE SPECIES OF *Uca* IN AUSTRALIA
(Guide to Characters: p. 616)

1. Front narrow: narrowest between eyestalk bases, its minimum breadth subequal to, rarely 1.5 times, basal breadth of erected eyestalk. 2

   Front wider: narrowest below eyestalk bases, its breadth between them twice or more basal breadth of erected eyestalk. *Tropical Australia.* .... *lactea* subspp.
(p. 292; subspp. Table 6)

2(1). No long, distinct furrow outside major dactyl; suborbital crenellations large & distinct; no tubercles on floor of orbit; female chelae never with enlarged teeth in gape. 3

   1 or 2 long furrows outside major dactyl (sometimes faint); suborbital crenellations small to absent; tubercles sometimes present

on orbit's floor; female chelae usually with at least 1 enlarged tooth on at least 1 side (female unknown in *demani australiae*). 4

3(2). Major dactyl strikingly broad & flat; oblique ridge on palm high with large tubercles; small cheliped in both sexes with fingers much longer than manus. Female: no pile posteriorly on carapace sides; a large protuberance beside gonopore. *Tropical & subtropical Australia.* ...... *vocans* subspp.
(p. 85; subspp. Table 3)

Major dactyl of conventional shape; oblique ridge on palm low, the tubercles small & irregular; small cheliped in both sexes with fingers shorter than manus, which is notably broad. Female: a patch of pile posteriorly on carapace sides; no protuberance beside gonopore. *Tropical Australia*: at least *east coast.* .... *tetragonon*
(p. 77)

4(2). Major dactyl with a long lateral furrow plus a subdorsal furrow about half or more dactyl's length; 4th ambulatory merus in female sometimes edged posteriorly with pile; gonopod flanges present. 5

Major dactyl with only one long furrow, the subdorsal furrow being proximal, short, sometimes vestigial; 4th ambulatory merus in female never edged posteriorly with pile; gonopod flanges absent, the tip tubular. 6

5(4). Poorly developed tubercles on orbital floor; major dactyl with subdorsal furrow about half dactyl's length; no single large tooth on dactyl or pollex, but a series of subdistal tubercles enlarged on each; 4th ambulatory merus moderately broad, its dorsal margin clearly convex; female unknown. Known from holotype only, *Broome.* .... *demani australiae*
(p. 41)

No tubercles on orbital floor; major dactyl with subdorsal furrow long, traceable almost to tip; a tubercle usually slightly enlarged near pollex's middle, but no subdistal series on it or dactyl; 4th ambulatory merus slender, its dorsal margin straight &, in female of *d. dussumieri* only, edged postero-ventrally with pile. *Tropical Australia.* .... *dussumieri* subspp.
(p. 32)

6(4). Major manus with outer tubercles large, largest at pollex base; dactyl usually with a predistal hook-like structure; pollex never

with a large projection but always with a lateral groove; 4th ambulatory merus broad, the dorsal margin clearly convex; female cheliped with dactyl's dorsal margin nearly naked. *Tropical Australia.* .... *coarctata* subspp.
(p. 52)

Major manus with outer tubercles small, largest near dorsal margin; dactyl never with a predistal hook-like structure; pollex always with a projection near middle, usually a large, tuberculate triangle, but sometimes a large tubercle; no pollex lateral groove; 4th ambulatory merus slender, the dorsal margin in middle almost straight; female cheliped with dactyl's dorsal margin conspicuously setose 7

7(6). Small cheliped in both sexes with large, triangular teeth, the opposing distal pair much the largest; major pollex with a long, low, triangular projection; female's 4th ambulatory carpus & manus with posterior patches of pile. *Tropical Australia.* .... *seismella*
(p. 70)

Small cheliped with teeth not strikingly large & triangular, the largest never distal; projection on pollex occupying less than half its length; no pile on carpus & manus of female's 4th ambulatory. 8

8(7). Palm with oblique ridge high, the tubercles subregular; no depression outside pollex base; orbital floor with a row of tubercles (exception: *bellator longidigita*); gonopore rim raised or uneven; crab size small. *Tropical & subtropical Australia.* .... *bellator* subspp.
(p. 64; subspp. Table 2)

Palm with oblique ridge low, the tubercles small & irregular; no tubercles on orbital floor; a faint depression outside pollex base; gonopore margin flat & smooth; crab size moderately large. *Tropical Australia.* .... *polita*
(p. 72)

5. KEY TO THE SPECIES OF *Uca* IN HONG KONG, MAINLAND CHINA, NORTHWEST TAIWAN AND JAPAN
(Guide to Characters: p. 616)

1. Front narrow: narrowest between eyestalk bases, its minimum breadth subequal to, rarely 1.5 times, basal breadth of erected eyestalk. 2

Front wider: narrowest below eyestalk bases,
its breadth between them twice or more basal
breadth of eyestalk. 6

2(1). No long, distinct furrow outside major
dactyl; pollex & major dactyl both flat;
suborbital crenellations large & distinct;
female chelae never with enlarged teeth in
gape. 3

1 or 2 long furrows outside major dactyl
(sometimes faint); suborbital crenellations
small & indistinct; female chelae usually with
1 or 2 enlarged teeth on at least one side. 4

3(2). Major merus with a cluster of small
tubercles distally on antero-distal margin;
in both sexes orbits scarcely or not at all
oblique and meri of 1st, 2nd, & 3rd
ambulatories strikingly broad; dorsal margin
of 4th convex in female.
Northwest Taiwan. .... *formosensis*
(p. 83)

Major merus with a large, sharp tooth distally
on antero-dorsal margin; in both sexes orbits
distinctly oblique & meri of ambulatories
slender with dorsal margin of 4th straight.
*Hong Kong, mainland China, & northwest
Taiwan.* .... *vocans borealis*
(p. 90)

4(2). 2 long furrows outside major dactyl; in
both sexes merus of 4th ambulatory slender,
its dorsal margin straight &, in female, edged
postero-ventrally with pile; gonopod tip
with a short, spinous projection. *Hong Kong
& mainland China.* .... *dussumieri spinata*
(p. 36)

1 long furrow outside major dactyl; in both
sexes merus of 4th ambulatory broad, its dorsal
margin convex; no short spinous or tuberculate
projection on gonopod tip. 5

5(4). Orbits almost straight, with antero-lateral
margins long; 4th ambulatory with ventral
margin clearly convex, in female not edged
postero-ventrally with pile; gonopod with
posterior flange large, anterior rudimentary;
crab size large. *Range as in key title.* .... *arcuata*
(p. 44)

Orbits oblique, with antero-lateral margins
very short; 4th ambulatory with ventral margin
almost straight, especially its distal half, & in
female edged postero-ventrally with pile (often
detached); gonopod with posterior flange

small, anterior large; crab size small.
Hong Kong & mainland China. .... *acuta acuta*
(p. 25)

6(1). A shallow, triangular depression outside
pollex base; oblique, tuberculate ridge on
palm moderate to low; 4th ambulatory
merus broad, its dorsal margin convex;
gonopod tip without flanges, tubular;
female's middle ambulatories with lower
margins of meri bent forward.
Hong Kong. .... *chlorophthalmus
crassipes*
(p. 101)

No triangular depression outside pollex base;
oblique, tuberculate ridge on palm strong;
4th ambulatory merus slender, its dorsal
margin straight; gonopod tip with large
flanges; female's middle ambulatories not with
bent lower margins of meri. *Range as in
key title.* .... *lactea lactea*
(p. 300)

6. KEY TO THE SPECIES OF *Uca* IN THE EASTERN PACIFIC
(Guide to Characters: p. 616)

1. Front narrow: narrowest between eyestalk
bases, its minimum breadth subequal to basal
breadth of erected eyestalk, its distal portion
slightly wider, spatuliform; tubercles of
lower, outer manus large. 2

Front wider: narrowest below eyestalk bases,
its breadth between them twice or more basal
breadth of erected eyestalk; tubercles of lower,
outer manus moderate to minute. 6

2(1). Pollex covered with conspicuous pits &
always clearly narrower than dactyl, which is
broadest distal to middle; dorso-lateral
margins of carapace & ventral margins of
ambulatory meri with large separated
tubercles or spines in female; at least traces
of similar armature in male. 3

Pollex without pits (exception: *stylifera* with
faint pits), subequal to or wider than dactyl
which is broadest at or proximal to middle;
no large separated tubercles on dorso-lateral
margin (except sometimes a small one in
female at posterior end) & none on ventral
margins of ambulatory meri. 4

3(2). Major merus margined with large,
separated tubercles antero-dorsally; male
without large tubercles on dorso-lateral
margins except for 1 at posterior end;

gonopod tip stout. Female with vertical lateral carapace margins granulate or tuberculate. *El Salvador—Peru* (? *Chile*). .... *maracoani insignis* (p. 147)

Major merus with a row of small, close-set tubercles & a single large, distal tooth; male with dorso-lateral margins more or less spinous. Female with vertical lateral margins vestigial, unarmed. *Central America to Peru.* .... *ornata* (p. 150)

4(2). Major merus with a large convex flange antero-dorsally; tip of gonopod thick, blunt; merus of 4th ambulatory in female with pile on postero-dorsal margin; female gonopore with a large tubercle. *Gulf of California to Peru.* .... *princeps* subspp. (p. 128)

Major merus with a low, straight-edged ridge ending distally in an abruptly higher tooth or small crest; tip of gonopod slender, tapering; merus of 4th ambulatory in female without pile on postero-dorsal margin; female gonopore without a large tubercle, its shape either crescentic or angular. 5

5(4). Male ocular stylet always present, much longer than cornea; major pollex faintly pitted; female chela with marginal setae short. *El Salvador to Peru.* .... *stylifera* (p. 140)

Male ocular stylet occasionally present, no longer than cornea; major pollex not pitted; female chela with marginal setae moderately long, stiff, close-set. *El Salvador to Ecuador.* .... *heteropleura* (p. 133)

6(1). Small chela: gape wide, in middle clearly wider than adjacent part of dactyl, usually much wider; opposing edges practically parallel at least in gape's proximal half & always with only the chela tips in contact; serrations absent or, rarely & maximally, few, minute & irregular, never in contact; carapace strongly arched. 7

Small chela: gape much narrower, ranging from about equal to width of adjacent part of dactyl to, in distal half, absent; serrations various; carapace arching various. 9

7(6). Eyebrow narrower than smaller dimension of adjacent part of depressed eyestalk; major pollex with a subdistal, tuberculate

tooth or crest; proximal palm without a stridulating ridge; 3rd to 6th abdominal segments in male with traces only of fusion. *Costa Rica & Panama.* .... *deichmanni* (p. 311)

Eyebrow clearly wider than smaller dimension of eyestalk; major pollex distally tapering; proximal palm with or without a stridulating ridge; 3rd to 6th abdominal segments in male almost completely fused. 8

8(7). Major cheliped with fingers shorter than manus; no proximal stridulating ridge on palm, its oblique, tuberculate ridge obsolescent to absent; 1st major ambulatory without anterior tubercles; anterior margin of front distinct. *Lower California (La Paz) to Ecuador.* .... *latimanus* (p. 319)

Major cheliped with fingers long; oblique ridge & a proximal stridulating ridge on palm both present; 1st major ambulatory with anterior tubercles on merus & carpus; anterior margin of front obsolescent. *Subtropical & tropical eastern Pacific.* .... *musica* subspp. (p. 314)

9(6). Small chela ending distally in long, thick brushes of stiff setae; major outer manus with proximal end projecting beyond distal part of carpus; front broad, contained about 3 times in carapace breadth; antero-lateral angles acute, produced anteriorly. *El Salvador to Peru.* .... *panamensis* (p. 158)

Small chela not ending in thick brushes; no proximal projection on major manus; front when broad not with antero-lateral angles projecting strongly forward. 10

10(9). Small chela with gape moderate, about equal to width of adjacent part of dactyl or slightly less, the serrations weak, not in contact; front never very wide, being contained clearly more than 3 times in carapace breadth; oblique tuberculate ridge on palm always strong; pollex normally slender & tapering, little or not at all deeper proximally than adjacent part of dactyl, never conspicuously triangular. 11

Small chela with gape narrow, diminishing distally, the serrations usually strong and, in distal half, often in contact or nearly so; front, oblique tuberculate ridge on palm & shape of pollex various. 14

1(10). Antero-lateral margins divergent, long & strong; major manus with dorsal margin strong, set off from upper, outer manus by a distinct, adjacent groove; gonopod with a long thumb; female sometimes with pile on posterior sides of carapace. *Southern California & subtropical Mexico.* .... *crenulata* subspp.
(p. 232)

Antero-lateral margins not divergent but short and usually poorly marked; major manus with dorsal margin indistinct, the adjacent groove absent or rudimentary; gonopod thumb various; female never with pile on posterior sides of carapace. 12

2(11). Orbits strongly oblique; major manus & triangle on proximal palm with minute tubercles in a reticulate pattern; gonopod thumb represented by a shelf. *Galapagos.* .... *helleri*
(p. 271)

Orbits straight to scarcely oblique; tubercles on major manus & proximal palm not in a reticulate pattern. 13

3(12). 1st major ambulatory with anterior tubercles on merus, carpus, & manus; suborbital crenellations abruptly larger in both sexes near antero-lateral angle; gonopod thumb represented by a shelf. ...... *dorotheae*
(p. 275)

No anterior tubercles on 1st major ambulatory; suborbital crenellations not abruptly larger near antero-lateral angle; gonopod thumb long. *El Salvador to Panama.* .... *limicola*
(p. 308)

4(10). Orbits extremely oblique; antero-lateral margins practically absent, with the dorso-laterals strongly convergent almost directly from antero-lateral angles; palm's oblique ridge absent. 15

Orbits little oblique (exception; moderately so in *zacae*); antero-lateral margins distinct, whether short or long; dorso-laterals variously convergent; palm's oblique ridge present or absent. 16

5(14). Major cheliped with fingers clearly shorter than manus, which is not unusually tumid; center of palm smooth; 4th ambulatory merus broad, its dorsal margin convex; 1st ambulatory on both sides with anterior tubercles on carpus. *Costa Rica to Colombia.* .... *argillicola*
(p. 220)

Major cheliped with dactyl subequal to manus which is strikingly tumid; center of palm coarsely tuberculate; merus of 4th ambulatory slender, its dorsal margin practically straight; 1st ambulatory without anterior tubercles. (Female unknown.) *Costa Rica to Colombia.* .... *pygmaea*
(p. 161)

16(14). Suborbital crenellations in both sexes absent to minute, with no gradual increase in size toward outer angle; close to outer angle 1 or more large crenellations present or absent; crab size small to very small; pollex much deeper proximally than major dactyl's base & therefore appearing triangular (exception: *batuenta*); gonopod thumb a shelf. 17

Suborbital crenellations well developed at least in outer half where they increase gradually in size toward outer angle; crab size, pollex shape, & gonopod thumb various. 20

17(16). Major cheliped with pollex proximally subequal to adjacent part of dactyl; a large, triangular projection in pollex' distal third; branchial region without pile; suborbital margin in male without a large isolated tooth near outer angle; female ambulatories with pile at least on 2nd carpus dorso-posteriorly. *El Salvador to Peru.* .... *batuenta*
(p. 244)

Major cheliped with pollex strikingly triangular, its proximal part being clearly wider than that of dactyl; pollex projection small or absent; rest of preceding characters various. 18

18(17). Carapace dorsally with at least 8 small patches of pile in both sexes (when abraded locations discernible through surface mounds or irregularities); 1st major ambulatory with anterior tubercles on merus & carpus; tubercles on palm's oblique ridge distally in a distinct row. *El Salvador to Peru.* .... *inaequalis*
(p. 254)

No patches of pile on carapace; no anterior tubercles on 1st major ambulatory; tubercles on palm's oblique ridge distally obsolescent to entirely absent. 19

19(18). Major dactyl with 2 large, triangular projections; ambulatories of usual proportions, with dorsal margins of 2nd & 3rd meri slightly convex. *El Salvador to Colombia.* .... *saltitanta*
(p. 247)

Major dactyl without projections; ambulatories in both sexes notably long & slender, with dorsal margins of meri straight. *Costa Rica & Peru.* .... *tenuipedis* (p. 258)

20(16). Front moderately narrow, contained 4 or more times in carapace breadth; gonopod flanges & thumb present or absent; thumb usually represented only by a shelf & always arising far down shaft; 1st major ambulatory with anterior tubercles present or absent. 21

Front wide, contained less than 3.5 times, usually less than 3, in carapace breadth; gonopod flanges and thumb always present, the thumb never represented by a shelf; 1st major ambulatory never with anterior tubercles. 29

21. Ambulatory meri very broad in both sexes, the middle ones with dorsal margins strongly convex; pile widespread dorsally on carapace, especially in females & young (caveat: sometimes largely abraded, persisting only in grooves), not confined to localized small patches; gonopod thumb present. 22

Ambulatory meri slender to moderate, the middle ones with dorsal margins at most slightly convex; pile when present confined to small patches in & near H-form depression, or to antero-dorsal region, or to postero-lateral region; gonopod thumb present or a shelf. 23

22. Crab size large; major carpus on inner side with a large tooth; palm's dorsal margin without a beaded edge or, if detectable, rudimentary & not curving downward; pile on ambulatories profuse, much of it persistent, especially in females & young; gonopod flanges present. *El Salvador to Panama.* .... *thayeri umbratila* (p. 113)

Crab size small; no large tooth on inner side of carpus but several minute tubercles present or absent; palm's dorsal margin with a strong beaded edge slanting downward along dorso-distal edge of carpal cavity; pile on ambulatories absent or practically so; gonopod flanges absent. *El Salvador to Peru.* .... *tomentosa* (p. 261)

23. Antero-lateral margins long & slightly to strongly diverging; no tubercles anteriorly on 1st ambulatory; no pilous depression outside pollex base; gonopod thumb present or a shelf. 24

Antero-lateral margins short & straight, concave, or converging; ambulatory tubercles & pollex base depression present or absent; gonopod thumb a shelf (exception: a nubbin in some specimens of *stenodactylus*). 25

24. Carapace dorsally with 2 to 6 patches of pile, conspicuous on branchial region; female without pile postero-laterally & on posterior sides; major chela short with pollex triangular, the pollex clearly broader proximally than dactyl base; no enlarged tubercles at gape base on outer manus; gonopod thumb a shelf. *El Salvador to Panama.* .... *oerstedi* (p. 251)

Carapace dorsally without pile in male; female with persistent pile postero-dorsally & posteriorly on sides; major cheliped with fingers strikingly long, $\pm$ 3 to 7 times manus length; a few enlarged tubercles on major manus outside gape base; gonopod thumb present, short. *El Salvador to Ecuador.* .... *festae* (p. 267)

25. Carapace dorsally in male usually with 6 curved patches of pile near middle, female with 2; pollex with a pilous depression outside base; 1st major ambulatory with anterior tubercles on carpus. *Ecuador & Peru.* .... *tallanica* (p. 264)

Carapace dorsally without pile or, at most, with traces in grooves; no pilous depression outside pollex base; anterior tubercles on 1st major ambulatory present or absent. 26

26. Palm with 2 predactyl ridges; small chela with serrations minute & similar; no longitudinal convexity on outer major manus. 27

Palm with only 1 predactyl ridge; small chela with serrations large & unequal (rarely worn or damaged in large specimens); a longitudinal convexity, broad & blunt, on outer major manus. 28

27. Gonopod with a protuberance on pore margin; 1st major ambulatory with anterior tubercles on merus distally, carpus & manus; female carapace without pile on sides posteriorly. *Panama to Peru.* .... *dorotheae* (p. 275)

Gonopod without a protuberance on pore margin (characteristics of 1st ambulatory unknown; female undescribed). *El Salvador.* .... *leptochela* (p. 274)

28. Tips of small chela obliquely truncate, meeting perfectly; 1st major ambulatory with tubercles anteriorly on carpus & manus; margins of middle ambulatory meri slightly convex in male; gonopod thumb an oblique shelf; female without pile on posterior sides of carapace. *Nicaragua to Peru.* .... *beebei* (p. 278)

Tips of small chela slender, pointed, not meeting perfectly; 1st major ambulatory with tubercles anteriorly on carpus only if at all; margins of middle ambulatory meri straight in male; gonopod thumb a transverse shelf or a nubbin; female with pile on posterior sides of carapace. *El Salvador to Peru (? Chile).* .... *stenodactylus* (p. 282)

29(20). Orbits oblique; antero-lateral margins short; dorso-laterals long & strongly converging; palm's oblique tuberculate ridge usually wholly absent, sometimes traceable through a few granules; female gonopore crescentic, its rim thickened externally but not unevenly raised, & without a tubercle. *El Salvador & Costa Rica.* .... *zacae* (p. 206)

Orbits little oblique to practically straight; antero-lateral margins long and convex; dorso-laterals short, little converging; palm's oblique tuberculate ridge vestigial (sometimes absent) to strong; female gonopore not crescentic. 30

30(29). Palm with oblique, tuberculate ridge vestigial, with apex low, to absent; pile in marbled pattern present over most of carapace dorsally (but often largely absent through abrasion); gonopod tip thickened by the large, truncate inner process; female gonopore with rim on three sides strong & unevenly raised. *Mexico to Peru.* .... *vocator ecuadoriensis* (p. 166)

Palm with oblique tuberculate ridge strong, the apex high; no pile dorsally on carapace; gonopod tip not appearing thick, the inner process being thin & tapering; female gonopore without a raised rim, but a marginal tubercle present or absent. 31

31(30). Palm without a down-curving beaded edge from dorsal margin; proximal ridge at dactyl base diverging from adjacent groove; pollex gape with a row of enlarged, separated teeth in distal half but no subdistal tuberculate crest; ambulatory meri strikingly slender,

even in female. *Lower California to Panama.* .... *brevifrons* (p. 180)

Palm with a strong down-curving beaded edge from dorsal margin; proximal ridge at dactyl base paralleling adjacent groove; no row of enlarged teeth on distal half of pollex gape but a subdistal tuberculate crest present on gape's outer edge. *El Salvador to Peru (? Chile).* .... *galapagensis* subspp. (p. 183)

7. KEY TO THE SPECIES OF *Uca* IN THE WESTERN ATLANTIC
(Guide to Characters: p. 616)

1. Front narrow: narrowest between eyestalk bases, its minimum breadth subequal to basal breadth of erected eyestalk, its distal portion slightly wider, spatuliform. Tubercles of lower, outer manus large. 2

Front wider: narrowest below eyestalk bases, its breadth between them twice or more basal breadth of erected eyestalk. Tubercles of lower, outer manus minute. 3

2(1). Major chela with fingers broad & flat, gape absent except proximally, dactyl widest in middle, pollex rough with pits; gonopod tip thick, its inner process tumid. Female without pile on 4th ambulatory carpus; chela with dactyl tip obliquely truncate. *Tropical Atlantic.* .... *m. maracoani* (p. 147)

Major chela with fingers tapering, in contact only distally, the gape normal; gonopod tip slender, tapering. Female with pile dorsally on 4th carpus; chela with dactyl tip pointed, its outer edge oblique but continuously convex, not truncate. *Tropical Atlantic.* .... *major* (p. 136)

3(1). Small chela: gape wide, in middle at least half width of adjacent part of dactyl, usually clearly more; opposing edges practically parallel in at least gape's proximal half & only the chela tips in contact; serrations absent or at most few, minute & irregular. Male abdomen with some segments partly fused. In both sexes 4th ambulatory merus slender, its dorsal margin straight. 4

Small chela: gape narrow, in middle clearly less than half width of adjacent part of dactyl, diminishing distally; opposing edges often almost in contact except near gape's base (exception: uncommon individuals of *pugilator* in Florida); serrations distinct & regular throughout middle section. Male abdomen with all segments distinct. 4th ambulatory merus various. 5

4(3). Major manus distal to carpal cavity without a strong, dorsal, tuberculate ridge and without an adjacent, pilous groove; 1st ambulatory in male without an anterior ridge on manus; female gonopore crescentic, the "horns" directed outward. *Tropical Atlantic south to southern Brazil.* .... *leptodactyla*
(p. 304)

Major manus with a dorsal, tuberculate ridge and adjacent pilous groove running almost full length of subdorsal margin; 1st ambulatory in male with an anterior, supraventral ridge on manus on both sides; female gonopore not crescentic. *South Atlantic (Rio de Janeiro to Buenos Aires).* .... *uruguayensis*
(p. 229)

5(3). No pile on ambulatories in either sex. 6
Ambulatory pile always present at least on 2nd & 3rd carpus & manus. (Exception: female *vocator*; see 10 below.) 8

6(5). Oblique tuberculate ridge on palm present; no lateral ridge on outer pollex; 1st major ambulatory with a row of anterior tubercles on carpus; gonopod tip tubular; flanges absent; 3rd & 4th ambulatory meri in female with margins convex; gonopore with marginal tubercle. *Tropical Atlantic.* .... *cumulanta*
(p. 240)

No oblique tuberculate ridge on palm; outer pollex with a lateral ridge; no row of anterior tubercles on carpus of 1st major ambulatory; gonopod tip not tubular; 3rd & 4th ambulatory meri in female with margins straight; gonopore without a marginal tubercle. 7

7(6). Carapace practically semi-cylindrical; gonopod tip thick, contorted, with inner process distally broad and appearing fringed; female gonopore large, oval, rim raised except antero-internally. *Southern United States (Texas) & northeast Mexico.* .... *subcylindrica*
(p. 209)

Carapace only moderately arched; gonopod tip not thick & contorted but relatively flat & narrow with 2 flanges & a tapering inner process; female gonopore not unusually large,

without a raised rim. *Eastern & southern United States; Bahamas.* .... *pugilator*
(p. 223)

8(5). Front narrow, contained at least 4.5 times in carapace breadth; palm with dorsal beaded edge above carpal cavity not curving down around cavity's distal margin. *Subtropical & tropical Atlantic.* .... *t. thayeri*
(p. 114)

Front wider, contained at most 3.5 times in carapace breadth, usually less; downward degree of curving of palm's dorsal beaded edge various. 9

9(8). Antero-lateral margins practically straight, posteriorly always sharply angled; palm's dorsal beaded edge slanting only slightly downward, usually with little or no curvature; crab size small. *United States: south Florida & Gulf of Mexico; Mexico; Cuba.* .... *speciosa* subspp.
(p. 236)

Antero-lateral margins convex, curving gradually into postero-dorsal margins; palm's dorsal beaded edge strong, curving distinctly downward along carpal cavity's upper distal edge. 10

10(9). Palm with oblique, tuberculate ridge vestigial to absent; pile in a marbled pattern present over most of carapace (but often largely absent through abrasion); 2nd & 3rd ambulatories without pile in females, present in males including on lower manus; gonopod tip thick, its inner process broad & truncate; female gonopore with edge unevenly raised, with 3 unequal tubercles. *Tropical north Atlantic.* .... *v. vocator*
(p. 166)

Palm with oblique, tuberculate ridge always distinct (exception: atypical *mordax*, usually large), although the tubercles are often in irregular rows or bands; pile on carapace absent or scanty, confined to H-form depression, &, rarely, other grooves or antero-lateral region, never in a widely distributed marbled pattern; 2nd & 3rd ambulatories always with pile on carpus & manus in both sexes at least dorsally; gonopod with inner process narrow, tapering; female gonopore with edge raised or not & with or without a single tubercle. 11

11(10). 2nd & 3rd ambulatories with pile on ventral as well as dorsal sides of carpus & manus. 12

Pile completely absent on lower sides of
ambulatories. 13

12(11). Major chela with proximal ridge at
dactyl base paralleling adjacent furrow;
eyebrow strongly inclined, almost vertical;
pile on ventral sides of ambulatory carpus &
manus scanty, fragile, confined to antero-
ventral margins. *United States: northwest
Florida to Texas.* .... *rapax longsignalis*
(p. 197)

Major chela with proximal ridge at dactyl base
clearly diverging upward from adjacent furrow;
eyebrow moderately inclined; pile on ventral
sides of ambulatory carpus & manus in both
sexes thick, covering entire surface, persistent.
*Tropical mainland Atlantic (Guatemala
to the Amazon).* .... *mordax*
(p. 173)

13(11). Proximal ridge at dactyl's base clearly
diverging upward from adjacent groove,
often either with an angle ventrally or with a
curve throughout; center of palm always
rough with tubercles of moderate size, not
fine granules; pollex tip never with an outer
subdistal crest but always with an enlarged,
subdistal tubercle in gape's median row;
meri of 2nd through 4th ambulatories
slender, the dorsal & ventral margins of 4th
straight in male, practically so in female;
female gonopore without a tubercle but with
the posterior part of edge clearly raised.
Subtropical & tropical Atlantic. .... *burgersi*
(p. 168)

Proximal ridge at dactyl's base straight, closely
paralleling adjacent furrow, or (*minax* only)
in upper portion minutely diverging from it;
center of palm various; pollex tip always with
an outer, subdistal crest at least indicated &
never with an enlarged, subdistal tubercle in
gape's median row; meri of ambulatories
various; female gonopore various. 14

14(13). Center of palm almost always finely
granulate, usually appearing almost smooth,
although exceptions occur; subdistal crest
on outer pollex almost always strongly
developed, the highest tubercle usually
proximal with several others diminishing
regularly toward tip; ambulatory meri broad,
the dorsal margins of 3rd & 4th clearly
convex at least on one side in both sexes;
apex of oblique tuberculate ridge on palm
high, the tubercles almost always continued
little or not at all upward around carpal
cavity; eyebrow only moderately inclined

& usually narrower than smaller dimension
of thickness of adjacent, depressed eyestalk;
female gonopore with a tubercle.
Subtropical & tropical Atlantic. .... *r. rapax*
(p. 196)

Center of palm almost always with large
tubercles that are sometimes flat; apex of
oblique ridge low, often lower than its median
section, the ridge continued or not upward
around carpal cavity; crest on outer pollex tip
highly variable within each species in strength
& form; ambulatory meri slender in males, the
dorsal margins of 4th scarcely or not at all
convex, broader in females; eyebrow various;
female gonopore with or without a
small tubercle. 15

15(14). Front extremely broad, clearly more
than one-third carapace breadth in both
sexes; eyebrow wider than smaller dimension
of adjacent, depressed eyestalk; oblique ridge
inside palm not continued upward around
carpal cavity; female carapace dorsally with
antero-lateral patches of conspicuous
tubercles; crab size large. *Eastern &
southern United States.* .... *minax*
(p. 176)

Front narrower, less than one-third carapace
breadth in males, about one-third in females;
eyebrow various; oblique ridge inside palm
continued to variable extents upward around
carpal cavity; female carapace dorsally without
antero-lateral patches of tubercles or, at most,
with a few small tubercles, highly variable;
crab size small to moderate. 16

16(15). Eyebrow almost always strongly
inclined, almost vertical, its breadth in males
narrower than smaller dimension of adjacent,
depressed eyestalk, in females subequal to it;
front always with distal margin's inner
edge normally rounded; female gonopore
with posterior edge slightly raised and
sometimes with a minute tubercle.
Eastern United States. .... *p. pugnax*
(p. 203)

Eyebrow weakly inclined, its breadth in males
subequal to smaller dimension of adjacent,
depressed eyestalk, in females broader; front
with margin's inner edge often, but not always,
appearing almost truncate in antero-dorsal view;
female gonopore with edge entirely flat and
without a tubercle. *Southern United States
(Mississippi to Texas) & northern
Mexico.* .... *pugnax virens*
(p. 203)

Appendix C. Tables

TABLE 1
Characteristics of the Subgenera

	Deluca	Australuca	Thalassuca	Amphiuca	Boboruca	Afruca	Uca	Minuca	Celuca
Range	Indo-Pacific →	→	→	→	America	Africa	America →	→	(plus 2 in Indo-Pacific)
No. of species	8	3	3	2	1	1	6	12	26
Front	Narrow →	→	→	Moderate →	→	→	Very narrow	Wide to very wide	Moderate to wide
Carapace convexity	Slight* →	→	→ *	Moderate	Slight	Moderate	Slight*	Moderate to great	Moderate to extreme
Eyebrows	Short; narrow; complete	Long; narrow to almost absent	Narrow to moderate; complete	Moderate; complete	Long; wide; complete	Moderate; lower edge absent	→	Moderate to wide; complete }	→
Eye diameter compared with stalk	Markedly greater →	→	Scarcely greater →	→	Moderately greater	Scarcely greater	→	Similar →	→
Suborbital margin	Rolled out →	→	Erect	Slightly rolled out	Erect →	→	→	→	→
Elevation(s) on orbital floor	Present or absent	Present	Absent →	→	→	Present	Absent →	→	**
Suborbital crenellations	Almost absent, except sometimes near angle }	→	Large	Absent or small except near angle	Moderate	Large	Moderate to large	Small except larger near angle	Almost absent to extremely large
Major merus Antero-dorsal margin: Armature	No large structures	Large distal crest	With or without a large tooth	With or without large crest	No large structures	Large distal tooth	Large structures	With or without moderate structure }	→
Major manus: Outer tubercles: General size	Large	Moderate to small	Large to moderate	Moderate	Small	Large	Large	Moderate or small (upper) & very small (lower)	Very small

TABLE 1 (*Continued*)

Characteristics of the Subgenera

	Deluca	*Australuca*	*Thalassuca*	*Amphiuca*	*Boboruca*	*Afruca*	*Uca*	*Minuca*	*Celuca*
Major manus: Location of largest tubercles	Pollex base (usually)	Dorsally	Pollex base	Dorsally →	→	Dorsally and near pollex base	Dorsally and near pollex base →	Dorsally	→
Palm: Oblique ridge tubercles continue upward when present?	No	→	→	→	Yes →	→	Usually →	→	Usually (18 spp.; sometimes weakly)
Palm: Beaded ridge along upper carpal cavity?	No →	→	→	→	Rudimentary No	→	→	Yes (1 exception)	Usually (15 spp; sometimes weakly)
Palm: Carpal cavity continued distally?	Yes →	→	→	Slightly →	→	Yes →	→	No (same exception)	Sometimes
Grooves on outer pollex and/or major dactyl	Present on both	Present on dactyl	Short on pollex; absent or trace on dactyl	Absent except for trace on dactyl	Absent	→	→	→	→ †
Pollex with ventral carina?	No	Sometimes	No	No →	Vestigial	Yes →		No	Sometimes
Gill on third maxilliped	Small to vestigial; no books	→	Large with books in many individuals	→	Vestigial	Large with books →		Vestigial	Small to Vestigial

TABLE 1 (Continued)
Characteristics of the Subgenera

	Deluca	Australuca	Thalassuca	Amphiuca	Boboruca	Afruca	Uca	Minuca	Celuca
Waving display (general type; moderate & high intensity)	Vertical	Vertical; semi-lateral trends →		Semi-lateral	Lateral-straight	Lateral, straight & circular	→	→	→
Construction activities	Chimneys	None	None	Chimneys	Chimneys	None	None	Chimneys; rudimentary hoods	Chimneys (rare); pillars; hoods
Down-pointing included in threats	No	→	→	→	→	Yes →	→	No	In one species
Special characters	Tubercles on ambulatory meri always simple	Gonopod tip always tubular	—	Depression outside pollex base; pre-dactyl ridges diverging	—	2nd maxilliped spoon-tips basally with spine	Eyes sometimes with styles	Antero-lateral margins long, curving behind (1 exception)	Palm-ambulatory armature in 13 spp.; gape sometimes wide; Dactyl crest in 5 spp.

*Except much stronger in one northern species or subspecies.
**Except for traces in *triangularis* subspp. & individual *lactea annulipes* females.
†Except for trace on pollex in individual *triangularis*.

TABLE 2

UCA BELLATOR: Principal Characteristics of the Subspecies

Characteristic		*bellator*	*longidigita*	*signata*	*minima*
General Range		*Philippines; East Indies*	*Near Brisbane, Australia*	*Tropical Queensland, Australia*	*Northwest Australia*
Gonopod:	Inner process	Straight; thick; appressed	Curved; appressed	Curved; appressed	Straight; slim; not appressed
	Thumb	Large; thick	Small	Absent	Absent
Gonopore:	Edge structure	Postero-external tubercle (moderate)	No tubercle, but external side of rim slightly raised	External tubercle (small)	Postero-external tubercle (moderate)
Orbital floor:	Tubercles or granules (both sexes)	Present	Absent	Present	Present in female; present or absent in male
Suborbital margin:	Crenellations (male)	Small, distinct, regular	Absent	Small, distinct, regular	Minute
Major merus:	Crest	Distally convex; tuberculate	Distally convex; weakly tuberculate or serrate	Straight, low, thick; tubercles irregular	Convex, large, smooth
Major pollex:	Ventral carina	Present	Absent	Present	Present
	Inner subdistal carina: tubercles or beading	Present	Present	Present	Absent
Minor chela:	1 pr. opposed, enlarged teeth near middle (adult male)	Present or absent	Present	Absent	Absent
Waving display:	Jerks present?	No	No	No	Yes

TABLE 3

UCA VOCANS: Principal Characteristics of the Subspecies

Characteristic		*borealis*	*pacificensis*	*dampieri*	*vomeris*	*hesperiae*	*vocans*	Apparent Hybrids *pacificensis* x *vomeris*	Apparent Hybrids *pacificensis* x *vocans*
General Range		*Hong Kong →Taiwan*	*Tropical west Pacific*	*Northwest Australia*	*East Australia & vicinity*	*East Africa, India, rarely →Singapore*	*Malaya →Nansei (Ryu Kyu) Is.*	*East New Guinea*	*South Philippines*
Gonopod:	Torsion	Absent	Slight	Moderate	Strong	Extreme	Extreme	Slight	Slight
	Inner process base tumid?	No	No	Yes	Yes	No	No	No	No
	Which flange broader?	Anterior	Anterior	Posterior (both narrow)	Posterior	Posterior (anterior minute)	Anterior	Anterior	Anterior
	Thumb very short?	No	No	Yes	No	Yes	No	No	No
	Anterior hollow	Absent	Absent	Absent	Moderate	Deep	Deep	Absent	Absent
Gonopore:	Posterior ridge	Moderate	Large	Large	Small	Minute→absent	Minute→absent	Large	Large
	Anterior ridge	Minute→absent	Minute→absent	Small	Large	Large	Large	Minute→absent	Minute→absent
Major manus:	Oblique ridge height	Low	High	High	High	High	High	High	High
	Outer tubercles	Large	Large	Moderate	Large	Large	Large	Large	Large
Major pollex:	2 projections sometimes present?	Yes, but weak	Yes	No	No	Yes	Yes	Yes	Yes
Major dactyl:	Notably deeper than pollex	Yes, in young	Yes	No	No	No	No	(Slightly)	(Slightly)
Carapace:	Dorso-lateral margins in adult male	Absent	Present	Absent	Absent	Weak→absent	Absent	Present, often short	Present, often short
Color:	Special pre-white phase?	No	Sometimes yellow	Blue spots	No	Sometimes yellow; often pale spots	Sometimes pale streaks	?	?
	Dactyl	Dull yellow→white	Violet→pink	Sometimes pink	Violet→pink (except yellow→white in New Guinea)	Yellow→white; rarely violet.	Yellow→white	?	?
Waving display:	Semi-lateral sometimes pronounced?	No	No	No (Maximally vertical)	No	Yes	No	?	?
	Seriality present?	No	No	Yes (high & medium intensities)	Sometimes (at medium intensity only)	No	No	?	?

638

TABLE 4
Comparison of Closely Related Forms in the Subgenus *UCA*

	m. maracoani Atlantic	*m. insignis* Pacific	*ornata* Pacific
Major merus with a row of long, separated tubercles	+	+	−
Major merus with small close-set tubercles and a large, distal tooth	−	−	+
Male: Postero-dorsal tubercles large only at margin's beginning & end	+	+	−
Male: Postero-dorsal tubercles large and separate*	−	−	+
Female: Postero-dorsal tubercles similar to those in male *maracoani*	+	−	−
Female: Postero-dorsal tubercles similar to those in male *ornata*	−	+	+
Female: Vertical lateral margins and their armature strong, not vestigial	+	+	−
Female: Strong tubercles across posterior carapace margin	−	−	+
Gonopod: not distally slender; inner process tumid	+	+	−
Gonopore: with tubercle large, depression deep	+	+	−

*One individual exception: see text.

TABLE 5
Geographic Distribution of Color Components in
UCA LACTEA
(Displaying males only)

KEY: Place names in *italics:* Component usual.
Place names in roman: Component occasional.
*Component present only in uncommon individuals suspected of being either *l. annulipes*, or hybrids between that subspecies and *l. perplexa*.
**Component present only in uncommon individuals suspected of being either *l. perplexa*, or hybrids between that subspecies and *l. annulipes*.

Color Components	l. lactea	l. perplexa	l. annulipes	l. mjobergi
Entire crab				
Polished white fully attained	*Taiwan*	Philippines	Singapore* Pakistan	—
Pale gray; pale buff; dull white	Hong Kong	*Philippines* *New Guinea* *Fiji*	Singapore East Africa (Massawa only)	N. W. Australia
Carapace: dorsal aspect				
Fine marbling: black and white; black and bluish; or brown and buff on white	—	Philippines *New Guinea* *New Caledonia* *Fiji* *Java*	*East Africa*	N. W. Australia
Transverse markings in bands or spots: black or blue on blue or white	—	Philippines** New Guinea (rare) (posterior band only)	Sarawak Singapore Ceylon S.W. India E. Africa (rare) (posterior bands only)	—
Major cheliped (often confined to outer manus)				
Polished white	*Hong Kong*	Java	Sarawak	—
Yellow present (buff to orange yellow)	Taiwan	*Philippines* *New Guinea* *New Caledonia* *Fiji* *Java*	Singapore*	N. W. Australia
Red present (pink to red)	—	Philippines**	Sarawak Singapore Ceylon S.W. India E. Africa	—
Anterior aspect: buccal & adjacent areas; minor chelipeds; ambulatories				
Yellow present	—	Philippines New Guinea Fiji	—	—
Red present (often on ambulatory meri only)	Hong Kong Taiwan	Philippines New Guinea New Caledonia	*Singapore* Ceylon *East Africa*	N. W. Australia

TABLE 6
UCA LACTEA: Principal Characteristics of the Subspecies*

Characteristic		mjobergi	annulipes	perplexa	lactea
	General Range	Northwest Australia	W. Africa to Singapore (occasionally to central Philippines)	Tropical W. Pacific (rarely to E. India)	Hong Kong to Japan
Gonopod:	Torsion	Slight	Absent	Strong	Slight
	Which flange longer & broader?	Anterior	Anterior	Posterior	Posterior
	Pore in deep notch?	No	Yes	No	Yes
	Thumb reaching beyond flange base?	No	No	Yes	Yes
Gonopore:	Tilting of marginal lip	Slight	Great	Slight	Moderate
	Edge colored corneous brown?	Yes	No	Yes	Yes
Major pollex:	Outer row of tubercles	Absent	Prevalent only in S.W. & Central Africa	Rarely present, weak	Absent
	Supramarginal keel & groove	Absent	Usually present & strong in S. W. & Central Africa; elsewhere very variable	Absent in east; sometimes present in west	Usually present
	Predistal triangular tooth	Small or, usually, absent	Small to moderate (excl. zones of mingling with perplexa; there sometimes large)	Moderate to large (excl. zones of mingling with annulipes; there sometimes small)	Small or absent
Major dactyl:	Dorsal convexity	Throughout	Throughout	Distally only	Throughout
	Central portion wider than adjacent part of gape (excluding smallest adults)?	Yes, except in rare leptochelous individuals	No	Yes	No
Color:	Display white	Absent	Usually absent	Present or absent	Strongly present
	Outer manus: usual display color	Yellow	Pink or red	Yellow to chrome	White
Waving display:	Diminishing waves	Present	Absent	Absent	Absent
Structures:	Hoods observed?	No	In Karachi, Pakistan	In Madang, New Guinea (rare & rudimentary)	In N.W. Taiwan

*In order to facilitate comparison between *annulipes* and *perplexa,* which sometimes intermingle and apparently hybridize, the systematic order followed in the text is altered in this table. See discussion, p. 294.

TABLE 7
Probable Limits of Present Distribution in UCA

	Approx. N. Lat.	Approx. S. Lat.	No. of Species
Indo-Pacific Region			
Red Sea: Sinai Peninsula	28°	–	2
Southeast Africa: Umtata River	–	32°	2
West Australia: Monte Bello Is.	–	21°	2
East Australia: Sydney	–	34°	1
Japan: Fukuoka*	34°	–	2
Mid-Pacific: Marquesas Is.	–	10° (long. 140° W)	2
(Absent from Hawaii)			
American Region			
Northeast Pacific: southern California	34°	–	1
Southeast Pacific: at least to Zorritos, Peru; early records to Concepción and Valparaiso, Chile**	–	3°	2
Northwest Atlantic: Cape Cod at Provincetown†	42°	–	1
Southwest Atlantic: Buenos Aires, Argentina	–	35°	1
Galapagos Is.	0°	0°	3
Eastern Atlantic Region			
Northeast Atlantic: South Portugal	37°	–	1
Southeast Atlantic: Baia dos Tigres, Angola	–	16°	1
(Absent from Mediterranean Sea)			

*Formerly north to east coast of Honshu (35° N).

**Unconfirmed old records of three species near Valparaiso, Chile (33° S); records between 1950 and 1960 as far as Puerto Casma, Peru (10° S); distribution reviewed by von Hagen, 1968.2: 457ff.; see present study, p. 438.

†Formerly slightly north to Boston Harbor.

TABLE 8

Subregional Distribution of Species and Subspecies of *UCA*: Indo-Pacific Region

NOTE: Names in parentheses indicate the occasional occurrence of species or subspecies atypical of the region. Usually, they are found in scattered populations, as shown on the maps. In the case of the allopatric species *forcipata* and *coarctata*, which are shown coinciding in subregions 5 and 6, the populations are separate in the known examples. When on the contrary, different subspecies of *vocans* or *lactea* occur in the same subregion, local populations sometimes intermingle and there is morphological evidence of hybridization.

	1. N.W. Australia	2. China; N.W. Taiwan; Japan	3. Pacific Islands	4. E. Australia; New Caledonia	5. Ryukyus→ Philippines→E. Indies Axis; New Guinea	6. N. Borneo; N. Sumatra; Malaya→E. India	7. W. India; E. Africa
DELT-UCA	*dussumieri capricornis* *demani australiae* *coarctata flammula*	*acuta acuta* *dussumieri spinata* *arcuata*	*coarctata coarctata*	*dussumieri dussumieri* *coarctata coarctata*	*dussumieri dussumieri* (*d. spinata*) *demani typhoni* *demani demani* *coarctata coarctata* (*forcipata*) *coarctata flammula*	*acuta rhizophorae* *rosea* *dussumieri spinata* *forcipata* (*c. coarctata*)	*urvillei*
AUSTRAL-UCA	*bellator minima* *seismella* *polita*			*bellator signata* *bellator longidigita* *seismella* *polita*	*bellator bellator*	(*bellator bellator*)	
THALASS-UCA	*vocans dampieri*	*formosensis* *vocans borealis*	*tetragonon* *vocans pacificensis*	*tetragonon* *vocans vomeris*	*tetragonon* *vocans vocans* (*v. pacificensis*) (*v. vomeris*)	*tetragonon* *vocans hesperiae* (*v. vocans*)	*tetragonon* *vocans hesperiae*
AMPHI-UCA		*chlorophthalmus crassipes*	*chlorophthalmus crassipes*	*chlorophthalmus crassipes*	*chlorophthalmus crassipes*	*chlorophthalmus crassipes*	*chlorophthalmus* *chlorophthalmus inversa inversa* *inversa sindensis*
CELUCA	*lactea mjobergi*	*lactea lactea*	*lactea perplexa*	*triangularis triangularis* *lactea perplexa*	*triangularis triangularis* *lactea perplexa* (*l. annulipes*)	*triangularis bengali* *lactea annulipes* (*l. perplexa*)	*lactea annulipes*

TABLE 9

Subregional Distribution of Species and Subspecies of *UCA*
Eastern Pacific and Western Atlantic Regions

	Eastern Pacific		Western Atlantic		
	1. S.W. U.S.A. and Mexico	2. Salvador→ N. Peru*	3. U.S.A.: (S. Florida)→ S. Brazil	4. U.S.A. and Mexico: Gulf of Mexico	5. U.S.A.: Massachusetts→ N. Florida
BOBORUCA		*thayeri umbratila*	*thayeri thayeri*		*thayeri thayeri*
UCA	*princeps monilifera*	*princeps princeps* *heteropleura* *stylifera* *maracoani insignis* *ornata*	*major* *maracoani maracoani*		
MINUCA	*brevifrons* *galapagensis herradurensis*	*panamensis* *pygmaea* *vocator ecuadoriensis* *brevifrons* *galapagensis herradurensis* *galapagensis galapagensis* *zacae*	*vocator vocator* *burgersi* *mordax* *rapax rapax*	*vocator vocator* *burgersi* *minax* *rapax rapax* *rapax longisignalis* *pugnax virens* *subcylindrica*	*minax* *pugnax pugnax*
CELUCA	*crenulata coloradensis* *crenulata crenulata* *musica musica*	*argillicola* *batuenta* *saltitanta* *oerstedi* *inaequalis* *tenuipedis* *tomentosa* *tallanica* *festae* *helleri* *leptochela* *dorotheae* *beebei* *stenodactylus* *limicola* *deichmanni* *musica terpsichores* *latimanus*	*speciosa speciosa* *cumulanta* *uruguayensis*** *leptodactyla*	*pugilator* *speciosa spinicarpa*	*pugilator*

*U. maracoani insignis, macrodactylus (name rejected), and stenodactylus were all described from Valparaiso, Chile, but not recently reported.

**U. uruguayensis is known only from Rio de Janeiro, Brazil to Buenos Aires, Argentina.

TABLE 10
Biotopes of *UCA*

(See Figs. 21–23)

NOTE: Species and subspecies are omitted if their habitats are unknown or known only in general terms.

Subgenus	Species and Subspecies	4. Stones with Sand, Shelly Conglomerate, or Coral Substrate (All with Some Mud)	5. Sand over Mud	6. Muddy Sand to Sandy Mud (Tide Levels: A. Upper; B. Lower)	8. Open Mud Flats near Stream Mouth	9. Muddy Sand to Mud with Partial Shade	11. Open Flats of Muddy Sand to Mud in Delta Lagoon	12. Muddy Stream Banks and Protected Flats, near Mouth, near Vegetation	13. Same as 12 but with Banks Steep	14. Same as 12 and 13, but Water Brackish	15. Upstream River Banks, Water Almost or Wholly Fresh	16. Subtropical and Temperate Marshes (Mangroves Absent)
Deltuca	acuta subspp.							+	+			
	rosea							+	+			
	dussumieri subspp.							+				
	demani demani						+	+				
	arcuata							+	+			+
	forcipata							+	+	+		
	coarctata subspp.							+	+	+		
	urvillei				+			+				
Australuca	bellator subspp.							+		+	+	
	seismella							+	+			
	polita	+	+						+			
Thalassuca	tetragonon	+	+									
	vocans subspp.			+B	+							
Amphiuca	chlorophthalmus crassipes		+	+A			+	+				
	chlorophthalmus chlorophthalmus							+			+	
	inversa subspp.						+	+				
Boboruca	thayeri subspp.							+	+	+		
Afruca	tangeri	+	+	+A,B	+	+	+	+		+		+
Uca	princeps princeps			+B	+		+					
	heteropleura			+B	+							
	major			+B			+					
	stylifera			+A,B	+							
	maracoani				+	+	+	+				
	ornata				+	+	+	+				
Minuca	panamenis	+										
	pygmaea										+	
	vocator subspp.					+		+		+		
	burgersi					+	+	+				
	mordax						+	+		+	+	
	minax									+	+	+
	brevifrons									+	+	
	galapagensis subspp.						+	+		+		
	rapax			+A,B		+	+	+				
	pugnax			+A,B				+				+
	zacae					+		+		+	+	

TABLE 10 *(Continued)*

Subgenus	Species and Subspecies	4. Stones with Sand, Shelly Conglomerate, or Coral Substrate (All with Some Mud)	5. Sand over Mud	6. Muddy Sand to Sandy Mud (Tide Levels: A. Upper; B. Lower)	8. Open Mud Flats near Stream Mouth	9. Muddy Sand to Mud with Partial Shade	11. Open Flats of Muddy Sand to Mud in Delta Lagoon	12. Muddy Stream Banks and Protected Flats, near Mouth, near Vegetation	13. Same as 12 but with Banks Steep	14. Same as 12 and 13, but Water Brackish	15. Upstream River Banks, Water Almost or Wholly Fresh	16. Subtropical and Temperate Marshes (Mangroves Absent)
Celuca	*argillicola*									+	+	
	pugilator		+	+A,B								+
	uruguayensis					+		+				
	crenulata crenulata											+
	speciosa speciosa			+B		+						
	cumulanta						+	+				
	batuenta				+		+	+				
	saltitanta				+							
	oerstedi			+B				+				
	inaequalis					+	+	+				
	tenuipedis							+				
	tomentosa (after von Hagen)		+	+	+							
	tallanica				+	+						
	festae									+	+	
	dorotheae						+					
	beebei			+B	+	+	+	+		+		
	stenodactylus		+	+A			+					
	triangularis subspp.					+			+	+		
	lactea subspp.			+A		+	+	+		+		+
	leptodactyla	+	+	+A		+						
	limicola							+	+	+		
	deichmanni	+	+	+A								
	musica terpsichores			+A								
	latimanus					+	+	+		+		

TABLE 11

Variation of morphological characters in populations
of *UCA RAPAX:* Mature males

NOTE: Individuals are counted as mature when, on the major cheliped, the dactyl is longer than the manus (that is, excluding pollex).

Country	Town	No. in Sample	Length Range (mm.)	Length/Breadth (%) Range	Length/Breadth (%) Mean	Front/Length (%) Range	Front/Length (%) Mean	Brow/Length (%) Range	Brow/Length (%) Mean
U.S.A.	Miami	28	10–15	59–71	65.2	40.0–50.0	44.5	5.1–7.2	5.8
Puerto Rico	San Juan	65	10–15	55–69	65.0	36.0–49.0	42.7	4.4–6.6	5.4
Puerto Rico	Ponce	60	10–15	59–73	65.3	40.9–47.0	43.5	4.7–6.9	5.6
Virgin Is.	St. Thomas	51	10–13	61–69	64.7	37.3–45.0	40.0	4.9–6.8	5.8
Guatemala	Puerto Barrios	23	10–14	60–68	63.6	42.2–50.0	45.1	4.0–6.6	5.6
Venezuela	Maracaibo	103	10–18	55–74	64.6	35.0–48.0	40.4	5.2–5.9	5.8
Venezuela	Puerto Cabello	7	10–13	62–68	64.2	39.2–41.9	40.6	–	–
Trinidad, W.I.	Port-of-Spain	71	10–17	59–71	65.7	–	–	4.8–6.6	5.5
Guyana	Georgetown	21	10–16	58–68	63.4	38.6–46.4	42.1	5.3–7.5	6.3
Brazil	São Luiz	35	10–15	59–67	62.7	37.6–43.0	40.6	5.9–7.5	6.8
Brazil	Fortaleza	4	16–17	64–65	64.3	38.2–40.0	39.3	5.4–6.4	6.0
Brazil	Recife	7	10–14	61–72	66.1	38.4–44.8	41.7	5.5–6.8	5.9
Brazil	Bahía	4	11–13	61–65	62.5	41.6–41.8	41.7	5.5–6.8	6.1
Brazil	Río de Janeiro	71	10–17	58–70	64.5	35.0–45.0	39.6	5.1–8.4	6.1

TABLE 12

Distribution of Acoustic Components in *UCA*

KEY: +—Component present.
×—Component probable.
? —Presence of component questionable (observation or film record dubious)

Columns 1–6 = *Type of Data (Present Study)*. Columns 1–15 = *Sound Produced by a Single Crab* [Stridulation: 1. Major-merus-rub, 2. Minor-merus-rub, 3. Minor-claw-rub, 4. Palm-leg-rub, 5. Leg-wag, 6. Leg-side-rub; Vibration Against Carapace: 7. Major-merus-drum, 8. Minor-merus-drum; Vibration Against Substrate: 9. Major-manus-drum, 10. Minor-chela-tap, 11. Leg-Stamp; Sound Associated With Respiration: 12. Bubbling, 13. Membrane-vibration] except 14. Claw-rub and 15. Claw-tap = *Sound Produced by Contact Between Two Crabs*.

Subgenus	Species	Tape	Heard field (vib. det.)	Heard crabbery/lab.	Film	Seen field	Seen crabbery/lab.	1	2	3	4	5	6	7	8	9	10	11	12	13	14	15
Deltuca	acuta				+	+						+										
	rosea				+	+						+	?									
	dussumieri	+	+		+	+	+	+				+				+		+			+	
	demani				+				+			?	?									
	arcuata				+			?														
	coarctata	+	+		+	+	+					+									+	
	urvillei				+	+		+				+			+			?			+	+
Australuca	bellator				+	+	+	+				+		+								
	seismella				+	+			+			+			+							
	polita				+	+		+														
Thalassuca	tetragonon	+	+		+	+		+				+									+	
	vocans	+	+		+	+		+				+									+	
Amphiuca	chlorophthalmus				+							+								+		
	inversa				+							+								+		
Boboruca	thayeri			+	+																	
Afruca	tangeri				+	+						+	+	+		+		+			+	
Uca	major					+		+														
	heteropleura				+							+										
	maracoani	+	+	+	+							+						+			+	
	ornata				+			+				+					+	+			+	
Minuca	panamensis															×						
	vocator	+	+	+	+	+	+									+			+			
	burgersi	+	+	+		+	+											+				
	mordax																					
	minax																					
	galapagensis				+			+	+			+										
	rapax	+	+	+	+	+	+					+						+			+	+
	pugnax		+		+	+	+					+						+			+	+
Celuca	pugilator		+		+	+		+			×	+				+		+			+	+
	uruguayensis										×											
	speciosa				+	+		?								+						
	cumulanta	+	+	+	+	+	+			?		+		+		+						+
	batuenta				+	+			+			+		+		?						
	saltitanta				+	+						+				?						
	oerstedi				+	+					?	+		+								
	inaequalis				+	+		+	+		+	+		+		?						
	festae				+			?	?			+		?								
	beebei				+	+			+		×	+			?							
	stenodactylus				+				+		×	+			?							
	triangularis	+	+																			
	lactea	+	+		+	+		+				+		+		+	+	?			+	+
	limicola				+	+						+										
	leptodactyla																					
	deichmanni				+	+		+		+	+	+		+			?	+			+	
	musica										×											
	latimanus							+	+	?		+										

Remarks	Locality	References	
		Previously Published	Page No. (Systematic Section)
	Singapore	–	28
	Malaya	–	30
5. Incl. rubbing against minor cheliped	Philippines, N. Caledonia	Crane, '66	34
5. Incl. rubbing against minor cheliped	Philippines	–	42
	Taiwan		46
	N. W. Australia, Fiji	Crane, '66	54
	Zanzibar	–	60
	Java, Philippines	–	67
	N. W. Australia	–	71
	N. W. Australia	–	73
	Red Sea, Zanzibar	Crane, '66	79
1. With circling motion	Red Sea, N. Caledonia, Fiji	Crane, '66	88
	Zanzibar	–	100
	Zanzibar, W. Pakistan	–	106
	Trinidad	von Hagen, '68.1	113
	Spain, Portugal, Angola	Altevogt, '59, '64; von Hagen, '62	121
	Trinidad	von Hagen, '68.1	137
	Panama (Pacific)	–	135
	Trinidad	Crane, '66; von Hagen, '68.1	146
	Panama (Pacific)		152
9. Inferred only from morphology	Panama (Pacific)	–	159
9. Synchronized with long roll	Trinidad	Crane, '66; von Hagen, '68.1; '70.3	165
	Trinidad, St. Thomas, Florida	von Hagen, '68.1; Salmon, '67	171
	Trinidad	von Hagen, '68.1	–
	East coast, U.S.A.	Salmon, '65	–
	Costa Rica	–	185
	U.S.A., W. Indies, Brazil	Crane, '66; Salmon, '67; Salmon & Atsaides, '68.1,2; von Hagen, '	194
	U.S.A., (east & Gulf coasts)	Salmon, '65, '67; Salmon & Atsaides, '68.1,2.	202
4. Inferred only from morphology	U.S.A. (east & Gulf coasts)	Early refs.: see, p. 480 Salmon & Stout, '62; Salmon, '65; Salmon & Atsaides, '68.2	225
4. Inferred only from morphology	Uruguay	–	229
	U.S.A. (Florida)	Salmon, '67; Salmon & Atsaides, '68.2	238
	Trinidad	Crane, '66; von Hagen, '68.1	241
	Panama (Pacific)	–	246
	Panama (Pacific)	–	249
	Panama (Pacific)	–	253
	Panama (Pacific)		256
	Ecuador		269
4. Inferred only from morphology	Panama (Pacific)	–	280
4. Inferred only from morphology	Panama (Pacific)	–	284
	N. Caledonia	–	289
	E. Africa; S.E. Asia; Taiwan; Fiji	Crane, '66	296
	Panama (Pacific)	Crane, '66	309
	Trinidad	von Hagen, '68.1	306
1. Performed antiphonally by 2 males	Panama (Pacific)	–	313
4. Inferred only from morphology	E. Pacific	Specializations pointed out first by Rathbun, 1914	314
	Costa Rica (Pacific)		321

TABLE 13
Organization of 84 Structures on *UCA* Claws and Their Relation to Ritualized Combat

Area (Figs. 42, 43, 44)	Structure No.	Structure Name	Type of Variation	Range of Variation	Combat Components Using Structure (cf. Table 14)	Structure No.
OUTER MANUS *Ventral Margin:* Length (plus any continuation on ventral pollex margin)	1	Enlarged tubercles	a. Occurrence b. Size c. Location of largest d. Distal end of row e. Arrangement f. Location of Keel	Present or absent Small to moderate Proximally or near pollex base Between pollex base and tip Tubercles usually in single line; rarely multiple, sometimes on low keel Full length, proximal or distal	Pollex-under-&-over slide Subdactyl-&-subpollex-slide Upper-&-lower-manus-rub Dactyl-submanus-slide	1
	2	Furrow beside tubercles on outer side	a. Occurrence b. Extent c. Depth	Present or absent Full length, proximal or distal Slight to moderate	Pollex-under-&-over-slide Dactyl-submanus-slide	2
Outer Surface: A. Entire region	3	Tubercles (excluding local specializations)	a. Size, general b. Size, regional c. Homogeneity d. Arrangement	Minute to small Ventral to dorsal size differences Rarely of different sizes in same region Rarely in reticulated pattern	Manus-rub Heel-&-hollow Heel-&-ridge Supraheel-rub (vertical)	3
B. ± Ventral half	4	Convexity of heel (= proximal ventral surface)	Range	Slight to moderate	Heel-&-hollow Heel-&-ridge	4
C. Pollex base	5	Depression	a. Size b. Tuberculation	Large, shallow and indefinite to small, deeper and distinctly bounded Present or absent, thick or sparse, similar or different in size from those of adjacent manus	Pollex-base-rub	5
	6	Keel below depression	a. Occurrence b. Extent proximally c. Extent distally d. Tuberculation	Present or absent Sometimes proximally to mid-manus; at least as row of tubercles From proximal part of pollex almost to tip Present or absent	Pollex-rub Pollex-base-rub	6
D. Cuff (= vertical convexity beside dactyl base)	7	Groove	Occurrence	Present or absent	Interlace	7
	8	Tubercles	Occurrence & arrangement	Linear to scattered to absent	Interlace	8
E. Central portion	9	Special longitudinal convexity	Occurrence	Rarely present	(Unknown)	9
	10	Tubercles	Size near gape	Sometimes abruptly larger	Interlace Pregape-rub	10

Upper-&-lower-manus-rub: some or all, 11-24-(+1)

(Also 17: Guide to heel in heel-&-ridge)

#	Feature	Degree	Slight to pronounced	#
F. ± Upper third				
11	Inward bending of surface		Slight to pronounced	11
12	Flattening of submarginal area	a. Degree b. Location of maximum	Absent to pronounced Throughout manus length or pronounced proximally, minimal to absent distally	12
13	Tubercles: proximal specializations	a. Occurrence b. Arrangement	Present or absent Vertical or oblique rows, reticulations, and/or rugosities	13
14	Tubercles: submarginal specializations	a. Occurrence b. Size (cf'd. #3b) c. Arrangement	Present or absent Secondarily reduced As in 13b	14
15	Submarginal longitudinal ridge	Occurrence	Rarely present	15
16	Submarginal longitudinal smooth area	Occurrence	Present or absent	16
17	Submarginal groove	a. Occurrence b. Extent c. Distinctness	Present or absent Throughout or proximal only Wellmarked to faint	17
Dorsal Margin:				
A. Proximal part (coincides with upper edge of carpal cavity)				
18	Beaded edge (see also #27)	a. Distinctness b. Extent	Beading strong or weak Distinct throughout or proximally only	18
B. Distal part				
19	Breadth	Degree	Narrow to wide	19
20	Flattening	Degree	Absent to strong	20
21	Inward tilt	Occurrence	Present or absent	21
22	Armature of outer edge	a. Distinctness b. Extent	Whether marked by a ridge and/or tubercles Edge distinct throughout or proximally only	22
23	Armature of marginal surface	a. Occurrence b. Tubercles unpatterned c. Tubercles in rows d. Tubercles in reticulations e. Rugosities	Roughness of some kind(s) present or absent; form range b – e Crowded to very sparse, surface being otherwise smooth Transverse and/or oblique, slanted proximo-externally Size of smooth spaces Degree of tuberculation	23
24	Demarcation between margin and upper palm	Occurrence	Present or absent	24

TABLE 13 (Continued)
Organization of 84 Structures on *UCA* Claws and Their Relation to Ritualized Combat

Area (Figs. 42, 43, 44)	Structure No.	Structure Name	Type of Variation	Range of Variation	Combat Components Using Structure (cf. Table 14)	Structure No.
PALM						
A. Upper, predistal area (= between upper distal part of carpal cavity, dorsal margin, & dactyl base)	25	General form	a. Relative size b. Shape	Small to large Narrow to broad	Unknown	25
	26	Distal extension of carpal cavity into predistal area	a. Occurrence b. Shape	Present or absent Tapering; narrow to broad		26
	27	Downward extension of carpal cavity's beaded, dorsal edge	a. Occurrence b. Separation of cavity from predistal area	None, slight or extensive None, partial or total		27
	28	Other separations between cavity & predistal area	a. Grouped tubercles b. Tubercles continued up from apex c. Higher predistal plane	Present or absent Present or absent Present or absent		28
	29	Separation by tubercles, differences of lower part of predactyl area from upper palm	Occurrence	Present or absent		29
	30	Depression in middle of area	Occurrence	Present or absent		30
	31	Armature of area	a. Occurrence b. Tubercles unpatterned c. Tubercles in row d. Tubercles in reticulations e. Rugosities	Range as in #23, a–e		31
B. Predactyl area (= upper distal palm, beside dactyl base)	32	Proximal row of tubercles (continued as inner row along pollex gape)	a. Occurrence b. Extent upward c. Tubercle characteristics	Present, few or absent Full length or ventral part only In row regular, irregular or multiple; small to large	Interlace	32
	33	Distal row of tubercles	a. Occurrence b. Tubercle characteristics c. Series stronger than proximal	Present, few, obsolescent, or absent In row single or multiple; minute to small (1 species): Tubercles larger and row longer	Interlace	33
	34	Intervening groove	Occurrence	Well-marked, indistinct or absent	Unknown (partial interlace)	34
C. Central palm	35	Convexity	Occurrence	Direction(s) of slope; degree	Pregape-rub	35
	36	Small depression near gape	Occurrence	Present or absent	Unknown	36
	37	Oblique depression, dorsal to oblique ridge	Occurrence	Present or absent	Unknown	37
	38	Tubercles	Occurrence	Varying local distributions & sizes	Pregape-rub	38
D. Oblique, tuberculate ridge	39	Ridge form between carpal cavity and lower palm at pollex base	a. General height b. Edge thickness c. Steepness of sides d. Extent distally	High to obsolescent Thick to thin Steep to very gentle Almost or entirely to pollex base	Heel-&-ridge	39
	40	Apex, at lower, distal end of carpal cavity	Height, cf'd. ridge	Higher, equal or lower	Heel-&-ridge	40
	41	Tuberculation, ridge, and apex	a. Occurrence of linear tubercles b. Location of largest c. Degree of irregularity	Present or absent On apex or more distally Slight to great	Heel-&-ridge	41
	42	Tuberculation continued from apex, upward around distal end of carpal cavity	a. Occurrence b. Extent c. Irregularity	Present or absent Slightly, moderately or merging with downturned dorsal margin Slight to great	Heel-&-ridge	42

No.	Feature	Aspect	Degree	Gradual to steep	Heel-&-ridge	No.
43	E. Carpal cavity	Distal slope		Gradual to steep	Heel-&-ridge	43
44	F. Lower, proximal triangle	Lower edge	a. Height b. Thickness c. Tuberculation	Low to moderate Thick to thin Present or absent	Unknown (in *triangularis* prob. heel-&-ridge)	44
45		Armature	a. Tubercles: occurrence b. Tubercles: general size c. Tubercles: location of largest d. Tubercles: where regionally absent e. Tubercles: in rows f. Tubercles: in reticulations g. Rugosities h. Oblique row of parallel striae: occurrence	Present or absent Minute to moderate Proximal end ventral to dorsal or distal Rarely from distal half, dorsal half or entire triangle except near oblique ridge Present or absent-location variable Present or absent-location variable Present or absent-location variable Present or absent	Unknown (excluding use in autostridulation with 1st ambulatory)	45
46	G. Depression at pollex base	Form	a. Extent b. Shape	Large or small Shallow with boundaries indistinct or relatively deep, subtriangular	Heel-&-hollow	46
47		Tuberculation	a. Occurrence b. Size	Present or absent Same as on adjacent palm or different	Heel-&-hollow	47
	POLLEX					
48	*Entire*	Length, cf'd. manus	Range	Longer or clearly shorter	Heel-&-ridge when long	48
49		Shape	Range	Approximately straight & slender; lower margin clearly convex, or straight with tip turned up; broad; or triangular	Heel-&-ridge when straight (?)	49
1	*Ventral Margin*	Tubercles (see under *MANUS*)	—	—	See Structure 1	1
2		Outer furrow (see under *MANUS*)	—	—	See Structure 2	2
6		Keel (see under *MANUS*)	—	—	See Structure 6	6
50	*Outer Surface* A. Lower part	Groove above keel	Occurrence	Present or absent	Pollex-rub Dactyl-along-pollex-groove	50
5	B. Outer surface as a whole	Distal end of depression at pollex base (see under *MANUS*)	—	—	See Structure 5	5
51		Tuberculation	a. Occurrence b. Extent	Present or absent General or near ventral margin and/or near prehensile edge	Manus-rub Pollex-rub	51
52	*Dorsal Margin* (= prehensile edge) A. Entire Margin	Curvature	a. Degree b. Location of concavity	Straight to strongly concave General or subdistal only	(Unknown)	52

TABLE 13 (Continued)
Organization of 84 Structures on *UCA* Claws and Their Relation to Ritualized Combat

Area (Figs. 42, 43, 44)	Structure No.	Structure Name	Type of Variation	Range of Variation	Combat Components Using Structure (cf. Table 14)	Structure No.
B. Proximal part	53	Tubercles: outer row	Occurrence	Present, weak or absent	1 or more of these rows in components: Pollex-under-&-over-slide	53
	54	Tubercles: median row	Occurrence	Present, weak or absent	Interlace	54
	55	Tubercles: inner row	Occurrence	Present, weak or absent		55
	56	Space between outer and median rows	Breadth	Narrow or wide	Interlace	56
	57	Space between median and inner rows	Breadth	Narrow or wide	Interlace	57
C. Median & distal parts	58	Tubercles: outer row	Occurrence & extent	Present, weak or absent, extended to tip or not	1 or more of these rows in components: Pollex-under-&-over-slide / Subdactyl-&-subpollex slide	58
	59	Tubercles: median row	Occurrence & extent	Present, weak or absent, extended to tip or not	Pollex-base-rub / Upper-&-lower-manus-rub	59
	60	Tubercles: inner row	Occurrence & extent	Present, weak or absent, extended to tip or not	Pregape-rub / Heel-&-hollow / Heel-&-ridge / Subdactyl-&-suprapollex-saw	60
	61	Large, tuberculate tooth	a. Occurrence b. Location c. Row(s) of origin	Present or absent / Submedian or subdistal / Median and/or outer	Unknown	61
	62	Keel: tip of median row	a. Occurrence b. Tubercles	Present or absent / Present or absent	Unknown	62
	63	Keel: tip of inner row	a. Occurrence b. Tubercles	Present or absent / Present or absent	Unknown	63
	64	Pollex tip appearing bifid or trifid	Occurrence	Present or absent	Unknown	64
Inner Surface	46	Distal end of depression inside pollex base (see under *MANUS*)	–	–	See Structure 46	46
	65	Tuberculation	a. Occurrence b. Extent	Present or absent / General, proximal only, or near prehensile edge only	Unknown	65
DACTYL						
Entire	66	Length compared with manus	Range	Shorter to much longer	Heel-&-hollow / Heel-&-ridge	66
	67	General shape	Range	Slender to broad; central portion broader or narrower than corresponding part of pollex	Heel-&-hollow / Heel-&-ridge	67
Ventral Margin (= prehensile edge)						
A. Entire margin	68	Curvature	Range	Arched throughout or distally only	Heel-&-hollow / Heel-&-ridge	68
B. Proximal part	69	Tubercles: outer row	Occurrence	Present, weak or absent	1 or more of these rows in components: Interlace	69
	70	Tubercles: median row	Occurrence	Present, weak or absent	Dactyl-submanus-slide	70
	71	Tubercles: inner row	Occurrence	Present, weak or absent		71

No.	Section	Feature	Aspect	Description	Component rows
72	C. Median & distal parts	Tubercles: outer row	Occurrence & extent	Present, weak or absent	1 or more of these rows in components: Dactyl-slide
73		Tubercles: median row	Occurrence & extent	Present, weak or absent	Upper-&-lower-manus-rub
74		Tubercles: inner row	Occurrence & extent	Present, weak or absent	Heel-&-hollow / Heel-&-ridge / Supraheel-rub (vertical) / Dactyl-along-pollex-groove / Subdactyl-&-suprapollex-saw
75		Large tuberculate tooth	a. Occurrence b. Location	Present or absent / Proximal, median or distal	Unknown
76		Keel: tip of median row	a. Occurrence b. Tubercles	Present or absent / Present or absent	Unknown
77		Keel: tip of inner row	a. Occurrence b. Tubercles	Present or absent / Present or absent	Unknown
78	*Outer Surface*	Tuberculation	a. Occurrence b. Extent	Present or absent / General or near dorsal margin, and/or near prehensile edge	Manus-rub
79		Lateral Groove	a. Occurrence b. Extent	Present or absent / Proximal only or extending beyond the middle	Unknown
80		Proximal subdorsal groove	a. Occurrence b. Extent	Present, weak or absent / Extremely short to moderately long	Heel-&-ridge (Guide to heel)
81	*Dorsal Margin*	Tuberculation	a. Density b. Extent	Close-set, sometimes continuing in subdorsal groove, to very sparse / Above and below subdorsal groove only, to continuing dorsally nearly to dactyl tip	Dactyl-slide / Upper-&-lower-manus-rub
82	*Inner Surface*	Tuberculation	a. Occurrence b. Extent	Present or absent / General, proximal only or near prehensile edge only	Unknown
83		Lateral groove or depression	a. Occurrence b. Extent	Present or absent / Proximal only or extending almost to tip	Unknown
84	*GAPE*	Breadth	Range	Middle part much narrower than to much broader than adjacent pollex (exclusive of any tuberculate tooth)	(when wide): Heel-&-ridge

TABLE 14*

Ritualized Combat in *UCA:* Distribution of Components

Component	Actor		Inactor	
	Instruments	*Structure Nos.*	*Contact Areas*	*Structure Nos.*
1. Manus-rub	Outer manus Outer pollex Outer dactyl	3 51 78	Outer manus, almost always ventral half Outer pollex Outer dactyl	3 51 78, 79
2. Pollex-rub	Lower, outer pollex	6, 50, 51	Lower outer pollex	6, 50, 51
3. Pollex-under- &-over-slide	a. Pollex: prehensile edge, median	58, 59, 60	aa. Pollex: ventral margin.	1, 2
	b. Pollex: ventral margin	1, 2	bb. Pollex: prehensile edge, median & proximal rows.	53, 54, 55 58, 59, 60
4. Subdactyl-&-sub- pollex-slide	Dactyl: proximal half dorsal margin	81	Dactyl: prehensile edge, median part	72, 73, 74
	Pollex: prehensile edge, median part	58, 59, 60	Pollex: proximal part ventral margin	1, 2
5. Pollex-base-rub	Pollex: prehensile edge, median	58, 59, 60	Outer manus: flat area at pollex base	5, 6
6. Dactyl-slide	Dactyl: prehensile edge, median	72, 73, 74	Dactyl: upper margin	81
7. Upper-&-lower manus-rub	Dactyl: prehensile edge, median and distal part	72, 73, 74	Dactyl: proximal dorsal margin. Outer manus: upper third, submarginal area & dorsal margin	81 11–24 incl. (some or all)
	Pollex: prehensile edge, distal part	58, 59, 60	& lower margin	1
8. Dactyl-submanus- slide	Dactyl: prehensile edge, median part	72, 73, 74	Manus: ventral margin	1, 2
9. Interlace	Dactyl: prehensile edge, proximal parts, inner, median and/or outer rows	69, 70, 71	*When actor engages from outer side:* Palm: predactyl area Pollex: prehensile edge. inner row, extreme proximal part	32, 33, 55
	Pollex: prehensile edge, proximal parts, inner, median &/or outer rows plus enlarged space between rows	53, 54, 55 56, 57	*When actor engages from inner side:* Outer manus: cuff Outer manus: central, distal portion Pollex: prehensile edge, outer &/or median rows, extreme proximal parts	53, 54
10. Pregape-rub (longitudinal)	Dactyl: prehensile edge, median part	72, 73, 74	Outer manus: central, distal area	10
	Pollex: prehensile edge, median part	58, 59, 60	Palm: central distal area	35, 38
11. Heel-&-hollow	a. Dactyl: prehensile edge, distal part; length, shape, curvature	72, 73, 74 66, 67, 68	aa. Outer manus: heel area	3, 4
	b. Pollex: prehensile edge, distal part	58, 59, 60	bb. Palm: Depression at pollex base	46, 47
12. Heel-&-ridge	a. Dactyl: prehensile edge, distal part; length, shape, curvature	72, 73, 74 ?76, ?77, 66, 67, 68	aa. Outer manus: heel area	3, 4
	b. Pollex: prehensile edge, median & distal parts; length & shape	59, 60 ?62, ?63, ?64 48, 49	bb. Oblique ridge inside palm & its upward extension Ventral edge, carpal cavity	39–43 incl. 44
	Indirect instrument gape: breadth	84		
			Guides to heel: Dactyl: subdorsal groove Upper manus: submarginal groove	80 17
13. Supraheel-rub (vertical)	Dactyl: prehensile edge, median & distal parts	72, 73, 74	Outer Manus: subdorsal proximal area	3, 13
14. Dactyl-along- pollex-groove	Dactyl: prehensile edge, distal part	72, 73, 74	Proximal outer pollex	50
15. Subdactyl-&- suprapollex-saw	Dactyl: prehensile edge, median-&-distal part	72, 73, 74	Pollex: prehensile edge, median part	58, 59, 60

*See also Table 13 and Figs. 42, 43, and 44.
**Plus a different component related to the numbered one listed; data being analyzed.

TABLE 14 (*Continued*)
Ritualized Combat in *UCA:* Distribution of Components

Known Occurrence of Component in Genus Uca						
Deltuca	Thalassuca	Amphiuca	Afruca	Uca	Minuca	Celuca
dussumieri coarctata urvillei	vocans	chlorophthalmus inversa	tangeri	maracoani	rapax pugnax	pugilator cumulanta inaequalis stenodactylus lactea**
					rapax	pugilator lactea**
						pugilator lactea
					rapax pugnax	lactea
						pugilator
dussumieri urvillei					rapax pugnax	pugilator cumulanta lactea** deichmanni
					rapax pugnax	pugilator lactea**
						pugilator
	vocans		tangeri		rapax pugnax	pugilator cumulanta lactea
						pugilator
urvillei	tetragonon vocans					lactea
					rapax pugnax	cumulanta lactea**
		chlorophthalmus inversa				pugilator
	vocans	inversa				
	vocans					

TABLE 15*

UCA RAPAX. Composition of 154 Combats Observed at Cocorite, Trinidad, October 13–17 and November 26–29, 1966

NOTE: Combats are listed only if observation is believed to have included the first component

Sequences of Components	Combats Between an Aggressive Wanderer and a Burrow Holder		Combats Between Two Burrow Holders		Total
	Homo-clawed	Hetero-clawed	Homo-clawed	Hetero-clawed	
Manus-push only	1	–	1	1	3
Manus-push + manus-rub	–	–	2	2	4
Manus-rub only	19	10	18	10	57
Manus-rub + dactyl-slide	1	13	3	6	23
Dactyl-slide only	1	3	1	1	6
Manus-rub + dactyl-slide + heel-&-ridge	1	1**	1	–	3
Manus-push + manus-rub + heel-&-ridge	–	–	1	–	1
Manus-rub + heel-&-ridge	9†‡	4**	4	2	19
Dactyl-slide + heel-&-ridge	1	–	1	1**	3
Heel-&-ridge only	5**‡	–	5†	1	11
Manus-rub + interlace	–	5†	–	–	5
Heel-&-right + interlace	–	4	–	–	4
Interlace only	–	2‡	–	2	4
Manus-rub + heel-&-ridge + interlace	1	–	–	–	1
Dactyl-slide + heel-&-ridge + interlace	–	1‡	–	–	1
Manus-rub + dactyl-slide + heel-&-ridge + interlace	–	3‡	–	1	4
Manus-rub + dactyl-slide + interlace	–	2	–	1	3
Dactyl-slide + interlace	1	1	–	–	2
Total	40	49	37	28	154

*From Crane, 1967: 56, Table II.
**1 heel-&-ridge component not followed by tapping
†2 heel-&-ridge components not followed by tapping.
‡1 combat with forceful ending included.

TABLE 16*

UCA RAPAX. Relative Frequency of Components in 154 Combats (From data in Table 15)

Component	Frequency (%)	
	In 77 Homoclawed Combats	In 77 Heteroclawed Combats
Manus-push	6	4
Manus-rub	78	77
Dactyl-slide	14	44
Heel-&-ridge	38	23
Interlace	3	29

*From Crane, 1967: 56, Table III.

TABLE 17*

U. RAPAX. Divisions of 104 Combats of Known Duration

Intensity	Duration Less than 20 secs.		Duration More than 60 secs.		Total
	With Force	*Fully Ritualized*	*With Force*	*Fully Ritualized*	
Low	10	35	0	0	45
High	4	46	4	5	59
Total	14	81	4	5	104

*From Crane, 1967: 66, Table VII.

TABLE 18*

UCA RAPAX. Behavior of Opponents. Following 148 Combats

KEY: AW—Aggressive wanderer larger than opponent
aw —Aggressive wanderer smaller than opponent
BH —Burrow-holder larger than opponent
bh —Burrow-holder smaller than opponent
M —Mutual component(s) clearly present
† —Burrow-holders dispossessed; waving resumption delayed more than 2 minutes

Result	Combat Class	General Combat Composition						Subtotal	Total
		Low intensity only		High intensity					
				With forceful end		No forceful end			
		With push	No push apparent	No taps	With taps	No taps	With taps		
Subsequent behavior apparently unchanged	AW & bh	1	5 (1M)	–	–	1	3 (1M)	10	
	aw & BH	–	15 (2M)	5 (1M)	–	23 (6M)	13 (10M)	56	
	BH & bh (BH = trespasser)	4 (1M)	7	–	–	7 (3M)	3 (2M)	21	
	BH & bh (bh = trespasser)	1	5 (3M)	1	–	4 (1M)	5	16	
	BH & bh (on boundary)	–	6 (2M)	–	–	3	–	9	
	Subtotals	6	38	6	–	38	24	112	112
Resumption of waving by burrow-holder delayed (less than 2 minutes except as noted)	AW & bh	–	4	–	2 (1M)†	3	2	11	
	aw & BH	–	2	–	–	1	–	3	
	BH & bh (BH = trespasser)	2	3	–	–	2	2 (1M)	9	
	BH & bh (bh = trespasser)	–	–	–	–	–	–	–	
	BH & bh (on boundary)	–	–	–	–	1 (1M)	2 (1M)	3	
	Subtotals	2	9	–	2	7	6	26	26
Wanderer's aggressiveness reduced	AW & bh	–	–	2	1 (M)	–	1 (M)	4	
	aw & BH	–	4	–	2	–	–	6	
	Subtotals	–	4	2	3	–	1	10	10
	Totals	8	51	8	5	45	31	148	148

*From Crane, 1967: 67, Table VIII.

TABLE 19

Waving Display in *UCA*: Counts and Timing Data from Film Analysis

NOTE: Characteristics of low-intensity waves excluded. See also p. 498.

KEY:　V Vertical　　L Lateral　　V/L Semi-lateral　　— No data
* Pause at apex of wave
† Secondary waves present, included in durations
= C No. of high-intensity courtship waves

Name	Locality	Chief Wave Form	Counts		Duration (sec.)						Pause between Waves (in Same Series)	Material Analyzed	
			Jerks Up	Jerks Down	Entire Wave — Moderate Intensity	Entire Wave — High-intensity Courtship	Up-strokes Only — Moderate Intensity	Up-strokes Only — High-intensity Courtship	Down-strokes Only — Moderate Intensity	Down-strokes Only — High-intensity Courtship		No. of Waves	No. of Crabs
DELTUCA													
acuta rhizophorae	Singapore	V	None	None	0.15-0.25	0.25-0.3	0.08-0.11	0.11-0.17	0.08-0.11	0.08-0.11	0.08-0.11	29(9 = C)	2
rosea	Malaysia: Penang	V	5-6	3	3.71-4.12	—	3.32-3.58	—	0.50-0.54	—	—	2	1
dussumieri capricornis	Australia: Broome	V	None	None	0.12-0.38	0.16-0.38	0.13-0.25	0.18-0.21	0.18-0.21	0.18-0.21	0.08-0.25	46(13 = C)	8
dussumieri dussumieri	Philippines: Madaum	V	1 or 0	None	0.38-0.50	—	0.29-0.46	—	0.08-0.13	—	0.29-0.58	8	1
dussumieri dussumieri	Philippines: Sasa	V	2-3	None	0.54-1.58	—	0.38-1.46	—	0.08-0.21	—	0.42-0.88	12	4
dussumieri dussumieri	Philippines: Zamboanga	V	2-3	None	0.71-2.04	0.80-1.08	0.58-1.84	0.58-0.88	0.13-0.25	0.08-0.42	0.08-1.34	39(11 = C)	6
demani	Philippines: Malalag	V	None	None	0.58-0.92	—	0.46-0.75	—	0.16-0.21	—	0.88-9.08	6	1
forcipata	Singapore	V†	None	None	0.21-0.25	—	0.13-0.17	—	0.08	—	0.08-0.13	5	1
cqarctata coarctata	Fiji Is.	V†*	2-4	None	0.84-1.25	0.84-1.08	0.46-0.96	0.42-0.63	0.29-0.54	0.34-0.54	0.50-1.54	10(5 = C)	3
coarctata coarctata	Philippines: Iling	V†*	0-4	None	1.71-3.38	0.96-1.29	Inapplicable	0.71-1.08	Inapplicable	0.21-0.34	0.08-0.75	14(3 = C)	2
coarctata flammula	Australia: Darwin	V†	3-5	None	1.63-2.58	1.69-2.58	0.92-1.8	0.96-1.42	0.38-0.96	0.63-0.80	0.08-0.29	12(4 = C)	2
urvillei	E. Africa: Pemba	V	None	None	1.00-1.21	0.75-0.88	0.42-0.63	0.38-0.67	0.38-0.80	0.21-0.38	0.04-0.75	7(4 = C)	2
AUSTRALUCA													
bellator bellator	Philippines: Manila	V	None	None	0.75-1.42	0.50-1.08	0.50-1.21	0.24-0.80	0.17-0.38	0.21-0.29	0.46-0.92	22(6 = C)	4
bellator bellator	Philippines: Iling	V	None	None	0.71-1.63	0.67-1.42					0.21-0.87	59(19 = C)	7
bellator bellator	Java: Semarang	V	None	None	0.71-0.96	0.58-1.42					0.21-0.54	40(32 = C)	6
bellator signata	Australia: Gladstone	V	None	None	0.63-2.50	—	—	—	—	—	0.67-2.50	20	8
bellator minima	Australia: Darwin	V	None	None	0.67 1.25	—	—	—	—	—	0.46-1.13	22	3
seismella	Australia: Darwin	V	None	None	0.08-0.20	—	0.04-0.12	—	0.04-0.12	—	0.04-0.08	16	1
THALASSUCA													
tetragonon	Ethiopia: Massawa	V	None	None	0.20-0.50	—	0.08-0.27	—	0.08-0.20	—	0.21-0.54	25	3
vocans borealis	Hong Kong	V	None	None	0.38-1.39	—	0.21-1.00	—	0.17-0.38	—	—	5	5
vocans pacificensis	Fiji Is.	V	None	None	0.38-2.25	—	0.17-2.00	—	0.17-0.54	—	0.21-2.54	43	6
vocans dampieri	Australia: Broome	V	None	None	0.21-0.75	—	0.13-0.46	—	0.13-0.42	—	0.04-0.96	51	3
vocans dampieri	Australia: Darwin	V	None	None	0.38-1.84	0.46-1.25	0.21-1.58	0.29-1.08	0.17-0.34	0.13-0.17	0.08-0.71	38(18 = C)	3
vocans hesperiae	E. Africa: Zanzibar	V	None	None	0.50-1.71	—	—	—	—	—	0.08-2.04	25	5
vocans vocans	Philippines: Madaum	V	None	None	1.04-1.17	1.34-1.88	0.88-0.92	0.92-1.50	0.13-0.25	0.21-0.42	2.67-> 3	13(8 = C)	1
vocans vocans	Philippines: Sasa	V	None	None	0.58-1.97	—	—	—	—	—	0.50-4.00	19	5
vocans vocans	Philippines: Puerto Princessa	V	None	None	—	1.29-1.67	—	0.96-1.00	—	0.25-0.42	1.6-4.8	6(6 = C)	2
AMPHIUCA													
chlorophthalmus crassipes	Philippines: Zamboanga	V/L	None	None	0.54-0.75	—		—		—	0.50-0.63	11	2
chlorophthalmus crassipes	Tahiti	V/L	None	None	0.38-0.88	—	0.25-0.58	—	0.08-0.17	—	0.29-0.80	45	6
chlorophthalmus chlorophthalmus	E. Africa: Pemba	V/L	None	None	0.42-0.96	—	0.29-0.63	—	0.08-0.25	—	0.04-0.80	37	3
inversa inversa	E. Africa: Zanzibar	V/L	None	None	0.46-1.54	—	—	—	—	—	0.75-2.29	32	3
inversa sindensis	Pakistan: Karachi	V/L	None	None	0.63-0.92	—	0.46-0.80	—	0.13-0.21	—	0.21-0.25	10	1

Genus / Species	Locality	V/L											
BOBORUCA													
thayeri thayeri	West Indies: Trinidad	V/L	2-3	None	1.29-2.4	—	0.63-1.50	—	0.54-0.92	—	3.13-7.0	6	2
thayeri thayeri	Brazil	V/L	2-3	None	1.17-1.84	—	0.80-1.42	—	0.34-0.42	—	—	3	1
AFRUCA													
tangeri	Portugal: Faro	L	None	None		0.75-1.00	0.54-0.80	—	0.54-0.92	0.20-0.25	0.34-1.80	12(12 = C)	1
tangeri	W. Africa: Angola	L	None	None		1.08-1.66	0.71-1.16	—		0.33-0.66	0.25-1.50	12(12 = C)	1
UCA													
heteropleura	Panama	L	None	None	0.71-1.20	0.42-1.04	0.54-0.87	0.25-0.79	0.25-0.29	0.16-0.25	1.17-5.1	19(9 = C)	2
maracoani maracoani	West Indies: Trinidad	L	None	None	1.41-3.54	1.04-1.54	1.00-1.58	0.63-0.92	0.34-1.96	0.34-0.42	0.58-1.88	37(22 = C)	4
ornata	Panama	L	None	None	0.50-1.66	1.08-1.83				0.34-0.42	0.67-4.34	19(6 = C)	2
MINUCA													
galapagensis herradurensis	Costa Rica: Golfito	L*	5-7+1	3-4	1.21-2.67	—	← Not relevant — Double peaks →			0.54-1.42	(Single wave)	5	2
rapax rapax	Florida: Miami (Tahiti B.)	L	12-20	2-4	6.20-6.54	4.04-5.87	5.0-5.29	3.29-5.20	1.20-1.25	0.54-1.42	0.08-0.50	9(5 = C)	3
rapax rapax	Florida: Miami	L	5-16	1-6	2.12-5.08		1.62-3.70		0.25-1.91	—	0.08-0.80	19	5
rapax rapax	Puerto Rico: San Juan	L	8-29	1-5	3.91-12.00		2.04-9.75		0.29-2.25	—	0.04-0.80	9	3
rapax rapax	West Indies: Trinidad	L	13-18	3	4.87-7.33		4.0-6.12		0.87-1.20	—	0.58-1.71	5	2
rapax rapax	Venezuela: Pedernales	L	14-15	4-7	5.98-7.29		4.29-4.88		1.17-2.42	—	1.08-1.17	3	1
rapax rapax	Brazil: São Luiz	L	8-22	2-5	3.67-10.37		3.00-8.70		0.67-2.58	—	1.25-2.71	6	4
rapax rapax	Brazil: Recife	L	11-17	2-7	4.70-8.79	3.75	3.62-6.08	3.62	0.91-2.70	0.12	0.13-0.17	10(1 = C)	5
rapax rapax	Brazil: Rio de Janeiro	L	8-34	1-8	5.2-12.87	4.2-5.6	3.67-11.70		0.91-4.04	—	0.17-0.96	50	15
pugnax pugnax	New York: Long Island	L	3-14	1-7	0.87-5.0		0.5-3.25		0.25-1.83	—	0.08-0.34	8	2
zacae	Costa Rica: Golfito	L*	2	None	3.66-5.16		0.16-0.29		0.20-0.33	—	—	3	2
CELUCA													
pugilator	Florida: St. Augustine	L*	None	None	0.79-0.91		0.29-0.46		0.33-0.42	—	0.96-2.34	4	1
pugilator	Florida: Miami	L	2-3	None	0.92-1.80	0.63-1.08	0.38-1.25	0.25-0.71	0.29-0.50	0.17-0.54	0.38-0.75	18(8 = C)	3
speciosa speciosa	Florida: Miami	L	None	None	0.38-0.50		0.21-0.42		0.13-0.17	—	1.29-2.29	8	1
cumulanta	West Indies: Trindad	L	None	None	1.00-3.4		0.58-2.21		0.46-1.46	—	0.80-3.00	17	3
batuenta	Panama	L*	None	None	0.34-0.58		0.08-0.34		0.08-0.17	—	0.42-1.34	25	4
saltitanta	Panama	L	None	None	0.29-0.89		0.16-0.54		0.08-0.38	—	0.50-1.58	21	2
oerstedi	Panama	L	None	None	1.12-2.84		0.91-2.00		0.20-1.29	—	0.75-2.17	29	5
festae	ex-Buenaventura, Colombia (crabbery)	L	None	None	2.63-4.71		2.21-3.83		0.42-0.88	—	0.88-3.17	16	2
beebei	Panama	L	None	None	0.29-0.46	0.29-0.50	0.16-0.33	0.20-0.37	0.08-0.20	0.08-0.13	0.13-0.96	27(17 = C)	2
stenodactylus	Panama	L	None	None	1.37-2.79		0.75-1.42		0.25-0.62	—	(1 only = 1.21)	4.	3
lactea annulipes	E. Africa: Pemba	L	None	None	0.42-1.17	0.58-0.88	0.29-0.58	0.67-1.00	0.34-0.58	0.25-0.42	0.98-3.63	47(9 = C)	10
lactea mjobergi	Australia: Broome	L†	None	None	0.6-1.25	0.3-2.3			0.08-0.17	—	0.80-0.96	35(15-C)	8
lactea lactea	Taiwan: Tamsui	L	None	None	0.58-1.46		0.50-1.34	0.04-0.13	0.17-0.50	0.04-0.08	0.25-2.08	41(16 = C)	5
lactea lactea	Hong Kong	L	None	None	1.04-1.63		0.75-1.46		0.17-0.34	—	0.67-2.38	5	1
lactea perplexa-form	Singapore	L	None	None	0.54-0.71	0.17-0.34	0.38-0.54	0.08-0.17	0.17-0.29	0.08-0.13	0.17-3.58	14(9 = C)	1
lactea perplexa	Fiji	L	None	None	0.42-1.00	0.13-0.25	0.38-0.71	0.04-0.13	0.13-0.25	0.08-0.17	0.04-3.5	94	12
lactea perplexa	Philippines: Madaum	L	None	None	0.38-1.04				0.08-0.17	—	1.25-5.5	62(31 = C)	14
leptodactyla	Brazil: Recife	L	None	None	0.87-1.42	0.67-1.00	0.54-1.00	0.46-0.83	0.29-0.62	0.16-0.46	0.50-2.46	21	3
limicola	Panama	L	2	None	1.62-2.46		1.04-1.87		0.37-0.75	—	2.38-4.41	5	1
deichmanni	Panama	L*	None	None	0.50-1.29	0.46-1.34		0.08-0.17	0.08-0.38	0.08-0.13	0.13-2.75	28(5 = C)	4
latimanus	Costa Rica: Golfito	L*	None	None	0.92-2.79	1.71-1.75	0.58-1.16	0.62-1.13	0.20-0.75	0.25-0.33	0.92-2.20	12(3 = C)	2

TABLE 20
Waving Display in *UCA;* Distribution of Components

KEY: + Component present.
× Component present but weak.
? Presence of component questionable (field observation or film record dubious).

Subgenus	Species	Wave Forms								Other Components						Timing Elements Résumé				(Drumming Incorporated. Sometimes Ritualized)
		1. Vertical	2. Jerking-vertical	3. Semi-unflexed	4. Lateral-straight	5. Lateral-circular	6. Jerking-oblique	7. Reversed-circular	8. Overhead-circling	9. Minor-wave	10. Leg-stretch	11. Prolonged-leg-stretch	12. Leg-wave	13. Curtsy	14. Herding	Pause at Peak	< 1 Sec.	1-2 Sec.	> 2 Sec.	
Deltuca	acuta	+															+			
	rosea	+	+																+	
	dussumieri	+	+								×						+			
	demani	+		+							+	+					+			
	arcuata	+		×							+						+	+		
	forcipata	+	+									+					+	+		
	coarctata	+	+	×							×						+	+		
	urvillei	+															+	+		
Australuca	bellator	+		+	+					+	+				+		+	+		
	seismella	+		+													+			
	polita			+	+					?	?						+			
Thalassuca	tetragonon	+		+						+	+						+			
	vocans	+		+						+	+				+		+	+		
Amphiuca	chlorophthalmus			+	+					+							+			
	inversa	+		+	+					+	×		+		+		+	+		
Boboruca	thayeri		+	+						+								+		
Afruca	tangeri				+	+				?		+		+			+	+		
Uca	princeps				+	+				+	+	+			+		+	+		
	heteropleura				+	+				+	+						+	+		
	major			+	+	+				+	+								+	
	stylifera			+	+					+		+							+	
	maracoani				+				+	+		+	+						+	+
	ornata				+				+	?		+	+				+	+		
Minuca	panamensis				+													+		
	vocator				+	+	+				+					+			+	
	burgersi				+	+				+		+	+						+	
	mordax				+	+				+		+	+						+	
	minax				+	+				+		+	+						+	
	galapagensis				+	+				+		+	+			+		+	+	
	rapax			+	+	+				+		+	+	+	+	+			+	
	pugnax				+	+				+		+	+	+	+	+			+	
	zacae			+		+						+	+						+	

TABLE 20 (*continued*)

Subgenus	Species	1. Vertical	2. Jerking-vertical	3. Semi-unflexed	4. Lateral-straight	5. Lateral-circular	6. Jerking-oblique	7. Reversed-circular	8. Overhead-circling	9. Minor-wave	10. Leg-stretch	11. Prolonged-leg-stretch	12. Leg-wave	13. Curtsy	14. Herding	Pause at Peak	< 1 Sec.	1–2 Sec.	> 2 Sec.	(Drumming Incorporated, Sometimes Ritualized)
		Wave Forms								*Other Components*						*Timing Elements Résumé*	*Usual Wave Duration (Moderate to High Intensity)*			
Celuca	pugilator				+		+	+		+	+					+	+	+		
	uruguayensis					+														
	crenulata				+												+			
	speciosa				+	+				+		+					+	+		
	cumulanta				+	+				+	+							+	+	
	batuenta					+				+	+					+	+	+		+
	saltitanta				+	+				+	+						+	+		+
	oerstedi				+	+				+		+						+	+	
	inaequalis				+	+				+	+							+		+
	tenuipedis				+	+				+	+					+	+			+
	tomentosa					+				+	+		+			+		+		
	festae				+	+				+		+			+	+	+	+	+	
	dorotheae				+	+				+	+		+		×	+		+		
	beebei				+	+				+		+	+				+			
	stenodactylus				+					+		+			+			+	+	
	triangularis	+		+		+								×			+	+		
	lactea	+			+	+				+	+	+	+	+	+		+	+		
	leptodactyla				+	+				+	+						+	+		
	limicola					+	+			+	+							+	+	
	deichmanni				+						+					+	+	+		+
	musica					+				+		+	?				+			
	latimanus				+	+	+					+			×	+	+	+	+	

TABLE 21

Reference List of Components in Four Categories of Social Behavior in *UCA*

(For explanation see text)

Ritualized Combat *(pp. 488–491; Table 14, p. 652)*	*Waving Display* *(pp. 494–496; Table 20. p. 658)*	*Agonistic Postures and Associated Motions* *(pp. 479–480; no table)*	*Sound Components* *(pp 482–484; Table 12, p. 644)*
1. Manus-rub	1. Vertical-wave	1. Raised-carpus	1. Major-merus-rub
2. Pollex-rub	2. Jerking-vertical-wave	2. Down-point	2. Minor-merus-rub
3. Pollex-under-&-over-slide	3. Semi-unflexed-wave	3. Frontal-arc	3. Minor-claw-rub
4. Subdactyl-&-subpollex-slide	4. Lateral-straight-wave	4. Forward-point	4. Palm-leg-rub
5. Pollex-base-rub	5. Lateral-circular-wave	5. Lunge	5. Leg-wag
6. Dactyl-slide	6. Jerking-oblique-wave	6. After-lunge	6. Leg-side-rub
7. Upper-&-lower-manus-rub	7. Reversed-circular-wave	7. Carpus-out	7. Major-merus-drum
8. Dactyl-submanus-slide	8. Overhead-circling	8. Flat-claw	8. Minor-merus-drum
9. Interlace	9. Minor-wave	9. Chela-out	9. Major-manus-drum
10. Pregape-rub	10. Leg-stretch	10. Lateral-stretch	10. Minor-chela-tap
11. Heel-&-hollow	11. Prolonged-leg-stretch	11. Creep	11. Leg-stamp
12. Heel-&-ridge	12. Leg-wave	12. Prance	12. Bubbling
13. Supraheel-rub	13. Curtsy	13. High-rise	13. Membrane-vibration
14. Dactyl-along-pollex-groove	14. Herding	14. Legs-out	14. Claw-rub
15. Subdactyl-&-suprapollex-saw	(Duration components omitted)		15. Claw-tap
			16. Interdigitated-leg-wag

TABLE 22

Coincident populations of subspecies in *UCA*

NOTE: Specimens from localities marked with asterisk were collected personally; the others were examined in museum collections. Sources in Appendix A, pp. 597f. and 611ff. Discussion in text, pp. 87, 294. See also maps 20, 21.

Area	Subspecies of U. VOCANS	Locality	Subspecies of U. LACTEA	Locality
New Guinea	*pacificensis, vomeris:*	Port Moresby	–	–
	pacificensis, vomeris:	Madang*	–	–
Philippines				
Sulu	*pacificensis, vocans:*	Joló*	*perplexa, annulipes:*	Tawi Tawi*
Mindanao	*pacificensis, vocans:*	Zamboanga*	–	–
G. of Davao	*pacificensis, vocans:*	near Davao*	*perplexa, annulipes:*	Malalag*
G. of Davao	*pacificensis, vocans:*	Madaum*	*perplexa, annulipes:*	Sasa*
Panay	–	–	*perplexa, annulipes:*	Iloilo
Indonesia				
Celebes	–	–	*perpexa, annulipes:*	Makassar, Para Pare
Java	–	–	*perplexa, annulipes:*	Madera, Besoeki, Djakarta
Borneo	–	–	*perplexa, annulipes:*	Pontianak
Malaysia				
Singapore	*hesperiae, vocans:*	Bedok*	*perplexa, annulipes:*	Kallong*
Malaya	–	–	*perplexa, annulipes:*	Port Dickson
India	–	–	*perplexa, annulipes:*	Pondicherry

TABLE 23

Comparisons of High Intensity Courtship Display Between Two Series of Allopatric Forms: A Possible Factor in Sympatric Coexistence (See p. 529)

Area	Allopatric Group	Display		Allopatric Group	Display	
	UCA DUSSUMIERI	*Jerks During Primary Wave*	*Secondary Waves*	*UCA, superspecies COARCTATA*	*Jerks During Primary Wave*	*Secondary Waves*
Fiji Is.	[absent]	–	–	*c. coarctata*	Present	Weak
Philippine Is.	*d. dussumieri*	Present	Absent	*c. coarctata*	Present	Strong
N. W. Australia	*d. capricornis*	Absent	Absent	*c. flammula*	Present	Weak
Malaysia	*d. spinata*	Absent	Absent	*forcipata*	Present	Weak
E. Africa	[absent]	–	–	*urvillei*	Absent	Absent

TABLE 24
Sites of Field Work on *UCA* 1953–1970*

Region	Locality	No. of Visits	Duration of Stay (Number of Day-time Low Tides)		
			1 to 2	3 to 7	More than 10
Eastern Atlantic					
Portugal:	The Algarve: Vila Real de Sto. Antonio to Sagres	1			+
Nigeria:	Lagos	1		+	
Angola:	Luanda	1		+	+
Indo-Pacific					
Ethiopia:	Massawa	2			+
Tanzania:	Pemba	1		+	
	Zanzibar	1			+**
	Dar-es-Salaam	1	+		
Mozambique:	Inhaca I.	1		+	
Aden		1	+		
Pakistan:	Karachi	2	+		
India:	Bombay	1	+		
	Ernakulam	1		+	
Ceylon:	Negombo	2			+
Malaysia:	Penang	1	+		
	Negri Sembilan: near Sungei Dua	1	+		
	Malacca	1	+		
	Sarawak: Santobong	1	+		
Singapore		1			+**
Indonesia:	Java: Semarang	1		+	
	Surabaja	1		+	
Australia:	Broome	1		+	
	Darwin	1		+	
	Gladstone	1		+	
	Shorncliffe: near Brisbane	1	+		
New Caledonia:	near Nouméa	1		+	
N. E. New Guinea:	near Madang	1			+**
Philippines:	Sulu: Tawi-Tawi	1	+		
	Joló	1	+		
	Mindanao: Zamboanga	1			+
	near Davao	1		+	
	Palawan: Puerto Princesa	1	+		
	Basilan R.	1	+		
	Luzon: near Manila	1	+		
Hong Kong:	Kowloon	1		+	
Taiwan:	Tamsui	1		+	
Japan:	Kyushu: Ariadne Bay	1		+	
Fiji:	Viti Levu	3			+
Tahiti		1			+
Bora Bora		1	+		

TABLE 24 (*Continued*)

Region	Locality	No. of Visits	Duration of Stay (Number Day-time Low Tides)		
			1 to 2	3 to 7	More than 10
Eastern Pacific					
Costa Rica:	Golfito	1		+	
Panama:	Panama City	2		+	
Western Atlantic					
U.S.A.:	Massachusetts: Cotuit	1			+**
	Florida: St. Augustine	2		+	
	Miami	2		+	
	Puerto Rico	3		+	
	St. Thomas	3		+	
Guatemala:	Puerto Barrios	1		+	
Guadeloupe		2		+	
Martinique		1	+		
Barbados		1	+		
Trinidad-Tobago		Many			+†
Guyana	Georgetown	2	+		
Surinam:	Paramaribo	2		+	
Brazil:	Belém	1	+		
	São Luiz	1		+	
	Fortaleza	1	+		
	Recife	1	+		
	São Salvador	1	+		
	Rio de Janeiro	1			+

*Before 1953 work was also done on *Uca,* without cinematography, on the coast of Venezuela and, in the eastern Pacific, from southern California to the Gulf of Guayaquil.

**Between one and two months.

†Sporadic work on *Uca,* including the use of crabberies, carried out at the William Beebe Tropical Research Station, Simla, Trinidad, West Indies, especially during 1957, 1958, and 1962–1966.

Appendix D. Field Methods and the Maintenance of Fiddler Crabs in Captivity

CONTENTS

INTRODUCTION

As in every other branch of zoology, each worker who begins to study fiddler crabs quickly develops his own methods of work. Accordingly the paragraphs below will be most helpful to newcomers. Little will be said of particular instruments, since those available on the market change rapidly, often with improvements useful to a biologist in the field, while the disadvantages of any model may be soon corrected.

On the other hand it seems worthwhile to include a number of suggestions for elementary aids to success. Many will seem obvious to a worker experienced in dealing with the minor crises that often loom when living animals, both delicate and strange, are the subjects of his study. Yet the simple solutions may not occur to him in time to avert a disaster to his efforts—whether these are attempts to fly a hundred healthy crabs across an ocean or merely to keep some fiddlers in a pail from kicking off their legs.

TIMING OF FIELD TRIPS

Many species in the tropics are socially active throughout the year, so that seasons may be disregarded in the making of plans. This attitude is reasonably safe when the trip includes territory in the equatorial zone, extending from about lat. 10° N to lat. 10° S. Even here local peculiarities of climate are to be expected, when drought or excessive rain may immobilize the crabs. It follows that the only reliable course is to learn as much as possible about the seasons of the area before selecting dates for the trip. In out-of-the-way places letters to the government meteorological service, to the nearest museum,

or to a school of agriculture should be sent, preferably months ahead, in order to allow time for follow-up correspondence. The enquiry should stress that average rainfall at the capital city or the average number of days on which rain falls during each month will not give the information desired; instead, detailed rainfall tables should be requested for the coastal zone over a period of years. In parts of the tropics which have long had stable governments, such as Singapore, there is of course no problem; in many other areas time and persistence are necessary.

In the tropics outside of the equatorial zone, a single dry season usually contrasts strongly with a single rainy season and here information on the seasons and their variations becomes essential. The species that occur farthest from the open sea are subject to desiccation during the dry season and these populations accordingly aestivate as need arises. Even when they do not do so, they show little or no social activity until after the start of the rains; on the other hand even species that breed throughout the year may show a peak of social behavior near the beginning of the rainy season, always in accordance with a characteristic lunar or semi-lunar cycle. For example, in the West Indies social behavior studies of the subgenus *Minuca* should be avoided from about the middle of January until May or June. Even though meteorological information is easily available in this region, annual variations in the arrival of the rains are sometimes striking and always unpredictable; if time or funds are short, it may prove helpful to ask a local field biologist to keep watch and cable when the crabs start waving; if such a correspondent does not exist, it would be almost equally helpful for an acquaintance or official to cable as soon as the first strong downpour—not the first shower—occurs.

In the subtropics temperature becomes a factor and, as in the temperate zone, most social activity is confined to the local spring and summer. The length of the season depends in general on the latitude, but adequate information has not yet been gathered.

An essential tool in the planning of any field work is the volume of tide tables appropriate to the area. A complete set is published annually in English by both the United States Department of Commerce and by the British Admiralty. The tables permit the selection for almost any suitable locality of appropriate periods for the visit; when only a short stay on a tropical shore is possible, such planning is fully as important as a knowledge of the seasons. If both the locality itself and the activity rhythms of the desired species are unfamiliar, it is best to arrive at least two days before the occurrence of a low tide around 0800; often this tide occurs several days after new or full moon. If time is limited to a week or so, optimal populations can be located during the first two days of the stay when most species—with the tide low around 0600 to 0700—are not socially at their most active. Then, if the work period can extend through the week to include the day on which low tide occurs about 1300, the chances are good that the major part of the social activity cycle will have been observed. At least in tropical species, when the tide is low during the afternoon the populations show minimal social activity even when heat is not excessive. In a few species waving display, but not combat, is resumed when the tide is sufficiently low between 1700 and dusk. In several species the most active periods of all fall between 1700 and 1800 as well as in the morning between 0700 and 0900. Nocturnal social activity when present apparently takes place chiefly during periods of low tide before midnight, corresponding to the high activity of corresponding morning lows; many further observations are needed on appropriate species, however, during parts of the cycle when low tide takes place between midnight and dawn.

Coasts having very irregular tides underline the need for special comparative studies of fiddler crab activity rhythms. Here during parts of the year the tide ebbs conspicuously only once every 24 hours and only during daylight or, as the tide tables put it, "low tide largely diurnal." At these times the diurnal highs and nocturnal lows are usually barely perceptible in the tables as very slight, brief reversals of flow direction that are virtually undetectable in the real world of a mud flat. A normal set of tides may or may not abruptly appear for a few days at or near the change of the moon; very often even these periods do not provide a diurnal low during the hours elsewhere found to be optimal for social activity in a particular species. Sometimes the tides are reversed,

being low only at night for long periods. Good examples of these and other irregularities appear on the Gulf coast of the United States, the north coast of Java, and the north central coast of New Guinea. All these conditions provide challenges to the investigator and also, it would seem, to the crabs. A short-term visitor can only aim for a morning or midday low occurring reasonably near full moon or new moon, and hope for the best.

SELECTION AND USE OF SITES

A vital factor in selecting for field work a particular stretch of shore is its history of pollution. Unfortunately, clean-looking small bays, margined ideally with mangroves or northern marshes, often turn out to be contaminated by runoffs from the inland use of agricultural pesticides, chemical fertilizers, and processing plants. Sometimes the health and behavior of the crabs appear to be unaffected. I found one happy example in Barbados, where a population of *burgersi* behaved characteristically, although the crabs shared a stream mouth with a sugar refinery; this plant was probably responsible for the vivid red that entirely suffused every individual in the population, a situation never found elsewhere. As we learn more of the effects on behavior and reproduction of pollution in other animals, however, the need for care in the investigation of local conditions becomes always more apparent. This is especially true before work on species in which the criteria and ranges of "normal behavior" are as yet unknown, or when the forms to be expected are unfamiliar in life to the investigator.

The importance of this factor in certain studies cannot be exaggerated. It becomes crucial, for instance, in comparative investigations of different populations of the same species. Another example will illustrate the uncertainties that can arise. Several summers ago on Cape Cod, Massachusetts, I concentrated on observing the social behavior of an apparently typical population of *pugilator*. The site appeared ideal and reliable sources reported that no aerial spraying against mosquitoes had taken place over the small cove for three years; in contrast to this certainty, the runoff patterns from tilled fields and cranberry bogs, prevalent inland, were unknown; the collection of shellfish for food from the cove and adjoining bay was permitted by the authorities, although areas a few miles away were blocked off as contaminated.

The purpose of the study was to compare the social behavior of the population here near the northern boundary of the species' range with that of populations in more southern localities. Individual components of waving behavior, courtship, threat, and combat proved to be entirely comparable to those previously observed in Connecticut and Florida; the crabs appeared to feed as energetically as usual in the genus; finally, that particular summer included a proportion of warm and sunny weather that was normal for the locality. Yet the amount of social activity of the crabs was far below that of more southern populations. In particular, bouts of waving and combat were rare and the periods devoted to these activities exceedingly short. I ended the season uncertain as to whether I had been watching an interesting effect of climate, or whether pollutants were after all at work, or both. Data accumulated on gonads and egg production have not yet been analyzed.

Fiddler workers will keep in mind a sharp division between artificial pollution and the presence of natural sewage, which gives one of the best guarantees of a rich and varied fauna of *Uca*. Such a favorable location is often easily spotted from the air in the tropics, as the plane circles for a landing, since it characteristically includes thatched huts, preferably on stilts, along the edge of a cove partly fringed with mangroves and flanking the mouth of the usual stream. Small fishing boats drawn up on such a shore almost insure good crabbing, except close to the frequent turmoil at the landing place itself.

In selecting such sites, it is useful automatically to remember three points. First, a close association often occurs here of species that are not usually sympatric; accordingly the assembly is in that particular atypical. Second, in future studies of the effects of crowding we shall very likely find that even conspecific members of such populations behave differently in some ways from aggregations with more space among the burrows. Finally, no able-bodied observer of fiddlers should avoid wading into the mud as usual and getting down to his customary crab's eye point of view.

Work in tropical mud deserves amplification. Aesthetically the experience practically never proves displeasing; regardless of their unsavoury reputation, mangrove swamps and flats frequented by thriving fiddlers smell only of good, flourishing vegetation and the fresh odor of tide-washed mud and salt air. We counteract the slight risk of infection after working near houses by scrubbing when back at the field base with soap and water to which we add a liquid disinfectant; the local pharmacy always carries some appropriate and usually familiar brand. As a further precaution we apply additional solution after washing to any area of skin that came in contact with the mud and allow it to dry there. Since modern workers take whatever prophylactic measures against disease that have been professionally recommended, most health risks are small. In the mud, especially, these include the danger of infection of small cuts, which should be well protected. Otherwise, nowadays most

indispositions are fortunately quickly curable, although they can be exasperating wasters of time and opportunity. In brief, only the local prevalence of a serious epidemic should keep a healthy worker out of the mud.

When a newcomer to a region is unfamiliar with the appearance of the local species in the field, he may not be able to recognize them in spite of preliminary work on preserved specimens. This difficulty may be partly because in each species the individuals show great variation in proportions due to allometric growth and to contrasts among males that are strongly leptochelous or brachychelous; almost all populations also show a striking range of color differences. Biologists who are not systematists sometimes feel understandably reluctant to become familiar with the morphological features of gonopods, preferring to concentrate on characters that do not need a lens, much less a microscope. Yet if a worker takes time before the trip briefly to investigate gonopods of expected species, he can catch sample males in the field and usually identify them in the hand with a pocket lens; when time in the locality is limited, this rapid certainty proves rewarding.

Finally, as will be amplified below, in work with unfamiliar species it seems best not to cut down on observation time in favor of operating cameras and other demanding instruments. Human eyes, binoculars, and patience give the best foundations.

EQUIPMENT AND ITS USES

The selection of instruments and supplies, and the divisions of time for their use, depends of course on the primary interests of the investigator, not to mention the size of his budget. The following remarks therefore provide rough guidelines only.

Binoculars. In my experience this important instrument's most desirable characteristics, aside from good optical quality, are light weight to aid prolonged use without shifting position, and adjustment to permit short-range focusing, preferably to less than 2 meters. The ideal magnification for me is $\times 7$. Unlike its effect in bird glasses, a narrow field is not a disadvantage.

Motion Picture Photography. For all serious motion picture work on fiddler crabs a 16 mm camera of professional caliber is essential. Lesser instruments at present do not seem sufficiently flexible for the demands of the work and in particular are inadequate for photographing in sufficient detail for ethological analysis many of the significant motions of individuals. The most useful lenses, all telephotos, have proved to be 63, 135, and 150 mm in focal length, respectively. The 63 mm lens gives the greatest magnification and good depth of field for small crabs that allow a close approach, since in the particular model used the lens can be racked out without removing it from the camera; extension tubes, which under field conditions are at the least inconvenient, become unnecessary. The 150 mm lens makes possible usable films of excessively shy populations. When air travel, combined with the need for taking other heavy equipment, makes it desirable to carry minimal cine gear, I take only the camera, two magazines, and the 135 mm lens along with its adapter; a camera case is omitted, since everything can be safely padded with clothing in a suitcase during travel; in the field plastic bags and an umbrella protect the instruments and film from mud and the weather. Styrotex picnic boxes are efficient insulators against heat and can usually be bought locally; otherwise the ubiquitous plastic pail serves well.

Exposure at 24 frames per second proves better than at 16 frames, not only because a sound track can then be added later if desired but because, since more frames cover a given action, inspection of fine details in projection or under a microscope is facilitated.

Color film, in spite of its relative slowness and the extra expense, is far better than black-and-white for observing and analyzing the motions of waving display, combat, and sound production, whether during ordinary projection or special analyses; on black-and-white film the crabs tend to merge with their backgrounds so that their morphological details can be frustratingly indistinct during attempts to distinguish, for example, their methods of stridulation. Since I have always used color film, I did not appreciate its advantages during analyses until good quality black-and-white prints were made to save the original from wear caused by repeated projections; the experiment was not a success, when judged even from the limited viewpoint of eyestrain alone. With care, projection does not hurt the original and, with specially important footages, an investment in color duplicates can and should be made.

No matter how limited the budget, ample film should be carried on a trip if at all possible; it should be exposed in quantity even if conditions are not optimal; many important insights into fiddler behavior, and suggestions for future work, were obtained through rerunning films made months and years later from film exposed for quite different reasons and often in bad weather.

For any behavioral study it is unsatisfactory to include many individuals in one frame in the hope of being able to analyze display and details of social interchanges in this way of an entire group. Details are invariably disappointing, and the use of long shots is, for serious film analysis, strictly limited. An exception that has not yet been tried will probably prove to be wide-angle photography directed vertically down on a population to record the progress of aggressive wanderers and of wandering females, or to determine the division of waving and non-waving time in individual burrow-holders. Because of the shortness of reels and the expense of film this project, when instruments improve in resolution, will probably be found to be a very suitable use for a television camera.

During film analysis, the duration of components and their parts are determined by counting the individual frames, either manually through a microscope or by means of the counter on a projector for time-motion studies. Accordingly, after every season in the field and more often after rugged use, the camera's motor should be checked to ensure the accuracy of its indicated speed.

Video Equipment. At this writing videotape is not yet available in color in portable television equipment. Since black-and-white videotape is not nearly equivalent in definition to color motion picture film, and because of other shortcomings, videotape images do not replace motion picture close-ups for ethological analysis. Nevertheless video equipment proves invaluable as a means of inexpensively recording large numbers of repetitions of the gross characteristics of waving displays; these sequences can determine reliably ranges in variation in such categories as duration of individual waves, height, and angle of the major cheliped, and elevations of the ambulatories during display. The equipment also serves excellently as a monitor close to the mouths of one or more individual burrows, since it can operate unattended for long periods. Finally video work is the only technique now available for proving that certain motions result in sound, the synchrony of sound and image being perfect. Related comments will be found in the following paragraphs.

Microphones, Tape-recorders, and Batteries. As far as is known, fiddlers do not make high-frequency sounds; accordingly the range in a good microphone for general use proves adequate. Future work may well show that certain kinds of stridulation produce high-frequency, airborne sounds, comparable to those of some orthopterans; if so, special equipment will of course be needed.

Tough, small, contact microphones, when pressed into the substrate close to the crab, are at present the most useful instruments for receiving fiddler sounds transmitted through the substrate. A variety of inexpensive hearing aids and guitar amplifiers have been used with success by others and by me.

Microphones capable of picking up airborne sounds produced by such components as leg-twiddling are expensive. I have had no success with any except a strongly directional instrument. Even the best must be used very close to the performing crab;

they are far more delicate than contact microphones and must be protected from direct contact with both substrate and moisture. For these microphones a small windscreen is essential and its size is important; although a large one may be more efficient, it may then constitute for the crabs a strange object of such importance that they need more than a few minutes to become habituated; sometimes a selected individual will not again during that particular low tide ever become active, or even reemerge from his burrow sufficiently close to the microphone for its operation.

Tests with two models of hydrophones were failures, since each instrument picked up subsurface sounds from, apparently, a number of neighboring fiddlers; not enough is yet known of fiddler sounds for the observer confidently to disregard all except a particular sound, much less to attribute each correctly to sex, phase, and species, or even usually to be certain that the producing animal is a fiddler. As soon as sufficient knowledge is accumulated, hydrophones, because of this very sensitivity, will certainly prove invaluable. For example they should serve well to detect social situations underground, to determine the presence and distribution of males and perhaps females when producing sounds deep in their burrows, and to detect antiphonies and choruses.

Tape-recorders of sufficient toughness for travel and for hard use in the tropics are now fortunately prevalent. In my experience their weak points remain their batteries. If rechargeable types are carried, along with a range of adapters for foreign electrical outlets, the problem is only partly solved. The best of them soon become weakened and hold their charges inadequately, even when they are fully recharged after each use. The trouble sometimes results from weak voltage in the local current; in this case charging for long periods—for example for 24 consecutive hours—sometimes will yield several needed hours of current in the field. More often the only reliable solution is to rent or buy a car storage battery. If power is needed daily at the same site for a week or more, and if the chosen spot is either sufficiently guarded, fenced, or isolated, the battery may be wrapped after use in a piece of plastic and left on a board above the level of high tide. Unfortunately, in perhaps most parts of the world likely to be visited by a short-term field worker the object would prove to be such a temptation to pilferers that its weight would be no deterrent. Nevertheless its dependable power is worth a large sum in car-hire money, perhaps not otherwise needed, to take it to and from the base.

Still Photography. The general methods and equipment employed in macrophotography of small living animals are altogether applicable to fiddler crabs, although the uses of such photographs are limited. Because of the fast and complex movements of socially active individuals, still photographs are almost always less useful for ethological study than in many other groups of animals. One needs to be able, for reconstruction of memories after field work and for comparative analyses, to review action patterns rather than simple postures. Obvious exceptions exist, such as various threat positions, the highest point a cheliped attains in waving display, the form of a structure beside a burrow, and the position held during a surface copulation; nevertheless these useful pieces of patterns are few. Again, color photographs show limitations in recording color changes. For scientific use, at least, a well-exposed sequence on color motion picture film, made at a suitably close distance, is worth much more than a series of stills in illustrating any aspect of fiddler behavior, and requires far less precious field time. The relatively poor quality of single frames for reproduction is counterbalanced by the wide choice of the moment to be illustrated.

Tripods. In most sequences it is extremely important to photograph the crabs close to ground level, as near as feasible to a crab's eye view. When taken from other angles the films may not show details vital to ethological analysis and correct interpretation. For example, a display sequence shot obliquely from above often cannot settle whether the major cheliped touches the ground in a drumming component, whether ritualized drumming occurs instead, or whether, in contrast, the major merus is vibrating against the suborbital region. On the other hand, occasional long shots and wide-angle views require normal heights for the camera.

A necessity therefore is a tripod with adjustable legs which when fully contracted measure less than 10 inches long. Such lengths are readily available in "table-top" tripods; unfortunately none of these models has the requisite strength to support the heavy, tilting, tripod head that must be added, plus the weight of a professional 16 mm motion picture camera, or of a television unit. If a light tripod is used, the film shows the effects of vibration, while a weak and unversatile head both fails to hold the camera at an angle and cannot cope with some of the work's demands. Finally, the design of the tripod's legs is important, since in spite of a large diameter they must be quickly sinkable, with the aid of strong distal points and a trowel, into firm substrates for further height reduction, while the locks on the joints must work easily for rapid extension at need. Frequent cleaning and oiling help avoid lost opportunities; as any photographer knows, few frustrations except a stuck camera are worse in a crisis than a balky tripod.

Although at least one adequate model used to be

available, it is no longer manufactured; at present a machinist must generally be asked to alter a heavy duty tripod to the desired specifications. While it is always possible to use a standard model sunk deep in a pit, the digging wastes time, proves impracticable in underlayers of coral or rock, slows position shifts, and, most important, messes up the habitat. It is far better to travel with a suitable instrument.

Collecting Tools. The best all-around tool for catching fiddlers is a gardening trowel. This implement is unknown in many parts of Asia and no substitute proves as useful, with the occasional exception of the human hand. Accordingly it is advisable to carry a heavy model of stainless steel; the handle and scoop should be forged in one piece; lighter designs will probably not last out the trip.

When crabs are wanted merely as preserved specimens, members of small species with shallow burrows can often be dug up efficiently by wielding the trowel as fast as possible and dropping the crabs into pails variously supplied with liquid, as described in a later section. Difficulties arise when the crabs are larger and live in deep burrows, when they are aestivating or hibernating in hard ground, when the substrate is laced with roots or rhizophorae, when stones are prevalent, or when the burrows extend into coral or creviced limestone. Under any of these conditions the collector must simply take whatever measures suggest themselves, from depending on difficult digging by hand, through the use of a shovel, to changing his activities to more cooperative terrain.

When large general collections are needed and no selection of individuals is required, the fastest method is to borrow a shovel and persuade a cooperative adult to wield it at a distance from the site of ethological observations; troops of enthusiastic small boys should be provided with jars or pails and rewarded for staying far away.

When particular individual crabs must be caught, whether as records following observations, after filming, or for transport alive to crabberies, a trowel still proves to be the most convenient tool. This procedure can never be hurried. The direction of the passage that slants downward inside the burrow mouth must first be determined. When the crab is in its burrow, the collector moves, with as little vibration of the substrate as possible, to a position on one side of the hole so that the trowel can be held vertically, close above the ground under which the passage lies and several inches from the hole itself. As the crab starts to emerge and before it becomes aware of strange objects close by, the blade thrusts swiftly into the ground, scooping underneath the crab and blocking its retreat. Although the subsequent capture and handling will shock the crab it recovers quickly. When the individual is destined for a crabbery it should be placed at once in a dish by itself, as described in the section on crabberies. If a trowel is not available, individual crabs may be similarly cut off from their burrows with any broad blade; butcher knives, machetes, and even a Malay kris have all worked.

One method seems most effective for seizing male fiddlers so that their appendages remain in place and their claws do not pinch human fingers. The system works even when a large male is sitting at arm's length in the bottom of a burrow. It consists of grasping the crab by placing your thumb firmly against the posterior part of his carapace and then using your fingers, with the first one bent above his major dactyl, to push his claw into the flexed rest position in front of his mouth region and to hold it there. If the crab is in a high intensity threat position above ground, with the claw "open," you must meanwhile force his major dactyl slowly down against his pollex with help from your other hand. Even females should be grasped similarly to prevent their losing some legs.

Marking Paint. Several opaque, fast-drying lacquers are available in artists' supply stores and hobby shops that are suitable for marking crabs. The brands selected by behavioral entomologists for use on bees and other insects are often appropriate. Providing the carapace and outer major manus of individual crabs are cleaned and dried before marking, and the marked crab kept in a dry pail for several minutes before releasing, the paint needs renewal in outdoor crabberies only about every four weeks, unless of course an individual molts.

I have not yet had success in marking crabs either in the field or in crabberies with spray paint. Out-of-doors air currents blow the spray even on calm days so that it either misses the target crab or gets in its eyes or soft areas connecting segments. When this happens the fiddler interrupts its activities to rub eyes or legs with other appendages; sometimes paralysis of an appendage occurs and sometimes captive crabs have soon died after showing one or more symptoms. I have watched similar effects on salticid spiders and butterflies when the paint by accident flowed over sensitive areas. An array of brands should be systematically tested in the home laboratory and perhaps a suitable formula concocted.

Sometimes particular crabs in the field can be touched with a brush fastened to the end of a dark-colored, slender, flexible rod of bamboo or other material. My chief difficulty in limited attempts has been that when lacquer dries fast enough to stick on the crab's next abrasive trip underground it also dries before the target crab can be touched, as I wait with the rod poised above its burrow. No matter how habituated it has become to an observer, it retreats almost inevitably before the paint-filled brush can be eased down to touch it; therefore the actual dab must await its emergence. Nevertheless some variation of this technique should be made eventually to work.

Color of Clothing and Equipment. Like birds and many other animals, fiddler crabs—easily startled by strange objects and abrupt motions—are keenly sensitive to moving objects that contrast with the substrate. Accordingly, in order to encourage their rapid habituation to a nearby human being, experienced observers avoid fast or sudden motions, white or pastel clothing, and large expanses of untanned European skin. Tennis hats can be tinted in coffee or tea, while a felt-tipped pen or crab-marking enamel will blacken the shine on a camera's chrome trim.

FIELD DATA

In the study of *Uca* it seems more important than in many groups that all notes, films, and recordings be accompanied by plentiful data on both meteorology and ecology. This necessity is caused by the importance of weather, circadian and other rhythms, and even microhabitats in the behavior of the crabs. The data include ideally the following information: exact geographical location, including any local names in dialect that might simplify the location of the study area by a later worker; date; time; weather during the period, as well as any special events, such as a typhoon in the recent past; temperature both at the surface and within a burrow; substrate; proximity and type of vegetation; associated species of *Uca* and other animals; size and degree of crowding of the population; relative numbers of displaying and non-displaying males; evidence of partial isolation of displaying males in a lek-like formation; prevalence of aggressive wanderers; number of wandering females; proportion of ovigerous to non-ovigerous females; presence and proportion of young; color phases and their distribution; information related to possible pollution. It is convenient when reviewing notes to find at the beginning of each day's data a note beside the date giving the time of low tide and the phase of the moon, such as "low 0850; 3rd day after full," even though this information, unlike the rest, is readily retrievable at home.

Especially important are records of precise display circumstances surrounding the observation or filming of particular episodes. All of these written details, particularly of short film sequences made on poorly known species, prove exceedingly useful when analysis is under way, months and sometimes years later. For example, helpful notes will include whether the filmed display appeared to be of high or low intensity, and whether the behavioral fragment was apparently elicited by the presence of a female beyond range of the lens or by another male making some particular motion.

When either a species, a locality, or the behavior on which I intended to concentrate was unfamiliar, it seemed essential to devote time for as long as possible—whether the first low-tide period or, in a long stay, a week or more—to observations and note-taking only, with photography and recording saved for later sessions. When this course was followed, the non-observational chores could all be planned and carried out far more intelligently. For me attempts to combine in a single session the use of camera and recorder with observation resulted in poor work in both areas. Odd notes scribbled inconspicuously, while waiting between instrument set-ups and their operation, sometimes of course prove invaluable, but they probably should be viewed as unexpected dividends; because of the division of attention, they may prove unreliable. As described in the preceding paragraph, exceptions are the descriptions of circumstances surrounding an episode, which should be written immediately after filming or recording.

A final exception to the postponement of the use of equipment seems to me to exist when, in a far-off place, a concatenation occurs that consists of very brief field time, uncertain weather, and rare or unexpected species; then it usually proves wise to photograph a few representative waving displays as rapidly as possible and so, in the photographer's phrase, "get some film under your belt."

DATA FOR ALLOMETRIC STUDIES

The growth characteristics of series of individual fiddlers were investigated in different populations of a number of species. In previous studies of the changing proportions of the major cheliped to the body, as in Huxley, 1924, the investigator selected the relation of the weight of the major cheliped to that of the total weight of the crab.

In the present study this procedure proved unsatisfactory and linear proportions were substituted. The relation of the carapace length to that of the propodus (major manus plux pollex) appeared to be the most useful and convenient. Efficiency in the use of the material required that at least these two measurements be made at the field base rather than at the home laboratory, since in spite of care some claws become detached during travel. In detailed growth studies of large collections all needed additional measurements should of course be made at the same time, to keep the measurements of individuals of the same length distinct. Since caliper measurements cannot be accurately made at less than about 10 mm, and because of the usual dearth of microscopes with micrometer scales at field bases, it is often necessary to take the time gently to flex the major cheliped against the front of the body of each small crab and tie it in place before packing for travel. Every measurement can then be taken with confidence at home.

This use of meristic relationships does not share several disadvantages characteristic of weight proportions, even when only ratios of major claw to body are investigated. First, preserved individuals do not lend themselves to any technique of weighing that makes the results acceptably comparable, in the absence of information on the effects of chemicals on the organs. Different kinds, strengths, and sequences of preservatives must, it seems, affect the weight of the material differently. Again, no formula has been devised to regulate the time to be given for draining off the liquid which would make the weights of crabs of different sizes fully comparable. Even if all specimens are thoroughly dried before weighing, the values obtained are probably undesirably artificial. Second, even if in contrast living specimens are weighed, each comparably dried and drained of water from the branchial chambers, the weighing must be done on a sensitive balance. Yet in most of the places where I made collections it was impossible to bring an adequate instrument and living crabs together. On tests in Trinidad, where we had a suitable balance, the total weight of a crab changed as much as 20 percent in either direction between weighing within two hours after capture and weighing again after one week in the crabberies. Significant variations in weight presumably occur also in crabs living under natural conditions. For example, when crabs in the dry season have not been feeding regularly, the general body weight may be expected to be less with respect to the major claw, the weight of which consists largely of the integument. It is obvious that linear proportions do not share these disadvantages.

PRESERVATION OF SPECIMENS

In preserving specimens of *Uca* for general study in the laboratory the following system has been adopted. It forms a compromise among several methods favored by various investigators. Its advantages for the needs of the present kind of study are the following: a minimum number of specimens lose their chelipeds; the internal organs are well preserved; the joints of the appendages are eventually left flexible enough for convenient manipulation. Although these intersegmental areas remain somewhat stiffer than in specimens placed sooner or solely in alcohol, the improved condition of the internal organs appears to give more than adequate compensation.

Jars or plastic pails are partly filled with 4 percent formalin made with fresh water in which crustaceans have already died. This solution, carried daily to the

collecting site, is used repeatedly and becomes increasingly effective. The animals in a well "ripened" solution succumb quickly with practically no struggle or autotomy, while the formalin prevents the internal decomposition which so often begins in the heat of a mud flat.

In the field base, whether laboratory or hotel room, the specimens are drained, rinsed gently in tap water, and covered to twice their depth with 4 percent formalin, previously unused, to which borax has been added in the proportion of one tablespoon per gallon. After seven days the crabs are changed to 70 percent grain alcohol and this, after another week, is changed once more. Crabs may be left longer than one week in formalin, but should be transferred to alcohol at the first opportunity. In many foreign countries pure ethyl alcohol (C_2H_5OH) is both prohibitively expensive and difficult to obtain. Here ethyl alcohol that has been slightly adulterated under government control, to make it unfit for drinking, is often both usable for crab preservation and readily available. Methyl (wood) alcohol (CH_3OH), however, should never be used except as a temporary last resort.

Sufficient full strength formaldehyde, with its bottle tightly sealed, padded and packed in a taped plastic pail, can safely be carried by air as checked baggage in sufficient quantity to last throughout a long field trip. In its place, emergency supplies can almost always be obtained in small tropical towns from the local undertaker, if not from a pharmacy; the only difficulty is the prevalence of holidays.

In an extreme emergency full-strength rum, whisky, arrack, or other hard liquor may be used full strength, but it should be replaced with a conventional preservative as soon as possible, since the specimens soften while the appendages gradually drop off. If a high-proof grade is available, it should of course be selected.

If formalin is not carried from the field base onto the mud flat and the crabs are alive on the return to headquarters, they may be placed in a refrigerator freezer for an hour, then well covered with 4 percent formalin and returned to the refrigerator overnight; at this time they should not be stored in the freezer compartment. After that washing and change of solutions continues as before. Or, on being brought from the field and in the absence of refrigeration they may be killed by covering to twice their depth with 10 to 15 percent formalin for 20 minutes. Washing, 4 percent formalin, and subsequent steps then continue as usual. The disadvantage to the latter technique is that while it prevents autotomy and struggle it leaves the legs permanently less flexible than any of the other methods. However in my opinion it is still preferable to the sole use of alcohol on either living or dead crabs, if both intact legs and well-preserved internal organs are desired.

Tweedie's suggestion (1950.2) for placing tropical grapsoids as soon as caught into wide-mouth thermos jars of ice water with floating ice is, as he recommends, unexcelled for small, delicate crabs, particularly of certain Indo-Malayan genera. The containers are, nevertheless, unwieldy and unnecessary for most *Uca*. A modern compromise is provided by styrofoam picnic boxes, in which ice can be heaped around plastic containers each holding a little ice water, in which the crabs can be dropped upon capture.

Well-preserved crabs travel successfully by air, when both the expense of weight and regulations against the transport of alcohol are factors, if the specimens are packed in plastic pails with only a small amount of preserving liquid and a layer of cotton or cloth soaked in weak formalin or water on top. Labeled packets of groups or individuals to be kept separate may be wrapped in porous material, such as cheesecloth, and tied with string. The pail's cover is sealed on with freezer tape. Needless to say, the crabs should be unpacked promptly on arrival and immersed in suitable preservative.

TRANSPORTATION OF LIVING CRABS

For ethological work in crabberies the only feasible system of transportation has proved to be the following. Individual crabs are selected in the field, collected as described above (p. 669), washed off by submersion in a clean plastic pail that has never been used for chemicals and is filled with seawater, and then placed at once, while the collector is still at the site, in a plastic refrigerator dish large enough to allow the crab ample space to move about. A small amount of the local seawater is added, not sufficient even fully to cover the bottom of the dish. The lid is taped on, preferably with freezer tape, since its adhesion is unaffected by water. Each dish is placed in the nearest patch of shade until time to return to the field base; it is never moved abruptly. If at any stage the journey involves transportation by car over a rough road, the dishes are protected by padding from both shock and engine heat; if styrofoam picnic boxes are available, they provide additional insulation for groups of the dishes; the car is driven slowly. The captures are preferably made one to two days before departure from the field base for the home laboratory. Local seawater, preferably taken at high tide from the shore where the crabs were captured, is brought to the local base in plastic pails that, as usual, have not been used for chemically preserved specimens. Each crab is submerged in a tumblerful of the water and allowed to move about freely at least once before traveling, while its dish is cleaned with

more of the seawater, thus removing mud and faeces. After each cleaning the crab is replaced in its dish with seawater only a few millimeters deep; the amount depends on the size of the crab and should be only enough to enable it to moisten its gills through the afferent apertures when it settles on the bottom. If the air is tropically humid, even less moisture is used in the final servicing before travel, since the moisture will condense at lower temperatures during flight, or during a cold-weather arrival in the north. For reasons still unknown, a fiddler's chance of survival is decreased when more than minimal water is provided; it may be that waste matter dissolved in the water acts as a poison. The boxes are sealed as usual with freezer tape. No food should be provided; fiddlers have been carried for seven days with no food and even without a change of water, providing they have been prepared for travel as just described. Ideally, as in a trip across the Pacific, a long voyage should be broken at, for example, Honolulu, and the containers cleaned, with the water replenished by freshly dipped, fully marine, unpolluted seawater, taken preferably at high tide and always in an uncontaminated plastic pail; no attempt should be made in the middle of a trip either to feed the crabs or to provide them with water from a local mud flat. When transporting crabs on long trips by sea the water may of course be provided from mid-ocean; even the crew on large liners seem to enjoy this kind of request, but must be warned to dip the water after leaving a polluted harbor and not to take it from near one of the ship's waste outlets. The crabs feed well on a trip of this kind on ordinary food for marine fish or for turtles; a few pellets should be provided once a day after cleaning, and removed within one hour. The dish should then be cleaned again and fresh seawater provided.

In the more usual trip by air, every attempt should be made to carry the crabs into the cabin and not ship them either as checked baggage or as air freight. In order to carry more than a small box or flight bag of specimens in the cabin, requests should be made well ahead to the local airline representative; in out-of-the-way places impressive letters of identification help, especially when stamped with institutional seals; in addition, polite persistence is usually essential. Sometimes the agent refers the matter to the pilot.

In case all persuasion fails and the crabs must be checked in the baggage compartment, the individual dishes should be packed in styrofoam boxes, which provide some degree of insulation, and marked clearly that the contents are live animals to be kept upright and away from heat and cold. An airline should not only be selected that is known to maintain the baggage compartments at cabin temperature and pressure, but the point should be rechecked locally before departure. "Approximately cabin temperature" sometimes turns out to be less than 15° C, which can kill a tropical crab. We do not yet know the effects of moderate but swift changes in pressure. I have had few occasions to carry crabs in small, unpressurized planes; the flights were always short and at low altitudes; no ill-effects were apparent.

If it is absolutely necessary to send crabs by air freight, they should be completely routed ahead of time, with space reserved for each leg of the voyage; day flights with stops at tropical airports should be avoided even when no transshipment is needed, since baggage compartments often heat up rapidly when the plane is on the ground; a transfer between airlines should also be avoided, unless it can be handled personally by an acquaintance, since in spite of the labels boxes can easily be left out-of-doors in freezing or torrid weather long enough to kill the entire contents; such shipments are always a gamble and the worst must be expected.

Two examples will emphasize the varied fortunes to be encountered during transportation. I have carried 92 living specimens of *Uca arcuata*, including adults and young of both sexes, from Japan to Trinidad without a single death; water was changed once, during an overnight stop in Honolulu; the entire trip took six days. On the other hand, another trans-Pacific collection of about 100 specimens of several species, including *arcuata*, was lost; I assembled, packed, and routed it in Hong Kong, with every safeguard dictated by experience, sending it by air freight; although the booking was followed in detail as shown by the papers delivered with the shipment, and although the trip only lasted three days, every crab was dead on arrival. More success has, fortunately, been achieved with other, shorter shipments of air freight, although nothing replaces the advantage of carrying the crabs by hand.

If a change of water is indicated en route at a point where no freshly collected seawater is available, a sufficient amount, taken at high tide as close to the open ocean as possible, should be brought with the traveler; modern plastic jugs with screw tops make the transport easy. The water may be safely checked in the baggage compartment; the point is that water from or close to the site where the crabs were collected should not be used except when absolutely necessary, and then taken at or close before high tide, because of the abundant animal and plant matter that may soon pollute the stored water. The transport of seawater for a change in mid-journey worked, for example, at a stopover in Paris on a trip from Ceylon to New York; the entire, hand-carried collection of 96 examples of *Uca lactea annulipes* came through alive and the great majority lived in excellent condition for several months in the crabberies at the New York Zoological Park; at that point the

work was terminated—but the crabs were still showing every sign of good health and going through cycles of social activity.

A common practice of commercial collectors in Florida and elsewhere is to ship local fiddlers by the hundred to laboratories, crowding the living specimens into cartons in a tangle of legs. Such a procedure does not work when specimens are wanted for behavior study, since the crabs' subsequent activity is affected; in any case it is suitable only for such relatively hardy forms as *pugilator*. Another system that does not work with *Uca* is the shipment of crabs in plastic bags filled with water and pumped-in air, in the method used so successfully in carrying and shipping small fishes.

CRABBERIES

For ethological work on fiddler crabs in captivity, only a few principles will be given. The details of successful construction and maintenance, and the suitability of various subjects for research, can vary within wide limits.

The essential points that need attention are provision for space suitable for the species to be kept; artificial tides; water appropriate for the species; a bank of substrate that, again, must be composed of material within the toleration of the species; an efficient filter system; and sunlight, whether real or an artificial facsimile. Almost equally important is care in the capture and transportation of specimens, as described above, and the provision of a population density resembling, at least initially, that found at the source of supply.

Regardless of the small size of the species under observation the smallest tank I have found useful measures 3 ft. by 5 ft. As has long been known, specimens of some *Uca* will live and even breed in fingerbowls, but for ethological work this sort of maintenance is virtually worthless. At the other extreme of size out-of-door crabberies 10 ft. by 15 ft. or more in dimensions are practicable and convenient, at least in the tropics. The minimum advisable depth of any crabbery seems to be with substrate, exclusive of filter, about 7 inches deep at its deepest part; more is of course preferable. Sometimes the tops of open-air crabberies need screening against predators, especially kingfishers; in the mountains of Trinidad these birds quickly learned of the new source of food. At the same station we had to fence the crabberies with wire netting to a height of about 8 inches even in installations so close to the house that the kingfishers did not trouble them; the reason was the attraction of crabbery water for marine toads (*Bufo marinus*) during the dry season; although they apparently did not seize the crabs, the amphibians jumped in and waddled about over the mud bank, knocking down burrow markers and generally disturbing the fiddlers.

The simplest method of simulating tides is to pour a few pails of seawater into the tank at the time of high tide and siphon it slowly out again, starting several hours after its introduction. Out-of-doors in their natural climate, all the tropical species of *Uca* kept in captivity, totaling about 28, responded with apparent health and appropriate behavior when provided with a single low tide in each 24-hour period, with the low arranged to occur at the same hour every morning. The caveats given at the end of this section on crabberies must be particularly heeded in the conduction of any ethological work under such an entirely artificial tidal regimen. Its great advantage is its convenience for indefinite maintenance of a population when minimum facilities and help are available.

In contrast, at the Zoological Park in New York three pairs of fiberglass tanks were maintained with success, seawater being pumped back and forth between the members of each pair with the time controlled automatically by a clockwork mechanism. The dials could be set so that the water flow would reverse any desired number of minutes later at each tide, in accordance with the schedule being followed. While this regimen provided a useful approximation to reality for several days, such as on holiday weekends, it became too quickly out of synchrony with the normal semi-lunar tidal periods to be used for longer periods under the only conditions when such precise schedules were needed. These occurred during particular observations when the crabs' activities had to be carried out under conditions as nearly approaching those in the field as possible. Accordingly, in the usual type of operation the clocks were set once a day during the regular maintenance period in strict accord with the tide table being followed, unless the tanks were under a simple maintenance regimen. At those times a morning or midday low tide was given, the choice depending on the natural period of highest activity for the species; since the clockwork mechanism was available, a low was also set to occur 12 hours later, at night. Just as in out-of-door tanks, the indoor population stayed in good shape indefinitely on this artificial schedule.

Between the extremes of the pail-and-siphon system and the clockwork controls in simulating tides, a number of intermediate systems have been used with success. For example, the two largest Trinidad crabberies were built on a slope, with a storage tank below each. To bring about low tide the tanks were slowly drained by gravity into the storage tanks. From these the water was pumped up again with regulated speed, in time to cover the bank of substrate at the appropriate hour. In this system only one inexpensive pump was needed for each tank; although it had to be started manually, it stopped automatically when the storage tank was almost empty. Whatever the system, just as with apparatus for any saltwater aquarium, the parts of the pumps and fittings that came into contact with the water had to be corrosion-proof and free of copper.

Whether indoors or outdoors it proved possible to use the same seawater for several months or more. Out-of-doors in the rainy season natural showers and downpours maintained the seawater in a naturally variable brackish state, the crabs being adaptable, in the species kept, to a wide range of salinity. In the dry season evaporation was so rapid that liberal additions of fresh water were needed daily from the garden hose; this water had not yet been chemically treated and so could be used from the tap. In the crabberies as in the field the salts in the water that stood in the crab holes during low tide were persistent, giving higher salinity in the burrows; these levels doubtless went far to counteract the wide swings in salinity we provided in the crabberies.

In the indoor crabberies in New York evaporation had also to be carefully watched, since the strong electric lights had an effect on the small tanks similar to that of the sun and wind out-of-doors. Again the daily addition of fresh water, along with frequent checks with a hydrometer, kept the salinity within optimal ranges. Care was taken to dechlorinate all tap water used with a material obtainable from any aquarium supply house; the plastic pails of treated water were allowed to stand at least 12 hours before use in the crabberies, as a safeguard, even though the directions sometimes state that the water is fit for use in ordinary aquaria at once. It was also important that the tap water reach a temperature comparable to that of the circulating "tides"; the fresh water was added to the storage tanks, or at least to the crabbery water at the time of the incoming flow, rather than poured suddenly onto the exposed bank itself. An exception which may, perhaps, be important was the daily simulation at the time of the diurnal low tide of a brief shower of rain, through dribbling dechlorinated water through a perforated plastic dish over the substrate. This procedure seemed to stimulate the crabs to a higher level of activity, just as does a true shower in the field in the middle of a hot and sunny day. Care was taken for some of the fresh water to get into the burrows from time to time, so that the salinity level did not build up excessively. Standing water from the burrows, either indoors or outdoors, was drawn out now and then by syringe and tested with a hydrometer, to be sure that the concentration had not risen above about 40°/00, the highest figure I have found in burrows in the field in Trinidad and the Philippines.

Crabs from special localities, such as the shores of the Red Sea, tolerate considerably higher peaks under their normal conditions of life, and probably should be maintained in captivity at similar levels.

Artificial seawater can always be used for short periods in crabberies, but it should probably not be used indefinitely, any more than such reliance is generally recommended for ordinary aquaria.

The substrate selected should be of the general kind in which the species to be kept most often occurs. For example, it obviously would not be wise to provide a mud-living crab such as *coarctata* or *maracoani* with a relatively sandy substrate that would be entirely suitable for *pugilator*. Nevertheless three of the most pleasant surprises resulting from the crabbery work were connected with substrate provisions, and emerged in New York. I had expected that it would be necessary to transport portions of their own substrate along with foreign crabs; this would have entailed great expense, moderate inconvenience, and an almost certain impasse at customs unless the soil were sterilized. Even if foreign crabs proved adaptable to local substrates, no doubt with the provision of appropriate (and unknown) kinds and quantities of added food, I foresaw that the artificial bank would have to be replaced at frequent intervals which, in temperate latitudes, would be difficult in the winter when microorganisms on local shores are largely inactive.

These apprehensions fortunately proved needless. Substrate from nearby Long Island was used without replenishment for periods lasting up to 6 months, and supported a variety of species from the same locality, from the West Indies and from Ceylon. We took great care to collect the substrate from a shore which had not been subjected as far as we could learn to chemical pollution and which remained the continuing habitat of large and apparently healthy populations of *Uca*. We were also careful to bring back to the laboratory only the top layer of substrate on which the crabs normally fed, and to layer the sections carefully in chemically uncontaminated plastic pails. Finally, we did not collect the material during the winter, November 1 of a mild autumn being the latest we gathered a load; there had been as yet no frost. We always included small tussocks of swamp weed which, although they died back during the winter, doubtless continued to provide organic richness and resprouted in the spring. Most surprising was the fact that substrate of this kind without replenishment and without provision to the crabs of supplementary food kept the captive populations in vigorous health. After our success the first winter—a full period of 6 months—we brought additional substrate from the same locality, but kept half of the original material to "season" the new sandy mud, and managed the addition with minimal upset to the crabs.

We feel sure that a most important part of the success indoors in New York was our provision for a filter system far more extensive than I can find is recommended for any home aquaria, either freshwater or marine. The systems used by marine biological stations and exhibition aquaria are not really comparable, because of the availability to them of circulating seawater and their lack of need for natural substrate for their specimens. The system that worked in New York I devised in consultation with experienced aquarists and a search of the literature. The result uses parts of several arrangements, as well as screens to prevent disruption of the filters by digging crabs, and, finally, the addition of the thick layer of substrate on top. The system includes a total of six different layers. It may prove in the future feasible to reduce or omit one or more of the ingredients, thus cutting down on their considerable cost; in the course of the work to date there has been no time to experiment along these lines.

The filter and substrate arrangement in each tank is as follows, with the layers listed from the bottom up. (1) Sheets of plastic, manufactured commercially for the purpose, their surfaces covered with parallel slits and their edges turned down so that each surface is raised slightly above the tank's bottom; several sizes fit against one another so that one side and a part of each end of a tank is covered, about one-third of the bottom being left bare; the siphons provided with the apparatus are discarded. (2) A layer of coarse, white, quartz gravel, at least two inches thick. (3) A similar layer of fine, coral sand. (4) A layer of powdered carbon. (5) A layer of fine, aluminum screening, to prevent the crabs from digging into the filter layers. (6) The substrate, arranged to form a gently sloping bank with its top extending along the tank's long dimension. Like the substrate the entire filter system can be reused with little replenishment when a tank is reorganized. An investigator familiar with aquaria will be reassured to find, on examining all levels of the filter, that they remain sweet, even though small crabs occasionally die in their burrows and dissolve without being found.

In both indoor and outdoor crabberies two additional safeguards are needed. First, the open, fourth side of the filter layers, adjoining the uncovered third of the tank bottom, are held approximately in place by a vertical strip of aluminum screening. This not only prevents sliding of the layers and digging by the crabs but also minimizes loss of material from the layers during drainage. Second, since most *Uca* are good climbers and since it is always desirable to have the top of the bank heaped as high as possible close to the lip of the tank, to provide maximum depth for the burrows, overlapping rectangles of window glass are thrust into the substrate all around the tank against its sides, except on the side bare of substrate; there it is not necessary provided the tank sides are very smooth and kept clean. The glass itself should be polished often. Once the crabs have settled down in the new quarters they normally do not attempt at all to climb out except during the wandering and aggressive wandering phases; then they literally climb the walls if measures have not been taken. The glass should project at least 5 inches above the substrate.

In outdoor tropical aquaria we have had success with only a thin layer of coarse sand as a filter, covering the portion of the tank beneath the substrate to a depth of several inches at most; we have also used instead a single, similar layer of coarse charcoal, such as is used in local stoves. In Trinidad, however, it was always possible to bring up frequent truckloads of mangrove mud and seawater from the shore, and because of this we changed the mud, seawater, or both on the average of once every 6 weeks during seasons when we were working seriously with the installations and not keeping them on a purely maintenance level. Again, the substrate remained completely sweet, even though the usual deaths sometimes occurred without our being aware of them, so that we failed to remove the casualties.

In New York the question of illumination gave us the most difficulty in design because of the need for an intensity sufficient to simulate sunlight. This strength seems to be of importance because in the field a population attains its highest levels of social activity when the sun is shining, even though at the peak of a display phase individuals often wave vigorously in dull weather and although some acoustic behavior is prevalent at night. At the end of our attempts we finally managed to achieve an intensity level approaching for short periods that of tropical sunlight, although overheating of the substrate remained a limiting factor and further improvement is needed.

Our first care was to select a group of 12 fluorescent lamps for suspension over each crabbery tank. Each storage tank also had its own group, so that the seawater would be exposed to light during a daytime period in that container; we hoped that the water's consequent exposure to a full quota of illumination daily would help to keep it in acceptably wholesome condition. Additionally, we sometimes used the storage tanks, on an alternate tidal schedule, to maintain crabs temporarily when we had more specimens than could be properly housed in the fully fitted crabberies, or when they housed large specimens being saved for physiological work in which maintenance conditions were not, it seemed, so demanding.

Each group of fluorescent lamps was composed of four kinds of tubes of known emission spectra, selected to resemble as a group the spectral composition of sunlight. Because of the limitations of their spectra, an exact reproduction could not of course be made. Fortunately, the crabs did well under this illumination alone. Nevertheless it seemed unsatisfactory that the level of intensity of the light reflected from the substrate, as measured in foot candles, was only equivalent to that reflected from a comparable substrate in the tropics on a cloudy day with imminent rain. The quality of the light furnished by the lamps was of course in some ways nearer to that of sunlight than was that of a tropical overcast, but the situation still called for improvement. Supplementary photofloods and similar lamps were unsatisfactory because of their short lives and high heat. We finally secured quartz halogen lamps and fitted them with heat absorbing glass filters that cut down the intensity only moderately. Limited tests with this setup indicate that their use during particular periods when observations, photographs, or experiments were under way did indeed stimulate the crabs to somewhat greater activity. The area struck by the beam had to be checked frequently for overheating; usually the crabs themselves gave the first warning, as they do on hot afternoons in the field, by dropping down their burrows and staying there—a response an investigator hopes fervently to avoid.

The large banks of fluorescent tubes were controlled automatically, as part of the clockwork system operating the pumps. When tropical species were kept, the lights were set to go on and off at 0600 and 1800 respectively, corresponding roughly to the times of sunrise and sunset in the tropics. Local hours were kept for temperate zone crabs. Since the laboratory admitted some daylight through a skylight, no attempt was made to keep crabs from overseas on their original circadian schedule; this particular biological clock in *Uca* is easily shifted without apparent effects on the behavior patterns to be investigated. The rhythm, as well as the tidal rhythm, had of course already been interrupted and upset by a long journey.

In conclusion, although fiddler crabs can be kept in health in captivity, and although surprising degrees and varieties of activities can be elicited, the use of crabberies for reliable ethological study is very strictly limited. I firmly believe that no descriptions of waving displays or combat behavior, for example, should be derived from observations made wholly in crabberies, whether indoors or out. A single example will illustrate the dangers. The usual waving rate of *festae* in the field is half again as fast as that shown by several individuals that lived for months in apparent health in a Trinidad crabbery out-of-doors and that went through the expected behavioral phases. Such a difference between a component in the field and in captivity is unusual, but it can serve as a warning. Nevertheless, to verify fine details of displays or of combat techniques, or to learn the kind of behavior that may yield specially interesting observations on future field trips, crabbery work is unexcelled.

The tanks also give excellent opportunities for certain kinds of photography and for sound recordings, including television work. Again, an appropriately designed crabbery is certainly the place where observations of crabs inside their burrows can best be made. Finally, various experiments, based safely on a foundation of familiarity in the field with the behavior of the species, can be done as well in crabberies, and some of them doubtless only under those conditions; the sensible possibilities for future work seem in fact to be unlimited. It seems very clear, nevertheless, that any observations or experiments that depend on quantitative aspects of the data obtained should be undertaken only with the greatest care; in particular they should rest on a solid basis of preliminary field work, accomplished on the particular subspecies, and preferably on members of the same population, from which the transplanted crabs are afterwards derived.

Appendix E. Conventions, Abbreviations, and Glossary

CONVENTIONS

! The specimens listed have been examined by the author of the present contribution. This convention is used in the systematic section under the headings Type Material and References and Synonymy.

( ) 1. A scientific name or initial enclosed in parentheses is that of a subgenus. *Examples. Uca (Deltuca) dussumieri, U. (D.) dussumieri*, or simply, in a list of species in the text, (*Deltuca*) *dussumieri*.

2. An author's name and date enclosed in parentheses after the name of a species indicate that when he described the species he placed it in a genus with a different name. *Example. Uca (Deltuca) dussumieri* (Milne-Edwards, 1852).

When no parentheses surround the author's name, his original description of the species placed it in the same genus in which it appears in the present study. *Example. Uca (Deltuca) demani* Ortmann, 1897.

[ ] A scientific name or initial enclosed in brackets is that of a superspecies. *Examples. Uca (Deltuca) [acuta] acuta* (Stimpson, 1858); *U. (D.) [a.] acuta rhizophorae* Tweedie, 1950.

Note. For general treatments of taxonomic practice, see Blackwelder, 1967, and Mayr, 1969. For technical details, see also the "International Code of Zoological Nomenclature adopted by the XV International Congress of Zoology," published for the International Commission on Zoological Nomenclature by the International Trust for Zoological Nomenclature; London, 1961. References to specific bulletins published by the International Trust are given in the present text.

Throughout this volume "Milne-Edwards," without an initial, refers to H. (Henri) Milne-Edwards; "A. Milne-Edwards" to his son, Alphonse. The hyphen between the two parts of the surname has often been employed in the past; its use is revived here as a possible aid to non-carcinologists who may only occasionally need to use the bibliography.

ABBREVIATIONS

AMNH—American Museum of Natural History; Central Park West at 79th St., New York, New York 10024, U.S.A.

Amsterdam—Zoölogisch Museum; Plantage Middenlaan 53, Amsterdam C, Netherlands.

Bishop—Bernice Pauahi Bishop Museum; Honolulu, Hawaii 96818, U.S.A.

BM—British Museum (Natural History); London, S.W.7, England.

Copenhagen—Universitetets Zoologiske Museum; København K, Denmark.

Frankfurt—Natur-Museum und Forschungs-Institut "Senckenberg"; Senckenberg-Anlage 25, Frankfurt am Main, Germany.

Göttingen—Zoologisch Institut der Universität; Göttingen, Germany.

Hancock—Allan Hancock Foundation; University of Southern California, University Park, Los Angeles, California 90007, U.S.A.

Leiden—Rijksmuseum van Natuurlijke Historie; Raamsteeg 2, Leiden, Netherlands.

MCZ—Museum of Comparative Zoology; Harvard University, Cambridge, Massachusetts 02138, U.S.A.

NYZS—New York Zoological Society; Bronx, New York 10460, U.S.A. (*Note.* Collections of *Uca* have been transferred to USNM; see p. 591.)

Paris—Muséum National d'Histoire Naturelle; Paris Ve, France. Address for correspondence on Crustacea: Laboratoire de Zoologie, 61 Rue de Buffon, Paris Ve.

Philadelphia—The Academy of Natural Sciences at Philadelphia; 19th St. and the Parkway, Philadelphia, Pennsylvania 19103, U.S.A.

Raffles—Raffles Museum, Singapore.

Torino—Istituto e Museo di Zoologia della Università di Torino; Via Accademia Albertina, 17, Torino (204), Italy.

UPNG—Department of Biology, University of Papua and New Guinea; Boroko, Territory of Papua and New Guinea.

USNM—Division of Marine Invertebrates, National Museum of Natural History; Smithsonian Institution, Washington, D.C. 20525, U.S.A.

Yale—Peabody Museum of Natural History; Yale University, New Haven, Connecticut 06520, U.S.A.

Yokohama—Faculty of Liberal Arts and Education; Yokohama National University, Yokohama, Kamakura, Japan.

GLOSSARY

INTRODUCTION

This glossary consists of terms that may be divided roughly into three categories. First are words used here in a restricted sense; many morphological terms are included for this reason. The second group comprises coined words and phrases, most of which are names for behavioral components in *Uca* or for structures on the gonopod. The final category includes terms that are widely used in particular fields of biology. Each appears in the glossary because the term may not be familiar to a worker outside the discipline it serves. When complex and sometimes controversial concepts are involved, recent definitions are quoted directly and references given to discussion by the authorities cited; in some cases an annotation comments on the term's use in the present study.

Abdomen. The segmented part of the body that is folded underneath the carapace, fitting into a depression in the sternum. It is narrow in males (Fig. 2) and broad in females. (P. 463.)

Actor. In combat, the individual at any particular moment performing the motions of a component. (P. 487.)

After-lunge. A feint by a burrow-holder directed toward a departing opponent. This activity is often associated with combat. (Agonistic component no. 6, p. 479.)

Agonistic. See *Behavior, agonistic.*

Alliance. A group of closely related species consisting both of allopatric forms and of one or more other species living sympatrically with one or more of the allopatric forms. This usage corresponds approximately to the definition given by Mayr (1969: 412) of superspecies used in the broader sense: ". . . entirely or *largely* allopatric species. . . ." (Italics mine.) The introduction of *alliance*, in an apparently new, informal usage of the word, provides a brief term to cover this wider situation, the term *superspecies* being reserved for a series of allopatric forms. Cf. *Superspecies.*

Allies. Members of an alliance. See above.

Allometry. "The study of proportion changes correlated with variation in size of either the total organism or the part under consideration. . . . The variates may be morphological, physiological or chemical . . ." (Gould, 1966: 629.) Also: "1.

growth of a part of an organism in relation to the growth of the whole organism or some part of it." (Random House Dictionary of the English Language, 1966.) (P. 449.)

Allopatry. The distribution of distinct populations of a species or of closely related species so that they occupy areas that with marginal exceptions are mutually exclusive. Adj.: *Allopatric.* In this study *allopatric* is used also as a n. (cf. patronymic, in accepted use as both adj. and n.). See also *Superspecies*; *Alliance*; cf. *Sympatry.* Discussions on pp. 432 and 527ff.

Ambulatory. One of the eight walking legs; arranged in four pairs, they are inserted behind the chelipeds. They correspond to the second to fifth pairs of legs or periopods of authors who count the chelipeds as the first pair of legs. The latter designation is of course phylogenetically exact, but in *Uca* the use of *chelipeds* and *ambulatories* is preferred for clarity and convenience. Where *leg* is used in this study, it is always for the sake of brevity in an unambiguous situation and refers only to ambulatories, as in a coined name for a behavioral component, such as *leg-wave.* (Figs. 1, 2.)

Angle, antero-lateral. The angle formed by the meeting of the anterior and side (antero-lateral) margins of the carapace. (Fig. 3.)

Antenna, pl. *antennae.* On the lower, anterior surface of the body, one of the outer pair of very short, flagellate appendages lying between the edge of the front and the buccal cavity. (Figs. 2, 28.)

Antennule. On the lower, anterior surface of the body, one of the inner pair of very short appendages lying between the edge of the front and the buccal cavity; it lies folded inconspicuously in a cavity. (Figs. 2, 28.)

Aperture, afferent branchial. The opening between the bases of the second and third ambulatories through which water is drawn into the branchial chambers.

Aperture, efferent branchial. The opening between the outer anterior margin of the buccal cavity and the outer anterior edge of the flexed palp of the third maxilliped. Through it water and bubbles are expelled from the branchial cavity.

Area, friction. In combat, a part of the inactor's claw correlated with one or more structures on the actor's claw during the performance of a component. See Table 14, "contact area."

Armature. Specializations of the integument consisting principally of ridges, tubercles, and grooves. Occurring on the carapace, chelipeds, and ambulatories, they are used in sound production or, when on the major claw, almost altogether in intermale combat.

Armed. Equipped with armature.

Autotomy. The casting off, by reflex action, of a cheliped or ambulatory that has been strongly stimulated, as when seized by a predator. The break is always cleanly made between the basis and ischium. The lost appendage is regenerated more or less perfectly, the process requiring a number of molts.

Basis, pl. *bases*. The second segment from the proximal end of a cheliped or an ambulatory. (Fig. 2.)

Beading. A row of similar, rounded tubercles, contiguous or nearly so and usually very small. See *Edge, beaded.*

Behavior, acoustic. Production of sound, along with associated activities.

Behavior, agonistic. Postures and motions indicating aggression, defense, submission, withdrawal, or flight; often ambivalent.

Behavior, conflict. Postures and motions indicating the simultaneous or alternate activation of two drives, such as combat and flight.

Behavior, displacement. Activities, inappropriate to the circumstances, which partially or wholly replace suitable behavior, such as the occurrence of feeding during conflict between drives to court a female and to threaten another male.

Behavior, post-combat. Activities, such as the after-lunge, that occur immediately following combat.

Behavior, precopulatory. Final stages of courtship immediately preceding copulation on the surface or presumed copulation underground. Characterized by behavior not found either in earlier stages of courtship or in agonistic behavior.

Behavior, social. Postures and motions that ordinarily serve as mutual stimulation among members of the same species.

Behavior, submissive. Postures and motions indicating unreadiness to engage in or to continue aggressive behavior. Cf. *Creep.*

Behavior, territorial. Postures and motions indicating readiness to defend from intruders the mouth of a burrow, or the mouth and its immediate vicinity; the behavior exhibited consists of waving display, threat components, and combat.

Belt, hybrid. "A zone of interbreeding between two species, subspecies, or other unlike populations; zone of secondary intergradation." (Mayr, 1969; 405.)

Biotope. A habitat characteristic of a group of animals, whether of a formal taxon, such as a species, or of a local population. Cf. *Niche.* (See also p. 445.)

Book. One of the contiguous layers of tissue in a gill. All lie perpendicular to each side of the gill's long axis. (Figs. 78, 81.)

Brachyura; adj. *brachyuran.* The order of Crustacea consisting of the true crabs, including *Uca.* In adults the abdomen is always relatively small and folds underneath the body, while the abdominal appendages are not used in locomotion.

Brachychelous. Individuals or species having the fingers of the major chela relatively short and broad in comparison with those of other members of the group. Cf. *Leptochelous.*

Bubbling. Emission of foam from the efferent branchial openings. Composed of air bubbles and moisture from the branchial regions, it apparently serves several functions, as a cleansing agent, heat regulator, aerator of eggs, and sound-producing mechanism. In its latter role it is described as component no. 12, p. 484. (General account on p. 472.)

Burrow-holder. A male in the display phase, centering his activities around a particular burrow, and defending its vicinity from intrusion by conspecific males through threat postures, combat, or both. (P. 487.)

Carapace. The crab's "shell," covering the dorsal and lateral parts of the body above the chelipeds and legs; anteriorly it includes the orbits and pterygostomian regions. The carapace is posteriorly truncate, the abdomen being bent underneath the body. (Figs. 1, 2, 3.)

Carpus. The fifth segment from the proximal end of a cheliped or an ambulatory. (Figs. 1, 2.)

Carpus-out. An agonistic posture in which a male, having descended his burrow, leaves the carpus of his flexed major cheliped projecting above the surface. (Agonistic component no. 7, p. 479.)

Cavity, buccal. The mouth area, lying between the antennae and antennules anteriorly and the chelipeds posteriorly. Covered, when not in use, by the third maxillipeds. (Fig. 2.)

Cavity, carpal. On a major cheliped, the depression in the proximal part of the palm. When the cheliped is flexed, bent into rest position, the carpus fits into the cavity. (Fig. 44.)

Chela. The grasping, distal portion of the claw on either of the two chelipeds. Each chela is formed by a distal extension of the sixth segment, the manus, and the opposing distal seventh segment, termed the dactyl. The two parts of the chela are referred to together as fingers; the lower, immovable finger as the pollex; and the pollex along with the rest of the manus as the propodus. (Figs. 1, 2, 42, 44.)

Chela-out. An agonistic posture in which a crab, otherwise completely withdrawn into his burrow, leaves his major chela's tip projecting. (Component 9, p. 479.)

Cheliped. One of the two appendages that end in a chela. Morphologically these appendages form the first pair of periopods. Each is bounded anteriorly by the suborbital region, externally by the vertical lateral margin, and posteriorly both by the sternum and by the coxa of the first ambulatory.

Cheliped, major. The large cheliped; confined to males. (Fig. 1.)

Cheliped, minor. The small cheliped in a male. (Fig. 1.)

Cheliped, small. Either of the two chelipeds in a female, or, when used plurally in simultaneous reference to both sexes, the chelipeds excluding the major.

Chimney. A wall made of substrate erected by an individual around its burrow. (P. 500.)

Chromatophore. A pigment-bearing body in the integment that influences the individual's color through its expansion and contraction.

Claw. The sixth and seventh segments of a cheliped, formed of the propodus (the specialized manus) and the dactyl. An informal term, but conveniently brief. See *Chela.*

Claw-rub. A general term for sound produced by rubbing of the claws of two males against each other during combat. (Sound component no. 14, p. 484.)

Claw-tap. A general term for sound produced by tapping or vibration of the claws of two males against each other during combat. (Sound component no. 15, p. 484.)

Cline. "A gradual and nearly continuous change of a character in a series of contiguous populations; a character gradient (cf. *Subspecies*)." (Mayr, 1969: 400.)

Clock, biological. See *Rhythm, endogenous.*

Close-set (adj.). Applied to tubercles in the same series which, while not widely spaced, are not continuous and hence cannot be termed beading.

Combat. A general term for any behavior between males in which the claws of the chelipeds come into contact. Because of the usually high degree of ritualization, it might be preferable to substitute the word *encounter*, thus avoiding the "loaded" words *combat* and *fight*. Since the behavior discussed undoubtedly has an aggressive base, frequently shows overtly forceful components, and probably often includes pushing elements effectively masked by ritualizations, it seems permis-

sible to use all three terms. In this contribution, therefore, *combat* is selected for general use; *encounter* appears occasionally in the discussion of fully ritualized combats; and *fight* is restricted to combats with overtly forceful components.

Combat, heteroclawed. In one opponent the claw on the right side is enlarged, in the other on the left.

Combat, high-intensity. Part of the claw of each opponent comes between the dactyl and pollex of the other, in other words enters into the gape. In forceful endings the claw tips may grip the opponent's claw; in fully ritualized encounters they do not do so.

Combat, homoclawed. Both opponents have the claw of the same side enlarged, whether right or left.

Combat, low-intensity. Contact between opponents is confined to the outer surfaces of mani and chelae.

Combat, mutual. In many ritualized combats both crabs perform one or more of the components, either in sequence or alternately. They can perform simultaneously only during manus-rubs.

Combat, ritualized. Encounters consisting of ritualized components and lacking the ingredient of irregular force.

Component. An activity that is a characteristic part of some aspect of social behavior and appears to the observer to be distinct from adjacent actions.

The most clear-cut examples are so distinct and so stereotyped that they may confidently be termed fixed-action patterns, in the sense developed by Lorenz and Tinbergen and now often used in ethological studies. These examples are characterized not only by distinctive motor patterns but in ritualized combat by the juxtaposition of specialized morphological structures.

Other activities, however, show considerable variability connected neither with intensity nor with transition to other components. All of these, instead, often appear instantly adaptable to the changing circumstances of an agonistic situation or a courtship. In combat, for example, the sequence of components used in a particular kind of encounter in a certain species is only moderately predictable, while force may be suddenly interjected in the midst of an otherwise ritualized combat at any time. Any or all of the components in the various classes of social behavior may need subdivision or other modification. Only further study with emphasis on comparative work within the genus can resolve the uncertainties.

Therefore it seems that the use of *fixed action pattern* would be at present a semantic disservice. The more general word *component* is adopted instead, in the same spirit shown by morphological taxonomists when they feel it premature to use a definite term such as *subgenus* and compromise on the noncommittal *group.*

Component, forceful. In low-intensity combat, manus-pushes; in high-intensity combat, grips, flings, and upsets. All are highly variable and irregular, and hence are here considered to be un-ritualized, since the cheliped is wielded variously and unpredictably in pushing, grasping, and lifting. Most or all ritualized components at times also are interrupted by the use of overt force which sometimes develops almost insensibly during the performance of the ritualized component, in the form of visible pressure. Exceptions to the above definition of forceful components are now being studied in *Uca lactea*, in which overt force itself is sometimes ritualized. (P. 494.)

Component, mutual. In combat, the same component is performed by both crabs, either in sequence or alternately. Only the manus-rub can be performed by both crabs simultaneously.

Component, ritualized. In combat, a component in which no pushing, grasping, or lifting motions are ordinarily included and which is distinguished by its predictability of form, and sometimes sequence, and by its association with a particular series of structures. See also *Component, forceful* and *Ritualization.*

Conglomerate. An assembly of stones, sessile marine organisms, or both, fastened together by natural deposits or secretions; the organisms often include corals, dead or alive, mollusks or their shells, and tube worms. Sometimes the conglomerate is in the form of lumps, large or small; sometimes it forms in itself a local substrate.

Cornea. See *Eye.*

Courtship. Behavior patterns in both sexes that, when fully elicited, are followed by copulation. They usually include high-intensity waving display, along with following of the female or her attraction to the male's burrow. Special display motions and sound production also sometimes are part of a species-specific pattern. See also *Behavior, pre-copulatory.*

Coxa, pl. *coxae.* The first segment of a cheliped or ambulatory, always short (Figs. 1, 2).

Creep. Method of locomotion adopted by non-aggressive individuals under certain conditions, the body being held close to the ground. Cf. *Behavior, submissive.* (Agonistic posture no. 11, p. 479; Fig. 84C.)

Crenellations. Tubercles along the suborbital margins of the orbit, often separated and truncate. (Figs. 2, 3, 26, 27.)

Crest. On a cheliped or walking leg, a thin ridge, often relatively high, with the edge either entire or tuberculate. Cf. *Ridge.*

Curtsy. In waving display and in high-intensity courtship, the crab's body rapidly lowers through bending the legs and is raised again. A bob. (Component no. 13, p. 496.)

Dactyl. The most distal segment, the seventh, of a cheliped or ambulatory. On a cheliped also termed the movable finger.

Dactyl-along-pollex-groove. In high-intensity combat the actor slides the tip of his dactyl along the narrow, longitudinal furrow on the inactor's outer pollex. (Ritualized component no. 14, p. 491.)

Dactyl-slide. In high-intensity combat the prehensile edge of one dactyl slides along the upper edge of the opponent's dactyl. (Ritualized component no. 6, p. 489.)

Dactyl-submanus-slide. In high-intensity combat the actor rubs his dactyl's prehensile edge along the lower margin of the inactor's manus, both claws being appropriately tilted. (Ritualized component no. 8, p. 489.)

Deme. "A local population of a species; the community of potentially interbreeding individuals at a given locality." (Mayr, 1969: 401.)

Dendrogram. A diagram, more or less in the form of a tree, designed to indicate apparent degrees of relationship.

Depression. An indentation on the integument, usually shallow and irregularly shaped.

Depression, H-form. On the carapace the design formed by the meetings of the furrows dividing the cardiac, intestinal, and branchial regions from one another. Often this roughly H-shaped result includes the only distinctly marked regional boundaries in a species of *Uca*, where the regions are in general weakly indicated. (Fig. 1.)

Display, visual. A general term that includes agonistic postures, their associated motions, and waving display.

Display, waving. A rhythmic motion of the major cheliped, along with any associated movements of other appendages. Used interchangeably with *waving.*

Display, waving: high-intensity. Waving characterized in general by the maximum tempo, precision of motions, amplitude of wave, brevity of pauses between waves, and sometimes additional motions or elision of motions characteristic of a species.

Display, waving, low-intensity. The wave is relatively slow, and often variable in tempo and amplitude, with the series widely spaced and pauses between waves often longer than at high intensity; feeding during and between waves is prevalent; characteristics may be present that are absent at high intensities, or the converse may be true. Sharp boundaries between the two intensities are rare, gradual change being the rule.

Down-point. A threat posture of high-intensity often leading to combat. Two opponents face each other, their major claws pointed straight downward. (Component no. 2, p. 479.)

Down-push. One crab is pushed down his own burrow by his opponent. An activity associated with combat. (P. 491.)

Drove. An aggregation of individuals, moving more or less in unison. (P. 478.)

Drumming. Sound production through repeated tapping of the major or minor merus against the carapace, or of the major manus against the ground. (Sound components 7, 8, and 9, p. 483.)

Duration. The length of a combat timed from the moment at which the two chelipeds come into contact to their separation immediately preceding the departure of one of the crabs. Associated activities, ranging from preliminary threat behavior to after-lunges, are not included. See also Duration, p. 493.

Ecology. The study of organisms in relation to their environment.

Edge, beaded. A structure sometimes present on the upper, major palm, between the dorsal margin and the carpal cavity. (Fig. 44.)

Edge, prehensile. On a cheliped, the dorsal margin of the pollex or the ventral margin of the dactyl.

Encounter. A fully ritualized combat. See also *Combat.*

Estuary. A tidal waterway in the delta region of a stream or river, usually running through mangrove or other kinds of swampland. See also *Lagoon.*

Ethology. "The science of the comparative study of animal behavior." (Mayr, 1969: 402.) In a wider sense, "the biological study of animal behavior." (Eibl-Eibesfeldt, 1970.)

Eye. The faceted structure at the tip of the eyestalk; the cornea.

Eyebrow. An elongate area along the dorsal margin of the orbits, varying in length, breadth and inclination; bounded by raised edges that are sometimes beaded or granulate.

Eyestalk. The peduncle supporting the eye.

Fight. A combat including components that are at once forceful, irregular in form, and unpredictable. See also *Combat.*

Finger. On a cheliped, either of the two distal elements forming the chela or pincer. Usually used in the plural, to designate in one word both dactyl and pollex.

Finger, fixed. The pollex.

Finger, movable. The dactyl of a cheliped.

Flagellum, pl. *flagella.* On each of the two antennae, the slender, tapering, distal portion composed of many segments. (Fig. 28.)

Flange. On the gonopod, a calcified wing near the tip, normally extending anteriorly or posteriorly but sometimes differently oriented because of torsion. (Fig. 58.)

Flat-claw. An agonistic posture in which a crab, having descended his burrow, leaves his major claw projecting and bent so that it lies flat on the surface. (Component no. 8, p. 479.)

Fling. In combat, a variable, unritualized component at the close of a forceful ending. One opponent is pushed backward in a skid or is partly overturned. (Forceful component no. 2, part, p. 488.)

Forward-point. A threat posture of moderate intensity. The major claw is directed forward, the fingers held apart. (Agonistic component no. 4, p. 479.)

Front. On the anterior part of the carapace, the middle section that projects forward and down between the orbits. (Figs. 1, 2.)

Frontal-arc. A threat pattern of low intensity; the major chela parallels the ground, fingers open, and moves forward and back. (Agonistic component no. 3, p. 479.)

Furrow. See *Groove.*

Gape. On a claw, the space between the dactyl and pollex when their distal portions are in contact or, in other words, when the claw is "closed."

Gills. Organs responsible for the extraction of oxygen from the water. Five principal ones, all elongate, distally tapering and divided into sections (*books*), are located in the posterior part of the branchial region; single gills, small to vestigial and variously shaped, are located proximally on the 2nd and 3rd maxillipeds. (Figs. 81, 82, 83.)

Gonopod. In males, one of the pair of anterior abdominal appendages, situated on the ventral, proximal part of the abdomen. When not in use, these slender, stiff appendages are bent forward, parallel to the similarly bent abdomen, fitting into an indentation in the sternum. Spermatophores are introduced into the female gonopores through the specialized tips. (Fig. 58.)

Gonopore. In females, one of the pair of genital openings. They are located on the third sternal segment near the midline. The term has often been applied also to the genital opening of the male; in this study the word *pore* is used for the male opening in order to avoid confusion.

Granule. An imprecise term for a tubercle that is considered by the observer to be particularly small in relation to other tubercles on the same individual.

Grip. In combat, the prehensile edges of one claw seize the claw of the opponent. (Forceful component no. 2, part, p. 488.)

Groove. A narrow depression or furrow in the integument.

Group. "A neutral term for a number of related taxa, especially an assemblage of closely related species within a genus." (Mayr, 1969: 404.)

Growth, allometric. Development in which one part

of an organism grows at a different rate than another; heterogony. (P. 449.)

Habitat. The general kind of environment in which a species usually lives. (P. 440.) (Cf. *Biotope; Niche.*

Heel. The proximal ventral portion of the major manus. (Fig. 42.)

Heel-and-hollow. In high-intensity combat, the pollex tip rubs or taps the depression on the palm lying at the base of the opponent's thumb, while the dactyl acts similarly against the outer manus. (Ritualized component no. 11, p. 490.)

Heel-and-ridge. In high-intensity combat, the pollex tip rubs or taps the oblique ridge of the opponent, while the dactyl curves around the opponent's heel or taps with its tip against that region. (Ritualized component, no. 12, p. 490.)

Hepatopancreas. A digestive gland, consisting mostly of two large lobes that underlie the paired hepatic regions of the carapace. Formerly often termed the *liver.*

Herding. During a waving display, a male maneuvers a female toward his burrow. (Component no. 14, p. 496.)

Heterogony. (Replaced in current usage by *Allometry.*)

High-rise. A posture, occurring in the threat behavior of both sexes and in the courtship of males, in which the crab raises its body high on the extended ambulatories. (Agonistic component no. 13, p. 480.)

Holotype. In taxonomy, "the single specimen, designated or indicated as 'the type' by the original author at the time of the publication of the original description." (Mayr, 1969: 404.)

Hood. A concave, arching structure erected by a male in display phase beside his burrow. (P. 500; Pls. 48, 49.)

Hybridization. "The crossing of individuals belonging to two unlike natural populations, principally species." (Mayr, 1969: 405.) See also *Hybridization, allopatric; Subspecies.*

Hybridization, allopatric. "Hybridization between two allopatric populations (species or subspecies) along a well-defined contact zone." (Mayr, 1969: 397.)

Inactor. In combat, the opponent holding his major claw temporarily motionless. (P. 487.)

Ischium. On a maxilla, maxilliped, cheliped, or ambulatory, the third segment from the proximal end; in each of these appendages it is distal to the basis. (Figs. 1, 2.)

Instigator. In combats between two burrow-holders, the crab that approaches his future opponent; except for this approach, he does not necessarily ever become an actor. In combats between an aggressive wanderer and a burrow-holder, the wanderer is apparently always the instigator. Because of the usual high degree of ritualization, the words *aggressor* and *attacker* are not used.

Instrument. In combat a structure on the actor's claw used during the performance of a component.

Intensity, low. (1) In threat postures and motions, a general term for slight motions of the major cheliped toward an encroaching crab, in which the fingers are separated only partly if at all and the body little raised on the ambulatories; often accompanied by feeding. (Fig. 84.) (2) In combat, a component in which only the outer surfaces of the claws come into contact. (3) In waving display, the motion of the major cheliped is relatively slow when compared with that of high intensity, the tip of the claw usually does not reach as high in the air, in lateral waves the waving motion may be straight rather than circular, and, finally, any special motions occurring in advanced courtship are absent; often accompanied by feeding, low-intensity display usually indicates relatively simple territoriality rather than threat toward a particular male or definite courtship of a female. There is no sharp dividing line distinguished between low and high intensities in any of the three uses.

Intensity, high. (1) In threat postures and motions, the major dactyl is raised so that the fingers of the claw are widely separated and the body is raised on the extended ambulatories. (2) In combat, a component in which a part of each claw enters the gape of the opponent. (3) In waving display, the major cheliped moves at the maximum speed attained by the species, and describes the widest arc; in both threat and courtship, but particularly in courtship, special motions of both the major cheliped and of the other appendages sometimes involved in sound production are often included. Actual feeding only very rarely accompanies high-intensity behavior of any kind, the motions of the minor cheliped that often occur at these times being incomplete; they are here interpreted as displacement feeding under conditions of conflict.

Interdigitated-leg-wag. The more distal segments of one or more ambulatories overlap those of a parallel individual with accompanying rubbing or vibration. (Sound component no. 16, p. 484.)

Interlace. In high-intensity combat, each manus lies within the gape of the opponent, the bases of the two chelae coming almost into contact; the proximal prehensile edge of one pollex then rubs along one or both predactyl ridges of the opponent. (Ritualized component no. 9.)

Isolation, reproductive. "A condition in which interbreeding between two or more populations is prevented by intrinsic factors." (Mayr, 1969: 410.) Cf. *Mechanism, isolating.*

Jerk. In waving display, one of a series of short motions of the major cheliped that lifts or lowers the appendage with short pauses between. In this contribution the period of motion alone is counted as the jerk. No wave is described as including jerks unless either the upward or downward sweep, or both, are broken by at least one pause; a wave with a pause only at its peak, however, is not considered to be a jerking wave.

Jerking-oblique-wave. In waving display, the major cheliped unflexes obliquely upward with jerks at least during its rise. (Component no. 6, p. 496.)

Jerking-vertical-wave. In waving display, the flexed major cheliped is moved up and down with jerks at least during its rise. (Component no. 2, p. 496.)

Keel. A sharp ridge armed or smooth, at or close to the dorsal or ventral margin of part of a cheliped or ambulatory.

Kick. See *Leg-wave.*

Lagoon. A body of seawater, sometimes brackish, usually elongate, and always largely cut off from the surf of the open ocean. Sometimes the barrier is a reef, which may be largely submerged; sometimes it is the upper part of a beach or a stretch of dunes. Communication with the open sea is often by a permanent channel, as on coral atolls; sometimes, on continents and larger islands, a connection occurs only at spring tides or during heavy rains. Direct connection with a stream, if any, is frequently intermittent. Except on small atolls, lagoons are often bordered by mangrove swamps. The word *lagoon* and its translations are often used imprecisely, with variable implications, in different parts of the world. The above remarks cover its use in the present contribution. Cf. *Estuary.*

Lateral-circular-wave. In waving display the major cheliped is completely unflexed, then raised and returned to rest position from above the eyes. (Component no. 5, p. 496.)

Lateral-straight-wave. In waving display, the cheliped is completely unflexed, then returned to the flexed position in the same plane. (Component no. 4, p. 496.)

Lateral-stretch. A threat posture in which at least the major cheliped is unflexed to the side. (Agonistic posture no. 10, p. 479.)

Lateral-wave. In waving display, a general term for a motion of the major cheliped in which the claw is unflexed to the side. See also *Lateral-straight-wave* and *Lateral-circular-wave.*

Lectotype. "One of a series of syntypes which, subsequent to the publication of the original description, is selected and designated through publication to serve as 'the type.' " (Mayr, 1969: 406.)

Left-clawed. The cheliped on the crab's left side is enlarged.

Leg. See *Ambulatory.*

Leg-side-rub. Stridulation involving the more posterior ambulatory meri and the sides of the carapace. (Sound component no. 6, p. 483.)

Leg-stamp. The turned-under dactyls of two or more ambulatories strike the ground. (Sound component no. 11, p. 484.)

Leg-stretch. In waving display the crab raises its body with each wave. (Component no. 10, p. 496.)

Leg-wag. Stridulation involving rubbing of the ambulatory meri. (Sound component no. 5, p. 482.)

Leg-wave. In waving display, two or more ambulatories are raised on a side, often in a kicking motion; their meri however do not make contact and therefore there is no stridulation. (Component no. 12, p. 496.)

Legs-out. Posture assumed by a non-receptive female; when descending into her burrow she leaves the ambulatories of one side projecting in the air. (Agonistic posture no. 14, p. 480.)

Lek. A restricted locality in which males in many species of birds and other animals congregate in the breeding season. Here they defend individual territories from encroaching males and court females. In a restricted sense, the word is used of associations in which the strongest males hold territories near the center and do most of the mating. Leks in the restricted sense are not yet known to occur in *Uca.*

Leptochelous. The pollex and dactyl of the major cheliped are unusually long and slender in comparison with most individuals in the same species of similar carapace size. Cf. *Brachychelous.*

Line, raised. A ridge, long and very low, not paralleling similar structures. Cf. *Stria.*

Lunge. A threat pattern of high intensity. One male, with claw pointed forward, makes a feint toward another. (Agonistic component no. 5, p. 479.)

Major. Adjective preceding either the noun *side* or the name of an appendage or one of its segments; the word indicates that the location of the area is on the same side of a male as the large cheliped.

Major-manus-drum. Vibration of the manus of the major cheliped against the ground. (Sound component no. 9, p. 483.)

Major-merus-drum. Vibration of the merus of the major cheliped against an anterior part of the carapace. (Sound component no. 7, p. 483.)

Major-merus-rub. Stridulation in which part of the major merus rubs against an adjacent part of the carapace. (Sound component no. 1, p. 482.)

Mandible. One of the two small jaws, heavily calcified and meeting in the midline, in the buccal cavity. In a ventral view of the crab they lie underneath the three pairs of maxillipeds and two pairs of maxillae. Morphologically the mandibles are anterior to the first maxillae. (Fig. 37.)

Mangrove. Any of a number of shrubby and arboreal

plants associated on sheltered coasts of the tropics and subtropics and in their brackish river deltas. Members of such a community are characterized most conspicuously by stilt roots. The community itself is called in English *mangrove* or *mangroves*. Macnae (1968) suggested substituting the word *mangal* to distinguish the forest community from its individual plants; the suggestion was published too late to be followed here, but its use in the future would certainly be sensible.

Manus, pl. *mani*. The sixth segment from the proximal end of a cheliped or ambulatory, it is simultaneously the predistal segment. On each of the two chelipeds its ventral distal portion is produced, forming the lower part of the chela, the pollex; the manus and pollex together are termed the *propodus* (*q.v.*). In discussions of the major cheliped, the word *manus* signifies its outer side, plus its dorsal and ventral margins. Cf. *Palm.* (Figs. 1, 2.)

Manus-rub. In low-intensity combat, the outer mani of the opponents rub longitudinally. (Ritualized component no. 1, p. 488.)

Margin, antero-lateral. On the dorsal part of the carapace, that portion of the side margins immediately behind the antero-lateral angle.

Margin, dorso-lateral. On the dorsal part of the carapace, a long raised line, smooth, beaded, or tuberculate, starting at the posterior end of the antero-lateral margin and continuing posteriorly and somewhat toward the center. (Figs. 1, 3.)

Margin, suborbital. On the antero-ventral part of the carapace the lower edge of the orbit; it is usually more or less crenellate at least near its outer angle. (Figs. 2, 3.)

Margin, vertical lateral. On the side of the carapace the long raised line extending upward from the ventral margin, directed toward or reaching the junction between the antero-lateral and dorso-lateral margins; often obsolescent at least in upper part. (Fig. 3.)

Maxilliped. Any one of six mouthparts, consisting of three pairs. The third pair, the outermost, includes broad, flat segments that fit closely over the buccal cavity. The second pair, in a ventral view of the crab, underlies the third pair and is external to, and overlying, the first pair. The first pair is external to and overlies the maxillae. Morphologically the maxillipeds are posterior to the maxillae. (Figs. 2, 33, 36.)

Mechanisms, isolating. "Properties of individuals that prevent successful interbreeding with individuals that belong to different populations." (Mayr, 1969: 405.)

Megalops. A post-embryological stage of crab development occurring after the zoea and before the crab molts into the first crab stage. In most crabs, including all *Uca*, both zoea and megalops are free-swimming. In *Uca* the few larvae so far identified are pelagic but most individuals of many species probably pass even their early stages in brackish water and a few far up tidal streams. Cf. *Zoea*.

Membrane-vibration. Air or air and water inside the epibranchial chambers vibrate against the membranes at the chelipeds' proximal ends. (Sound component no. 13, p. 484.)

Merus, pl. *meri*. On a third maxilliped, the more distal of the two large segments that cover the buccal cavity; on a cheliped or ambulatory, the fourth segment from the proximal end; on each of these appendages the merus is distal to the ischium. On chelipeds and ambulatories it is the most proximal large segment; it is always longer than broad and is followed distally by a short carpus. (Figs. 1, 2, 33.)

Microhabitat. A local subdivision of a habitat.

Mill, gastric. See *Stomach*.

Minor. Adjective preceding either the noun *side* or the name of an appendage or one of its segments; it indicates that the area is on the same side of a male as the minor cheliped. Cf. *Major*.

Minor-chela-tap. The tip of the claw on the minor cheliped strikes the ground several times. (Sound component no. 10, p. 484.)

Minor-merus-rub. Stridulation in which the merus of the minor cheliped rubs against an adjacent part of the carapace. (Sound component no. 2, p. 482.)

Minor-claw-rub. Stridulation in which the claw of the minor cheliped rubs the suborbital crenellations. (Sound component no. 3, p. 482.)

Minor-merus-drum. Vibration of the merus of the minor cheliped against the anterior part of the carapace. (Sound component no. 8, p. 483.)

Minor-wave. In waving display, the minor cheliped moves similarly to the major. (Component no. 9, p. 496.)

Mound. A low elevation with rounded top on the carapace or an appendage.

Mud. "The result of the reaction and interaction of alluvium and living organisms." (Macnae, 1956; 40.)

Mud flat. Expanses of muddy land left uncovered at low tide. Most occur near river mouths that often open into sheltered bays. In the tropics the flats are often partly bordered with mangroves.

Neotype. "A specimen selected as type subsequent to the original description in cases where the original types are known to be destroyed or were suppressed by the [International Commission on Zoological Nomenclature]." (Mayr, 1969: 407.)

Niche. "The precise constellation of environmental factors into which a species fits or which is required by a species." (Mayr, 1969: 407.) Cf. *Habitat*; *Biotope*.

Orbit. One of the pair of elongate trenches extending along most of the anterior part of the carapace, in which the eyestalk and eye lie when not in use. (Fig. 3.)

Osmoregulation. See *Regulation, osmotic.*

Ovary. In non-breeding *Uca* females, each of the two egg-bearing glands is confined in the coelomic cavity to a region between the heart and the posterior branchial region. In breeding individuals each ovary extends conspicuously far forward and laterally and they are in part joined in the midline. (Fig. 77.)

Overhead-circling. In waving display a component in which the cheliped does not return to the flexed rest position during a series of waves but describes aerial circles above the crab's body. (Component no. 8, p. 496.)

Palm. On a cheliped, the inner surface of the propodus, proximal to the pollex; in other words, the inner side of a claw proximal to the fingers. (Figs. 2, 44.)

Palm-leg-rub. Stridulation resulting from rubbing of the major palm against the anterior side of the first ambulatory. (Sound component no. 4, p. 482.)

Palp. The segments of a maxilliped distal to the merus.

Paratype. "A specimen other than the holotype which was before the author at the time of the preparation of the original description and was so designated or indicated by the original author." (Mayr, 1969: 408.)

Pattern, fixed action. See *Component.*

Pereiopod. See *Ambulatory.*

Phase. A temporary state that is characterized in males by one of a number of general behavior patterns. (P. 505.)

Phase, aggressive wandering. A male moves apparently at random through a population that includes displaying males, punctuates his passage with threats toward them, engages them in combat, makes superficial burrow explorations, and attempts unsuccessfully to mate.

Phase, display. In males the temporary condition characterized by waving display, burrow-holding, threat, combat, and courtship.

Phase, non-aggressive wandering. A male moves through a population or forms part of a drove, feeding near the tide's edge. He does not threaten or enter into combat, often passes close to displaying males in a crouching posture, does not hold a burrow during low tide, does not wave, and does not court.

Phase, territorial. A condition intermediate between the aggressive wandering phase and the display phase, often or usually very brief to absent. At this time a male holds a burrow and threatens other males but does not wave or court.

Phase, underground. A male remains underground in his burrow throughout one or more low-tide periods; the period of submersion is usually several days unless the crab is undergoing a period of hibernation or aestivation.

Phenetics, numerical. "The hypothesis that relationship of taxa can be determined by a calculation of an overall, unweighted, similarity value." (Mayr, 1969: 408.) See also *Taxonomy, numerical.*

Pheromone. An external secretion selectively affecting the behavior of conspecifics.

Phylogeny. "The study of the history of the lines of evolution in a group of organisms; the origin and evolution of the higher taxa." (Mayr, 1969: 409.)

Pile. A group of short, thickly set setae, often occurring in patches on the carapace and appendages; the appearance is reminiscent of velvet or fur.

Pillar. A tower-like structure built of substrate raised by a male in display phase beside his burrow. (P. 500.)

Pit. An indentation in the integument, small to minute and more or less rounded.

Pleopod. One of the paired appendages arising on the ventral sides of certain abdominal segments. In males there are only two pairs, one on each side of the first two segments; the first of these consists of the two genital organs or gonopods. Females have one pair on each of the first five segments; they are modified for holding the egg-mass.

Pollex; pl. *pollices.* On a cheliped's claw, the fixed finger, which is the ventral, distal extension of the propodus. (Figs. 1, 2, 42, 44.)

Pollex-base-rub. In high-intensity combat a component performed when the inactor has partly descended his burrow, leaving the claw projecting. The actor rubs the prehensile edge of his pollex against the outer pollex base of his opponent. (Ritualized component no. 5, p. 489.)

Pollex-rub. In low-intensity combat the outer sides of the pollices rub longitudinally. (Ritualized component no. 2, p. 488.)

Pollex-under-and-over-slide. In high-intensity combat, one pollex slides in turn along the ventral and dorsal edges of the opposing pollex. (Ritualized component no. 3, p. 488.)

Population, local. "The individuals of a given locality which potentially form a single interbreeding community." (Mayr, 1969: 409.)

Pore. The genital opening of the male, located on the distal end of the gonopod. (Fig. 58 *D, E.*) Cf. female's *Gonopore.*

Posing. A rigid posture adopted occasionally by individuals of both sexes in which the chelipeds are usually unflexed and the carapace tilted toward the sun. (P. 506.)

Post-megalops. The first littoral stage or instar; also known as the first crab stage. Cf. *Zoea* and *Megalops.*

Prance. A motion in which a male walks stiffly on the bent-under dactyls of the ambulatories. Possibly sounds are produced. (Agonistic component no. 12, p. 479.)

Pregape-rub. In high-intensity combat the tips of the dactyl and pollex rub longitudinally along the distal outer manus and palm, respectively, of the opponent. (Ritualized component no. 10, p. 490.)

Process, inner. A structure on the distal end of the gonopod arising on the inner side but often displaced anteriorly by torsion. Its shape varies from a large and tumid projection to a slender transparent spine. (Fig. 58; p. 464.)

Prolonged-leg-stretch. In waving display the crab holds his body high off the ground throughout a series of waves. (Component no. 11, p. 496.)

Propodus. In crustaceans generally, the predistal segment of a leg-like appendage; in this contribution, confined to either of the two chelipeds in *Uca*; it there consists of the manus and its distal extension, the pollex. The propodus on the ambulatories in this contribution is termed the *manus* (*q.v.*). (Fig. 2.)

Pubescence. See *Pile.*

Raised-carpus. A threat posture of low intensity. The major carpus is raised while the claw points obliquely down. (Agonistic component no. 1, p. 479.)

Rap; rapping. Terms used by Crane (1941.1ff.) in previous descriptions of waving display; they were equivalent to the present general term *drumming.* The behavior formerly called *rapping* was found to include manus-drums, merus-drums, and ritualizations of both. (P. 483.)

Region (of carapace). An area, usually convex, separated from its neighbors by narrow grooves.

Region, branchial. (Paired.) A large lateral region overlying the branchial cavity. (Fig. 1.) The region includes two other areas with names used in this contribution, the epibranchial and metabranchial regions (see below).

Region, cardiac. (Unpaired.) The central area on the carapace bounded anteriorly by the mesogastric, laterally by the branchial, and posteriorly by the intestinal regions. (Fig. 1.)

Region, epibranchial. (Paired.) The anterior part of the branchial region. It is bounded anteriorly and externally by the margins of the carapace and internally by the hepatic region. (P. 471.)

Region, hepatic. (Paired.) An antero-lateral area bounded anteriorly by the orbital region, internally by the mesogastric region and other gastric areas, and both externally and posteriorly by the branchial region. (Fig. 1.)

Region, intestinal. (Unpaired.) The most posterior of the central areas of the carapace, bounded anteriorly by the cardiac region, laterally on each side by the branchial region and posteriorly by the carapace margin. (Fig. 1.)

Region, mesogastric. (Unpaired.) The central area of the carapace that is bounded posteriorly by the cardiac region and laterally by the hepatic and branchial regions. (Fig. 1.) Anteriorly lie other gastric areas.

Region, metabranchial. (Paired.) The posterior, larger part of the branchial region, it is bounded anteriorly by the epibranchial part of the branchial region and the hepatic region, and internally by the mesogastric, cardiac, and intestinal regions; it overlies the cavity containing the large gills. (P. 469.)

Region, orbital. (Paired.) The area immediately behind the eyebrow and adjacent parts of the upper margin of the orbit. (Fig. 1.)

Region, pterygostomian. (Paired.) On the antero-ventral part of the carapace, the area external to the buccal cavity. (Fig. 2.)

Region, suborbital. (Paired.) On the antero-ventral part of carapace, the area immediately behind the suborbital margin. Paired. (Fig. 2.)

Regulation, osmotic. "Osmotic regulation may be defined as the regulation of the total particle concentration of such fluids at levels different from those of the external medium." (Robertson, 1962: 323.)

Reversed-circular-wave. In waving display as in the usual lateral-circular-wave, except that the cheliped is unflexed at a high point and flexed into rest position at a lower level. (Component no. 7, p. 496.)

Rhythm, circadian. An endogenous rhythm with a cycle roughly 24 hours in length.

Rhythm, endogenous. A cycle under physiological control that continues to operate for a time on its previously established schedule in the absence of the appropriate external stimuli.

Rhythm, lunar. An endogenous rhythm with a cycle roughly 28 days in length.

Rhythm, semi-lunar. An endogenous rhythm with a cycle roughly 14 days in length.

Rhythm, tidal. An endogenous rhythm with a cycle roughly 12.4 hours in length, its particular temporal characteristics changing continually in accordance with the tidal schedule prevailing locally.

Ridge. An elongate, narrow elevation of the integument with a more or less sharp edge that is sometimes tuberculate or otherwise armed. Cf. *Stria*; *Line, raised*; *Rugosity*; *Crest*.

Ridge, distal at dactyl base. The more distal ridge, usually tuberculate, on the upper distal part of the major palm. (Fig. 43.)

Ridge, oblique; *ridge, oblique tuberculate.* On the upper, distal part of the major palm, a raised area with its projecting edge usually sharply ridge-like but sometimes blunt; usually with tubercles extending from the carpal cavity to the vicinity of the proximal ventral part of the pollex. (Fig. 43.)

Ridge, proximal at dactyl base. The more proximal ridge, usually tuberculate, on the upper distal part of the major palm; vertical or nearly so in its upper part, its lower portion curves distally to merge with the tubercles of the inner row on the dactyl's prehensile edge. (Fig. 42.)

Right-clawed. The cheliped on the crab's right side is enlarged.

Ritualization. In the evolution of an animal's behavior, changes in a movement and certain associated structures so that they come to serve as a signal in communication, or as a different signal, or to function in some other social capacity. (P. 519.)

Rugosity. A roughness of the integument, usually in the form of short, blunt, irregular ridges, non-tuberculate, irregularly shaped, and arranged in groups, the alignment of the individual rugosities being roughly parallel.

Sac, pericardial. (Paired.) A water-absorbing organ in the posterior part of the branchial cavity. (Fig. 78.) It aids in controlling the maintenance of moisture.

Salinity. The proportion of dissolved materials in seawater. Measured in parts per thousand (°/00). Normal seawater ranges, depending on latitude, between 34°/00 and 37°/00.

Sand. "A material consisting of comminuted fragments and water-worn particles of rocks (mainly siliceous) finer than those of gravel; often *spec.* as the material of a beach, desert, etc." (*The Shorter Oxford English Dictionary*, 3rd ed.)

Semi-unflexed-wave. In waving display the cheliped is partly unflexed as it is raised. (Component no. 3, p. 496.)

Series. (1) "In taxonomy, the sample which the collector takes in the field or the sample available for taxonomic study." (Mayr, 1969: 411.) (2) In waving display a group of waves and their associated motions performed by an individual in an uninterrupted sequence.

Serration. One of a series of somewhat compressed tubercles occurring on the margins of appendages; each is more or less triangular with the apex directed toward the distal end of the segment.

Seta, adj. *setose.* An unjointed appendage, either soft and hair-like or a stiff bristle. Setae vary greatly in microscopic characteristics and many are doubtless sensory in function; they occur plentifully on most parts of the body and jointed appendages.

Seta, spoon-tipped. On a mouthpart, especially on the merus of the second maxilliped, a seta ending in a more or less concave expansion, usually lobed or pectinate. (Figs. 36, 37.)

Sonagram. "A graphic representation of the vocalization of an animal." (Mayr, 1969: 411.)

Speciation. "The splitting of a phyletic line; the process of the multiplication of species; the origin of discontinuities between populations caused by the development of reproductive isolating mechanisms." (Mayr, 1969: 412.) Cf. *Speciation, Allopatric*; *Speciation, sympatric.*

Speciation, allopatric. "Species formation during geographic isolation." (Mayr, 1969: 397.)

Speciation, sympatric. "Speciation without geographic isolation; the acquisition of isolating mechanisms within a deme." (Mayr, 1969: 412.)

Species. "Groups of actually (or potentially) interbreeding natural populations which are reproductively isolated from other such groups. . . ." (Mayr, 1969: 412.) Cf. *Subspecies*; *Population, local*; *Isolation, reproductive*; *Mechanisms, isolating.*

Species-specific. An attribute characteristic of a species; often used in connection with behavior patterns or their components.

Spermatophore. A capsule containing spermatozoa transferred through the male gonopod into the female gonopore.

Spine. A long, pointed tubercle.

Sternum. The segmented, ventral surface of the body lying between the proximal segments of the two chelipeds and of each pair of ambulatories. (Fig. 2.) In males the narrow abdomen folds forward into a groove that runs longitudinally down the sternum's midline. In adult females the broad abdomen, folded similarly forward, almost covers the sternum.

Stomach. A general term for the central part of the gut. Its most conspicuous component is the so-called gastric mill, a muscular sac with hard, internal ridges located between the lobes of the hepatopancreas. (Fig. 80.)

Stria. In this contribution used only of a ridge or raised line, very small to minute in all dimensions—length, height, and thickness; usually more or less parallel to other such structures; it may be unarmed, beaded, or tuberculate. This meaning of the word is included among the definitions given, for example, in *The Shorter Oxford English Dictionary*, 3rd ed.; there it appears as follows: "2.

Chiefly in scientific use. A small groove, channel, or ridge." Cf. *Ridge*; *Line, raised*.

Stria, postero-lateral. A stria on the postero-lateral part of the carapace, behind and usually external to the postero-lateral end of the dorso-lateral margin; when fully developed there are two, more or less parallel to each other, on each side. (Figs. 1, 3.)

Stridulation. Sound production by a single individual performed by rubbing, tapping, or vibrating one part of the body against another, either or both of which are suitably armed. The parts involved may be those of an appendage against the carapace or of two or more appendages against one another.

Style. A slender projection of the eyestalk beyond the distal end of the eye (cornea).

Subdactyl-and-subpollex-slide. In high-intensity combat, the upper edge of one dactyl moves to and fro longitudinally along the prehensile edge of the opponent's dactyl, while the prehensile edge of the pollex moves similarly along the lower edge of the opponent's pollex. (Ritualized component no. 4, p. 489.)

Subdactyl-and-suprapollex-saw. In high-intensity combat, one dactyl's prehensile edge moves transversely across part of the prehensile edge of the opponent's pollex. (Ritualized component no. 15, p. 491.)

Subgenus. An optional taxonomic category, consisting of a group of species the members of which appear to be more nearly related to one another than they are to other members of the genus.

Subspecies. "A geographically defined aggregate of local populations which differs taxonomically from other such subdivisions of the species." (Mayr, 1969: 412.)

Substrate. A general term for the terrestrial components of a biotope, composed partly or largely of inorganic matter, such as mud or sand.

Superspecies. "A monophyletic group of entirely or largely allopatric species." (Mayr, 1969: 412.) In the present contribution the term is used as follows: a group of allopatric species, each of which appears to be more closely related to its neighbors within the superspecies than to other members of the genus, with the occasional exception of sympatric members of its alliance. The term has no official taxonomic standing. In headings the name is enclosed in brackets, as suggested by Amadon 1966: 245). Cf. *Alliance*; *Allopatry*.

Supraheel-rub. In high-intensity combat, the dactyl rubs vertically against the heel of the opponent's manus, the pollex meanwhile lying against his palm. (Ritualized component no. 13, p. 491.)

Sympatry. "The occurrence of two or more populations in the same area; more precisely, the existence of a population in breeding condition within the cruising range of individuals of another population." (Mayr, 1969: 413.) In the present contribution, use of the word is restricted as follows: in the broad sense sympatry indicates the occurrence of more than one species of *Uca* in the same area, within sight of one another and without physical barriers between the populations; in the restricted sense it indicates such an occurrence of members of the same subgenus, particularly of those showing the closest interspecific relationship, as among members of an alliance (*q.v.*). Adj.: *Sympatric*; in this study used also as a n. (cf. allopatric).

Synonym. "In nomenclature, each of two or more different names for the same taxon." (Mayr, 1969: 413.)

Synonymy. "A chronological list of the scientific names which have been applied to a given taxon, including the dates of publication and the authors of the names." (Mayr, 1969: 413.)

Syntype. "Every specimen in a type-series in which no holotype was designated." (Mayr, 1969: 413.)

Systematics. "The science dealing with the diversity of organisms." (Mayr, 1969: 413.)

Tapping. In combat of both low and high intensities, part of the prehensile edge of the dactyl, the pollex, or both tap against part of the opponent's dactyl or propodus. Tapping sometimes forms a part of a number of ritualized components. (P. 491.) Cf. *Vibration*; *Claw-tap*.

Taxon, pl. *taxa*. "A taxonomic group that is sufficiently distinct to be worthy of being distinguished by name and to be ranked in a definite category." (Mayr, 1969: 413.)

Taxonomy. "The theory and practice of classifying organisms." (Mayr, 1969: 413.)

Threat. Aggressive postures and motions confined in males to individuals in the aggressive wandering, territorial, and display phases; in females they appear when approached by an aggressive wanderer, or when in a non-receptive phase. Although in males threats usually precede combat, the behavior appears far more often than combats. See agonistic postures and motions, and discussion; pp. 478, 516.

Thumb. (1) In older literature the fixed finger or pollex on a cheliped. (2) On the male gonopod, a subdistal structure (Fig. 56). Equivalent to the *palpus* of von Hagen, 1962.

Tomentum. See *Pile*.

Tooth. A tubercle, larger than the other tubercles in the same series and usually of a different shape.

Topotype. "A specimen collected at the type-locality." (Mayr, 1969: 413.) Such specimens are not necessarily given official taxonomic status.

Torsion. A characteristic of the male gonopod in some species in which part of the distal end is twisted so that one or more of its structures are displaced from their more usual positions. (P. 464.)

Tubercle. A small projection on the integument, either blunt or more or less conical, but of widely diverse shapes and sizes. Usually in series or groups. When very small in comparison with adjacent armature, the structure is here termed a granule, the name being imprecise. If larger and, particularly, different in shape from its neighbors, it is called a *tooth.* See also *Serration, Spine, Crenellation,* and *Edge, Beaded.*

Type. "A zoological object which serves as the base for the name of a taxon." (Mayr, 1969: 413.)

Type-locality. "The locality at which a holotype, lectotype or neotype was collected. (Cf. *topotype.*)" (Mayr, 1969: 414.)

Type-species. The species in a genus or subgenus which was designated as its type.

Upper-and-lower-manus-rub. In high-intensity combat the prehensile edges of one claw rub along the upper and lower margins of the opponent's manus. (Ritualized component no. 7, p. 489.)

Upset. In combat, an unritualized final component in which one crab is turned completely upside-down. The rarest of all combat components. (Forceful component no. 2, part, p. 488.)

Vertical-wave. In waving display, the flexed cheliped moves up and down. (Component no. 1, p. 496.)

Vibration. In sound production and in combat the rapid tapping by an appendage against the carapace, another appendage, or the substrate, or, in ritualized waving display, certain motions in the air. For present practical purposes, the word is confined to motions repeated at rates faster than 50 per second. At this rate, a blur on the film results when motion picture film is exposed at 24 frames per second, giving, when the frames are viewed through a dissecting microscope, a convenient means of distinguishing in combat between tapping and vibration. Vibration is here used synonymously with *drumming,* as employed in the names of certain behavioral components. Cf. *Tapping.*

Walking leg. See *Ambulatory.*

Wanderer. (1) A male in a non-aggressive phase, characterized by lack of attachment to a particular burrow. See *Phase, non-aggressive wandering.* (2) A receptive female at least in the subgenera *Minuca* and *Celuca,* as she moves through a displaying population.

Wanderer, aggressive. See *Phase, aggressive wandering.*

Wave. A general term for the rhythmic raising and lowering of the major cheliped during waving display. (P. 494.)

Wave, diminishing. In waving display, a wave lower than its predecessor in a single series.

Wave, primary. In waving display, the first and highest wave in a series.

Whitening, display. Temporary color lightening during the display phase of males and less often of females; it is characterized by the expansion of white chromatophores at least on the carapace. (P. 466.)

Withdrawal. Behavior associated with combat, in which a burrow-holder descends partway or entirely into his own burrow. (P. 491.)

Zoea, pl. *Zoeae.* One of the aquatic larval stages of crabs. Its outstanding morphological features are the occurrence of one or more large spines on the carapace. See also *Megalops.*

Bibliography

Note: For abbreviations of serial publications "The World List of Scientific Periodicals" has been used. In the few cases where a particular periodical appeared to be unlisted, I have abbreviated it in this bibliography in a form consistent with that of the "World List," except that unusual words are spelled out.

References of importance for this study that were received too late for incorporation in the text are listed in the Addendum, p. 715.

Adams, A., and A. White
1848.   Crustacea. *In* The zoology of the voyage of H.M.S. Samarang; under the command of Captain Sir Edward Belcher, during the years 1843-1846, ed. Arthur Adams. Published under the authority of the Lords Commissioners of the Admiralty. Reeve, Benham, & Reeve, London. (Crustacea: viii + 66 pp.)

Alcock, A.
1892.1.   On the stridulating apparatus of the red *Ocypode* crab. Ann. Mag. nat. Hist. (6)10: 336.
1892.2.   On the habits of *Gelasimus annulipes*. Edw. Ann. Mag. nat. Hist. (6)10: 415-16.
1900.   Materials for a carcinological fauna of India. No. 6. The Brachyura Catometopa or Grapsoidea. J. Asiat. Soc. Bengal 69: 279-486.
1902.   A naturalist in Indian seas, *or* Four years with the royal Indian marine survey ship "Investigator." London. xxiv + 328 pp.

Alcock, A., and A. R. Anderson
1894.   Natural history notes from H.M. Indian marine survey steamer "Investigator," Commander C. F. Oldham, R.N., commanding. Series 11, No. 14. An account of a recent collection of deep-sea Crustacea from the Bay of Bengal, Laccadive Sea. J. Asiat. Soc. Bengal 63: 141-85.

Allee, W. C.
1931.   Animal aggregations. A study in general sociology. University of Chicago Press.
1938.   The social life of animals. Norton, New York. 293 pp.

Allee, W. C., A. E. Emerson, O. Park, T. Park, and K. P. Schmidt
1949.   The principles of animal ecology. W. B. Saunders Company, Philadelphia and London. xii + 837 pp.

Altevogt, R.
1955.1.   Beobachtungen und Untersuchungen an indischen Winkerkrabben. Z. Morph. Ökol. Tiere 43: 501-22.
1955.2.   Some studies on two species of Indian fiddler crabs, *Uca marionis nitidus* (Dana) and *U. annulipes* (Latr.). J. Bombay nat. Hist. Soc. 52: 702-16.
1956.1.   Der Mechanismus der Nahrungsaufnahme bei Winkerkrabben. Naturwissenschaften 43 (4): 92-93.
1956.2.   Neue Untersuchungen an indischen Winkerkrabben. *In* "Verhandlungen der Deutschen Zoologischen Gesellschaft in Hamburg 1956." Akademische Verlagsgesellschaft Geest & Portig K.-G., Leipzig: 148-50.
1957.1.   Untersuchungen zur Biologie, Ökologie und Physiologie indischer Winkerkrabben. Z. Morph. Ökol. Tiere 46: 1-110.
1957.2.   Beiträge zur Biologie und Ethologie von *Dotilla blanfordi* Alcock und *Dotilla myctiroides* (Milne-Edwards) (Crustacea Decapoda). Z. Morph. Ökol. Tiere 46: 369-88.
1957.3.   Text associated with FILM. Zur Biologie Indischer Winkerkrabben. Wissenschaftl. Film D 756 des Instituts für den Wissenschaftlichen Film. Göttingen. [Not seen.]
1959.   Ökologische und ethologische Studien an Europas einziger Winkerkrabbe *Uca tangeri* Eydoux. Z. Morph. Ökol. Tiere 48: 123-46.
1962.   Akustische Epiphanomene im Sozialverhalten von *Uca tangeri* in Südspanien. Verh. dt. zool. Ges. supplement to Zool. Anz. 22: 309-15.
1963.1.   Wirksamkeit polarisierten Lichtes bei *Uca tangeri*. Naturwissenschaften 50: 697-98.
1963.2.   Lernversuche bei *Uca tangeri*. Zool. Beitr., NF. 9: 447-60.
1964.1.   Ein antiphoner Klopfkode und eine neue Winkfunktion bei *Uca tangeri*. Naturwissenschaften 51: 644-45.
1964.2.   FILMS. Encyclopaedia Cinematographica, Göttingen. [Not seen.]

E 691. *Uca tangeri* (Ocypodidae). Nahrungsaufnahme. (Duration: 3 minutes.)

E 692. *Uca tangeri* (Ocypodidae). Drohen und Kampf. (Duration: 3 minutes.)

E 693. *Uca tangeri* (Ocypodidae). Klopfen und Winken. (Duration: 9 minutes.)

1965.1. *Uca tangeri* (Eydoux, 1835) in der *Terra typica*. Crustaceana 8: 31-36.

1965.2. Lichtkompass- und Landmarkendressuren bei *Uca tangeri* in Andalusien. Z. Morph. Ökol. Tiere 55: 641-55.

1968. Text associated with FILMS (1964.2, above). Encyclopaedia Cinematographica, Göttingen: Publ. Inst. Wiss. Film 2a (3).
E 691: 277-84.
E 692: 271-76.
E 693: 259-69.

1969.1. Ein sexualethologischer Isolationsmechanismus bei sympatrischen *Uca*-Arten (Ocypodidae) des Östpazifik. Forma et Functio 1: 238-49.

1969.2. Das "Schaumbaden" brachyurer Crustaceen als Temperatur-Regulator. Zool. Anz. 181: 5-6.

1970. Form und Funktion der vibratorischen Signale von *Uca tangeri* und *Uca inaequalis* (Crustacea, Ocypodidae). Forma et Functio 2: 178-87.

Altevogt, R., and R. Altevogt
1967. FILMS. Encyclopaedia Cinematographica, Göttingen. [Not seen.]
E 1268. *Uca stylifera*—Balz.
E 1269. *Uca princeps*—Balz.
E 1288. *Uca insignis*—Balz.
E 1289. *Uca beebei*—Balz.
E 1290. *Uca mertensi*—Balz.
E 1291. *Uca rapax*—Balz.
E 1292. *Uca batuenta*—Balz.
E 1293. *Uca terpsichores*—Balz.

Altevogt, A., and H.-O. von Hagen
1964. Über die Orientierung von *Uca tangeri* Eydoux im Freiland. Z. Morph. Ökol. Tiere 53: 636-56.

Amadon, D.
1966. The superspecies concept. Syst. Zool. 15: 245-49.

Anderson, A. R.
1894. Note on the sound produced by the ocypode crab (*Ocypoda ceratophthalma*). J. Asiat. Soc. Bengal 63: 138-39.

Aurivillius, C.W.S.
1893. Die Beziehungen der Sinnesorgane amphibischer Decapoden zür Lebenweise und Athmung. Nova Acta R. Soc. Scient. upsal. Nyt. Mag. Naturvid. ser. 3: 1-48.

1898. Krustaceen aus dem Kamerun. Gebiete. Bih. K. svenska Vetensk Akad. Handl. 24, afd. 4 (1): 1-31.

Azrin, N. H., R. R. Hutchinson, and R. McLaughlin
1965. The opportunity for aggression as an operant reinforcer during aversive stimulation. J. exp. anal. Behavior 8: 171-80.

Baerends, G. P.
1950. Specializations in organs and movements with a releasing function. Symp. Soc. exp. Biol. 4: 337-60.

Baerends, G. P., and J. M. Baerends-van Roon
1950. An introduction to the study of the ethology of cichlid fishes. Behaviour, Suppl. no. 1: 1-242.

Balss, H.
1921. Über stridulations Organe bei dekapoden Crustaceen. Eine susammenfassende Übersicht. Naturw. Wschr. Jena, neue Folge 20: 697-701.

1922.1. Östasiatische Decapoden. IV. Die Brachyrhynchen (Cancridea). Archiv. f. Naturgesch., A (11): 94-166.

1922.2. Crustacea VII; Decapoda Brachyura (Oxyrhyncha und Brachyrhyncha) und geographische Übersicht über Crustacea Decapoda. Beitr. Kennt. Meeresfauna Westafr. Herausg. von W. Michaelsen Hamburg, 3 (3): 71-110.

1924. Expedition S.M. Schiff "Pola" in das Rote Meer 1895/6-1897/8. Zool. Ergeb. XXXIV. Decapoden des Roten Meeres. III. Die Parthenopiden. Cyclo- und Catometopen. Denkschr. Acad. Wiss. Wien 99: 1-18.

1938. Die Dekapoda Brachyura von Dr. Sixten Bocks' Pazifik-Expedition, 1917-1918. Göteborgs K. Vetensk.-O. Vittern Samh. Handl. (5B), 5 (7): 1-85.

Barnard, K. H.
1950. Descriptive catalogue of South African decapod Crustacea. Ann. S. Afr. Mus. 38: 1-837.

Barnwell, F. H.
1963. Observations on daily and tidal rhythms in some fiddler crabs from equatorial Brazil. Biol. Bull. 125 (3): 399-415.

1966. Daily and tidal patterns of activity in individual fiddler crabs (genus *Uca*) from the Woods Hole region. Biol. Bull. 130 (1): 1-17.

1968.1. Comparative aspects of the chromatophoric responses to light and temperature in fiddler crabs of the genus *Uca*. Biol. Bull. 134 (2): 221-34.

1968.2. The role of rhythmic systems in the adaptation of fiddler crabs to the intertidal zone. Am. Zool. 8: 569-83.

Barrass, R.
1963. The burrows of *Ocypode ceratophthalma* (Pallas) (Crustacea, Ocypodidae) on a tidal wave beach at Inhaca Island, Moçambique. J. Anim. Ecol. 32: 73-85.

Bate, C. S.
1866. Vancouver Island Crabs. *In* "A Naturalist in Vancouver Island and British Columbia"; J. Keast Lord. London. Vol. ii: 262-84.
1868. Carcinological gleanings. No. 3. Letter of Dr. R. Cunningham concerning Brazilian Crustaceans. Ann. Mag. nat. Hist. 1868, 1: 442-48.

Baudouin, M.
1903. Autotomie et repoussé des pinces chez le *Gelasimus tangieri* Eyd. Bull. Mus. Hist. nat. 9: 341-42.
1906. Le *Gelasimus tangeri*, crustacé d'Andalousie. Annls Sci. nat. 3: 1-33.

Beebe, W.
1928. Beneath tropic seas. G. P. Putnam's Sons, New York. xiii + 234 pp.
1944. The function of secondary sexual characters in two species of Dynastidae (Coleoptera). Zoologica, N.Y. 29 (2): 53-58.

Beer, C. G.
1959. Notes on the behaviour of two estuarine crab species. Trans. R. Soc. N.Z. 86: 197-203.

Bliss, D. E.
1968. Transition from water to land in decapod crustaceans. Am. Zool. 8: 355-92.

Bliss, D. E., and L. H. Mantel
1968. Adaptations of crustaceans to land: A summary and analysis of new findings. Am. Zool. 8: 673-85.

Bliss, D. E., and L. H. Mantel, organizers and eds.
1968. Terrestrial adaptations in Crustacea: a symposium organized for the Division of Invertebrate Zoology, American Society of Zoologists, 1967. Am. Zool. 8: 307-685.

Bolau, H.
1878. Neue oder sonst bemerkenswerthe Bewohner des Aquiereums im zoologischen Garten zu Hamburg. Zool. Gart. Frankf. 19 (5): 149.

Boone, L.
1927. The littoral crustacean fauna of the Galapagos Islands. Part 1. Brachyura. Zoologica, N.Y. 8 (4): 127-288.
1930. Scientific results of the cruises of the yachts "Eagle" and "Ara," 1921-1928, William K. Vanderbilt commanding. Crustacea: Stomatopoda and Brachyura. Bull. Vanderbilt mar. Mus. 2: 5-228.

Borradaile, L. A.
1900. On some crustaceans from the South Pacific. Part 4. The crabs. Proc. zool. Soc. Lond. 568-96.
1907. The Percy Sladen Trust Expedition to the Indian Ocean in 1905, under the leadership of Mr. J. Stanley Gardiner. No. 3—Land and freshwater Decapoda. Trans. Linn. Soc. Lond., ser. 2, Zool. 12: 63-68.
1910. On the land and amphibious Decapoda of Aldabra. Trans. Linn. Soc. Lond., ser. 2, Zool. 13 (3): 405-409.

Bosc, L.A.G.
1802. Histoire naturelle des crustacés, contenant leur description et leurs moeurs; avec figures dessinées d'après nature. Deterville, Paris: Vol. i: 258 pp.

Bott, R.
1954. Dekapoden (Crustacea) aus El Salvador. 1. Winkerkrabben (*Uca*). Senck. Biol. 35: 155-80.
1958. Dekapoden von den Galapagos-Inseln. Senck. Biol. 39: 209-11.

Bouvier, E. L.
1906. Sur les crustacés decapodes marins recueillis par M. Gruvel en Mauritanie. Bull. Mus. Hist. nat., Paris 12: 185-87. Reprinted in "Mission des pêcheries de la côte occidentale d'Afrique." Paris, no. 7: 95-97; and Act. Soc. linn. Bordeaux 61: 198-200.
1915. Decapodes marcheurs (Reptantia) et stomatopodes recueillis a l'île Maurice par M. Paul Carié. Bull. scient. Fr. Belg. Paris 48: 178-318.

Bovbjerg, R. V.
1960. Behavioral ecology of the crab, *Pachygrapsus crassipes*. Ecology 41: 668-72.

Boyce, D. R.
1924. The calling crabs of Durban Bay, *Uca annulipes* (Milne-Edwards). S. Afr. J. nat. Hist. 4: 250-52.

Brocchi, M.
1875. Récherches sur les organes genitaux mâles des crustacés decapodes. Annls Sci. nat. Zoologie (6) 2 (2): 73-74.

Brown, F. A., Jr.
1944. Hormones in the Crustacea: their sources and activities. Q. Rev. Biol. 19 (1): 32-46; (2): 118-43.
1961. Physiological rhythms. *In* The physiology of Crustacea. T. H. Waterman (ed.). Academic Press, New York and London. Vol. ii: 401-30.

Buitendijk, A. M.
1947.   Zoological notes from Port Dickson, III. Crustacea Anomura and Brachyura. Zool. Meded., Leiden 28: 280-84.

Burkenroad, M. D.
1947.   Production of sound by the fiddler crab *Uca pugilator* Bosc, with remarks on its nocturnal and mating behavior. Ecology 28: 458-62.

Cameron, A. M.
1966.   Some aspects of the behaviour of the soldier crab, *Mictyris longicarpus*. Pacific Science 20 (2): 224-34.

Cano, G.
1889.   Crostacei brachiuri ed anomuri raccolti nel viaggio della "Vettor Pisani" intorno al globo. Boll. Soc. Nat. Napoli, 3: 79-106 and 169-269.

Carlson, S.
1935.   The color changes in *Uca pugilator*. Proc. natn. Acad. Sci. U.S.A. 21 (9): 549-51.
1936.   Color changes in brachyuran crustaceans, especially in *Uca pugilator*. K. Fysiogr. Sallsk, i Lund Forhandl. 6: 63-80.

Chace, F. A., Jr.
1942.   Scientific results of a fourth expedition to forested areas in Eastern Africa. III. Decapod Crustacea. Bull. Mus. comp. Zool. Harvard 91: 185-233.

Chace, F. A., Jr., and H. H. Hobbs, Jr.
1969.   The freshwater and terrestrial decapod crustaceans of the West Indies with special reference to Dominica. Bull. U.S. natn. Mus. 292: 1-258.

Chapgar, B. F.
1957.   On the marine crabs (Decapoda Brachyura) of Bombay State. Part 2. J. Bombay nat. Hist. Soc. 54 (3): 503-49.

Chilton, C., and E. W. Bennett
1929.   Contributions for a revision of the Crustacea Brachyura of New Zealand. Trans. & Proc. New Zealand Inst. 59: 731-78.

Chopra, B., and K. N. Das
1937.   Further notes on Crustacea Decapoda in the Indian Museum. Rec. Indian Mus. 39: 377-434.

Colosi, G.
1924.   Crostacei raccolti nella Somalia dalla missione della R. Società Geographica 1924. Boll. Musei Zool. Anat. comp. R. Univ. Torino 39 (32): 1-4.

Coventry, G. A.
1944.   The Crustacea. *In* "Results of the 5th George Vanderbilt Expedition (1941). (Bahamas, Caribbean Sea, Panama, Galapagos Archipelago and Mexican Pacific Is-

lands)." Monograph no. 6 of the Academy of Natural Sciences of Philadelphia: 531-44.

Crane, J.
1941.1.   Eastern Pacific Expeditions of the New York Zoological Society. XXVI. Crabs of the genus *Uca* from the west coast of Central America. Zoologica, N. Y. 26 (3): 145-208.
1941.2.   Eastern Pacific Expeditions of the New York Zoological Society. XXIX. On the growth and ecology of brachyuran crabs of the genus *Ocypode*. Zoologica, N.Y. 26 (4): 297-310.
1943.1.   Eastern Pacific Expeditions of the New York Zoological Society. XXXI. *Uca schmitti*, a new species of brachyuran crab from the west coast of Central America. Zoologica, N.Y. 28 (6): 31-32.
1943.2.   Crabs of the genus *Uca* from Venezuela. Zoologica, N.Y. 28 (7): 33-44.
1943.3.   Display, breeding and relationships of fiddler crabs (Brachyura, genus *Uca*) in the northeastern United States. Zoologica, N.Y. 28 (23): 217-23.
1944.   On the color changes of fiddler crabs (genus *Uca*) in the field. Zoologica, N.Y. 29 (3): 161-68.
1957.   Basic patterns of display in fiddler crabs (Ocypodidae, genus *Uca*). Zoologica, N.Y. 42 (2): 69-82.
1958.   Aspects of social behavior in fiddler crabs with special reference to *Uca maracoani* (Latreille). Zoologica, N.Y. 43: 113-30.
1966.1.   A discussion on ritualization of behaviour in animals and man, organized by Sir Julian Huxley: Combat, display and ritualization in fiddler crabs (Ocypodidae, genus *Uca*). Phil. Trans. R. Soc. Lond. B 772; 251: 459-72.
1966.2.   Comparative aspects of social behavior in fiddler crabs of the world (Ocypodidae, genus *Uca*). Proc. Symposium on Crustacea [at Ernakulam, Kerala, S. India, 1965, under auspices of Bureau of Fisheries]. *In* Symp. Ser. Marine Biol. Assoc. India No. 2 1965 [1966], Part 1, p. 28. [Contribution consists only of a brief summary of contents of a film prepared for the symposium; film not available.]
1967.   Combat and its ritualization in fiddler crabs (Ocypodidae) with special reference to *Uca rapax*. Zoologica, N.Y. 52 (3): 49-75.

Crosnier, A.
1965.   Faune de Madagascar XVIII: Crustacés decapodes Grapsidae et Ocypodidae. Centre

National de la Récherche Scientifique et de l'Office de la Récherche Scientifique et Technique Outre-Mer: 1-143.

Cuvier, G.
1817.   [See Latreille, 1817.1.]
1836-   Le règne animal distribué d'après son
1849.   organisation. Edition accompagnée de planches gravées, réprésentant les types de tous les genres par une réunion de disciples de Cuvier. [See Milne-Edwards, H., ? 1836.]

Dakin, W. J. assisted by I. Bennett & E. Pope
1954.   Australian seashores. Angus & Robertson, London. xii + 322 pp.

Dana, J. D.
1851.   Conspectus crustaceorum quae in orbis terrarum circumnavigatione, Carolo Wilkes e classe reipublicae foederatoe duce, lexit et descripsit. Proc. Acad. nat. Sci. Philad. 5: 247-54.
1852.   Crustacea. *In* "United States Exploring Expedition . . . during the years 1838, 1839, 1840, 1841, 1842 . . . under the command of Charles Wilkes, U.S.N." 13 (1); 8 (1, 2); atlas.

Darling, F. F.
1937.   A herd of red deer: A study in animal behaviour. Oxford University Press, New York and London. viii + 215 pp.
1938.   Bird flocks and the breeding cycle. Cambridge University Press, Cambridge. x + 124 pp.
1952.   Social behaviour and survival. Auk 69: 183-91.

Darwin, C.
1871.   The descent of man and selection in relation to sex. 1st ed. [Not seen.]
1874.   The descent of man and selection in relation to sex. 2nd ed. revised and augmented; authorized. D. Appleton & Co.: printing of 1901. xvi + 688 pp. [Refs. to *Gelasimus* on pp. 259, 272, 274, 275.]

Day, J. H., and J.F.C. Morgans
1956.   The ecology of South African estuaries. Part 8. The biology of Durban Bay. Durban Mus. Novit. 8 (3): 259-312.

De Geer, C. [= K.]
1778.   Mémoires pour servir a l'histoire des insectes. Des crabes. De l'Imprimerie de Pierre Hesselberg, Stockholm, 7: 409-32.

De Kay, J. E.
1844.   Natural history of New York. Zoology of New York *or* The New York fauna. Pt. 6. Crustacea. Carroll & Cook, Albany, 70 pp.

Dembowski, J.
1925.   On the "speech" of the fiddler crab, *Uca pugilator*. Pr. Inst. M. Nencki. 3 (48): 1-7.

1926.   Notes on the behaviour of the fiddler crab. Biol. Bull. mar. biol. Lab., Woods Hole 50: 179-200.

Démeusy, N.
1957.   Respiratory metabolism of the fiddler crab *Uca pugilator* from two different latitudinal populations. Biol. Bull. mar. biol. Lab., Woods Hole 113 (2): 245-53.

Desbonne, I., and A. Schramm
1867.   Crustacés de la Guadeloupe. See Schramm, A.

Desmarest, A.-G.
1817.   *In article on* "Crustacés fossiles," p. 505, no. 14. *In* "Nouveau dictionnaire d'histoire naturelle, appliquée aux arts, à l'agriculture . . . etc. par une societé de naturalistes et d'agriculteurs. . . ." Deterville, Paris: Edition 2, Vol. VIII.
1822.   Les crustacés proprement dits. *In* "Histoire naturelle des crustacés fossiles sous les rapports zoologiques et géologiques." A. Brongniart and A.-G. Desmarest. F.-G. Levrault, Paris. vii + 154 pp.
1825.   Considérations générales sur la classe des crustacés et description des espèces de ces animaux, qui vivent dans la mer, sur les côtes, ou dans les eaux douces de la France. Paris. xix + 446 pp.

Doflein, F.
1899.   Amerikanische Dekapoden der k. bayerischen Staatssammlungen. Sitzungsber. Alm. K. bayer. Akad. Wiss. 29: 177-95.

Dorf, E.
1959.   Climatic changes of the past and present. Contrib. Mus. Paleont. Univ. Michigan 13: 181-210.

Dumortier, B.
1963.   Morphology of sound emission apparatus in Arthropoda. *In* "Acoustic behavior in animals," ed. R.-G. Busnel. Elsevier Press, New York. Pp. 310-15.

Durham, J. W.
1950.   Cenozoic marine climates of the Pacific coast. Bull. Geol. Soc. Am. 61: 1243-64.

Durham, J. W., and E. C. Allison
1960.   The geologic history of Baja California and its marine provinces. (Contrib. to Symposium: The biogeography of Baja California and adjacent seas.) Syst. Zool. 9: 47-91.

Edmondson, C. H.
1925.   Marine zoology of the tropical central Pacific (Tanager Exped. Publ. 1): Crustacea. Bull. Bernice Pauahi Bishop Mus. 24: 3-62.
1933.   Reef and shore fauna of Hawaii. Crustacea. Spec. Publs Bernice Pauahi Bishop Mus. 22: 191-271.

1946. 2nd ed. of above: pp. 219-315.

Edney, E. B.
1960. Terrestrial adaptations. *In* "The physiology of Crustacea"; T. H. Waterman (ed.). Academic Press, New York and London. Vol. I: 367-93.
1961. The water and heat relationships of fiddler crabs (*Uca* spp.). Trans. R. Soc. S. Afr. 36: 71-91.
1962. Some aspects of the temperature relations of fiddler crabs (*Uca* spp.). *In* "Biometeorology," ed. S. W. Tromp. Pergamon Press, Oxford, London, New York, Paris. Pp. 79-85.

Eibl-Eibesfeldt, I.
1970. Ethology, the biology of behavior. Holt, Rinehart and Winston, New York . . . , Montreal, Toronto, London, Sydney. xiv + 530 pp.

Eisner, T., and F. C. Kafatos
1962. Defense mechanisms of arthropods. x. A pheromone promoting aggregation in an aposematic distasteful insect. Psyche 69 (2): 53-61.

Ekman, S.
1953. Zoogeography of the sea. Sidgwick and Jackson, London. v + 417 pp.
1967. 2nd ed. of above.

Estampador, E. P.
1937. A check list of Philippine crustacean decapods. Philipp. J. Sci. D. Gen. Biol. Ethnol. Anthrop. 62: 465-559.
1959. Revised check list of Philippine crustacean decapods. Natural & Applied Sci. Bull. Coll. Liberal Arts, Univ. of Philippines 17 (1): 100-103.

Eydoux, F.
1835. Nouvelle espèce de Gélasime. Mag. de Zool. 5, cl. 7; 4 pp. (not numbered).

Fabricius, J. C.
1775. Systema entomologiae, sistens insectorum classes, ordines, genera, species. . . . Flensburgi et Lipsiae in officina Libraria Kortii. 832 pp.
1787. Mantissa insectorum, sistens eorum specie nuper detectas. Adiectis characteribus genericis, differentiis specificis, emendationibus, observationibus. Hafniae: Impensis Christ. Gottl. Proft. Vol. I: xx + 348 pp.
1798. Supplementum entomologiae systematicae. Hafniae: Proft & Storch. 572 pp.

Feest, J.
1969. Morphophysiologische Untersuchungen zur Ontogenese und Fortpflanzungsbiologie von *Uca annulipes* und *Uca triangularis* mit Vergleichsbefunden an *Ilyoplax gangetica*. Forma et Functio 1: 159-225.

Filhol, H.
1885. Considérations relatives à la faune des crustacés de la Nouvelle Zélande. Paris. Hautes Études Bibl. 30 (2): 60 pp.

Fingerman, M.
1957. Relation between position of burrows and tidal rhythm of *Uca*. Biol. Bull. 12 (1): 7-20.

Forest, J., and D. Guinot
1961. Crustacés décapodes brachyoures de Tahiti et des Tuamotu. *Extrait de* "Expedition française sur les récifs coralliens de la Nouvelle Caledonie." Paris. xi + 195 pp.
1962. Remarques biogéographiques sur les crabes des Archipels de la Societé et des Tuamotu. Cah. Pacif. 4: 41-75.

Fourmanoir, P.
1953. Notes sur la faune de la mangrove dans la région de Majunga. Naturaliste malgache 5 (1): 87-92.

Fowler, H.
1912. The Crustacea of New Jersey. Ann. Rept. New Jersey State Museum for 1911. Part 2: 29-650.

Freycinet, L. de
1825-
1829 Voyage autour du monde entrepris par ordre du roi . . . execute sur les corvettes l'Uranie et la Physicienne pendant les années 1817, 1818, 1819 et 1820. Historique. Vol. I; vol. II, parts 1-3. Pillet ainé, Paris.

Garth, J. S.
1946. Littoral brachyuran fauna of the Galapagos Archipelago. Allan Hancock Pacif. Exped. Ser. 1, 5 (10): 341-601.
1948. The Brachyura of the "Askoy" Expedition with remarks on carcinological collecting in the Panama Bight. Bull. Am. Mus. nat. Hist. 92 (1): 1-66.
1957. Reports of the Lund University Chile Expedition 1948-1949. No. 29. The Crustacea Decapoda Brachyura of Chile. Lunds Univ. Arsskrift., n.s., section 2 53 (7): 1-128.
1960. Distribution and affinities of the brachyuran Crustacea. (Contribution to Symposium: The biogeography of Baja California and adjacent seas.) Syst. Zool. 9 (3): 105-123.
1965. The brachyuran decapod crustaceans of Clipperton Island. Proc. Calif. Acad. Sci. ser. 4, 33 (1): 1-46.
1966. On the oceanic transport of crab larval stages. Proc. Symposium on Crustacea [at Ernakulam, Kerala, S. India, 1965, under auspices of Bureau of Fisheries]. *In* Symp. Ser. Marine Biol. Assoc. India No. 2 1965 [1966], Part 1. Pp. 443-48.

Gee, N. G.
1925.   Tentative list of Chinese decapod Crustacea including those represented in the collection of the U.S.N.M. with localities at which collected. Lingnaam agric. Rev. 3 (2): 156-66.

Gerlach, S. A.
1958.1. Die Mangroveregion tropischer Küsten als Lebensraum. Z. Morph. Ökol. Tiere, 46: 436-530.
1958.2. Beobachtungen über das Verhalten von Winkerkrabben (*Uca leptodactyla*). Z. Tierpsychol. 15: 50-53.

Gmitter, T. E., and R. M. Wotton
1953.   Crabs from the island of St. Thomas. Proc. Pa. Acad. Sci. 27: 261-72.

Göldi, E. A.
1885.   Studien über neue und wenig bekannte Podophthalmen Braziliens. Zool. Anz. 8: 662-63.
1886.   Studien über neue und weniger bekannte Podophthalmen Braziliens. Arch. Naturgesch. 50: 19-46.

Goodbody, I.
1961.   Mass mortality of a marine fauna following tropical rains. Ecology 42: 150-55.

Gordon, H.R.S.
1955.   Displacement activities in fiddler crabs. Nature, Lond. 176: 356-57.
1958.   Synchronous claw-waving of fiddler crabs. Animal Behav. 6: 238-41.

Gordon, I.
1931.   Brachyura from the coasts of China. J. Linn. Soc. Zoology 37 (254): 525-58.
1934.   Crustacea Brachyura. Res. sci. du voyage aux Indes orientales Néerlandaises de LL. AA. RR. le Prince et Princesse Léopold de Belgique. Vol. III (15): 1-78.

Gould, A. A.
1841.   A report on the invertebrata of Massachusetts, comprising the Mollusca, Crustacea, Annelida, and Radiata. Folsom, Wells, and Thurston; Cambridge, Mass. xiii + 373 pp.

Gould, S. J.
1966.   Allometry and size in ontogeny and phylogeny. Biol. Rev. (Cambridge Phil. Soc.) 41: 587-640.

Grant, F. E., and A. R. McCulloch
1906.   On a collection of Crustacea from the Port Curtis District, Queensland. Proc. Linn. Soc. N.S.W. 31: 2-53.

Grant, V.
1963.   The origin of adaptations. Columbia University Press, New York and London. x + 606 pp.

Gravely, F. H.
1927.   [See Raj, B. S. *et al.*]

Gravier, C.
1920.   Sur une collection de crustacés recuellis á Madagascar par M. le lieutenant Décary. Bull. Mus. natn. Hist. nat., Paris: 465-72.

Gray, E. H.
1942.   Ecological and life history aspects of the red-jointed fiddler crab, *Uca minax* (Le Conte), region of Solomon Island, Maryland. Contr. Chesapeake biol. Lab., Publ. No. 51: 3-20.

Gray, E. H., and C. L. Newcombe
1938.1. Relative growth of parts in the blue crab. Growth 2: 235-46.
1938.2. Studies of moulting in *Callinectes sapidus* Rathbun. Growth 2: 285-96.

Gray, I. E.
1957.   A comparative study of the gill area of crabs. Biol. Bull. 112 (1): 34-42.

Griffin, D.J.G.
1968.   Social and maintenance behaviour in two Australian ocypodid crabs (Crustacea: Brachyura). J. Zool. Lond. 156: 291-305.

Guérin-Méneville, F.-E.
1829.   Atlas. Crustacés, arachnides et insectes. *In* "Voyage autour du monde, execute par ordre du Roi, sur la corvette de sa Majesté, la 'Coquille,' pendant les années 1822, 1823, 1824 et 1825"; L. I. Duperrey. Paris. Crustacean pl. no. 1. [Plates of crustaceans, 1-5, were all published during 1829-1830; see Holthuis, 1961.]
1838.   Texte. Crustacés, arachnides et insectes. *In* "Voyage autour du monde, execute par ordre du Roi, sur la corvette de sa Majesté, la 'Coquille,' pendant les années 1822, 1823, 1824 et 1825"; L. I. Duperrey. Paris. Zool. vol. II, pt. 2, div. 1: xii + 319. Crustaceans: Livr. 28, pp. 1-47.
1829-
1843.   Iconographie du règne animal de G. Cuvier. Représentation d'après nature de l'une des espèces les plus remarquables et souvent non encore figurées de chaque genre d'animaux. Baillière, Paris. Vol. II, Planches; vol. III, Texte explicatif.

Guinot, D., and A. Ribeiro
1962.   Sur une collection de crustacés brachyoures des Iles du Cap-Vert et de l'Angola. Trab. Cent. Biol. Piscatoria Lisboa: Mem. Junta Invest. ultram. ser. 2, no. 40: 9-89.

Guinot-Dumortier, D.
1959.   Sur une collection de crustacés (Decapoda Reptantia) de Guyane Française. Bull. Mus. natn. Hist. nat., Paris 31 (5): 423-34.

Guinot-Dumortier, D., and B. Dumortier
1960.   La stridulation chez les crabes. Crustaceana 1 (2): 117-55.

Gunther, H.-J.
1963. Untersuchungen zur Verbreitung und Ökologie von *Uca tangeri* an der SW-iberischen Küste. Z. Morph. Ökol. Tiere 53: 242-310.

Haan, W. de
1835. Crustacea. *In* "Fauna Japanica," P. F. von Siebold. Lugduni Batavorum: 1833-1850. Part as follows: Decade II; Sheets 7-16: pp. 25-64; pls. 9-15, 17 C, D.

Hagen, H.-O. von
1961. Nächtliche Aktivität von *Uca tangeri* in Südspanien. Naturwissenschaften 48: 140.
1962. Freilandstudien zur Sexual-und-Fortpflanzungsbiologie von *Uca tangeri* in Andalusien. Z. Morph. Ökol. Tiere 51: 611-725.
1967.1. Nachweis einer kinasthetischen orientierung bei *Uca rapax*. Z. Morph. Ökol. Tiere 58: 301-20.
1967.2. Preliminary report. Klopfsignale auch bei Grapsiden (Decapoda Brachyura). Naturwissenschaften 7: 177-78.
1968.1. Zischende Drohgerausche bei westindischen Krabben. Naturwissenschaften 3: 139-40.
1968.2. Studien an peruanischen Winkerkrabben (*Uca*). Zool. Jb. Syst. 95: 395-468.
1968.3. *Gelasimus macrodactylus* H. Milne Edwards & Lucas, 1843 (Crustacea, Decapoda): Proposed suppression under the plenary powers Z.N.(S.). Bull. zool. Nom. 25 (1): 60-61.
1969. Stärlinge und Karpflinge als Eiräuber bei der Winkerkrabbe *Uca vocator* (Herbst). Z. Tierpsychol. 26: 1-6.
1970.1. Verwandtschaftliche Gruppierung und Verbreitung der Karibischen Winkerkrabben (Ocypodidae, Gattung *Uca*). Zool. Meded. Leiden 44 (15): 217-35.
1970.2. Zur Deutung langstieliger und gehörnter Augen bei Ocypodiden (Decapoda, Brachyura). Forma et Functio 2: 13-57.
1970.3. Die Balz von *Uca vocator* (Herbst) als ökologisches Problem. Forma et Functio 2: 238-53.
1970.4. Anpassungen an das spezielle Gezeitenzonen-Niveau bei Ocypodiden (Decapoda, Brachyura). Forma et Functio 2: 361-413.

Hallam, A., ed.
1973. Atlas of palaeobiogeography. Elsevier Scientific Publishing Co., Amsterdam. xii + 531 pp.

Hartnoll, R. G.
1969. Mating in the Brachyura. Crustaceana 16: 161-81.

Haswell, W. A.
1882. Catalogue of the Australian stalk-eyed and sessile-eyed Crustacea. Sydney: 326 pp.

Hay, W. P., and C. A. Shore
1918. The decapod crustaceans of Beaufort, N.C., and the surrounding region. Bull. Bur. Fish., Wash. 35: 369-475.

Heberer, G.
1930. Am Mangrovestrand von Ekas. *In* "Eine biologische Reise nach den Kleinen Sunda-Inseln"; B. Rensch. Borntraeger, Berlin. xii + 236 pp.

Hedgpeth, J. W.
1953. An introduction to the zoogeography of the northwestern Gulf of Mexico with reference to the invertebrate fauna. Publs Inst. mar. Sci. Univ. Tex. 3: 107-224.
1957.1. Concepts of marine ecology. *Chap. 3 in* "Treatise on marine ecology and paleoecology," vol. I. Mem. geol. Soc. Am. 67: 29-52.
1957.2. Marine biogeography. *Chap. 13 in* "Treatise on marine ecology and paleoecology," vol. I. Mem. geol. Soc. Am. 67: 359-82.

Hediger, H.
1933.1. Beobachtungen an der marokkanischen Winkerkrabbe, *Uca tangeri* (Eydoux). Verh. schweiz. naturf. Ges. 114: 388-89.
1933.2. Notes sur la biologie d'un crabe de l'embouchure de l'Oued Bou Regreg *Uca tangeri* (Eydoux). Bull. Soc. Sci. nat. Maroc. 13: 254-59.
1934. Zur Biologie und Psychologie der Flucht bei Tieren. Biol. Zbl. 54: 21-40.

Heller, C.
1862. Neue Crustaceen ges. während der Weltumseglund der K. K. Fregatte "Novara." Verh. zool.-bot. Ges. Wien. 12: 519-28.
1863. Die crustaceen des südlichen Europa. Crustacea Podophthalmia: 333 pp.
1865. Reise der österreichischen Fregatte "Novara" um die Erde in den Jahren 1857-58-59 unter den Befehlen des Commodore B. von Wüllerstorf-Urbair. Zool. 2 (3): Crustaceen. Vienna: 280 pp.

Henderson, J. R.
1893. A contribution to Indian carcinology. Trans. Linn. Soc. Lond. 2 (10): 325-458.

Henschell, A.G.E.T.
1833. Clavis Rumphiana botanica et zoologica. Accedunt vita G. E. Rumphii, Plinii Indici, specimenque materiae medicae amboinensis. Vratislaviae xiv + 216 pp.

Herbst, J.F.W.
1782-1804. Versuch einer Naturfeschichte der Krabben und Krebse. Nebst einer systematischen Beschreibung ihrer verschiedenen Arten. J. C. Fuessly, Zurich. References given appear in the following sections: 1782-1790:

1 (2): 71-274; pls. 1-21. 1804: 3 (4): 1-49; pls. 59-62.

Herklots, J. A.
1851.   Additamenta ad faunam carcinologicam Africae occidentalis . . . in littore Guineae. 1-28.

Herrick, C. L.
1887.   Contribution to the fauna of the Gulf of Mexico and the South. List of the fresh-water and marine Crustacea of Alabama. Mem. Denison sci. Ass. 1 (1): 1-56.

Herrnkind, W.
1966.   The ability of young and adult sand fiddler crabs, *Uca pugilator* (Bosc), to orient by polarized light. Summer meeting, Am. Soc. Zool., 1966. Am. Zool. 6 (3): author's abstract.

1967.   Development of celestial orientation during ontogeny in the sand fiddler crab *Uca pugilator*. Summer meeting, Am. Soc. Zool., 1967. Am. Zool. 7 (4): author's abstract.

1968.1.   Ecological and ontogenetic aspects of visual orientation in the sand fiddler crab *Uca pugilator* (Bosc). Ph.D. thesis, Univ. of Miami, Florida.

1968.2.   The breeding of *Uca pugilator* (Bosc) and mass rearing of the larvae with comments on the behavior of the larval and early crab stages (Brachyura, Ocypodidae). Crustaceana: Suppl. 2: 214-24.

1968.3.   Adaptive visually directed orientation in *Uca pugilator*. Am. Zool. 8: 585-98.

Hess, W.
1865.   Beiträge zur Kenntniss der Decapoden-Krebse Öst-Australiens. Arch. Naturgesch: 127-73.

Hesse, R., W. C. Allee, and K. P. Schmidt
1951.   Ecological animal geography. 2nd American ed. John Wiley and Sons, by arrangement with the University of Chicago Press. ix + 715 pp.

Hiatt, R. W.
1948.   The biology of the lined shore crab, *Pachygrapsus crassipes* Randall. Pacif. Sci. 2: 135-213.

Hilgendorf, F.
1869.   Crustaceen *in* "Reisen in Öst-Afrika," by v.d. Decken; 3: 69-116; also p. 147.

1879.   Die von Hrn. W. Peters in Mozambique gesammelten crustaceen. Abh. dt. Akad. Wiss. Berl. for 1878: 782-851.

1882.   Carcinologische Mittheilungen. Sber. Ges. naturf. Freunde Berl.: 22-25.

Hinde, R.
1970.   Animal behaviour: a synthesis of ethology and comparative psychology. 2nd ed. Mc-Graw-Hill Co., New York. xvi + 876 pp.

Hoffmann, C. K.
1874.   Crustacés et echinodermes de Madagascar et de l'Ile de la Réunion: Catalogue des crustacés recueillis par mm. Pollen et van Dam a Madagascar et ses dépendences. *In* "Récherches sur la faune de Madagascar et de ses dépendences d'après les decouvertes de François p. 1. Pollen et D. C. van Dam." Part 5, livr. 2: 1-58.

Hoffmann, K.
1971.   Biological clocks in animal orientation and other functions. Proc. int. Symp. circadian Rhythmicity (Wageningen, 1971): 175-205.

Holmes, S. J.
1900.   Synopsis of California stalk-eyed Crustacea. Occ. Pap. Calif. Acad. Sci. vii + 262 pp.

1904.   On some new or imperfectly known species of West American Crustacea. Proc. Calif. Acad. Sci. 3 (12): 307-28.

Holthuis, L. B.
1954.1.   On a collection of decapod crustaceans from the Republic of El Salvador (Central America). Zool. Verhand. Leiden No. 23: 1-43.

1954.2.   Observaciones sobre los crustaceos decapodos de la Republica de El Salvador. Comun. Inst. trop. Invest. cient. S. Salv. 3 (4): 159-66.

1958.   Crustacea Decapoda from the northern Red Sea (Gulf of Aqaba and Sinai Peninsula). II. Hippidea and Brachyura (Dromiacea, Oxystomata, and Grapsoidea). Bull. Sea Fish. Res. Sta. Israel Bull. 17 (9): 41-54.

1959.1.   Notes on pre-Linnean carcinology (including the study of *Xiphosura*) of the Malay Archipelago. Chap. 5 in "Rumphius Memorial Volume": 63-125.

1959.2.   H. E. van Rijgersma—a little-known naturalist of St. Martin (Netherlands Antilles). Stud. Fauna Curaçao Caribb. Isl. 9: 69-78.

1959.3.   The Crustacea Decapoda of Suriname (Dutch Guiana). Zool. Verh., Leiden 44: 1-296.

1961.   On the dates of publication of the crustacean plates in Duperrey's "Voyage autour du monde . . . sur . . . la 'Coquille.' " Crustaceana 3 (2): 168-69.

1962.   Forty-seven genera of Decapoda (Crustacea); proposed addition to the official list. Z.N.(S.) 1499. Bull. zool. Nom. 19 (4): 232-53.

1967.   On a new species of *Uca* from the West Indian region (Crustacea, Brachyura, Ocypodidae). Zool. Meded., Leiden 42 (6): 51-54.

Hombron, J.-B.
[See under Jacquinot, H., and under Jacquinot, H. and H. Lucas.]

Horch, K. W., and M. Salmon
1969. Production, perception and reception of acoustic stimuli by semiterrestrial crabs (genera *Ocypode* and *Uca*, family Ocypodidae). Forma et Functio 1: 1-25.

Hubbs, C.
1960. The marine vertebrates of the outer coast. (Contrib. to Symposium: The biogeography of Baja California and adjacent seas.) Syst. Zool. 9: 134-47.

Hughes, D. A.
1966. Behavioural and ecological investigations of the crab *Ocypode ceratophthalmus* (Crustacea: Ocypodidae). J. Zool. London 150: 129-43.

Hult, J.
1938. Crustacea Decapoda from the Galapagos Islands collected by Mr. Rolf Blomberg. Ark. Zool. 30A (5): 1-18.

Huxley, J. S.
1914. The courtship-habits of the great crested grebe (*Podiceps cristatus*) with an addition to the theory of sexual selection. Proc. zool. Soc. Lond. p. 491.

1924. Constant differential growth-ratios and their significance. Nature, Lond. 114: 895-96.

1942. Evolution, the modern synthesis. Harper, New York.

1966. A discussion on ritualization of behaviour in animals and man. Introduction. *In* "A discussion on ritualization . . ." J. S. Huxley *et al.*: 249-71. (See below.)

Huxley, J. S. *et al.*
1966. A discussion on ritualization of behaviour in animals and man, organized by Sir Julian Huxley. Phil. Trans. R. Soc. B. 772; 251: 247-526.

Huxley, J. S., and F. S. Callow
1933. A note on the asymmetry of male fiddler-crabs (*Uca pugilator*). Wilhelm Roux Arch. Entw. Mech. Org. 129 (2): 379-92.

Huxley, J. S., and G. Teissier
1936. Terminology of relations growth. Nature (Lond.) 137: 780-81.

Hyman, O. W.
1920. The development of *Gelasimus* after hatching. J. Morph. 33: 485-501.

1922. Adventures in the life of a fiddler crab. Rep. Smithsonian Instn. for 1920: 443-60.

Ives, J. E.
1891. Crustacea from the northern coast of Yucatan, the harbor of Vera Cruz, the west coast of Florida and the Bermuda Islands. Proc. Acad. nat. Sci. Philad. for 1891-1892, 43: 176-80.

Jacquinot, H.
1852. Zoologie, Atlas Crustacés: pls. 1-9. *In* "Voyage au Pôle Sud et dans l'Océanie sur les corvettes 'L'Astrolabe' et 'La Zelée' pendant les années 1837-1838-1839-1840 sous le commandement de M. Dumont d'Urville, Capitaine de Vaisseau, publié par ordre du Gouvernement et sous la direction superièure de M. Jacquinot, Capitaine de Vaisseau, Commandant de la 'La Zelée.' Zoologie." Gide et J. Baudry, Paris: 1842-1853. [Notes: Publication date of crustacean plates *fide* Forest & Guinot, 1961. In some bibliographies and library catalogues Hombron, a surgeon on the expedition, appears as the senior or junior co-author; his name is absent from the title page of this volume and from that of its section made up of the crustacean plates, but present on the title page of the text, cited below. In some catalogues the entire series of volumes on the expedition is entered only under the name of "Dumont d'Urville," or "Urville, J. Dumont d'."]

Jacquinot, H., and H. Lucas
1853. Zoologie, Texte Crustacés: 3: 1-107. *In* "Voyage au Pôle Sud. . . . M. J. Dumont d'Urville . . . Gouvernement sous la direction superièure de M. Jacquinot . . . Zoologie, par MM. Hombron et Jacquinot." Gide et J. Baudry, Paris.

Jansen, P.
1970. Physiologisch-ökologisch Untersuchungen zum "Posen" von *Uca tangeri*. Forma et Functio 2: 58-100.

Johnson, M. E., and H. J. Snook
1927. Seashore animals of the Pacific coast. Macmillan, N.Y.

Johnston, H.
1906. Liberia. 2 vols. ɪɪ: Appendix 8: 860-1116. Crustacea: 861-62.

Kamita, T.
1935. On the Brachyura of the west Korean waters (Yellow Sea). Zool. Mag., Tokyo (= Dobutsugak u zasshi) 47: 61 and 69. (Japanese with English res.)

Kellogg, C. R.
1928. Crustacea of Fukien Province. Lingnan Sci. J. 5 (4): 351-56.

Kemp, S.
1915. Fauna of the Chilka Lake. Crustacea Decapoda. Mem. Indian Mus. 5: 199-325.

1918. Zoological results of a tour in the Far East; N. Annandale (ed.). Part 5. Crustacea Decapoda and Stomatopoda. J. Proc. Asiat. Soc. Beng. 6: 217-97.

Kingsley, J. S.
1878. List of decapod Crustacea of the Atlantic coast, whose range embraces Fort Macon. Proc. Acad. nat. Sci. Philad. 316-30.
1880.1. Carcinological Notes. Proc. Acad. nat. Sci. Philad. for 1879: 34-37; II. Revision of the *Gelasimi*: 135-52.
1880.2. On a collection of Crustacea from Virginia, North Carolina and Florida, with a review of the genera of Cragonidae and Palaemonidae. Proc. Acad. nat. Sci. Philad. for 1879: 383-427.
1888. Something about crabs. Am. Nat. 22: 888-96.

Kirk, T. W.
1881. Notice of new crustaceans. Trans. N.Z. Inst. 12: 236-37.

Kleinholz, L. H.
1942. Hormones in Crustacea. Biol. Rev. 17 (2): 91-119.

Knopf, G. N.
1966. Observations on behavioral ecology of the fiddler crab, *Uca pugilator* (Bosc). Crustaceana 11: 302-306.

Kohli, G. R.
1924. Brachyura of the Karachi Coast. Proc. Lahore phil. Soc. 3: 81-85.

Korte, R.
1966. Untersuchungen zum Sehvermögen einiger Dekapoden, insbesondere von *Uca tangeri*. Z. Morph. Ökol. Tiere 58: 1-37.

Kossmann, R.
1877. Zoologische Ergebnisse einer Reise in die Kustengegenden des Rothen Meeres. III. Crustacea. Leipzig. [Not seen.]
1878. Kurze Notizen über einige neue Crustaceen. Arch. Naturgesch. 44 (part 1): 251-56.

Krauss, F.
1843. Die Süd-Afrikanische Crustaceen. Eine zusammenstellung aller bekannten Malacostraca . . . E. schweizerbartische verlagsbuchhandlung. Stuttgart. 68 pp.

Kummel, B.
1961. History of the earth. An introduction to historical geology. W. H. Freeman and Co., San Francisco and London. 610 pp.

Lamarck, J.B.P.A. de
1801. Système des animaux sans vertèbres, . . . Deterville, Paris. viii + 432 pp.
1818. Histoire naturelle des animaux sans vertèbres . . . Paris. Vol. v: 612 pp.

Lanchester, W. F.
1900.1. On a collection of crustaceans made at Singapore and Malacca. Part 1. Crustacea Brachyura. P. zool. Soc. Lond.: 719-70.
1900.2. On some malacostracous crustaceans from Malaysia in the collection of the Sarawak Museum. Ann. Mag. nat. Hist. (ser. 7) 6: 249-64.
1902. On the Crustacea collected during the "Skeat" Expedition to the Malay Peninsula, together with a note on the genus *Actaeopsis*. Part 1. Brachyura, Stomatopoda and Macrura. Proc. zool. Soc. Lond. 1901 (ii): 534-74.

Langdon, J. W.
1971. Shape discrimination and learning in the fiddler crab, *Uca pugilator*. Dissertation. Ph.D. Florida State University, 1971. University Microfilms, Ann Arbor, Michigan: 102 pp.

Latreille, P. A.
1802- Histoire naturelle, générale et particulière,
1803. des crustacés et des insectes. Paris. Vol. vi: 391 pp.
1817.1. Les crustacés, les arachnides et les insectes. *In* "Le règne animal, distribué d'après son organisation, pour servir de base a l'histoire naturelle des animaux et d'introduction a l'anatomie comparée"; G.L.C.F.D. Cuvier. Deterville, Paris. Vol. iii.
1817.2. Gélasime, *Gelasimus* (Buffon). *In* "Nouveau dictionnaire d'histoire naturelle, appliquée aux arts, à l'agriculture . . . etc. par une societé de naturalistes et d'agriculteurs . . ." Deterville, Paris: Edition 2, Vol. xii: 517-20.
1818. Crustacés, arachnides et insectes. Explication des planches: pp. 1-38; pls. 269-397. *In* "Tableau encyclopédique et méthodique des trois règnes de la nature." Agasse, Paris. Part 24.

Laurie, R. D.
1906. Report on the Brachyura collected by Professor Herdman at Ceylon in 1902. London Rep. Pearl Oyster Fish. 5: 349-432.
1915. Reports on the marine biology of the Sudanese Red Sea. 21. On the Brachyura. J. Linn. Soc. 31: 407-75.

Leach, W. E.
1814. Crustaceology. *In* "The Edinburgh encyclopaedia"; Brewster. Vol. vii: 383-437.
1815. A tabular view of the external characters of four classes of animals, which Linné arranged under Insects. . . . Trans. Linn. Soc. Lond. 11: 306-400.

LeConte, J.

1855. On a new species of *Gelasimus*. Proc. Acad. nat. Sci. Philad. 7: 402-403.

Lenz, H.

1905. Östafrikanische Dekapoden und Stomatopoden. Gesammelt von Herrn Prof. Dr. A. Voeltzkow. Abh. Senckenb. Ges. 27 (4): 341-92.

1910. Crustaceen von Madagaskar, Östafrika und Ceylon. *In* "Reise in Östafrika v. A. Voeltzkow." Stuttgart. Vol. ii: 539-76.

Lenz, H., and F. Richters

1881. Beitrag sur Krustaceenfauna von Madagascar. Abh. Senckenb. Ges. 421-28.

Lin, C. C.

1949. A catalogue of brachyurous Crustacea of Taiwan. Q. J1 Taiwan Mus. 2 (1): 10-33.

Linnaeus, C.

1758. Systema naturae . . . Ed. 10. Vol. i: iii + 824 pp.

1767. Systema naturae . . . Ed. 12. Vol. i (part 2): 533-1327.

Linsenmair, K. E.

1967. Konstruktion und Signalfunktion der Sandpyramide der Reiterkrabbe *Ocypode saratan* Försk. (Decapoda Brachyura Ocypodidae.) Z. Tierpsychol. 24: 403-56.

Lockington, W. N.

1877. Remarks on the Crustacea of the west coast of North America with a catalogue of the species in the museum of the California Academy of Sciences. Proc. Calif. Acad. Sci. 7: 94-108 and 145-56.

Lorenz, K. Z.

1941. Vergleichende Bewegnungstudien an Antinen. J. Ornithol. 83: 137-213 and 289-413.

1965. Evolution and modification of behavior. University of Chicago Press, Chicago and London: 121 pp.

1966.1. On aggression. (Translated by Marjorie Kerr Wilson.) Harcourt, Brace & World, New York. xiv + 306 pp.

1966.2. A discussion on ritualization of behaviour in animals and man, organized by Sir Julian Huxley: Evolution of ritualization in the biological and cultural spheres. Phil. Trans. R. Soc. Lond. B 772; 251: 273-82.

Luederwaldt, H.

1919.1. Os manguesaes de Santos. Rev. Mus. Paul. 11: 311-407.

1919.2. Lista dos crustaceos superiores. Rev. Mus. Paul. 11: 429-35.

1929. Resultados de umo excursão scientifica a Ilha de São Sebastião em 1925. Rev. Mus. Paul. 16: 1-79.

Maccagno, T.

1928. Crostacei decapodi. Le specie del genere *Uca* Leach conservate nel Regio Museo Zoologico di Torino. Boll. Musei Zool. Anat. comp. R. Univ. Torino 41 (11): 1-52.

MacGinitie, H. D.

1958. Climate since the late Cretaceous. *In* "Zoogeography," ed. C. L. Hubbs. Am. Ass. Advmt Sci. Publ. 51: 61-79.

MacKay, D.C.G.

1943. Relative growth of the European edible crab, *Cancer pagurus*: iii. Growth of the sternum and appendages. Growth 7: 217-26.

MacKay, D.C.G., and F. W. Weymouth

1934. The growth of the Pacific edible crab, *Cancer magister* Dana. J. Biol. Bd. Toronto 1: 210-11.

MacLeay, W. S.

1838. On the brachyurous decapod Crustacea brought from the Cape by Dr. Smith. *In* "Illustrations of the Annulosa of South Africa" by MacLeay, *which is in* Vol. v of "Illustrations of the Zoology of South Africa; consisting chiefly of figures and descriptions of the objects of natural history collected during . . . 1834, 1835, and 1836"; Andrew Smith, M.D. Invertebratae, pp. 53-71.

Macnae, W.

1956. Aspects of life on muddy shores in South Africa. S. Afr. J. Sci. 53: 40-43.

1957. The ecology of the plants and animals in the intertidal regions of the Swartkops estuary near Port Elizabeth, South Africa. J. Ecol. 45: 113-31 and 361-87.

1963. Mangrove swamps in South Africa. J. Ecol. 51: 1-25.

1966. Mangroves in eastern and southern Australia. Aust. J. Bot. 15: 67-104.

1968. A general account of the fauna and flora of mangrove swamps and forests in the Indo-West-Pacific region. *In* "Advances in marine biology," ed. F. S. Russell and M. Yonge. Academic Press, London and New York, 6: 73-270.

Macnae, W., and M. Kalk

1962. The ecology of the mangrove swamps at Inhaca Island, Moçambique. J. Ecol. 50: 19-34.

Macnae, W., and M. Kalk (eds.)

1958. A natural history of Inhaca Island, Moçambique. Witwatersrand University Press, Johannesburg. v + 153 pp.

Maki, M., and H. Tsuchiya

1923. A monograph of the decapod Crustacea of Formosa. Rep. Govt Res. Inst. Dep. Agric.

Formosa 3: 1-195; 14 pls. [Wholly in Japanese. Not seen.]

de Man, J. G.
1879.   Notes on some new or imperfectly known Podophthalmous Crustacea of the Leyden Museum. Notes Leyden Mus. 53-73.
1880.   On some species of *Gelasimus* and *Macrophthalmus*. Notes Leyden Mus. 2: 67-72.
1887.1. Report on the Podophthalmous Crustacea of the Mergui Archipelago. J. Linn. Soc. 137 and 138: 1-176.
1887.2. Bericht über die im indischen Archipel von Dr. J. Brock gesammelten Decapoden u. Stomatopoden. Arch. Naturgesch. 53: 215-600.
1891.   Carcinological studies in the Leyden Museum No. 5. Notes Leyden Mus. 13: 1-61.
1892.   Decapoda des indischen Archipels. *In* "Zoologische Ergebnisse einer Reise in Niederländisch Öst-Indien"; M. Weber. Brill, Leiden. Vol. II: 265-527.
1895.   Bericht über die von Herrn Schiffscapitan Storm zu Atjeh, an den westlichen Kusten von Malakka, Borneo und Celebes sowie in der Javasee gesammelten Decapoden und Stomatopoden. Zool. Jb. Abteilung für Systematik Ökologie und Geographie der Thiere 8: 485-609.
1902.   Die von Herrn Professor Kükenthal im Indischen Archipel gesammelten Dekapoden und Stomatopoden. Abh. Senckenb. naturforsch. Ges. 25: 467-929.

Manton, S. M.
1952.   The evolution of arthropodan locomotory mechanisms. Part 2. General introduction to the locomotory mechanisms of the Arthropoda. J. Linn. Soc. (Zool.) 42: 93-117.
1959.   Habits of life and evolution of body design in Arthropoda. J. Linn. Soc. (Zool.) 44: 58-72.

Marcgrave, G., de Liebstad
1648.   Historia rerum naturalium Brasiliae. *In* "De medicina Brasiliensi . . ." by G. Piso. Libro Octo: Quartus des piscibus (Crustacei Pisces): 182-89. Lugdun Batavorum et Amstelodami.

Marler, P. R., and W. J. Hamilton III
1966.   Mechanisms of animal behavior. John Wiley and Sons, New York, London, Sydney. xi + 771 pp.

Martens, E. von
1869.   Südbrasilische Suss-und-Brackwasser-Crustaceen nach den Sammlungen des Dr. Reinh. Hensel. Arch. Naturgesch. 35: 1-37.

Matthews, L. H.
1930.   Notes on the fiddler crab, *Uca leptodactyla*,

Rathbun. Ann. Mag. nat. Hist., ser. 10, 5: 659-63.

Mayr, E.
1963.   Animal species and evolution. Harvard University Press, Cambridge, Mass. xiv + 797 pp.
1969.   Principles of systematic zoology. McGraw-Hill Co., New York. xi + 428 pp.
1970.   Populations, species, and evolution. An abridgment of "Animal species and evolution." Harvard University Press, Cambridge, Mass. xv + 453 pp.

Mayr, E., E. G. Linsley, and R. L. Usinger
1953.   Methods and principles of systematic zoology. McGraw Hill Co., New York. ix + 328 pp.

McNeill, F. A.
1920.   Studies in Australian carcinology. No. 1. Rec. Aust. Mus. 13: 105-109.

Miers, E. J.
1879.1. On a collection of Crustacea made by Capt. H. G. St. John in the Corean and Japanese Seas. Proc. zool. Soc. Lond.: 18-61.
1879.2. An account of the petrological, botanical, and zoological collections made in Kerguelen's Land and Rodriguez during the Transit of Venus Expedition in the years 1874 and 1875. Crustacea. Phil. Trans. R. Soc. 168: 485-96.
1880.   On a collection of Crustacea from the Malayasian region. Part 2. Ann. Mag. nat. Hist., series 5, 5: 304-17.
1881.   On a collection of Crustacea made by Baron Hermann Maltzam at Gorée Island, Senegambia. Ann. Mag. nat. Hist. series 5, 8 (26): 259-81.
1884.   Crustacea. *In* "Report of the zoological collections made in the Indo-Pacific Ocean during the voyage of H.M.S. 'Alert,' 1881-1882." London. Pp. 178-322 and 518-75.
1886.   The Brachyura collected by H.M.S. "Challenger" during the years 1873-76. "Challenger" Rep. Zool. Vol. XVII: 412 pp.

Miller, D. C.
1961.   The feeding mechanisms of fiddler crabs, with ecological considerations of feeding adaptations. Zoologica, N.Y. 46: 89-100.
1965.   Studies of the systematics, ecology and geographical distribution of certain fiddler crabs. Doctoral Dissertation, Duke University. University Microfilms; Ann Arbor, Michigan. 240 pp. Diss. Abstr. No. 26: 3545.

Miller, D. C., and F. J. Vernberg
1968.   Some thermal requirements of fiddler crabs of the temperate and tropical zones and

their influence on geographic distribution. Am. Zool. 8: 459-69.

Milne-Edwards, A.

1868. Description de quelques crustacés nouveaux provenant des voyages de M. Alfred Grandidier à Zanzibar et à Madagascar. Nouv. Archs Mus. Hist. nat., Paris 4: 69-72.

1873. Récherches sur la faune carcinologique de la Nouvelle-Caledonie, 9 (2): 156-332.

Milne-Edwards, H.

?1836. Les Crustacés. *In* "Le Règne animal distribué d'après son organisation"; Georges Cuvier. Edition accompagnée de planches gravées, réprésentant les types de tous les genres . . . par une réunion de disciples de Cuvier. Paris. Vol. XVII (texte): 278 pp.; vol. XVIII (atlas). [Exact date of issue of these two volumes is apparently uncertain.]

1837. Histoire naturelle des crustacés, comprenant l'anatomie, la physiologie et la classification de ces animaux. Librairie Encyclopédique de Roret, Paris. Vol. II: 531 pp.; + separate atlas to Vol. II: 32 pp.

1852. Observations sur les affinités zoologiques et la classification naturelle des crustacés. Annls Sci. nat. Zoologie, series 3, 18: 109-166.

1854. Notes sur quelques crustacés nouveaux ou peu connus, conservés dans la collection du muséum d'histoire naturelle. Arch. Mus. natn. Hist. nat., Paris 7: 145-88.

Milne-Edwards, H., and H. Lucas

1843. Crustacés. *In* "Voyage dans l'Amerique méridionale (le Brézil . . . l'Uruguay, la République Argentine, la Patagonie . . . Chili . . . Bolivia . . . Pérou), exécuté pendant . . . 1826-33"; D. d'Orbigny. Bertrand and Levrault, Paris and Strassbourg. Vol. VI, part 1: pp. 1-52.

Miranda y Rivera, A. de

1933.1. Ensayo de un catalogo de los crustaceos decapodos marinos de España y Marruecos Español. Notas Resúm. Inst. esp. Oceanogr. 2 (67): 1-72.

1933.2. Notas carinológicas. Notas Resúm. Inst. esp. Oceanogr. 2 (68): 1-9.

Miyake, S.

1936. Reports on the Brachyura of the Riukiu Islands collected by the Yaéyama expeditions during the years 1932-34. I. Note on a new and some rare crabs from Iriometeshima: 494-505. II. A list of the known species of the Brachyura from Ishigakishima: 506-513. Annotnes zool. jap. 15.

1938. Notes on decapod crustaceans collected by Prof. Teiso Esaki from Micronesia. Annotnes zool. jap. 17: 107-12.

1939. Notes on Crustacea Brachyura collected by Professor Teiso Esaki's Micronesia Expeditions 1937-1938 together with a check list of Micronesian Brachyura. Rec. ocean. Wks Japan 10 (2): 168-247.

1961. A list of the decapod Crustacea of the Sea of Ariake, Kyushu. Rec. ocean. Wks Japan, special no. 5: 165-78.

Monod, T.

1923. Sur la biologie de l'*Uca tangeri* Eydoux. Rev. gen. Sci. (Paris). 34 (5): 133.

1927. Crustacea. 4. Decapoda (excl. Palaemonidae, Atyidae et Potamonidae). *In* "Contribution a l'étude de la faune du Cameroun; T. Monod." Faune Colon. fr. 1 (6): 593-624.

1932. Sur quelques crustacés de l'Afrique Occidentale. Bull. Com. Étud. hist. scient. Afr. occid. fr. 15: 456-548.

1956. Hippidea et Brachyura ouest-africains. Mem. I.F.A.N. 45: 1-674.

Moore, H. B.

1958. Marine ecology. John Wiley and Sons, New York. Chapman and Hall, London. v + 493 pp.

Moreira, C.

1901. Contribuiçoes para o conhecimento da fauna Braziliera. Crostaceos do Brazil. Archos Mus. nac., Rio de J. 11: 1-153.

Müller, F.

1869. Facts and arguments for Darwin. Translated from the German by W. S. Dallas. John Murray, London. 144 pp.

1881. Farbenwechsel bei Krabben und Garnelen. Kosmos 8: 472-73.

Murphy, R. C.

1939. The littoral of Pacific Colombia and Ecuador. Geog. Rev. 29: 1-33.

Musgrave, A.

1929. Life in a mangrove swamp. Aust. Mus. Mag. 3 (10): 341-47.

Nemec, C.

1939. Carcinological notes. Publs Field Mus. nat. Hist. Zoological series 24 (9): 105-108. (Publ. 451.)

Newell, R. C.

1970. Biology of intertidal animals. Elsevier Press, New York. 555 pp.

Nicolet, H.

1849. Crustaceos. *In* "Historia fisica y politica de Chile, Zoologia"; C. Gay. Santiago and Paris. Vol. III: 547 pp.

Nobili, G.

1897. Decapodi e stomatopodi raccolti dal Dr. Enrico Festa nel Darien, a Curaçao, La Guayra, Porto Cabello, Colon, Panama ecc.

Boll. Musei Zool. Anat. comp. R. Univ. Torino 12 (280): 1-3.

1899.1. Intorno ad alcuni crostacei decapodi del Brasile. Boll. Musei Zool. Anat. comp. R. Univ. Torino 14 (355): 1-6.

1899.2. Contribuzioni alla conoscenza della fauna carcinologica della Papuasia, delle Molucche e dell' Australia. Annali Mus. civ. stor. nat. Giacomo Doria (ser. 2a, 20) 40: 230-82.

1899.3. Decapodi e stomatopodi Indo-Malesi. Annali Mus. civ. stor. nat. Giacomo Doria (ser. 2a, 20) 40: 473-523.

1901.1. Note intorno ad una collezione di crostacei di Sarawak (Borneo). Boll. Musei Zool. Anat. comp. R. Univ. Torino 16 (397): 1-14.

1901.2. Decapodi raccolti dal Dr. Filippo Silvestri nell America meridionale. Boll. Musei Zool. Anat. comp. R. Univ. Torino 16 (402): 1-16.

1901.3. Viaggio del Dr. Enrico Festa nella Republica dell Ecuador e regione vicine. Decapodi e stomatopodi. Boll. Musei Zool. Anat. comp. R. Univ. Torino 16 (415): 1-58.

1903.1. Contributo alla fauna carcinologica di Borneo. Boll. Musei Zool. Anat. comp. R. Univ. Torino 18 (447): 1-32.

1903.2. Crostacei di Pondichéry, Mahé, Bombay, &c. Boll. Musei Zool. Anat. comp. R. Univ. Torino 18 (452): 1-24.

1906.1. Decapodi della Guinea Spagnuola. Madrid. Mems R. Soc. esp. Hist. nat. 1: 297-321.

1906.2. Mission J. Bonnier et Ch. Perez (Golfe Persique 1901). Crustacés decapodes et stomatopodes. Bull. scient. Fr. Belg. 40: 13-159.

1906.3. Faune carcinologique de la Mer Rouge. Decapodes et stomatopodes. Annls Sci. nat. Zoologie 4: 1-347.

1907. Ricerche sui crostacei della Polinesia. Decapodi, stomatopodi, anisopodi e isopodi. Memorie Accad. Sci. Torino, ser 2, 57: 351-430.

Nobre, A.
1931.1. Crustáceos decápodes de Portugal. Anais, Fac. Sci. Porto 16 (3): 134-86.

1931.2. Crustáceos decápodes e stomatópodes marinhos de Portugal. Imp. Portuguesa, Porto, iv + 5-307 pp.

1936. Fauna marinha de Portugal. IV. Crustáceos decápodes e stomatópodes marinhos de Portugal. Cia Ed. Minho, Barcelona. viii + 213 pp.

Nutting, C. C.
1919. Barbados-Antigua Expedition. Narrative and preliminary report of a zoological expedition from the University of Iowa to the Lesser Antilles under the auspices of the Graduate College. Stud. nat. Hist. Iowa Univ. 8: 72-79; 180-87.

Oliveira, L.P.H. de
1939.1. Contribuição ao conhecimento dos crustaceos do Rio de Janeiro. Genero *Uca* (Decapoda: Ocypodidae). Mems Inst. Oswaldo Cruz 34: 115-48.

1939.2. Observações sobre a biologia dos adultos do genero *Uca* Leach 1814. Liv. Hom. Profs. A. e M. Ozorio de Almeida Rio de J.: 490-97.

1939.3. Alguns fatores que limitam o habitat de varias especies de caranguelos do genero *Uca* Leach. Mems Inst. Oswaldo Cruz 34 (4): 519-26.

Olivier, M.
1811. Encyclopédie méthodique. Histoire naturelle. Insectes. H. Agasse, Paris. Vols. VI and VIII.

Ono, Y.
1959. The ecological studies on Brachyura in the estuary. Bull. biol. Stn Asamushi 9 (4): 145-48.

1962. On the habitat preference of ocypoid crabs. I. Mem. Fac. Sci. Kyushu Univ., ser. E (Biology), 3 (2): 143-63.

1963. First report of the Kyushu University Expedition to the Yaéyama Group, Ryukyus. The ecological distribution of ocypoid crabs in the Yaéyama Group, the Ryukyus. Rep. Comm. foreign sci. Res. Kyushu Univ. No. 1: 49-60. [In Japanese, with summary and captions in English.] [Not seen.]

1965. On the ecological distribution of ocypoid crabs in the estuary. Mem. Fac. Sci. Kyushu Univ., ser. E (Biology), 4 (1): 1-60.

Orr, P. R.
1955. Heat death. I. Time temperature relationship in marine animals. Physiol. Zoöl. 28: 290-93.

Ortmann, A. E.
1894.1. Crustaceen. *In* "Zoologische Forschungsreisen in Australien und dem malayischen Archipel"; R. Semon. Vol. V. Denkschr. med.-naturw. Ges. Jena. 8: 1-80.

1894.2. Die Decapoden-Krebse des Strassburger Museum. Zool. Jb. Abteilung für Systematik, Ökologie und Geographie der Thiere 7: 683-772.

1897. Carcinologische studien. Zool. Jb. Abteilung für Systematik, Ökologie und Geographie der Thiere 10: 258-372.

Osorio, B.
1887. Liste des crustacés des possessions portugaises d'Afrique Occidentale dans les collections du Muséum d'Histoire Naturelle de Lisbonne. Jorn. Sci. math. phys. nat. 11: 220-31.
1888. Liste des crustacés des possessions portugaises d'Afrique Occidentale dans les collections du Muséum d'Histoire Naturelle de Lisbonne. (Suite.) Jorn. Sci. math. phys. nat. 12: 186-91.
1889. Nouvelle contribution pour la connaissance de la faune carcinologique des Iles Saint Thome et du Prince. Jorn. Sci. math. phys. nat. ser. 2, 1: 129-39.
1891. Note sur quelques espèces des crustacés des Iles S. Thome, du Prince et Ilheo das Rolas. Jorn. Sci. math. phys. nat. 2, 2: 45-49.
1906. Uma nova lista de crustaceos africanos. Jorn. Sci. math. phys. nat. ser. 2, 7: 149-50.

Owen, R.
1839. Crustacea. *In* "The zoology of Captain Beechey's voyage; compiled from the collections and notes made by Captain Beechey, the officers and naturalist of the expedition during a voyage to the Pacific and Behrin's Straits performed in His Majesty's ship 'Blossom,' under the command of Capt. E. W. Beechey, R.N., F.R.S., &c. in the years 1825, 26, 27 and 28." London. Pp. 77-92.

Panikkar, N. K., and R. G. Aiyar
1937. The brackish-water fauna of Madras. Proc. Indian Acad. Sci. 6 (B 5): 284-337.

Parisi, B.
1918. I decapodi giapponesi del Museo di Milano. VI. Catometopa e Paguridea. Atti Soc. ital. Sci. nat. 57: 90-115.

Passano, L. M.
1960. Low temperature blockage of molting in *Uca pugnax*. Biol. Bull. 118 (1): 129-36.

Pearse, A. S.
1912.1. A new Philippine fiddler-crab. Philipp. J. Sci. 7: 91-95.
1912.2. The habits of fiddler-crabs. Philipp. J. Sci. 7: 113-33.
1914.1. On the habits of *Uca pugnax* (Smith) and *U. pugilator* (Bosc). Trans. Wis. Acad. Sci. Arts Lett. 17, part 2: 791-802.
1914.2. Habits of fiddler crabs. Rep. Smithson. Instn 1913 (1914): 415-28.
1916. An account of the Crustacea collected by the Walker Expedition to Santa Marta, Colombia. Proc. U.S. Nat. Mus. 49 (2123): 531-56.
1928. The ecology of certain estuarine crabs at Beaufort, N.C. J. Elisha Mitchell scient. Soc. 44 (2): 230-37.
1932. Observations on the ecology of certain fishes and crustaceans along the bank of the Malta River at Port Canning (India). Rec. Indian Mus. 34 (3): 289-98.
1936. The Ganges delta. Scient. Mon., N.Y. 42: 349-54.

Pearson, J.
1907-1908. Memoir on *Cancer*, the edible crab. L.M.B.C. Memoirs, no. 16. Liverpool biol. Soc. 22: 198-406.

Pesta, O.
1913. Crustacea. I. Decapoda Brachyura aus Samoa. *In* "Botanische und zoologische Ergebnisse einer wissenschaftlichen Forschungsreise nach den Samoa-Inseln, dem Neuguinea-Archipel und den Salomonsinseln"; K. Rechinger. Denksch. Acad. Wiss., Wien 88 (for 1911): 36-56; 57-65.
1931. Ergebnisse der Österreichischen biologischen Costa-Rica-Expedition 1930. I. Crustacea Decapoda aus Costa-Rica. Annln naturh. Mus. Wien 45: 173-81.

Peters, H. M.
1955. Die Winkgebärde von *Uca* and *Minuca* (Brachyura) in vergleichend-ethologischer, ökologischer und morphologischanatomischer Betrachtung. Z. Morph. Ökol. Tiere 43: 425-500.

Petiver, J.
1713. Aquatilium animalium Amboinae, &c. icones & nomina. Containing near 400 figures, engraven on copper plates of aquatick crustaceous and testaceous animals; as lobsters, crawfish, prawns, shrimps, sea-urchins, eggs, buttons, stars, couries, concks, perywinkles, whelks, oysters, muscles, cockles, frills, purrs, scallops, with divers other sort of sea and river shell-fish; all found about Amboina, and the neighboring Indian shores, with their Latin, English, Dutch and Native Names. Printed for Mr. Christopher Bateman in Paternoster Row, London. Vol. I: 1-4; pls. 1-22.
1767. Jacobi Petiveri opera, historiam naturalem spectantia: containing several thousand figures of birds, beasts, fish, reptiles, insects, shells, corals, and fossils; also of trees, shrubs, herbs, fruits, fungus's, mosses, seaweeds, etc. from all parts, adapted to Ray's History of plants . . . to which are now added seventeen curious tracts. . . . The additions corrected by the late Mr. James Empson, of the British Museum, etc. . . . John Millan, London. Vol. I.

Pfeffer, G.

1889.　Uebersicht der von Herrn Dr. F. Stühlmann in Aegypten, auf Sansibar und dem gegenüberliegenden Festlände gesammelten Reptilien . . . und Krebse. Jb. hamb. wiss. Anst. 6: 1-36.

Ping, C.

1930.　Zoological notes on Amoy and its vicinity. Bull. Fan meml Inst. Biol. 1 (8): 126-42.

Piso, G.

[See under Marcgrave, G.]

Pocock, R. I.

1903.　Crustacea: Malacostraca. In "The natural history of Sokotra and Abd-el-Kuri . . . ," ed. H. O. Forbes. Bull. Lpool Mus. I. The Decapoda of Sokotra: 212-13; 4 figs. II. Decapod and sessile-eyed crustaceans from Abd-el-Kuri (Brachyura): 216-32.

Porter, C. E.

1913.　Sinopsis de los Ocypodidae de Chile. Bol. Mus. nac. Chile 5: 313-18.

Raben, K. von

1934.　Veranderungen im Kiemendeckel und in Kiemen einiger Brachyuren (Decapoden) im Verlauf der Anpassung an die Feuchtluftatmung. Z. wiss. Zool. 145: 425-61.

Raj, B. S., et al.

1927.　Crustacea. In "The littoral fauna of the Krusadai Island in the Gulf of Manaar . . . Other Decapoda and Stomatopoda," ed. F. H. Gravely. Bull. Madras Govt Mus. new ser. 1 (1): 135-55.

Rankin, W. M.

1898 and 1910.　The Northrop collection of Crustacea from the Bahamas. Ann. N.Y. Acad. Sci. 11 (12): 225-58; 7 pls. Reprinted in: A naturalist in the Bahamas. J. I. Northrop memorial volume. New York, 1910: 69-96.

Rao, K. R.

1968.　The pericardial sacs of Ocypode in relation to the conservation of water, molting, and behavior. Am. Zool. 8: 561-67.

Rasa, O.A.E.

1971.　Appetence for aggression in juvenile damselfish. Zeits. Tierpsychologie Suppl. 7: 1-70.

Rathbun, M. J.

1893.　Descriptions of new genera and species of crabs from the west coast of America and the Sandwich Islands. Proc. U.S. natn. Mus. 16: 223-60.

1897.1.　A revision of the nomenclature of the Brachyura. Proc. biol. Soc. Wash. 11: 153-67.

1897.2.　List of the decapod Crustacea of Jamaica. Rep. Inst. Jamaica 1 (1): 1-49.

1898.1.　[Type description of Uca leptodactyla.] P. 227 in "The Northrop collection of Crustacea from the Bahamas"; W. M. Rankin. Ann. N.Y. Acad. Sci. 11 (12): 225-58. [See also reprint data under Rankin.]

1898.2.　The Brachyura collected by the U.S. Fish Commission steamer "Albatross," on the voyage from Norfolk, Virginia, to San Francisco, California, 1887-1888. Proc. U.S. natn. Mus. 21 (1162): 567-616.

1900.1.　The decapod crustaceans of West Africa. Proc. U.S. natn. Mus. 22 (1199): 271-316.

1900.2.　Synopses of North-American invertebrates. XI. The catometopous or grapsoid crabs of North America. Am. Nat. 34: 583-92.

1900.3.　Results of the Branner-Agassiz Expedition to Brazil. I. The decapod and stomatopod Crustacea. Proc. Wash. Acad. Sci. 2: 133-56.

1902.1.　The Brachyura and Macrura of Porto Rico. Bull. U.S. Fish Comm. for 1900, part 2: 1-137. [Pre-print date: 1901.]

1902.2.　Crabs from the Maldive Islands. Bull. Mus. comp. Zool. Harv. 39 (5): 123-38.

1902.3.　Papers from the Hopkins Stanford Galapagos Expedition 1898-1899. Brachyura and Macrura. Proc. Wash. Acad. Sci. 4: 275-92.

1904.　Descriptions of three new species of American crabs. Proc. biol. Soc. Wash. 17: 161-62.

1905.　Fauna of New England. V. List of the Crustacea. Pap. Soc. nat. Hist. Boston 7: 1-117.

1907.　Reports on the scientific results of the expedition to the tropical Pacific in Charge of Alexander Agassiz, by the U.S. Fish Commission steamer "Albatross," from August 1899 to March 1900, Commander Jefferson F. Moser, U.S.N., commanding. IX. Reports . . . to the Eastern Tropical Pacific . . . Lieut.-Commander L. M. Garret, U.S.N., commanding. X. The Brachyura. Mem. Mus. comp. Zool. Harv. 35 (2): 21-74 and 91.

1909.　New crabs from the Gulf of Siam. Proc. biol. Soc. Wash. 22: 107-114.

1910.　The Danish Expedition to Siam 1899-1900. V. Brachyura. K. danske Vidensk. Selske. Skr. 7 Raekke 5 (4): 303-367.

1911.　The stalk-eyed Crustacea of Peru and the adjacent coast. Proc. U.S. natn. Mus. 38: 531-620.

1913.　Descriptions of new species of crabs of the family Ocypodidae. Proc. U.S. natn. Mus. 44: 615-20.

1914.1. Stalk-eyed crustaceans collected at the Monte Bello Islands. Proc. zool. Soc. Lond. 653-64.

1914.2. New genera and species of American brachyrhynchous crabs. Proc. U.S. natn. Mus. 47: 117-29.

1918.1. The grapsoid crabs of America. Bull. U.S. natn. Mus. no. 97: xxii + 461 pp.

1918.2. Decapod crustaceans from the Panama region. Contributions to the geology and paleontology of the Canal Zone, Panama, and geologically related areas in Central America and the West Indies. Bull. U.S. natn. Mus. no. 103: 123-84.

1919. West Indian Tertiary decapod crustaceans. *In* "Contributions to the geology and paleontology of the West Indies," prepared under the direction of T. L. Vaughan. Publs Carnegie Instn 291 (5): 157-84.

1920. Stalk-eyed crustaceans of the Dutch West Indies. *In* "Rapport betreffende een voorloopig onderzoek naar den toestand van de Visscherij en de Industrie van Zeeproducten in de Kolonie Curaçao ingevolge het Ministerieel Besluit van 22 November 1904." J. Boeke, Netherlands, 2: 317-49.

1921.1. The brachyuran crabs collected by the American Museum Congo Expedition, 1909-1915. (Ecological and other notes by H. Lang.) Bull. Am. Mus. nat. Hist. 43: 379-468.

1921.2. New species of crabs from Formosa. Proc. biol. Soc. Wash. 34: 155-56.

1923. The brachyuran crabs collected by the U.S. Fisheries steamer "Albatross" in 1911; chiefly on the west coast of Mexico. Bull. Am. Mus. nat. Hist. 48 (8): 619-37.

1924.1. Brachyuran crabs collected by the Williams Galapagos Expedition 1923. Zoologica, N.Y. 5: 153-59.

1924.2. Results of Dr. E. Mjoberg's Swedish scientific expeditions to Australia 1910-1913. Brachyura, Albuneidae and Porcellanidae. Ark. Zool. Uppsala 16 (23): 1-33.

1924.3. Brachyuran crabs collected at Curaçao. Bijdr. Dierk. 23: 13-22.

1924.4. Expedition of the California Academy of Sciences to the Gulf of California in 1921. Crustacea (Brachyura). Proc. Calif. Acad. Sci. (Ser. 4) 13: 373-79.

1926.1. Brachyuran crabs from Australia and New Guinea. Rec. Aust. Mus. 15: 177-82.

1926.2. The fossil stalk-eyed Crustacea of the Pacific slope of North America. Bull. U.S. natn. Mus. no. 138: viii + 155 pp.

1933. Brachyuran crabs of Porto Rico and the Virgin Islands. *In* "Scientific survey of Porto Rico and the Virgin Islands." N.Y. Acad. Sci. 15 (1): 1-121.

1935.1. Preliminary descriptions of six new species of crabs from the Pacific coast of America. Proc. biol. Soc. Wash. 48: 49-52.

1935.2. Scientific results of an expedition to rain forest regions in Eastern Africa. II. Crustacea. Bull. Mus. comp. Zool. Harv. 79 (2): 23-28.

Raut, M. R.
1943. Whimbrel and fiddler crabs. J. Bombay nat. Hist. Soc. 44 (2): 300.

Reese, E. S.
1964. Ethology and marine zoology. *In* "Annual review of oceanography and marine biology," ed. H. Barnes. Allen and Unwin, London: 455-88.

Richardson, L. R.
1949. A guide to brachyrhynchous crabs. Tautara, Wellington, N.Z. 2 (1): 29-36.

Richters, E.
1880. Decapoda. *In* "Beitrage zur Meeresfauna der Insel Mauritius und der Seychellen"; K. Möbius, E. Richters, and E. von Martens. Gutmannschen, Berlin: 139-78. [Not seen.]

Robertson, J. D.
1960. Osmotic and ionic regulation. *In* "The physiology of the Crustacea," ed. T. H. Waterman. Academic Press, New York and London, vol. I: 317-39.

Rochebrune, A. T. de (fils)
1883. Diagnoses d'arthropodes nouveaux propres à la Senegambie. Crustacea. Bull. Soc. philomath. Paris 7 (7): 167-75.

Rossignol, M.
1957. Crustacés décapodes marins de la région de Pointe Noire. *In* "Mollusques, crustacés, poissons marins des côtes d'A. E. F. en collection au Centre d'Océanographie de l'Institute d'Etudes Centrafricaines de Pointe Noire"; J. Collignon, M. Rossignol and C. Roux. Vol. II: 75-136.

Rouch, J.
1953. Les explorations des océans de 1815 a nos jours. *In* "Histoire universelle des explorations," publiée sous la direction de L. H. Parias. Nouvelle Librairie de France, Paris. Chap. 1. (Publication date approximate.)

Roux, J.
1917. Résultats de l'expedition scientifique néerlandaise à la Nouvelle Guinea. Crustacés (Expedition de 1903): Nova Guinea. Paris. Vol. V: 589-621.

1927. Note sur une collection de crustacés decapodes du Gabon. Bull. Soc. vaud. Sci. nat. 56 (218): 237-44.

Rumphius [= Rumpf], G. E.

1705.   D'Amboinsche Rariteitkamer, Behelzende eene Beschryvinge van allerhande zoo weeke als harde Schaalvisschen, te weeten raare Drabben, Kreeften, en diergelyke Zeedieren, als mede allerhande Hoorntjes en Schulpen, die men in d'Amboinsche Zee vindt. . . . T'Amsterdam, Gedrukt by François Halma, Boekverkoper in Konstantijn den Grooten. 1st ed. In 3 parts: 28 pp., 340 pp., 43 pp. 60 pls. Other eds., not seen: 1711, 1739, 1740, 1741.

Rüppell, E.

1830.   Beschreibung und Abbildung von 24 Arten Kurz Schwanzigen Krabben als Beitrag zur Naturgeschichte des rothen Meeres. Brönner, Frankfurt am Main. 28 pp.

1834.   Description de 24 espèces de crabes pour servir à l'histoire naturelle de la Mer Rouge. Férussac, Bull. Sci. nat. 24: 100-104 [= abstract of 1830 contribution].

Ryan, E. P.

1966.   Pheromone: evidence in a decapod crustacean. Science, 151 (Jan. 21): 340-41.

Sakai, T.

1934.   Brachyura from the coast of Kyushu, Japan. Sci. Rep. Tokyo Bunrika Daig. Sect. B. 1 (25): 281-330.

1936.   Report on the Brachyura collected by Mr. F. Hiro at Palao (Pelew) Islands. Sci. Rep. Tokyo Bunrika Daig. (B) 2: 155-77.

1939.   Studies on the crabs of Japan. IV. Brachygnatha, Brachyrhyncha: 365-741. (Publ. by author, Tokyo.)

1940.   Bio-geographic review on the distribution of crabs in Japanese waters. Rec. oceanogr. Wks Japan 11: 27-63.

Salmon, M.

1965.   Waving display and sound production in *Uca pugilator* Bosc, with comparisons to *U. minax* and *U. pugnax*. Zoologica, N.Y. 50: 123-50.

1967.   Coastal distribution, display and sound production by Florida fiddler crabs (genus *Uca*). Anim. Behav. 15: 449-59.

Salmon, M., and S. P. Atsaides

1968.1. Behavioral, morphological and ecological evidence for two new species of fiddler crabs (genus *Uca*) from the Gulf coast of the United States. Proc. biol. Soc. Wash. 81: 275-90.

1968.2. Visual and acoustical signalling during courtship by fiddler crabs (genus *Uca*). Am. Zool. 8: 623-39.

1969.   Sensitivity to substrate vibration in the fiddler crab *Uca pugilator* Bosc. Anim. Behav. 17: 68-76.

Salmon, M., and J. F. Stout

1962.   Sexual discrimination and sound production in *Uca pugilator* Bosc. Zoologica, N.Y. 47: 15-21.

Sandeen, M. I.

1950.   Chromatophorotropins in the central nervous system of *Uca pugilator*, with special reference to their origins and actions. Physiol. Zoöl. 23 (4): 337-52.

Sankarankutty, C.

1961.   On Decapoda Brachyura from the Andaman and Nicobar Islands. 1. Families Portunidae, Ocypodidae, Grapsidae and Mictyridae. J. mar. Biol. Ass. India 3 (1 & 2): 101-119.

Saussure, H. de

1853.   Description de quelques crustacés nouveaux de la côte occidentale du Mexique. Rev. et Mag. Zool. Paris ser. 2, 5: 354-68.

Say, T.

1817-   An account of the Crustacea of the United
1818.   States. J. Acad. nat. Sci. Philad. (ser. 1) 1: 57-64, *65-80*, 97-101, 155-69, 235-53, 313-26, 374-401, *423-58*. [Sections in italics include remarks on *Ocypode pugilator*.]

Schäfer, W.

1954.   Form und Funktion der Brachyuren-Schere. Abh. senckenb. naturforsch. Ges. No. 489: 1-65.

Schenkel, E.

1902.   Beitrag zur Kenntnis der Dekapodenfauna von Celebes. Verh. schweiz. naturf. Ges. Basel 13: 485-585.

Schmitt, W. L.

1921.   The marine decapod Crustacea of California. Univ. Calif. Publs Zool. 23: 1-470.

Schöne, H.

1961.   Complex behavior. *In* "The physiology of the Crustacea," ed. T. H. Waterman. Academic Press, New York and London, vol. II: 465-620.

1968.   Agonistic and sexual display in aquatic and semi-terrestrial brachyuran crabs. Am. Zool. 8: 641-54.

Schöne, H., and I. Eibl-Eibesfeldt

1965.   FILM. Encyclopaedia Cinematographica, Göttingen. E 599. *Grapsus grapsus* (Brachyura). [Not seen.]

Schöne, H., and H. Schöne

1963.   Balz und andere Verhaltensweisen der Mangrovekrabbe *Goniopsis cruentata* Latr. und das Winkverhalten der eulitoralen Brachyuran. Z. Tierpsychol. 20: 641-56.

Schramm, A.

1867.   Crustacés de la Guadeloupe d'après un manuscrit du Docteur Isis Desbonne comparé avec les échantillons de crustacés de sa collection et les dernières publications de MM. Henri de Saussure et William Stimpson. Première partie, brachyures. Imprimérie du gouvernement, Basseterre, ii + 60 pp.

Schroff, K. D.
1920. Notes on some land and marine crabs and field-snails which are pests in Burma. Rept. Proc. 3rd entomological meeting Pusa for 1919, 2: 689-94.

Schwartz, B., and S. R. Safir
1915. The natural history and behavior of the fiddler crab. Cold Spring Harb. Monogr. 8: 1-24.

Seba, A.
1758 and 1761. Locupletissimi rerum naturalium thesauri accurata descriptio et iconibus artificiosissimis expressio per universam physices historiam . . . Amstelaedami. Vol. III: 22 pp. + 212 pp. 116 pls. Publ. in 1758 "apud Janssonio—Waesbergios" and in 1761 "apud H. K. Arksteum & H. Merkum, et Petrum Schouten." [1761 printing not seen.]

Sendler, A.
1912. Zehnfusskrebse aus dem Wiesbadener Naturhistorischen Museum. Jb. nassau. Ver. Naturk. 65-66: 189-207.
1923. Die Decapoden und Stomatopoden der Hanseatischen Südsee Expedition. Abh. Senckenb. naturforsch. Ges. 38: 21-47.

Serène, R.
1937. Inventaire des invertébrés marines de l'Indochine (1re. liste). Notes Stn marit. Cauda No. 30: 65-78.

Sewell, R. B.
1913. Notes on the biological work of the R. I. M. S. S. "Investigator" during survey seasons 1910-1911 and 1911-1912. J. Proc. Asiat. Soc. Beng. 9: 329-90.

Shaw, G., and E. R. Nodder
1802. The naturalist's miscellany. London. Vol. XIV: Pl. 588 and 2 pp. (not numbered) preceding it.

Shen, C.
1932. The brachyuran Crustacea of North China. Zool. Sinica, ser. A. Invertebrates of China 9 (1): 1-300.
1937.1. On some account of the crabs of North China. Bull. Fan meml Inst. Biol. 7 (5): 167-85.
1937.2. Second addition to the fauna of brachyuran Crustacea of North China, with a check list of the species recorded in this particular region. Contr. Inst. Zool. natn. Acad. Peiping 3: 277-312.
1940. The brachyuran fauna of Hongkong. Journ. Hongkong Fish. Res. Stn 1 (2): 211-42.

Silas, E. G., and C. Sankarankutty
1960. On the castle building habit of the crab *Cardisoma carnifex* (Herbst) (Family Geocarcinidae), of the Andaman Islands. J. mar. biol. Ass. India 2 (2): 237-40.

Simpson, G. G.
1953. The major features of evolution. Columbia University Press, New York. xx + 434 pp.
1961. Principles of animal taxonomy. Columbia University Press, New York. xii + 247 pp.

Sivertsen, E.
1934. Littoral Crustacea Decapoda from the Galapagos Islands. (Norwegian Exped. to the Galapagos Is. 1925, VII.) Nyt Mag. Naturvid. 74: 1-23.

Sloane, H.
1725. A voyage to the islands Madera, Barbadoes, Nieves, St. Christophers, and Jamaica; with the natural history of the herbs and trees, four-footed beasts, fishes, birds, insects, reptiles &c. of the last of those islands. . . . Printed for the author, London. Vol. II: xviii + 499 pp. 125 pls.

Smith, S. I.
1869.1. Notice of the Crustacea collected by Prof. C. F. Hartt on the coast of Brazil in 1867. Trans. Conn. Acad. Arts Sci. 2: 1-42.
1869.2. A fiddler-crab with two large hands. Am. Nat. 3 (10): 557. Issue of December, 1869; bound vol. published 1870.
1870. Notes on American Crustacea. No. 1. Ocypodoidea. Trans. Conn. Acad. Arts Sci. 2: 113-76. [Issue of March 1870.]
1871. Thirty-two species of Crustacea collected by J. A. McNiel at west coast of Central America, Nicaragua, and Bay of Fonseca. Rep. Peabody Acad. Arts Sci. for 1869: 87-98. [For publication data, see note and dates under Director's & Treas. Reports, Dec. 31, 1870, & Jan. 15, 1871.]

Stebbing, T.R.R.
1905. South Africa Crustacea. Part 3. Marine investigations in South Africa. Cape of Good Hope Dept. of Agric., Cape Town; 4: 21-123.
1910. General catalogue of South African Crustacea. (Part 5 of S.A. Crustacea, for the marine investigations in South Africa.) Ann. S. Afr. Mus. 6 (4): 281-593.
1917. The Malacostraca of Natal. Ann. Durban Mus. 2: 1-33.
1921. Some Crustacea of Natal. Ann. Durban Mus. 3: 12-26.

Stephensen, K.
1921. Nogle Traek af Strandkrabbernes, saerlig VinkeKrabbernes, Biologi. Contributions to

the biology of the crabs, especially of *Uca pugilator*. Naturens Verd. 5: 458-65.

1946. The Brachyura of the Iranian Gulf, with an appendix: The male pleopoda of the Brachyura. Dan. scient. Invest. Iran., part 4: 57-237. [Note: Appendix lists references in which gonopods in species of *Uca* are figured; no original work on this genus is included.]

Stimpson, W.

1858. Prodromus descriptionis animalium evertebratorum quae in expeditione ad Oceanum Pacificum Septentrionalem, a Republica Fed. . . . Pars. 5. Crustacea Ocypoidea. Proc. Acad. nat. Sci. Philad. 10: 93-110.

1859. Notes on North American Crustacea, No. 1. Ann. Lyc. nat. Hist. N.Y. for 1859, 7: 49-93.

1860. Notes on North American Crustacea in the museum of the Smithsonian Institution for 1860, no. 2. Ann. Lyc. nat. Hist. N.Y. for 1860, 7: 176-246.

1907. Report on the Crustacea (Brachyura and Anomura) collected by the North Pacific Exploring Expedition 1853-1856. (With introductory note, and edited by M. J. Rathbun.) Smithson. misc. Collns 49 (1717): 1-240.

Stossich, M.

1877. Sulla geologia e zoologia dell'isola di Pelagosa. Boll. Soc. Adriatica Sci. nat. Trieste, 1877: 184-92.

Streets, T. H.

1872. Notice of some Crustacea from the island of St. Martin, W.I., collected by Dr. van Rijgersma. Proc. Acad. nat. Sci. Philadelphia, 24: 131-34.

1877. Contributions to the natural history of the Hawaiian and Fanning Islands and Lower California. Bull. U. S. natn. Mus. No. 7 and Smithson. misc. Colln 13. Crustacea: 103-141.

Studer, T.

1882. Verzeichniss der während der Reise S. M. S. "Gazelle" an der Westküste von Afrika, Ascension, und dem Cap der Guten Hoffnung gesammelten Crustaceen. Abh. dt. Akad. Wiss. Berl. 32 pp.

Sumner, F. S., R. C. Osburn, and L. J. Cole

1913. A biological survey of the waters of Woods Hole and vicinity. Part 2. Sect. 3. A catalogue of the marine fauna. Bull. Bur. Fish., Wash. 31 (2): 669-75.

Suvatti, C.

1938. A check-list of aquatic fauna in Siam (excluding fishes). Bureau of Fisheries, Bangkok; B.E. 2480. 116 pp.

Sverdrup, H. U., M. W. Johnson, and R. H. Fleming

1942. The oceans. Their physics, chemistry, and general biology. Prentice-Hall, Inc., Englewood Cliffs, New Jersey. x + 1087 pp.

Symons, C. T.

1920. Notes on certain shore crabs. Spolia zeylan. 306-13.

Takahasi, S.

1935. Ecological notes on the ocypodian crabs (Ocypodidae) in Formosa, Japan. Annotnes zool. jap. 15: 78-87.

Tashian, R. E.

1958. The specific distinctness of the fiddler crabs *Uca pugnax* (Smith) and *Uca rapax* (Smith) at their zone of overlap in northeastern Florida. Zoologica, N.Y. 43: 89-92.

Tazelaar, M. A.

1933. A study of relative growth in *Uca pugnax*. Wilhelm Roux Arch. EntwMech. Org. 129 (2): 393-401.

Teal, J. M.

1958. Distribution of fiddler crabs in Georgia salt marshes. Ecology 39: 185-93.

1959. Respiration of crabs in Georgia salt marshes and its relation to their ecology. Physiol. Zoöl. 32 (1): 1-14.

Tesch, J. J.

1918. The Decapoda Brachyura of the "Siboga" Expedition. I. Hymenosomidae, Retroplumidae, Ocypodidae, Grapsidae and Gecarcinidae. Siboga-Expeditie, uitkomsten . . . H. M. "Siboga." Brill, Leiden. Vol. xxxix C: 148 pp.

Thallwitz, J.

1892. Decapoden Studien, insbesondere basiert auf A. B. Meyer's Sammlungen im östindischen Archipel, nebst einer Aufzahlung der Decapoden und Stomatopoden des Dresdener Museums. Abh. Ber. K. zool. anthrop.-ethn. Mus. Dresden. No. 3: 1-55.

Thompson, T. I.

1963. Visual reinforcement in Siamese fighting fish. Science 141: 55-57.

1964. Visual reinforcement in fighting cocks. J. Exptl Anal. Behavior 7: 45-49.

Tinbergen, N.

1952. "Derived" activities; their causation, biological significance, origin and emancipation during evolution. Q. Rev. Biol. 27: 1-32.

1953. Social behaviour in animals. Methuen and Co., London. 150 pp.

Trewartha, G. T.

1954. An introduction to climate. 3rd ed. McGraw-Hill Book Co. vii + 402 pp.

Troll, C., and K. H. Paffen

1963. Weltkarten zur Klimakunde / World maps

of climatology. Heidelberger Akademie der Wissenschaften; Springer Verlag, Berlin, Göttingen, and Heidelberg.

Tweedie, M.W.F.
1937. On the crabs of the family Ocypodidae in the collection of the Raffles Museum. Bull. Raffles Mus. 13: 140-70.
1950.1. Grapsoid crabs from Labuan and Sarawak. Sarawak Mus. J. 5 (2): 356-67.
1950.2. The fauna of the Cocos-Keeling Islands. Brachyura and Stomatopoda. Bull. Raffles Mus. 22: 105-48.
1950.3. Notes on grapsoid crabs from the Raffles Museum. II. On the habits of three ocypodid crabs. Bull. Raffles Mus. 23: 317-24.
1952. Two crabs of the sandy shore. Malayan Nature J. 7: 3-10.
1954. Notes on grapsoid crabs from the Raffles Museum Nos. 3, 4, and 5. Bull. Raffles Mus. 25: 118-27.

Umbgrove, J.H.F.
1949. Structural history of the East Indies. Cambridge University Press, Cambridge.

Vatova, A.
1943. I. Decapoda della Somalia. Thalassia 6 (2): 1-37.

Vernberg, F. J.
1959.1. Studies on the physiological variation between tropical and temperate zone fiddler crabs of the genus Uca. II. Oxygen consumption of whole organisms. Biol. Bull. 117 (1): 163-84.
1959.2. Studies on the physiological variation between tropical and temperate zone fiddler crabs of the genus Uca. III. The influence of temperature acclimation on oxygen consumption of whole organisms. Biol. Bull. 117 (3): 582-93.

Vernberg, F. J., and J. D. Costlow, Jr.
1966. Handedness in fiddler crabs (genus Uca). Crustaceana 11: 61-64.

Vernberg, F. J., and R. E. Tashian
1959. Studies on the physiological variation between tropical and temperate zone fiddler crabs of the genus Uca. 1. Thermal death limits. Ecology 40 (4): 589-93.

Vernberg, W. B., and F. J. Vernberg
1968. Physiological diversity in metabolism in marine and terrestrial Crustacea. Am. Zool. 8: 449-58.

Verrill, A. E.
1873. Report upon the invertebrate animals of Vineyard Sound and the adjacent waters, with an account of the physical characters of the region. In "The report of Professor S. F. Baird, Commissioner of Fish and Fisheries, on the conditions of the sea-fisheries of the south coast of New England in 1871 and 1872"; Washington: Govt. Printing Office, Part 8: 295-747.

Verrill, A. E., and S. I. Smith
1874. [Same as Verrill, 1873, above, extracted and reprinted with new pagination (vi + 478 pp.), old pagination appearing additionally on each page; table of contents and index added.]

Verwey, J.
1930. Einiges über die biologie öst-indischer mangrovekrabben. Treubia 12 (2): 167-261.

Vilela, H.
1939. A tragédia de um caranguejo (Uca tangieri). Naturalia 3: 177-81.
1949. Crustaceos decapodes e estomatopodes da Guine Portuguesa. Anais Jta Invest. colon. 4: 47-70.

Volz, P.
1938. Droh- und Warn-signale bei zehnfussigen Krebsen. Forschn Fortschr. 14: 284-86.

Walker, A. O.
1887. Notes on a collection of Crustacea from Singapore. J. Linn. Soc. 20: 107-17.

Ward, M.
1928. The habits of our common shore crabs. Aust. Mus. Mag. 3 (7): 242-47.
1939. The Brachyura of the Second Templeton Crocker—American Museum Expedition to the Pacific Ocean. Am. Mus. Novit. No. 1049: 1-15.
1941. New Brachyura from the Gulf of Davao, Mindanao, Philippine Islands. Am. Mus. Novit. 1104: 1-15.

Warner, G. F.
1969. The occurrence and distribution of crabs in a Jamaican mangrove swamp. J. Anim. Ecol. 38: 379-89.
1970. Behavior of two species of grapsid crabs during intraspecific encounters. Behavior 36: 9-19.

Waterman, T. H. (ed.)
1960. The physiology of Crustacea. Vol. I. Metabolism and growth. Academic Press, New York and London. x + 670 pp.
1961. The physiology of Crustacea. Vol. II. Sense organs, integration, and behavior. Academic Press, New York and London. v + 681 pp.

Werner, F.
1938. Ergebnisse einer zoologischen Forschungsreise nach Marokko. 7. Insekten, Arachnoiden und Crustaceen. Sber. Akad. Wiss. Wien 1, 147 (3): 111-34.

White, A.
  1847.   List of the specimens of Crustacea in the
          collection of the British Museum. Printed
          by order of the trustees. Edward Newman,
          London. viii + 143 pp.
  1848.   Short descriptions of some new species of
          Crustacea in the collection of the British
          Museum. Proc. zool. Soc. Lond. Part xv for
          1847: 84-86.
Whitelegge, T.
  1898.   The Crustacea of Funafuti. The Atoll of
          Funafuti. Pt. 2. Mem. Austral. Mus. 3:
          127-51.
Wright, H. O.
  1968.   Visual displays in brachyuran crabs: Field
          and laboratory studies. Am. Zool. 8: 655-
          65.

Wynne-Edwards, V. C.
  1962.   Animal dispersion in relation to social be-
          haviour. Hafner Publishing Co., New York.
          v + 653 pp.
Yerkes, R. M.
  1901.   A study of variation in the fiddler crab
          Gelasimus pugilator Latr. Proc. Am. Acad.
          Arts Sci. 36 (24): 417-42.
Young, C. G.
  1900.   The stalk-eyed Crustacea of British Guiana,
          W. Indies and Bermuda. London. xix +
          514 pp.
Zehnter, L.
  1894.   Crustacés de l'Archipel malais—voyage de
          M.M. Bedot et Ch. Pichtet dans l'Archipel
          malais. Revue suisse Zool. 2 (1): 135-214.

ADDENDUM TO BIBLIOGRAPHY

Aspey, W. P.
  1971.   Inter-species sexual discrimination and ap-
          proach-avoidance conflict in two species of
          fiddler crabs, *Uca pugnax* and *Uca pugila-
          tor*. Anim. Behav. 19: 669-76.
Bright, D. B., and C. L. Hogue
  1972.   A synopsis of the burrowing land crabs of
          the world and list of their arthropod sym-
          bionts and burrow associates. Contrib. in
          Science, Los Angeles County Mus. No. 220:
          58 pp.
Hagen, H.-O. von
  1972.   Text associated with FILM E 1421 / 1971.
          Uca leptodactyla (Ocypodidae). Balz. En-
          cyclopaedia Cinematographia, Göttingen:
          Publ. Inst. Wiss. Film: pp. 3-20.
  1972.   Text associated with FILM E 1423 / 1971.
          Uca maracoani (Ocypodidae). Balz. En-
          cyclopaedia Cinematographia, Göttingen:
          Publ. Inst. Wiss. Film: pp. 3-20.
Hogue, C. L., and D. B. Bright
  1971.   Observations on the biology of land crabs
          and their burrow associates on the Kenya
          coast. Contrib. in Science, Los Angeles
          County Mus. No. 210: 10 pp.

Salmon, M.
  1971.   Signal characteristics and acoustic detection
          by the fiddler crabs, *Uca rapax* and *Uca
          pugilator*. Physiol. Zool. 44: 210-24.
Salmon, M., and K. W. Horch
  1972.   Acoustic signalling and detection by semi-
          terrestrial crabs of the Family Ocypodidae.
          In "Behavior of marine animals: current
          perspectives in research," H. E. Winn, ed.
          Plenus Publishing Corporation, New York.
          Vol. i: 60-96.
Selander, R. K., W. E. Johnson, and J. C. Avise
  1972.   Biochemical population genetics of fiddler
          crabs (*Uca*). Abstract. Biol. Bull. (Marine
          Biol. Lab.) 141 (2): 402.
Vannini, M., and A. Sardini
  1971.   Aggressivity and dominance in river crab
          Potamon fluviatile (Herbst). Monitore Zool.
          ital. (n.s.) 5: 173-213.
Zucker, N.
  1972.   Shelter building in the tropical fiddler crab
          Uca terpsichores. Abstract. Bull. Ecol. Soc.
          Am. 53 (4): 22.

Indexes

Index to Scientific Names

Entries in **boldface** are the names of subgenera, species, and subspecies of the genus *Uca* which in this contribution are considered valid. Page numbers in boldface refer to the principal treatment of each of these entries, in accordance with the topics listed and described on pp. 10–13.

Several of the subheadings provided for the longer entries serve only as general guides. For example, the subheading "evolution" may include references to apparent phylogeny, morphological responses to ecological pressures, and possible derivations of behavioral components.

Acanthoplax, 20, 147, 149, 153, 323, 602

acuta, 25–28; distribution, 433, 437; keys, 623, 625; material, 592; morphological cfs., 22–23, 30, 33, 49, 466; nomenclature, 29, 31, 39, 44, 51, 53. *See also* subspecies, below

acuta (superspecies), 24, **25**, **27–29**, 40, 432, 528

acuta acuta, 25–28, **28**, 36, 45–46; key, 625; material 592

acuta rhizophorae, 25–28, **27**, 30, 33, 35, 42, 51, 524; key, 623; material, 592

acutus, 25, 27–28, 31, 47, 51, 61

affinis (of Guérin), 81, 171, 172, 322

affinis (of Streets), 171, 172, 175, 322

Afruca, 116–17, 118–24; behavior, 128, 479, 481, 484, 500–502, 522; color, 467–68; evolution, 19, 127, 432, 459, 532–33; key, 619; material, 600; morphological cfs., 125, 145, 151, 213, 454–55, 459, 464, 466, 469; size 17, 449

Agama (as predator), 446

albimana, see annulipes var. *albimana*

amazonensis, 101, 103, 322, 599

Amphiuca, 96–97, 98–108; behavior, 481, 497, 500, 502; color, 467–69; distribution, 432; evolution, 19, 163, 532–33; key, 619; material 598–99; morphological cfs., 86, 125, 454, 462, 464, 469; size, 449; sympatrics, 79, 297

angustifrons, 66, 69

annulipes, 188, 292, 298–99, 301–302, 326, 611. *See also annulipes* var. *albimana, annulipes* var. *lacteus, annulipes* var. *orientalis*, and **lactea annulipes**

annulipes var. *albimana*, 298, 302

annulipes var. *lacteus*, 303

annulipes var. *orientalis*, 298, 303, 612

arborescens, 446

arcuata, 44–47; behavior, 50, 474, 500, 513, 673; distribution, 176, 433, 437; ecology 58, 442, 452; evolution, 24, 452; key, 625; material, 594; morphological cfs., 21, 25–26, 30, 33, 36, 49, 53, 84; nomenclature, 28, 57; sympatrics, 27, 84

arcuatus, 28, 47, 57, 61, 594

argillicola, 161, 186, 211, 213, 217, **220–22;** key, 627; material, 607

armatus, 147, 149, 602

aspera, 486

australiae, *see* **demani australiae**

Australuca, 62–63, 64–74; behavior, 461, 497, 500, 502; color, 467: distribution, 295, 432; evolution, 18, 295, 436–37, 531–33; key, 619; material, 595–96; morphological cfs., 40, 46, 53, 59, 86, 218, 454, 459, 460, 462, 464; nomenclature, 327; size, 449

Avicenna, 445

basipes, 322

batuenta, 244–46; behavior, 63, 248–49, 254, 256, 259, 262, 266, 306, 313, 461, 480; evolution, 218; key, 627; material, 609; morphological cf., 248, 461; nomenclature, 609; size, 17, 449; sympatrics, 249, 253, 256, 280, 282 (cf. only), 309

beebei, 278–81; behavior, 277, 283–84, 311, 461, 480, 498, 500–501, 504, 507; color, 468–69; distribution, 282; evolution, 217–18, 268, 275; key, 629; material, 610; morphological cfs., 134, 211, 216, 275–76, 282–83, 287, 461, 464, 480; sympatrics, 135, 253, 256, 282, 284, 309, 313, 317, 321

bellator, 64–69; behavior, 72, 497, 500; distribution, 326–27; evolution, 63; keys, 623–24; material, 595–96; morphological cfs., 40, 62–63, 73; nomenclature, 62, 323, 326; sympatrics, 35, 71. *See also* subspecies below

bellator bellator, 64–66, **66**, 69, 595

bellator longidigita, 64, 66, **68**, 69, 596, 624

bellator minima, 64–66, **68**, 596

bellator signata, 66, **67**, 69, 596

bengali, *see* **triangularis bengali**

Boboruca, 109–11, 112–15; behavior, 497, 500, 502; color, 467; evolution, 18–19, 145, 163, 436, 532–33; key, 619; material, 600; morphological cfs., 463; size, 449; sympatrics, 146, 195

borealis, *see* **vocans borealis**

brasiliensis, see **pugnax brasiliensis**

Brasiliensis, see Uka una, Brasiliensis

brevifrons, 154–56, 164–65, 174, 176–77, **180–82,** 184, 222, 233, 461, 472, 602; key, 629; material, 604–605

brevifrons var. *delicata*, 182, 605

brevipes, 47, 594

Bufo marinus, 673

burgersi, 168–72; behavior, 156, 174, 443, 498–99; color, 665; distribution, 326–27; ecology, 665; evolution, 157, 176; key, 631; material, 603–604; morphological cfs., 154, 156, 164, 173–74, 177, 180–81, 192–93, 196–97, 237; nomenclature, 165, 174, 322, 324–25, 327; sympatrics, 165, 173, 175, 195–96, 230, 306

Callinectes, 479

Cancer, 20, 75, 81, 89, 93, 138–39, 148, 165, 167, 172, 198, 204, 227, 324, 326, 455, 495

Cancridae, 486

capricornis, *see* **dussumieri capricornis**

Cardisoma, 479

Carpilius, 479

Celuca, 211–19, 220–321; behavior, 100, 460–61, 479, 481–83, 485, 487, 494, 498–502, 516, 518, 522, 524; color, 466–68; descriptive methods, 11; ecology, 442, 444, 455; evolution, 18–20, 157, 436, 497, 524, 530, 532–35; key, 620; material, 607–614; morphological adaptations to combat, stridulation, 457, 458, 462; morphological cfs., 10, 17, 63, 96, 114, 154–56, 161, 177, 180, 207, 451, 453–54, 459, 463–66; nomenclature, 18, 323, 327; size and allometry, 449–50; sympatrics, 65, 88, 113, 134–35, 146, 186, 195, 200, 203

ceratophthalmus, 434

chlorophthalmus, 98–104; behavior, 105, 500, 503, 513; color, 468–69; distribution, 77, 80, 428, 432, 437, 528; evolution, 528, 532; heat tolerance, 442; keys, 620, 622; material, 598–99; morphological cfs., 23, 27, 59, 78, 96, 105–106, 163, 294, 464, 469; nomenclature, 108, 322, 324; sympatrics, 60, 77, 297. *See also* subspecies below

chlorophthalmus chlorophthalmus, 99–102, **102**, 103, 322, 599

chlorophthalmus crassipes, 79, 100–101, **101**, 102, 291, 324, 623, 625, 598–99

Ciccaba (predator), 188

cimatodus, 122, 123

General Index

This index includes geographical localities, names of persons, and subjects.

Geographical Localities. Only major areas are listed, such as names of countries and certain islands. Page references to the species recorded from an indexed locality are given under the subheading "recorded spp.", followed by the first page of each pertinent species treatment in the Systematic Section (for example, "recorded spp., 77ff, 284ff"). These lists include species recorded from the locality in print, on specimen labels in material examined, or both. Some lists include records considered by the present author to be erroneous; the most questionable are also referred in the entry to additional page numbers; minor queries are discussed within the species treatments. The locality indexed, such as "India," as well as names of smaller places included

within the area, occur chiefly in these treatment topics: Introduction, Range, Field Material, Type Material and Nomenclature, and References and Synonymy, as well as in Material Examined (Appendix A). Incidental references to indexed localities occurring in non-systematic sections are omitted when the geographical information is not directly important to the context, as in locality identification of behavioral observations.

Names of Persons. Both collectors and donors of material, as given on the labels of occasional preserved specimens, are listed as donors. Page numbers in boldface refer to citations of a writer only as the author of a scientific name. When further material involving the same person appears on the same page, the page number appears also in roman.

504; courtship behavior, 502–503; herding of young by males, 496, 498, 503; morphological characters, summaries, 17, 470; receptivity, 503, 523; as taxonomic topic, 11; territoriality, 511; wandering 502–503, 517. *See also* text: Systematic Section, recurrent topic "Female"; Chap. 3, structure accounts, 451–69

field trips, planning 664–66

fight, 681, 683. *See also* combat

Fiji Is. (Pacific O.): material, 595, 597, 599, 613; nomenclature, 324; recorded spp., 52ff, 77ff, 85ff, 98ff, 292ff; zoogeography, 21, 434, 437

Filhol, H., 323, **323**

finger, 683

Fingerman, M., 228

fixed action pattern, 681

flagellum: of antenna, 454, 683; of 3rd maxilliped, eye-cleaning, 455

flange (of gonopod), 463–64, 683

flat-claw (agonistic component), 479, 522, 660 (Table 21), 683

Fleming, R. H., 431

fling (in combat), 488, 683

Flinn, M., 483

Flores (Indonesia): recorded sp., 98ff

Florida, *see* United States of America

Fonseca, Gulf of (eastern Pacific): distribution and zoogeography, 432, 438; material, 601–602; recorded spp., 133ff, 140ff, 143ff, 158ff, 251ff, 282ff. *See also* El Salvador, Nicaragua

food, 472, 534

Forest, J., 82, 99, 101–104, 303

Formosa, *see* Taiwan

forward-point (agonistic component), 479, 495, 522, 660 (Table 21), 683

fossils, 11, 127, 157, 324, 435

Fourmanoir, P., 61, 103, 303

Fowler, H., 204, 227

Frankfurt, *see* institutions

French Equatorial Africa, *see* Africa, west

French Guiana, *see* Cayenne

Freycinet, L. de., 4

friction (contact) area, 459–60, 487, 652 (Table 14), 680

Friendly Is. (Pacific O.): material, 596, 599, 613; recorded spp., 77ff, 85ff, 98ff, 292ff

front, 452, 455, 527, 683

frontal-arc (agonistic component), 479, 522, 660 (Table 21), 683

Fukien, *see* China

furrow: frontal, 452. *See also* groove

Gabon, *see* Africa, west

Gaimard, P. (donor), 60, 101

Galapagos Is.: material, 605, 610; recorded spp., 183ff, 271ff, 274f

Gambia, *see* Africa, west

Ganges (delta), *see* India

gape (of claw), 683; major cheliped, 458–60, 651 (Table 13); small cheliped, 460–61

Garth, J. S., 160, 162, 182, 186, 188, 222, 235, 272–73, 285, 321, 431, 433, 435, 438

Gay, Mr. (donor), 147

Gee, N.G., 28, 37, 47, 94, 103, 303

Gélasime, 20

Genard, Mr. (donor), 592

genetics, 9, 525–26

Georgia, *see* United States of America, Atlantic coast north of Florida

Gerlach, S. A., 114, 148, 175, 198, 231, 307

Gerstaecker, A., **323**. *Ref.:* Archiv. f. Naturgeschichte, 1856, 22 (1): 138

Gilbert Is. (Pacific O.): material, 596; recorded sp., 77ff

gills, 469, 683

glossary, 679–90

Gmitter, T. E., 198

Göldi, E. A., 322

gonopod, 463–65, 504, 527, 530–33, 683. *See also:* text, Systematic Section, recurrent topic

Gordon, H. R. S., 87, 303, 518, 520

Gordon, I., 38, 47, 57, 69, 95, 103, 290, 302, 324

Göttingen, *see* institutions

Gould, A. A., 204, 227

Gould, S. J., 449–50, 679

Grant, F. E., 38, 57

Grant, V., 526

granulation, *see* granule

granule, 454, 683

Gravier, C., 82

Gray, E. H., 179, 495

Greeley, A. W. (donor), 113

Griffin, D. J. G., 486, 495

grip (in combat), 488, 683

groove, 460, 683

growth, *see* development, post-larval

growth, allometric, 449–50, 455, 670, 683

Gruner, H.-E., 165

Guadeloupe, *see* West Indies

Guam: material, 597; recorded sp., 85

Guatemala, Atlantic coast: material, 600, 603–605; recorded spp., 112ff, 163ff, 168ff, 173ff, 190ff

Guayaquil, Gulf of, 438

Guérin, see Guérin-Meneville

Guérin-Meneville, F. E., **80**, 81, 103, 171, **171–72**, 306, **307**, 307, **322**, 322, 324, 325, **596**, 600

Guinea, *see* Africa, west

Guinea, Gulf of, *see* Africa, west

Guinea, Spanish, *see* Africa, west

Guinot, D., 82, 99, 101–104, 124, 146, 303, 325

Guinot-Dumortier, D., 148, 480, 482, 487

Gunther, H.-J., 124

Guyana: material, 602–604, 606, 609; recorded spp., 143ff, 163ff, 173ff, 190ff, 240ff

Haan, W. de., **28, 44, 47**, 47, **57, 292**, 292, **298, 300,** 301, **324, 594, 611–12**

habitat, 440, 442, 683; use of term in topic, "References and Synonymy," 13. *See also* biotope habits, use of term in topic, "References and Synonymy," 13; behavior

Hadramaut: material, 599; recorded sp., 105ff

Hagen, H.-O. von, 113–14, 118, 120–22, 124, 128, 132, 135, 138–40, 142, 146, 148–49, **153**, 153, **166**, 166, 167, 171–72, 175, 185–89, 194–95, 198, 218, 239, 243, 246, 255, 257, 259, 260–66, **264, 267, 275, 277**, 281, 283, 285, 306–307, 318, 323, 444, 446, 452, 455, 468, 474, 480–81, 484–485, 495, 498, 500–502, 504, 506–507, 511, 601, 604, **609–610**, 639 (Table 7), 643 (Table 10), 645 (Table 12), 690

Haight, A. L. (donor), 606

Haiti, *see* West Indies

Hallam, A., ed., 431

Halmahera, *see* Molucca Is.

Hamilton, W. J., III, 471, 478

Hamlin, H. (donor), 157

Hancock, Hancock Foundation, *see* institutions

Hartnoll, R. G., 500, 503

Haswell, W. A., 57, 69, 93

Hawaii, 80, 82, 323–24, 327, 597, 599

Hay, W. P., 179, 204, 227

Haygood, B. (donor), 613

Hedgpeth, J. W., 431, 440–41

Hediger, H., 124, 495

heel (on major manus), 684

heel-and-hollow (in combat), 490, 646–51 (Table 13), 652 (Table 14), 660 (Table 21), 684

heel-and-ridge (in combat), 460, 489–90, 646–51 (Table 13), 652 (Table 14), 654 (Table 15), 660 (Table 21), 684

Heller, C., 57, 81, 93, 102, 291, 301, **326**

Henderson, J. R., 291, 301

Hendriksen, Mr. (donor), 596

Henschell, A.G.E.T., 93

hepatopancreas, 684

Herbst, J. F. W., **20, 75, 77, 80–81,** 80–81, **132, 136,** 138–39, **138, 163,** 165, 165–67, 167, **171–72, 198,** 326, **596, 601, 602**

herding (in waving display), 202, 496, 498, 503, 658 (Table 20), 660 (Table 21), 684

Herklots, J. A., **122,** 123

Herrick, C. L., 198

Herrnkind, W., 228, 450, 472, 474, 478, 510

Hess, W., 66, **66–67, 69,** 69, **80,** 81, 323, **596**

Hesse, R., 440, 445

hibernation, 44, 441, 452, 471

hierarchies, dominance, 506

high-rise (agonistic component), 480, 523–24, 660 (Table 21), 684

Hilgendorf, F., 61, 81, 93, 103, 123, 301

Hinde, R., 478

Hispaniola, *see* West Indies

Hobbs, H. H., Jr., 114, 138–39, 167, 172, 198, 239, 243, 306–307

Hoffmann, C. K., 35, 61, 81, 93–94, **105–108,** 106–108, 301, 593, **599**

Hoffmann, K., 448

Holmes, S. J., 182, 235, 325

Holthuis, L. B., 20, 66, 82, 85, 89, 93, 94, 108, 114, 135, 138–39, 142, 148, 160, 165, 167–68, **168,** 170–72, **171–72,** 175, 182, 198, 208, 243, 281, 285, 289, 306, 310, 593, 596, 598–99, **603,** 613

Library of Congress Cataloging in Publication Data

Crane, Jocelyn.
   Fiddler crabs of the world (Ocypodidae: genus *Uca*)

   Bibliography: p.
   1. Fiddler crabs.   I. Title.
QL444.M33C7   595′.3842   73-16781
ISBN 0-691-08102-6